First E.C. Conference on Solar Heating

Commission of the European Communities

First E.C. Conference on

Solar Heating

Proceedings of the International Conference held at Amsterdam, April 30-May 4, 1984

Edited by

C. DEN OUDEN

Institute of Applied Physics TNO-TH, Delft, The Netherlands

D. REIDEL PUBLISHING COMPANY

A MEMBER OF THE KLUWER ACADEMIC PUBLISHERS GROUP

DORDRECHT / BOSTON / LANCASTER

Library of Congress Cataloging in Publication Data
Main entry under title:

E.C. Conference on Solar Heating (1st : 1984 : Amsterdam, Netherlands)
First E.C. Conference on Solar Heating.

At head of title: Commission of the European Communities.
Includes index.
1. Solar heating-Congresses. I. Ouden, C. Den. II. Commission of the European Communities. III. Title.
TH7413.E25 1984 621.47. 84–18139
ISBN-13: 978-94-009-6510-2 e-ISBN-13: 978-94-009-6508-9
DOI: 10.1007/978-94-009-6508-9

Organization of the conference by
Commission of the European Communities
Directorate-General Science, Research and Development, Brussels

in co-operation with the
- International Energy Agency (IEA)
- Dutch National Solar Energy Programme
- Netherlands Organization for Applied Scientific Research (TNO)

Publication arrangements by
Commission of the European Communities
Directorate-General Information Market and Innovation, Luxembourg

EUR 9437

Softcover reprint of the hardcover 1st edition 1984

Published by D. Reidel Publishing Company
P.O. Box 17, 3300 AA Dordrecht, Holland

Sold and distributed in the U.S.A. and Canada
by Kluwer Boston Inc.,
190 Old Derby Street, Hingham, MA 02043, U.S.A.

In all other countries, sold and distributed
by Kluwer Academic Publishers Group,
P.O. Box 322, 3300 AH Dordrecht, Holland

PREFACE

Contributors to this Conference have shown the wide range of active and passive solar heating systems which have been researched, installed and monitored in recent years throughout western Europe and elsewhere. Yet much remains to be done if solar heating is to reach its full potential. The Conference Committee hopes that this record of the proceedings will provide a basis for the further development of these systems.

Many difficulties have been surmounted in arriving at today's position. The foundations of the growing confidence of architects and engineers are to be found in the concerted programmes of research and development mounted by two of the sponsors of the Conference - the European Community and the International Energy Agency. Some of the more tangible products of these programmes have been reported here: component and system behaviour has been subjected to rigorous scientific study; new test facilities have been founded; test procedures devised; simulation methods developed and evaluated; design rules formulated and checked against measured performance. It has been apparent here that the willingness to exchange information and experiences, which has always been a feature of the solar energy scene, remains as strong as ever. A further information-sharing initiative was noted on the part of another sponsor, UNESCO - the setting-up of the European Cooperative Network on Solar Energy, involving countries from both eastern and western Europe.

The pressures of oil price and supply which brought solar energy into prominence a decade ago have temporarily eased, making this a difficult time for the industry. As it strives to become more firmly established in the market-place, hard-pressed governments are applying the strictest criteria in assessing continued support for the R&D that is vital for the industry's well-being. Governments are not yet ready to attach values to the very characteristics that distinguish solar energy from conventional sources: that it is perpetual, secure, non-polluting, distributed, accessible and so on. One of the most notable features of the Conference has been the realistic way in which this situation has been accepted. In every Session there have been presentations and discussions on cost-effectiveness. New products and ideas for others have been described. In spite of the difficulties, I sensed a feeling of optimism that the goal of competitiveness could be reached if the partnership which has been established between industry, research centres and governments is sustained.

Brian Brinkworth

Brian Brinkworth
CONFERENCE CHAIRMAN

Dr. A. Strub, Director of non-nuclear energy R&D programmes,
Commission of the European Communities.

CONFERENCE COMMITTEE

General Chairman
Prof. B.J. Brinkworth
University College Cardiff,
Solar Energy Unit,
United Kingdom.

Technical Programme Management
Dr. E. Aranovitch
Joint Research Centre
CEC, Ispra, Italy

Mr T.C. Steemers
CEC, DG-XII
Brussels, Belgium

Ir. C. den Ouden
Institute of Applied Physics
TNO-TH Delft, The Netherlands.

Publication
Mr D. Nicolay
CEC, Luxembourg

Conference Secretariat
Mrs E.L.S. Janssen
Institute of Applied Physics
TNO-TH P.O. Box 155
2600 AD Delft
The Netherlands

Local Organisation
TNO Corporate Communication Department
Mr H. van den Berg
P.O. Box 297
2501 BD The Hague
The Netherlands

Exhibition
De Boer en Van Teylingen
Mr B.W. Bartstra
Oranjelaan 60
2281 GG Rijswijk
The Netherlands

Scientific Committee

Drs. A. Strub
CEC, Brussels

Dr. W. Palz
CES, Brussels

Ir. P.F. Sens
Stichting Energieonderzoek
Centrum Nederland, Petten
The Netherlands

Dr. G. Beer
PHOEBUS-ENEL
Catania, Italy

Dr. G. Beghi
Joint Research Centre
CEC, Italy.

Prof. C. Boffa
Torino, Italy.

Mr E.N. Carabateas
Ministry of Research and
Technology, Athens, Greece

Mr A. Debosscher
Katholieke Universiteit Leuven,
Belgium

Prof. A. Dupagne
Université de Liège,
Belgium

Prof.Dr.Ir. W.L. Dutré
Katholieke Universiteit Leuven
Belgium

Mr R. Ferraro
Energy Conscious Design,
London, United Kingdom

Dr. F.J. Friedrich
Kernforschungsanlage Jülich
Federal Republic of Germany

Ir. E. van Galen
Institute of Applied Physics
TNO-TH, Delft, The Netherlands

Dr. W.B. Gillett
Sir William Halcrow & Partners
Swindon, United Kingdom

Prof. Dr.-Ing. E. Hahne
Universität Stuttgart,
Federal Republic of Germany

Prof. Ir. C.J. Hoogendoorn
Delft University of Technology,
The Netherlands

Drs. K. Joon
Stichting Energieonderzoek
Centrum Nederland, Petten,
The Netherlands

Mr P. Kesselring
Eidgenossen Institut für
Reaktorforschung,
Würenlingen, Switzerland.

Prof. V. Korsgaard
Technical University of
Denmark Lyngby, Denmark.

Mr R. Lebens
Ralph Lebens Associates,
London, United Kingdom.

Mr J.P. LePoivre
Agence Française pour la
Maîtrise de l'Energie
Valbonne, France.

Mr J.O. Lewis
University College Dublin
Dublin, Ireland

Prof. S. Los
Bassano Del Grappo, Italy

Mr E. Öfverholm
Swedish Council for Building
Research Stockholm, Sweden

Mr G. Olive
Ingénieur ECP
Paris. France

Mr K. Robinson
National Board for Science and
Technology, Dublin, Ireland

Prof. H. Schreck
Technische Universität Berlin
Berlin, Federal Republic of
Germany

Prof. G. Schepens
Facultés Universitaires
Notre-Dame de la Paix, Namur
Belgium

Mr S. Svendsen
Technical University of Denmark
Lyngby, Denmark

Dr. M. Tsamparlis
University of Athens
Department of Physics,
Division of Mechanics
Athens, Greece

Conference chairman Prof. B.J. Brinkworth welcomes Dutch Minister of Economic Affairs Drs. G.M.V. Van Aardenne

Delegation of Conference Committee informs Dutch Minister of Economic Affairs about details of conference

CONTENTS

SESSION I - PERFORMANCE OF SOLAR HEATING SYSTEMS

SESSION II - CORRELATION METHODS AND DESIGN TOOLS

SESSION VI - SOLAR ENERGY APPLICATIONS IN AGRICULTURE AND INDUSTRY

SESSION VII - SOLAR HEAT STORAGE, SHORT TERM AND INTERSEASONAL STORAGE CONCEPTS

OPENING SESSION

Opening speech by
A. STRUB, Commmission of the European Communities, Brussels, Belgium

Opening speech by
W.A. DE JONG, The Netherlands Organization for Applied Scientific Research

The Dutch National Solar Energy and the International Energy Agency's solar heating and cooling programme

The status and future of the active use of solar energy

Solar architecture

PANEL DISCUSSION

A report on the panel discussion
T.C. STEEMERS, Commission of the European Communities, Brussels, Belgium

CLOSING ADDRESS

by G.M.V. VAN AARDENNE, Minister of Economic Affairs, The Netherlands

Dutch Minister of Economic Affairs addresses conference participants.

Ir. W.A. de Jong, President of the Netherlands Organisation for Applied Scientific Research TNO addresses conference participants.

OPENING SPEECH

Albert STRUB
Commission of the European Communities
Brussels

It is a great honour and pleasure for me to welcome you here on behalf of the Commission of the European Communities. As you know, the European Communities are involved in energy research, development and demonstration both through a number of multi-annual R&D programmes and through regulations providing financial support to energy relevant demonstration projects.

It is therefore no surprise to me to meet at this Conference many participants and, in particular, many authors and speakers who are linked to our work in solar heating and cooling. I am particularly pleased, however, to hear that this gathering has also attracted many scientists and engineers from throughout the rest of the world. May I extend my warmest welcome to you all.

You are certainly aware that the "calm on the oil front" and the recent development (or non-development) of the oil price has somewhat deviated the public interest, away from energy matters to other subjects which also determine our economic or general well-being. It goes without saying that this is strategically rather dangerous. It is widely known that these trends have led to particularly painful reductions in public funding for those relatively young disciplines which aim at exploiting renewable energy sources. The Europeans seem to be still in the midst of this downward movement, whereas in the US there appears to be a stabilization or even a new impetus. Japan, we know, has always maintained its level of effort.

I think it is not sufficient just to complain and to wait for a possible change in this situation. We should rather use this state of affairs to critically assess the work done up to now and to design a strategy for the future under these new circumstances. A difficult situation such as this is also a challenge and an occasion to think things over.

This Conference can help us to carry out such an exercise, and it has to fulfil a very important task: it must not only aim at presenting the state of the art, but it must also identify achievements, weaknesses and gaps, and give us guidance on where to concentrate our future efforts. In drawing our conclusions, we must keep in mind, however, that the utilisation of solar energy for heating and cooling (in buildings, and also in other applications) cannot exclusively be judged upon its potential contribution to reduce our energy bill or our dependence on oil. There are many other positive or beneficial aspects to be considered: be they in the field of environmental pollution, in the field of generally improving our conditions of life and well-being, and maybe also the general (educational) impact on our way of life. In addition, we should not forget that preparing and keeping open the technical options for alternatives to oil, indigenous energy sources have a high strategic value. Who knows what the oil price situation would be today, if you and your colleagues had not undertaken all the efforts made to exploit or to prepare the future exploitation of solar energy?

Very often the techniques of active and passive use of solar energy are tightly linked with conventional building techniques. And you all know that the "building people" are very critical of and reluctant to adopt new methods if their reliability and economic interest is not fully demonstrated. The real potential of solar techniques depends strongly therefore on the existence of reliable components, purpose oriented testing and performance measuring methods, design tools, and the sort of information resulting from research and development which you are dealing with at this Conference. Maybe the papers presented here will allow a better and more realistic assessment of the crucial issue of heat storage (short term or seasonal); and, maybe, those who are working in collateral fields will receive some further guidance on what they could contribute by developing better materials, more appropriate control systems and the like.

Ladies and Gentlemen, the weather is with us, the organisation is fine and we are celebrating the birthday of Her Majesty the Queen of this beautiful country, (my congratulations go to all Dutch participants at this occasion). Can you imagine a better setting for a conference?

I therefore wish you a fruitful Conference in this sense and a pleasant stay in Amsterdam.

OPENING SPEECH

Ir. W.A. de Jong, President of
the Netherlands Organization for Applied Scientific Research

Mr Chairman, Ladies and Gentlemen,

It is a pleasure to welcome so many of you from so many countries to this first European Community Conference on Solar Heating, on behalf of the Netherlands Organization for Applied Scientific Research.
This conference - as I see it - comes after approximately ten years of research into the application of solar energy. The initiative of the principal organizer, the Commission of the European Communities, to hold this conference, is welcomed by my Organization as an effective means of assessing what has been achieved in this field in the past decade. The participation of the co-sponsors, the International Energy Agency, the Dutch National Solar Energy Programme and UNESCO is, I think, evidence that these organizations also welcome the opportunity of strengthening applied research and development in the solar energy field.

This conference comes at a time when it is believed by many that for the next 3 to 5 years or so, there will be relatively little difficulty in meeting energy demands. This is not only due to the progress in introducing new energy conservation methods, but, of course, also to the stagnant economy.
Contribution by new and renewable energy resources have played a minor role. However, the upward trend of the economies of many industrial countries is making itself felt in the total consumption of energy. This in conjunction with the strategic importance of an adequate and sufficiently idependent supply of energy, makes it necessary to continue the search for economically viable sources of energy, including renewable sources of decentralised energy supply.
For my own Organisation this implies that further participation in solar energy research, development and demonstration is foreseen. These efforts date back from about 10 years ago, when the Institute of Applied Physics at Delft began its work in this field. The scientists who started research in converting solar radiation into useful energy for various applications, belonged to a group which was involved mainly in designing buildings in such a way that overheating by the sun did not occur.
Their progress can be read from the programme of the present conference. By participating in the Dutch National Solar Energy Programme, the Solar R & D Programme of the Commission of the European Communities and a similar programme of the International Energy Agency, several studies were carried out to assist industry in improving the various solar energy system components, such as collectors and heat storge systems. Furthermore, various research tools were developed to improve solar system performance and a new laboratory was set up to test solar installations and newly developed components.

The question should be answered where we now stand as far as actual application and transfer of the results to industry is concerned. Such stock-taking of our present position is necessary, if we are to set the course for future work in the solar energy field. To do this, I will mainly refer to the present state of the art in this country, and to a lesser extent to the Conference programme which is about to begin.

It seems to me that considerable progress towards the economic use of solar energy has been made in the past three or four years. Although the Netherlands are certainly not ideal from the point of view of its climate, I believe that the next three to five years will show much increased use for certain applications, in particular in combined heating systems, even here.
In this context the increased interest of selected industries indicates that practical application has come closer. In the past years or so several Dutch firms, large and small, have formed small units dedicated to solar energy usage. The efforts of industry and research institutions should therefore be continued, and if possible, intensified, the objective being to prepare ourselves adequately for future increases in the price of fossil fuels which sooner or later will occur.

Although I expect that industry's share in the cost of solar energy programmes is likely to rise, a sizeable part of the costs of such programmes should continue to be borne by national governments or international agencies. I must express my sincere hope that the member states of the European Communities will soon succeed in approving the research programme on solar energy for the next budgetary period. Similarly, I hope that the current discussion on some of our national energy-related programmes will lead to adequate funding. I am convinced that the authorities involved are fully aware of the strategic (and the economic) importance of these programmes.

I must apologize for make these comments on national and international science policies during an opening session of what I hope will be a stimulating and fruitful Conference. But I do think that such remarks must be made - in view of the importance and the long-term nature of the subject. The recent invention by the Dutch Government of a process for stimulating the efficiency of research by appreciable cuts in subsidies and funds has caused me to make these remarks.

Ladies and Gentlemen,

I sincerely hope that the formal sessions of your Conference as well as the informal discussions of the next few days will provide you with new ideas and information for your own research. Also, that the initiative to organize this Conference will lead to closer cooperation between the many organizations working on solar energy in the various countries represented here.
And, finally, I wish you an enjoyable conference and a pleasant stay in Amsterdam.

THE DUTCH NATIONAL SOLAR ENERGY AND THE INTERNATIONAL ENERGY AGENCY'S SOLAR HEATING AND COOLING PROGRAMME

P.F. Sens
Project Office for Energy Research, ECN

Summary

A short overview is given of the Dutch and the IEA Solar energy R and D programmes, serving as a sketch of the framework for a number of contributions to this conference.

1. INTRODUCTION

The oil crisis, as from 1974, generated an enhanced interest in the question of how to make an efficient use of the world's oldest, most conventional, most reliable, most abundant and most nuclear energy source: the Sun. In many countries national solar energy programmes were initiated and, as from 1976, an international liaison between those programmes in a number of OECD-countries took place under the auspices of the International Energy Agency.

2. THE DUTCH NATIONAL SOLAR ENERGY PROGRAMME

The programme started in 1978 as a government sponsored, multidisciplinary and multi-party programme under the management of the Project Office for Energy Research.
The objectives of this programme were:
- justifiable introduction
- expansion of application
- sound industry and market
- low cost.

The programme was divided in two phases of which the first (1978-1981) has been completed and evaluated. It has led us to the following conclusions:

- boilers
  . computer programmes are available on an international scale;
  . air and water systems are well developed;
  . no need for further system development;
  . cost reduction required a.o. by integration of components;
  . economic applications with low temperature water in agriculture and in swimming pool heating, with higher temperature water in food preparation for cattle and in dwellings (in competition with electric boilers);
  . field experiments have provided installation and operation experience, initiated measurement campaigns and presented some non-technical problems.
- space heating
  . computer programmes are under development in international cooperation;

	. test methods concerning efficiency, durability and reliability should be further developed a.o. in IEA and EC cooperation; . short term storage is feasible applying stratification and phase change materials; . field experiments with first generation systems are in progress; . second generation systems, including new concepts like evaluated tube collectors, should be developed to reach economic viability; . emphasis should be on systems development.
- storage	. no perspective for moisture absorption; . storage in an artificial lake presented too high cost in connection with water tightness; . subsurface storage of solar heat is subject of a pilot project at Groningen and offer perspective of 60% coverage. The storage is ready and operates to expectation. 96 dwellings with evacuated tube collectors, to be connected to the storage, are under construction; . also aquifer storage offers good perspectives and will be subject of another pilot project.

At present, the 2nd phase of the programma (1982-1985) is in progress and contains the following subprogrammes:

- conversion to thermal energy
  - . general supporting activities
  - . components development
  - . system studies and modelling
  - . seasonal storage
  - . passive solar
  - . non-technical aspects
  - . field experiments
- photovoltaics, etc.
  - . basic research
  - . system development

The general level of government funding for the national solar programme has been of the order of D.Fl 7 million per year. In addition, financial contribution have been obtained from participants in the programme, such as building corporations, and the European Communities.

3. THE IEA SOLAR HEATING AND COOLING PROGRAMME

The Interational Energy Agency was founded in 1974 under the auspices of the Organization for Economic Cooperation and Development with the aim to coordinate energy policy and planning. Twenty-one countries are member of the IEA and in a number of activities also the European Communities are participating. One element of the IEA programme is the enhancement of cooperation in R and D of alternative energy sources, such as solar energy.

In 1976 an agreement has been concluded between a number of countries to implement a programme to develop and test solar heating and cooling systems.

The objective of this programme has been to conduct a programme of cooperation activities which would contribute to the development of solar technology and supplement research programmes carried on at a national level. With the present decline in solar energy budgets, experienced by many countries, international collaborative efforts which share cost and expertise take on even more importance.

The programme is carried out under the management of an Executive Committee. It is subdivided in tasks and each task is put under the responsability of an Operating Agent.

The following tasks are in progress or have been completed:

I Performance of solar heating and cooling systems (completed)
II Coordination of R and D on solar heating and cooling
III Performance testing of solar collectors
IV Insolation handbook and instrument package (completed)
V Use of existing meteorological information for solar energy applications (completed)
VI Performance of solar systems using evacuated collectors
VII Central solar heating with seasonal storage
VIII Passive and hybrid solar low energy buildings
IX Solar radiation and pyranometry studies.

Participation in the various tasks is as follows:

Implementing Agreement	Task								
Signatories	[I]	II	III	[IV]	[V]	VI	VII	VIII	IX
Australia			X			X			
Austria		X	X		X		X	X	X
Belgium	X	X	X	X	X			X	
Canada			X	X	X	X	X	X	(X)
Denmark	(X)	X	X		X		X	X	X
Germany	X	X *	(X)	X	X	X	X		
Greece		X	X						
Italy	X	X	X	X	X			X	X
Japan	X	(X)	X			X			X
Netherlands	X	X	X	X	X	X	X	X	X
New Zealand			X *					X	
Norway		X						X	X
Spain	X	X	X	X				X	
Sweden	X	X	X		(X)	X	(X)	X	X
Switzerland	X	X	X	X	X	X	X	X	X
UK	X		(X)***	X	X	X *	X **	X	
USA	X	X	X	(X)	X	(X)	X	(X)	X
CEC			X	X	X	X	X		
Total	11	11	16	10	12	9	9	13	10

(X) = Operating Agent
* Former Participant
** Phase I Participant
[ ] = Completed Task
***Operating Agent as from April 1984

Additional activities have been undertaken outside the structure and scope of the formal tasks. Examples are workshops on ground-coupled solar-assisted heat pumps and air heating systems. A workshop on thermal losses from large collector arrays will be held in June 1984.

A list of 29 reports, issued in the past and based on the various task, has been made available to all participants to this conference and should be considered as the main source of information on the results obtained so far. At regular intervals the programme isuues a newsletter "up-date" to provide recent information on the task-activities. During this confernce a number of papers will be presented, orally or as poster, that have their basis in the IEA cooperative solar heating and cooling programme.

THE STATUS AND FUTURE OF THE ACTIVE USE OF SOLAR ENERGY

General keynote speech
by
C. den Ouden

INTRODUCTION

As the first 'scientific technical' keynote speaker in this conference I feel honoured to present the status of 'low temperature' thermal applications of solar energy. As this conference is devoted to both active and passive use of solar energy I will restrict myself to the 'active¡ part of solar energy applications, whereas my colleague and friend Theo Steemers will give an introduction of the 'passive' solar topics.The content of my speech will be of a more general nature, because during the conference a lot of scientific and technical achievements will be discussed in the various technical sessions. It will give you an indication - at least I hope so - of where we stand now and as I think it is entirely justified, this speech will be a plea to use solar energy more than is the case today.

I hope I will be able to convince you that solar energy can and must be used for many more applications than the 'general public' is aware of today.

In this speech I will make a few statements that may be somewhat provocative. I will do that deliberately to stimulate the discussions during this conference. When I state that governments should do more to stimulate the use of solar energy I blame myself and many of you that we have not made a greater effort to convince the decision makers!

WHAT IS THE VALUE OF SOLAR ENERGY AND WHAT CAN WE DO WITH IT?

In the Netherlands (and other parts of north-western Europe) an average amount of solar energy per m^2 per year of approximately 1000 kWh (3.6 MJ) comes from the sun. When we could use all this energy for those applications where we now use electricity to heat water, the value of this amount of energy would be 1000 times twenty-five cents, which amounts to 250.- Dutch guilders (see figure 1).

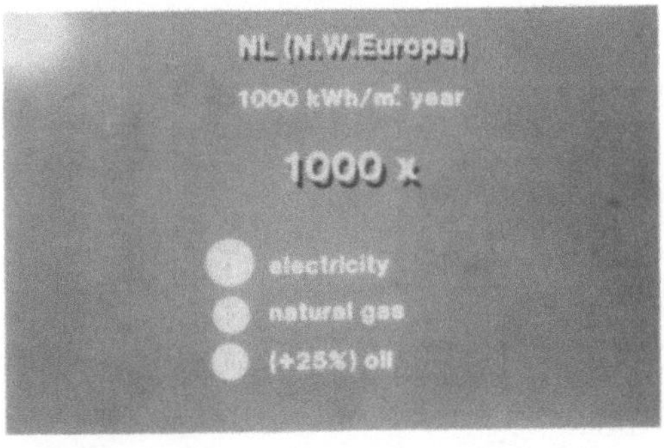

Figure 1: Value of solar energy in N.W. Europe.

When using this energy for those applications where nowadays natural gas is used to heat water, the value of this amount of energy would be 1000 times ten cents, which amounts to 100.- Dutch guilders. When using this energy for those applications where we use oil to heat water, the value of this amount of energy would be about 125.- Dutch guilders.
This stream of money comes to us at random (see figure 2). Today we use between 65 and 25% of this freely available clean form of energy, depending on the application and the system efficiency of the solar energy installation.

Figure 2: Maximum value of solar energy if competing with electricity.

In the Netherlands the 1984 status for using solar energy is such that swimming pool heating with well-designed 2nd generation systems is cost effective now! Using a solar domestic hot water system to preheat the tapwater currently heated by electricity is another cost-effective application. For certain agricultural and industrial applications, especially when using well-designed systems in processes asking continuously around the year warm water up to about 60 °C solar energy can be used nearly cost-effective. This is the situation for only a small part of Europe. Much more favourable situations for solar applications exist. For instance in France, solar energy has even more value because of the higher solar radiation (see figure 3).

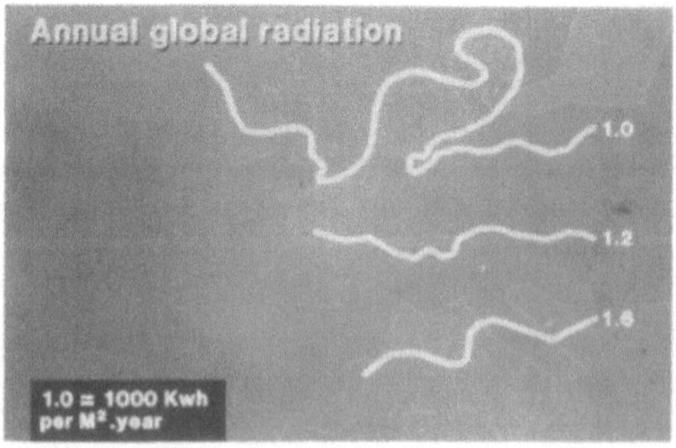

Figure 3: Annual global radiation in Europe.

When looking at the world's potential for solar we can easily see that for the heating of water for many applications, indeed much more favourable conditions exist, not only because of the higher solar radiation conditions (see figure 4) but also because of the higher energy costs for traditional energy souyrces such as electricity, oil and gas. Moreover, we should not forget to mention the many rural areas where some forms of energy today are not available at all.

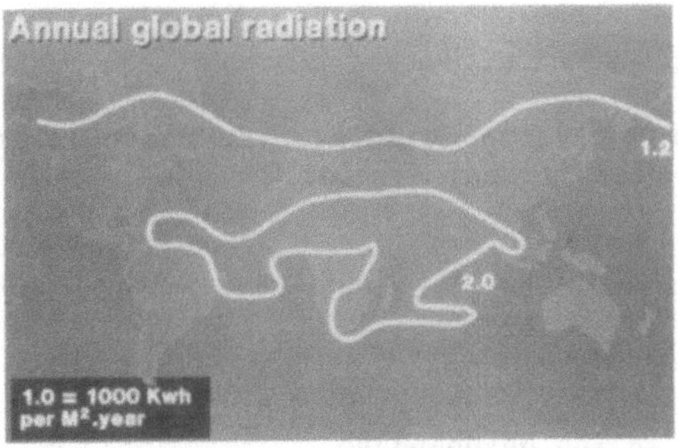

Figure 4: Annual global radiation (world).

WHY DON'T WE COLLECT MORE ENERGY FROM THE SUN TODAY?

Why are 'governments' so 'afraid' of stimulating the use of solar heating installations (figure 5).

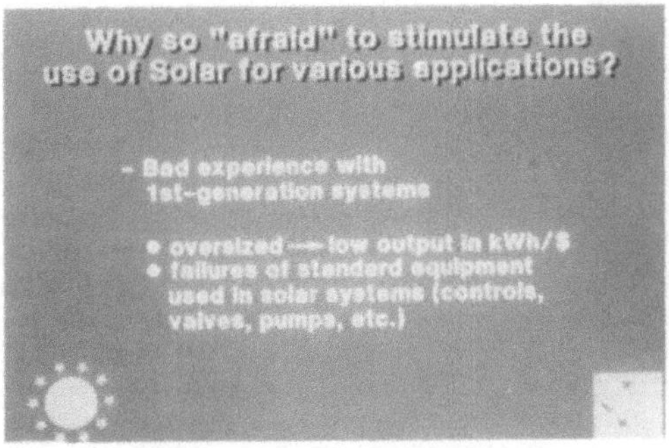

Figure 5: Reasons for not stimulating solar.

Since it was almost a decade ago that the Research & Development was taken up somewhat more seriously than before and as a result of that, rather optimistic predictions were made which could in many cases not be proven by the 1st generation 'demonstration projects', several decision making agencies became less believing in the future for solar heating applications.
In my opinion many of the so-called 'demonstration projects' were in fact R & D projects to learn from and many mistakes were made although they were unavoidable.

On top of that came the bad economic situation at the end of the seventies, continuing in the beginning of the eighties, resulting in less money for house-building etc. This resulted in a drastic cut of solar collector sales etc. Today the economic situation is improving again and hopefully the solar community can also profit from this situation. Compared to the tradiational energy sources we have an unfavourable position, solar energy for heating purposes has principally a decentralized collecting and consuming nature and can, therefore, less easily be 'controlled' like is the case with:
- electricity production and distribution
- oil production and distribution
- gas production and distribution

Why couldn't we consider to change the attitude of the decision-making authorities in such a way that:
- electricity, gas and oil companies have to stimulate the use of solar energy if it is competitive with 'traditional' fuels.

This could already be done for:
- swimming pool heating
- preheating of tapwater currently heated by electricity
- some agricultural application of solar, etc.

HOW HAS DOMESTIC HEATING DEVELOPED IN THE NETHERLANDS

I want to use the last 10 minutes of my speech to show you how domestic heating has developed over the years in this country. The aim of doing this is that I believe that even for domestic heating with solar energy there is a potential in our unfavourable climate. Before 1960 the majority of the houses in this country had only a heating facility in the living room, as shown in figure 6.

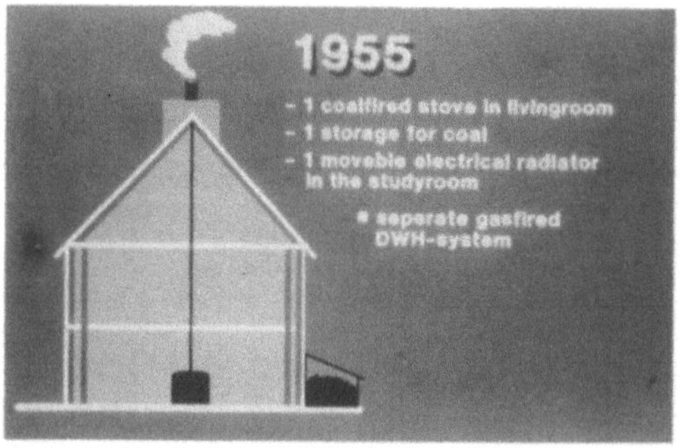

Figure 6: House heating situation in the Netherlands before 1960.

When the large natural gas reservoir in the north of this country came into exploitation all new houses were equipped with gas-fired central heating installations (see figure 7). The components of a central heating system are summarized in the same picture.
The total plumbing work is indicated in figure 8. In figure 9 you see the costs for such a total system including labour costs for installing.
Some of you may think that it is almost impossible to do all this at such low cost, especially when I give you an impression in figures 10, 11, 12 and 13 of what is needed with respect to hardware and labour to heat a so-called terraced house in Holland.

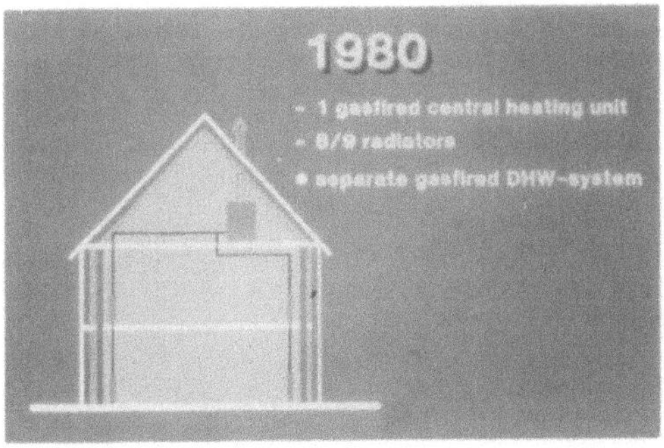

Figure 7 : Standard gas-fired central heating systems (1980).

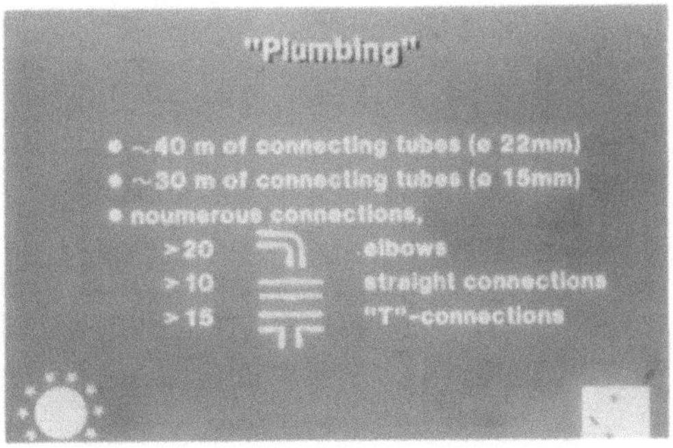

Figure 8 : Overview of plumbing work.

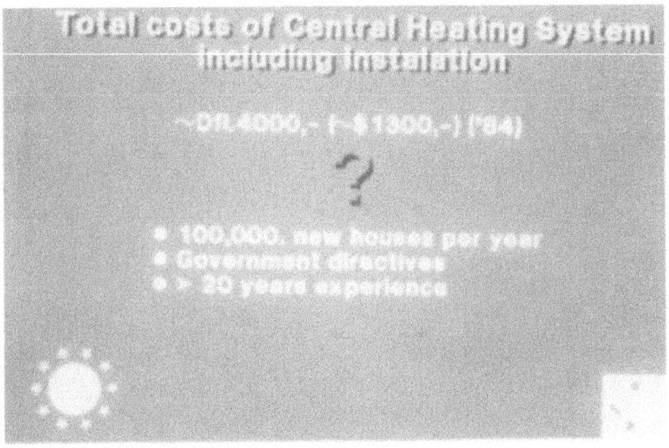

Figure 9 : Total costs of central heating system.

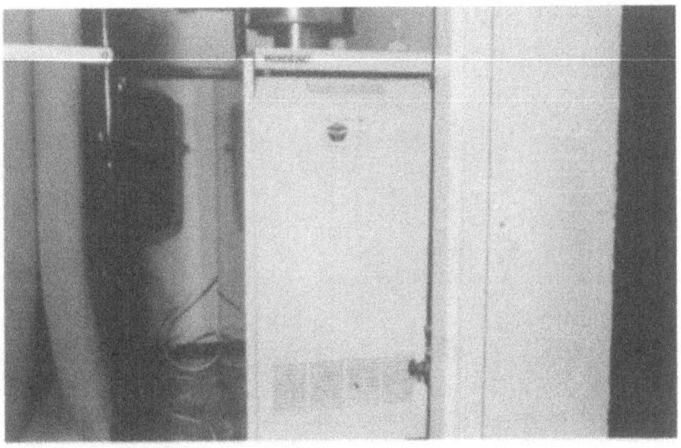

Figure 10 : Gas-fired heating unit.

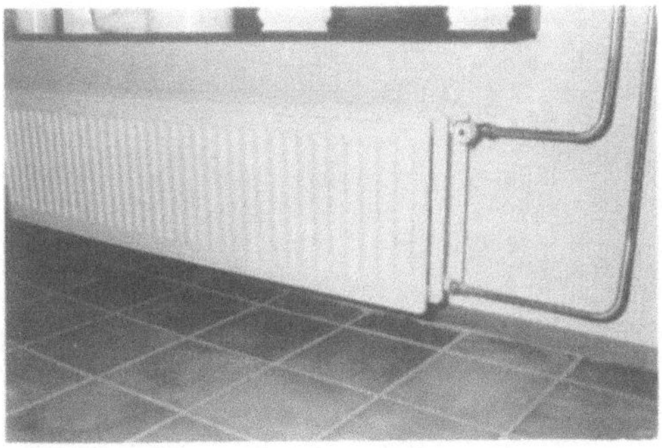

Figure 11 : Radiator in living room.

Figure 12 : Radiator in bathroom.

Figure 13: Impression of the plumbing work.

In these figures you see a gas-fired central heater, containing a pump, a gas-supply control unit; various types of radiators and an impression of the plumbing work. Finally several appliances such as expansion tanks and safety devices are included. All this equipment is only used for heating, for the tapwater heating another gas-fired heater is n ormally used (see figure 14).

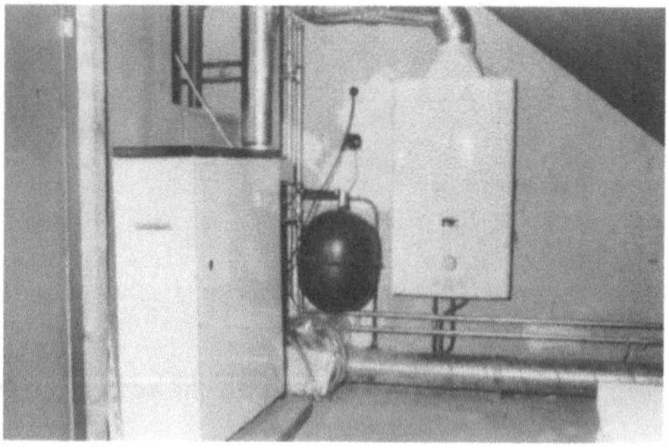

Figure 14: Gas-fired tapwater heater.

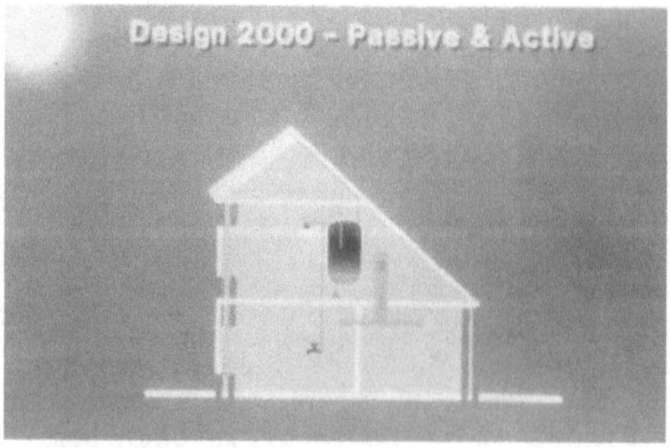

Figure 15: Future solar house.

Knowing that in a well-designed future house (see figure 15) the heat requirement will be much less than it is today, *I expect that an integrated solar energy system for domestic hot water and space heating can be installed in future houses for marginal extra costs, at least for those costs that can be justified by the savings on the energy bill.* An impression of such a system is given in figure 16.

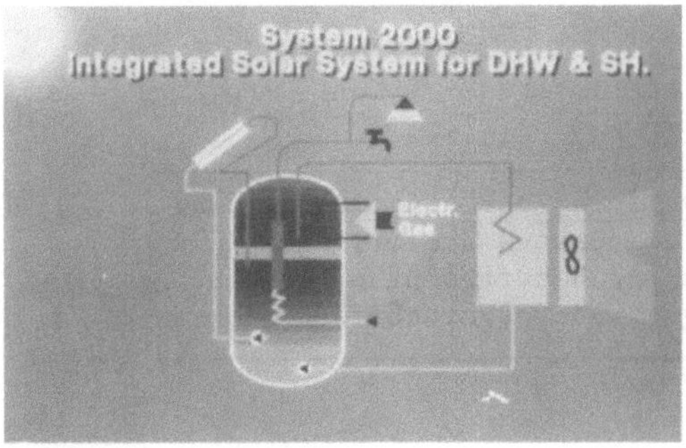

Figure 16: 'System 2000'

CONCLUSIONS

Mr Chairman, ladies and gentlemen, I hope to have succeeded in showing you that a sunny solar future can be expected and I conclude my speech by expressing the hope that the R & D on Solar Heating as well as the 'Industrial Development' in Europe will continue, but above all it is my belief that many applications are worthwhile stimulating today.

SOLAR ARCHITECTURE

General keynote speech

by

T.C. STEEMERS

Definition of a passive solar building: A building in which the building components are arranged in a manner which maximizes the benefits of solar heat without reducing standards of comfort

This simplified definition of a passive solar building already indicates that the introduction of passive solar energy will have an impact on the art and science of building construction; in other words on architecture.

The 1970's/1980's might be recognized later as the decades in which climate-responsive building revived again since it had been abandoned in the 19th century due to the introduction of mechanical technologies for heating and environmental control in general.

However, since then, standards of comfort have risen considerably and a more subtle balance between building and environment is necessary these days as compared to the last century. We need therefore to develop a more precise basis for design than that which sufficed in the past in order to reduce the auxiliary energy without reducing standards of comfort.

Before we deal with the important aspect of comfort, we should remind ourselves of the four principal passive solar concepts, which are:

- direct gain; where the low angle rays of the winter sun are admitted into the living space through windows in the south façade (see figure 1);

- sunspace; where an attached greenhouse functions as a solar collector (see figure 2);

- Trombe wall; which consists of a south facing wall with the external surface glazed to reduce heat loss (see figure 3); and

- roof space collector; where part of the roof is glazed to allow solar heat to be collected in the roof space (see figure 4).

Combinations and refinements of these concepts give the designer a variety of design possibilities. All these concepts, however, lead to an increase of the glazed area in the south façade of the building.

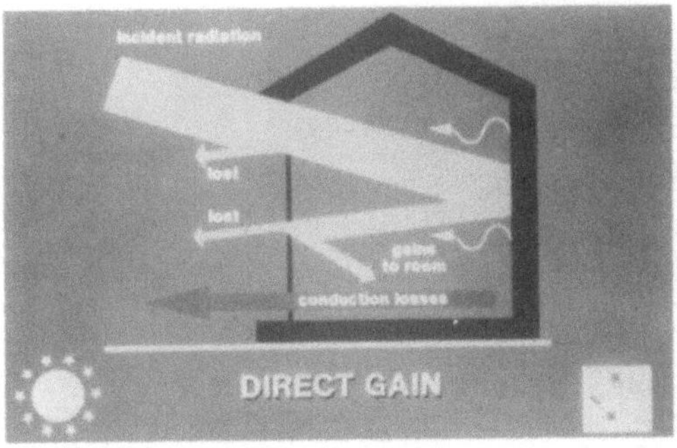

Figure 1: Direct gain.

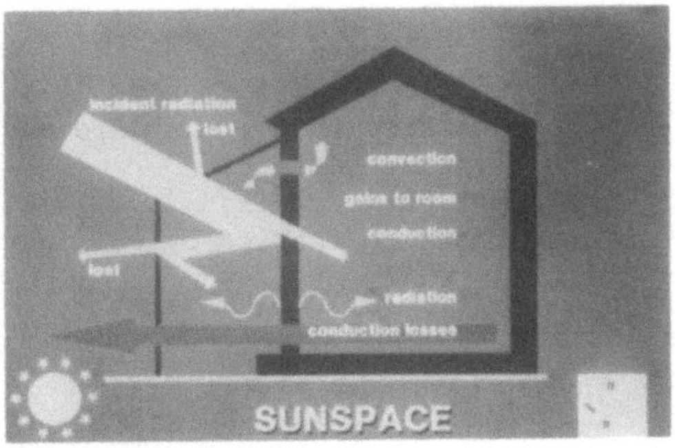

Figure 2: Sunspace

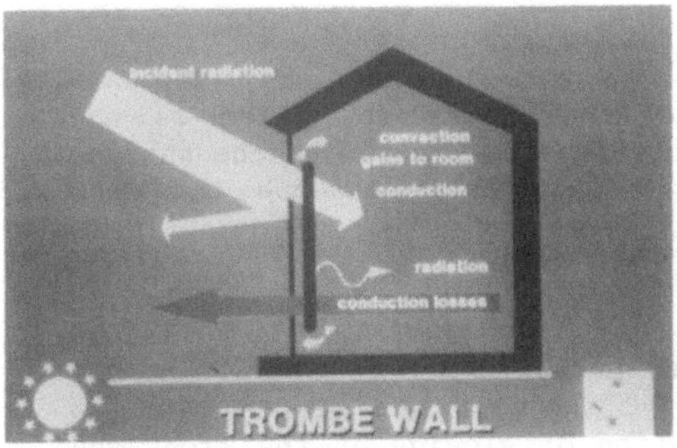

Figure 3: Trombe wall.

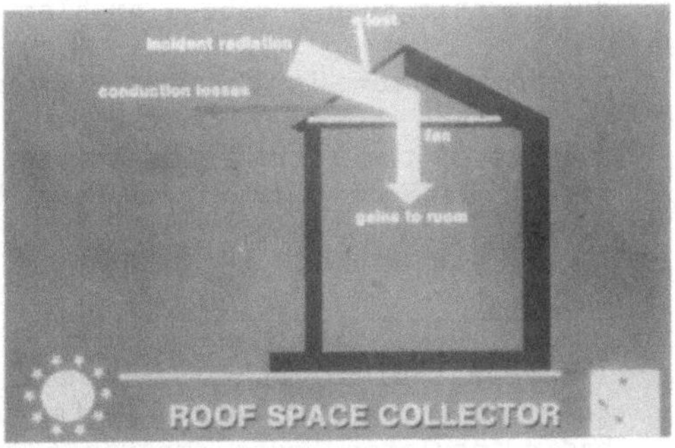

Figure 4: Roof space collector.

Comfort

The standards of comfort which can as a consequence easily be affected in passive solar design are :

- the room temperature;
- temperature differences in a room;
- visual comfort i.e. glare and strong light contrasts;
- acoustic comfort;
- ventilation rate;
- damage to fabrics by sunlight.

The temperature

The temperature limits within which one feels comfortable depend on humidity, air movement, clothing and the metabolic rate. The latter is in turn determined by the activity level of the person(s) concerned and therefore by the purpose of the room.

A lower temperature limit is controlled by the thermostat setting of the auxiliary heating. The upper limit should be restricted by proper passive solar design.

Summer overheating is often identified as a major comfort penalty with passive buildings; however, with proper shading and natural ventilation design this need not be a problem.

A complication is that the human body senses not so much the air temperature but rather the mean radiant temperature, which is the average of the surrounding surface temperatures weighted by the surface areas concerned. Devices have been developed to measure the mean radiant temperatures.

Temperature differences

The mean radiant temperature does not, however, detect discomfort due to large temperature differences. Open fire heating in a badly-insulated house creates a situation of glowing at the front and freezing at the back, which might be romantic but which is in conflict with modern standards of comfort.

Visual comfort

Large glazing areas increase the chance of glare and strong light contrasts. Daylight calculations should assess the visual comfort of a passive solar design before construction.

Large glazed areas can also cause acoustic problems and loss of privacy. Internal room layout should prevent this.

Passive solar design is in principle very simple, but in practice extremely complex due to the boundary conditions imposed by standards of comfort.

Many of the early passive solar builders designed on a rather intuitive basis, hoping for the best. However, those who are willing to experiment with an investment as large as a building, are small in number. The vast majority of builders are quite conservative and insist on reliable performance predictions, well-founded in experimental observations. It goes without saying that the first requisite is of course cost-effectiveness.

The question of cost-effectiveness has been dealt with in a European-wide assessment study on solar energy appplications to dwellings. The conclusions are particularly encouraging for passive solar :

- direct gain systems and retrofit sunspaces show almost immediate cost-effectiveness;

- sunspace (newbuild) will become cost-effective from 1990 onwards.

In addition to comfort and cost-effectiveness, there are other factors which are important for the introduction of passive solar techniques in the building industry. These are mainly :

- the familiarity of the architect with passive solar design techniques;

- the availability and reliability of design tools; as well as

- the availability on the market of reliable passive solar components.

European Passive Solar Programme; Achievements

It was only since 1980 that in the European Energy R&D programme, attention was directed towards these factors and to passive solar research in general.

Two passive solar design competitions for architects and building engineers were organized consecutively. They were successful in the sense that over 2400 applications for documents were received and about 500 entries were registered. The entries were - according to a group of international judges - of a high standard.

From 1981 to 1983 a European Passive Solar Working Group (PSWG), with experts from all the member states, has been active within the framework of the CEC Solar Energy R&D programme. A major task has been completed, i.e. the drafting of a European Passive Solar Handbook, a passive solar design manual for architects and building engineers which contains the state of the art in the field concerned. The handbook will be published later this year.

The draft handbook forms the main teaching material at the European Passive Solar Workshops which are being organized by the members of the PSWG in each of the member states.

The modelling subgroup of the PSWG has thoroughly examined the passive solar design aids in Europe and elsewhere and checked them against a long list of selection criteria such as suitability for passive solar design, easiness to commission, user friendliness, documentation and availability.

In the section "main frame and mini-computer models", ESP is recommended. ESP is a sophisticated simulation model developed at Strathclyde University with the financial aid of the Commission.

At the other end of the scale, Method 5000 was recommended as a manual method. Method 5000, developed by Claux-Pesso-Raoust in Paris, is well documented, very transparent and well orientated towards passive solar design, especially direct gains, sunspaces, and Trombe walls. It calculates a monthly utilization factor and a monthly indoor temperature (without auxiliary heating) in the dwelling. This last parameter can be used for an assessment of overheating. Method 5000 is fully described in the European Passive Solar Handbook.

Another important aspect and a prerequisite for the introduction of passive solar techniques in the building industry is the availability of reliable passive solar components. The Commission has supported research in this area by concluding 20 contracts with industry, universities and research organizations. Industrialization of production processes of some of these components is now being considered. Most results have now been published in final reports and performance figures have been quoted. However, it is difficult to compare the performance results of even the same type of components because different test procedures were used to assess performance and the way results were reported was not uniform.

Future Passive Solar Research

In the next European Solar Heating programme, the development and validation of the passive simulation programme, ESP, and the passive design aid, Method 5000, will receive a high priority. Validation is perhaps too ambitious but in any case a concerted European action will be started to improve the confidence in ESP and Method 5000.

It is not expected that sufficient experimental data for such an exercise can be obtained from monitoring passive solar houses. Besides, it would be too costly and also interference on the part of occupants would cause too much scatter.

An investigation into this validation problem has shown that an advanced type of passive solar test cell would be the best facility for obtaining experimental data on different passive systems under different climatological conditions. At the same time, these passive solar test facilities could be used for the development of test procedures for passive solar components, which, as we have seen, are badly needed.

Progress has been made in starting such a concerted action, i.e. a pre-design study has been completed and a design brief is ready to start the design phase as soon as the Council of Ministers ratifies the budget for the next programme.

A prototype of the test facility will be built at the Joint Research Centre, Ispra, to learn the teething troubles in the construction of such a facility.

Eventually, some six passive solar research centres will participate in this concerted action: developing and testing ESP and Method 5000 and developing test procedures for passive solar components.

Quite a number of specific passive solar topics need further investigation. Partly because they have been investigated for non-European conditions only and partly because they have been investigated only superficially.

Typical examples of these topics are:

- comfort (thermal, visual, acoustic, etc.);
- heat transfer and air movement;
- auxiliary heating and controls;
- day-lighting;
- planning and urban design;
- user response in passive buildings.

An update of the European Passive Solar Handbook in which the new findings are included should then be considered.

The number of architects in Europe is in the order of 200.000 and it is the architects who are possibly the most important link in the introduction of passive solar techniques into the building industry. The information transfer from the passive solar research community to the architectural profession should therefore remain a key area for the future. The actions which will be considered are:

- the provision of instruction material for courses and workshops and the provision of teaching aids in general;

- the stimulation and organization of workshops;

- the organization of passive solar competitions;

- the organization of passive solar conferences.

It can be said in conclusion that passive solar energy offers great opportunities for energy conservation in buildings provided it is supported by a well-directed R, D & D programme.

I really do hope that the national and international authorities will have the wisdom to provide the necessary funds for such research programmes.

A REPORT ON THE PANEL DISCUSSION.

T.C. Steemers, CEC, DG XII.

Panel : - W. Palz, CEC, DG XII, Chairman
- T.C. Steemers, CEC, DG XII, Rapporteur
- C. den Ouden, Institute of Applied Physics TNO, Secretary
- E. Aranovitch, CEC, Joint Research Centre Ispra
- P. Sens, Energiecentrum Nederland
- Prof. Korsgaard, Thermal Insulation Laboratory, Denmark
- A. Boettcher, Federal Ministry for Research and Technology, Germany
- Prof. H. Schreck, Professor in Architecture, Berlin
- Prof. G. Schepens, Professor in Economy, Belgium
- N. Baker, Energy Conscious Design, UK
- W. Kaut, CEC, DG XVII
- Prof. W.A. Beckman, University of Wisconsin, USA
- J. Michel, Architect, France
- Prof. B. Brinkworth, University College Cardiff, UK

The panel chairman, Mr. W. Palz, emphasised in his introductory remarks that a readjustment of the research strategy in solar heating is requested since we have been working intensive in this area for 10 years, and especially now since research budgets in Europe have been decreased and the solar market as a whole has not expanded since three years. He asked that more attention be given to solar architectural topics.

The chairman subsequently invited each of the panel members to make a short statement on the general topic, i.e. solar heating, or on a particular subject dealt with at this conference.

Mr. den Ouden expressed his personal belief that if research budgets were doubled and 5 to 10 % were spent on industrial development, solar energy would have a greater impact within a couple of years.

Mr. Aranovitch highlighted three tasks for the future i.e. :

- the demonstration of the reliability and durability of solar heating systems; to start with solar water heaters;

- the further development of seasonal heat storage;

- the accentuation of the cost effectiveness of certain passive solar technologies.

Mr. Sens reiterated the different tasks in the IEA solar heating and cooling implementing agreement. Regarding future tasks in solar heating research in general, he emphasised the importance of information transfer, seasonal storage, testing of components and solar air heating systems.

Prof. Korsgaard pointed out that to obtain a big solar market it is not sufficient to stress cost effectiveness; subsidies of governments are, however, necessary. The advantages for the authorities would be both, a reduction of employment problems as well as foreign currency problem (less oil import !).

He stressed that it can be shown that subsidies on solar heating systems do not cost the government a penny even if subsidies up to 75 % were given.

Prof. Boettcher expressed his opinion that the lack of success in solar heating marketing was mainly due to the high cost of the systems and their installation.
His advice : make systems simple and suitable for the do-it-yourself market.

Prof. Schreck was strongly in favour of the construction of more passive solar buildings so that user response can be studied. He favoured the exchange of experience with colleagues of member states. We should, however, not forget that energy was only one element in the design process.

Prof. Schepens stressed the fact that energy research in agriculture is rather fragmented due to the large variety of applications. Besides the aim was not only to save energy but also to qualitatively and quantitatively improve the crop.

Mr. Baker emphasised the importance of the transfer of know-how from the research community to the design professionals. This could best be done by demonstration and technical analysis of the examples demonstrated. In this manner intuitive design could be encouraged.

Mr. Steemers reminded the audience of the negative propaganda for solar heating resulting from the badly performing, overdimensioned first generation active solar heating systems. Due care should be taken, he said, to prevent the same happening in passive solar energy as the consequences of a badly performing passive solar system are even worse.

Mr. Kaut supplied some facts and figures on the CEC demonstration programme. In the five year programme from '78 - '83, 3 calls for tenders were published in the field of solar energy applications; 84 proposals were accepted. The subsidy from the Commission to these projects was approx. 40 % in general.
For the three-year programme '83 - '85, 265 million ECU subsidy will be spent on projects in the field of alternative energies, energy conservation, etc. In 1983, 37 solar projects have been selected for financial support from the Commission. Of these projects 27 were concerned with solar thermal applications.

Prof. Beckman drew attention to the differences in European and American research in solar heating. Little attention was paid to long term storage in the U.S. as opposed to Europe, whereas the collective storage domestic hot water systems obtained little interest in Europe as opposed to the U.S.

Mr. J. Michel emphasised the importance of the involvement of architects in the development of components. He further stressed that energy is only one of the considerations in building design. Mr. Michel ended with the statement : "Build for people, not for science".

After these statements of the members of the panel the questions which had been previously submitted were discussed in turn. Before asking the panel members to reply, each question was first put to the floor.

Question 1. (Prof. Dutré, KUL, Belgium).

The poor performance of many of the systems being studied nowadays appears to be caused by considerable losses on all levels of the system following a very efficient capture of energy by the collectors. Does the panel feel that the present state of the art is such that demonstration projects will solve these problems or that more basically-oriented research with respect to the complete system concept and design should be intensified.

Ir. E. van Galen (TNO, The Netherlands). The losses mentioned in the question are due to bad system design and bad performance of the heat storage. These problems cannot be solved by demonstration programmes. This view was supported by the panel secretary, Ir. C. den Ouden.

Question 2. (Dr. H. Tabor, SRF, Israel).

How are the costs for a solar heating system distributed over the collectors, storage, installation, maintenance, etc. The purpose of the question is to find out where cost reductions could be obtained. Future R & D should be concentrated on these items.

Mr. M. Neves (Shell, Portugal) was of the opinion that cost reduction in installation could be obtained by more and better courses for installers.

Prof. Löf (Colorado State University) informed the audience that, based on the experience in the US, the cost of selling solar water heaters was 40 % of the total cost.

Mr. C. den Ouden answered that the experience in Holland was different in the sense that the selling cost was much less. The answer for the US could possibly be to sell in smaller regions with regional selling and servicing centres.

A participant from Canada put forward that integrated collectors in roofing systems could save money.

Question 3. (G. Löf, Colorado State University).

I wish to comment and make a proposal concerning a uniform basis for reporting performance of active and passive systems for space heating with solar energy.

Prof. G. Löff.
In active and passive system performance one does not speak the same language. One should, in comparing systems, use a comparable basis e.g. a building of conservative design.

Mr. W. Palz.
We will put more emphasis on this point as soon as the oil prices have risen by 1 dollar.

Question 4. (W.B. Gillett, Sir William Halcrow and partners).

We seem, with respect to sunspaces, to have conflicting advice : ECD-Partnership says these are cost-effective now and France says no.

Mr. N. Baker.
Cost-effectiveness is dependent on the cost of the component, how the sunspace is used and how it is fitted to the building.
In the assessment study, to which this question refers, this was all well-defined. It certainly was not stated that any sunspace is cost-effective.

Mr. C. den Ouden finally summarized the panel statements and discussion as follows : "We are optimistic regarding the possibilities of active solar, whereas in passive more R & D is needed".

"Let us learn from the mistakes made in active solar and prevent the construction of bad first generation systems in passive by careful design".

"We have gone through a learning process in the active solar energy area; we have learned that solar systems should not be made up from separate components plumbed together without consideration of the interaction between the components and the effect on the system performance".

"Fresh thinking from collector to tap is required". "It is important to design systems with high kWh/ECU-value". A cost reduction of DHW systems of 20-40 % below the values quoted at this conference must be possible".

CLOSING ADDRESS

Drs. G.M.V. Van Aardenne,
Minister of Economic Affairs,
The Netherlands

Mr. Chairman, Ladies and Gentlemen, it is my pleasure to be able to say a few words at the closing of this conference, especially as it is you yourselves who have provided me with the material to do so during this past week.

The conference has covered many subject matters. Subjects in the field of active as well as passive utilization came to order, as well as storage systems, test facilities, models and application possibilities worked out to the finest details. This general approach to the subject matter made it possible for the participants not only to take part in the presentations and discussions of their own subjects, but also to actively take part in the discussions and thoughts of researchers in other subject matters. The more objective and more unconventional approach of an outsider can very often trigger off new ideas and solutions, especially when both the expert and the outsider have the same objective in view.

There is an interesting comparision to be made between the technical development of solar energy systems and the development of the government's budgetting for energy research. 10 years ago it was as yet unclear to us which systems offered the best technical and economic perspective for the use of solar energy. There were certainly enough funds available for the preliminary investigations into the possibilities. The research done during the past years has provided us with techniques which we can most profitably use for particular applications in particular locations. We have acquired some new techniques, but have also learned that some of them offer absolutely no perspective at all. This makes it possible for us to put the, unfortunately, ever decreasing funds to the most effective use for the further development of those techniques offering the most perspective. It is now necessary to get an even greater return from our investments in research in order not to delay the necessary technical and scientific development, despite the decreasing budget available.

I have just mentioned that the choice of system for an optimal solar energy system is very much dependent upon the geographical location of the system. This applies particularly if we review the area of the European Community. In the Mediterranean area more water can be warmed up by cheaper systems than in more northerly areas with much more expensive systems. This great climatic dependence could easily have led to a form of isolationism, as a southerner would be interested in different aspects to a northerner. However, both receive advantage from co-operation in their development work as by uniting the best aspects of both ideologies it becomes possible for Europe (I shall keep to the same example) to have at its disposal an efficient and cheap system to produce warm water.

Even in this country where the weather is not so sunny as in the southern European countries, many swimming pools are already warmed by solar energy. It is, therefore, very gratifying that such isolationism has not appeared, and that co-operation exists which extends even further than our own countries or even neighbouring countries. And, at this particular stage, I would very much like to mention my appreciation of the driving force which - in energy research in general - flows out from the European Commission. The many projects which have been realized in member countries in the past years with the support of the European Community have contributed greatly to the development of this research field. Resulting from this flow to all member countries, the contribution has been much larger than when each project had been a national one.

However, European co-operation has also its limits. As soon as the technical development reaches the stage of commercial production, the willingness to co-operate internationally (but also nationally) decreases. The participants have then to each decide which form of involvement is best for them. This can mean that for certain countries the priorities slide from international scientific co-operation to national or even sub-national system or component development. It can already be noticed that there is a tendency on a European scale for less strenuous efforts to be made in the field of solar energy in favour of an increase on a European scale in the field of photo voltaic solar cells.

It is the task of the researchers to continue on an international scale to further work on fundamental innovations; such work earns the support of the European Community and of the national governments.

As you are now our guests here in Holland, I shall give you a brief resumé of what the Dutch government is doing in the field of solar energy. An important consideration in the Dutch policy in the field of solar energy is the large theoretical potential; even in our densely populated country, the yearly total amount of solar energy coming from the sun is easily fifty times larger than the national primary use of energy. Obviously there are many reasons as to why only a small proportion of this energy can in fact be utilized, but, nevertheless, the amount which can still be utilized is quite considerable. At this particular time, and from a financial-economic point of view, it is not yet an attractive enough proposition to use solar energy on a large scale - the energy produced is, in general, still more expensive than the rival forms of energy. Even so, there are already solar energy systems in certain sectors - and in this connection I have already mentioned collectors for swimming pools - which are producing energy at a comparable or even at a lower price than conventional systems; but in many other sectors it is still not viable.

Our policy is in particular directed at achieving simple, cheap and energy saving systems which could provide good industrial prospects. For such systems, the customary marketing machine takes care of the introduction, after which further work can be done on product development, e.g. to produce a higher return. From this point of view we have built up a comprehensive and structured research programme in the field of solar energy, partly in combination with energy-saving techniques. To this end we spend about fifteen million guilders yearly (about six million ECU). Projects are executed in both passive and active systems. These include partly practical experiments and partly projects relative to the development of components. These efforts have already led to various industrial products from cheap solar boilers to good heat pump systems and - to come back to today's theme - various excellent solar energy systems, both active and passive. Tomorrow, on the occasion of the 'study tours' you will be given the opportunity to make acquaintance with some field projects. One of the largest field projects - unfortunately also the farthest away from Amsterdam - is the project in Groningen, in which the European Community also participates. In this project 96 homes are warmed by evacuated tubual collectors. The warmth which is collected in the summer is stored in the ground by means of a 14 kilometer long folded tube which is brought into place by means of a giant 'stapling machine'. However, I shall not go into details about this as you have already heard about it in detail this morning.

Mr. Chairman, Ladies and Gentlemen, I would finally like to express the wish that all your further efforts may flourish and that the international co-operation, of which this conference is a perfect example, may quickly come to fruition whereby the whole international community may benefit. This clean source of energy, which despite constant use remains inexhaustable, is worthy of all your efforts to make it available at a low cost so that many generations hereafter may literally and figuratively bathe in the warmth of the sun.

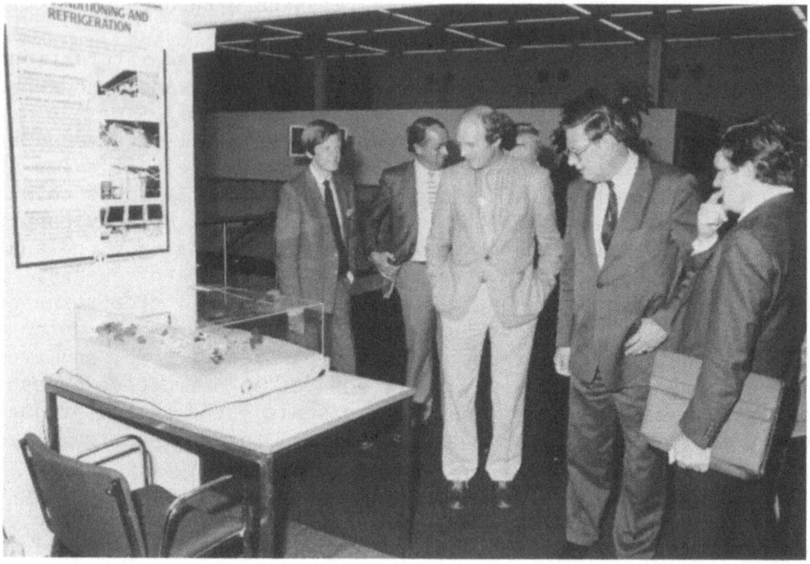

Dutch Minister of Economic Affairs visits the conference exhibition

SESSION I - PERFORMANCE OF SOLAR HEATING SYSTEMS

Simulation and monitoring of solar domestic hot water systems

Solar heating test design facility for bulk PCM storage

Low energy passive solar houses at Milton Keynes, United Kingdom

THE EUROPEAN SOLAR PILOT TEST FACILITIES, A POWERFUL TOOL FOR MODEL VALIDATION

W.L. DUTRE
Afdeling Toegepaste Mechanica en Energieconversie
Katholieke Universiteit Leuven
Celestijnenlaan 300 A, B-3030 Heverlee, Belgium

Summary

Detailed solar system simulation models constitute the basic tool for system development and optimisation as well as for the development of simplified methods for solar system calculations. The validation of simulation models is a task of utmost importance to arrive at sufficiently reliable design and development tools for solar systems. The Solar Pilot Test Facility project and the European Modelling Group for Solar Systems, two concerted actions of the solar energy research program of the Commission of the European Communities, were established to achieve this goal. The basic principles of the Solar Pilot Test Facilities operation, in view of reliable data production and of the methodology applied by the European Modelling group, are closely linked and requires intensive collaboration of both groups. Most of the validation results to date are related to the reference systems of the Solar Pilot Test Facilities such that generalised conclusions with respect to model accuracy for different system types are not yet available. For the cases considered thus far, no systematic model errors are detected and in spite of simplification in the simulated system configurations and the physical model of the system components, the model predictions are in good agreement with experimental performance data. The continuation of the validation task will be devoted to the second solar system of each of the Solar Pilot Test Facilities.

1. The use of simulation models and the need for validation

Solar systems for domestic hot water production, space heating, cooling or combinations of these are conceptually different from classical systems used in these applications. A classical system is basically designed and dimensioned such that it can satisfy the peak demand, whereas solar systems in most cases contribute only partially to the heating or cooling demand. The basic concern in designing and dimensioning a thermal solar system is therefore not related to the peak value of the demand, but rather to the total useful energy output of the system over its entire operation period. This quantity depends on the solar system characteristics, component and regulation properties, the history of the local climatic conditions and the dynamic thermal behaviour of the building or a time dependent load profile. As the economic feasibility of a solar system depends strongly on its energy savings, these must be predictable with a sufficient degree of accuracy.

Since the overall useful energy output of a solar system constitutes only a very global evaluation of the considered system, it does not permit judgements to be made regarding the modification which should be introduced into the system in order to increase its performance or to improve the system concept and design. A clear understanding of how a system reacts to changes in its design parameters, the system concept, component properties and regulation criteria, the heating or cooling demand and its distribution strategy, can be obtained only from more detailed considerations, including the dynamic energy balance of every component and the response of the system to the continuously occuring variations in its operating conditions. This level of understanding of the system behaviour permits a detailed improvement of the system, based on the sensitivity of its performance to the various parameters involved. Such detailed theoretical predictions can be obtained only from a correspondingly detailed calculation of the system behaviour, requiring the use of a detailed mathematical simulation model. The kind of information which can be obtained from such calculations performed for system design and optimisation, depends of course on the degree of sophistication of the model being used. Generally, the use of simulation models involves a big calculational effort on a powerful computer such that the cost is usually quite high. Simulation models are therefore mainly considered as research and development tools in solar engineering, rather than for use in the daily practice of solar engineers and architects for overall dimensioning of every individual domestic solar system they are dealing with. For the latter type of calculations, simplified methods are considered to be more appropriate. The development of reliable simplified methods for various types of solar systems, either as a directly applicable correlation for the system performance or as a simplified energy balance equation combined with corrective algorithms is on the use of a well developed simulation model.

In every simulation model, many simplifying physical and mathematical approximations are used with regard to heat transfer and fluid dynamics in the various system components, the energy demand calculation and the interaction between different components. The accuracy of theoretically predicted system behaviour can therefore be estimated only from comparisons of theoretical results with experimental data obtained from measurements on separate system components and complete systems. Based on such, usually long term experience, a computer simulation model can be gradually validated and improved in order to obtain a reliable design and development tool for solar systems. Validation of a theoretical model implies the investigation of the model accuracy for a sufficiently wide range of the system parameters, component properties and climatic conditions and to estimate the limits of its applicability. For generalised computer programmes which can handle a large variety of solar systems, this information should be obtained for the different types of systems of interest to the user. Validation of a simulation model can thus be defined as the progressive investigation of the level of confidence one can attribute to the theoretical model.

Recognising the importance of the role reliable mathematical models can play in the development of solar systems, the Commission of the European Communities decided to include two major concerted actions in its four year solar energy research programme 1979-1983, one to stimulate the development and validation of simulation models and simplified methods, known as the European Modelling Group for Solar Systems (EMG) and one to provide reliable experimental data to validate the models which became known as the Solar Pilot Test Facility Group (SPTFG).

2. The Solar Pilot Test Facilities

Eight Solar Pilot Test Facilities have been constructed, each facility consisting of two full scale solar systems : a space heating system used as a common reference and a second system of which the design and the system

components represent a particular interest of the participating country or organisation. From the reference systems, being nearly identical in all facilities, experimental data over a wide range of climatic conditions are obtained for the same type of system. The test facilities are designed such that operating conditions as in occupied solar houses are closely approximated without being coupled to a real building.
The principle of these installations, to be situated somewhere between laboratory experiments and detailed monitoring of solar houses, is illustrated in figure 1. Each facility consists of the real solar system to be studied, with collectors sited outdoors subject to the prevailing weather conditions, the house and its heat distribution and emission system being replaced by a computer controlled physical simulator. The simulator imposes a heat demand that can account for any theoretical house its characteristics, occupancy, lighting usage and heat distribution and emission, while being fully interactive with the prevailing weather conditions and the status of the solar system.

In comparison with the monitoring of a complete solar system installed in a house, the use of these test facilities has several advantages. Its flexibility with respect to the heat demand and the heat distribution strategy, only requiring software changes, is of particular importance to investigate the performance of the system for different house heat distribution and emission systems and/or a different type of house. The results of the on-line simulation of this part of the real system are continuously used to control the heat extraction from the solar system as though it were installed in an actual house. This, together with the possibility to change the size of the solar collector array, nominal flow rates and regulation criteria, constitute the main differences between the pilot test facilities and occupied solar houses. Also substantial modifications have been carried out during the period 1981-1983. It follows from this flexibility that the Solar Pilot Test Facilities offer extensive possibilities for experimental system studies and data production for model validation, under more favourable conditions than those usually encountered in actual houses.

The measurements performed during operation of the test facilities must be sufficiently detailed in view of the validation of the simulation model. These measurements include three groups of physical quantities : meteorological quantities, status variables of the solar system and the variables related to the heat withdrawal from the system by the physical simulator. From the measurements, performed at one or five minute time intervals, hourly averages are evaluated which are stored for subsequent use by the European Modelling Group for the validation of simulation programs and for direct experimental system development studies.

Each solar system of the Pilot Test Facilities is described by a set of parameters, such as geometrical parameters, heat loss coefficients, thermal capacities, heat exchanger characteristics, regulation criteria,... giving a complete description of the system and the considered operating sequence. These descriptive parameters are determined from estimations, calculations and independently performed measurements.

A scheme of the reference system is shown in figure 2, in which the different subsystems considered in the description of the installation are indicated. A description of the other systems of the Solar Pilot Test facilities, the operating procedures and the various experimental sequences considered is given in the references 1, 2 and 3.

3. Validation Methodology

Basically, the methodology of the validation is that data obtained from the SPTF on actual performance of a solar system is compared with the performance predicted from the model of the system operating under the same wheather conditions and the same load conditions as the SPTF-system. In the validation procedure applied by the EMG since 1982, the comparison is performed on the level of integrated energy flows. Previous experience indicated that, in spite of an apparently satisfactory agreement of temperature and power profiles, the discrepancies observed for energy transfers can be much more significant. Integrated energy flows are obviously more sensitive to accumulation of errors and therefore constitute a more severe validation criterion. The observed discrepancies are due to the combined and accumulated effect of measurement errors on the experimental data, systematic errors on thephysical installation description parameters, unknown malfunctioning of system components and shortcomings of the physical model as well as the mathematical integration algorithm. The different levels at which errors or reasons for discrepancy enter in the validation calculations are indicated in figure 3.

Considering the uncertainty margin to be associated with each variable and parameter used as an input quantity of the simulation and considering the error margins of the experimentally determined energy flows, it should not be surprising that a good agreement between experimental and theoretical values may be the result of a fortunate cancellation of errors. The uncertainty margins to be associated with each experimental and theoretical output quantity being considered, as well as with their discrepancies, are therefore evaluated. The uncertainty margin of the experimentally determined energy flows can be calculated from the accuracy of the instruments being used and the measured values of flow rates and temperature differences. The estimation of the theoretical uncertainty is based on a linear error propagation assumption and the calculation of the sensitivity of each of the considered output quantities to each of the input data and input variables used. In the calculation of the uncertainty margin of the discrepancy, allowance is made for the fact that the experimental and theoretical values are not entirely independent, as some experimental energy flows are evaluated by using measurements which are also used as input data for the theoretical calculation, such that their error margins are correlated.

Validation calculations can be performed according to different representations of the simulated system heat removal, as indicated in figure 4 by the three different data flow paths from the actual SPTF-system operation to the simulation model in as far as the storage heat removal is concerned [3, 5]. Each method offers particular advantages but is also subjected to some shortcomings and the results of the calculations are not necessarily identical. Method 1 can only be applied succesfully when the physical simulator of the heat distribution and emission system functions properly. In method 2, the validation of the model is necessarily limited to internal energy flows, as the measured system energy output is applied to the model. Method 3 eliminates the main disadvantages of the methods 1 and 2. The methods 1 and 3 have been used by the EMG for validations related the SPTF-reference system.

4. The simulation program being validated

From previous experience of the EMG it was realised that the validation of a simulation model is a very slowly progressing process and that all the effort should be concentrated on the validation of one simulation program. A modular simulation program has been developed taking account of the shortcomings of the models of individual EMG participants used in the first phase of the validation work. A first version of the program, called EMGP1 [4], has been considered for validation work, using validation method 1 of figure 4 since it did not include the possibility to apply method 3 [3, 5] .

An improved and generalised modular simulation program, called EMGP2 [5, 6], has since been developed. As compared to EMGP1 it allows for many additional features in view of general studies in solar system development and optimisation. EMGP2 can handle a large variety of different systems, including the various second systems of the SPTF-group participants. For the SPTF reference systems several validation calculations for EMGP2, based on validation method 3, have been performed.

5. Validation Results to Date

The validation of EMGP1 and EMGP2 with respect to the SPTF-reference system has been concerned only with the validation of a simplified configuration of the considered system. In this simplified representation, the entire collector array is represented by only one element, the thermal capacity of the collector loop piping as well as the storage loop piping is not explicitely accounted for, the heat exchanger thermal capacity and its heat losses are neglected and the storage tank is assumed to be unstratified. The effect of these simplifications depends on the operating conditions during each of the individual system operating sequences and is not taken into account in the uncertainty calculations. It may result in systematic deviations or increase the scattering of the discrepancies when many experimental sequences are considered.

Validation method 1 has been applied for a number of selected operaing sequences in which the physical simulator appeared to function with sufficient accuracy. A comparison of the calculated and experimental values of the integrated storage energy input and the integrated storage energy output is shown in the figures 5 and 6 respectively, for sequences with a length from 7 to 44 days, totalising 345 days of operation.

Validation method 3 could applied to a larger number of operation sequences since no conditions are imposed with regard to the accuracy of the physical simulator operation. The results of these validation calculations, for 33 experimental sequences with a length from 7 to 45 days and totalising 512 days of operation, are shown in the figures 7 and 8.

Although the uncertainties of the experimental as well as the calulated energy values represent an essential result of the validation calculations, they could not be indicated in the figures 5 to 8. The resulting undertainty domains would in fact be to much overlapping. For the same data sequences as represented in figure 8, the differences between the calculated and experimental values of the storage energy output, together with the associated total undertainty margins (3 σ-intervals), are shown in figure 9.

6. Conclusions

Although experimental data production and subsequent model validation may appear to be an obvious and relatively task, numerous problems are usually encountered with regard to measurements of system characteristics, reliability of the data production during actual operation and the validation procedure to be applied. The SPTF- and EMG-projects, developed as concerted actions in the CEC-research program, have developed powerful experimental facilities and the theoretical tools to contribute to this

important field of solar engineering. The validation procedure includes the evaluation of the expected range of deviations as a result of various sources of errors.
From the validation results obtained for the SPTF-reference system, several problems areas, mainly related to averaging of data, the accuracy of system descriptive parameters and sensor locations, could be identified. Improvements in these various aspects of the data production, together with the use of more detailed configurations in simulating the considered systems, is expected to reduce the statistical scattering of the data points of model validation as well as the differences in systematic deviations observed for different installations and to narrow the statistical uncertainties involved in the comparison of theoretical and experimental results. The results presently available show that uncertainty margins of 5 to 10 % are obtained, which are usually larger than the observed discrepancy, such that systematic model errors can not be detected. For the considered sequences, obtained from different SPTF-reference systems and a variety of operating conditions, the correlation coefficient of the linear regression of the calculated and measured storage energy output is better than 99 % and most of the discrepancies are within the total uncertainty interval.

REFERENCES

[1] OLIVES, G. (June 1982). Solar Energy Applications to Dwellings. Proceedings of the EC-Contractor Meeting. Vol. 2, p. 59-66, D. Reidel Publ. Co.
[2] OLIVES, G. (June 1983). Solar Energy Applications to Dwellings. Proceedings of the EC-Contractor Meeting. Vol. 4, p. 55-64, D. Reidel Publ. Co.
[3] SPTFG-Report (June 1983). The Solar Pilot Test Facility, a powerful tool for Model Validation and Solar System Development.
[4] DUTRE, W. (June 1982). Solar Energy Applications to Dwellings. Proceedings of the EC-Contractor Meeting. Vol. 2, p. 41-46, D. Reidel Publ. Co.
[5] DUTRE, W. (June 1983). Solar Energy Applications to Dwellings. Proceedings of the EC-Contractor Meeting. Vol. 4, p. 30-42, D. Reidel Publ. Co.
[6] DUTRE, W. "EMGP2 : A Transient Simulation Model for Solar Systems" (to be published).

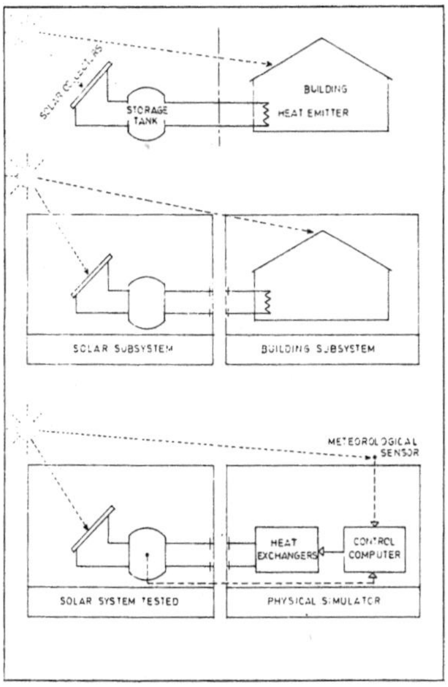

FIGURE 1 : PRINCIPLE OF THE SPTF-OPERATION

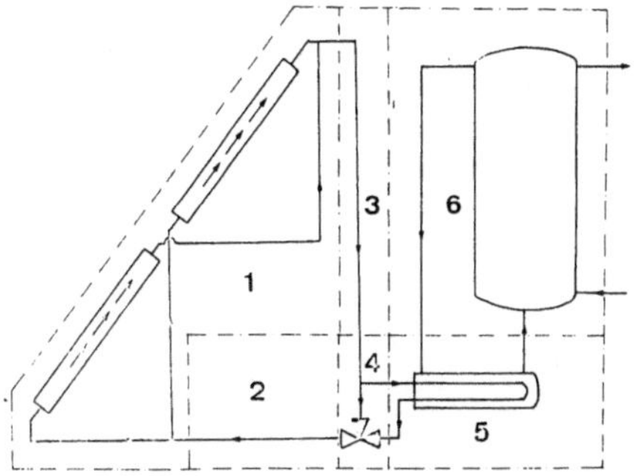

FIGURE 2 : SPTF-REFERENCE SYSTEM

SUBSYSTEM 1 : COLLECTOR ARRAY AND INTERCONNECTING PIPING
SUBSYSTEM 2 : COLLECTOR LOOP COLD LEG
SUBSYSTEM 3 : COLLECTOR LOOP HOT LEG
SUBSYSTEM 4 : BYPASS AND SWITCHING VALVE
SUBSYSTEM 5 : HEAT EXCHANGER
SUBSYSTEM 6 : STORAGE TANK AND STORAGE LOOP

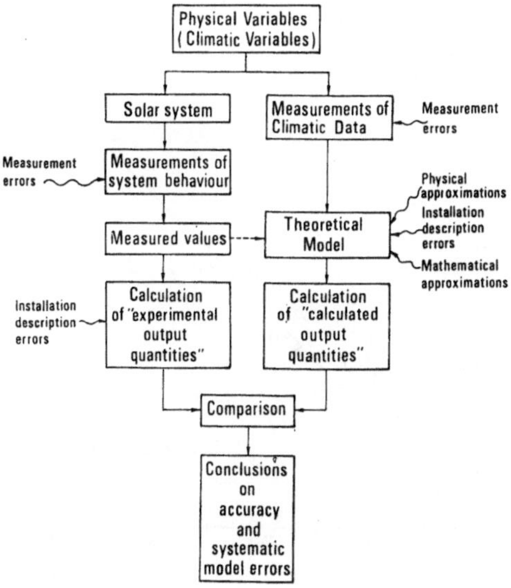

FIGURE 3 : ERROR SOURCES IN VALIDATION CALCULATIONS

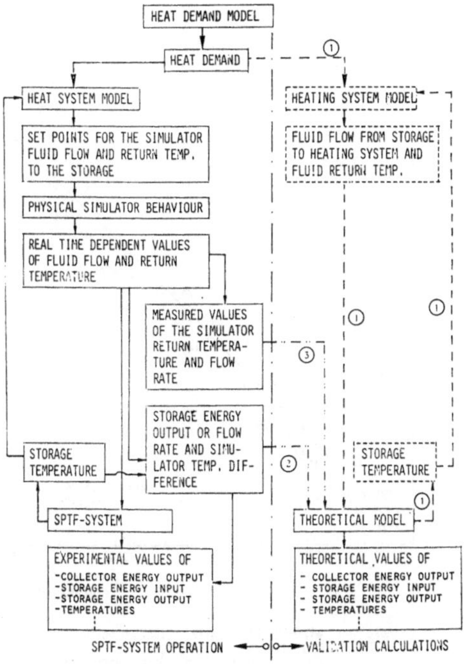

FIG. 4 DIFFERENT APPROACHES FOR THE REPRESENTATION OF THE STORAGE HEAT REMOVAL IN VALIDATION CALCULATIONS

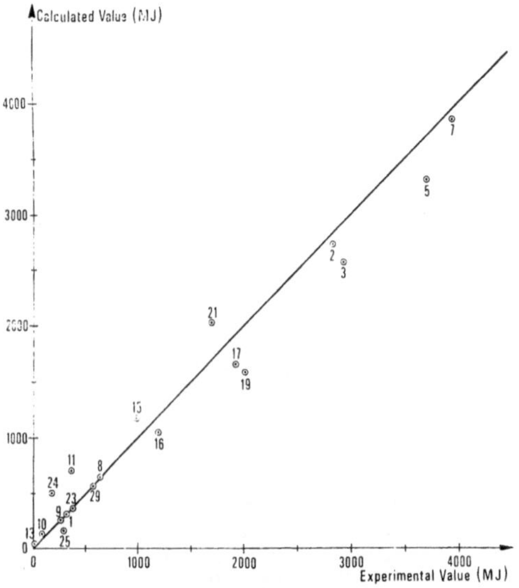

FIGURE 5 : EMGP1-VALIDATION WITH VALIDATION METHOD 1
SPTF-REFERENCE SYSTEM
CALCULATED VS MEASURED STORAGE ENERGY INPUT

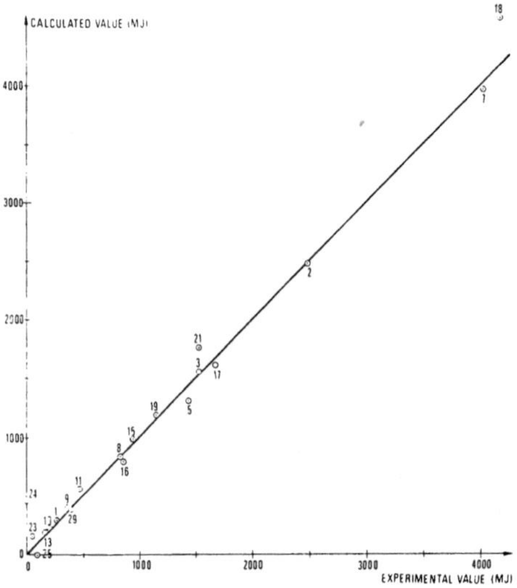

FIGURE 6 : EMGP1-VALIDATION WITH VALIDATION METHOD 1
SPTF-REFERENCE SYSTEM
CALCULATED VS MEASURED STORAGE ENERGY OUTPUT

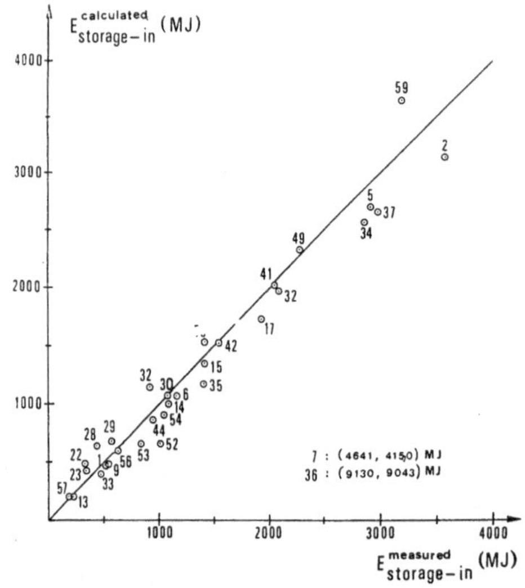

FIGURE 7 : EMGP2-VALIDATION WITH VALIDATION METHOD 3
SPTF-REFERENCE SYSTEM
CALCULATED VS MEASURED STORAGE ENERGY INPUT

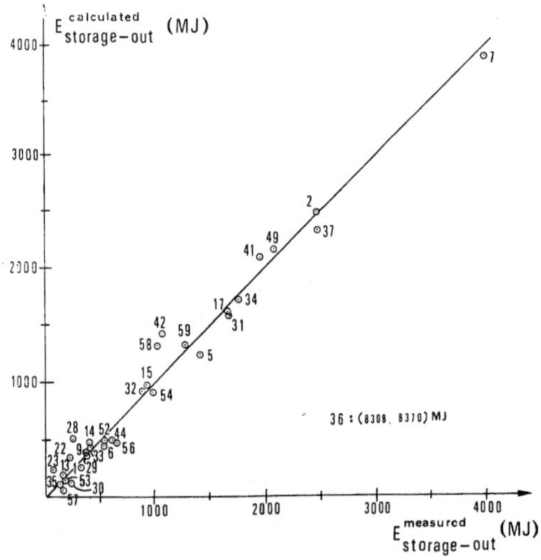

FIGURE 8 : EMGP2-VALIDATION WITH VALIDATION METHOD 3
SPTF-REFERENCE SYSTEM
CALCULATED VS MEASURED STORAGE ENERGY OUTPUT

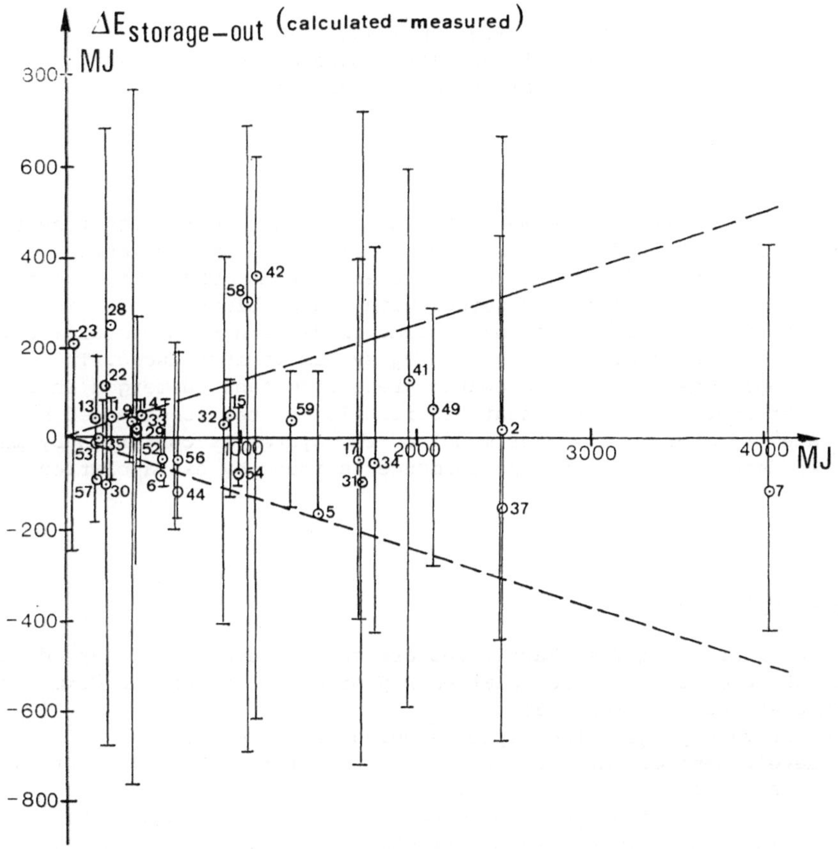

FIGURE 9 : EMGP2-VALIDATION WITH VALIDATION METHOD 3
SPTF-REFERENCE SYSTEM
STORAGE ENERGY OUTPUT DISCREPANCY
AND ITS TOTAL UNCERTAINTY INTERVAL (± 3σ)
VS MEASURED STORAGE ENERGY OUTPUT

COMBINED SOLAR SYSTEM FOR SPACE HEATING AND DOMESTIC HOT WATER SUPPLY: PERFORMANCE MEASURING DATA FROM THE DANISH PILOT TEST FACILITY

N.B. Andersen
Thermal Insulation Laboratory
Technical University of Denmark

Summary

This paper describes some of the most satisfying results from the Danish Pilot Test Facility: the development and improvement of combined space heating and domestic hot water systems. The Danish history of evolution is presented together with the problems of the implementation of solar space heating systems in dwellings. A certain well functioning combined space heating system is introduced. The solar storage consists of a hot water preheating tank submerged in a bigger outer storage, and the heat distribution system is extended by a separate radiator. A computer model of the system is confirmed by measured data and the yearly performance is found by extrapolation.

1. INTRODUCTION

The Solar Pilot Test Facilities are one of several "concerted actions" within the project A of the solar energy programme under the Commision of the European Communities (CEC).

Under this project the topics studied are in general:

- the development of European test procedures for solar components (collectors, heat storage systems, etc.)
- the development of uniform approaches for total system design, monitoring and evaluation of total system performance and validation of computer simulation models for different solar systems, both for active and passive solar designs, etc.

The activities as listed above are in general directed towards design, development and improvements of solar systems (passive and active) in order to stimulate the implementation of these systems within Europe.

In Denmark this work has been partly funded by the CEC and partly by the Danish Ministry of Energy.

The SPTF is a test rig (1,2) on which you mount the solar system including solar collectors, storages, back-up systems, etc., and measure the overall performance of the system under well defined conditions.

This rig has been a valuable tool in the development of solar space heating systems because of the possibility of making changes quickly and because of the measuring accuracy of the rig. These advantages would not have been achieved if the solar system had been mounted in a real occupied house.

2. EVOLUTION AND DEVELOPMENT OF SPACE HEATING SYSTEMS

In Denmark, during the latest 10 years a lot of solar systems have been installed, but the expectations of the consumers and the manufacturers have not been satisfied in all cases, especially not concerning the space heating systems. Often technical problems have occured with the constructions, problems that was not easy to solve with the very few experiences obtained. Concerning the domestic hot water systems the problems have been solved a few years ago (4,5), and thereby the systems installed after 1981 are well-functioning.

Regarding the space heating systems it is a fact, that it is more difficult to cover the requirements of the consumers because of the differences in the heat distribution systems of the dwellings. One of the problems has been, that for a dwelling with an ordinary water based heat distribution system, it is necessary to obtain a solar storage temperature, which is higher than the return temperature in the distribution system, to draw out energy from the storage.

It is well known, that the distribution of solar insulation over the year do not fit the heat demand distribution very well under the Danish and similar climates. Therefore most of the savings caused by a space heating system will be obtained in the spring and in the autumn.

Regarding the great amount of money the space heating system is representing and the uncertainty of the output from the solar system, very few house-owners have done their investments in solar space heating devices.

The system described in section 3 seems to justify, that it is possible to get a reasonable high solar output for a reasonable low price. Because of the separate radiator included in the system it is possible to utilize even low accessible storage temperatures.

3. DESCRIPTION OF THE SYSTEM

The Energy consumption of the house, which the system is designed for, is about 48.800 MJ for the space heating and 11.450 MJ (200 l/day) for the domestic hot water supply.

The DHW storage tank (0.2 m^3 of water) is placed inside the SH storage tank (1.0 m^3 of water) and has nearly the same height as the SH tank. The solar collector system is liquid based, and it is connected to the storage with a pipe heat exchanger placed in the bottom part of the SH storage.The inlet of the cold water is placed at the bottom of the DHW-tank and the outlet is placed at the top. This storage configuration provides higher efficiences of the primary circuit, because the cold inlet will cool down the bottom of the storage and give higher temperature differences between the fluid in the primary circuit and the water in the storage tank.

Solar collectors: 10 elements with selective surface on aluminium, 1 layer of glass, total transparent area: 15,7 m^2, liquid: propylene glycol flow: about 0,111 kg/s (~0,4 l/min/m^2). Control: differential thermostat with separate setting of the start- and stop difference. Storage: hot water tank: 200 l, storage (excl. hot water) 1000 l. Heat exchanger: 20 meters of 16 mm/18 mm copper tube. Electric water heater: 3,0 l and 1.2 kW.

Back up: As seen on the diagram of the system configuration the system is backed up with a boiler. In summer the no-load loss of the boiler would have been of considerable magnitude in comparison with the demand for DHW. Therefore the boiler is cut off when there is no heat demand and the DHW supply is backed up with an electric heater, which has only little no-load loss.

Separate radiator: The space heating system is equipped with a separate radiator for the solar space heating.This separate heater is placed in the largest room of the dwelling, which has f.ex. 25% of the total demand for space heating, and it will make better conditions for using low storage temperatures.

Summer operation: the boiler is cut off by the three-way valve. The demand for space heating is satisfied by the existing radiators and the

Fig. 1: System configuration

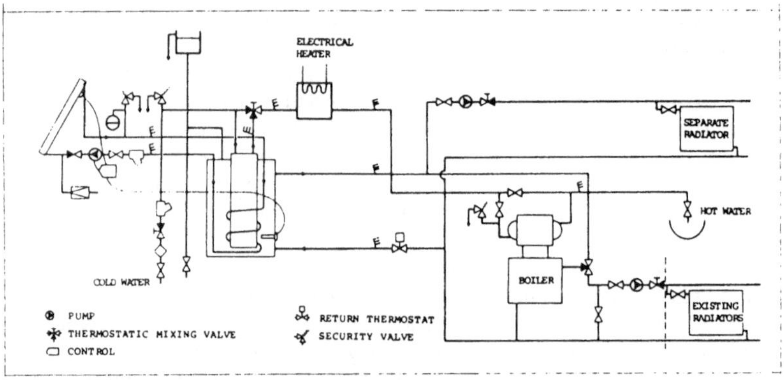

separate radiator. The demand for hot water is satisfied by the content of the submerged hot water storage. Winter operation: the boiler is operating on the existing radiators but is separated from the rest of the installation by the three-way valve. Water from the outer space heating storage is used for the separate radiator as often as the storage temperature makes it possible (ca. 25°C). The hot water is preheated in the submerged storage and is brought up in temperature in the domestic hot water device connected to the boiler.

4. MEASUREMENTS

The measuring system has been based on a 80-channel datalogger system from Hewlett-Pachard, 3025A. This system consists of the following components: scanner, voltmeter, disc-calculator (controller), multiprogrammer. All measuring points are scanned every twenty seconds and the mean values of these measurements together with mean values of the energy flows computed over a period of half an hour are stored. The solar system has not been installed in a real house, but the demands have been simulated by the controller and the connected control system. The tapping of hot water was controlled by a watch with a daily cycle and the heat removal in the interface system is controlled by the computer. The return temperature from the interface is calculated and set by the computer according to the actual heat demand. The accuracy of the measurements: 0.1°C (temperatures), 3% (flows).

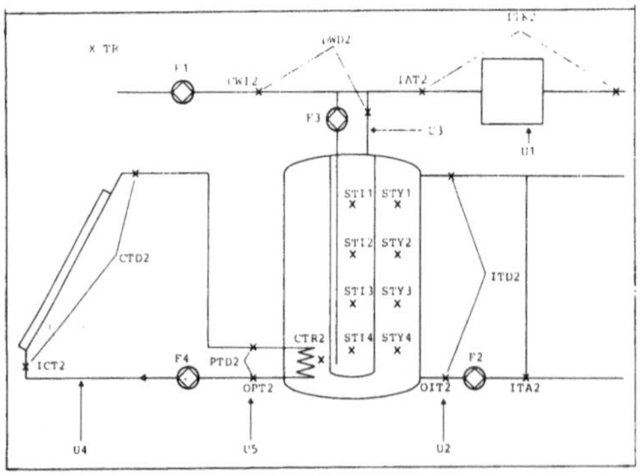

Fig. 2: The system connected to the interface- most of the measuring points are shown

5. PERFORMANCE

During several periods of operation measurements have been obtained and compared with an DMGP2 computer simulation (6) for the periods respectively and a reasonable fair agreement was found, which reflected a confirmation of the simulation and modelling parameters. The stratification of storage temperatures, heat- and capacity losses in the primary curcuit and the influence of the primary flow on the collector output are taken into account.

Fig. 3: Example of measured and computed energy

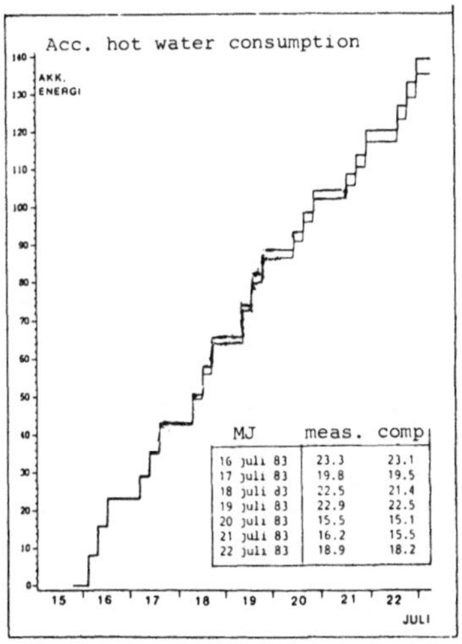

5.1 Yearly performance

By use of the validated modelling parameters, the yearly performance of the system has been computed in accordance with the test reference year giving the results shown on table 1. If the electrical heater is implemented in the system as suggested in section 3. It would be possible to cut off the boiler during summer and save the idle heat loss for nearly five months and the annual savings would be increased by 25%.

Table 1: Yearly performance of the space heating system

	Solar Output		Heat demand		Solar fraction		
	DHW MJ	SH MJ	DHW MJ	SH MJ	DHW %	SH	SH+DHW
Jan	320	242	972	8734	33	3	6
Feb	393	700	878	7738	45	9	13
Mar	471	906	972	6756	48	13	18
Apr	656	1799	941	4295	70	42	47
May	787	2034	972	2544	81	80	80
Jun	941	0	941	0	100	-	100
Jul	972	0	972	0	100	-	100
Aug	972	52	972	41	100	100	100
Sep	752	1445	941	1484	80	97	91
Oct	527	1055	972	3823	54	28	33
Nov	380	451	941	5811	40	8	12
Dec	373	353	972	7606	38	5	8
Tot	7545	9038	11450	48833	66	19	28

5.2 Stratification of storage temperatures

As expected the temperature stratification of storage temperatures was good in the way that at the same time the storage could keep a high temperature in the top f.ex. 60°C and a low temperature in the bottom, and this stratification was nearly undisturbed by the tapping of hot water and by the circulating water in the space heating curcuit. When there is a solar input to the storage the lower part of the storage will be disturbed, of course, up to at certain level and then the volume under this level will have a common mean temperature, which is still lower than the temperature in the upper part.

5.3 Efficiencies of the system

The computations of the yearly performance gave these additional results, which are mean values.

Table 2: Efficiencies of the system

System efficiency:				
	$\frac{\text{System output}}{\text{Global radiation}}$	=	$\frac{16583 \text{ MJ}}{62265 \text{ MJ}}$	= 27%
Collector efficiency:				
	$\frac{\text{Heat withdrawal}}{\text{Global radiation}}$	=	$\frac{22519 \text{ MJ}}{62265 \text{ MJ}}$	= 36%
Pipe efficiency:				
	$\frac{\text{Storage input}}{\text{Heat withdrawal}}$	=	$\frac{20631 \text{ MJ}}{22519 \text{ MJ}}$	= 92%
Storage efficiency:				
	$\frac{\text{System output}}{\text{Storage input}}$	=	$\frac{16583 \text{ MJ}}{20631 \text{ MJ}}$	= 80%

5.4 Dimensioning of the system

The yearly performance of the system for other sizes of the storage has shown that the annual output of the system would only decrease by 5% with a 0,7 m^3 storage (0.2 m^3 hot water storage). Concerning the other parameters they seem to be appropriate.

6. CONCLUSIONS

- the system was well-functioning and in accordance with the expectations, though the storage was bigger than needed.
- a good agreement between measured and computed data was found justifying the use of the validated model in an extrapolation of the yearly performance.
- a good storage temperature stratification was obtained providing a more efficient storage.
- 28% of the total demand is delivered by the sun.
- Assuming a mass production and regarding the economic situation for a Danish consumer, tax rates, fuel prices etc. the system design seems to be interesting, as there is a positive balance between savings and costs even during the first year of operation if the back-up heating is provided by electricity. If the solar output is substituting oil or gas though, the savings will just counter balance the costs.
- Anyway, further development especially of the system components will reduce the costs and hereby make the system profitable substituting electricity and/or gas.

REFERENCES

1. Olesen, O.B. (1981). Solar Pilot Test Facility, Final Report - Denmark. Thermal Insulation Laboratory, Technical University of Denmark.
2. Ellehauge, K., Olesen, O.B. and Andersen, N.B. (1983). Solar Pilot Test Facility, Final Report - Denmark. Thermal Insulation Laboratory.
3. Olesen, O.B. and Andersen, N.B. (1984). Kombineret solvarmeanlæg på systemprøvestanden. Laboratoriet for varmeisolering.
4. Ellehauge, K. (1982). Solar systems for domestic hot water. The development of an efficient solar system for domestic hot water. Thermal Insulation Laboratory.
5. Papers related to posters P5.13 and P5.14 of the conference.
6. Dutre, W.L. (1983). The EMGP2-computer program. Laboratorium voor Warmteoverdracht en Reaktorkunde, Belgium.

MEASURED PERFORMANCE OF EXPERIMENTAL AND COMMERCIAL SOLAR HEATING SYSTEMS

G.O.G. Löf and S. Karaki
Solar Energy Applications Laboratory
Colorado State University
Fort Collins, Colorado 80523

F.H. Morse
Office of Solar Heat Technologies
Conservation and Renewable Energy
U.S. Department of Energy
Washington, D.C. 20545

Summary

Selected results of a multi-year, multi-organization program of solar heating system development based mainly on experimental and demonstration projects supported by the Office of Solar Heat Technologies of the U.S. Department of Energy are presented. The capabilities of various types of solar water heating and space heating systems are compared, the low performance of most systems in practical use is contrasted with that of well designed experimental systems, and opportunities for improving the average efficiency of solar heating are shown There is a trend toward considerably better average performance and toward higher efficiencies of the best systems in practical use. It is concluded that even without major design changes or system cost increases, there are excellent prospects for significant increases in average performance of solar heating systems.

1. INTRODUCTION

Although thousands of solar heating systems of many types are in use, reliable performance data are available on only a limited number, most of which were installed several years ago. A few systems in R&D centers have also been monitored. The results generally show that systems in routine residential and commercial use have not performed as well as expected nor as well as systems in R&D sites. The differences are mainly due to the soundness of the design, the quality of the installation, and the extent of maintenance required and performed.

There have been significant increases in system performance at the research sites, primarily through system design improvements based on experience. Further increases are possible, through better component integration, close attention to thermal loss reduction, and changes in operating and control strategy.

Efficiencies of solar collection and utilization in practical systems have shown a significant increase in recent years. The average system efficiency of monitored solar space heating systems in the National Solar Data Network has increased from ten percent in the 1980-81 season to over 15 percent in 1981-82, but there are numerous systems that continue to operate poorly.

2. SOLAR WATER HEATERS

Two types of performance data have been collected and assembled. The output of solar water heaters has been measured under closely controlled conditions so that the only major variable was the system type and specific design. This laboratory testing has provided data which permit comparison of one system type with another and, within a particular type, one design or brand with another.

The other type of performance testing has been conducted in the field in dwellings and commercial buildings provided with solar water heaters. The results of these monitoring programs show typical performance levels and operational characteristics of a large variety of systems, but because of highly different climatic and operating conditions, close comparisons of one design with another cannot be reliably made.

Tests by the Tennessee Valley Authority (1) summarized in Figure 1 show a wide variation in performance of 25 systems supplied by 14 manufacturers. Although several system types are represented, variation in factors other than type, such as use of double or single tanks, selective or non-selective absorbers, single or double glazing, glass or plastic covers, effectiveness of heat exchanger (if used), fluid circulation rate, may mask differences due to system type. Daily heat output, expressed as MJ/m^2 of collector, has been computed from the actual test results on each collector, at an average solar radiation level of 14.4 MJ/m^2 day and atmospheric temperature of 15°C. Average daily efficiencies of converting solar radiation to energy in heated water are also shown.

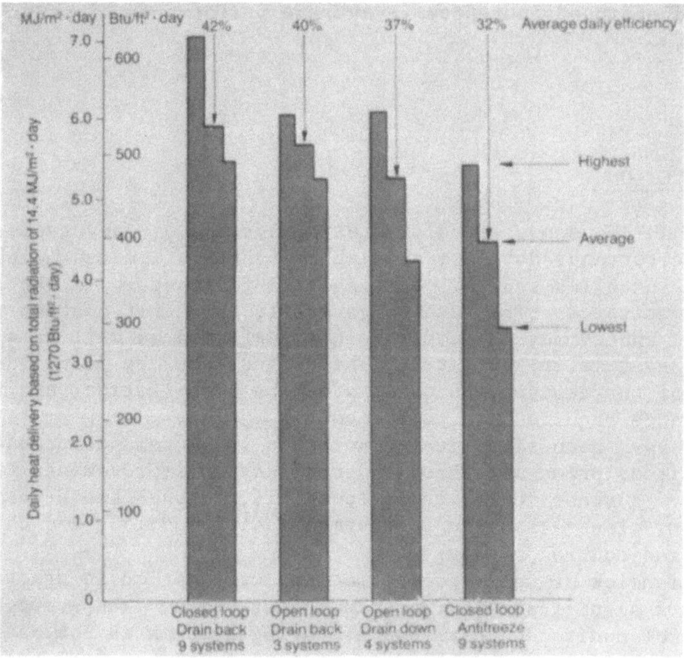

Figure 1 Performance of solar water heating systems tested by TVA (Ref 1)

The results show that solar hot water delivery from one system can be almost twice that from another of equal size. It is seen that the drain-down and drain-back systems employing water in the collector loop had higher efficiencies than systems employing glycol solutions and non-aqueous media in the collector circuit.

Although the foregoing tests show substantial differences in the average performance of these types, the number of systems may not be large enough for broad generalization, particularly when component variations are also present. But since the lowest output of the nine closed loop drain-back systems equals the highest output of the ten anti-freeze systems, a significant performance advantage of the drain-back system is evident. The somewhat lower outputs of the drain-down and open loop drain-back systems may not be significant differences, however, and the absence of heat exchange requirements should place their performance at levels comparable to that of the most efficient systems.

The results of field monitoring of residential solar water heating systems in Pennsylvania are shown in Figure 2 (2). A wide range of outputs is observed, the best systems delivering an average of 5 MJ/day m^2 collector. Ten percent of the systems provided an insignificant amount of solar heat, and another 30 percent supplied less than half the output of the best systems.

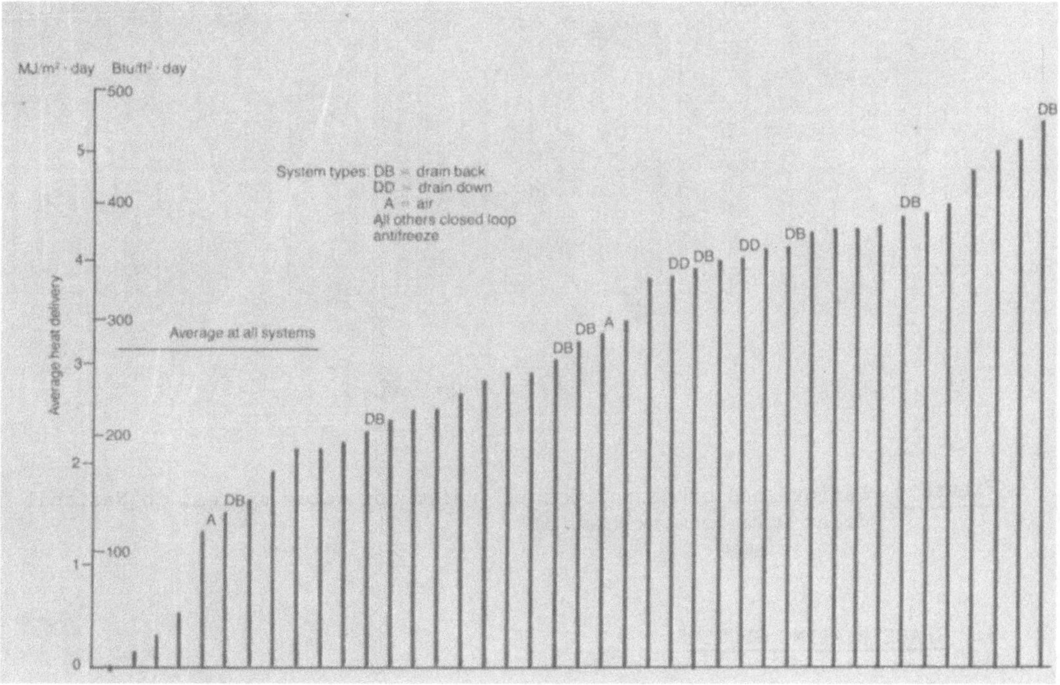

Figure 2 Performance of 41 domestic solar water heaters in Pennsylvania Field tests (Ref 2)

The performance of solar hot water demonstration systems in commercial buildings, shown in Figure 3, indicate a wide range of efficiencies and operational problems (3). Collection and delivery of 30 to 40 percent of the incident solar energy as hot water, seen in the best systems, indicates good design, construction, and operation. More than a ten-fold variation in system efficiency and hot water output per square foot of collector is seen, the best performers showing system efficiencies of 35 to 40 percent and average daily energy deliveries above 5 MJ/m^2, in sunny climates. Since system type does not appear to be a strong determinant of performance, variations are due mainly to quality of installation, control, and maintenance.

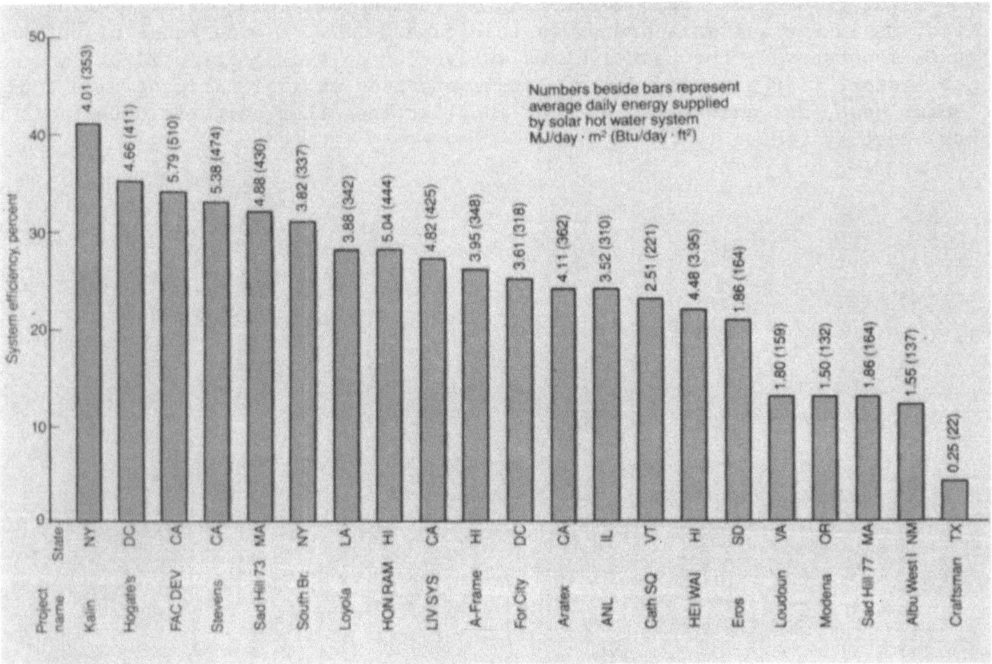

Figure 3 Performance of nonresidential solar hot water systems in National Solar Data Network (Ref 3)

3. SPACE HEATING SYSTEMS

Performance data on solar heating systems are from two principal sources. One is the monitoring program in the National Solar Data Network, in which over 100 installations in occupied new and converted buildings have been monitored for several years. Data on approximately 50 solar heating systems have been procured and reported during the past three years.

The other source of performance data is a much smaller group of installations in buildings located in a few institutions where research, development, and detailed data procurement have been under way. Much of this information comes from a 10-year program involving about 25 systems at Colorado State University.

From the extensive reports of the NSDN program (4,5), Tables I and II have been adapted. They show large differences in solar heat supply and in the percentage of available solar radiation actually collected and used in widely distributed residential and commercial buildings. Because of wide differences in climate, designer skills, quality of the installation, diligence of the building owner or systems operator, and other variables, system types cannot be quantitatively compared. In four installations having system efficiencies above 25 percent, air collectors were used in two, and liquid collectors with an anti-freeze solution and heat exchangers in two others. Clearly indicated is the relatively poor performance of many systems in comparison with the best. In virtually all cases, the systems were intended to operate at seasonal solar delivery efficiencies of 25% to 40%, so it is evident that faults in design, construction, and operation have often been encountered.

Average system performance was substantially better in 1981-82 than in the previous year. There were fewer poor systems, the average was half again better, the best system efficiency increased from a value of 25 percent to nearly 35 percent, and one fourth rather than about one tenth of the systems had efficiencies above 20 percent. A significant increase in average solar heat output and use is also evident. The best daily average levels increased from 3.7 to 4.7 MJ/m^2 collector.

The lack of generic superiority of any one type of system is also evident, although air systems finally appear to have demonstrated high efficiency in practical use as they had previously shown in experimental installations. The several liquid types appear to be of comparable capability.

Figure 4, based on performance of systems at CSU (Ref 6), shows seasonal solar collection and utilization efficiency of heating systems designed, installed, and operated by solar specialists in a closely controlled measurement program. The reduction and elimination of several constraints encountered in the NSDN sites makes possible (a) the achievement of performance approaching the maximum capability of the systems, and (b) reliable performance comparison between different types of solar heating systems.

Typical monthly and seasonal efficiencies of solar collection and delivery, December through March, are 30% to 40% of the incident solar energy. Energy required for system operation (for pumps and fans) has not been subtracted from the solar delivery values, so comparison with "system efficiencies" in Tables I and II requires a small adjustment. Seasonal operating energy for the air system was 7.3 percent of the solar energy supplied and used, and 10.0 percent in the flat-plate liquid system. The <u>net</u> solar usage (solar used minus operating energy) directly comparable to the values in Tables I and II is therefore 0.3 to 0.5 MJ/day m^2 lower than shown in Figure 4, corresponding to efficiencies about 2.5 to 3.5 percentage points below those indicated. The fraction of seasonal heating and hot water demand met by solar was in the 60 to 65 percent range.

Table I. Performance of Solar Heating and Hot Water Systems - 1980-1981 Season

Site	State	Collector Area, m²	System Type	System Efficiency %(1)	Net Energy Saved MJ/m² day (1)
Saddle Hill Lot 36	MA	29	HX-G	25	3.7
Oakmead Industries	CA	188	HX-G	21	3.8
Montecito Pines	CA	88	DB	18	2.9
RHRU Clemson	SC	36	Air	13	2.0
John Byram	KS	29	Air	13	2.0
Scattergood	IA	232	Air	13	1.8
M.F. Smith	RI	48	DD	12	1.9
Summerwood M	CT	36	DD	12	1.8
Design Construction	MT	74	DD	12	1.4
Lawrence Berkeley	CA			11	1.8
J.D. Evans A	MD	33	DD	11	1.6
First Manufactured #9	TX	27	Air	10	1.7
J.D. Evans E	MD	33	W	10	1.4
Helio Thermics #8	SC	39	DD	8	1.1
First Manufactured #10	TX	27	Air	7	1.2
Perl Mack	CO			7	1.1
Summerwood G (Attic)	CT	42	A-W	6	.9
Sir Galahad	VA	59	HX-G	5	.8
Matt Cannon	FL	55	W	5	.7
Helio Thermics #6 (Attic)	SC	39	Air+W	5	.6
Maulden Corporation	IN	65	Air	3	.4
Washington Natural Gas	WA	55	Air	3	.2
Brookhaven	NY	232	HX-G	2	.2
Average				10	1.5

(1) Based on total solar usage minus solar operating energy used
Note: All systems included solar preheating of domestic water
Air = air collection, pebble bed storage sytem
W = Water, no freeze protection
DD = Drain down, water collection
DB = Drain back, water collection
HX-G = Heat exchange, glycol solution collection

Table II. Performance of Solar Heating and Hot Water Systems - 1981-1982 Season

Name of Building	State	System Type	Collector Area, m²	System Efficiency, %(1)	Solar Load Fraction, %	Net Energy Saved, MJ/m² day (1)
Contemporary Systems	NH	Air	74	34.1	60	4.2
Oakmead Industries*	CA	HX-G	244	32.5	46	4.5
First Manufacturing home #9*	TX	Air	27	26.8	59	4.7
Karasek home*	MA	W-T	59	21.1	31	2.3
El Toro Library	CA	W-R	133	18.1	49	2.7
Cushing home*	UT	DB	42	17.7	42	2.2
Williamson home*	MA	Air	31	16.9	48	1.7
Honeywell-Salt River Project	AZ	HX-G	763	16.1	73	2.4
First Manufacturing home #10*	TX	Air	27	15.7	60	2.7
La Quinta Motor Inn*	NV	HX-G	111	14.5		2.7
Wood Road School*	NY	DB	2047	11.3	27	1.8
Crown Realty #1	CO	HX-G	84	7.6	20	1.3
Carter County Visitor Center*	KY	HX-G	100	7.4	5	.6
San Anselmo School	CA	W-R	347	6.5	25	.8
Scottsdale Courthouse	AZ	W-R	253	2.1	7	.2
Permalloy Corp.	UT	Air	332	1.9	8	.1
Average				15.6		2.2

(1) Based on total solar usage minus solar operating energy used
* includes solar preheating of domestic water
Air = air collector, pebble bed storage system
HX-G = heat exchange, glycol solution collection
W-T = water collector, trickle type
DB = water collection, drain-back storage
W-R = water collection, recirculated freeze protection

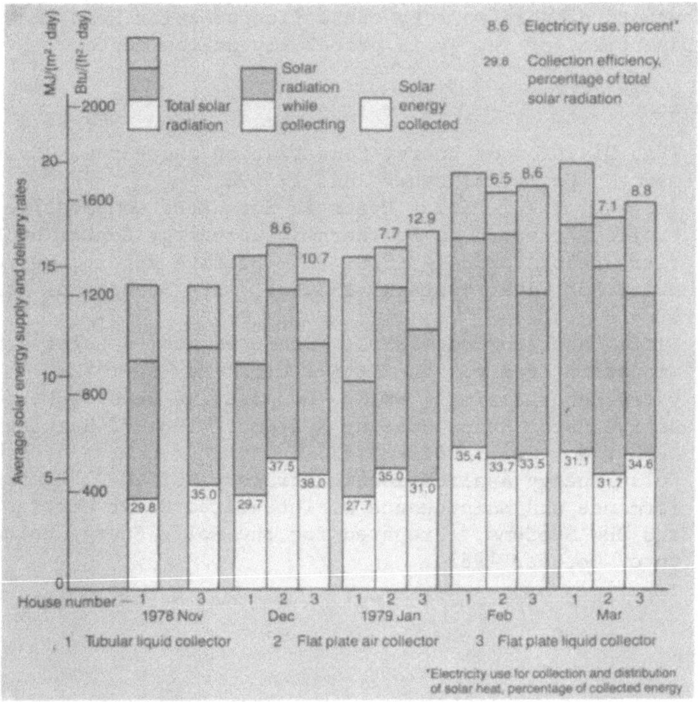

Figure 4 Performance of solar space heating systems in three solar houses at Colorado State University (Ref 6)

Seasonal efficiencies in an air system with pebble bed storage and in a liquid system with glycol collection and water storage are essentially identical. These measurements confirm the predicted similarity in performance of air and liquid solar heating systems of comparable design and construction quality. The evacuated tube collector used in House I was of a type which failed to show the higher performance attainable with other types of evacuated collectors.

Comparisons of performance measured in closely controlled systems and in systems constructed earlier in the demonstration program show that substantial improvements in the latter are possible. It also shows that attaining such performance requires engineering and construction skills and good system maintenance. Improvements that have been observed in the performance of practical systems designed, built, and operated most recently (1981-82 data), the best actually equaling the performance of closely controlled and monitored systems at CSU, indicate the potential of good solar design and operation. The results emphasize that differences between the several types of well designed and installed systems are considerably less than the differences in systems of the same type but of good and poor quality.

4. ACKNOWLEDGEMENT

The support of this work by the Office of Solar Heat Technologies of the U.S. Department of Energy is gratefully acknowledged.

5. REFERENCES

1. TVA, Division of Energy Conservation and Rates, "System Performance," Bulletin issues June 1, 198
2. Aungst, W.K., "Solar Domestic Hot Water Comparative Analysis Project," Report to Southern Solar Energy Center by Pennsylvania.
3. Vitro Laboratories - NSDN, "Comparative Report, Performance of Solar Hot Water Systems, 1980-81," DOE Report No. Solar/0024-82/41.
4. Vitro Laboratories - NSDN, "Draft of Active Solar Program Documentation Effort," VL-ESB-T-118(82) May 5, 1982.
5. Vitro Laboratories - NSDN, "Comparative Report, Performance of Active Solar Space Heating Systems," 1981-82 Heating Season, DOE Report No. Solar/0025-83/42.
6. Solar Energy Applications Laboratory, C.S.U., "Operation, Performance and Maintenance of Integrated Solar Heating, Cooling and DHW Systems," Prepared for the Solar Energy Research Institute, October 1981.

THE ACTIVE SOLAR BUILDING OF BOURGOIN-JALLIEU LESSONS FROM 5 YEARS OF MEASUREMENTS (1)

G. KUHN and P. PATAUD
Centre de Recherches sur les Très Basses Températures,
C.N.R.S., BP 166 X, 38042 Grenoble-Cédex, France.

Summary

General purpose of this work was to measure actual performances of a big active solar heating system and to test detailed computer models. Numerous and continuous measurements are not easy to achieve for an inhabited building and many improvements were necessary to get complete and precise energy schedules. The Bourgoin-Jallieu Solar Building is a very good example of first generation active solar systems, which has proved at least the reliability of the components and the interest of heating floors associated to solar collectors. The mean productivity of the vertical collectors is about 160 kWh/m^2. The thermal mass of the building, increased by the heating floors, provides good thermal inertia (one week effects) leading to a small auxiliary energy consumption. Sensitivity study to the size of the main solar components was also performed, showing as expected big dependence of collector area, relatively small influence of storage volume and no interest to use phase change material in this type of systems.

1. THE BOURGOIN-JALLIEU SOLAR BUILDING AND THE SOLAR SYSTEM

The project consists of a three storey block containing twelve flats, located in Bourgoin-Jallieu between Lyon and Grenoble (France). The plan is symmetrical with six flats in each half of the building. Total area of the flats is 1145 m^2. The South Face of the building is a curtain wall with forty bands of steel collectors, single glazed, with a total area of 305 m^2. The solar and auxiliary heating systems are also independent for each half. Artful design of internal balconies in the South Wall provides a good natural lighting and a consequent passive solar contribution in winter.

The solar system (fig. 1) is a classical first generation type. Heat storage is provided in six 10 m^3 tanks, three in each half, which can be used, or not. The stored solar energy is distributed to heating floors and a solar DHW preheating tank.

After six years, very few problems were encountered with the system. About 4 glazings (1 m^2) were broken by childrens. 1 collector (2 m^2) had to be changed in 1980 due to corrosion. Presence of glycol and lateral collector inlets are probably responsible, and this could now be avoided with more recent absorbers. No overheating occurs in the summer due to the vertical position of the collectors.

2. INSTRUMENTATION

The building is completely instrumented for weather data and energy balance measurements. Weather data include 3 Kipp Solarimeters (Global horizontal, Diffuse Horizontal, and Global Vertical), relative humidity, wind speed and direction and external temperature. 14 diaphragm flowmeters and

28 Pt 100 Ω thermometers serve for energy balances. Other Pt resistances follow storage tank or technical room temperature. A 64 channel data acquisition system can read or integrate the data for each scrutation.

First, data were stored on punched tape, but rapidly this shows his unreliability and big magnetic tape had to be used, giving 2 weeks of autonomy with six minutes measurement intervals.

Even these short intervals are in fact too long when precise analysis has to be made, for instance for energy balances in part of the solar circuits, as energies were not integrated and no information was available on pump working inside the intervals. Repeated calibrations were also necessary for the Pt sensors, as the temperature differences in solar circuits are often small. 1 to 4 d°C difference in collector circuits is common. Measurement currents have to be inverted in the Pt resistance circuit to take into account thermoelectric forces. Automatic battery auxiliary power supply had also to be installed, specially for power cuts due to summer storms.

To diminish measurement intervals, we use now an Apple II microcomputer, which analyses the data every 15 seconds, and mean or integrated results are stored every quarter of hour.

As the laboratory is 70 kms far from Bourgoin, a MODEM system was also installed and the results are available by phone ; so, rapid intervention can be done if necessary. Without daily control, sensors failure for instance, could only be detected after analysis of the magnetic tape, that means after some weeks.

To follow the global thermal behaviour of the building, we measured first auxiliary energies only weekly. This was no satisfactory and since the beginning of 1983, we measure, on cassettes, electric auxiliary energy every 10 minutes for all the apartments of one half building and global electrical input for the two sides. Internal temperatures are also measured in a central point of each apartment.

Very good information on what and how measure in solar buildings is now available (2) but this was not the case some years ago. Nevertheless precise continuous measurements remain expensive and laborious.

3. THERMAL BEHAVIOUR OF THE BUILDING AND INHABITANTS

Volumetric heat loss coefficient G was calculated and it was found G= 0.93 W/m^3K. Forced ventilation losses were included with about 0.8 air change per hour. This value was confirmed by experience.

We reported global energy used for space heating, for three years and the two half buildings in function of external temperature (Fig. 2). Very economical behaviour of inhabitants is observed in winter for cold periods and where generally small solar contributions are furnished. The theoretical curve corresponds to the energy necessary for the calculated G value, admiting a non heating temperature of 16°C.

Auxiliary energy for S.H. can be very different for the apartments, varying from quite zero to two or three times the mean value. But the sum for each half building are very similar, i.e. that a global value is significative for the six flats.

The experimental G value was tested on cold periods, without sun, and if we take at least 3 or 4 days, never values of G over 0.7 W/m^3K were found. More, after a sunny and mild period, a one week effect is observed, where no or few auxiliary heating is used, the building cooling down various degrees (fig. 3).

Even if we admit that the temperature measured in the apartments are not well representative of the coldest part, the low G values are difficult to explain. We will try to study this in a next future with the heavy buil-

South side of the building

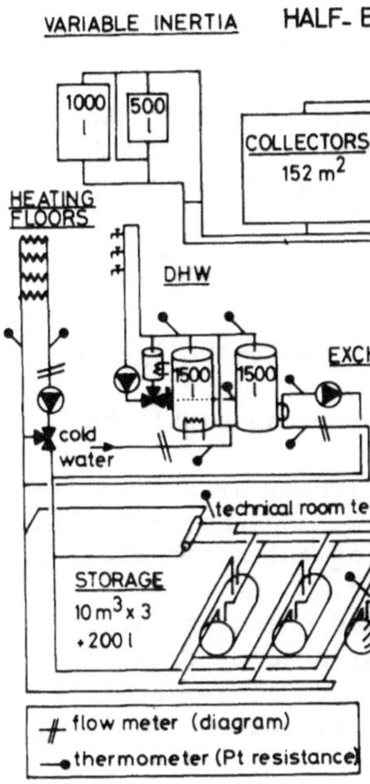
VARIABLE INERTIA HALF- BUILDING
1000 l
500 l
COLLECTORS 152 m2
HEATING FLOORS
DHW
HEAT EXCHANGER
1500 l
1500 l
cold water
technical room temperature
STORAGE 10 m3 x 3 + 200 l
flow meter (diagram)
thermometer (Pt resistance)

Figure 1

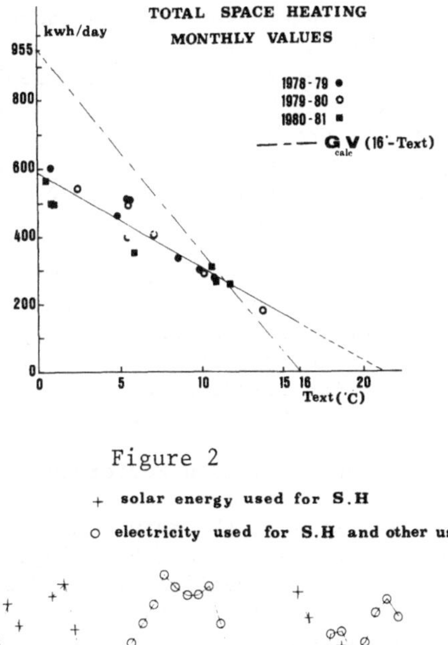
TOTAL SPACE HEATING
MONTHLY VALUES
kwh/day
955
800
600
400
200
0
1978-79
1979-80
1980-81
GV calc (16°-Text)
0
5
10
15 16
20
Text (°C)

Figure 2

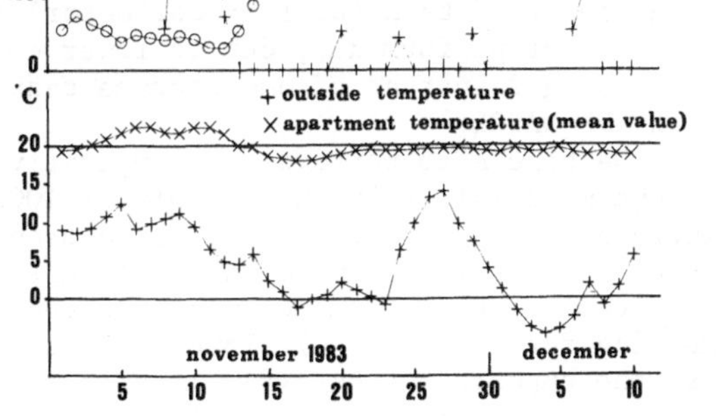
solar energy used for S.H
electricity used for S.H and other uses
kwh
300
200
100
0
°C
outside temperature
apartment temperature (mean value)
20
15
10
5
0
november 1983
december
5
10
15
20
25
30
5
10

Figure 3

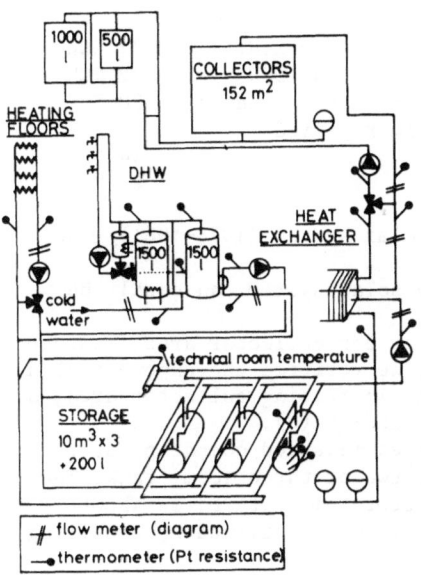

ding thermal load computer model CLIM of EDF. We have as yet used this model with climate of Lyon and fixed internal temperatures (3). We will now introduce measured meteorological data and internal temperatures. It is sure that simple energy need calculations, of linear type like GV (T_{int}-T_{ext}) cannot follow detailed variations of consumption, but even mean values seem to be well over-estimated in cold season.

4. PERFORMANCES OF THE SOLAR SYSTEM

Detailed results and comments have already been published (4,5,6). Mean performances for three years of measurements are given in table I . Productivity of the collectors are typical for the first generation solar systems. Vertical position of the collectors is not optimal, but lowers certainly installation cost, at least for multistorey buildings. Solar contribution to DHW is small, as during the heating season temperatures of the storage are generally below 30°C.

Many characteristics of system response have been identified, when modeling is undertaken. For instance a threshold incident energy has to be taken into account.

Incident solar energy has really to be measured in the plan of the collectors as big scattering exists for vertical global flux, even during nice days where horizontal flux follows astronomical previsions.

Solar system modeling was performed with TRNSYS and Electricité de France MASOL program. Main problems appeared with pipe losses, pipe inertia, and storage stratification (7).

. Solar Energy Used per Unit Collector Area for S.H.	131 kWh
. Solar Energy Used per Unit Collector Area for DHW	32 kWh
. Efficiency of the solar system for the heating season	30 %
. Efficiency of the solar system in summer period	5 %
. Solar fraction for S.H.	44 %
. Solar fraction for DHW	25 %

Table I : Mean Performances of the solar system (1978-1981)/year

The size of the main components of the solar system can easily be changed on the two half buildings.

The most significative studies were done for collector area, storage volume or material. For the collector area, we verified the exponential decrease of additional solar contribution with increasing collector area, as admitted in various simulation models, like TRNSYS (8).

Influence of the storage volume, has been measured comparing storage with one 10 m^3 tank (65 l/m^2 collector) and two or three 10 m^3 tank. The increase of performance, due to lower collector temperature, is about 10 % when using 2 instead of one storage tank. This was confirmed by modelling with EDF-MASOL.

We have also performed a comparative study of heat storage with phase change material (Polyethylene glycol 1000). The quantity of material (6 T), introduced in a storage tank, was chosen to have the same total energy storage capacity (between 20 and 60°) that the original water tank. Although improvement of performance was observed when working temperature was in the melting region (30-35°C), lower heat capacity, in solid or liquid phase, has the opposite effect (9). If this study was interesting on an academical point of view, it seems doubtful to recommend use of this phase change material in the condition of working of the Bourgoin-Jallieu Solar Building.

After six years, the Bourgoin-Jallieu Solar Building has proved the reliability of active solar systems. Performance of the system, in actual conditions of use, have been measured. The heating floors, associated to solar systems, proves their particular interest and inertia effects obtained are very favourable to low global energy consumption.

Many lessons were drawn for measurements and how to improve active solar systems. For instance, is intermediate storage in water useful or should we store energy directly in the heating floors ? This is one point we will study the next heating season.

REFERENCES

1. Main Participants :
   - Financing of solar system and experiment :
     Plan Construction, Ministère de l'Urbanisme et du logement, Avenue du Parc de Passy, 75016 Paris-Cédex, Tel. (1) 524.52.34. CNRS-PIRSEM, 282 Bd St-Germain, 75007 Paris, Tel. (1) 705.77.15.
   - Construction : S.A. d'HLM de Voiron et des Terres Froides, 6 rue Genevoise, 38503 Voiron, Tel. (76) 05.40.66.
   - Architects : Audrain and Geneve, 6 rue Beyle Stendhal, 38000 Grenoble, Tel. (76) 54.09.71.
   - Thermal System Conception :
     CET (M. Chavin) 47 Chemin de la Taillat, 38240 Meylan, Tel. (76) 90.62.18.
2. Monitoring Solar Heating Systems (A Practical Handbook). Commission of the European Communities (CEC), Pergamon Press 1983.
3. Solar Houses in Europe - How They Have Worked, CEC, Pergamon Press 1981.
4. Solar Space Heating, CEC, Doc. N° EUR 8004, 1982.
5. G. KUHN, P. PATAUD, Trois Ans de Bilans à Bourgoin-Jallieu - Les performances expérimentales des installations solaires à capteurs plans, Colloque ESIM, Sophia-Antipolis, Dec. 1981.
6. B. RUEL et G. KUHN, CLIM et MASOL à Bourgoin-Jallieu - Bilans simulés et mesurés, Bâtiment-Energie N° 27, June 1983.
7. A. CORDIER, A. CLEMENT, F. FLOUQUET, N. GALANIS, G. KUHN and P. PATAUD Simulation of the Bourgoin-Jallieu Solar Building with TRNSYS. International Congress on Building Energy Management, Portugal (1980), Pergamon Press.
8. K. CHANG and A. MINARDI, Solar Energy 24 (1979) 99-103.
9. C. AUBERT-DASSE, G. KUHN, P. PATAUD, J.L. CHEVALLIER, P. EURIN, S. SALLEE Experimental and Theoretical Comparison between Large Phase Change Material Storage and Water Storage, ICBEM, Portugal (1980) Pergamon Press.

PERFORMANCE MONITORING OF SOLAR HEATING SYSTEMS

R FERRARO and R GODOY
(on behalf of the CEC Performance Monitoring Group)
Energy Conscious Design
11-15 Emerald Street
London WC1N 3QL
England
(Tel: 01-405 3121)

MEMBERS OF THE PERFORMANCE MONITORING GROUP (1982/83)

Co-ordinator	- Energy Conscious Design, London, UK
A Debosscher	- Katholieke Universiteit Leuven, Belgium
P E Kristensen	- Thermal Insulation Laboratory, Lyngby, Denmark
G Kuhn	- CNRS - CRTBT, Grenoble, France
U Luboschik	- IST Energietechnik GmbH, Germany
J Owen Lewis	- University College Dublin, Ireland
F Cecchi Paone	- Fidimi Consulting Spa, Rome, Italy
D Brethouwer	- Institute of Applied Physics (TNO), Netherlands

Persons responsible at the Commission: T C Steemers and W Palz
Scientific adviser to the Commission: C den Ouden

Summary

The work has been carried out in three phases by a coordinated group of European experts, called the Performance Monitoring Group (PMG), 100% funded by the CEC. The aim has been to collect data from monitored solar heating systems in Europe and draw conclusions about system performance, and this has resulted in a series of publications including data from over 100 monitored solar projects (1,2,3,4,5,13,15). To achieve this, standard Reporting Formats have been prepared and distributed to allow meaningful comparisons between different monitored projects (7,8,9,10,11,12). Finally, a detailed understanding of system design and construction has been obtained, enabling improved designs of 'second generation' systems to emerge. Data from suitable monitored solar projects has been collected in each European community, via the local PMG representatives, and analysis work has been carried out by the coordinator, under the scrutiny of the Group. In the analysis, attention has been paid not only to overall performance achievement, but also to more detailed aspects of design, construction and cost. In the case of space heating systems (active and passive) attention is also given to the design of the houses themselves. Considerable work has been done on the improvement of monitoring techniques, and detailed guidelines with recommendations for standard 'levels' of monitoring (i.e. degrees of complexity) have been published (6).

The results show that significant improvements are possible in system design and construction, likely to result in higher outputs per unit collector area and/or lower capital costs. This formed the basis of a related work programme carried out by the PMG and presented separately in session 5B (14).

1.0 INTRODUCTION

The work of the CEC Performance Monitoring Group (PMG) has been executed under Project A of the CEC Solar Energy Research and Development Programme. Within Project A there have been a number of 'concerted actions' undertaken since the start of the programme in 1979, related to solar energy applications in dwellings. This work forms one such action.

The PMG has drawn conclusions about the likely benefits to be gained from incorporating solar heating systems in different parts of the Community. At the outset, work was initiated whereby performance data and detailed design information could be collected and analysed from existing active and passive monitored solar projects (1,2,3,4,5,13,15). These data collection exercises result in a comprehensive overview of different system types, system configurations, specifications, integration, performance aspects and costs, leading to the establishment of Design Groups capable of producing system designs with improved performance and reduced overall costs. (N.B. This latter aspect is covered in a related paper in Session 5B. See also ref. (14)). The present paper deals with the collection and analysis of data and information from built projects monitored to date. At the same time, the PMG undertook the task of documenting detailed recommendations on the subject of monitoring procedures, practice and degrees of complexity (6).

1.1 Aims and Objectives

In summary, four main objectives were established concerning performance monitoring aspects. These were:

- To produce Reporting Formats to enable uniform recording of system parameters, costs and performance results from different types of system.
- To report on the performance of domestic solar heating systems in the European Community including active space heating, passive space heating, and domestic water heating systems.
- To produce guidelines for monitoring solar heating systems to ensure uniformity of approach, reliability and accuracy of results.
- To establish criteria for the selection of better optimised field trials and to make recommendations on the suitability of different system types based on performance results.

1.2 Method and Phases of Work

Throughout the project, the work has been undertaken collaboratively by a group of European experts, one institute per country, with the Coordinator responsible for constructing the work programme and reporting to the Commission. The work has been undertaken in three phases.

Phase 1: (79-80):

Collection and analysis of data from 20 projects, mostly active solar space heating systems.

Phase 2: (80-81):

The experience gained in Phase 1 was used to considerably improve the methods of reporting and analysis, and the exercise was extended. Data from some 60 projects was reported, with a greater emphasis on solar assisted solar water heating systems.

Phase 3: (82-83):

The process of collecting and analysing data from monitored solar heating projects was continued in this phase of the work, with special emphasis on passive solar space heating projects. Some 17 projects were reported, including five passive projects, several of which involved multiple housing units. Selected projects were reported in greater depth.

During Phase 3 of the work, the PMG tried to limit the number of projects to those which were particularly promising in terms of performance or technical design. This aim could only be achieved in part, because few better optimised projects are yet being monitored; this is partly due to the long time periods required for a project to yield reliable data.

Each member of the PMG was responsible for identifying suitably well-monitored projects within their own country and arranging for the data to be supplied. The projects were to be monitored at a suitable level for supplying the correct amount of data and were also to be capable of supplying monthly performance figures, preferably for one year of operation but, at the very least, six months.

All the projects reported have been monitored by individual research groups, nearly always with financial support from their respective National Governments.

2.0 BACKGROUND TO PERFORMANCE MONITORING

It should be appreciated that few experimental solar heated buildings existed in Europe prior to 1974. The first systems were generally over-complicated and over-sized, and little attempt was made in the early stages to optimise the designs from either technical or economic points of view. Monitoring of projects was carried out to different standards and with varying degrees of success, there being a tendency to collect excessive quantities of data. Data was often not processed or totalised into useful quantities until months or even years had passed after collection. This frequently resulted in wastage of money and time, as system malfunctions that could have been rectified often went unchecked. Similarly, improvements in performance that could have resulted from relatively minor adjustments or improvements were also frequently not realised, as experimenters struggled with their complex monitoring equipment and data. All of these factors contributed to a slow process of learning and development, which characterised the field in the late 1970s, and in which context the work of the Performance Monitoring Group began.

2.1 PMG Reporting Formats

Over the period of its work, the PMG has produced a set of six published Reporting Formats for recording the performance of monitored solar heating systems (7,8,9,10,11,12). These have been designed to obtain data in a directly comparable form from projects monitored by different teams in different countries. The Reporting Formats are in two sets of three: one set for small projects with high level monitoring (e.g. individual houses); one set for projects involving multiple housing, monitored at low level (e.g. pilot field trial projects). Each set covers three distinct solar system types:

- Active Space Heating and Water Heating Systems (combined)
- Domestic Water Heating Systems
- Passive Space Heating Systems

The Reporting Formats contain sections for description of the project, climate, the building, solar heating systems, system operation, monthly performance, annual performance, detailed performance data, cost information and conclusions. Additionally, there is a summary sheet in each case.

The Reporting Formats have been used successfully by the PMG to report data from monitored projects and they are available to individuals and organisations involved in the monitoring of solar heated buildings.

Indirectly they can be used to help at the planning stage of a monitoring programme, as minimum data requirements have been carefully thought out. Each Reporting Format includes a comprehensive User Guide.

(N.B. Translations in French of the PMG Reporting Formats for projects with high level monitoring have been prepared by the Agence Francaise pour la Maitrise de l'Energie (AFME)).

2.2 Improvements in Monitoring

To provide back-up to the Reporting Formats and to provide guidance in the monitoring process, the PMG has produced an easily used book on the subject, covering detailed aspects of planning, data requirements, instrumentation, data recording and transmission, data treatment and data storage (6).

This Monitoring Handbook sets out the process of defining and setting up a monitoring system from its inception to completion and commissioning, and covers solar assisted active space heating, passive space heating and domestic hot water systems. It is generally assumed that monitoring will occur in occupied dwellings. Three major stages in monitoring are identified and discussed: a pre-monitoring diagnostic stage, the actual performance monitoring stage, and post-monitoring requirements. In addition, 'one-time' measurements are identified and methods for obtaining them described.

The pre-monitoring period is to discover if the installed system components and the control equipment are functioning as designed. The monitoring equipment must be adequately checked and on-site calibration may have to be carried out. During the actual performance monitoring period, the objectives are to find out how the system functions in practice in comparison with its predicted behaviour, and to evaluate variations. During this period adequate and prompt data treatment is essential to discover faults without delay (faults of the data acquisition system or of the solar installation) and to understand the operation of the solar heating system. Practice has shown that it is essential to avoid backlogs of untreated data, which can become impossible to analyse after memories of events have subsided. The post-monitoring phase is to regularly check the reliability of the system and to establish the long-term behaviour of major components.

3.0 RESULTS

Over the last four years, the PMG has collected and analysed data from over one hundred monitored solar heating projects. Half of these are active space heating systems, including most types of designs, the remainder being solar water heaters (34%) and, more recently, passive space heating systems (16%). Apart from obvious selection criteria such as the need for a project to produce sufficient reliable data, and more recently improved performance, the proportions of different project types selected have relied very much on availability. It is noteworthy that there has been a recent increase in the number of passive projects available for study. The projects extend from 38° to 57°N and include most European climates, ranging from maritime to more continental areas.

The resulting information, which has been published by the CEC is thought to represent a comprehensive appraisal of existing solar heating systems and a base for the development of improved system designs.

3.1 Performance, Cost and Cost-effectiveness

The main performance figure relating to a solar heating installation is the amount of solar energy used per annum per unit collector area, and is shown in Fig. 1.

Overall system efficiency, defined as the ratio of solar energy used to that incident on the collector over the same period, is another important performance parameter. Performance figures show efficiencies of up to 30% (active), 64% or 35% for passive systems (depending on whether back losses through the collector are discounted from the gains or not) and 56% for solar water heaters.

MJ p.a./m^2			AVERAGE	Maximum	Minimum
PHASE III:	ACTIVE SPACE HEATING		571	1,288	126
	PASSIVE SPACE HEATING				
		Direct Gain (1)	835	1,096	494
		Direct Gain (2)	202	655	228
		Trombe Wall (1)	496(3)	na	na
		Trombe Wall (2)	505(3)	na	na
		Sun Space (1)	279(3)	na	na
		Sun Space (2)	206(3)	na	na
	DOMESTIC HOT WATER		1,070	2,234	387
PHASE II:	ACTIVE SPACE HEATING		560	1,195	291
	PASSIVE SPACE HEATING:				
		Direct Gain	na	na	na
		Trombe Wall (1)	290(3)	na	na
		Sun Space	na	na	na
	DOMESTIC HOT WATER		820	1,551	249

(1): measured useful solar conbtribution
(2): estimated useful energy balance of solar collector, adjusted for missing days data
(3): one project only; max and min not applicable (na).

Fig. 1 Annual performance summary: phases II and III

The main cost figure relating to a solar heating installation is the total capital cost per unit collector area, and information collected indicates that prices have not changed significantly. These are on average 618, 400 and 50 to 300 ECU/m^2 for active, hot water and passive solar heating systems respectively. It must of course be remembered that the projects studied were designed several years ago, and work underway at the present time indicates that significant cost reductions are to be expected.

Most passive designs provide other benefits in addition to energy savings, which occupants may value as much or more than the auxiliary energy savings. This will affect the way in which the costs and benefits of passive solar heating are calculated.

Measured hot water load (approx. 27 to 45 litres per day per person) is lower than normally assumed for design purposes, and this has important implications for system design.

Significant developments are being made in the cost effectiveness of solar heating systems, and up to 2.01, 9.39 and 6.11 MJ of useful solar energy delivered annually per unit cost (ECU) are reported for active, passive and solar water heating systems respectively. The average performance-to-cost (PTC) ratio has improved from earlier projects, but further improvements are required and could be achieved to arrive at shorter pay-back periods.

3.2 Factors affecting performance

- The main climatic factors affecting the performance of solar heating systems are: Incident solar radiation (DHW, A-SH and P-SH systems); and Degree days (A-SH and P-SH systems).
- The performance of solar heating systems increases with both quantities, if all other variables remain unchanged.
- The annual load per unit collector area is the main design parameter and all solar heating systems should be designed in proportion to the load. Large values (greater than approx 2,000 MJ pa/m^2) indicate undersized collector arrays and result in large savings per unit area of collector, better cost effectiveness, but lower total energy savings.
- Heat storage and piping should be well insulated to avoid excessive heat losses, and pipework should be kept as short as practicable. Measurements prove that actual losses are often two or three times bigger than indicated from manufacturers figures, although storage losses as low as 3 to 5 W/m^3K have been reported.
- Some mechanical components tend to break-down or fall out of adjustment and therefore complex systems are more prone to develop problems. Simple systems should be used where possible. Many existing components for space heating systems are oversized and inadequate for use in low energy houses.

4.0 CONCLUSIONS

The following summarises the overall conclusions and recommendations that have resulted from the PMG data collection exercises:

- Significant improvements are being made in designing more cost effective systems and the best projects reported are cost effective now, particularly in areas where lower cost fuels are un-available. Where low cost energy is available, further developments are required to improve cost-effectiveness. (N.B. Other benefits derived from solar heating projects, such as improved standards of comfort, are recognised by designers and occupants, but are difficult to evaluate in economic terms.)
- The design and development of better optimised solar heating systems should be accelerated, with a view to obtaining practical designs of solar heating projects which achieve annual savings of 5 to 7 MJ/ECU (approx. 1.4 to 1.9 kWh/ECU). It is essential that existing experience, as reported by the PMG, be taken into account in this process, so that mistakes are not repeated, and more rapid development occurs.
- Committed companies with appropriate track records in the field, should participate in this process, so that economies of scale can be achieved. Without these economies, cost-effectiveness will not be achieved so rapidly. Joint work by research organisations and industry, undertaken under an adequate organisational structure, may be particularly productive in this area.

(N.B. PMG work in the development of more cost-effective systems is reported in Session 5.B).

References

1 Solar Houses in Europe: How They Have Worked. Commission of the European Communities, edited by W Palz and T C Steemers. EUR 7109. ISBN 0 08 026744 0. 1981.

2 Solar Houses in Europe (Slide Package). Commission of the European Communities. EUR 7344. 1981.

3 Performance Monitoring of Solar Heating Systems in Dwellings: Executive Summary and Recommendations. Commission of the European Communities. EUR 8002. 1982.
4 Solar Water Heating: An analysis of design and performance data from 28 systems. Commission of the European Communities. EUR 8003. 1982.
5 Solar Space Heating: An analysis of design and performance data from 33 systems. Commission of the European Communities. EUR 8004. 1982.
6 Monitoring Solar Heating Systems: A Practical Handbook. Commission of the European Communities. EUR 8005. ISBN 0 08 029992 X. Published for the CEC by Pergamon Press Ltd. 1983.
7 Reporting Format for Solar Heating Systems (for projects with high level monitoring): Active Space Heating and Water Heating Systems (combined). Commission of the European Communities. EUR 7785. 1981.
8 Reporting Format for Solar Heating Systems (for projects with high level monitoring): Domestic Water Heating Systems. Commission of the European Communities. EUR 7786. 1981.
9 Reporting Format for Solar Heating Systems (for projects with high level monitoring: Passive Space Heating Systems. Commission of the European Communities. EUR 7787. 1981.
10 Reporting Format for Solar Heating Systems (for projects with low level monitoring): Passive Space Heating Systems. Commission of the European Communities. EUR 8944. 1983.
11 Reporting Format for Solar Heating Systems (for projects with low level monitoring): Active Space Heating and Water Heating Systems (combined). Commission of the European Communities. EUR 8945. 1983.
12 Reporting Format for Solar Heating Systems (for projects with low level monitoring): Domestic Water Heating Systems. Commission of the European Communities. EUR 8946. 1983.
13 Solar Heating: Performance of Recent Systems. Commission of the European Communities. EUR 8947. 1983.
14 Solar Heating: Performance and Cost Improvements by Design. Commission of the European Communities. EUR 8948. 1983.
15 Solar Heating: Performance Monitoring 82/83 - Executive Summary and Recommendations. Commission of the European Communities. EUR 8949. 1983.

FIELD MEASUREMENT ON 65 SINGLE FAMILY SDWH SYSTEMS THROUGHOUT FRANCE

A. FILLOUX - D. BIENFAIT - S. SIINO
CENTRE SCIENTIFIQUE ET TECHNIQUE DU BATIMENT
BP. 21 06562 VALBONNE - FRANCE

Summary

An investigation of the performance of Solar Domestic Water Heaters (SDWH) was conducted on the overall French territory.
A total of 65 SDWH were examined for this study. Nine of these were selected for in-depth evaluation because of the availability of long term data.
The systems were monitored at 15 minutes time interval over a period of 9 months and their performances analysed.

1. INTRODUCTION

Among the thermal Solar Energy application of interest in the near future, Solar Water Heaters are very likely the most important for European Countries.
The objective of the work wich was supported by the "Agence Française pour la Maîtrise de l'Energie" was to determine the thermal performances of presently manufactured Domestic Solar Water Heaters (DSWH) in order to schedule the ability of the systems and to settle a data bank for simulation programs.
Two major types of data were collected (Figure 1)

1) overall data, collected from heat meters and flow meters wich were monthly monitored. (1) (2)

2) fine data, from temperature sensors, power meters and flow meters, monitored with magnetic tape recorder. Every 15 minutes, the magnetic tape records pulses from pulse generating meters and electric measurements from temperature sensors.

The weather data, mainly insolation, was obtained by the French Meteorology Agency from specific sites.
The solar hot water systems selected for this paper were installed in 1982.
They consist on solar water heaters with forced primary circulation and electric auxiliary.
The collector areas are 3 and 4 m² with respectively 200 l and 300 l storage tanks.
The collector are south tilted and their inclination angle is 45 degres.

2. PERFORMANCES

Three factors have been considered essential for the performance evaluation of Domestic Solar Water Heaters.(3)

They are :

a) The solar fraction of total water load
b) The conventional energy saved by unit collector area
c) The conventional energy consumption required for water heating.

Others parameters are :

- Hot water load
- Average hot water load temperature

2.1 Solar Fraction of total water load

The solar fraction of hot water load (F) is calculated by means of the booster consumption. It can be equated in the following manner.

$$F = 1 - \frac{\text{C Auxiliary (kWh)}}{\text{C electric (kWh)}}$$

Where C electric is the electric energy required to operate the water heater when the solar fraction F is equal to 0. The basic approach is first to measure the total booster energy and to calculate the electric energy consumed by the solar system to provide a given hot water load at a given temperature level, when there is no energy transfered from the collector to the tank.

C electric has been computed by the calculation program "CODE OSOL" wich has been developped by the C.S.T.B. for the simulation of solar water heaters. (4)

2.2. Conventional energy saved by unit collector area

The conventional energy saved by unit collector area P is related to the Auxiliary consumption as follow :

$$P = \frac{\text{C electric (kWh) - C Auxiliary (kWh)}}{\text{Collector area (m}^2\text{)}}$$

Where C electric and C Auxiliary are defined as previously.
P represents the energy saved by the water heater, including the losses by the pipes and the tank, the collector efficiency and the heat exchanger effectiveness.

2.3. Conventional energy consumption required for water heating

The conventional energy consumption can be related to the hot water load by a Coefficient of Performance C.O.P.
The C.O.P. is the ratio of the energy consumed in the water load to the conventional energy entering the system.

$$\text{C.O.P.} = \frac{\text{C water (kWh)}}{\text{C electric (kWh)}}$$

2.4. Results

The result of the calculation concerning 9 Solar Hot Water System installed in Cannes is presented in Table I.

The figure 2 presents the variation of the Auxiliary consumption with the total water load. It clearly appears that the electric energy consumed for water heating is related, with some deviation to the water load.

But in a general case, one can notice from figure 3, that larger water loads favour higher productivities.
A particular exemple is number 9.
The bad performances observed during the monitoring are mainly due to the behaviour of the inhabitants and the high set point of the electric temperature set control (75°C).

An illustration of the seasonal variation in performances is given in table 2. It clearly appears that the solar energy contribution in the solar hot water is larger in the summer.

Thermal performances simulation

The recorded data has been correlated with calculated values.
The program (CODE OSOL) allows the simulation of the thermal performances of a solar water heater for a one year period, and a one hour time interval.
The program input datawere the technical and site characteristics of the installations with a conventional insolation data and a conventional draw-pattern as driving functions.

For every installation , the monthly water load was based on the montlhly data recorded from the flow-meters. Figure 4 presents the comparative results of the simulation for the most representative installations, according to the duration of the monitoring.

Taking into account the accuracy of the measurement, procedure and correlation, the results fit quite well with the theoretical calculation method.

As an average, the actual measured performances are 10 per cent lower than the calculated performances ; this is due to the controller efficiency, the size of the sample (9 SDWH) and the conventional insolation data and draw-pattern.

In a further work, these differences will be reduced by entering the simulation program with insolation data and draw-pattern corresponding to the conditions of the on-site monitoring.

3. CONCLUSION

Working solar energy hot water systems have been monitored. Nine of them were scanned with a short time interval.
Most solar water heaters have significant energy saving. In general, saving to load ratios were 0,42 to 0,86, and coefficients of performances were found to be into the range of 1,4 to 4 (for the south of France).

The performances are inhabitants dependant. A higher level temperature set control leads to lower performances.

For a nine months period, including summer time, the mean productivity can be related with some deviation to hot water load.
Attempt to predict the installation performances was successfull. The main factors limiting correlation are insolation data and corresponding draw-pattern.

	Collector area m²	F	COP	P(kWh/m²) (nine months)
DWH 1	4	0,82	3,4	656
DWH 2	3	0,73	2,2	500
DWH 3	4	0,85	3,8	342
DWH 4	4	0,84	4,0	363
DWH 5	3	0,86	3,1	387
DWH 6	3	0,42	1,4	239
DWH 7	3	0,71	1,9	569
DWH 8	3	0,58	1,7	655
DWH 9	3	0,18	0,9	95

TABLE I : Thermal performances of nine DSWH monitored for a 9 months period

	DSWH 1	DSWH 2	DSWH 3	DSWH 4	DSWH 5	DSWH 6	DSWH 7	DSWH 8	DSWH 9
MARCH	0,76	-	-	-	-	-	0,61	0,62	0,40
APRIL	0,82	0,56	0,89	-	0,83	-	0,57	0,36	0,00
MAY	0,92	0,70	0,82	0,93	0,86	0,54	0,73	0,63	0,22
JUNE	0,93	0,81	0,99	0,96	0,98	0,74	0,81	0,69	0,26
JULY	0,99	0,84	0,63	1,00	0,99	0,95	0,92	0,88	0,28
AUGUST	0,97	0,99	1,00	1,00	0,99	0,96	0,92	0,88	0,43
SEPTEMBER	0,90	0,98	1,00	0,99	0,97	0,86	0,87	0,77	0,31
OCTOBER	0,68	0,89	0,99	0,97	0,91	0,61	0,76	0,64	0,00
NOVEMBER	0,69	0,47	0,78	0,74	0,70	0,44	0,59	0,42	0,18
DECEMBER	0,64	0,49	0,74	0,62	0,67	0,28	0,45	0,16	0,00

TABLE II : Monthly variation of the Solar Fraction of total water load

REFERENCES

1. K.A.REED - Instrumentation for thermal performance measurements : striving for measurement assurance in solar collector testing SOLAR ENGINEERING - 1982 p. 337

2. SOLAR WATER HEATING - An analysis of design and performance data from 28 systems - CEC - 1981

3. J.T. PYTLINSKI - Method of on-site testing for performance rating of Solar Water Heating Systems.

4. D. BIENFAIT - Modélisation thermique de chauffe-eau solaires - CODE OSOL - Document CSTB - Décembre 1982

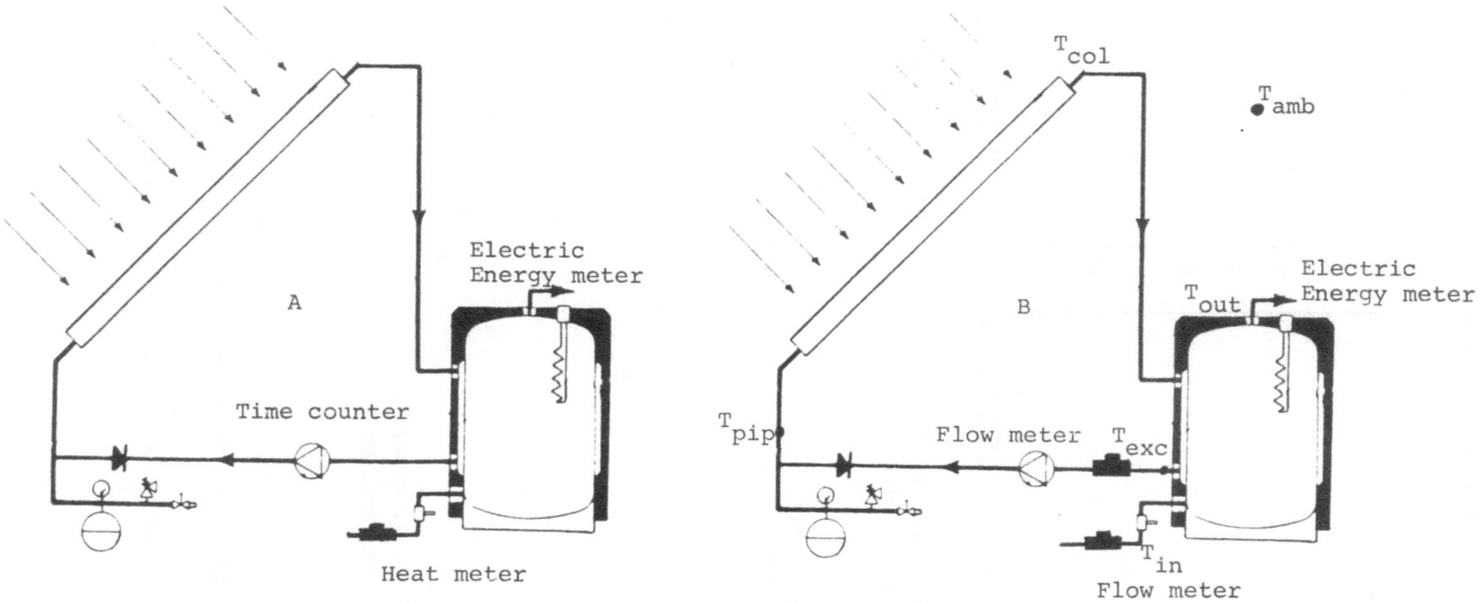

FIGURE 1 : Installation Instrumentation
(A) global instrumentation
(B) Analytical instrumentation

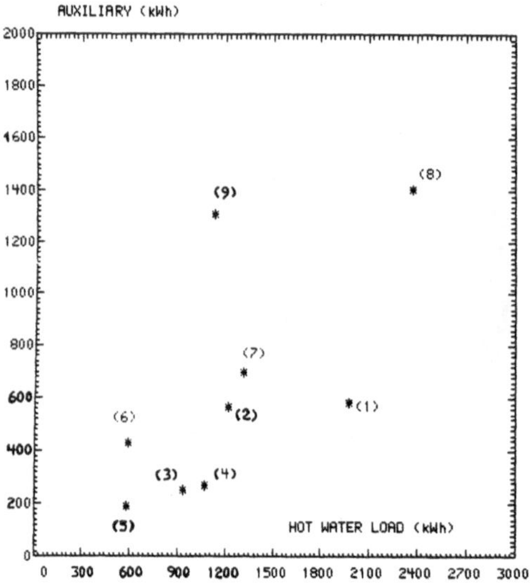

FIG 2: Auxiliary consumption versus total water load

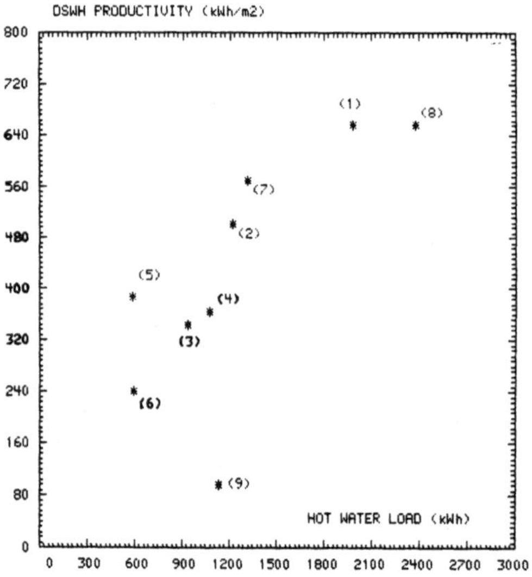

FIG 3: DSWH productivities versus total water load

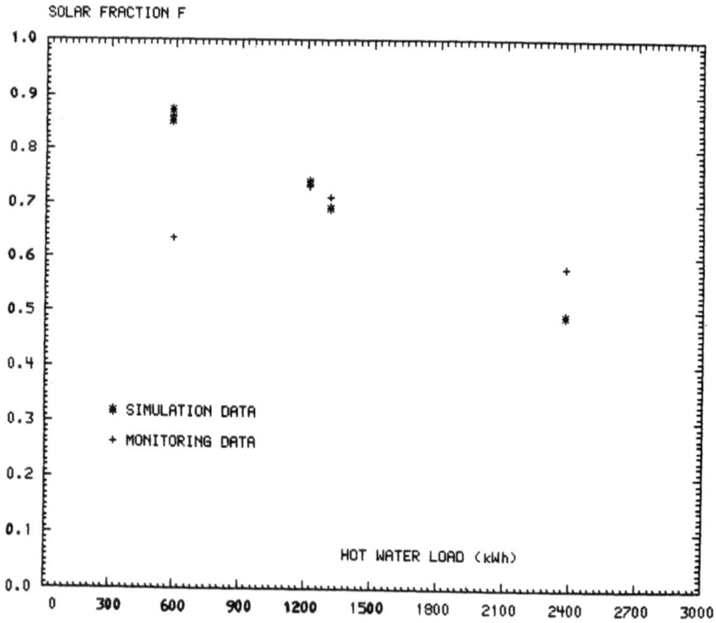

.FIG 4: DSWH thermal performances simulated by CODE OSOL

EUROPEAN PASSIVE SOLAR TEST FACILITY

Baker N V
Energy Conscious Design
11-15 Emerald Street
London WC1N 3QL

Summary

The paper describes the main considerations and conclusions drawn from a pre-design study for a European Passive Solar Test Facility. The main conclusions are that there is a need for an outdoor test facility providing standardised and easily specified, but realistic test environments. Two main purposes are identified; firstly to provide data sets for mathematical model validation and development, and secondly, for the performance testing of passive solar components.

1. INTRODUCTION

At a meeting of the Passive Solar Working Group (PSWG) in December 1982, it was suggested that there is the need for a physical test facility for passive solar components and systems, to be coordinated at a European level. This would be partly for the evaluation of passive solar components such as those being currently developed with CEC funding, and partly to assist in the development and validation of a European Model for passive systems.

The next four year Solar Energy R & D programme for the European Community will commence in 1984. Experience in the previous programme with the setting up of the Pilot Test Facilities (PTFs), to test active solar components at a number of European sites, had indicated that projects of this type require major organisational and technical effort. Thus it was considered advantageous to have plans well advanced before the commencement of the next four year programme.

For the purpose of discussion, an outline proposal for a test cell was made by DGXII and comments from members of the PSWG were invited. At the same time a pre-design study was commissioned.

The study involved consultation with experts on both physical testing and the use of test data. The results of that study have been published in full as internal report to DGXII (1). The paper presented here summarises the main conclusions of the study, and illustrates some of the more important considerations which lead to these conclusions.

We would like to thank all those people with whom we have discussed this project. The range of views about physical testing has prevented us from explicitly incorporating many of their views in our proposal. However, their comments have contributed much to the overall balance of the proposal.

2. MAIN CONCLUSIONS

The main conclusion of the study was that there is a need for highly standardised passive solar test cells in Eúrope, designed to provide a realistic and clearly definable test environment for systems and components.

Most experts consulted agreed with this broad conclusion, but there was and probably still is some divergence of opinion of the form that the test facility should take.

We identified two main purposes for the physical testing of passive components and systems. Firstly, to provide reliable data sets for the purpose of validating mathematical models. These models could then be used directly as design tools or may subsequently produce generalised solutions which would be used as design guidelines. Secondly, physical testing could provide direct information on the performance of passive solar components (ie physical elements such as glazing, shutters, storage systems controls etc).

As well as the two main purposes, we identified two main categories of experimental work. Firstly, that which takes place in highly specific test environments where boundary conditions are grossly simplified in order to have a small range of variables. These facilities are likely to be indoor, since the realistic and complex combination of weather parameters is not required. Typical examples of such facilities are hot boxes, convection study rooms, wind tunnel models etc. An important function of this kind of testing is the measurement of component parameters which will form the input data to mathematical models of systems incorporating these components.

The second category of physical testing is where far more complex boundary conditions are allowed, but at the same time these have to be easy to specify with a high degree of accuracy. Where comparisons have to be made such as in component performance testing, then the parameters of the test environment should be highly standardised. This category would generally require outdoor 'real' weather conditions, since the simulation of all relevant weather parameters at large scale is not economically feasible.

These two inter-linking streams of work are illustrated in Fig. (1).

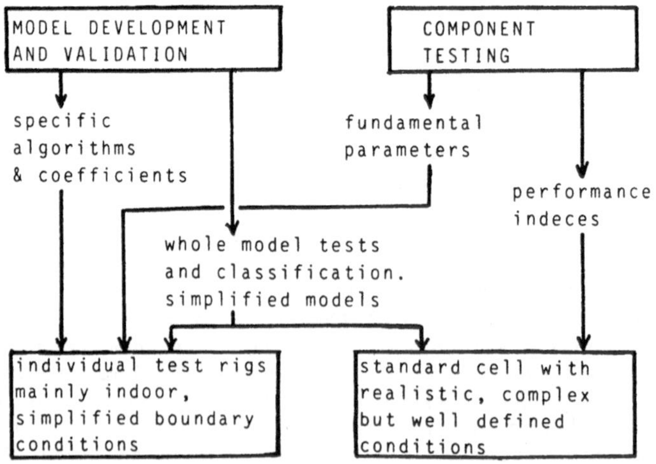

1 *Physical testing strategy*

3. PHYSICAL CONSIDERATIONS

In the study we took some time reviewing the important physical processes occuring in a passive solar building. This helped us to make recommendations for the actual physical form of the test facility, the most important of which we outline here.

Two classes of physcial processes in passive solar buildings can be identified - (i) strongly climate-sensitive processes and (ii) weekly climate-sensitive processes. Processes which occur between zones are generally of the second kind, and can be studied with indoor test rigs. Thus outdoor cells can generally be single zone, provided certain technical measures are adopted, except where an unheated sunspace is being tested.

The study of the physical processes in a passive building, also drew attention to the importance of convective exchange within zones. Convective flow is highly influenced by geometry and dimensions and this suggested that realistic room-sized dimensions should be adopted. Indeed it is unlikely that scaling down would make significant cost savings since the actual cell itself is not likely to be the major cost element.

It is well established that the absolute performance of a solar system is strongly dependent upon load. Thus it is important to make the heating load of the test cell as realistic as possible by providing the appropriate fabric losses and ventilation losses, and by heating the cell to a realistic comfort condition.

It is clear that as buildings become better insulated so the heat loss becomes dominated by ventilation loss and loss via glazing, usually related to the solar component This lead us to propose a cell where the net loss through the five surfaces, other than the south facing wall, were kept to a minimum by adopting high levels of insulation.

All dynamic effects are provided by mass located within the highly insulated envelope. Time varying heat flows in and out of room surfaces can still take place but net losses are predominantly via the solar component and ventilation. Errors in calculating the small losses through the highly insulated walls would have only a small effect upon a calculated heat balance.

The emphasis on ventilation as the main heat loss mechanism, lead us to propose a precision method of handling this. The cell would be sealed to a high standard and natural ventilation would be 'simulated' be a mechanical system under the control of a computer. Relevant wind data, ie direction and strength and temperature difference would be transformed to a ventilation rate via a standard alogorithm. This control would also give the opportunity for introducing a realistic occupancy dependence.

Radiant energy exchange (both visible and IR) are relatively unaffected by scaling at practicable levels, but are influenced by geometry and the surface characteristics of the room and its contents. This suggested that realistic geometries should be adopted together with realistic finishes, ie. materials and colours. Since in direct gain systems, much radiation falls upon room contents, it is proposed that the test cell includes a set of standard objects of appropriate physical characteristics which correspond to furnishings. These will not only influence radiant processes but also, due to their short time constant, influence the thermal capacity of the 'room air'.

4. MEASUREMENT AND ASSESSMENT

An important question in test cell work is 'what is to be measured'? To a great extent this is linked with the nature of the test cell. One approach would be to use a cell in which all possible heat fluxes were measured and the performance of the system calculated. This necessitates the use of a large number of heat flux meters, or a calormetric method using for example, water-cooled heat exchangers. These approaches, simple in principle, are likely to be difficult and expensive in practice and the elimination of artifacts from the data, difficult to guarantee.

We have adopted a much simplier approach. Since ultimately we are concerned with displacing auxiliary heating energy, this is what we propose to measure. However, since using real weather we cannot repeat a given sequence of days, it is necessary to make a similtaneous comparison between a heated test cell incorporating the component or system under test, and a standard or reference cell subjected to the same weather, and the same internal comfort requirements.

A Performance Index could then be defined by relating in a simple way the relative quantities of axuiliary heating energy to provide the same conditions in the test and reference cell.

Unfortunately, this simple concept has to become more complicated. If we are to run successive (or serial) tests for different components and systems, how can we be sure that the weather conditions over the successive test periods are similar enough to allow comparisons between these successive tests? One solution would be to have an array of test cells as great in number as the number of components to be tested, for a one-time test, i.e. a parallel test. This carries enormous cost penalties and is not consistant with the concept of an on-going facility. We suggest that a compromise between parallel and serial testing, i.e. an array of 6-8 cells one of which acts as a reference, be adopted.

The directly measured performance indeces of successive test runs are corrected to what they would be if the test had taken place over the standard weather day sequence. This correction is done by a mathematical model of the test cell and though the correction factor may contain errors due to uncertainties of the model, the weather sequence can be chosen to keep these errors to second order. We appreciate that this area is uncertain and needs further study.

Passive components may serve other functions than reducing auxiliary heating. For example it may be required to test a shading device whose function is to reduce overheating. Or, a glazing system may be tested in an unheated space, corresponding to a conservatory or atrium, where we are only interested in the floating temperature. Thus we propose other performance indices corresponding to these test situations.

Finally we must ask, 'to what use will these indeces be put?' We see two kinds of use. Firstly performance indices will be useful to the designer and specifier in helping him to choose between options. The numerical values of performnce indices will gradually become familiar to designers, in much the same way that other quantitative indices associated with properties such as fire resistance, noise control etc, which are used in non-analytical way.

Secondly performance indices could also form the basis of simplified calculations methods, where the indices are used analytically to predict the energy consumption of a proposed design.

5. RECOMMENDATIONS

The study concluded with the following recommendations:
A Pilot Project be implimented as a prototype array of test cells, adopting the design principles described here (and in more detail in the report (1)). The pilot project would have a number of distinct components under the two broad categories of theoretical and experimental work.

1. Drawing up of a detailed engineering design from the design brief, and specification of monitoring systems.

2. Development of computer code to handle the normalisation procedure, (to compare sequential test runs over different weather-day periods), and the criteria for selection of suitable runs. This work could proceed on a purely theoretical basis, or by using existing test cell data, before results from the new test cell were available.

3. Development of procedure and computer code to control test cell hardware and data acquisition.

4. Further appraisal of the Performance Indices already defined, and investigation of feasibility of developing simplified calculation method based upon these indices.

5. Building and commissioning an array of four cells on a single site. Nominating and commissioning logging and sensors.

6. Operation of cells for at least one fully operational year.

7. Evaluation of all aspects of test cell operation and purpose. Distribution of preliminary Performance Indices and data sets to experts.

6. FUTURE WORK

On completion of the Pilot Project, the appraisal ((7) above) should be considered with a view to setting up a substantive European Test Cell Project.

We feel that at least four European test sites should be set up, one of which would develope on the site of the Pilot Project. These four sites would represent the main climatic regions - Mediterranean, mid European (coastal), mid European (continental) and North European.

One of the sites, probably the orignal Pilot site, would have a special role in providing data sets for model validation. This function would be of two distinct types. Firstly the provision of data sets for conventional or standard passive systems. This function would not be continual since once a range of data sets had been produced they would not (in principle) have to be produced again. The second role would be the provision of data sets from test cells incorporating new and innovative components and systems. This role would be necessary to validate and develope component algorithms used in the large models.

All sites would have the main role of providing performance indices for new components and systems. The minimum of four sites would provide broad coverage of climatic types and ensure that communication and transportation was not too inconvenient for manufacturers and designers who might wish to visit the test sites.

REFERENCES

1. Baker N V. Passive Solar Test Facility. A report for the CEC DG XII, Energy Conscious Design, London, 1983.

THE PERFORMANCE OF AIR COOLED COLLECTOR SYSTEMS IN DUTCH FIELD TESTS

D.E. Brethouwer
Technisch Physische Dienst TNO-TH
(Institute of Applied Physics TNO-TH)
P.O. Box 155
2600 AD DELFT
The Netherlands

Summary

Since 1975 the Technisch Physische Dienst TNO-TH in Delft has monitored five solar assisted heating systems that have air cooled collectors. The data gathered is used to record the performance and also to validate a computer simulation model. With this model the results have been evaluated. In general the performance of the systems was disappointing. The annual output per m^2 of collector area ranged from 35 to 300 MJ (10 to 80 kWh). On the one hand this was caused by incorrect assumption in the design phase and on the other hand many small technical failures led to reduced net output of the collector. With this experience we have defined recommendations for the design of such solar heating systems in present Dutch housing. As an example of a system in which some of these recommendations have been brought into practice, the double-flow solar heater of Isomur is presented.

INTRODUCTION

From 1974 till 1981 about a dozen buildings in the Netherlands were equipped with solar-assisted space heating system with air-cooled collectors. Five of these were the subject of detailed monitoring programmes, carried out by the Technisch Physische Dienst TNO-TH in Delft.

These programmes were funded by the Dutch National Solar Energy Programme. The purpose of the monitoring activities was to measure and analyse the performance of the installations and to evaluate the results. Apart from a description of the five installations and their performance figures, there are given some general conclusions and is tried to formulate recommendations for the design of air-cooled solar heating systems.

DESCRIPTION

Of the 5 project investigated, 4 are single family houses. The remaining one at Schagen is an office building.
The most important characteristics are listed in table 1.

project name	collector area (m^2)	collector type	cover	storage type	storage capacity (kJ/K)
OSS	25	concrete	double	concrete	13860
ZOETERMEER	35	flat plate	single	water	8400
ZUNDERT	40	concrete	evacuated tubes	concrete	9280
STAPHORST	30	flat plate	double	bricks	11300
SCHAGEN	168	flat plate	uncovered	water	12600

Table 1: The most important characteristics of the five installations.

As space is restricted here it is impossible to show the installation schemes, but the most important features are given below:

OSS : - No solar support to DHW.
- Combined collector and storage system in a slab of concrete.

ZOETERMEER: - Collector plate assembled from juxtaposed quadrangular aluminium profiles with a spectral-selective coating.
- Storage of heat in a water tank via an air to water heat exchanger.
- Preheat of DHW in a copper tube heat exchanger in the top of the tank.

ZUNDERT : - Combined collector and storage system in a slab of concrete.
- Heat extraction in quandrangular alumimium profiles in the top layer.
- Cover consisting of evacuated tubes locked up between two panes.

STAPHORST : - Heat storage in bricks under the floor of the garage.
- Separate pre-heat tank for DHW, fed by an air to water heat exchanger.

SCHAGEN : - Uncovered plastic (Lexan) collectors.
- Ambient air used as heat transport medium.

For further information the reader is referred to references (1) to (5).

METHODOLOGY

Our general approach towards field tests is to start with a diagnosis of the functioning of the system(s) in question, using chart recorders. If the actual behaviour of the installation differs from the design, it is tried to adjust the imperfections. At this stage a data acquisition system is installed to be able to start with the actual monitoring as soon as failures, if any, have been adjusted.

The data is gathered on magnetic tape and transported to the institute for further processing by computer. The results are analysed thoroughly and long term performance figures are produced. In most cases evaluation of the results is based on computer simulation.

During the monitoring period a computer model is developed and validated with the aid of the data gathered. Afterwards the influence of several options is calculated, in order to formulate general conclusions concerning comparable installations.

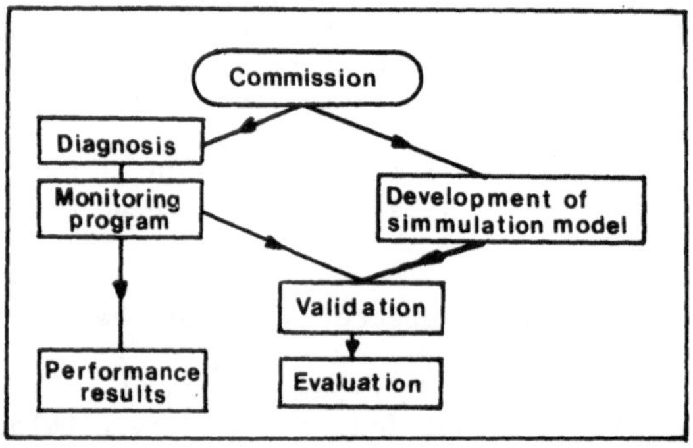

Figure 1: Schematic representation of our approach towards field tests.

RESULTS

In general the performance of the solar heating systems was disappointing. The annual output per m^2 of collector area ranged from 35 to 300 MJ (10 to 80 kWh). On the one hand this was caused by incorrect assumptions in the design phase and on the other hand many small technical failures led to reduced net output of the collectors. The performance figures of the projects and the main reasons for the reduced output are:

OSS :
- Solar output: 70 MJ/m^2.an (20 kWh/m^2.an).
- Very high storage losses, due to the combination of collector and storage.
- Heat exchanger area too small.
- Strong influence of leakage of collector/storage system and valves on net outoput.
- No solar contribution in DHW-production.

ZOETERMEER:
- Solar output: 285 MJ/m^2.an (80 kWh/m^2.an).
- The net output in the "direct use" mode is relatively low, because this mode interferes with "passive" measures.
- The solar installation is overdimenioned with respect to an optimized design.
- Heat losses of storage tank relatively high.

ZUNDERT :
- Solar output: 200 MJ/m^2.an (60 kWh/m^2.an).
- Reduced output because of leakage of collector/storage system and valves.
- Very unfavourable proportion of net and gross collector area.
- Heat exchange area of the DHW-tank too small.

STAPHORST :
- Solar output: 220 MJ/m^2.an (65 kWh/m^2.an).
- Relatively long air ducts between the various components through unheated spaces.
- Poor performance of the heat storage in bricks (placed outside the envelope of the house.

SCHAGEN : - Solar output: 40 MJ/m^2.an (10 kWh/m^2.an).
- The use of uncovered collectors.
- The use of ambient air as heat transport medium.
- The profile of the load in an office building is less suitable for solar heating.
- Very low load for water heating (4 MJ/day).

CONCLUSIONS

From this practical experience the following recommendations can be derived for the design of solar heating systems, using air cooled collectors:

- Pay a lot of attention to solar water heating, because the load for space heating decreases due to high insulation standards and increasing application of passive measures (heat loss coefficient of the houses: 180-300 W/K).

- Leakage of air into the collector can be limited by a design, in which the pressure difference between absorber and ambient is minimized.

- Reduce the use of valves to a minimum.

- Take precautions to avoid convective flows.

- Reduce the length of air ducts and pay a lot of attention to the insulation of the heat storage to avoid large heat losses.

- Minimize the mutual effect between the solar and the auxiliary heating by designing the system and selecting or developing of controllers suitable for their task. Use the heat collected on the lowest possible temperature level (heating of ventilation air).

Furthermore the development of short term heat stores for air systems should be emphasized. Apart from the storage capacity, the heat exchange capacity and the heat loss coefficient are of paramount importance for the adequate functioning of solar heating systems.

Therefore, a final word about a new development in the Netherlands: the double-flow heat storage and the double-flow collector.

THE DOUBLE-FLOW AIR HEATING SYSTEM

These and other experiences from practice actuated the firm Isomur in Gouderak to develop the so-called double-flow collector and heat storage. With these components a solar assisted system can be built that has no valves and of which the price is relatively low (see figure 2). The flows in each of the three circuits are fully separated. The top layer of the absorber heats the DHW-tank, whereas the heat extracted from the lower one is fed into the heat storage. Besides storage of heat it is possible to exchange heat to the third circuit, in which the air from the house is circulated. The storage is built up of juxtaposed profiles, between which the storage material is placed (see figure 3). Because of the simplicity of the design use can be made of standard controllers.

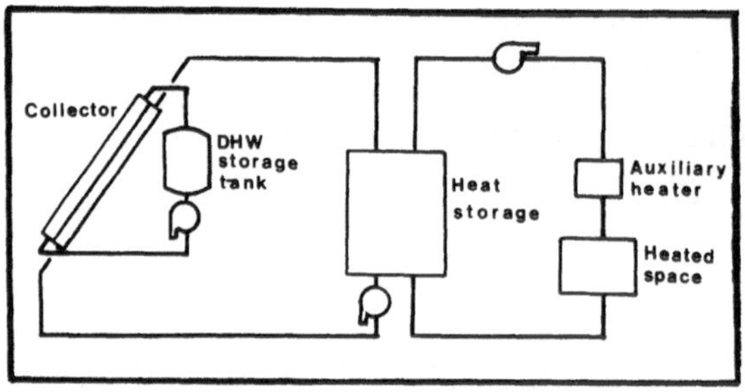

Figure 2: The double-flow air heating system.

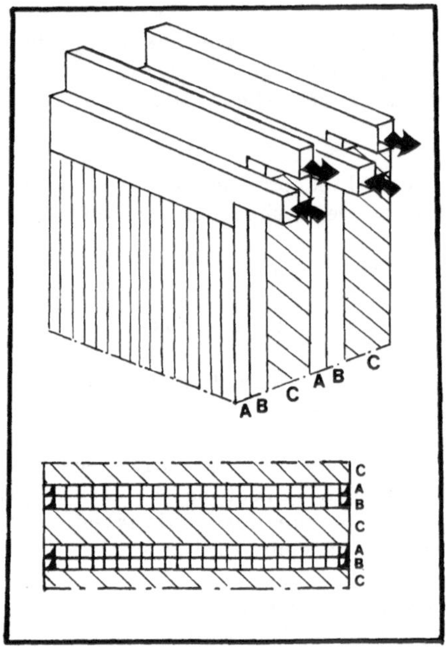

Figure 3: Cross section of the double-flow heat storage.

REFERENCES

1. Brethouwer, D.E. and Den Ouden, C. (1977): "Praktijkonderzoek naar het effect van de energiebesparende maatregelen, zoals toegepast in een aantal experimentele woningen te Oss". TPD-report no. 403.246.
2. Den Ouden, C. and Brethouwer, D.E. (1982): "Meet- en evaluatieonderzoek: Project Zoetermeer". TPD-report no. 803.259.
3. Verdonschot, J.K.M., Brethouwer, D.E. and Den Ouden, C. (1981): "Meetresultaten van de zonneverwarmingsinstallatie te Zundert en berekeningen aan een collector met geïntegreerde warmte-opslag". TPD-report no. 803.211.
4. Brethouwer, D.E. and Verdonschot, J.K.M. (1983): "Meet- en evaluatieonderzoek: Project Staphorst". TPD-report no. 003.215.
5. Brethouwer, D.E. and Verdonschot, J.K.M. (1984): "Meet- en evaluatieonderzoek: Project Schagen". TPD-report no. 103.204.

PERFORMANCE OF 12 PASSIVE AND/OR ACTIVE SOLAR HOUSES IN BASSENS-FRANCE

B. BOURRET et R. JAVELAS
I.N.S.A. - U.P.S., Laboratoire
de Génie Civil, équipe
de recherche associée au C.N.R.S.
Av. de Rangueil
31077 TOULOUSE CÉDEX - FRANCE
TEL : 16 (61) 55.96.68

L. GIOL
Centre d'Etudes
Techniques de l'Equipement
(C.E.T.E.) du Sud-Ouest
Groupe Construction
B.P. 57 - 33019 BORDEAUX CÉDEX - FRANCE
TEL : 16 (56) 47.14.24

Summary

Located near Bordeaux this project was founded by the Town and Housing Ministery, the Plan-Construction, the French Agency for Energy Conservation and the Aquitaine Regional Council. It consists of 12 approximately similar houses classified into 5 thermal categories according to the arrangement of passive and active heating and insulation systems. Unoccupied houses heat consumption and thermal confort were measured from April 1981 to May 1982. C.E.T.E. developed thermal balances and compared them with prediction methods and I.N.S.A. specifically studied some thermal subsystems and transitory evolutions.

1 • BRIEF DESCRIPTION OF THE PROJECT AND VARIOUS MEASUREMENTS

1 • 1 • Description

- Passive systems : direct gain. South facing double glazed windows are normal size for D and E types and very large size for A, B and C types (in this last case, fitted with internal insulated wooden shutters).

- Active systems (A and D - types) : 14,6 m² of single glazed black painted collectors (south facing; 75 degrees/horizontal plane) are connected with 70 m² of underfloor heating coils at ground and first floor levels, without any storage tank. A pumped circuit through the hot water cylinder provides solar preheating (used in A 15 house only).

- Insulation : 8 cm polystyrene walls insulation is internal for C, D and E - types (medium thermal mass) and external for A and B -types (heavy thermal mass).

- Common features. Net floor area : 89 m²; volume : 251 m³; two storeys; concrete block walls; concrete floors; double glazed windows with shutters electric convectors; mechanical ventilation system.

1 • 2 • Measurements

Occupancy was simulated during 232 days of 1981/82 winter (see table I). 106 various meters were readen daily. A high performance microcomputer controlled data acquisition system worked out on 2 houses A 16 and D 7 (28 000 measurements performed and 3 800 data recorded daily). Various chart recorders and small data acquisition systems were used for the control of the 10 other houses thermal confort.

House Type	E		D		C				B		A	
House Number	5	6	7	8	9	10	11	12	13	14	15	16
South facing window net area m^2	6,5	6,5	6,5	6,5	16,0	16,0	16,0	16,0	16,0	16,0	16,0	16,0
Active Heating System			×	×							×	×
Walls insulation I = Internal E = External	I	I	I	I	I	I	I	I	E	E	E	E
Thermal Mass M = Medium H = Heavy	M	M	M	M	M	M	M	M	H	H	H	H
Heat Loss Coeff. W.K-1	235	228	247	247	250	231	250	250	238	238	248	248
Set Point Temp degrees Celsius	18	18	18	16	18	18	18	16	18	20	18	18
Shutters Opening/Shutting Time	8-19	0	8-19	8-19	8-19	0	8-19	8-19	8-19	8-19	8-19	8-19
D.H.W. l/day	0	0	0	0	0	0	0	0	0	0	150	0

Table I - Comparison chart of thermal features and occupancy simulation.

1 • 3 • Climatic data during measurement period (and long term averages)

	Oct 1981	Nov	Dec	Jan 1982	Feb	Mar	Apr	May (20 days)	Winter (232 days)
External Temperature	14,0	8,4	7,8	9,0	8,0	8,8	12,3	14,4	10,2
degrees Celsius	(13,5)	(8,7)	(6,4)	(5,7)	(6,5)	(9,0)	(11,6)	(14,7)	(9,3)
Degree-Days	135	287	318	279	281	286	185	98	1 870
(at 18° c)	(140)	(279)	(362)	(387)	(331)	(280)	(198)	(74)	(2 051)
Sunshine	118	152	47	82	96	175	294	142	1 106
hours	(163)	(87)	(67)	(82)	(110)	(170)	(199)	(152)	(1 030)
Horiz. Plane	2,17	1,89	1,01	1,31	1,92	3,37	5,76	5,20	2,72
Global Irradiation kwh/m².day	(2,65)	(1,41)	(0,97)	(1,18)	(2,02)	(3,27)	(4,65)	(5,65)	(2,59)
South Facing Vertical Plane kwh/m².day	2,23	3,33	0,98	1,79	2,33	3,18	4,19	2,85	2,59
Solar Collectors Plane kwh/m².day	2,55	3,54	1,07	1,94	2,59	3,71	5,22	3,60	3,00
Average Wind Speed m/sec	2,4	1,5	3,5	2,3	2,1	2,6	2,4	2,2	2,4

Table II - Climatic data (and long term averages)

2 • THERMAL BALANCES AND COMPARISON WITH PREDICTIONS

(L. GIOL, C.E.T.E. du Sud-Ouest)

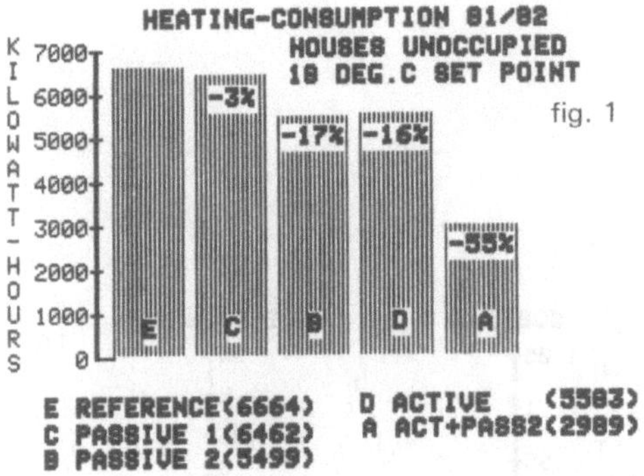

fig. 1: fairly small differencies in heat consumption between reference E and passive houses C and B and active houses D: the difference is large for passive + active house A

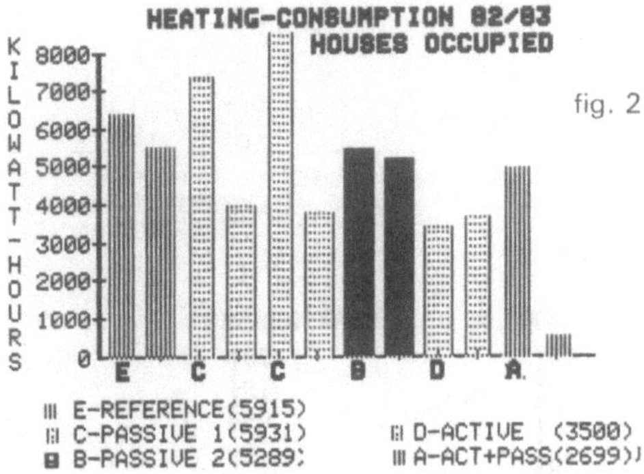

fig. 2: during winter 82/83 when the houses were occupied, large dispersions are predicted, and average results going the same way as previously

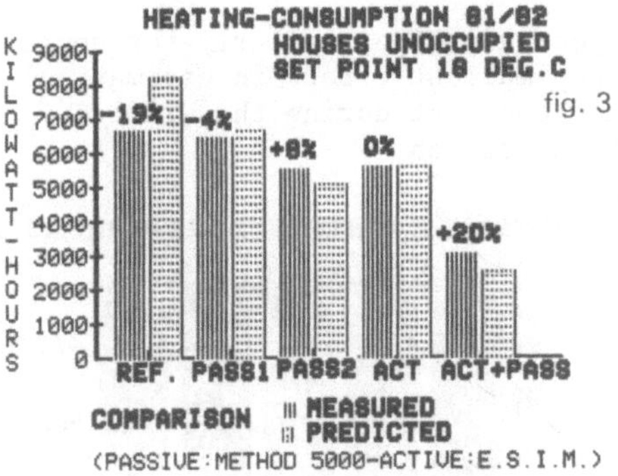

fig. 3: reference E has clearly used less energy then predicted, which upsets direct comparison

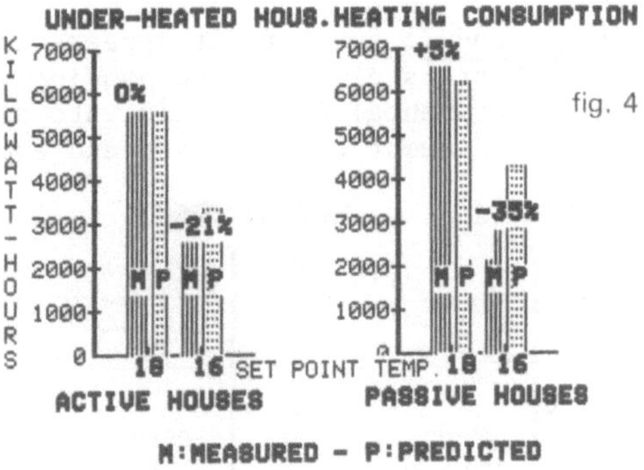

fig. 4: the 2 under-heated houses (16°C) also used much less energy than predicted

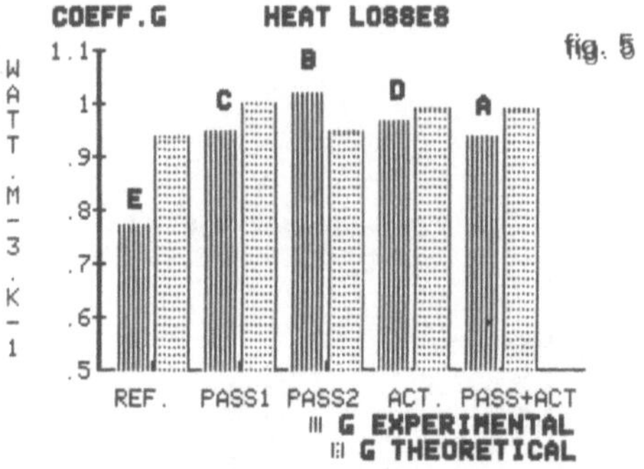

fig. 5: the actual heat losses of reference E are much less on average than the theoretical losses (1)

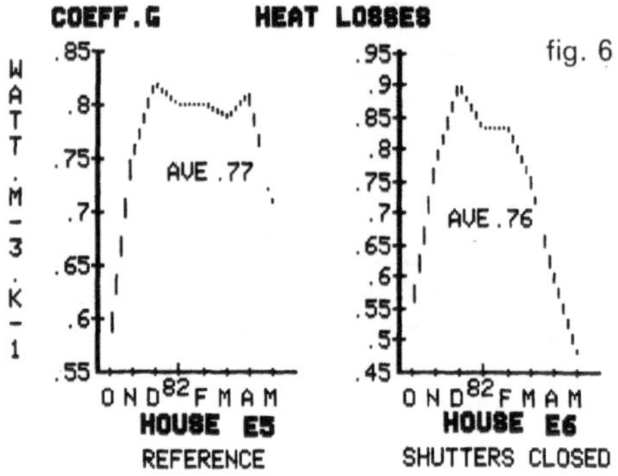

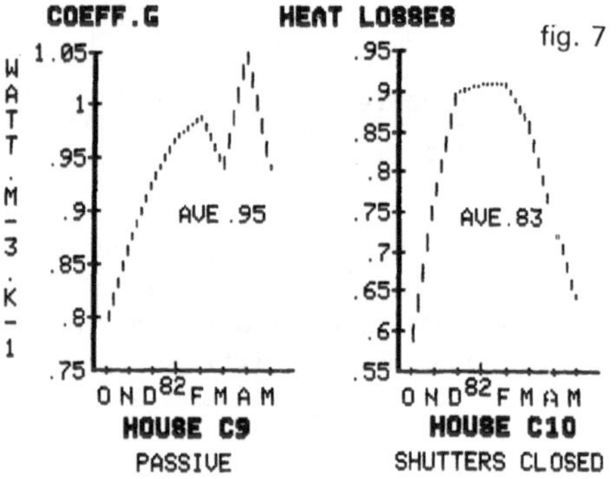

fig. 6 and 7: further, it appears that generally (particularly for houses with shutters closed corresponding to the smallest errors in estimation) the experimental G - coefficients are not constant during the winter: they are lowest at the start and end of the season

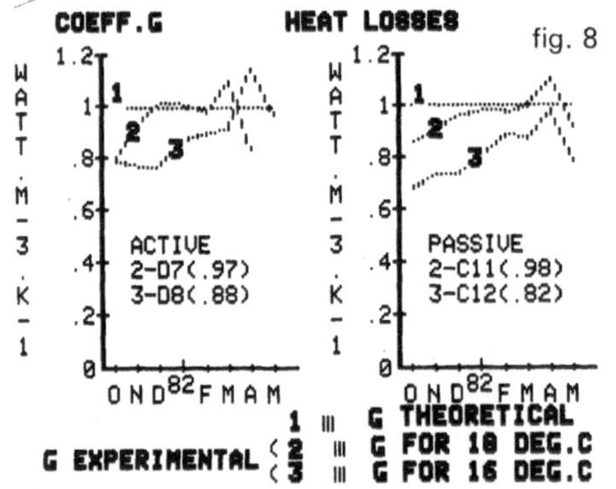

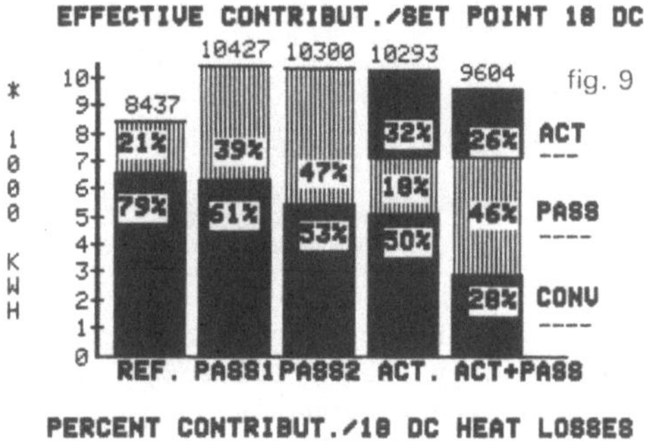

fig. 8: note also that the apparent experimental G-coefficient of the houses under-heated at 16°C is lower than the similar ones normally heated at 18°C

fig. 9: the estimation of the effective energy contribution given the required room temperature uses the experimental knowledge (difficult) of the actual temperature and the effective required temperature as well as the assumption of linearity between heat losses and temperature differencies

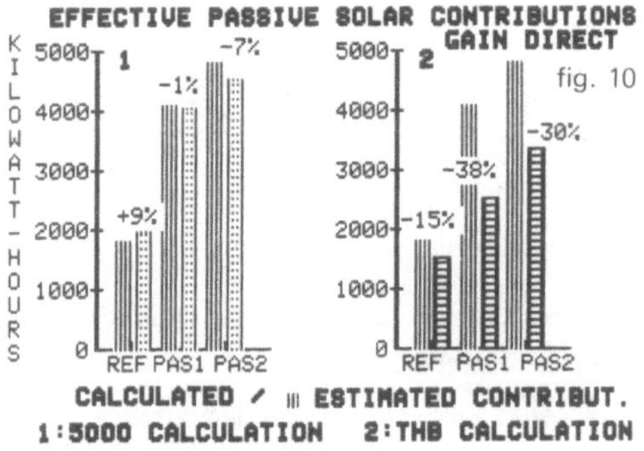

fig. 10: the effective passive contribution seems to be close to the predicted results of the 5 000 method (2) and greater than the ThB method (3) conceived for occupied houses (20% reduction of theoretical passive gains)

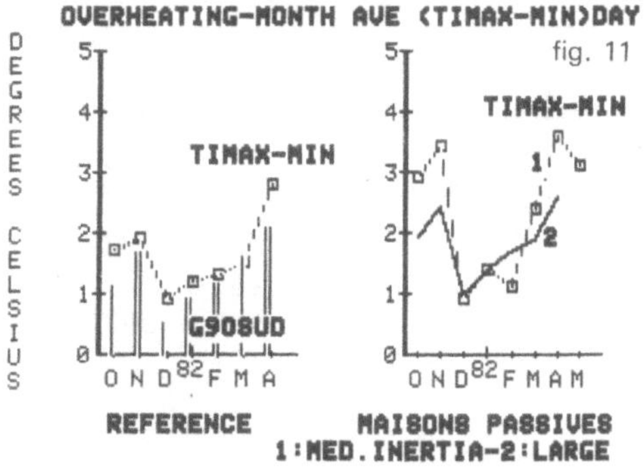

fig. 11: the over heating of the passive houses (which is well correlated to the south facing solar irradiation) is reduced by the large inertia

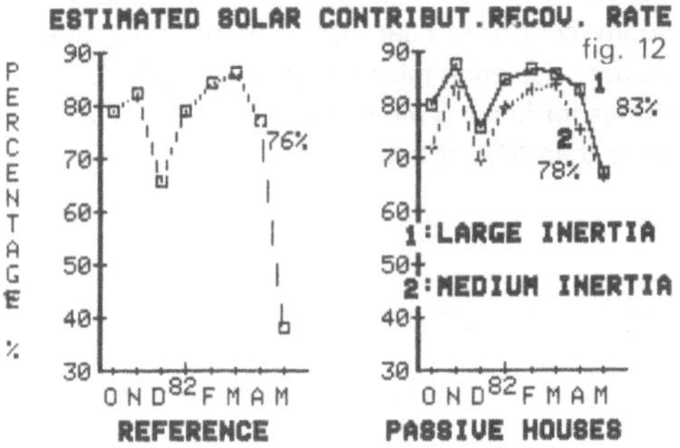

fig. 12: the average recovery levels of the passive solar contributions approach 70 to 80%

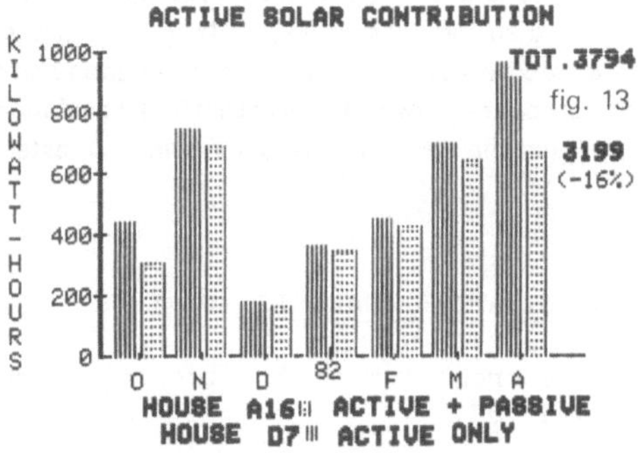

fig. 13: the total active solar gains through the solar floors are considerably larger for the active only houses than for the active + passive houses

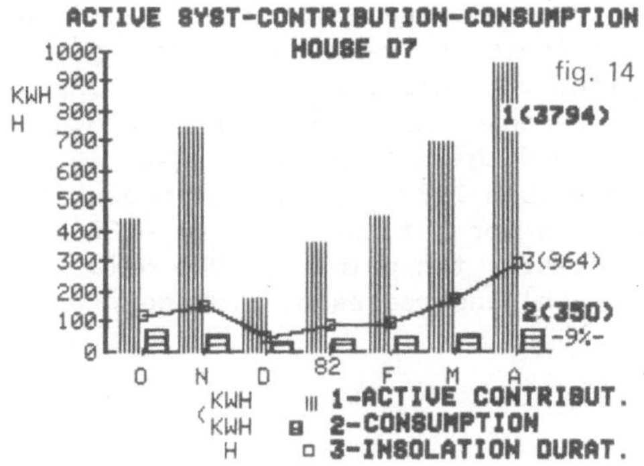

fig. 14: the electrical consumption of the active system (9% of the gains) was excessive

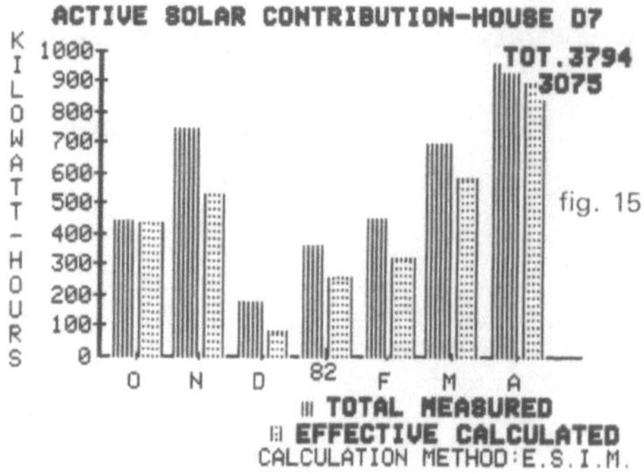

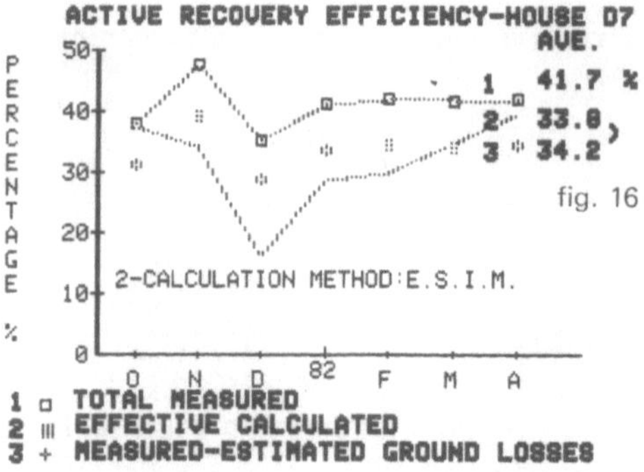

fig. 15: globally, the total active gains minus the losses from the lower floor (18%) are well estimated by the E.S.I.M. method (4)

fig. 16: conversely, in evolutionary terms, the measured efficiencies during the winter are much more stable than predicted by the E.S.I.M. method

Our conclusions on the second section are : the active houses of Bassens are better performers in every day use than the passive houses by direct gain with poorly controlled heat losses. In all cases, solar energy recovery is easier than its conservation and the thermal insulation must be carefully studied. This idea can be transposed to the simplified calçulation methods for heating requirement : paradoxecally it seems that the calculation of solar gains is mastered better than that of heat losses.

3. QUALITATIVE ANALYSIS (Laboratoire INSA-UPS - Equipe de Recherche Associée CNRS n° 998)

We present here the thermal behaviour of two sub-systems in two houses "A" and "D".

3.1 Experimental method

Our main objective was to study the transient thermal behaviour of these houses. So we needed to measure three kinds of parameters :
- meteorological data,
- parameters which allowed us to calculate thermal balances (active and passive solar gains, auxiliary energy),
- parameters required to study the envelope (surface temperatures, flux).

We chose the house "A" as reference (96 measuring points) because it has the maximum of interesting characteristics (outside insulation, south glass surface : 16 m^2, active solar heating system). A less important measuring programme was realized in the house "D" (32 measuring points). A data acquisition system controlled by a computer allowed the data processing of 128 measuring points from April 1st 1981 to May 20th 1982. Out of 232 measuring days, 33 were lost : - 8 due to monthly stops for data processing. - 25 due to failures of the acquisition system (specially the printer). The data bank is available in 16 floppy disks (256 K Ø each) and represents 199 days of hourly measures.

3.2 Results

We describe the thermal performance of the insulating shutter and the heating floor, then we give some details about the thermal comfort.

* The insulating shutter (figure 17) is less efficient than expected.

We have calculated the hourly values of the ratio $p = (TI-TL)/(TI-TE)$ which represents the insulating shutter effect on the wall thermal resistance. The figure 18 represents the variations of p with the difference TI-TE for different wind speeds. We can note that p is all the more low since TI-TE is high.

It can be thought inside warm air goes down in the air slide for TL is too high. This natural convection movement increases with the difference TI-TE and decreases with the wind speed.

The thermal performances of this wall can't be described by the simple notion of the heat transfert coefficient. Yet, a mean value of this parameter can be included between 1 and 1,2 W/m^2K (in theory : 0,4 W/m^2K).

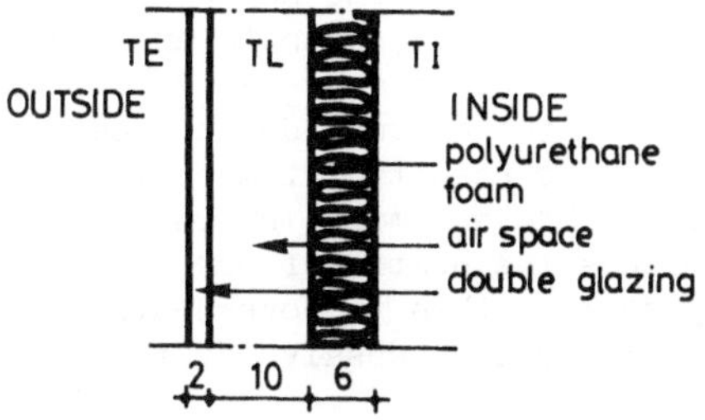

FIG. 17 - Insulating shutter

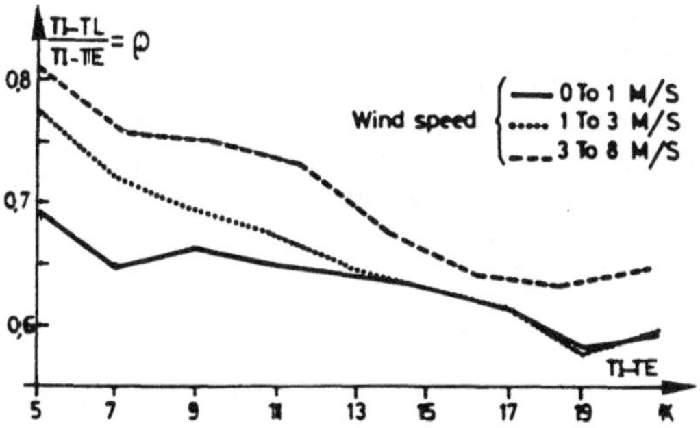

FIG. 18 - Insulating shutter thermal behaviour

* The ground heating floor has a complex behaviour. It has been very difficult to measure accurately all the thermal losses. So, we only want to point out what happens between the soil and the slab. The figure 19 shows the variation of the heat fluxes at this place for two successive periods. During the first period of ten sunny days, if the energy stored up is about 100, the energy released is about 12. But in the second period of ten cloudy days, the energy released is more important than the stored one.

The soil under the house can really be considered as a storage ; but the inadequate accuracy, with which its efficiency can be known, led us to continue this study in our laboratory. An experience has been realized and the theorical approch is made by a finite element programme.

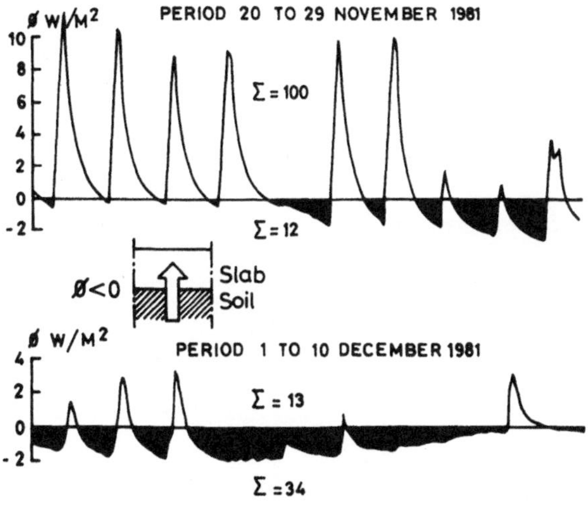

FIG. 19 - Thermal flux between the soil and the slab

* The thermal comfort of the two houses can be described by the cumulative frequency curves of temperature (figure 20).

In the house "D" (no passive system) the curves of ambiant and resultant temperatures are almost similar : there is no overheating owing to the active system (no temperature above 24°C). In the house "A", (large south glass), we note an important difference above 21°C : if the ambiant temperature don't exceed 25°C, the resultant temperature can reach 27°C and more. During these periods the heating floor is storing up energy, then the overheating risk is only due to passive system.

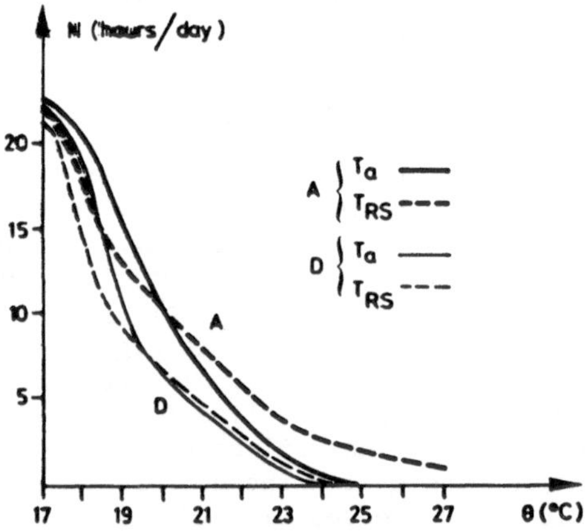

FIG. 20 - Cumulative frequency curves of ambiant (Ta) and resultant (T_{RS}) temperatures

Conclusion :

- this "in situ" experimentation allowed us to constitute a very full and reliable data bank which can be used now by all the research staffs.
- We have pointed out some interesting subjects which must be studied in laboratories (natural convection in insulating shutter -behaviour of a ground heating floor- thermal comfort).

4. REFERENCES

(1) C.S.T.B., Règles de calcul TH-G 77.

(2) CLAUX, FRANCA, GILLES. Méthode 5000. Pyc Editions.

(3) C.S.T.B., Règles Th-B 82.

(4) E.S.I.M., Calcul des planchers solaires directs. Edisud.

SUMMARY OF PERFORMANCE RESULTS FROM THE U.S. RESIDENTIAL CLASS B MONITORING PROGRAM

M.J. HOLTZ, J.N. SWISHER, D.J. FREY, R.C. BISHOP
Architectural Energy Corporation
8753 Yates Drive, Suite 105
Westminster, Colorado 80030 USA

Summary

The thermal performance of 70 passive solar homes is summarized. The homes were monitored as part of the U.S. Residential Class B Passive Solar Performance Evaluation Program. The Class B program represents the largest sample of passive solar homes in the world monitored in a consistent manner. The Class B performance evaluation methodology is briefly described. Some general conclusions and recommendations are made based on an analysis of the aggregrate performance results.

1. INTRODUCTION

The Residential Class B Passive Solar Performance Evaluation Program is a low cost program that evaluates the thermal performance of a large number of residential passive buildings throughout the United States (1). The goal of this program is to provide a consistent measure of the thermal performance of different types of passive buildings/systems in different climates. The methodology for the Class B program emerged from an effort led by the authors to develop a low cost approach to monitoring the performance of passive solar buildings (2). Instrumentation is limited to that needed to calculate the monthly building energy balance, separating passive solar heating from the other building energy flows. Thermal storage and other individual components are not instrumented, and no attempt is made to determine a building's thermal processes in detail.

Research and development on the Class B program was initiated at the Solar Energy Research Institute (SERI) in late 1978 with funding from the U.S. Department of Energy. Selection and instrumentation of passive buildings began in 1981. Over 70 passive buildings have been instrumented under SERI supervision. An additional 21 passive buildings have been evaluated using the Class B methodology without SERI or DOE support. To date, 94 passive homes have been monitored using the Class B approach, representing over 874 months of collected performance data.

Performance data from 70 passive solar homes are reported in this paper; 55 from the SERI Class B Program and 15 from the Bonneville Power Administration (BPA) Class B Program. The results are for the 1981-1982 and the 1982-1983 heating seasons. Only an aggregate performance summary is presented.

Detailed monthly performance calculations, hourly raw measurement values, and complete site handbooks that describe the monitored buildings and measurements are available through SERI (3), BPA (4) or the present Class B coordinating organization, NAHB Research Foundation (5).

2. PERFORMANCE EVALUATION METHODOLOGY

The Class B performance evaluation approach features on-site data processing and display/reporting of performance results using a standardized microprocessor-based data acquisition system (6). Data are received on up to 22 channels every 15 seconds, and channel averages or totals are stored on cassette tape every hour. The system calculates 53 daily and monthly performance factors including the building's major energy flows, in real-time, and then prints them on a daily basis. Monthly the data is transcribed from the cassette tape to a mainframe computer system for post analysis. Physical and thermal characteristics of the building, such as furnace efficiency and solar aperture area, are measured at the beginning of the monitoring period and stored in the microprocessor software for use during the real-time processing.

Building thermal performance calculations are based on a monthly building energy balance for a single control-volume that includes the air of the living zone, defined as conditioned space and space that always remains within 6°C (10 F) of conditioned space temperature. This criterion usually excludes attics and garages and includes basements and sunspaces only when these spaces are conditioned.

The passive solar heat actually used by the building is calculated by subtracting the measured heat delivered to the building by auxiliary and internal sources from the building heat loss, calculated from measured temperature differential. This "subtractive" method of performance calculation is easier to implement than an "additive" approach to determining passive solar heating because the quantities that are difficult to measure, such as venting of overheated air and heat transfer to and from mass walls, are not required.

Living-zone heat loss, Q-loss, is calculated using a single living zone temperature (averaged from multiple sensors) and a living-zone loss coefficient. The loss coefficient is measured using electric coheating. Infiltration measurements are used to correct the loss coefficient for wind speed and other effects (8,9).

Delivered auxiliary heating, Q-auxh, is continuously monitored using furnace burner on-time, with a single measurement of delivery efficiency, for gas or oil fired systems, or measured power for electric systems. The fuel delivery efficiency is measured during the coheating experiment and may be corrected periodically for part-load effects. No wood stoves or fireplaces are allowed, and air-source heat pumps must be operated only in a resistance heating mode. Delivered internal heat gains, Q-int, include the space heating effects of domestic hot water, water heater tank losses and miscellaneous electric appliances. Heat gains from domestic hot water and tank losses are estimated based upon a continuous measurement of domestic water heater burner on-time for gas or oil fired systems or measured power for electric water heaters. Heat from other electric appliances and lights is included in the measurement of total electrical power delivered to the building.

From these values the passive solar heating used by the living zone, Q-pash, is calculated as:

$$\text{Q-pash} = \text{Q-loss} - \text{Q-auxh} - \text{Q-int}$$

The heating can be divided among three sources:

PHR = passive heating ratio = Q-pash/Q-loss

AHR = auxiliary heating ratio = Q-auxh/Q-loss

IHR = internal heating ratio = Q-int/Q-loss

PHR calculated in this manner is not the same as the Solar Savings Fraction (SSF) used in the Solar Load Ratio monthly analysis technique (10). SSF is based on a building load that excludes the contribution of internal gains, that maintains the building at a uniform temperature, and that excludes the steady-state thermal losses through the solar glazing. The building load in the Class B methodology includes internal gains and solar aperture losses and allows the building temperature to float above the thermostat set point.

3. SUMMARY OF RESULTS

Figure 1 shows the location of each monitored passive

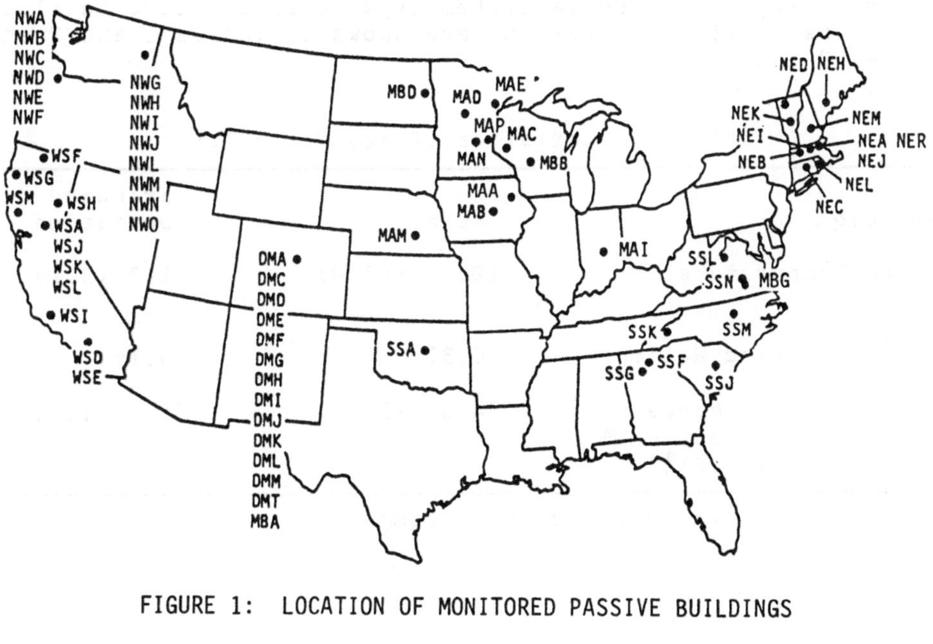

FIGURE 1: LOCATION OF MONITORED PASSIVE BUILDINGS

solar home summarized in this paper. The buildings are identified by a three-letter site code. The first two letters denote the region of the U.S. and the third letter denotes the specific building within the region (DM = Denver, Colorado, MA = Mid-America, NE = Northeast, SS = South, WS = California, NW = Northeast). An additional site code prefix, MB, identifies buildings in the SERI/DOE Passive Solar Manufactured Buildings Program (MBA is in Denver, MBB is in Wisconsin, and MBD is in North Dakota).

Figure 2 shows the overall heating season performance for 70 passive homes in the Class B Program. Each bar represents the total building heating load divided into passive, auxiliary and internal heating components. The energy quantities are normalized per unit floor area and degree-day (based on measured indoor-outdoor temperature difference). The buildings are ordered from left to right according to total purchased heating energy (auxiliary) plus internal). This value is defined as the Building Performance Index (BPI). At the top of each bar, the passive system type (DG = direct gain, SS = sunspace, TW = Trombe wall, WW = water wall) and the south glazing to floor area ratio are indicated.

The 70 passive solar homes included in this report represent a wide variation in building and passive system size, configuration and operation. The thermal performance of these buildings also varies considerably. To fully appreciate and understand the practical relevance of the results requires the study of the individual building summaries. However, nine general observations can be made regarding the overall performance of the monitored buildings.

The following conclusions are based on interpretations of the results in Figure 2 and a simple statistical analysis of the results. The statistical analysis involves averaging the results and grouping or plotting them according to various criteria such as passive system type solar aperture to floor area. The statistical results are shown in Tables 1 and 2 and Figures 3 and 4.

Table 1: Statistics of Key Variables

Variable	Mean	Standard Deviation
Inside Temperature °C (°F)	19.4 (67.0)	1.7 (3.1)
Passive Heating Ratio	0.37	0.16
Building Performance Index (KJ/°C - day - m^2 (Btu/°F - day - FT^2)	74.8 (3.66)	26.1 (1.28)

Note: 70 passive solar homes in sample.

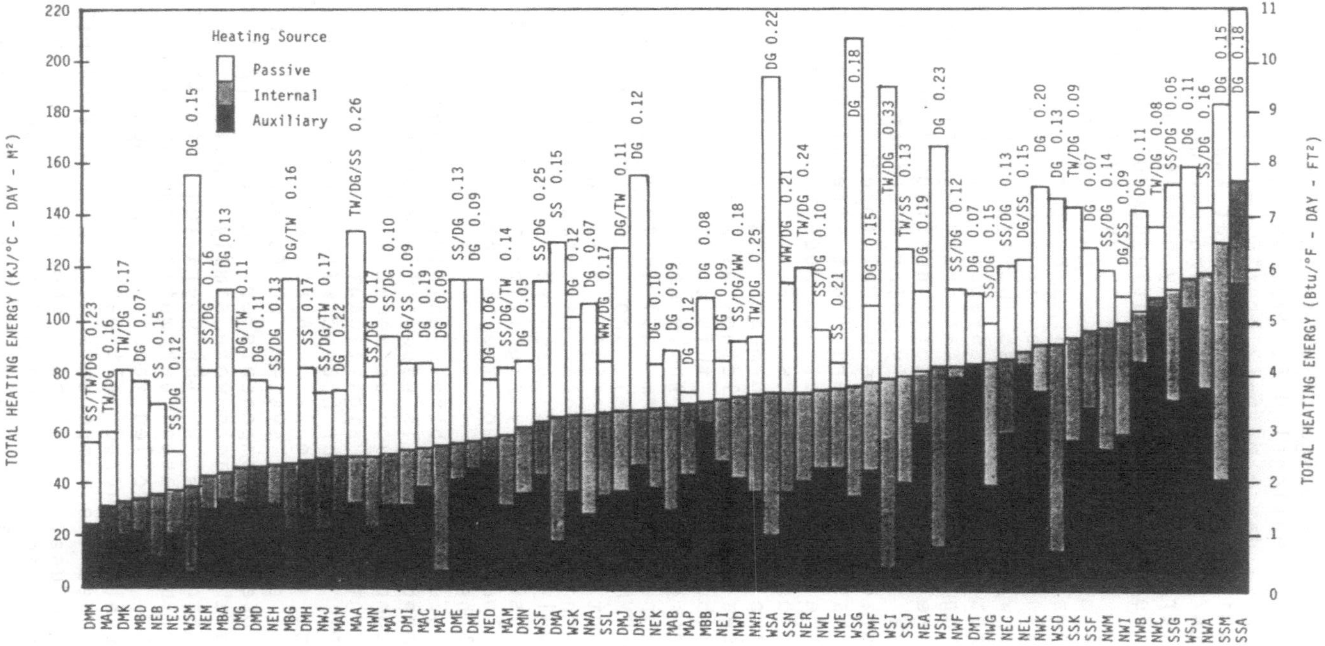

FIGURE 2. NORMALIZED HEATING SEASON ENERGY SUMMARIES FOR 70 PASSIVE RESIDENCES

First, these passive solar homes have low auxiliary heating requirements. Figure 2 shows that the auxiliary heating is generally less than 60 KJ/°C - day - m^2 (3 Btu/°F - day - FT^2) with an average of 43 KJ/°C - day - m^2 (2.1 Btu/°F - day - Ft^2). Conventional U.S. residences typically use from 120-240 KJ/ °C - day - m^2 (6-12 Btu/°F - day - FT^2). The average BPI, which includes internal heat was 74.8 KJ/°C - day - m^2 (3.66 Btu/° F - day - Ft^2), indicating that internal heating was nearly equal to auxiliary heating on the average.

Second, the energy saving influence of insulation and weatherization are critical for passive system performance. Buildings with modest passive solar aperture area on thermally tight buildings generally use less heat than building with larger passive solar aperture area on thermally loose buildings. Low heat losses also increase the fraction of the heating load met by the passive system. In Figure 2, those buildings using the most auxiliary heating (right side of the graph) have relatively high total heating loads with the exception of a few very heavily solar - driven buildings in Denver, Colorado and Northern California. Some of the smallest purchased energy users (left side of the graph) were buildings in the Northeast that had modest passive heating and very low heat-loss coefficients.

Third, the solar performance is extremely variable. Figure 2 shows several buildings with large passive solar heating contributions, but this does not necessarily mean significant energy savings. While many buildings with a large passive solar aperture perform well in terms of auxiliary heat (sites DMK, DMM, NEB, NEM, and WSM for example), some buildings do not (sites NEA, SSA, SSM, and WSJ). However, regardless of this passive solar performance variability, we can conclusively say that passive solar heating systems contribute a statistically significant portion (37%) of the total heating load, or 55% of the net heating load (total load minus internal heat).

Fourth, occupant habits significantly influence passive system performance. This is especially true regarding the use of operable components such as movable insulation, sunspace doors, and vents. No significant statistical difference exists in BPI or PHR for houses with night insulation and those without (See Table 2). A review of buildings with disappointing passive system performance typically had problems with operable components or occupant participation. In many cases the occupants were simply inattentive or were not given instructions on the operation of the building. However, in many cases automatic components, such as thermostatically controlled fans, did not operate properly. Several occupants ignored altogether the use of operable components because the components were not sufficiently simple or convenient to use. Designers should continually strive to improve the simplicity and convenience of manual components and select only those automatic components that will perform reliability.

Fifth, the three passive system types - direct gain, sunspace and thermal storage wall - all had comparable performance. No one system type performed significantly better or worse than the others (see Table 2). Several of the sunspace designs used very little auxiliary heating (DMM, NEB, and NES for example) but did not have especially large passive heating contributions. This results speaks to the value of the sunspace as a thermal buffer, for reducing heat losses, as well as a solar collection component.

Table 2: Thermal Performance Mean Estimates by Category

Category	Number in Sample	As/Af	BPI KJ/day-m (Btu/ F-day-Ft)	PHR	Horizontal Solar Radiation Btu/day-FT
System Type					
Direct Gain	37	0.13	77.9 (3.81)	0.36	-
Sunspace	22	0.15	69.7 (3.41)	0.33	-
Mass Wall	11	0.19	74.9 (3.67)	0.41	-
Location					
Denver	15	0.13	59.0 (2.89)	0.46	11,460 (1010)
Northeast	11	0.14	67.2 (3.29)	0.32	7,650 (674)
Mid-America	11	0.14	58.7 (2.87)	0.35	8,450 (746)
South	9	0.14	99.9 (4.90)	0.33	8,800 (776)
California	9	0.19	80.7 (3.95)	0.51	10,740 (947)
Northwest	15	0.14	89.2 (4.36)	0.25	6,610 (583)
Night Insulation					
Yes	26	0.14	72.5 (3.55)	0.35	-
No	44	0.14	76.2 (3.73)	0.38	-

Sixth, passive system performance varied signficantly by location. The PHR means for the California and Denver buildings were distinctly higher than for the other regions (See Table 2). Greater measured solar radiation levels account for some of this difference; however the BPI mean for California is comparable to the south and the northwest. Three reasons seem to explain this apparent contradiction: (1) the buildings in the colder regions tend to be more heavily insulated; (2) venting of internal gains is greater in the warmer regions; (3) degree day normalization may introduce a small bias in favor of colder climates.

Seventh, passive solar homes provide comfort comparable, and perhaps greater, than in conventional homes. The average indoor temperature for all the homes was 19.4°C (67.0°F). However, assuming a higher mean radiant temperature on the massive surfaces of the home, comfort may be better than a conventional home. A globe temperature measurement is not included in the Class B approach so only subjective occupant reaction is available to evaluate this effect.

Eight, total purchased heat does not correlate to the ratio of solar aperture to floor area (See Figure 3). Even within an individual region, buildings with a similar solar aperture to floor area ratio can have significantly different auxiliary heating. Approximately one-half of the buildings had a passive solar aperture between 10-15% of total floor area.

Nine, total purchased energy tends to correlate to a building's heat loss coefficient (see Figure 4). Obviously, the tighter a building is thermally, the lower its total purchased energy requirement. This seems to be the case regardless of the passive solar aperture to total floor area

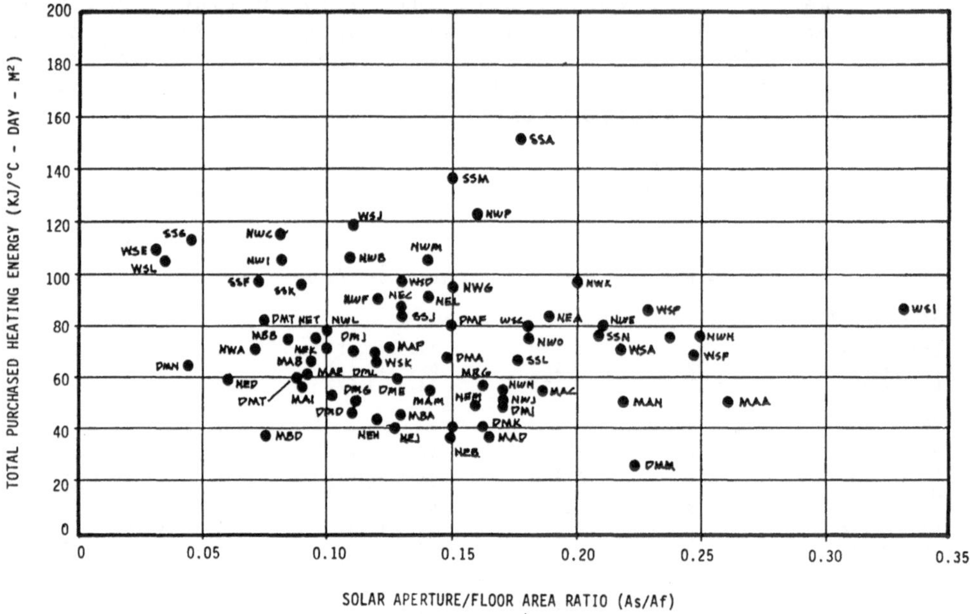

FIGURE 3: TOTAL PURCHASED HEATING ENERGY VS. SOLAR APERTURE - FLOOR AREA RATIO

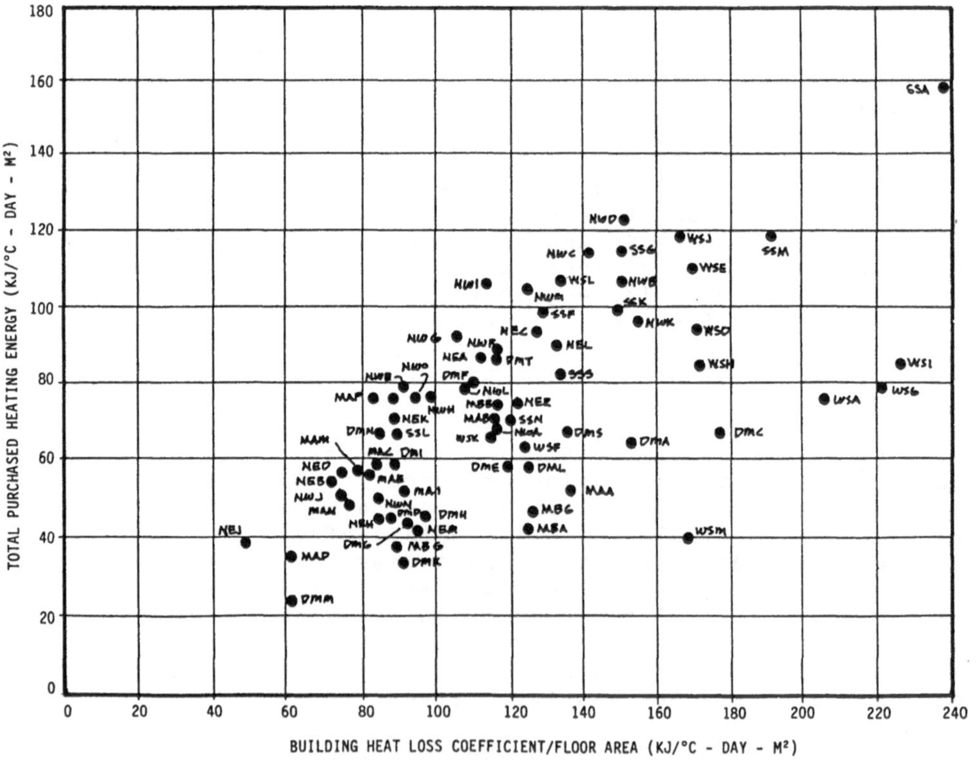

FIGURE 4: TOTAL PURCHASED HEATING ENERGY VS. BUILDING HEAT LOSS

ratio. Buildings with large As/Af ratios such as DMM and MAD have a relatively small building loss coefficient. This fact speaks to the need of selecting high quality windows and proper construction detailing and installation practices.

4. LESSONS LEARNED

Based on the analysis of the performance results from these 70 passive solar homes, what recommendations can one make for the design and construction of passive solar homes in general. Key recommendations are briefly stated below.

- Good energy conservation practices must come first. Thermally tight houses perform better and are more comfortable.

- A large passive solar aperture does not guarantee low auxiliary heating. The solar aperture must be sized for year round performance. Overheating can easily occur during both the heating and cooling season unless effective solar control and venting strategies are employed. Most passive designs are overglazed.

- The simplier the passive design from an architectural integration, mechanical integration and occupant interaction perspective the better the performance is likely to be. The last 10 years of passive design and construction has taught us that simplier passive designs work better. Movable insulation does no good if not used. Design for fail safe operation.

- The choice of passive system types has as much to do with architectural and aesthetic considerations as it does with energy saving considerations. Daylighting, construction costs, performance, reliability, creating tempered usable space, and greater comfort all must be carefully weighed when selecting the appropriate passive system.

5. ACKNOWLEDGEMENTS

The Class B Program was conceived, developed and implemented by tens of people. Without their dedication and support, we would not now have the rich data base of passive system performance from which to learn and improve. We would like to specifically thank the following individuals: Michael Maybaum, Fred Morse, Blair Hamilton, Larry Palmiter, Ralph Beckman, George Yeagle, John Duffy, and Suk Mahajan.

REFERENCES

1. Holtz, M. Hamilton, B., et al., "Program Area Plan: Performance Evaluation of Passive/Hybrid Solar Heating and Cooling Systems: SERI/PR-721-788. October 1980.

2. Palmiter, L., Hamilton, B., and Holtz, M., "Low Cost Performance Evaluation of Passive Solar Buildings." SERI/RR-63-223. October 1979.

3. Swisher, J., and Cowing T., "Passive Solar Performance: Summary of 1981-1982 Class B Results, SERI/SP-281-1847, Solar Energy Research Institute, Golden, Colorado (1983). Available from USGPO, Superintendent of Documents, Washington, DC.

4. Bonneville Power Administration -KTT, P.O. Box 3621, Portland, Oregon 97208.

5. NAHB Research Foundation, 8753 Yates Drive, Suite 105, Westminster, Colorado 80030, USA.

6. Henderson, J., Holtz M., "Evaluation of Prototype Low-Cost Data Acquisition Systems for Performance Monitoring of Passive Solar Buildings." SERI/RI-721-869. March 1981.

7. Sonderregger, R., et at., "In-situ Measurements of Residential Energy Performance Uisng Electric Co-Heating." LBL-10117. January 1980. Lawrence Berkeley Laboratory, Berkeley, California.

8. National Bureau of Standards, "A Low-Cost Method of Measuring Air Infiltration Rates in a Large Sample of Dwellings," NBSIR 79-1728, (1979), National Bureau of Standards, Washington, DC.

9. Sherman, M. and Grimsrud, D., "Measurement of Infiltration Using Fan Pressurization and Weather Data," LBL-10852 (1980), Lawrence Berkeley Laboratory, Berkeley, California.

10. Balcomb, J., et at., "Passive Solar Design Handbook, Vol. 2," DOE/CS-0127/2 (January 1980), Low Alamos Scientific Laboratory, Los Alamos, New Mexico.

MEASUREMENT OF PASSIVE SOLAR GAIN WITHIN THE PROJECT "LANDSTUHL"

D. Oswald
Fraunhofer-Institut für Bauphysik
Nobelstr. 12
7000 Stuttgart-80
Federal Republic of Germany

Summary

In the past, detailed investigations into solar contribution to the energy requirements of one-family houses have been made in laboratory or experimental houses only. In the Landstuhl project, however, the investigations take place with people living in the houses. The 26 buildings contain both active and passive solar components, but the investigations concentrate on the passive solar gains. The most important aim of the investigations is to determine the annual heating energy consumption and the solar contribution. The efficiency of the individual components -that is, of the passive and active solar devices and of the conventional heating systems- is also of interest. Finally the influence of the behaviour of the inhabitants on heating energy consumption and use of solar gain is evaluated.

The experimental investigations are accompanied by computer simulations. These simulations can be used, on the one hand, to provide a back-up for the measurements, for instance, by verifying the data, and, on the other hand, to optimize the building, particularly in terms of its orientation, and the design of the passive components. Two houses have been built and people now live in them. The measuring devices have been installed in one house. The computer simulation shows an annual heating energy consumption of 75 kWh/m^2 a and a solar contribution of 54 %.

1. INTRODUCTION

In the Landstuhl project the most important passive solar components are large south-facing windows, which form the distinctive feature of 10 of the buildings. The solar radiation transmitted through them provides the living area with a direct gain.

Greenhouses also help reduce the heating energy required. On the one hand they operate as a thermal buffer zone with regard to transmission and ventilation losses from adjacent living areas. On the other hand, direct solar radiation can enter the living areas. Finally, a further energy saving effect can be achieved by storing or distributing the heated greenhouse air. 10 buildings in the Landstuhl project are provided with greenhouses. Air conduction into the northern sited rooms takes place in 3 houses, heat storage in stone storage devices in 2 houses.

In order to be able to compare solar and conventional houses, the Landstuhl project contains 6 houses of conventional design. They are equipped with the same measurement devices as the solar houses. This relatively large number of conventional houses within a solar project is necessary because of the lack of detailed measurement data on conventional houses, and because a certain range of variation in the energy consumption data is expected due to the influence of the users.

2. METHOD OF APPROACH

2.1 Correlation analysis

Correlation analysis is one method of evaluating the measurement data. In this method, the dependence of auxialiary heating energy on climatic or user parameters like indoor and outdoor temperature or ventilation behaviour is investigated during specified periods. The minimum useful period is one day as the climatic and operating conditions recur periodically day by day. Sometimes a period of one week may be better since the influence of the storage capacity of the building decreases over a longer period of time. The dependence of auxiliary heat requirements on outdoor temperature reveals the quality of the thermal insulation of the building. The contribution of individual components and building design to passive solar gain can be determined from the dependence on solar radiation.

With multiple regression analysis, the measurement data can be used to determine the influence of several parameters on the auxiliary heating energy at the same time:

$$Q_{aux} = a\,(\theta_i - \theta_o) + b\,S + c\,w\,(\theta_i - \theta_o) + \dots \qquad (1)$$

with Q_{aux} = auxiliary heating energy (within the chosen period)
S = solar radiation " " " "
w = wind speed (mean value within the chosen period)
θ_i = indoor temperature " " " " " "
θ_o = outdoor temperature " " " " " "
a,b,c = constants to be determined

Only a few measured quantities are needed when evaluating the measurement data with correlation analysis. These quantities are indoor and outdoor temperature, auxiliary heating demand, solar radiation and wind speed. The accumulation of measurement data in certain regions of outdoor temperature and sun radiation is disadvantageous as the experimental investigations in the other regions then become less valid. Furthermore, the climate parameters are also interdependent. Finally one needs a large number of measured data (e.g. weekly mean and integral values) in order to get sufficiently accurate results.

2.2 Energy balance

Another method of data processing involves making a complete balance of all energy flows in a building at anyone point in time. This method requires more complex measuring techniques, but yields more detailed results. It allows evaluation of the influence of the inhabitants' behaviour as well as of the effect of short-term outdoor climate fluctuations. Using the energy balance method, the passive solar gain of a building can be defined in the following way:

$$Q_{pass} = Q_{loss} - Q_{aux} - Q_{int} \qquad (2)$$

with

$$Q_{loss} = Q_{trans} + Q_{vent} \qquad (3)$$

and

$$Q_{trans} = \sum_n A_n\,u_n\,(\theta_{i_n} - \theta_o) \qquad (4)$$

The heat losses Q_{loss} include the heat flow from the interior to the outside resulting from the transmission losses Q_{trans} and ventilation losses Q_{vent}.

The transmission losses are determined by measuring the outdoor temperature, and the indoor temperature in each room. The areas A_n are taken from building plans, whereby the u-values u_n follow from material data and the construction of walls, ceilings, roofs and windows.

Opening windows for ventilation and leakage caused by window joints are the reasons for ventilation losses. Measuring these losses presents special problems. Firstly the losses are determined by a calibration procedure based on tracer gas methods. Then they are determined by continuous measurement of temperature distribution, frequency and duration of window opening and wind direction and speed.

The auxiliary heat requirement Q_{aux} is the conventional energy delivered by oil or gas. It has to be measured at the point where the energy is delivered to the heated volume of the house. People in the house and domestic appliances produce heat too, the internal gains Q_{int}. The resulting passive solar gain Q_{pass} includes the gain from all the passive components together.

2.3 Components

Special measurement points have been set up to evaluate the efficiency and reliability of individual components or parts of the system. On average, there are 40 measurement points in a house; the precise number and arrangement of the points depends mainly on the heating system and the active components. Thus the measurement data allow us to carry out correlation analysis, to work out the energy balance, and to investigate the efficiency and mode of operation of the components. An example of the most important measuring points or quantities installed in the house described above (shown in Fig. 1 from the south) is given in Table I.

sector	subsector	measured quantity
outdoor climate	solar radiation	global, horizontal south, vertical north, vertical (diffuse)
	wind	direction speed
	temperature	outdoor temperature
heat losses	transmission	indoor temperature (8 x) greenhouse temperature use of night insulation
	ventilation	opening of windows, doors
auxiliary heating	gas boiler	delivered heating energy domestic hot water supply energy gas consumption
	solid-fuel stove	waste gas temperature
internal gains	domestic appliances cooking	electrical energy gas consumption
components	stone storage device	heat charge and discharge

Table I: The most important measured quantities in the house taken for example.

2.4 Computer simulation

The theoretical evaluation of solar energy use is carried out by computer simulation. This computer program, for instance, calculates indoor temperatures, and the auxiliary heating energy required, including the dependence on climatic data or storing capacity of the building, especially of the inner walls and ceilings. It is thus possible to

- compare measurement data with calculated ones (indoor temperature, auxiliary heating energy)
- eliminate the influence of the different locations by using the same climatic data for all buildings (test reference year)
- evaluate the energy gains from individual components
- optimize the passive solar gains of the whole building.

Fig. 1: The house taken for example from the south.

3. RESULTS

The measuring devices have been installed in the house shown in Fig. 1. The characteristic features of this house (living area 180 m^2) are:

- large greenhouse devided into 3 parts, facing south-east, south and south-west, area 68 m^2
- night insulation between greenhouse and living area, u-value 0,65 W/m^2 K
- stone storage device under the living room, charged with greenhouse air,. volume 12 m^3
- heating system containing gas boiler, floor heating system and solid-fuel stove

The computer simulation was carried out using weather data from Stuttgart in 1981. The most important results are (see also Fig. 2):

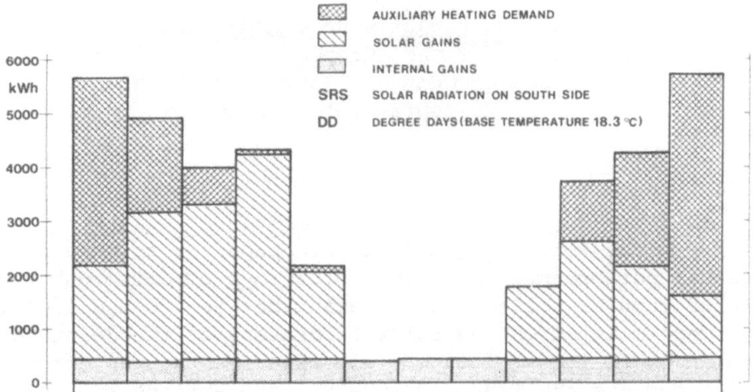

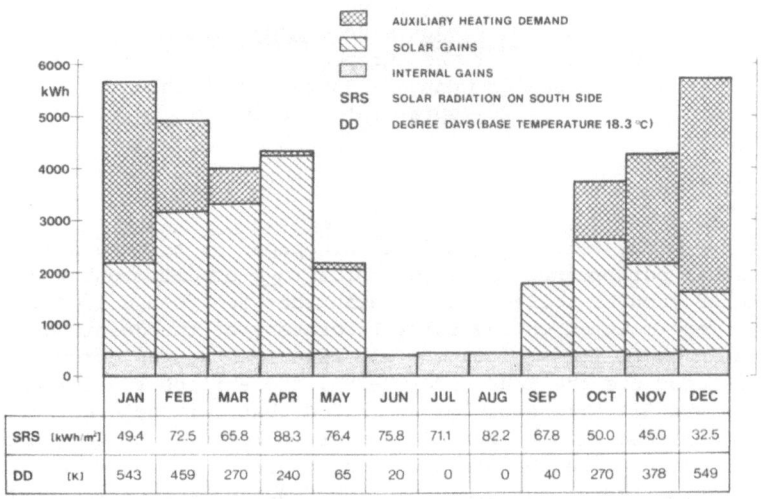

	JAN	FEB	MAR	APR	MAY	JUN	JUL	AUG	SEP	OCT	NOV	DEC
SRS [kWh/m²]	49.4	72.5	65.8	88.3	76.4	75.8	71.1	82.2	67.8	50.0	45.0	32.5
DD [K]	543	459	270	240	65	20	0	0	40	270	378	549

Fig. 2: Calculated auxiliary heat demand, solar gains and internal gains in the described house.

- The annual auxiliary heat demand amounts to 13400 kWh or 75 kWh/m^2 a. This is a low value compared with energy consumption in conventional buildings. The solar contribution during the heating period from the middle of September to the middle of May (with 2834 degree days taken a base temperature of 18.3 °C) amounts to 54 %.
- The auxiliary heat is practically needed from October to March only. In the rest of the year the losses are compensated by internal and solar gains.
- The useful solar gains reach maximum values in the transition period, when both heat losses and solar radiation are comparitively high.
- In the cold months December and January with low solar radiation, the solar gains are proportional to the radiation available at south-facing sides. The ratio of available radiation of January to December amounts to 1.52; the ratio of useful solar gains in the two months 1.49.

4. CONCLUSIONS

The use of solar energy can reduce the conventional energy consumption, provided the components are working with sufficient efficiency, economy and reliability. Because of the variety of heating systems, components and architectural conceptions within this project, it will be possible to estimate in general the solar energy contribution to the energy supply of single-family houses under German climatic conditions.

THREE YEARS OF ARRAY PERFORMANCE OF DIFFERENT TYPES OF EVACUATED TUBULAR SOLAR COLLECTORS

D. van Hattem, P. Actis Dato and P. Tebaldi
Commission of the European Communities
Joint Research Centre - Ispra Establishment
21020 Ispra (Va) - Italy

Summary

The scope of the work presented in this paper is to make a detailed experimental performance assessment of arrays of different types of evacuated tubular collectors operating between 80 and 100°C. The arrays were mounted on the Solar Laboratory of the Joint Research Centre in Ispra for various cooling experiments. The evacuated collectors used are the Sanyo STC-CU-250, the Philips VTR 261 and the Philips VTR 361. The experiments showed that very good efficiencies may be obtained with evacuated tubular collectors. During summer conditions daily collector efficiencies of around 50% have been measured, at operating temperatures of about 60°C above ambient temperature. The results demonstrate that evacuated tubular collectors could be very promising for applications where high operation temperatures are required.

1. INTRODUCTION

Because of the relatively modest levels of insolation which characterize most of Europe, high performance collectors, such as evacuated tubular collectors, may increase significantly the yearly performance of active solar systems for space heating, the production of domestic hot water, etc. Their capability of operating in the range of 100°C with still a good efficiency make them also interesting for other applications specially when higher temperatures are required.

It is estimated that the energy needed to produce one m^2 of such a collector is of the order of 660 kWh /1/. This means that the collector can have a short "energy pay-back time" and that the cost can become acceptable provided that the market permits a large scale production.

The scope of the work presented in this paper is to make a detailed experimental performance assessment of arrays of different types of evacuated collectors operating between 80 and 100°C.

2. EXPERIMENTS

The experiments presented in this paper are part of a general study concerning the technical and economical feasibility of combined solar heating and cooling in Mediterranean climates. Computer simulations have shown that the combination of solar heating, cooling and DHW, can be an economically valid solution for certain climates /2/.

In order to evaluate the performance of such a system with more sophisticated collectors, a solar cooling system equipped with different types of evacuated tubular collectors, has been monitored for several years. This cooling system was built in the Solar Laboratory of the Joint Research Centre in Ispra (Fig.1). The Solar Laboratory is a small office building (ground surface 20x8 m) which was specially built for the study of active solar heating and cooling systems /3/. The cooling system is used to provide air-conditioning to the building during the summer months. The heating of the building is done with a separate heating system. The results of this installation are presented within Task VI of the IEA Solar Heating & Cooling Program.

System Description

The cooling system is schematically shown in Fig.2. It consists essentially of:

- four arrays of evacuated tubular collectors:
  - A) 6 modules of the Sanyo STC-CU 250 L collectors(Sanyo I collectors);
  - B) 6 modules of the Sanyo STC-CU 250 L collectors with reflectors in front of them (Sanyo II collectors);
  - C) 5 modules of the Philips VTR 261 collectors;
  - D) 3 modules of the Philips VTR 361 collectors.
- a $LiBr/H_2O$ absorption chiller, Arkla WF36.
- a small (0.5 m^3 of water) heat storage.
- a large (50 m^2 of water) chilled water storage.
- fan coil units.

In Table I the dates on which the different components were installed are given.

TABLE I - Installation dates of the system components

April 1981	:	Installation Sanyo collectors
April 1982	:	Installation Arkla chiller
June 1982	:	Installation Philips VTR 261 collectors
May 1983	:	Installation Philips VTR 361 collectors

The specifications of the collectors are given in Table II.

TABLE II - Specifications of the evacuated tube collectors

Parameter	Specification		
Collector	Sanyo STC-CU	Philips VTR 261	Philips VTR 361
Tubes per module	10	19	14
Aperture area per module	2.45 m^2	2.22 m^2	2.27 m^2
Selective surface	Nickelbase	Cobalt sulfide oxide	Cobalt sulfide oxide
$F'\alpha\tau$	0.68	0.65	0.64
$F'U_L$	2.1 W/m^2C	1.5 W/m^2C	1.37 W/m^2C
Heat pipe fluid	xx	neopentane	water

Operation and Control

The Sanyo collectors are divided over two arrays, one with reflectors and one without reflectors. The four arrays of collectors can be operated in two groups, independently from each other. The two Philips arrays operate together and the two Sanyo arrays also. This configuration enables a performance comparison of the two collector types.

The climate of Ispra is such that during the month of May and part of June, there is a fair amount of sunshine with rather low outdoor temperatures. On the other hand, during July and August there is a high cooling load with less solar irradiation than in the previous months. Since there was a large underground storage available in the Solar Laboratory, it was decided to try and store the overcapacity of cooling available at the end of the Spring and use it at the end of the Summer. This solution with a large chilled water storage enables the chiller to work without auxiliary energy.
The room temperature is regulated in function of the outdoor temperature.

3. RESULTS

The collector-array performance for 1983 is given in Table III. This table shows the monthly and seasonal collector efficiencies for the four arrays. The table shows also the collection losses, that is the incident energy which is not collected, as a percentage of the incident energy, when the collectors are on and when they are off.

TABLE III - Array performance for 1983

1983		June	July	August	Sept.	Summer	
		%	%	%	%	MJ	%
	collected	30	32	29	27	9163	30
Sanyo I	on losses*	50	55	48	43	15592	50
	off losses**	20	13	23	30	6199	20
	collected	38	41	36	33	11672	38
Sanyo II	on losses	42	46	41	37	13082	42
	off losses	20	13	23	30	6199	20
	collected	48	49	50	48	11523	49
Philips	on losses	43	45	38	33	9823	41
VTR 261	off losses	9	6	12	19	2318	10
	collected	48	51	50	49	7173	49
Philips	on losses	43	43	38	32	5923	41
VTR 361	off losses	9	6	12	19	1422	10
	collected	39	42	39	37	39530	40
Total of	on losses	45	48	42	37	44422	44
4 arrays	off losses	16	10	19	26	16138	16

* Energy not collected while the collector pump is on.
** Energy not collected while the collector pump is off.

Remarkable are the low off losses for the Philips collectors. Also remarkable is the almost equal performance of both Philips arrays. Only in the month of July there is a difference in collector performance. This is probably due to too high operating temperatures on some very sunny days. Under these conditions the working fluid in heat pipes could have been heated to temperatures above its critical temperature. In that case, all the liquid is transformed in vapour and the tube can deliver only very little energy through conduction and convection.

The operating temperatures which are almost the same for all collectors, were on an average about 60°C above the ambient temperature. In Fig; 3 the daily output of the Philips VTR 361 is given as a function of the daily solar irradiation. The figure shows that the collectors have a very low threshold for collection and that even at low daily insulation levels (e.g. 10 MJ/m^2) there is still a good array efficiency (n). This is remarkable considering the relatively high operating temperatures.

Table IV shows the array performance measured over different years. The Sanyo I collectors showed first a decrease in performance but improved in 1983. In 1983 one collector module was exchanged for a new one, this could explain some performance increase but not all of it.

TABLE IV - Collector array efficiency measured over three years

	1981		1982		1983	
	No. of days	Array effic. %	No.of days	Array effic. %	No.of days	Array effic. %
Sanyo I	68	31	42	28	132	30
Sanyo II	68	37	42	38	132	38
Philips VTR 261	-	-	42	54	132	49
Philips VTR 361	-	-	-	-	132	49

The Philips VTR 261 showed some performance decrease, reasons for this could be:

- decrease of glass transmittance and reflectance of reflector due to outdoor exposure. In 1982 the collectors were installed immediately before the measurements;
- the collectors were accidently stagnated, this could have caused some outgassing of methane;
- a large number of tubes have been exchanged, this could have increased the thermal resistance of the condensor block.

However, the performance decrease is not due to the above mentioned cut-off, since it was also noticed at low irradiation levels.

Diurability

The three years of operation gave also some indications concerning the diurability of evacuated tubular collectors. With the help of infrared thermo vision equipment it was established that e.g. some Sanyo collectors had a temperature gradient along the glass tube. This is probably due to a bad bonding of the absorber plate to the heat transfer fluid tube.

During a very heavy hailstrom with stones with a diameter of 4 cm and more, a large number of tubes from the Philips VTR 261 collector were broken. Though this storm was certainly exceptional, there are regions where one should reckon with it.

4. CONCLUSIONS

Evacuated tubular collectors have proven to perform very well at relatively high operation temperatures (85-95°C). This makes this type of collectors very well suited for applications where higher operation temperatures are required, e.g.:

- solar heating systems with seasonal storage;
- district heating applications;
- industrial heat;
- combined solar heating and cooling.

Some diurability problems were identified. However, these problems do not look unresolvable and should not inhibit the further development of this kind of collectors.

REFERENCES

1. DE GRIJS, J.C., BLOEM, H. and DE VAAN, R. (1981) "Evacuated tubular collector with two-phase heat transfer into the system", Solar World Forum, Brighton, August 1981.
2. AXAOPOULOS, P. and VAN HATTEM, D. (1983) Simulation of a combined solar heating and cooling system for a middle size building in Greece. Int. Conf. on Passive and Low Energy Architecture, Chania, Greece.
3. VAN HATTEM, D. and ACTIS-DATO, P. (1980) Description and performance of an active solar cooling system using a LiBr-H_2O absorption machine. Commission of the European Communities, EUR 6864EN.
4. VAN HATTEM, D. et al. (1983) Solar climatization experiments with evacuated tube collectors in Northern Italy. Solar World Congress, Perth, Australia.

Fig.1 The Solar Laboratory of the Joint Research Centre in Ispra.

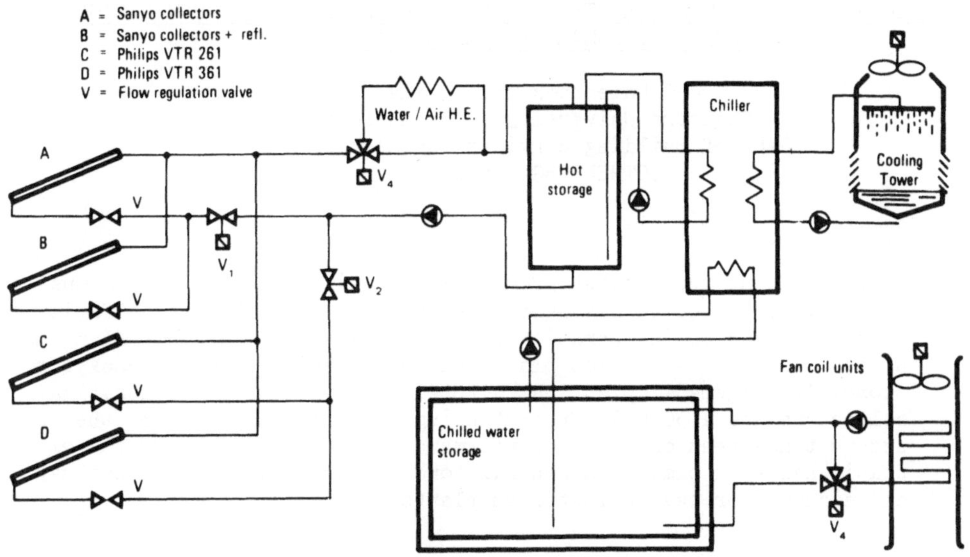

Fig. 2 Schematic view of the solar cooling system.

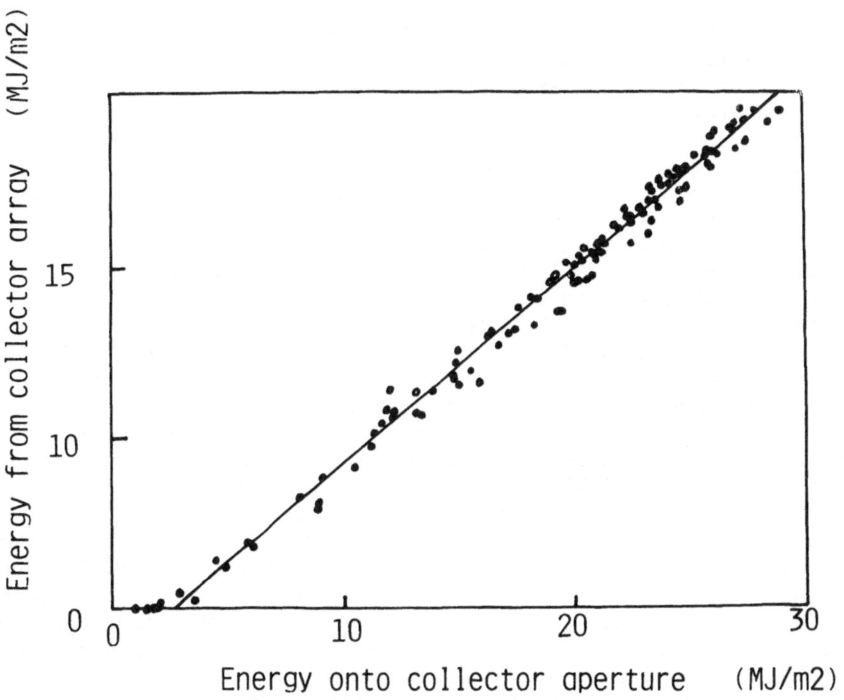

Fig. 3 Daily input/output diagram for the Philips VTR 361 collector.

SWEDISH SOLAR HEATING PLANTS WITH SEASONAL STORAGE - SYSTEM DESIGN INFLUENCE ON THERMAL PERFORMANCE AND ECONOMY

J-O Dalenbäck and T Jilar
Chalmers University of Technology
Dept of Building Services Engineering
GOTHENBURG, Sweden

Summary

At the present two Swedish group solar heating plants with seasonal storage, connected to residential areas with about 50 single-family houses, have been in operation for four years. The Lambohov Plant has a total of 2700 m^2 of flat plate collectors and a 10000 m^3 rock pit store. The Ingelstad plant has a total of 1300 m^2 of concentrating collectors and a 5000 m^3 free-standing tank. The emphasis of the assessment has been on system performance as a whole. Most of the emphasis has been placed on general conclusions concerning technology and economy for seasonal storage plants.

1. INTRODUCTION

One of the aims of the Swedish national energy research and development programme is to find realistic ways and prepare for the introduction of solar heating systems and energy storage in Sweden.

At the present time three Swedish group solar heating plants with seasonal storage are in operation. The first and the second one to be erected were the Ingelstad and the Lambohov group solar heating plants. These two plants may be regarded as half-scale sized on behalf of their connection to residential areas with about 50 single-family houses. The last one to be erected was the Lyckebo plant which is connected to about 500 households and thus represents a full-scale concept in this context.

The erection work as well as the research work has been financed by governmental grants from the Swedish Council for Building Research (BFR). The Council has allocated a total investment during the period 1978-1985 of about SEK 400 million (1 SEK ≈ 0,13 US$) in 1978 current prices for the Solar Heating System and Energy Storage sub-programme.

In 1979 and in 1980 solar heating plants connected to newly-built residential areas at Ingelstad, outside the city of Växjö, and at Lambohov, outside the city of Linköping, were taken into operation.

The energy performance measurements, follow-up and assessment elements of the research program have been carried out by the Building Services Engineering Department of Chalmers University of Technology in Gothenburg.

The emphasis of the follow-up work has been on system performance as a whole. When processing the recorded results and appraising the experience gained, most of the emphasis has been placed on the more general conclusions concerning solar energy technology and economy that can be drawn from the results, rather than on the plant installations as such.

2. SYSTEM DESIGN AND PERFORMANCE

The Lambohov plant has a total of 2700 m^2 of flat plate solar collectors mounted on the roofs of the houses. The heat store contains 10000 m^3 of water. Heat pumps are installed in order to increase the storage capacity. The system concept is shown in Fig 1. The Lambohov plant was designed to meet 100 % of the annual energy demand for space heating and domestic hot water of 55 terraced houses.

The solar collectors are mounted on the roofs of the houses in the form of elements integrated into the external roof cladding. They are inclined at 55°, i.e. the same as the slope of the roof.

The heat store is as excavated rock pit. Wall thermal insulation is provided by thick layers of cement-bound lightweight sintered clay granules and lightweight concrete. The water seal is provided by a sheet of butyl rubber. The water temperature in the store ranges between approximately 10 °C and 65 °C.

Each block of houses is equipped with a hot air unit for space heating. The design operating water temperature is 55 °C at an outdoor temperature of -20 °C.

The total plant cost amounted to about SEK 22 million in 1985 price levels. The cost for the heat store was SEK 7,9 million, the solar collector system SEK 4,6 million and the central equipment involving heat pumps SEK 5,5 million.

Concerning thermal performance the as-measured 1983 annual results for the system as a whole are:

- 2900 MWh (1072 kWh/m^2) of global irradiation on the plane of the solar collectors and 1070 MWh (397 kWh/m^2) of solar energy yield, i.e. 37 % seasonal efficiency
- 395 MWh heat store losses and 675 MWh net solar energy distribution
- 280 MWh total electrical energy consumption

The performance is quite close to predictions except for the great heat store loss which is owing to wet underground thermal insulation.

The Ingelstad plant has concentrating, solar tracking collectors mounted on the ground. The heat store is a cylindrical free-standing concrete tank. The system concept is shown in Fig 1.

The solar heating plant was designed to meet 50 % of the total annual energy requirement of 52 detached houses. The rest of the energy requirement was expected to be met by fossil fuelling.

The solar collector array consists of 35 groups of collectors with a total of 1320 m^2 of collcetor area. The heat store contains 5000 m^3 of water, and has a thick thermal insulation in the walls and tank top. The water temperature in the store ranges between approximately 45 °C and 70 °C. The distribution system has a design operating water temperature of 80 °C at an outdoor temperature of -20 °C.

The total plant cost amounted to about SEK 16,4 million in 1985 price levels. The cost for the tank was SEK 5,4 million, and the collector system SEK 6,0 million.

Concerning thermal performance the as-measured 1982 annual results for the system as a whole are:

- 833 MWh (670 kWh/m^2) of direct irradiation on the tracking plane of the solar collectors and 247 MWh (200 kWh/m^2) of solar energy yield, i.e. 30 % seasonal efficiency
- 120 MWh heat store losses and 130 MWh net solar energy distribution, i.e. 14 % energy coverage

The performance is far below predictions in the main owing to low collector efficiency and great heat store losses.

3. ASSESSMENT METHOD - DEFINITIONS

As far as system performance aspects are concerned, assessment work has been concentrated on calculation methods that can be used to enable the performance of larger sub-systems to be simulated accurately. Methods of calculation for determining the thermal yield from solar collector arrays of different types of solar collector, and for determining the heat losses from heat store types used in Ingelstad and Lambohov, have been considered. Measurement data for the existing plants have been used for validations of the simulations.

The annual thermal performance for the whole plant has been calculated for the following cases:

The Lambohov concept - flat plate solar collectors having as-measured efficiency in combination with a 10000 m^3 rock pit having up to standard or as-measured underground thermal insulation. The Ingelstad concept - concentrating solar collectors having as-measured or up to standard efficiency alternatively efficient flat plate collectors in combination with a 5000 m^3 free-standing storage tank having up to standard thermal insulation.

The solar thermal yield for the average year has been calculated providing a constant store volume and varying collector area.

Economic calculations have been made using a present value method. The life of the heat store has been assumed to be 40 years, while the life of other parts of the installation have been assumed to be 20 years.

The cost of heat production, in SEK/kWh, has been calculated by setting the present value of all investment and maintenance costs against the present value of the total amount of solar heating energy produced during the 40-year life of the installation. A real rate of interest of 4 % and a real rate of annual energy price increase of 2 % has been assumed. 1985 has been taken as the reference year for all cost discussions.

4. PRINCIPAL RESULTS

Technical assessment

The calculations for the Lambohov concept show that (Fig 2):
- The heat store thermal insulation performance has a weak influence on the solar energy yield and the electrical energy consumption. This is owing to the fact that the collectors work is more efficient at low temperatures and thus increased heat store losses are compensated to some extent.
- A collector area lower limit is set by the risk of freezing the heat store water and an upper limit is set by the temperature durability of the heat store water sealing.

The calculations for the Ingelstad concept show that (Fig 2):
- A solar collector system with somewhat over 1300 m^2 of concentrating collectors having as-measured efficiency should produce about 260 MWh of solar heat, i.e. 25 % solar heat coverage.
- To achieve the design objective of 50 % solar heat coverage would require either 1700 m^2 of an efficient concentrating solar collector, 2300 m^2 of the Ingelstad concentrating collectors or 1900 m^2 of efficient flat plate collectors.

Economics and development potential

At present, the cost of heat produced by the Ingelstad and Lambohov plants is about SEK 1,90:-/kWh and SEK 1,10:-/kWh respectively based on the initial capital costs.

If solar heating installations of this kind are to become economically attractive for residential space heating purposes, it will be necessary to bring down the cost of heat from them in the fairly near future to the same level as that of heat from oil firing. A reasonable time perspective in this context is perhaps 20-30 years. If a 2 % real annual increase in the price of oil is assumed, the cost of heat from oil at the end of that period will be between SEK 0,43:- and SEK 0,53:-/kWh.

The question therefore resolves into one of which parts of the plant are capable of producing heat for the stipulated cost. The lowest heating costs in systems operating on the same principle as in Ingelstad and in Lambohov is obtained with about 2500 m^2 and about 2000 m^2 of solar collectors respectively.

An important factor, if the cost of the solar collectors and the heat store is to be realistic, is that the cost for the ancillary piping and water-side equipment should be much lower than is the case in Ingelstad and Lambohov. By experience from installations in the last few years it seems possible to hold the cost down to a level of less than 50 % of the initial costs.

At present the Swedish market cost is about SEK 700:-/m^2 for the type of flat plate collector which is used in the Lambohov plant. These assumptions result in the following maximum cost levels for the heat store in the Lambohov concept (Fig 3):

- Excavated pit store: SEK 100-300:-/m^3

Using data from a number of Swedish investigations into seasonal heat storage costs, it is possible to assume a present cost of SEK 350:-/m^3 for a pit store. For the Ingelstad concept a discussion similar to the one above results in the following maximum cost levels for the collectors (Fig 3):

- Efficient flat plate collectors: SEK 800-1300:-/m^2
- Efficient concentrating collectors: SEK 1000-1500:-/m^2

In this case the solar collector cost includes the cost of the collectors themselves, the supporting structure, foundations and land.

5. CONCLUSIONS

- Solar collector plants fitted with flat plate collectors are more likely to become competitive with oil firing than plants fitted with concentrating collectors for residential heating in Sweden.
- Low cost ancillary piping and water-side equipment must be used in solar heating plants with seasonal storage aimed to be competitive with oil firing.

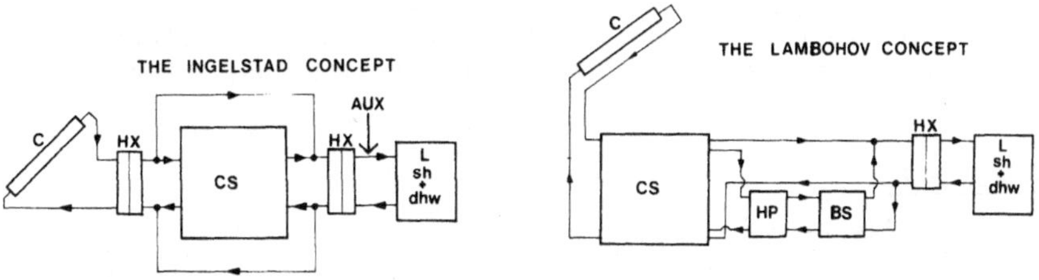

Figure 1 System concepts

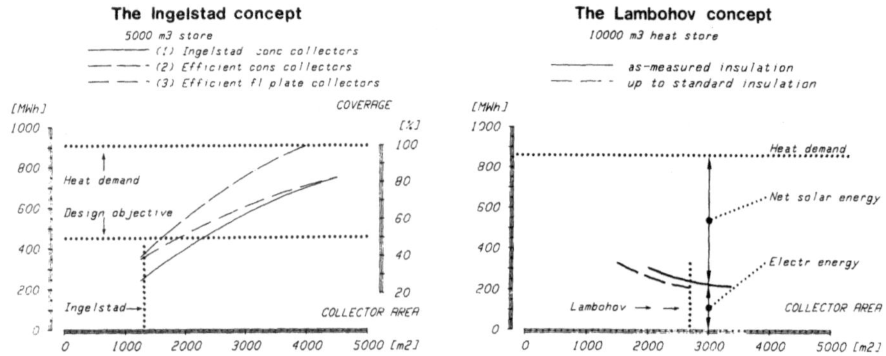

Figure 2 Calculated annual solar energy yield versus solar collector area

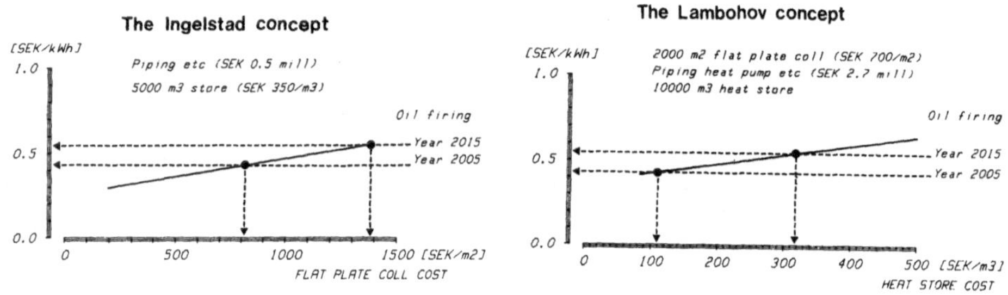

Figure 3 Maximum cost levels (in 1985 terms) for the heat store in the Lambohov concept and for the collectors in the Ingelstad concept to compete with fossil heat costs

REFERENCES

Norbäck, K, Hallenberg, J, 1980. A Swedish solar heating plant with seasonal storage: a technical and economic description of the Lambohov project. (Swedish Council for Building Research.) D36:1980. Stockholm.

Finn, L, 1979. A Swedish solar heating plant with seasonal storage. The Ingelstad project: design and construction stage. (Swedish Council for Building Research.) D14:1979. Stockholm.

SOLAR HEATING PLANT AT MOUNT ZUGSPITZE

H.J. Stein and M. Köhnen
Institut für Kernphysik
Kernforschungsanlage Jülich GmbH

Summary

Significant contributions of solar energy for room heating purposes without using long-term storage can only be expected in climatic regions with high insolation combined with low air temperatures. Middle Europe has a moderate climate with many cloudy days during the winter time. This situation may be different in mountain regions where the air temperature is low also during the summer months, and where the insolation all over the year is higher than in the lowland. The research project under study here is a solar collector based heat pump room heating system for the radio transmission station of the German Federal Post at Mount Zugspitze, 2964 m. The plant has been operational since February 1982. A sophisticated monitoring technique has made possible a detailed study of the technical feasibility of using solar energy under these circumstances. The results obtained over a period of two years show that the solar contribution can be remarkably high. However, it seems to be almost impossible to design a system which would be economically attractive. Nevertheless, the practical experiences obtained on monitoring techniques and on the long-term behaviour of system components are very useful for many other applications.

1. INTRODUCTION

The German Federal Post has installed a new radio transmission station at Mount Zugspitze/Garmisch. The service building of this station has been equipped with a solar collector based heat pump system for room heating purposes. The solar system was financed by the German Minister for Research and Technology (1) and constructed by Brown, Boveri & Cie. on a commercial basis. The authors have undertaken the task of designing and running a research oriented monitoring programme, in order to get reliable data on system performance as a basis for an optimum operational strategy and possible technical improvements.

2. BUILDING AND HEATING SYSTEM

The construction housing the radio transmission systems consists of a weather protecting cover made of aluminium under which the three storey well-insulated windowless service building is situated. The total floor area and the volume to be heated are about 300 m^2 and 900 m^3, respectively. Acrylic glass windows are integrated in the lower part of the weather protecting cover. Behind these windows single-glazed black-paint collectors have been placed, Fig. 1. The entire solar system, Fig. 2, consists of an array of 48 collectors inclined 72° and oriented south-east (field aperture area 53 m^2), two storage vessels (2,3 m^3 each), and an electric heat pump (3 to 4 kW compressor power). Due to safety reasons the whole system is filled with antifreeze 70% glycol/30% water. Heat distribution occurs via a conventional air circulation system including the option of regenerative

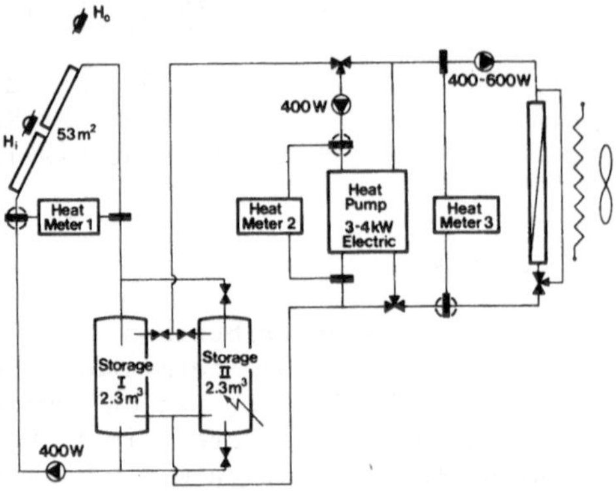

Fig. 1. Outside view of the radio transmission station building.

Fig. 2. Scheme of the solar heating plant with basic instrumentation.

heat exchange with fresh air. The air system was chosen in order to make use of internal loads generated by the telecommunication electronics. The solar system is regulated in such a way that above 30 °C storage temperature heat is directly fed into the air circulation system (direct solar mode). Below 30 °C the heat pump is switching on, operating down to a storage temperature of -5 °C (heat pump mode). Back up energy is provided by electric resistor heaters in the air distribution system. Use of low tariff electricity is possible by loading one of the two storage vessels with an electric heater. It should be noted that all system components are located in the "cold area" outside the heated rooms.

3. MONITORING SYSTEMS

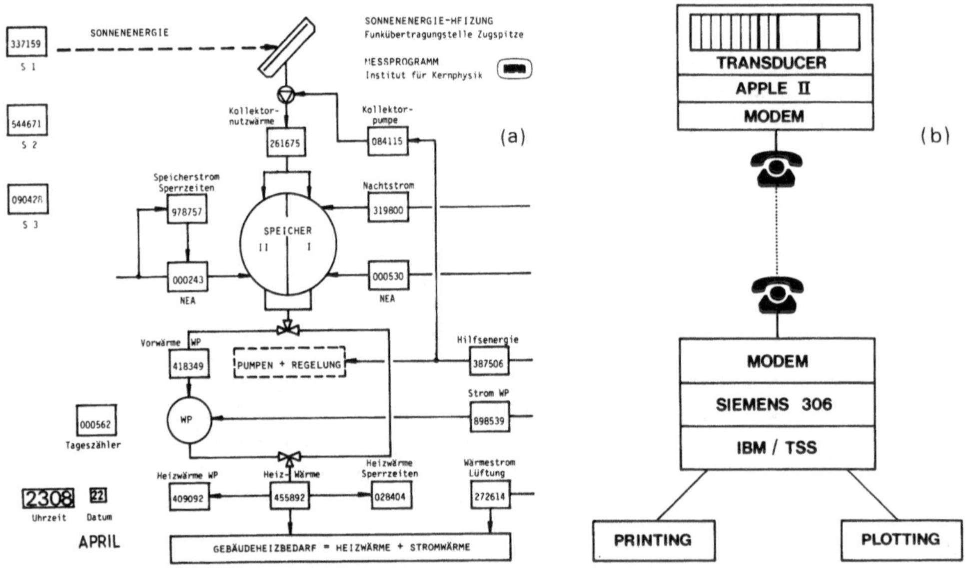

Fig. 3. Central telltale board with numerical registers for automatic photographing (a) and scheme of the computer based data acquisition system (b).

Data taking for monitoring the solar system is performed with two systems which are completely independent from each other. With the first system, the INTEGRAL system, only the main energy fluxes in the system like solar irradiation, heat fluxes and electric energy inputs are detected and integrated with numerical registers placed in a central board showing the energy flux diagram of the system, Fig. 3a. Readings are automatically photographed every day before midnight. Energy balance diagrams on a daily, weekly, or monthly basis can easily be made and enable a simple and quick evaluation of the performance of the system and the main components.

In the second system, the DIFFERENTIAL system, important temperatures and status information is recorded in addition to the energy fluxes by a specially developed low-cost data acquisition system based on the personal computer APPLE II (2). Data are taken every 20 sec., integrated or averaged over 10 min., stored on flexible disk, and retrieved daily via the public telephone network to a central computer at the KFA Jülich, Fig. 3 b. Here, plotting and printing of the results is automatically performed the next day. The data acquisition system has a capacity of 32 temperature, 16 analogue, 32 status, and 16 counter channels. The calculation of the three important heat fluxes of collector output, heat pump input, and total heat output is performed on line with the APPLE II computer independently from the heat meters used in the integral system. A status signal initiates differentiation between direct solar and heat pump mode. In Fig. 4 is shown a selection of representative computer plots.

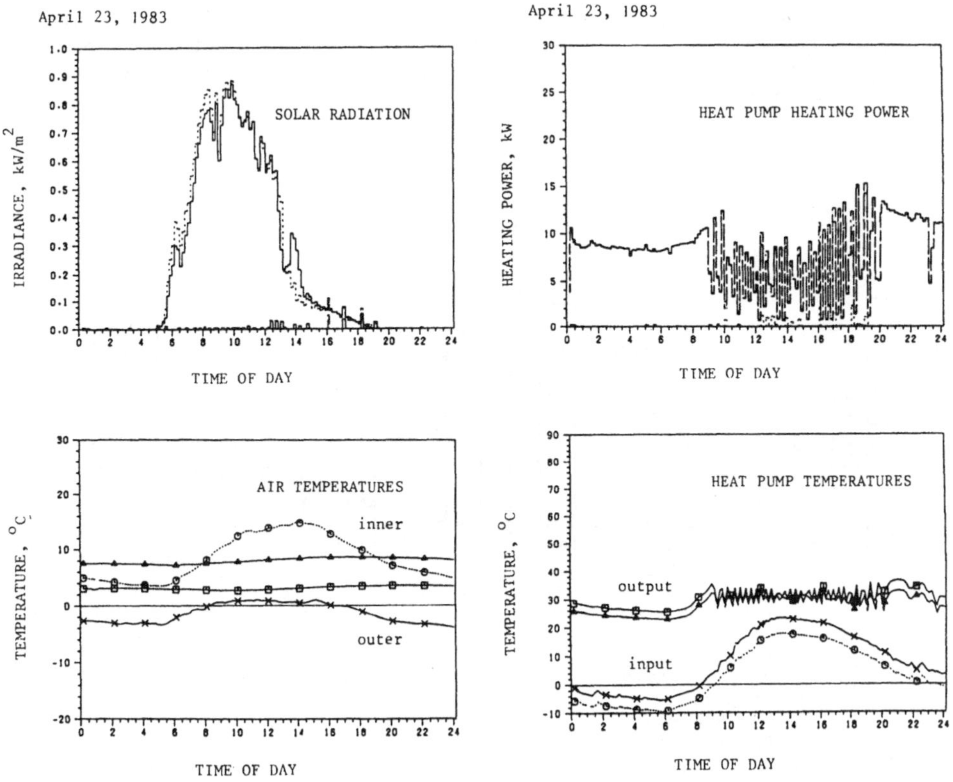

Fig. 4. Selection of automatically generated computer plots for a typical day. Note the difference of air temperatures inside and outside the weather protecting cover.

4. RESULTS

The solar system and the integral monitoring system have been in full operation since April 82. After some troubles with the telecommunication line, the computer data acquisition has been reliably functioning since October 1982. In Fig. 5 is shown the result of energy balance measurements with approximately monthly time resolution. Substracted from the mean daily heat load are the mean values of back up, compressor, auxiliary, stored electric, and useful collector energy. The white bars above the zero line define the savings against 100% direct electric heating. Negative values describe thermal losses which are predominantly storage losses. The numbers below give the solar irradiation inside and outside the weather protecting cover related to 53 m^2 collector field area. Due to shadowing about 30% are lost for the collectors.

In the period from April to October 1982 savings were 40%. In the following winter months a drastic reduction in system performance was caused by loading too much electrical heat into the twin storage system. With the help of the computer data a malfunction of the electronic regulation and a hydraulic coupling between the two storage vessels were found. As a consequence, the mean temperatures in both vessels were too high causing losses of electrically generated heat and low collector efficiencies. After closing an open valve in an expansion pipe and limiting the temperature in the storage vessel II, the system was again working properly. Year-long savings in the period from February 1983 until March 1984 were 43% of a heating load of 52.000 kWh. The contribution of high temperature collector heat used in the direct solar mode was only 5%. The heat pump was working with a coefficient of performance between 3.0 and 3.8 with a mean value of 3.4.

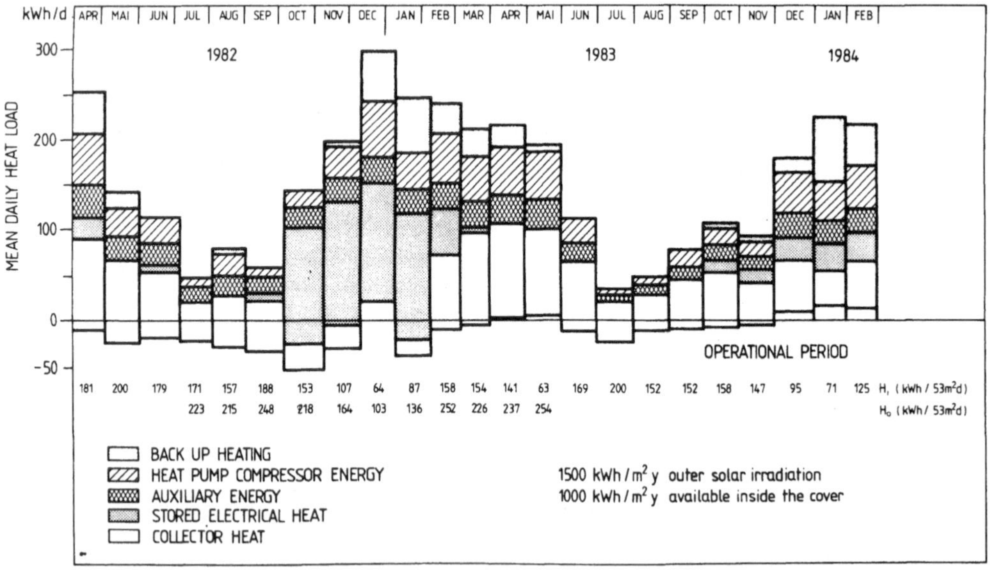

Fig. 5. Result of energy balance measurements on a monthly basis. Only 5% of the collector heat were directly used without the heat pump.

5. CONCLUSIONS

The investigated plant is essentially a solar collector assisted heat pump system. It can be fully characterized by the measured energy balances. Differential analysis was necessary to find out faulty operation modes and to propose measures for repair. The heating load of the building is lower than expected, because forced ventilation with fresh air is not applied at present. Passive solar gains through the weather protecting cover additionally reduce the heat load during daytime. In the summer time there is excess solar energy during sunny periods. 30% of solar radiation available outside the cover is lost by shadowing. This reduces system performance during cloudy periods and during the winter time. Mounting the collector in the open air, however, would have implied possible damage due to extreme climatic loads. Since the air temperature inside the cover is strongly correlated with the solar irradiance, simple absorbers combined with the heat pump would result in a better system performance. Due to the limited area, a system using high-efficient collectors without a heat pump would not yield higher savings. Economic aspects could only be discussed if it were possible to achieve system costs in the range of 1000,-- - 2000,-- DM/m^2 collector area.

ACKNOWLEDGEMENTS

The authors would like to express their thanks for the excellent cooperation with all participating parties, as Oberpostdirektion München, Brown, Boveri & Cie., and the staff of Fernmeldeamt Garmisch.

REFERENCES

(1) BMFT, Annual Report 1980 on New Sources of Energy, Project No. ET 4315 Z/2, 4.1, p. 1616.

(2) Made by INTRO, Ing.-Büro GmbH, Nideggen, Germany, according to the directions prescribed by "Zentralstelle für Solartechnik", KFA Jülich.

MEASUREMENT AND EVALUATION TECHNIQUES FOR SOLAR HEATING SYSTEMS

Per Holst, Bengt Perers, Heimo Zinko, Leif Eriksson
Studsvik Energiteknik AB, Sweden

Summary

Since 1982 the solar district heating demonstration plant, at Södertörn, has been in operation with 7 different solar collector systems, 100 - 200 m^2 each. The plant has been subject to a detailed monitoring program carried out by the group for solar - and measurement techniques at Studsvik.

The dataacquisition was carried out using a computerbased measurement system with a software system developed at Studsvik.

The software was given a structure that could solve a number of problems involved with continuous data collection from processes with different operating statuses and fits to a software system of similar structure for evaluation of data and selection of reporting formats.

By use of generalized baseprograms controlled by parameters specifying the dataprocessing general purpose computers can effectively be used to monitor and evaluate energy systems.

The handling of great numbers of data require a standardized process and well tested software for data acquisition, datareduction, data storage and preliminary operational reports.

Results from system comparisons tests are presented in the paper. The highest system efficiency shown was 40 % for Philips VTR 141 system at 60 $^\circ$C operation temperature. The second generation of flat plate collectors represented by Scandinavian Solar HT gives system efficiencies comparable to evacuated tube collectors.

1. INTRODUCTION

The Södertörn plant serves to gain operating experience with larger solar collector fields with different collector types. For the moment, 7 subgroups are under operation, the size of each unit being between 120 and 216 m^2.

Each unit is separately connected to the district heating return pipe by means of a heat exchanger which results in operating temperatures of 60 - 80 $^\circ$C for the collectors.

2. THE MEASUREMENT SYSTEM

All seven solar collector units are monitored and partially evaluated on site by means of a computerbased system, consisting of a HP-3497 scanning voltmeter and a HP-85 desk top computer. The final evaluation is performed off-line at a HP-9845 computer in connection with a Winchester disc memory at Studsvik evaluation center.

The system is based on a data acquisition program with functions for measurements, calculation of theoretical predicted variables, data reduction and storage.

The measurement software system was developed to achieve:

- A generalized parameter controlled software system for a number of measurement projects.
- Possibilities to include special calculations.
- A conditional datareduction scheme that increases information quality in reduced data and enables the selection of data reduction from special operating conditions.
- On-line simulation of the collectorsystem based on measured instantanous values.
- Possibilities to include control functions for slow varying control applications as strategy selection, solar tracking and so on.

The measurement parameters, PARF, that control the handling of data in the measurement computer are defined and checked off-line in a parameter generation program, PGEN.

The measurement parameters together with parameters for evaluation functions and formats also control the evaluation process.

2.1 Measurement and data collection program

The measurement software system is shown as one part in Figure 1. The system includes the measurement program as a module.

The main purpose of the measurement program is to control the data acquisition, to calculate theoretical values and at certain time intervals store the reduced data (mean values, accumulated totals, integrals, etc) on the storage media (cartridge or flexible disc).

The periodic measurement sequence consists of the following phases:

- sampling of measurement data and time
- calculation of conditions and calculation variables
- examining the connected condition for data reduction and if the status is true conduct the specified data reduction namely:
- accumulation
- integration
  - mean value
  - weighted mean value
  - momentary value
  - maximum, minimum value
  - variance
  - min-, maximum period/day
- store specified variables connected to the storage modes which are active
- activate or inactivate storage modes (1-3) depending on time of day and the status of the specified condition.

The possibilities to select data for data reduction according to the status of other processvariables and to calculate mean values based on energy weighted measurement values has been of special importance for the interpretation of differences between the solar collector systems.

2.2 Evaluation software system

The measurement system enables the user to select data reduction and storage functions that results in primary data covering different periods of the day for different measured and calculated variables.

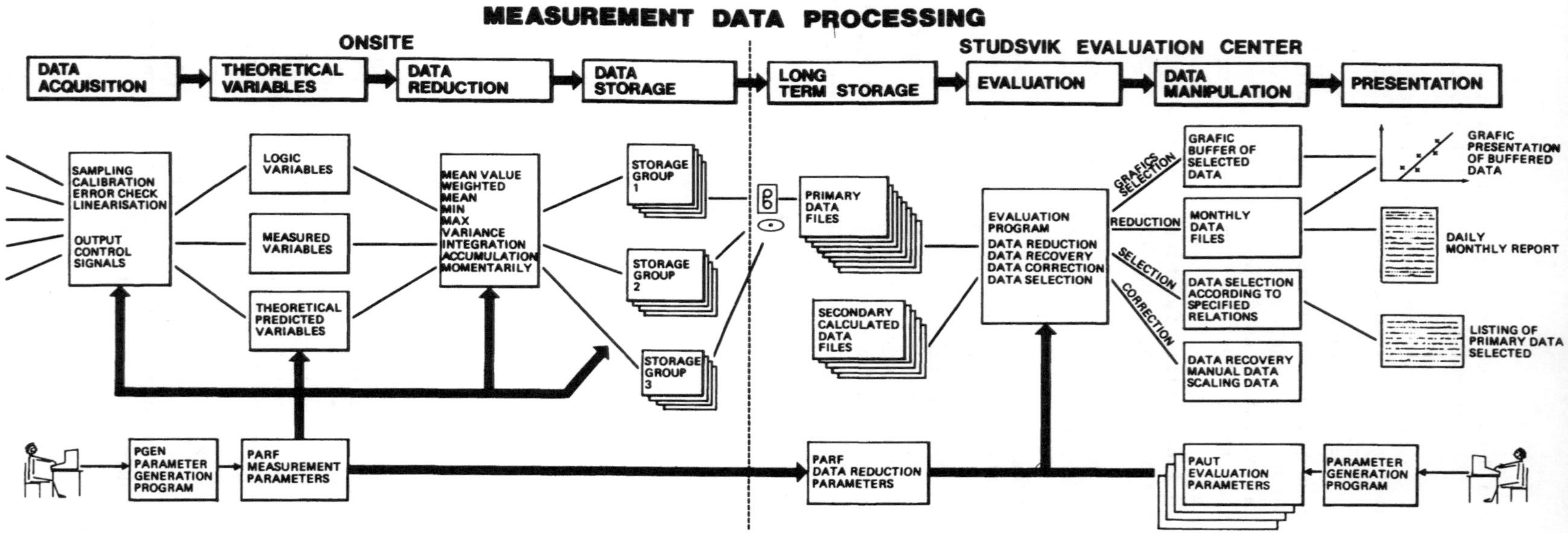

Figure 1 Measurement and evaluation flow chart

To get a flexible system for handling this primary data and put them together in daily or monthly reports a desktop computer HP-9845 with a whincester disc memory of 16.7 Mb has been used.

The evaluation software system was developed following the same principles used in the measurement software system and is shown in Figure 1.

The evaluation process is controlled by the same parameters, PARF that specified the data processing in the measurement system together with additional parameters, PAUT, specifying the evaluation calculations, reporting formats, etc.

The evaluation program RUTV contains the following functions:

- calculation and storage of secondary calculation variables using standard algorithms implemented in the program
- generation of a monthly file containing 24 h values put together from primary data using the same data reduction principles as in the measurement program.
- selection of data records according to a number of conditions specified and buffering of selected variables from these data records in a buffer for later presentation in graphical form.

The archival data files of primary data also contain identification records with data reduction parameters and data storage information.

The evaluation and data processing programs use these archival data files for input data together with evaluation parameters PAUT to perform calculation of secondary data and data reduction to 24 h values and monthly values.

According to a number of conditions defined in the evaluation program data records can be selected from special operating conditions for storage in a buffer for later graphical presentation.

In order to be able to present monthly values even if some data is lost, the evaluation program enables the user to calculate corrected monthly values by replacing 24 h values lost or incomplete with other 24 h values.

The measuring process then results in the following presentation:

- monthly report
- primary data reports at selected operating conditions
- graphical presentations of primary data or reduced/calculated data selected from several different evaluation periods
- secondary calculated values based on calculation algorithms defined in the evaluation program.

3. PRESENTATION OF RESULTS

During the period presented June 1982 - Dec 1983 two different control modes has been tested. 60°C outlet temperature to the district heating network and 80°C outlet temperature.

When comparing the collectors it is important to keep in mind the deviation in slope for the Philips collectors, 60° compared to 40° for the other collectors. This results in about 10 % less yearly solar insolation to the front aperture of the Philips collectors.

The Philips collectors show the highest performance in spite of the infavourable slope.

The Scandinavian Solar HT shows the second best performance just 10 % lower than the Philips collectors for the 80°C operation. Partly this is due to a special control system which gives a lower average

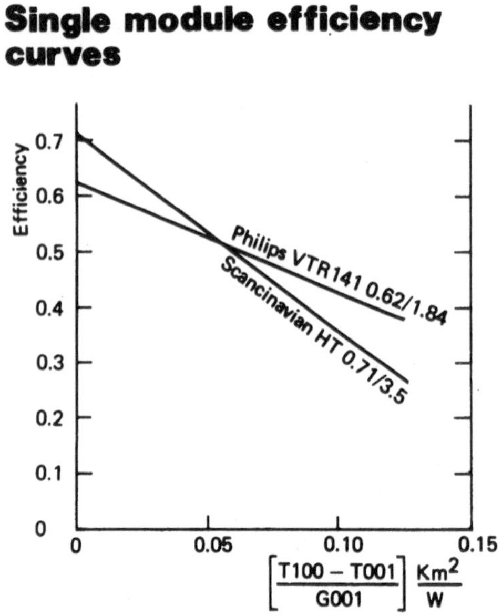

Fig. 2 Collector efficiency curves for Scandinavian HT and Philips VTR 141

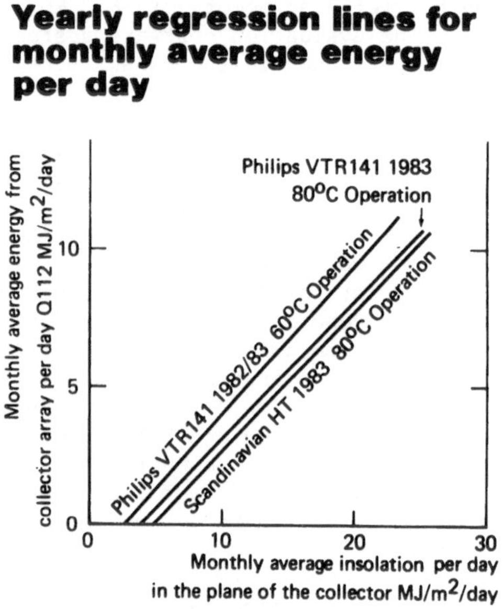

Fig. 3 Long term regression lines for the input/output relationship

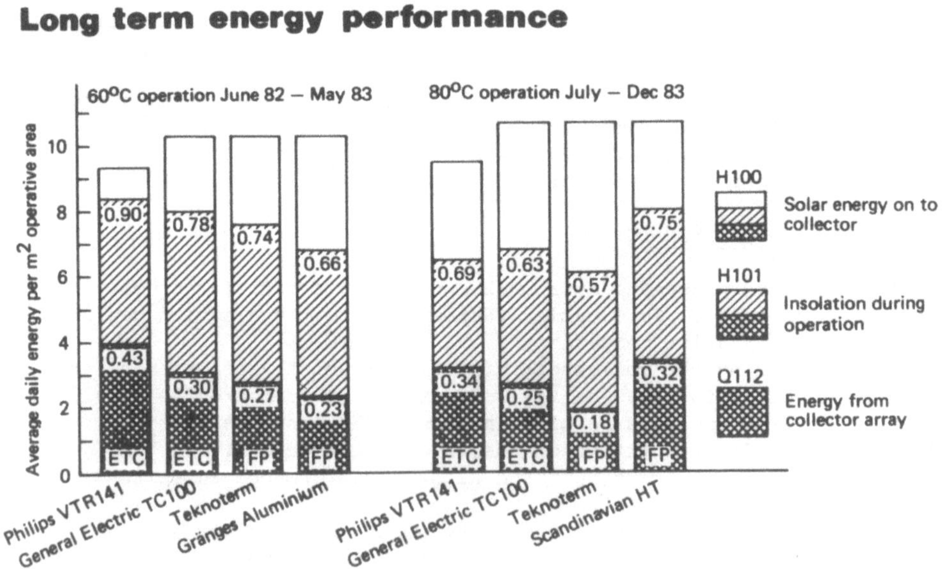

Fig. 4 Long term energy bar chart for 60°C and 80°C operation

operating ΔT ($9^{o}C$ lower than for the Philips array) with nearly the same output temperature to the network.

The other flat plate collector arrays show much lower performance. 30 - 40 % lower energy output in spite of a more favourable slope, compared to the Philips collectors.

In Figure 2 the collector efficiency curves for Scandinavian HT and Philips VTR 141 are shown for comparison.

Figure 3 show the long term regression lines for monthly average insolation and collector array output over a full year both for ETC and FP collectors.

The collected energy and insolation over a period with operation at $60^{o}C$ (June 1982 - May 1983) and a period with operation at $80^{o}C$ (July 1983 - Dec 1983) is shown in Figure 4.

4. CONCLUSIONS

- An advanced general data collection and processing system has been successfully developed for desktop HP-computers and has proven to give high measuring availability even for full year periods.
- The use of conditional data reduction has proven to increase the information quality and enables the selection of data reduction from special operating conditions.
- The parameter controlled software system allows flexible data processing according to parameters specification.
- Large module size, anticonvection glazing and improved control system can increase flat plate collector performance significantly.
- Well designed high efficiency flat plate collector systems can be successfully operated even at $80^{o}C$.
- On-line simulation enables fast identification of measurement and operational malfunctions and reduce data storage requirements to 1 hour values.

PERFORMANCE RESULTS FROM A LARGE SOLAR HEATING AND COOLING SYSTEM OPERATIONAL SINCE 1981 NEAR MARSEILLE (FRANCE)

P. KOZOULIA and R. LEDUC
COMPAGNIE MERIDIONALE D'EQUIPEMENTS TECHNIQUES (COMETEC)

Summary

Since 1981 a solar assisted thermal equipment supplies CTRCE (CENTRE TECHNIQUE REGIONAL DES CAISSES D'EPARGNE) heating and cooling needs. To obtained maximum energy saving and occupants confort, computer simulations were used to design the solar heating and cooling system. In account of the previous results and of the importance of such plant, for solar energy used development, 50 % of solar equipements over cost was financed by A.F.M.E (Agence Française pour la Maîtrise de l'Energie formerly COMES Commissariat à l'Energie Solaire). Since 1982 an on site microprocessor - based data acquisition system is used to analyse the real all system and components, behavior and performance. This field testing, performed by COMETEC and financed by A.F.M.E, have allowed, to rectify some installation failures, and to compare the performance and energy saving predictions generated by computer simulation to the measured values.

1. THE BUILDING

Established in 1973, the CTRCE provides data processing services to his whole menbers divided on 20 establishments in Bouches-du-Rhône, Languedoc and Roussillon. At the time of the building extension, the Caisses d'Epargne have shown their desire to built a building functional but saving energy. The local sunshine conditions were particulary conductive to a solar solution to provide a part of thermal needs for space heating and air conditionning. The CTRCE extension is a crown of offices connected by a large central space around the old building wich represents (central space + offices) 1.400 m2. The building roof plane, 30° inclined on south direction, supports the solar collectors used for heating and cooling.

2. WORK OUT PROJECT

-A reasonable returns on investment (less than 10 years) was obtained through an optimised system components size, after computer studies on different architectural and system scenarios. The low pay back obtained is due to the installation facilities to use up all over the year the whole solar energy collectors production. Computer simulations results for the choisen building and system are :

- thermal needs for space heating : 130.702 kWh/year
- cover rate for space heating needs by solar energy : 63 %
- thermal needs for space cooling : 23.418 kWh/year
- cover rate for space cooling needs by solar energy : 98 %
- thermal solar energy available for room computers cooling : 129.057 kWh/year

A test plant to scale 1/6 of CTRCE plant project was tested by the Compagnie des Lampes MAZDA (Vaccum tube collectors constructor) in collaboration with COMETEC at the S.M.E.H (Station Mediterranéenne d'Essais Heliothermiques). On this solar laboratory, with a complete instrumentation, different sort of regulation and safety device were tested as well as installation performance was measured and compared to computer simulation. Built in 1981 CTRCE building extension devices a solar assisted thermal equipment to provide thermal needs for space (1.400 m2) heating and cooling.

3. PLANT DESCRIPTION

The solar thermal energy provides from 304 m2 vaccum tube collectors 30° inclined. For space heating, the thermal energy distributed on fan coil units network is sampled in priority order, on solar hot water storage tank through a flat plate heat exchanger, after, on compressor chiller water condensor, and at last, on existing gaz boiller through a multitubular heat exchanger. Cold water (9-13°C) for space cooling is produced by two lithium - bromide absorption machines (2 x 35.000 fg/h) supplied with solar hot water at 80-85°C. Temperature of hot water boiller inlet is restricted at 85°C by a mixing - valve and the solar thermal energy excess is stocked in the solar hot water storage (8,5m3). Cooling absorption chiller output is absorbed on two cold water storages (2 m3 each) below utilization. In rescue, cooling needs are satisfied by the compressor chiller, ordinary used for room computers air conditionning. Recirculation hot storage water is used for the field collectors freeze protection. Two hight temperature securities are available on field collectors. In case of stagnation, field collectors is drain down by two differential thermostat controlled electro-valves. Up to 110°C on running field collectors water, an aerotherm dispatches solar thermal energy excess.

4. RESULTS

The solar field collectors performance evaluated from data measurment is satisfying (52 % efficiency). Results obtained on absorption chiller show a significant difference (about 20 % less) between constructor and measured performance, but still correct according to this type of solar installation state working. Both, heating and cooling, measured building requirement exceed computer simulation expectation, about 75 % more for heating and 200 % more for cooling. These results proceed essencially from an intermittently mismanagement. Also, the cover rate, respectively for heating and cooling building needs by solar energy, are not higher than 22 % and 55 %. The relatively bad performance in heating derives from three heat sources mismanagement, and field collector freeze protection heavy on solar energy.Lessons drawn from last three years system testing have been used to improve energy managemert and field collectors freeze protection. Last values, measured during the just starling new testing period, after installation modifications, still show that predicted computer plant performance data could be reached rapidely. The solar collectors productivity, wich had been evaluated on first year at about 800 kWh/m2 year, could reach, in this conditions 980 kWh/m3 year as computer simulation. Finaly the feasibility of saving cooling and heating energy at low pay-back, in using vaccum tube collector and lithium bromide absorption machine has been demonstrated.

PERFORMANCE MONITORING OF APPARTMENT BUILDINGS IN BERLIN

R. Hanitsch, G. Valentin
Technische Universität Berlin
Institut für Elektrische Maschinen
Einsteinufer 13-15, 1000 Berlin 10
Federal Republik of Germany

Summary

In order to demonstrate the application of solar energy for domestic hot water production in Berlin on ten appartment buildings large flate plate collector systems were installed between 1978 and 1982. Five of these buildings with six solarsystems have been equiped with measurement systems. Measurements were carried out by the Technical University Berlin by order of the German Federal Ministry for Research and Technology. The purpose of this program is to collect information about the energy balances of all systems.

The measured data analysed so far show that the system efficiency is rather different from system to system. We obtained values between 5 % and 38 %. During the monitoring of the systems we could improve the control strategy of most of the systems and simultanously the system efficiency.

1. INTRODUCTION

The experiments on solarsystems for domestic hot water production in Berlin are integrated in a project ("ZIP"), sponcered by the German Federal Ministry for Research and Technology. The solarsystems were added to conventionel DHW-systems in existing appartment buildings. The design and calculation of dimensions of all components are made by the manufacturer or the fitter service of each solarsystem. Six of these solarsystems have been equiped with measurement systems and were carried out by the Technical University Berlin. The general description of these solarsystems is summarized in the following table:

solarsystem	collector-area m^2	storage volume m^3	hot water concumption m^3/d
1	82	12	2,7
2	172	25	1,8
3	72	2	1,1
4	66	10	2,4
5	98	11	1,0
6	128	8	3,5

T. 1 Main values of the 6 solarsystems in Berlin

2. PERFORMANCE MONITORING

In the six projects, a mixture of monitoring levels is used. One project is highly instrumented and controlled for assessing the performance of the system. The other projects are well instrumented. The measured data are temperatures, counting rates of flow meters, solar radiation, windvelocity, operating time of pumps and electricity consumption. Compared with other experiences, measurements on solarsystems need a data acquisition for a long period. The values of all the sensors are stored in counters or in read-out devices for weekly reading off and in addition one data acquisition system with a memory recorder was used. The measured data are analysed on a central computer.

3. RESULTS IN DIAGRAMS

The measured data analysed so far show that the system efficiency is rather different from system to system. Figure 1 shows the obtained values of the six solarsystems from April to September 1983. System 2 and 5 had beakdowns during the measure period, distinguished in both low efficiency and delivery rate.

A summary of this efficiency results shows the following table

solarsystem	collectorsystem efficiency %	solarsystem efficiency %	delivery rate %
1	37,8	21,5	33,8
2	5,5	4,9	18,3
3	20,5	14,3	54,1
4	28,8	23,0	35,9
5	27,7	12,2	50,0
6	31,4	20,5	23,1

The energy balance of all systems shows Figure 2. To compare the six systems, the reference value for all systems is the hot water consumption.

The system efficiency gs and delivery rate D for one of the best system is shown in Figure 3. Figure 4 shows the Energy Flow Diagram for the same system.

4. CONCLUSIONS

During the monitoring of the systems we could improve the control strategy of most of the systems and simultanously the system efficiency.

On the basis of this practical experience the following preliminary results can be defined:
- Insert enough air exhausters on the top of the collector field.
- Minimize the thermal capacity of all devices in the collector circuit.
- Calculate the heat exchanger more carefully in an optimum design.
- Use a simple and reliable controll system for simple solar systems, a microcomputer based controll system for sophisticated solar systems.

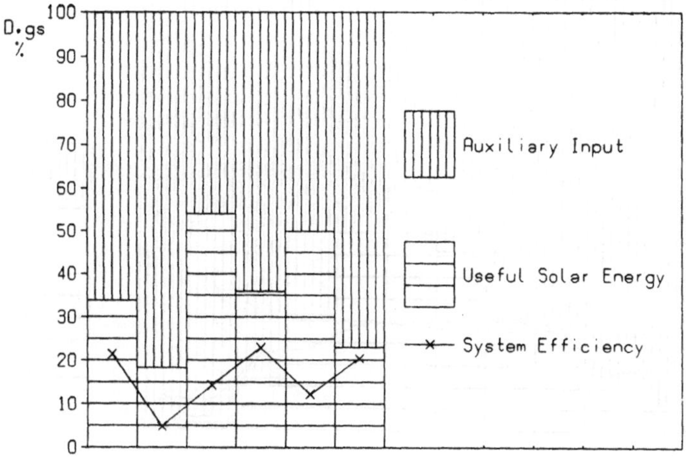

Fig.1 System efficiency gs and delivery rate D of six different solar systems in Berlin. Period: April to September 1983.

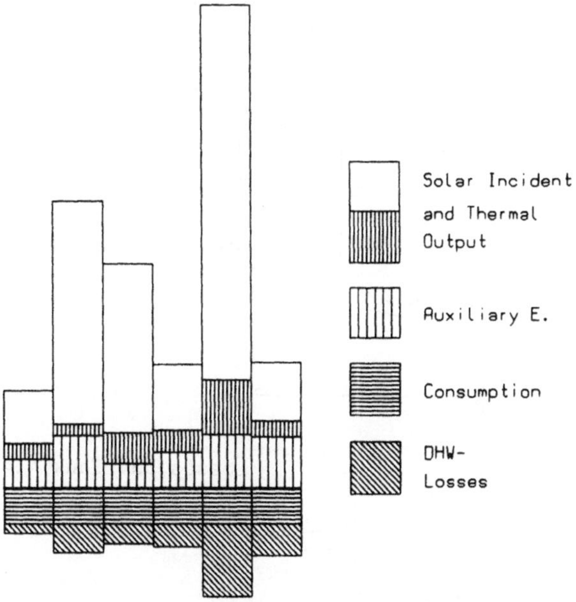

Fig.2 Energy balance of the six different solar systems from April to September 1983. Reference value: hot water consumption.

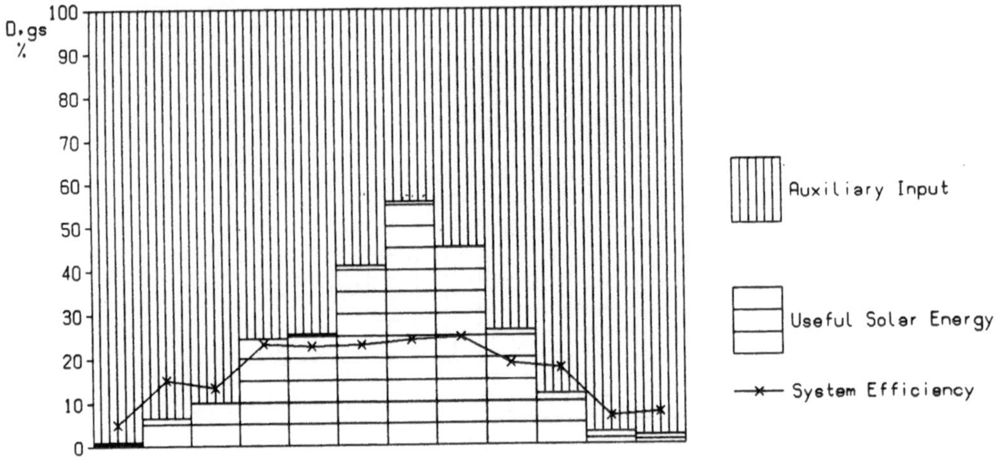

Fig.3 System efficiency gs and delivery rate D 1983 for solar system 4.

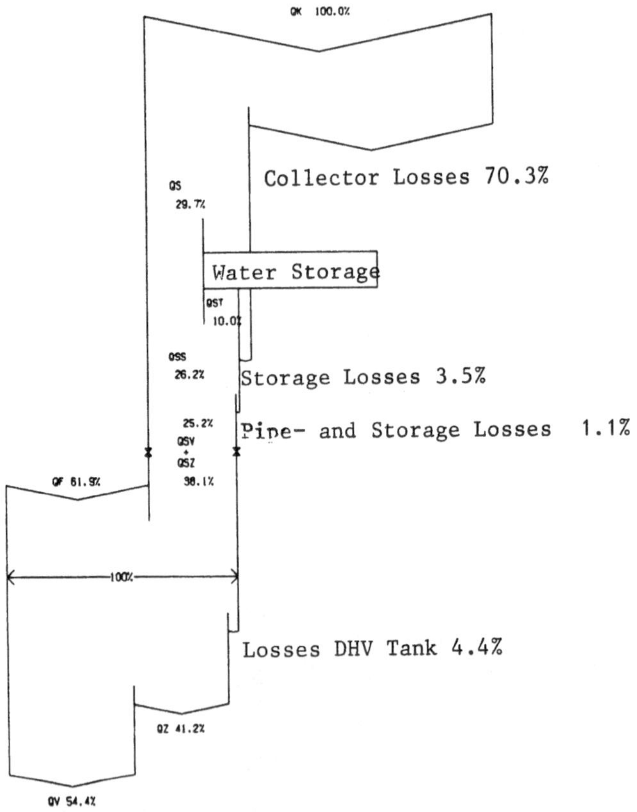

Fig.4 Energy flow diagram, April to September 1983 Solar system 4

DESCRIPTION AND EXPERIMENTAL LONG RANGE EVALUATION OF A SOLAR ELECTRIC PLANT INSTALLED IN A RESIDENTIAL BUILDING IN SOUTHERN ITALY

F. Parrini - R. Viadana
ENEL (Italian Electricity Generating Board) CRTN - Milan
N. Viti
ENEL (Italian Electricity Generating Board) SPT - Rossano Calabro
A. Biondo - C. Filinceri
PHOEBUS S.p.A. - Catania

Summary

From winter season 1979/'80 ENEL installed and monitorized an SHS and a DHW solar system in a six flats building located near the Rossano Calabro (Cosenza) thermal power plant. The main purpose of the installation was to verify the effective solar contribution to the winter building load in a climatically favourable site with a plant formed of non selective solar collectors and of a daily thermic storage. The solar plant is composed of a 50 m^2 conventional type solar collector field, located on the roof, and of an 80 m^2 collector field integrated in the South facing wall. An automatic data acquisition and recording system allows the evaluation of the energetic bahaviour and the reliability of the plant and of its main items. The long period running of the plant (five years) allows besides to estimate the performance degrade of solar collectors caused from aging processes.
The measurements runs during five winter seasons performs a value of solar fraction about 20÷25% and an average collecting efficiency about 20÷25%. The reliability of the plant and particulary of solar system has been very satisfactory. None extraordinary operation was necessary. Solar collectors efficiency, after four years, is reduced about 2%. Data acquised was utilised for the validation of mathematical simulation models of energetic building-plant systems.

1. INTRODUCTION

From winter 1979'80 ENEL realized and made operating a SHS and DHW experimental plant for a six flats civil building located near the Rossano Calabro (Cosenza) thermal power plant.

2. PURPOSE OF THE WORK

The purpose of the above realization was to verify the effective solar contribution by the use of a non selective solar collector plant integrated with a daily thermic storage system, in a site favourable climate (see table I).

Location	: Rossano Calabro (Cosenza)	Altitude s.l.m.: 10 m
Latitude	: 39,5° North	Longitude : 16,7° East
Typical climate	: Temperate zone	
Degree day	: 1500°C day (base 19°C)	
Ambient temperature	: 15°C (annual average)	9°C (dec-jan)
Global irradation horizontal plane	: 3,78 kWh/m^2	(annual average)
" " " "	: 2,5 kWh/m^2	(dec-jan)
Wind speed	: 2 m/s	
Relative humidity	: 70%	

Table I : Climate

Another purpose of this enterpise was to verify, in operating conditions, the realiability of non conventional installed items and the occupancy plant agreement.

The long running period (5 years) has besides allowed to estimate the energethic performance degrade os solar collectors due to aging processing.

The plant is equiped with a measurement, collecting and data treatement system whose purpose is to determine the energetic behaviour of its main items measurement campaigns worked out in 5 years; collected informations permitted to validate simulation mathematical models of the whole system energetic behaviour.

The collectors on the South wall facing are another interesting aspect of this realization; the above collectors, are in fact integrated with the wall and the whole building and, besides forming an additional thermal insulation, have olso good aesthetical characteristics.

3. BUILDING-PLANT SYSTEM

On table II you can see the main constructive characteristics of the building, the global heat loss coefficient, the nominal heating power and the main energy conservation measures adopted.

Heated foor area	: 780 m^2	
Building height	: 12,5 m (three flats)	
Heat loss coefficient	: 3120 W/K	
Design temperature	: 2°C (ambient)	20°C (internal)
Nominal power rating	: 60 KW	

Energy conservation measures:

a - Insulated roof (15 cm light weight concrete)
b - Insulated South wall (100 mm rockwool)
c - Double glazing, insulated shutters
d - Individual thermostat controls in each room

Table II: Building

The plant, whose semplified diagram with the main measure points is reported inf fig 1, is formed of:

a- a vertical collector field integrated on the South facing wall, with an absorber area of about 80 mq. (N. 4 collectors of about 10 m^2 absorber area);

b- a conventional collector field located on the roof 45° tilted, with an absorber area of 50 m^2 about (N. 32 roll bond solar collectors of 1,5m^2 absorber area);

c- a daily solar storage tank of 7,5 m^3.

Auxiliary system is formed by:

a- electric night heaters (resistance)

b- daily storage tank of 10 m^3.

The purpose of the auxiliary storage tank is to displace electric energy demand to off-peack hours (night) in which there is a low utilization coefficient of power plants.

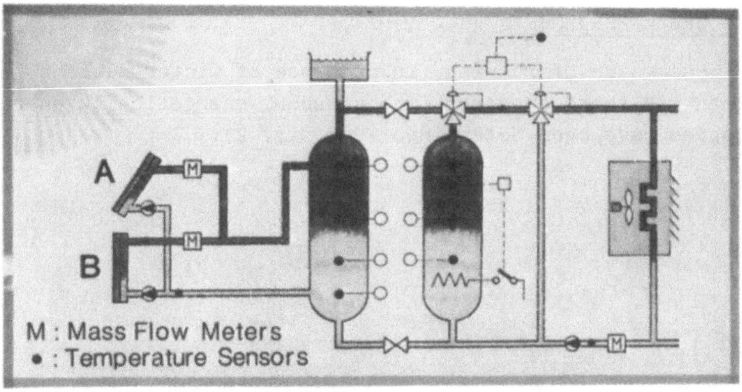

Fig. 1: Simplified system diagram

On table III you can see the main technical characteristics of the whole plant.

Solar section: a- solar collectors

	ROOF	FACING WALL
- absorber	Al roll bond	Al extruded
- treatment	black non selective	black non selective
glazing	single	single
back insulation	50 mm rock wool	100 mm rock wool
coolant	water	water
antifreeze	=	=
absorber area	50 m^2	80 m^2

(table III cont.)

tilt	45°	90°
design Δt	10°C	10°C
	Solar section: b- daily storage	
Volume	7,5 m^3	
Design temperatures	60°C (max)	40°C (min)
Thermal capacity	0,6 x 10^6 kJ	
	Auxiliary system	
Nominal electric power	80 kW	
Heating period	from 8 p.m. to 7 a.m.	
Daily storage volume	10 m^3	
Design temperatures	85°C (max)	45°C (min)
Thermal capacity	1,9 x 10^6 kJ	

Table III: Plant

4. RESULTS OF THE EXPERIENCE

During these years of running, in presence of winter seasons less favourable than the those expected, the seasonal energetic fluxes reported in Sankey diagram have been determined (see fig. 2).

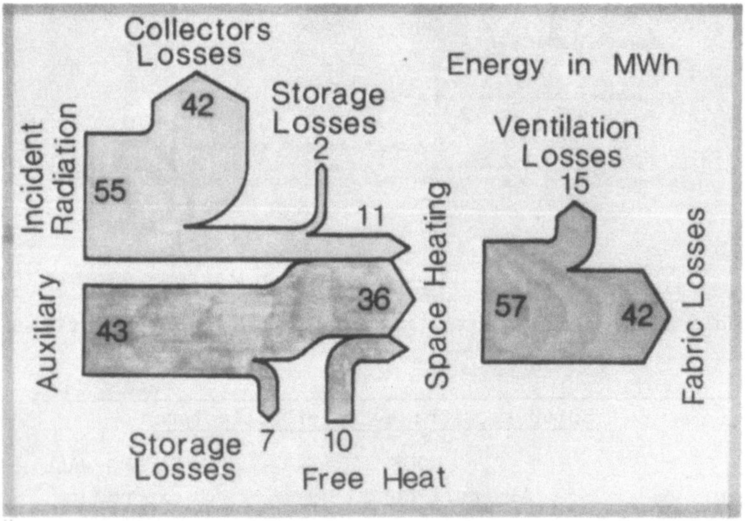

Fig. 2: Sankey diagram

The average plant performances are:

- solar fraction intended as the rate between het exploited solar energy and global energy supplied to the building: 20÷25% ;
- seasonal efficiency of solar collectors intended as the rate between solar energy supplied to the plant and incident energy on the collectors: 20÷25%;
- global storage efficiency: 80÷85%.

The building winter requirement, whose available flats have been only half occupied, is about 47.000 kWh.

In 1983, two roof collectors, one running for about 3 years; the other spare in the plant, was subjected to a comparative test for the determination of the instantaneans efficiency.

The results (see fig. 3) tell us that the four years running collector has given an average efficiency lower than the other about 2%.

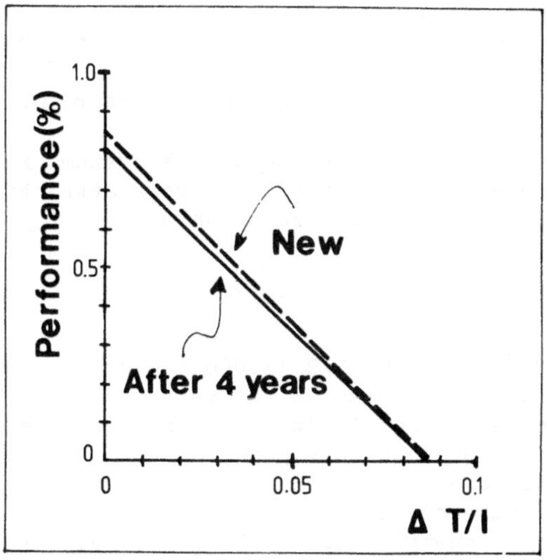

Fig. 3: Tilted (roof) solar collector performances

5. CONCLUSIONS

Running period data and the operation effected during the five years running, have pointed up what following:

a- the best plant reliability proved by an absolute regular running and a lack of damages on collector fields;

b- high acceptation degree from occupancy;

c- solar collectors energetic performances lower than those initially estimated;

d- solar collector characteristics have remained steady during the years in spite of the proximity of the shore;

e- good precisation of the installed measure system.

SIMULATION AND MONITORING OF SOLAR DOMESTIC HOT WATER SYSTEMS

G. Brouwer
Van Heugten Consulting Engineers
P.O. Box 305, 6500 AH NIJMEGEN
The Netherlands

Summary

The use of solar energy for domestic hot water production seems to be very attractive due to the fact that there is a demand during the complete year. In the period of the last 10 years about 1000 Domestic Hot Water solar systems are installed. Experiences with existing installations and measured results are analyzed in serval projects taking part of the National Research Program in Solar Energy in the Netherlands.
These results are compared to predicted results from computer simulation models.
There is ample correlation between monitored results and simulation models.
Though the number of installed solar DHW systems is quite small, the price-performance of Dutch systems is very attractive. The prospects for mass production look very favourable.

1. INTRODUCTION

In the past 10 years about 1/3 (3.000 m²) of the total installed collector area in the Netherlands is for solar water heating in dwellings. Experiences and monitoring results about the performance of these systems are presented in this paper.
The actual daily hot water consumption of an average household is about 100 ltr at a temperature of 60 °C.
Variations in water consumption of 4 : 1 are not uncommon.
Therefore solar systems for average households are designed with a 3 m² collector (spectral selective absorber and single glazing) and a 150 ltr heat storage, so that anually approximiately 50% is saved on the fuel bill for water heating.
The installed system-costs are currently without grants in the order of 1200 ECU (about Dfl. 3360).
With the present gasprice of 0,0135 ECU/MJ heat demand, (about 13,5 ct/kWh), the predicted annual saving is nearly 50 ECU (140 Dfl.).
The equivalence of predicted and measured results of the performance of solar systems on yearly and monthly basis is very good. This can be used for analysing improvements and for operational feedback of installed systems.
Based on the measured results and the gathered experiences from a number of subsidized projects the aim of the study is to develope most promising routes for the introduction of solar water heating systems.

2. SHORT DESCRIPTION OF 2 PROJECTS

1. Project Holendrecht, Amsterdam
   86 one-family dwellings from which 12 solar systems have been monitored realisation september 1981
   - Pumped/drain down system, transfer of heated fluid (water) from the collector to the pre-heated tank by means of forced circulation
   - Collector : 2,7 m² area, spectral selective absorber (α = 0,92, Σ = 0,24), single polycarbonate cover, Collector flow rate : 90 l/h.m²
   - Heat storage : 160 ltr water, one heat exchanger (single wall)
   - Operating modes circulation pump : +8,5, +2,5 °C
   - Orientation SSE, tilt 45°
   - Domestic hot water consumption (design) 100 l/day, 60 °C.

2. Project Indische Buurt, Amsterdam
   127 dwellings with grouped solar systems installed on appartment buildings, from which 12 solar systems have been monitored (for 48 dwellings), realisation, febr. 1982
   - Pumped system, (anti freeze)
   - Collector : 8 m² area, spectral selective absorber (α = 0,92, Σ = 0,24), single tempered low iron glass, collector flow rate : 90 l/h.m²
   - Heat storage : 500 ltr water, double heat exchanger
   - Operating modes circulation pump : +8,5, +2,5 °C
   - Orientation : S, tilt 45 °
   - Domestic hot water consumption (design) 320 l/day, 60 °C (4 dwellings)

3. PREDICTED RESULTS AND PARAMETER STUDY

The expected performance of a solar DHW system depends to a large extent on the climatical conditions. On a monthly basis the analysis of long term weather data shows a spread between 475 MJ/m² month in summer and 70 MJ/m² month in winter. The average monthly ambient temperatures range from 2 to 18 °C (annual 9,5 °C). Solar DHW systems are, due to the fact that demand is continuous during the whole year, very attractive in summer.

The heat gain as well as the validation accuracy of the results during the 4 month winter period are very small.

Mathematical modelling and system evaluation are used as a frame work for carefully balanced design of Solar systems.

The used computer programs to predict efficiences and to design systems are very detailed, they are executed at our computing centre.

The first step in the design process is to implement the instantaneous efficiency of the measurements into the simulation model of the collector.

Secondly the simulation model of the total solar DHW system is used for predicting the hourly, daily or monthly heat-gain of the system with reference weather data. The predicted monthly results for project 1 are shown in figure 1.

Heat-gain Solar DHW 1125 kWh

Solar fraction 32%

Wind speed 5m/s.

Control Δt 0,5 °C

Flow 100 l/h,m²

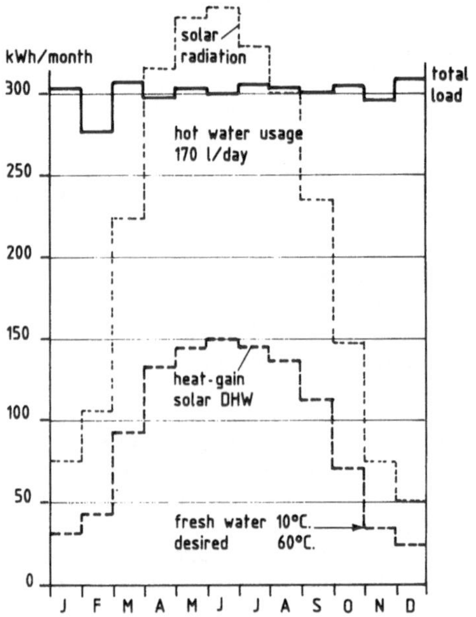

fig. 1

Very interesting are the influences of some parameters on the heat gain. Figure 2 shows the solar fraction of a solar DHW system as a function of the hot water demand on an average per day (a) and the collector flow rate (b).

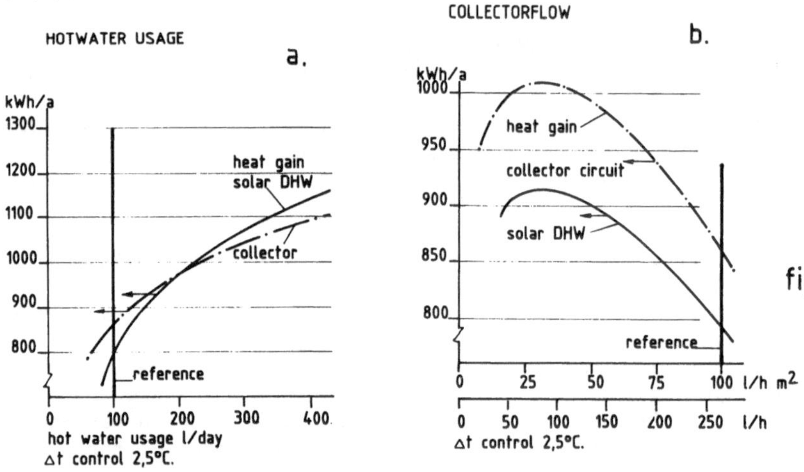

fig. 2

From the results of parameterstudies optimal conditions of the systems can be derived.

Decrease of the flow-rate in the collector circuit and decrease of the temperature difference of the out operating mode of the circulation pump contribute to about 15% higher heat gain for this type of solar DHW system.

On the results of these parameter studies with the simulation models easy readable graphics are carried out for energy gain estimation, system dimensioning and economic evaluation.

4. THE MONITORING PROCESS OF SOLAR DHW SYSTEMS

The data of the projects are obtained with relatively sophisticated data processing systems (manufacturer Acurex type Autodata 10/5). The sensors are more simple, a solarimeter (Dirmhirn) in the collector plane, fluid (turbine) flow meters, temperature sensors (4 leads platinum resistance elements) and a wind velocity meter. The scanning of analogous and digital inputs, mostly every 20 seconds and the programmable arithmetic functions of the data acquisition system make it possible to obtain instantaneous and long term performance data of the solar system.

The scheme of project 1 is given in figure 3.

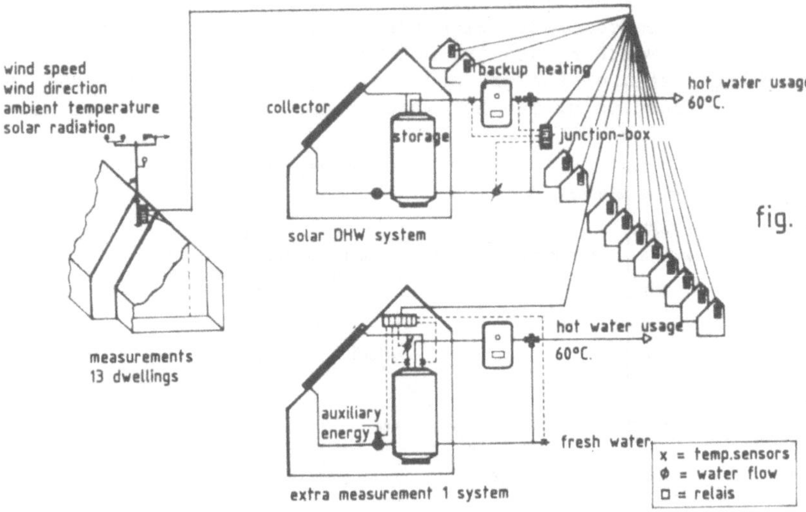

fig. 3

SOLAR DHW SYSTEMS 'HOLENDRECHT' -AMSTERDAM

These analysed data are stored on a recording medium (cassette). Programming and transfer of data with data recording is also possible at the computer analysis centre by means of 2 modems and a telephone line ; the modem on site has an automatic answer function.

In this way more than one monitoring system can be controlled, programmed, changed and recorded at the computercentre of our company.

In the datalogging system the complete Performance Data Sheet of the EC Reporting format can be programmed.

5. THE MONITORING RESULTS

For different types of Solar DHW systems (for detached houses, project 1 and apartment buildings project 2) measured data in a large extent are analysed for the year 1983.

The hot water demand for the one family dwelling 50 - 325 ltr/day of a temperature of 60 °C varies considerably (not quite equivalent with the average Dutch households).

The yearly average is about 160 ltr/day.

The water consumption in the apartment buildings is much lower, about 70 ltr/day.

In summer the fresh water temperature is much higher as was predicted (10 °C). In 1983 this temperature varies during the year from 5 to 20 °C. Therefore the solar fraction decreases considerably. Heat losses of not insulated appendages and sensors slightly decrease the heat gain of the solar systems. On sunny days during the summer the storage temperature is high enough for consuming. In most dwellings the post heaters (including the watch flames) are frequently not in operation.
These consumer attitudes stand for an indirect saving of the solar DHW system.
Absence of the inhabitants during vacations also diminishes the heat gain of the solar system.
The measured instantaneous efficiency data of the actual collectors are compared with the predicted results.

The measuring results of the yearly heat gain of the Solar DHW systems can also be comprared with the results from the simulation model using the actual weather data and hot water demand.
For the solar system of the grouped dwellings figure 4 shows the solar fraction as a function of the hot water demand per day on average.

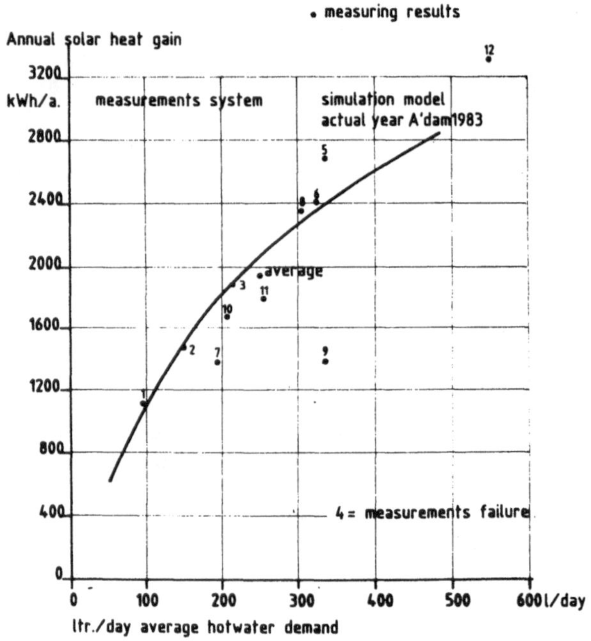

fig. 4

The before mentioned analysis is executed for each month of the year. From these results the following conclusion is made : The functioning and the yearly performance of actual solar DHW systems can be obtained by actual monitoring data for a relatively short period and by the use of reliable simulation models.
The total solar radiation during this monitoring periode is a measure of the accuracy of the performance prediction.

6. CONCLUSIONS

Based on the experiences and stimulated by the sophisticated know-how of consultants the Dutch market offers solar energy products of a guaranteed quality and at competitive price levels.
The price prospects for mass production look very favourable.

The improvement of system efficiency of actual solar DHW systems occurs by :

- optimizing the collector flow rate
- decreasing temperature difference of the control of the circulation pump
- better insulation of storage, pipes and components.

The system efficiency for collective solar boilers is lower than for individual solar boilers for dwellings due to the double heat exchanger prescribed by the Government. A high heat transfer capacity of the heat exchangers is necessary.

There is ample correlation between monitoring results and simulation models. With the use of simulation models and with short term monitoring results (weeks) the functioning and yearly performance can be easily and accurately determined.

The heat gain of solar DHW systems for 1983 amounts to (excluding failing systems) :

- 3600 MJ for one-family dwellings (project 1, 160 ltr hot water/day, 60 °C)
- 2050 MJ for apartment dwellings (project 2, 70 ltr hot water/day, 60 °C).

If indirect savings of back up heating (not in operation) are included the savings will be about 20% higher, resp. 56 ECU (Dfl. 158,--) and 32 ECU (Dfl. 90,--). The gasprice for 1983 is 0,013 ECU/MJ (13,2 ct/kWh).

SOLAR HEATING TEST DESIGN FACILITY FOR BULK PCM STORAGE

P. ACHARD - B. AMANN - D. MAYER
Ecole Nationale Supérieure des Mines de Paris
Centre d'Energétique - Sophia Antipolis
06565 VALBONNE Cedex
FRANCE

Summary

This experimentation, conducted by the "Centre d'Energétique de l'ENSMP", was designed to analyze the interest of bulk PCM storage centralized in a real water active solar heating system consisting of a low temperature distribution by floor heating. The main difficulty concerning the use of phase change enthalpy of melting bodies resides in the transfer of energy between the fluid carrier and its materials and vice versa. Adequate exchangers should be elaborated to permit maintainance of a given power after a certain time of charge-discharge. The research for an optimum running system lies, therefore, on specific dimensionning of the exchange surface as well as the storage volume.

The analysis of the interest of a PCM storage is carried out by comparison with a classical water storage. The principle of the regulations of the loop is to permit the loop to satisfy the maximum needs as directly as possible with the solar gains. The analysis shows

- the variability of
  - . the power absorbed or delivered by the storage
  - . the required capacity of the storage (daily, monthly...)
- the behaviour of the storage as a piston type as long as the first rotation of the volume of the storage is not achieved.

1. INTRODUCTION

Cette installation (outil de recherche pour le Centre d'Energétique de l'ENSMP à Sophia Antipolis) a été réalisée par l'office public des HLM de Cannes qui en a confié la maîtrise d'oeuvre à son bureau d'études ENERSCOP.

L'Agence Française pour la Maîtrise de l'Energie (AFME) et la direction de la construction du Ministère de l'Urbanisme ont décidé de soutenir conjointement cette expérimentation. Celle-ci conduite par le Centre d'Energétique de l'ENSMP doit permettre de situer l'intérêt d'un stockage par chaleur latente dans un système de chauffage solaire actif comprenant une distribution à basse température (16 m2 de capteurs, 70m2 de plancher chauffant). L'analyse de cet intérêt est effectué par comparaison avec un stockage classique en eau (800 l). Un système d'acquisition de données autonome a été élaboré spécifiquement, il est piloté par un micro-ordinateur et comporte 30 voies de mesures. Il permet d'acquérir, de visualiser et de traiter les données sur place. La saison de chauffe 83-84 a permis une analyse fine du fonctionnement du système avec le stockage en eau. Ont été mis en évidence en particulier: le rôle de la régulation sur une dalle mince, les séquences de fonctionnement piston du stockage, la variabilité des sollicitations qu'il subit en puissance et de la capacité mise en oeuvre.

2. ANALYSE DU FONCTIONNEMENT

La gestion de l'ensemble est prévue de façon à privilégier une satisfaction des besoins aussi directe que possible. Le circuit de captation et celui de distribution fonctionnent à même débit constant. La régulation en température de la distribution se fait par vanne trois voies pilotée sur la température extérieure. Le stockage, couplé en parallèle entre le circuit capteur et la distribution reçoit un débit variable dans un sens ou dans le sens inverse.

3. ANALYSE DES RESULTATS OBTENUS

Cette analyse du système avec stockage en eau est effectuée grace à l'enregistrement de 16 voies de mesures:

. prises de température (9)
. prises de débit (3)
. mesure ensoleillement (1)
. états de vannes (2)
. mesure de l'appoint électrique (1)

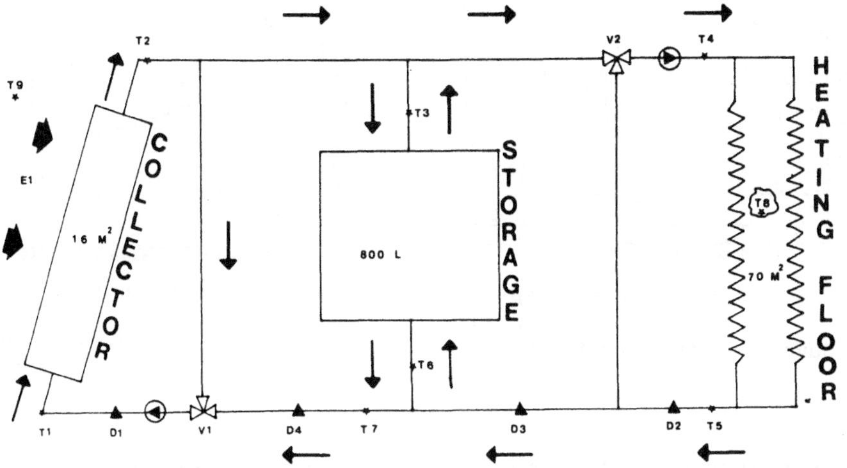

Une séquence de deux journées d'hiver est présentée:

. la première (belle journée froide) montre une couverture à 70% des besoins sur 24 h
. la seconde (avec passage nuageux) montre une satisfaction à 50%

Les courbes enregistrées mettent en évidence un fonctionnement piston du stockage en charge et surtout en décharge tant que le volume du stockage n'a pas effectué une rotation complète. Tout au long de ces séquences de fonctionnement, les rendements sont aussi élevés que possible, (notamment celui de captation), la température d'entrée restant constante. Dès que cette séquence est achevée:

. la température d'entrée du circuit capteur croit très vite (en captation)
. la température de sortie du stock s'élève (en captation) ou s'écroule (en distribution) pour devenir "égale" à celle d'entrée

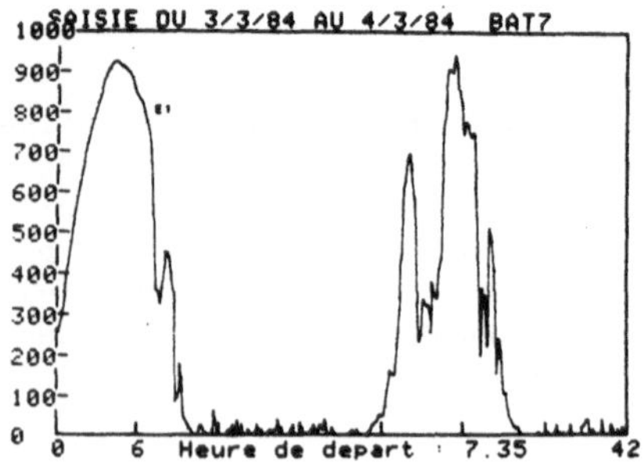

Ensoleillement: E1
(W/m2)

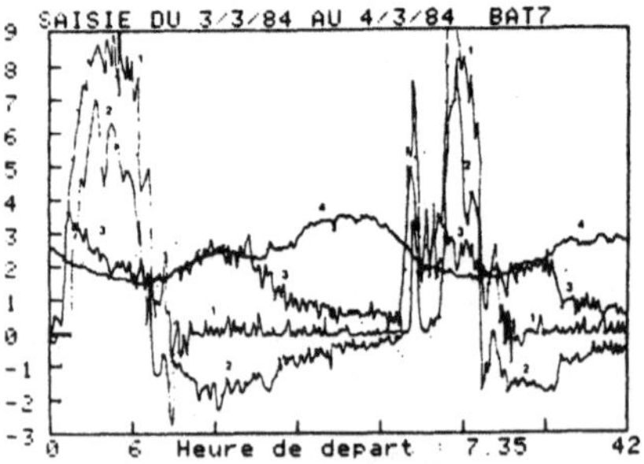

Puissance: 1 captation
(kWh) 2 stockage
3 distribution
4 déperdition

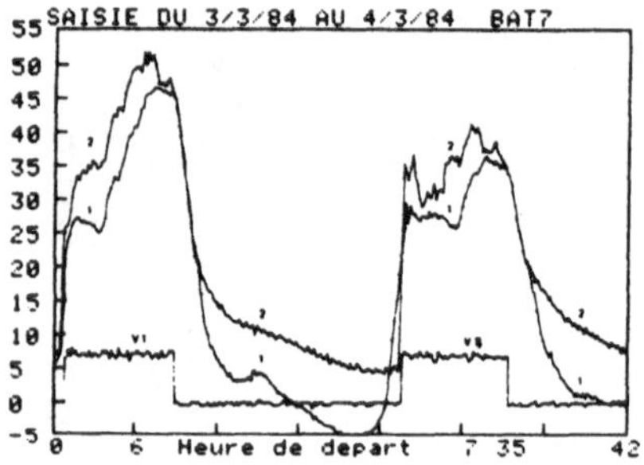

Température: 1 entrée capteurs
(°C) 2 sortie capteurs
Ouverture vanne capteurs: VI

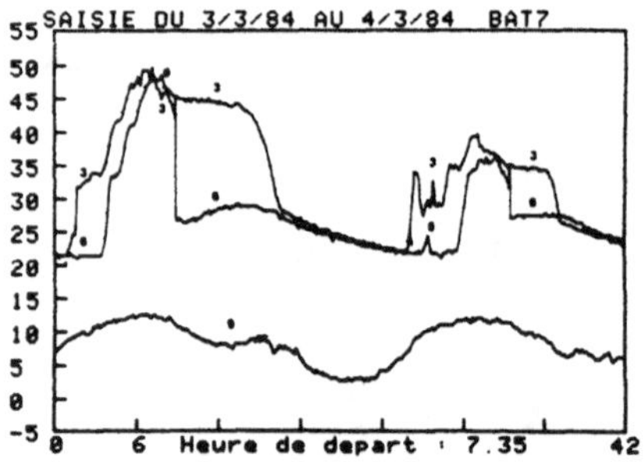

Température: 3 entrée cuve
(°c) 6 sortie cuve
9 extérieure

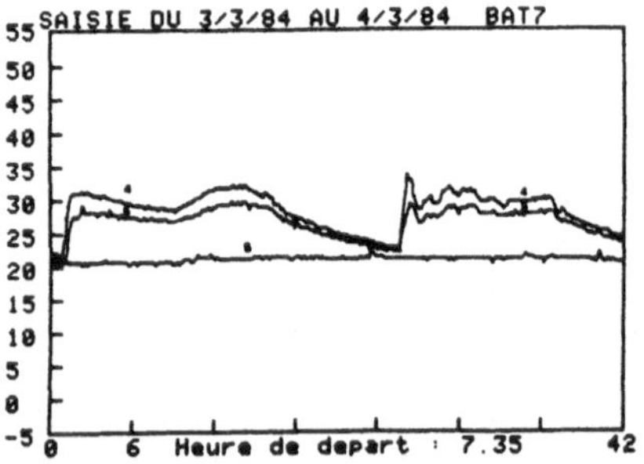

Température: 4 entrée plancher
(°C) 5 sortie plancher
8 appartement

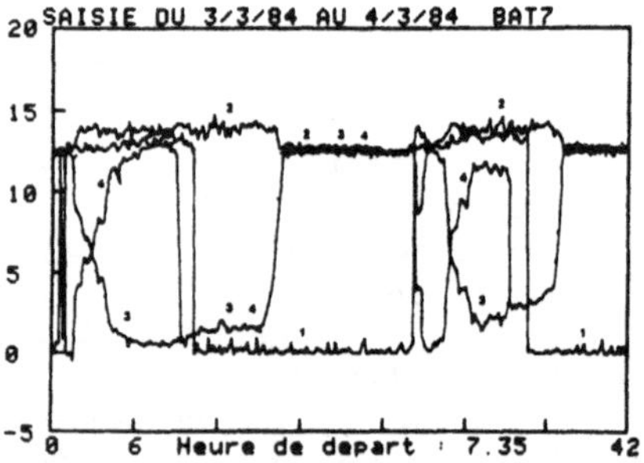

Débit: 1 captation
(l/mn) 2 distribution
3 retour
4 stockage

Ces résultats ainsi que ceux d'autres séquences explicitent la variabilité des sollicitations que subit le stockage:

. variabilité de la puissance absorbée ou délivrée (à débit variable)
. variabilité de la capacité demandée pour les différentes séquences météo

4. RECHERCHE EN COURS

Le stockage idéal devrait fonctionner en "piston" avec une capacité infinie. Il semble a priori qu'un stock latent puisse fonctionner en piston pour ce qui est du maintien de la température de sortie à une valeur quasi-constante, mais pas pour n'importe quelle puissance. Le problème à résoudre consiste à dimensionner un échangeur de manière à avoir un fonctionnement du stock latent aussi près que possible du "modèle piston" pour les gammes de puissance mises en oeuvre. Cette expérimentation va donc permettre de valider les règles de dimensionnement élaborées par le Centre d'Energétique (1). La saison de chauffe étant finie, les tests de deux géomètries de cuve (l'une parallèlipédique, l'autre cylindrique) munies de structure d'échange à faisceau tubulaire légèrement différente et de $CaCl_2$, $6H_2O$ comme matériau à changement de phase , sont effectuées sur le banc d'essai de l'ENSMP. Les résultats obtenus permettant de choisir la configuration qui sera mise en place lors de la prochaine saison de chauffe pour des essais in situ. Ultérieurement, d'autres types de stockage travaillant avec l'eau comme fluide caloporteur pourront être expérimentés.

5. CONCLUSION

Cette expérimentation opérationnelle depuis Décembre 1983, permet d'étudier le comportement d'unité de stockage court terme dans un système de chauffage solaire actif, de caractériser leur fonctionnement du point de vue transfert thermique, ainsi que de valider les méthodes de dimensionnement d'un stockage par changement de phase soumis à des conditions d'entrée-sorties réelles.

REFERENCES

/1/ D. LECOMTE - D. MAYER
"Sizing a latent heat storage with shell and tubes heat exchanger a new simplified method"
First EC Conference on Solar Heating - Amsterdam April 30 May 4 1984

LOW ENERGY PASSIVE SOLAR HOUSES AT MILTON KEYNES, UK

S. FULLER
Milton Keynes Development Corporation
Milton Keynes

J. DOGGART
Energy Conscious Design
London

Summary

Two housing developments at Milton Keynes with a total of 185 houses have been monitored over the last three years, in a collaborative venture between Milton Keynes Development Corporation and the Open University. Both the house design and housing layout incorporate simple passive design techniques to increase the solar contribution to space heating, and both are highly insulated.

Interim results are now available and show that:

1. A package of insulation and solar measures have reduced gas usage by 11,000 kWh/yr compared to houses built to current UK building standards, a reduction of nearly 50% and worth over £100/yr.

2. The solar contribution was substantial, with 25% (3500 kWh/yr) of the average space heat requirement of four houses being met by passive direct gain measures.

3. Modelling the effects of solar radiation on heat demand shows that the house acts as if the south window area is replaced by an unglazed aperture which is 56% of the actual double glazed area.

4. The house heat loss coefficient of 225W/°C is close to the predicted value, although the heat loss through ventilation rate was about half that predicted and floor heat losses about double that predicted.

5. All the conservation measures were very cost effective, with solar measures having a simple payback 2.6 years. Insulation measures have a payback of 1.6 years and the total conservation package has a payback of 2 years.

Introduction

Milton Keynes is a new city, designed for 200,000 people and situated 80 km North West of London; the current population is 110,000. In 1967 the Milton Keynes Development Corporation was set up to coordinate the design and development of the city, and as a result of the oil crisis of 1974 became concerned at the impact of future energy costs and availability on the design of the new city. As a result of this concern, the Corporation sponsored several energy saving projects, with the aim of assessing their costs and energy savings.

Two of the housing projects, Pennyland and Linford, are major studies, involving 185 houses in all. The projects commenced in 1979 with the extra over costs being funded by the Department of Energy and the Department of the Environment. Both groups of houses are insulated to a high standard, and incorporate the same passive solar features. Pennyland consists of 177 three bedroom houses with 90 m^2 floor area; the majority of houses facing south $\pm$ 45° and are laid out to avoid overshading. House plans are designed so that major rooms are on the south side, with circulation, storage and bathrooms on the north side. This plan arrangement allows window area on the north facade to be transferred to the south facade, to enhance solar contribution.

Fig 1. Typical south aspect house plans at Pennyland

About half the houses have been insulated slightly better than current UK standards, the other half to about double the current standards. This involves double glazing and draught stripping, 100mm insulation in walls, 150 mm in the roof and perimeter insulation of the ground floor slab. Fuel meters, average room temperature and heat meters in the radiator circuit were monitored weekly and compared with 20 control houses on the adjacent Neath Hill estate which have been built to current thermal standards.

The Linford project consists of 8 four bedroom houses for sale. Houses face south and have the same insulation standards as the highly insulated Pennyland group; the house plans are like Pennyland but have a slightly larger floor area of 110 m^2. Up to 60 sensor positions were monitored at any one time including hourly room temperatures, fuel meters, heat flux sensors in building elements and heat meters in radiator and hot water heating circuits. Additionally ventilation rate measurements were carried out on a heated but unoccupied test house over extended periods.

Fuel savings

Fuel savings for the two projects were determined in different ways. For Pennyland which has large numbers of houses in the sample groups, it was statistically satisfactory to compare annual fuel bills. Gas usage in the highly insulated group is 14,280 kWh/yr, 11,340 kWh less than the control group at Neath Hill. Electricity usage is comparable between the groups, implying approximately equal socio-economic occupancy, and average internal temperatures are also very comparable.

Group	Elec use	Gas use	Total use	Standard deviation	Error on mean ±
Normal (Neath Hill)	2890	25620	28510	3296	824
Area 1 (highly insulated)	2600	16823	19423	3449	438
Area 2	2660	14280	16940	3357	685

Table 1. Comparison of fuel bills at Pennyland Areas 1 and 2 with control group.

For Linford, which has only eight houses, it was not possible to establish a statistically significant control group with which to compare fuel bills. The approach therefore was to establish a reference house design which represented what would have been built in a typical housing development. The fuel bills of such a house were then calculated using SUNCODE, a computer program which was validated using measured field trial data from Linford. The fuel use of the reference house was then compared with the fuel use of the as-built houses. The reference house was assumed to be built to current thermal building standards, to be randomly oriented, overshaded, and to have windows equally distributed between the principal facades.

This approach showed that the as built house gave a saving of 9,300 kWh/yr of purchased fuel compared with the reference house, worth £104 at current prices. This compares with the Pennyland figures which gave a larger £127/yr saving, although Pennyland has a somewhat smaller floor area. One of the reasons for this is that the Pennyland figures include the savings resulting from a more efficient boiler, while Linford does not. When this factor is taken into account the figures are comparable.

Solar Contribution and Solar Aperture

The solar contribution was calculated in two ways. The first method was to calculate the balance of heat requirement and energy supply in four intensively monitored Linford houses. The difference between demand and supply represents the solar contribution, after incidental gains have been allowed for.

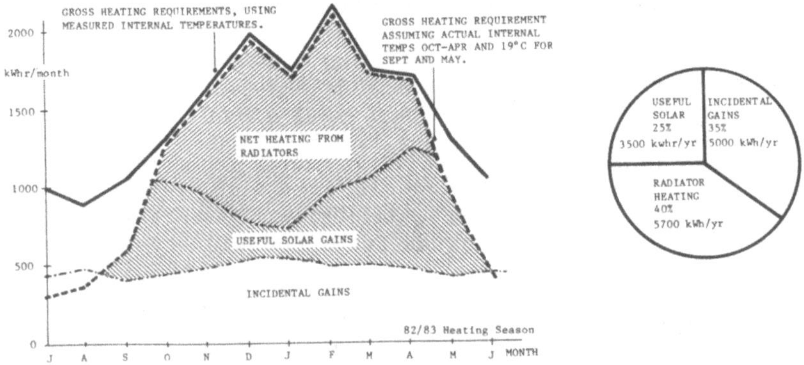

Fig 2. Space heating energy balance for a typical Linford house, and average solar contribution for four houses.

The useful solar contribution, ie that which displaces auxiliary heating requirement varies between the four houses from 2,900 to 4,000 kWh/yr, representing between 16% and 35% of the total space heat demand; the average is 3,500 kWh/yr, or 25% of useful space heat demand.

The second calculation method involves multiple regression analysis of data collected during constant heating temperature tests carried out on the Linford test house. In this method, the house is heated to a constant internal temperature and heat requirement (Q) is plotted against south face vertical solar radiation (S) and inside-outside temperature difference (ΔT). In practice, it is convenient to plot Q/ΔT against S/ΔT, which reduces a three dimensional plot to two dimensions, and to make separate allowance for heat losses through floor (F) and for ventilation rates (Cv).

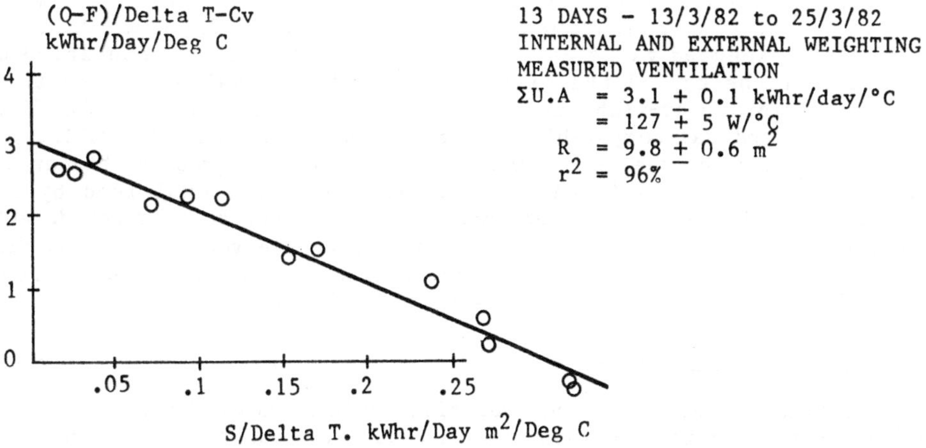

Fig 3. Relationship of heat requirement, inside-outside temperature difference and solar radiation for Linford test house.

Results from plotting daily values gives value to the heat balance equation $(Q-F)/\Delta T - Cv = \Sigma AU - RS/\Delta T$, where R is the apparent solar aperture. The values vary according to the allowances made for thermal mass, but best estimates suggest an apparent solar aperture of 9.8 ± 0.6 m^2, or 56% of the 17.6 m^2 of south facing glass area. This figure would give a solar contribution of 3500-4500 kWh/yr, depending on the assumed length of the heating season, and compares reasonably with the 3500 kWh/yr calculated from the heat balances of four houses described above.

Of this 3500 kWh/yr it is calculated that 2,000 kWh/yr would be available from the windows of a normal house; thus the extra energy saving due to the additional solar measures is 1,500 kWhrs/yr, displacing 2,300 kWhr/yr of purchased fuel if the boiler efficiency is 65%.

Fabric and ventilation losses

Analysis of heat flux measurements showed that much of the fabric losses were as expected, and the heat loss through walls, roof and windows appear to be close to predicted values. The major differences were in ventilation rates and floor heat losses.

Ventilation rate measurements were made continuously during two extended periods from February to April 1982 and November 1982 to March 1983. A large amount of data has been analysed, and shows ventilation rates to dependent on wind direction, inside-outside temperature difference and wind shielding by adjacent buildings.

Wind direction and building shielding can vary infiltration rates considerably; thus on the test house which was shielded on the W.S.W side ventilation remained independent of wind speed at around 0.4 airchanges/hr, while from the unshielded E.N.E direction it varied from 0.5 to 0.7 airchanges/hr. In all cases however, the infiltration was less than the predicted rate of 1 airchange/hr with a measured average of 0.5 airchange/hr. This discrepancy between predicted values was probably due to well fitting and sealed windows and doors and sealed cavity wall construction, but was surprising since low infiltration rates are not normally reported in UK construction.

The second major variation from expected values was in the floor heat losses. Heat flux measurement showed U values of 0.9 W/m^2°C, nearly double the expected value of 0.5 W/m^2°C. After 50 mm of insulation was placed over the top of the slab the equivalent U value reduced to 0.4 W/m^2°C. Before over-slab insulation the heat losses through the floors was 50 W/°C and was comparable to the losses through all the walls, even through the edge of the slab was insulated. These figures are higher than other reported values in the UK, and may be partly explained by soil conductivity; however, a fuller understanding is yet to be reached. When the overall heat loss coefficient is calculated the ventilation and floor heat loss factors tend to cancel each other out. The predicted value was 220 W/°C, and the actual value appears to be 225 W/°C.

Cost and Cost effectiveness

Analysis of the costs and fuel bill savings shows the measures to be highly cost effective. The changes in insulation compared with present UK standards resulted from double glazing with draught stripping, 100 mm wall cavity insulation replacing 50 mm and 150 mm loft insulation replacing 100 mm. This cost an extra £350, but was offset by smaller and cheaper boiler and radiators which could be fitted due to the smaller heat load (£156).

The major cost for the solar measures was to use dense concrete blockwork for the inner walls to act as a thermal store (£68). The other measures, such as transferring glazed area from the north to the south facade, orienting the building correctly and avoiding overshading, appear to carry no cost penalty.

INSULATION		SOLAR	COMBINED
1. Increase wall insulation from 50 to 100 mm		1. Increase density of internal walls.	
2. Increase roof insulation from 100 to 140 mm		2. Transfer glass from North to South elevation	
3. Insulate edge of ground slab		3. Orient house to South	
4. Double glaze (including draughtstripping)		4. Avoid overshading	
COST	£126 +	£68	= £194
FUEL COST SAVING	£78/yr +	£26/yr	= £104/yr
SIMPLE PAYBACK	1.6years	2.6 years	= £1.9 years

Table 2. Cost, fuel savings and cost effectiveness of measures.

The cost effectiveness of the measures was calculated from the costs and the fuel savings reported earlier. The package of insulation measures pays back in 1.6 years, the solar measures repay in 2.6 years and the combined package in 1.9 years. Thus all the measures appear highly cost effective.

SESSION II - CORRELATION METHODS AND DESIGN TOOLS

DESIGN AND VALIDATION OF SIMPLIFIED METHODS FOR SIZING ACTIVE SOLAR SPACE HEATING SYSTEMS

J. ADNOT[+], B. BOURGES[+], B. PEUPORTIER[+], W. DUTRE[++], T.C. STEEMERS[+++]

\+ Centre d'Energétique, Ecole des Mines de Paris, 60, bld St Michel 75272 PARIS CEDEX 06 FRANCE

++ Afdeling Toegepaste Mechanika en Energie Konversie, K.U.LEUVEN, B 3030 Heverlee BELGIUM

+++ Commission of the European Communities, 200, rue de la loi, B 1049 BRUXELLES BELGIUM

Participants in the concerted action : A. DEBOSSCHER (Katholieke Universiteit Leuven, BE), O. BALSLEV-OLESEN (Technical University of Denmark, DK), J. REICHERT (Fraunhofer Institut für system technik und Innovations-forschung, GE), M. FRANK (Trinity College, IR), A. BIONDO (Phoebus SpA, IT), E. VAN GALEN (TNO-TH, NE), R. La Fontaine (Faber Computer Operations, UK), D. BOYD (Polytechnic of Central London, UK) S. GROVE (Plymouth Polytechnic UK), J. LEE (J. LEE Computing, UK), B. ROGERS (University College, UK), Pr O'CONNELL (University College, IR), W. DUTRE (Coordinator).

Financial Support : Commission of the European Communities (Directorate General XII), Agence Française pour la Maîtrise de l'Energie (Direction de la Recherche et du Développement).

Keywords : Active Solar Heating, Simplified Methods, Utilisability.

Summary : One of the activity of the European Modelling Group (coordinated by Prof. DUTRE of the Katholieke Universiteit Leuven, Belgium) is the design and validation of Simplified Sizing Methods; This activity of the Group is being coordinated by Ecole des Mines de Paris and the final results of the first phase (1981-1983) deal with :

1 - The validity of Existing Simplified Methods for the European climates and Heating Practice ;
2 - The design of a more general and more accurate method : ESM1 ;
3 - Its adaptation to the climates and Heating Practice of European Countries and making it available to users (under two forms : version A for Hand Computations, version B for micro-computers).

The accuraty obtained (standard deviation) is 50 % better than that of the best preexisting method (f-charts) on a yearly basis and four times better on a monthly basis. With the new method (ESM1) more parameters are taken into account, a fact users will certainly appreciate.

1 - VALIDITY OF EXISTING SIMPLIFIED METHODS

1.1 - Selection of Existing Methods

A review of the existing methods gave us sixteen references. We rejected half of them due to the lack of some steps (meteorological data not available e.g.) or restrictions of use (for a given country or a given collector). Among the remaining methods with some generality, three different approaches are found : analytical (utilizability-based) physically meaningful but biased, empirical (like f-charts) easy to use but accurate on an annual basis only for a limited range, hybrid (corrective step added to an analytical approach).

Dealing with Space Heating only we selected seven methods with meteorological data available for all European Climates and one method for France

(CSTB).

METHOD	AUTHOR	COUNTRY	TYPE
f-charts	Beckman et al.	U.S.A.	empirical
φ-f-charts	Klein et al.	U.S.A.	hybrid
Lunde 2	Lunde	U.S.A.	hybrid
CFC 2	Adnot et al.	FRANCE	hybrid
CFC 3	Bourges	FRANCE	analytical
ESIM	Chateauminois et al.	FRANCE	empirical
SEU	Kenna	U.K.	hybrid
CSTB	Delcambre et al.	FRANCE	empirical

They are all applicable to the system shown on the following figure. Standardized versions with the same nomenclatures and the same steps were edited.

1.2-Accuracy of the methods in common cases

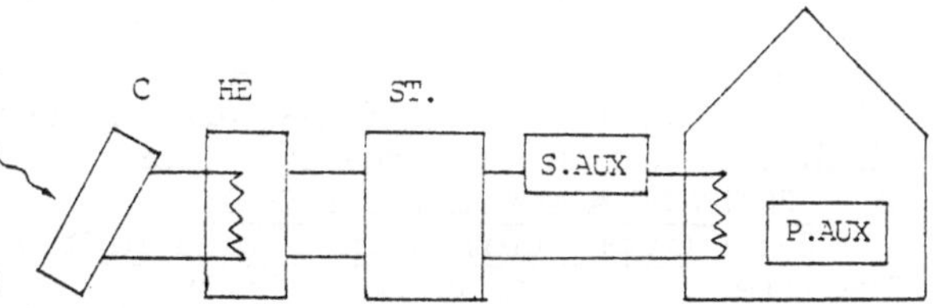

Each method has been compared extensively to results obtained with the Standard European Detailed Model EMGP2 (K.U. Leuven) with and without the perturbations created by the parameters not taken into account usually by Simplified Methods : pump power dissipation, tank losses, stratification, etc ... The reason for this choice is that we wanted to show the main physical phenomena and to check the validity in the common range (which may be obtained even with physically incorrect methods).
The results consist mainly of graphs like the examples given : f-charts versus EMGP2, ESIM versus EMGP2, etc ...A first statistical analysis was performed on the population of the ratios f_{SM}/f_{EMGP2} (fraction given by the simplified method, SM, divided by the same given by the Detailed Model) either on a yearly or monthly basis.

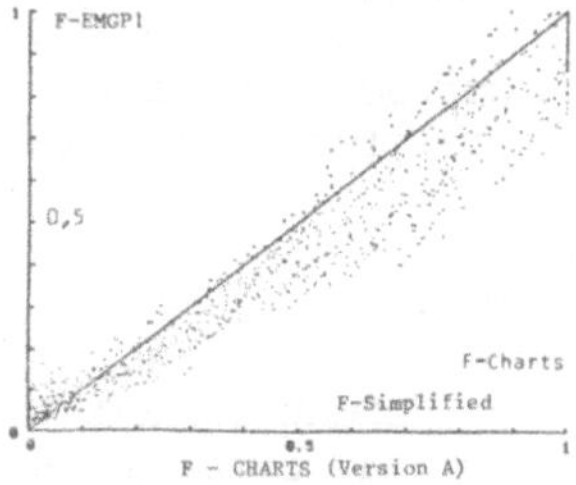

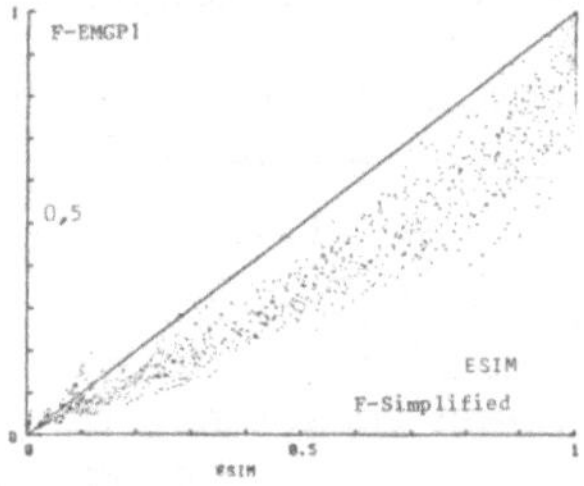

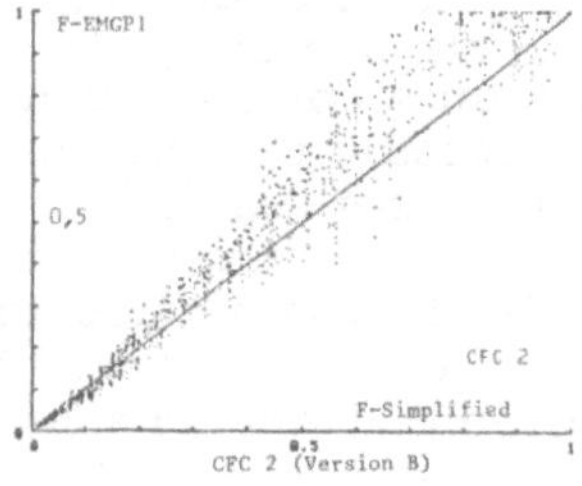

For the actuel situations of Solar Heating in Europe the results are the following, m being the bias and σ the standard deviation of the ratios :

METHOD	ANNUAL FRACTION m	ANNUAL FRACTION σ	MONTHLY FRACTION m	MONTHLY FRACTION σ
f-charts	1.023	0.0864	0.878	0.3899
φ-f-charts	1.179	0.0676	1.090	0.2043
Lunde	1.109	0.0393	1.079	0.1017
CFC2	0.910	0.0643	0.905	0.0931
CFC3	1.248	0.1407	1.162	0.2726
ESIM	1.262	0.0723	1.112	0.3843
SEU	1.286	0.2376	1.360	0.5925
	(Dimension of the sample : 27)		(Dimension of the sample:216)	

The general tendencies are consistent with what we mentionned previously about the three types of methods :

- the analytical method (CFC3) is clearly biased like all the other availability-based methods without corrective term ;
- the empirical methods (f-charts and ESIM) are accurate in a given range (southern climate and low-temperature distribution) and only for a yearly prediction ;
- the hybrid methods can provide both a small bias (accurate fitting of the corrective term for storage capacity) and a small standard deviation.

The best existing method for the prediction of the yearly solar fraction in European conditions (and the only one with acceptable results) is f-charts with $m = 1.02$ and $\sigma = 0.086$; 95.4 % of the ratios f_{SM}/f_{EMGP2} may be found in the range 0.85 - 1.20. We shall see that the proposed European Method gives a smaller range of possible values.

A special validation has been performed for France, including the CSTB method. The results are consistent with those for Europe. Using f-charts, 95.4 % of the ratios f_{SM}/f_{EMGP2} are in the range 0.90-1.18. Using CSTB method, they are in the range 0.45-1.29

2 - DESIGN OF A NEW METHOD : ESM1

2.1 - Computation of the Available Energy

The first step is to compute the Available Energy without inertial losses (collector in steady state, hour after hour) and with the differentials of the controller set to 0.

Let us consider the Hottel-Whillier relationship as a steady state model :

$$Q_u = AF_R \left[(\tau\alpha) I - U_L (T_i - T_a) \right]^+$$

We can write for a month (N days) :

$$E = N\ A\ F_R (\tau\alpha) \bar{H}\ \varphi$$

($\bar{H}$ is the daily average, φ is the utilisability)

From the systematical study of the Cumulative Frequencies Curves (CFC, in Project F of the CEC Research Programme), we know that we can compute simply φ from the following parameters : $\bar{H}$, t_d (day duration), I_{max} (maximum radiation intensity), B_1, B_2, B_3, B_4. These last parameters may be read in tables (for 15 european locations) or estimated from $\bar{H}$ in the other places ($\bar{H}$ may be read in the CEC Meteorological Atlas for the considered tilted plane and location).

The values computed with the described procedure may be corrected to take into account the inertial losses of the collector. A general formulation of these losses was found :

$$\text{Loss} : (\tau\alpha) F_R (a\bar{H} - b\varphi\bar{H})\, \tau_c^{\,1.40}$$

(τ_c is the characteristic time of the collectors, in hours).

The relative losses of 1008 individual computations (with EMGP2 of the Available Energy in three European Locations, for different collectors and two operating temperatures appear relatively close in general from the above mentionned average (see fig.)

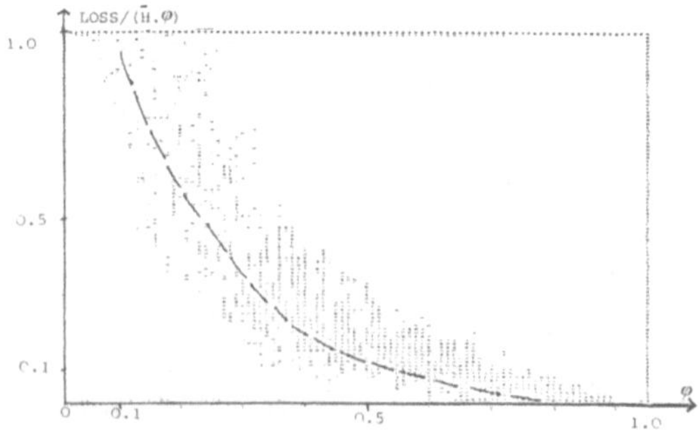

In fact the bias of the whole process (reduction from hourly files to CFC, fitting, inertia,...) is +1 % and the standard deviation ± 2.5 %.

A corrective term was also introduced to take into account the effect of both the starting and the stopping differentials ; in the usual range (1 to 5K) it reprodu-

ces the very small effect of the differentials (up to 3 %) with an accuracy better than 0.1 %).
Another set of corrective terms allows for a good evaluation of the consumption of the pumps, of the recovery of the pumping power by the fluid. It is possible also to consider separately the back and front losses of the collectors, namely in the case where the collectors are integrated in the building and to account for the effect both on efficiency and on the load.
The base temperature depends on the load, on the effectiveness of the Heat Emitters ε_2 (with their mass flow-rate $\dot{m}_2 C_2$) on the heating duration t_e and, obviously on the temperature inside the heated space T_c. It depends also on the type of control of distribution (Thermostat, ideal or non-ideal, law of temperature, etc ...)

2.2 - Storage Size Correction

We took the following definition of the normalized storage size :

$$C_s = (MC) / (75850 \, A \, F'_R \, U_L)$$

with (MC) in $J.K^{-1}$
Over a special set of EMGP2 runs with varied storage capacities but fully-mixed behaviour the following corrective term was defined and allows for the transformation of SF_o : (= Available Energy / LOAD) into SF (Solar Fraction in terms of Useful Energy) :

$$SF = 1 - \exp(-C_1 SF_o (1 + C_2 SF_o))$$

with

$$C_1 = 1.0205 \, (1-0.47 \exp(-1.785 \, C_s))$$

$$C_2 = 0.205 + 0.1515 \ln C_s$$

This empirical approach of the storage size correction gives almost similar results for all locations and shows the effect of the sequences of good and bad days (excess in some days, lack of Energy in the others).

2.3 - Corrections for other perturbations

The formulae by Beckmann giving the effect of piping losses on collection efficiency were tested and chosen.
The effect of flow rate and heat exchanger effectiveness on collection efficiency given by F_R and F'_R was assumed correct.
When the storage tank is likely to be stratified a corrective factor is applied to the available energy and gives the positive effect of stratification (from a few percent to 10-16 %) :

$$C_{str} = 1 + (F'_R \, A \, U_L) / (2 \, \dot{m}_1 \, C_1)$$

(with a very large influence of the flow rate on collection side $\dot{m}_1$).
The storage losses were studied closely. The usual procedure (by Klein) is to compute the losses as :
$UA(T_B - T_a) \times 24.N$ (T_a, outside temperature, may be substituted by the ambient temperature around the storage). In fact in winter months the tank is most of the time under T_B and in mid-seasons over T_B. A carefull study gave origin to a new formula with better monthly and yearly accuracy, as may be seen on the graphes (next page).

3 - VALIDATION AND DISTRIBUTION OF ESM1 - CONCLUSIONS

3.1 - The two versions (A) and (B) of ESM1

We took advantage of the best of each simplified methods and of some elements we derived (tank losses, general inertial correction, pumps and differentials...) to build a Simplified Method adequate for European practice and European Climates : ESM1 (European Simplified Method of Space Heating). It has two versions : (A) for hand computations or pocket calculators with very few parameters taken into account (collectors, storage, heat emitters) and (B) with all the refinements that a Consultant or an Architect could need : pumps consumption and dissipation, effect of the differentials, stratification and so on ...

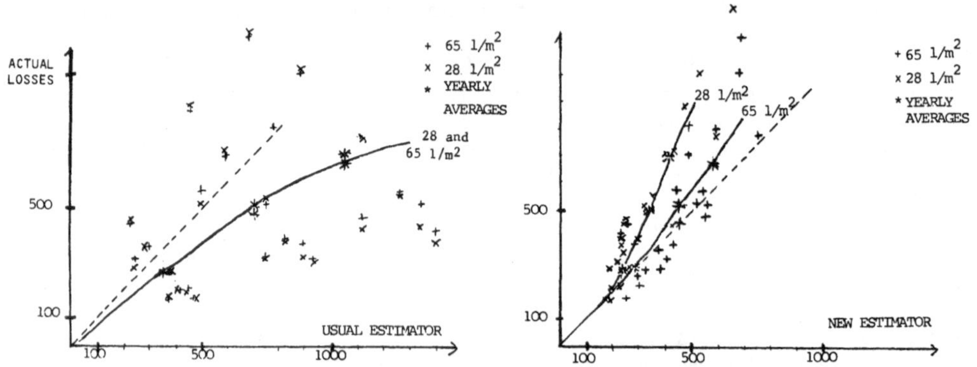

(Unit = MJ) * yearly average + monthly value

Version (A) for hand computations (with charts or calculator) was simplified as far as possible. The meteorological data will be included in its Handbook (reading of curves). In the case of version (B) all the possible parameters were included. The use of a micro-computer in this case is obviously needed and more accurate meteorological data may be used. A true interactive software may be built around the method but in the first step only the listing of the computationnal steps in BASIC and FORTRAN will be available.

3.2 - Accuracy comparison with pre-existing Methods

The following graph was obtained with 216 monthly runs of EMGP2 corresponding in European Heating Practice ; it represents the statistical distribution of f_{SM}/f_{EMGP2} : bias and interval of two standard deviations (95.4% of the yearly or monthly predictions) ; both version (A) and (B) appear equally accurate on a yearly basis with the advantage for ESM1 (B) of giving the rignt sensivity to numerous phenomena ; on a monthly basis f-charts cannot be used, as it is known.

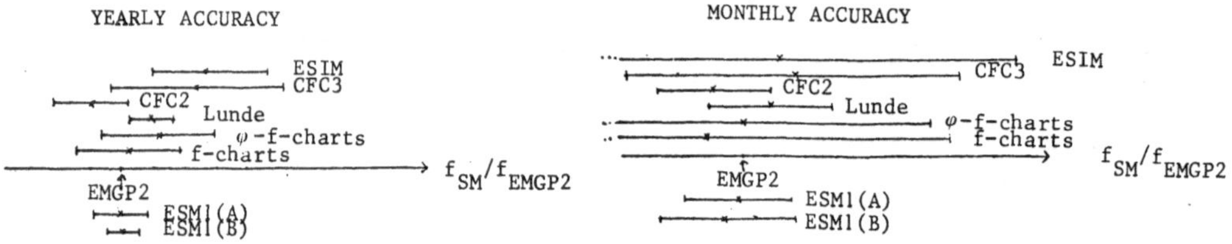

3.3 - Conclusions

Existing Simplified Methods show unsatisfactory accuracies ; with the best existing method (f-charts) the ratio between the estimator given and the actual value has only a probability of 95 % of falling on the range 0.84 - 1.20. Refinements introduced in the newly-built European Method ESM1 allow for :

- an annuracy better by 50 % (0.90 - 1.10) on a yearly basis and four times better on a monthly basis (version A) ;
- the reproduction of the sensitivity of the performance to a large number of parameters not usually at the disposal of the designer (version B).

F-CHART IN EUROPEAN CLIMATES

B.L. Evans, W.A. Beckman and J.A. Duffie
Solar Energy Laboratory
University of Wisconsin-Madison
1500 Johnson Drive
Madison, WI 53706 USA

Summary

The f-chart design method, and the design methods based on utilizability concepts ($\overline{\phi}$ and $\overline{\phi}$,f-charts) were developed for meteorological data of North America. These methods have been built into two design programs, FCHART 3 and FCHART 4, which have been widely used to predict the performance of standard types of systems.

Comparisons of design method predictions, detailed simulation and experimental measurements in various U.S. climates indicate that agreement is satisfactory in all U.S. climates. The differences are greatest using the f-Chart method for Seattle, a station with cloudy weather. This paper includes a discussion of FCHART predictions compared to TRNSYS simulation results in European climates.

The f-Chart design method used in FCHART 3 predicted solar system performance to within ±4.7% (RMS error of the solar fractions for the seven cities) for air space heating and ±2.2% for domestic hot water systems. The f-Chart method tended to underpredict solar system performance for space heating systems, with the differences greatest for Ukkel, Belgium and Copenhagen, Denmark. This appears to be similar to the underprediction encountered in cloudy U.S. climates. The $\overline{\phi}$,f-Chart method, used for liquid based systems in the FCHART 4 program gave better predictions, predicting solar system performance to within ±4.2% of the simulation results for liquid space heating and 1.1% for domestic hot water systems.

1. INTRODUCTION

The f-Chart (1,2) and $\overline{\phi}$,f-Chart (3) design methods for active solar heating systems have been widely used to design solar systems throughout North America. They have been shown to provide reasonable estimates of the long term average performance of system types which have a configuration similar to the systems for which the design methods were developed. Both design methods were developed using weather statistics for North America. There is, therefore, some concern regarding how well they will apply to locations which have somewhat different weather patterns, such as Europe. This paper briefly reviews comparisons of the design methods with both simulation results and measured system performance in North America, and then explains in more detail comparisons of f-Chart and $\overline{\phi}$,f-Chart predictions with simulation results in Europe.

A number of studies have been conducted to compare f-Chart and $\overline{\phi}$,f-Chart system performance predictions with simulation results. Klein, et al. (1,2) in the original publication of the f-Chart methods reported agreement between f-Chart predicted solar fraction and simulation results of approximately 2% on an annual basis. The $\bar{\phi}$,f-Charts gave very similar results, but provided better predictions for cloudy climates such as Seattle, WA (3). More recent studies (4) have confirmed the tendency of the f-Charts to underpredict in the 0.5-0.7 solar fraction range for cloudy climates.

Klein, et al. (1) also compared f-Chart performance predictions with experimental results obtained from the MIT solar House IV (5). Two years of data were compared with f-Chart predictions. f-Chart overpredicted performance by 8% and 5% for these two years, which may have been due to difficulties in operating the MIT system during its first two years. (Engebretson of MIT stated that "Solar House IV could utilize from 4 to 10% more of the available radiation than it does" (5)).

Mitchell and Duffie (6) conducted a more extensive survey after more experimental data had been accumulated on several different systems. Approximately 30 different systems were included in the study. The overall conclusion from their paper is that the f-Chart method gave adequate estimates of solar system performance, provided that the systems being analyzed are close in configuration to the systems for which the f-Charts were developed.

2. METHOD OF APPROACH

The f-Chart and $\bar{\phi}$ f-Chart design methods were examined for European climates in a similar manner to that which was used for North American climates. Detailed simulations using the simulation program TRNSYS 12.1 (7) were run for three different types of solar heating systems: liquid based DHW; liquid based space heating; and air based space heating. The simulations used hourly weather data which was available for seven locations, predominantly in northern Europe. The cities used in this study are listed in Table I. The design methods used monthly average weather data values calculated from the hourly data.

TABLE I

City	Country	Latitude	$\bar{K}_T$
Carpentras	France	44.1	0.62
Trappes	France	48.1	0.44
Valentia	Ireland	52.0	0.44
Hamburg	Germany	53.5	0.43
Ukkel	Belgium	50.8	0.38
London	England	51.1	0.38
Copenhagen	Denmark	55.7	0.46

The first system to be analyzed is a liquid based domestic hot water system. This system is a closed loop system with a liquid collector, optional collector to storage heat exchanger, preheat storage tank and either a conventional auxiliary water heater or a demand auxiliary heater. The water being delivered to the load is tempered so that it will not exceed the set temperature. Specific system parameters used in this study are listed in Table II. Simulations of this type of solar domestic hot water system can be compared with results from the f-Chart and $\bar{\phi}$,f-Chart methods.

The solar space heating systems used in this study are broken down into liquid based systems and air based systems. The liquid systems use water (or an antifreeze solution) as the circulating fluid and store energy in a fully mixed, insulated storage tank. Energy is then delivered to the load via a liquid to air heat exchanger. The air based space heating system uses a pebble bed for the thermal storage media.

TABLE II

DHW System Parameters

Collector Area	1-20m^2
$F_R(\overline{\tau\alpha})$	0.70
F_RU_L	3.89 W/m^2-°C
Tilt	Latitude
Storage Capacity	315 KJ/°C-m^2
Load	300 ℓ/day
Set Temperature	60°C

Consequently, there is no need for a load heat exchanger as the room air can be directly circulated through either the collectors or the storage and returned to the building. System parameters for the air and liquid space heating system are summarized in Tables III and IV. Liquid based systems can be analyzed using either the f-Chart or the $\overline{\phi}$,f-Chart methods and compared with simulation results, while only the f-Chart method is applicable to air based systems.

The comparisons between the design method predictions and the simulation results were done using two computer programs, FCHART 3 (8) and FCHART 4.1 (9). FCHART 3 uses the f-Chart method to analyze all three types of systems. FCHART 4.1 uses the $\overline{\phi}$,f-Chart method for liquid systems and the f-Chart method for air systems.

TABLE III

Liquid System Parameters

Collector Area	10-120 m^2
$F_R(\overline{\tau\alpha})$	0.70
F_RU_L	3.89 W/m^2-°C
Tilt	Latitude
Storage Capacity	315 KJ/m^2-°C
UA (Building)	270 W/°C

TABLE IV

Air System Parameters

Collector Area	10-120 m^2
$F_R(\overline{\tau\alpha})$	0.80
F_RU_L	4.50 W/m^2-°C
Tilt	Latitude
Storage Capacity	315 KJ/m^2-°C
UA (Building)	270 W/°C

3. RESULTS

Comparisons of TRNSYS simulations with the design method predictions are summarized for each of the three system types in Figures 1-3. Each figure plots the solar fraction predicted by the design method as a function of the solar fraction from the associated TRNSYS simulation. Ideally, each point would lie on a 45 degree line. Deviation from this line indicates an over or under prediction of solar fraction by the f-Chart or $\overline{\phi}$,f-Chart methods. These plots are for solar fraction on an annual basis and somewhat more deviation will be found on a monthly basis. The intent of the design methods, however, is to predict yearly, not monthly energy consumption, hence the annual prediction accuracy is of greater importance.

Solar domestic hot water system performance was quite adequately predicted by either the f-Chart or $\overline{\phi}$,f-Chart methods. Only slight differences could be noted between the predictions from the two design methods. The f-Chart predictions, shown in Figure 1 as the symbol, □, had an rms error in solar fraction of 0.022 on an annual basis, with a -0.009 bias towards underpredicting. There appears to be a slight tendency to underpredict at low solar fractions and overpredict at solar fractions approaching 1.0. The $\overline{\phi}$,f-Chart predictions are shown on Figure 4 as the + symbol. The error in the $\overline{\phi}$,f-Chart predictions had virtually no dependence upon solar fraction, and had an rms error in solar fraction of 0.011. The bias error for the $\overline{\phi}$,f-Charts for domestic hot water systems was a 0.0068 tendency to overpredict solar fraction.

Liquid based space heating system comparisons are shown in Figure 2. This type of system showed more of a difference between the f-Charts and $\overline{\phi}$,f-Charts. The f-Chart method had a tendency to underpredict solar fraction, with the greatest underprediction in the .60-.80 solar fraction range. This is consistent with previous studies (4) of f-Chart predictions for space heating systems in predominantly cloudy climates (six of the seven cities used in this study were cloudy in comparison to typical North American cities). The rms error for the f-Charts was 0.061, with an bias error of -0.047. The $\overline{\phi}$,f-Chart method gave significantly better accuracy in performance predictions than the f-Chart method. The bias error in solar fraction was 0.034 (overprediction) with an rms error of 0.042. The error for $\overline{\phi}$,f-Charts was, however, more evenly distributed than for the f-Charts. There was only a slight tendency for the mid-range solar fraction predictions to have greater errors.

Figure 3 shows a comparison of simulation and design method results for air based space heating systems. The air f-Chart predictions have trends similar to the liquid f-Chart predictions; a tendency to underpredict solar fraction, particularly in the 0.6-0.8 solar fraction range. The predictions are statistically better than the liquid system f-Chart predictions, with a rms error in solar fraction of 0.047 and a bias error of -0.034. This is of comparable accuracy to the $\overline{\phi}$,f-Chart method for liquid space heating systems.

4. CONCLUSIONS

The comparisons presented in this paper are encouraging in that the f-Chart and $\overline{\phi}$,f-Chart solar system design methods seem to provide reasonable estimates of solar fraction in European climates. The design method performance predictions for domestic hot water system were within an rms error of 2.2% of simulation results for both design methods. This is similar to the results found for North America. Space heating system performance predictions in Europe have somewhat larger errors than the corresponding systems have in North America. System performance can be estimated to within rms error of 4.2% for liquid systems and 4.7% for air systems using the $\overline{\phi}$,f-Chart, and f-Chart methods respectively.

5. REFERENCES

1. KLEIN, S.A., BECKMAN, W.A. and DUFFIE, J.A., (1976). "Design Procedure for Solar Heating Systems," Solar Energy, 18, 113.
2. KLEIN, S.A., BECKMAN, W.A. and DUFFIE, J.A., (1977). "A Design Procedure for Solar Air Heating Systems," Solar Energy, 19, 509.
3. KLEIN, S.A. and BECKMAN, W.A., (1979). "A General Design Method for Closed Loop Solar Energy Systems" Solar Energy, 22, 269.
4. EVANS, B.L., KLEIN, S.A. and DUFFIE, J.A., (1983). "A Correction Factor of f-Chart Predictions of Active Solar Fraction in Active-Passive Heating Systems,"ASES/ISES conference proceedings, Minneapolis.
5. ENGEBRETSON, C.D., (1964). "The Use of Solar Energy for Space Heating: MIT Solar House IV," Proceedings of the UN Conference on New Sources of Energy, 5, 159, United Nations, New York.
6. DUFFIE, J.A. and MITCHELL, J.W., (1983). "f-Chart: Predictions and Measurements," Journal of Solar Energy Engineering," 105, 3.
7. TRNSYS 12.1: A Transient Simulation Program, (1984). Report 38-12, University of Wisconsin-Madison.
8. FCHART 3.0, (1978). EES Report 49, University of Wisconsin-Madison.
9. FCHART 4.1, (1980). EES Report 50, University of Wisconsin-Madison.

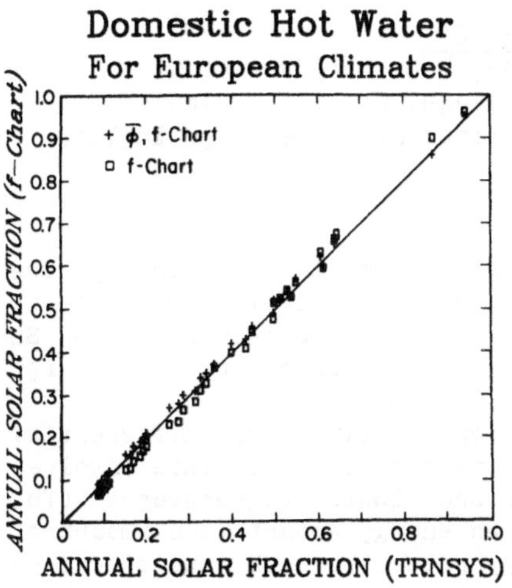

Figure 1. Comparison of f-Chart and $\bar{\phi}$,f-Chart performance predictions with TRNSYS results for solar domestic hot water systems. The rms error for the f-Chart method is 2.2%, and the rms error for the $\bar{\phi}$,f-Chart method is 1.1%.

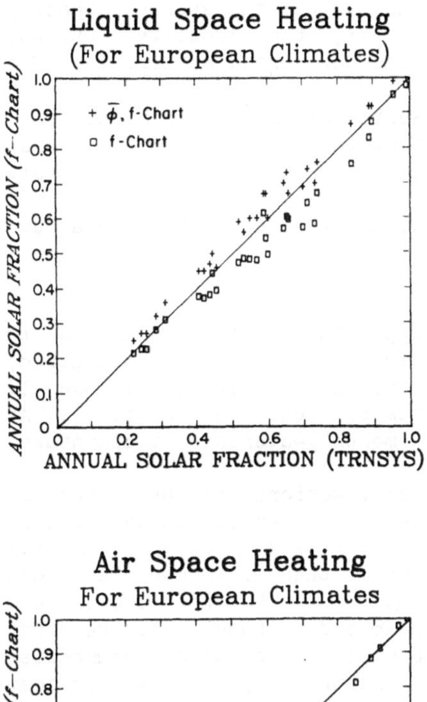

Figure 2. Comparison of f-Chart and $\bar{\phi}$,f-Chart performance predictions with TRNSYS results for liquid based solar space heating systems. The rms error for the f-Chart method is 6.1%, and the rms error for the $\bar{\phi}$,f-Chart is 4.2%.

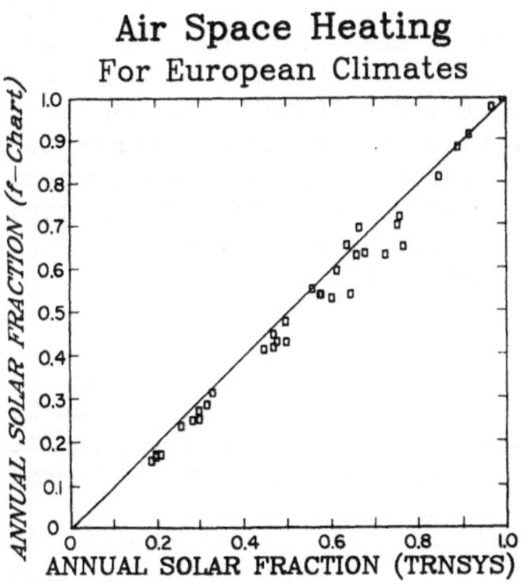

Figure 3. Comparison of f-Chart performance predictions with TRNSYS results for air based solar space heating systems. The rms error is 4.7%.

THE SEU DESIGN METHOD

R.H. Marshall*, J.P. Kenna**, P.B. Lloyd*
* Solar Energy Unit, University College, Cardiff, Wales
**Intermediate Technology Power Ltd., Mortimer Hill, Reading, England

Summary

The SEU Design Method (SEUDM) is an algorithm for predicting the performance of liquid-based solar-heating systems. It is applicable to domestic hot-water (DHW), space-heating (SH), and combined DHW and SH systems, and it is intended for use as an aid to architects and design engineers in the rapid assessment of design options.

The derivation of the method is based on putting the differential equation for the heat-store into a non-dimensional form. This involves introducing five reference quantities: upper and lower temperatures (Thi and Tlo), a time interval (t* = 1 day), an energy amount (the amount of energy collected in a day, H*), and an energy rate (the peak rate of energy collection in a day, G*). The quantities that enter into the differential equation are grouped into six non-dimensional parameters, which represent the major influences on the performance of the system: the storage capacity (R), the total amount of available energy (M), the 'quality' of the available energy (K), the store loss (Ls), the DHW demand pattern (P), and the SH demand pattern (Q).

The performance of the system is represented by the 'solar fraction', F, defined to be the fraction of the heating load that is met with solar energy. In SEUDM, the solar fraction is computed as a function of the six non-dimensional parameters.

For a detailed exposition of the method, see the papers by Kenna and Marshall (1-3).

1. Introduction

In what follows, we consider the application of the method to space-heating (SH) systems. The non-dimensional parameters are discussed, and two case studies of European sites are considered. Tables I and II give information about the sites (Carpentras and Uccle respectively). At the top of each table are the location, the monthly-averaged temperature, and the monthly space-heating loads.

At the bottom of the tables are the solar fractions as predicted by EMGP2 and SEUDM. In the middle of the tables are intermediate values that arise in the calculations performed in SEUDM. For each step, alternative values are given to illustrate the sensitivity of these intermediate values to the choice of prediction formula. Near the top are two rows showing two ways of finding the radiation components (beam and diffuse). In the first row, the global irradiation is that measured with an unshaded pyranometer and the diffuse fraction is that found by subtracting the measured beam irradiation. In the second row, the global irradiation is found by summing the measured beam and diffuse irradiations. Appreciable differences are found in the Carpentras data but not in the Uccle data.

2. The non-dimensional parameters

Of the six parameters, P (representing the DHW demand pattern) is inapplicable here, Q (representing the SH demand pattern) is fixed, and Ls (representing store loss) has no effect as the store loss is assumed to be negligible. We shall consider the other parameters, M,K,R, in turn. These are formally defined in papers giving full details of the method (1-3).

The parameter M represents the amount of solar energy that is available to the system. The 'usable' part of the solar energy is that which falls above the cut-off irradiance level, Gc(Tlo), which is determined by the collector and air temperatures:

Gc(Tlo) = (U/no)(Tlo-Ta)

The energy available to the system over any period is the time integral of the usable irradiance. If the irradiance does not rise above Gc then there is no energy available to the system; if, in the other extreme, the ambient temperature Ta rises above Tlo then the system gains ambient heat.

The parameter M is then defined as

M = (energy available to the system)/(monthly heating load)

The user of the method is expected to know the irradiation on a horizontal surface but not the irradiation on a tilted surface or the usable fraction of it. These two quantities must be estimated from the horizontal global irradiation using empirical formulae.

The first step toward finding M is to split the global irradiation into its beam and diffuse parts. A number of formulae have been published for this, and they usually express the diffuse fraction D as a function of the global transmission K, which is the ratio of global to extraterrestrial irradiation (4). Analysis of British meteorological records has shown that (a) the random scatter swamps the effect of geographic differences in hourly and daily data; (b) there are non-negligible differences between sites in the monthly data; (c) the seasonal dependence is negligible; (d) non-linearity in the correlation is negligible. For monthly data in Britain, D=1-K was found to be a good fit, with an RMS error of about 0.06. Two curves lie outside the scatter of British data: Liu & Jordan's (5) and Collares-Pereira & Rabl's (6). (Note: LJ gave only a hand-drawn curve, so Klein's fit was used.) CPR (6) and Erbs (4) showed that the severe under-prediction of LJ's curve was due to the latter's use of data uncorrected for the shading ring. CPR's curve is highly non-linear and, although it fitted their data, it fails with our data. Both Erbs and CPR used values of the monthly transmission K between about 0.4 and 0.7, but the range typical of Britain (and Uccle) is about 0.2 to 0.5. In Tables I and II, the linear SEU fit for the diffuse fraction is seen to predict the largest amounts of diffuse radiation. It works well for Uccle but less well for Carpentras. Exceptions are the very low diffuse irradiations in midsummer in Carpentras, which are so low that LJ's curve is applicable. These may be due to measurement errors; see above.

Second step. The diffuse irradiation on the tilted surface is computed by assuming an isotropic diffuse field. This seems a reasonable approximation.

Third step. LJ (5) computed the beam on the tilted surface by assuming that the irradiance normal to the solar beam is constant over the day. Theilacker & Klein (7) assumed CPR's (6) formula for the diurnal variation in irradiance and derived a more complex expression for converting beam irradiations to tilted surfaces. As is seen in the Tables (rows "LJ/tr" and "TK/tr") the later formula is significantly better in winter.

The combined effect of putting estimates of the monthly beam and diffuse irradiations into Theilacker & Klein's formula is seen in the next four rows ("TK/S" etc.). The prediction of tilted-surface irradiation becomes much worse when one uses estimated beam and diffuse irradiations.

The next parameter, K, is defined as

K = (mean daily-peak usable irradiance)/(all-day mean irradiance)

The numerator is computed by taking the highest hourly irradiation of each day and forming the average over the month. SEUDM estimates this mean peak from the monthly total irradiation. As a first approximation, CPR's (6) formula for the diurnal variation in irradiance is used to predict the mean noon-hour irradiation, and this is assumed equal to the mean peak.

Finally, the R parameter is defined as

R = (store capacity for Tlo to Thi)/(daily load)

The reference temperature-interval is arrived at from several considerations (3). In the present case (radiators in series with the solar system),

Thi = Trm + (L/t*)/mCpE and Tlo = Trm

in which Trm is the room temperature, E is the radiator effectiveness, and mCp the capacitance flowrate. Here, the store comprises 750 kg of water, and the radiators have a high performance (E=0.7) and a large capacitance flowrate, mCp=1000 W/K. The R value is hence fixed as 0.052, which is near to the lowest value of R for which SEUDM is applicable.

3. Comparison with EMGP2

The European Modelling Group (EMG) has recently tested seven simplified methods for predicting the performance of solar-heating installations. The tests involved comparing the predictions of the simplified methods with the predictions of the detailed simulation program EMGP2. The inputs were meteorological data that came from widely spread sites in Europe and covered 27 station-years. The tests showed up clear differences between the methods.

The study reported here has also compared the SEU Design Method (one of the seven used by the EMG), with EMGP2. This comparison involved only two years of data and is therefore more limited than the EMG's tests. Nontheless, the study indicates that the SEU Design Method predicts more accurately than is suggested by the EMG's results. Further investigations are in hand, using a larger set of meteorological records. The origin of the difference, however, cannot be explained until the EMG's tests are reported more fully.

This study has shown that there is a continuing problem in estimating input data for system models from meteorological records.

References.

1. Kenna, J.P. (1982), "A Study of Solar Heating System Performance and Collector Design", PhD thesis, Dept. Mech. Eng., U. College Cardiff.
2. Kenna, J.P. (1982) "The SEU Design Method for Solar Heating Systems", Helios 15, pp. 8-11; corrections, Helios 17, p. 9; publ. SEU, U. College Cardiff.
3. Marshall (1982), "Evaluation of thermal storage for solar heating systems", Proc. EC Contractors' Meeting, Meersburg, 14-16 June 1982, 'Solar Energy Applications to Dwellings', Ser. A, Vol. 2, pp. 363-373, eds. W. Palz & T. Steemers, publ. Reidel.
4. Erbs, D.G. (1980), "Methods for estimating the diffuse fraction of hourly, daily and monthly-average global radiation", MSc thesis, U. Wisconsin-Madison.
5. Liu, B.Y.H. (1960) "The interrelationship between and characteristic distribution of direct, diffuse and total solar radiation", Solar Energy 4(3), 1-19.
6. Collares-Pereira, M., and A. Rabl (1979), "The average distribution of solar radiation - correlations between diffuse and hemispherical and between daily and hourly insolation values", Solar Energy 22, 155-164.
7. Theilacker, J.C., and S.A. Klein (1981), "An algorithm for calculating monthly-average radiation on inclined surfaces", ASME J. Solar Engineering 103, 29-33.

Table I. Carpentras (latitude 44.05 deg. N, longitude 5.03 deg. E)

		Jan	Feb	Mar	Apr	May	Jun	Jul	Aug	Sep	Oct	Nov	Dec	J	F	M	A	M	J	J	A	S	O	N	D
Air temp (C)	Ta	5.9	6.7	10.2	11.9	19.1	22.2	24.6	25.5	22.0	14.6	11.1	7.3												
Load (Wh)	L	9806	8417	6222	4583	0	0	0	0	0	2306	5028	7472												
		Below: monthly-averaged global irradiations (Wh/m2)												Below: diffuse fractions (%)											
Measured H	unshd	1673	2746	4202	5309	6048	*	*	*	*	*	*	*	45	37	40	39	42	34	34	38	34	39	42	44
	shade	1664	2707	4010	5119	5829	6743	6782	5993	4732	3012	1945	1583	44	36	37	36	39	35	34	38	34	39	42	44
Estimated D	SEU	-	-	-	-	-	-	-	-	-	-	-	-	54	46	42	43	44	42	39	39	40	48	51	50
	Erbs	-	-	-	-	-	-	-	-	-	-	-	-	47	39	35	37	38	35	33	33	34	41	44	43
	CPR	-	-	-	-	-	-	-	-	-	-	-	-	45	40	37	38	38	37	35	35	36	41	43	42
	LJ/K	-	-	-	-	-	-	-	-	-	-	-	-	40	34	31	32	33	31	29	29	30	35	38	37
Hrly est. Ht	iso	2918	4211	5078	5375	5311	5838	6046	5950	5551	4337	3374	2996	23	22	28	33	40	38	36	36	27	25	22	22
Mhly est. Ht	LJ/tr	3140	4418	5300	5533	5571	5911	6090	5920	5632	4445	3538	3187	22	21	26	31	38	37	36	36	27	24	21	20
	TK/tr	3011	4275	5221	5545	5589	5896	6083	5939	5603	4321	3399	3055	23	21	26	31	38	37	36	36	27	25	22	21
	TK/S	2743	3994	5184	5489	5585	5923	6091	5929	5464	4056	3103	2869	30	30	31	39	44	44	41	37	32	33	29	25
	TK/Er	2955	4211	5351	5573	5595	5898	6082	5981	5597	4258	3324	3096	24	24	26	33	38	38	35	31	27	27	24	20
	TK/CP	3002	4203	5313	5559	5594	5904	6085	5962	5557	4262	3353	3118	23	24	27	34	39	39	37	33	29	27	23	20
	TK/LJ	3144	4377	5473	5634	5603	5881	6075	6017	5694	4417	3510	3283	20	20	21	28	33	33	31	27	23	23	20	17
Mean noon H	true	490	627	744	741	690	741	809	779	779	670	541	528												
	CPR	501	660	760	760	727	758	790	798	796	654	552	518												
		Below: solar fractions (%)																							
Solar fract.	EMGP2	25.3	40.5	63.0	77.8	100.	100.	100.	100.	100.	88.2	50.5	34.2	year: 46.9											
	SEUDM	23.3	40.1	67.1	87.6	100.	100.	100.	100.	100.	100.	49.8	33.2	year: 48.3											

unshd=unshaded pyranometer, shade=shaded pyranometer
iso=tilted-surface irradiation Ht computed assuming isotropic diffuse field.
D='diffuse fraction'=ratio of diffuse irradiation to global irradiation.
Diffuse-fraction models: Er=Erbs; CPR,CP=Collares-Pereira & Rabl; LJ=Liu & Jordan; S=SEU correlation.
Models for tilted-surface irradiation Ht: LJ=Liu & Jordan; TK=Theilacker & Klein
Note: LJ/K = Klein's fit to Liu & Jordans curve; TK/tr = Theilacker & Klein's model with true beam and diffuse irradiations; TK/S = Theilacker & Klein's model with SEU correlation for the diffuse fraction; similarly TK/Er, TK/CP, TK/LJ.
* - The values of global irradiation on the horizontal surface in the EMG tape were invalid for these months.

Table II. Uccle (latitude 50.8 deg. N, longitude 4.35 deg. E)

		Jan	Feb	Mar	Apr	May	Jun	Jul	Aug	Sep	Oct	Nov	Dec	J	F	M	A	M	J	J	A	S	O	N	D
Air temp (C)	Ta	4.6	4.0	8.0	10.5	13.2	18.2	17.3	18.6	17.9	12.0	6.9	3.5												
Load (Wh)	L	10806	7833	7694	6028	3750	0	0	0	889	4944	8667	11416												
		Below: monthly-averaged global irradiations (Wh/m2)												Below: diffuse fractions (%)											
Measured H	unshd	635	1321	2090	3143	4131	5184	4046	3782	3178	1551	707	458	79	69	66	66	67	57	69	62	51	69	77	82
	shade	632	1312	2081	3132	4118	5164	4039	3773	3166	1545	706	456	79	69	66	65	67	57	69	62	51	69	76	82
Estimated D	SEU	-	-	-	-	-	-	-	-	-	-	-	-	74	66	66	64	61	54	63	60	55	67	75	77
	Erbs	-	-	-	-	-	-	-	-	-	-	-	-	72	61	61	58	54	47	57	53	48	62	74	78
	CPR	-	-	-	-	-	-	-	-	-	-	-	-	63	56	56	53	51	46	53	50	46	56	64	66
	LJ/K	-	-	-	-	-	-	-	-	-	-	-	-	66	54	54	51	47	41	50	46	42	55	68	72
Hrly est. Ht	iso	871	1663	2455	3027	3642	4413	3504	3564	3681	1822	939	626	50	47	49	59	66	59	69	57	39	51	50	51
Mhly est. Ht	LJ/tr	922	1825	2428	3096	3716	4539	3564	3611	3736	2007	1023	676	46	43	49	58	64	57	68	57	38	46	45	47
	TK/tr	857	1730	2379	3123	3769	4567	3618	3650	3719	1920	955	628	50	45	50	57	63	57	67	56	38	48	49	51
	TK/S	962	1802	2369	3145	3791	4566	3627	3669	3624	1953	980	702	42	42	51	56	58	55	62	54	43	46	47	43
	TK/E	993	1918	2465	3212	3815	4563	3635	3728	3771	2057	995	694	40	37	45	50	52	48	56	48	36	41	46	44
	TK/CP	1166	2044	2569	3264	3828	4562	3641	3757	3813	2186	1179	876	30	32	40	45	49	46	52	45	35	35	34	30
	TK/LJ	1110	2078	2598	3293	3841	4559	3644	3790	3907	2209	1103	782	33	30	39	43	46	42	50	41	31	34	38	37
Mean noon H	true	173	272	330	394	449	531	379	437	525	285	171	121												
	CPR	175	297	359	419	462	549	432	474	532	316	185	136												
	Below: solar fractions (%)																								
Solar fract.	EMGP2	5.0	11.2	25.3	34.3	59.7	100.	100.	100.	100.	28.6	6.4	2.6	year: 17.2											
	SEUDM	5.9	11.8	22.0	37.5	69.0	100.	100.	100.	100.	32.0	7.2	0.0	year: 17.8											

unshd=unshaded pyranometer, shade=shaded pyranometer
iso=tilted-surface irradiation Ht computed assuming isotropic diffuse field.
D='diffuse fraction'=ratio of diffuse irradiation to global irradiation.
Diffuse-fraction models: Er=Erbs; CPR,CP=Collares-Pereira & Rabl; LJ=Liu & Jordan; S=SEU correlation.
Models for tilted-surface irradiation Ht: LJ=Liu & Jordan; TK=Theilacker & Klein
Note: LJ/K = Klein's fit to Liu & Jordans curve; TK/tr = Theilacker & Klein's model with true beam and diffuse irradiations; TK/S = Theilacker & Klein's model with SEU correlation for the diffuse fraction; similarly TK/Er, TK/CP, TK/LJ.

COMPARISON OF THE REM HEATING ENERGY CALCULATION PROCEDURE TO MEASURED DATA

M.J. HOLTZ, R.G. DERICKSON, F.D. TELLER
Architectural Energy Corporation
8753 Yates Drive, Suite 105
Westminster, Colorado 80030 USA

Summary

1. INTRODUCTION

The Residential Energy Manual (REM) is a flexible, accurate and easy to use residential energy design and analysis procedure developed to serve the needs of homebuilders, home designers and consumers (1). The REM encourages the intelligent blending and balancing of energy conservation and passive solar features through the use of quantitative design guidelines and heating and cooling calculation procedures. The REM calculation procedures are applicable to the continuum of residential design - conventional, suntempered, passive and superinsulated. Also, the basic REM approach is appropriate to both new and retrofit single and multi-family construction.

The REM is organized into two major sections: (1) design/construction guidelines; and (2) heating and cooling calculation procedures. The first section provides the needed background to familiarize the homebuilder with energy-efficient design principles and techniques. It identifies those techniques providing the greatest energy savings and includes a procedure for selecting the most effective combination of techniques. Critical construction details are also presented so that the energy efficient design can be turned into a practical reality.

The second section presents the calculation procedures used to accurately estimate the home's annual heating and cooling energy consumption. The procedures allow for the integrated treatment of energy conservation and passive solar design, calculates heating and cooling loads and includes the effects of arbitrary amounts of thermal mass, ventilation and shading strategies, multiple window orientation, movable insulation, internal gains, thermal zoning and mechanical system efficiency.

The REM has been developed to be a true design and analysis tool. The design guidelines provide sufficient quantitative and qualitative information to enable the homebuilder and designer to develop a near-optimum design solution for both new and retrofit applications without the need of calculations. Only when the design has been developed and adequately specified are the calculation procedures needed and then used to show compliance with energy codes, size mechanical equipment, establish an energy rating or to fine tune the design to improve its energy performance.

2. TECHNICAL BASIS OF REM CALCULATION PROCEDURES

The REM heating and cooling calculation procedures are based on computer generated seasonal "load curves" for specific geographic locations in which the building load coefficient (BLC) for each curve is held constant over variation in solar aperture area and internal gains (2). The approach of an invariant BLC with increasing solar or internal gains eliminates the need to establish fixed assumptions such as number or type of glazing panes, rate of internal gains and so on. For the REM Heating Calculation Procedure, these load curves are used to produce a Utilization Table for specific geographic locations. The Utilization Table expresses the fraction of solar and internal gains used in offsetting auxiliary heating energy consumption. Also, it indicates the maximum potential auxiliary heat reduction achieved by adding thermal mass above the "free" mass (gypsum board, furniture, fixtures, etc.) present in all buildings. For each geographic location, separate tables representing component heating loads, solar gains and internal gains are used to determine the heating load index and the gain index needed to enter the Utilization Table.

The REM Cooling Calculation Procedure uses an Overload Table to express the cooling loads generated by window solar gains and internal gains. The overload value is then added to the envelope conductive gains to represent the building's total sensible cooling load. The effects of shading, ventilation and thermal mass are included in the procedure. Geographic specific tables for component loads, solar and internal gains are used for determining cooling season performance.

3. SOFTWARE VERIFICATION OF REM HEATING CALCULATION PROCEDURE

Numerous software verification tests have been made between the REM heating calculation procedure and detailed building energy analysis simulations (3). Figure 1 shows a comparison of REM hand calculations and BLAST 3.0 simulations for 32 house cases in Denver, Colorado. Each point represents a house with a different combination of window area and orientation, levels of thermal mass and insulation, and infiltration rates. The figure shows excellent agreement over a wide range of design variables; therefore, we are confident that the thermophysics of the REM method are correct and that trends of building energy performance due to parametric changes are accurate.

4. COMPARISON OF REM HEATING CALCULATION TO MEASURED DATA

4.1 Description of Monitored Passive Solar Home

A passive solar home, located in Golden, Colorado and monitored under the SERI Residential Class B Passive Solar Performance Monitoring Program, was selected as the basis of comparison to a REM calculated heating energy performance prediction. The home has 271.4 square meters (3015 square feet) of conditioned floor area on two levels. The primary passive heating systems are direct gain windows and a large two story sunspace. The total passive solar aperture area is 33.9 square meters (365 square feet) or 12 percent of the total conditioned floor area. Thermal mass is provided in the form of a quarry tile covered 10 cm (4 inch) thick concrete slab floor and 30cm (12 inch) thick brick veneered concrete block wall in the sunspace. The house is insulated to RSI-5.28 (R-30) in the roof, RSI-2.82 (R-16) in the exterior walls and RSI-1.76 (R-10) on the exterior below grade walls. The lower level is below grade on the north and partially below grade on the east and west. All windows are double glazed and have full solar access. Auxiliary heating is provided by a gas-fired forced air heating system. Two small ceiling fans in the solarium help destratify and distribute solar heated air. The south windows are fitted with manually operated insulating shutters. The main entry on the north is through a vestibule.

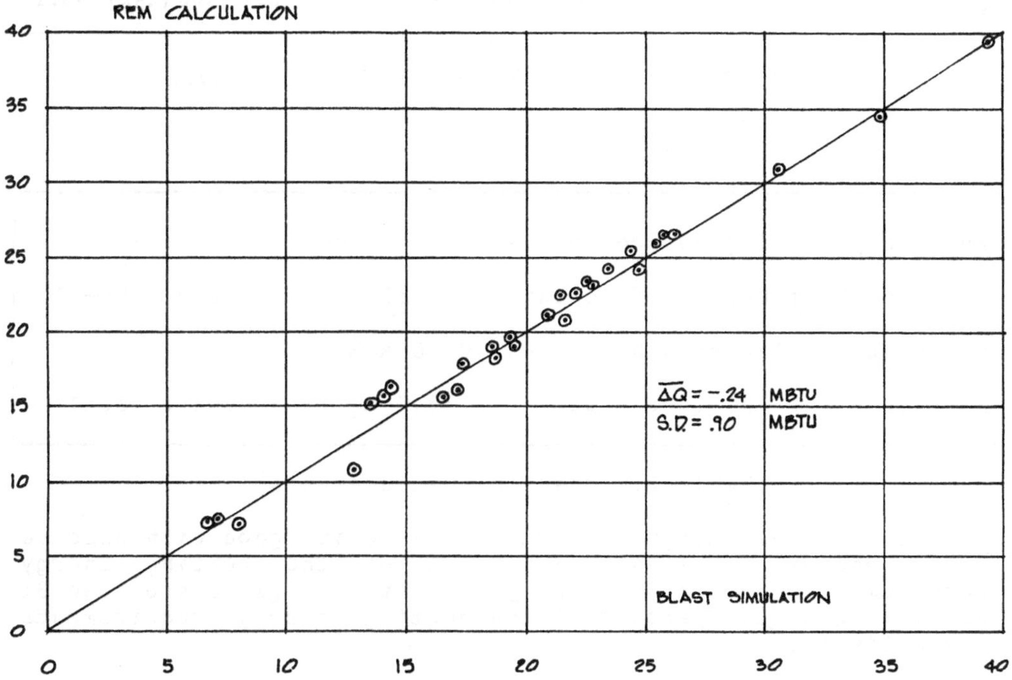

Figure 1: Comparison of REM Prediction to BLAST Simulation

4.2 <u>Measured Performance Results: 1982-1983 Heating Season</u>

The building was monitored as part of the U.S. Residential Class B Passive Solar Monitoring Program (4). The revised Class B Performance Evaluation Methodology was employed. An anometer was used during the 1982-83 heating season to measure wind speed as a basis of calculating the building's infiltration rate using an effective leakage air and the LBL simplified infiltration algorithm.

Table 1 presents the performance results for the 1982-1983 heating season. The key one-time measurements are shown in Table 2.

SI Units

Number of Days of Complete Data	Radiation (MJ/m^2-day)		Average Temperature (°C)		Degree Days (18.3°C) (°C-day)	Heating Energy Balance (MJ/day)				Specific Purchased Heat (kJ/°C-day-m^2)	
	Horizontal	Incident	Outdoor	Indoor		Total Load	Passive	Auxiliary	Internal	Purchased	Auxiliar
Nov 30	9.1	15.4	1.72	20.4	498	501	(48%)	129 (26%)	133 (26%)	44.8	22.(
Dec 25	7.3	14.5	0.17	20.0	563	556	(42%)	199 (37%)	121 (21%)	52.5	32.8
Jan 23	8.2	14.5	1.66	20.5	515	520	(44%)	179 (34%)	110 (22%)	50.1	30.9
Feb 28	12.1	15.5	4.88	20.9	376	436	(54%)	94 (21%)	108 (25%)	40.1	18.7
Mar 31	13.9	13.6	2.27	19.9	497	473	(41%)	170 (36%)	105 (23%)	50.7	30.9
Apr 30	19.1	8.2	4.33	20.4	419	80	(31%)	29 (36%)	26 (33%)	54.6	35.(
Average	11.6	15.5	2.5	20.4	428	193	(45%)	133 (31%)	100 (24%)	49.1	28.7

Table 1: Summary of Performance Results in 1982-1983

Auxiliary Heating System Delivery Efficiency	- 0.65
Building Heat Loss Coefficient	- 316 W/°C (600 Btu/hr- F)
Measured Infiltration Rate	- 0.40 ACH
Passive Solar Aperture Area	- 33.9 m (365 square feet)

Table 2: One-Time Measurements

The homes energy performance was quite good with passive solar energy supplying 45 percent of the heating energy requirement. Auxiliary energy and internal gains provided 31 percent and 24 percent of the heating energy requirements respectively.

4.3 Comparison of Calculated Performance to Measured Performance

Input to the REM Heating Calculation Procedure was developed from the construction drawings, on-site inspection, one-time measurements and the Class B site handbook. Because the monitoring period was shorter than the defined heating season assumed in the REM method, 181 days compared to 227 days, a degree day adjustment of the calculated load was made. Also, solar gains and the utilization factor were adjusted to compensate for the shortened monitored heating season. These adjustments are somewhat artitrary, and as will be discussed later, diminish the ability to compare measured to calculated energy performance for this building.

The comparison between measured and calculated energy performance is shown in Table 3.

Measured - Nov. 1982 - April 1983		24.05 MJ
Calculated - Nov. 1982 - April 1983		28.90 MJ
Difference		4.84 MJ or 17%

Table 3: Summary of Heating Performance Results - Measured and Predicted

4.4 Discussion of Results

Although the agreement between measured and calculated energy performance appears to be good, it will be shown that in this case it is probably a coincidence. Numerous uncertainties in climate, building characteristics, and measurement, eliminate any valid and useful comparison between measured and calculated performance.

Tables 4, 5 and 6 show a comparison of measured and long term climate data. Denver TMY was used in generating values for the REM. The monitored project is 32 kilometers west of Denver in the foothills of the Rocky Mountains. Sufficient variation exists between the measured and long term climate data to create significant differences in the comparison.

Month	Measured	Long Term	Deviation
Oct	-	9.49	-
Nov	1.72	3.49	-1.77
Dec	0.17	-0.76	+0.88
Jan	1.66	-1.69	+3.35
Feb	4.88	-0.60	+5.48
Mar	2.27	3.56	-1.29
Apr	4.33	9.31	-4.98
May	-	14.04	-
Average	2.5	2.22 (6) mo.	+0.28

Table 4: Climate Data: Average Ambient Temperature C

Month	Measured	Long Term	Deviation
Oct	-	15.2	-
Nov	9.1	10.0	-0.9
Dec	7.3	8.4	-1.1
Jan	8.2	9.5	-1.3
Feb	12.1	12.3	-0.2
Mar	13.9	18.3	-4.4
Apr	19.1	21.7	-2.6
May	-	25.0	-
Average	11.6	13.4 (6 mo.)	-

Table 5: Climate Data: Total Horizontal Solar Radiation (MJ/M -Day

Month	Measured	Long Term	Deviation
Oct	-	291.3	-
Nov	497.8	445.8	+ 52.0
Dec	562.8	590.5	- 27.7
Jan	515.5	620.6	-105.1
Feb	376.1	530.1	-154.0
Mar	497.2	462.3	+ 34.9
Apr	419.4	280.7	+138.7
May	-	165.9	-
Total	2868.8	2930.0 (6 mo.)	- 61.2

Table 6: Climate Data: Heating Degree Days - 18.3°C Base

Table 7 shows a comparison of measured overall UA of the building to an ASHRAE steady-state calculation using assumed "U" values and building surface areas. The difference is almost a factor of two. Given that the conductive load could possibly be off by a factor of two and the uncertainty of the infiltration rate, the basis of comparison between measured and calculated energy performance is further called into question.

Measured UA	214 W/°C
Calculated UA	410 W/°C
Difference	195 W/°C

Table 7: Measured Verus Calculated Building UA

A comparison of the construction drawings to the constructed house show major discrepancies. All the window sizes are different; material selection was different; location and possibly the existance of insulation is different; and thermal mass placement is different, and areas of the envelope surfaces varied considerably from the drawings. Conversation with the homebuilder failed to clarify all of these discrepancies.

4.5 Conclusions and Findings

The primary conclusions and findings from this attempt to compare measured energy performance to calculated energy performance are:

1) any close agreement between the measured and calculated energy performance of this passive solar home is purely coincidential;

(2) differences of climate data between those during the monitoring period and those long term values used in the calculation procedure make it difficult to compare performance results;

(3) too many uncertainties exist in the input to the calculation procedure and in the measurements to allow for an accurate comparison;

(4) the REM calculation procedure does show good agreement over a range design variations when compared to detailed simulation model results;

(5) unless all input values are accurately known and the base load is the same, there is little validity and value in comparing measured energy performance to calculated energy performance; and

(6) simplified design tools should not be evaluated/verified against validated building energy analysis simulations.

The important thing to remember is that design tools are used as a basis of design. The concern is for relative accuracy and the correct trends of performance due to design parameter changes and not absolute accuracy and fixed performance analysis.

REFERENCES

1. R.G. Derickson and M.J. Holtz, The Residential Energy Manual (REM), International Solar Architecture Conference, Cannes, France, December 1982.

2. R.G. Derickson and M.J. Holtz, Report on the Scientific Basis of a Simplified Residential Energy Design and Analysis Procedure, forthcoming technical report from the Solar Energy Research Institute, Golden, Colorado, May 1984.

3. R.G. Derickson and M.J. Holtz, The Residential Energy Manual: An Example of Use and Comparison to Results to Detailed Computer Simulation, Proceedings of the 8th National Passive Solar Conference, ASES, September 1983.

4. D.J. Frey, M. McKinstry, J. Swisher, Installation Manual for the SERI Class B Passive Solar Data Acquisition System, Solar Energy Research Institute, Golden, Colorado, November 1982.

A MICRO-COMPUTER AIDED DESIGN TOOL : CASAMO

P. BREJON, D. CAMPANA, F. NEIRAC, G. WATREMEZ
Centre d'Energétique
Ecole des Mines de Paris
60, bld St Michel 75006 - PARIS

Summary

CASAMO is a micro-computer aided design tool, specific for architects and designers which allows easy optimization of solar energy use in housing : energy saving and comfort.
The flexibility of use and conversational procedure allow various systems or configurations (hyper insulation, passive systems, active systems...) to be checked and optimized.
CASAMO is used, since 1982, on many micro computers : APPLE-2, CBM 8032, VICTOR-S1, IBM-PC ...
CASAMO calculates monthly heat losses of the building and monthly heating loads - taking into account passive solar gains and internal gains-. The method can analyze the effects of direct gains, TROMBE walls, sunspaces, and also active solar systems. CASAMO has obtained the aproval of the French Government to check the conformity of buildings to thermal regulations.
A new version of CASAMO performs an analysis of the internal building comfort. In particular, this new version predicts the possibility of overheating by calculating the hourly mean radiant temperature evolution, evaluated over a period of several typical days. The model, based on the finite difference method, was successfully validated by comparison with MINERVE code simulation results. MINERVE is a detailed code which was elaborated by centre d'Energetique of Ecole des Mines and is validated by experiments on a solar test cell built in Sophia-Antipolis. The adequate agreement between CASAMO and MINERVE is shown on figure 2.

Résumé

CASAMO est un outil d'aide à la conception thermique des bâtiments disponible sur micro-ordinateur. Depuis 1982, sa première version est diffusée en milieu professionnel sur APPLE-2.

Il permet de répondre aux préoccupations thermiques suivantes:

. comment implanter un habitat sur site de façon à utiliser au mieux le potentiel solaire disponible ?

. à combien s'élèvent les déperditions thermiques , les besoins nets de chauffage et les consommations d'énergie pour le chauffage des locaux et la production d'ECS ?

. quelles économies peuvent apporter des systèmes solaires "actifs" ?

. y-a-t-il risque de surchauffe du fait des configurations solaires passives retenues ?

Une fois élaborées les méthodes de calcul permettant de répondre à ces diverses questions, la 2ème étape du travail a consisté à les intégrer dans des logiciels aisément accessibles aux concepteurs de projet.
La règlementation thermique des locaux d'habitation en vigueur en France depuis 1982, autorise l'usage de différentes méthodes de calcul des besoins de chauffage, parmi lesquelles la méthode informatisée CASAMO.

Depuis 1979, le Centre d'Energétique de l'Ecole des Mines étudie et réalise des outils d'aide à la conception thermique des bâtiments, disponibles sur micro-ordinateurs. L'année 1982 a constitué la fin d'une première étape avec le début de la diffusion en milieu professionnel - architectes, bureaux d'études ... - du logiciel CASAMO, sur le matériel informatique APPLE-2 ; il s'agissait, bien évidemment, d'une première version de ce logiciel que la poursuite des études permet d'enrichir progressivement. L'objet du présent texte est de faire le point de l'état actuel des logiciels CASAMO dont les objectifs à moyen terme dépassent, nous le verrons, ceux des origines, à savoir la seule "Conception Architecturale Solaire Assistée par Micro-Ordinateur".
La démarche qui soustend ce projet est double : d'une part développer des méthodes de calcul "portables" sur les micro-ordinateurs 8 et 16 bits les plus courants ; d'autre part mettre en forme ces méthodes et structurer leur mode d'utilisation de façon à atteindre l'objectif annoncé d'aide à la conception. Dans un premier temps il ne s'agira que des bâtiments d'habitation, en climats tempérés. Les préoccupations d'ordre thermique et énergétique sont les suivantes :

. comment implanter un projet d'habitat sur site, de façon à tirer le meilleur parti du potentiel solaire ?
. quelles sont les déperditions thermiques de l'habitat étudié et, en conséquence, la puissance de chauffage à installer ?
. à combien s'élèvent les besoins nets de chauffage une fois déduite des déperditions, l'énergie gratuite récupérée ?
. quelles sont les consommations d'énergie des 2 usages, chauffage et eau chaude sanitaire (ECS) ? notamment dans le cas où l'on recourt à une installation solaire "active" ?
. n'y-a-t-il pas risque de surchauffe, l'été ou en mi-saison, dans certaines pièces très exposées ?

Quant à l'aspect "logiciel", il induit des qualités minimales que les futurs utilisateurs sont en droit d'en attendre à quatre niveaux :

. la conversationnalité - langage clair, non codé ... -
. l'entrée des données - forme accessible, droit à l'erreur, redondance minimale ... -
. les sorties, à l'écran et sur papier - rédaction et présentation soignées, sorties graphiques, résultats intermédiaires optionnels ...-
. la structure d'ensemble - modularité, possibilité d'étudier des variantes, des sensibilités paramétriques ... -

Nous mettrons l'accent dans les pages qui suivent sur 3 des aspects thermiques cités plus haut ; l'implantation sur site, les besoins de chauffage, les conditions de confort.

1 - <u>Implantation d'un projet d'habitat sur site</u>

Pour l'essentiel, il s'agit de donner au concepteur les moyens d'intégrer dans l'ensemble des contraintes d'implantation d'un projet, des préoccupations thermiques, et de lui permettre de tirer le meilleur parti du site climatique et de l'environnement.
La première question est donc de savoir quelles sont les données météorologiques disponibles. La méthode de calcul retenue pour les besoins de chauffage suppose connues les valeurs moyennes mensuelles des 2 variables:

. fraction d'insolation (ou éclairement global sur plan horizontal) ;
. degrés-jours en base quelconque.

Dans le contexte français, le recours à ces 2 seules variables permet de disposer de données pour 88 sites répartis d'une façon qui autorise, dans la plupart des cas, une bonne approche climatique du projet étudié.

La deuxième question est relative à l'affaiblissement du potentiel solaire compte tenu de la présence de masques liés à l'environnement du projet ou au projet lui-même (fenêtre en retrait ...). La méthode de calcul adoptée consiste à suivre la course du soleil, avec un pas de temps de l'ordre de l'heure, pour une journée moyenne de chaque mois, et, à calculer un coefficient d'affaiblissement énergétique moyen mensuel liés aux différents masques affectant les plan récepteurs du projet étudié.
A cette approche numérique, s'ajoute pour les masques lointains un traitement graphique à l'aide du tracé du diagramme solaire (fig. 1).

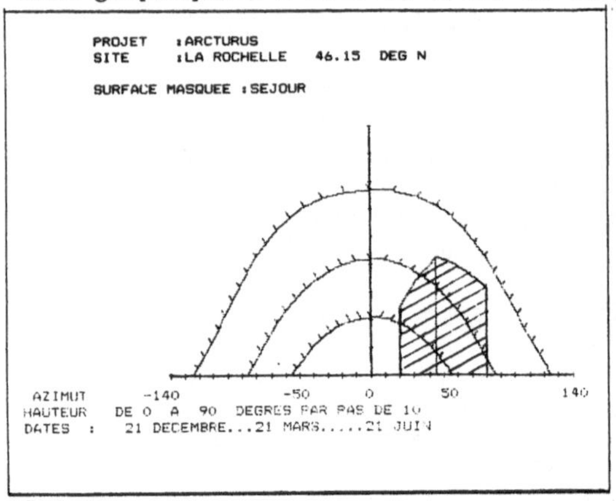

Fig. 1 : Tracé d'un diagramme solaire avec représentation d'un masque lointain (zône hachurée).

Dans un repère angulaire azimut/hauteur, le programme superpose au tracé des 3 courbes illustrant la course du soleil aux solstices et aux équinoxes, l'image du masque considéré (surface hachurée). Le soleil n'est "vu" du point d'observation que lorsqu'il se trouve "au-dessus" des bâtiments portant ombrage. Les graduations des courbes en demi-heures permettent d'apprécier de façon qualitative les effets d'ombre portée à n'importe quel instant de l'année. L'architecte trouve là une information supplémentaire relative à la qualité de l'éclairage naturel.

2 - Les besoins de chauffage

Il s'agit des besoins nets de chauffage, nécessaires au maintien du volume chauffé à une température de consigne, comprise entre 12 et 21°C ; ainsi on déduit, des déperditions thermiques calculées pour cette température de consigne, la quantité récupérée d'apports énergétiques gratuits. Ceux-ci, on le sait, sont de 2 ordres : des apports internes qui sont, dans la méthode CASAMO, forfaitisés à une valeur brute de 0,13 kWh/jour/m^2 et des apports solaires transmis par différents types de composants passifs: vitrages, vérandas, murs TROMBE.
La méthode de calcul que nous avons développée, repose sur une loi de corrélation issue d'une confrontation des travaux concordants de D.BALCOMB et R. GILLES menés sur des modèles détaillés. Cette loi permet d'obtenir la fraction (f_x) des déperditions (D) couvertes par les apports gratuits - solaires (AS) et internes (AI) - à partir d'une variable caractéristique du projet étudié. Cette variable (X), calculée mois par mois, s'exprime de la façon :

$$X = \frac{AS + AI}{D}$$

La loi de corrélation s'acrit : $f_x = 1 - \exp(-0,9 \times (1 + 0,3x))$
Il vient alors : $B = D(1 - f_x)$, où B désigne les besoins nets de chauffage du mois considéré.
La sommation sur les 8 mois de la saison de chauffage permet d'obtenir, en kWh, la valeur des besoins de chauffage annuels (B_a). Signalons enfin que la règlementation thermique relative aux locaux d'habitation, en vigueur en FRANCE depuis 1982, impose le calcul d'un coefficient volumique

de besoins de chauffage qui s'exprime sous la forme : $B = \frac{B_a}{24\ V_H\ D_j}$ en $W/m3.°C$

avec V_H = volume habitable chauffé, en m^3

D_j = degrés-jour pour une température de consigne de 18 ou 19°C, en °C.Jour

Plusieurs méthodes de calcul de ce coefficient règlementaire ont été "agréées" par le Ministère de l'Urbanisme et du Logement parmi lesquelles CASAMO.

3 - Les conditions de confort

Doter les concepteurs d'un outil permettant d'apprécier les qualités d'un habitat en termes de confort, constitue une nécessité afin que la recherche de la performance énergétique ne s'effectue pas au détriment de la notion de confort thermique de l'occupant. Ce type d'outil, encore très peu usité dans la pratique professionnelle, est appelé à se développer du fait de l'évolution actuelle des normes de construction : très forte isolation thermique associée parfois à d'importants apports solaires passifs.

Cette notion de confort fait intervenir un grand nombre de facteurs relatifs à l'occupant lui-même et à l'ambiance intérieure. Nous avons admis que sous les climats tempérés, pour les types d'habitations "européennes", les facteurs qui déterminent le niveau de confort d'une ambiance intérieure sont principalement la température d'air et la température radiante.

Nous avons enfin pris le parti de nous intéresser aux risques de surchauffe qui nous semblent être la principale source d'inconfort des ambiances: aussi raisonnons-nous sur des journées à fort ensoleillement et sur les conditions d'ambiance d'une pièce - la plus exposée - d'un logement.

Sur le plan méthodologique, nous avons recours à une simulation en régime variable effectuée selon le principe des différences finies ; la résolution numérique s'effectue selon un schéma de Crank-Nickolson. La discrétisation en espace et en temps résulte d'un compromis entre les 2 objectifs suivants :

. finesse du maillage pour rendre compte de façon satisfaisante du comportement thermique du local étudié

. limitation du nombre de noeuds pour autoriser une implantation sur des microordinateurs 8 bits du type de l'APPLE-2, sans conduire à des temps de calcul trop longs.

Ces derniers stériliseraient en effet l'objectif d'aide à la conception qui se traduit par une incitation à tester diverses variantes d'un même projet.

Nous admettons des écarts sur les niveaux de température maximale obtenus par la méthode CASAMO et le modèle détaillé, MINERVE, utilisé comme référence, inférieurs à 1°C, pour des sollicitations extrêmes (fig. 2). Nous considérons par ailleurs, que la simulation d'une séquence de 3 jours ne doit pas excéder quelques minutes - de 2 à 5 selon qu'il y a ou non présence d'une serre -. Ces deux conditions nous ont conduit à :

. un nombre de noeuds compris entre 10 et 20 selon l'importance inertielle des différentes parois, l'éventualité d'une serre et le choix de l'utilisateur entre temps de calcul et finesse des résultats

. un pas de temps des calculs de simulation d'une demi-heure.

L'originalité d'une telle méthode ne réside bien sûr pas dans le mode de résolution retenu mais plutôt dans la mise en forme d'un logiciel sur micro-ordinateur, conversationnel, directement utilisable par un architecte ou un thermicien. Les entrées du logiciel se limitent en effet à :

. la description détaillée des éléments constitutifs des différentes parois - avec recours possible à des bibliothèques de parois et de matériaux qui rende aisée la saisie -,

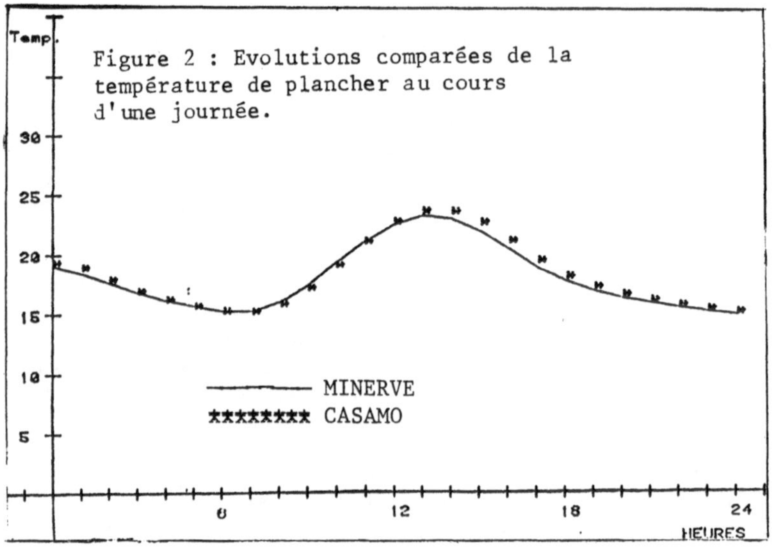

Figure 2 : Evolutions comparées de la température de plancher au cours d'une journée.

. la description du site, analogue à celle proposée lors du calcul des besoins de chauffage - latitude, albédo, masques, orientation des vitrages...,
. les températures extérieures quotidiennes minimales et maximales - le programme génère une évolution sinusoïdale entre ces 2 valeurs -,
- le mois de l'année et s'il se trouve en période de chauffage, la température de consigne et le mode d'émission de la chaleur - plancher chauffant ou convecteur -.

En sorties, affichées à l'écran ou imprimées, le concepteur obtient les évolutions des différentes températures pour la séquence étudiée : air intérieur, parois, température résultante, ainsi que les courbes d'irradiation solaire transmise et de température extérieure.

4 - Conclusion

Les logiciels CASAMO constituent d'ores et déjà des outils d'aide à la conception thermique des bâtiments utilisés en milieu professionnel depuis la fin de l'année 1982 (*).

Cette diffusion atteste d'un souci des pouvoirs publics qui ont partiellement financé ce projet - Ministère de l'Urbanisme et du Logement, Agence Française pour la Maîtrise de l'Energie - et de l'Ecole des Mines de favoriser la mise à disposition des professionnels, de résultats de recherche.

On ne saurait toutefois conclure sans mentionner les limites actuelles de ces outils :
- sur le plan de l'optimisation économique, encore inexistante
- sur le plan thermique : non prise en compte de l'intermittence de fonctionnement des locaux
- sur le plan climatique, limité aux zones tempérées.

Ce sont ces lacunes qui nous ont conduit à définir un plan de recherche pour 1984-1986 axé sur :
. la mise au point de méthodes simplifiées adaptées aux conditions particulières des locaux du tertiaire
. le développement de méthodes de calcul et des logiciels associés, adaptés aux conditions climatiques des pays tropicaux et à leurs besoins spécifiques
. la réflexion et la mise en forme d'outils intégrés de CAO traitant conjointement les aspects dessin, métré, évaluation thermique et optimisation énergétique.

(*)disponibles aujourd'hui sur APPLE-2, CBM-8032, VICTOR-S1, IBM-PC et la plupart des micro-ordinateurs utilisant les systèmes d'exploitation CP/M ou MS-DOS

ANALYTICAL MODEL FOR THE INPUT/OUTPUT ENERGY RELATIONSHIP

Bengt Perers, Heimo Zinko, Per Holst
Studsvik Energiteknik AB, Sweden

Summary

The purpose of the work is to analyse the relationship between daily energy collected and daily total solar insolation in the plane of the collector observed in so called input/output diagrams. As a first approximation linear regression lines are used to present this relationship.

In this paper we describe a way how to take into account the variation of the daily and seasonal distribution of the irradiation, which influence the daily energy output significantly.

A model has been developed for analytical calculation of the average input/output relation with high accuracy.

1. INTRODUCTION

Input/output regression lines are often used for comparing solar collector array performance from different systems and climates. So is done within the IEA Solar Heating and Cooling task VI for evacuated collectors.

Usually for well performing solar collector systems a very linear relationship is experienced between daily solar irradiation H and daily energy collected Q (see ref 1) and Fig 7. In general this relationship can be expressed by

$$\boxed{Q = a\,H - b} \qquad (1)$$

were a give the slope of the curve and b represent the daily threshold energy for the system.

2. THEORY

An analytical expression for the daily energy flows can be derived from a short term energy balance for the solar collector system.

The collectors are modelled with the Hottel Whillier Bliss equation including angle modifier, dynamic losses, piping losses and pumping energy dissipation.

When integrating this equation for 24 hrs we get a very simple analytical expression for the daily energies:

$$\boxed{Q = A\,H - B\,\langle \Delta T \rangle_{24\,h} - C} \qquad (2)$$

A is the all day zero loss efficiency and B is the collector array heat loss factor. $\langle \Delta T \rangle 24h$ is the 24 hrs average temperature difference between collector mean and ambient temperature. C is a correction term where parasitic energies as pumping energy dissipation and piping losses are taken into account. In a well designed system C is almost zero. But in a test system C can be significant.

By dividing this equation with H we get an expression for the daily efficiency. This equation has been tested with experimental data showing that A and B really can be approximated by constants. See Fig 1.

In practice H and <ΔT>24h are not independent variables as shown in Fig 2. There is a clear tendency that the <ΔT>24h increases with H due to a systematic variation of the average time distribution of H.

This means that the resulting input/output relation for constant operating ΔT is composed of daily results along individual lines for constant <ΔT>24h according to equation 2. See Fig 3.

This explains why the regression line has a slightly lower slope than the all day zero loss efficiency A. The <ΔT>24h dependence on the time distribution of the insolation during the day explains the variation of Q by season and also the scatter for individual days. For example the shorter winter days give a lower <ΔT>24h for the same H and operating ΔT. This results in a higher Q compared to summer operation.

For the purpose of system modelling the coefficients A, B and C as well as <ΔT>24h have to be derived analytically.

This can be done by establishing a system energy balance according to Fig 4. From this energy balance a substitution for B<ΔT>24h can be derived and insertet into equation 2. For simple analytical treatment however some approximations are necessary. These approximations are explained by means of Fig 5 which shows a typical daily irradiation. A x H is the thermal energy input to the system and B<ΔT>24h is the energy losses from the collector array.

The total 24 hr energy losses from the array B<ΔT>24h is substituted by analytical expressions for the four loss areas M, R, E and D. R is derived from the operating ΔT requirement and the operating time. M and E are derived from the difference between daylength and operating time and the shape of the limiting curve represented by an empirical formfactor f. D is the all day capacitance energy which is leaving the system as heat losses after the operating period.

M, R and E have expressions with the operating time involved. The operating time increases with H in a systematic way and corresponds to the previously observed increase in <ΔT>24h with H. Longterm data analyses for both operating time and irradiation time-distribution show a very linear average relationship, see Fig 6. This has been used to eliminate the operating time from M, R and E.

When inserting the expression for <ΔT>24h into 2 we get a model where all the important system parameters are in explicit form and the input/output average relations according to the regression lines can be analytically predicted. C is not shown in explicit form as it normally can be neglected.

$$Q = H(A-U\cdot\Delta T(1-f)d)+U\cdot\Delta T\cdot s-U\cdot\Delta T\cdot f\cdot(S_{day}+s)-\Delta T\cdot C_s-C$$

Q	= Daily energy from collector array	MJ/m^2 day
H	= Daily insolation in the plane of the collector	MJ/m^2 day
A	= All day zero loss efficiency for the collector array	-
U	= Array heat loss coefficient (= B/24/3600)	W/m^2 K
ΔT	= Average temperature difference between mean collector temperature and ambient during operating time	K
f	= form factor for energy losses outside operating time (∿0.30)	-

d = operating time gradient with H (longterm average) (∿0.0015) s/J

s = threshold operating time = S_{day} d f_1 ΔT U/n. s

f_1 = form factor for energy losses for days with threshold insolations (∿0.25) -

n = all day zero loss efficiency (100% diffuse insolation) -

S_{day} = Daylength s

C^s = Collector array capacitance $J/K/m^2$

C^s = Correction term for parasitic energy flows in the system (same C as in 2) MJ/m^2 day

3. RESULTS

The model has been tested against monthly linear regression lines from different Swedish installations to evaluate the accuracy. See Figs 7 and 8. The best accuracy is achieved for systems with low threshold values. These systems also show the highest correlation coefficient for the measured regression lines. When analysing the diagrams one should keep in mind that the analytical model gives the expected average monthly line for a system in full operation whereas single monthly experimental lines can be slightly shifted by climatic and operational effects.

4. CONCLUSIONS

- The linear relationship between the daily energy collected Q and daily insolation H at constant operating ΔT can be explained and predicted by a simple model.
- The relationship between Q and H is theoretically just linear for constant 24 hr ΔT. But there exists an average relation between 24 hr ΔT and H which causes a linear relation between Q and H also for constant operating ΔT. This line has a lower slope than those for constant 24 hr ΔT.
- The time distribution of the insolation during the day influences the energy output significantly and causes a seasonal shift of the curves as well as a scatter along the monthly regression lines. This effect is correctly taken into account by the model.
- The time distribution dependence increases with increasing threshold value for the system. Systems operating near their maximum temperature level therefore show a high scatter in the input/output diagrams.
- Collector array performance parameters can be evaluated directly from daily values of Q and H when the 24 hrs ΔT between collector mean and ambient temperature is known.

REFERENCES

1. DUFF, William, S. Eight Evacuated Collector Installations. Interim report for the IEA task VI on evacuated collectors. Nov 1982 Colorado State University, USA.

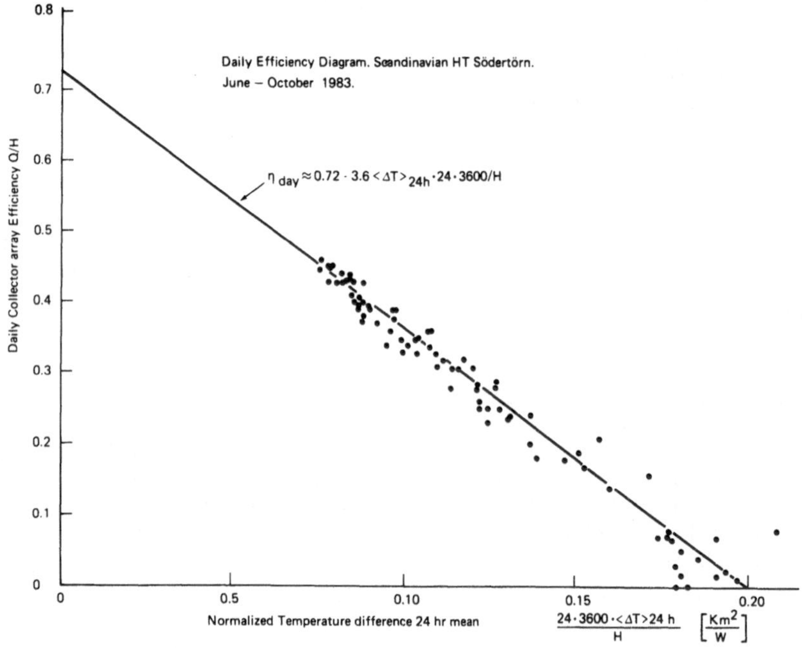

Figure I - Daily efficiency diagram

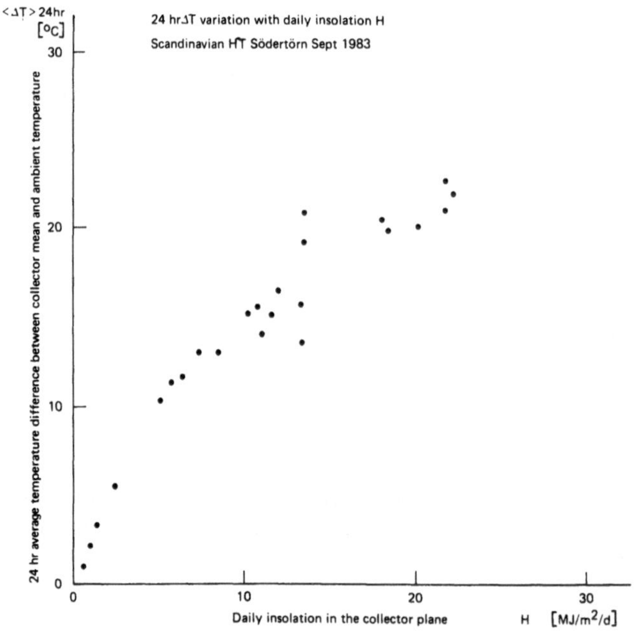

Figure II - Variation of 24 hrs ΔT with daily isolation H

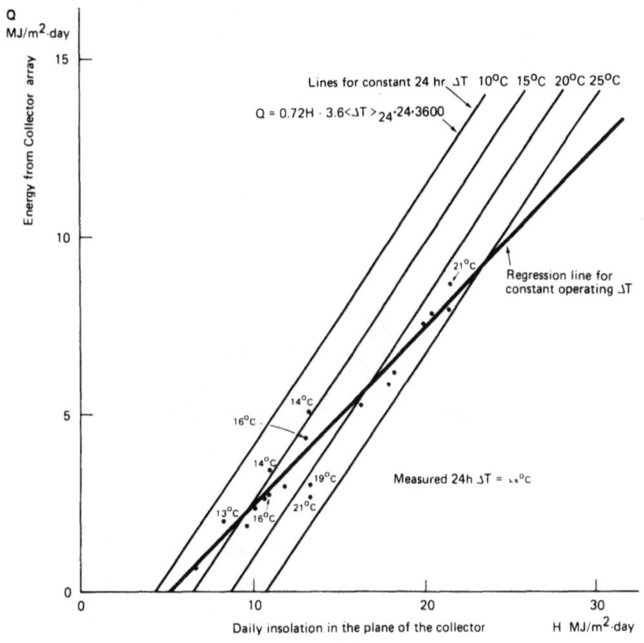

Figure III - Daily lines for constant 24 hr ΔT and resulting monthly regression line for constant operating ΔT

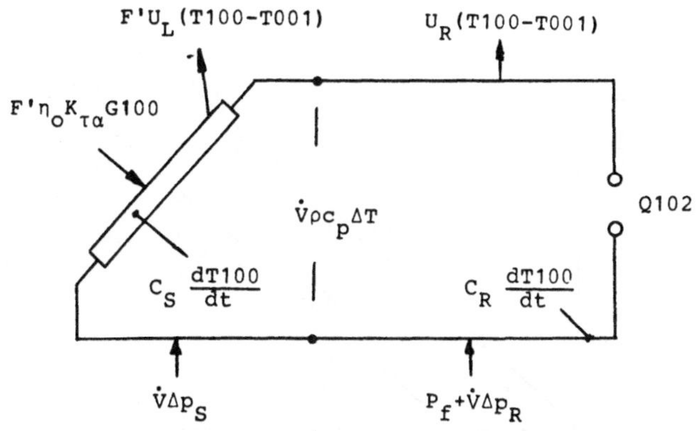

Figure IV - Energy flow in the collector subsystem

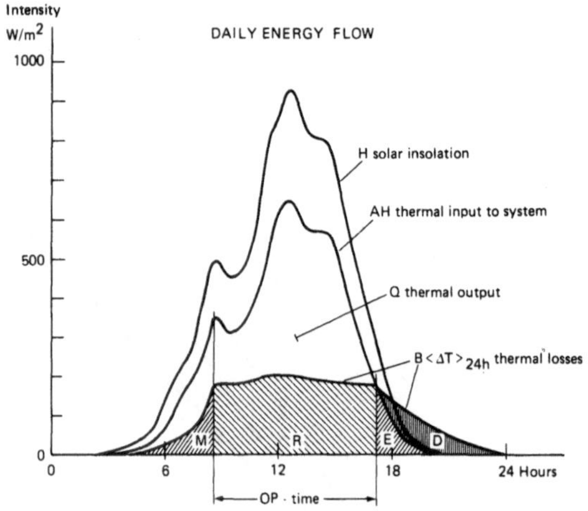

Figure V - Daily energy flow in the collector system

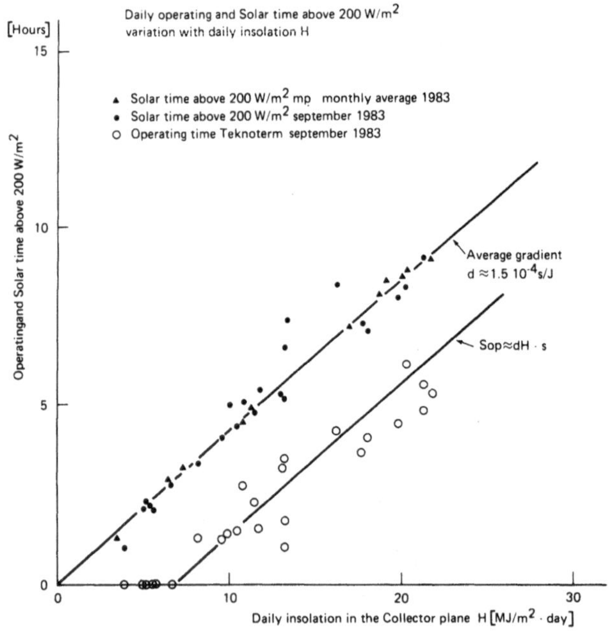

Figure VI - Daily operating time and solar time above 200 w/m^2
dependence of daily insolation in the collector plane

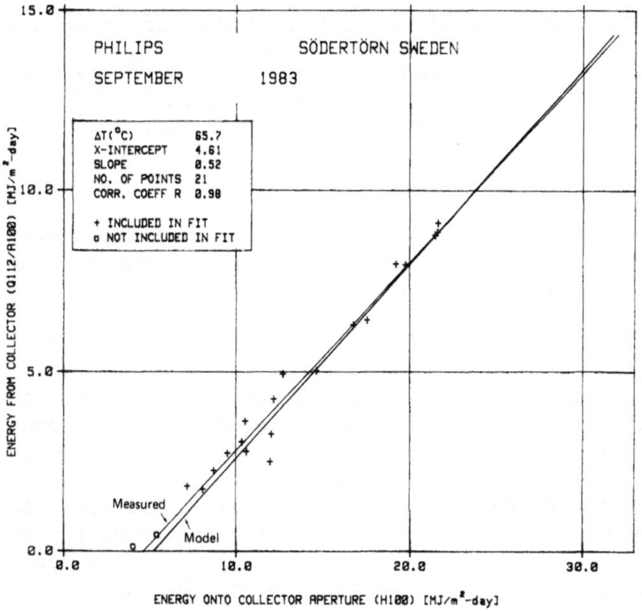

Figure VII - Input/output diagram with analytical and experimental linear relationship

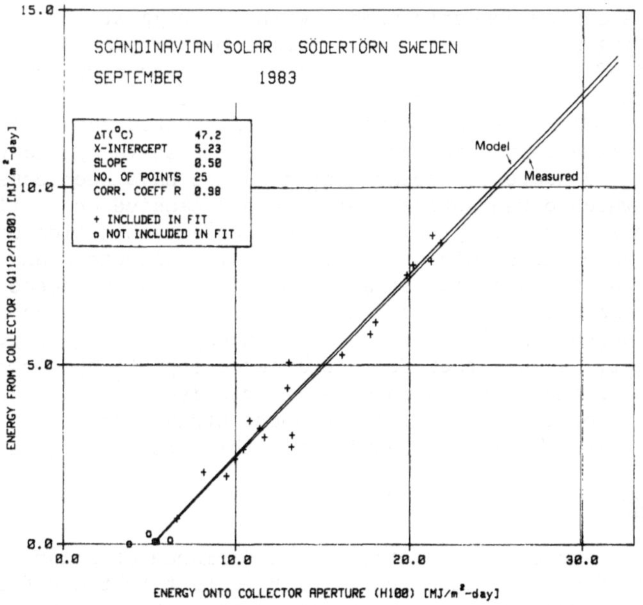

Figure VIII - Input/output diagram with analytical and experimental linear relationship

PERFORMANCE AND COST EVALUATION OF SOLAR PASSIVE BUILDINGS IN GREECE

C. Stambolis*, M.Santamouris, M.Valindras, R.Rigopoulos.
Physics Laboratory B, Dpt of Physics, University of Patras, Patras, Greece.
* Heliotechnic (Hellas) Ltd, 106 71, Athens, Greece.

Summary

The objectives of this paper is to examine the thermal performance of different type solar passive systems, for the heating of buildings, in different parts of Greece and also to look into the economic aspects in terms of required capital investment and simple payback ratios. The present investigation is narrowed down to two passive systems, namely Trombe walls and Direct Gain. The un-utilizability design method for collector-storage walls is used in the case of Trombe walls while a method developed by the authors is used for the calculation of energy gain in the case of Direct-Gain systems. The results of this analysis are summarised in several tables corresponding to the different types of houses used, its location and the chosen solar passive system. These tables can be directly used by the architect or engineer involved in solar passive design to quickly assess the anticipated performance of the passive system under study.

Background

Although Greece claims the largest solar water heater market in the European Community with more than 400,000 m² of installed solar collectors, very little has been done in the field of solar space heating whether active or passive. As a significant part of the country's national energy budget is spend on domestic needs, with the biggest component going for space heating needs (approx. 4.6 m.t.o.e. or 33% of the total energy budget), and or 88% of the country's oil is imported. The need for energy conservation in the domestic sector is apparent. It is further estimated that approximately 100,000 of new houses are constructed every year thus presenting an excellent opportunity for solar passive applications. So far only a limited number of solar heated houses have been constructed in Greece of which 8 use active systems and 7 passive ones. (1) Most of these houses are single storey buildings and use a combination of passive systems such as Trombe walls, direct gain, wall-air collectors and greenhouses. The potential for applying solar passive heating in Greece is enormous due to the good climatological conditions (high solar radiation, small number of degree days) and the notable house building activity of modern Greek society. It was therefore considered important to carry out a first assessment of the performance of solar passive houses in Greece for different areas of the country. The present study narrows its investigation to houses only and reports results for three climatological areas.

Methodology

In order to assess the thermal performance of solar passive buildings we selected four reference houses with respect to their floor area, their insulation characteristics and their south wall solar passive exposure. The four types of reference houses range from a single storey small house with an 80 m² floor area to a two storey 300 m² dwelling. The construction characteristics of all houses are the same and they asume the standard reinforced concrete framework construction which is typical of house buildings in Greece. The main characteristics of these reference houses are given in Table 1 while Fig. 1 and Fig. 2, 3 show floor plans and sections of three houses. For each reference house we considered two possible versions for the total wall surface glass area for the Trombe and the Direct Gain system. Each version represents what we determined to be an upper

TABLE 1

Floor Area (m^2)	80	130	180	300
Height (m)	3	3	3	6(3+3)
Heat Loss Coefficient				
(W/m^2) Salonica	0.62	0.62	0.62	0.67
Athens	0.79	0.79	0.79	0.83
Chania	1.07	1.07	1.07	1.10
Total surface area F				
(m^2)	270	400	530	700
Volume (m^3)	240	390	540	900
F/V	1.12	1.02	0.97	0.77
Trombe Wall Area (m^2)(11%)	9,18	(22.5) 15,30	20,40	30,60
Direct Gain (m^2)	9,18	15,30	20,40	30,60

and lower limit of useful glass surface in terms of practical south wall coverage. Within the upper and lower glass surface area limits lies a whole range of possible glazed areas. However, for the purpose of this study we only considered the upper and lower limit. It is the authors' intention to complete their investigation by examining energy gains, both for the Trombe and Direct Gain systems, for a much larger number of glazed areas, say at increments of 2 m^2, within the range set by the upper and lower limits given in Table 1.

To calculate the total energy gain from the Trombe Wall in each of the reference houses considered we used the Un-Utilizability Design Method for collector-storage walls, which is described in reference (2). In the case of Direct Gain systems we used a method developed by the authors as this is described in reference (3). Throughout the calculations for determining the various energy gains we assumed that night insulation is being used of $R=1.59\ m^2.C^o\ W^{-1}$. We further assumed house designs that enable the removal of overhangs during the winter months. Therefore no shadow effects, resulting from overhangs, have been considered. Another important assumption is that for the reference houses we have used thermal insulation as recommended by Greece's thermal insulation code (4). Solar radiation data for Greece was used as described in reference (5).

Performance Results

A summary of the obtained results are presented in Tables 2, 3 and 4 each corresponding to the three locations we selected for our investigation, namely. Salonica, Athens and Chania. Each city located in one of the three distinct climatological zones of the country. Looking at the variation of solar collector south wall surface area, in both the Trombe and Direct Gain systems, in each of the reference houses considered we discern a considerable difference in the monthly total energy gain between the upper and lower glazed area limits. So far example in the version of the 80 m^2 reference house located in Athens we observe that the total energy gain for January is almost double, i.e. 1.60 GJ/m for the 18 m^2 glazed surface compared to 0.81 GJ/m for the 9 m^2 surface. i.e. the solar load ratio in the first version is .3 compared to .16 in the second version. Using a relatively large Trombe wall area, but still practical from a construction point of

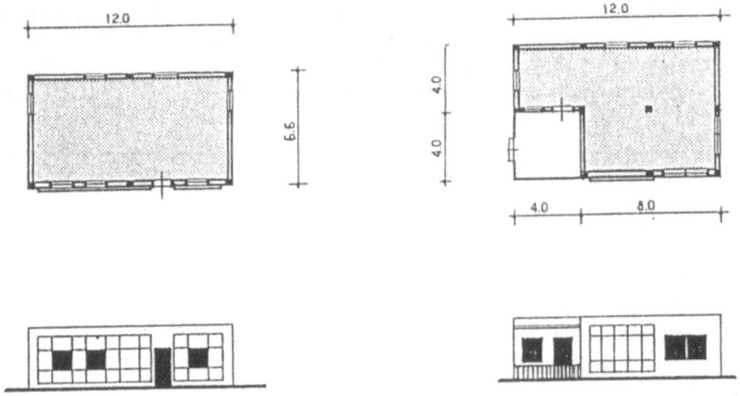

<u>Fig.1:</u> the 80 m² reference house

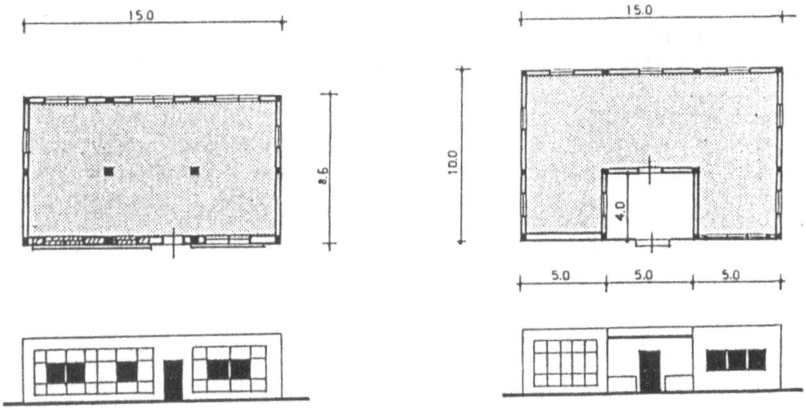

<u>Fig.2:</u> The 130 m² reference house

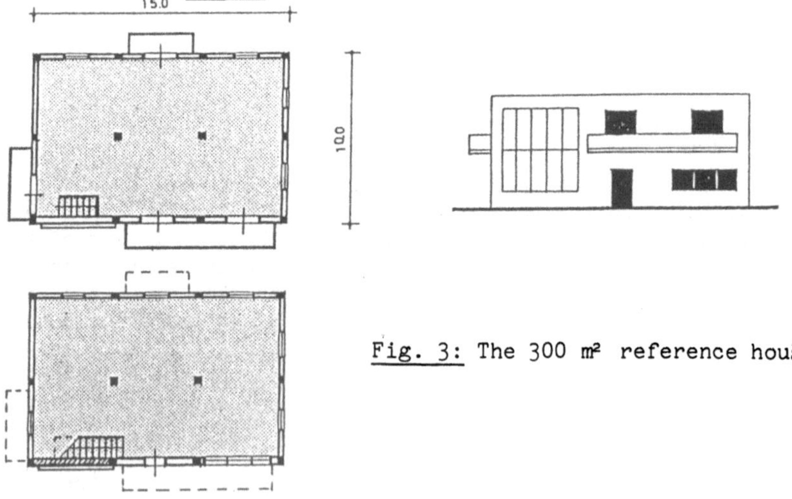

<u>Fig. 3:</u> The 300 m² reference house

TABLE 2

ATHENS (37° 58´)	J	F	M	N	D
FLOOR SURFACE: 80 m²					
Passive Wall: 9 m²					
QLOS	5.11	4.34	3.79	1.86	4.00
QTRB	0.81	0.86	0.91	0.99	0.94
QDG	1.11	1.09	1.25	0.80	1.19
Passive Wall: 18 m²					
QLOS	5.35	4.59	3.97	1.94	4.18
QTRB	1.60	1.67	1.75	1.66	1.80
QDG	1.72	1.62	1.79	2.23	1.65
FLOOR SURFACE: 130 m²					
Passive Wall: 15 m²					
QLOS	7.62	6.46	5.65	2.77	5.94
QTRB	1.35	1.43	1.51	1.58	1.56
QDG	1.81	1.75	1.99	1.22	1.89
Passive Wall: 30 m²					
QLOS	8.02	6.81	5.96	2.91	6.26
QTRB	2.63	2.71	2.81	2.62	2.90
QDG	2.69	2.53	2.77	1.37	2.55
FLOOR SURFACE: 180 m²					
Passive Wall: 20 m²					
QLOS	10.10	8.57	7.50	3.67	7.88
QTRB	1.80	1.90	2.00	2.07	2.07
QDG	2.40	2.33	2.65	1.62	2.51
Passive Wall: 40 m²					
QLOS	10.64	9.03	7.90	3.87	8.30
QTRB	3.48	3.58	3.70	3.38	3.81
QDG	3.58	3.36	3.68	1.82	3.39
FLOOR SURFACE: 300 m²					
Passive Wall: 30 m²					
QLOS	14.12	11.94	10.48	5.13	11.02
QTRB	2.70	2.84	2.98	2.99	3.08
QDG	3.53	3.41	3.86	2.29	3.64
Passive Wall: 60 m²					
QLOS	14.94	12.67	11.09	5.43	11.65
QTRB	5.12	5.25	5.38	4.78	5.56
QDG	5.16	4.82	5.27	2.57	4.84

QLOS = Total heat loss per month GJ/m
QTRB = Energy gain from Trombe Wall per month GJ/m
QDG = Energy gain from Direct Gain per month GJ/m

TABLE 3

CHANIA (35° 30′)	J	F	M	N	D
FLOOR SURFACE: 80 m²					
Passive Wall: 9 m²					
QLOS	4.55	3.78	3.46	1.21	2.98
QTRB	0.84	0.83	0.97	0.97	0.91
QDG	1.16	1.19	1.25	0.17	1.30
Passive Wall: 18 m²					
QLOS	4.73	3.92	3.59	1.25	3.09
QTRB	1.64	1.60	1.71	1.25	1.71
QDG	1.69	1.73	1.72	0.18	1.59
FLOOR SURFACE: 180 m²					
Passive Wall: 20 m²					
QLOS	8.99	7.46	6.34	2.39	5.88
QTRB	1.86	1.83	1.96	1.96	2.01
QDG	2.47	2.54	2.62	0.35	2.54
Passive Wall: 40 m²					
QLOS	9.38	7.78	7.13	2.49	6.13
QTRB	3.53	3.41	3.83	2.49	3.57
QDG	3.49	3.56	3.38	0.37	3.23

TABLE 4

SALONICA (40° 33′)					
FLOOR SURFACE: 80 m²					
Passive Wall: 9 m²					
QLOS	6.05	4.78	4.09	2.85	5.17
QTRB	0.76	0.81	0.86	0.93	0.82
QDG	0.94	0.95	1.09	1.12	0.95
Passive Wall: 18 m²					
QLOS	6.42	5.07	4.34	3.03	5.48
QTRB	1.60	1.61	1.69	1.73	1.62
QDG	1.51	1.52	1.62	1.44	1.45
FLOOR SURFACE: 180 m²					
Passive Wall: 20 m²					
QLOS	11.98	9.47	8.10	5.55	10.22
QTRB	1.79	1.80	2.03	2.03	1.83
QDG	2.02	2.07	2.00	2.33	2.05
Passive Wall: 40 m²					
QLOS	12.79	10.11	8.65	6.04	10.92
QTRB	3.51	3.70	3.60	3.58	3.52
QDG	3.17	3.18	3.35	2.93	3.03

view, we may cover 85% of our heating requirements in November and 43% in December in Athens, for the 80 m² reference house. Concerning the direct gain systems here we obtain solar load ratios generally bigger than for the Trombe wall. This is in agreement with the findings of reference (2). A general conclusion that can be arrived at by examining the results given in Tables 2, 3, 4 is that solar passive systems, both Trombe and Direct Gain, can be used successfully in different parts of Greece, from North to South, to cover substantial parts of houses heating loads. More important is the observation that solar load ratios in the range of .3 to .45 can be achieved without departing from existing and traditional building construction methods and without requiring any serious extra investment from the part of the owner.

Economics

The net total cost per m² for the construction of a Trombe wall in Greece is approx. 10,000 drachs (i.e. 100 US $/m²), while for a Direct Gain the cost drops to approx. 6,000 drachs including night insulation. For the 130 m² reference house with a 30 m² solar wall area the total Trombe wall cost would reach 300,000 drachs. Since the total construction cost for the 130 m² house could reach 5,2 million drachs (40,000 drachs/m²) the cost for the construction of the Trombe wall corresponds to just 5.7% of the total construction budget. In the version of the 300 m² reference house the upper limit Trombe wall would cost 600,000 drachs and the total construction cost of the house could reach 12 million drachs. In this case the Trombe-wall construction would correspond to only 5% of the total construction budget. According to currently available figures regarding the cost for the construction of an equivalent solar active system (in terms of annual heat gains) solar passive systems cost half and many times less than half cost of active systems. The payback ratio P/A for the passive systems under study, (upper limit) and for current oil prices in Greece, is 6.

References

1. A.TOMBAZIS, Solar and Wind Energy Applications in Greece, in Proceedings of Inaugural meeting, Greek Section of ISES, Athens March 84, (In Greek).
2. W.A.MONSEN, S.A.KLEIN, W.A.BECKMAN, The un-utilizability design method for collector storage walls. Solar Energy, 29, 421-429, (1982).
3. R.RIGOPOULOS, M.SANTAMOURIS, C.STAMBOLIS, M.VALINDRAS, Windows in Greece: Possibilities for Solar Direct Gain Utilisation, Windows in Building Design and Maintenance Conference Proceedings, Gothenburg, 13-15 June, 1984.
4. THERMAL INSULATION CODE, Government Gazette, Hellenic Republic Athens, July 4, 1979, No 362.
5. D.P.LALAS, D.K.PISSIMANIS, V.A.NOTARIDOU, Methods of Estimation of the Intensity of Solar Radiation on a Tilted Surface and Tabulated Chronica Scientific Journal of the Technical Chamber of Greece-Section B, Vol. 2, No 3.4 (1982).

UPPER LIMIT OF THE USEFUL SOLAR HEAT ACHIEVABLE IN THE CENTRAL EUROPEAN CLIMATES WITH DHW AND HEATING SYSTEMS. IMPORTANT TECHNICAL FACTORS DETERMINING THE USEFUL SOLAR HEAT OF A SOLAR SYSTEM

J.M. Suter, J. Keller, B. Schläpfer and T.H. Schucan
Swiss Federal Institute for Reactor Research
Solar heating project
CH-5303 Würenlingen

Summary

An improved technical assessment of active solar heating - space heating and domestic hot water (DHW) with hydronic systems - in the various climatic zones of Switzerland and, hence, of Central Europe is presented and the identified important technical factors determining the amount of solar heat which can be effectively used in a system are described. We used a combined experimental and theoretical approach: (i) We measured the thermal performance of nearly 100 different collectors on our test facility and we monitored the thermal performance of approximately 20 commercial solar energy systems located at various places in Switzerland; (ii) We developed and validated a simplified model to predict the all-day performance of a collector in a given climate under known operating conditions. The resulting upper limits for the useful solar heat have been found to vary by a factor of 4. This points out the essential role of system optimization in order to achieve the breakthrough of active use of solar energy. We further conclude that no global assessment of the potential of this technology is possible in Central Europe; a detailed analysis is required.

1. Introduction

A few years ago, the Swiss Federal Institute for Reactor Research (EIR) started systematic measurements on about 20 solar systems throughout Switzerland(1, 2). Microprocessor controlled data acquisition units are used to monitor the energy flows and the related physical quantities (meteorological and operating parameters) in the solar systems under actual service conditions. Technical aspects of the systems are studied and comparisons are made between the different systems. This activity supplements the thermal performance tests of the collectors of the Swiss market carried out on the EIR test facilities (3); in the last years these tests led to improvements of the available collectors. A better assessment of the solar heating technology in Switzerland is now possible. Domains requiring future developments to enable the breakthrough of the active use of solar energy in Switzerland have been identified.

Following the statistical approach described in Ref. (4) we divide our analysis into two subtasks by distinguishing between the heat collection and the transport and storage of heat.

2. Gross Heat Output (GHO)

A model has been developed at EIR to calculate the average gross heat output ("all day performance") of a glazed collector with given parameter values under given operating and weather conditions (5). For this purpose meteorological data of the 3 stations Zürich-Kloten, Davos and Locarno-Monti have been used. Table I summarizes some features of the three main climates in Switzerland. These values reflect typical conditions for Central Europe.

Table I: some meteorological data (mean values) of the 3 typical stations

	Zürich-Kloten		Davos		Locarno-Monti	
Region	Midlands		Alps		South of the Alps	
Altitude (m)	436		1588		366	
	G	T	G	T	G	T
Year	1183	8,5	1530	3,1	1579	11,7
March	95	3,7	144	-2,6	142	7,6
June	154	15,8	148	10,6	163	18,8
December	26	-0,9	76	-6,0	82	3,2

G Global radiation in a south-facing plane of 45° tilt angle (kWh/m^2)
T Ambient dry bulb temperature (^{o}C)

Table II: Sensitivity analysis of the collector gross heat output

a) Reference case: single glazed, selective collector with optical efficiency for beam radiation at normal incidence η_o = 0,75 and thermal loss factor U_m = 3,75 W/m^2K, at Kloten, south facing, tilt angle = 45°, average operating collector temperature T_m = $40^{o}C$.

	March	June	December	Year
Gross heat output (kWh/m^2)	32,6	74,0	3,1	498,2
Average efficiency (%)*	34,2	48,1	12,0	42,1

*) 100 % = incident radiation on collector aperture

b) Sensitivity analysis: only one parameter is varied at a time; others as in reference case.

Parameter	Parameter variation	Gross heat output (kWh/m^2) March	Gross heat output (kWh/m^2) Year
Location	Kloten ⟶ Davos	56,5 (+73%)	644,0 (+29%)
	Kloten ⟶ Locarno	67,1 (+106%)	773,9 (+55%)
Optical efficiency η_o	0,75 ⟶ 0,85	37,5 (+15%)	560,5 (+12%)
	0,75 ⟶ 0,55	22,9 (-30%)	374,5 (-25%)
Thermal loss factor U_m	3,75 ⟶ 3,0 W/m^2K	37,1 (+14%)	540,8 (+9%)
	3,75 ⟶ 5,0 W/m^2K	26,4 (-19%)	436,3 (-12%)
Temperature T_m	$40^{o}C$ ⟶ $60^{o}C$	21,1 (-35%)	346,4 (-30%)
Tilt angle α	$45^{o}C$ ⟶ $30^{o}C$	31,2 (-4%)	516,2 (+4%)
Azimuth γ	0^{o}(S) ⟶ 30°	28,3 (-13%)	467,6 (-6%)
	0^{o}(S) ⟶ 45° (SW)	25,0 (-23%)	440,6 (-12%)

The results obtained from our model calculations have been **experimentally** validated (5, 6) by measurements at our test facilities as well as by the field measurements mentioned in the introduction. From this agreement we conclude that the link between the collector and the storage subsystem can be adequately described by one single parameter, i.e. the monthly average operating collector temperature. The dynamic effects play a secondary role. A sensitivity analysis of the GHO is given in Table II. It is found to depend strongly on the season, the location, the collector parameters and the operating temperature. However, the heat output is much less sensitive to the collector tilt angle and azimuth and to the fluid flow rate (as long as the latter is kept above $\sim$ 20 kg/m^{2}h). It is concluded that the designer of a solar energy system can improve the collector heat output by a factor of 2 by choosing the most suitable collector and the appropriate operating conditions. Our published Tables (5) of the GHO can be used for this purpose.

3. Net Heat Output (NHO)

The net heat output or useful solar heat is defined as the gross (collector) heat output minus the losses during the heat transfer from the collector to the user (line and storage losses). However, this definition is somewhat arbitrary in two ways: (i) Some part of the "losses" may contribute to space heating. (ii) The separation of the solar losses from the conventional losses - those related to the unavoidable back-up energy (oil, gas, electricity, etc) - is not always uniquely defined, as both the solar heat and the heat produced using back-up energy may be stored - and partly lost - in the storage subsystem. In order to sketch the main features of solar DHW-systems we first present the results for one of the houses investigated in our measuring campaign and subsequently discuss some of the important factors influencing the performance of such systems.

3.1. Example: DHW for a single family house in the Midlands

The system described in Fig. 1 can be considered as a modern one and is located near Berne. The collector area (aperture) is 6,9 m^2 and the storage volume 1200 l. The upper part of the storage is kept at 60°C all the year round by means of the oil burner. Because of the costs, only the two boiler inserts are made of stainless steel. The largest part of the storage medium is heating water with low oxigen concentration.

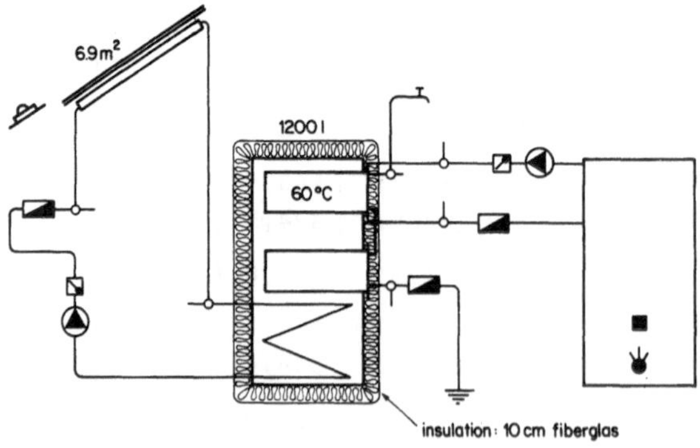

Fig. 1
Hydraulic scheme of DHW solar system in a single family house in the Swiss Midlands. The measurement points are indicated (radiation, flow, temperature).

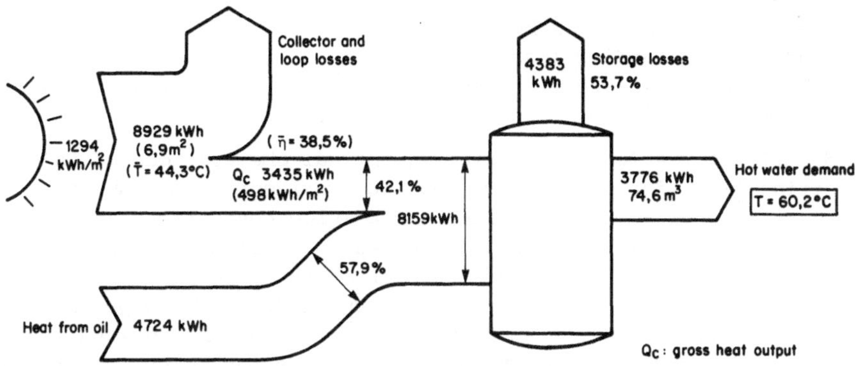

Fig. 2: Energy balance April 1982 - March 1983 of the system described in Fig. 1. A reduction of the storage losses would be a good thing and is feasible.

The measured energy balance (Fig. 2) indicates a high collector gross heat output of about 500 kWh/m^2 y. However, the storage losses are so high that the oil burner must be used also in summer to keep the temperature at the tap at the required level. If the storage losses would be drastically reduced (e.g. from 54 to 30%), the net heat output and the yearly solar fraction would reach the values of 350 kWh/m^2 and 64 %, respectively. Such an improvement is feasible (see below), but it requires an additional overheating protection for the nice weather periods in summer.

3.2. Thermosyphon Systems

Our measurements show that thermosyphon systems (systems with natural convection flow induced by gravity in the collector loop) can be recommended. They have net heat outputs comparable with those of systems with forced convection and they are very simple, thus have less troubles: no pump, no control device, no sensors, no non-return valve, no auxiliary energy demand!

3.3. Components of Solar Systems

In 8 solar systems, we measured the heat losses of the collector loop. They amount to 10 to 30 % of the yearly gross heat output, depending on insulation quality and operation mode. In large systems, it is worth transferring the heat remaining in the collector loop each evening to a cold storage tank, if such a tank or part of tank is available. For this purpose a bypass valve may be installed at the exit of the collector array.

The largest heat losses of solar energy systems occur in the storage subsystem. Losses of 20 to 75 % of the heat input to storage have been recorded (yearly averages). However, it must be distinguished between the necessary losses that happen in summer at high temperature (excess heat) and the unnecessary losses reducing the useful heat. There are many reasons for the latter to occur. One can group them in two categories:

- too long storage time: the heat produced by the collector stays unnecessarily long in the storage subsystem before reaching the consumer. Possible reasons for this are unclever storage management and pipe connections, heat production at a too low temperature level (compared to user's requirements) or unwanted mixing of hot and cold liquids in the storage tanks (perturbations of the vertical stratification in the tanks) so that

available heat can no longer be distributed to the user.
- unexpectedly high standing losses: it turned out that the standing losses of a storage tank observed in practice are 2 to 5 times higher than calculated from insulation data and geometry (7).

In detailed investigations including optical observations of the natural convection flow within a storage tank (8) we identified insulation failures and improper (although quite common) pipe connections as the main causes of the standing losses. As a consequence a permanent natural convection flow occurs within the cross section of the connected pipes and heat is transported from the tank to the colder part of the pipe. Independently a chimney effect is induced between the tank and the surrounding insulation layer due to air gaps. The standing losses may be characterized by the time constant τ describing the time in which the mean temperature difference between the content of the hydraulically isolated tank and its surroundings falls to 37 % (= 1/e) of its original value. We measured τ-values between 50 and 300 hours. Evidently the time lag between loading and unloading the storage (for both solar and conventional heat) should be small compared with τ for a reasonably designed system. The high storage losses in our example (Fig. 2) are caused by the fact that this condition is violated in this case.

In Switzerland, circulation loops to keep warm the DHW distribution line to the taps are often installed even in single family houses. This leads to a substantial additional heat demand; in a single family house, for example, to the double of the demand for DHW itself. A lot of heat can be saved by carefully designing and installing the circulation loops, or even omitting them. The cost effectiveness of a solar energy system strongly depends on the matching of heat production (i.e. the size of the installed collectors) and heat demand (4). Hence, it is particularly important, at design time, to estimate correctly also the heat required by a circulation loop, if any.

The back-up energy savings of a solar energy system are strongly influenced by the quality of the integration of the back-up device, so that it interacts properly with the solar part. In retrofit solar systems, the existing conventional part has to be thoroughly checked when designing the solar extension; mostly, the non-solar part must be adapted to the new situation; otherwise, one gets poor solar net heat outputs.

There are also trivial factors which determine the NHO of a solar energy system; they are well known by the experienced companies. Our measurements confirm their importance: complete degasing of the collectors, homogeneous flow through them, use of non-ageing plastic parts, very good design and construction of the tracking devices for focussing collectors (accuracy!), heat exchangers with large enough heat transfer area, suppression of unnecessary heat exchangers, preventing the heat to flow back from storage to collectors, etc.

4. Determination of Upper Limits

By combining all our results we find that a typical solar DHW-system (solar fraction $\sim$ 100 % in July, water temperature at the tap 60°C) in the Swiss Midlands built according to present technical standards could operate with a collector efficiency of 40 %, collector loop losses of 10 % and storage losses of 25 %. The first number if confirmed by the example presented above, while the other two have been achieved, e.g., in the DHW-system of the restaurant of our institute (2). For a yearly (global) irradiation of 1250 kWh/m^2 the resulting collector and net heat outputs are 500 and 350 kWh/m^2y, respectively.

If, by increasing the role of the auxiliary heating system, the DHW-system is designed to cover a lower solar fraction all year round (e.g. large systems used in hospitals), the collector efficiency is increased due to the lower operating temperature. The produced heat is used faster, hence the storage losses are reduced. In this case an upper limit of 450 kWh/m^2y results for the net heat output. For a similar system installed in a region with better insolation in winter it should be possible to achieve a value of 800 kWh/m^2y due to considerable increase of the beam irradiation.

If a domestic hot water system is combined with space heating, larger collector arrays have to be installed. The resulting excess heat in summer cannot be put to use and consequently the net heat output of the combined system is reduced. A typical upper limit for the net heat output in this situation is 200 kWh/m^2y for a single family house in the Swiss Midlands.

5. Conclusion

A solar energy system has to be designed seriously, realized properly and optimized. This is reflected by the fact that we measured a net heat output as low as 70 kWh/m^2y for a system built in the beginning of solar energy use in Switzerland. This is one fifth of the standard upper limit value of 350 kWh/m^2y reported above. On the other hand, even for a well designed solar heating system built according to today's technical standards, the amount of useful solar heat may vary by a factor of two (upwards or downwards) from that limit.

Many of the sensitive factors enumerated in this paper are taken care of very well today in solar energy systems built by experienced and reliable companies. However, certain domains still need R + D efforts. Especially the future heat management philosophy should include storage durations as short as possible, and adequate simplified design and test methods for complete systems should be developed (if not yet available) and systematically used.

In the present analysis the question of the costs was nearly completely ignored. However, it is clear that not only energetical considerations but also acquisition and maintenance costs as well as the lifetime of a system have to be taken into account. Only in this way one may speak of a true optimum. Today the active use of solar energy with liquid heat transfer medium is reaching the limit of cost-effectiveness in several cases under Swiss conditions. Two examples are provided by swimming pool heating with unglazed plastic solar absorbers and by DHW in systems with a high demand in summer. The latter use of solar energy is especially profitable with large systems (e.g. hotels, tourism). More details about the cost effectiveness of solar energy can be found in Ref. (9), (10).

As solar energy systems nearly reach the cost-effectiveness limit in the favourable cases, it is particularly important to optimize the future systems to ensure the breakthrough of this technology. An efficient tool for this optimization is provided by our model for the calculation of the gross heat output or collector all-day performance described in Section 2.

We believe that our paper explained clearly enough how carefully the potential of the active use of solar energy has to be assessed in Switzerland. Global lump estimates must not be made for this purpose. A detailed analysis considering the local meteorological parameters and the possible applications is required as large variations exist between various situations.

Acknowledgements

The authors are very grateful to their colleagues of the project "Solar heating and technical systems of buildings" of the EIR who were largely participating to the measurements, data evaluation and interpretation.

Part of the reported work belongs to the framework of the "Solar heating and cooling programme" of the International Energy Agency, and has been partly supported by the Swiss National Energy Research Foundation (NEFF). We express our gratitude to the NEFF and to the coordinating authority, the Swiss Federal Office of Energy.

References

1. J.M. Suter and P. Kesselring, "Measurement of Performance and Efficiency of Solar Energy Systems", contribution to the ISES 1979 International Congress on Solar Energy, May 28 - June 1, Atlanta (Georgia), USA, Pergamon Press, Oxford 1979.
2. J.M. Suter, B. Schläpfer, "How and how much the useful solar heat can be enhanced in a solar energy system for domestic hot water and/or space heating. Results of a large measurement programme on commercial systems", contribution to the ISES 1983 Solar World Congress, August 14 - 19, 1983, Perth, Australia; Pergamon Press, Oxford 1983.
3. J. Keller and V. Kyburz, "Wärmeerträge und Kenngrössen von Sonnenkollektoren", EIR Würenlingen, 1983.
4. P. Kesselring and A. Duppenthaler, "The layout of solar hot water systems, using statistical meteo- and heat demand data", contribution to the ISES 1979 International Congress on Solar Energy, May 28 - June 1, Atlanta (Georgia), USA, Pergamon Press, Oxford 1979.
5. P. Ambrosetti, "Das neue Bruttowärmeertragsmodell für verglaste Sonnenkollektoren", EIR Würenlingen, 1983 (ISBN 3-85677-012-7).
6. B. Schläpfer and P. Ambrosetti, "Ueberprüfung zweier Methoden zur Vorhersage bzw. Kontrolle der Erträge von Sonnenkollektoren (Schlussbericht)", EIR Würenlingen, 1983 (ISBN 3-85677-011-9).
7. H. Weber, M. Brack and J.M. Suter, "Energetische Optimierung eines Warmwasserspeichers in Theorie und Praxis: Wärmeverluste, Schichtung und Auswirkungen der Zirkulation", Contribution to the 2nd status seminar "Wärmeschutzforschung im Hochbau", ETH Zürich 1982
8. H. Weber, M. Brack and J.M. Suter, "Ein Wildbach unter Wasser/Zur freien Konvektion in Warmwassertanks", Contribution to the 4th Symposium "Recherche et Développement en Energie Solaire en Suisse", EPF Lausanne, 1983.
9. H. Rüesch, "Sonnenenergie 1983 - Der Stand aus der Sicht der Schweizer Sonnenenergie-Industrie (ökonomisch, politisch, technisch)", Heizung Klima 10 (1983) 56.
10. G. Bergmann, "Solare Warmwasserbereitung und Heizung von Gebäuden in Mitteleuropa", Contribution to the Topical Meeting on "Bewertung der Wirtschaftlichkeit regenerativer Energien" organized by the German section of the ISES, Munich 1982.

EXPERIENCE IN MODELLING AND EXPERIMENTING OF THERMOSYPHONIC SOLAR WATER HEATERS WITH HEAT EXCHANGER

E.Vazeos: B.P. Greece, Solar Department, Research Assistant, Thermal Eng. Section, Nat. Techn. Univ. Athens.
P.Canellopoulos: B.P. Greece, Solar Department.
Dr. C.D.Rakopoulos: Dipl. Ing., M.Sc., D.I.C., Ph.D. (Lon) Assistant Professor, Thermal Eng. Section, Mech. Eng. Dpt., Nat. Techn. Univ. Athens, Greece.

Summary

A theoretical model has been developed to predict the behaviour of thermosyphonic solar water heaters with heat exchanger. The governing equations describing the operation of the system are numerically treated and a computer program is written to implement the analysis. Input data include the hour by hour meteorological conditions and various system design parameters. At the same time experimental work has been carried out to check the validity of the model. The results show that the analysis predicts correctly the performance of the system, under various operational conditions, as evidenced by the agreement of the theoretical and experimental results.

1. INTRODUCTION

The higher efficiency (by 5÷45%) of the thermosyphonic solar water heaters over those with pump circulation is today well established. Furthermore their use in conjuction with a heat exchanger in the storage tank (indirect heating) has resulted in a very popular design configuration (antifreeze protection ability) [1]. Mertol et al [2] presented a review and a theoretical model of a solar thermosyphon with heat exchanger which remained to be validated experimentally. A work by the present authors [3] developed a comprehensive model which was validated experimentally for different design characteristics (heights, tank volumes, collector areas) for the nodraw case. The present work is an extension of the previous one with a view to validating the model for different operating conditions for the actual draw case.

2. THEORETICAL MODEL

The assumptions and equations for the model are outlined.

Collector parameters are based on actual data for a typical one cover, non-selective flat plate collector with negligible heat capacity (see Fig. 1). Actual hourly solar radiation, air temperature T_α and draw profile data are used. From the measured total solar radiation I on a horizontal plane, the hourly clearness index is computed [1], then the diffused part [4] and finally the total solar radiation on the tilted collector I_T [5].

The variation of water properties with temperature is taken into account by fitting second order polynomials (density ρ, thermal conductivity k, kinematic viscosity ν, specific heat capacity c_p, thermal expansion coefficient β and Prandtl number $Pr = \nu \cdot \rho \cdot c_p / k$).
The useful solar energy Q_u (in a time internal Δt) is given by [1],

$$Q_u = A_c F' \left[I_\tau(\tau\alpha) - U_L(T_m - T_\alpha)\ \Delta t \right] \quad (1)$$

where $(\tau\alpha)$ is the transmittance-absorptance product, U_L the overall collector loss coefficient and F' the collector efficiency factor. $F'(\tau\alpha)$ and $F'U_L$ are connected with the test values $F_R(\tau\alpha)_n$ and $F_R U_L$ (see Table I) through simple relations [1]. The mean water temperature in the collector is $T_m = 0.5\ (T_1 + T_2)$ (2)

Also, $Q_u = \dot{m}\ c_p\ (T_2 - T_1)\ \Delta t$ (3)

The heat exchanger is composed of a horizontal tube inside the tank (see Table I). The heat loss from the connecting pipes (riser-downcomer) is neglected, so that T_2, T_1 are the inlet/outlet temperatures for the tube exchanger. The tank stratification is neglected [1,2]. The internal heat transfer coefficient h_i is calculated from [6],

$$Nu = 1.75 \left[Gz + 0.012(Gz \cdot Gr^{0.33})^{1.33} \right]^{0.33} \left[(\nu\rho)_w / (\nu\rho)_m \right]^{-0.14} \quad (4)$$

where the properties in the last brackets are evaluated at wall temperature and T_m respectively. The external coefficient h_o is calculated from,

Laminar flow $(10^4 < Gr \cdot Pr < 10^9)$ $Nu = 0.47\ (Gr \cdot Pr)^{0.25}$ (5a)

Turbulent " $(Gr \cdot Pr > 10^9)$ $Nu = 0.10\ (Gr \cdot Pr)^{0.33}$ (5b)

The overall heat transfer coefficient U_e in the heat exchanger is [6],

$$U_e^{-1} = h_i^{-1} + (d_i/2\ k_w) \cdot \ln\ (d_o/d_i) + d_i/d_o\ h_o^{-1} \quad (6)$$

Therefore a heat balance in the heat exchanger gives

$$Q_u = U_e\ (\pi\ d_i\ l_e)(T_m - T_s)\ \Delta t \quad (7)$$

It is assumed that the variation of density in the collector and exchanger tube is linear with distance. The application of the momentum equation around the entire thermosyphon loop takes the form: driving (buoyancy) pressure difference = frictional pressure drop. Therefore according to the assumptions and the fact that the friction coefficient for laminar flow is $f = 64/Re$, we take (see Table I)

$$0.5\ g(\rho_1 - \rho_2)(Z_1 + 2Z_2) = 64\nu l \dot{m}/2A_h d_h^2 + J\ \dot{m}^2/2\rho A_h^2 \quad (8)$$

and if $\rho = aT^2 + bT + c$ (a,b,c = const), eq.(8) becomes $B_2\dot{m}^2 + B_1\dot{m} + B_o = 0$ (9)

with $B_2 = J/2\rho A_h^2$, $B_1 = 32\nu l/A_h d_h^2$, $B_o = 0.5g(T_2 - T_1)(2aT_m + b)(Z_1 + 2Z_2)$

The heat balance in the storage tank gives in finite difference notation

$$T_{new} = T_{old} + (mc_p)_s^{-1} [Q_u - L - (UA)_s(T_s - T_\alpha)\Delta t] \quad (10)$$

where L = energy removal to load $\dot{m}_d\ c_p(T_s - T_d)\Delta t$ (11)

T_d = tank temperature supply and $\dot{m}_d$ = actual draw profile

3. COMPUTER PROGRAM PROCEDURE

Given the geographical and meteorological data, I_τ is calculated from measured values of I for every hour. A value of T_m is selected to be subsequently corrected. Then Q_u is found from eq. (1). The system of eqs(3) and (9) can now be solved for $\dot{m}$ and $(T_2 - T_1)$. The validity of eq. (7) is checked and the value of T_m is corrected. After T_m is stabilized, the tank temperature T_s is computed for each hour from eq. (10) using Eulers method, $\dot{m}_d$ being known from measurements. The time step is conveniently taken as Δt=1hr.

4. COMPARISON WITH EXPERIMENTAL RESULTS

Fig. 2 shows I_τ, T_α and $\dot{m}_d$ measured (every hour) for one case (day). For the measurements the following instruments were used: a) solarimeter KIPP & ZONEN b) temperature sensors (resistance bridge) c) turbine flow meter. For the same case, Fig. 3 shows the measured and calculated cumulative value of load energy removal E_T, together with the computed circulating flow rate $\dot{m}$. The agreement as can be seen is satisfactory, being of the same order as for the other cases. In Fig. 4 the experimental and theoretical cumulative values at the end of each day E_{T24} are plotted in the usual way. The agreement is better than ±10%.

5. CONCLUSIONS

It is concluded that the present analysis predicts satisfactory the performance of the solar water heater under various design and operating conditions despite its low computer time run requirements. Therefore it can be proved a very useful tool for the designer and the user of the system.

TABLE I (System design data)

COLLECTOR
A_c (area) = 2.12 m^2, s (tilt angle) = 45^0, $\dot{m}$ (mass flow rate)
$F_R(\tau\alpha)_n$ (heat removal factor x ($\tau\alpha$) for normal radiation) = 0.83
F_RU_L (" " " x overall loss coefficient) = 10 W/m^2 ^{0}C
TANK
m (mass) = 143 kg
$(UA)_s$ (tank loss coefficient x area) = 1.165 W/^{0}C
HEAT EXCHANGER
d_i/d_o (tube inside/outside diameter) = 0.108 m/0.114 m
l_e (tube length (horizontal)) = 1.75 m
k_w (wall thermal conductivity) = 48 W/m ^{0}C
COLLECTOR CIRCUIT
d_h (hydraulic diameter) = 0.012 m, l (equivalent length) = 1.37 m
J (local friction loss coefficient) = 2.04, $A_h = \pi d_h^2/4$
Z_1 (height of collector plate) = 1.00 m
Z_2 (vertical distance between collector top-exchanger axis) = 0.33 m

DIMENSIONLESS NUMBERS

Reynolds number $(R_e) = \rho \cdot u \cdot d/\nu$	Grashof number $(Gr) = \beta \cdot g \cdot \Delta T \cdot d^3/\nu^2$
Nusselt " $(Nu) = h \cdot d/k$	Graetz " $(Gz) = \pi d \cdot Re \cdot Pr/4 \cdot l$

REFERENCES

1. Duffie, J.A. and Beckman, W.A., "Solar Engineering of Thermal Processes", John Wiley, New York, 1980.
2. Mertol, A., Place, W. and Webster, T., "Detailed Loop Model Analysis of Liquid Solar Thermosyphons with Heat Exchangers", Solar Energy, Vol. 27, pp. 367-386, 1981.
3. Vazeos, E., Rakopoulos, C.D., Canellopoulos, P.T. and Kammenos, G., "A Digital Computer Simulation of Solar Thermosyphons with Heat Exchanger and Comparison with Experimental Results". Proc. First National Conference on "Mild Energy Forms", Salonika Univ., 1982.
4. Orgill, J.F., and Hollands, K.G.T., "Correlation Equation for Hourly Diffuse Radiation on a Horizontal Surface", Solar Energy, Vol. 19, p. 357, 1977.
5. Liu, B.Y.H. and Jordan, R.C., "The Interrelationship and Characteristic Distribution of Direct, Diffuse and Total Solar Radiation", Solar Energy, Vol. 4, No 3, 1960.
6. Kreith, F., Black W.Z., "Basic Heat Transfer", Harper and Row, New York, 1980.

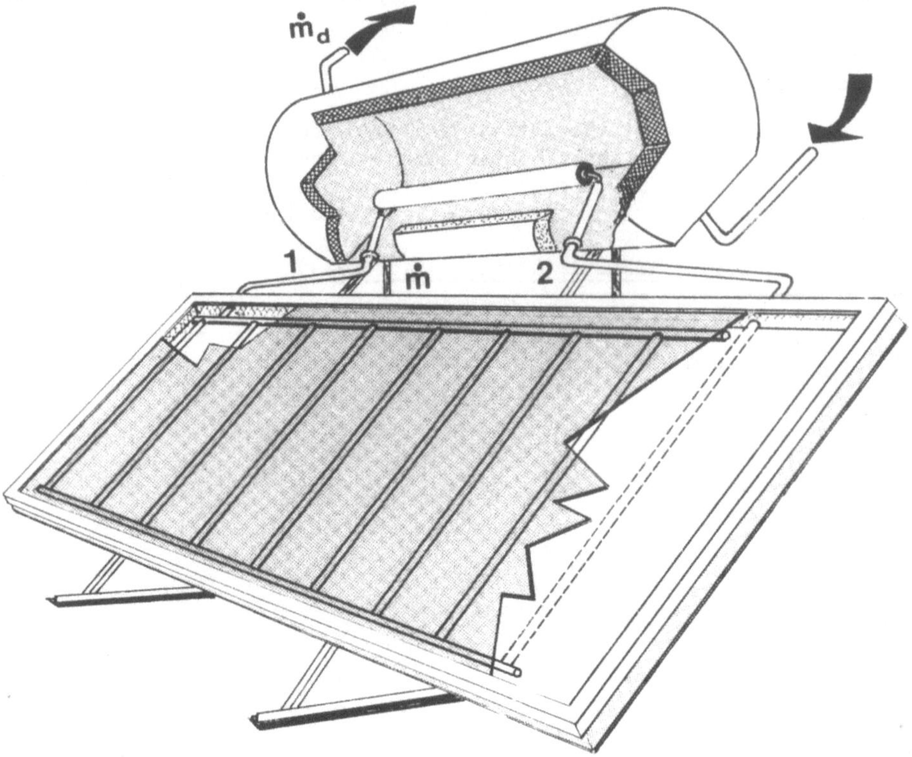

Fig 1 Thermosyphon solar heater with heat exchanger

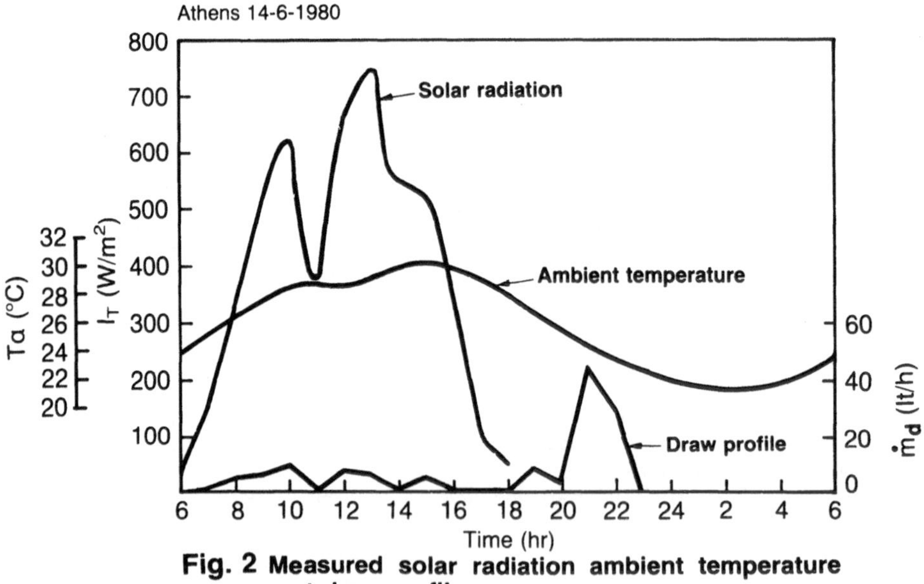

Fig. 2 Measured solar radiation ambient temperature and draw profile

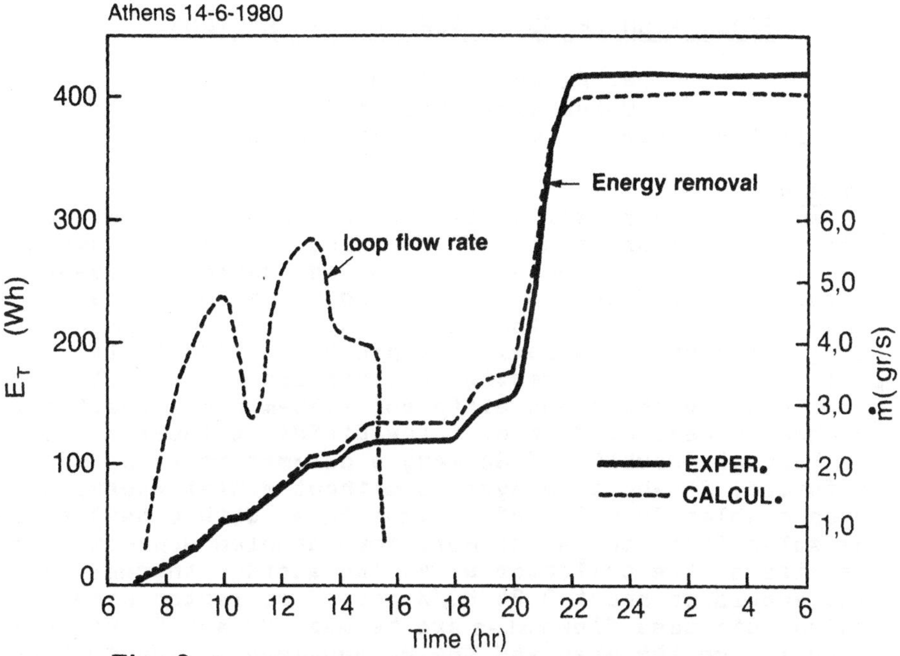

Fig. 3 **Cumulative energy removal and loop mass flow rate**

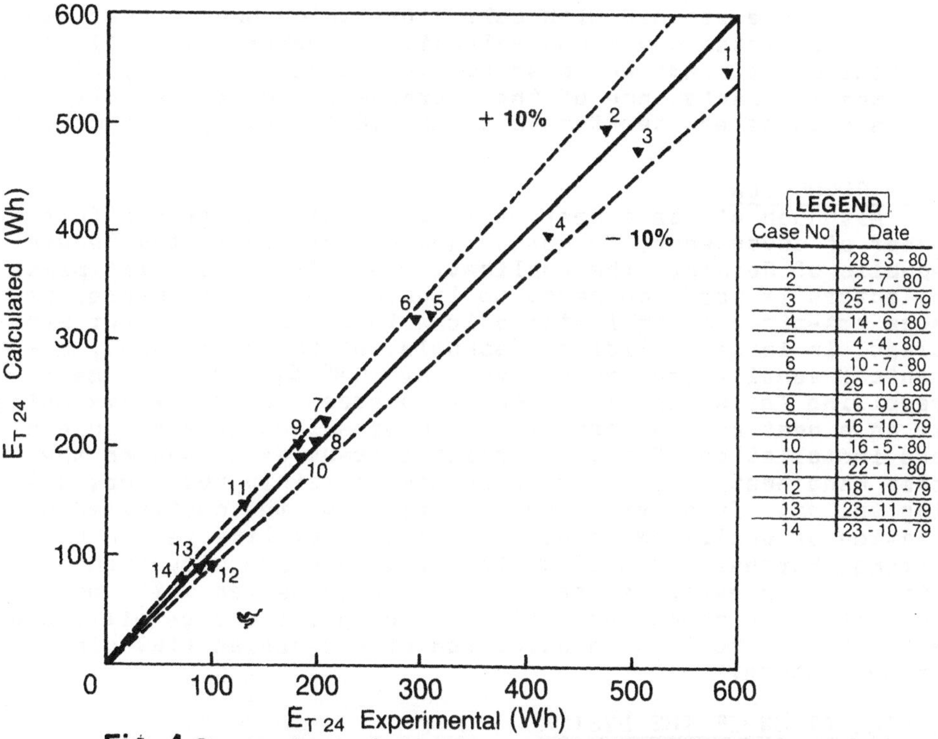

Fig. 4 **Comparison of experimental and theoretical results for daily energy removal for various days (cases) in Athens**

SIMULATION OF SOLAR AIR-HEATING SYSTEMS

N. Fisch and E. Hahne
Universität Stuttgart
Institut für Thermodynamik und Wärmetechnik

Summary
The application of air collectors and pebble-bed storage bins in the solar heating of dwelling-houses and domestic hot water preparation is investigated. Based on measured meteorological data and fixed building parameters and component parameters, the solar fraction is determined using a computer program. This program ("SIMUL") is capable to consider presently four different air collector constructions and eight different system configurations. For hourly weather data of three different locations in the Federal Republic of Germany a parametric study was performed. Air-heating systems without a heat store, only reach a solar fraction of maximum 10 %. With a heat store, the solar fraction can be more than doubled depending on the size of the collector area. The storage thermal capacity should be about 0.24 MJ/K per m^2 collector area. The optimal air mass flow rates are between 20 and 50 $kg/(h \cdot m^2)$ depending on the area and the arrangement of the collector field. The thermal capacity of the collector should be smaller than about 25 $kJ/(K \cdot m^2)$. Similar solar fractions are reached with collectors which have either two cover plates and a non-selective absorber ($\alpha^*_A = \varepsilon^*_A$) or one cover plate and a selective absorber ($\alpha^*_A / \varepsilon^*_A > 5$). The thermal resistance of the storage- and duct insulation should be greater than 2.0 and 0.6 $(m^2 \cdot K)/W$, respectively.

1. INTRODUCTION

Although air as a heat carrier in solar systems offers a number of advantages, in Europe and especially in the Federal Republic of Germany, the application of air-cooled flat plate collectors is poor, compared to liquid-cooled collectors, because hot-air-heating installations in dwellings are not widespread. In larger buildings (schools, office buildings), mechanical ventilation and air vent by HVAC-System has been prevailing for a long time. For this area, solar systems using air as a heat carrier are highly interesting, because on one hand installations for ventilation already exist and on the other hand heating of such buildings is mainly necessary during the day. In private houses, till now, the controlled ventilation of dwellings is not dominant, just like return-air-heating. Further intensification of insulation directions in this area, however, is useful only, if connected with controlled ventilation of rooms. An active use of solar energy will necessarily bring about the application of air-cooled flat plate solar collectors.

2. SIMULATION OF THE SYSTEM

The thermal behaviour of solar systems for room heating is influenced by numerous parameters. Meteorological conditions and consumers' habits always cause time-dependent performance. Therefore, the investigation of such systems requires a detailed numerical modelling of all single components of the building ("HAUS") and of the solar system ("SIMUL"). In figure 1, the total model is presented schematically.

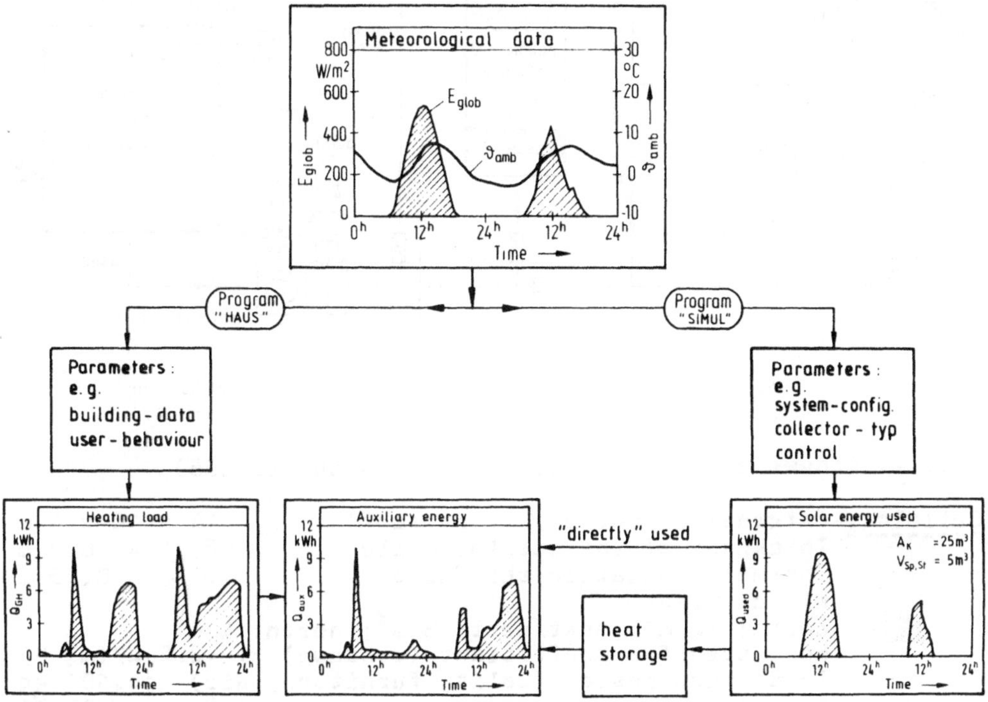

Fig. 1: The simulation problem

At the Institute of Thermodynamics and Heat Transfer (ITW) of Stuttgart University, a program has been developed especially determined for simulation of solar systems using air as a heat carrier ("SIMUL"). The advantage of "SIMUL" is its relatively simple applicability and its special adjustment of solar installations to HVAC-System . At present, there is a choice between eight different system configurations (SYS01 - SYS08) and four different collector types (type 1 to type 4) /1/. Figure 2 shows the scheme of a relatively complicated solar system (SYS08) with a closed primary circuit, a pebble-bed store and a domestic water store. Disadvantageous are the relatively long computing times (CP-Time on a CD 6600 approx. 50 - 250 seconds) and the relatively high storage demand (CM approx. 150 KByte). The program "HAUS" applied for calculation of transient behaviour of the building is described in /2/.

2.1 Constant parameters for investigation

For simulation, hourly weather data are used. The essential design parameters ("standard data") for the building and the solar system are:

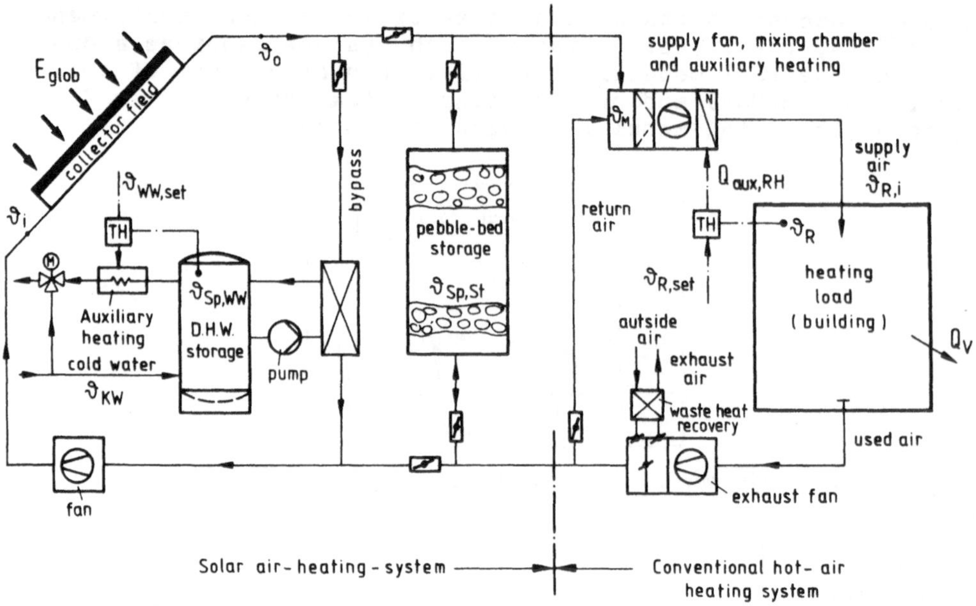

Fig. 2: Scheme of a solar air-heating system (SYS08)

Building: External dimensions (flat roof) 10 x 10 x 4 m
Thickness of the outside walls δ_{Wa} = 0.3 m
Thermal insulation thickness $\delta_{Is,H}$ = 0.05 m
Window areas,
south: 20 m^2; east/west: 8 m^2; north: 4 m^2
Heat transfer coefficient (windows) k_F =2.4 W/(m^2 K)
Total mass inside (walls, furniture, air) ca.10^3 kg
Room temperature $\vartheta_{R,set}$ = 20 °C

Solar system: SYS08

Collector field	Collector field area	A_K	= 40 m^2
	Aspect ratio of collector field	l_K / b_K	= 0.4
	Air flow rate	$\overline{\dot{m}}_L$	=25 kg/(h·m²)
	Collector type 1, two covers	n	= 2
	Absorptivity of the absorber	α^*_A	= 0.97
	Emissivity of the absorber	ε^*_A	= 0.97
	Gap of the duct in the collector	s	= 0.01 m
Pebble-bed store	Storage volume (3.54x3.54x5 m)	$V_{Sp,St}$	= 10 m³
	Porosity	χ_{Sp}	= 0.43
	Thermal insulation thickness	$\delta_{Is,Sp}$	= 0.1 m
DHW system	Storage tank volume	$V_{Sp,BW}$	= 0.2 m³
	Rated temperature	$\vartheta_{BW,set}$	= 45 °C
	Hot water demand	$\dot{m}_{BW}$	= 200 kg/d
Ducts	Total length of the flow lines	l_P	= 20 m
	Thermal insulation thickness	$\delta_{Is,P}$	= 0.05 m
	Thermal capacity	C_P	= 2000 J/(K·m)

The most important parameters - especially those deviating from "standard data" - are given in the figures. The transient behaviour of the building as well as "passive" energy gains from window areas are accounted for in the calculation of the heating load in the following chapter.

3. RESULTS

3.1 Effect of meteorological data

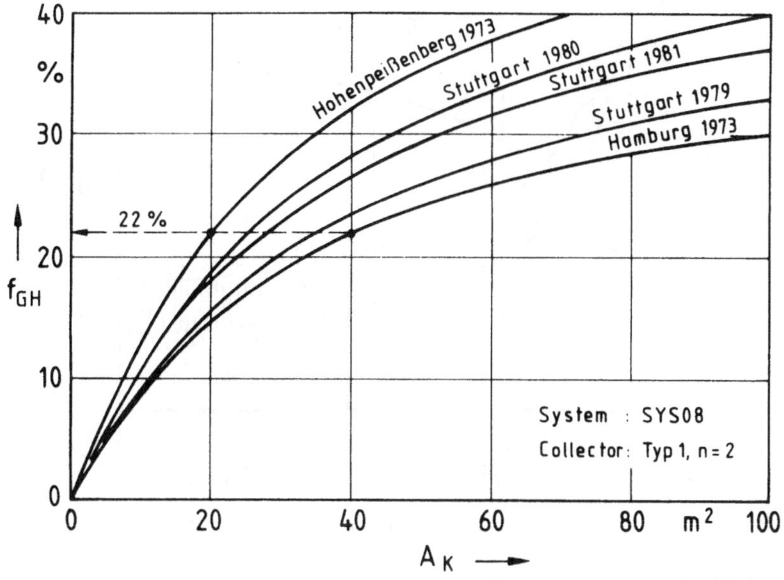

Fig. 3: Effect of the collector area (A_K) on the solar fraction (f_{GH}) for locations in West Germany

In figure 3 the performance of the solar fraction is presented for different weather data in dependence on the size of collector area. The growth of the solar fraction, of the used quantity of heat, respectively, decreases with increasing collector area. In figure 3 it is shown that Hohenpeißenberg is the "best" of the locations investigated for the use of solar energy and Hamburg is the "worst". In Hohenpeißenberg, e.g., 22 % of the solar fraction can be obtained with only half of the collector area necessary in Hamburg. Dependent on the location, the solar fraction steeply rises for collector areas up to 40 and 60 m^2; beyond this area the increase is only small. For the Stuttgart location (weather data of 1979) e.g., an extension of the collector area from 20 to 40 m^2 brings forth an increase of the solar fraction twice as high as an extension from 40 to 60 m^2.

3.2 Effect of the thermal insulation of the building

Figure 4 indicates that higher solar fractions are received by an improved thermal insulation of the building. For the building without additional thermal insulation ($\delta_{Is,H}$ = 0), the solar fraction only amounts to 12.7 %. The solar fraction increases strongly (from 13 to 24 %) with a thermal insulation thickness up to about 50 mm; then, to about 100 mm, the increase becomes smaller (up to 27 %) and reaches for $\delta_{Is,H}$ = 200 mm nearly 30 %.

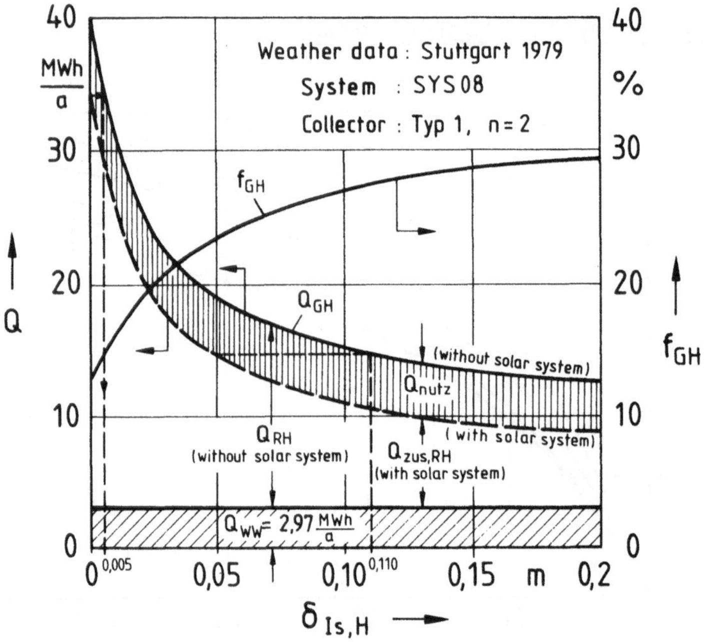

Fig. 4: Effect of the thermal insulation thickness ($\delta_{Is,H}$) on the total heating load (Q_{GH}) and the solar fraction (f_{GH})

Compared to the house which has no additional thermal insulation ($\delta_{Is,H}$ = 0), a 50 mm thick insulation helps to decrease energy consumption by approximately 55 %; a further decrease of 11 % is obtained with another 50 mm insulation layer ($\delta_{Is,H}$ = 100 mm). Figure 4 shows that the same saving of heating energy can be obtained, if an insulation of 5 mm is applied to the not additionally insulated house - instead of a solar heating system (SYS08, A_K = 40 m²). For a building with a solar heating system and 50 mm thermal insulation, the useful heat supplied by the solar system can be compensated by a thermal insulation layer of 110 mm thickness. Another result is derived from figure 4: For thermal insulation layers thicker than 100 mm, significant reductions can be achieved only with a solar heating system; by reasonably thick insulation layers a compensation cannot be obtained any more.

3.3 Heat capacity of the pebble-bed store

Figure 5 describes the effect of the heat capacity of a pebble-bed store on the solar fraction for differently sized collector areas. The store capacity was varied up to 60 MJ/K. This corresponds to a storage volume of approximately 48 m³ with ρ_{St} = 2.6 kg/dm³; c_{St} = 0.835 kJ/(kg·K) and X_{Sp} = 0.43

$$C_{Sp,St} = \rho_{St} \cdot c_{St} \cdot (1 - X_{Sp}) \cdot V_{Sp,St} \tag{1}$$

For these investigations, the store was designed as a cube. For a given collector area, the solar fraction increases first with the store capacity and then remains approx. constant within the region investigated here. From the results presented in figure 5, an empirical equation for the dimension of a pebble-bed store can be given as:

store capacity = collector area $\cdot C_2$; $\quad C_{Sp,St} = A_K \cdot C_2$ (2)

with $C_2 \approx 0.24 - 0.28$ MJ/(K·m^2).

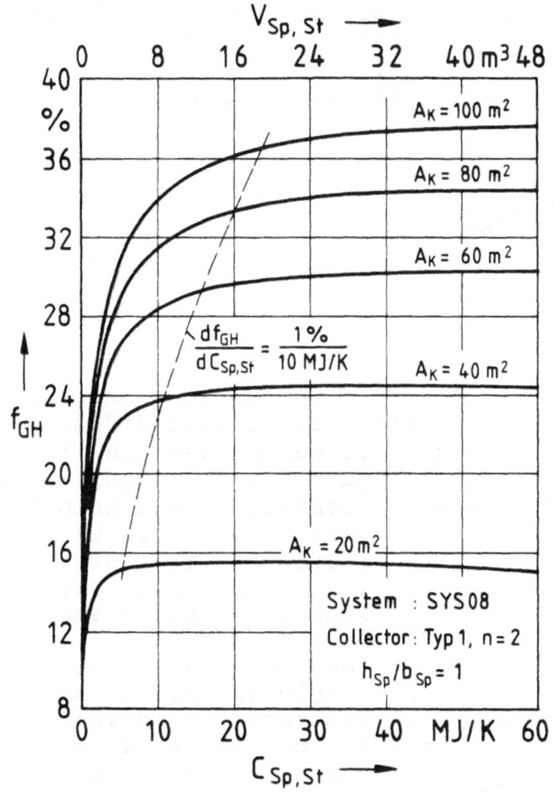

Fig.5: Effect of the thermal capacity of a pebble-bed storage ($C_{Sp,St}$) on the solar fraction (f_{GH}) for different collector areas (A_K)

For the determination of C_2, that heat storage capacity was taken from figure 5 and for each collector area, which yielded an increase of the solar fraction by only 1 % if the capacity was increased by 10 MJ/K (dash-dotted line). For a collector area of 40 m² a heat storage capacity of $C_{Sp,St} \approx 10$ MJ/K is received according to eq.(2) and this corresponds to a volume of 8 m³, in this case. For a very large storage volume, the solar fraction slightly decreases. This can be noticed only for A_K = 20 m² in the region of storage volume presented in figure 5. For larger collector areas, 40 m², e.g. and a storage volume of 100 m³, the calculation results in a solar fraction of approx. 23.5 %.

3.4 Length, heat capacity and thermal insulation layer of ducts

Figure 6 shows the influence on solar fraction affected by the thickness of the thermal insulation layer and by the heat capacity of ducts in dependence on the total length of flow lines. The heat capacity is of minor importance in regard of the thermal insulation. The influence of thermal insulation rises correspondingly to duct length. Figure 6 also explains that the increase of solar fraction gets smaller according to the thickness of the thermal insulation layer; e.g. for l_P = 20 m and C_P = 4000 J/(K·m), the solar fraction increases from approx. 16.5 % for the non-insulated duct to 22 % for a layer thickness of 2.5 cm; a doubling of the thermal insulation layer up to 5 cm, only brings about a growth of absolute 1 % in the solar fraction.

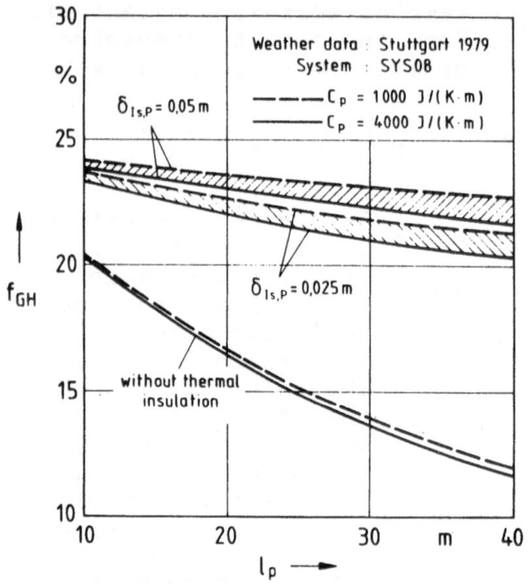

Fig. 6: Effect of the duct length (l_P) on the solar fraction (f_{GH}) for different thermal insulation thickness ($\delta_{Is,P}$) and heat capacity (C_P)

3.5 Air mass flow

Figure 7 presents the solar fraction versus the mass flow rate based on the collector area; various areas with different aspect ratios are given as parameter. It can be seen in figure 7 that larger solar fractions are obtained working with longer collector fields (large l_K/b_K). Besides, figure 7 indicates that a smaller mass flow rate should be taken when area and length (l_K/b_K) of the collector field increase. For an area of e.g. 40 m^2 and an aspect ratio of 0.05 - this corresponds to l_K=1.4 m and b_K=28 m - an optimal air flow rate is obtained at approx. 35 kg/(h·m^2). For the same collector area and an aspect ratio of 2.5 (l_K= 10 m, b_K= 4 m) an air flow rate of approx. 25 kg/(h·m^2) only, has to be chosen.

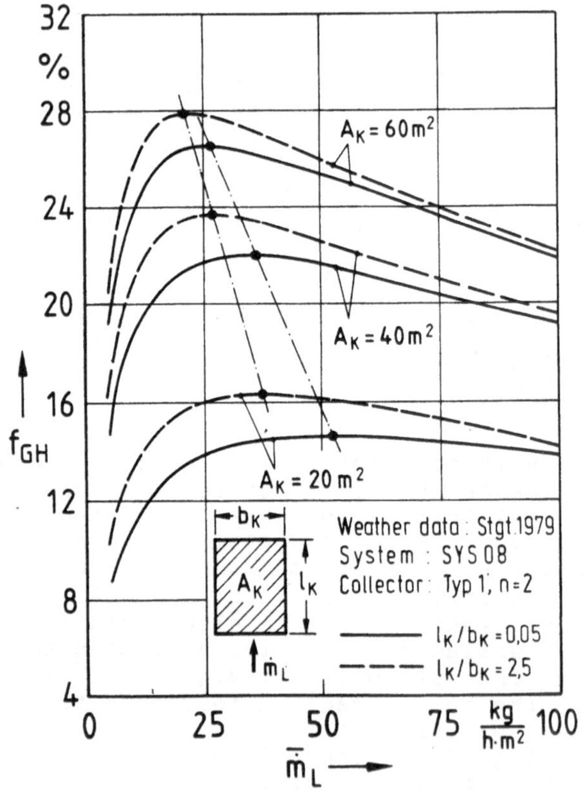

Fig. 7: Effect of the air flow rate ($\bar{\dot{m}}_L$) on the solar fraction (f_{GH}) for three collector areas (A_K) and two field geometries (l_K/b_K)

3.6 Additional significant results of the parametric investigation

- The arrangement of collectors was investigated indirectly by the length/width ratio of the collector area. For collectors with "two-pass flow" /1/ optimal aspect ratios were received between 0.025 and 0.1 (A_K = 40 m^2). For the "one-pass flow" /1/, the solar fractions rise correspondingly to the collector field length and remain nearly constant for $l_K/b_K>1$.
- The collector field of a solar heating system should be orientated towards south and tilted against the horizontal line under an angle of $\gamma = \phi + 10^o$ (ϕ= degree of latitude), e.g. for Stuttgart $\gamma \approx 60^o$. In comparison to this optimal orientation, a south-orientated vertical field in Stuttgart yields a solar fraction which is smaller by 11 % and a west-orientated and by 48.8^o tilted ($\gamma = \phi$) collector field yields solar fractions smaller by 32 %.
- The heat capacity of the absorber should not exceed approx. 25 kJ/(K·m^2).
- By a second cover plate additional reductions of heating energy between 25 and 45 kWh/(m^2·a) are obtained for a collector with a usual "black" absorber ($\alpha_A^* = \varepsilon_A^*$) depending on the collector type.
- Similar reductions are realized with one cover plate, if either a selective absorber or an IR-reflecting layer on the cover bottom are used.
- In the case of two cover plates and a usual "black" absorber, essential increases in the solar fraction are not obtained, neither by a second cover plate, nor by a selective absorber, or by an IR-reflecting layer on the cover.
- The width of the flow duct should be between 5 and 15 mm depending on the collector type.
- The thermal resistance of the thermal insulation should be higher than 2.0 (m^2·K)/W for the storage volumes investigated.

Acknowledgment: The investigations have been financed by the Federal Ministry of Technology under ET 4045 A; the authors gratefully acknowledge this support.

REFERENCES

1. FISCH, N. and HAHNE, E. (1984). Nutzung der Sonnenenergie in Heizungsanlagen mit Luft als Wärmeträger. Int. Journal of Solar Energy, Vol. 3
2. FISCH, N., MÜNDER, J. and HAHNE, E. (1984) Untersuchung über Maßnahmen zur Reduzierung des Wärmeverbrauchs eines Wohngebäudes. Bauphysik, Heft 2, 55-62

ACCURACY AND LIMITATION OF THE LESO-A SIMPLIFIED METHOD

Ch. Eriksson and A. Faist
Ecole Polytechnique Fédérale
Bâtiment LESO CH - 1015 Lausanne

Purpose of the work

A very simple and straighforward method of calculation called LESO-A has been proposed to compute the energy balance of solar houses. In the following paper, the results from the LESO-A simplified method are compared to those obtained by detailed computer simulations in order to test LESO-A's accuracy and its limitations.

Method of approach

A passive solar house can be characterised by the following basic parameters :

- H_o, the global energy loss coefficient : to calculate the yearly (or monthly) losses, H_o is multiplied by the corresponding "degree-days" (cumulated temperature differences).
- S, the equivalent collecting area also called solar aperture : to calculate the solar gains, S is multiplied by the corresponding, yearly (or monthly) solar radiation on the collecting surface
- the yearly (or monthly) wild gains

The LESO-A simplified method allows the quick calculation of H_o, S and the yearly (or monthly) auxiliary energy and solar fraction.

Using the computer program, PASSIM-3 [1] detailed simulations have been performed on four solar houses : two direct gain, one Trombe wall and one sunspace design.

The following work shows some interesting results from these simulations and then compares them with the results of the simplified method LESO-A.

Ground temperature

The variation of the ground temperature from fall to spring is smaller than the corresponding outdoor air temperature variations. It is also shifted in time : about 1 month. Thus when using the indoor - outdoor degree-days to compute the heat losses, the overall heat loss coefficient of the house is not constant but varies smoothly from fall to spring (fig. 1).

$$H_{ext} \cdot (T_{int} - T_{ext}) + H_{ground} (T_{int} - T_{ground}) = H_L (T_{int} - T_{ext})$$

$$H_L = H_{ext} + H_{ground} \cdot \left(\frac{T_{int} - T_{ground}}{T_{int} - T_{ext}} \right)$$

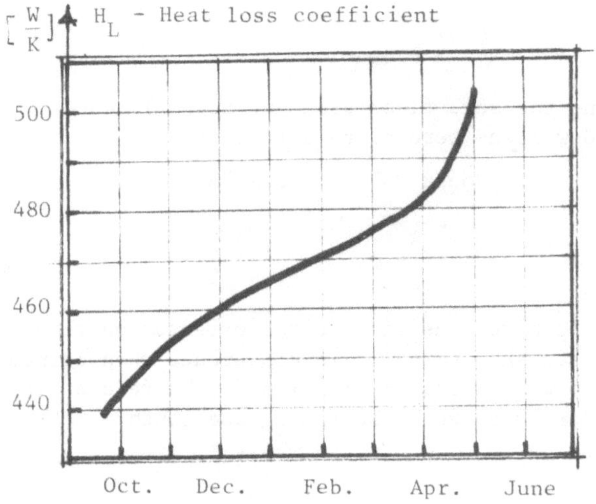

<u>Fig. 1:</u> Yearly variation of the overall heat loss coefficient

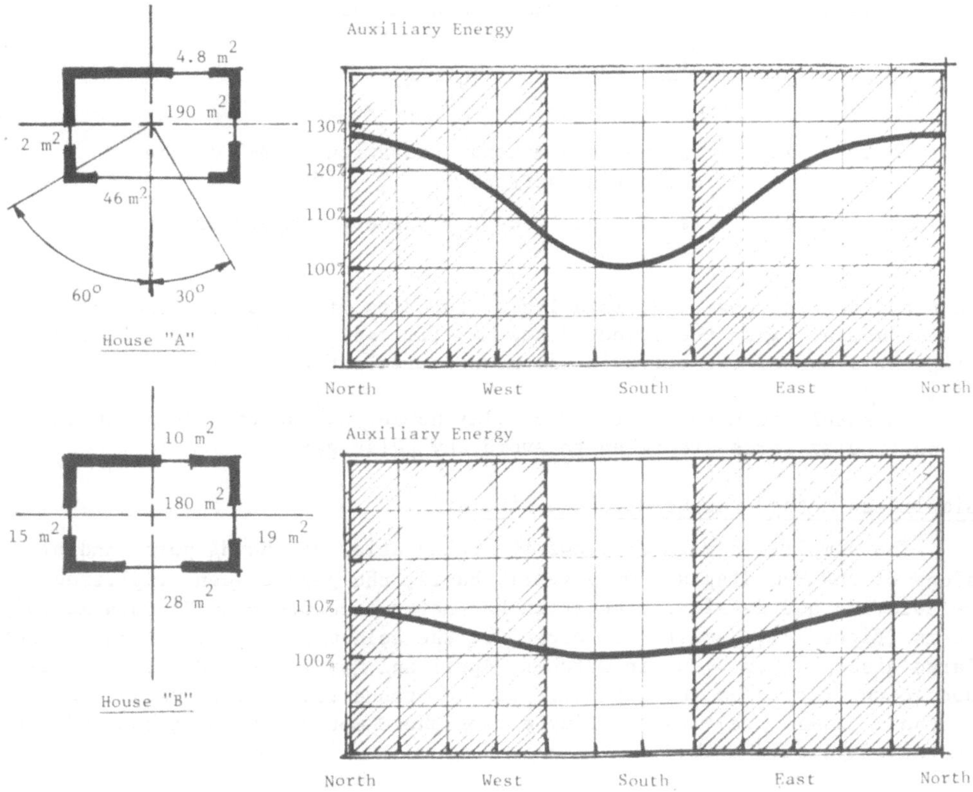

<u>Fig. 2:</u> Effect of a change of orientation
House "A" is more sensitive than house "B"

Small one storey houses are more sensitive to this effect than high rise or large buildings.

Orientation

Two direct gain houses having the same floor area and thermal heat losses were simulated. Their window areas were quite different :

		House "A"	House "B"
Window area :	South	46 [m^2]	28 [m^2]
	East/West	3.2/2 m^2	19/15 m^2
	North	4.8 m^2	10 m^2

As a consequence, house "A" is more sensitive to the orientation than house "B" (Figure 2) but both houses show very little difference when turned from south - south-east (- 30°) to west - south-west (+ 60°). Therefore, it is sufficient, in most cases, to analyse the house facing due south.

The H-m diagram

At each time of the year the house is characterised by a particular net heat loss coefficient, H, and a particular equivalent collecting area, S, (or solar aperture S).

If we define the meteorological index m as

$$m = \frac{\bar{I}}{\overline{\Delta T}} \qquad [W/m^2K]$$

where $\bar{I}$ is the mean monthly radiation [W/m^2] on the collecting surface

$\overline{\Delta T}$ is the mean indoor - outdoor temperature difference [K]

then the heating power, H, is a linear function of m during the cold winter months (when the rejected heat is limited). In this particular diagram the slope of the curve is the mean equivalent collecting area, S (figure 3).

It should be noted that for a solar house, H_o can only be measured if particular cares are taken to avoid any solar gain.

Simplified method versus detailed simulation

The simplified method allows the calculation of the H_o point and of the slope of the H-m diagram. On a yearly basis, the gap between simplified calculation and detailed simulation has never exceeded ± 10 %. On a monthly basis (figure 4), results are close for the cold months (± 10 %) but rather large discepencies occur in October, April and May (up to 100 %). This is due to the fact that the influence of the ground temperature variation and of the high heat rejects occuring during these months is not accounted for in the simplified method.

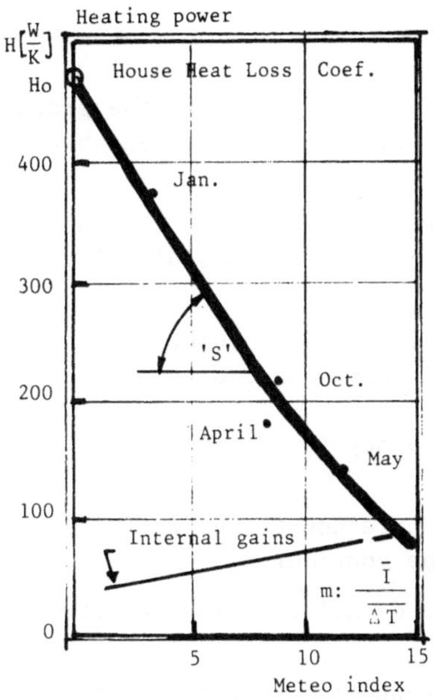

Fig. 3: The H-m diagram

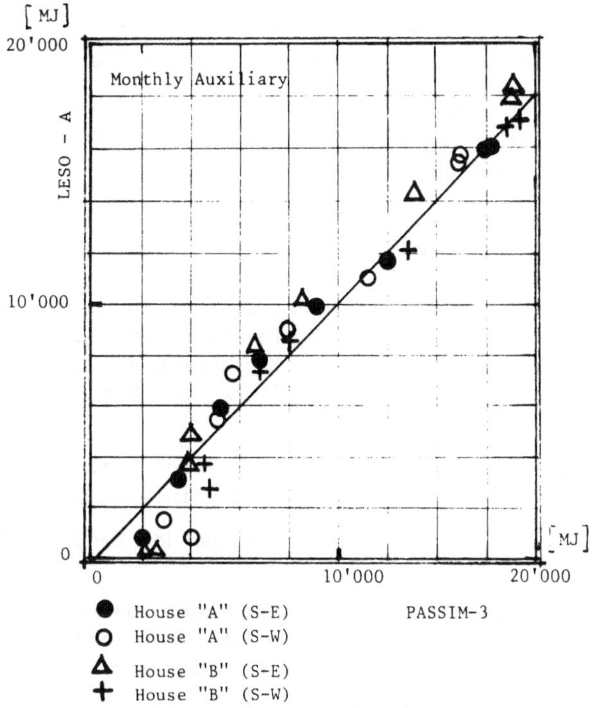

Fig. 4A: LESO-A versus PASSIM-3

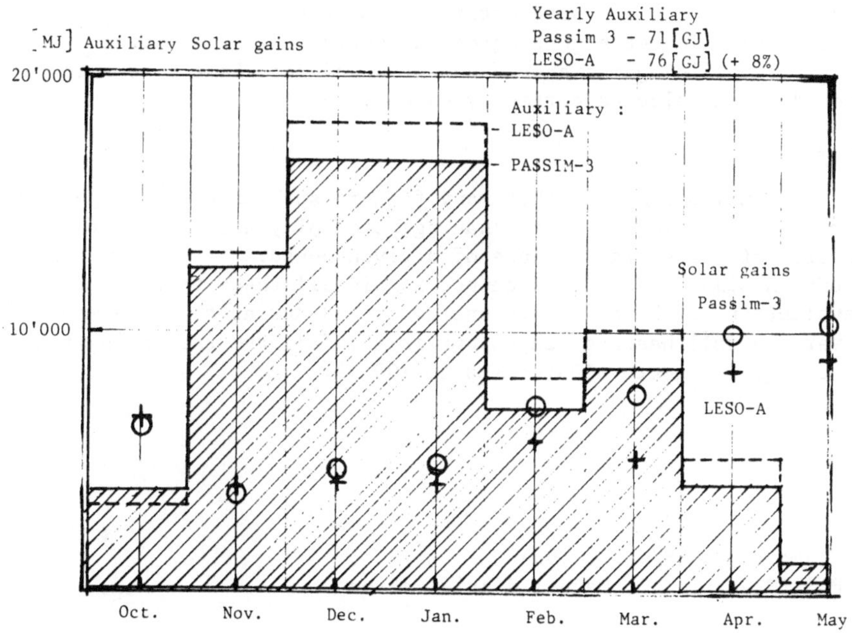

Fig. 4B: LESO-A versus PASSIM-3 House "B"

Significant results

Detailed simulations have shown that :

- A significant part of the losses of the houses are going to the ground. Thus the apparent overall heat loss coefficient is not constant but increases from fall to spring.
- The equivalent collecting area (solar aperture) varies firstly, because of shading effect and secondly, (if one considers the useful solar energy) because of heat rejects.
- In the H-m diagram, the heating power, H, is a linear function of m for $m < 7\ W/m^2K$. The slope of the curve is the mean equivalent collecting area (mean solar aperture), S.
  Because of heat rejects, October, April and May ($m > 7$) do not lie on the straight line.
- When establishing a detailed heat balance of a house, the north oriented windows solar gains should also be considered.

Comparison of simplified method results with detailed simulation shows that :

- The computation of the heat loss to the ground is critical. A bad choice of the equivalent heat loss coefficient to the ground can bring an error in the yearly auxiliary energy demand as large as 10 %.
- The effect of shading varies very slowly when the orientation of the building is changed. The due south analysis of shading may be used for S-E to S-W orientations without significant change in the auxiliary energy consumption (5 to 10 %).
- A very good agreement with detailed computation is obtained during the cold months. To get a good agreement during fall and spring, rejected solar (and wild) gains should be accounted for. Heat losses to the ground should also be correctly calculated.

Conclusions

Detailed simulations show that many factors such as groundtemperature, indoor temperature, rejected heat, shading, etc. play a significant role in the evaluation of the energy balance of a given house.

The LESO-A simplified method cannot handle all these factors. It gives a precise evaluation of the auxiliary energy demand during the cold months as well as over the full heating season but in the fall and the spring the auxiliary energy is underevaluated (up to 100 %).

Reference

1 N. Morel GRES/EPFL, 1984
PASSIM version 3, Manuel d'utilisation

A SIMPLE DESIGN METHOD FOR SOLAR HEATING SYSTEMS WITH IN-GROUND LONG TERM STORAGE

J.L. SALAGNAC, B. DELCAMBRE, M. RUBINSTEIN
Centre Scientifique et Technique du Bâtiment
BP 21
06562 VALBONNE Cédex
France

Summary

The aim of this work is to provide a simple design tool that avoids the use of large simulation models and is then in position to be used by designers that cannot afford the help of computer centers. Numerous runs of a detailed simulation model have been performed and input parameters have been set in such a way that simple correlations can be found between the solar fraction and functions of the input parameters (characteristics of the collector array, of the storage units - short term and in-ground long term-, of the distribution pipes, of the heat demand and meteorological datas). The agreement between the detailed computer model and the simple method is better than 10 % for small values of the solar fraction and results are presented in such a way that it is easy to perform parametric studies so as to help to the determination of an optimum technico-economical solution.

1. PURPOSE OF THE WORK

The aim of this work is to provide an easy-to-use design tool that avoids the costly computer time needed by large simulation models.

The considered system (see figure 1) consists of water solar collectors, a fully mixed short term water storage including the solar collectors heat exchanger, an in-ground long term stratified water storage, water distribution pipings, an auxiliary heater, and the heat load.

We only consider the heat demand during the heating season, excluding domestic hot water.

The inputs of this calculation method are :

- the sizes and main characteristics of the components,
- the annual meteorological data of the system location.

The main output is the annual solar fraction.

The ambition of this manual calculation method is to give easily an order of magnitude of the annual solar fraction.

2. METHOD OF APPROACH AND RESULTS

2.1 Detailed simulation program

The correlations used in this method come from the results of a detailed simulation program describing the behaviour of the previously mentionned system.

This program is based on an hour time step and takes into account the transient thermal behaviour of the different parts of the system (solar collectors, in-ground long term water store, ...).

The energy management is pre-determined and gives priority to solar collectors, then to heat storages (short term first and then long term) and to the auxiliary heater.

Under special circumstances, this strategy might not be optimal but the expected gain from such an optimal energy management should be smaller than the final accuracy on the solar function.

The heat emitters are radiators, the inlet temperature of which is calculated hourly in the detailed program depending on the heat demand and on the emitters characteristics.

At the starting up time, the ground surrounding the long term storage is considered to be at the annual average ambient temperature of the location.

Due to storage heat losses, this temperature varies with time and calculations are stopped when a stationnary state is reached (this requires generally three to four years depending on the storage insulation and shape and on the ground characteristics).

3. CORRELATIONS

For a given set of parameters, the annual solar fraction

$$F = \frac{\text{Annual heat demand - Annual Auxiliary Energy}}{\text{Annual heat demand}}$$

is a continuous function of the ratio (see figure 2) :

$$\frac{\text{maximum annual collected solar energy}}{\text{annual heat demand}}$$

The numerous results of the detailed simulation program can be presented by such curves but, in order to simplify the use of such drawings, we have tried to describe all these curves starting from one particular set of parameters, that is chosen as a reference (see figure 2).

The other parameters values are put in the method by means of corrective factors corresponding to the different parts of the system in the following way :

$$F = F_{ref} * C_{SE} * C_{ST} * C_{SC} * C_{RES} \quad (1)$$

where :

F : annual solar fraction

F_{ref} : annual solar fraction in the reference case (given by figure 2)

C_{SE} : corrective factor for the long term storage

C_{ST} : " " the short term "

C_{SC} : " " the collectors

C_{RES} : " " the distribution piping.

The corrective factors are given by curves where characteristic parameters of the considered part of the system are found (see example of C_{SE} in figure 3).

The agreement between the F factor given by (1) and the corresponding factor given by the detailed simulation is fairly good mainly for small values (< 0,5) (see figure 4).

This method needs then a limited number of curves and replaces advantageously the use of a computer.

A complementary work is being done to give a simple way to evaluate the volumetric heat loss coefficient of the in-ground storage (see corresponding work in the present volume , poster P. 7.21).

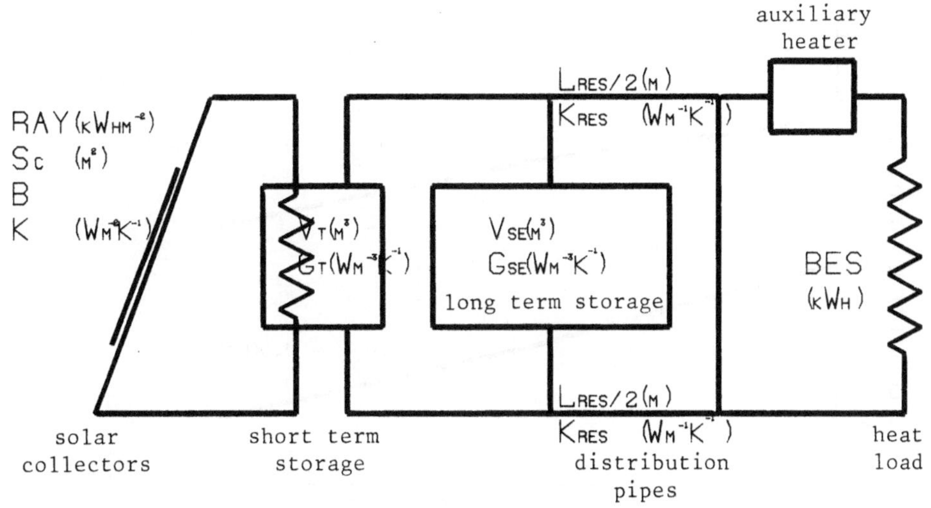

Figure 1 : SOLAR HEATING SYSTEM

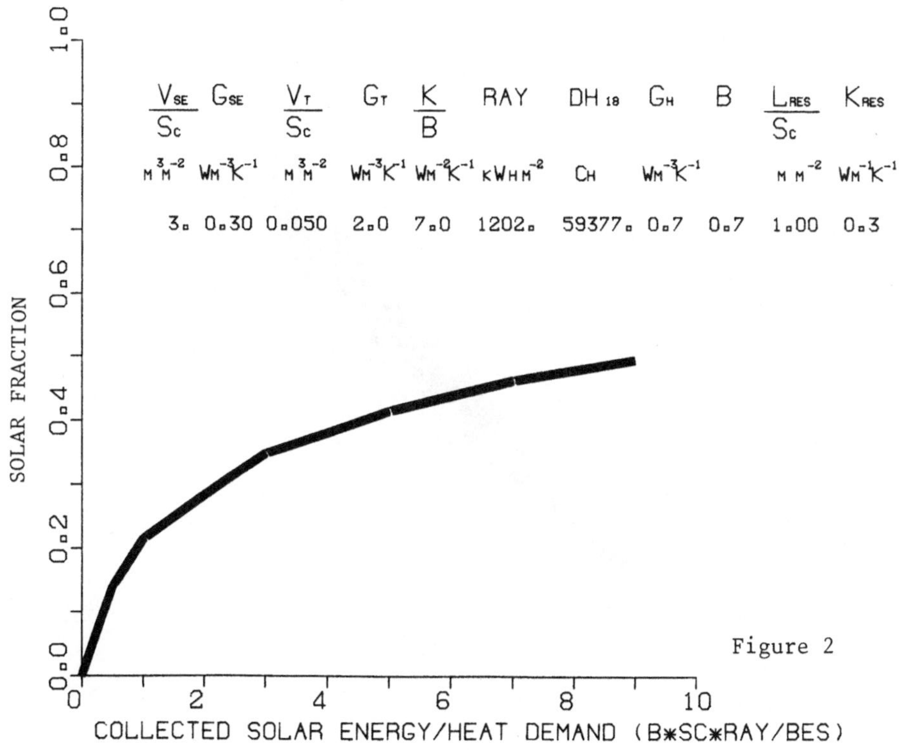

Figure 2

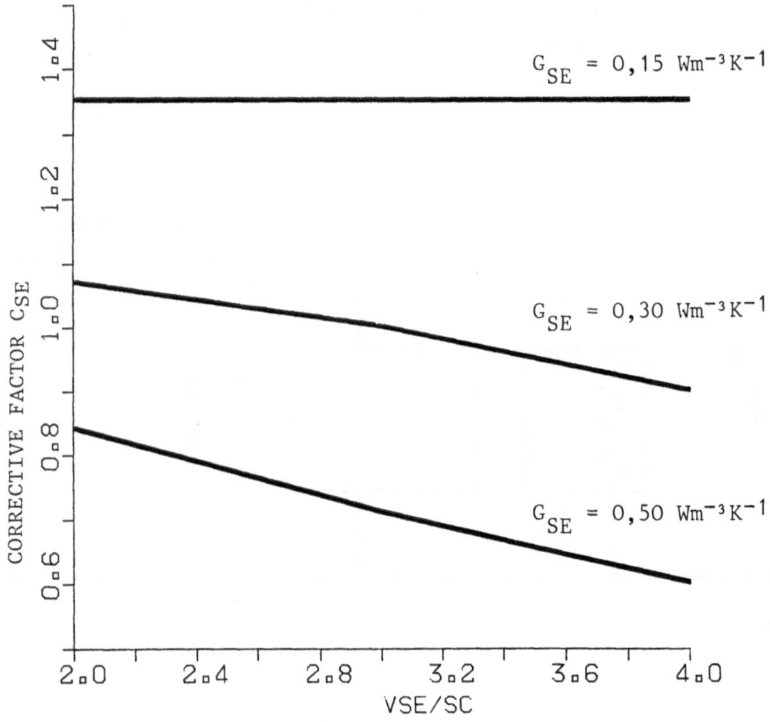

Figure 3

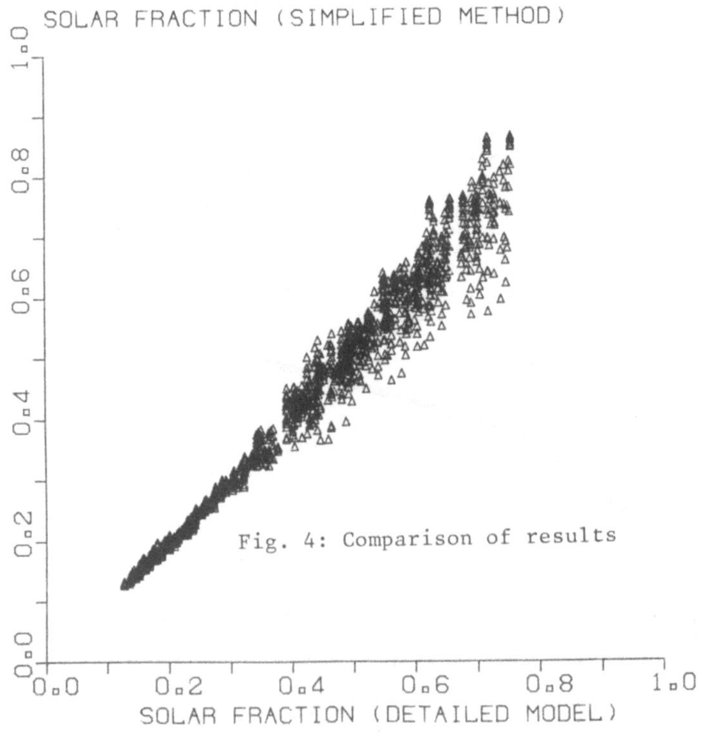

Fig. 4: Comparison of results

EASY-TO-USE METHODS FOR PERFORMANCE EVALUATION OF ACTIVE SOLAR SPACE HEATING SYSTEMS

L. BOURDEAU, B. DELCAMBRE, M. RUBINSTEIN
Centre Scientifique et Technique du Bâtiment
BP 21
06562 VALBONNE Cédex
France

Summary

The easy-to-use calculation methods presented here concern heating systems using air or liquid solar collectors connected to a packed bed or liquid short term storage unit. These methods have been prepared within the framework of the implementation of a French standardized performance specification - the 'Label Solaire' - concerning new dwellings. They permit to calculate the consumption of the auxiliary heating system, taking into account the energy provided by the solar system and the various efficiency factors. The paper describes the main characteristics of these annual methods and of the detailed computer simulations which the methods are based on.

1. INTRODUCTION

One of the main tasks of the Centre Scientifique et Technique du Bâtiment in the field of thermal design is to develop simplified calculation methods in order to provide users with easy-to-use but accurate tools for evaluation of the thermal behavior of a building. Thus, can be mentioned :

- a method to calculate the thermal losses through the walls of a building : the 'Règles Th-K 77'(1) ;
- a method to calculate the thermal load of a building : the 'Règles Th-G 77'(2) ;
- a method to calculate the heating needs of a building : the 'Règles Th-B 82'(3) ;
- a method to calculate the energy consumption (gas, fuel-oil or electricity) to meet the heating needs and satisfy the domestic hot water demand: the 'Méthode LHPE-LS'(4).

This last method has been prepared within the framework of two French standardized performance specifications - the 'Label Haute Performance Energétique' and the 'Label Solaire' - concerning new dwellings. It contains a chapter about Solar Domestic Hot Water Systems described in another paper of the Proceedings of the Conference.

The present paper deals with the chapter of 'Méthode LHPE-LS' concerning active solar space heating systems with short term storage. It can be noted that a calculation tool for systems with in-ground long term storage, not yet included in 'Méthode LHPE-LS', is also described in another paper.

2. DESCRIPTION OF THE CALCULATION METHODS FOR ACTIVE SOLAR SPACE HEATING SYSTEMS

2.1. Types of systems

The methods are applicable to :

- water heating systems consisting of liquid collectors connected to a storage tank by a water loop with or without a submerged or separating heat exchanger ;
- air heating systems consisting of air collectors connected to a packed bed by an air loop. Heat transfer is insured by forced air circulation from collectors to house, collectors to storage or storage to house, independently of the ventilation system of the house.

The auxiliary system can in both cases be connected or not to the solar system.

2.2. Principle of the methods

The aim of the project was to provide designers with easy-to-use but accurate manual calculation methods that could take into account all the parameters likely to have a significant influence on the final result.

The principle of the methods is defined by the following characteristics :

- the calculation is made in one time step on an annual basis ;
- the userscan, at many points (determination of the piping losses, of the heat exchanger efficiency, of the pump or fan consumption, of the efficiency of the space heating distribution system, of the heat generating system efficiency) either perform a full calculation considering all the influencing parameters or choose a fixed value depending just on one or two main parameters ;
- the methods first give the result for a reference system. A corrective factor is then applied if the characteristics of the system are different from those of the reference system.

2.3.Determination of the gross annual solar contribution factor

Figure 1 shows a flow chart describing the calculation process of the methods. First, a gross solar contribution factor, F°, is calculated. This factor corresponds to the reference systems defined by :

- a ratio of the storage capacity to the collector area equal to 0.05 m^3/m^2 (water systems : WS) 0.15 m^3/m^2 (air systems : AS) ;
- a ratio of the storage unit heat losses to the storage capacity equal to 2 $W/m^3\,°C$(WS) or 1 $W/m^3\,°C$ (AS) ;
- a storage length (AS) equal to 2.5 m ;
- an air flow rate (AS) equal to 50 m^3/h per unit of collector area

2.3.1. Fraction of incident solar energy available at distribution level

This fraction ω is computed by a formulation taking into account :

- the area and thermal characteristics (linear model) of the collectors ;
- the pipe or duct losses calculated from the loop characteristics or by a given linear function of the collector area. This function assumes a middle length of pipes or ducts and a rather low level of insulation ;
- the position of the auxiliary heating system (connected or not connected with the solar system) and the average temperature of the heat emitters ;
- the efficiency of the heat transfer from the collectors to the storage (WS only). This efficiency depends on the characteristics of the collectors, the pipe losses, the mass flow rates and specific heat capacities of the thermal fluids, the effectiveness of the control system, and the thermal conductance of the heat exchanger. It can be determined from appropriate formulations given for the main configurations or taken equal to a fixed low value (0.68).

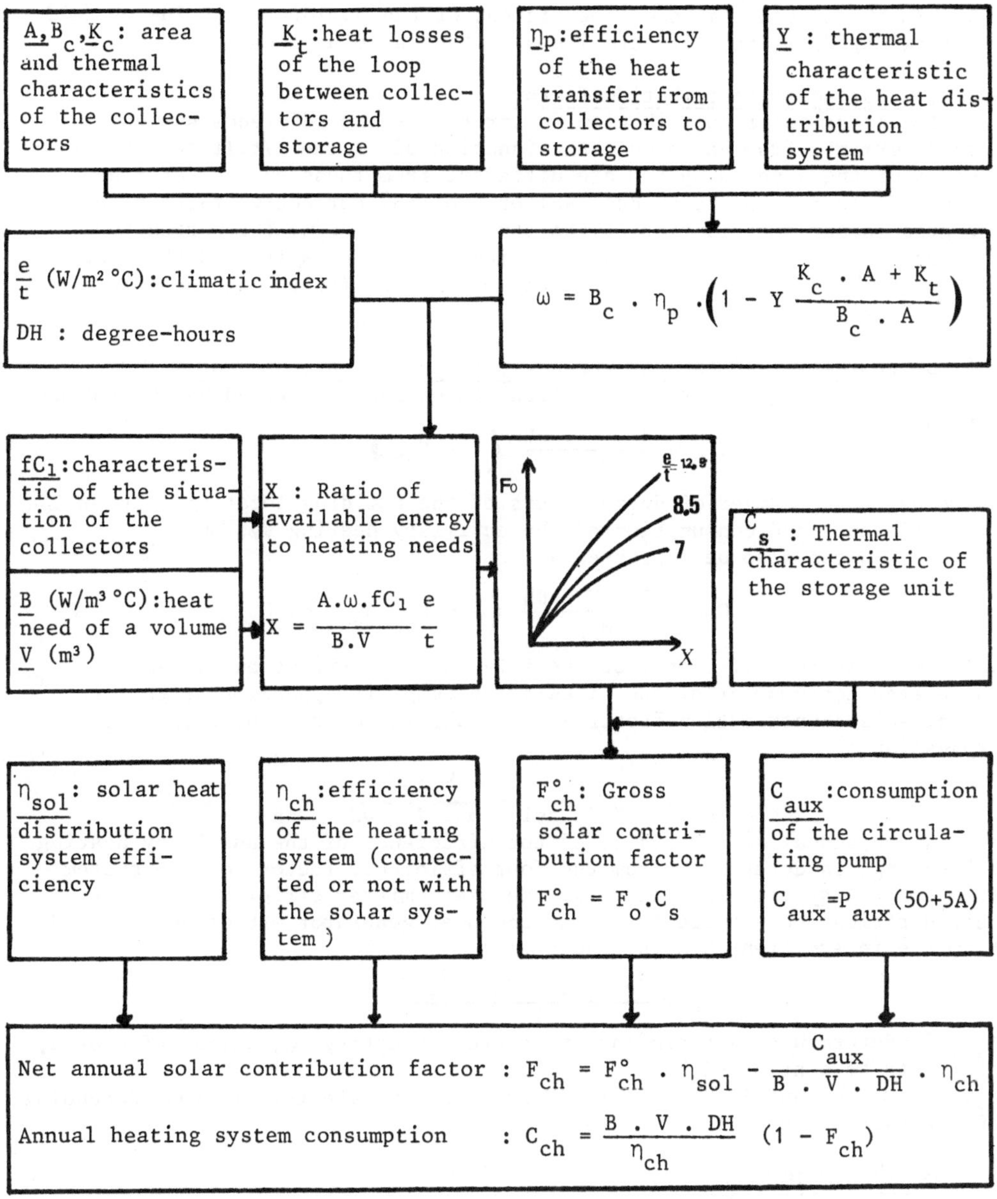

Figure 1 : Calculation process of the methods

2.3.2. Ratio of the mean available solar energy to the heating needs

This ratio X is expressed as :

$$X = \frac{A \cdot \omega \cdot f \cdot C_1}{B \cdot V} \cdot \frac{e}{t}$$

where :

- A (m²) is the area of the collectors ;
- ω is the fraction calculated as mentioned above ;
- fC_1 is the characteristic of the shading, orientation and tilt of the collectors ;

- B (W/m³ °C) is the heat need coefficient of the building of volume V (m³) ;
- e/t (W/m² /°C) is the climatic index of the building location.

2.3.3. Gross solar contribution factor

The gross solar contribution factor F° for the reference system is then determined from an abacus as a function of X and e/t. If the characteristics of the studied system are different from those of the reference system, F° is multiplied by one (WS) or two (AS) corrective factors to get the gross solar contribution of the studied system F°_{ch}. The corrective factors can be read in tables as functions of the capacity and the heat losses of the storage (WS and AS), and of the bed length and the mass flow rate (AS only).

2.4. Determination of the net annual solar contribution factor

The net annual solar contribution factor F_{ch} is defined by the formulation :

$$C_{ch} = \frac{B \cdot V \cdot DH}{\eta_{ch}} \left(1 - F_{ch}\right)$$

where DH is the number of degree-hours of the building location, η_{ch} et C_{ch} the efficiency and consumption of the auxiliary heating system.

As C_{ch} can also be expressed as

$$C_{ch} = \frac{B \cdot V \cdot DH}{\eta_{ch}} \left(1 - F^{\circ}_{ch}\,\eta_{sol}\right) + C_{aux}$$

where η_{sol} is the efficiency of the solar heat distribution system and C_{aux} the energy consumption of the circulating pump, it appears that the net annual solar contribution factor F_{ch} is calculated from the following expression :

$$F_{ch} = F^{\circ}_{ch} \cdot \eta_{sol} - \frac{C_{aux}}{B \cdot V \cdot DH} \cdot \eta_{ch}$$

η_{ch} is the annual average value of the efficiency of the auxiliary heating system. η_{ch} is determined from the four efficiency factors of the system (the control system efficiency υ_r, the heat emitter efficiency υ_e, the heat distribution efficiency υ_d and the heat generator efficiency υ_g) and from the intermittent heating factor I :

$$\eta_{ch} = \frac{\upsilon_r \cdot \upsilon_e \cdot \upsilon_d \cdot \upsilon_g}{I}$$

η_{sol} is determined by a similar procedure, excepting υ_g, which, of course, is not to be considered here.

υ_r, υ_e, υ_d, υ_g and I are calculated from approximate formulations depending on the type of the heating system.

3. BASE OF THE METHODS

The calculation procedures of 'Méthode LHPE-LS' have been elaborated from theoretical studies (analytical approach and computer simulations) supported by numerous experiments (field measurements and equipment testing).

Thus computer simulations of the solar heating systems have been made using actual climate data sets and pertinent models of the different components :

- solar collectors are represented following the basic derivation by HOTTEL, WHILLIER and BLISS ;
- liquid store is supposed to be unstratified ;
- packed bed store is described by a finite difference model ;
- collector loop and distribution loop are characterized by their lengths and thermal fluid-to-outside conductances ;

. system operation-collection, storage, distribution- is controlled by numerical components simulating the differential temperature controllers in the loops ;

. heat distribution to the house is modelled either by a heat exchange between a heater and a temperature controlled environment or by forced-air heating ;

. heating needs of the house are evaluated by a model accounting for passive solar gains, casual gains and building thermal mass.

Thousands of computer simulations have been conducted over complete heating seasons and for several French locations. The analysis of the results has allowed to quantify the respective and mutual influences of the design parameters of the solar heating systems. Then, the simplified analysis methods presented above, amenable to hand calculation have been developed, giving results well correlated to computer simulations.

REFERENCES

1. C.S.T.B. : "Règles Th-K 77", Cahiers du C.S.T.B., novembre 1977.
2. C.S.T.B. : "Règles Th-G 77", Cahiers du C.S.T.B., novembre 1977.
3. C.S.T.B. : "Règles Th-B 82", Cahier du C.S.T.B. n°1767, avril 1982.
4. C.S.T.B. : "Méthode LHPE-LS", Cahier du C.S.T.B. n°1849, mai 1983.

A SIMPLE METHOD FOR PERFORMANCE EVALUATION OF SOLAR DOMESTIC HOT WATER SYSTEMS

D. BIENFAIT
CENTRE SCIENTIFIQUE ET TECHNIQUE DU BATIMENT
BP. 21 - 06562 VALBONNE FRANCE

Summary

A simple method intented to predict the annual thermal performance of individual and collective SDHW systems has been developed by CSTB. This method which addresses different kinds of design and back-up energy is derived from results of the detailed, one hour step, computer program OSOL.
It is now being validated by comparison with field measurement on SDHW systems.

1. REFERENCE COMPUTER PROGRAM

The FORTRAN written computer program, called OSOL, is intended to simulate the thermal behaviour of a solar domestic hot water system during a one year period. The main features of this program are the following :

a) Main assumptions

. Hot water demand :

The pattern of domestic hot water demand throughout the day has been chosen by reference to statistical data available for France ; it is given by figure 1

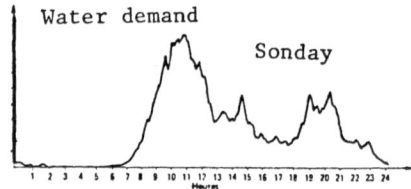

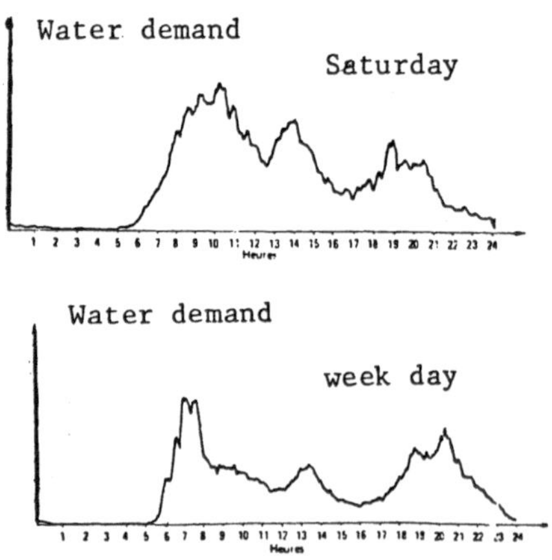

Fig. 1: Pattern of domestic hot water demand

according to the same statistical data, the average volume of hot water demanded depends on the month considered as follows :

Janv.	Fév.	Mars	Avr.	Mai	Juin	Juil.	Août	Sept.	Oct.	Nov.	Déc.
1,25	1,20	1,10	1,05	1,00	0,8	0,5	0,6	0,9	1,05	1,15	1,40

The cold water temperature is assumed to be different for each month of the year.

. Climatic data :

The climatic data are the actual records of air temperature, direct irradiation and diffuse irradiation on an horizontal plan. These date have been provided by the meteorological services for different towns with a 1 hour time-step for one year or more. The values of hourly irradiations on tilted plans is derived from these data by trigonometric calculations. The ground albedo is taken equal to 0,2.
The diffuse irradiation is assumed to be isotropic.

. Solar collector efficiency, η :

The instantaneous solar collector efficiency is assumed to be given by the following relation :

$$\eta = \eta_o . f(i) - a_1 . \frac{\Delta T}{E}$$

Where ΔT is the difference between the mean liquid temperature and the air temperature, E is the solar irradiance calculated on the collector plan, i the incidence angle of the beam, f(i) is the angle modifier factor assumed to have the same expression for all collectors, η_o and a_1 are input data defined below.

. Heat storage :

The tank is assumed to be fully stratified and is divided into 5 superimposed zones. The solar heat exchanger is located in the lower zone. A back-up electric resistance may be placed in one of the upper zones.

b) Input data

The main input data are the following :

- tilt angle and azimuth of the solar collector
- η_o and a_1 : collector efficiency parameters
- A (m^2) : collector area
- Ke (W/°C) : heat exchanger heat transfer coefficient
- Q (W/°C) : solar loop energy flowrate
- K_t (W/°C) : heat loss coefficient of the solar loop pipes
- K_b (W/°C) : tank heat loss coefficient
- M (kg) : tank thermal mass
- M_e (kg) : thermal mass of the part of the tank heated by back-up energy (may be equal to zero)
- geographical location : latitude and meteorological data
- B (m^3/year) : domestic hot water needs

c) Output data

The main output data are :

- F (solar fraction on a yearly basis)
- C (consumption of back up energy)
- T_m (mean temperature of the DHW supplied, in order to verify that the installation sizing is sufficient)

. SIMPLIFIED CALCULATION OF THE SOLAR FRACTION

A simplified calculation method of the solar fraction F has been derived from the computer program OSOL :
The program OSOL has been used to calculate the solar fraction F over a large range of input data values (nearly one thousand computer runs have been performed) and the results have been analysed by least square methods in order to obtain the following emperical method :

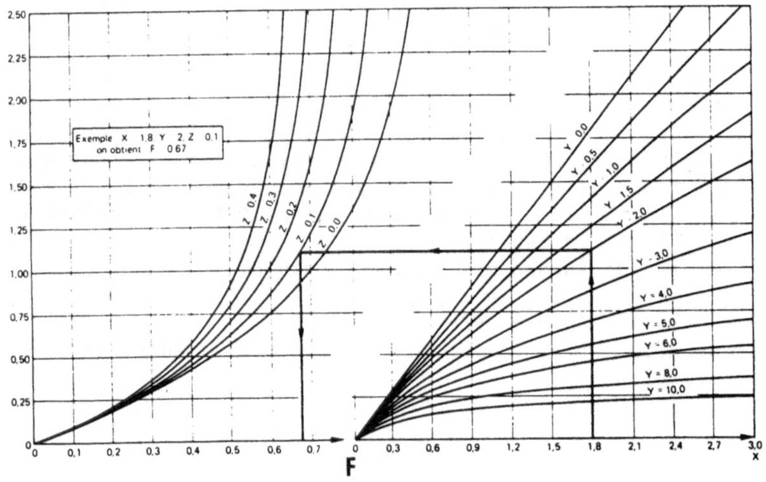

FIG 2 : Graphic determination of solar fraction F as a function of non dimensional parameters X, Y and Z.

The solar fraction F is given by figure 2 as a function of three non dimensional parameters, X, Y, Z whose expressions are :

$$X = \sqrt{p} \cdot \frac{A . \eta_o . f . E}{B + \alpha K_b} = \frac{\text{solar energy input}}{\text{thermal load}}$$

$$Y = \frac{(A . a_1 + K_t) . \alpha}{A . \eta_o . f . E} = \frac{\text{solar loop heat loss coefficient}}{\text{solar energy input}}$$

$$Z = \frac{B + \alpha K_b}{\beta . M} = \frac{\text{thermal load}}{\text{tank capacity}}$$

Where α and β are numerical coefficients
- f is a corrective coefficient which holds for shadow, tilt and azimuth influences
- E (kWh/m^2.year) is a value depending on the geographical region under consideration (see figure 3)

$\sqrt{p}$ is an analytical function which accounts for the influence of K_e (heat exchanger) and Q (flowrate)

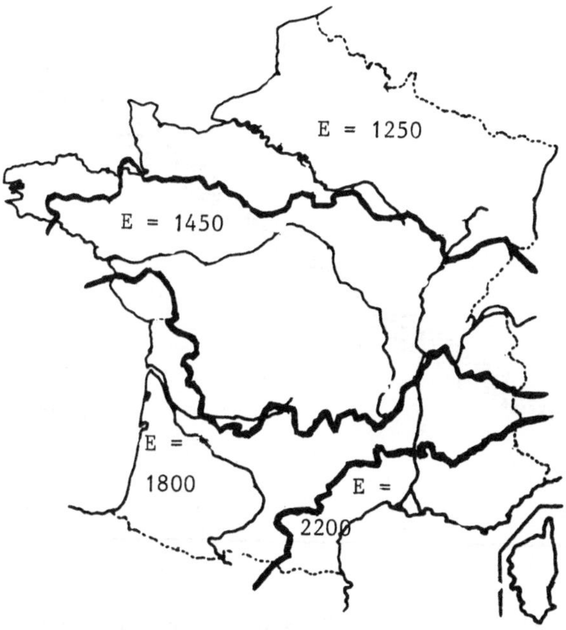

FIG 3 : values of parameter E for the four french geographical areas

3. ACCURACY OF THE SIMPLIFIED METHOD

The correlation between the results given by the reference program and the simplified method for a wide range of input parameter variations is shown on fig. 4

It can be seen that in most cases the agreement between the reference program and the simplified method is better than 5 %.

4. CALCULATION OF BACK UP ENERGY CONSUMPTION C

The ultimate goal of the method is to calculate the yearly back-up energy consumption, C :

This consumption depends on the kind of design (2 examples of design are shown in fig 5). It is given for each design as a simple algebraic function of solar fraction F, heat loss coefficient of back-up tank, heat loss coefficient of DHW supply pipes, and burning efficiency when gas or fuel is used.

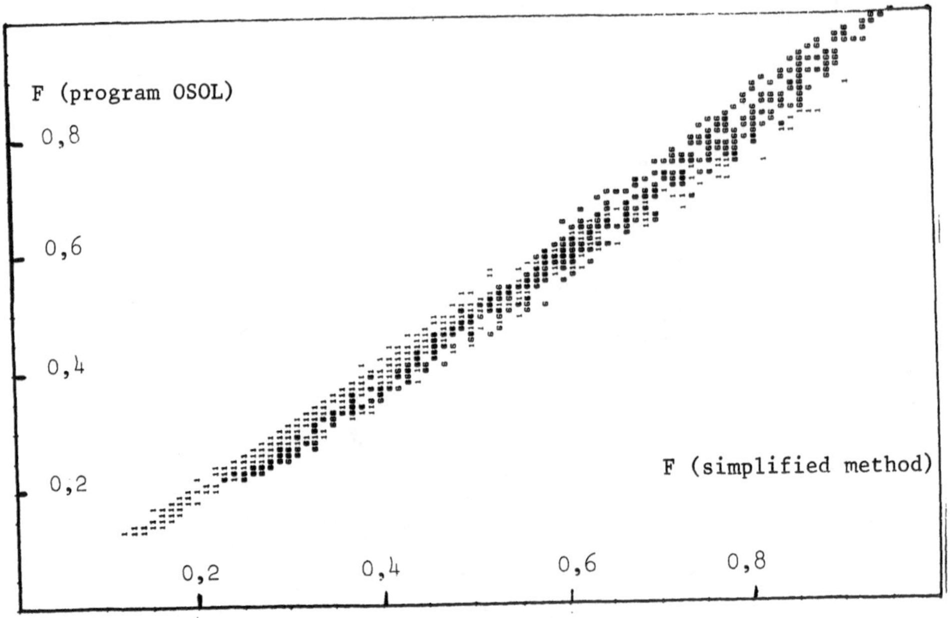

FIG 4 : plot of F values calculated with the reference program versus F values calculated with the simplified method over a large range of input data.

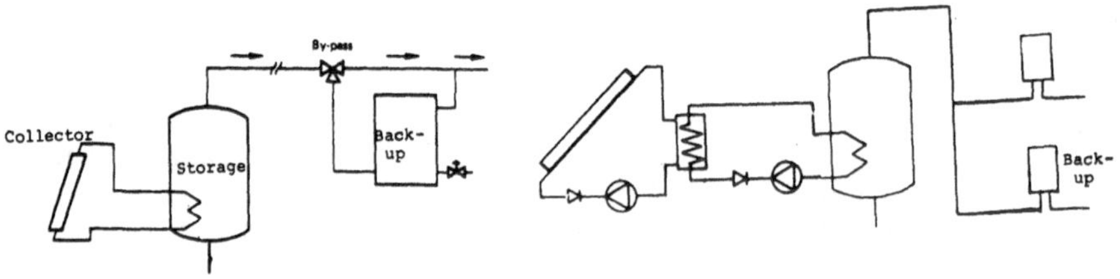

Single family SDHW with distant back-up

Collective SDHW with back-up energy individually provided in separate tanks

Fig 5. : examples of installation schemes that can be processed by the method

DESIGN TOOL FOR BURIED HORIZONTAL HEAT EXCHANGER

JM. CARDI - P. NOLAY
Ecole Nationale Supérieure des Mines de Paris
Centre d'Energétique Sophia Antipolis
06560 VALBONNE Cedex
FRANCE

Summary

In the first part of this paper the objective of the researches is defined. The aim of the study is to obtain a design method for heating systems using buried horizontal heat exchangers. Then, we present the simulation code SCHUSS (simulation code for horizontal under ground source system) and describe each component model in particular:

- the models of building
- the model for heating slab
- the model of heat pump
- . the thermal model of soil around the heat exchanger

The last part of this paper presents some results calculated by the computer program. Some of them concerning the thermal behavior of the installation are detailled. The results are compared with experimental data from a monitoring of a system located in Nice area. Now the program SCHUSS permits to compare thermal efficiency of an installation and also investment and functioning costs for different designs.

1. INTRODUCTION

Le but de la recherche menée par l'Ecole des Mines est de développer une méthode de calcul pour les systèmes utilisant des échangeurs horizontaux enterrés et en particulier pour les systèmes centralisés (lotissement de plusieurs maisons par exemple).

A l'heure actuelle très peu de projets importants ont été réalisés. Une raison de ce fait est le manque d'outils de conception et de méthode de dimensionnement pour ce type de systèmes. En effet les méthodes utilisées actuellement sont essentiellement empiriques et conduisent généralement à un surdimensionnement.

2. PRESENTATION DU CODE DE CALCUL

Le logiciel de simulation SCHUSS, développé par le Centre d'Energétique de l'Ecole des Mines de Paris à Sophia Antipolis et dont on trouvera la description dans les pages suivantes permet l'étude des systèmes énergétiques présentés ci-dessous.

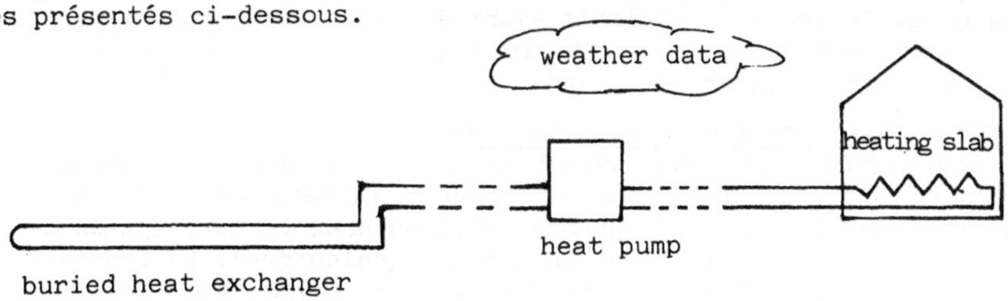

L'utilisation de SCHUSS peut conduire pour ces systèmes à:

- une analyse plus fine des comportements du système à des pas de temps supérieurs ou égaux à 5'
- l'établissement de bilans énergétiques sous forme de tableaux ou de graphes montrant leur évolution en fonction du temps
- l'analyse de l'intérêt économique des systèmes à l'étude

Ce programme est écrit en FORTRAN IV et nécessite 8 mn de temps de calcul sur un DPS 7 de CII pour un cycle annuel et dans une configuration normale.

3. DESCRIPTIONS DE MODELES DE COMPOSANTS

3.1 Modélisation de l'habitat

SCHUSS permet de choisir entre deux modèles:

. un modèle où l'on décrit le bâtiment paroi par paroi et où l'on effectue pour chacunes d'elles le bilan entre les apports passifs et les déperditions en utilisant les températures équivalentes des vitrages et des parois. On en déduit alors le flux global passant par les parois du bâtiment. Ce modèle prend en compte l'inertie du bâtiment.

. le second modèle considère uniquement le bilan entre les apports passifs et les déperditions en utilisant le coefficient G des déperditions volumiques de la surface sud équivalente de l'habitat.

3.2 Modélisation de la restitution de chaleur

La restitution de la chaleur est assurée par un plancher chauffant à eau. Le modèle utilise en entrée un vecteur de commande fourni par la régulation associée (régulation sur la température extérieure par exemple) Le vecteur de commande comprend la température d'attaque du plancher par le fluide ainsi que la température de la dalle au pas de temps précédent.

A partir de ces données on peut calculer la température du fluide à la sortie du plancher

$$TS = TE \times e^{-\frac{hs}{\dot{m}\, cp}} + (1 - e^{-\frac{hs}{\dot{m}\, cp}}) \times T\ dalle$$

et donc la puissance distribuée dans la dalle ainsi que la température de surface T plancher. Par conséquent le bilan sur l'air peut s'écrire en fonction des apports internes E int, des apports solaires passifs Easol et de la distribution.

$$\rho\, cp\, \frac{\dot{T}\ air}{t} = Eint + Easol + h\ (T\ plancher - \dot{T}\ air)$$

3.3 Modélisation de la pompe à chaleur

Le modèle utilisé dans ce logiciel ne rend pas compte physiquement du fonctionnement de la pompe. Pour cela il faut, à l'aide des données du constructeur, formuler une expression du coefficient de performance et de la puissance électrique consommée par la pompe à chaleur qui lissent parfaitement les valeurs mesurées pour différentes températures d'entrées à l'évaporateur de sortie au condenseur, et pour des pompes de diverses puissances. Ce travail préalable ayant été fait, le logiciel peut à l'aide des expressions du C.O.P et de la puissance, calculer à chaque pas de temps, les performances de la PAC.

3.4 Modélisation de l'échangeur enterré

Pour décrire le comportement thermique du sol, le programme utilise un modèle bidimensionnel aux différences finies. La maillage est conçu de façon à laisser une très grande latitude dans la description de l'échangeur (grand nombre de tubes - emplacement quelconque). La condition à la surface prend en compte la convection, le rayonnement terrestre et le rayonnement solaire .

Les résultats fournis par le programme sur le comportement du sol sont illustrés par les deux courbes ci-dessus. La première représente le profil de température du sol à la verticale du tuyau, la seconde la trajectoire de la température du sol à proximité du tuyau au cours de l'année.

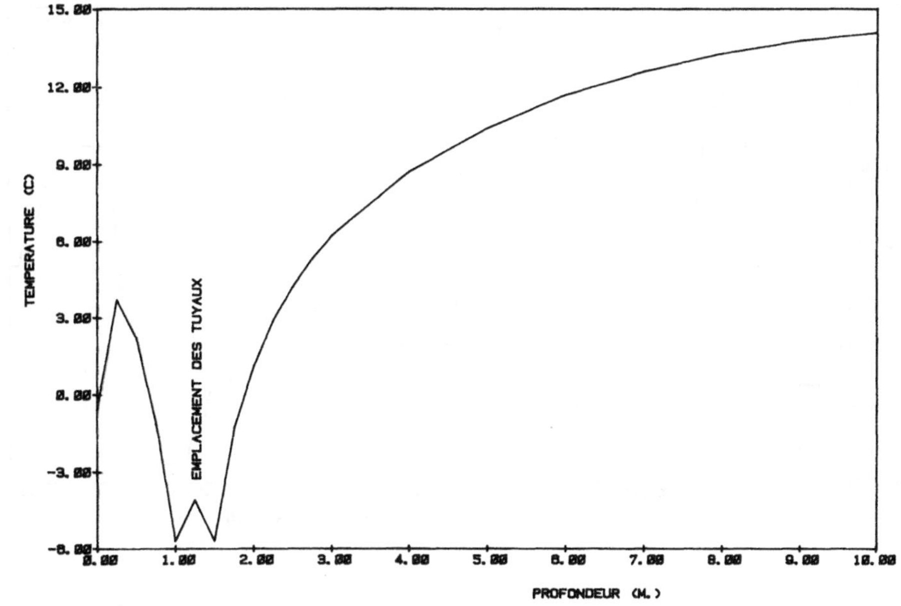

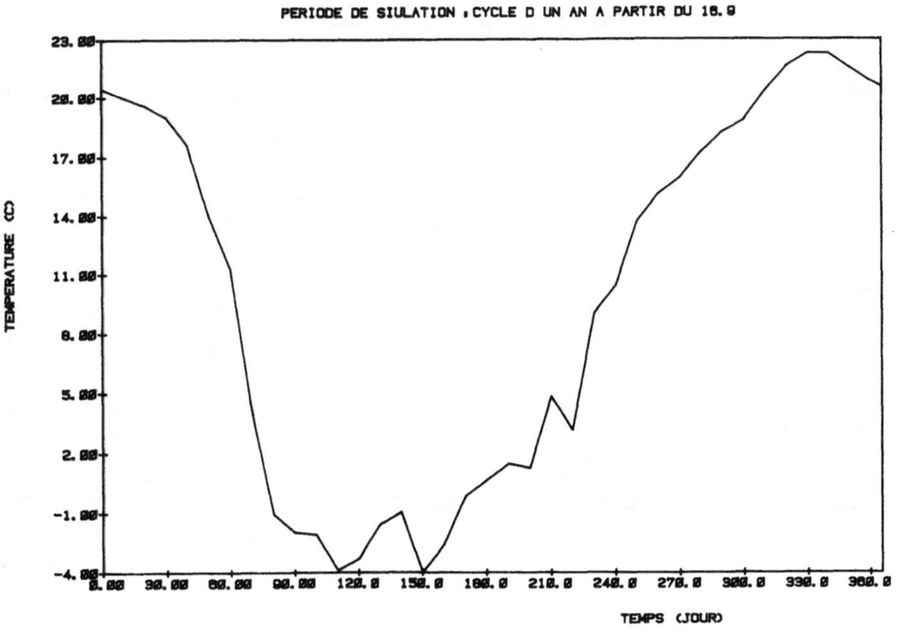

4. RESULTATS SUR LE FONCTIONNEMENT GLOBAL DE L'INSTALLATION

Les résultats présentés ici, décrivent le comportement thermique d'une installation située dans la région de Nice.

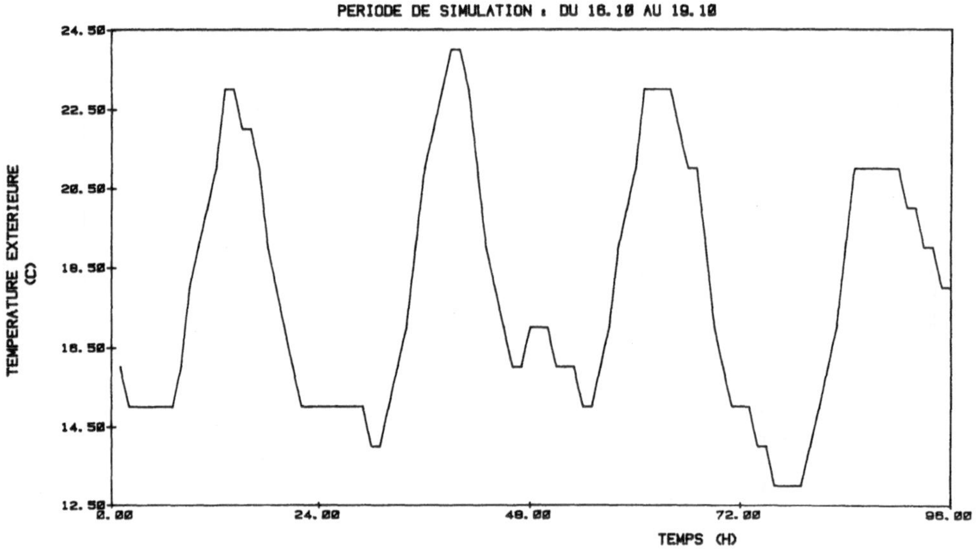

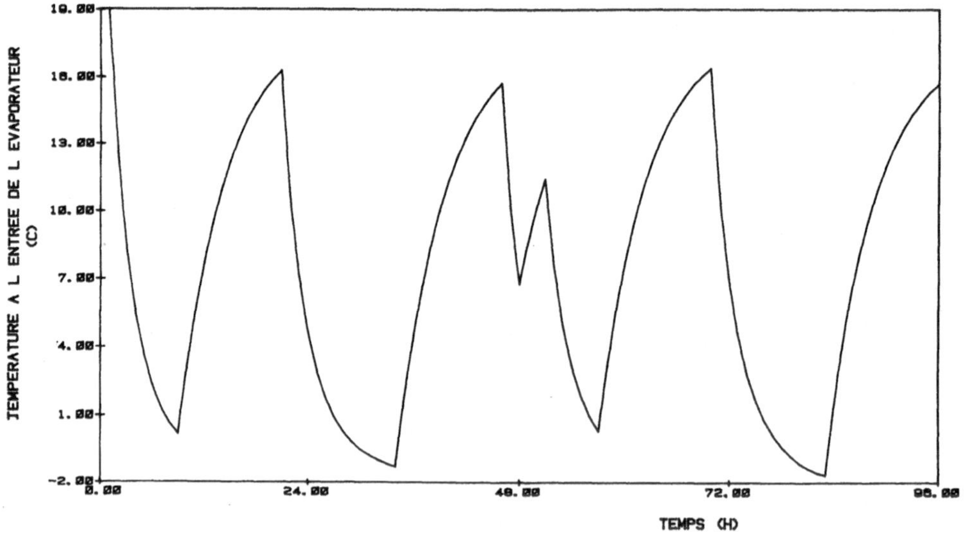

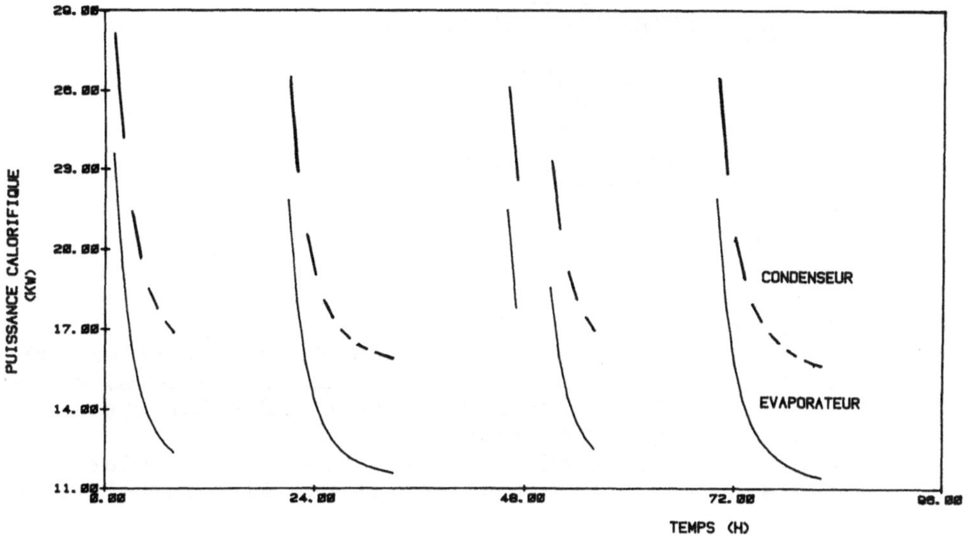

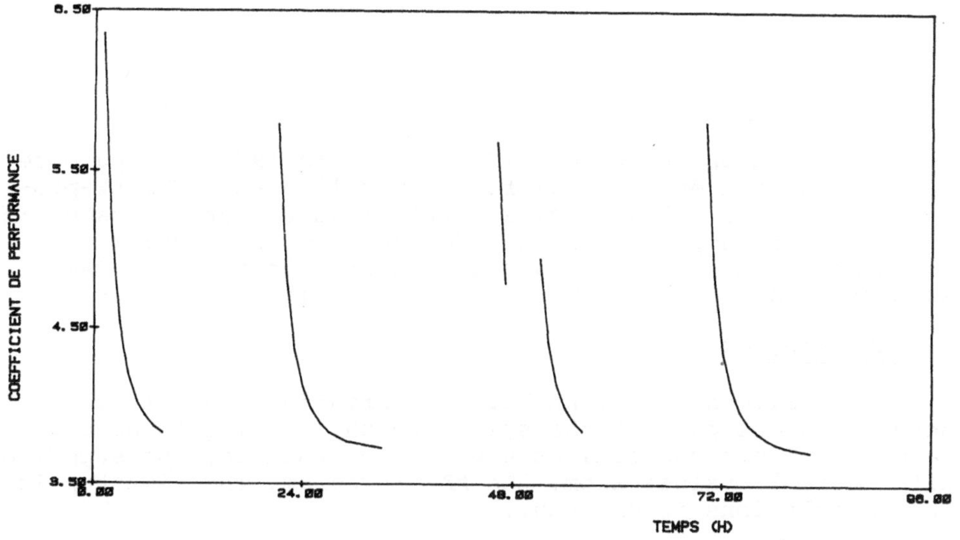

5. <u>CONCLUSION</u>

L'accord entre les résultats des modèles présentés et les mesures effectuées sur une installation permet d'utiliser le modèle d'échange de chaleur dans le sol, bidimensionnel purement conductif pour comparer les différentes configurations pour l'échangeur enterré (géomètrie, profondeur...) et conduira à une optimisation économique afin de réduire le temps de retour de tels systèmes. La diffusion dans des systèmes de grande taille nécessite le développement de méthode de conception. En effet, même si le C.O.P des grosses pompes à chaleur est plutôt bon pendant tout l'hiver, le COP du système est fortement pénalisé par l'énergie nécessaire aux pompes de circulation: ainsi le code de simulation SCHUSS permet de comparer l'efficacité thermique de l'installation et aussi l'investissement et le coût de fonctionnement pour différentes configurations.

SUNSYST -
THE COMPUTER PROGRAM FOR SOLAR ENERGY

S. OLSSON
AB Andersson & Hultmark
Box 24135
400 22 Göteborg
Sweden

Summary

In order to create a design tool the development of SUNSYST started in 1979. With SUNSYST it is possible to simulate systems with flat plate solar collectors, heat-pumps, vertical duct storage systems and solar systems with a watertank as storage.
SUNSYST have been used within the big Swedish solar projects during the last years. It has also been involved in works within the IEA.

1. INTRODUCTION

The development of SUNSYST started in 1979. The purpose was to create a design tool for plants like the SUNCLAY-project with solar collectors, a vertical duct storage system in clay and heatpumps. During the following years SUNSYST was supplemented in order to make it possible to describe DAW-systems with stratified water as storage in a tank.

2. DESCRIPTION

SUNSYST contains a number of subroutines one of each describing one part of the system which is going to be calculated on. These subroutines are put together in, and supplied with variable values from, the MAIN-program. The timestep in the calculations is one hour.

Here follows a short description of each subroutine, (see fig 1 on next page):

HDSOL1 - Contains statistical data for outdoor temperature, cloudiness and windspeed.

VAEDER - Calculates the outdoor temperature, the cloudiness and the windspeed for every specific hour. The outdoor temperature is determined from statistical data on every day's maximum and minimum temperature.

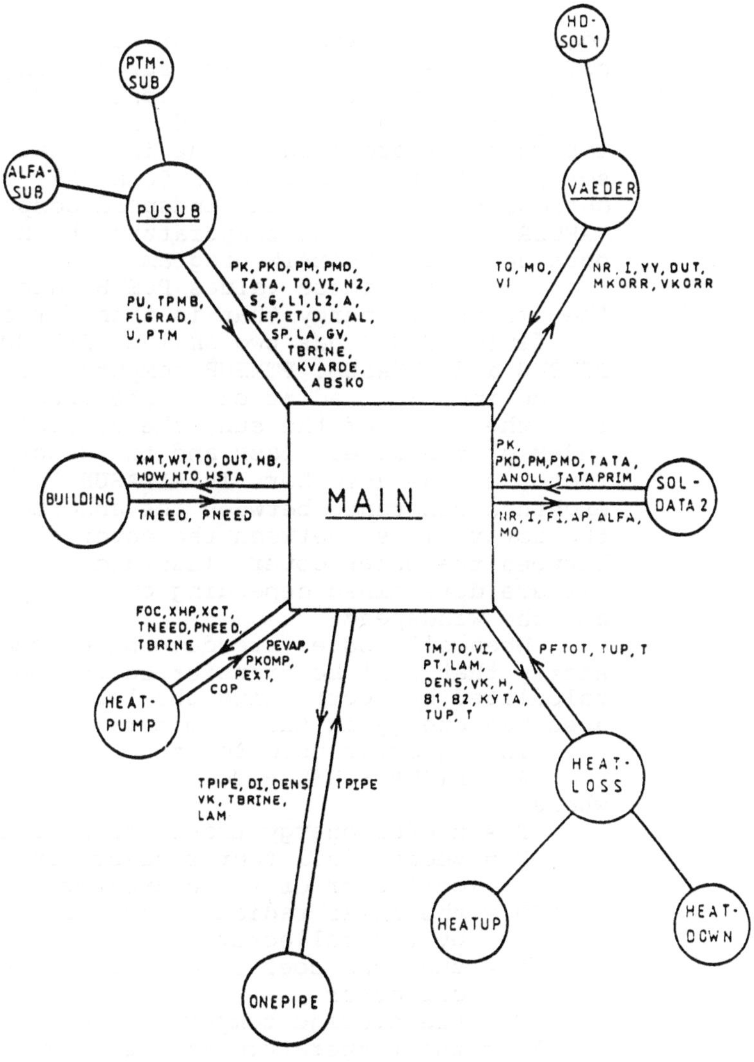

Figure 1 SUNSYST

<u>SOLDATA2</u> - Determines the solar radiation and the angle of incidence of the solar radiation against a solar collector placed in any direction. Consideration is taken to the cloudiness given by VAEDER.

<u>PUSUB</u> - Models a flat plate collector with or without cover glasses or convection surpressing device. To proceed with the calculations PUSUB requires the geometry and material of the solar collector which is provided in the input data, the solar radiation computed in SOLDATA2, the outdoor temperature and windspeed computed in VAEDER and the brine temperature which is obtained from the MAIN program.

Utilizing these values PUSUB calculates the amount of energy that is transferred to the fluid. PUSUB has two smaller SUBROUTINES PTMSUB and ALFASUB. PTMSUB computes how much of the incoming radiation is absorbed considering the angle of the sun, the refractive index of the cover glass and the absorption factor of the absorber. In ALFASUB the thermal exchange constants between the absorber and the cover glass, between the cover glasses and between the outer cover glass and the ambient air are determined depending on the spacing and the windspeed.

When all losses are determined the useful energy that will be transferred to the fluid is calculated by subtracting the losses from the absorbed energy in the absorber.

The equation used for this is:

P = F(PTM - U(T - TO))

where

P = useful energy transferred to the fluid
F = geometrical factor describing the collector or the convector
PTM = the solar radiation that is absorbed by the collector
U = the loss coefficient of the solar collector
T = the average temperature of the fluid
TO = the temperature of the surroundings

The computation for P is an iterative process.

<u>HEATLOSS</u> - This subroutine calculates the heat losses from the ground storage using the division of the surrounding ground and the temperature in the storage from the MAIN program.

The heat transmittance of the ground surface is calculated in HEATUP considering any insulating material on the surface and influence of the solar radiation.

HEATDOWN is another subroutine that determines the losses downwards and sidewards.

The heat losses are calculated with a normal heat transmission equation for the element closest to the storage.

Both these subroutines use a finite difference method to calculate the temperature field in the surrounding ground.

When calculating the heat losses from HEATUP and HEATDOWN consider the storage as a box in the ground. The surrounding ground is divided into siz loss-zones which are further divided into thin elements. The thickness of these elements depends on the thermal properties of the ground. Figure 2 shows this division.

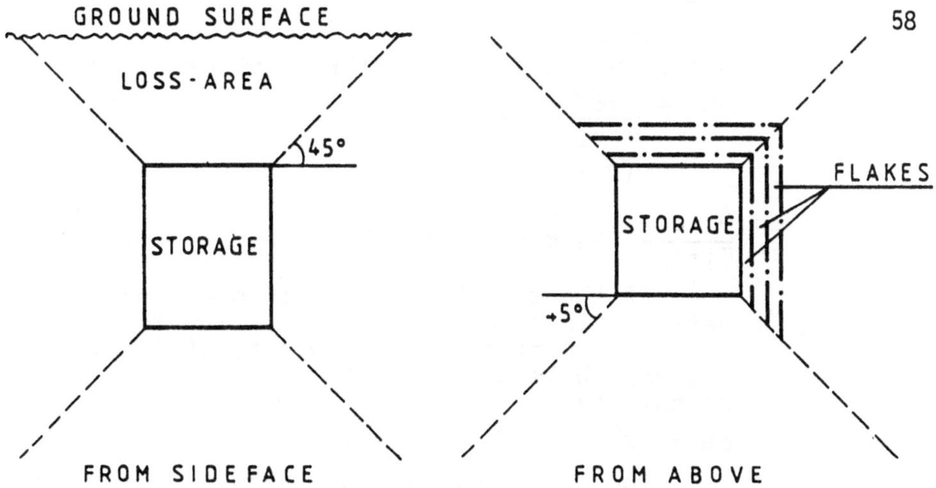

Figure 2 Subdivision of ground around a storage system.

The thickness of the elements R can be determined as follows:

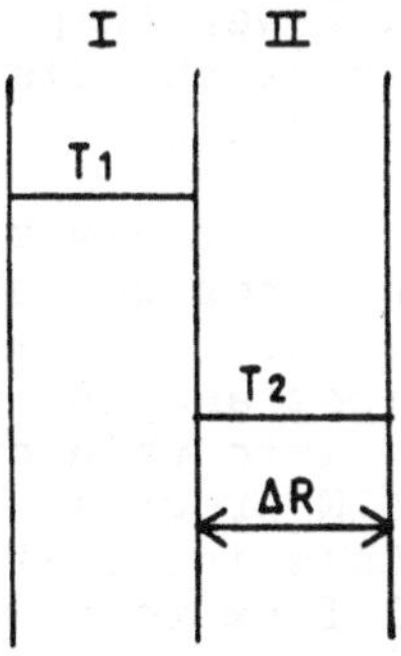

Figure 3 Elements I and II

The flux between element I and II (Figure 3) is

$$P = k \cdot A \cdot (T_1 - T_2) \text{ where } k = \frac{\lambda}{\Delta R}$$

and λ = thermal conductivity

$$P = \frac{\lambda}{\Delta R} \cdot A \cdot (T_1 - T_2)$$

If P is multiplied with 3600 s (1 hour) and set equal to $m \cdot c \cdot \Delta T_{II}$, where m is the mass of element II and ΔT_{II} is the temperature rise of element II, and if a unit area is considered,

$$3600 \cdot \frac{\lambda}{\Delta R}(T_1 - T_2) = \Delta R \cdot p \cdot c \cdot \Delta T_{II}$$

where

p = density (kg/m^3)
c = specific heat capacity (J/kg K)

$$\Delta R^2 = \frac{3600 \cdot \lambda}{\frac{\Delta T_{II}}{(T_1 - T_2)} \cdot p \cdot c}$$

$\frac{\Delta T_{II}}{(T_1 - T_2)}$ is defined as the stability limit.

If the stability limit is set then R is defined. To obtain a good stability and significant element sizes a stability of 0.2 is chosen, while the size of elements will depend on this factor.

$$\Delta R = \sqrt{(3600\ )/(0.2 \cdot p \cdot c)}.$$

The first step in the calculations is to raise the temperature in element II according to the stability limit. Changes in the temperature of element I can be determined depending on the relative volume of elements I and II. In this manner calculations proceed hourly always distributing energy from a higher temperature to a lower temperature regime independent of whether the distribution is from or towards the storage.

ONEPIPE - The local thermal distribution around the pipes in the storage is calculated in this subroutine. The temperature of the brine inside the pipes and the division of the surrounding ground into thin cylinders is obtained from the MAIN program.

With this information ONEPIPE calculates the thermal flux to or from the pipe. Using the thermal conductivity, temperatures at different distances from the pipe can also be determined. A finite difference method is used in these calculations.

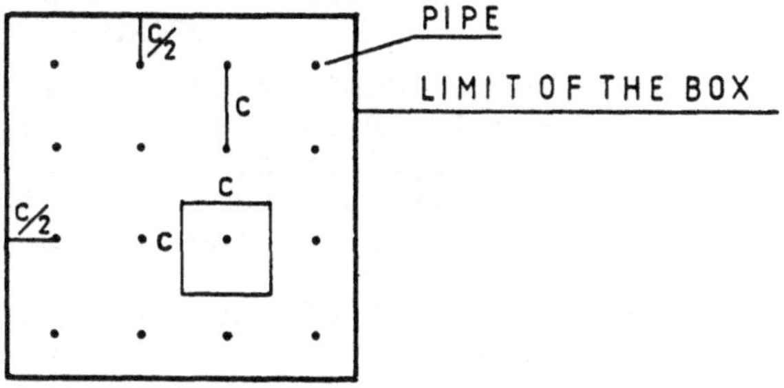

Figure 4 Plan view of a vertical pipe storage system.

Figure 4 shows an example of a vertical duct storage system in plan view.

The distance between two pipes is c. The limit of the box is assumed at a distance c/2 from the outer pipes. In this manner one pipe will affect an area c·c. In the calculations this square is transformed into a circle with the same area (figure 5).

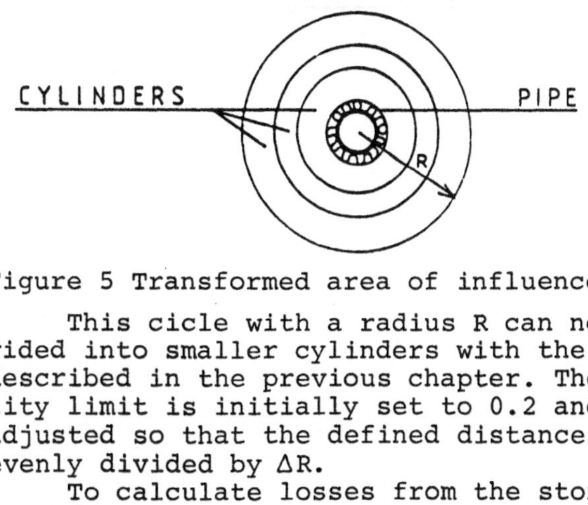

Figure 5 Transformed area of influence

This cicle with a radius R can now be divided into smaller cylinders with the method described in the previous chapter. The stability limit is initially set to 0.2 and the adjusted so that the defined distance R is evenly divided by ΔR.

To calculate losses from the storage the temperature at the radius R is taken as the boundary condition.

<u>BUILDING</u> - Describes the energy demand of a building. The factor KxA (W/k) for the building is provided as a input data.

HEATPUMP - In this subroutine an ordinary heat pump is described. Initially the COP of Carnot is determined with the following formula:

$$\text{COP of Carnot} = \frac{\text{TCOND}}{\text{TCOND} - \text{TEVAP}}$$

where TCOND is the condensation temperature and TEVAP is the evaporation temperature of the regrigerant in Kelvin degrees.

It is assumed that the smalles temperature difference in the condenser and evaporator is five degrees.

Part of the input data to HEATPUMP is the efficiency of Carnot (FOC) defined as FOC = $\varepsilon_{act}/\varepsilon_c$, where ε_{act} is the actual factor of efficiency and ε_c is Carnot factor of efficiency.

ε_{act} = heat from condensor/power to compressor drive unit

ε_c = condensing temperature/(condensing temperature-evaporating temperature)

To get a representative COP for a typical heat pump the Carnot heat factor ε_c is multiplied by FOC.

The power demand of the compressor can be determined by dividing the power demand of the building by COP.

IEA TASK VIII: PASSIVE AND HYBRID SOLAR LOW ENERGY BUILDINGS

M.J. HOLTZ
Operating Agent on Behalf of the U.S. DOE
Architectural Energy Corporation
8753 Yates Drive, Suite 105
Westminster, Colorado 80030 USA

Summary

This paper discusses the background and results achieved to date of an International Energy Agency Solar Heating and Cooling Program Task on Passive and Hybrid Solar Low Energy Buildings. Task objective, approach and participants are presented as well as research results from each of the five major subtasks.

1. INTRODUCTION

The International Energy Agency Solar Heating and Cooling Program undertakes cooperative research, development, demonstration and exchanges of information in order to advance the activities of member countries in the field of solar heating and cooling (1,2). Ten research tasks have been approved by the Executive Committee since December 1976 with seven tasks currently active. In October 1981 the Executive Committee approved Task VIII entitled Passive and Hybrid Solar Low Energy Buildings. Table 1 lists the participating countries.

TABLE 1: Task VIII Participating Countries

Austria	Italy	Sweden
Belgium	Netherland	Switzerland
Canada	New Zealand	United Kingdom
Denmark	Norway	United States
F.R. Germany	Spain	

2. TASK OBJECTIVE

The overall objective of Task VIII is to <u>accelerate the development and use of passive and hybrid heated and cooled low energy buildings in the participants'countries</u>. This will be done, in general terms, by:

(1) Increasing the understanding of the design and performance of buildings using active and passive solar and conservation technologies, the interactions of these technologies and their effective combination in various climatic regions; and

(2) Verifying that passive and hybrid solar low energy buildings can substantially reduce the building load and consumption of non-renewable energy over that of conventional buildings while maintaining acceptable levels of year-round comfort.

The present Task scope includes only new residential single family and multi-family buildings. The participants may, at a later date, choose to expand the scope to include commercial buildings. The current schedule calls for the work to be completed over four years.

A unique aspect of Task VIII is its emphasis on integrated active and passive solar systems, so called hybrid systems, combined with energy conservation techniques and advanced mechanical and electrical systems into optimized energy efficient dwellings. The underlying intent of Task VIII is to gather, organize, evaluate, apply and share the current body of knowledge on passive and hybrid solar low energy building design and construction. As the Task is structured, it combines applied research in performance testing and modeling with design tool development and building design, construction and evaluation.

New research on integrated systems analysis and design will also be conducted during the course of Task VIII activities. The research results are designed to provide the homebuilding industry in each participants' country with the understanding, experience and tools to apply current knowledge of passive and hybrid techniques to current design and construction problems.

3. APPROACH

Task VIII is organized into five major Subtasks each coordinated by a lead country (Lead Country):

Subtask 0: Technology Baseline Definition (U.S.A.)
Subtask A: Performance Measurement and Analysis (Denmark)
Subtask B: Modeling and Simulation (Denmark)
Subtask C: Design Methods (U.S.A.)
Subtask D: Building Design, Construction and Evaluation (Netherland)

Subtask "0" established a technology baseline in terms of cost and performance of a conventional reference building from which to compare the results (improvements) obtained from designs developed during Task VIII research. Also, candidate passive and hybrid solar and energy conservation techniques and cost/performance goals are identified in Subtask "0." Subtask "A" collects, analyzes and reports performed data on passive and hybrid solar techniques for comparison against simulation results (validation studies of Subtask "B") and for characterizing system performance parameters. Common measurement procedures for monitoring passive and hybrid solar low energy buildings are to be developed within this Subtask.

Subtask "B" will evaluate the capabilities, accuracy and limitations of current building energy analysis simulations and conduct parameter studies to understand the design and integration issues of passive and hybrid systems. The results of this Subtask are used in preparation of the design handbook. Subtask "C" will survey and evaluate available passive/hybrid design tools, develop new or improved existing design tools and prepare a Passive and Hybrid Solar Low Energy Building Design Handbook. Subtask "D" will design, construct and evaluate a state-of-the-art passive and hybrid solar low energy building(s) based upon the research results and information obtained from Subtasks "O," "A," "B," "C."

The primary results of the Task are to be:

(1) IEA design and construction handbooks for passive and hybrid solar low energy buildings;
(2) Exemplary passive and hybrid solar low energy buildings designed, constructed and evaluated; and
(3) Technical reports on performance measurement, design tools, simulation models, validation tests, parameter sensitivity studies, etc...

4. RESULTS ACHIEVED TO DATE

The Task was formally initiated in February 1982 when the first Task coordination meeting was held. Since this time, three experts meetings have been held to review the research accomplished since the previous meeting and to plan future subtask research activities. Results are summarized for each Subtask.

4.1 Subtask O: Technology Baseline Definition

Three activities have been accomplished under this subtask. First, reference buildings have been documented by each Participant to be the basis of comparison for passive and hybrid solar low energy building designs developed during the Task. In this manner, improvements in energy efficiency and energy consumption of the passive and hybrid building can be evaluated against the conventional reference building. The reference building is to be "typical" of standard design and construction practice and represent the largest housing type of this country.

Second, the climates of representative cities within each participating country have been compared through the use of a graphic/analytical technique that evaluates key climatic variables. This heating climate similarity index, developed by Derickson (3), is based on the ratio of solar radiation on a south-facing vertical surface to degree day heating load during the heating season. Figure 1, illustrates the concept for three cities: Madison, Wisconsin; Ottawa, Ontario and Venice, Italy. Assuming a time averaged ratio of VS/HDD for the middle 60% of the heating season. A heating climate similarity based on ±5% range in CSI between cities at the extremes of a group and a mean value for the group as shown in Table 2.

Third, candidate energy saving techniques have been defined by all Participants. These techniques, spanning the full range of energy conservation and active, passive and hybrid solar measures, will be the focus of research in the other Subtasks.

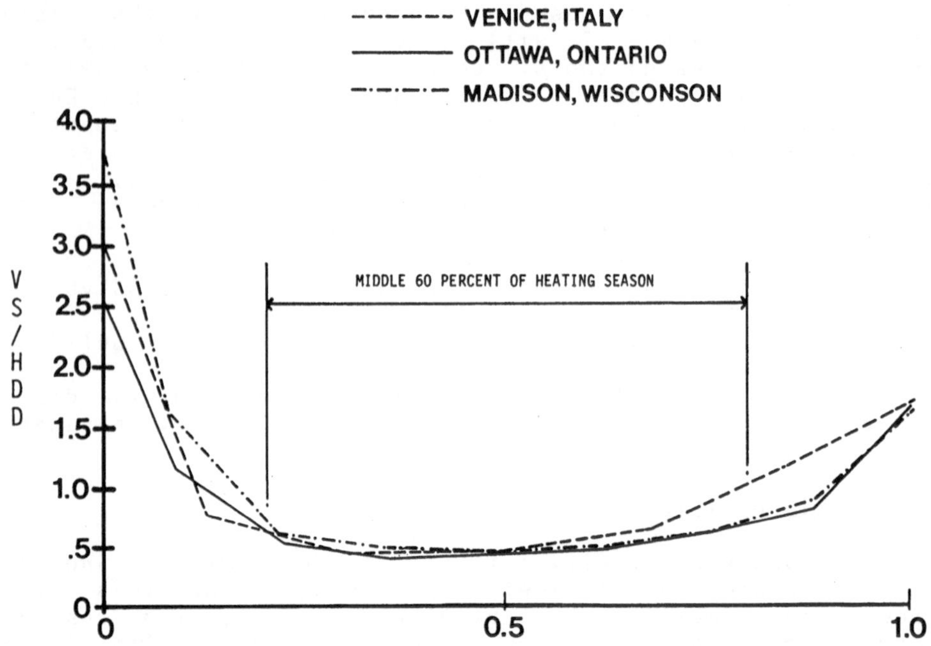

FIGURE 1: VERTICAL SOLAR HEATING DEGREE DAYS VS. NORMALIZED HEATING SEASON FOR VENICE, OTTAWA, MADISON

TABLE 2: CLIMATE SIMILARITY INDEX

CITY	HDD	VS	CSI_H	CLIMATE GROUP
MILAN	1382	457	.34	1
VIENNA	2096	758	.37	1
STOCKHOLM	3008	1114	.37	1
COPENHAGEN	2535	1055	.43	2
DEBILT	1788	758	.43	2
UCCLE	2203	945	.44	2
BERGEN	2803	1205	.44	2
KEW	1933	852	.45	2
SALZBURG	2721	1268	.48	3
ZURICH	2452	1143	.48	3
WINNIPEG	4556	2315	.52	4
OTTAWA	3722	1902	.52	4
VENICE	1266	660	.52	4
VANCOUVER	2059	1092	.54	4
GENEVA	1802	941	.54	4
MADISON	3400	1852	.55	4
EDMONTON	4564	2636	.63	5
WASHINGTON	1370	998	.73	6
ROME	964	944	.98	7
MADRID	1100	1081	.99	7
DENVER	1978	2045	1.04	7
ALBUQUERQUE	1389	1840	1.33	8

ALL VALUES ABOVE CORRESPOND TO TOTALS FROM THOSE MONTHS CORRESPONDING MOST CLOSELY TO 60% OF THE HEATING SEASON FOR A PARTICULAR LOCATION.

HDD: HEATING DEGREE DAYS (18.3°C BASE)

VS: SOLAR RADIATION INCIDENT ON A VERTICAL SURFACE (MJ/M2)

CSI: TIME AVERAGED RATIO OF VS/HDD FOR THE MIDDLE 60% OF THE HEATING SEASON

4.2 Subtask A: Performance Measurement and Analysis

Sources of performance data have been identified and documented for over 200 passive and hybrid solar low energy buildings and test facilities. A summary document is presently being prepared by the Lead Country that includes a description of the building, its energy systems, measured quantities, and data availability. The Participants have also developed a common passive/hybrid system performance evaluation procedure (See Figure 2). The purpose of this procedure is to determine with a relatively high degree of accuracy the energy saved by the passive/hybrid building over a conventional reference building of similar size.

4.3 Subtask B: Modeling and Simulation

Over thirty detailed building energy analysis simulations have been identified and documented during a survey of Task Participants (6). The most significant findings of this survey were: (1) few of the computer programs can model integrated residential energy systems using active and passive solar features; and (2) other than direct gain windows, validation of these programs is extremely limited.

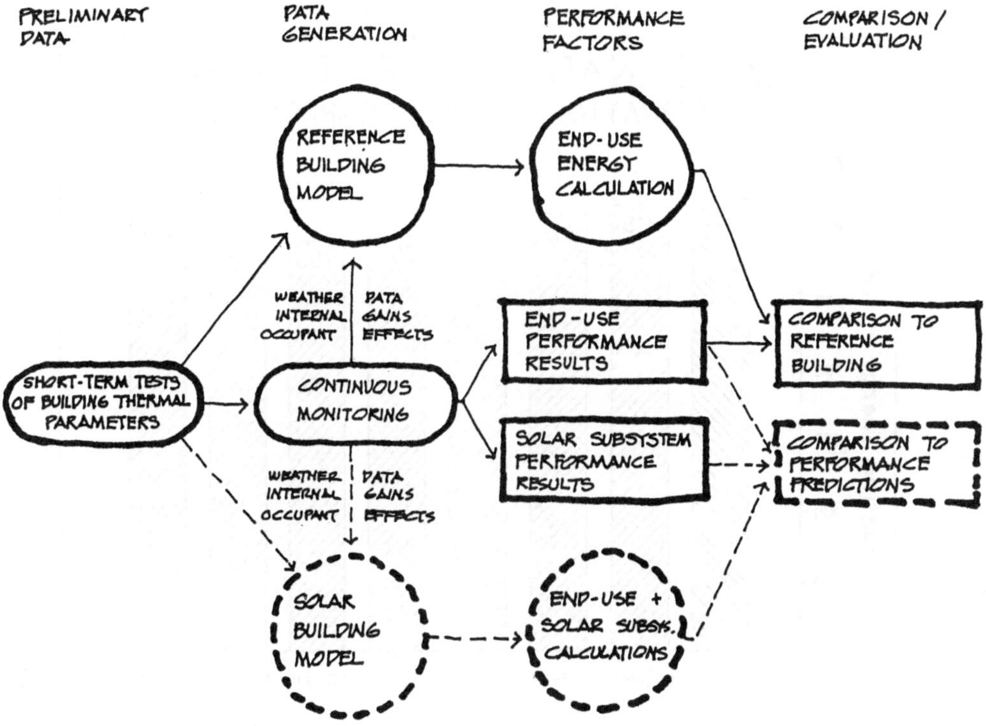

Figure 2: Passive/Hybrid Building Performance

Approximately ten programs were selected for evaluation/validation. Two evaluation approaches are used: (1) comparison to empirical data; and (2) model to model comparison. The first approach involves three data sets - a direct gain house in Ottawa, Ontario, a Trombe wall test room in Switzerland and a sunspace test room in New Mexico, USA. Figure 3 presents results the direct gain house. Measured heating energy consumption of the unoccupied house during the test period was 323 KwH. The seven programs evaluated all came within 10 percent of the measured value with the average of all programs within 6 percent. Comparison of measured versus predicted temperature fluctuations was also made with extremely good results.

The ultimate purpose of the model evaluation/validation effort is to establish a "reasonable" level of confidence in the simulation models to accurately predict changes in performance based on changes in building/system design. Models that achieve this level of accuracy/confidence will be used to perform parametric sensitivity studies on key passive/hybrid design features. Planning of the parametric study has begun. Results are to be available for incorporation into the Design Handbook of Subtask C by the Fall of 1984.

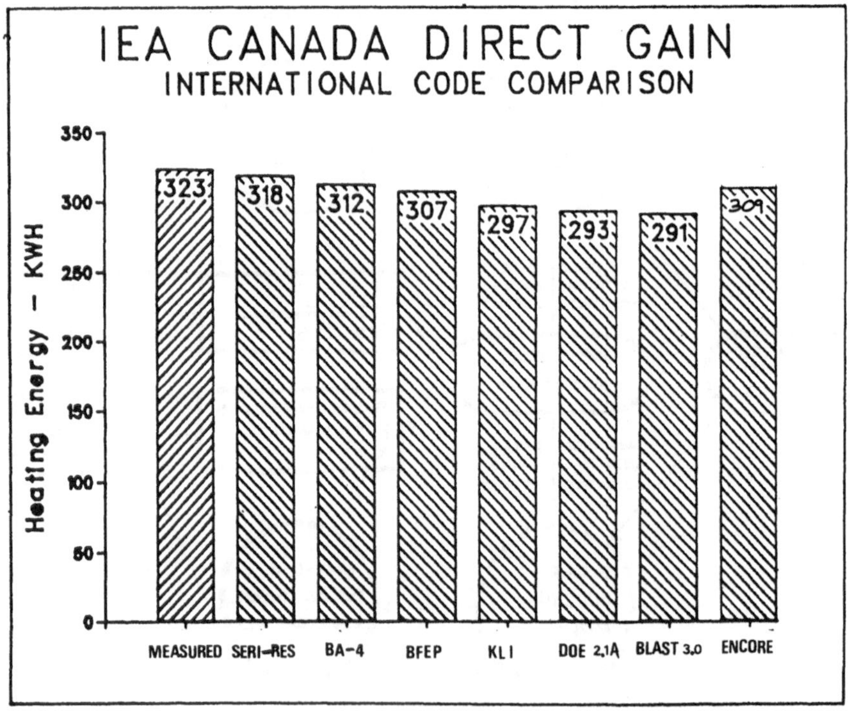

Figure 3: Model Comparison to Measured Data from the Canadian Direct Gain Test Facility

4.4 Subtask C: Design Methods

Over two hundred simplified residential energy design tools have been identified and documented during a survey of Task Participants (7). The survey questionnaire was designed to obtain an overview of each design tool, including phase of design for which the tool is suitable, initial investment, operating costs, ease of use, range of applications, calculation method used, and type/number of inputs required.

The most significant findings from the survey were: (1) a general lack of verification of design tool against detailed models and measured data; (2) very few simplified calculation procedures for cooling analysis; (3) growing trend to use microcomputers as equipment of design tool; (4) very few procedures that use quantitative design guidelines as part of design tool procedure; and (4) great range in cost, complexity and ability to analyze full range of energy conservation and solar measures.

Approximately 25 design tools were selected by the Participants for evaluation/verification. The evaluation is to concern the user characteristics of the design tool as well as its technical accuracy. Two test homes were documented and analyzed for three climates (Denver, Colorado; Geneva, Switzerland; and Copenhagen, Denmark). These data packages were distributed to the Participants along with standard evaluation forms. The design tools were applied to these data packages and the results recorded on the evaluation forms. Preliminary results are available and are undergoing analysis. Consequently, no findings or conclusions can be reported at this time.

The preparation of a Passive and Hybrid Solar Low Energy Building Design Handbook has been initiated. The Design Handbook wil be a combination of international/universal principles and concepts and of national/specific design guidelines and evaluation procedures. The intent is to prepare a Handbook that addresses residential energy design integration questions and does not duplicate available reference documents, such as the CEC Passive Design Handbook.

The results from the parametric sensitivity study of Subtask B will be included in the handbook as quantitative design guidelines. Also, new or improved design tools emerging from the design tool evaluation/development activities will be part of or referenced in the handbook. The current schedule calls for having a very rough first draft of the Handbook available in the Fall of 1984.

4.5 Subtask D: Building Design, Construction, and Evaluation

This Subtask is to be the culmination of the ideas and research results from the other four Subtasks. Based on the new state of the art of passive and hybrid solar low energy buildings, exemplary integrated residential energy designs are to be developed, constructed and evaluated. Currently work is in progress in seven countries to develop preliminary designs. The results from parametric sensitivity study will directly benefit the designs conceived under this Subtask. The projects range in size from single family detailed homes to low rise apartment buildings. Careful documentaion of the design and construction process is being maintained so that the lessons learned from this project can be summarized in the Design Handbook, a Construction Handbook or separate technical reports.

The evaluation procedure developed in Subtask A will be used to evaluate the performance of the completed projects. The thermal and economic performance of the passive and hybrid solar buildings will be better understood at the completion of this Subtask.

6. CONCLUSION

IEA Task VIII represents a bold, innovative experiment to link, both nationally and internationally, solar researchers with design practitioners for the purpose of developing and accelerating the use of an important energy technology. To date, the program has been very successful and this researcher-designer partnership has proved of benefit to both parties. It is vital to the long term energy savings of all participating countries that this partnership remain after the completion of this Task.

7. ACKNOWLEDGEMENTS

The author is indebted to Dr. Fred Morse, Mr. Robert Holliday, Ms. Margaret Jenior and Ms. Janet Neville of the U.S. Department of Energy for their support of this program.

8. REFERENCES

1. Implementing Agreement, Solar Heating and Cooling Programme, International Enrgy Agency, December 1976.
2. Holtz, M.J., International Energy Agency Solar Heating and Cooling Programme: Task VIII, Passive and Hybrid Solar Low Energy Buildings, Proceedings of International Solar Architect Conference. Cannes, France; December 1982.
3. Derickson, R.G., and Claridge, D.E., "Combinations of Passive Solar and Energy Conservation Measures That Provide Balanced Heating and Cooling Performance in Residential Design: A Look at the Full Range of U.S. Climates," Proceedings of 6th Miami International Conference on Alternative Energy Sources, Miami Beach, Florida, December 12-14, 1983.
4. Jones, R.W. (Editor) Passive Solar Design Handbook: Volume Three, Los Alamos National Laboratory, April 1982.
5. Derickson, R.G., Simplified Home Energy Design Manual, unpublished research results, July 1981.
6. Jorgensen, O., Analysis Model Survey Thermal Insulation Laboratory, Technical University of Denmark, Bldg. 118, DK 2800 Lyngby, Denmark, , 1983.
7. Burt Hill Kosar Rittelmann Associates, Design Tool Survey, Unpublished Working Document, October 1982.

STATISTICAL METHODS FOR THE STORAGE SIZE OPTIMIZATION IN SOLAR HOT WATER SYSTEMS

T.H. Schucan and P. Scherer
Swiss Federal Institute for Reactor Research
CH-5303 Würenlingen, Switzerland

Summary

The optimal size of the short term storage in a solar hot water system depends significantly on meteorological data. The crucial link connecting these data with the properties of the collectors, the heat demand and the storage size, is the amount of backup energy per year which has to be secured from secondary sources. We propose that this connection can be formulated in the most concise way by the introduction of statistical concepts. In a first step we derive a simple analytical formula relating the backup energy per month to the storage size and to the mean values and variances of the daily net heat input characteristic for this month. The meteo data of 8 years have been used to derive those quantities and the daily energy demand for hot water has been assumed constant. In the second step the values of these backup energies summed over the whole year have been validated by comparison with those obtained from a day-to-day numerical simulation. An agreement within $\leq$ 6 % has been found in most cases. By introducing the autocorrelation functions describing the memory of the weather, we have further improved this agreement.

1. INTRODUCTION

The efficiency of solar heating systems is depending on various properties of the weather. Whereas the gross heat output of a collector is related to the mean value of the incident radiation, the behaviour of a short term storage device is determined by the variances of both incident radiation and heat demand representing the day-to-day variation of the meteorological conditions and of the users behaviour. It has been suggested in Ref. (1) that these connections should be described with statistical methods. In this paper we demonstrate the efficiency of this approach by applying it to the size optimization of a solar domestic hot water system.

2. STATEMENT OF THE PROBLEM

The heat balance of a storage system on day i is given by the difference s_i between the gross heat output p_i of the collectors and the heat demand d_i. The heat content of the storage changes by this amount until it happens to be full or empty. In these cases excess heat cannot be stored or backup energy from other sources has to be provided to satisfy the demand. The backup energy needed per year thus depends both on the daily fluctuations of the net heat input s_i as well as on the storage size K. The optimal size is then determined by the relation between the storage cost and the backup energy price.

In this paper we focus the attention on the connection between the backup energy per year and the daily fluctuations of the gross heat input in to the storage tank which are caused by meteorological data and by collector parameters. We base our calculation on global irradiation data measured over 8 years (1963 - 1970) at three locations in Switzerland (2) and on the corresponding values obtained for the gross heat output (3) with a whole range of collector parameters (optical efficiency, thermal loss factor, mean collector temperature, azimuth and tilt angle). Since a similar set of data for the daily demand of hot water is not yet available to us, we approximate it by a constant value d. Thus the two main questions to be answered are the following:

1) How does the backup energy per year depend on the meteorological data, on the collector parameters and on the storage size?
2) Is it possible to establish a simple formula describing this connection with a very limited set of parameters?

Starting from the data mentioned above the first of these questions can be answered by straightforward day-to-day bookkeeping calculations over the whole period of eight years for any desired combination of collector parameters and storage sizes. It is by providing the answer to the second question that the statistical methods described in the next chapter display their full force.

3. THE STATISTICAL APPROACH

3.1. Description of the concepts for a storage with infinite capacity

The statistical concepts used in this paper have been described in Ref. (1). We recall some of the basic notions in this section. The first of these is the function f (E, t) representing the probability that the heat content of the storage at the time t is given by E. For a storage with infinite capacity a diffusion equation for this function can be obtained from the energy balance condition. The solution of this equation is given by a simple Gaussian function. Since we want to establish a connection between the function f and the meteorological data we replace the continuous description by a discrete one. From the laws of combination of probabilities we obtain the relation

$$f\,(E,\ i+1) = \int_{-\infty}^{+\infty} f(E',\ i)\ \phi\ (E-E')\ dE' \qquad (1)$$

between the probability functions for two consecutive days. With $\phi(E)$ we denote the probability that the heat content at the storage is increased by E in one day.

In the next step we assume that the daily net heat inputs s_i are a sample of a stationary stochastic distribution. We have verified that this is a reasonable assumption provided that the data for each given month are averaged over the whole period of observation. The corresponding probability distribution can thus be given as

$$\phi(E) = \frac{1}{\sqrt{2\ \pi\sigma^2}}\ e^{-(E-m)^2/2\sigma^2} \qquad (2)$$

where the mean values m and the variances σ^2 are characteristic numbers for each month. Inserting this function into the recursion relation (1) we find that the probability distributions f (E, i) are given by Gaussian functions with mean values and variances increasing by m and σ^2, respectively, on each day.

3.2. Stationary solution for finite size storage

The boundary conditions for a storage with capacity K can be taken into account by replacing the recursion relation (1) by

$$f(E,\ i+1) = a_i\ \phi(E) + b_i\ \phi(K-E) + \int_0^K f(E',i)\ \phi(E-E')\ dE' \tag{3a}$$

with $a_i(b_i)$ denoting the probabilities for the storage to be empty (full) on the day i. These probabilities are given by

$$a_i = \int_{-\infty}^{0} f(E,i)dE \qquad\qquad b_i = \int_K^{\infty} f(E,i)dE \tag{3b}$$

and the backup energy needed on the same day is obtained as

$$E_B(i) = - \int_{-\infty}^{0} E\ f(E,i)\ dE \tag{4}$$

Starting with an initial distribution the time evolution of the probability function f(E,i) can again be obtained from these recursion relations. In contrast to the case of the infinite storage these functions approach a stationary distribution after some time. This distribution does not depend on the initial conditions and once it is reached it will not change any more by further application of the recursion relations (3). It is determined by the parameters m and σ of the daily input distribution (2) and by the storage capacity K.

By numerical solution of the recursion relations (3) for a large variety of values m and σ we found that the resulting stationary distributions can be very well approximated by

$$f^{stat}(E) = N \cdot e^{m(2E-K)/\sigma^2} \left[\Phi\ (\frac{E+\alpha}{\beta}) + \Phi\ (\frac{K-E+\alpha}{\beta})\right] \tag{5}$$

The exponential function is the solution of the diffusion equation mentioned above in the stationary limit and the error functions represent the influence of the finite size of the storage. The form of eq. (5) guarantees that all the appropriate symmetries and limiting conditions are satisfied. It contains two additional parameters α and β.

An approximate value for the backup energy per day can be obtained by inserting the stationary distribution (5) into eq. (4). The parameter β may be eliminated from the resulting expression with the aid of the very simple condition that the backup energy has to be equal to the daily demand on a day without any sunshine. In addition we found that $\alpha = 0.581$ is an excellent choice for all cases which we have considered. Combining these results we find the simple expression

$$E_B^{stat}(m,\ \sigma) = m \cdot n \cdot \left[e^{2m(1.162\sigma+K)/\sigma^2} - 1\right]^{-1} \tag{6}$$

for the backup energy needed in a month with n days. The results obtained with this formula agree extremely well (deviations less than 0.1 % in all cases) with those obtained from the numerically determined stationary distributions. As this formula connects the statistical parameters of the daily input distribution in a very simple way with the storage size K, it is essentially the answer to the second of the questions raised in chapter 2.

4. RESULTS

In order to check the applicability of the stationary approximation described above, we have compared the monthly backup energies obtained with eq. (6) with the results of the day-to-day numerical simulations mentioned at the end of chapter 2 for the three locations Zurich-Kloten, Locarno-Monti and Davos and for a variety of collector parameters and storage sizes. In Table I we display an example of the results obtained for Kloten in a typical case. It is seen by comparing the results given in the first two blocks that the backup energies per year obtained with the two methods agree with each other within 6 %. For storage sizes larger than those given in the table the agreement is better as a consequence of the right asymptotic behaviour built into eq. (5).

Table I: Backup energies in kWh for storage sizes K = 1 - 3 (measured in multiples of the daily demand)

K	Jan	Feb	Mar	Apr	Mai	Jun	Jul	Aug	Sep	Oct	Nov	Dez	Year
Numerical simulation													
1	368	197	117	55	21	16	4	19	26	128	285	404	1640
2	368	189	91	29	8	7	0	9	7	106	278	404	1496
3	368	188	78	21	3	2	0	5	2	88	272	404	1431
Statistical calculation from m and σ													
1	365	181	93	46	20	6	3	17	22	104	281	404	1542
2	365	176	63	20	5	1	0	4	6	77	281	404	1402
3	365	175	48	9	1	0	0	1	2	64	281	404	1350
Statistical calculation in checking 2-day autocorrelations													
1	365	185	116	72	35	15	8	31	36	124	281	404	1672
2	365	177	80	37	12	3	1	10	14	92	281	404	1476
3	365	175	61	20	5	1	0	3	5	75	281	404	1395

These results have been obtained for a typical collector array ($A_o = 0.8$, $K_o = 3.5\ W/m^2K$, $T_m = 40^oC$, $\gamma = 0^o$, $\alpha = 45^o$) in the meteorological conditions of Zurich-Kloten and for a daily energy demand for hot water of 14.5 kWh.

A further improvement of the agreement between the simulation results and those of obtained with the statistical concept is obtained if the latter is extended to include the effects connected with the memory of the weather. As shown in Ref. (1) this may be achieved by introducing the autocorrelation functions and by a corresponding modification of the variances. With this extension the same formula eq. (5) can again be used to obtain improved expressions for the yearly backup energies. The improvement obtained with two-day autocorrelations is shown in Table I as well.

In calculations performed for many different cases we found a convenient parametrization for the autocorrelation functions. In particular we found that they can be represented by multiplying the squared gross heat input with a correction factor depending on the location only but not on the values of the collector parameters. This is a considerable simplification for practical purpose. The results of this investigation and their application to practical situations are reported elsewhere (4).

5. CONCLUSIONS AND OUTLOOK

By using statistical concepts we have been able to find an answer to the two central questions raised in chapter 2. A simple connection has been established between the backup energy per year, the storage size, and the mean values and variances of the net heat input. In the example of a hot water system with constant daily demand considered in this paper these variances equal those of the gross heat input. By using the parametrization mentioned at the end of the last chapter the calculation of these variances can be considerably simplified and the autocorrelations may be easily incorporated. The resulting backup energies have been successfully compared with those obtained from numerical simulations. The optimal storage size can then be determined by comparing the backup energy costs with the investment costs needed for the construction.

The method used in this work can be extended to include the case of variable heat demand and to solar energy systems used for both hot water and space heating. In both cases the variances of the heat demand have to be added to those of the gross heat output. In the second case the cross-correlation between gross heat output and heat demand has to be considered in addition, which is mainly due to the strong link between those quantities through the outdoor temperature.

ACKNOWLEDGEMENTS

The authors are very grateful to Dr. P. Kesselring for the stimulation of this work and for many helpful discussions. The work has been performed under the auspices of the Swiss Solar Energy Committee (KNS) and financed by the National Energy Research Foundation (NEFF) and by the Federal Energy Office (BEW). We would like to thank all those institutions for their support.

REFERENCES

1. P. Kesselring and A. Duppenthaler, "The layout of solar hot water systems, using statistical meteo- and heat demand data", contribution to the ISES 1979 International Congress on Solar Energy, May 28 - June 1, Atlanta (Georgia), USA.
2. We are grateful to the Swiss Meteorological Institution (SMA) Zurich for providing these data.
3. P. Ambrosetti, "Das neue Bruttowärmeertragsmodell für verglaste Sonnenkollektoren", EIR Würenlingen, 1983 (ISBN 3-85677-012-7).
4. P. Scherer, P. Kesselring and T.H. Schucan, "Einfache Auslegung von Kurzzeitspeichern solarer Anlagen auf Grund statistischer Methoden", EIR Würenlingen, 1984

SESSION III - RESEARCH AND DEVELOPMENT OF PASSIVE SOLAR COMPONENTS

INNOVATIVE PASSIVE/HYBRID SOLAR DESIGN CONCEPTS EMPLOYED IN THE U.S.

J. Mewshaw and S.B. Blum: International Planning Associates
5010 Sunnyside Avenue, Beltsville, MD USA, 20705
M.J. Holtz: Architectural Energy Corporation
8753 Yates Drive, Westminster, CO USA, 80030

Summary

A survey of leading U.S. solar designers and a review of recent literature conducted for the International Energy Agency's Passive Solar Task revealed that designers are experimenting with a variety of new techniques and designs such as creative thermal mass, active collection with passive discharge, novel distribution and control devices, and unconventional building configurations. From the whole building point of view, designers are giving careful attention to the successful integration of all aspects of the energy system. However, a recent survey conducted by the National Association of Homebuilders indicated that conventional builders tend to view passive/hybrid design as separate marketable features. More effective technology transfer between designers and builders; better performance data on solar low energy features and residences; and simplified design tools are needed to help passive/hybrid solar low energy homes enter the mass housing market.

1. INTRODUCTION

The past few years have witnessed an explosion of ideas and creativity in passive and hybrid low energy residential designs in the U.S. Leading solar architects and designers have produced a second generation of passive/hybrid solar homes which encompass both innovative concepts and refinements of standard passive features.

This paper provides an overview of some of these exciting passive/hybrid solar design directions. It also briefly explores the implications of these kinds of innovative approaches as well as other more common designs for the mass housing market. Moreover, the differences between design professionals and conventional homebuilders' approaches to passive solar design and construction is examined.

2. SIGNIFICANT FINDINGS

Perhaps the area which has seen the most exciting innovations is that of thermal storage techniques. Architects are experimenting with creative designs for mass walls which enhance their aesthetic character while providing daylighting and views.(1) For example, a "floating mass wall" employed in the Lucas residence utilizes glass blocks within concrete slump blocks. The storage mass wall in the Kairath residence incorporates a combination of clear and stained glass which allows some direct gain.

A unique louvered mass wall in the Franta residence consists of four rectangular two-story columns with their horizontal axes pointing southeast. This design combines direct and indirect gain systems by allowing early morning sun to penetrate into the living areas while the afternoon sun is absorbed by the columns and serves as a source for night radiant heating.

A special insulated glazing, "Heat Mirror^R," is installed between the louvers to minimize heat loss during the night.

Although many designers question the effectiveness, cost and long-term durability of passive phase change storage, substantial progress has been made towards its successful application. One innovative example, the Brooks residence, employs a motorized array cylinder containing phase change rods which are positioned under a large skylight in the south-facing solarium. Heat is directed to both the front and back sides of the rods with an aluminum reflector and, when needed, is drawn from the storage rods and distributed into three heating zones via the duct system. In the summer, rods function as venetian blinds: the heat absorbed by the reflector travels up the ceiling into the cupola where it exhausts outside.(2)

Water tube thermal storage is becoming an increasingly popular alternative to mass wall storage as designers are integrating the transparent water tubes as structural and visually pleasing elements of the architectural design. In the Dewinkle residence retrofit project, water tubes and a planting bench are incorporated as a divider between the sunroom and living space. The water tubes store the heat while registers in the planting bench help convey it to the living space.(3)

Water storage tubes were used in the Baker residence retrofit in Berkeley, California where there was no opportunity to integrate structural thermal mass. Mass was provided by two 10-ft high (3 m), 18-inch (45.5 cm) fiber glass tubes filled with blue colored water. The translucent tubes are almost luminescent and visitors have mistaken the installation for sculpture.

An interesting variation of the water storage wall is the thermal diode wall utilized in the Sun/Tronic demonstration house. One inch (2.5 cm) copper heat pipes attached to an absorber plate transfer energy to 12 inch (30.5 cm) copper water storage tubes where it is radiated to the living space. An insulated curtain can be drawn over the tubes to reduce heat transfer and the tubes can be partially drained to reduce thermal capacitance.(5)

One of the most innovative approaches to thermal storage in the U.S. is a hybrid system which combines active collection with passive delivery. An example is the Trident System in which water is heated by roof-mounted collectors and is stored in the solar domestic hot water tanks. When the temperature of the floor slabs drops below the thermostat setting, water from storage is distributed to the slab through embedded polybutelene tubing. The slab discharges heat to the room by radiation and convection.(6)

Passive solar designers are also experimenting with a number of unique movable insulation concepts. In the above-mentioned Kairath residence, the Trombe walls are automatically insulated by an exterior sensor with self-inflating curtains consisting of several layers of Mylar^R film. When the curtains are lowered, air naturally rises between each layer, fluffing them apart, thus creating an insulative barrier of several air spaces.(7)

Lightweight translucent movable insulating panels in the Parham residence are used to overcome excessive losses through the south glazing. The uncommon translucency of the double-skinned fiberglass panels offers a tight weatherskin while allowing light to enter the space.(8)

Although this discussion has focused on specific passive/hybrid features, it is important to remember that these features are only parts of the total energy system; their performance depends on the successful integration of all the architectural elements of the building construction. Following are three projects which illustrate the close interrelationship of the total building energy system.

The Earth Lodge, a 140 m^2 single family home located in Connecticut, illustrates the innovative integration of passive, active and auxiliary systems. The passive features include a solar oriented, earth-bermed construction with south-facing windows and patio doors as well as skylights with insulating and shading controls. A rock-bed plenum floor is utilized for storage and radiant heating and cooling. The second floor is constructed with truss-joists, thus opening the flooring for use as an air distribution plenum which heats the second floor through a combined radiant/convection effect. The active system not only provides a majority of the domestic hot water (DHW), but also supplements the passive heating system. Two thermosiphon arrangements are utilized: one was created by installing the hot water tank above the solar DHW collectors; and another by placing a water jacket around the auxiliary stove, thus permitting heat recovery from the wood stove flue. Low-temperature heat in the range of 32-38°C from the hot water tank and woodstove heats the second floor by radiation/convection.(9)

Architect Richard Crowther's 631.5 m^2 solar house and research facility in Denver consists of interactive thermal zones which may be isolated due to the unique cellular nature of the plan. The key thermal zone is the central gallery which collects direct and reflected sunlight and serves as the heat "accumulator" for the house. Excess heat is destratified by a fan-and-duct system that delivers it to the thermal mass storage room in the basement. The home office and the living, dining, breakfast and kitchen areas are heated by direct gain strategies and may be theromodynamically isolated from the rest of the house, if necessary. The bedrooms are heated by thermal energy collected and stored in the central gallery, a solar-heated water tank, a sauna, and south-facing windows. A concrete-covered earth berm and a west-facing sunroom act as climatic buffers for the bedrooms. A "solar chimney" in the master bedroom is used in the summer to induce ventilation and exhaust hot air. The lower-level thermal mass storage room and the sauna act as primary, well-insulated heat sinks for the building. A south-facing greenhouse heats the recreation spaces adjacent to the solar-heated pool which, in turn, helps heat the living room located above.(10)

The Van Teeckelenburgh residence in New Jersey represents an exeptional deviation from the conventional passive site plan because it has two interacting solar orientations (southeast and southwest), permitting increased solar gain to offset the onsite shading and enhance the living space with two views. The 162 m^2 home tracks the path of the sun throughout the day with alternative time-clock thermal sensors at each orientation which automatically insulate the glass wall not receiving significant solar gain. Phase change material mounted on the ceiling provides heat storage (up to 74,000 Btu/day) for cloudy day carryover situations and also reduces the average summer interior temperature. All interior living spaces are designed to include both natural and fan-assisted cross ventilation.(11)

3. DISCUSSION

What are the prospects for these advanced, innovative designs in the mainstream housing market? How well do they perform? Are these approaches which can be easily replicated by conventional builders? Will some of these concepts become part of future passive "prototype" designs?

Definitive answers to these questions are not yet available. First of all, there is insufficient performance data and a lack of effective design tools for most of these advanced concepts. Despite the fact that designers and their clients report excellent performance, the actual cost effectiveness and net energy contribution of the low energy features remain unclear since their selection and sizing are generally made on an intuitive basis and their performance has not been formally monitored. Builders, however, need detailed, reliable performance data and guidance on designing and building good passive homes.

The prospects for replicability of any type of effective passive design is also affected by the difference in approach of builders and designers to passive solar buildings. Experienced design professionals tend to focus on an integrated system approach to passive and hybrid low energy residential design. Their aim is to balance all elements of the building and its total energy system, taking into account the climatic impacts to which the building is exposed throughout the year as well as client preferences and aesthetics. Since designers are generally employed to produce custom plans for expensive homes, they have the freedom to pay considerable attention to system integration and quality.

Conventional homebuilders, on the other hand, seldom employ designers and architects. In responding to the growing market demand for energy efficient and passive solar homes, builders often treat solar as a collection of marketable features, with little understanding of their effect on heating and cooling loads. Thus, many builders indiscriminately add such features as large expanses of south-facing glass without adequate storage, distribution, or control mechanisms. The result, of course, is poor performance, a disappointed owner, and a bad name for passive solar. The challenge facing designers and builders, then, is to fully integrate energy-related design features with current builder practices from both the technical and marketable perspectives.

The situation is further complicated by the absence of a "prototype" passive house which would serve as guidance to builders who want to construct passive houses on a widespread scale. There is no single optimum design approach due to the complexity of designing for different climates, sites, building styles, and market conditions. If good passive solar design is to enter the conventional housing market on a widespread basis, the designs must be reasonably consistent with current builder practice. Builders need information which focuses on those designs which are likely to have the highest chances for success in the mainstream market, i.e., those which are simple, easy to implement, cost-effective, and marketable.(12) Thus, the applicability of some of the designs discussed earlier is questionable since their cost and complexity may make them unlikely candidates for the mass housing market.

Indeed, many of the prominent solar architects and designers surveyed have come to believe that passive will move into the mass market with simple, modest, inconspicuous features such as suntempering, radiant release slabs, and mass incorporated as part of the decor. A frequently heard theme was that "glass and mass" is evolving to "light and tight"--superinsulation and other energy conservation features combined with suntempering. Many designers indicated they are avoiding features they consider too gimicky or exotic, such as phase change materials, rock beds, and double envelope house design. These designers believe that passive design will make the transition into the mass market only if the solar technologies are unobstrusive and place few demands on the occupants. They are therefore emphasizing tried and true passive features combined with good whole building design.

4. CONCLUSIONS

- Many innovative and advanced passive/hybrid residential design concepts have been developed in the U.S., especially in the areas of thermal storage mass, distribution, and control devices. The suitability of these innovations for the mass housing market remains unclear.
- The natural focus of passive solar commercialization falls upon specific products or building components, but it is the successful integration of all aspects of a building's energy system (passive/hybrid features, conventional HVAC, auxiliary systems and climatic conditions) which results in low energy, cost effective residences.

If passive/hybrid residential design is to enter into the marketplace, certain steps need to be taken:

- There must be more effective transfer of information between the researchers, designers and homebuilders on how to design and build economical, well-integrated passive/hybrid solar low energy homes.
- Performance data must be collected and analyzed to help the researchers, designers and homebuilders understand the overall energy performance of passive/hybrid homes and the interaction of the components.
- Simplified, user-friendly tools for the design and construction of passive/hybrid low energy residences must be developed.
- Designers and builders must work together to develop regional prototype low energy designs which take into account factors such as climate, indigenous construction materials, standard construction techniques, and simple inconspicuous passive features which can be easily incorporated into the home.

ACKNOWLEDGMENTS

Special thanks to the following designers for providing information on their projects: Gregory Franta, ENSAR Group, Lakewood, CO (Lucas residence, Franta residence); Peter Dobrovolny, Sun-Up Ltd, Snowmass, CO (Kairath residence); Robert Hicks, Grand Rapids, MI (Brooks residence); Don Schramm, Prado, Madison, WI (Dewinkel residence retrofit); David Baker, David Baker and Assocs., Berkeley, CA (Baker residence retrofit); Barry Berkus, Berkus Group Architects, Santa Barbara, CA (Sun/Tronic house); Michael Anderson, Trident Energy Systems, Davis, CA; John Alt, North Carolina (Parham residence); Donald Watson, FAIA, CT (Earth Lodge); Richard Crowther, Crowther Solar Group, Denver, CO (Crowther residence); M. Stephen Zdepski, NJ School of Architecture, NJ (Van Teeckelenburgh residence).

REFERENCES

1. Gregory Franta and Steven Hogg, "Creative Interior Designs for Thermal Storage Walls," Proceedings American Solar Energy Soc., 8th National Passive Solar Conference, 1983, pp. 519-521.
2. Phone conversation with Robert Hicks, designer of Brooks residence, July 29, 1983.
3. David Wright and Dennis Andrejko, Passive Solar Architecture: Logic and Beauty (New York: Van Nostrand Reinhold 1984), pp. 100-101.
4. Wright and Andrejko, p. 194.
5. Robert Hedden, "CDA Project Showcases Energy Saving Options," Heating/Piping/Air Conditioning, October 1980, p. 4.
6. Joe Kohler, "Active Collection with Passive Delivery," Solar Age 8, January 1983, p. 26.
7. Wright and Andrejko, p. 161.
8. Wright and Andrejko, p. 72.
9. Phone conversation with Donald Watson, FAIA, designer of Earth Lodge, July 13, 1983.
10. Richard Rush, "This Building is Loaded," Progressive Architecture, April 1980, pp. 150-154.
11. M. Stephen Zdepski, "The Van Teeckelenburgh Residence," Proceedings American Solar Energy Soc., 8th National Passive Solar Conference, 1983, pp. 1009-1013.
12. Paul Kando, "How Builders View Passive Housing," Solar Age 8, September 1983, pp. 16-17.

BUILDING COMPONENTS WITH INTEGRATED LATENT HEAT STORAGE

G. Pellegrini and R. Colombo
Commission of the European Communities
Joint Research Centre - Ispra Establishment
21020 Ispra (Va) - Italy

Summary

In many climates the fluctuation of indoor air temperature in passive solar buildings can be reduced by an appropriate design of the thermal mass. Consequently, there is a potential for the reduction of the annual energy consumption. An optimization of the thermal mass can be achieved using advanced storage concepts like latent heat salts and implying that the storage component is an integral part of the building envelope, as e.g. part of the ceiling, floor and walls. Latent heat storage materials are being considered particularly attractive because of their relatively high storage capacity and because of the isothermal nature of the storage process which improves living comfort. This opens new economic possibilities for the climatization technology. Here an overview of the most recent developments concerning the design and the construction of building components with integrated latent heat storage is presented.

1. INTRODUCTION

Architecture should aim at fulfilling the basic human requirements for comfort by providing an indoor environment which mitigates climatic extremes at reasonable costs. This is achieved by utilizing optimally the natural renewable energy available for heating and cooling. Research and testing is going on in this area along two lines:

- the study of a proper design based on the concept of integrated passive systems for heating and cooling of the building;
- the improvement of thermophysical properties of building materials.

1.2 Integrated Systems

The term integrated means that various elements of the building such as for example the roof, walls, window, porches, participate in the various functions of the energy system, namely the collection, storage, transport and distribution of thermal energy. In this way the whole building itself is considered as an energy system and the architectural and structural details become integral parts of it. The main classes of integrated natural energy systems in buildings are: the direct heating by penetration of solar radiation, sun-porches and greenhouses as solar collectors, southern walls as collector-storage components, roofs for solar absorption and storage and southern slopes with gravel or water storage /1,2/. Concerning the improvement of the building material properties, three thermophysical features are of primary importance for energy conservation and living comfort, namely: the thermal resistance (R), the heat capacity (C) and the solar absorptivity (A) of the surface. High thermal resistance of the building envelope minimizes the heat flow by conduction, while a thermal capacity has the effect of stabilizing the indoor day- and nighttime temperature for large passive solar systems, especially in mild continental climates.

2. INTEGRATED LATENT HEAT STORAGE

A problem common to all integrated passive systems is that of the storage. Integrated wall or roof storage systems operate at low temperatures, since the collection temperature is limited by technical as well as by human comfort requirements. As a result the temperature range of the thermal storage within the structural mass is very limited, a factor which limits the storage capacity as well as the overall storage efficiency.

This problem can be minimized using latent heat storage. Latent heat storage materials are interesting because of their high storage capacity (up to 3-4 times that of water) over a relatively narrow range of temperatures. Furthermore because of the isothermal nature of the heat storage and the release process, the living comfort in buildings is better. The calorimetric data of a typical charge-discharge process in a salt hydrate mixture is shown in Fig.1 /3/. The narrow temperature range, in which most part of the storage process takes place, and the high storage capacity per unit of weight are clearly indicated on the diagram. The high thermal capacity of the latent heat storage material combined for example with a high thermal resistance of its envelope, yields the high thermal time constant (TTC) suitable for continental climates.

The major drawbacks for an expanded use of these materials are the relatively high costs of the encapsulation, the degradation of the storage process, corrosion problems in the container material and cycling lifetime.

2.1 Selected Latent Heat Storage Materials

The transition temperature of the phase change material is determined by the desired application /4,5,6/. Suitable temperature regions are: 35-55°C for hot air, 20-35°C for wall panels and 5-20°C for cold storage. For thermal energy storage in passive systems and for climatization purposes, the temperature range of interest is 20-35°C and preferably 20-25°C. This restriction limits the available choice to a few inorganic salt hydrates and some organic systems such as e.g. $CaCl_2.6H_2O$ (29°C), $Na_2SO_4.10H_2O$ (32°C), decanoic acid (31.4^{o}C), octanoic acid (15°C) and eutectic mixtures of the above. The organic materials are inflammable, and more expensive than the inorganic salts. However they exhibit less supercooling and are not corrosive.

2.2 Encapsulation

Since the most promising way to store thermal energy is in the building structure, encapsulation techniques are being explored which will incorporate the latent heat storage directly into the construction material (e.g. concrete) or into building components for floors, ceilings, roofs, structural walls and partition walls.

Polymer concrete blocks and tiles: Concrete has a low average specific heat, less than one fourth that of water (0.22 cal/gr.^{o}C) /7/. The incorporation into concrete, e.g. of 20% of an active phase change material with a latent heat of fusion of 35 cal/gr and a transition temperature of 25°C, raises the heat capacity from 2.2 cal/gr to 9.0 cal/gr over the temperature interval 20-30°C. Polymer concrete is a concrete in which ordinary cement is replaced by a polymer compatible with the phase change material. Highly porous foamed glass beads impregnated with the phase change material constitute part of the concrete aggregate. Polymer concrete blocks with incorporated latent heat storage have been designed as structural elements in the basements of buildings with central heating or air conditioning, as bricks for passive solar south walls, as floor blocks for the storage of off-peak electric heat and as ceiling tiles in rooms with southern exposures /8,9,10/.

Macroencapsulation: Various methods are being explored for packaging phase change material in a sealed container, which itself acts as the heat exchanger. Size and shape of the sealed container vary depending on the storage concept and application. Examples are: translucent fiberglass-reinforced plastic pods /11/, plastic coated aluminium pouches /12/, polypropylene spheres /13/, plastic rods /14/, blistered polyethylene sheets /15/ and lattices of plastic capsules /16/.

2.3 Wall Structures

The incorporation of latent heat storage modules into southern walls has in general the double function of increased storage capacity of solar energy and of mitigating the temperature oscillations of the back-coupled sun space, thus providing improved space conditioning.

Transparent fiberglass-reinforced plastic pods operating at 27°C are utilized in solar warehouses along the south wall /17/. A latent heat storage system integrated in a south wall and consisting of a stack of plastic coated Al-pouches filled with Glauber's salt mixtures, is being studied in a test cell for air conditioning (Fig.2) /18/. Wall panels consisting of a lattice of plastic spheres filled with a phase change material are being developed for ventilated wall structures (Fig.3) /16/. The use of spherical particles containing a phase change material has been proposed for ameliorating the natural air circulation in modified Trombe Wall-structures /19/.

2.4 Ceiling- and Floor Components

Polyester concrete tiles containing two thin layers of phase change material have been designed to replace the sheetrock or the acoustic tiles normally used for ceilings /10/. A relatively cheap (but mechanically weak) package of the phase change material is that of plastic coated aluminium pouches, 20 cm large and 2 cm thick, developed and tested at the MIT and Cabot Corporation (USA).

Special attention is now being devoted to the storage of "cheap" off-peak electric heat during the night in floor structures. Structures of this type have to combine a high strength with good heat exchange perperties. An example of the most recent development of this is shown in Fig.4 /16/.

3. CONCLUSIONS

Phase change materials with a transition temperature between 20 and 35°C have received particular attention for passive heat and cold storage in building structures, especially for climatization purposes. Important efforts have been made to encapsulate these products into construction materials (bricks and tiles) and storage components to be incorporated in building structures such as south walls, partition walls, ceilings and floors.

Considerable progress has been also made in the design of building components with integrated absorption/storage systems to be produced on an industrial scale.

There is, however, a need of improved and more extensive testing in order to evaluate the thermal performance of such building components, particularly concerning the long term efficiency in real field conditions.

4. REFERENCES

1. GIVONI, B. (1981) Conservation and the use of integrated passive energy systems in architecture. Energy and Buildings 3, 213-228.
2. ARANOVITC, E. Editor (1979) Design and technology of solar heating and cooling systems for buildings. Comm. of the European Communities, Ispra Courses.
3. REITER, F. (1982) Commission of the European Communities, JRC Ispra Establishment, unpublished work.
4. LANE, G.A. and GLEW, D.N. (1975) Heat-of-fusion systems for solar energy storage. Proc. Workshop on Solar Energy Storage Subsystems for the Heating and Cooling of Buildings, pp.43-55, Charlottensville, Virginia.
5. ABATH, A. (1983) Temperature latent heat thermal energy storage: heat storage materials. Solar Energy 30, 4, 313-332.
6. SWET, C.J. (1980) Phase change storage in passive solar architecture. Proc. 5th Nat. Passive Solar Conf., pp282-286, Amhearst, Massachussets.
7. NEVILLE, A.M. (1975) Properties of concrete. John Wiley & Sons, New York, p.431.
8. Brookhaven National Laboratory (1978) Process Science Division, Department of Energy and Environment, private communication.
9. Suntek Research Associates, 506 Tamal Vista Blvd., Corte Madera, Colorado 94925, USA.
10. JOHNSTON, T.E. (1976) New building materials and components for passive heating of buildings. Symp. on Passive Solar Heating Systems, Albuquerque.
11. Solar Components Corporation, 88 Pine Street, P.O.Box 237, Manchester, NH 03105, USA.
12. Cabot Corporation and Massachussets Institute of Technology, USA.
13. Cristopia, Place J. Bermond, Sophia Antipolis, F-06560, Valbonne, France.
14. PSI, Thermal Energy Storage Division, Fento, Mo., USA. Energy Materials Inc., Englewood, Co., USA.
15. Thermoform, Lugano, Switzerland.
16. ANTONINI, G. and PAIN, J.P., Division Génie des Transferts et Energétique, Département de Génie Chimique, UTC, B.P.233 (60206) Compiegne Cédex, France, private communication.
17. KELLER, M.B. et al. (1978) Passive solar heated warehouse. Proc. 2nd National Passive Solar Conf., Philadelphia, Pa., March 16-18, 1978.
18. COLOMBO, R. and PELLEGRINI, G. (1984) A hybrid collector-latent heat storage wall for air conditioning. 1st EC Conf. on Solar Heating, Amsterdam, April 30 - May 4, 1984, Poster Session P3.
19. FLAMM, J. (1983) Solar energy collector with integrated heat storage and and radiator. Commission of the European Communities, JRC Ispra Establishment, 21020 Ispra (Va), Italy. EUR-PAT.Ref.2021, UK Nr.2, 103, 783.

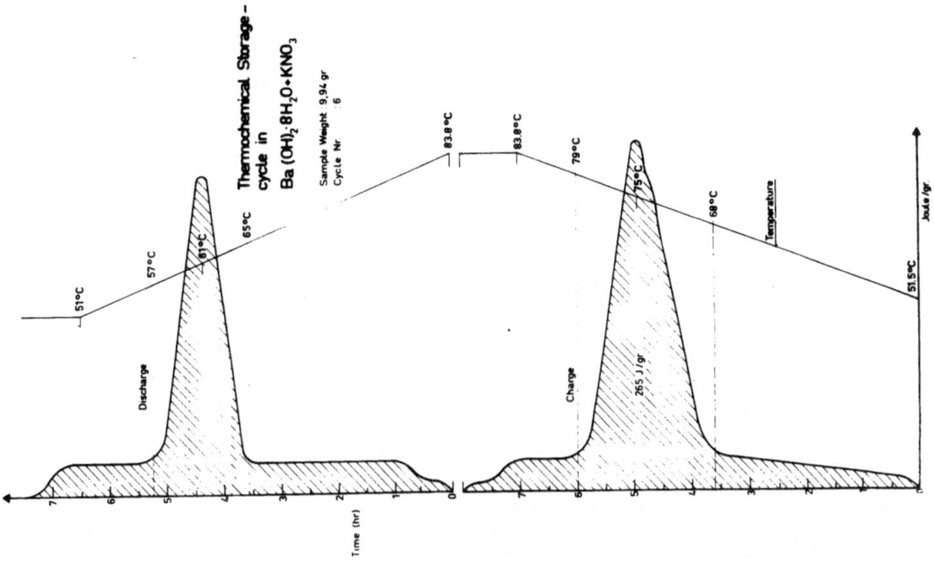

Fig.1 - Calorimetric diagram of a heat storage cycle in $Ba(OH)_2 \cdot 8H_2O + KNO_3$-mixture /3/.

Fig.2 - Latent heat storage system for space conditioning integrated into a southern wall /18/.

Fig.3 - Latent heat storage panel consisting of a lattice of elastomeric capsules containing PCM /16/.

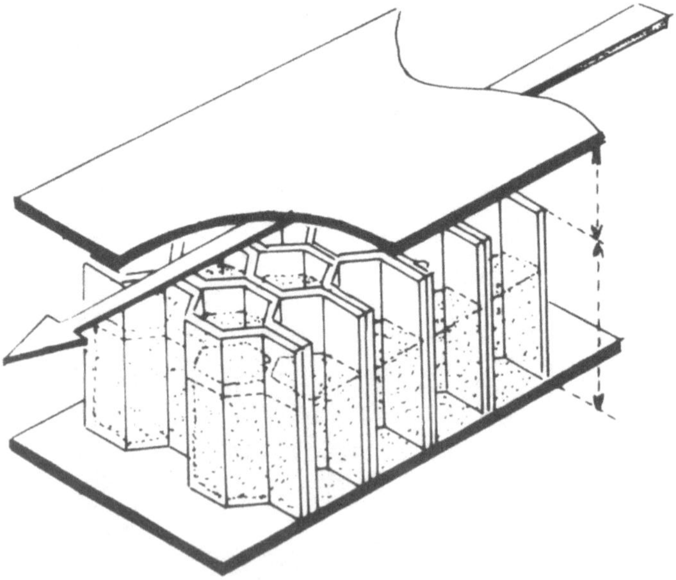

Fig.4 - Honeycomb structure for latent heat storage in the floor /16/.

ENERGY-SAVINGS DUE TO THE INTEGRATION OF A THERMOGENIOUS ELEMENT IN PREFABRICATED FACADES .

R. VAN DE PERRE, J.P. DEMOOR, M. LIBERT
FREE UNIVERSITY OF BRUSSELS
DEPARTMENT OF SOLAR ENERGY.
Pleinlaan, 2 -B-1050 Brussels (Belgium)

Summary

Energy savings induced by the integration of passive components (combination of direct and indirect gain) are discussed. Results were obtained from a measurement-campain during Februari 1984. They are compared with some static calculations (real climate and standard climate conditions). As we compare semi-experimental results each to one another, we have an idea of the netto-heat-gain for the described components.

1. INTRODUCTION

The considered building ("kultuurkaffee") is located on the campus of the Free University of Brussels. The building consists of three main parts: a pub on the ground floor (south-west), some offices (closed and open landscape) -also situated on the ground floor (north-east), and an exposition-room, with some studios (pottery, photography, ...), on the first floor.

The pub is not considered in this study, for reasons of the important, but unknown casual gains, and losses due to occupancy. For the offices on the ground floor, their will be no global energy balance either. Only some part (exposition room) of the first floor will be discussed in detail (cfr. Fig.1), several components will be compared each to one another.

The building is SE orientated, with a buffer-zone (closed offices and studios) on the NW and the NE side. The exposition room (without studios) has a floor area of 179 m2, and a volume of 465 m3. The total floor area on the first floor equals 321 m2. The roof is flat and ventilated (K roof:1.82) The SE façade on the first floor has an area of 104 m2 (76 for the exposition room) (K wall:2.25), with 35% double glazed windows, and with 30% opaque passive wall collector. The SW façade on the same level has an area of 33 m2 (K wall:1.31), with 60% opaque passive wall collector and 8% window. The NW wall has the same configuration as the SE wall, but here are the collectors replaced by an insulation board (K wall:1.9). No windows has been placed in the NE wall (K wall:0.58). The K value of the floor between ground and first floor equals 3.22 W/m2°C.

The building is prefabricated. The outside walls are composed of repetitive concrete frames (important thermal bridges). Windows, passive components or insulation boards are fixed in each concrete frame. Inside walls are light weight structures.

Several configurations of passive componants are under study. There are:

- the "parapet-type" (combination of direct and indirect gain, 54% window fraction)
- the "entire-wall-type" (only indirect gain).

There is a possible choice between the kind of glass-layer placed in front of the opaque wall, there are several components with different orientations, and there are different ways of natural convection control. The opaque thermogenious component consists (from inside to outside) of a black painted wall (18cm)-without possibility for natural convection through the wall , an air layer (2cm), and a glass layer -hold in a wooden frame. At the inside, a permanent insulation (8cm) is placed in front of the opaque wall, to avoid excessive heat-losses during cold and cloudy periods. Manually controlled natural convection is possible between the ambiant room-air and the enclosed air-gap (12cm) -between insulation and concrete, during cold and sunny periods (after sunset). The window-sill consists of a wooden valve which can be opened to allow natural convection. In the same way wooden valves can be placed at floor- and/or at ceiling level.

The components which are discussed here in detail are: (cfr. Fig 2)

- the "parapet-type" with valves, they are SE orientated. On the first floor a double glazing was placed in front of the collector, versus a single glazing on the ground floor.
- the "entire-wall-type", SW orientated, single glazed, with only one valve on floor-level.

For the considered zone we have 39 m2 double glazed windows (27 m2 SE), 23 m2 opaque double glazed (SE), and 20 m2 opaque single glazed mass-walls (SW). The basic idea (concept developped by prof. J. Van Loeij) was an attempt to make a performant combination between direct and indirect gain. The Belgian climate (intermittant and medium intensity sunshine) does not allow the integration of a classical Trombe Wall in (residential) buildings. The only way to diminish the heat losses through a mass wall is to place an insulation. There are but two possibilities:

- insulation at the outside (diode wall: air-collector directly coupled to the storage-wall. Disadvantages are: difficult heat-exchange between air and mass-wall, necissity of a fan.
- insulation at the inside (VUB-wall). Disadvantages are: screening of the radiation from the wall to the inside, elevated losses from wall to outside.

A comparison (1) , shows that the first solution is probably more performant than the second one. On the other hand, it is clear that the second solution has an important lower investment- and maintenance cost. But first of all, we have to know the real performance of the considered components.

2. MODELLISATION.

A first attempt to model the behaviour of the described components in a real building geometry, proved to be incorrect, due to the difficulties to estimate the varying border conditions, the effect of ventilation losses, and the coupling of different zones. An electrical network is now under study, but no results are available today.

3. EXPERIMENTAL SET-UP.

Calorimeters are placed on each circuit of the auxiliary heating system, and each radiator can be turned off individually, from a central control board. 42 thermocouples (Pt100,CuCt,NiCrNi) are placed all-around, to measure air-temperatures (inside, air-gaps, outside) and surface temperatures (inside, outside). Solar data is registrated with a solarimeter. All this data is scanned each quarter of an hour, and is registrated on cassette, which can be treated on a mini-computer.

In this paper we analyse 9 cold but sunny February days (09-17/02)

mean outside T : -0.31 °C	normal :	3.9 °C	
mean solar radiation (H) : 109 W/m2	normal :	52 W/m2	(24 hours)
mean inside T upstairs : 14.21 °C			
mean real DG : 405 (monthly basis)	normal :	295	(15/15)

We had the possibility to maintain low ambiant temperatures during this period, and to turn off the auxiliary heating system in the studios for all the time, and in the exposition room for some days. At the first day of the period, there was a large overheating (up to 24 °C). At the end of the period, we did start again the auxiliary heating, in order to have one cycle in which the termal inertia of the building plays in both directions.

4. CALCULATIONS FOR THE EXPOSITION ROOM.

Heat losses to the outside, and to the not-heated adjacent studios, (transmission and ventilation), are calculated in a static way, using the real temperature differences -each quarter of an hour. Heat-gains through the floor are taken into account, as well as heat gains through the roof (using a sol-air temperature), penetrating diffuse radiation from the NE-side and some casual gains 77MJ/day. As a final balance we obtain:

$$Q\ aux1 + Q\ floor + Q\ roof + Q\ diff + Q\ cg + Q\ comp = Q\ out1$$

We know the auxiliary energy input, so we calculate Q component for the considered period. In our static evaluation we over-estimate reality for Q out, but we do the same for Q floor and Q roof. As we compare Q component with Q solar incident, we obtain the efficiency of the considered components. Over 9 days this efficiency was equal to 0.57.

When we replace the passive components by a perfect insulation, and we assume the same internal and external conditions, our final balance becomes:

$$Q\ aux2 + Q\ floor + Q\ roof + Q\ diff + Q\ cg = Q\ out2$$

As we can calculate Q out2, we can easily obtain Q aux2.

- Q aux2 − Q aux1 represents the netto saved energy, compared to a perfect insulated component. We found 328 MJ/day
- (Q aux2 − Q aux1)/Q incident represents the solar fraction which was effectively used. We found 0.35
- (Q aux1 + Q roof + Q floor)/(Q aux2 + Q roof + Q floor) represents the saving fraction of auxiliary energy. We found 0.56

If we want to have a very rough estimation of the behaviour of those components in a statistical mean Februarimonth,we can make the assumption that the components efficiency is invariant for the considered climatological change. In fact we assume that the very high solar intensity during the measurement-campaign, compensates the very low outside air-temperature, during the same period. If we recalculate Q component, Q roof, Q diff and Q aux1,2 , we find respectively: 39 MJ/day , 0.09 , 0.93 , (calculations made for 15/15 degree - days, T out = 3.9 , I solar = 52).

5. COMPARISON BETWEEN DIFFERENT COMPONENTS.

Although a comparison between the first floor and the ground flour is rather dangereous (other inside-temperature-variation, other surface-temperature distribution), the different behaviour of the single glazed versus the double glazed component is considerable. (The higher downstair's inside ambiant air temperature is due to a bigger Q aux). We can conclude that the time for which Tsi (inside wall-temperature) is higher than Tse (external wall temperature) is only an indication, but no correct estimation, of the heat-loss/heat-gain time. This is shown by a simple comparison with the gradient of air-temperature in the air-gap (negative means downstream cooling air) . We notice a discrepancy of about 14% . We see that the double glazing elevates the mean Tse with about 2 °C, while the maximum reached external surface temperatures are almost the same in both cases.

	Ti	T gap	gr.gap T.	gr.gap > 0	Tsi > Tse	Ti > Tgap	Tse - Te
double	14.2	13.9	3.37	18	48	59	16.4
single	15.9	10.7	1.55	62	52	96	15.6

A detailed description of the behaviour is given in (2) .

6. CONCLUSIONS.

We found that the described passive components have an efficiency of 0.57 for the considered test-period. Compared to a perfect insulation, the induced energy savings are 328 MJ/day during the test period, and 39 MJ/day during a standard climatological Februarimonth. The effective used solar fraction equals 0.35 during the test period and 0.09 during the standard month. The relative savings, compared to a perfect insulation represent 0.44 for the test-period and 0.07 for the standard month. The comparison with the standard month is a very rough estimation.

We proved that those passive components do give a real energetic contribution to the building, even in severe winter-months. The experiments show that those components are well fitted to maintain a minimum air-temperature of 10 to 15 °C, in severe climate conditions. A detailed description (2) showsthat therefore a combination between direct and indirect gain is necessary. The considered VUB-wall is a valid alternative for low temperature heated spaces, in which natural daylight is needed. In order to reduce the heat-losses from the mass-wall to the outside, double glazing is a must. When large surfaces will have to be covered with this, a financial problem may arise.

7. REFERENCES.

1. Qualitative investigation towards the use of materials and components in passive solar energy applications for the Belgian climate.
   R. VAN DE PERRE , J. VAN LOEIJ report E3/III/2.3/1
   (dutch language) January '84 Belgian Ministry of Science

2. Energy-savings due to integrated passive components in te "kultuur-kaffee".
   R. VAN DE PERRE , J.P. DE MOOR , J. VAN LOEIJ internal VUB-report
   (english language) June '84

8. NOTES.

Q aux1 represents the auxiliary heating in the considered building (measured value)

Q aux2 represents the auxiliary heating in the considered building, where all solar components are replaced by a perfect insulation (calculated value)

Q floor represents the energy penetrating through the floor, due to measured temperatures upstairs and downstairs; in 0 out1,2 the floor was considered as perfect insulated

Q roof represents the extra energy penetrating through the roof due to the solar radiation; in Q out1,2 the roof was only sollicitated by the external temperature

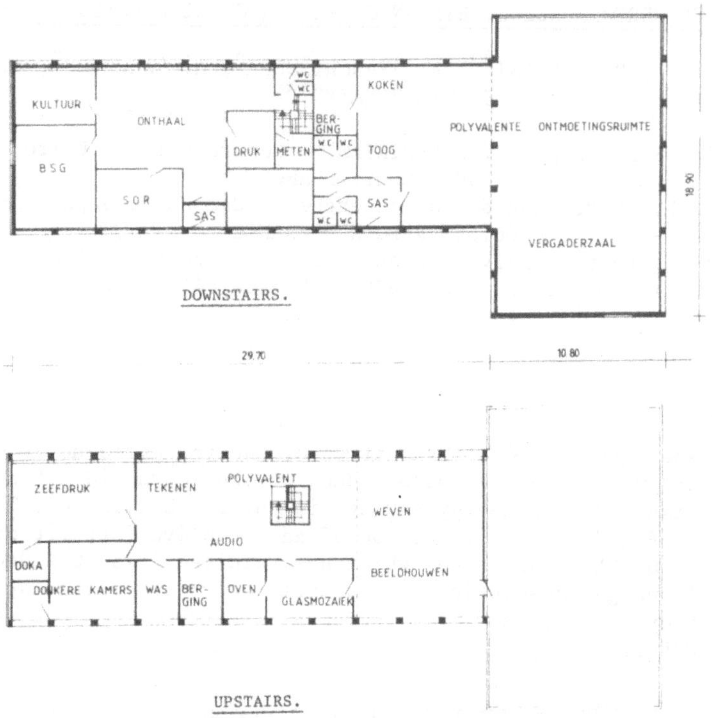
KULTUUR
ONTHAAL
B.S.G
S.O.R
SAS
DRUK
METEN
BER-GING
WC
KOKEN
TOOG
SAS
POLYVALENTE
ONTMOETINGSRUIMTE
VERGADERZAAL
DOWNSTAIRS.
29.70
10.80
18.90
ZEEFDRUK
TEKENEN
POLYVALENT
WEVEN
AUDIO
DOKA
DONKERE KAMERS
WAS
BER-GING
OVEN
GLASMOZAIEK
BEELDHOUWEN
UPSTAIRS.

FIG. 1

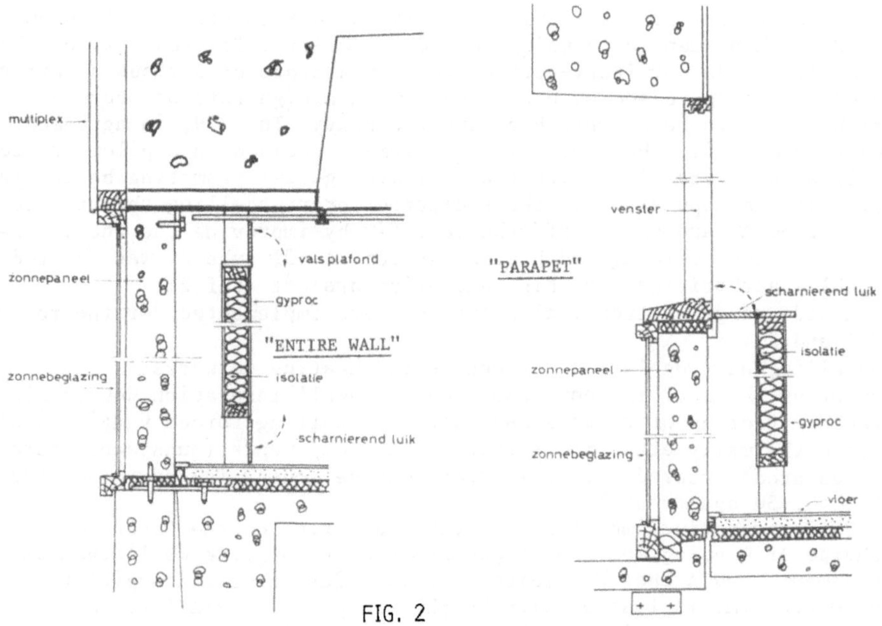
multiplex
vals plafond
zonnepaneel
gyproc
"ENTIRE WALL"
zonnebeglazing
isolatie
scharnierend luik
venster
"PARAPET"
scharnierend luik
isolatie
zonnepaneel
gyproc
zonnebeglazing
vloer

FIG. 2

SUNSPACE PERFORMANCE WITHIN AN OVERALL HIGH RISE BUILDING RETROFIT

L.M. CHOUNET*, D. CLEMENT**, R. FRANCHISSEUR*, A. LAHELLEC*, J. NOPPE*, F. WINKELMANN***

(*) Equipe R.A.M.S.E.S. (CNRS-PIRSEM) - Centre Universitaire, Bât. 208 91405 ORSAY - FRANCE

(**) A.F.M.E. (Direction de la Recherche) - Campus de Sophia Antipolis 06565 VALBONNE - FRANCE

(***) Building Energy Analysis Group - Lawrence Berkeley Laboratory;Blg.90 University of California - BERKELEY - CA.94720 - U.S.A.

Summary

This paper gives a first evaluation of the thermal benefit due to a ventilation pre-heat sunspace, added on the south side of an apartment (approximately 25 % energy savings in the Paris climate). This study is part of a detailed evaluation of an extensive retrofit of 593 apartments at DREUX (PARIS area) which has resulted in a 50 % decrease in overall energy consumption. The principles of the method applied in this study are presented : measurements and computer code simulations (CALECO/DOE2).

1.PRESENTATION OF THE "CITE DU LIEVRE D'OR" RETROFIT - DREUX (PARIS AREA)

Built in 1966,about 80 km West of PARIS, "Cité du Lièvre d'Or" is a set of 19 buildings (593 apartments) belonging to DREUX O.P.H.L.M.(Office Public d'Habitations à Loyers Modérés). The architecture of this estate which houses 2000 inhabitants is typical of the town planning policy implemented in France in the fifties and sixties to meet the strong demand for housing due to the population influx related to industrial development. DREUX O.P.H.L.M. manages housing for 5000 families. Ten years after they were built, these dormitory-estates were in a state of serious deterioration contributing to social uneasiness, with a high rate of occupant turnover; heating overheads were becoming too heavy. In 1979, in agreement with DREUX town council, the O.P.H.L.M. management decided on a pilot project for "Cité du Lièvre d'Or" with the following goals: promoting better integration of the population in their district by remodelling the architectural environment and paying off the retrofit by improving the thermal performance of the building and heating system. A 40% subsidy was granted by the public authorities for this renovation project (ref.2).

A wide range of technical solutions were implemented for the retrofit (Ref.1 and 2):

- installing new regulation devices on the heating networks;
- improving the building envelope (roof and wall insulation,double glazing)
- draught-proofing of windows and doors; installing forced ventilation;
- adding thermally effective architectural components (sunspaces,increased glazed areas, glazed wall collectors on the south side of respectively 130, 40, 30 apartments).

The prospective importance of this retrofit led the Government bodies in charge of housing and energy conservation (Ministère de l'Urbanisme et du Logement, and Agence Française pour la Maîtrise de l'Energie) to request DREUX O.P.H.L.M. in liaison with RAMSES group (CNRS-ORSAY) to make a detailed thermal evaluation of the retrofit (Ref.2). After two years of monitoring and simulation work it can be said that the objectives of the thermal and architectural retrofit are fully achieved, and that energy saving prospects are enormous for two million dwellings under comparable climatic conditions in FRANCE.

2. THE OVERALL THERMAL EVALUATION

The heating system consists of a centralized gas-fired furnace which delivers hot water to the apartment radiators via 8 circulation/regulation groups. In terms of gas consumption alone, the gross result shows a redution of 48% (see fig.2). Energy-meters were installed to provide continuous recording of the consumption of 15 out of the 19 buildings on the estate. Monthly values per m^3 and per K(base 18°C) for 1983 are given in figure 3. The result of the retrofit can thus be added to the BECA graph (see ref.3): 47 GJ/100m^2 for 2550 degree-days, base 18.3°C. At this stage of the analysis, it is not possible on the basis of energy measurements alone to estimate the performance of the individual solutions implemented.

3. DETAILED ANALYSIS OF A PASSIVE SOLAR SOLUTION

A more fundamental approach to the retrofit component analysis consisted of monitoring unoccupied apartments while the meteorological data were recorded on site.

A conventionally retrofitted test-apartment was monitored continuously from 1981 to 1983 (winter and summer) with 60 measurement channels recorded at 15-minute intervals.

Shorter measurement campaigns were made on thermally characteristic apartments when released from occupancy: three 2-month periods (40 channels recorded at 15-minute intervals).

The analysis of the recorded data is based on the use of the computer program CALECO/DOE2 originally developed by the Energy Efficient Building Group at Lawrence Berkeley Laboratory(California), and since 1979 with the collaboration of RAMSES Group (Ref.4 and 5). We have also developed a computer tool, connected to CALECO, for hour-by-hour comparison between calculated and measured data. This combined measurement/simulation device was used to calibrate a realistic parametric model for the various components of the retrofit.

After having presented the comparison results of a conventionally retrofitted building at the CANNES International Conference in December 82 (Ref.2), we shall now show the first results of our analysis of the architectural solution adopted for 7 apartments of one of the buildings, whereby a sunspace providing preheat ventilation was added on the south side of two rooms per apartment.

The results of a rough calibration are illustrated on Figs.4 and 5; the worst and the best of a 40-day fit sequence are shown. A parametric study then led us to estimate that 2100 KWh/year/apartment were saved due to the sunspace, for a consumption of 8600 KWh, had the sunspace not been added.

For a more basic understanding of the behaviour of this sunspace, several thermal effects can be separated by varying the parameters given in the simulation model used to estimate the overall amount of energy saved (North-South rotation of the apartment with and without sunspace, with and without forced pre-heat ventilation).

The results are collected in Figure 6 showing that the buffer effect accounts for 15% of the saving achieved, then the dynamic insulation effect (preheat of admission air recovering conductive losses) provides an extra 9% saving, and lastly that the collector effect adds nothing, which may be explained by the fact that this architectural device throws shade on the windows of the apartment as well as collecting solar radiation. An important consequence is that this sunspace built onto the North side of the building would pay off just as quickly as on the South side.

4. CONCLUSIONS

The detailed analysis of the retrofit, to which this paper is but one contribution, confirms the overall success of the project. The measurement and simulation methods adopted have provided us the means for a critical study of the retrofit components, the ultimate goal of which is to rationalize the choice of later retrofit solutions. The computer program CALECO/DOE2 has proved to be an invaluable tool for establishing simulation models and has convinced us of its physical relevance and realistic predictions.

Besides the success of the thermal retrofit in terms of energy consumption, we should also state that the social results can be measured in terms of the decrease in occupant turnover (210 families departed in 1978 against 60 in 1982).

ACKNOWLEDGMENTS

We are happy to thank Jean-Pierre HAMON, Manager of DREUX O.P.H.L.M. who initiated and directed the retrofit of Cité du Lièvre d'Or and gave us constant and friendly support throughout monitoring operations, and Professor Arthur H. ROSENFELD who aroused our interest towards thermal retrofit analysis during many fruitful discussions. We emphasize also the strong support giver to us during this study by A.F.M.E.(Agence Française pour la Maîtrise de l'Energie), M.U.L. (Ministère de l'Urbanisme et du Logement), and C.N.R.S. (Centre National de la Recherche Scientifique).

REFERENCES

1. RAOUST, M. (1980). Opération du Lièvre d'Or: étude des serres et des murs-capteurs. Rapport final d'étude pour l'O.P.H.L.M. de DREUX (mars 1980). Internal report of DREUX-O.P.H.L.M.
2. CLEMENT, D. et al. (1982). Réhabilitation de la Cité du Lièvre d'Or à DREUX. Proceedings of "Conférence Internationale sur l'ARCHITECTURE SOLAIRE", held at CANNES from 13 to 16 December 1982. Published in 1983, at "Techniques et Documentation-LAVOISIER" Editions (PARIS).
3. FLOUQUET, F. et al. (1984). Monitored European solar and American low-energy houses - performances analysis. First CEC - Conference on Solar Heating, held in AMSTERDAM in May 1984. Proceedings published by CEC in 1984.
4. LOKMANHEKIM, M. et al. (1979). DOE-2: a new state-of-the-art computer program for the energy utilization analysis of buildings. Paper presented at the Second C.I.B. Symposium on Energy Conservation in the Built Environment, COPENHAGEN, DENMARK, May 1979.
5. Equipe R.A.M.S.E.S. (1979). Présentation du code CALECO-DOE2: Colloquium "CNRS/Plan Construction" on thermal simulation for buildings, held at POITIERS in 1979. Internal report of RAMSES group.

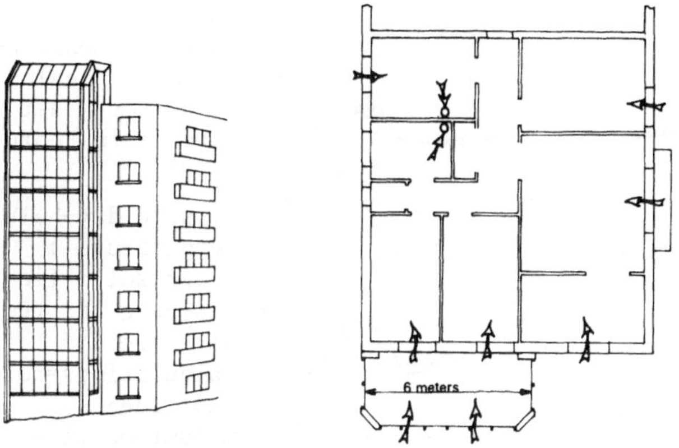

FIGURE 1

Cité du Lièvre d'Or at DREUX : South view of building 19 (left), and plan view of the 5th floor apartment which was monitored.

* Flow. Total extraction Rate = 0.6 vol/hour

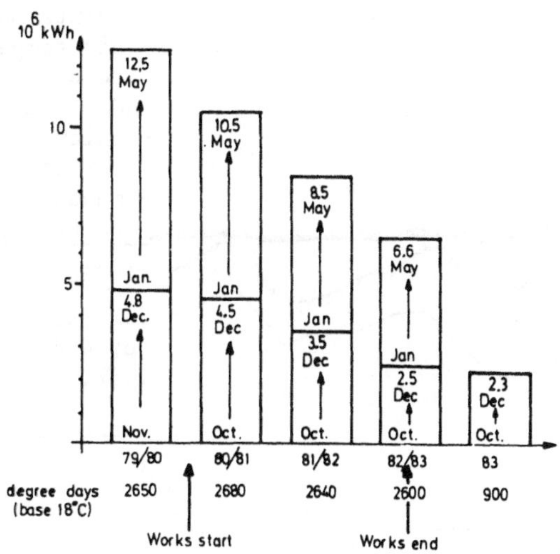

FIGURE 2

Cité du Lièvre d'Or : overall gas consumption billed from November 79 to December 83.

ENERGY CONSUMPTION PER UNIT VOLUME: W/M^3.°C

BUILDING	RETROFIT FEATURE	JAN	FEB	MAR	APR	MAY	OCT	NOV	DEC	TOTAL
1	SUNSPACES SOUTHSIDE	1.11	1.04	0.89	0.86	0.72	0.85	1.00	1.01	0.98
2	NON SOLAR	1.16	1.12	1.04	1.06	0.87	1.02	1.17	1.09	1.10
4	NON SOLAR	1.06	0.99	0.83	0.78	0.63	0.84	0.97	0.96	0.93
5	NON SOLAR	0.95	0.95	0.89	0.88	0.69	0.79	0.87	0.87	0.89
6	SUNSPACES SOUTHSIDE	1.05	1.03	0.98	0.97	0.77	—	—	—	1.00
7	NON SOLAR	0.95	0.93	0.88	0.91	0.72	1.01	0.93	0.93	0.92
8 + 9	WALL COLLECTORS	0.99	0.94	0.86	0.86	0.73	0.74	0.96	0.98	0.92
10	NON SOLAR	0.93	0.88	0.79	0.79	0.63	0.68	0.89	0.94	0.86
11	SUNSPACES SOUTHSIDE	0.83	0.79	0.72	0.68	0.60	0.58	0.78	0.77	0.75
13	SUNSPACES SOUTHSIDE	0.88	0.82	0.76	0.72	0.65	0.64	0.85	0.84	0.80
14 + 15 + 16	SUNSPACES SOUTHEND	0.86	0.79	0.73	0.75	0.63	0.60	0.73	0.76	0.76
19	SUNSPACES SOUTHEND	0.84	0.75	0.68	0.66	0.54	—	—	—	0.73
TOTAL		0.95	0.89	0.81	0.80	0.67	0.73	0.88	0.88	0.86
DEGREE-DAYS (BASE 18 °C)		357	439	333	272	45	140	324	434	2344

FIGURE 3 Energy consumption coefficient for the different buildings of Cité du Lièvre d'Or (DREUX) during the year 1983.

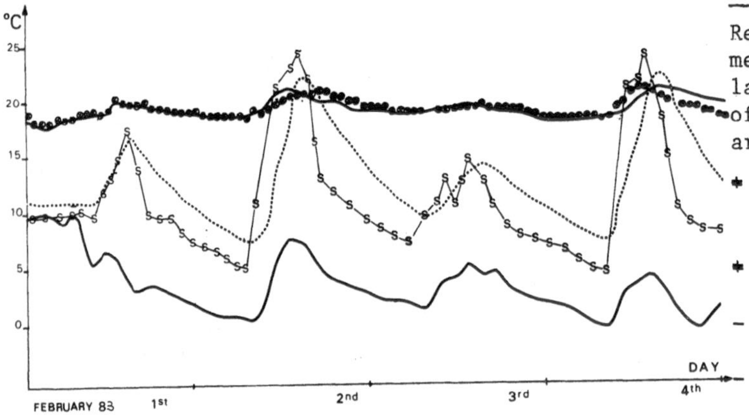

FIGURE 4

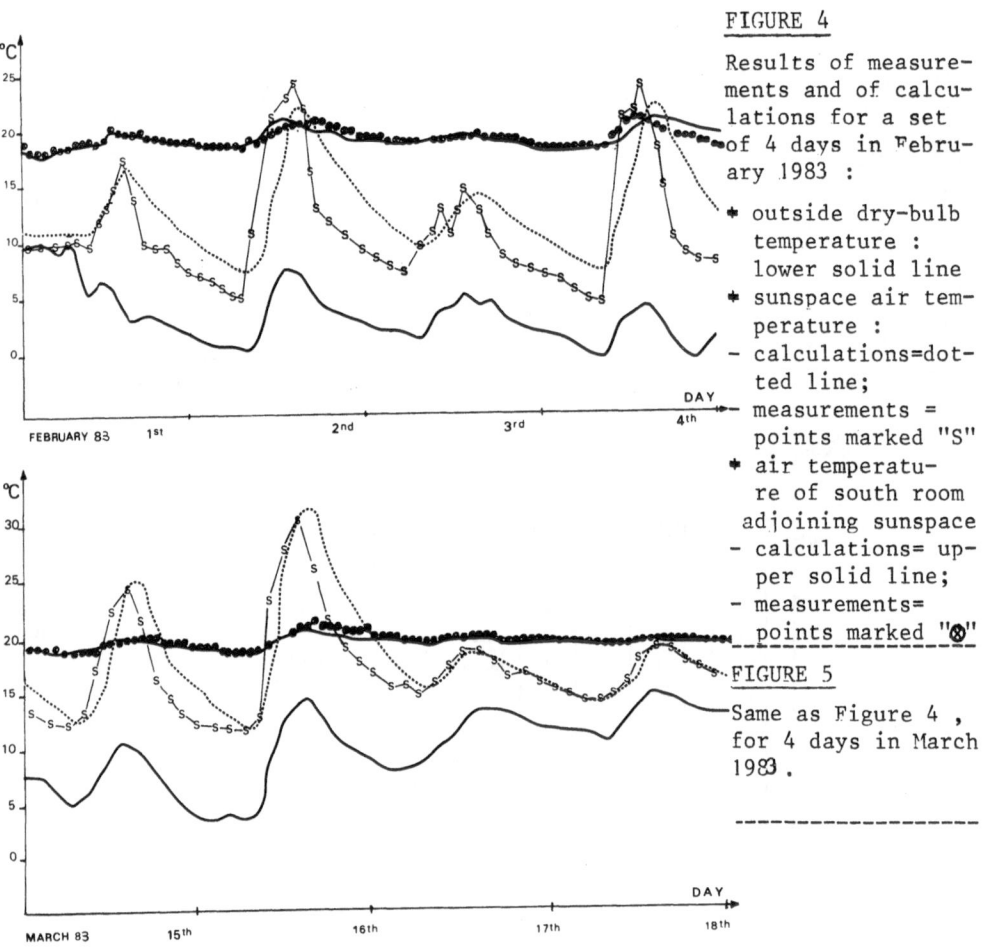

Results of measurements and of calculations for a set of 4 days in February 1983 :

- outside dry-bulb temperature : lower solid line
- sunspace air temperature :
  - calculations=dotted line;
  - measurements = points marked "S"
- air temperature of south room adjoining sunspace
  - calculations= upper solid line;
  - measurements= points marked "⊗"

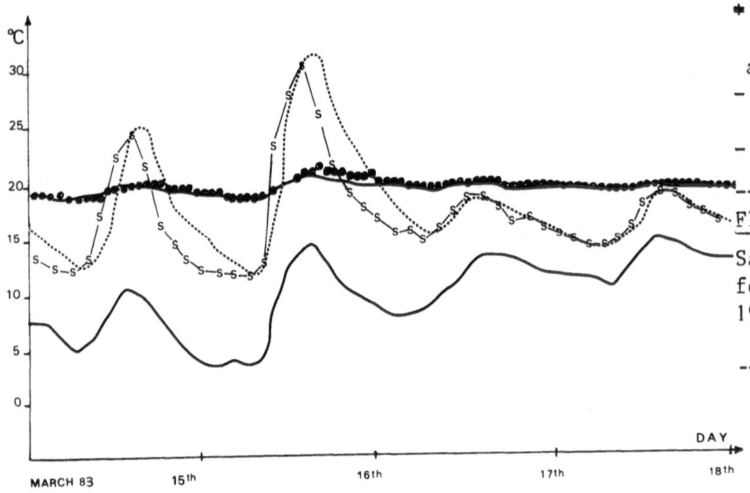

FIGURE 5

Same as Figure 4 , for 4 days in March 1983 .

	Energy consumption/year/appartment		Cumulated energy saving	Partial effect added
	appartment without sunspace	appartment with sunspace		
On non-retrofitted building				
Adding sunspace on Northside	9150	7710	1440(16%)	1440(16%) Buffer effect
With pre-heated ventilation	9150	6960	2190(24%)	750(8%) Exchange effect
Rotating North/South	8640	6520	2120(25%)	-70(-) Collector effect
On retrofitted building				
Adding sunspace on Northside	6790	5860	930(14%)	930(14%) Buffer effect
With pre-heated ventilation	6700	5100	1600(24%)	670(10%) Exchange effect
Rotating North/South	6200	4900	1300(21%)	-300(-) Collector effect

FIGURE 6: Individual thermal benefits due to sunspace (Kwh)

USE IN PASSIVE SOLAR ARCHITECTURE OF SUNLIGHT DOUSERS WITH A SEASONAL EFFECT

F. SCAMONI, U. FLISI, U. MONACO
Montepolimeri C.S.I.
Viale Lombardia, 20 -I 20021 Bollate (Mi)
ITALY

Summary

This research was undertaken with the aim of verifying the practical validity of a theoretical idea postulating the possibility of controlling the sunlight admission into dwellings by using PMMA prismatic sheets as solar dousers. The optical properties of the prismatic sheets have been theoretically calculated with a computerized calculation programme and experimentally investigated in different ways with light transmission apparatuses. Light transmittance data were thus obtained under different angles, and the experimental curves gave the same trend as the theoretical ones. The practical performance of the solar dousers assembled in a real-size architectural component, has been tested by means of test cells. Smaller test boxes have also been built in order to compare the behaviour of solar dousers with the shading effect of overhangs. Incident sunlight data and temperature measurements of test cells and boxes were worked out in order to obtain the daily mean transmittance of the solar dousers in comparison with that of a standard glazing. The results of such an exercise, have shown that the dousing effect, expressed as a reduction of light transmittance, is of the order of 30% in summertime and decreases to zero in the Fall.

1. INTRODUCTION

The idea of controlling sunlight transmission through transparent sheets in accordance with the position of the sun appeared several years ago in some patents and publications . However, the means generally used for this purpose were different from the one proposed here. In fact, in those cases which concern entirely "passive" system, the shape of these is quite different from the present proposal, which appears simpler and more efficient than all of them.

2. THEORETICAL BACKGROUND

The invention concerning this research (8), patented by G.V. Nardini (1-4), is based on the phenomenon of "total internal reflection", which occurs in any transparent material, such as glass or polymethylmethacrylate (PMMA), having a refractive index greater than 1.0 and an appropriate shape of prismatic sheet, with such a profile as shown in fig. 1. Owing to this particular shape, the prismatic sheet will allow a light beam to pass through, when it is hit with a small angle, (fig. 1a), whereas it will totally reflect the same beam, when it is reached with an incidence angle greater than a critical value (fig. 1b). If such sheets are assembled in an architectural component and installed in a vertical position facing

South, they will admit sunlight during the winter period and will reject it away during the summer months. The period of total reflection depends, at a given latitude, on the refractive index of the material and on the amplitude of angle 1 of fig. 1. This period can thus be defined a priori by using an appropriate material and by choosing a particular value of angle 1, so that the total internal reflection will occur when the sun has a given incidence angle at noon. This angle will set the limit between the summer and winter behaviour.In this piece of research, (PMMA) prismatic sheets have been used with the following internal angles:angle 1 = 16°30' angle 2 = 33° 20' angle 3 = 130° 10'.The critical incidence angle thus becomes 56°, corresponding to the height of the sun at noon on April 22nd, at the latitude of 45° 30' N, where the experiments have been carried out. From this date to August 20th, direct sunlight will be rejected for the whole day by vertical douser facing South.A computerized calculation (7), carried out for the above conditions and using the statistical solar radiation data for Milano-Linate (which is close to the laboratory where tests were run), gives the ratio of direct, diffuse and reflected radiation passing through the solar dousers. These theoretical data were plotted, together with those of a flat sheet as a reference, versus the hours of the day in the diagrams of fig. 2 and versus the month of the year in fig. 3. The same data of fig. 3 were also worked out to get the optical transmittance of the two glazings averaged on the month, (fig. 4)

3. EXPERIMENTAL PART

Prismatic sheets of PMMA, having a profile like that described above, have been prepared, in order to verify with experiments the results of our calculations. Their behaviour has been studied in four different ways with:

- a light transmission apparatus, operating with artificial light,
- a light transmission apparatus, mounted on a solar tracker, operating with sunlight,
- real-size test cells built and instrumented for testing the architectural solar passive component,
- test boxes, similar to the real-size test cells built in order to compare the behaviour of solar dousers with that of overhangs.

3.1. Preparation of the prismatic sheets

The prismatic sheets used for the tests have a square shape (fig. 5), with the side 30 cm long, and a thickness of 3 mm. They have been prepared by compression molding of flat PMMA sheets, with a mold designed and built for this purpose. Internal angles of the prism have the values given in § 2.

3.2. Light transmission apparatuses

The light transmission apparatus (5), operating with artificial light ,is 5 m long and consists of a light source, a number of path defining baffles and a spherical integrator (fig. 6). The light source is a 1 KW halogen lamp inserted in a parabolic light concentrator. The spherical integrator, has a diameter of 83 cm. Inside the sphere, in a lateral position facing the interior, a thermopile sensor is placed, which measures the radiation passing through the prismatic sheet placed on an opening of the sphere (fig. 7). The sheet can rotate on an axis perpendicular to its plane and this movement combined with that of the sphere, pivoting on its vertical axis, allows the sheet to assume various angles on respect to the light beam. The sunlight transmission apparatus consists of a spherical integrator identical to the previous one supported by a solar tracking device, placed on the top of a scaffold and oriented to the South (fig. 8). Measurements of light trasmission coefficient with both transmission apparatuses, carried out under several incidence and orientation angles, gave the information needed to describe the optical behaviour of the prismatic sheets.

3.3. Real-size test cells

Two test cells have been set up, one for testing the glazing made of prismatic sheets and the other one for testing a plane standard glazing of PMMA as a reference sample. The cells are made of steel-polyuretane-steel sandwich panels, 12 cm thick, assembled to give a 3 m side cube. Inside this cube and near the face of the glazing, a set of vertical polypropylene pipes has been installed (fig. 9). These black pipes, filled with water,represent the energy storage of the cells. The differential warming up of these storages and the temperatures of air inside the cells have been elaborated in order to calculate the optical transmittance of the two glazings.

3.4. Reduced-size test boxes

Six boxes made of expanded polystyrene panels, 12 cm thick, with an 80 cm side, have been built to test glazings of reduced size (60 cm x 60 cm). For modelling these boxes, the room thermal conductance and storage thermal capacitance have been normalized to the area of the glazing: the glazing area will be in the same "ratio" for the test boxes as for the real-size cells. These boxes contain plastic pipes filled with water, with the same diameter than those of real-size cells, acting as heat accumulator. Two groups of three units have been set up on the top of real-size cells (fig. 10); the two groups are at different inclination and each of them consists of one box with solar dousers glazing, a second one with standard glazing and a last one equipped with on overhang, whose depth is 50% of the wall height. This overhang is considered here in order to compare its shading effect with the dousing effect of the prismatic sheets.

4. RESULTS

4.1. Optical properties of the prismatic sheets

The light transmittance of prismatic sheets, obtained with the apparatuses described in paragraph 3.2., has been plotted versus the incidence angle "i" for different values of orientation angle "a" . The angle "a" is that between the projection "p" of the sunlight vector on the plane of the sheet and the vertical to the prism axis (fig. 11). The experimental values obtained with artificial light are reported in fig. 12, while the theoretical curves, calculated by the mathematical model are given in fig. 13. The experimental values obtained with the solar tracker are not reported here, since they have the same trend of those of fig. 12, only being a little higher. In order to make a comparison between these results, only the curves at angle "a" = 180°, corresponding to noon, were reported in fig. 14, where the experimental behaviour of the prismatic sheet is compared with the theoretical one (curve 4) and with that of a flat sheet (curve 1), tested with the same apparatus. It can thus be seen that the experimental curves 2 and 3 deviate from the theoretical one, turning down earlier than expected (especially curve 3), and what is most important, showing a minimum at about 60° after which they raise again. This behaviour may be due to phenomena of internal reflection subsequent to the first one, depending on the thickness of the sheet related to the dimensions of the prisms.To conclude this paragraph, it must be said that curve 2 of fig. 14were obtained with the solar tracker and contain both the direct and the diffuse components of solar radiation, whereas curve 3, were obtained with only direct rediation.

4.2. Performance of the solar douser glazing

Storage temperature, external temperature and incident radiation data for the test cells and boxes are reported in fig. 15 and 16 versus the time of the day. Fig. 15 well describes the dousing effect of the prismatic glazing in comparison with the standard one: at the begining of the test in fact, both real-size cells have been warmed up to the same temperature (left side of the figure) after which they have been let running their natural way. It can seen that after the second day, the water storage temperature of the first cell (curve E) becomes progressively lower than thatof the reference one (curve D) thus indicating the dousing effect of the prismatic glazing. Fig. 16 refers to the test-boxes in summertime: the shading effect of the overhang (curve D) is more appreciable than that of the solar douser (curve E), but this was expected, owing to the high depth/height ratio of the overhang. This shading effect however lasts until the late Fall, which is undesirable. Incident radiation and temperature data of the test cells and boxes have been worked out in order to get the optical transmittance of the solar douser and to compare it with that of the standard glazing. For this computation it has been necessary to calculate the energy balance between the solar energy admission on one hand and the thermal storage and losses on the other. Daily mean transmittance values have been thus obtained and those referring to the real-size cells are reported in fig. 17. It can be seen that the optical transmittance of the douser is 30% lower than that of the flat glazing in August and increases in September almost reaching the reference one at the end of the month, as it was foreseen by the theory. The same results were also obtained for the test-boxes.

5. CONCLUSIONS

The sunlight dousing effect of PMMA prismatic sheets has been well evidenced by both optical and performance measurements, but the extent of such an effect is in practice lower than expected by the theoretical model. This lower efficiency may be due either to physical defects of the prismatic sheet or to phenomena of internal reflection subsequent to the first one depending on the sheet thickness ratio to the dimensions of the prisms, which seems not the most proper one.

REFERENCE

1) G.V. Nardini - Italian Patent No. 24943 A/79 (1979)
2) G.V. Nardini - French Patent No. 8017364 (1980)
3) G.V. Nardini - Spanish Patent No. 494002 (1980)
4) G.V. Nardini - U.S. Patent No. 173146 (1980)
5) C.S. Herrick - Solar Energy 28 (1), 5 (1982)
6) D.P. Grinner - Solar Energy 22 343 (1979)
7) M. Chalon - Graduation Thesis - Milan (1983)
8) U. Flisi , F. Scamoni - M. Chalom - U. Monaco - Proc. of the EC Contractors' Meeting Brussels 1 - 3 June 1983 p. 220

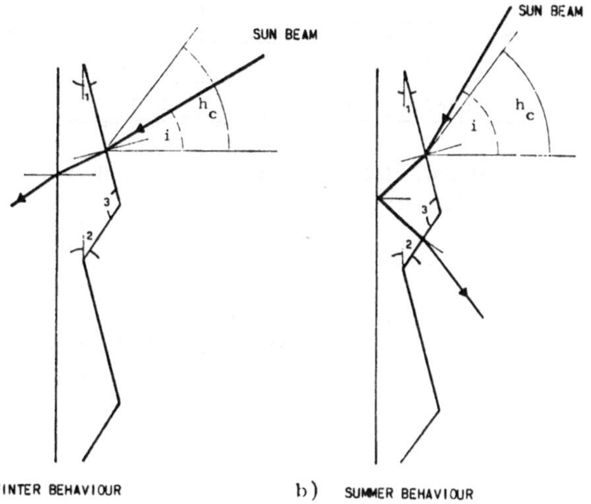

Fig. 1 - Profiles of prismatic plates, showing mechanism of light transmission (a) and of total internal reflection (b). i = incidence angle; h_c = critical height of the sun

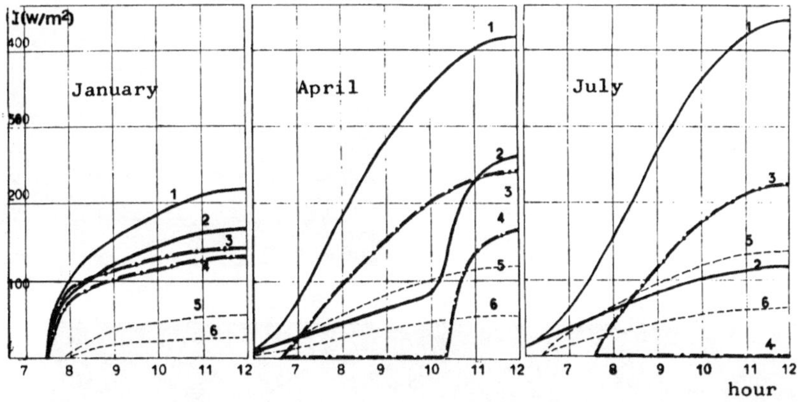

Fig. 2 - Daily theoretical data for Milano-Linate: 1: total incident radiation; 2: total radiation passing through the douser; 3: direct incident radiation; 4: direct radiation passing through the douser; 5: diffuse incident radiation; 6: diffuse radiation passing through the douser

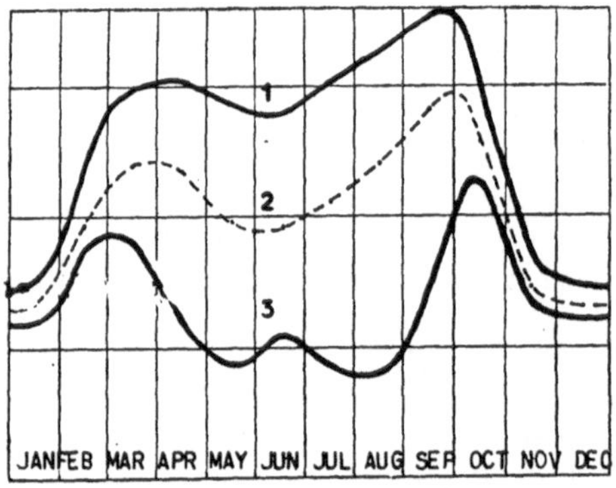

Fig. 3 - Monthly theoretical data of radiation for Milano-Linate:
1: total incident radiation
2: total radiation passing through the flat sheet
3: total radiation passing through the douser

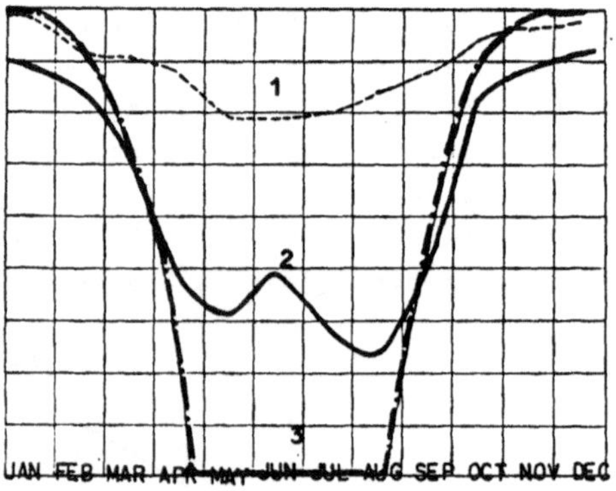

Fig. 4 - Monthly theoretical transmittance
1: total radiation transmittance of a flat sheet
2: total radiation transmittance of the douser
3: direct radiation transmittance of the douser

Fig. 5 - Front view of a prismatic sheet made of PMMA

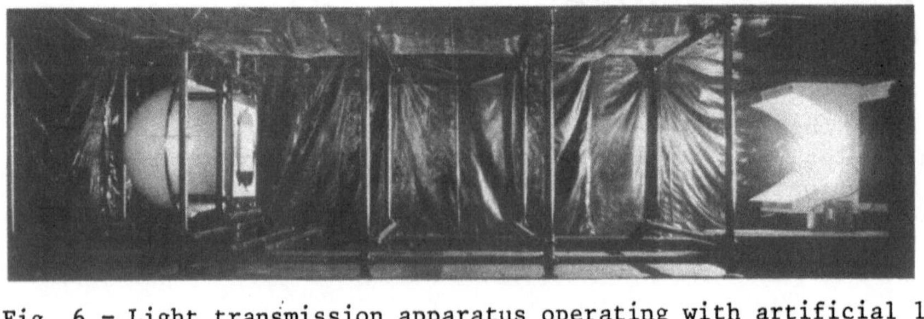

Fig. 6 - Light transmission apparatus operating with artificial light

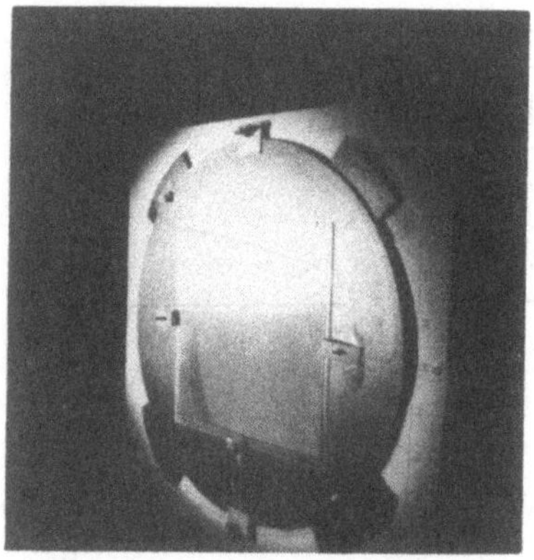

Fig. 7 - Prismatic sheet placed on the opening of the sphere

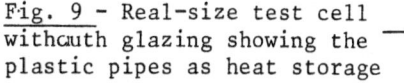

Fig. 8 - Sunlight transmission apparatus supported by a solar tracker apparatus

Fig. 9 - Real-size test cell withouth glazing showing the plastic pipes as heat storage

Fig. 10 - View of the two groups of test boxes on the top of real-size cells

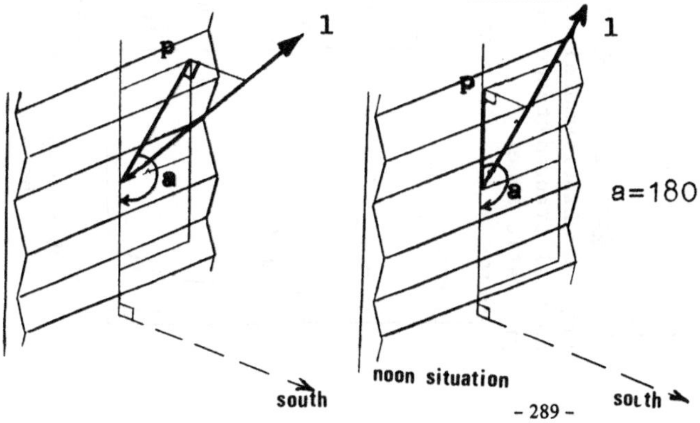

Fig. 11 - Orientation angle "a" between the projection "p" of the sunlight vector "1" on t on the plane of the sheet and the vertical to the prism axis

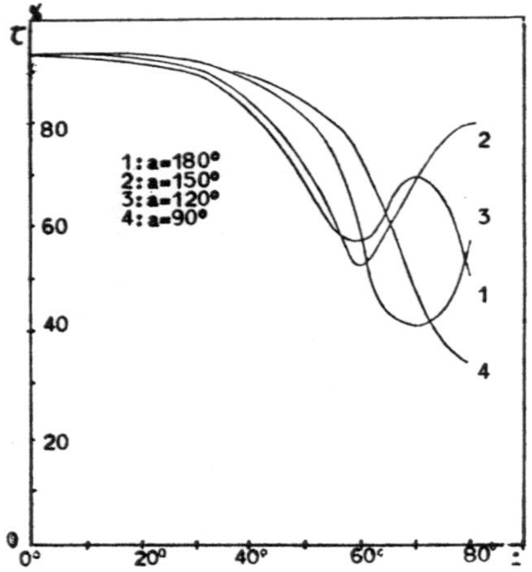

Fig. 12- Experimental curves of optical transmittance of the prismatic sheet

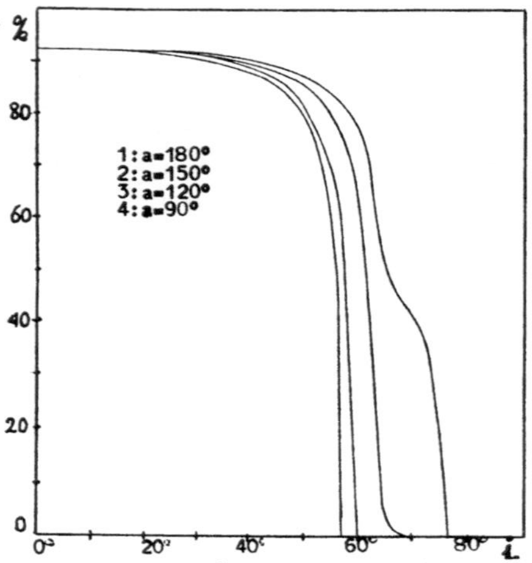

Fig. 13- Theoretical curves of optical transmittance of the prismatic sheet

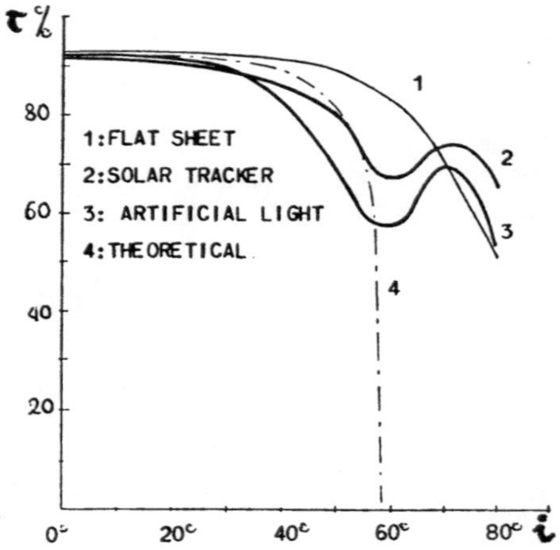

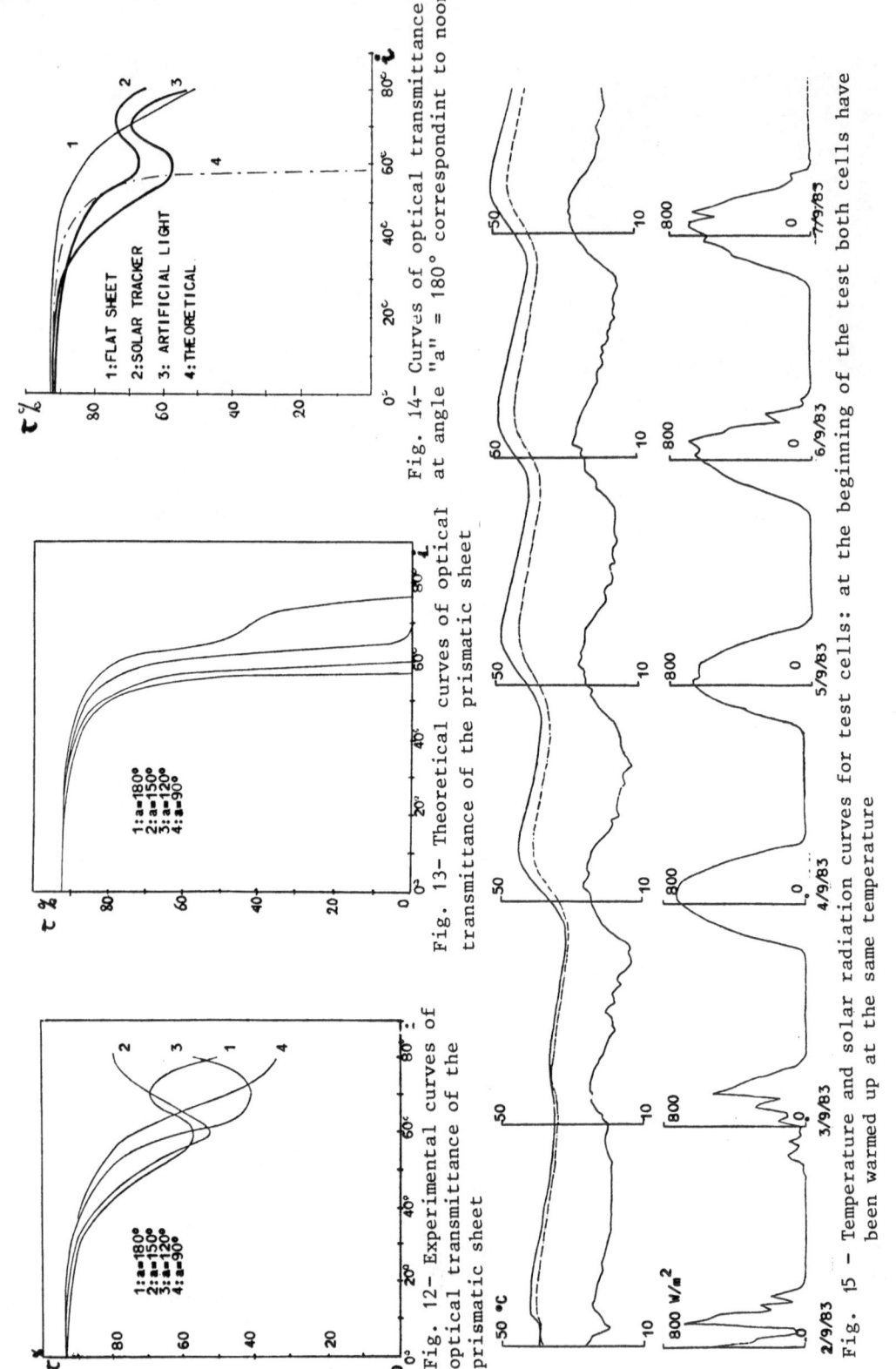

Fig. 14- Curves of optical transmittance at angle "a" = 180° correspondint to noon

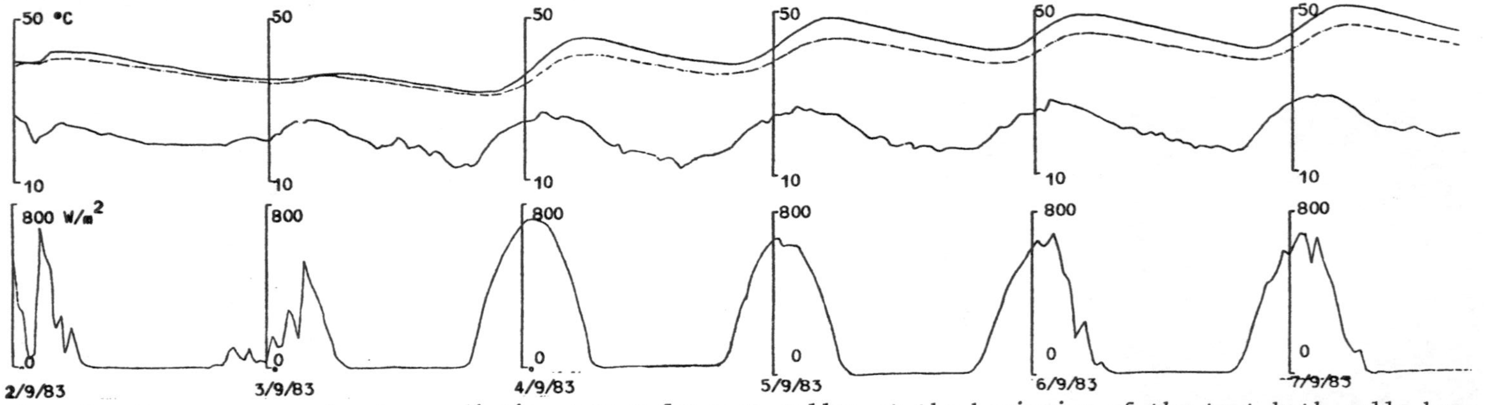

Fig. 15 - Temperature and solar radiation curves for test cells: at the beginning of the test both cells have been warmed up at the same temperature

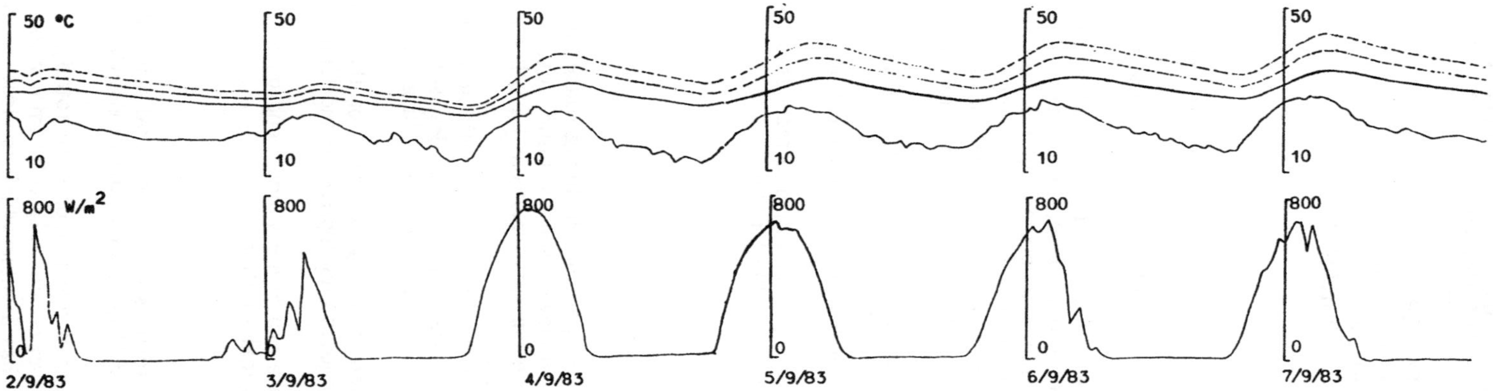

Fig. 16 - Temperature and solar radiation curves for the test boxes

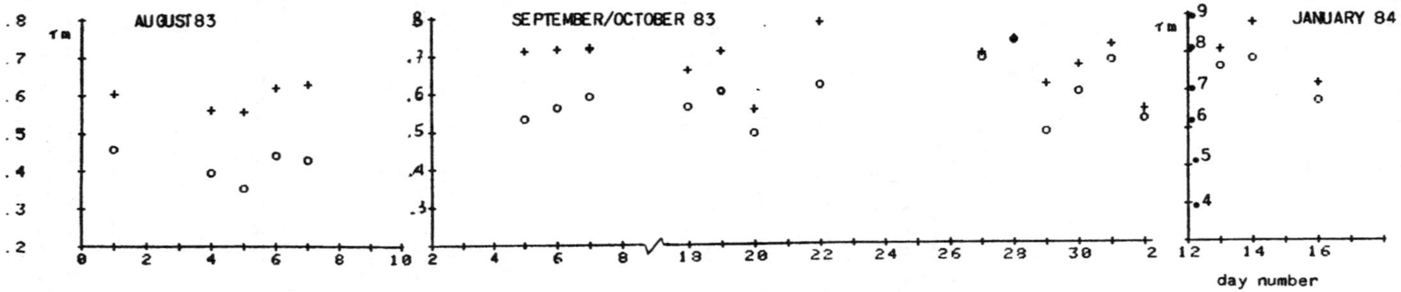

Fig. 17 - Daily mean transmittance values referring to the real-size cells at different months

o : dousers + : standard glazing

OPTICAL MEASUREMENTS ON A VARIABLE TRANSPARENCY WINDOW USING SELECTIVE CATADIOPTRIC REFLECTION

R. GUICHERD and D. ROYER
CEA - Centre d'Etudes Nucléaires de Cadarache
Service d'Etudes Energétiques
B.P. 1 - 13115 SAINT PAUL LEZ DURANCE
FRANCE

Summary

Research at the Commissariat à l'Energie Atomique on the means to control heat inputs in large glazed buildings, has led to the construction of a variable transparency window. This window uses the selectively obtained optical reflection on wet or dry catadioptric reflectors.
The window consists of three glazed panes with a plane of methyl polymetacrylate cat's-eye reflector in between. In reflecting mode, direct sun rays in the proper admittance angle are reflected back, thus protecting the inside space from overheating and reducing the summer cooling loads. Provision is made in the space between two panes of glass to drain the reflector with water. This has the effect to suppress the reflection property, and to allow for desirable direct solar gains in winter.
A prototype of theses panels has been used to caracterize the optical properties of the system. The transmittance of the window is measured in both reflection and transmission modes and the influence of the incidence angle of the direct sun rays is investigated.
The experimental results are used to evaluate the annual performances of the window in terms of heating and cooling charges of the building.

1. INTRODUCTION

The most common way to regulate ambiant temperature in living spaces is the control of incoming energetic fluxes through windows : shades, blinds, etc... In the past few years, original adaptations of these classical solutions have been proposed and sometimes successfully industrialized.

The case of large glazed areas, e.g. in office buildings, is not satisfactorily solved. Inside or outside solutions, for instance venetian blinds or shades, suffer from either thermal or mechanical drawbacks : thermal dissipation, failures to wind, etc...

In the search for new solutions to the problem of energy control for the large glazed areas, a catadioptric solution has been proposed and applied to the CISI building at the International Milan Fair (1) (2).

We report here on optical measurements on a prototype window and on a comparison of performance with other more classical solutions.

2. PRINCIPLE OF THE WINDOW

For the short description which follows, we focuses only on general principles and results, without details on the calculations and Milan Fair realizations which can be found in references (1) to (3).

As is well known, the main property of the cat's-eye mirror is to totally reflect incident beam backwards. The reflection can be suppressed by immersion of the reflector in a medium whose optical index has a value which is close to the reflector index. A representation of the catadioptric grating is shown in Fig. 1.

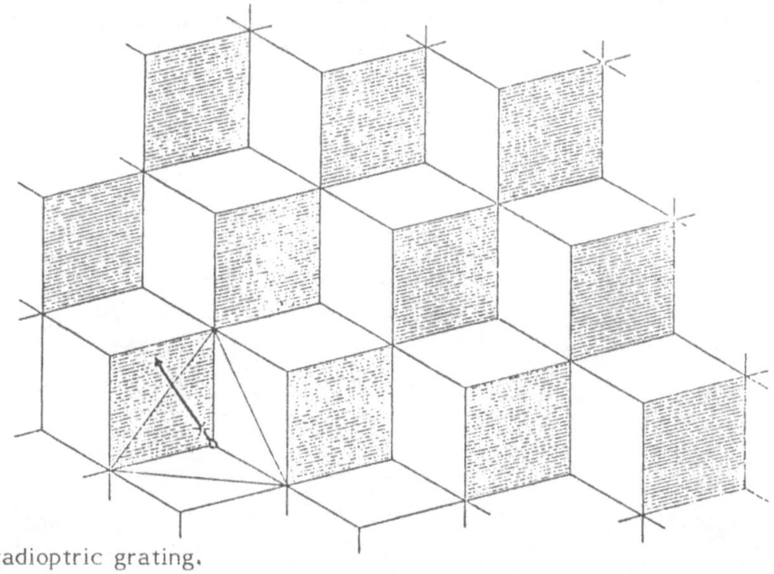

Fig. 1 : Catadioptric grating.

An additional property is that reflection takes place only when the incident beam is within an acceptance angle which can be described in the reference frame depicted in Fig. 1. In this figure, the trihedral axis has its trace at the center of the base triangle, the angular coordinates of the axis (height above horizontal plane and azimuth) beeing choose by construction. In this frame, the exclusion (transmission) angles are visualized as sectors centered at the triangle's summits and the direct sun rays traces as trajectories labeled with time (Fig. 2). In the case of Fig. 2, for a south vertical window at latitude 45°, with a trihedral element pointing 45 degrees above horizontal, the acceptance (reflection) mode takes place from april to august.

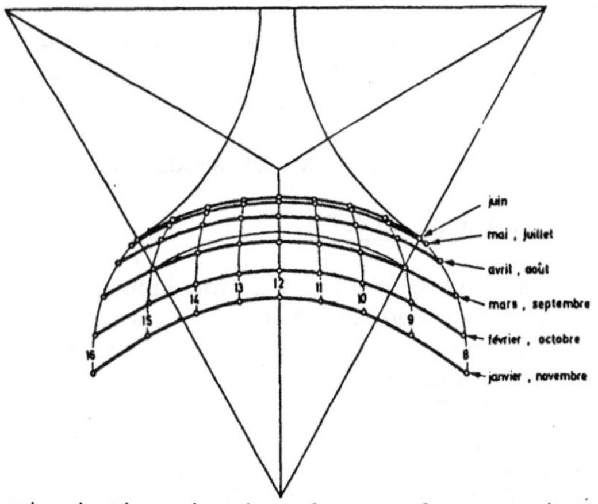

Fig. 2 : Sun rays trajectories in the triangle reference frame and acceptance sectors (see text).

The unit window, realized by Boussois S.A. for the CISI building is a three tempered glass panes panel 1,73 x 0,98 m wide. The catadioptric reflector is made out of polymethylmetacrylate (PMMA) and is set up between the two rear panes. The third outside glass pane is provided for additional winter insulation. On Fig. 3, a photograph of the reflector unit shows the overall dimensions and size of the trihedral element. The units are assembled together to form a continuous plane.

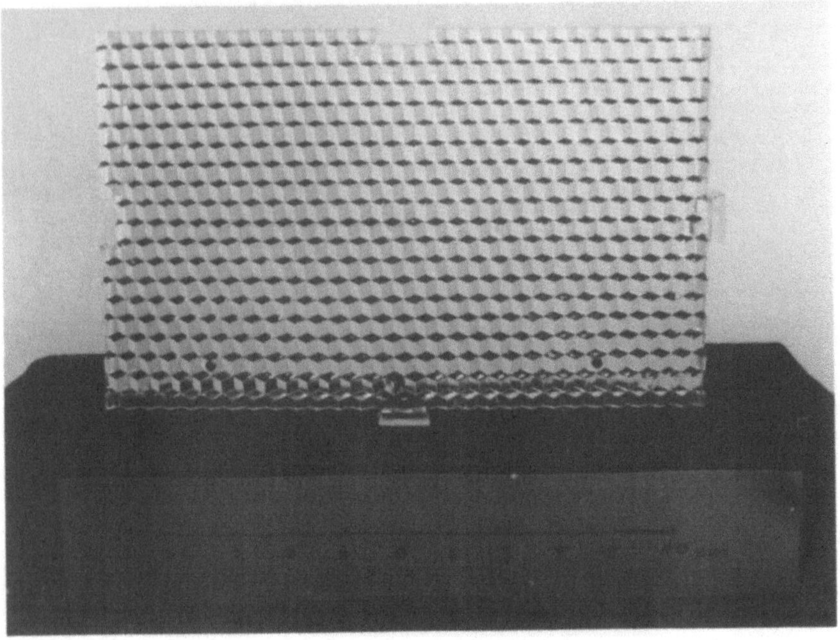

Fig. 3 : PMMA reflector element of the CISI building.

The design of the reflector is especially suited for the CISI building location and orientation (latitude 43° and south-east orientation).

The transmission/reflection mode of the window is selected by admitting distilled water in the space containing the reflector. For the operation, partial vacuum on the upper part of the window allows the water to be pumped up from a small water storage below the lower level of the window. In this way, the total hydrostatic pressure in the window is maintained below atmospheric pressure. The hydraulic circuit is common for a raw of windows.

3. THE OPTICAL MEASUREMENTS

Beside the sample measurements at the laboratory scale, the complete caracterization of the window, requires optical and thermal measurements in test cells with natural sun conditions, to fully take into account the behavior of the window for direct and diffuse radiations.

We report here on optical measurements which have been conducted in Cadarache in august-september 1983.

The window unit is enclosed as south face in a test cell of inside dimensions : 1 m x 1 m x 1,8 m made out of polystyrene blocks, whose reflectivity has been measured to be 0,92 in the wavelength band 550 to 1600 nm.

The instrumentation consists of :
- outside the cell, a Kipp and Zonen pyranometer (0,3 - 2,7 μm) and a Se cell (visible band) which are positioned in the window plane,
- inside the cell, a Cimel pyranometer (0,3 - 2,7 μm) and a Se cell (visible band) are vertically positioned on the internal face of the window, viewing all the internal reflective surface of the cavity. In addition, a pyranometer with a 0,4 - 0,72 μm filter is used.

The experiment is conducted by alternatively filling and draining the window in 15 minutes intervals from sunrise to sunset. The results are shown in Fig. 4 to 7 and table 1.

The qualitative observations we can made are the following :
- As is seen on Fig. 4 for August 3, the sun rays trajectory escape the acceptance angle between sun hours 11 a.m and 1 p.m. The transmission curves, Fig. 4, exhibits a clear enhancement from ~ 10 a.m to 1 p.m in the reflective mode (R mode). Instead, for the same time interval, the transmitted beam is attenuated in the transmission mode (T mode). This effect, which is not completely expected, remains to be explained.
- The diffuse light transmission is fairly constant all the day along both in T and R modes (Fig. 6).
- For the equinox data (Fig. 7) the total (0,3 - 2,7 μm) and visible (0,4 - 0,72 μm) transmissivities in T mode behaves differently, the former beeing peeked around noon, while in R mode, both total and visible values remains constant.

The mean transmissivity values extracted from these experiments are shown in table 1. They are not corrected for the effect of the absorptivity of the cavity walls. Nevertheless, the ratio of the transmissivities for the two modes should not be influenced by this effect.

	diffuse / global	0,3 - 2,7 μm		0,4 - 0,72 μm		Figures
		T mode	R mode	T mode	R mode	
23/08/83	> 0,90	0,45 - 0,50	0,30			6
equinox	< 0,11	0,55 - 0,70	0,30	0,35 - 0,40	0,20	4, 7

Table 1 : Transmissivities for diffuse and direct sunlight.

4. OVERALL PERFORMANCE EVALUATION

With the preceeding mean transmissivity values, a performance evaluation has been conducted for comparison of the variable transparency window with more classical solutions.

For temperate and hot climates (latitudes 41° and 33°) we have evaluated the solar gains and heat losses (or gains) for a south faced wall of single, double (without and with reflective indoor summer shade) and variable transparency glazings. The method used is described in ref. 4.

The assumed degrees-day values for the heating and cooling loads are the following (°C x day) :

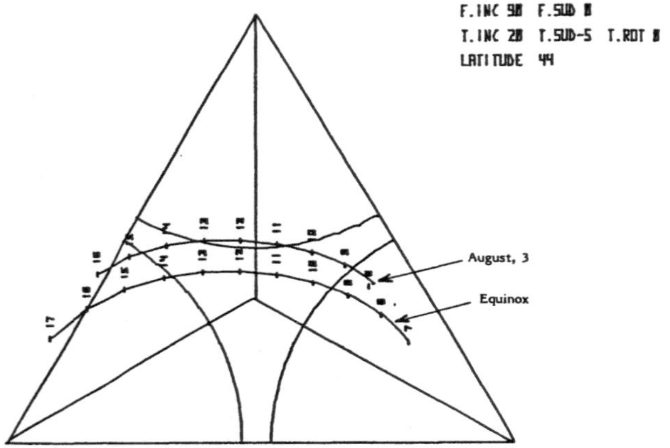

Fig. 4 : Experimental acceptance conditions

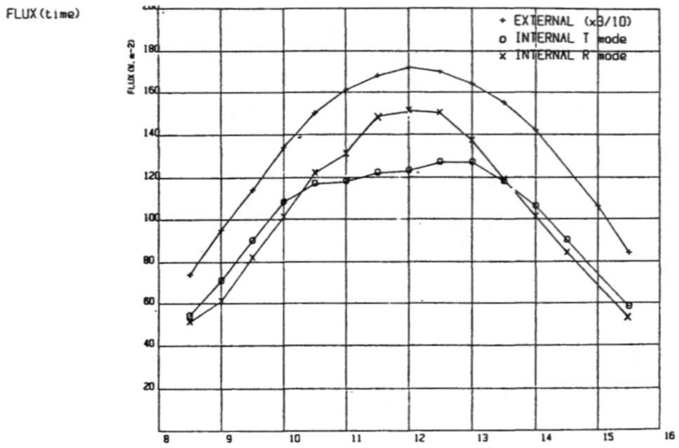

Fig. 5 : Effect of the acceptance angle (3 Août 1983)

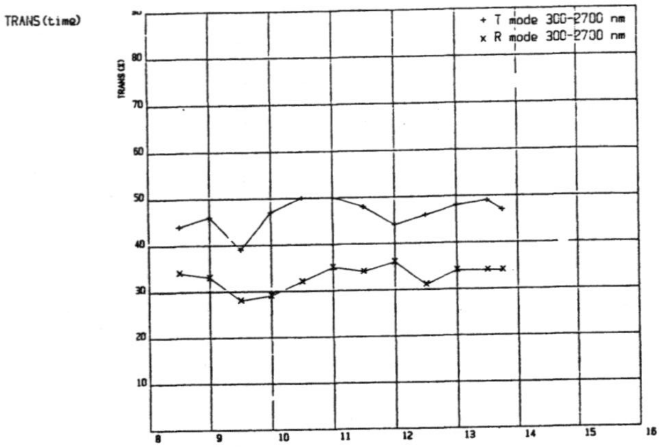

Fig. 6 : Transmissivities for diffuse light (23 Août 1983)

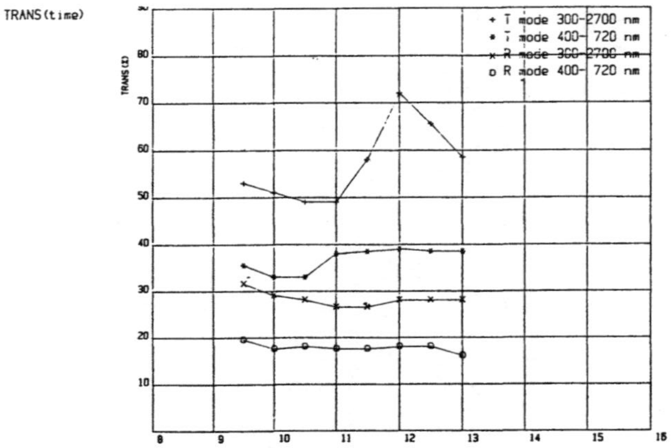

Fig. 7 : Transmissivities for direct light (equinox)

	heating	cooling
Latitude 41°	2620	0
Latitude 33°	1245	550

The thermal transfer coefficients U and optical attenuation coefficients τ (relative to single glazing) of the windows are taken as follows (ref. 4 and this study) :

	Winter		Summer	
	U W/m². k	τ	U W/m². k	τ
1 Single	6,4	1	6	1
2 Double	3,1	0,85	3,0	0,85
3 Double + indoor reflective shade (ϱ = 0,70)	3,1	0,85	3,0	0,3
4 Variable transparency	2,8	0,75	1,7	0,38

These data are used to calculate net energy balances for winter and summer seasons, which are summarized in table 2 :

	Latitude 41°			Latitude 33°		
	Winter	Summer	Total	Winter	Summer	Total
1 Single	3,4	- 7,4	- 4,0	11,6	- 13,7	- 2,1
2 Double	8,2	- 6,3	1.9	12,4	- 10,7	1,7
3 Double + reflective shade	8,2	- 2,2	6	12,4	- 4,7	7,7
4 Variable transparency	7,1	- 2,8	4,9	10,9	- 4,9	6

Table 2 : Net energy balances through south vertical windows (Savings on auxiliaries if positive, expenses if negative). Units 10^8 J/m^2 .

5. CONCLUDING REMARKS

Compared in the way above with solution 3, the variable transparency window performs unfavorably. This judgement should be weighted by the following remarks:
- Ultimate performance of the variable transparency window have not been reached and room exists for improvements, for instance by changing the reflector's material PMMA (n = 1,50) to polystyrene or polycarbonate (n = 1,60) which will have the effect to increase the acceptance angle.
- The thermal behavior of the various devices can be different as is their spectral selectivity (Fig. 7).

- Comfort aspects, such as diffuse light transmissivity are in favor of the variable transparency window.

These criteria should be taken into account for a detailed comparison. Most of them, the thermal performance of the V.T. window remains to be investigated in test cell experiments.

REFERENCES

1. B. DEVIN, D. ROYER. Fenêtres et capteurs catadioptriques à transparence variable. Principes et conditions d'emploi ; International Conference. Ente Fiera Milano, April 18,20 1980. Italy
2. J. MICHEL. La façade solaire à transparence variable ; ibid.
3. Les Cahiers Techniques du Bâtiment. Le Moniteur des Travaux Publics n° 45 Mai 1982
4. Efficient use of energy, American Institute of Physics Proceedings n° 25 New York 1975

RECHERCHE ET DEVELOPPEMENT D'ARCHITECTURE CLIMATIQUE ET DE COMPOSANTS SOLAIRES

Jacques MICHEL, Architecte, M.A. Harvard 58
14, rue des Poissonniers - B.P. 32
92204 NEUILLY-sur-SEINE CEDEX - FRANCE

Sommaire

Sachant que le verre est transparent à la lumière directe et diffuse du soleil, un pan de glace est placé devant les murs d'une maison. La lumière traverse le vitrage et au contact de la surface du mur peint en noir se transforme en rayonnement infrarouge réchauffant la lame d'air située entre le vitrage et le mur. Cette approche de la morphologie architecturale climatique perçue au cours de ma collaboration avec le Corbusier dans les années 1955, poursuivie avec Félix Trombe en 1966, a permis la réalisation des façades solaires telles que les maisons solaires de CHAUVENCY-le-CHATEAU en 1968, ODEILLO en 1972, plus récemment les façades des bâtiments du Centre A.F.P.A. de BEZIERS et du gymnase de SAINT-PERAY, etc... Ces projets conduisent aux définitions actuelles et prévisionnelles de l'auteur en matière de formes architecturales et d'enveloppes climatiques liées aux besoins et aux progrès de la Technologie.

Purpose of the work

Since 1962 more than 100 realisations of the author's solar architecture have been equipped or designed with solar components (circular house at LISSEY, first public Trombe-Michel wall house at CHAUVENCY-le CHATEAU in 1968, three solar houses at ODEILLO 1972, solar houses at SAINT-CHERON, ORSAY, A.F.P.A. Center at BEZIERS, solar gymnasium at SAINT-PERAY, Vitrac building, Demirsac building at ISTAMBUL, economical houses of french program 5000 solar houses, design of the mexican station of BAHIA de Los ANGELES with the use of passive and active systems at low average and high temperature, building envelopes with variable transparency of the CISI Palace at MILAN Fair etc...) The majority of these projects have been the subject of detailed monitoring programmes, carried out by the CNRS, EDF, CEA, CETIAT, private laboratory...

PRINCIPES CLES

Méthodes de construction

Procédés de construction liés aux composants solaires permettant la production en série du "paquet héliothermique" adaptable aux contenants habitables de divers volumes par les enveloppes de façade et de couverture comprenant les insolateurs, l'asservissement, le stockage thermique - intérêt scientifique et technique, vulgarisation sans imposer aucune contrainte architecturale.
Réalisation de 4000 m2 de bâtiments à BEZIERS, immeuble DEMIRSAC à ISTAMBUL Turquie.

Concepts de développement

Programme gouvernemental français des 5000 maisons solaires et insertion architecturale des brevets TROMBE-MICHEL des technologies passive et active (lotissements d'habitations à loyer modéré).

Etablissements humains

Dévelopment en 4 phases de la station Mexicaine de BAHIA de LOS ANGELES en Basse Californie en matière d'équipement portuaire, hotelier et des habitations en partant d'une urbanisation solaire et de l'emploi de systèmes passif et actif à basse, moyenne et haute température.

Réhabilitation des bâtiments anciens

Bloc fenêtre à 4 fonctions climatiques et phoniques équipé d'un store à lamelles orientables, l'une des faces est traitée en corps noir, l'autre avec un colorant sélectif émettant dans l'infrarouge.

Contrôle des enveloppes

Façade à transparence variable du Palais CISI à la Foire de Milan. Contrôle du flux solaire entrant dans une surface vitrée ou réflexion vers l'extérieur. En combinaison avec des surfaces noires, le système permet de disposer de capteurs solaires qui seront à volonté producteurs de chaleur ou réflecteur de l'énergie incidente vers l'extérieur.

Possibilité d'utilisation du rayonnement terrestre nocturne pour le rafraichissement du bâtiment.

1. INTRODUCTION

Exemples d'applications

Fig. 1 Maisons solaires de FONT-ROMEU (ODEILLO)

Vue de la façade Sud en cours de construction, sur un piton rocheux situé non loin du four solaire. L'Architecture tient compte de l'environnement, des conditions climatiques et des orientations ; les façades comportant des capteurs solaires sont situées en orientations Sud, Est et Ouest ; les ouvrants forment une unité avec les capteurs solaires. Une architecture personnalisée, tenant compte de programmes différents, démontre la souplesse d'emploi des capteurs verticaux dans un plan libre. La construction est réalisée en matériaux traditionnels. L'opération a obtenu une aide officielle et est considérée comme l'image de marque des maisons solaires du CNRS.

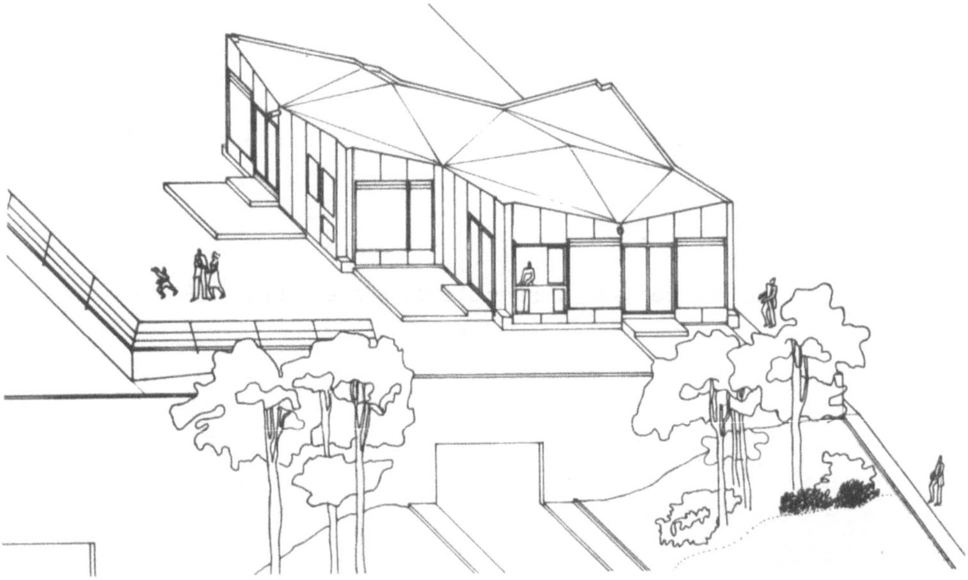

Fig. 2 Prototype de maison solaire industrialisée.

Au printemps 1973, le Comité des Expositions de Paris mettait à ma disposition, au Village de France de la Foire de Paris, un emplacement pour exposer une maison solaire. Avec l'aide bénévole d'un groupe d'industriels et d'entreprises, j'ai présenté une architecture volontairement prémonitoire découlant des recherches en cours en ce qui concerne l'application de matériaux secs permettant de réaliser des habitations à bon marché et rapidement montées. Cette maison était opérationnelle et comportait les nouveaux capteurs légers à masse thermique liquide, soit statique soit avec transfert après échauffement solaire dans un dispositif accumulateur calorifugé celui-ci situé dans une pièce de l'habitation non exposée au rayonnement solaire permet le chauffage de cette pièce par ventilation forcée.
A chaque volume correspondant à la trame de 3,60 m x 3,60 m de l'ossature tubulaire, était associée une façade comportant deux panneaux de 1,80 m de largeur. L'un de ces panneaux était constitué par un pan de verre assurant l'éclairement ; l'autre comportait le vitrage de la serre verticale, dont la surface de captage était calculée pour chauffer le volume compris dans la trame de 3,60 x 3,60 m.
Les conditions ci-après étaient parfaitement assurées :
- thermocirculation naturelle, (chauffage et ventilation d'été)
- renouvellement d'air,
- capacité de réserve thermique et restitution,
- extraction des airs viciés,
- chauffage électrique d'appoint.

Fig. 3 Vue perspective d'ensemble du projet AFPA à BEZIERS

Fig. 4 Façade solaire d'un atelier

La figure 3 montre la vue perspective d'ensemble de l'Association Nationale pour la formation Professionnelle des Adultes, réalisée à BEZIERS en 1978. Outre le bloc administratif, restauration, foyer, alimentés par préchauffage solaire grâce à un champ de collecteurs laminaires à eau disposés en terrasse et reliés à un stockage avec échangeurs sur circuit chaufferie, une première tranche de 11 ateliers de 4.500 m2 de surface est totalement "solarisée" au moyen des murs capteurs "F. Trombe-J. Michel", assurant la température de base de 12 à 14 °C pendant l'hiver et la ventilation d'été par thermocirculation naturelle d'air frais pris en façade Nord. Les 1.500 m2 de façades solaires assurent aux ateliers l'indépendance énergétique sans chauffage d'appoint.
Economie annuelle 45 à 50 tonnes de fuel.

Circuit hiver

1. capteur chaud
2. capteur froid
3. cheminée d'extraction
4. ventilateur
5. bouche de soufflage
6. bouche de reprise
7. bouche de reprise
8. stockage
9. admission d'air en façade nord
10. sonde de température de gaine
11. thermostat d'ambiance
12. sonde de température de stockage
13. convecteur d'appoint
V volet d'obturation
R registre 2 positions
-- chauffage direct
— — charge du stockage
..... décharge du stockage

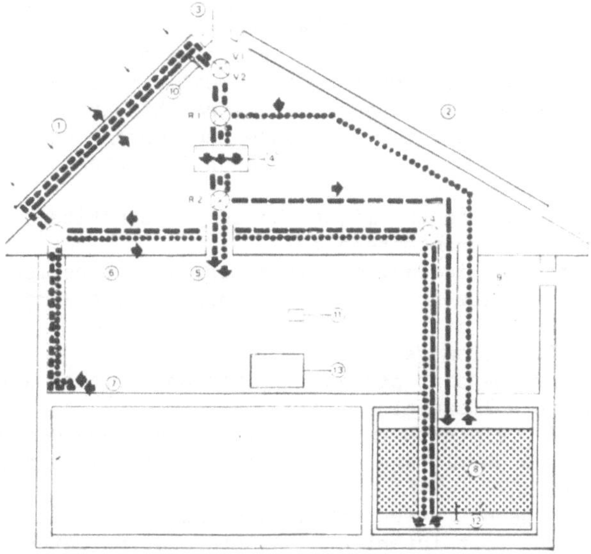

Fig. 5 Circuit hiver

Circuit été

1. capteur chaud
2. capteur froid
3. cheminée d'extraction
6. bouche de reprise
7. bouche d'arrivée air froid
9. admission d'air en façade nord
V volet d'obturation
-- refroidissement diurne
— — refroidissement nocturne

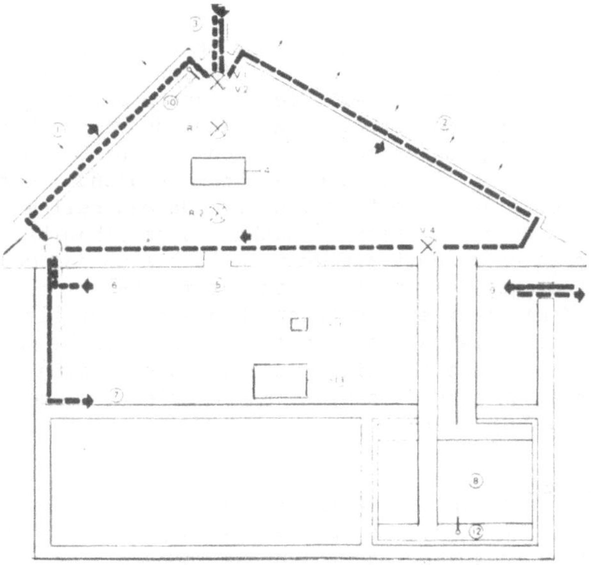

Fig. 6 Circuit été

Il est fait usage du brevet "Jour-Nuit" Trombe-Michel, utilisant les toitures Sud et Nord avec un procédé de chauffage solaire actif à air, distribution d'air,,refroidissement nocturne en été et stockage dans un lit de roches.

En ce qui concerne le rafraichissement nous nous référons aux travaux de J. FOURNIER et Félix TROMBE.

L'étude des composants proposés constitue, à la connaissance des auteurs du brevet, les premières réalisations en France de ce dispositif de chauffage solaire associé au rafraichissement nocturne, complément intéressant en été, notamment en zone méditerranéenne.

L'habitation comportera deux enceintes thermiques inclinées formant tout ou partie de la toiture.

L'enceinte thermique chaude est constituée d'un ou plusiers capteurs. Chaque capteur comprend :

- un matelas isolant en laine minérale fibre longue,
- une face absorbante en acier noir de faible masse thermique,
- une lame d'air,
- un vitrage en verre épais sélectif,
- un couvre-joint permet la jonction des bacs entre eux et assure l'étanchéité à l'eau et à l'air.

L'enceinte thermique froide ne diffère de la "chaude" que par le remplacement du vitrage par une plaque de tôle d'acier peinte avec un colorant sélectif.

L'une est orientée du côté exposé au soleil et est délimitée par un élément de toit intérieur constitué par un matériau capteur de masse thermique très faible, absorbant les rayonnements solaires de longueur d'onde comprise entre 0,25 et 4 microns, et un élément de couverture extérieur, transparent ou non, étanche à l'eau et à l'air et également de faible masse thermique. Cette enceinte thermique chaude est reliée d'une part à sa partie supérieure à un conduit formant extracteur, et à un conduit de circulation débouchant dans l'habitation et d'autre part, à sa partie inférieure à un conduit dirigé vers le bas et débouchant dans l'habitation et le stockage.

L'autre enceinte thermique (enceinte froide) est orientée du côté non exposé au soleil et est délimitée par un élément de toit intérieur et un élément de couverture extérieur étanche à l'eau et à l'air, essentiellement constitué par un matériau émetteur, de masse thermique très faible, se comportant comme un corps opaque non absorbant pour les rayons solaires et émetteur dans l'infrarouge de longueur d'onde comprise entre 4 et 40 microns.

Cette enceinte thermique froide est reliée à sa partie supérieure à un conduit de circulation débouchant dans l'habitation, et à sa partie inférieure à un conduit dirigé vers le bas et débouchant dans l'habitation et le stockage. Cet ensemble est automatisé et équilibré avec le renouvellement d'air neuf et l'extraction des airs viciés.

Une variante économique du dispositif de captage peut consister à créer une enceinte thermique unique entre un plan de verre (ou de tôle) extérieur et un plan de couverture inférieur, ce dernier étant bien entendu isolé thermiquement.

Régulation

L'installation héliothermique est automatisée par un système de régulation. Cette régulation se fait au moyen d'un thermostat d'ambiance à deux étages avec asservissement à une sonde placée dans le capteur ; le thermostat étant réglé à + 30° C, si la température d'ambiance descend au-dessous de cette valeur le ventilateur se met en route, à condition que la température de capteur soit au moins de 30° C.

Si l'ambiance descend au dessous de 19° C (insuffisance de chauffage solaire de base), le deuxième contact assure la mise en route d'un chauffage d'appoint.

Dans ces conditions, la priorité de fonctionnement est toujours donnée au chauffage solaire et interdit tout chevauchement inutile des deux systèmes.

Schémas de fonctionnement :

- Régime hiver - chauffage direct diurne.
- Régime été - rafraichissement nocturne.

a) circuit direct : capteur - habitation,
b) circuit de stockage : capteur masse accumulatrice,
c) circuit de restitution : masse accumulatrice - habitation.

Un système comprenant un thermostat d'ambiance, sonde de gaine, sonde de stockage et registres motorisés, permet le fonctionnement de l'un ou l'autre de ces trois circuits ou du chauffage d'appoint, en fonction des trois variables : température de capteur, de stockage et d'ambiance.

Points de départ théoriques ou scientifiques

Les mises en oeuvre actuelles ne traitent que l'aspect calorifique. Néanmoins les antériorités nous permettent toutes adaptations pour un rafraichissement.

On peut citer comme exemple le refroidissement très rapide du sable et de l'air au coucher du soleil au Sahara. Le sol constitué par du sable de faible masse thermique et de faible conductibilité calorifique prend très vite son équilibre de température.

En une demi-heure, le sable peut passer de 50° C à 10° C ou 15° C par rayonnement sur le ciel et il descend de plus en plus à mesure que l'air se refroidit. Référence : J. Fournier, Congrès AFEDES. UNESCO : Le soleil au service de l'homme, Paris 1973.

Des expériences au Chili ont montré qu'on obtient 20 Kg de glace par m2 et par nuit, grâce au rayonnement sur le ciel et à l'évaporation de l'eau dans l'air.

Ces deux phénomènes interviennent à peu près à égalité (50% rayonnement, 50% évaporation). Cela donne 10 Kg/m2/nuit pour le rayonnement.

La glace pour se former dégage 80 calories par Kg. Il faut donc fournir, pour 10 Kg, 800 calories sous forme de froid, soit à peu près 1 KWH (860 calories)

On dispose donc, dans de bonnes conditions, de 1 KWH par nuit sous forme de froid.

Pour une toiture rayonnante de 10 m x 5 m = 50 m2, cela donne 50 KWH par cycle journalier pour climatiser. Cela est une limite et on produira moins dans un ciel moins clair que celui du Nord Chilien, mais il s'agit là pour le Chili, d'une moyenne. J. FOURNIER a obtenu certains jours 2,5 fois plus.

Ajoutons qu'un toit rayonnant sur le ciel distribue son refroidissement sur l'air extérieur qui "coule" selon le principe du patio (exemple de la maison de MONTELIER TENSARD, brevet F. TROMBE-J. MICHEL)

On peut également constituer le versant en double peau de sorte que l'air refroidi "coule" à l'intérieur sans subir les turbulences atmosphériques.

Etat d'avancement

Les bâtiments équipés de ces composants reçoivent une toiture dont l'inclinaison Sud est située à 45° ou 60° sur l'horizontale avec montage d'insolateurs à air sans inertie pour un dispositif de chauffage de type "actif", fonctionnant avec des moyens mécaniques et transmission de l'énergie et une accumulation dissociée du captage.

L'originalité consiste à séparer les générateurs chaud et froid ; avec ce dispositif on peut choisir les meilleurs angles de captage pour le soleil et le froid et utiliser pour ce dernier, les surfaces sélectives (peintures blanches à l'oxyde de titane par exemple). On aura ainsi une toiture noire ou de valeur sombre côté soleil et blanche à l'opposé, avec un angle de 30° sur l'horizontale, extraction sur les 2 versants du faîtage, côté soleil pour chauffer ou aspirer de l'air dans le bâtiment, et côté froid avec gaine verticale d'écoulement d'air dirigée vers le bas. Le stockage sera constitué par un lit de galets calibrés, aménagé dans le sous sol de la maison. Une centaine de maisons solaires a été réalisée dans le cadre du Concours National des 5000 Maisons Solaires utilisant un paquet solaire "Jour-Nuit".

2. LES PROPOSITIONS

Description :

La rénovation récente de la façade du Palais CISI à MILAN nous a donné l'occasion d'oeuvrer dans cet esprit.

Moduler la quantité de lumière passant à travers un vitrage solutionnant ainsi le contrôle des surfaces vitrées avec également une production d'énergie.

On réalise une transmission variable du rayonnement solaire en utilisant les propriétés réfléchissantes des structures catadioptriques composées de pyramides trirectangles (utilisées dans l'industrie automobile) et l'on module ou supprime la réflexion totale de la lumière sur les faces des pyramides en introduisant un liquide approprié au contact du réseau de pyramides.

J'ai proposé à la Foire de MILAN ce contrôle de la lumière mis au point par le laboratoire des techniques solaires du C.E.A.

La façade du Palais CISI, d'une surface générale de 470 m2 est composée de 250 vitrages de 5 types différents, dimensions 1,73 m x 0,983, produits par la Société BOUSSOIS.

L'ensemble est intégré dans une structure métallique tubulaire conçue pour ces vitrages.

Monté dans un profil aluminium, les 3 vitres sont en verre Float de 4 mm trempé thermiquement, les espaces d'air sont de 12 mm pour la lame extérieure et de 8 mm pour la lame intérieure. La conception du vitrage le garantit contre la condensation grâce à l'étanchéité et aux dessicateurs intégrés et renouvelables.

La lame d'air de 8 mm est occupée par un catadioptre en métacrylate et sa périphérie est pourvue d'un dispositif de remplissage et vidange d'eau distillée.

Quand le film d'eau est mis en place (par aspiration) le relief du catadioptre s'estompe et la lumière solaire est transmise à l'intérieur du bâtiment - transmission lumineuse de l'ensemble plein d'eau : environ 85%.

Compte tenu de la différence d'indice entre l'eau et le métacrylate, la vision optique est floue.

Quand le vitrage est vide, le catadioptre joue son rôle de réflecteur la lumière est renvoyée vers le soleil dans une proportion de 90%.

La disposition de l'axe des trièdres du catadioptre, adaptée pour la région de MILAN doit permettre de réfléchir la lumière solaire en Avril et Octobre de manière efficace, entre 7 H 30 et 11 H 30 (heure solaire vraie)

L'étude à conduire sur le plan scientifique, comporte essentiellement le choix des heures de la journée auxquelles la transparence variable doit être utilisée, afin de choisir la structure catadioptrique en conséquence.
Références : Trombe F. et Michel J. (1970) Brevet Français n° 2 144.066 et

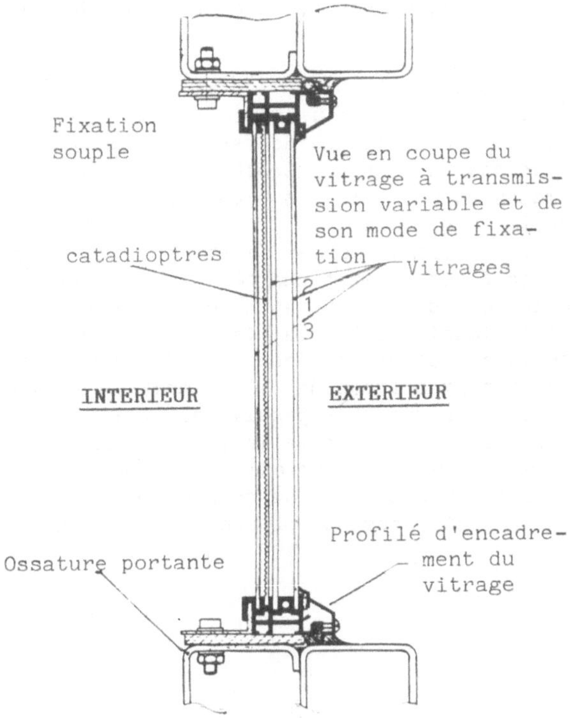

Fig. 7 Vue en coupe du vitrage à transmission variable

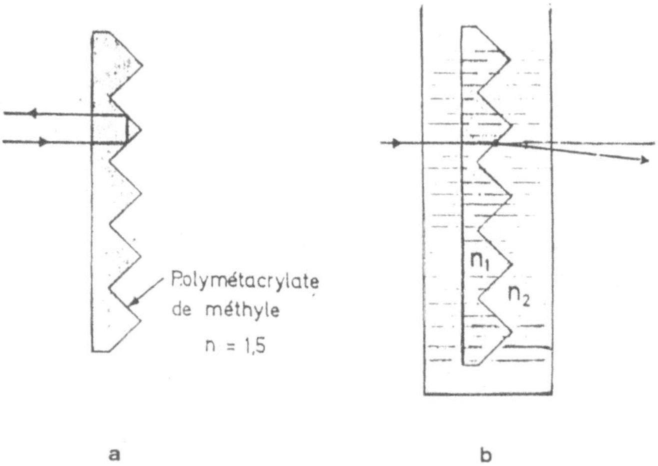

Fig. 8 Principe de la fenêtre à transmission variable

a) Le catadioptre renvoie la lumière
b) Le catadioptre immergé n'a plus d'effet, la lumière passe

et 2.189.686 voir aussi U.S. patent n° 3.832.992 - Trombe F. et MICHEL J. (1975) "Jour-Nuit Climatisation" brevet français n° 75 32921 - International Conference - Alternative Energy and the offer of the sun and the sea - Milan Avril 1980.

3. CONSEQUENCES ATTENDUES

Ce nouveau traitement de la peau vitrée des bâtiments ouvre des possibilités fantastiques à l'architecture de verre et aux murs rideaux traditionnels puisque l'on pourra ainsi contrôler la lumière, donc l'énergie transmise aux bâtiments et utiliser les plans vitrés favorables en orientation pour la production d'énergie.

D'autre part, la perception de ces quelques exemples montre que l'équipement solaire passif des habitations et des bâtiments à bas besoins énergétiques sans chauffage d'appoint, apporterait la survie en cas de pénurie de combustibles classiques par la mise hors gel hors humidité des espaces habitables abaissant d'autant l'escalade du chiffre des besoins tels que nous les voyons et permettant ainsi l'intégration de composants solaires actifs sophistiqués sur une base passive réduisant les besoins classiques, l'importance et le coût des installations solaires.

Ceci est valable également en période d'été pour la recherche du confort grâce à la ventilation naturelle de jour par effet de cheminée des capteurs à air et le refroidissement nocturne des toitures rayonnant sur le ciel.

SIMPLE AIR COLLECTORS FOR PREHEATING FRESH AIR

J.L.M. Hensen, M.H. de Wit
group FAGO-TNO-THE
Eindhoven University of Technology

Summary

In dwellings with mechanical ventilation systems the fresh air can easily be preheated by means of simple solar air systems. These can be an integral part of the building facade or roof and the costs are expected to be low. By means of computer experiments a large number of systems were evaluated. It was found that these systems can provide interesting energy gains providing the system is well dimensioned and has a storage. For this massive airducts are more effective than a massive absorber.

1. Introduction

In dwellings with mechanical ventilation systems the fresh air can easily be preheated by means of a simple air collector. Figure I shows an outline of such a system. This could be an integral part of the building structure and does not need any control mechanism. Therefore the costs are expected to be low. The system can be equipped with a massive absorber or remote storage (e.g. hollow concrete floor) to improve the overall performance and to reduce the temperature swings of the outlet air.

Figure I Outline of the system.

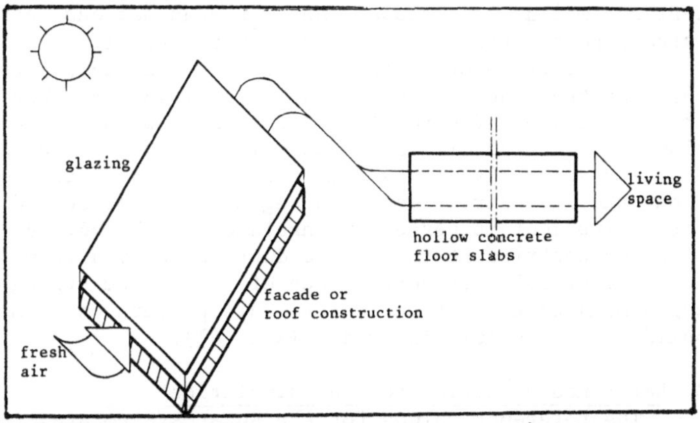

The influence of several parameters on the performance was studied by means of computer simulation. Section 2 describes the simulation model. The reference situation is outlined in section 3. The simulation results and the conclusions are described in sections 4. en 5.

2. Simulation model

The collector without massive absorber can be represented by a thermal conductorenetwork as shown in figure II. With the assumptions that there is only one-dimensional heat flow in the glazing and the absorber (perpendicular to air flow) and no heat loss from the absorber to the back, the efficiency of the collector can be computed as indicated in figure II. A detailed description of the different conductances is beyond the scope of this paper. The convective heat transfer coefficients were derived from (1,2). To be able to account an enlarged absorber area (e.g. corrugated sheet) a distinction is made between convective heat transfer coefficients on either side of the gap.

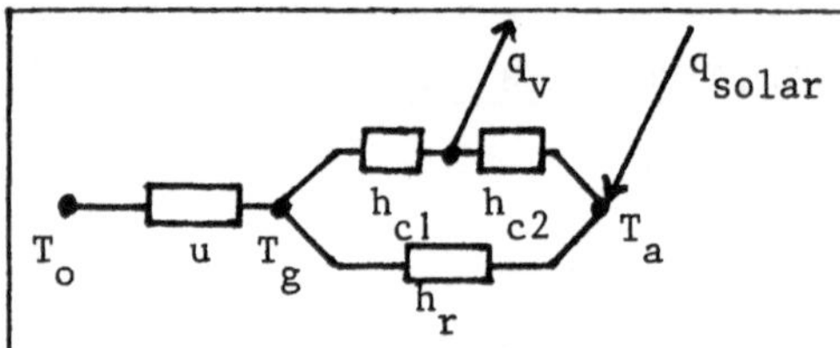

$$q_v = \eta_v \cdot q_s$$

$$\eta_v = \frac{\rho.c.V/A}{k1+k3+\frac{k1.k3}{k2}}\left(1-\exp\left\{\frac{-(k1+\frac{k2.k3}{k2+k3})}{\rho.c.V/A}\right\}\right)$$

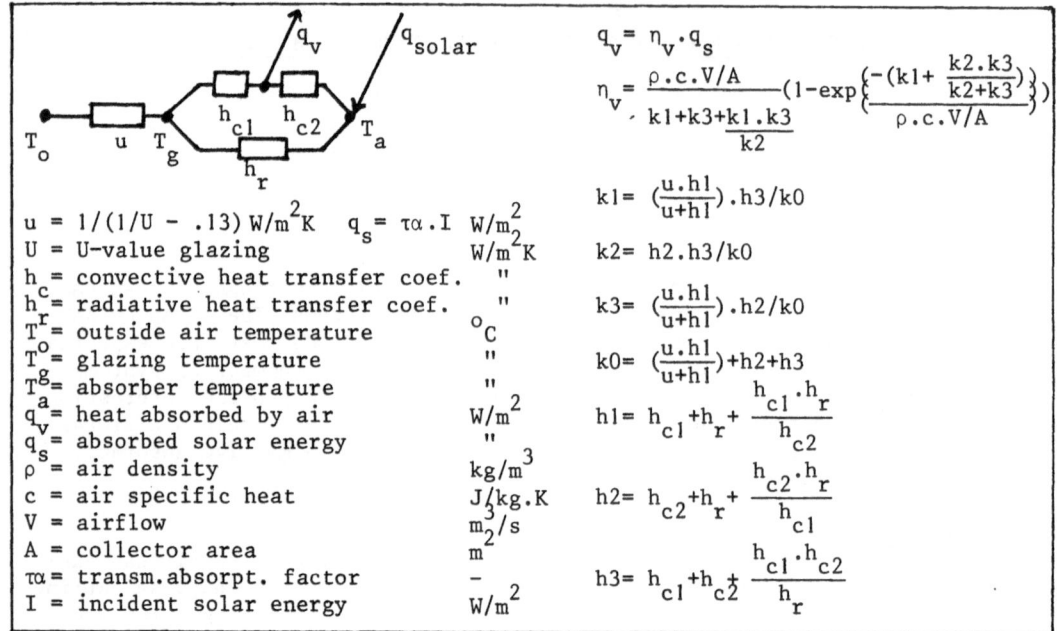

$u = 1/(1/U - .13)$ W/m^{2}K	$q_s = \tau\alpha.I$	W/m^2
U = U-value glazing		W/m^{2}K
h_c = convective heat transfer coef.		"
h_r = radiative heat transfer coef.		"
T_o = outside air temperature		^{o}C
T_g = glazing temperature		"
T_a = absorber temperature		"
q_v = heat absorbed by air		W/m^2
q_s = absorbed solar energy		"
ρ = air density		kg/m^3
c = air specific heat		J/kg.K
V = airflow		m^3/s
A = collector area		m^2
$\tau\alpha$ = transm.absorpt. factor		-
I = incident solar energy		W/m^2

$$k1 = \left(\frac{u.h1}{u+h1}\right).h3/k0$$

$$k2 = h2.h3/k0$$

$$k3 = \left(\frac{u.h1}{u+h1}\right).h2/k0$$

$$k0 = \left(\frac{u.h1}{u+h1}\right)+h2+h3$$

$$h1 = h_{c1}+h_r+\frac{h_{c1}.h_r}{h_{c2}}$$

$$h2 = h_{c2}+h_r+\frac{h_{c2}.h_r}{h_{c1}}$$

$$h3 = h_{c1}+h_{c2}+\frac{h_{c1}.h_{c2}}{h_r}$$

Figure II Thermal conductance network representing the collector

The radiative heat transfer coefficient is evaluated at mean surface temperatures. When this coefficient would be a constant the efficiency is not dependent on ambient conditions. Then the mean seasonal efficiency could be calculated directly from the equations in figure II.

A massive absorber is simulated by subtracting the heat flow into the absorber from the absorbed solar energy. The absorber heat flow is computed from an implicit finite difference scheme with one hour time steps and .02 m place steps perpendicular to the air flow. The absorber is perfectly insulated.

A massive airduct is simulated by dividing the duct in segments (.50 m) along the flow direction. Each segment is represented by one resistance/capacitor. The duct is perfectly insulated on the outside. The convective heat transfer coefficient inside the duct is derived from (3).

The overall net performance of the system strongly depends upon the heating demand pattern of the house. This pattern was computed by means of a large building load simulation model (KLI (4)).

3. Reference situation for the simulation

The reference values for the solar system are:

Airflow per area		V/A	25	50	100	200	m^3/h.m^2
Area for total flow = 200 m^3/h		A	8	4	2	1	m^2
Lenght and absorber-glazing dist.		l, d	2	,.04			m
Glazing:	single	U	6				W/m^{2}K
	solar transmission	τ	.85				-
	emissivity	ε	.90				-
Absorber:	absorptivity = emiss.	$\alpha=\varepsilon$	.90				-
	density	ρ	2500				kg/m^3
	specific heat	c	840				J/kg.K
	south oriented; tilt		45				degrees
Ducts:	diameter inside (outs.)	d_i (d_o)	.10	(.20)			m
	density	ρ	2500				kg/m^3
	specific heat	c	840				J/kg.K

The reference dwellings are taken from (5). The dwellings are relatively small (in a row); only the ground floor level is heated (with night set-back); approx. 4.5 m^2 south oriented double glazed window per story. The following distinctions in type of dwelling were made:

1. Standard insulation (.04 m); heavy construction (brickw.); Qaux 7400 kWh/a
2. Standard insulation (.04 m); light construction (timber); Qaux 6500 "
3. Increased insulation (.10 m); heavy construction (brickw.); Qaux 3000 "
4. Increased insulation (.10 m); light construction (timber); Qaux 2400 "

The reference ambient conditions are a standard reference year (6). The total solar irradiation per heating season is 510 kWh/m^2 for a vertical south plane and 610 kWh/m^2 for a 45 degr. tilted south plane.

4. Simulation results

The results are presented as a function of the parameter V/A because this is the main influencing parameter on the collector efficiency (fig.II). The influence of airspeed (determined by the aspect ratio l/d) and glazing type is shown in figure III and IV. There is a negligable difference between varying l or d because the Reynolds-number depends both on airspeed and d.

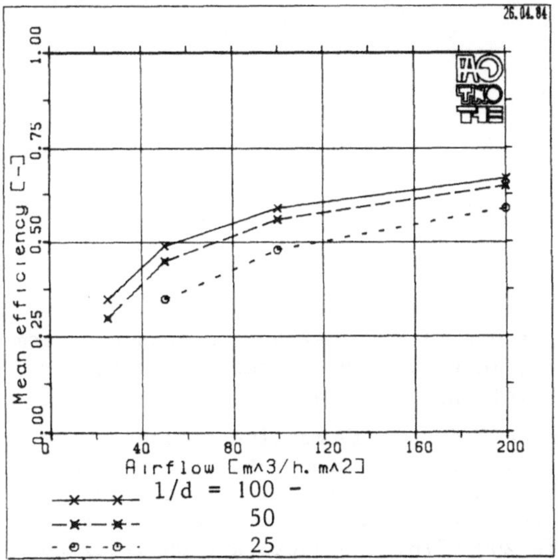

Figure III Mean heatingseason collector efficiency and aspect ratio

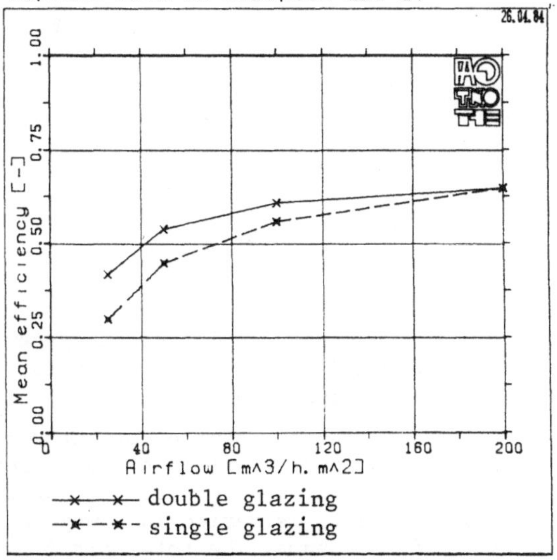

Figure IV Mean heatingseason collector efficiency and type of glazing

Higher aspect ratios (longer) result in a better efficiency. Double glazing is favourable at low airflow rates. A higher flow rates the difference with single glazing becoms negligable.

Selective surfaces were not considered because these would make the system more expensive and problems of degradation (dust etc.) could arise.

If the mean (output weighted) radiative heat transfer coefficient is known, the mean collector efficiency can be calculated from the equations in figure II. It was found that this radiative heat transfer coefficient varied from approx. 5.5 W/m^2K (V/A = 25) to approx. 4.5 W/m^2K (V/A = 200).

Only part of the collector output is useful because of the simultaneous input of solar energy by the windows and the solar collector. The mean utilization factor for a total flow of 200 m^3/h varied from .85 (A = 1 m^2) to .75 (A = 8 m^2) for a vertical south collector and dwelling 1. respectively from .55 to .40 for dwelling 4.. The mean utilization factor will be higher when an adequate storage is provided. For this there are two possibilities: a massive absorber or massive airducts coupled to the collector. The absorber could be made of e.g. concrete or a brickwork facade. Figure V shows the mean the mean net efficiency for a collector with a massive absorber and dwelling 4. The total air flow is 200 m^3/h. The efficiency increases only at low flowrates. The effect of the concrete layer is very small. It was found that a large part of the collected solar energy is released during the night-time set-back when there is no heating demand. This reduces the benefit of the heat released during evening hours in such a way that the two effects compensate each other. Warmer air during the evenings and reduction of the temperature swings during daytime have a positive effect on the thermal comfort conditions indoors.

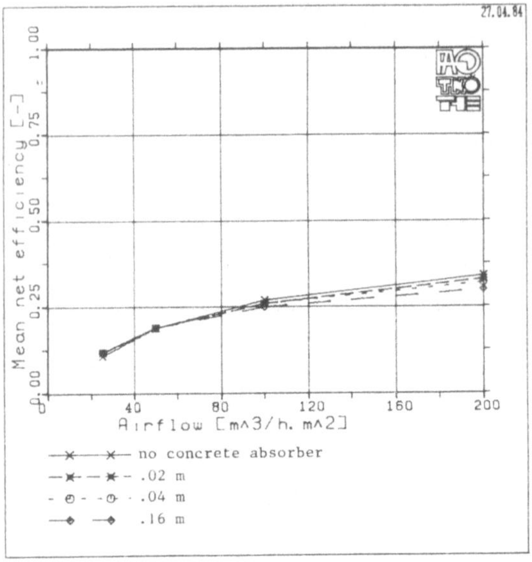

Figure V Mean net heatingseason system efficiency and thickness of concrete absorber for dwelling 4.

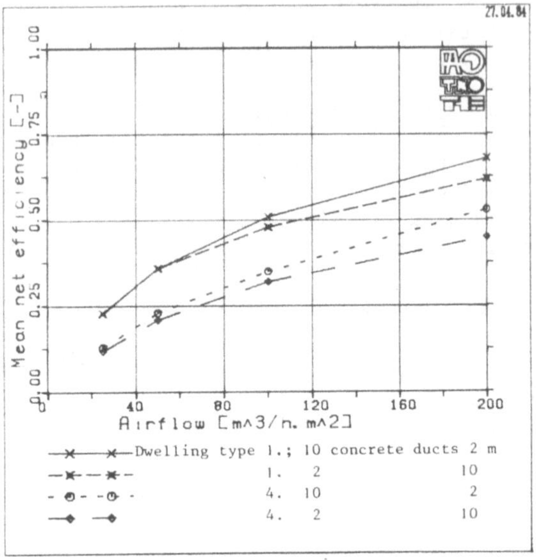

Figure VI Mean net heatingseason system efficiency with concrete airducts for dwelling 1. and 4.

The outlet air of the collector can be led through massive airducts like e.g. hollow concrete floor slabs. Figure VI shows the mean net efficiency of the system with massive airducts for two types of dwelling. The lenght and number of ducts is varied in a way that the total volume of concrete (approx. .5 m^3) is the same. The total air flow is 200 m^3/h. The ducts are parallel to each other. Increasing the number of ducts decreases the air speed in the ducts and increases the system performance. Increasing the number of ducts also reduces the total pressure loss of the ventilation system and thus the energy demand of the fan. The transfer of heat into the storage is more direct in a massive absorber. The heat exchange area is however much smaller. The system with massive airducts can also have much more effective storage material per collector area (max. about .10 m^3/m^2 for a massive absorber). This enables the system to distribute the collected solar energy over a longer period resulting in a higher utilization factor.

5. Concluding remarks

The simulation results lead to the conclusions:

- this type of solar system can provide an interesting energy gain at probable relatively low costs
- because the fresh air need in a dwelling is relatively small, the collector should not be large to achieve a good performance per m^2
- the performance of the system increases with increasing flow rate per collector area. The total airflow should however not be above the fresh air need because the extra heat output of the solar system is always less than extra heating demand due to the increased ventilation losses
- the efficiency of the collector increases with increasing aspect ratio l/d
- single glazing is sufficient providing the flow rate per m^2 is high

- the performance of the system is much better when an adequate storage is added to the collector. For the considered types of dwelling massive air ducts are more effective than a massive absorber
- adding storage to the system improves the thermal comfort conditions indoors by reducing the air temperature swings during daytime and providing warmer ventilation air during evenings
- the system will cause an extra pressure loss for the ventilation system resulting in a higher energy demand for the fan. This aspect still has to be researched.

REFERENCES

1. VDI-WARMEATLAS (1977). VDI-verlag, Düsseldorf.
2. DUFFIE, J.A. and BECKMAN, W.A. (1980). Solar engineering of thermal processes. John Wiley & Sons, New York.
3. SPRENGER, E. (1979). Taschenbuch für heizung und klimatechnik. R.Oldenbourg, München.
4. BRUGGEN, R.J.A. van de (1978). Energy consumption for heating and cooling in relation to building design. Thesis Eindhoven University of Technology.
5. KOK, J.W.M. e.a. (1982). Referentiewoning, bouwtechnische omschrijving en kostenbegroting. Rapport no. 7975 Bouwcentrum, Rotterdam.
6. Verkort referentiejaar voor buitencondities (1983). ISSO-publicatie 12, ISSO, Rotterdam.

TRANSPARENT INSULATION SYSTEM FOR PASSIVE SOLAR ENERGY UTILIZATION IN BUILDINGS

A. Goetzberger, J. Schmid and V. Wittwer
Fraunhofer-Institut für Solare Energiesysteme, 7800 Freiburg, West-Germany

Summary

A new concept for the passive use of solar energy, transparent insulation, is described together with the first experimental results. Transparent insulation material has the property of being transparent or translucent to solar radiation while at the same time acting as heat insulation. Elements made of this material can be attached to the walls of buildings and thus permit the utilization of solar energy for heating. Relations are given for the dependence of heat flux and conversion efficiency of radiation into useful heat on the thermal resistance of the components. Calculations using meteorological data show that with material-parameters achievable with present technology not only south but also west/east and possibly even north orientations can lead to significant contributions to heating. In order to avoid overheating in summer, control of radiation must be provided. Experiments with two different materials on two buildings showed promising results. In sunny periods in wintertime a heat flux into the house was measured for south and west orientated walls. The mean heat flux is reduced drastically depending on the material and orientation used.
The beneficial effect of masonry walls with regard to heat storage and damping of temperature fluctuation was also demonstrated.

1. INTRODUCTION

The passive use of solar energy has attracted much attention in recent years. For instance, it has been demonstrated that windows, particularly those oriented towards the south,can produce noticeable heat gains even at low global irradiation (1, 2). Other concepts, such as the Trombe wall or the greenhouse have been investigated both theoretically and experimentally (3). In this contribution the principles and first experimental results are reported of a very simple concept for passive solar energy use, namely transparent heat insulation. The principle is shown in Figure 1. A layer of insulation having transparent or translucent properties is arranged in front of the facade of a building. The surface of the wall has the properties of a light absorber, thus the incoming radiation is converted into thermal energy (4, 5, 6). Depending on the heat insulation efficiency of the transparent layer, more or less of the solar energy will flow into the building, thus contributing useful heat. The quantitative relations will be described below.

The advantages of the concept are the following:

- Possible utilization of solar radiation on all four sides of a building.
- Easy retrofit of existing structures.
- The heat storage effect of massive walls is employed.

In connection with the last point, it should be mentioned that use of the storage effect of walls offers advantages compared to windows because practically no overheating occurs during the winter. Thus the entire radiative energy can be converted to thermal heating energy. On the other hand, because of the flow of heat in two opposing directions, only part of the energy absorbed is converted into useful heat.
Another advantage of this kind of heating may be found in the favourable radiation ambient within the room. The level of comfort is increased by radiation from the walls.
As well as the advantages just mentioned, there are also a number of disadvantages and problems which appear, however, to be surmountable. The most serious problem is the risk of overheating during summer. A wall covering capable of contributing to the winter heating load by conversion of radiation will tend to produce overheating during summer. Thus shading devices or other means of removing the unwanted heat are needed. A seasonal adjustment may be sufficient in some cases but automatic control would be most desirable. These devices have yet to be developed.

2. BASIC PRINCIPLES

In this section simple relations are derived for the heat gain from radiation through transparent insulation systems. Because of the thermal inertia of a heavy wall combined with short-term fluctuations of outside and inside temperature, the instantaneous temperature and heat flow conditions can only be obtained by numerical computation. However, if only average values of temperature and heat flow over longer periodes, such as months, are considered the relations can be drastically simplified. In particular, all heat storage effects can be neglected since they are of importance only during shorter periods. The thermal equivalent circuit for this case of the combination transparent insulation - wall is sketched in Figure 2.
The total efficiency for the conversion of the incoming radiation into thermal heat is given by

$$\eta_c = \tau \, \alpha \, / \, \left(1 + \frac{k_I}{k_W}\right) ,$$

where τ is the transparency of the element and α the absorption of the wall.
Good efficiencies are obtained with good transparency and low k-value of the insulation, and high k and α of the wall. Since old building stock often has badly insulated walls, such buildings are particularly amenable to retrofitting with transparent insulation.
First theoretical estimates are given in reference (7). These calculations show the principal difference between conventional insulation and transparent insulation: conventional insulation can only minimize energy loss, while transparent insulation can also deliver energy gain to the interior of the building.

3. EXPERIMENTAL RESULTS

In order to test the potential of the ideas advanced here, experiments on two single family dwellings in Freiburg were carried out. Material 1, a PMMA foam, was tested at a south-, west- and north-facade in the first months of 1983. Material 2, a capillary structure, was tested at a west-facade beginning in January 1984. The cross-section through the experimental elements attached to the buildings are given in Figure 3a and 3b. The test area in the case of the south-facade was about 9 m² and 12 m² and 8 m² for the west- and northern-facade.

The characteristic data of both elements are given in Table 1.

Table 1 Characteristic data fo the elements

	transmission	heat losses coefficient
PMMA-foam	0.5	1.8 W/m²K
Capillary-structure	0.6	1.0 W/m²K

From these data the mean heat flux average values for each week and month as well as the total system efficiency were calculated.
In Figure 4 mean fluxes in W/m² for the south-facade are plotted on a weekly basis.
The upper curve represents the uninsulated wall, including the effect of irradiation on the exposed surface, which is rather small. The curve below shows the change due to the heat insulation alone (k_W = 1.3 W/m², k_I = 1.8 W/m²K) and the lowest curve shows the measured heat flux dominated by radiative gain. The results for the south orientation are summarized in Table 2

Table 2 Results for the south facade

	Global radiation (W/m²)	Heat flux (W/m²)	ΔT °C	Conversion efficiency	Heat gain (Wh/m²d)	Heat demand T_{in} = 20°C (Wh/m²d)
Mean values Jan. 1983	55	-4	7.6	0.17	222	362
16.-23 Febr. 83 (sunny period)	175	-21	10.6	0.16	671	334

The results show, that in sunny periods there is no need of an additional heating system on a south facade in wintertime with material 1. Even in cloudy periods and on west and north orientated walls large amounts of energy could be saved. More details are given in reference (7).
The capillary structured material 2 is tested on the west-facade since January 1984. Figure 5 shows the mean fluxes in W/m² on a monthly basis in a similar way as done in Figure 4 for the south facade on a weekly basis (K_I = 1 W/m²K, K_W = 1.3 W/m² K). Once more one can see, that the energy demand is reduced drastically. In February and March the mean heat flow of the total system is negative even at the west facade. A summary of the results is shown in Table 3:

Table 3 Results for the west facade

	Global radiation (W/m²)	Heat flux (W/m²)	T °C	Conversion efficiency	Heat gain (Wh/m²d)	Heat demand T_{in} = 20°C (Wh/m²d)
January 1984	14.0	2.2	13.1	0.37	125	224
February 84	32.7	-0.18	14.8	0.26	205	249
March 1984	62.8	-7.64	13.0	0.24	360	217

4. CONCLUSION

The results reported here pertain to an early phase of the development of transparent insulation. The main objective was to show that the effects are as expected and that solar radiation can be used in the manner described. Thus considerable energy savings in buildings of all kinds can be expected with further development of this concept. Future planned work will be directed towards improving the transparency and heat insulation properties of materials, development and testing of radiation control devices, and questions related to practical use in architecture.

ACKNOWLEDGEMENTS

We are greatly indebted to F. J. Kamerewerd and W. Platzer for installing and operating the data acquisition system and to W. Stahl and A. Pflüger for measurements of thermal material properties and helpful discussion. In addition we have to thank IMC-Acrylguss and Diveno-Kapillar for providing the insulation material.

REFERENCES

1. WERNER, H.; Wärmedurchgang durch Fenster und Wand unter Berücksichtigung der Sonneneinstrahlung, Haustechnik Bauphysik und Umwelt 102, 121 (1981)
2. GERTIS, K., HAUSER, G., KÜNZEL, H., NIKOLIC, V. ROUVEL, L., and WERNER, H. Energetische Beurteilung von Fenstern während der Heizperiode, DAB 12, No. 2, 201 (1980)
3. PALZ, W. and DENOUDEN, Solar Energy Applications to Dwellings, Solar Energy R & D in the European Community Series A. Vol 2, D. Reidel Publishing Company, Dordrecht (1982)
4. HAUSER, G. Passive Sonnenenergienutzung durch Fenster, Außenwände und temporäre Wärmeschutzmaßnahmen, HLH 34, No. 4,144, No. 5,200, No. 6, 259 (1983)
5. OLSEN, L. Transparente Insulation for Thermal Storage Walls, Solar Energy R & D in the European Community Series A, Vol. 2,127, D. Reidel Publishing Company, Dordrecht (1982)
6. BERTSCH, K., BOY, E., FRANGOUDAKIS, A. and HEIM, U., Transparent insulation materials - a possibility for reducing the consumption of fossil fuels in households, 1st EC-Conference on Solar Heating
7. GOETZBERGER, A., SCHMID, J., WITTWER, V., Transparent Insulation System for Passive Solar Energy Utilization in Buildings, International Journal of Solar Energy , to be published.

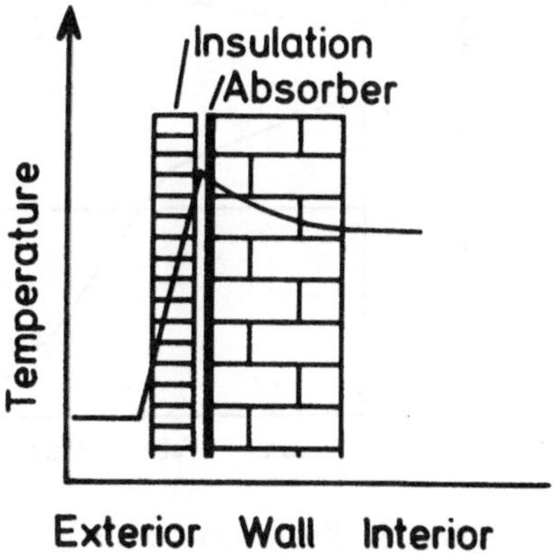

Figure 1: Principle

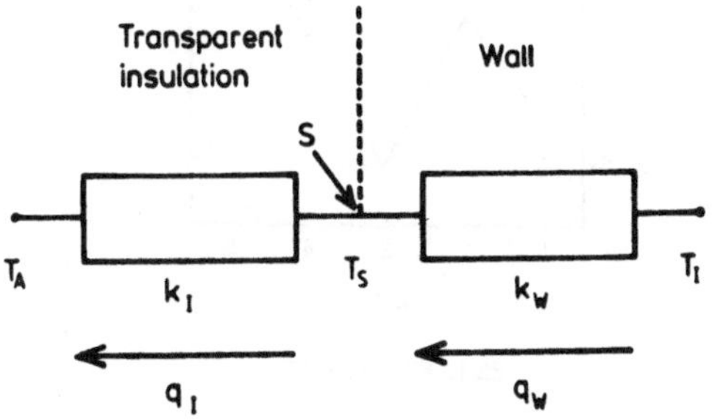

Figure 2: Thermal equivalent circuit
T_A, T_S, T_I temperatures
K_I, K_W heat transfer coefficients
q_I, q_W heat flows
S irradiation

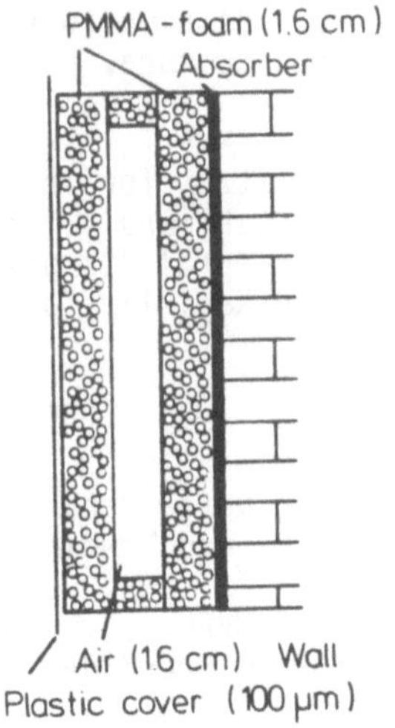

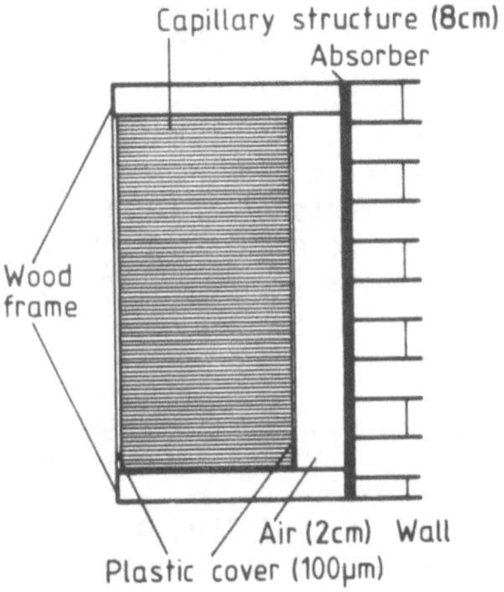

Figure 3a:
Cross section of the transparent insulation with the PMMA-foam

Figure 3b:
Cross section of the transparent insulation with the capillary-structure

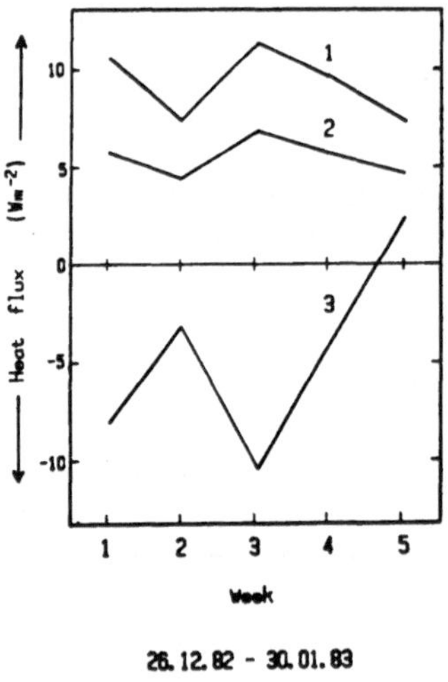

Figure 4: Mean heat flux through south facade K_W=1.3W/m² K, K_I=1.8W/m² K , PMMA-foam, indoor temperature variable

1 without insulation
2 with opaque insulation
3 with transparent insulation

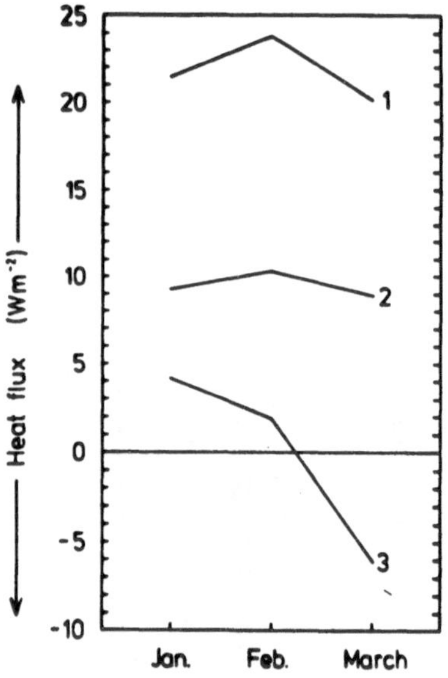

Figure 5: Mean heat flux through west facade K_W=1.3W/m² K, K_I= 1 W/m² K, capillary structure, indoor temperature 20°C

1, 2, 3 see Figure 4

COMPOSANT SOLAIRE PASSIF A LAMES MOBILES
ET A STOCKAGE PAR CHALEUR LATENTE

P. ACHARD - D. MAYER - M.D WHITE*
Ecole Nationale Supérieure des Mines de Paris
Centre d'Energétique - Sophia Antipolis
06565 VALBONNE Cedex
FRANCE

*Mechanical Engeneering Dept.
Colorado State University
FORT COLLINS CO.80523
USA

Summary

This component has been designed to permit collection storage and distribution of solar energy. It is a kind of green house effect wall that uses phase change materials (PCM) as storage medium. The system is composed of flat vertical or horizontal blades rotating on an axis behind a fixed glazing. The blades are for half composed of a container filled with PCM and for half of an insulating material.The aim of the project was to solve the main problems of the green house effect walls looking for an improvement of the global efficiency. This passive building component can be presented as the complement of the window - i.e the component of direct gain - on a sunny building face.

. it allows a solar energy gain
. it allows to distribute the gain when useful
. it can be fitted in the building enveloppe as a direct gain component
. it allows to vary the proportion direct gain/out of phase gain
. it allows to let in the day-light and to vary it
. it allows visual contact with the outside
. it supports its own thermal insulation and protects the building from housebreaking

The performances of this component seem to be promising. The system can be mounted either in new buildings or on retrofit, in non residential or residential buildings, individual or collective ones. It participates in thermal comfort, light comfort, visual comfort and esthetic.

1. INTRODUCTION

Ce composant est étudié par le Centre d'Energétique de l'Ecole des Mines en collaboration avec Cégédur-Péchiney, SGEC, le groupe ABC, Enerscop. Il est conçu pour assurer la captation, le stockage et la distribution de l'énergie solaire en utilisant des matériaux à changement de phase. Il a été élaboré dans le but de pallier les inconvénients des murs à effet de serre tout en recherchant des performances thermiques élevées.

1.1 Présentation

Dans un mur Trombe classique, les masses mises en oeuvre sont importantes et par la suite il n'y a pas ou peu de commande possible. La restitution d'énergie est fortement déphasée et les pertes vers l'extérieur sont importantes. L'avantage majeur de ce mur est sa simplicité au plan technologique. Pour diminuer la masse, il a été proposé d'utiliser des matériaux à changement de phase. Ont alors été étudiés des murs à effet de serre contenant des MCP. Les problèmes de déphasage et de pertes notamment nocturnes restent importants. Nous sommes arrivés à la conclusion qu'un système utilisant des lames mobiles contenant un MCP permettait de résoudre efficacement ces problèmes. Il permet de plus de donner la priorité au gain direct lorsque cela s'avère intéressant.

Ce système se présente alors comme le complément de la fenêtre sur une façade ensoleillée:

- il permet un gain d'énergie décalé dans le temps
- il s'installe comme un composant-fenêtre
- il permet de moduler le rapport gain direct/gain déphasé
- il permet la vue sur l'extérieur, l'occultation est variable
- il est porteur de sa propre isolation thermique et permet une occultation complète

1.2 Description du système

Il est constitué d'un ensemble de lames plates verticales ou horizontales tournant sur un axe derrière un vitrage fixe. Ces lames sont constituées d'une structure conteneur réalisée en aluminium et d'un isolant. Elles sont dimensionnées pour permettre lorsqu'elles sont en position intermédiaire des gains directs importants, et une large vue sur l'extérieur. Leur capacité de stockage est dimensionnée pour permettre l'accumulation des apports captés correspondant à une belle journée d'hiver.

La face absorbante des lames subit une excursion en température limitée dans la phase captation du fait de l'emploi des MCP. Les pertes en captation sont d'autant limitées. L'arrière de cette face peut être ailettée pour augmenter les transferts thermiques dans le matériau à changement de phase. Il paraît hors de question d'utiliser un revêtement sélectif pour la face absorbante; en effet lorsque le système fonctionne en restitution, cette même face doit avoir les meilleures caractéristiques possibles en émission, donc un coefficient d'émission I.R aussi grand que possible.

Ce système comporte les deux caractères actifs et passifs:

actif:
- une commande est possible
- une motorisation et une régulation peuvent être installées

passif:
- il n'y a pas de fluide caloporteur
- le système intervient pour moduler des gains directs et apporte une inertie et une isolation à un composant de gain direct

DISPOSITIF DE MANOEUVRE PAR VIS SANS FIN

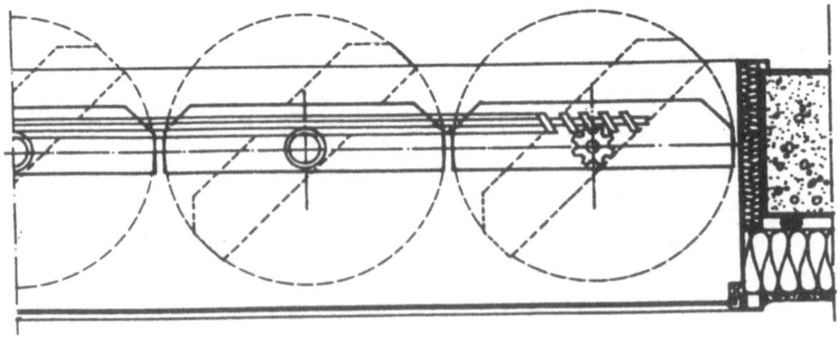

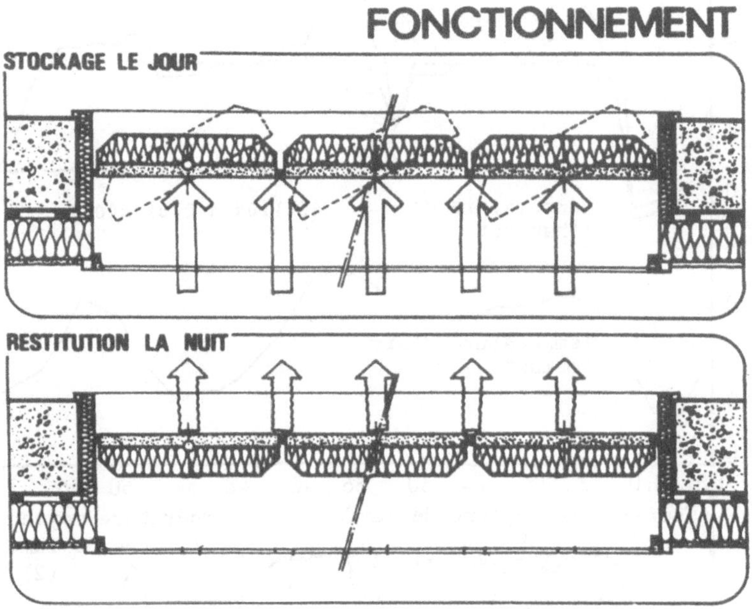

2. ANALYSE DES PERFORMANCES

Des résultats d'expérimentations effectuées en ensoleillement naturel sont données sur les courbes ci-dessous. Ce sont des mesures faites en extérieur sur des lames échelle 1 soumises au rayonnement solaire derrière un vitrage, celui-ci étant retiré pour la restitution nocturne. (cela simule la rotation). Deux MCP sont utilisés, l'un ayant une fusion à 26°C, l'autre à 36°C. On peut remarquer la constance de la puissance de restitution, de l'ordre de 150W/m2 pour le MCP à 26°C, dans une ambiance à 8°C.

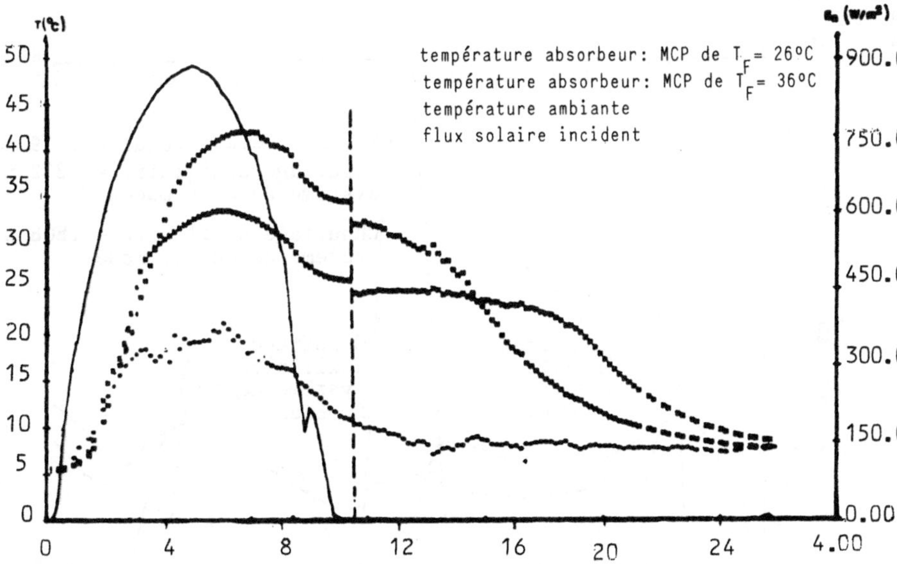

RESULTATS DE SIMULATION

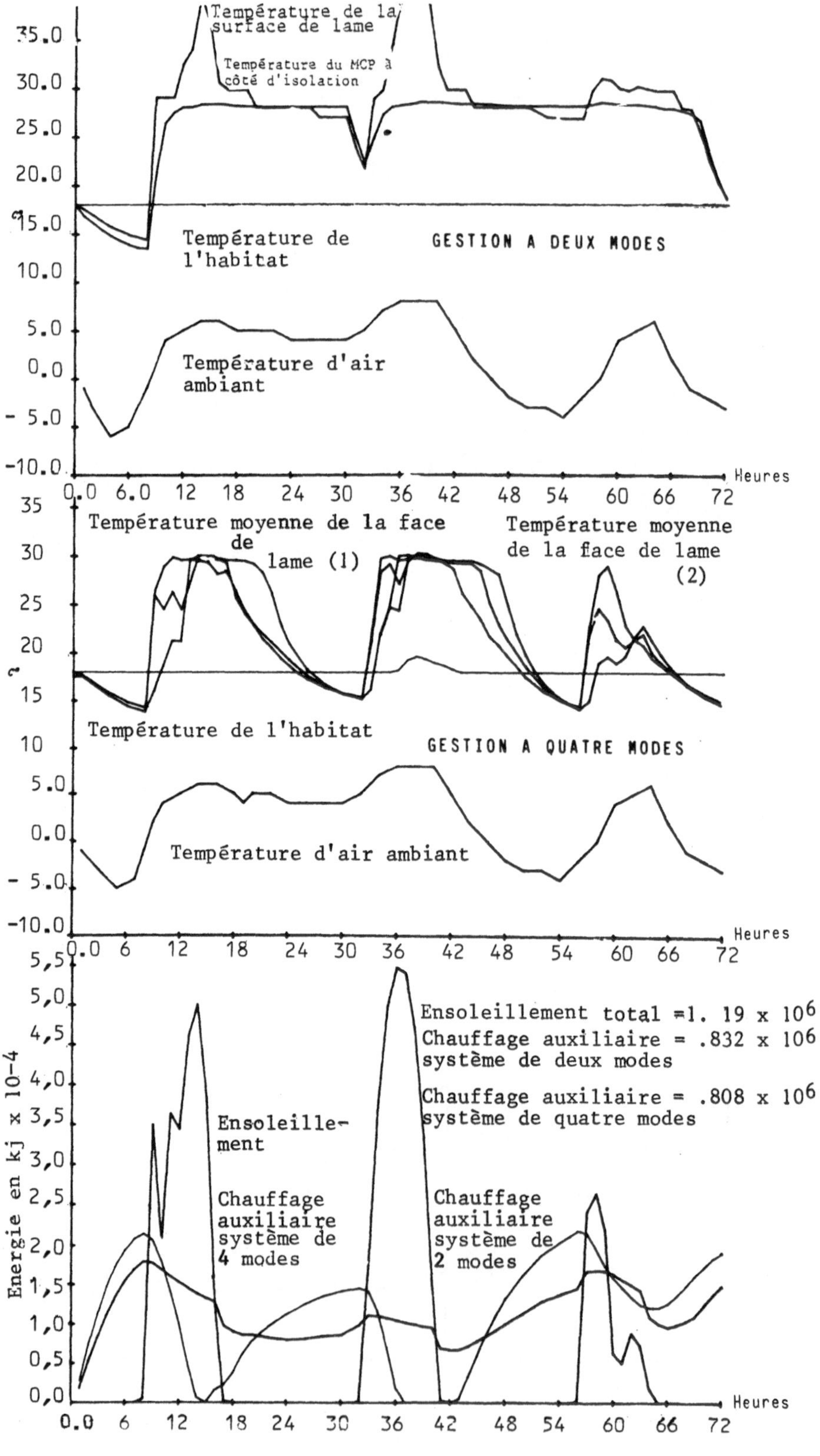

Température de la surface de lame
Température du MCP à côté d'isolation
Température de l'habitat
GESTION A DEUX MODES
Température d'air ambiant
Heures
Température moyenne de la face de lame (1)
Température moyenne de la face de lame (2)
Température de l'habitat
GESTION A QUATRE MODES
Température d'air ambiant
Heures
Energie en kj x 10-4
Ensoleillement total =1. 19 x 10^6
Chauffage auxiliaire = .832 x 10^6
système de deux modes
Chauffage auxiliaire = .808 x 10^6
système de quatre modes
Ensoleillement
Chauffage auxiliaire système de 4 modes
Chauffage auxiliaire système de 2 modes
Heures

Un modèle a été construit, pour simuler le fonctionnement du composant dans une cellule d'habitat type, avec différentes gestions. Les transferts thermiques au sein du matériau à changement de phase sont calculés par des équations aux différences finies. Deux gestions automatiques sont simulées:

. l'une, tout ou rien, permet de faire passer le système d'une position captage à une position restitution lorsque le flux thermique s'inverse. (gestion à deux modes).

. l'autre donne la priorité aux gains directs lorsqu'ils peuvent satisfaire directement des besoins. (gestion à 4 modes)

Les résultats de simulation montrent que les bilans d'énergie journaliers de l'habitat sont peu influencés par la régulation. En fait chaque gestion engendre une répartition propre de la puissance d'appoint.

La priorité au gain direct ne grève donc pas le fonctionnement global. Elle permet de plus de minimiser la capacité de stockage du composant.

Des corrélations du Solar Load Ratio (SLR) en fonction du Solar Saving Fraction (SSF) ont été recherchées pour quatre villes: Carpentras, Nice, Trappes et Stockholm. On peut remarquer que Carpentras et Nice montrent une bonne corrélation, ainsi que Trappes et Stockholm.

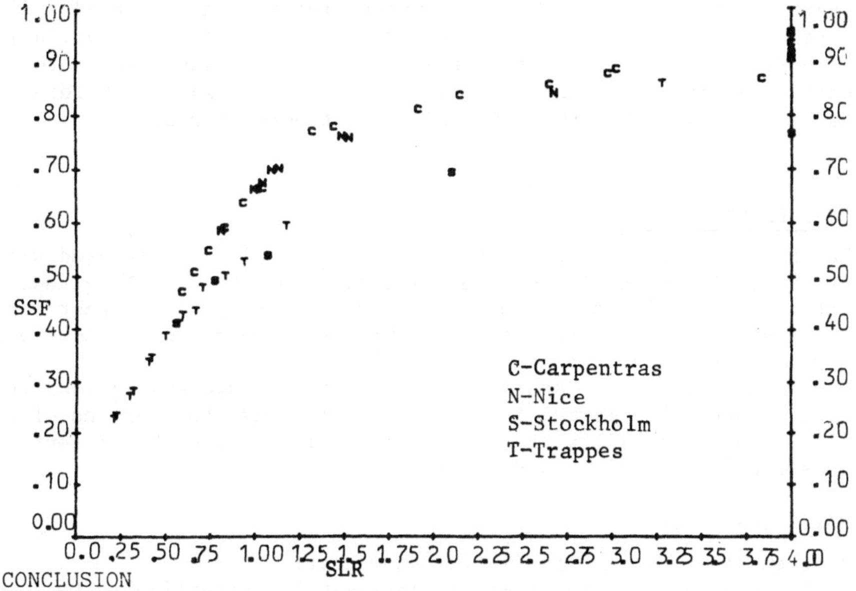

3. CONCLUSION

D'après les corrélations étudiées pour d'autres composants passifs par le groupe solaire du "Los Alamos Scientific Laboratory", il semble que ce système, lorsqu'il est conduit avec l'une des gestions automatiques présentées, ait des performances au moins égales à celles d'un mur à eau à isolation dynamique. Par rapport à ce dernier il offre de plus l'avantage de permettre le gain direct, la vue sur l'extérieur en créant une variabilité de la transparence de l'enveloppe du bâtiment. Ce composant est susceptible d'une grande diffusion dans l'habitat et le tertiaire, en construction neuve ou en réhabilitation. En effet il s'intègre au bâtiment en contribuant au confort thermique, au confort lumineux, au confort visuel ainsi qu'à l'esthétique. Il permet de moduler l'ambiance intérieure quant à ses quatre paramètres, il participe à l'animation de l'espace.

THE DESIGN OF A LOW-COST HOUSING ESTATE WITH THE USE OF THERMAL INERTIA IN A SOLAR ROOF, UNDER THE CLIMATIC CONDITIONS OF BOLIVIA.

M. LIBERT, civ.eng. & W.C. VILLAGOMEZ, arch.
Free University of Brussels (V.U.B.)
Pleinlaan 2 - B- 1050 Brussel - Belgium

Summary

The principle intention of the study was to develop a scheme of passive solar energy approach, maximizing the use of available resources for the Bolivian climate. Starting from the hourly values of external temperatures and the solar radiation a low-cost passive solar estate was designed and tested on its dynamical behaviour. The calculations have been performed with the use of Fourier Series to represent the contacttemperatures of each surface. Damping and phase-lag due to the thermal inertia were calculated with the results of the second Fourier Law for structures without internal complex heatflux. Resulting comforttemperatures have been calculated starting from the internal air temperatures and the weighted radiation temperatures. The results, for each month of the year, are valid in non steady regime. They showed that the use of a very simple solar roof, using only local materials and a traditional and local way of building, resulted in a 4°C gain for the internal comforttemperatures.

1. GENERAL DESCRIPTION

The project site is located in La Paz city. La Paz is situated within the intertropical zone of 16°30' south latitude and 68° west longitude. La Paz shows three units of urban morphology, El Alto - the highest part, La Cuenca and El Bajo. Due to its high altitude, 4100 m above sea-level, El Alto has extremely severe climatic conditions.

A passive solar energy approach of a housing estate should make it possible to reduce the diurnal temperature variations. For reasons of cost, flexibility and adaptibility the estate was designed with the present and local construction materials and know-how.

2. CLIMATIC CONDITIONS

Measurements performed in El Alto show that:

- Eastern and western winds are dominant in respectively the summer and the winter period. The eastern wind is warm and comes from the Atlantic zone. The western wind is cold and comes from the mountains. Figure I relates the monthly preponderant directions with their mean velocities.

The data concerning the outdoor temperatures and the solar radiation are statistical values computed over a period of 25 years (see figure II). As one can see the maximum temperatures show few variations (16°C to 19°C). The average temperatures have a similar behaviour. On the contrary, the minimum temperatures follow the season cycle with a maximum in January and a minimum in July. Examples of these values are given in figure III.

The climatic conditions of La Paz can be resumed as follows:

- The temperature variations can be considered as continuous over the whole year except for the minimum temperatures.

- In general La Paz climate is cold.
- There are large diurnal temperature variations.
- For an internal design temperature of 18°C 2717 degree-days are to be taken into account.
- The coldest period in the year is that one with the largest insolation.
- The sun altitude graphics show a maximum insolation on horizontal surfaces.

Out of the daily solar radiation figures the hourly values, needed for dynamical analysis, have been computed using the method of Liu and Jordan (1960). The found values have been compared with measurements of a particular meteorological station in Bolivia and showed to be very similar and certainly acceptable for horizontal surfaces.

3. CALCULATIONS

The outdoor temperatures and the solar radiation are represented in Fourier series. These series have been found after a curve fitting proces and showed to be very near to the real values measured in situ. Starting with these values the external contacttemperature of each wall, including the roof, has been computed. For mathematical use a new curve fitting proces was necessary. Each surface temperature was represented by a new Fourier serie. It showed to be necessary to take 5 terms for structural elements without solar radiation and 7 terms for those with. The general form of the Fourier serie is as follows:

$$\theta(t) = \frac{a_0}{2} + a_1 \sin\frac{2\pi t}{T} + b_1 \cos\frac{2\pi t}{T} + a_2 \sin\frac{4\pi t}{T} + b_2 \cos\frac{4\pi t}{T} + a_3 \sin\frac{6\pi t}{T} + \ldots$$

θ : temperature

T : period of the wave

a_i and b_i : coefficients to be found out of curve fitting

t : time in hours

So far the outdoor climatic data are all represented. This allows one to design the different structural elements taking into consideration the different parameters for a maximum free energetic input. Once all the external values are known the classical Fourier analysis delivers the damping and phase lag and thus the internal contacttemperatures of the different walls.

$$\theta'(t) = \frac{a_0}{2} + a_1' \sin\frac{2\pi(t-\phi_1)}{T} + b_1' \cos\frac{2\pi(t-\phi_1)}{T} + a_2' \sin\frac{4\pi(t-\phi_2)}{T} + \ldots$$

a_i', b_i' : amplitudes of non-steady parts of the internal contacttemperatures (for i=1,2,3,...). These values represent the external a_i and b_i multiplied with the damping factor found with the second Fourier Law.

All calculations were performed sticking as near as possible to the reality. Particularities like the absence of solar radiation on the northern wall* (southern hemisphere!) have been taken into account as well as the variation of the sun path during the seasons. The changing in transmission through windows during the day (hour by hour), the area of the different surfaces as well as their nature (wall or floor) struck by the sun have been calculated for every hour of a type day (the 21st ,arbitrarely choosen) of each month.

(* only during the months November, December and Januari due to lattitude)

The results of those calculations are the surface temperatures for each surface in the building and this for every hour of the choosen standard days. The different values can be found in figure IV. This values are function of the construction which is described in the following chapter.

4. CONSTRUCTIVE ELEMENTS

As mentionned in the general description only local materials and a traditional and local way of building have been considered. The thickness of the different materials that were used have been determined out of the phase lag, the damping and the moment on which maximum and minimum external temperatures occur.The selected structures for roof, internal and external walls are represented in the figures V, VI and VII.

5. CONCLUSION

A internal temperature rise of 3 à 4°C, depending of the season, is possible under the climatic conditions of Bolivia using very simple passive solar techniques.

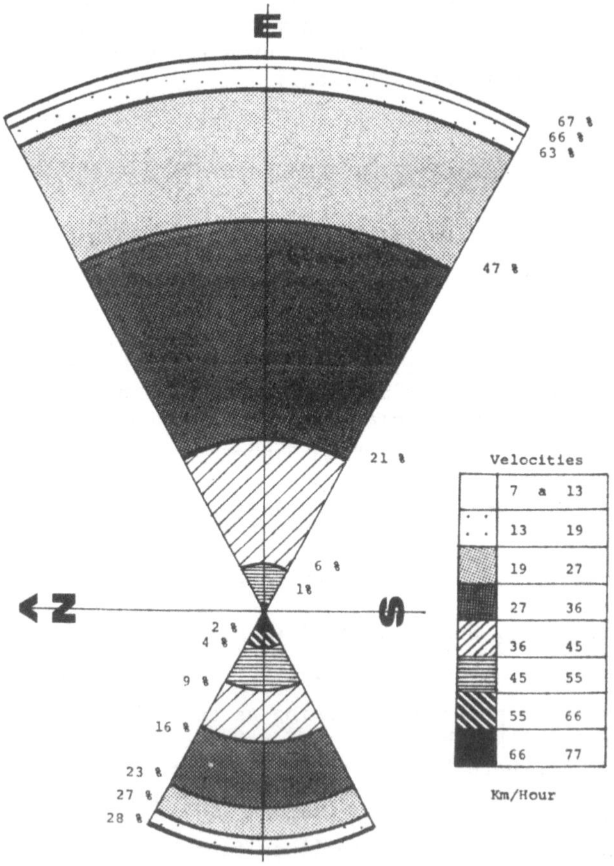

Figure I. Monthly preponderant wind direction and mean velocities classification

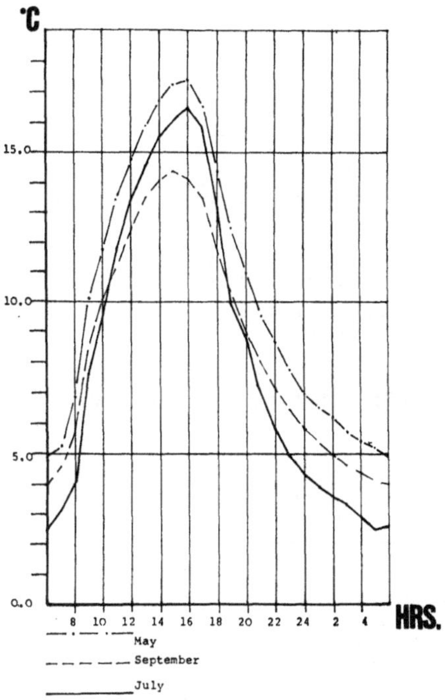

Figure II.
Hourly temperature variations
(period 1981)

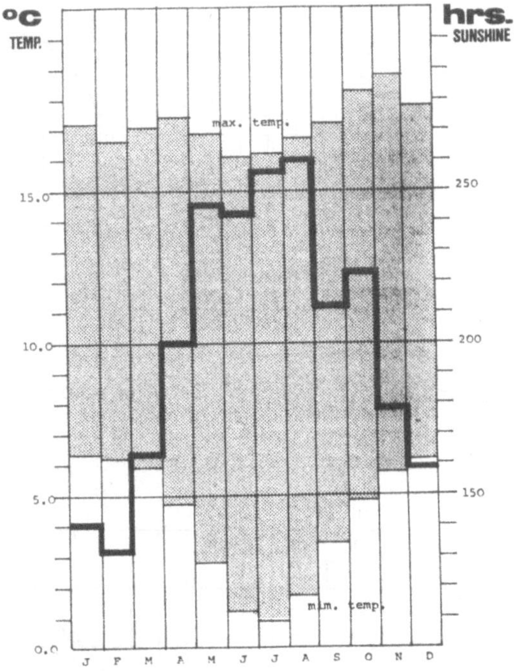

Figure III.
Yearly temperature variations in comparison with sunshine hours
(period 1951-1975)

Different Mean Temperatures

Period : 1951-1975
Station : S. Calixto
(values in °C)

Averages	N	D	J	F	M	A	M	J	J	A	S	O
Maximum	18.8	17.8	17.2	16.6	17.1	17.4	16.9	16.1	16.2	16.7	17.2	18.3
Medium	12.3	11.3	11.2	10.8	11.1	10.8	09.8	08.5	08.3	09.3	10.4	11.7
Minimum	05.7	06.1	06.3	06.2	05.9	04.7	02.8	01.2	00.9	01.7	03.4	04.8

Annual masimum temperature (average) = 17.2 °C
Annual medium temperature (average) = 10.5 °C
Annual minimum temperature (average) = 04.1 °C

Source : Urban Development Plan, La Paz, 1977
BCEOM-BRGM-PCA Consultors.

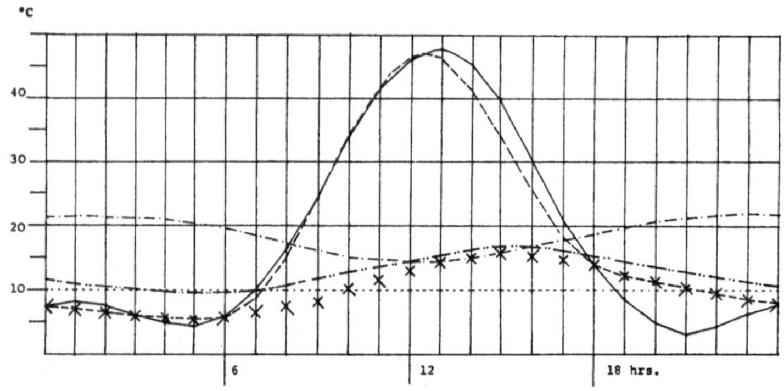

Figure IV.I. Temperature variations in function of time January

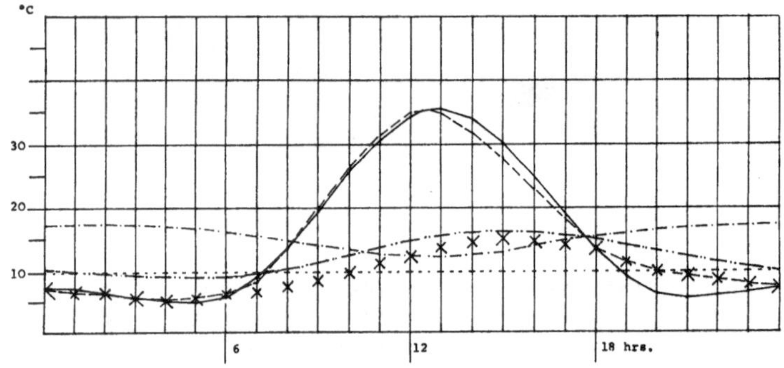

Figure IV.II. Temperature variations in function of time February

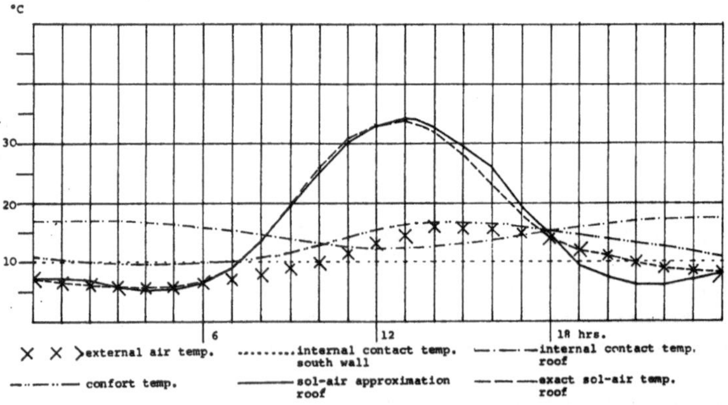

Figure IV.III. Temperature variations in function of time March

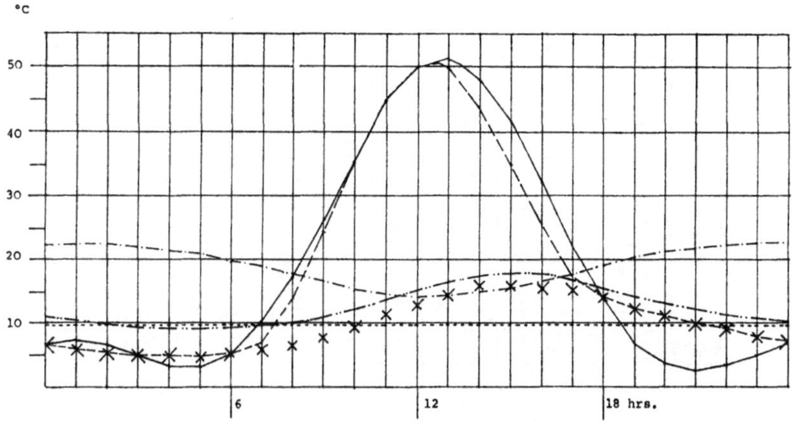

Figure IV.IV Temperature variations in function of time April

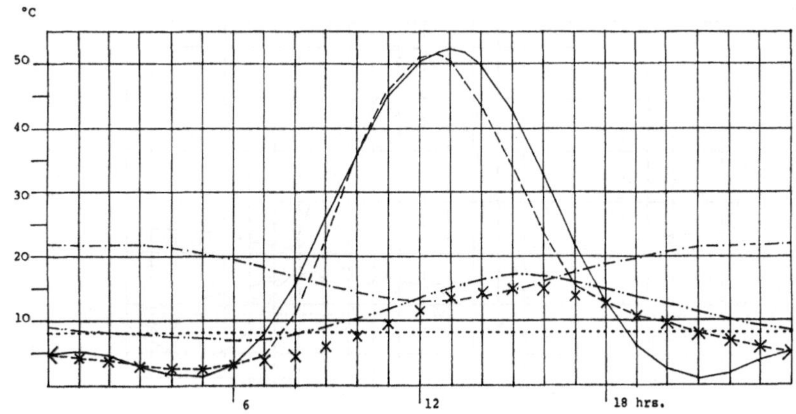

Figure IV.V Temperature variations in function of time May

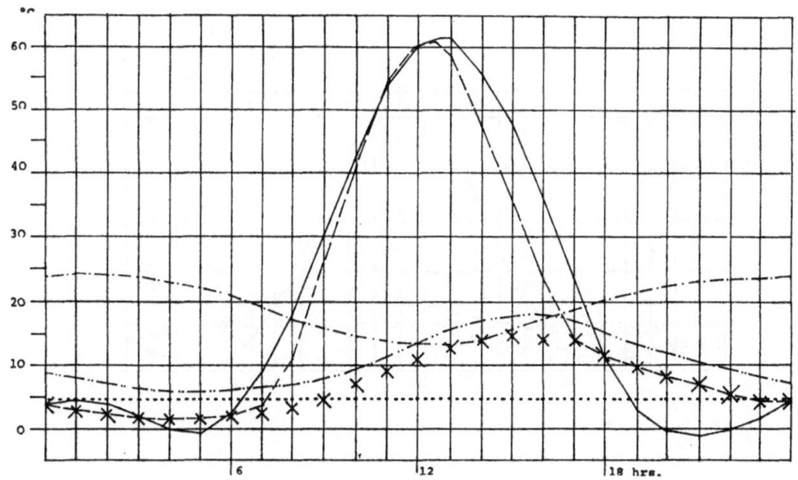

Figure IV.VI Temperature variations in function of time June

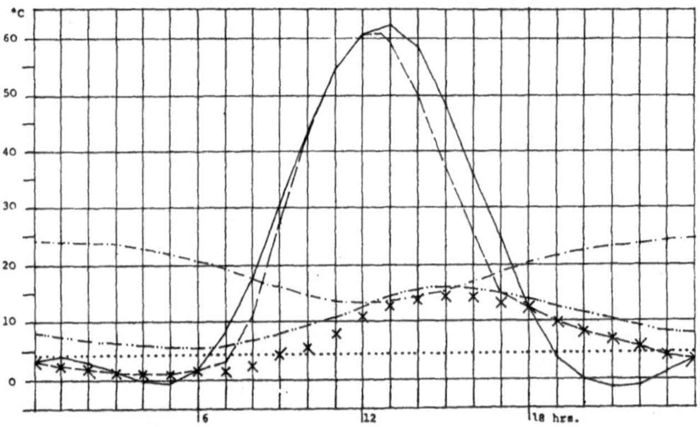

Figure IV.VII Temperature variations in function of time July

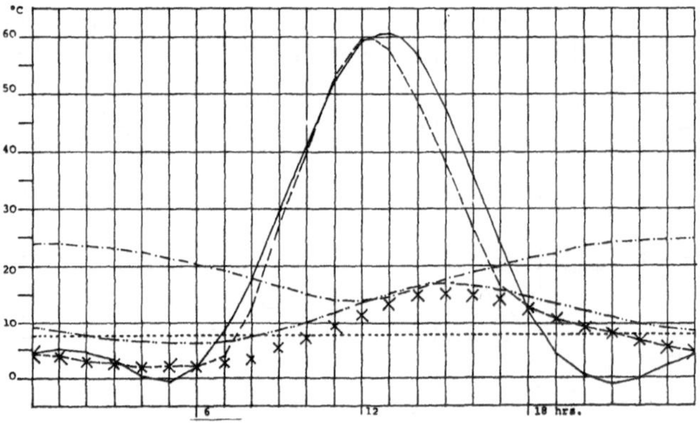

Figure IV.VIII Temperature variations in function of time August

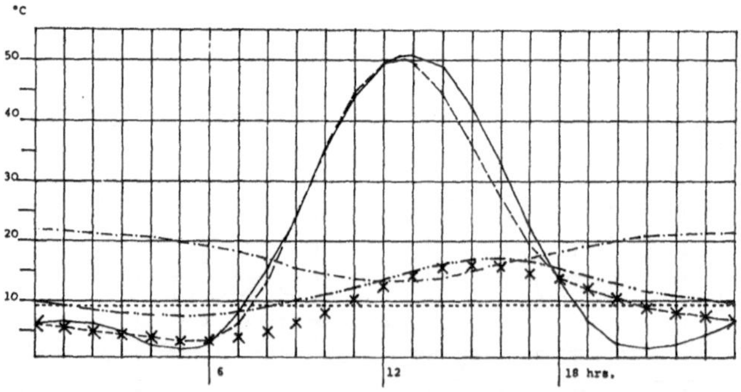

Figure IV.IX Temperature variations in function of time September

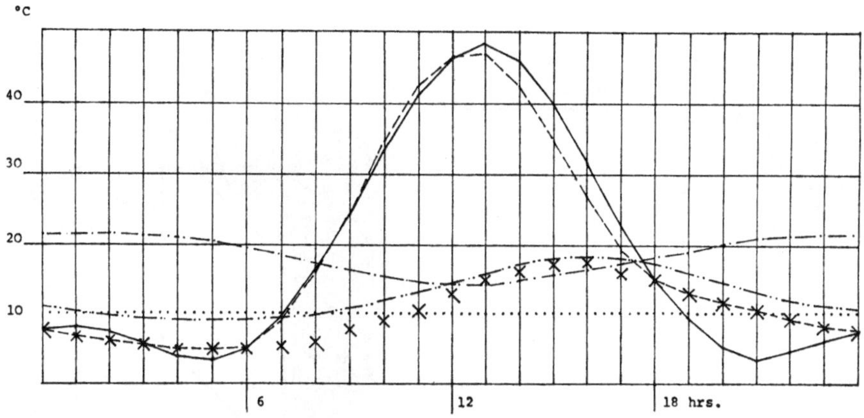

Figure IV.X Temperature variations in function of time October

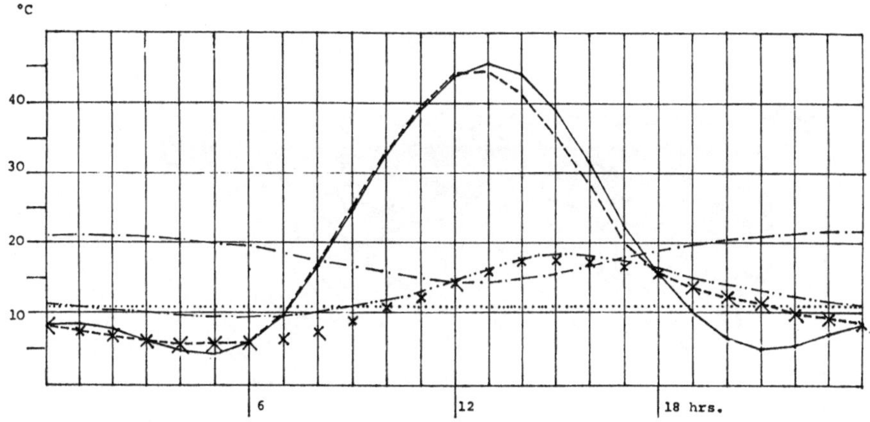

Figure IV.XI Temperature variations in function of time November

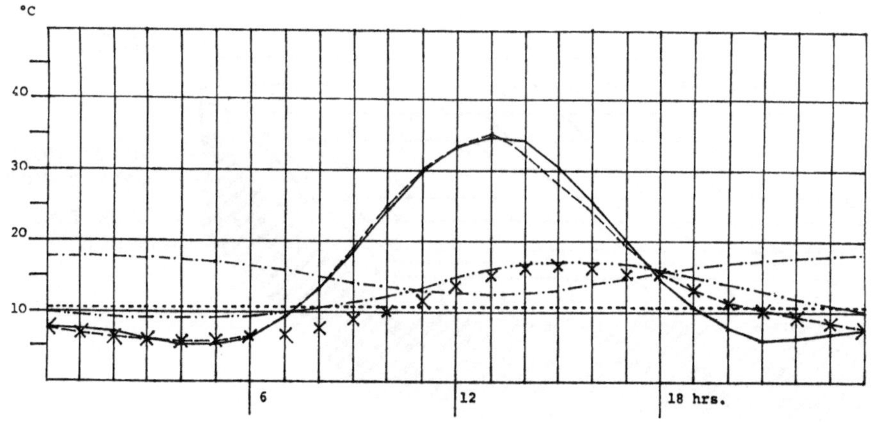

Figure IV.XII Temperature variations in function of time December

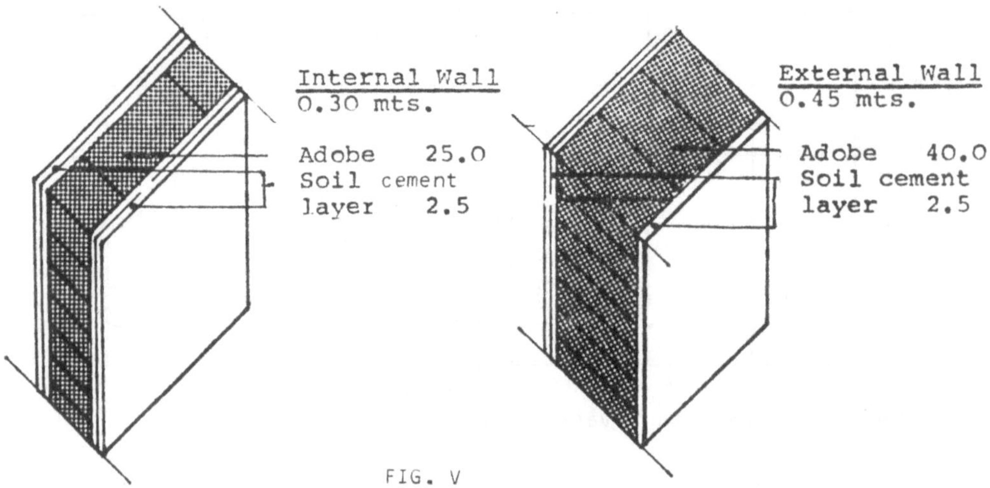
Internal Wall
0.30 mts.
Adobe 25.0
Soil cement
layer 2.5
External Wall
0.45 mts.
Adobe 40.0
Soil cement
layer 2.5

FIG. V

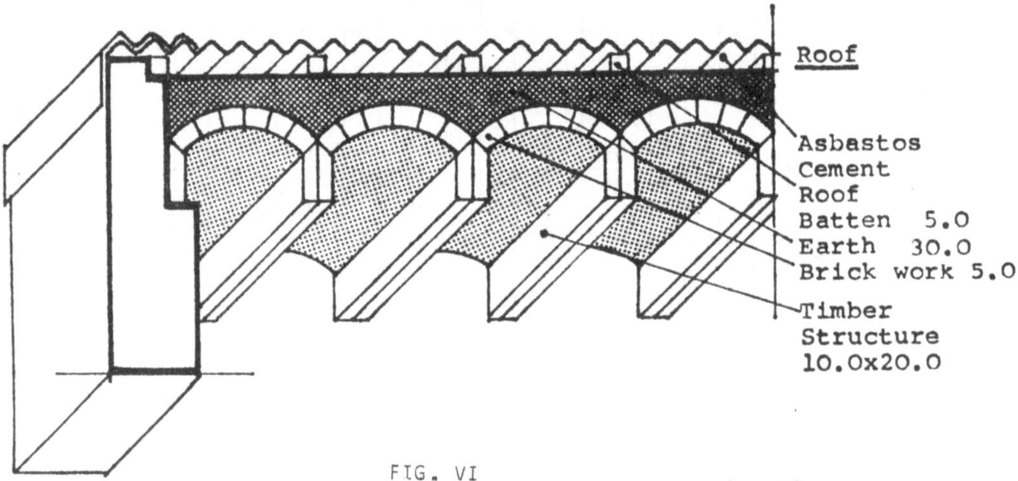
Roof
Asbastos
Cement
Roof
Batten 5.0
Earth 30.0
Brick work 5.0
Timber
Structure
10.0x20.0

FIG. VI

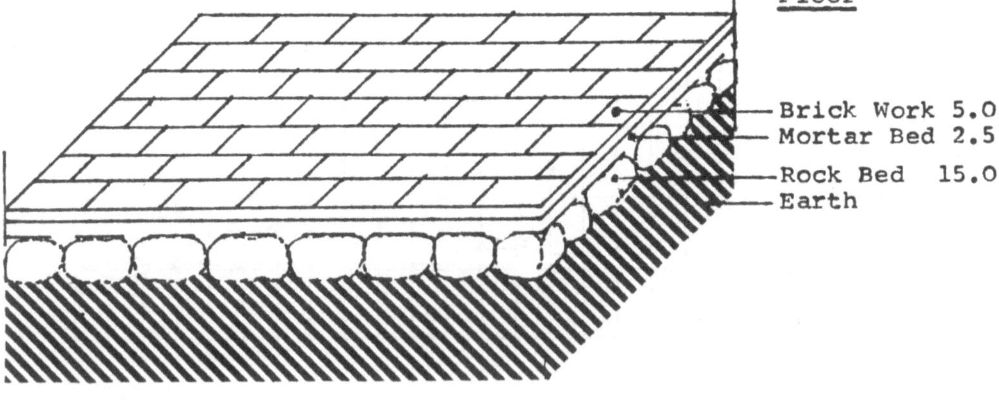
Floor
Brick Work 5.0
Mortar Bed 2.5
Rock Bed 15.0
Earth

FIG. VII

HYBRID SOLAR LOW ENERGY BUILDINGS IN DENMARK

O.C. JØRGENSEN
Thermal Insulation Laboratory
Technical University of Denmark

Co-Author: J. ANDERSEN, Architect maa
Royal Academy of Fine Arts, Copenhagen

Summary

Southfacing sloping windows, massive internal structures, outside insulating and reflecting shutters, building integrated, innovative solar water heaters, heat recovery on ventilation air, improved insulation levels in walls, roof, floor and footings and attached sunspaces are the passive solar and energy conservation concepts of a new building project which will be built in 1984 at a building site close to Elsinore. The results of the integration of these techniques in the building design is a calculated energy consumption for space heating and hot water heating of approx. one third of that of conventional houses built in Denmark at present.

The 55 single-family dwellings, designed by J. Andersen, were developed as part of the Danish contribution to Task VIII, Passive and Hybrid Solar Low Energy Buildings of the IEA Solar Heating and Cooling Programme. The houses will be monitored for a period of 18 months after erection.

1. INTRODUCTION

Energysaving is not a goal in itself for very many people other than a few idealists in this world. However, if it can be done economically feasible, resulting in saving of money, it becomes attractive to everybody. This is the underlying philosophy of one of the objectives of the IEA project Passive and Hybrid Solar Low Energy Buildings. This objective is to erect a number of buildings in the member countries to verify that passive and hybrid solar low energy techniques can be employed as cost-effective means to substantially lower the energy consumption for space heating and domestic hot water heating loads as compared to that of houses conventionally built.

The present project, designed by architect J. Andersen, has been developed as the Danish IEA-building project. The building design is a development of our contribution to the 2nd CEC Passive Solar Architecture competition in 1982.

2. INTEGRATION

The fundamental philosophy of the design process for this project has been that the techniques employed should be thought out into the finest detail in their integration into the building as a whole, not only as energy conservation techniques but as building elements.

3. APPLIED TECHNIQUES

3.1 Southfacing windows and massive structure

The windows are primarily oriented towards south. One large window slopes at an angle of about 60 degrees for maximum sunlight in the heating season. This window is placed in a special aluminum framing for maximum aperture area allowing as much sunlight as possible to be transmitted to the two heavy concrete consoles, projecting into the living room just behind the window. Apart from these consoles, the internal mass of the house has been increased by the use of heavy leca-concrete partitions and back walls.

3.2 Shutter

The large, sloping, southfacing window has been equipped with an outside roller shutter which encompasses two functions. During winter it can be used as an insulating shutter improving the U-value from 3 W/m^2K to 2 W/m^2K. In summertime the shutter can be pulled down completely or partly in order to reflect sunlight and thus prevent overheating of the house during warm, sunny periods.

3.3 Solar water heater

Previously it has always been a problem that the solar collectors had to be placed on top of outside walls or roofs causing flashing problems and increased installation costs. Further, the heat storage tank took up too much space and was therefore difficult to locate in the buildings.

In this project an innovative solar hot water system has been integrated in the south facade next to the large, sloping window and using the same aluminum frame. The absorber of approx. 3.1 m^2 and the 12 cm thick box-shaped storage tank of 140 litres of water have been combined into a unit or wall element which can be mounted just behind the glass. Suitably covered, it appears just like an ordinary outside wall when seen from the inside of the building. The system is developed based on an idea of Peder Vejsig Pedersen at the Thermal Insulation Laboratory of the Technical University of Denmark.

3.4 Heat recovery on ventilation air

In a well-insulated house the air infiltration and ventilation losses make up a relatively large part of the total heat losses. Consequently an attempt has been made to minimize air infiltration and an air-to-air countercurrent heat exchanger has been installed to recover the heat of the ventilation air. This type of heat exchanger has a very high efficiency (80%) and the make chosen is easy to install at a central place in the house. This makes it possible to use a few short air channels, thus keeping installation costs at a minimum.

3.5 Attached sunspace

At one end of the south facade a sunspace has been attached to the main building. The end walls of the sunspace are pre-fabricated leca-elements and the roof of the house is prolonged to exceed 1.2 m over the sunspace. Thus only the lower part of the roof, the south wall of the sunspace and a door in one of the end walls are glazed. The primary energy saving function of the sunspace is that it reduces heat losses from the windows of the adjoining room.

3.6 Improved insulation levels

The transmission losses of the house have been considerably reduced by using thicker layers of insulation material in all the surfaces exposed to ambient air temperature as well as in the floor and footings. U-values range from .13 to .16 W/m^2 K. Except for the sloped window with the outside insulating shutter, all windows are triple glazed. According to the current Danish building regulations U-values ranging from .2 to .4 W/m^2 K and double pane windows are sufficient.

4. CONCLUSION

The net result of the integration of these techniques in the building design is an anticipated energy consumption for space heating and domestic hot water heating of one third of that of conventional houses built according to the Danish building regulations.

5. ACKNOWLEDGEMENTS

The Danish participation in the IEA project is sponsored by the Danish Ministry of Energy. A special grant has been given by the Danish Building Industry Development Board to support building design, construction and monitoring.

MASS PRODUCED PASSIVE COMPONENTS FOR LOW-COST MULTISTORY BUILDING.

Gianni Scudo

Dipartimento di Programmazione Edilizia - Politecnico di Milano, Via Bonardi 3, Milano Italy.

Summary

The design and building processes of 40 solar passive flats in Marostica (Vicenza, Northen Italy) gave the opportunity to develop a mass produced low-cost passive component, the Barra-Costantini system, which is now produced by an italian Industry. One interesting feature of the buildings is the low pay-Back time (8 years) of the passive system which provides an Heating Solar Fraction of 50%.

1. INTRODUCTION

The main constraints to the large scale diffusion of multistory passive buildings in Italy are:

- the lack of integration between passive systems and current building systems;
- the lack of morphological control of innovative solar passive technologies appropriate to local dwelling cultures.

The passive solar settlement of Marostica (Fig.1,2) here presented was

Fig.1 View of the solar passive settlement

designed to overcome the above mentioned constraints. More specifically the research program is based on the following steps:

a) design of mass-produced and innovating passive components at acceptable cost for pubblic housing standards;

b) the integration of the passive components with the commonly uses Design Process. Innovating desing steps guarantee the penetration of "passive conscious design" into the professional practice;

c) design and realization of low cost and low energy profile multistory passive houses (with 60% energy saving mainly due to the solar contribution).

Fig.2 The sud facade with solar Chimneis.

1.2 THE BARRA-COSTANTINI SYSTEM

It is really inexplicable the amount of research work devoted through the world to improve low-efficiency passive system (Trombe, Direct Gain,etc...) and the few resources devoted to diffuse highly efficient ones (i.e. solar Chimneics which had been largely studied and experimented (Barra et al. 1980, Bilgen, 1981).
The Barra-Costantini system is a solar chimey integrated into the building structure. In winter days the air heated in the chimney (small thermal capacity air collector) flows through ducts plugged in the concrete structure of the ceiling (storage), enters in the room space and is convected back to the chimney by means of a natural convective loop. During night a dumpter is closed and stored heat is radiated from ceiling to room space (Fig.3).

The B-C systems eliminates the drowbacks of the most diffused passive systems (expensive night insulation, overheating, single facing typology etc..) because it has been designed to be tightly integrated with the most diffused building components and systems.

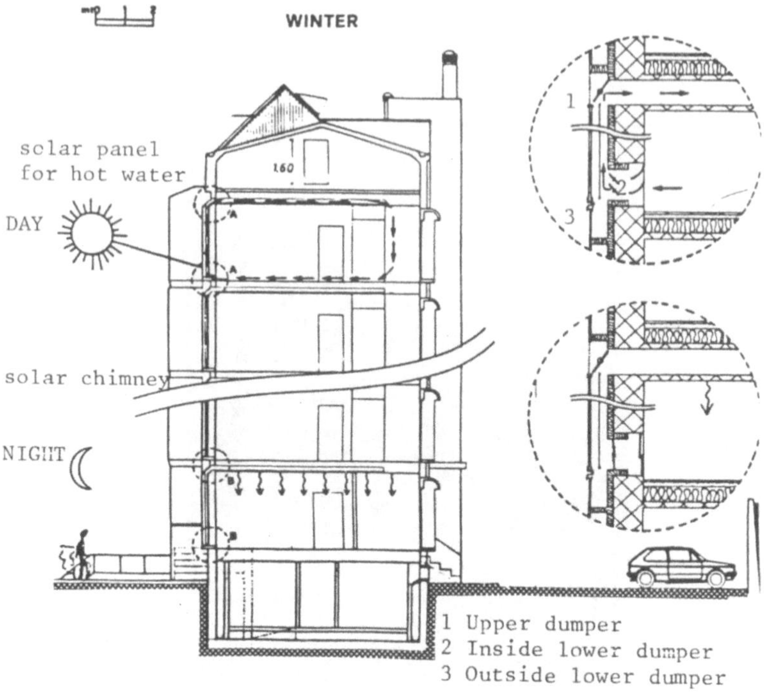

Fig.3 Winter operation of the B-C system

1.3 THE PASSIVE SETTLEMENT IN MAROSTICA

The settelment is made by one multistory and three row-house buildings.
The site is a foothill, lightly sloping ground with typical continental climate (2340 DD; 6.37 MJ/mq average dayly winter solar radiation). The four buildings face south (main axis East-West).

The multistory building (16 flats in four stories, 5045 mc) has a Surface/Volume ratio of 0.45 and a volumic coefficient of thermal losses of 0.72 w/mc/°C. The heated volume/south windows ratio is ceiling- is 600 KJ/°C mq of thermal ceiling. Secondary heat storage - in the walls - is 420 KL/°C/mq of floor.
Yearly Solar Heating Fraction is 50%. Global extra cost of the passive system is 5.5 million Lire, about 8% of the total building cost. (Scudo, 1984)

1.4 INDUSTRIAL PRODUCTION

The cooperative work carried out by designers (1) and Industry (2) is the most innovating festure of the passive/building system integration.

This system fits for industrialized and semi-industrialized building technology ("banches et predales" in the passive buildings here presented) and for mass-production.

The innovating components of the passive systems are:

- a new air flow regulation system based on a compact butterfly valve e-element easy to be installed and to operate;
- the reduction of return air openings to only one for single chimney: this reduces the passive system constraints on inside space,i.e. the arrangement of the air return opening is located under the fan-coil of the auxiliary heating system;
- the vertical-horizontal connection between the solar chimney and the storage distribution unit (the modified reinforced concrete ceiling) is done by extracting a special joint from the steel ducts of the ceiling (Figgs. 4,5,6,7) once the ceiling structure is put over the walls; this allows a very good fitness between concrete structure and solar chimney as well as a low cost installation.

Fig.4 The concrete ceilingwith steel ducts set up on the walls.

Fig.5 The "Banches et predales" Building technology

Fig.6 The joint from steel ducts to wall, before extraction

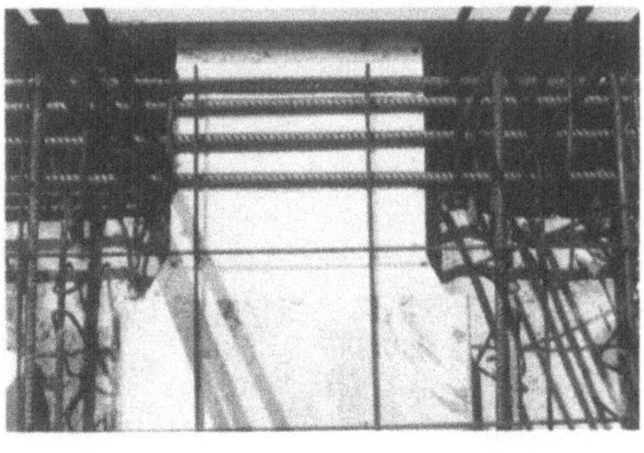

Fig.7 The joint after extraction and before concrete casting.

1.5 MONITORING AND EXPERIMENTAL PROGRAMME

Four flats - two of the multistory and two of the row-house buildings are going to be monitored by ENEA (3) in the next weeks. The main aim of the monitoring programme is to evaluate the congruity between designed energy performance and mesured energy performance. The thermal efficiency of the Barra-Costantini system under different behaviour patterns will be tested.

Attention will be given also to the critical problems of noise diffusion through air ducts and thermal losses through window operation.

REFERENCES AND NOTES.

Barra O.A. et al. (1980) Barra-Costantini solar passive system:experimental performances, BUILDING ENERGY MANAGEMENT, Proceeding of the International Congrees, Povoa de Varzim, Portugal, 12-16 may 1980, Pergamon Press.

Bilgem E., Chaabam M. (1982), Solar heating-ventilating system using a Solar Chimney, solar energy vol. 28 n.3 1982.

Scudo G. (1984). La progettazione termo-edilizia: 40 alloggi solari passivi a Marostica, Modulo n.2 - febbraio 1984.

(1) Ing. T. Costantini, Casa Solare, Salisano (RI), Italia
Cooprogetto, Via Calderari 9, Vicenza, Italia.

(2) Industrie Secco S.p.A., Preganziol (TV), Italia.

(3) Comitato Nazionale per la ricerca e per lo sviluppo dell'Energia Nucleare e delle Energie Alternative, Strada Anguillarese, Casaccia (Roma), Italia.

OPEN-LOOP THERMOSYPHON SOLAR-ENERGY SPACE HEATING

B. NORTON and S.D. PROBERT
Applied Energy Group
School of Mechanical Engineering
Cranfield Institute of Technology
Bedford MK43 OAL
Great Britain

Summary

The behaviours of solar-energy collectors for use in open-loop thermosyphons for space heating in institutional and commercial buildings have been studied experimentally. Almost identical instantaneous collector efficiencies were achieved by collectors whose base surfaces (which were orthogonal to the direct insolation) had fins which ran parallel or perpendicular to the mean convecting air flow direction in the collector. The temperatures of those surfaces with fins perpendicular to the flow took longer to respond to changes in insolation than surfaces with fins parallel to the airstream. The latter arrangement exhibited an almost equal response period to that achieved by a plane surface. It was also demonstrated that a cheap "letter-box" type flap-valve effectively prevented nocturnal reverse circulations.

1. INTRODUCTION

Many commercial and institutional buildings are only regularly occupied during day-light hours. Solar-energy space-heating systems, if they are to be used with such buildings, should exhibit the following attributes:-
(a) Minimum delays should ensue between the solar-energy collection and the input of the harnessed heat into the room(s).
(b) Only low rates of heat loss from the room(s) are permitted to occur during "cold" conditions.
(c) Overheating of the building to which the heater is attached must not happen during the summer months.
(d) The initial capital cost should be low.
(e) The construction should be robust, mass-producible and lightweight and incur low installation costs: these presently contribute up to 30% of the installed cost of solar collectors.
(f) Only minimal maintenance is likely to be needed.
(g) The whole system should be aesthetically acceptable when installed.

An open-loop natural-circulation solar-energy air-heater of the type illustrated in figure 1 could satisfy these criteria. During the period of the year when heating is required, air warmed in the collector enters the room under the action of buoyancy forces. During periods of high insolation, overheating is prevented by opening the external pivot window, so allowing the "solar chimney" effect to ventilate the collector space (1).

2. DESIGN AND CONSTRUCTION OF THE EXPERIMENTAL UNIT

Experiments were undertaken on an open-loop thermosyphon solar-energy air-heater located at Cranfield Institute of Technology, Great Britain. The site was not overshadowed at any time, and the local conditions were not significantly affected by industrial airborne pollution. The 1.4 m^2 area front-pass collector, inclined at an angle 60° to the horizontal, was glazed with a film of fluorinated ethylene-propylene, and the back of the collector was insulated by a 7 cm thick slab of expanded polystyrene. The bottom inlet and top outlet ducts were 8 cm across and extended the width of the collector. The flow channel depth was 4 cm. The absorber configuration could be changed by removing the cover, then substituting the new collector surface and replacing the cover. Copper/constantan thermo-junctions measured the air-stream temperature at the inlet and outlet and at five equally-separated vertical positions along the centre-line of each absorber plate. A 0.14 m^3 capacity wooden box, similar to that which held the collector, simulated the heated (admittedly very small) room: it was lined with a 5 cm thick layer of expanded polystyrene and covered externally with reflective foil. Centrally positioned along the major axis of the chamber was a length of thin wooden dowelling, to which were attached four equally-separated copper/constantan thermo-junctions. Heat could be extracted from the air by a spiral coil of 1 cm external diameter copper tubing, placed at the base of the cell and connected to a cold water supply. The insulant, its reflective coating and the copper cooling-coil facilitated the control of the working-space temperature. The system enclosing the environment to be heated was placed upon a supporting framework.

3. RESULTS

The relative merits of three absorber surfaces were studied. Each was manufactured from 1.6 m thick light aluminium alloy. The surfaces when in use were (i) plane, (ii) horizontally-finned and (iii) vertically finned. The fins were 1.6 mm thick, spaced 3 cm apart and in both cases, protruded 2.5 cm proud of the base surface. Each absorber surface was primed and coated with matt black paint.

The data recorded in the experiments included the temperature-time histories at various heights in the model room as well as the variation of the ambient temperature and the total radiation incident on the collector surface.

At a constant airflow rate, a plot of the efficiency of the collector, η_c, versus

$$\left(\frac{T_m - T_a}{IA}\right)$$

should be a straight line of gradient FU with a positive intercept of $F\tau\alpha$ on the efficiency axis (2). Experimental results for the considered three absorber systems are plotted in this manner in figure 2 for a steady natural -circulation airflow rate of 0.112 kg s^{-1}. Surprisingly little difference in instantaneous collector efficiencies was observed as a result of different alignments of the fins to the naturally-convecting air-stream. However, the addition of fins did provide a significant increase in the collector efficiency compared with that for a plane surface at all the considered collector temperatures.

An investigation was undertaken to determine the thermal response periods to changed in the applied conditions. Unlike forced circulation collectors, such response periods also include a contribution due to the initiation of the thermosyphonic flow. For a step decrease in the insolation, a time constant was defined as the period taken for the temperature difference between the absorber plate and the ambient environment to fall to 0.368 (i.e. 1/e) of its initial value. For a step increase in the insolation, there was not a single datum temperature towards which the temperatures would approach asymptotically for all three absorber surfaces. Thus it was not possible to obtain time constants in a similar way to that employed for temperature decreases. An alternative time constant was thus defined as the period that elapsed from exposure to the insolation until the temperature difference between the absorber plate and the airstream at the outlet became constant. The insolation level used was 400 Wm^{-1} and the time constants thus determined are shown in Table 1.

Absorber plate configuration	Response period for the plate to a step increase from zero to 400 Wm^{-2} in the insolation (s)	Relaxation period of the plate to a step decrease of 400 Wm^{-2} in the insolation (s)
Flat Surface	600 ± 16	1380 ± 30
Finned surface with fins aligned parallel to the convecting flow.	660 ± 18	1320 ± 33
The same finned surface but with the fins perpendicular to the convecting flow.	960 ± 18	1560 ± 35

TABLE 1. Behaviours of different collectors when subjected to step-changes in the insolation.

A simple "letter-box" type flap-valve was incorporated into the unit to prevent heat losses from the "room" due to nocturnal reverse thermosyphonic circulations. Figure 3 shows the profiles of the temperature variations with height in the system through two nights during which approximately similar ambient conditions prevailed. Both profiles are for the collector with a plane absorber surface, the only difference being the presence of a flap-valve in the flow circuit. As can be seen, the simulated room was maintained at a higher temperature when the flap-valve was in the flow circuit.

4. DESIGN RECOMMENDATIONS

The responses of the flat surface and the finned surface if the fins were aligned to the convecting air-stream, were almost similar. However, the response of the surface with the fins perpendiculat to the flow was markedly slower. It is thus preferable to employ the latter type of surface in collectors on buildings in climates experiencing rapid insolation changes in order to ameliorate their effects on the internal environment. The simple flap-valve was found to perform effectively. However, before it is adopted commercially, a study of its long-term reliability and durability should be undertaken.

In buildings in temperate climates, most energy is consumed in the provision of space-heating, hot-water, lighting and for operating electrical devices. A system such as the one described in this paper can meet a significant proportion of the energy required for space-heating, and natural-circulation solar-energy water heaters (3) can satisfy that for hot-water. The use of photovoltaic cells to provide electricity could complete a package of systems for the provision from the sun of a large portion of all the energy needed in a building, whilst not incurring the costs associated with mechanical systems and their sensors and controls. To date, except for

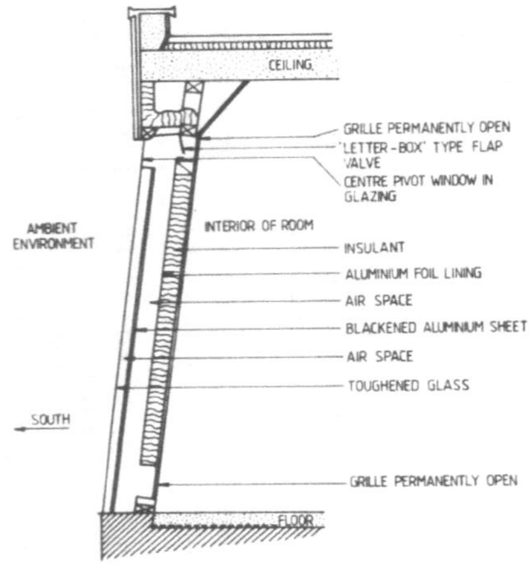

FIG.1
Conceptual diagram of a passive solar-energy air heater for schools or offices during the period of solar energy gain, the flap valve is open as shown.

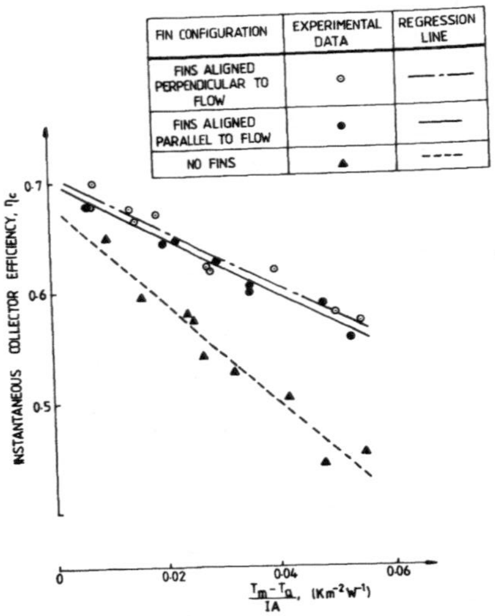

FIG.2
Collector characteristics for different types and configurations of absorber surface.

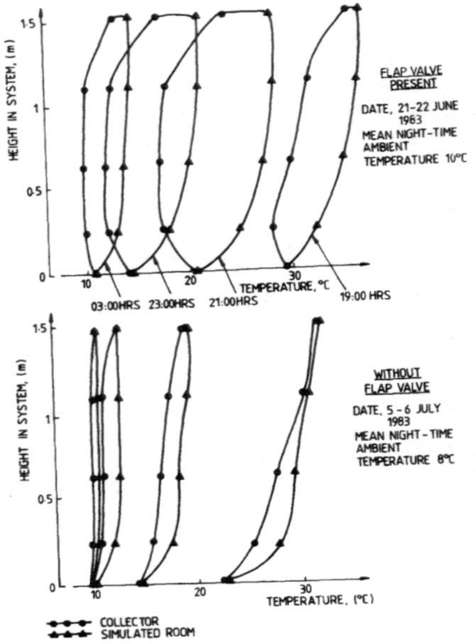

FIG.3
Nocturnal variations of system temperature profiles.

applications in extremely isolated locations, the cost of photovoltaic units has been prohibitive. However, in 1985, the Sanyo Electric Company in Japan intend to commence commercial production of photovoltaic roof-tiles each of about ·15 m^2 area and are producing typically of the order of 2 W of electrical power. Their relatively low intended cost (≃ £6.00 per tile) will make the realisation of a building in which solar energy economically satisfies an appreciable part, if not the whole, of each energy demand, both thermal and electrical, a practical proposition.

5. ACKNOWLEDGEMENTS

Financial support from the Waterloo Grille Company, Benfleet, Essex and the Science and Engineering Research Council, Swindon, Wiltshire, U.K. is greatly appreciated.

NOMENCLATURE

- A Aperture area of air-heating solar collector m^2
- F Collector heat removal factor
- I Global insolation on the plane of the collector, Wm^{-2}
- T_a Dry-bulb ambient temperature, K
- T_m Mean air-stream temperature in the collector, K
- U Total steady-state heat-loss coefficient for the collector, $Wm^{-2}K^{-1}$
- α Absorptivity for solar energy by the absorber plate
- n_c Instantaneous collector efficiency, i.e. rate of useful heat gain to the total insolation rate on the collector.
- τ Transmissivity of collector glazing with respect to solar energy.

REFERENCES

1. NORTON, B. and PROBERT, S.D., (1984), Solar-Energy Stimulated Open-Looped Thermosyphonic Air Heaters. Applied Energy, 17, 217-234.
2. BERNIER, M. (1984), Studies on Air Collector Testing at the NRCC. Proc. of the First European Community Conference on Solar Heating, Amsterdam, Netherlands, April-May.
3. NORTON, B. and PROBERT, S.D. (1983), Recent Advances in Natural-Circulation Solar-Energy Water Heater Designs, Applied Energy, 15 15-42.

STUDY OF NATURAL CONVECTION WITHIN A LARGE CAVITY - FIRST RESULTS

J.J. VULLIERME, F. YGUEL
Laboratoire d'Energétique Solaire - C.N.R.S.
C.E.A.T. - 43, route de l'Aérodrome - 86000 POITIERS, France

1. SUMMARY and INTRODUCTION

In the thermal analysis of building, one of the points of discrepancies between computed and measured values results from a rough estimate of the convective heat transfer. Presently, the numerical studies on natural convection heat transfer are restricted to the problems where the flow is laminar : Rayleigh number, Ra, less than 10^8, and consequently to small dimensions cavities. On the other hand, these numerical works allow to deal only with simple boundaries conditions, like a vertical face cold, its opposite warm, a linear temperature distribution on the four others or still four adiabatic walls. At the present time, only experimental works can provide qualitative and quantitative informations on this heat transfer, for Rayleigh number greater than 10^8, i.e. for large cavities similar to these existing in dwelling. The cavity under investigation, 3.1m width, 3.1m depth, 2.5m high, permits us to reproduce any temperatures distributions on its six walls. It is made indeed of 544 temperature-regulated elements on which a precise thermal balance can be carried out, and gives access to 380 values of the natural convection heat transfer coefficient within the cavity. Before to introduce complex temperatures boundaries conditions, similar to those existing in dwelling, we have worked on temperatures distributions identical to those used in the numerical works. We will present those first results.

2. DESCRIPTION DU MONTAGE EXPERIMENTAL

544 éléments fluxmétriques constituent les six faces de la cavité parallélépipèdique. Cette dénomination se justifie par le fait que ces éléments permettent d'imposer les températures de surface souhaitées et d'effectuer à l'équilibre thermique un bilan thermique précis aboutissant aux valeurs des densités de flux convecté ($\dot{Q}_{conv}$) :

$$\dot{Q}_{conv} = Pu - \dot{Q}_{cond} - \dot{Q}_{ray}.$$

Chacun d'eux se compose en effet, en partant de la face avant, d'un circuit imprimé bakelite-cuivre (ép. : 0.015m, λ = 0.254 $W\ m^{-1} K^{-1}$) qui constitue l'élément chauffant du fluxmètre, d'un matériau isolant rigide (ép. = 0.03m, λ = 0.029 $W\ m^{-1} K^{-1}$) et d'un radiateur qui permet de maintenir la face arrière du fluxmètre à une température constante et uniforme [1].

Les densités de flux conduit ($\dot{Q}_{cond}$) sont calculées à partir des mesures fournies par 2 thermocouples Chromel-Alumel (ϕ = 0.2mm), le premier placé sur le circuit de cuivre et le second sur le radiateur, les densités de flux injecté (Pu) sont obtenues à partir des mesures de tension et d'intensité aux bornes de la résistance chauffante. Les densités de flux rayonné ($\dot{Q}_{ray}$) proviennent d'une part des températures de surface (calculées à partir des mesures de la température sur le circuit de cuivre) et d'autre part du calcul des coefficients de couplage radiatif par la méthode des flux nets.

544 thermocouples Chromel-Alumel (ϕ = 0.2mm) placés à 0.1m devant chaque élément et 30 autres thermocouples disposés dans les plans de symétrie de la cavité, nous donnent accès à 280 températures d'air à l'intérieur de la cavité.

Le système de régulation est constitué d'un minicalculateur scrutant toutes les cinq minutes environ, 700 voies de mesures de température et 640 voies de mesure haut niveau pour les tensions et intensités. A la fin de chaque séquence d'acquisition, les densités de flux convecté sont calculées, et par extrapolation dans le temps, on affecte une nouvelle puissance de chauffe sur chacun des éléments ; ceci de façon à atteindre le plus rapidement possible, le régime stationnaire.

3. RESULTATS EXPERIMENTAUX

Les quatre expériences que nous avons effectuées jusqu'à présent, ont montré que le régime stationnaire était obtenu après une durée de l'ordre de 12 heures : la somme des flux de chaleur fournis est alors égale à la somme des flux conduits, la somme des flux rayonnés à celle des flux convectés.

Avant d'entreprendre l'étude de configuration de températures complexes, nous avons cherché d'abord à nous placer dans le cas très théorique et proche du "window problem" utilisé dans les études numériques, une face verticale chaude (Nord) est maintenue à température uniforme, ici 34°C ; les 5 autres faces ne sont pas régulées, mais la partie arrière des éléments fluxmétriques est maintenue à une température uniforme de 6°C. Les champs de températures de surface obtenus sont présentés sur la photographie 1, qui résulte d'une interpolation linéaire à partir des mesures.

On observe dans ce cas, la présence d'un gradient de température important, voisin de 9°K entre la face Nord et les éléments des faces verticales et horizontales proches de cette face. Ce gradient est induit par les conditions aux limites qui ne conduisent qu'à une faible génération de puissance dans la cavité : 950 W.

Les répartitions de températures d'air à 0.1m des parois sont présentées sur la photographie 2. La stratification de l'air observée à cette distance nous permet, en raison des répartitions de températures sur le plafond et le plancher, de conclure que l'air à l'intérieur du volume est lui aussi stratifié à des niveaux voisins de ceux qui sont obtenus à 0.1m. Ceci est confirmé par les mesures de température d'air dans les deux plans de symétrie de la cavité (figure 3). Le gradient de température vertical $\partial T/\partial z$, sur les cinq axes, évolue peu sur chacun d'eux, entre les cotes z=0.25m et z=2.0m. Il reste voisin de 1.5 K m^{-1}. Par ailleurs, la température au centre de la cavité est plus faible que la demi-somme des températures des faces Nord et Sud : 28.9°C ; il en résulte sans doute un accroissement des forces d'archimède génératrices du mouvement le long de la couche limite ascendante devant la face Nord, qui conduit à des vitesses plus fortes dans cette zone.

On trouve reproduites sur la photographie 4, les iso-densités de flux convecté interpolées à partir des mesures. On note que celles-ci sont faibles sur le plancher et fortes sur la face nord (les flux sortants sont comptés négativement). Sur le plafond les densités de flux convecté sont plus importantes que sur les faces Sud, Est et Ouest. Celles-ci sont sur les faces Est et Ouest, plus faibles à la même cote que celles sur la face Sud ; la différence est d'environ 2 W m^{-2}. Les densités de flux croissent par ailleurs sur les 3 faces verticales, avec la hauteur.

Les densités de flux rayonné évoluent sur les cinq faces froides, à l'inverse des densités de flux convecté. En dehors de la face Nord, elles sont les plus fortes sur le plancher (13 à 17 W m^{-2}) et les plus faibles sur le plafond (4 à 7 W m^{-2}).

4. RESULTATS NUMERIQUES

Il peut paraître étonnant d'entreprendre une étude numérique de la convection naturelle dans une grande cavité, puisque l'objet premier de ce travail de recherche est de pallier par l'expérience, à la déficience des méthodes numériques dans ce domaine. Il est évident que les solutions numériques obtenues à des nombres de Rayleigh aussi élevés seront sans doute approximatives en raison de la difficile prise en compte des phénomènes turbulents. Néanmoins, elles présentent un intérêt certain car elles nous aideront à analyser les phénomènes observés expérimentalement en nous fournissant des indications sur les grandeurs physiques à l'intérieur du volume.

Le modèle numérique employé est un schéma aux différences finies basé sur une approche volume de contrôle. Il utilise un maillage 38x33 à pas non constant [2].

Avant d'entreprendre une comparaison sur une cavité du type de celle présentée au § 2, nous avons d'abord comparé les solutions numériques obtenues dans une cavité de hauteur C= .272m, de largeur A= .408m, isolées sur les faces haute et basse, pour différentes valeurs de l'écart de température Δt entre les faces chaude et froide [3]. On a pu ainsi montrer que l'écart maximum sur le coefficient d'échange par convection, h calculé à partir de Δt, restait pour $\Delta t \leqslant 60°C$ inférieur à 15%. Pour ce dernier écart, nous présentons figure 5, l'évolution du coefficient h sur la face chaude avec la cote adimensi onnelle z/C.

CONCLUSION

Les résultats expérimentaux présentés sont les premiers d'une série de mesures relatives à différentes répartitions de température de paroi. La détermination de coefficients d'échange par convection naturelle qui demeure l'un des buts essentiels de cette recherche ne pourra être effectuée que lorsque nous disposerons d'un nombre d'expériences suffisamment important pour nous permettre de définir une température de référence.

En ce qui concerne la simulation numérique, les temps de calcul extrêmement importants (20 heures d'Unité Centrale d'un ordinateur Norsk Data 100 par configuration) ne nous autoriseront à étudier que quelques configurations. Nous espérons toutefois à partir de ces simulations, obtenir des informations tout au moins qualitatives sur la nature de l'écoulement à l'intérieur de l'enceinte.

BIBLIOGRAPHIE

[1] - F. YGUEL et alii, "Action de recherche coordonnée sur la convection naturelle dans l'habitat", Compte-rendu de la Conférence Internationale sur l'architecture solaire, Cannes, Décembre 1982

[2] - P. LE QUERE, F. YGUEL, "Comparaison de solutions numériques d'un problème de convection naturelle en cavité", rapport interne L.E.S., Novembre 1983

[3] - D. BAIRI, J.J. VULLIERME, "Etude expérimentale de la convection naturelle dans des cellules parallélépipèdiques inclinées", Communication personnelle.

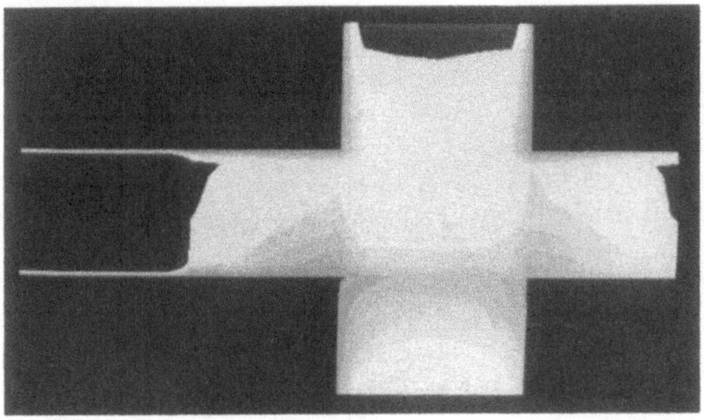

Températures de surface
- Tracé des isothermes
(faces Nord-Est situées à l'extrémité gauche de la figure et le plafond à l'extrémité haute)

- PHOTOGRAPHIE 1 -

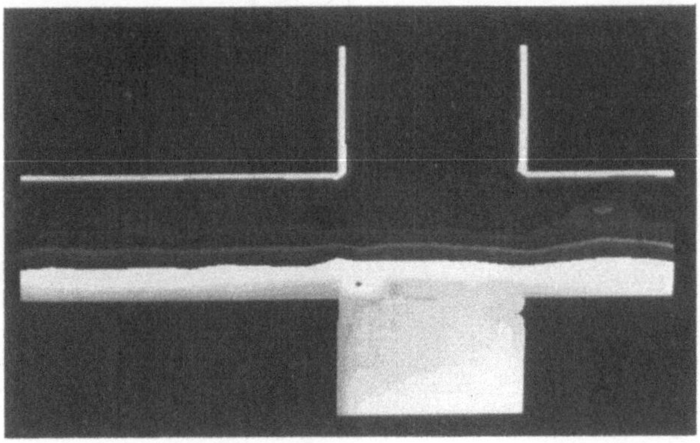

Température d'air à 10cm des parois
- Tracé des isothermes

- PHOTOGRAPHIE 2 -

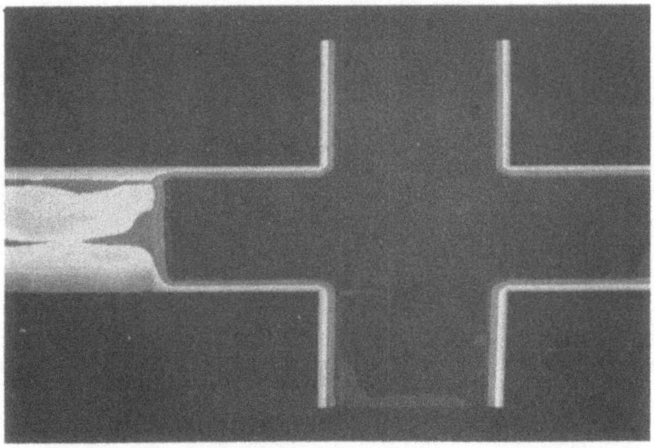

Isodensités de flux convecté

- PHOTOGRAPHIE 4 -

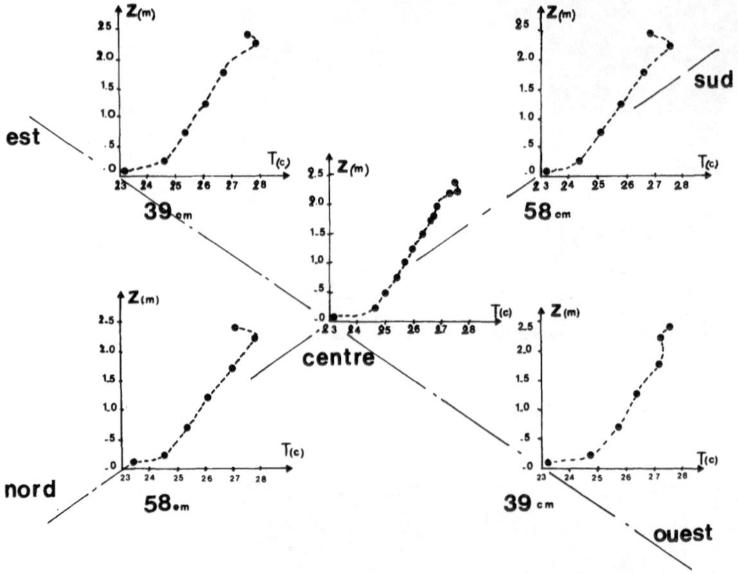

- FIGURE 3 -
Température de l'air sur 5 axes verticaux à l'intérieur de l'enceinte

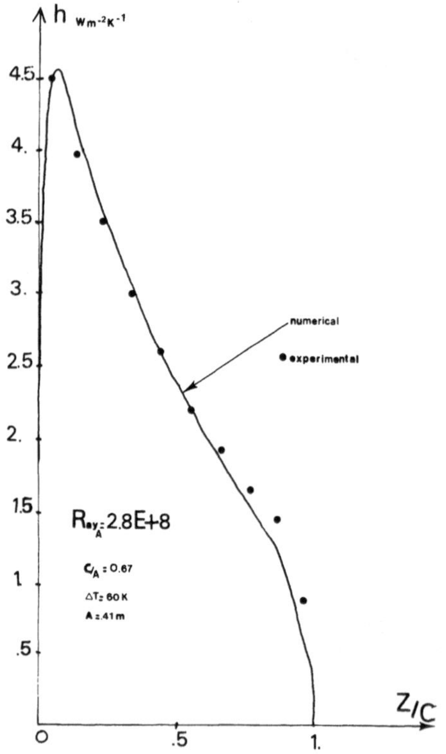

- FIGURE 5 -

PASSIVE SOLAR HOUSES WITH ENERGY EFFICIENT SOLAR WALLS CONTAINING PARAFFIN

P. DUBOIS, D. SINOBAD and T. VERHAEGE
Laboratoires de Marcoussis,
CR de la Compagnie Générale d'Electricité

Summary

We have investigated the possibility to design and build solar walls with a good efficiency and a low enough cost to be used in passive solar houses. The reported work includes :

. research on wall design including the solar heat collection and release.
. the building and test on an experimental house at Marcoussis of three 1.5 m2 paraffin solar walls, one brick colored wall with a selective coating developped at Marcoussis standing between two black non selective walls. The thermal behaviour of the walls and the space heating thermal regulation were thoroughly investigated. The efficiency of walls of this type compares favourably with the efficiency of solar collectors and separate heat store. The same solar efficiency was obtained for black walls and for a brick colored wall with a selective coating developped at Marcoussis, which is very promising from an architectural point of view.
a computer program study of the space heating of a 100 m2 house by south facing solar walls and electrical convectors.

0. INTRODUCTION

We have investigated the possibility to design and build solar walls with a good efficiency and a low enough cost to be used in passive solar houses. The reported work includes :

. research on wall design including the solar heat collection and release.
. the building and test on an experimental house at Marcoussis of three 1.5 m2 paraffin solar walls, one brick colored wall with a selective coating developped at Marcoussis standing between two black non selective walls. The thermal behaviour of the walls and the space heating thermal regulation were thoroughly investigated.
. a computer program study of the space heating of a 100 m2 house by south facing solar walls and electrical convectors.

1. SOLAR WALLS DESIGN

We have chosen a design which combines a solar effect and a dynamic insulation effect (cf. figure 1).

The solar heat is collected in wall elements containing paraffin, their absorbing surface being covered with a selective or a non selective coating. The elements containing paraffin can be made of metal or cement and plastics. The solar walls built at Marcoussis were made of metal casings containing 19 kilos of paraffin with an absorbing surface of 1 x .5 m2. Each metal casing is divided into six horizontal channels so as to improve convection heat exchanges in liquid paraffin.

The fresh air admitted in the house flows first in the front air channel between the solar collecting surface and a double glazing and then in the back air channel between the paraffin containing wall elements and the thermal insulation.

The solar heat collected and stored in the paraffin is transfered with a significant time lag to the fresh air, which is thus preheated before entering the house. After the stored solar energy has been completely released, the fresh air flowing through the wall saves energy through a dynamic insulation effect.

2. BUILDING AND TEST AT MARCOUSSIS OF THREE 1.5M2 PARAFFIN SOLAR WALLS

2.1 Building of three paraffin solar walls

Three 1.5 m2 paraffin solar walls have been built and tested at Marcoussis on an electrically heated experimental house with a floor area of 14.6 m2. The experimental house has poor thermal insulation so that the space heating needs of this experimental house are about one third of the needs of a well insulated 100 m2 house.

The design and construction of the walls was based on results we had acquired over the years on the behaviour of paraffin to thermal cycles, on the selection of encapsulating material and on the kinetics of heat exchange and mass transfer within the paraffin.

The absorbing surfaces of two of the walls consist of a commercially-available non-selective black paint. The absorbing surface of the third wall consists of commercially-available 10 x 20 cm2 enamelled brick colored tiles which were rendered selective by the application of a thin coat of tin oxide and antimony fired at 600°C. This coating, developped at Marcoussis, increases the surface absorption coefficient from 0.56 to 0.82 and at the same time reduces its low temperature infrared emission coefficient from 0.82 to 0.25. The tiles were glued to the metal of the paraffin-filled casings using a neoprenebased contact adhesive giving good thermal contact and good temperature stability.

2.2 Tests of the solar walls on an experimental house at Marcoussis

We have plotted on figure 2 results obtained during the tests of the three 1.5 m2 solar walls on an experimental house at Marcoussis.

- on days without sunshine or with very little sunshine, the solar walls provide several hundred watt-hour per day and per square meter of wall. This is due to a dynamic insulation effect.
- the heat provided to the house, including partial recovery of wall losses by dynamic insulation, is larger when fresh air flows continuously through the wall than when fresh air flows only from 5 p.m. to 8 a.m., that is when there will be no solar energy coming into the house through the windows. The solar efficiency of the walls goes up from 36 % with air flowing from 5 p.m. to 8 a.m. to 52 % with air flowing continuously through the walls.
- there is no significant difference between the thermal efficiency of a black non selective wall and the thermal efficiency of a brick colored wall with the selective coating developped at Marcoussis. This result is important from an architectural point of view because it shows that brighter colored solar walls are possible.

2.3 Space heating regulation

As indicated above, the solar efficiency of a paraffin wall is higher (52 %) with fresh air flowing continuously through the wall than with fresh air flowing through the wall only from 5 p.m. to 8 a.m. (36 %). It means that, in order to obtain the higher possible efficiency, the collected solar energy must be used as soon as possible.

The results obtained during the test of the solar walls at Marcoussis show that, even with fresh air flowing continuously through the walls, there is a significant time lag between solar energy collection and release. On a day with nearly 4 kWh/m2 of incident solar energy, only 30 % of the energy supplied to the house by the solar walls was delivered between 12 noon and 4 p.m.

For an electrically heated house with solar walls (cf fig. 4), it should be possible to obtain both comfort and a high thermal efficiency with a space heating thermal regulation which includes one blower and two shutters by applying the followings rules which gives priority to solar energy over electrical heaters :

When the temperature in the house is

- . lower than θ_0 : the electrical heaters are used
- . lower than $\theta_0 + \delta$: the fresh air flows into the house through the solar walls (V_1 open, V_2 closed)
- . higher than $\theta_0 + \delta$: the fresh air flows directly into the house (V_2 open, V_1 closed)

3. SPACE HEATING FOR A 100 M2 HOUSE WITH SOLAR WALLS

We have made a computer study of the space heating of a 100 m2 house by 15 m2 of south facing paraffin solar walls and electrical convectors. The space heating thermal regulation is operated according to the criteria indicated above (cf 2.3) with a reference temperature for the solar walls 3°C higher than for the electrical heaters.

For a house situated in the Paris area and with a space heating need of 200 Watt/°C, we have obtained the following results :

- in March, on sunny days, the fresh air flows most of the time but not continuously through the solar walls and there is a significant time lag between solar heat collection and release (cf fig. 5).
- over the full space heating period, from October to May, we have computed (cf fig. 6) an electricity comsumption of 6.438 kWh and a solar wall contribution of 2.584 kWh, nearly 30 % of the total.

4. CONCLUSION

The reported results show that paraffin solar walls can generate a significant time lag in the restitution of solar energy and their efficiency for space heating compares favourably with that of solar collectors associated with a separate heat store. There is no significant difference between the thermal efficiency of a black non selective solar wall and the thermal efficiency of a brick colored wall with the selective coating developped at Marcoussis. This last result is important from an architectural point of view because it shows that brighter colored solar walls are possible.

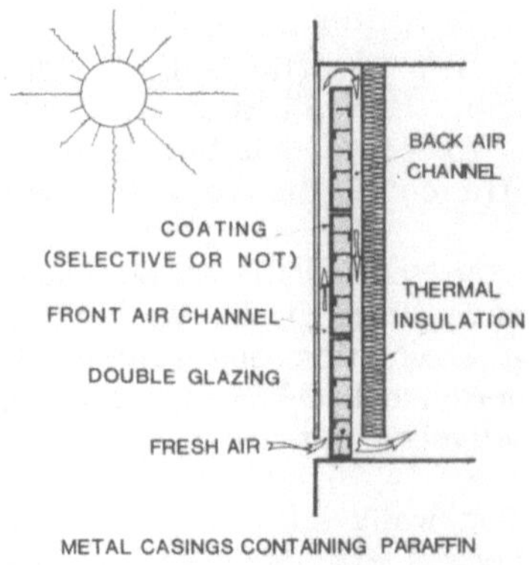

Fig. 1 PARAFFIN SOLAR WALLS DESIGN

Fig. 2 TEST ON AN EXPERIMENTAL HOUSE AT MARCOUSSIS - SOLAR WALLS EFFICIENCY

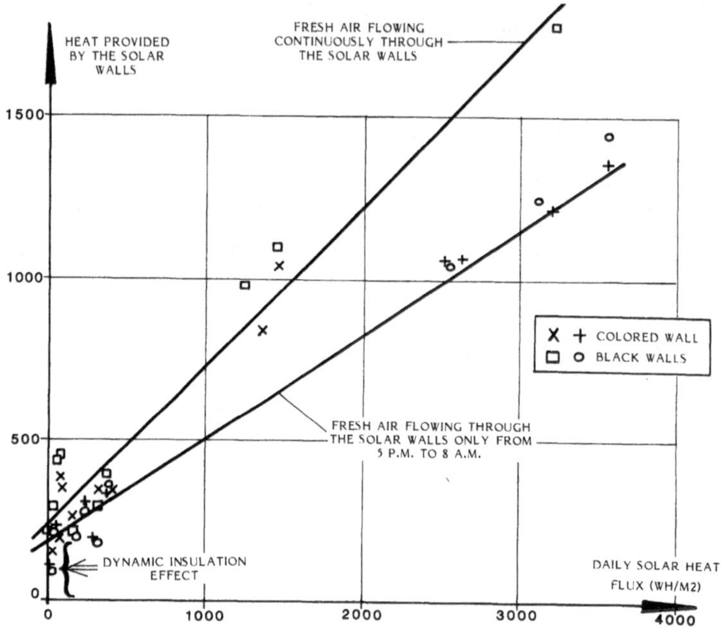

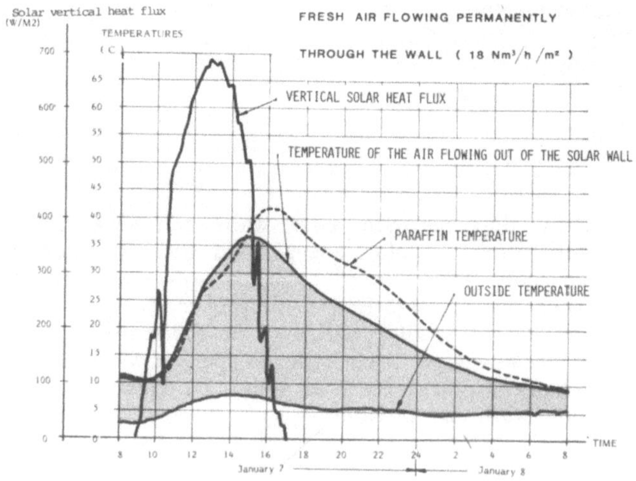

Fig. 3
TEST AT MARCOUSSIS-
SOLAR HEAT COLLECTION
AND RESEASE VERSUS
TIME

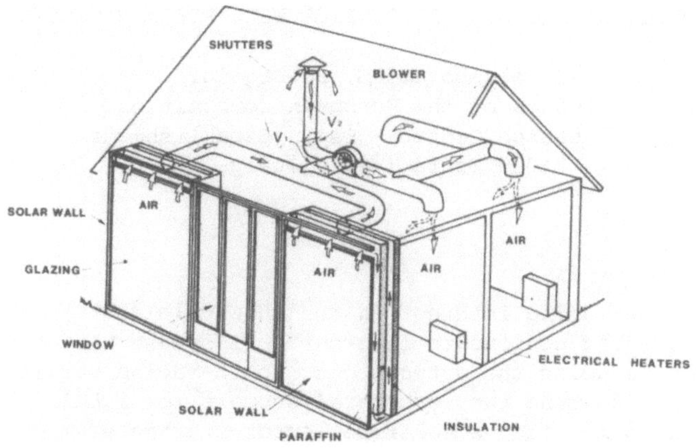

Fig. 4 SPACE HEATING REGULATION FOR AN ELECTRICALLY HEATED HOUSE WITH SOLAR WALLS

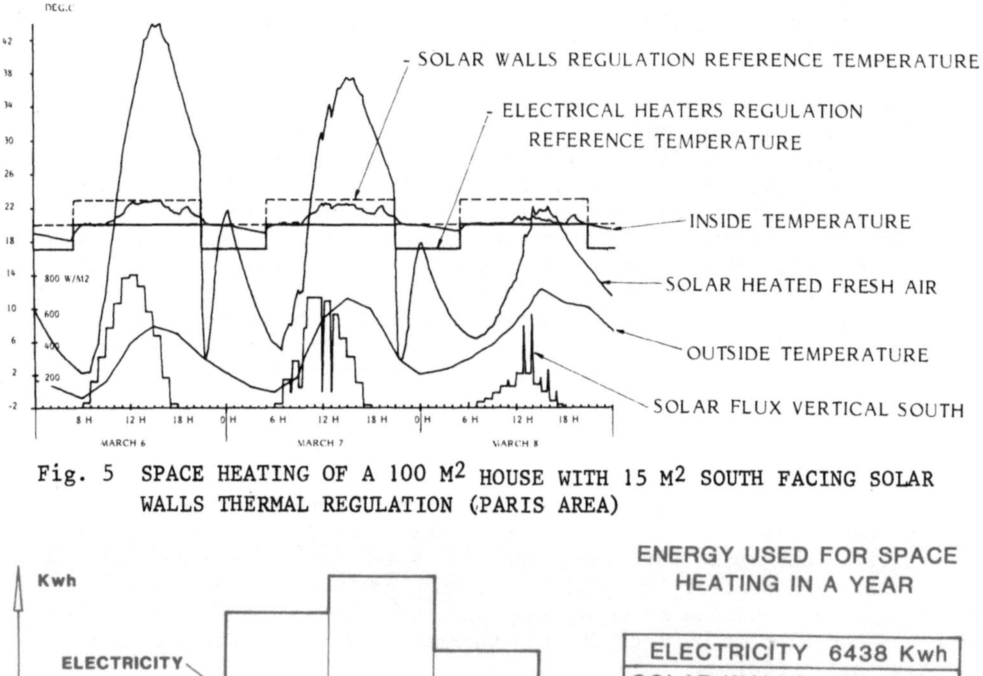

Fig. 5 SPACE HEATING OF A 100 M^2 HOUSE WITH 15 M^2 SOUTH FACING SOLAR WALLS THERMAL REGULATION (PARIS AREA)

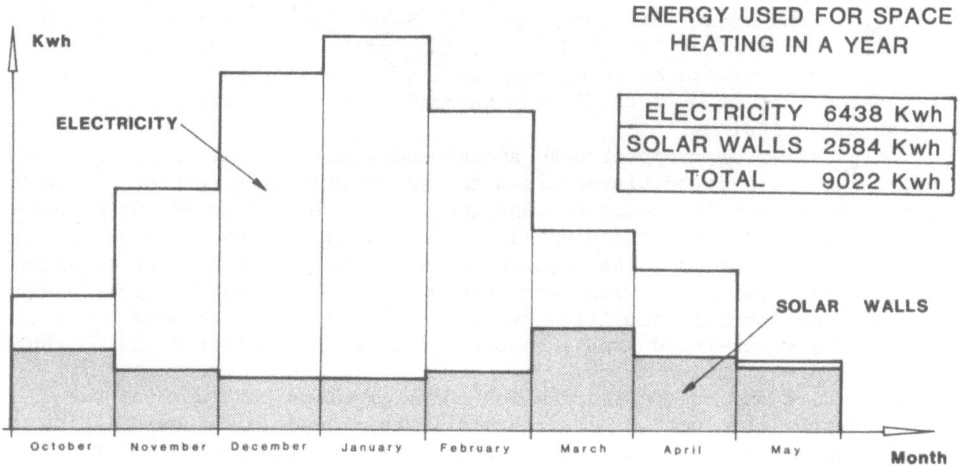

Fig. 6 SPACE HEATING OF A 100 M^2 HOUSE WITH 15 M^2 SOUTH FACING SOLAR WALLS - ENERGY BALANCE (PARIS AREA)

A HYBRID COLLECTOR-LATENT HEAT STORAGE WALL FOR AIR CONDITIONING

R. Colombo and G. Pellegrini
Commission of the European Communities
Joint Research Centre - Ispra Establishment
21020 Ispra (Va) - Italy

Summary

There is an increasing interest in the industrialization of thermally efficient building components in order to increase the solar energy contribution reducing the actual costs. South-facing vertical walls (Trombe Walls) forming the structural support for a building and acting at the same time as solar collector-storage elements are being evaluated theoretically and experimentally in various laboratories. Fully passive Trombe Walls, however, suffer in general from a series of negative aspects such as the slow thermal response, the high thermal losses during night, the poor thermocirculation characteristics and the limited useful utilization time covering only the cold period of the year.

Here the characteristics and the design of a latent heat storage curtain wall for air conditioning integrated in a hybrid solar building are presented and discussed from the point of view of the local climatological data and the energy saving potential. The wall combines in a single technological module the functions of collecting, storing and releasing solar heat at ambient temperature through a fan-assisted air circulation. The module is dimensioned in a way as to meet the daily energy demand for climatization during the summer and to supply part of the space heating requirements during the winter.

Thermal performance tests on the collector-storage wall coupled to the front of a 3x3x6 m test cell are in progress.

1. INTRODUCTION

Space heating and cooling requirements from both household and commercial sector represents about 24% of the total energy consumption in the EC-countries and about 18% in the USA /1/. Space heating and cooling systems utilize in general low grade heat. Important savings can be made in virtually every climate by using the passive solar approach to space heating and cooling, or by retrofitting thermally efficient structures to existing old buildings.

The passive solar approach to space conditioning usually implies the use of large south-faced glazed areas for an optimum exploitation of the locally evailable solar radiation and the utilization of an adequate thermal mass, suitably distributed within the building structure, to reduce temperature oscillations. The major problems to this approach are the high thermal losses associated with large window surfaces, sensible local overheating in the interior and large temperature differences between the north-faced and the south-faced areas, requiring important modifications to the actual heat distribution system.

Definitive and cheap remedies for these problems have not yet been found. Nevertheless, noticeable progress is being made with new building materials and components for passive heating and cooling, such as transparent insulation panels, specialized windows and louvers and building components with integrated heat storage /2/.

1.2 Southern walls

The concept of south-facing vertical walls (Trombe-walls) forming the structural support for a building and acting at the same time as solar collector-storage elements is being evaluated theoretically and experimentally in various laboraotires /3,4,5,6/. The experimental studies carried out so far show that considerable improvements in the efficiency of a Trombe-wall may be obtained using fan-assisted air circulation /7/. Moreover, further advantages are expected from the development of wall elements incorporating latent heat storage /8/.

Latent heat storage materials are of interest to passive solar technology because of their relatively high storage capacity per unit of volume and because of the isothermal nature of the storage process, which favours living comfort and which offers interesting applications in the climatization technology.

2. EXPERIMENTAL

The characteristics and the design of a latent heat storage curtain wall for space conditioning to be integrated in a hybrid solar building are presented. The wall combines in a single module the functions of collecting, storing and releasing solar heat at ambient temperature through a fan-assisted air circulation. The module is dimensioned in such a way as to meet the daily energy demand for climatization during the summer and to supply part of the space heating requirements during the winter.

2.1 Wall Structure

The design is the industrialized approach to thermally-efficient buildings components (e.g. the vertical facade of single rooms in a new passive apartment block or students residence).

The wall is a composite structure consisting of an outer transparent surface-double glass or polycarbonate- and an inner sandwich. This has high-strength insulation panels containing an air collector, the latent heat storage elements and a fan to force air from outside into the interior through the collector-storage system (Fig.1).

The air collector is integrated into the lower part of the vertical wall. The latent heat storage system consists of a stack of plastic coated aluminium pouches (40x15x2 cm) supported by an Al-frame allowing air to flow through the stack. Two types of phase change material are used, namely Glauber's salt melting at 32°C, and a mixture of Glauber's salt and NaCl melting around 22°C.

The size of the collector-storage system conforms to the architectural features and to the energy requirements of the building in which it is integrated (Fig.2).

2.2 Operation Mode

During a cold sunny day fresh air is sucked from outside by means of a fan into the air collector after being partially mixed with exhaust air coming from the interior. Hot air leaving the collector at temperatures between 30 and 70°C is forced through the stack of Al-pouches containing the PCM, where it is climatised around 20-22°C and then conveyed into the room.

During the night the collector is by-passed and the fresh air flows directly through the storage, where it is heated up to ambient temperature.

During warm summer days the outside air temperature may be sufficient to charge the storage system during the day. In that case the air collector is by-passed and the exhaust air is not recycled.

3. THERMAL PERFORMANCE TESTS

A prototype of 3x3 m size is mounted on the south-facade of a test cell and tested under the climate conditions of Ispra in Northern Italy (45°48'11"N) (Fig.3). The stabilization of the indoor temperature as recorded during a series of subsequent winter days, respectively winter nights, is shown in Fig.4. Preliminary energy balance evaluations indicate that an energy saving of around 30% is achieved during the cold winter months February-March, while the mean day-night indoor temperature oscillations were contained between 18 and 22°C.

Thermal performance tests for the evaluation of the air conditioning potential during the transition spring months April-May, and during the summer have been initiated. The thermal performance of the collector-storage wall will be compared with that of a conventional facade installed on a reference test cell climatized at ambient temperature (21°C).

4. CONCLUSIONS

An attempt has been made in order to improve the thermal performance and to promote the industrialized construction of south-facing vertical walls acting as passive solar components. The concept has been developed to substitute the conventional fully passive Trombe-wall with a hybrid wall containing an air collector and latent heat storage for air conditioning. The collector storage system is sized so as to meet the daily energy requirements for climatization of the back-coupled space during the summer and to supply part of the energy demand for fresh air heating during the winter.

Preliminary performance tests in real climatic conditions indicate that a good stabilization of the indoor temperature is achieved, and that important energy savings can be obtained by the use of such systems for space cooling.

5. REFERENCES

1. Eurostat Energy Statistics, Year Book, 1979.
2. JOHNSTON, T.E. (1976) New building materials and components for passive heating of buildings. Symp. on Passive Solar Heating Systems, Albuquerque, pp.288-292.
3. TROMBE, F., ROBERT, J.F., CABANAT, M. and SESOLIS, B. (1976) Some performance characteristics of the CNRS solar house collectors; Passive Solar Heating and Cooling Conf. Workshop and Proc., Albuquerque, New Mexico, 18-19 May, pp.201-202.
4. CASPERSON, R.L. and HOCEVAR, C.J. (1979) Experimental investigation of Trombe-wall passive solar systems. Proc. 3rd Nat. Passive Solar Conf., San José, California, pp.231-237.
5. BALCOMB, J.D., HEDSTROM, J.C. and MAC FARLAND (1977) Passive solar heating of buildings; Los Alamos Scientific Lab., Rep. La Ur 77-11-62.

6. OHANESSIAN, P. and CHARTERS, W.W.S. (1978) Thermal insulation of a passive solar house using a Trombe-Michel wall structure. Solar Energy, Vol.20, pp.275-281.
7. AKBARZADEH, A., CHARTERS, W.W.S. and LESSLIE, D.A. (1982) Solar Energy, Vol.28, 6, pp.461-468.
8. PELLEGRINI, G. and COLOMBO, R. (1984) Building components with integrated latent heat storage; Proc. 1st EC Conf. on Solar Heating, Amsterdam, April 30 - May 4, 1984.

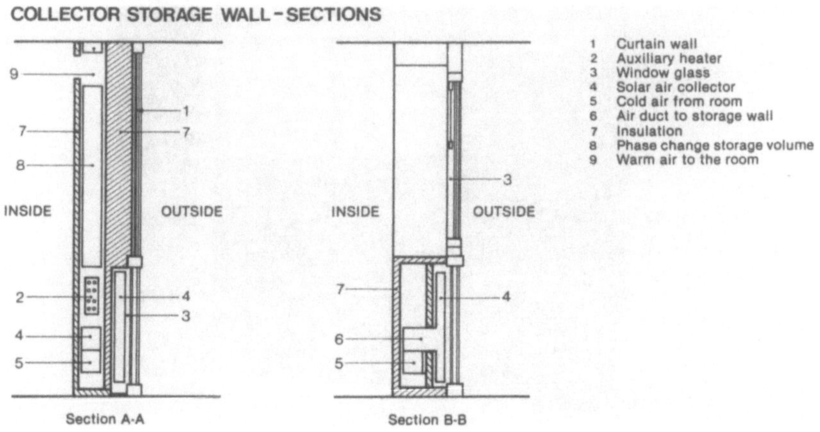

Fig.1 Schematic view of the structure of the collector-latent heat storage wall for air conditioning

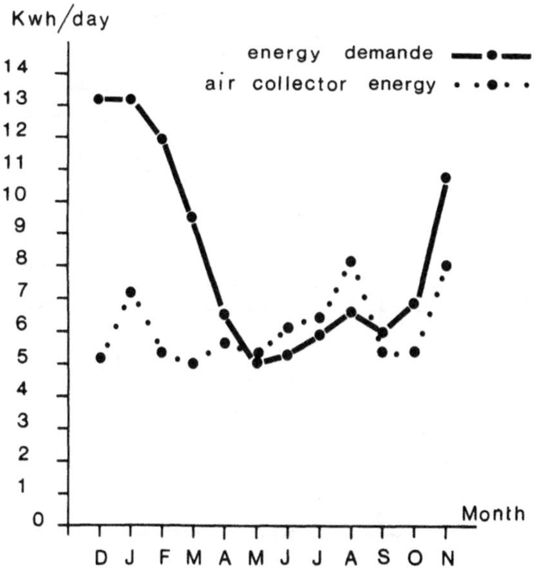

Fig.2 Calculated daily energy demand for climatization and mean daily energy available from the air collector

Fig.3 - The hybrid collector-storage wall mounted on the south-facade of a test cell at the JRC Ispra for performance evaluation tests.

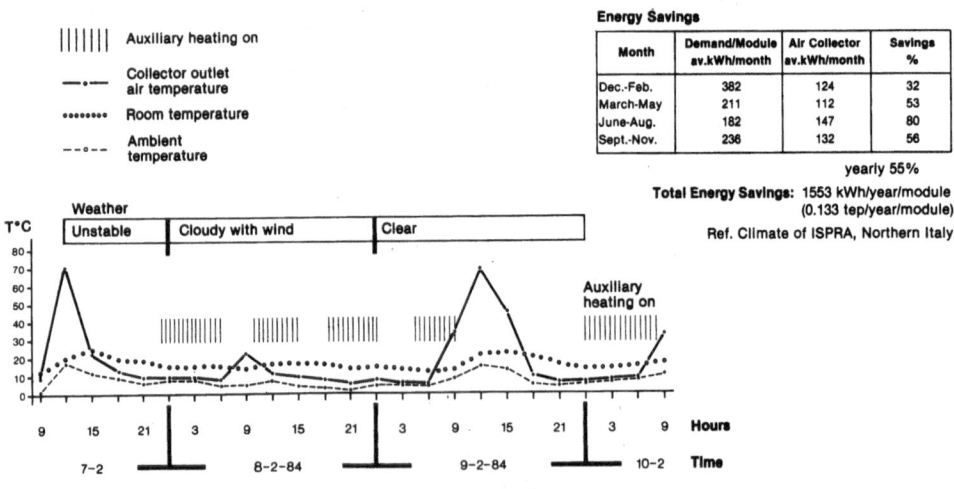

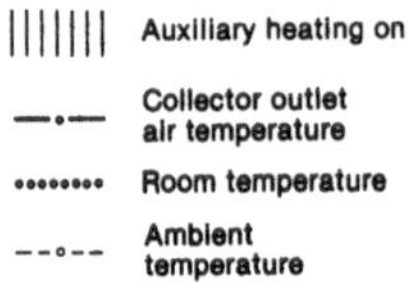

Energy Savings

Month	Demand/Module av.kWh/month	Air Collector av.kWh/month	Savings %
Dec.-Feb.	382	124	32
March-May	211	112	53
June-Aug.	182	147	80
Sept.-Nov.	236	132	56

yearly 55%

Total Energy Savings: 1553 kWh/year/module (0.133 tep/year/module)

Ref. Climate of ISPRA, Northern Italy

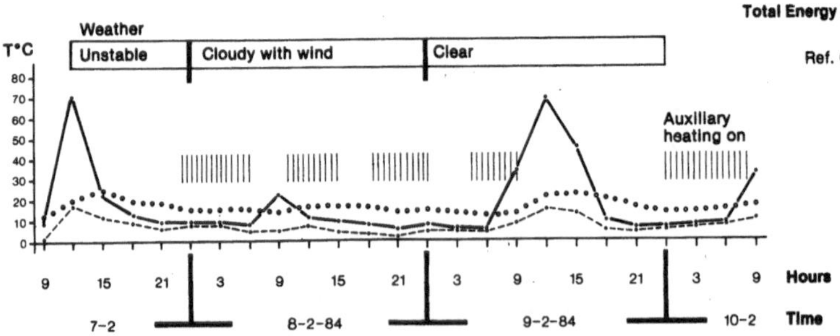

Fig.4 - Stabilization of the indoor temperature as observed during a series of subsequent winter days, respectively night.

EFFECTIVE 'U' VALUE

P.G.T. OWENS, BA, FCIBS
GROUPEMENT EUROPEEN DES PRODUCTEURS DE VERRE PLAT

Summary

The paper explains the philosophy of combining losses with solar gains through the modification of the conventional 'U' value used for windows. The definition of the modified value , called the Effective 'U' value is given in the form of an equation and a simple proof is given. The underlying research using detailed computer analysis of the energy requirements of buildings is discussed. Such computer analysis leads to the establishment of a methodology to determine Effective 'U' values for real situations. The methodology based on standard meteorological data and the thermal characteristics of glazing is explained and a program listing provided which enables the calculation to be made for any location and windoow design.

1. INTRODUCTION

Before the advent of more detailed computer analyses of energy consumption in buildings various methods based on the degree day concept were used. In these methods the specific design heat loss from the buildings is multiplied by the number of degree days to form a product proportional to the total energy requirement for the heating season. The influence of incidental gains, such as solar radiation, is dealt with by correcting the internal average temperature or base temperature in some relatively arbitrary way(1).

The consequence of this procedure for evaluating the contribution of the windows to the overall energy requirement is to dissociate the solar gain through the window from the heat lost through the window. This simplistic approach has undoubtedly resulted in a total misunderstanding of the role of the window in the energy balance of buildings by the majority of designers.

The more detailed calculations of energy use made practical by computers enables the windows and their influence in energy consumption to be more properly evaluated. However, for most building designs, such computer analyses are not undertaken for a variety of reasons. It seems that a simple method should be devised which can identify in a single figure the influence of the window on the total energy use.

To this end the Effective 'U' value is introduced. It is simply the conventional 'U' value modified to take account of solar radiation falling on the window during the course of the heating season.

2. DISCUSSION

There have been many studies done to evaluate the influence of solar radiation falling on the windows of buildings. Most of these have been done with computer models examining the dynamic response of the building to the continually varying weather data and solar radiation.

This discussion reports one such examination, however, all the calculations show the same trend and are in good agreement one with the other. Moreover such practical measurements in real buildings that have been published validate the findings of computer analysis.

Some computer models do not undertake a totally dynamic analysis and the criticism was made that simulations based on averaging methods of analysis were not realistic. More detailed models were then developed to add support to the growing body of evidence, based on computer analyses.

The computer model used here simulates the dynamic response of the building and its heating and control system to continually varying weather data. The weather data is in the form of hour by hour recorded weather at Kew, principally the temperature, direct and diffuse solar radiation.

It was set up to analyse the energy used for heating living rooms (day rooms) and bedrooms in a typical domestic dwelling.

The variables introduced were :

1) Windows single or double glazed.
2) Window area (up to 40 per cent of the external wall).
3) Windows with and without curtains.
4) Orientation of the dwelling.

Typical occupancy patterns were assumed.

Two room thermostat settings were taken : 20°C or 18°C in living rooms and 17°C or 15°C in bedrooms. The living rooms were assumed to be occupied from 1700-2300 and the bedrooms from 2300-0800.

The heating was a conventional boiler-radiator system with a boiler margin of 50 per cent and 0.5 hour time constant. Characteristic boiler part load efficiencies were taken and room thermostats with 1°C dead bands. The heating system was controlled by a time clock with the following schedule :

on : 0600 off : 1000 on : 1600 off : 2200

No heating was provided outside these scheduled periods which was used throughout the heating season.

Ventilation rates were 2 air changes per hour between 0600 and 2200 and 1 air change per hour between 2200 and 0600. The living room curtains were drawn at sunset and opened at 0800, the bedroom curtains were drawn at 2200 and opened at 0800.

If the temperature due to excess solar gains, rose at any time above 24°C the windows (i.e. increasing the ventilation rate) were opened and closed again when the temperature fell to 21°C.

The whole schedule of possibilities and results is too extensive to report here but a typical example is shown in Fig. 1 which characterises the results. The energy consumption curves, with increasing window area, are depressed due to solar radiation compared with the situation where no solar radiation is taken into account. The curves may be depressed to the extent that energy consumption is reduced with increased window area and in this case the window may be seen acting as a solar collector.

If we define U_{FE} an effective 'U' value it is given by :

$$U_{FE} = U_F \frac{(L_S - L_{WF})}{(L_O - L_{WF})} \qquad \text{(Appendix 1)} \qquad (1)$$

where :

$(L_O - L_{WF}) = U_F \, . \, D \, . \, C \, . \, A_W$

$(L_O - L_{WF})$ is the energy lost through the window during the heating season. Where C is 24/1000 if $(L_O - L_{WF})$ is in kWh, D is the number of degree days, and A_W the window area.
Equation (1) can be re-arranged.

$$U_{FE} = U_F - \frac{(L_O - L_S)}{D.C.A_W}$$

where $(L_O - L_S)$ is the amount of useful radiation.
If S is the amount of solar radiation falling per m^2 window in kWh for the heating season.

$$(L_O - L_S) = S.f.T_G.T_S.A_W$$

where :

f = utility factor (to be determined)
T_G = Glass transmission factor due to over shadowing, dirt or other causes.

whence :

$$U_{FE} = U_F - \frac{42.f.T_G.T_S.A_W}{D} \qquad (1/C = 42)$$

3. METEOROLOGICAL DATA

S is the available sun on the vertical surface. Generally speaking data for this is not available from meteorological stations.

The data available are mean monthly solar radiation totals on the horizontal surface (sometimes direct and diffuse data are available) or failing that sunshine hours or percentages of total possible sunshine hours, from which it may be possible to derive estimates for the mean monthly levels using the Angström correlations (3,4).

For this study the method proposed by Berlage has been used (5). This assumes that the radiation received on a horizontal plane on any single day is a constant proportion of that which would be received under clear sky conditions. It is implicit that there is no absorption of solar radiation in the atmosphere due to haziness however Basnett has shown there is little systematic deviation between predictions using the Berlage method and measurements on vertical walls (6).

Comparison of the cases studied show that for mean monthly averages of solar radiation for the months September through April inclusive the utility factor f lies in the region 0.6 - 0.7 in the colder climates.

For example making a comparison with Table I the utility factors for the south orientation we obtain the figures in Table II.

A computer program listing, which incorporates the Berlage method for deriving the vertical surface irradiation, to calculate effective 'U' values for any location is given (7).

U_F is the conventional 'U' value
L_S is the total energy used with sun on the window

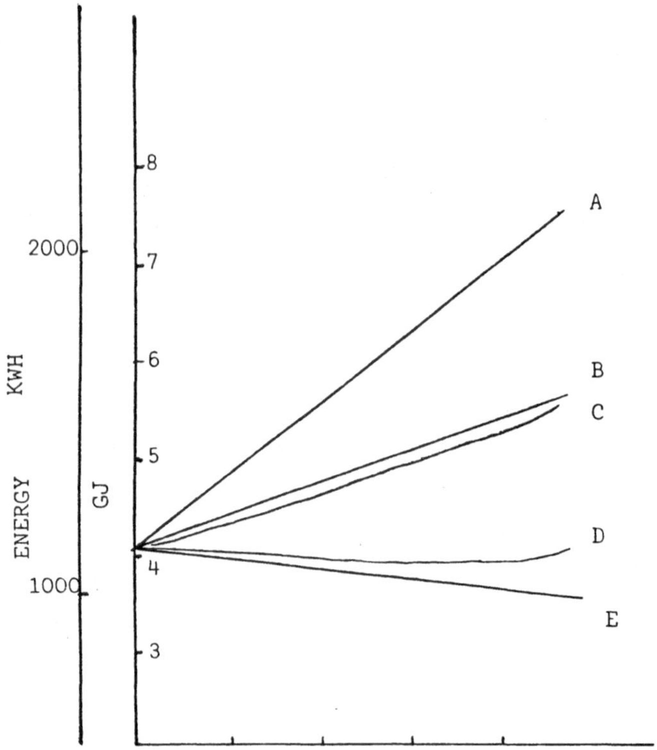

Fig. 1. Energy consumption curves, south facing room at Kew. (A) No sun single glazing, (B) No sun double glazing, (C) With sun single glazing, (D) With sun double glazing, (E) Total walls, ventilation, internal gains.
(BSRIA Research Report 1 Contract 3026).

L_O is the total energy load assuming no sun on the window.
L_{WF} is the total energy load due to ventilation, roof, wall, etc.

The effective 'U' values for the example in Fig. 1 are shown in Table I.

Table I. Effective 'U' values W/m^{2}K

Orientation/ per cent glazing		Single Glazing U = 5.7	Double Glazing U = 2.8
North	10	3.92	1.28
	40	3.98	1.36
South	10	2.57	0.13
	40	2.77	0.42

Table II. Utility factors

Per cent Glazing	Single Glazing	Double Glazing
10	0.66	0.65
40	0.62	0.58

APPENDIX 1

L_O : Total load with no sun on window
L_S : Total load with sun on window
L_{WF} : Load due to ventilation, roof, floor and externall wall at window area.

Therefore :

$$(L_O - L_{WF}) = U_F \, \Sigma A_W . \Delta T.HS$$

$$(L_O - L_{WF}) = U_{FE} \, \Sigma A_W . \Delta T.HS$$

$$\frac{L_S - L_{WF}}{L_O - L_{WF}} = \frac{U_{FE}}{U_F}$$

$$U_{FE} = U_F \frac{(L_S - L_{WF})}{(L_O - L_{WF})}$$

where :

U_F and U_{FE} are the normally used and effective thermal transmittances.

$\Sigma A_W . \Delta T.HS$ is the summation of the product : window area, average temperature difference, and heating season.

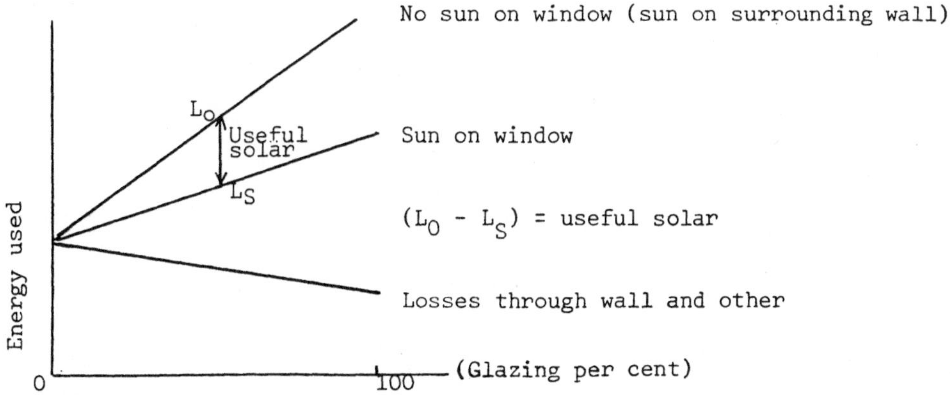

Fig. A1. Energy consumption curves showing derivation of Effective U value.

References :

(1) Dufton, A.F., Degree days, JIHVE, 2, 83-85 (1934).
Knight, J.C. and Cornell, A.A., Degree days and Fuel Consumption for Office Buildings, JIHVE, 26, 309-326 (1959).
Hitchin, E.R., Degree days in Britain, BSER&T, 2, (2) 73-82 (1981).

(2) Hamilton, G., BSRIA Report Contract 3026 Report No. 1 (March 1980).

(3) Ångström, A., Report to the International Commission for Solar Research in Actiometric Investigations of Solar and Atmospheric Radiation. Quarterly Journal Royal Meteorological Society 50, 121-125 (1924).

(4) Black,J.N., et al, Solar Radiation and Duration of Sunshine, QJRMS, 80, 231-235 (1954).

(5) Berlage, H.P., Zur Theorie der Beleuchtung einer Horizontalen Flache durch Tageslicht. Meteorlogische Zeitschrift, 174-180 (May 1928).

(6) Basnett, P., Estimation of Solar Radiation Falling on Vertical Surfaces from Measurements on a Horizontal Plane, ECRC/M846 (September 1975).

(7) Computer Program Listing. See P.G.T. Owens' Paper on Effective U Value, Building Services Engineering Research & Technology, 3, No. 4, 1982.

HIGH PERFORMANCE PASSIVE SOLAR HEATING SYSTEM WITH HEAT PIPE ENERGY TRANSFER

M.H. de Wit, J.L.M. Hensen
group FAGO-TNO-THE
Eindhoven University of Technology

H.A.L. van Dijk
G.J. van den Brink
E. van Galen
Institute of Applied Physics TNO-TH Delft

Summary

The aim of the project is to develop a passive solar heating system with a higher efficiency (regarding accumulation and transfer of solar heat into dwellings) than convential concrete thermal storage walls and with restricted extra costs for manufacturing the system. This is to be achieved by the introduction of three special components:

a. a heat pipe as a thermal diode tube for efficient transfer of collected solar heat from the absorber to the back of an insulation layer
b. a heat storage section with water or phase-change material
c. an extra insulation sheet with vents between the storage and the room for controlled transfer of heat into the room.

Additional advantegeous characteristics of such a system are that the maintenance costs are negligable and the performance is not affected by inadequate control. Series of measurements on single components and computer experiments have been carried out and thes have led to the expectation that a high performance can be achieved. The predicted costs, however, are still too high to be cost effective.

1. Introduction

The aim of the project is the development of a high performance passive solar heating system that collects, stores and distributes solar energy and can be installed as an integral part of the facade. In other words: an improved Trombe wall. Without multi-glazing or movable insulation the Trombe wall is hardly compatible with a good thermal insulation. It can be a heating burden in mid-winter and can easily cause overheating in summer. The reasons are:

1. Thermally the storage is very badly protected against the ambient climate.
2. The wall has a very high heat loss coefficient.
3. The temperature is not uniform in the concrete wall.
4. The transfer of heat into the building can hardly be controlled.

The first two problems can be solved by using an insulation sheet in front of the storage and by heat pipes for the transfer of energy from the absorber to storage. The advantage is clear: without movable parts or control equipment an effective insulation of the storage is achieved during periods in which the absorber plate temperature is lower than the storage temperature.

The third problem can be solved by using water or a phase-change material instead of concrete. The distribution of the heat can be done by means of a thermosiphonic loop behind the storage. This loop can be made by an insulation sheet with vents at some distance away from the storage.

Figure I shows an outline of the design. The prediction and computer experiments required a knowledge of all property values of the system. These were determined experimentally. In section 2 a simulation model of the system and in section 3 the influence of both the different system parameters and heating demand patterns on the performance is dealt with.

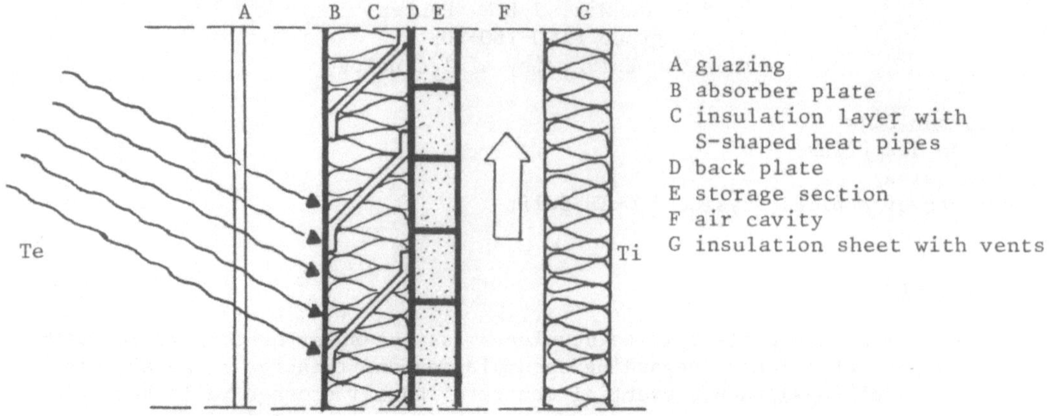

Figure I Outline of the system

2. Thermal analyses

A detailed description of the equations we derived from measurements (1) and literature is beyond the scope of this paper. We will only summarize the results. The simulation model of the system can be represented by a resistance network (figure II).

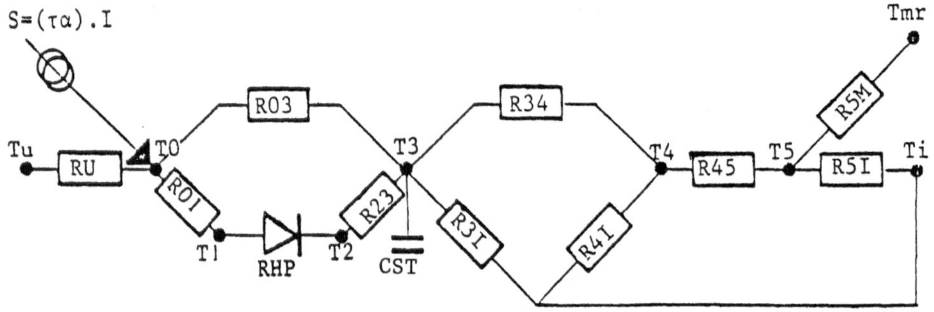

Figure II Resistance network

In figure II the thermal resistances (and one capacity) represent:

RU — Top-loss coefficient (3)

R01,R23 — Fin effect and bond resistance between heat pipe and absorber plate respectively storage unit

R03,R45 — Thermal resistance of insulation layers

RHP — The heat pipe resistance. The resistance can be derived from figure III This figure represents a heat pipe of plain copper filled with water. After testing of several types of heat pipes this one turned out to be sufficiently effective (1)

CST — Storage capacity. The properties of different storage materials were measured and tested under conditions simular to those in the element. The results are given in figure IV

R34,R3I R4I — Heat transfer by radiation and convection in the cavity. Many measurements have been carried out in a cavity with a configuration similar to that of the element. The result for a 4 cm cavity is shown in figure V.

R5I,R5M Heat transfer by radiation and convection to the room
($\alpha\tau$) Transmittance-absorptance product
Te Ambient air temperature
Ti Indoor air temperature
Tmr Indoor mean radiant temperature for facade

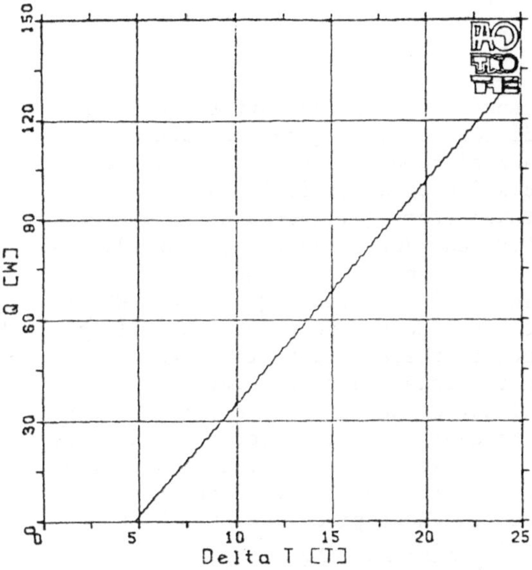

Figure III The heat flow as a function of the temperature difference over the heat pipe (Copper-water heat pipe; lenght .40 m, diameter 15 mm)

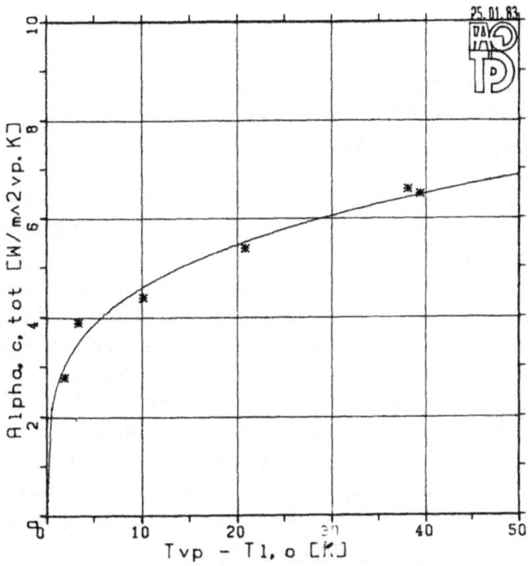

Figure V The heat transfer coefficient of the heated plate as a function of the tem-difference of the plate and the indoor air (cavity depth .04 m, height 1.80 m)

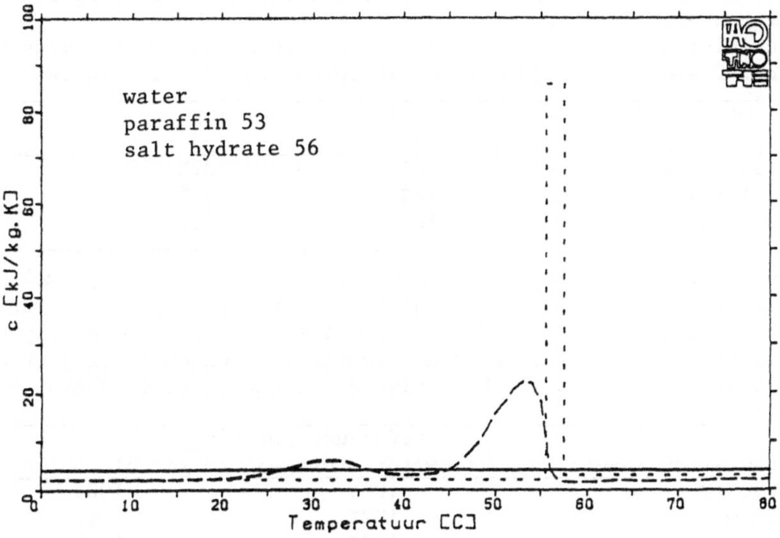

Figure IV Relation between specific heat capacity and temperature

The thermal analysis resulted in the following specifications of the system:
Dimensions: panel width 2.00 m ; panel height 1.00 m; number of glazings 1
Absorber plate: emissivity 0.01; absorptivity 0.95; thickness 5 mm
Heat pipes: number 10/m^2; length on absorber plate 0.20 m; insulation layer .15 m
Storage: material paraffin 53; mass 15 kg/m^2; storage layer 0.025 m
Cavity: depth 0.04 m; insulation layer 0.01 m; height vents 0.04 m
This system will be used as the reference unit.

3. The performance of the system

With the help of a digital computer the sensitivity of the performance to deviations from the reference unit and to different heating demand patterns was studied. For the climatic condition a standard reference year (5) was used. In a heating season the total solar irradiation of a vertical south plane is 470 kWh/m^2. The reference dwellings are taken from (4). The main properties:
no.1 normally insulated house; occupants present during daytime; Qaux 7350 kWh
no.2 well insulated house; occupants present during daytime; Qaux 2930 kWh
no.3 well insulated house; occupants absent during daytime; Qaux 3300 kWh
The nos. 1 and 3 represent extreme situations. The reference control will be "ideal"; that means the vents are only open if there is a heating demand and if the storage temperature is higher than the indoor air temperature.
For no. 1 a storage in a unit of 2 m^2 will not be necessary and this is very convenient to study the influence of the parameters that affect the transport of heat from absorber to storage and the heat exchange rate of the cavity. A better heat pipe (± 20%) turned out to have little influence on the performance (± 1%). The thermal resistance between the heat pipes and the absorber is much more crucial, as can be seen from the table below:

	number of heat pipes per m^2		
absorber plate thickness	5	10	15
2.5 mm	60%	91%	104%
5.0 mm	76%	100%	107%

By wing fins in the cavity a two times better heat transfer coefficient can be achieved. This results in a 10% better performance. With a 5 times better heat transfer coefficient (fan) the improvement was about 20%. However without the energy needed for the fan.

The sensitivity to storage and control of vents was studied with the dwellings nos. 2 and 3. The results for different control strategies are:

control of vents	reference dwelling	
	1	3
reference case (ideal)	100%	100%
0-18 closed, 18-24 ideal	87%	97%
0-24 open	80%	19%

It is obvious that in dwelling no.3 (occupants absent during daytime) control is necessary. Due to the good insulation of the heat pipe section a simple control (e.g. 0-18 closed, 18-24 open for no.3) is already expected to give good results. As heat will also be transferred (uncontrolled) through the insulation sheet of the cavity, the influence of the sheet thickness has also been studied:

	reference dwelling	
thickness insulation sheet	1 (water storage)	3 (double volume water)
reference case (0.01 m)	100%	100%
0.05 m	102%	119%
0.10 m	103%	125%

This table shows the importance of controlling the distribution to the room if day-time set-back of the thermostat is used. The same is expected in no.12 where due to the high insulation and direct gain the heating demand during daytime will be very low.

With an "ideal" control and an insulation sheet of 0.10 m the dependance of the performance on storage volume and material is shown in figure VI (2).

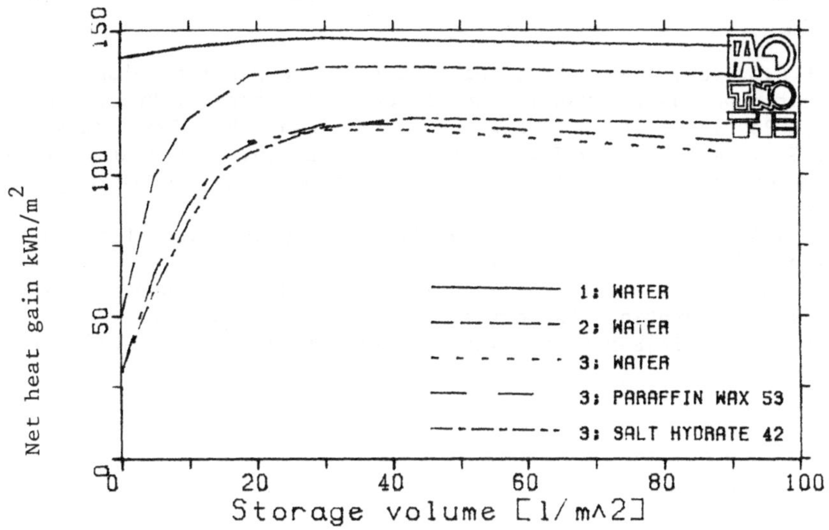

Figure VI Dependance of performance on storage volume and material

For dwelling no.1 storage is not important as was to be expected. The use of phase-change material instead of water does not give an important improvement. With a salt hydrate the storage volume can be smaller (appr. 30%). The difference between the curves for dwellings 2 and 3 is very small. This supports the assumption that for dwelling 12 (well insulated, occupants present) the heating demand during daytime is small.

4. Concluding remarks

- The maximum solar heat gain of the reference unit is about 160 kWh/m^2. Compared with a normally insulated wall the gain is even more because of the heat loss of this wall.
- Improvements of the performance can be achieved by a better thermal contact between heat pipes and absorber respectively storage. This could lead to fewer heat pipes per m^2.
- Further improvements are possible with fins in the cavity. A fan is not needed
- Good control of the distribution will be important in low-energy dwellings. This implies an insulation sheet of at least 5 cm at the room-side of the cavity and dampers on the vents which should not leak.
- The advantage of a phase-change material compared with water is too small to be cost effective. For special applications the volume reduction of 30% achieved by a salt hydrate might be interesting.
- In cases of a high auxilliary heating load or a small direct gain contribution the storage will not be needed. A much cheaper panel will be the result

The manufacturing costs of the first design for a proto-type are too high to get a favourable cost-benefit ratio. Still it will be important to realize a prote-type to get more insight in the possibilities to reduce the costs and to improve the performance of the proposed reference unit.

5. Acknowledgements

This project is financed by the Commission of the European Communities and by the Dutch National Solar Energy Programme.

References

1. DIJK, H.A.L. van, e.a. (1983). High performance passive solar system with heat pipe energy transfer and latent heat storage. Proc. 8th National Passive Solar Conference ISES-USA, Santa Fe, New Mexico.
2. BRINK, G.J. van den, HENSEN, J.L.M. (1984). Latent heat storage in a high performance passive solar heating system. Proc. Symposium on Heat Storage, Stuttgart.
3. DUFFIE, J.A. and BECKMAN, W.A. (1980). Solar engineering of thermal processes. John Wiley&Sons, New York.
4. KOK, J.W.M. , e.a. (1982). Referentiewoning, bouwtechnische omschrijving en kostenbegroting. Rapport nr. 7975 Bouwcentrum, Rotterdam.
5. Verkort referentiejaar voor buitencondities (1983). ISSO-publicatie 12, ISSO, Rotterdam.

SESSION IV - IMPORTANT ACTIVE AND PASSIVE SOLAR TOPICS

An assessment of energy performances of the housing stock in Italy: experimental data from 1000 dwellings

An assessment of the prospects for solar thermal application in Europe

Energy balance and thermal comfort in passive solar housing

Passive solar educational support for European architects

A graphical method to assess passive solar gains in the existing housing stock of the United Kingdom

Evaluation and development of thermal analysis models within the IEA solar heating and cooling programme, task VIII - Passive and hybrid solar low energy buildings

Comfort aspects in overglazed environments

Transparent insulating materials: a possibility for reducing the consumption of fossil fuel in households

The Swedish solar heating programme

Integrated design exemplified by solar cavity wall houses

Experimental study of free convection in a room model using a collector storage wall

Performance of passive solar systems in Dutch low-rise housing

The potential for solar heating in central urban areas: a case study

Air solar collecting roof for a low-energy sports-hall building

A review of the use of selective absorbers in foil form for passive solar collectors

Passive solar design in non-domestic buildings

The use of passive solar gains for the pre-heating of ventilation air in housing

The Sophia Antipolis passive test facility

Solar radiation and pyranometry studies for solar energy applications: an overview of IEA task IX

The design and establishment of a solar energy research centre in Greece

Solar heated swimming pools in Europe - An EC Demonstration Programme

AN ASSESSMENT OF ENERGY PERFORMANCES OF THE HOUSING STOCK IN ITALY: EXPERIMENTAL DATA FROM 1000 DWELLINGS

C.Boffa
CNR-Progetto Finalizzato Energetica
Politecnico, Corso Duca degli Abruzzi, 24
10129 Torino, Italy - Tel. 11-547859 telex 22046 POLITO

Purpose of the work

Solar research has been carried out extensively on national and international level since more than a decade. The research topics, originally focussed on single components of solar plant, have been extended first to solar systems and then to building behaviour, where factors as external climate, envelope, solar and conventional HVAC plant, and user's habits are considered as factors interacting with each other at a system level.

Our feeling is that now, in 1984, the knowledge available on single components of plants and on single buildings is far more advanced that the knowledge which is available, on a quantitative basis, of the real behaviour of the entire building stock of a country, from the energy and economical point of view.

Overall, quantitatively defined, statistically meaningfull data on the real behaviour of building stocks are indispensable in order to re-assess real prospects of applications of new technologies and to define new strategies for interventions, which should be suited to the changes, often drastic, which have already taken place in the standard building construction in various countries.

Therefore the purpose of the work reported in this paper is to obtain a quantitatively defined data base on the thermal and energy behaviour of a national housing stock, the Italian one, in order to assess the effectiveness of codes and regulations on energy conservation and on renewable energy sources utilization, and in order to define new building strategies and to assess their energy savings potentials and their cost to benefit ratios.

Method of approach

The work has been carried out in 1983 by interdisciplinary research teams, comprizing people from research laboratories, universities, utilities, and public and private companies already involved in fuel supply and HVAC control and management in housing stocks.

The work was sponsored by the Second National Energy Project (PFE2) (1983-1988), jointly sponsored by CNR and ENEA with 210 Million E.C.U., of which about 40 Million E.C.U. dedicated to final use of energy in buildings, including renewable energy sources, conservation, rational use of energy both in building envelope and HVAC plant.

On the basis of a statystical analysis of the Italian housing stock (table 1 and 2 show some of its basic overall characteristics), a set of over 1000 dwellings has been examined: data have been collected on the building architecture, on envelope characteristics and on heating plant characteristics, as well as on costs and on fuel consumptions.

The parameter which has been choosen to characterize the energy consumption of the dwellings is the overall seasonal specific energy consumption, referred to the heated volume and to the number of degree-days of the place where the building is located and of the heating season in which the seasonal energy consumption bas been recorded.

TABLE 1 - HEATING PLANT STRUCTURE IN ITALIAN HOUSING STOCK

	N° of plants	Dwellings	
		N°	%
Centralized heating plant	640.000	5.323.000	30,4
Autonomous single or duplex flats heating plant	3.710.000	4.587.000	26,2
Without fixed heating plant	-	7.599.000	43,4
	4.350.000	17.509.000	100,0

TABLE 2 - FUEL CONSUMPTIONS CHARACTERISTICS

	Oil	Gas	Other
Centralized heating plants	75,2%	19,1%	5,7%
Single apartment heating plants	44,8%	45,5%	9,7%

Significant results

The results give:

- a precise picture of the overall characteristics of the Italian housing stock (both architecture and HVAC plant);
- a complete set of data on seasonal energy consumption per unit volume and per degree-days as a function of climate, type of building, construction date, type of plant etc.;
- a complete set of data on cost to benefit ratios for several retrofitting interventions;
- a complete set of data on cost, performances and user's comfort on the existing solar (active and passive) houses in Italy.

Of the thousands of data now available on the existing standard Italian housing stock, only two sets are presented here, in table 3 and 4, as a mere example.

In the tables the average value of the specific energy consumption is reported together with the value at $\pm \sigma$. Moreover, for each value, the number of buildings to which it refers is indicated.

The data reported in tab. 3 show that the seasonal specific energy consumptions of the buildings constructed after 1976 are clearly smaller than the ones of the buildings constructed previously, and in particular between 1971-1975, and that the consumptions of buildings constructed after 1980 are even smaller. If we keep into account that several years pass between the design of a building and its construction and use, we can see that the data in tab. 3 indicate that in Italy a positive and effective reaction to the energy crisis of 1973 has taken place as far as new building construction is concerned.

Table 4 show that the specific energy consumption decreases with the number of the degree days. This is due to the fact that in the Italian climate the climatic zones with lower number of degree-days have a stronger insolation: thus solar gains in traditional houses reduce strongly the need of conventional energy: existing standard housing stock makes, already, a good use of solar energy!

TABLE 3 - SEASONAL SPECIFIC ENERGY CONSUMPTION AS A FUNCTION OF CONSTRUCTION YEAR

Construction year	Energy consumption /10^{-6} TEP/m^3 GG/ Value at $-\sigma$	Average value	Value at $+\sigma$	N° of examined buildings
ant. 1945	0,81	1,41	2,01	97
1946/60	1,0	1,54	2,08	167
1961/70	1,04	1,52	2,0	391
1971/75	1,09	1,59	2,09	143
1975/80	0,81	1,41	2,01	84
post.1980	0,71	1,24	1,81	26

TABLE 4 - SEASONAL SPECIFIC ENERGY CONSUMPTION AS A FUNCTION OF DEGREE DAYS

Degree days/GG/	Energy consumption /10^{-6} TEP/m^3 GG/ Value at $-\sigma$	Average value	Value at $+\sigma$	N° of examined buildings
less than 900	0,784	1,225	1,666	138
900÷1.400	0,827	1,285	1,743	71
1.400÷2.100	0,945	1,400	1,855	283
more than 2.100	1,155	1,600	2,145	417

Together with the above described assessment of energy consumption on 1000 existing Italian buildings, an extended cost to benefit analysis of retrofitting interventions has been performed.

Within the 1000 examined buildings, a group of 30 has been identified, representing the average behaviour of the various building typologies.

On this group of buildings a more detailed field analysis of the envelope, heating plant characteristics, and user's habits has been performed. The thermal and energetic behaviour of this buildings has been then evaluated using the simulation procedures developped in Italy in the First National Energy Project, (PFE1) and tested in various international programs (I.E.A., E.E.C.). These programs allow for a detailed simulation of the time dependent overall real building behaviour, at a system level, and keep into account envelope and heating plant characteristics, together with external climate and user's habits.

By using the above mentioned procedures with input data deriving from the detailed field analysis of the buildings, the seasonal energy consumption of each building has been calculated.

The data have been compared with the experimental ones and a good agreement has been found (differences were below 2%), which demonstrates, once more, that the available simulation codes are valid, when the input data are correct.

Then, a set of retrofitting interventions, both on building envelope and heating plant, has been defined, comprizing upgrading of the plant controls (changing of the calibration curve of the control and inserting thermostatic control in every room) changing of the boyler, insulating, double glazing, reducing of air infiltrations.

For each intervention the actual costs have been evaluated and the seasonal energy consumption in the real conditions of use have been determined. In this way for each intervention the cost to benefit ratio has been evaluated.

An example of the now available data in given in table 5.

TABLE 5 - COST TO BENEFIT RATIOS OF RETROFITTING INTERVENTIONS ON "AVERAGE" BUILDINGS

	Specific Energy Savings /10^{-6} TEP/m^3 GG/		Extra costs (ECU/m^3)
	1400÷2100/GG/	2100÷3000/GG/	
Upgrading of the controls: -changing the calibration curve	0,3		-
-inserting thermostatic control in energy room	0,4		0,7
Installing a new high efficiency boyler existing boyler efficiency 80%	0,14÷0,39	0,11÷0,30	2,3÷3,1(1,0÷1,5)*
" " " 85%	0,80÷0,21	0,06÷0,14	2,3÷3,1(1,0÷1,5)*
" " " 90%	0,07÷0,09	0,06÷0,08	2,3÷3,1(1,0÷1,5)*
Insulating external walls	0,19÷0,43	0,17÷0,37	4,6÷8,4(2,0÷2,6)**
Double gazing	0,028÷0,047	0,026÷0,042	2,3÷2,6(1,9÷2,5)**
Reducing of air infiltration	0,9	---	0,2÷1,5

* extra costs if old boyler has to be changed anyway

** extra cost if maintenance (decorating) of external walls has to be done anyway

The data reported in table 5 confirm that the calibration of the central controls of heating plant is usually not optimized and that its optimization yields substantial energy savings (≃ 7%) with practically zero cost.

Installing of thermostatic valves on the heating elements yields even larger energy savings with payback times in the order of 2÷3 years.

Heavy retrofitting of heating plants are often convenient: in particular if the existing boyler is substituted by a high efficiency one the payback time can be shorter than 3 years if the old one has reached the end of its life.

The double glazing on the "average" buildings considered here has a payback time much longer (around 10 years).

It is however a retrofitting intervention which the users like, also because of the higher thermal comfort (higher internal surface temperature of the the transparent surfaces) and acoustic comfort, which the double glazing yield, together with the comparatively small disturbance to the inhabitants that the installing of double glazed windows produces.

The insulating of external walls has a favorable cost to benefit ratio if it can be easily performed in suitably designed cavity walls; if the insulating is done by adding an external insulating coat the payback time is generally reasonable (less than 10 years) if maintenance (decoration etc.) of external walls has to be done anyway.

The reduction of air infiltration from window frames in Italian building yields usually large energy savings. Its costs are very variable: the do it-yourself techniques can be very cheap, while,if done by professionals it can be,sometimes,very expensive. This is why the cost indicated in the table has a very wide range.

Finally it must be remarked that the insulating and double glazing interventions show very long payback time because they are referred to "average" buildings.

The cost to benefit ratio might improve drastically if suitable buildings or suitable surfaces a building are choosen.

However the payback time on average buildings indicated in table 5 must be carefully considered, particularly by the people who are in charge of defining laws, regulations and incentives for energy conservation in buildings.

As before said, together with the above described assessments on the energy consumption of 1000 existing buildings and on cost to benefit ratios of retrofitting, an analysis of the bioclimatic buildings existing in Italy has been initiated.

The purpose of the study is to obtain a set of homogeneous data, comparable with each other and with the ones of existing standard buildings, which allow quantitative conclusions on the results actually achieved so far. The collected data refer to building and plant characteristics, energy consumption, actual overall building costs, acceptance by the users.

An example of the obtained results is given in table 6.

Conclusion

A detailed and quantitatively defined data base on energy performances versus costs of existing buildings (new, retrofitted, non retrofitted, bioclimatic, with single family, centralized and district heating plants) is now available in Italy on a nationwide basis. It is an indispensable, basic tool not only for planners, civil engineers and architects to define strategies for interventions and to assess real possibilities of practical diffusion of new technologies both for architecture and HVAC plant, but also for law-makers to define incentives and codes for rational use of energy in buildings.

Some peculiar considerations can be derived from these data.

TABLE 6 - EXAMPLE OF ENERGY CONSUMPTIONS AND COSTS OF EXISTING BIOCLIMATIC HOUSES IN ITALY

Buildings	Volume /m^{-3}/	N° of apartments	Standard energy consumption lex 373 /10^{-3} TOE/m^3 YEAR/	Energy savings /%/	Standard building costs lex 457 / ECU / m^3 /	Overall costs of building /ECU / m^3 /	Overall extra cost of building /ECU / m^3/	Cost of saved TOE/YEAR /10^3 ECU/TOE/YEAR/
1	6100	24	3,13	56	108	132	24	13,69
2	2786	12	4,43	64	121	155	34	11,99
3	9380	45	2,96	50	111	147	36	24,32
4	5134	24	4,25	59	119	163	44	17,54
5	2662	12	5,95	76	121	205	84	18,57
6	3688	18	4,55	71	121	207	86	26,60
7	2455	12	4,55	81	121	276	155	42,05
8	918	4	5,11	67	121	312	191	55,78

BUILDINGS CHARACTERISTICS

1) Trombe wall (thickness 25 cm) + direct gain - 46 m^2 flat plate collectors (sanitary water)

2) the buildings is correctly exposed to sun and highly insulated
   - 52 m^2 flat plate collectors (sanitary water)

3) sun spaces + direct gain
   - 180 m^2 flat plate collectors (sanitary water + space heating)

4) the buildings is highly insulated
   - 140 m^2 flat plate collectors (sanitary water + space heating)

5) solar air-to-water heat pump
   - 70 m^2 flat plate collectors (sanitary water + space heating)

6) direct gain
   - 72 m^2 flat plate collectors
   - heat pump

7) Trombe wall (thickness 25 cm)
   - 52 m^2 flat plate collectors

8) direct gains(mornings), indirect gains (afternoon)
   - 16 m^2 flat plate collectors (sanitary water)

Also in absence of incentive laws (the one now existing, lex 308, is operative only since 1983) an effective spontaneous reaction to the price increase of energy has taken place. As a conseguence a new building stock of about one million flats now exists which have a low energy consumption, much lower than what was generally thought and often even lower than the value fixed by the law.

The existing stock has reached spontaneously a qualified standard of low consumption, which makes it difficult for new generalized programs on energy conservation to be effective in further energy savings with reasonable costs.

This is shown also by the results of the studies on retrofitting interventions. If they are performed on an "average" building many of these interventions show prohibitive pay-back times.

Of course different results are obtained if the interventions are applied to buildings which have very poor energy performances, which still exist in our housing stock, as the spread of the results in tables 3 and 4 shows.

Similar considerations apply to the data on energy consumptions and costs of the existing bioclimatic houses.

Prohibitive costs of the "saved" TOE (as shown in the last column of table 6), which range from 15 to 50 times higher than the cost of TOE from oil, indicate that major changes in the research strategies and in the design and construction approach of bioclimatic buildings are highly necessary.

This appears even more clearly if we remember that we are now in 1984 and more than a decade of intensive research has been dedicated to these topics, and if we compare these results with the substantial improvements in energy savings which the standard buildings have attained in the same period with negligible cost increase, immediately absorbed by the marked.

AN ASSESSMENT OF THE PROSPECTS FOR SOLAR THERMAL APPLICATION IN EUROPE

Turrent D and Baker N
Energy Conscious Design
11-15 Emerald Street
LONDON WC1N 3QL

Summary

This paper describes the results of a study carried out in 1982/83 to assess the prospects for solar thermal applications in Europe up to the year 2020. The study addressed itself to the current state of solar technology both in terms of cost and performance, to the potential applications in domestic and non domestic buildings and to the likely energy savings resulting from significant market penetration in the future. The purpose of the study was to draw together the practical conclusions drawn so far in solar thermal research and development and to establish the priorities for future R & D in this area. The main conclusion was that most of the basic research has now been done and its technical viability has been established. In particular passive solar space heating appears to be a cost effective technology in northern Europe, and solar water heating also shows considerable promise. Active solar space heating systems are awaiting significant performance improvement and cost reduction before they can be taken up in the market place, although various second generation systems show encouraging results. In terms of future energy saving the study has shown that solar thermal energy could make a significant contribution to Europe's primary energy needs in the medium to long term, providing in excess of 50 million tonnes of oil equivalent per annum by the year 2020.

1. INTRODUCTION

The solar thermal assessment study was carried out during 1982 by Energy Conscious Design, assisted by a group of experts nominated by each member country of the community. Additionally, the views and opinions of some 50 leading researchers in Europe were taken into account through a series of structured interviews and meetings. The resulting assessment therefore refects a consensus view of the current state of the art and future prospects for solar thermal energy in Europe. It is addressed to policy makers and building professionals as well as to members of the research community, and is also intended as a guide for the developing solar industry.

The output of the work was a book published in 1983 by Reidel and entitled 'Solar Thermal Energy in Europe - An Assessment Study'. The book is structured into four sections; the context for solar energy research and development, the state of the technology, applications and assessment of the resource, and finally a profile of the developing industry.

2. THE CONTEXT FOR R & D

The second phase of the EEC Solar R & D programme had a total budget of some 46m ECU, of which 8.3m ECU was devoted to Project A 'Solar Energy Applications in Dwellings'. The programme has tended to support research areas which benefit from a co-ordinated approach, including collector testing, system testing, system modelling, performance monitoring and heat storage. Historically the main emphasis was on active space heating but towards the end of the programme passive solar energy became an increasingly important topic for research. In the next phase it is likely that passive solar energy will occupy a central position in the programme, together with the development of more refined design tools and the design of better optimised 'second generation' active systems.

3. STATE OF THE TECHNOLOGY

During the course of the study every aspect of solar thermal technology, including collector design, short and long term heat storage, active and passive systems, heating and cooling was investigated. On the basis of published research documents and discussions with individual researchers, priority areas for future R & D were drawn up. Some examples are given below:

Collector Technology

- Investigate and report on long-term durability and reliability of currently available components.
- Carry out further technical improvements in flat collector design to improve optical efficiency and reduce heat losses.
- Study potential applications of evacuated tube collectors and their integration in buildings.
- Investigate the potential for 'cladding type' collectors for use in industrial and commercial buildings.

Short Term Heat Storage

- The study and modelling of thermal stratification.
- Demonstration of better designed storage systems as a component of a total solar heating installation.
- Developing of a wider range of suitable materials, especially at higher temperatures.
- Research on durability and reliability of materials used in systems and long-term system behaviour.

Long-Term Heat Storage

- Basic research on more seasonal storage approaches especially aimed at low cost solutions.
- Development of computer simulation models for system-optimisation and cost optimisation purposes.
- Construction and monitoring of full scale experimental stores.

Solar Water Heating Systems

- Design of technically and ecomomically optimised systems.
- Cost reduction and performance improvement studies.
- Simple performance test for installed systems.

Active Space Heating Systems

- Design of technically and economically optimised '2nd generation' systems.
- Development of low temperature heat emitters.
- Feasibility studies of grouped systems with long-term storage.

Passive Solar Space Heating

- Production of simplified design tools for use by building designers.
- Studies in the field of dynamic thermal comfort, and user control.
- Detailed studies of heat transfer, air movement and heat storage in test cells.
- Design studies to produce exemplars for use in the design professions.

4. APPLICATIONS AND ASSESSMENTS

Clearly, there are technical limitations to the contribution that solar energy can make. The most obvious is the availability of solar radiation, which varies widely over Europe. Further technical limitations relate to the building stock: freedom from overshadowing, orientation and so on. Conservation itself influences the contribution solar energy can make, the smaller the annual heating load the smaller the absolute size of the solar contribution. For the purposes of this study a quantitive assessment method was developed to predict the potential for solar thermal applications in the European building stock. (Fig 1).

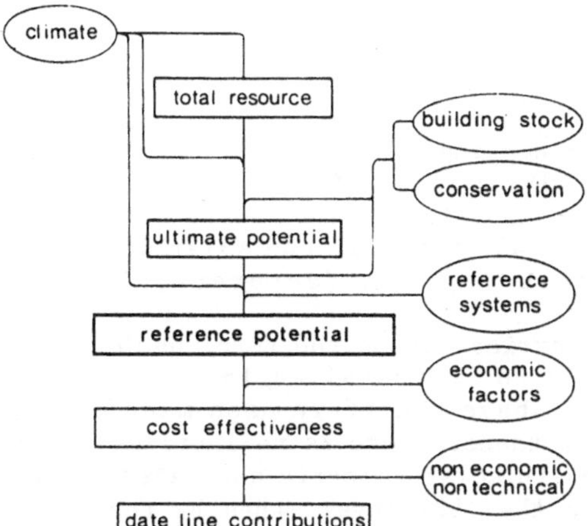

Fig (1) Assessment Method

Firstly, the method is used as a framework to discuss the various technical factors such as orientation, levels of conservation and system performance. These factors are then used to establish a technical limitation on the size of the potential solar contribution. For the purpose of calculation, the method involves the uses of Reference Systems. These are based upon current state-of-the-art practice, but with the option of improving performance and/or reducing costs in the future. Reference systems are defined for solar water heating, active space heating, and passive space heating, and they are applicable in both new and existing buildings. For every European country, the Reference System performance is modelled using appropriate climate data, to give an Annual Reference Performance. This is then applied to the suitable fraction of buildings in that country, to which the Reference System could be applied. This leads to a National Reference Potential. These Potentials are calculated for 1980, 2000 and 2020, taking account of the rates of new build and demolition.

The potential contribution shows a steady increase with successive time periods. This is due to the increased suitability of new buildings, where the opportunity for optimum orientation and freedom from overshadowing can be taken. The potentials for 1980, 2000 and 2020 are 25.9, 37.4 and 48.8 mtoe respectively for passive space heating, and 25.8, 35.0 and 44.55 mtoe for active space heating and solar water heating.

The Assessment Method explores the cost effectiveness of the systems by the application of discounted payback criteria. Future cost reductions are also considered. Various fuel cost scenarios and discount rates are employed in the analysis, and the importance of these parameters, when considering longer term investments, is discussed.

The 4 m^2 solar water heating system shows immediate cost effectiveness only in South Europe. Cost reductions by a factor of two, (reducing system costs to 1000 ECU), would also make systems cost effective in North Europe when displacing oil. For space heating, passive systems show almost immediate cost effectiveness for retrofit sunspaces. Due to the zero or very small extra cost, direct gain systems in new build are also immediately cost effective. Active systems are less attractive, their high capital cost more than offsetting their small improvement in performance over passive systems. Active systems are not likely to become generally cost effective until the period between 2000 and 2020. However, if a satisfactory solution to interseasonal storage can be found, then at the assumed costs, cost effectiveness could be brought forward to the 1990's.

5. PROFILE OF THE SOLAR INDUSTRY

The total number of solar water heating systems installed in dwellings in the EEC member states up to 1981 was 220,000, 30% of which were installed in 1981. The value of the industry in 1981 was 105 million ECU. There are currently between 400 and 450 firms in the solar domestic hot water industry in the EEC, including manufacturers, importers, distribution and installers.

No official studies have been carried out on the total number of solar domestic hot water systems likely to be installed in each country in Europe in the future. Projections have therefore been made based on a number of growth rates. If, for instance, there is no increase in growth and the area of collector installed in solar domestic hot water systems in each country continues each year on the 1981 level, then by the year 2000 around 5 million m^2 of collectors will have been installed in 1.5 million systems (about 1% of the total dwelling stock) in the EEC member states.

With an increase of 10% a year about 3% penetration would be achieved giving a total of 4 million installations and a 25% a year growth rate would give nearly 20 million installations, about 15% penetration. We have assumed that around 70% penetration is the maximum which can be achieved technically. With a 50% growth rate this year 70% penetration would be reached by the year 2003 for each of the EEC member states giving a total of nearly 90 million systems.

For 10% a year growth the annual value of the new installation work in the EEC would be nearly 650 million ECU in 1981 money in the year 2000. For 25% annual growth rates the annual value of the industry would be 5000 million ECU (1981 money) in the year 2000, representing a very considerable business potential.

6. CONCLUSIONS

The main conclusion drawn is that most of the basic research on solar thermal energy has now been done and its technical viability established. In particular the study has shown that the potential for solar thermal energy in Europe is considerably greater than previously thought. By the year 2020 it is estimated that solar energy could provide in excess of 50 million tonnes of oil (mtoe) per annum, as illustrated in Fig. ii.

The diffusion curve is based on the assumption that the market will respond when individual systems become cost-effective and a saturation value of about 60% of the full potential is anticipated. The combination of passive solar space heating and solar water heating shows a particularly promising potential contribution, and the slope of the diffusion curve steepens in the 1990's when these measures become cost effective.

Apart from technical factors, the future exploitation of solar thermal energy in Europe will depend strongly upon factors such as the development of public awareness and more vigorous marketing. Incentives in the form of government loans and subsidies are now being introduced in several countries. Exhibitions will also help to arouse public interest and create a demand. Once a healthy market is established the essential cost reductions are likely to follow. Just as there is a need for the public to be reassured of the 'respectability' of solar technology, so too the professions will have to be convinced that the transfer of the technology to their design skills is worthwhile. Governments can greatly assist in all these areas and many researchers and industrialists believe that government intervention is an essential ingredient if the full potential of solar energy is to be realised.

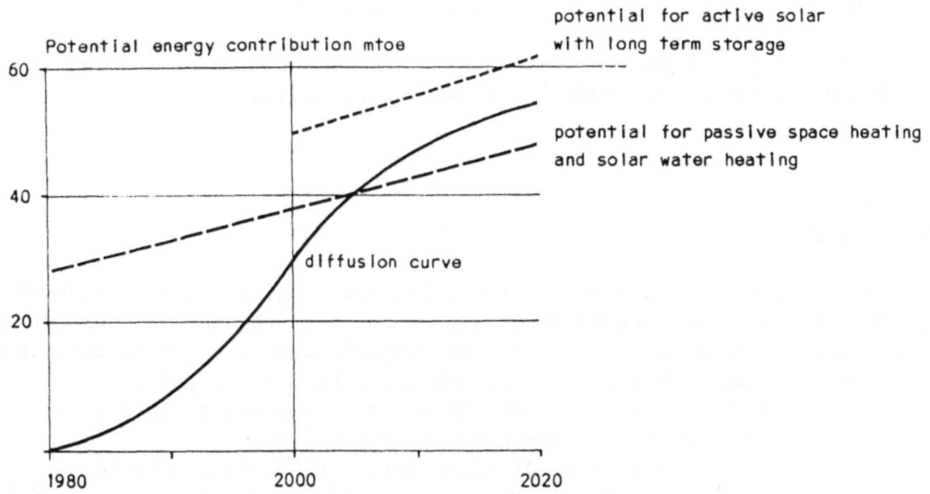

Fig (ii) Scenario for solar contributions from all systems

ENERGY BALANCE AND THERMAL COMFORT IN PASSIVE SOLAR HOUSING

K. Alder, Ch. Eriksson, A. Faist and N. Morel
Ecole Polytechnique Fédérale de Lausanne
Bâtiment LESO - CH 1015 LAUSANNE

Summary

To evaluate the performance of different passive solar dwellings it is necessary to consider not only the thermal performance but also the "comfort performance" of the system.
In order to achieve this goal the comfort has been described according to O. Fanger's theory. The Predicted mean vote (PMV), Predicted percentage of dissatisfied (PPD) and the Temperature index are computed hourly for an inhabitant located at different places in the room.
A complete set of parametric studies has been performed on two direct gain houses, one Trombe wall design and a sunspace house. In each case the thermal performance have been evaluated together with the comfort performance. The following parameters have been investigated.
For the direct gain houses :
- orientation of the building - position and shape of the windows
- effect of thermal mass
- effect of night insulation and night temperature setback

For the Trombe wall design :
- constructive optimisation - thickness of the wall
- convective gains of the Trombe wall - night insulation

And for the sunspace design :
- suppression of the sunspace glazing - thermal mass
- night insulation - night temperature setback
- strategy of window opening.

In this paper we show how the design and orientation of a direct gain house influence the thermal and comfort performance.

INTRODUCTION

To evaluate the performance of a solar dwelling, it is necessary to consider not only the thermal performance, but also the thermal comfort performance of the system. According to Fanger's work a clear description of the comfort can be given by three different parameters [1] .
- the predicted mean vote (PMV) on a - 3 to + 3 scale (cold or hot) where the zero vote is the comfort neutrality.
- the predicted percentage of dissatisfied (PPD) on a 0 to 100 % scale where a 5 % vote expresses the comfort neutrality and a 10 % vote the beginning of a noticable discomfort.
- the temperature index T_{index} which is the result of a ponderation of the air temperature T_{air} and the mean radiation temperature T_{mrt}.

Different solar houses have been modelized and simulated with the computer code PASSIM. The routine COMFORT attached to PASSIM allowed the calculation of T_{index}, PMV and PPD for an inhabitant located near the south wall

(1 m. distance) or in the middle of the room.

Direct gain houses

Two direct gain houses named "A" and "B" are compared. House "A" has an important south oriented glazing area which is shaded in summer by an overhang and a balcony. House "B" has got windows facing east, south and west (see table 1). Both houses have about the same heat loss coefficient.

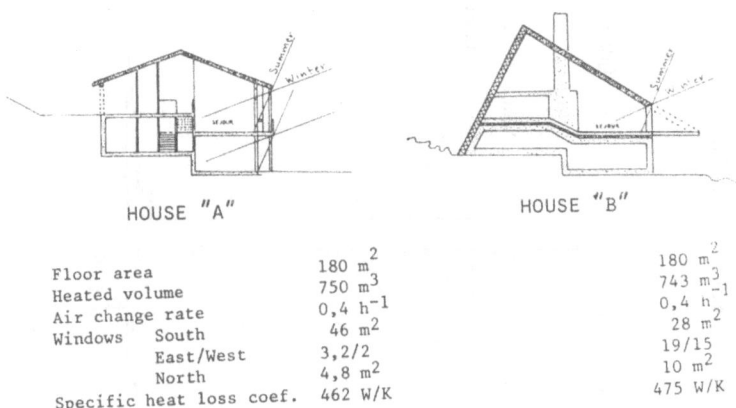

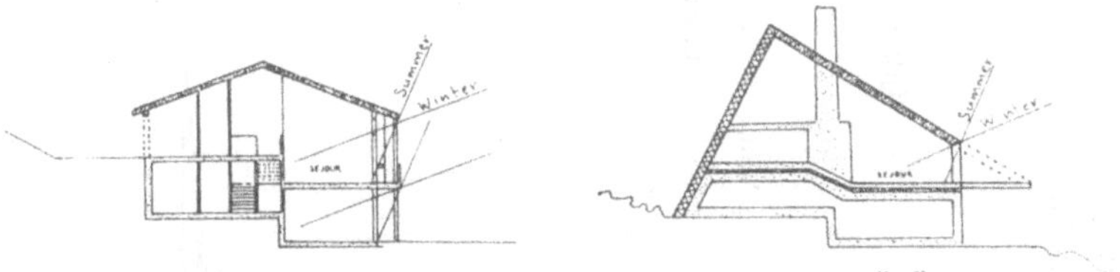

HOUSE "A" HOUSE "B"

		House "A"	House "B"
Floor area		180 m^2	180 m^2
Heated volume		750 m^3	743 m^3
Air change rate		0,4 h^{-1}	0,4 h^{-1}
Windows	South	46 m^2	28 m^2
	East/West	3,2/2	19/15
	North	4,8 m^2	10 m^2
Specific heat loss coef.		462 W/K	475 W/K

Table 1 : Principal characteristics of direct gain houses "A" and "B"

Influence of the house design

If we compare the temperature index felt by an inhabitant standing in March 1 m. away from the south glazing, we can see, in house "A" the favourable effect of the shading of the balcony and overhang : house "B" is slightly warmer and overheatings, up to 30 °C are experienced (figure 2). Thus the comfort conditions in house "B" are slightly worse than in house "A" : the PPD of 5 % is reached 94 % of the time in house "A" and only 80 % of the time in house "B".

A first question is : during what fraction of the time is there some real discomfort (PPD > 10 %) ?

The 10 % cut off is the answer : house "A" is always comfortable (10 % cut-off = 0 %) and house "B" shows some discomfort 11 % of the time. This difference is mainly due to the specific window pattern of the two houses and also to the 30° east orientation.

Second question is : what is the PPD that is not overrun more than 5% of the time ? The 95% point is the answer : PPD's of 6% for house "A" and 19% for house "B" are not overrun more than 5 % of the time.

For this particular orientation (30 °E) both houses show the same energy consumption but comfort in house "A" is better owing to the window distribution.

Effect of the orientation

Turning both houses from 30 °E to 15 °W leads to a limited auxiliary energy reduction : 7 % for house "A" but only 2 % for house "B". Solar gains are enhanced and occur later in the day. This leads to frequent overheatings as can be seen on figures 4 and 5 (compared to 1 and 2).

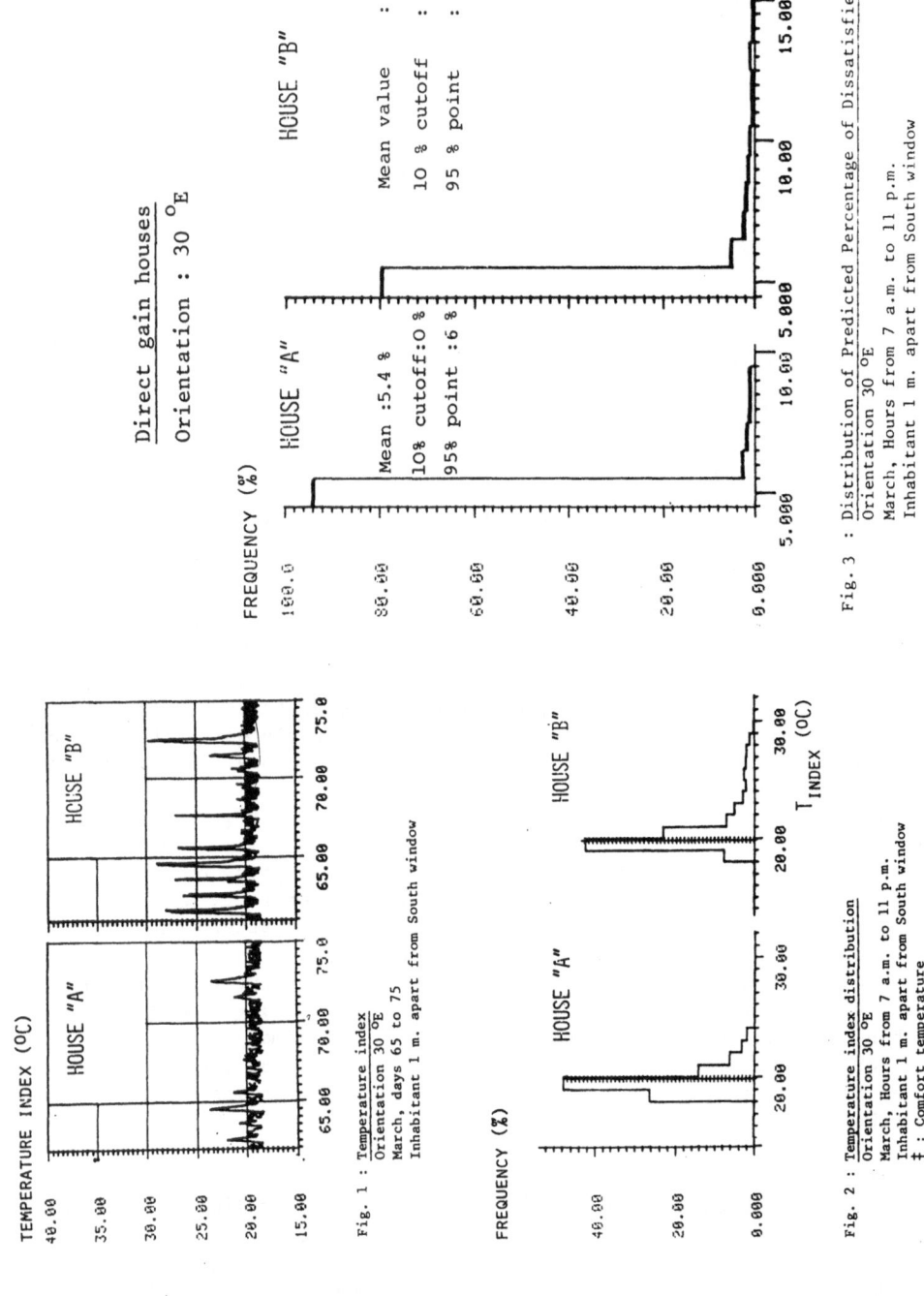

Direct gain houses

Orientation : 30 ^{o}E

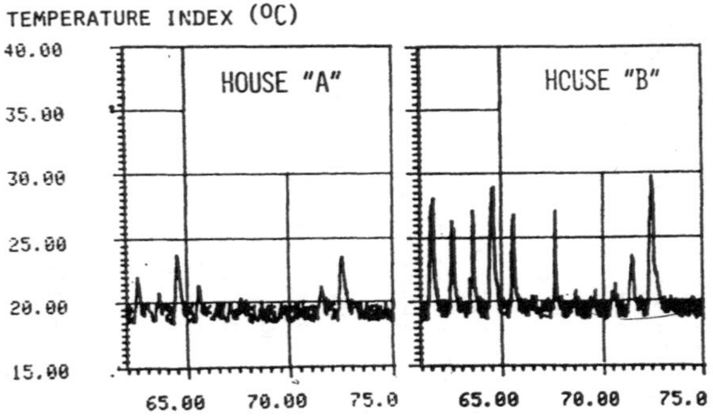

Fig. 1 : Temperature index
Orientation 30 ^{o}E
March, days 65 to 75
Inhabitant 1 m. apart from South window

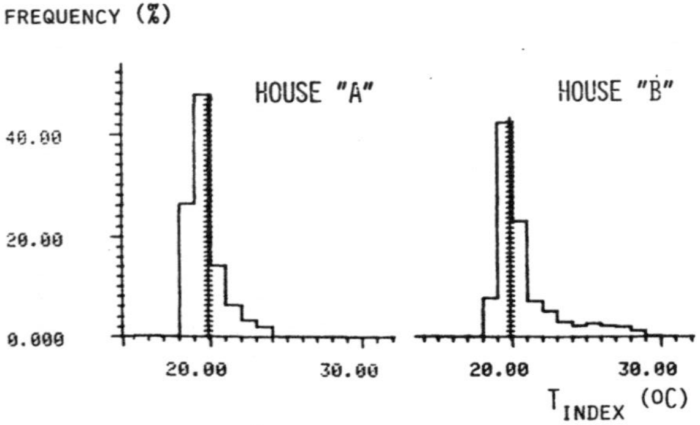

Fig. 2 : Temperature index distribution
Orientation 30 ^{o}E
March, Hours from 7 a.m. to 11 p.m.
Inhabitant 1 m. apart from South window
‡ : Comfort temperature

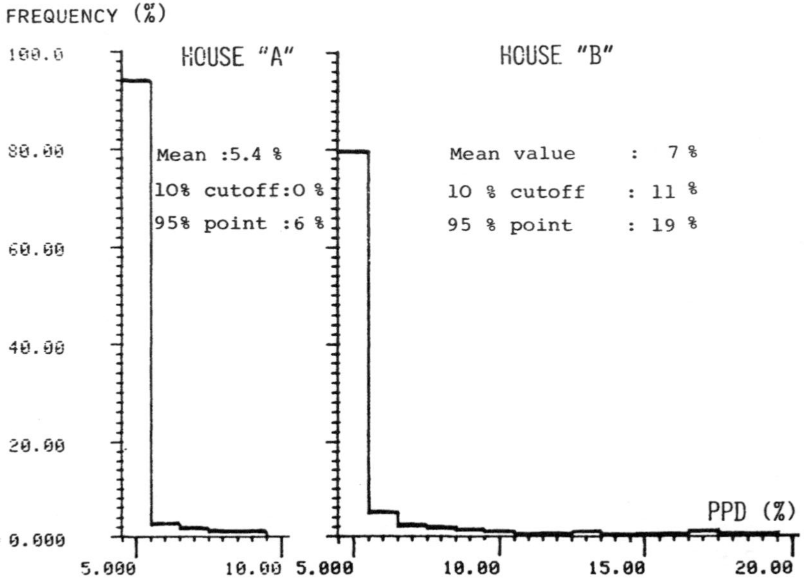

Fig. 3 : Distribution of Predicted Percentage of Dissatisfied (PPD)
Orientation 30 ^{o}E
March, Hours from 7 a.m. to 11 p.m.
Inhabitant 1 m. apart from South window

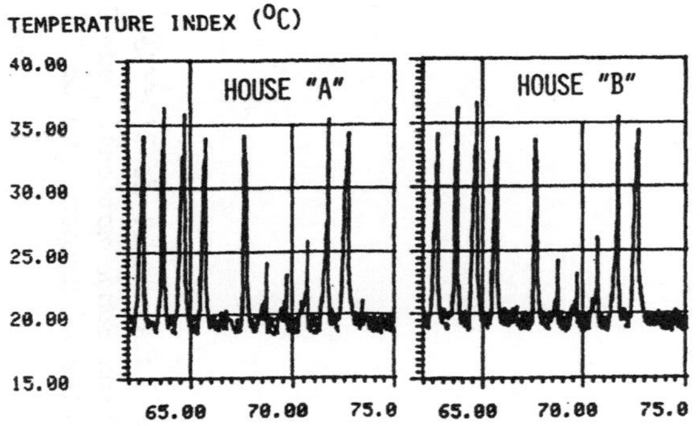

Fig 4 : Temperature index
Orientation : 15 °W
March, days 65 to 75
Inhabitant 1 m. apart from South window

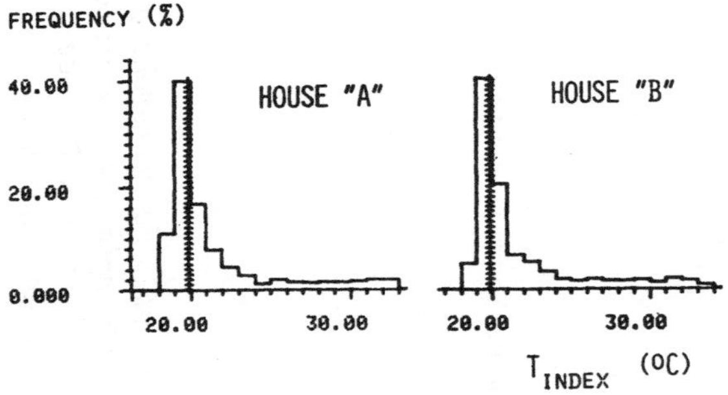

Fig. 5 : Temperature index distribution
Orientation : 15 °W
March, Hours from 7 a.m. to 11 p.m.
Inhabitant 1 m. apart from South window
‡ : Comfort temperature

Direct gain houses

Orientation 15 ° W

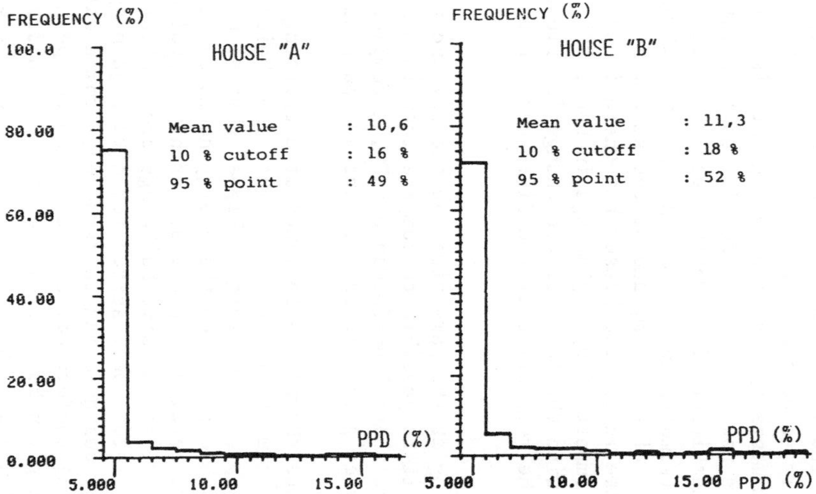

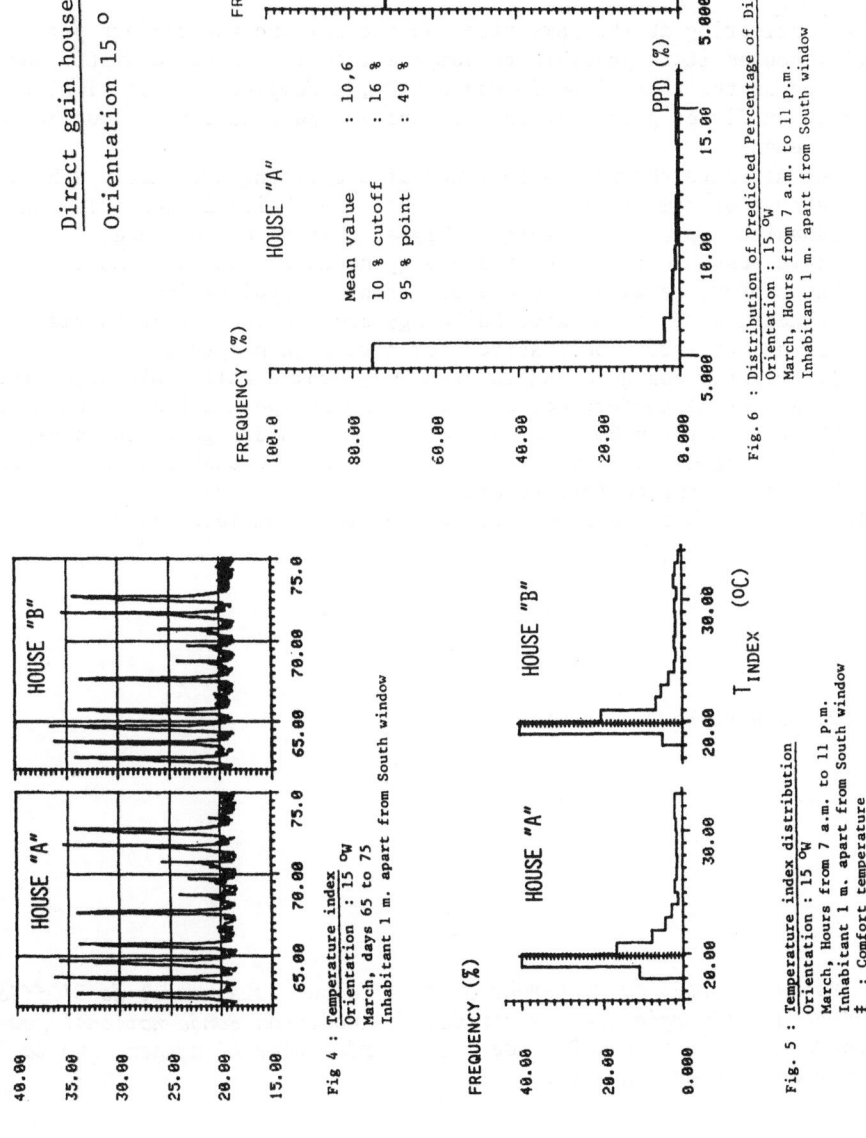

Fig. 6 : Distribution of Predicted Percentage of Dissatisfied (PPD)
Orientation : 15 °W
March, Hours from 7 a.m. to 11 p.m.
Inhabitant 1 m. apart from South window

The PPD distribution (figure 6) shows that some discomfort will be experienced by an inhabitant standing 1 m. away from the window.

In house "A" the 10 % cut-off is now 16 % : during this fraction of the time discomfort is to be expected (18 % in house "B").

At certain moments in the afternoon great discomfort may even occur: the 95 % point is now 49 % for house "A" and 52 % for house "B". In this situation the inhabitant will probably use sunshades in order to reduce the discomfort (simultanously reducing the thermal performance).

In the middle of the room, and at the same time, the comfort conditions are about the same in the two houses but there is still some negative effect due to the important west glazings in house "B".

In January, with the low sun, the change in orientation and the change of position in the room bring little difference in the comfort conditions.

CONCLUSIONS

By considering at the same time the thermal and the comfort performance of solar houses it is possible to compare them in a more realistic way than by looking at the thermal performance only. A complete set of simulations performed on direct gain, Trombe wall and sunspace designs allow the following statements :

- The ultimate thermal performance of a building are rarely achieved because of comfort limitations. When the choice is possible, the inhabitant prefers comfort to higher thermal performance.
- It is possible to reduce the energy needs and at the same time raise the comfort level by appropriate architectural design.
- Standard european massive buildings provide sufficient thermal mass and are therefore well suited for direct gain systems.
- If properly designed and constructed, Trombe wall could offer interesting thermal comfort performance under european climatic conditions.
- Changing a given house from the direct gain design to the sunspace design results in a moderate reduction of the auxiliary heat demand but raises the comfort conditions inside the house.

The whole work is to be published soon in a complete paper.

REFERENCE

1 O. Fanger, Thermal comfort

Le présent travail de recherche a été effectué sous mandat de l'Office fédéral de l'énergie avec le soutien financier du Fonds national pour la recherche énergétique NEFF. Les thèses présentées n'engagent que la responsabilité de l'auteur.

PASSIVE SOLAR EDUCATIONAL SUPPORT FOR EUROPEAN ARCHITECTS

J Owen Lewis
Energy Research Group
School of Architecture
University College Dublin

Summary

Education is highlighted as having a key role in advancing the use of passive solar energy in building design. The different requirements of undergraduate and mid-career education are considered and certain pioneering work is reported. Problems arise in integrating technological issues in the design process. Support already provided throughout Europe is described, specifically two architectural competitions and a series of professional workshops. It is concluded that some progress is being made in providing the necessary support to building designers to implement passive solar techniques in the European building industry.

1 INTRODUCTION

Technological feasibility does not in itself ensure the widespread implementation of new energy sources. The main non-technical factors which impede change include a lack of trained people, inadequate information, social and cultural inhibitions of all kinds, institutional intractibilities, problems of financing, and (last but not least) economic viability. Education in the broad sense has a key role to play in minimising many of these barriers. Throughout Europe programmes of energy education and training are being introduced in universities and technician training institutions, and diverse formal and informal programmes have been established which range from courses for policymakers and energy planners to activities designed to improve the understanding and acceptance of new technologies by the general public.

Architects can have a major influence on the efficiency with which energy is used in the servicing of buildings. At present architectural thinking in Europe is in a state of flux, and the broad consensus which obtained about the precepts of the Modern Movement is being reevaluated and partly replaced by Post-Modernism, Neo-Classicism, and the Romantic Vernacular, among many stylistic concerns. Some of the most distinguished examples of 'Modern architecture' made profligate use of energy, and the glass and steel buildings of the International Style frequently exaggerate the fluctuations of the outdoor climate; they depend for the provision of a comfortable environment on the consumption of large amounts of fuel in lighting, heating, ventilation and cooling systems. However, one of the architectural trends now discernable is towards a more energy-efficient architecture, and there is renewed interest in climate-sensitive design.

Passive solar energy is of particular importance among the new sources because of its economic attractiveness and its rich design potential. As the fascination with a symbolic functionalism and technology gives way to a more differentiated and site-oriented architecture, and as operating costs become a more significant factor in the design of buildings, there is a need and an opportunity to provide educational support so that passive solar techniques may be implemented in the European building industry. This paper is concerned with the provision of such support for the building design professions.

2 EDUCATION

Quite different educational requirements arise from the needs of students following professional undergraduate programmes compared with those of experienced designers.

There is of course substantial variation among the systems of architectural education throughout the European Community, reflecting differing traditions and professional circumstances. However the most common model has the design studio as a focus of the programme, with various theoretical, historical and technical courses. Lecture and seminar-based teaching methods are usually associated with the courses while studio projects demand that the student prepares design proposals in the context of the social technical environmental and formal issues which influence architectural design.

2.1 Teaching

An important American curriculum study (1) has generated a set of educational materials on passive solar design appropriate to these common teaching methods. Some parts provide simple course outlines for specific teaching situations, with clear objectives, assignments, reading lists and so on. Others include laboratory exercises, design exercises, and analytical procedures. Still others are general resource packages for specific passive topics which may be used in a variety of teaching settings from studios to lectures, in the circumstances of the USA. While these course materials are very useful, the original intention of the study - to encourage new teaching ideas on the fundamental issue of how technical questions are introduced into architectural design - was largely unaddressed.

A number of European architecture schools have done pioneering work in introducing energy studies into the architectural curriculum, including for instance Ecole d'Architecture de Marseille/Luminy and the Architectural Association in London. A short study (2) of certain schools with particular interest in passive solar energy, namely those at Liège, Berlin, Venice, Cambridge, and Dublin, identified some possibilities of exchanging experiences and an interest in establishing more formal links. Such a network has yet to be realised though some progress is being made at the bilateral level.

Passive solar and climate-sensitive design can be viewed as forming part of the established area of architectural science in which the building is conceived as an environmental filter between man and climate (for example, 3). Some have gone further and argued for the establishment of a new discipline of bioclimatic architecture (4). But it would appear that these principles and concepts are not being generally implemented and only appear in special projects. This is typical of the experience of very many schools of architecture that information taught in lecture or seminar courses is rarely applied or reinforced in the studio.

2.2 Technology in design

The elements of passive solar design cannot be considered only in their technical dimensions, as of their nature passive systems have profound architectural implications. A criticism which may be fairly levied at some early solar buildings is their diagrammatic nature, in that practically all other considerations would seem to have been made subservient to energy collection. However, the history of modern architecture includes many significant buildings in which energy technology has been an important generator of form (5). One of the clearest examples is the Central Pompidou by Piano and Rogers in Paris. Instead of exploiting building services hardware as architectural elements, passive buildings seek to provide comfort by their use of form fabric and space, and 'natural' energy flows.

To most designers the perception of the thermal and luminous implications of traditional elements such as walls and roofs is more difficult and less familiar than concepts such as architectural space and structure. Modern service systems tend to mask the direct experience of the adequacy of a building's environmental response to climate. Vernacular architecture often displays exemplary appreciation of the exigencies of local climate, but through a period of cheap energy building designers seem to have lost the skills of designing in harmony with climate. Designers usually must rely on abstract mathematical modelling to evaluate performance.

Both technical and formal considerations have general significance for energy in architecture. Hence the importance of developing teaching methods which integrate technological issues in the studio design process. A corollary is that it is necessary to increase the number of teachers who have the analytical insight, the design expertise and the pedagogical skills to attempt this integration. The introduction of outside 'experts' to the studio often serves only to reinforce the idea that technical analysis is separate from design.

The implications of some aspects of these matters is not restricted to the academic sphere.

3. PROFESSIONAL DEVELOPMENT

Architectural competitions are a well-established means of introducing new ideas and promoting development in building design among both practising architects and students of architecture. Several countries have organised regional and national competitions on passive solar design, and to date two such competitions have been sponsored by the Commission of the European Communities (CEC).

3.1 European competitions

The aim of the First European Passive Solar Competition in 1980 was explicitly educational in seeking to disseminate the principles of passive solar design to the European building design profession. A condensed guidelines handbook in cartoon format was specially prepared, and competitors were required to complete simplified performance calculators. The competition was open to architects and students of architecture resident in any of the EC countries. Ideas were sought for the application of passive principles either to a new construction or to the rehabilitation of an existing building. Three seperate categories of entries were invited: multi-storey housing, clustered housing and single dwellings. A total of 1019 applied for the competition documents. The 223 entries were judged in two stages: firstly by technical assessors from all over Europe, and secondly, the 107 designs which passed the technical assessment were examined by a panel of international architects with experience in passive solar design. Prizes totalling 30,000 ECU were awarded and the judges reported that the standard of entries was very good at both the technical and creative levels.

As a result of the response to the first competition and the interest shown in the winning schemes when they were exhibited in various places throughout the Community, the Commission held a second competition in 1982. This time two categories for entries were defined, high density low rise housing, and retrofit and rehabilitation of dwellings. Schools of architecture were encouraged to include the competition in their design studio work, and 20 schools applied for the competition documents. In addition, applications from 1395 individuals were received. A total of 249 entries was submitted and again following a two-stage assessment 30,000 ECU in prize money was awarded.

Both competitions were organised on behalf of the CEC Directorate General for Science Research and Development by Ralph Lebens Associates, London. Two books have been published which illustrate the results of the competitions in full and serve as valuable sources of design ideas (6,7).

3.2 Workshops

Action has been taken under the auspices of the CEC to advance and reinforce the process begun by these competitions in Europe. The Passive Solar Working Group (PSWG) was composed of one participant from each Member State and was co-ordinated by Ralph Lebens Associates. The PSWG was part of the Commissions 1979-'83 energy research and development programme, and one of its areas of activity was in the provision of educational support in passive solar energy.

The initial effort was concentrated on the organisation of a series of mid-career workshops throughout the Community and directed primarily towards building design practitioners. As far as possible the requirements of teachers in schools of architecture and building would also be met.

A two-day workshop was designed which would introduce the principles of passive solar design, illustrate a number of built examples, and provide various design and analytical tools so that participants might use this approach in practice. The draft European Passive Solar Handbook, the subject of another paper to this conference (8), was used as the basis of the workshops. Participation at each workshop was limited to 25 to 30 people, as those attending were asked to undertake a practical design exercise and apply simple manual and micro-computer based calculation methods to analyse their proposals.

At the planning stage contact was made with international professional organisations to foster co-operation and ensure compatibility with existing continuing professional development programmes. Individual PSWG members undertook local organisation with the support of national institutes, relating the workshop to national activities and making modifications as necessary to suit regional circumstances.

The first, prototype workshop took place in April 1983 in London. The finalised draft European Passive Solar Handbook was introduced at a Workshop in Dublin in December 1983. Further workshops were held in Copenhagen and Rennes in March 1984 and in Athens in April. The series is due to conclude with workshops in Venice, Brussels, Stuttgart and Eindhoven during May.

This initial series has been sponsored by the CEC and arrangements have been co-ordinated by J R Stammers, London. To date it is reported that all workshops have been successful, with very favourable reactions from participants. The widespread publicity achieved has had the further benefit of helping improve professional and public awareness of passive solar energy.

4. CONCLUSIONS

There is more than one aspect to the provision of passive solar educational support. On the one hand the issue is how best to provide the technical support needed by those interested in energy conscious design. Useful progress is being made, and the European Passive Solar Handbook is important in this context together with the European workshops. The Handbook is limited of course to the present state-of-the-art, and significant areas exist in which further R&D is needed. Further teaching aids and materials are needed, and the exchange of ideas and experience between both teachers and practitioners should be encouraged. Design and analytical tools require development and particularly the former.

But energy is or should be a design issue in all buildings. The inclusion in a few architectural journals of an energy assessment for each building reviewed is a useful advance. New or modified creative approaches and pedagogical models may be required if adequate account is to be taken of the technical and formal dimensions of energy in architecture, and if new information and insights are to be accommodated.

REFERENCES

1. PROWLER, D. and FRAKER, H. (1981). Teaching Passive Design in Architecture. University of Pennsylvania for US Department of Energy.
2. LEWIS, J.O. (1981). Changes in Energy Availability and Curriculum Development in Certain European Schools of Architecture. Unpublished report, Energy Research Group, University College Dublin.
3. MARKUS, T.A. and MORRIS, E.N. (1980). Buildings, Climate and Energy. Pitman, London.
4. DEBAT, R. (1982). A New Discipline: Architectural - Bioclimatology. Towards an Interdisciplinary Research Program. Energy and Buildings 5.
5. BANHAM, R. (1969). The Architecture of the Well-Tempered Environment. Architectural Press, London.
6. LEBENS, R.M. (Ed.) (1981). Passive Solar Architecture in Europe. Architectural Press, London.
7. LEBENS, R.M. (Ed.) (1983). Passive Solar Architecture in Europe 2. Architectural Press, London.
8. LEBENS, R.M. (1984). The European Passive Solar Handbook: Design Strategy and Tools. Proceedings, 1st E.C. Conference on Solar Heating.

A GRAPHICAL METHOD TO ASSESS PASSIVE SOLAR GAINS IN THE EXISTING HOUSING STOCK OF THE UNITED KINGDOM

F.A. PENZ
The Martin Centre for Architectural and Urban Studies
University of Cambridge, 6 Chaucer Road, Cambridge CB2 2EB, England

Summary

The Martin Centre carried out a survey of 413 dwellings representative of the city of Cambridge, as part of a project for the Department of Energy. The purpose of the study was to determine the potential of the existing housing stock to exploit passive solar energy for space heating. An energy model of the city of Cambridge was built and in order to perform a range of passive solar scenarios a graphical method was used to assess the gains of a large number of suitable retrofit candidates. A finite difference thermal model was used to construct the graphs which correlate total house-specific heat loss with solar gains. Patterns of occupancy were also taken into account as well as orientation and overshadowing.

Introduction

Since the 1973 'oil crisis' research and development in renewable energies throughout the world have demonstrated the potential of these energy sources to provide an alternative to scarcer fossil fuels in the long term future. In the UK a number of passive solar buildings have been built in recent years but clearly the interest lies with the existing stock which represents over 21 million units in the domestic sector. In an attempt to assess the potential of the domestic sector to exploit passive solar gains, the Martin Centre carried out a detailed survey of Cambridge, and built an energy model of the domestic sector of a town in order to extrapolate the results at the National level.

The Cambridge survey: the main results

It was found that 61.7% of the Cambridge housing stock sample had a potential for either a conservatory, a Trombe wall or a roof collector for a $\pm 45^{\circ}$ of south criterion. This figure was reduced to 41.2% of a $\pm 35^{\circ}$ if south criterion was applied. Semi-detached houses appeared to have the most potential. Two out of three could accommodate a passive solar retrofit. 74% of the houses built during the 1919-1939 period were south facing and in broad terms the 1919-1969 period seems to have been more favourable to southerly orientation than any other. Also, more local authority houses had a south facing side than owner-occupied houses. A rather interesting finding was that suitably orientated dwellings suffered only a 10% reduction in solar collection due to overshadowing. But the fact that 61.7% of the sample was suitable for a passive solar retrofit meant that a great number of houses would have to be assessed, thus the development of a simple graphical method to cope with this.

The development of the graphical method

Three typical houses were selected from the Cambridge sample: terraced, semi-detached and detached. They represent in terms of house types 82% of the dwelling stock and cover a wide range of thermal approaches with respectively 2,3 and 4 facades exposed to the environment. Each house was retrofitted with a conservatory, a Trombe wall and a roof collector, and the thermal performances were evaluated using a finite difference thermal computer model (1) which had been previously developed and tested at the Martin Centre. In each case passive solar gains were assessed as a function of the level of insulation, thermostat setting and pattern of occupancy. It was then possible to correlate on a graph the house-specific heat loss against passive solar gains for each passive system (as shown below for conservatories).

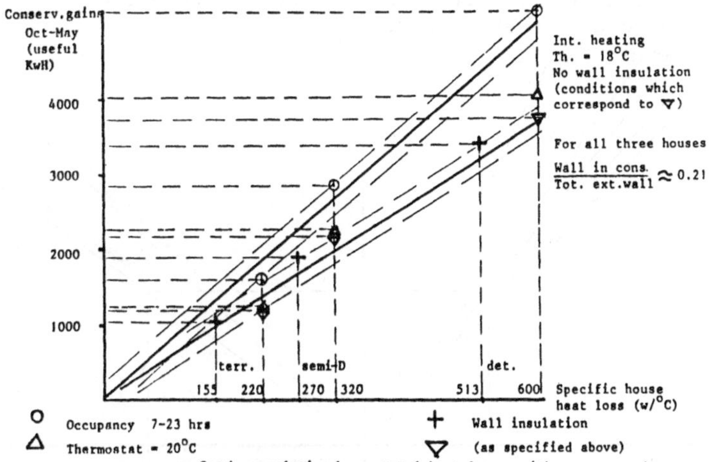

Graph correlating house total heat losses with conservatories gains for the three houses

Most of the results, as in the case of the conservatories, fit along a straight line and gave a high correlation coefficient. It was found quite interesting that houses very different in terms of wall construction, window area and specific heat loss could fit along the same straight line and this constituted the basis of the graphical method. Clearly these results are a direct consequence of the ratio of wall covered by the conservatory to total external wall area which is consistently around 0.21 for all three types of houses.

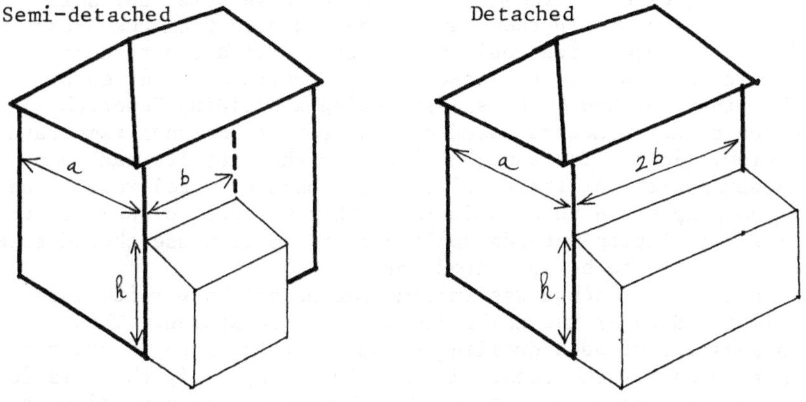

Both houses have the same ratio of wall enclosed in the conservatory to total external wall area (on average r=0.21)

It should also be stressed that the correlation graph would not make sense if the wrong size of conservatory was used, that is, if the conservatory fitted on to a mid-terrace house was added to a large detached house, the correlations shown previously would then be meaningless.

The same type of correlations were obtained for Trombe walls and roof collectors. In addition, a number of tests were performed to assess the effect of orientation and obstruction on passive solar gains.

Finally, the completed graph below brings together the main factors influencing passive system performances and shows specific house heat loss (or level of insulation), pattern of occupancy, the effect of orientation, and the influence of overshadowing. (2)

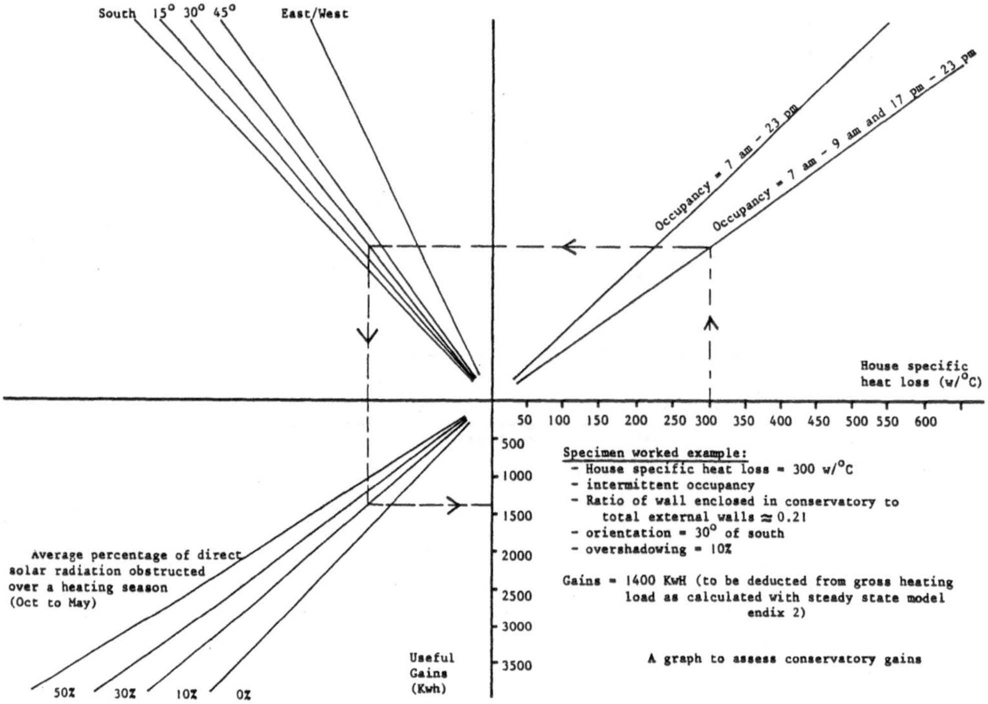

A graph to assess conservatory gains

The results expressed in the graph are the passive solar gains in useful kWh. These gains would then have to be deducted from the gross heating load of the house which could be calculated with any type of steady state calculations. In the case of the Cambridge survey each house space heating load had been assessed using a Building Research Establishment calculation method based on a survey of internal temperatures in housing (Uglow, 1981). The BRE calculation method results had been previously compared with the finite difference computer model results and consequently the graphs had to be adjusted. This was due to the fact that the two thermal calculation methods applied to the same house showed some discrepancies which had to be accounted for.

The BRE calculation method was implemented on the University main frame where all the data of the Cambridge survey were stored. This allowed the assessment of each dwelling's space heating load in order to build the energy model of the twon. As the first step, only the existing state of the domestic sector was assessed. Then each suitably orientated house ($\pm$ 45° of south) was retrofitted with the most appropriate type of passive system out of the three (conservatory, roof collector or Trombe wall).

The following results were obtained with respect to the suitability of the three basic types of passive solar systems for the domestic sector,

Conservatory	40 %
Roof collector	13 %
Trombe wall	8.7%
Not suitable for any system	38.3%
Total dwellings	100%

The potential gains for each suitable house were assessed through the graphical method and each result was fed back into the University main frame as input data. The total house-specific losses had been previously calculated for each dwelling as an input to the BRE method and the percentage of obstruction had been previously worked out by a separate program. Orientation and pattern of occupancy were obviously known from the survey data which made the process of going through the graphs very fast, although there were a large number of houses to assess.

Finally, it was possible to perform a range of passive solar scenarios in order to assess the inpact of such techniques in reducing the UK primary energy consumption.

Conclusions

The development of the graphical method was mainly intended to be a tool for assessing passive solar gains in the case of a large number of houses such as in the Cambridge Survey. It was found very practical and could also possibly be used as a 'back of the envelope' method at the early stage of a retrofit design. It was certainly worthwhile and easier to develop such a tool in England where the range of plans in housing is limited (compared to other European Countries) and where the method's 'built-in conditions' would apply to most cases.

References

Basnett, P.(1975). Estimation of solar radiation falling on vertical surfaces from measurements on a horizontal plane. The Electricity Research Centre, Capenhurst, Chester.

Hawkes, D. and Souza, C. (1981). Passive Solar Heating in Existing Housing. A Martin Centre Report, University of Cambridge.

Penz, F.A. (1983). Passive Solar Heating in Existing Dwellings. Ph.D Thesis, University of Cambridge.

Uglow, C. (1982). Energy use in dwellings: An exercise to investigate the validity of a simple calculation method. Building Services Engineering Research and Technology, Vol. 3, No. 1.

Notes

1. The thermal modelling used an hourly time step and the solar input was done through horizontal global data which were then split into direct and diffuse radiation with the Berlage equations (Basnett, 1975). The thermal calculation used a finite difference analysis with a matrix inversion to solve the diffusion equation (Penz, 1983).
2. The overshadowing is expressed as a percentage of the received radiation (taking into account obstruction) compared to the available radiation (if no obstruction) over the heating season.

EVALUATION AND DEVELOPMENT OF THERMAL ANALYSIS MODELS WITHIN THE IEA SOLAR HEATING AND COOLING PROGRAMME, TASK VIII PASSIVE AND HYBRID SOLAR LOW ENERGY BUILDINGS

O.C. Jørgensen
Thermal Insulation Laboratory
Technical University of Denmark

Summary

Subtask B, Modelling and Simulation is one of five subtasks that constitutes the international cooperative project: Passive and Hybrid Solar Low Energy Buildings within the Solar Heating and Cooling Programme of the International Energy Agency (1). The objectives of the subtasks are to increase the capability of the participants to accurately predict and analyse the performance of Passive and Hybrid Solar Low Energy Buildings, to provide a sound basis for the evaluation of design tools and to support the development of innovative designs. The work commenced in mid-1982 by surveying the existing simulation models in the member countries, their analysis capabilities and the validation experience of each of them. A total number of 31 models was included in the survey, which was published in November 1983. Based on this survey, a validation strategy was worked out and agreed to. The results of the models were compared to measured data from three different test cells and mutually on yearly simulations of the same three test cells. Finally, results were compared for yearly simulations of real buildings utilising the same three basic passive techniques. For some of the models involved, this was a first time validation effort; however, in most cases good agreement was obtained.

1. INTRODUCTION

Four major accomplishments were defined at the outset of the work for subtask B of the 8th Task of the IEA Solar Heating and Cooling Programme: Passive and Hybrid Solar Low Energy Buildings. These accomplishments were defined partly to fulfil general needs from the solar energy research world and partly to fulfil very specific needs from other subtasks of the Task itself. The four accomplishments are:

1. Survey of existing simulation models.
2. Validation of a selected number of models.
3. Parameter sensitivity study for handbook guidelines.
4. Analysis of subtask D designs.

The following paragraphs deal with these points separately.

2. SURVEY OF EXISTING SIMULATION MODELS

When the work commenced, one or more computer programs for a detailed thermal analysis of Passive and/or Hybrid Solar Low Energy Buildings had been developed in almost all the member countries of the IEA Programme.

However, it was felt that generally very little was known about the analysis capabilities and other important features of the models. Therefore it was decided to undertake a survey of these models which could serve two main objectives:

1. Assess the state of the art, and
2. create an overview of available models.

The survey was completed in 1983 and published in an IEA report entitled "ANALYSIS MODEL SURVEY" (2). The conclusions of this survey of 31 models were the following:

1. Hybrid systems can only be simulated by very few programs.
2. Even when written in FORTRAN most programs are very computer-mark dependent and therefore not easily transferable.
3. Most of the programs have been developed for research purposes by researchers; in general they do not represent energy analysis tools useful to building designers.
4. Validation experience with these models was extremely limited.

The recommendations thus derived from this survey are that future model developments should aim at user-friendly, computer-independent, design-oriented programs, capable of handling hybrid systems.

3. VALIDATION OF SELECTED NUMBER OF MODELS

Actual validation studies in the sense of comparing the results of the computer analysis models to measured data from test cells, test buildings or standard houses were surveyed at a working group meeting in July 1982.

The validation experiences with the programs were surprisingly limited. Even for Direct Gain test cells only a few studies were reported. The primary reason for this was claimed to be lack of adequate performance data. On the basis of this survey a validation strategy was formulated and agreed to by the participants.

The strategy is a three-step approach on the three basic passive designs (Direct Gain, Trombe-wall and Attached Sunspace). Second step is model-to-model comparisons on the same test cells simulated over one year in three different climates. Third step is model-to-model comparisons on the same three designs incorporated in real buildings and simulated under the same three weather conditions.

3.1 Test cell validation

Three different test cells were selected to provide data for the validation of the models. For Direct Gain a test building monitored by the National Research Council of Canada was chosen. The test building is a 3 by 9.5 m massive building with two rooms and a door and a fan between the rooms. The aperture area of the south window is 2.6 m^2.

The Trombe-wall test cell is located in the southern part of Switzerland. It is a 2.9 by 2.5 m cell with an aperture area of 5 m^2.

The Attached Sunspace test cell chosen is one of the test cells of the U.S. Los Alamos National Laboratory. This cell has a 5 m^2, 60° sloping glazing pane in front of two small rooms, each of 3.5 m^2. There is a door for natural convection between the sunspace and one of the rooms and another door is connecting the two rooms.

The Direct Gain test cell has been simulated by 11 different programs, the Attached Sunspace test cell by 4 and the Trombe-wall test cell by 4 programs. Results are expected from 3 more models on the latter case.

Good agreement was obtained for most of the programs simulating the direct gain test building.

Except for the preliminary results from USA, all models have predicted heating loads well within 10% of the measured value. Also the dynamic behaviour of the building was tracked very well by the models. Fig. 1 and fig. 2 show comparison plots of measured and predicted south room temperatures and auxiliary heating power. These plots have been produced by Sherif Barakat from Canada, but represent plots produced by other participants very well.

The attached sunspace test cell was a more difficult simulation object. The primary reason for this was that heat was transferred between the three zones by natural convection, which can only be handled by one of the models. Another reason was that the operation mode of the cell was changed during the period. Taking these difficulties into account, the results obtained showed reasonably good agreement. The problems with the simulation of natural convection heat transfer gave raise to a special side-activity trying to establish an overview of existing algorithms for modelling this phenomena.

The Trombe-wall simulations also showed fine agreement with the measured data. An interesting outcome of an attempt to model this test cell by the U.S. participant using Blast 3.0 revealed a number of bugs in the Trombe-wall subroutine of this program.

3.2 Model-to-model comparions

Following the validations of the model against the test cell data, the models were run on one-year weather data files using exactly the same input data file as for the validation work. Thus the results of the models could be mutually compared for a longer time period.

To date results have been presented from five programs on the Direct Gain case showing agreement to within 10% on the yearly heating load. Three programs have been used to calculate the Attached Sunspace using Denver weather data. Two programs agree to within 4%; the third, however, calculated the heat load to be 33% higher. More results are expected before this activity is closed.

3.3 Real building simulations

The third step of the validation strategy was model-to-model comparisons on real buildings. This activity was combined with the creation of a "Design Tool Evaluation Package" generated to spot-check the results obtained using design-tools against results obtained using detailed simulation models. The American program SERI-RES was used to generate the data for the package in which two single-family and two terrace house constructions (light and heavy mass), designed for either Direct Gain, Trombe-wall or Attached Sunspace solar utilization, were simulated using three different sets of yearly weather data (Copenhagen, Denver and Geneva). The other simulation programs could then be used to spot-check the SERI-RES results.

Until now results have been presented by the Canadian participant using the program Encore-Canada, to simulate the Direct Gain single-family cases. The results obtained agreed to within 5% on yearly heating load. The Direct Gain terrace house cases have been simulated by the author using the Danish program BA-4; the agreement obtained was better than 3% on yearly heating loads. Preliminary results from other participants show larger discrepancies, which primarily can be explained by different interpretations of the data package provided with the building descriptions.

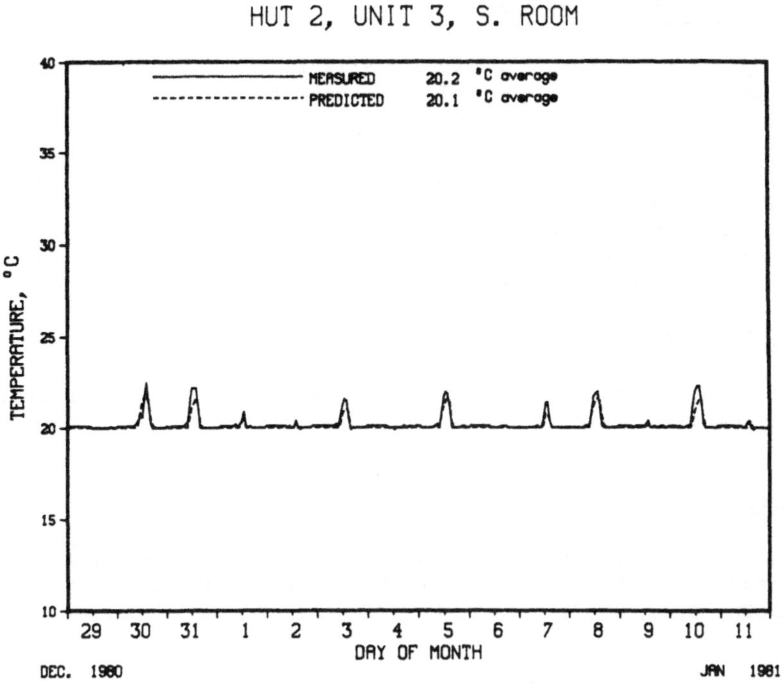

Fig. 1. Measured and predicted temperatures of Canadian test building.

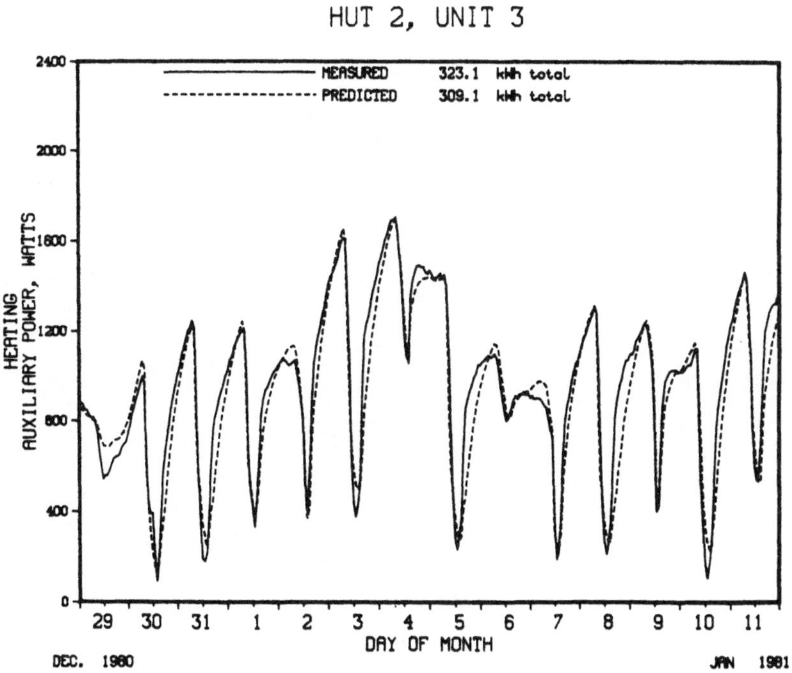

Fig. 2. Measured and predicted auxiliary heating power for Canadian test building.

4. ANALYSIS OF SUBTASK D DESIGNS

The objective of subtask D of Task VIII is to verify that passive and hybrid solar techniques can be used in a cost effective way to substantially reduce the heating load compared to conventionally built houses (1). The goal is to accomplish this objective by actually building one or more houses in each member country employing passive and hybrid techniques.

The simulation models will be used to analyse these designs which, in some cases, are rather innovative, thus calling upon the development of new routines for the models. The model developments and model studies will be documented and compiled into a report which will be published in 1986.

5. ACKNOWLEDGEMENT

The results of subtask B are produced by the subtask representatives of all the member countries. Thanks are due to their hard work and engagement.

6. REFERENCES

1. HOLTZ, M. (1982). IEA Task VIII: Passive and Hybrid Solar Low Energy Buildings. International Solar Architectur Conference, Cannes.
2. JØRGENSEN, O.C. (1983). IEA Task VIII, Analysis Model Survey.

COMFORT ASPECTS IN OVERGLAZED ENVIRONMENTS

J.C.DEVAL, X.BERGER and M.SCHNEIDER
C.N.R.S., Laboratoire d'Ecothermique Solaire
B.P.21, 06561 Valbonne Cedex. France

Summary

A review of the numerous works concerning human thermal losses has lead to a new human thermal comfort model. A set of diagrams has been obtained. We discuss here the opportunity of a norm laid down severely (too much rigid). The variability of the different parameters, mainly activity, and clothing insulation may be associated with other combinations of air and mean radiant temperatures than for example TAIR = 19°C and MRT = 19°C. The interest of the radiative field is pointed out on the diagrams. It seems that a thermal ambiance with 17°C for the air temperature and 26°C for the mean radiant temperature is one of the best to be proposed. To support that, some experiments have been performed in overglazed environments.

1. THE THERMAL COMFORT IN UNBALANCE VOLUMES

Radiatively out of balance volumes are the main characteristics of overglazed environments. Consequently, living in such conditions requires a knowledge of the human thermal behaviour.

The mean radiant temperature (MRT), accounting for blood circulation and clothing, is a well fitted parameter for a correct evaluation of the radiative exchanges between man and its environment. Its determination is generally a difficulty. To solve it, we built a code, named ECORAD, which takes into account solar radiation, emissivity of the walls, specular and diffuse reflections... It proceeds with a Monte-Carlo method (Ref.1). Figure 1 gives an example of its application. In a few cases the man is affected by asymetric radiative fields and this effect has to be considered. A cold or a hot radiative exchange between the ceiling and the head or shoulders is felt very unpleasant (Ref.2) ; for the head, the reasons lie in this usually uncovered part of the body, and in its relatively higher temperature. The spherical shape of the head and of the shoulders, the situation in the convective flux at the top of the body make them the most sensitive part (Ref.3). The commonly assumed limit for the amplitude of the radiant temperature vector is 6 celcius degrees. But this limit might be higher as long as the average temperature remains in the comfort zone.

Then, especially in sunny countries, the greenhouses will not have glazed-roofs. For large vertical windows facing the human body, this limit reaches 8 to 10 celcius degrees, a cold facing surface beeing better tolerated than a back one. These results of experiments have to be taken into account for the realization of overglazed environments. But, may be, they are too severe in regard to what is usually met.

Six parameters enter the expressions which describe the various thermal exchanges concerning human body : air temperature (TAIR), mean radiant temperature (MRT), air velocity, air humidity, activity and clothing insulation. But a few more are also important, as for instance, clothing selectivity, or clothing mean diffusivity to air and water vapour. We reviewed more than 200 papers of the last 40 years dealing with physiology and bioengineering of thermal comfort, we discussed them, and built numerical codes which lead to diagrams related to human feeling in various situations (Ref.4,5,6). The comfort diagrams point out three main aspects related to unbalance environment :

. <u>The optimal comfort zone</u>

The situation given by TAIR = 17°C, MRT = 26°C is one of the most comfortable for human body in standard conditions (Ref.5). Moreover, Fig.2 shows that the optimal comfort zone is not plotted on the line TAIR = MRT and demonstrates the interest of radiative ambiances. Therefore, the standard TAIR = MRT = 19°C, usually quoted for insulated volumes heated with convectors, is not the best fit. The stimulation of the organism requires gradients for thermal sensitiveness. The comfort feeling requires both null energetic balance and relative time and space temperature gradients. In association with the permanent variations of our conditions of life, and considering the difference between skin and breathing exchanges, these gradients are pleasant. The radiators, the fireplace of the stove which heat a cold volume, correspond to a convenient solution because they associate thermal comfort with pleasure. That is why direct solar gains and passive storages should be very good systems. But they are often rejected because of a wrong sizing of the system or a lack of control of the radiative ambiances. In overglazed environments, the TAIR = MRT = 19°C solution is no more than a possible one, not often met. An application of these criteria about comfort is given with the "Opio" solar house experiment.

. <u>The convective cooling zone</u>

This zone, where a cold air is compensated by a higher temperature radiative field, is plotted in the stable comfort zone. There, the clothing emissivity, and even the clothing itself, have a minor influence because this zone is delimited by the two lines corresponding to the relations :

TCLO = TSKIN (TCLO = mean temperature of outer surface of clothing
TSKIN = mean skin temperature)

TCLO = MRT

This convective cooling zone, which fits well with diurnal overglazed environments, is very suitable for human body. The control of the radiative field must be assumed to avoid overlighting and too large space gradients. Solar fluctuations must be compensated by immediate active systems such as air fans and radiative set-power sources. The P.C.M. venetian blind is an illustration of a solar system (european patent 0019573) to control the radiative ambiance.

Another example is the "Tourrettes" solar house for which a research in architecture has been assumed concerning large glazings. The aim was to get the necessary and sufficient energy at any time in winter as well as in summer during the most part of the day for comfort with the minimal action of the set-up power. Maximum solar gain has not been searched but a compromise between the aesthetic of the house, a large opening on the wonderful landscape and the optimal solar heating and lighting the whole year. The "Tourrettes" house, drawn by the architect CH.PETITCOLLOT, is the outcome of three previous experiments : "Grimaud", le "Tignet" and "Biot" solar houses. Overglazing which caracterizes this house, is associated with comfort by a research of the best orientations (south-east and south-west simultaneously for a natural control of the lighting), a selected insulation of the walls and of the glazings to avoid cold radiative fields, condensations and to get benefit of the solar contribution.

. <u>The radiative cooling zone</u>

This zone where the hot air is compensated by a cold radiative field is unstable and delimited by the evaporative limit (which is also a condensative limit) and the line corresponding to the relation TAIR = TCLO. The influence of the air motion is low near this line, and any deviation of one parameter from comfort situation immediatly produces an unpleasant feeling (a too high humidity, as frequently met in maritim countries during summer, can lead to condensations). Therefore, comfort in this zone is reached with a limited activity associated with low air velocity. A good example of this is shown by the numerous people sleeping on the attics in dry tropical regions. Other applications could be found for climatisation of offices. The cooling effect is mainly concerned by evening situations during summer time. It is often produced by large glazings (a cold source cannot be punctual) or during the day by massive walls refreshed during the night. During winter time, this zone is not normally reached, but it might be approached and experimented as an undesirable effect in overglazed environment when very high air temperatures are not able to compensate a very low radiative field. The cold radiation of the glazings after sunset is very often compensated by the radiative set-up power source or eliminated by the presence of curtains. As a matter of fact, it is mainly the unstable comfort which is unpleasant, and not the cold radiative field. The experiment of the new rectorship building in Nice (French riviera) is an example of natural radiative cooling in overglazed environments.

2. THE "OPIO" SOLAR HOUSE

This house has been the basis of several experiments. The integration of the solar collector and the complementarity between solar energy and electricity set-up power are described elsewhere (Ref.7). The diurnal thermal evolution of the living-room ambiance may be considered as a first success about the simultaneous control of TAIR, air velocity and MRT. It is an example of what could be done in solar houses : a permanent control of the varying parameters according to the solar radiation and the use of the room. Three situations are successively met during a clear winter-day :

. At breakfast, the room, which has not been heated during the night, has its walls and air cold (14°C). But the south-east orientation of the glazings, facing the sun at that time, permits the light to enter deeply (MRT = 23°C) and therefore, to produce a pleasant environment.

. At lunch, the sunlight inside the dining-room is reduced to a diffuse illumination. A hot air (25°C), coming from the solar collector on the roof, is propelled at 0.5 M/S. The walls are still fresh and this comfortable situation is mainly due to the air motion.

. At dinner, the air is still and, as the walls, at about 18°C. The closed shutters and curtains prevent from the external low radiative field. Again the thermal comfort is felt. Thus during daytime, going from one situation of comfort to another, produces a sensitive pleasant feeling (Fig.3).

3. THE P.C.M. VENITIAN BLIND

Before beeing an energy collector, the main thermal function of the blind is to provide and maintain the habitability of radiatively out of balance volumes (Ref.8). It is the reason why the blind has multiple actions:

. Diffusion of a part of the incident energy to insure a uniform lighting and to compensate for a cool air.

. Avoidance of overheating with storage of the excess solar energy, without impeding the outside view.

. Suppressing of the cold radiative influence of the glazing at night or through bad weather. Time lag restitution of the stored energy, mostly by radiative exchange.

. Increase of the effective living area, comfort improvement, and extension of the duration of habitability in intermittent living zones. The other functions of this patented system are energy saving, privacy protection and antieffraction shutting. Fig.4 shows the comfort cycle as measured with our experimental cell and with the industrial prototype/ This cycle remains near the wished theoretical comfort line. On this aspect the P.C.M. venitian blind may be considered as the complementary element of overglazing.

4. NATURAL RADIATIVE COOLING IN THE OVERGLAZED NICE RECTORAT BUILDING

This building was the support of a double action for summer cooling (Ref.9) :

. The first one concerns the choice of the glass : reflective glasses might appear convenient for sunny countries. Because of their reflective deposit deposit on one face, they have generally a non negligible absorptance. In winter it is always an error to refuse solar wins. In summer, because of the height of the sun at that time, only the neighbourhood of the glazings is sunlighted. This, and the radiation coming from the hot glazing, as from a radiator, make this part of the room uncomfortable. A too large gradient exists between the glazing and the opposite side of the room. The correct way is to refuse the sun outside the glazing in summer, and inside in winter (Ref.9). Fig.5 shows the comfort temperature in two cases of high reflective glazing, and of low reflective glazing with external blind in addition (the adopted solution).

. After limiting the solar wins to a minimum, which is the first aim to reach, the second action was a slight cooling of the rooms. A clear sky is correlated to a large daily temperature amplitude. We took advantage of this correlation to get a refreshment of the massive internal walls at night. During the diurnal period, the air is a little refreshed in the same time as a cold radiative field is present. Consequently, the comfort temperature is got under the outside air temperature and overheating is reduced to only a few days for the whole summer.

Of course such a simple system is not so efficient than an electrical cooler, but, in summer in Nice, it is sufficient for offices partially occupied. Fig.6 shows the histograms of the thermal situation which certifies the efficiency of this natural process.

5. CONCLUSION

If overglazing is an attractive proceeding to realize solar houses, it has to be associated with a severe control of the radiative field. Human body is conceived to live in unbalance ambiances, but that needs to take into account the possible effects of the various parameters which enter in the expressions of the thermal exchanges. In such ambiances, the compensation of the effect of a parameter by the effect of another one may be obtained mainly in the situations concerned by convective cooling. Radiative cooling needs more precautions. The reference TAIR = TMRT = 19°C has not to be searched in any way. Overglazing create radiative gradients which may be felt as a pleasantness if their importance is not inordinate. Finally, living in overglazed environments is an art in consequence of the complexity of the human alive machine.

REFERENCES

1. SCHNEIDER, M. and BERGER, X. (1980). Etude d'ambiances de rayonnement dans les systèmes solaires passifs. Colloque solaire international, Nice.
2. FANGER, P.O., BANHIDI, L., OLESEN, B.W. and LANGKILDE, G. (1980). Discomfort caused by overhead radiation. 7th International Congress of Heating and Air Conditioning "Clima 2000", Budapest.
3. McINTYRE (1980). Indoor climate. Applied Science Publishers, London.
4. DEVAL, J.C. and BERGER, X. (1983). Ambiances de confort. Applications du thermoconditionnement à la thermique du bâtiment, Colloque S.F.T., Lyon.
5. DEVAL, J.C. (1984). Le confort en climat tempéré. Revue de Physique Appliquée. (In press).
6. DEVAL, J.C. and BERGER, X. (1983). Le confort en climat chaud. Rapport intermédiaire Contrat REXCOOP, Plan Construction, A.F.M.E.
7. BERGER, X. (1983). Maison à Opio. Une expérience solaire dans les Alpes-Maritimes. Rapport sur la politique départementale de l'énergie, Conseil Général des Alpes-Maritimes, p. 77.
8. DEVAL, J.C. and BERGER, X. (1983). Latent heat venetian blind,in passive and low energy architecture. Proceedings of the second international PLEA conference (Rethimnon, Crete). Pergamon Press, p. 409-414.
9. BERGER, X. and DESLANDES, J. (1980). La climatisation estivale. Cahiers Techniques du Bâtiment, publication du Moniteur, n° 28.

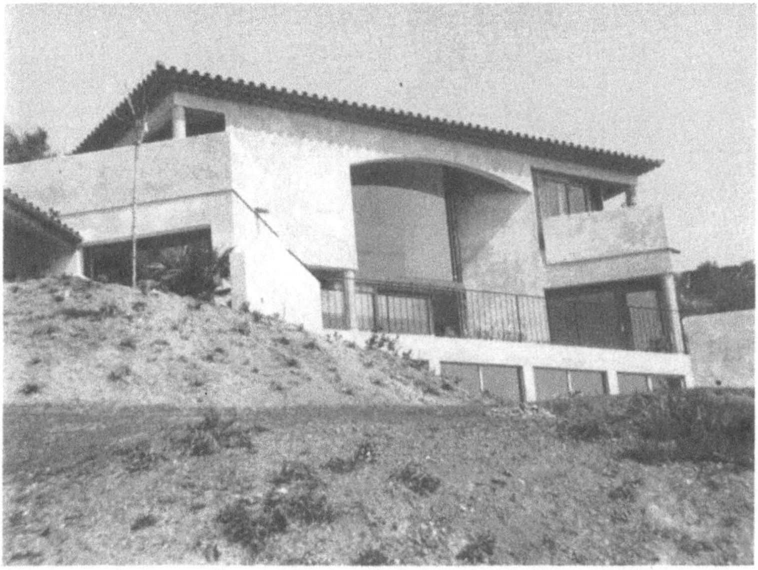

The "Tourrettes" solar house

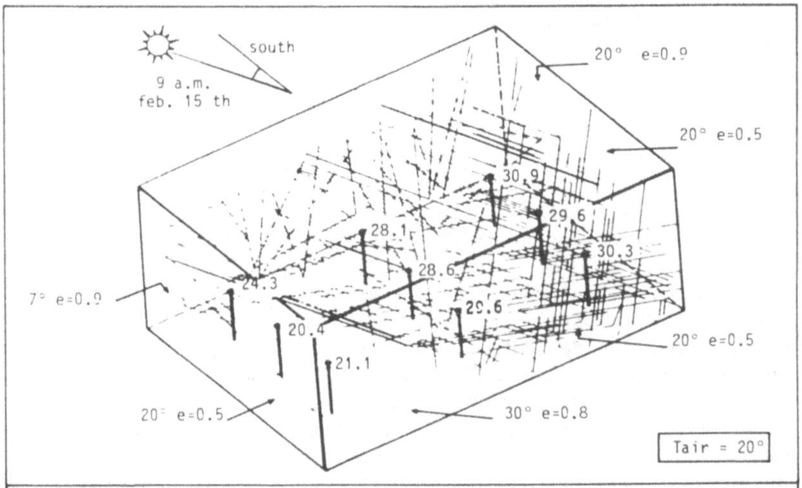

AN APPLICATION OF ECORAD CODE : SPATIAL DIAGRAM OF THE COMFORT TEMPERATURE IN A ROOM; 150 RAYS ARE PRESENTED AMONG 4000 COMPUTED ; e IS THE OPTICAL EMISSIVITY

Fig.1 - Monte Carlo methods

Fig.2
Comfort zones

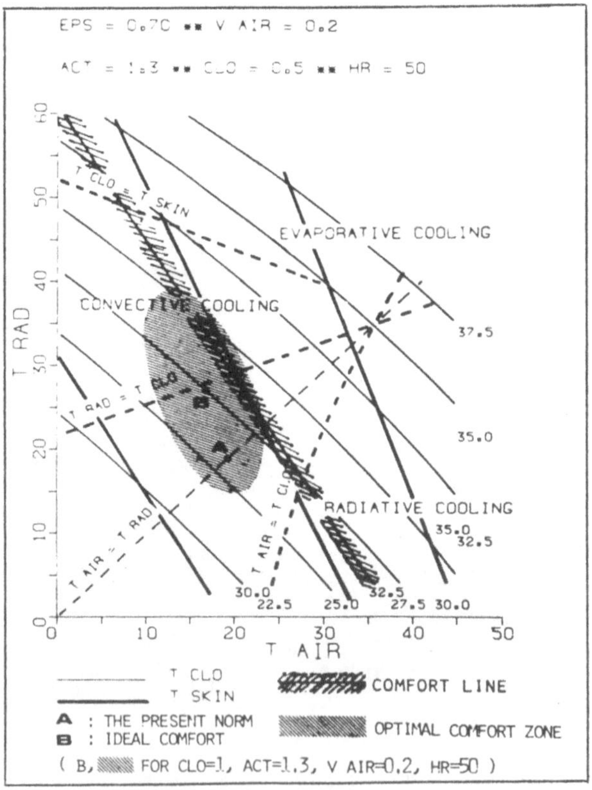

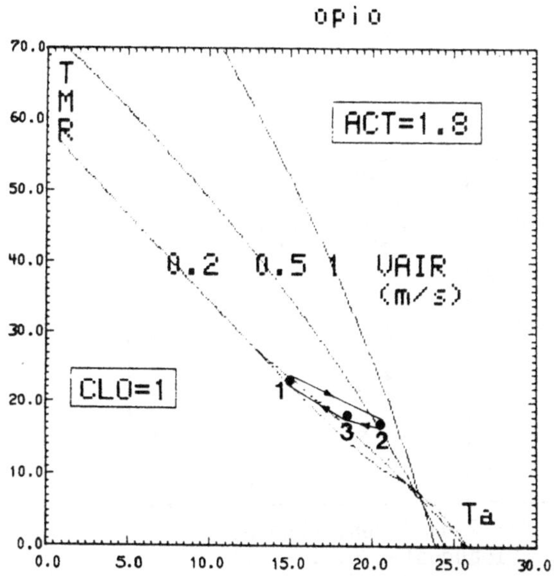

Fig.3

Diurnal thermal evolution of the living-room ambiance in "Opio" house.

1 : 9 a.m.
2 : 12 a.m.
3 : 8 p.m.

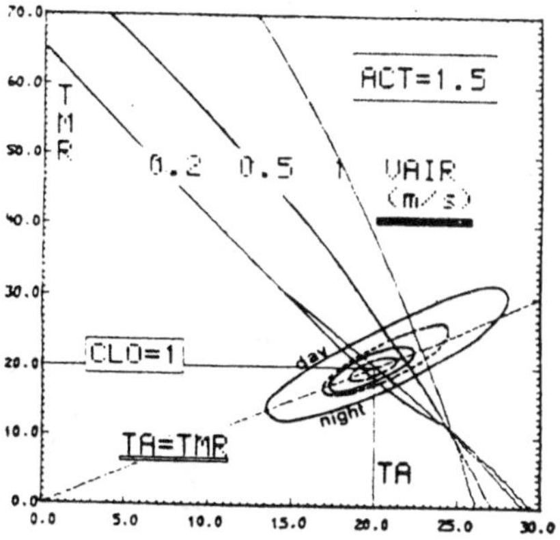

Fig.4

The comfort cycle as measured with our experimental cell (photo), in comparison with that of an ordinary well insulated volume. This point out the comfort effect of the PCM venetian blind.

An internal view of experimental cell with the comfort meter in front of the PCM venetian blind

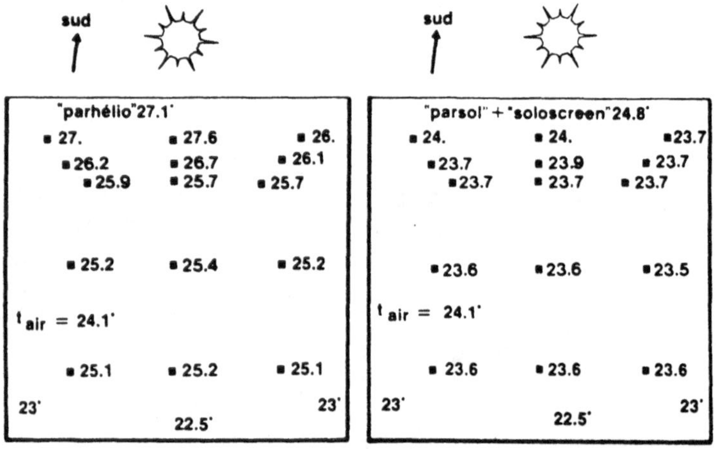

comfort temperatures for 2 different absorptive glasses (39% and 8%) ; 15august- 2p.m. - 1.1m high

Fig.5 - Nice rectorat building

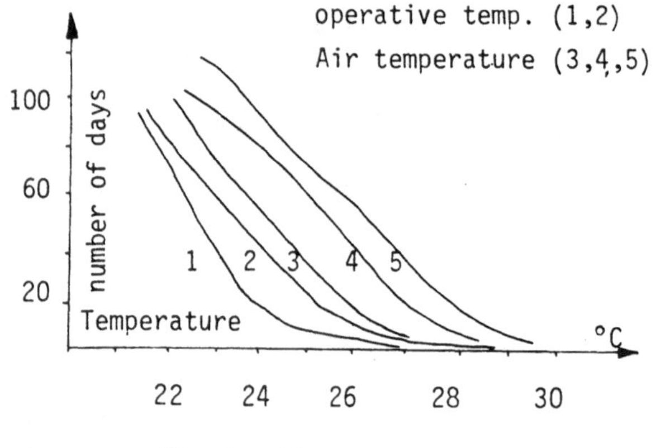

Fig.6 - histogram

1 : all day ventilation ; diurnal climatisation from july to september
2 : all day ventilation
3 : internal air temperature (nocturnal ventilation)
4 : external air temperature
5 : internal air temperature (without nocturnal vent.)

TRANSPARENT INSULATING MATERIALS - A POSSIBILITY FOR REDUCING THE CONSUMPTION OF FOSSIL FUEL IN HOUSEHOLDS

K. Bertsch, E. Boy, A. Frangoudakis und U. Heim

Fraunhofer-Institut für Bauphysik, Stuttgart
Nobelstrasse 12, D-7000 Stuttgart 80
Bundesrepublik Deutschland

Summary

Transparent insulating materials with transmission maxima in the visable range of the solar spectrum allow heat transfer inversion or at least temporary heat transfer inversion in outside wall constructions during the heating period.

If such transparent insulating material is attached to the outer surface of walls with a high thermal capacity (e.g. concrete or brick walls) day-time energy storage in the wall and reduced heat loss by night can be achieved. And this basically with only passive components.

Considering the application of transparent insulating materials in the building trade several factors have to be considered. Straight forward application could have many undesirable side-effects. Shielding the treated walls from direct sunlight in summer might be necessary for example.

Field measurements on test cells in natural climate show the enormous potential for energy saving as well as numerical calculations for different wall constructions and for complete houses.

The most important result concerns the comparision between conventional and transparent insulation materials. Problems that require detailed research and development work have been identified.

Technical problems standing in the way of immediate practical applications undoubtly exist, but they are not thought to be insuperable.

LICHTTRANSPARENTE WÄRMEDÄMMUNG - EINE MÖGLICHKEIT ZUR DRASTISCHEN REDUZIERUNG DES VERBRAUCHS AN FOSSILEN ENERGIETRÄGERN IM HAUSHALTSBEREICH

Zusammenfassung

Wärmedämmschichten mit einem Strahlungstransmissionsmaximum im sichtbaren Spektralbereich der Solarstrahlung erlauben während der Heizperiode die Umkehrung oder zumindest die zeitweilige Umkehrung des Transmissionswärmestromes in der Gebäudehülle von aussen nach innen.

Bei speicherfähigen Aussenwänden ist eine Wärmeeinspeicherung während der Sonneneinstrahlzeit und die Nutzung während strahlungsfreier Zeiten möglich, und das im wesentlichen mit passiven baulichen Mitteln. Bei der Anwendung dieser Dämmschichten im Bauwesen spielen viele Faktoren eine Rolle. Vor der unkritischen, nicht dimensionierten Anwendung muss dringend gewarnt werden. Während der Übergangszeit und in den Sommermonaten können zum Beispiel Verschattungsvorrichtungen erforderlich werden.
Messungen an Testzellen in natürlichem Aussenklima bestätigen das riesige Energieeinsparpotential das zuvor durch numerische Simulation verschiedener Wandkonstruktionen und ganzer Wohnhäuser ermittelt wurde. Für Zeitabschnitte mit direkter Sonneneinstrahlung stimmen die Ergebnisse der rechnerischen Simulation gut mit den Messergebnissen überein. Die vergleichenden Untersuchungen wurden im wesentlichen mit einer konventionellen Thermohaut-Aussendämmung als Referenz und einer transparenten Dämmung etwa gleichen Dämmwertes aus Acrylglasschaum durchgeführt. Problemstellungen die einer eingehenden Forschungs- und Entwicklungsarbeit bedürfen wurden identifiziert. Probleme für die Anwendung lichtdurchlässiger Dämmschichten bestehen zweifellos noch, wie zum Beispiel die Möglichkeit der Überhitzung, mechanische Beanspruchung durch starke Temperaturwechsel, brandschutztechnische Fragen oder Fragen der optischen Akzeptanz. Diese Probleme sind nach unserer Meinung jedoch wirtschaftlich lösbar.

1. EINLEITUNG

Die wirtschaftliche Nutzung der Solarenergie zur Raumheizung und zur Brauchwassererwärmung ist für die unterschiedlichen Klimaverhältnisse in Europa davon abhängig, wie die nur zeitweilig zur Verfügung stehende Strahlungsenergie gesammelt und kurz- oder langfristig gespeichert werden kann.

Transparente Wärmedämmschichten mit einem Strahlungstransmissionsmaximum für Globalstrahlung und einem Transmissionsminimum für längerwellige Wärmestrahlung lassen eine wirtschaftliche Solarenergienutzung erwarten, weil die Anwendung eine rein passive bauliche Massnahme ist und weil durch die Anwendung an der Aussenseite der Gebäudehülle, unabhängig von der Solarstrahlung auch bei Nacht oder während langer strahlungsarmer Perioden, wie bei der konventionellen Wärmedämmung der Wärmedurchgangskoeffizient (k-Wert) verringert wird. Die Aussenwandoberfläche kann als Strahlungsabsorber und die Aussenwand als Wärmespeicher dienen. Bei Anordnung der Dämmschicht mit Luftspalt vor der Aussenwand, wie in Bild 1 schematisch dargestellt, ist die Wärmenutzung für ein Luftheizsystem oder ein Lüftungssystem möglich. Mit der Anwendung lichtdurchlässiger Wärmedämmschichten kann der Transmissionswärmestrom in der Aussenwand stark reduziert und zeitweilig sogar von aussen nach innen umgekehrt werden. Eine drastische Reduzierung des Heizenergiebedarfes ist die Folge.

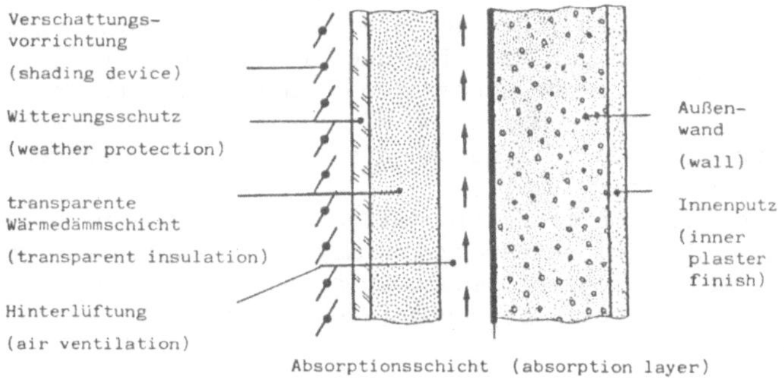

Bild 1: Schematische Darstellung des Anwendungssystems lichtdurchlässige Wärmedämmung.

2. ANWENDUNG LICHTDURCHLÄSSIGER WÄRMEDÄMMSCHICHTEN

2.1 Numerische Voruntersuchungen

Durch numerische Simulation von drei Aussenwandkonstruktionen mit jeweils einer konventionellen Aussendämmung (Thermohaut), einer Glasabdekkung (Trombe Wand) und einer lichtdurchlässigen Dämmschicht aus Acrylglasschaum, wie in Bild 2 skizziert, wurden mit gemessenen Aussenklimarandbedingungen die Innen- und Aussenoberflächentemperaturen und die Wärmestromdichten an der Innenwandoberfläche bei Südorientierung der Wände für den Standort Holzkirchen bei München berechnet.

Die für die Simulation verwendeten Abmessungen und Stoffdaten der Wände sind in Tabelle I und die Daten der Dämmschichten in Tabelle II angegeben. Die Simulationsmethode ist in (1) beschrieben. Dort sind auch die Klimadaten angegeben. Tabelle III enthält einige der wesentlichsten Ergebnisse der Simulation.

Material	Dicke (m)	Wärmeleitfähigkeit (W/m K)	Rohdichte (kg/m³)	spezifische Wärmekapazität (J/kg K)
Schwerbeton	0.175	2.10	2400	950
Ziegelmauerwerk	0.240	0.68	1600	950
porosierte Leichtziegel	0.240	0.33	800	950
Innenputz	0.010	0.35	1200	950

Tabelle I: Abmessungen und Stoffdaten der Wände

In der Zeitperiode vom 2. bis 30. Januar 1980 wurden Aussenlufttemperaturen zwischen -20°C und +4°C gemessen. Auf der Wandinnenseite wurden konstante Wärmeübergangsbedingungen und tagsüber eine konstante Raumtemperatur von 20°C und für die Nachtzeit eine Absenkung auf 15°C angenommen. Die auf der vertikalen Südfläche gemessene Gesamtstrahlungsintensität erreichte an vier sonnigen Tagen Werte um 900 W/m². Insgesamt lagen in dieser Periode nur 8 sonnige Tage. Trotzdem ist die energetische Überlegenheit der lichtdurchlässigen Wärmedämmung enorm, wie aus Tabelle III deutlich hervorgeht. In der Kombination mit der Betonwand werden die besten Ergebnisse erzielt. Die Glasabdeckung schneidet hier dagegen noch ungünstiger ab, als die Thermohaut.

Entsprechende numerische Simulationen für Perioden in der Übergangszeit und in den Sommermonaten ergaben natürlich Ergebnisse die auf Überhitzungsprobleme schliessen lassen. Aufgrund der überaus positiven Ergebnisse während der Heizperiode wurden auch praktische Untersuchungen durchgeführt.

Material	Dicke (m)	Wärmeleitfähigkeit (W/m K)	Rohdichte (kg/m^3)	spezifische Wärmekapazität (J/kg K)
konventionelle Dämmschicht	0.030	0.04	40	1460
Luftschicht	0.024	temperaturabhängig	1.3	1006
Klarglas	0.006	0.81	2500	920
Acrylglasschaum	0.030	0.04	30	1460
Plexiglas	0.001	0.19	1200	1460

Material	Strahlungsphysikalische Kennwerte für senkrechten Strahlungseinfall		
	Transmissionsgrad τ (-)	Reflexionsgrad ρ (-)	Absorptionsgrad α (-)
Klarglas	0.81	0.08	0.11
Acrylglasschaum	0.693	0.137	0.10
Plexiglas			0.07
konventionelle Dämmschicht	0	0.4	0.6

Tabelle II: Abmessungen, Stoffdaten und Kennwerte der Dämmschichten

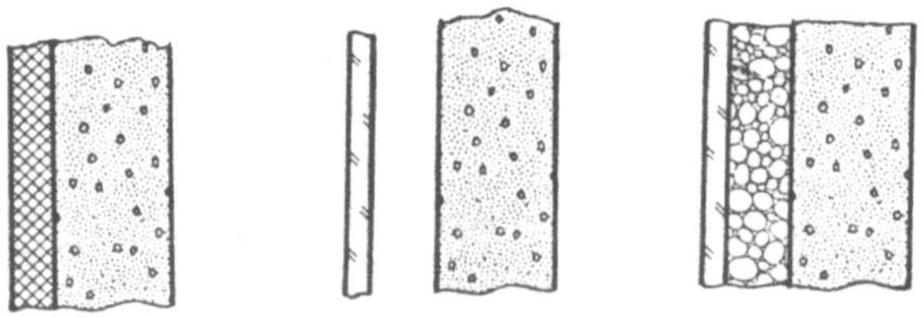

a) konventionelle Wärmedämmung (Thermohaut) b) Glasabdeckung (Trombe-Wand) c) lichtdurchlässige Wärmedämmung

Bild 2: Schematische Darstellung der bei der numerischen Simulation verwendeten Aussenwandkonstruktionen

Dämmsystem	Gesamttransmissionswärmestrom durch die Innenoberfläche der Aussenwand im Januar 1980					
	Beton		Ziegel		Poroton	
	kW/m²	%	kW/m²	%	kW/m²	%
konventionelle Wärmedämmung	-16.4	-100	-13.0	-79	-10.1	-62
Glasabdeckung	-19.0	-116	-11.5	-70	- 7.3	-45
lichtdurchlässige Wärmedämmung	+11.8	+ 68	+ 9.1	+55	+ 7.3	+45

Tabelle III: Ergebnisse der numerischen Simulation verschiedener süd-orientierter Aussenwandkonstruktionen

2.2 Praktische Untersuchungen an unbeheizten Versuchszellen im natürlichen Aussenklima

Auf dem Freigelände unseres Institutes in Stuttgart wurden zwei Fertiggaragen als Testzellen für erste praktische Grundsatzuntersuchungen installiert. Eine Testzelle wurde als Referenzzelle mit einer konventionellen Dämmschicht versehen, die andere mit einer lichtdurchlässigen Wärmedämmschicht etwa gleichen Wärmedurchlasswiderstandes.

In der Heizperiode 82/83 wurden an diesen Messzellen in jeweils unbeheiztem Zustand Parameterstudien durchgeführt. Das verwendete lichtdurchlässige Dämmmaterial war Acrylglasschaum. Die Montage der Dämmschicht erfolgte teilweise direkt auf der 50 mm dicken Beton- und Gasbetonwand und teilweise mit unterschiedlichen Abständen. Als Witterungsschutz wurde Plexiglas verwendet. Der Luftspalt zwischen Dämmschicht und Wand war zeitweilig verschlossen.

In Bild 3 sind die Raumlufttemperaturen der unbeheizten Messzellen während einiger Wintertage und in Bild 4 für einige Sommertage angegeben. Bei Aussenlufttemperaturen zwischen ca. - 5°C und + 12°C stellen sich in der unbeheizten Messzelle mit der lichtdurchlässigen Wärmedämmung Raumtemperaturen bis über 20°C ein. Die Aufzeichnungen aus den Sommertagen zeigen, dass zwar eine Verschattung erforderlich ist, dass bei offener Hinterlüftung die Überhitzungsprobleme jedoch geringer als erwartet ausfallen.

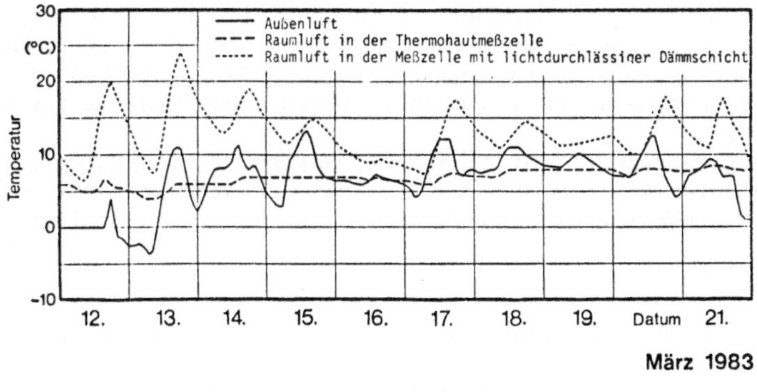

Bild 3: Vergleich der Raumlufttemperaturen in den unbeheizten Messzellen während einiger Tage in der Heizperiode 1982/83.

Bild 4: Vergleich der Raumlufttemperaturen der beiden Messzellen während einiger Sommertage 1983

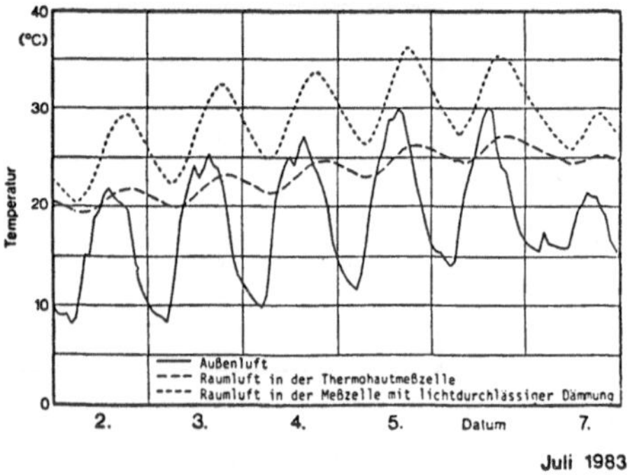

2.3 Praktische Untersuchungen an beheizten Versuchszellen im natürlichen Aussenklima

Während der Heizperiode 83/84 werden die Testzellen auf eine konstante Raumtemperatur von 20°C beheizt. Die Montage der lichtdurchlässigen Dämmschicht wurde modifiziert. Als Witterungsschutz wurde Glas verwendet. Der Luftspalt zwischen Aussenwand und Dämmschicht ist gegen die Umgebung verschlossen. Zur witterungsgeschützten Integration von Verschattungsvorrichtungen ist zwischen Glasabdeckung und Dämmschicht ebenfalls ein Luftspalt. In Bild 5 sind relative Heizenergieverbräuche der Messzelle mit der lichtdurchlässigen Dämmung bezogen auf den Verbrauch der Referenzzelle für einen kurzen Zeitausschnitt angegeben.

Bild 5: Relativer Heizenergieverbrauch der Messzelle mit der lichtdurchlässigen Wärmedämmung im Vergleich mit der Referenzzelle

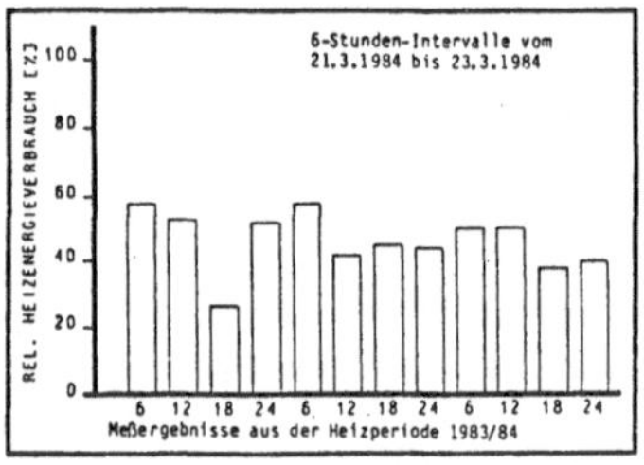

2.4 Ausblick

Die messtechnischen Untersuchungen mit lichtdurchlässigen Wärmedämmschichten bestätigen die theoretischen Untersuchungen weitgehend.

Für die Anwendung im Hochbau sind noch einige Probleme zu untersuchen und zu lösen. Zum Beispiel ist der Langzeiteinfluss der mechanischen Beanspruchung der Aussenwand infolge der starken Temperaturwechsel oder das brandschutztechnische Verhalten derzeit verfügbarer Materialien und anderes zu klären.

Das Energieeinsparpotential das die Anwendung lichtdurchlässiger Wärmedämmschichten bietet fordert verstärkte Entwicklungsanstrengungen. An Wirtschaftlichkeitsüberlegungen wird diese Entwicklung sicher nicht scheitern.

REFERENZEN:

1. FRANGOUDAKIS, A. Lichtdurchlässige Dämmschicht, SA 04/82 IBP Stuttgart
2. BOY, E., BERTSCH, K. und FRANGOUDAKIS, A. Lichtdurchlässige Dämmschicht, SA 04/83 IBP Stuttgart
3. BERTSCH, K., BOY, E. und FRANGOUDAKIS, A. Lichtdurchlässige Dämmschicht, SA 06/83 IBP Stuttgart

THE SWEDISH SOLAR HEATING PROGRAMME

E. ÖFVERHOLM
Swedish Council for Building Research

Summary

The paper covers the goals for, and the experience from the solar programme from 1976 and onwards.
The solar programme for space heating is managed by the Swedish Council for Building Research.Initially the programme covered a broad area of research and development.However a large amount of the resources where used for seasonal storage projects as seasonal storage of solar energy was believed to be the only possibility for large scale introduction of solar energy. Later on ,1979, a target oriented programme "SOLAR 85" was introduced with the main purpose of establishing a sound fact and information basis on which an energy policy decision concerning the introduction of solar heating could be taken by 1985.With this programme the funding was increased and secured until 1985.Both seasonal storage concepts and solar collector development have been succesful and energy costs for: domestic hot water in multifamily buildings,solar collector fields connected to small district heating networks, and solar seasonal storages are now almost competitive to oil fired plants. The "Solar 85" evaluation consists of two parts: the evaluation of research as well as planning of future research and identification of the solar potential in the future.
At the same time and in the same way heat pumps and enegy conservation is evaluated.

Background

The main goal of the for the swedish energy policy is to reduce the oil dependence. This has so far been achieved by energy conservation, nuclear energy introduction, use of other fuels and heat pumps. The oil reduction has been more rapid than expected. See fig.1. Due to Swedens northerly position, extending from latitude 56° N to 64° N, about 40% of the total consumtion is used for space heating purposes. The energy is supplied in a number of ways: individual oil fired burners, block centrals, district heating plants and individual electric heating units. District heating plants supply about 25% of the space heating load at present.
As a result of a referendum and parliament decision nuclear power is to be gradually phased out until the year 2010. Therefore, Swedens interest is now directed towards R&D on domestic energy sources.

The solar programme

Solar insolation amounts to about 1000 kWh/m^2,year on a horizontal surface, i.e. about the same as in central Europe. However the distribution of this energy throughout the year is less favourable than in many other countries.Irradiation on a horizontal surface in Stockholm is about 15 times less in December than in June, and in the far north the insolation is zero during some winter months. Further limitations on solar energy are imposed by the proportion of diffuse radiation, which amounts to about 50% of global radiation, and the number of days with alternating sun and cloud cover.

The Swedish solar energy programme was started in earnest in the middle of the 1970s. Initially the Swedish Council for Building Research concentrated on investigating the feasability of solar heating.In the national research and experimental programme , priority was given to work on solar heating systems that incorporated some form of seasonal heat storage capacity.
A target-oriented solar energy programme, known as the Solar 85 programme, was established at the end of the 1970s. Apart from R&D the programme contained a market goal of solar contribution by 1990 in the range of 1-3 TWh. The R&D part of this work is administered by the council.

Ojectives for the solar heating programme:

-component development:	Increased performance and reduced costs, better durability
-solar heated domestic hot water:	Development of low cost systems for multi-family housing units and monitoring of systems in detached houses.
-solar collectors in combination with district heating and block centrals:	Establishment of full scale systems and monitoring of systems already in operation
-basic research:	Development of insolation models, system models and durabilty test procedures,etc

The first stage of the programme was directed towards the development and evaluation of a variety of system concepts. During the second stage, from 1982 and onwards, a few interesting system types were developed towards better cost efficiency, see fig. 2. The programme is now in its third stage which also includes evaluation of the programme and the technology.

The market

Several small industries developed during 1974 - 1980.Most of them concentrated on domestic hot water systems. Production reached its peak in 1980, more than 20.000 m^2 of collectors sold. After that sales decreased rapidly. In all some 60.000 m^2 of collectors have been installed during the last ten years. See fig.3. Favourable subsidys have been offered to buyers of solar equipment. This year ,1984, 50% of the approved instalation cost is subsidized. The low price of electricity has been the mayor obstacle for solar introduction.
For solar energy in district heating systems the present prices of peat, coal, biomass and electricity summertime prohibits solar applications in the near future.

The programme evaluation and future of solar energy

A number of evaluation groups freestanding from the council have been following different parts of the programme. Apart from the evaluation they have contributed with input data for the solar 85 model. See fig.4.
The solar 85 model, see fig.5, has now been implemented. A tendency from the first runs is that oil is almost phased out of the energy system by 1990. Instead electricity is introduced. Solar energy and heat pumps can take a part of the market if they are subsidized: However these are very preliminary results and final runs of the model including sensitivity analysis might show another result.

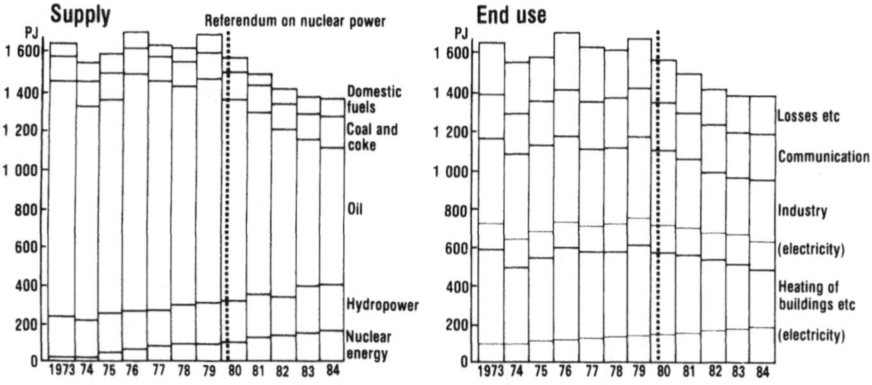

fig 1 Supply and the use of energy in Sweden 1973—1984

- Energy conservation has been succesful.
- Sweden's oil dependence has been drasticly reduced.
- Electricity from nuclear power has increased, and two more plants are yet to come.
- Some hydro power will also be added to the electric system
- Consumer prices for electric energy are among the lowest in Europe.
- Special price in summertime.
- Coal and biomass are being introduced in district heating systems.

It is tough for solar energy to compete.

Source: Ny Teknik 1984 no. 16

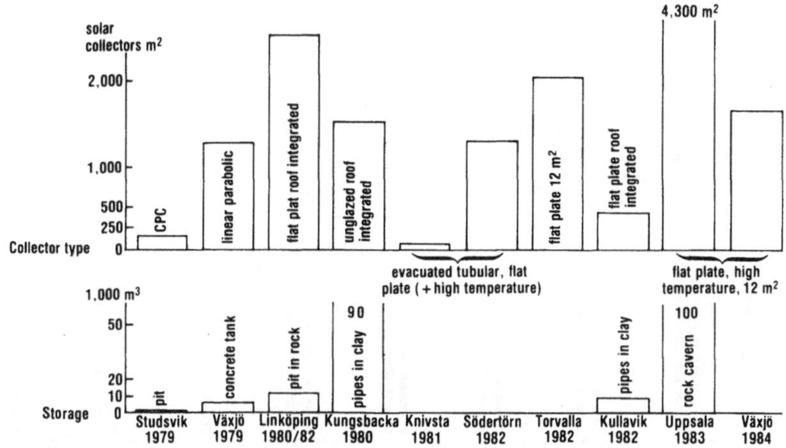

fig 2 Presentation of some seasonal storage and collector field projects

- Most types of collectors and seasonal storage have been tested in full scale plants.
- Experience so far:
  - — concentrating collectors require to much maintenance
  - — the new flat plate high temperature collectors outperform (cost/performance) other collectors in seasonal storage and district heating applications
  - — storage losses and storage costs can be increased drastically if the design is not made with utmost care
  - — store is very dependent on scale

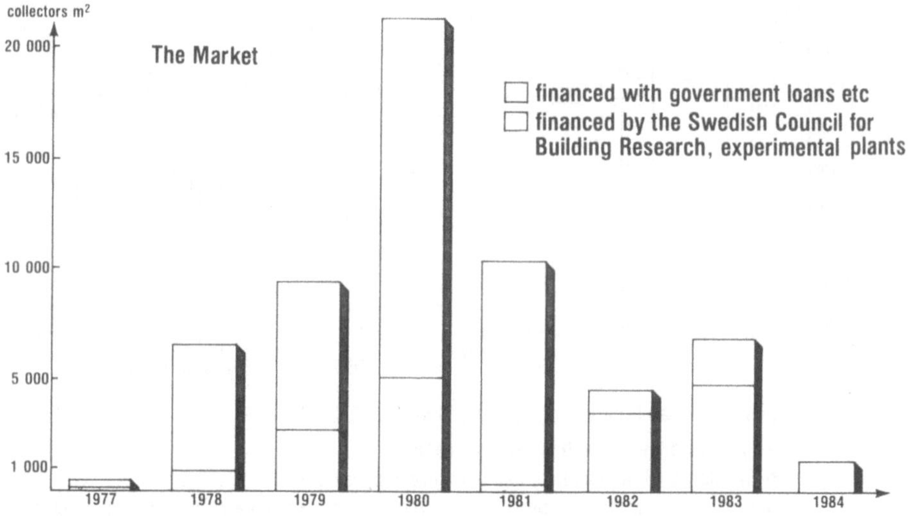

fig 3 The Market

- **The market reached its peak shortly after the second oil crisis and the referendum on nuclear energy.**
- **From 1982 and on many house-owners have changed from oil to electricity.**
- **Systems financed by the Council are mainly for seasonal storage, district heating and domestic hot water in multifamily houses.**

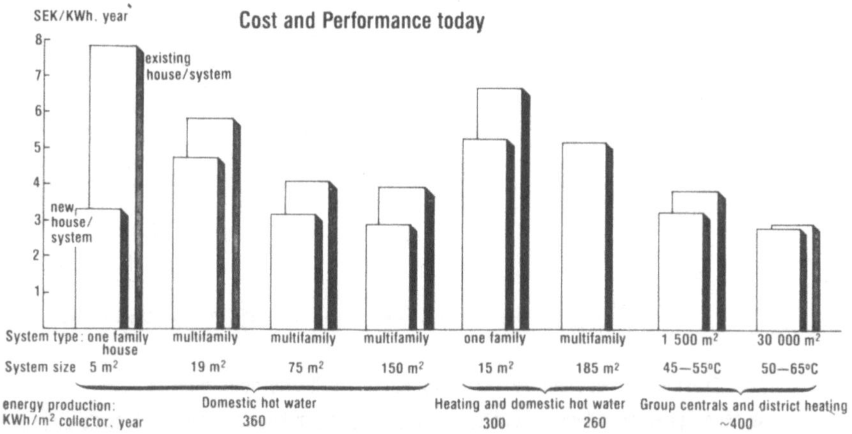

fig 4 Cost/performance for installed systems. (Investment cost per yearly produced KWh.).

- **These data are input to the solar 85 model, see fig 5.**
- **The data represent the best systems today.**
- **A significant decrease in cost and increase in performance has been experienced the last few years.**
- **Further improvements are expected, especially for large solar fields.**

The future

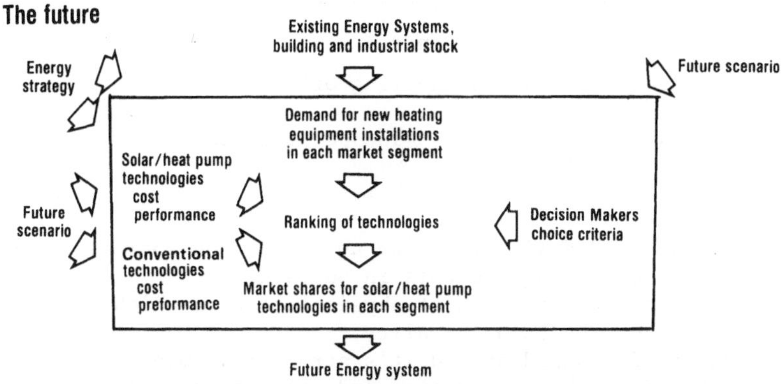

fig 5 The solar 85 model for prediction of solar storage and heat pump contribution to the Swedish solar energy system

- The model is basically a market model developed by RPA (Resource Planning Associates). — The best system wins (Ranking of technologies).
- The ,,market'' (the building stock) is divided into thousands of segments.
- The ,,competition'' between energy systems is performed within each market segment.
- The future is divided in two parts: the part which the government can control — strategies (taxes, district heating, incentives to home-owners etc) and the part which the government cannot control — scenarios (mainly international fuel prices and economic growth).
- Choice criteria for decision makers (house-owners) are: first cost, pay-back period and life-cycle cost.

INTEGRATED DESIGN

EXEMPLIFIED BY SOLAR CAVITY WALL HOUSES

J. Kristinsson

Architectural Engineering Office
Noordenbergsingel 10
7411 SE Deventer The Netherlands

Summary

"Integrated design" deals with (but not only) the optimalisation of all energy consuming elements in buildings.
It deals with a total complex of physical factors which have been transformed into a building reality.
The structure of a building and it's structure components need a thermal build-up, heattransport, and storage capabilities that are also desirable from a users point of view.
Hydrical behaviour, air-circulation and ventilation should be predictable. Only the lateral is known in the form of forced ventilation.
It is also possible to solve problems with acoustic, sound insulation and fire prevention in the design stage. The choise of daylight or electric lighting (and colouring) have a major effect on the appereance of a building. The human appriciation towards the architecture environment and psychological factors are to be considered.
Orientation to the sun, wind, traffic (noise) and planting are the last main points in this paper. An integrated disigned building is friendly to man and the human environment with a very long useful time of life. The "integrated design" will be exemplified by 6 experimental "solar-cavity-wall-houses" in Leiderdorp (western part of Holland) which were build at the end of 1983. They have an energy saving of about 85 to 90% related to normal Dutch building code.

1. INTRODUCTION

The designing of minimum energy buildings is still in an early stage. On one hand this is explaned because of the new generation of (non hygroscopical) insulating materials which have been recently introduced on the market in a time in which fossile fuel is very cheap; on the other hand because insulation doesn't fit in traditional building structures and therefore the insulating techniques can't be found in normal handbooks. By the dimensioning of the technical installation the minimum limits were often forgotten.
To project all oversized installation is easier and without risks. "Integrated design" sets new standards for environmental protection.

Summarizing we mention five secundary goals:

- Decrease heat-transmission
- Decrease ventilation losses
- Increase solar energy effect
- Use internal heat-production
- Minimize (heat supplation with), fossil energy forms

The projected long useful service-time of buildings is chained to a low fossile energy consumption for each building in it's own specific situation.

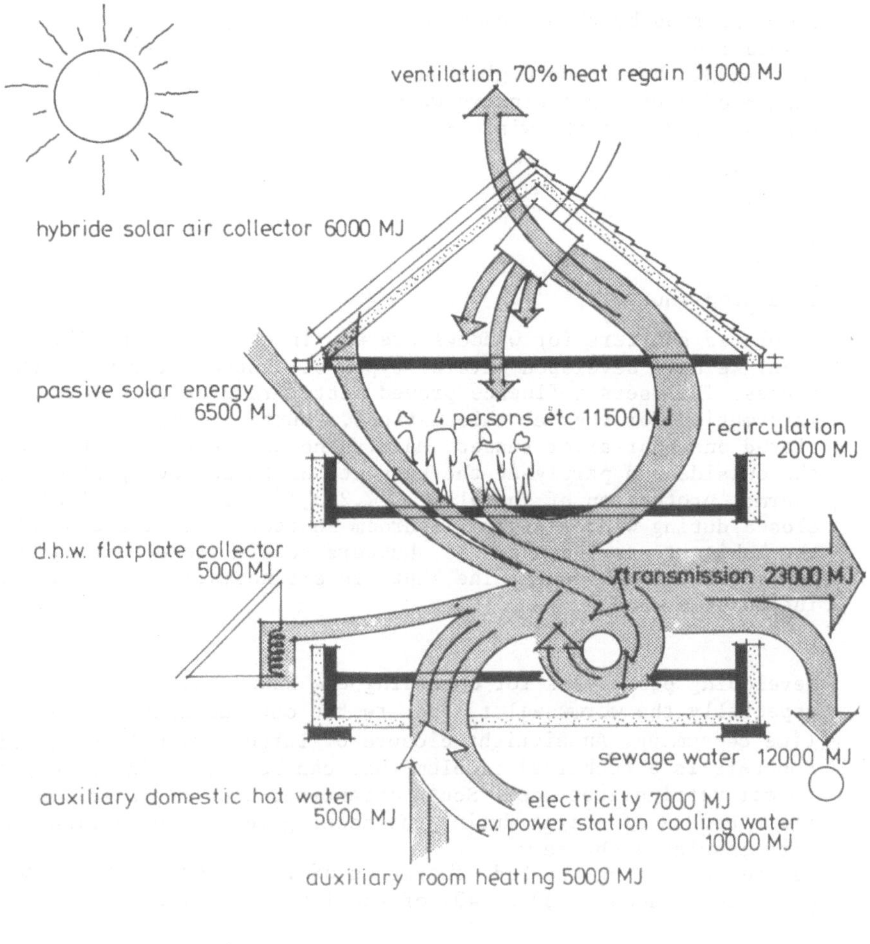

,thermal balance'
integrated design of solar cavity wall houses

1.1 <u>Decrease heat-transmission.</u>

	λ	R	Dutch b. code
The thermal insulation is made all around the house. The outside walls are filled with 180 mm mineral wool	λ= 0,04	R= 4,85 m2^u/w	1.3
adding coldbridge free-ties it increases to app.	λ= 0,4	R= 5,25 m2^u/w	
One part of the roof has tiles, the other half the build-in collector.			
a) tile-roof 150 mm mineral wool	λ= 0,04	R= 4,15	1.3
b) collector + 150 mm mineral wool		R= 4,85	
groundfloor un-insulated		R= 0,6	1.3
crawl-space is insulated with 100 mm polystyreen	λ= 0,03	R= 3,35	
The structural walls and groundfloor are supported by 65 mm foamglas insulation	λ= 0,05	R= 1,15	-
The windows have double glazing		R= 0,31	
Insulated shutters + windows which are used 2/3rd of the winter-time			0.25
- R theoretical		R= 1,5	
- R practical			

Insulated shutters.

Insulated shutters for windows are new in the dutch building tradition. We have developed several types of shutters for a few hundred houses. The users influence proved better results than were used as information for thermal calculations. Shutters: open 7.30 a.m. and closed one hour after sunset. Mainly the shutters are situated at the outside and partly at the subsections to achieve part wise thermal protection of the glas. 3/4,2/3,1/3 or 1/4 of 1/1 can be closed during winter-nights. Bedroom shutters are closed by all of the habitants an 3/4rd of the shutters remain closed during the day in december and january. The shutters are normally operated from inside by a handle.

Developing prototypes for operating the inner shutters and especially the wormwheel to operate the outside shutters, has been time consuming. An airtight closure of large (1 m2) insulated shutters is a technical problem that can be solved in the way of a normal refrigerator-door. South orientated double glazed windows are a source of energy during the heating season in combination with insulated shutters.
We are developing the 2nd generation of insulated shutters and the price is reduced to 30 or 40% of the 1st generation.

	transmission double-glazing	solar gains	netto gains	transmission with shutters	effect of shutters	double south-glazing with shutters
jan.	- 44	+ 23	- 21	-22	+ 22	+ 1
febr.	- 42	+ 38	- 4	-21	+ 21	+ 17
mar.	- 36	+ 49	+ 13	-17	+ 19	+ 32
apr.	- 28	+ 44	+ 16	-14	+ 14	+ 30
oct.	- 22	+ 49	+ 27	-10	+ 12	+ 39
nov.	- 33	+ 23	- 10	-17	+ 16	+ 6
dec.	- 38	+ 14	- 24	-18	+ 20	- 4
	-243	+ 240	- 3	- 119	+124	+ 121 kWh/m² jaar

- Passive solar energy gain - by using insulated shutters for double-glased south-windows at night - kWh/m^2 y -

Foundation insulation.

The bearing foundation insulation is developed by the manufacturer. It consists of a heavy load bearing layer of foamglas. The expectation towards the implementation of this new building material is that it will be introduced in the market at the end of 1984 for export to northern countries.

1.2 Decrease ventilation losses.

Optimal ventilation means four different things. First in relation to health, second humidity and third energy conservation. When we add to it heating capabilities for houses and buildings then things get complicated. There is a lack of experience and scientific documentation. The air-tightness of well insulated building is more important for energy conservation, then an increased thermal resistance. As a consequence of the achieved air-tightness the indoor-climate is not good enough to maintain a healthy atmosfer. We introduce forced ventilation to tackel this problem. The system consists of a unit for balanced ventilation, heat-regain, space-heating and the necessary pipe-work. The air-to-air heat-exchanger is a compact aluminium cross-plate-exchanger with an thermal efficiency of about 70%.
Warm, humid and used air is sucked out and passes the cold-fresh-air supply in the aluminium crosscurrent heat-exchanger.
Here the air-currents cross and heat from the exhaust-air is transferred to the supply-air.
The preheated fresh air is heated up (if needed) and blown in as ventilation-space heating-air with a capacity of 1.5 - 5 kW/h.
The ventilation/space-heating system has a very low sound level, also because the total air-volume doesn't exceed 80 - 240 m3/h.
In this way all rooms are ventilated and heated.

1.3 Increase solar effects.

The use of solar energy has three aspects in these houses: passive, hybride active (solar-cavity-wall) and domestic hot water.

. passive:
The passive solar energy, in which the influence of the greenhouse is not included, has transferred into a major source of energy because of the insulated shutters.
The windows on the south are not very large. The northern windows however are fairly big, but this has no large influence. The result is not that large because of the insulated shutters by which the bare glass surface can be reduced to 1/4th or even total closed.

. solar-cavity-wall:
The solar cavity-wall air-heating-system is the most technical advanced part of the houses and it gives the name to the experiment. After one (real) winter the Delft Technological University will publish the results of the tests. As a hybride part a small fan is installated in the crawl-space. The main characteristic is a direct solar-heating-system in combination with an indirect heating-system. The houses are equipped with integral build-in solar-collectors. The gained heat is transported to the internal heavy parts of the builded environment.
The heat distribution is purely achieved by convection and a slow heat-transmission goes through the groundfloor and the walls.
The main source of space-heating is heat-radiation.
The building envelope acts as an indirect heating-system. It's an active air-solar-collector-system which charches itself with the solar preheated air and flows through the collector, ducts, crawl-space, the tie-free cavity-wall (= solar-cavity-wall) and it's return ducts. All of it is controlled by a microcomputersystem.

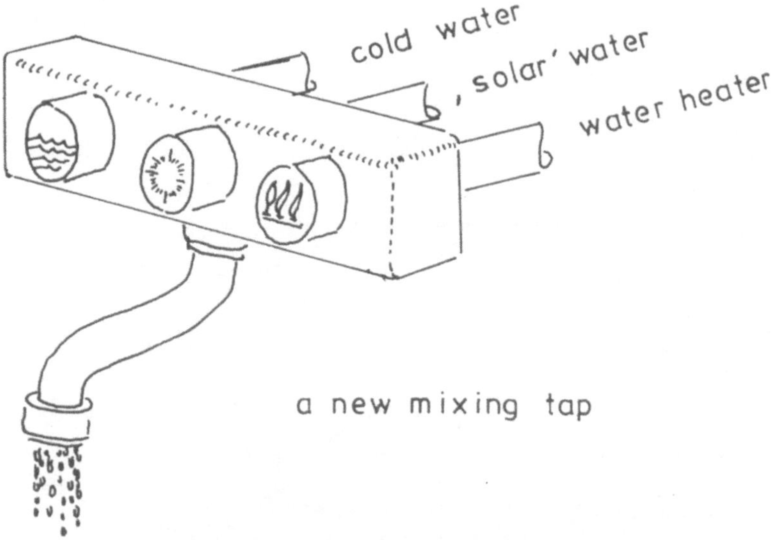

a new mixing tap

domestic hot water:
The third low caloric aspect is the domestic-hot-water. There has been many publications in this field so we add only few lines. Domestic hot water is (pre)heated in a steel-flatplate-collector which is integrated in the build-in solar-air-collector.
From the outside it's invisible.
Notwithstanding this rather unusual solution the bare collector works almost as normal. The short-term-storage is placed in the kitchen near to the tap.

1.4 Use internal heat-production

Because of the excellent insulation and a controled ventilation the internal heat-production can play a major roll.
Almost 3000 kWh/year is produced by a four-person household. This is almost the same amount as the total ventilation losses.
By an average temperature in the winter (of 4°C) the auxiliary heating limit is tremendously affected by the internal heat-production which varies between 500 and 1500 W.
The internal heat-production will decrease relatively a lot in the next 20 years. Nowadays the hot water production, cooking, washing, refrigerator and artificial light will exceed human heat-production of almost 100 w/pp 2 to 3 times.

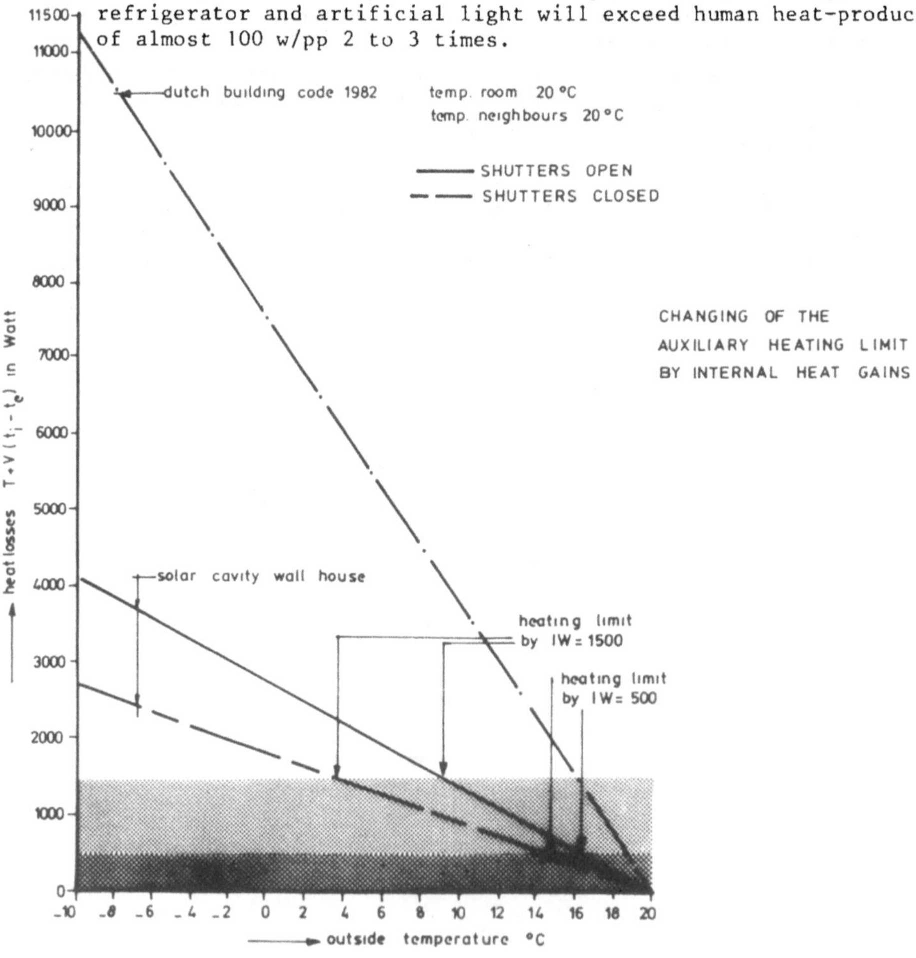

1.5 Minimize fossil energy forms.

By all mentioned measures the energy consumption is already very low but not yet at it's optimum.
The natural gasstove for domestic hot water and auxilery additional space-heating, is stand-by all year around. It is relatively inefficient in minimum energy houses.
The pilot flame in a boiler or a stove uses 175 and 225 m3 natural gas/year or 1600 and 2000 kWh/year.
The developement of a "Twin" boiler, for domestic hot water and space-heating, joined with a ventilation-unit has reduced this energy loss to zero because of it's electrical ignation. The fan for the solar cavity-wall must be "tempered" to use a minimal of electrical power. The optimal control of this fan and flap sections are evidently important for on optimal result of the hybride passive solar-heating-system. The selection of suitable lighting equipment (tube instead of bulbs) and light coloured walls and ceiling contribute to a decrease of electrical power consumption. The cooking equipment and cooking habits in most households are in extreem contrast with the latest development in hospital or institutional kitchens.

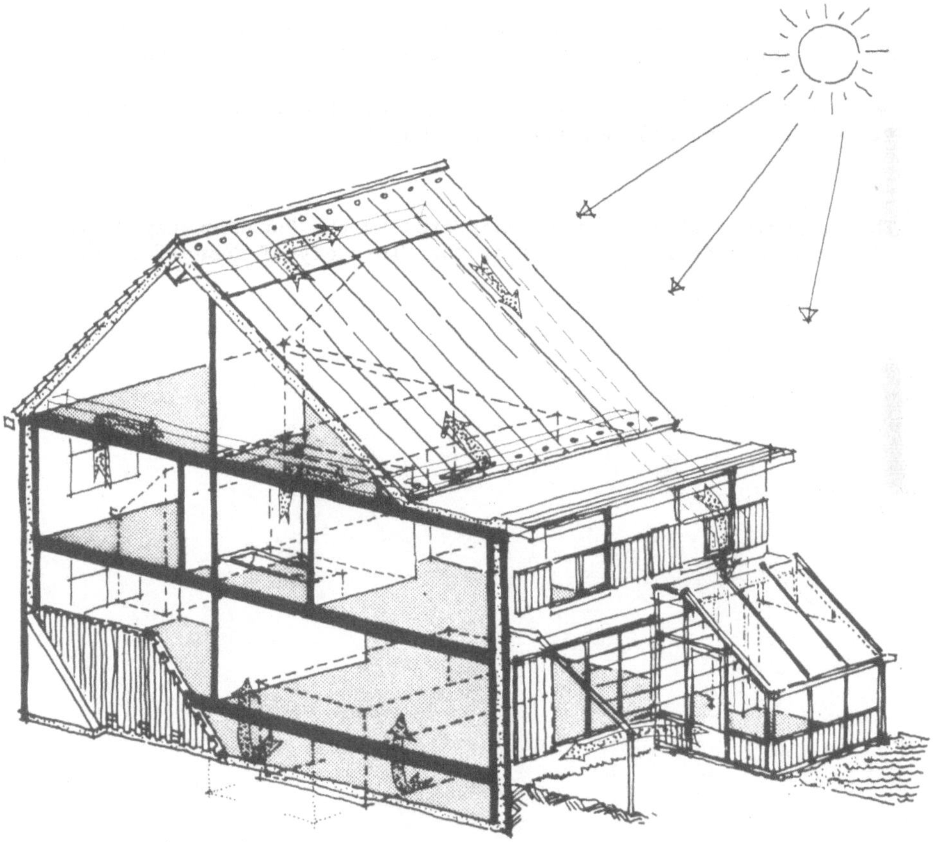

, system schematic '
solar cavity wall houses

CONCLUSION

For the time being we can conclude, that the solar-cavity-wall houses satisfy requirements.
The second and third generation of these houses are in a far stage of developing or are already in production.
New building materials are introduced because of these houses.
A further "spin-off" is the balanced ventilation Unit with a integrated heat-regain, space-heating etc. and the "Twin"-water heater without a pilot flame and equipped with an electronic ignition. This is also in production. If is was only for these two last reasons, we can consider the experiment to be a succes.

references:

Passive Solar Architecture in Europe 2 , edited by Ralph M. Lebens'82 the Commition of the European Communities, the Architectural Press.

Preformance characteristics of an indirect air based heating system connected to a solar collector, G.L.M. Augenbroe e.o. paper 2°World Solar Conference at Perth'83 , Delft University of Technology, Department of Civic Engineering, Building Physics Group,

Integraal ontwerpen,Zonnespouwwoningen te Leiderdorp, J.Kristinsson e.o. '83 Delft University of Technology, Department of Architecture.

De Architect monthly, 10th of March'83 , W. Schuringa, Ervaringen van adviseurs,. Uitgeverij ten Hagen, the Hague .

- This experimental project is supported by the Dutch Energy Projects
- Office at Apeldoorn (PBE-TNO) of the Ministery of Economical Affairs.

EXPERIMENTAL STUDY OF FREE CONVECTION IN A ROOM MODEL USING A COLLECTOR STORAGE WALL

P. CALVET and P. MILLAN
ONERA-CERT : Département de Mécanique et Energétique des Systèmes
2, Avenue Edouard Belin B.P. 4025 - 31055 TOULOUSE CEDEX
FRANCE

Summary

A small size room model using a Trombe wall system has been designed to make a qualitative as well as quantitative approach of convection and heat transfer aspects. The air flow is supposed to be under steady state and laminar conditions. From the following experimental measurements obtained in the model (i.e).
- velocity profiles mesured by Doppler laser Anemometer
- temperature distributions.

Different results were deduced, such as :
- 1) air flow motion inside the room
- 2) influence of each wall on the thermal balance
- 3) convective heat transfer inside the system.

Coupling between the room and the Trombe wall structure is therefore caracterized. Taking these results into account, we developped a modelisation based on a network model and implemented it on a microcomputer.

1. INTRODUCTION

In order to save energy and improve confort in solar buildings, it becomes necessary to study with the utmost care, natural convective heat transfers in enclosures.
The thermal interaction between buildings and the passive solar systems they contain (Trombe walls, grenhouse,...) has rarely been analysed.
This paper presents an experimental study of free convection in a room model using a collector storage wall.
Convective heat transfer can be divided into two different aspects : one, in the heated space, and the otherone in the room.
Natural convection phenomenous which are generated in the heated space can be considered as the flow between two vertical planes; theorical and experimental work has been carried out this field, as we can see from references 1, 2.
On the contrary, in the room we are confronted with two kinds of convection which act on each other; they can be described as follows :
- on one hand, the convective heat transfer which takes place in small aspect ratio room (A = H/L = 1).
- on the other hand, the convective motion crated by warm air blowing in a cavity, which interferes with previous one.

It is very difficult to solve such a problem, - i.e : to predict thermal performances of these enclosures -, unless hypotheses are made to simplify it. For lower values of Ra number, studies performed on simple geometric configurations, enables us to point out the existence of two regimes in the laminar zone of the free convection flow ; i.e. the convection one and the

boundary layer one.
In the convective regime, heat exchanges are mainly conductive and the flow is steady. As regards the boundary layer regime, a vertical gradient of temperature can be observed inside the room ; moreover, flow velocity by the wall is high by comparison with the core of the cavity.
For Rayleigh numbers, greater than 10^7, computer programs, which solve the full Navier Stockes equations, have been developped, using a K-E numerical scheme (3, 4).
Little fundamental research has been undertaken to study dynamic processes of natural convection in a building. In this work, a laser anemometer has been used to measure air velocities in the free convection flow.

2. HYPOTHESES OF THE STUDY (5)

The aim of this research is to predict from the dynamic field, flow rates distributions and air motion inside the room, and to determine the thermal field in order to quantify the influence of each wall on the thermal balance of the system.
However, such a work, can only be considered as an approach of the study of convection, because similitude is not respected, since we choose a simple geometric configuration and laminar air flow.

3. EXPERIMENTAL APPARATUS

The model used in these experiments in entirely transparent. The thermoconvective channel is 0,8 cm thick. The wall is a parrallelepipedic radiator made of brass (27 x 30 x 1 centimenters). The side which represent the wall exposed to solar radiation is four millimeters thick. In order to obtain a uniform temperature at the surface this temperature is induced by circulation of hot water, and controlled by thermocouple wires.(Cr. Al). (the cavity is insulated from the radiator). The room is a cube of 30 cm. This very little size gives a laminar boundary layer flow (Ref. 6). The walls temperature of the model can vary freely and their surface temperature is known from 27 thermocouple wires. The whole system has a vertical plane of symmetry.

To analyse the dynamic field in the two parts of the model, the laser anemometer works with back scattering light. The laser is a He Ne laser of 15 mW. Several preliminary experiments showed that the privileged component of the flow is the only one that can be measured because the anemometer only detects velocities greater than 4 mm/s. For instance : in the heated channel, we measure the vertical component and in the cavity, the boundary layer flow. In heated channel the thermal distribution is given by thermocouple wire set up on micrometric support. In the cavity, 7 thermocouples wires indicate the air temperature.

4. EXPERIMENTAL RESULTS AND THEIR CONSEQUENCES

The dynamic and thermal fields have been analysed in the heated channel and in the cavity for three different values of the surface temperatures. They are represented by :

$\Delta T_1 = 18°$ C, $\Delta T_2 = 28°$ C and $\Delta T_3 = 38°$ C with $T_i = T_p - T_a$

where : T_a is the ambiant temperature, T_p is the surface temperature of the wall.

4.1. The Trombe wall channel

A) Experimental results

The profile is fully developped and has a parabolic shape which tends to prove an important interaction of the developped boundary layers on the glasses and the warm wall. The evolution of temperature distribution along the channel (Ref.5) does n't show any flat zone. The different thermal boundary layers are overlapped.

B) Interpretation and discussion

The mass flow is calculated from velocity and temperature measurement in the different sections, this flow is conserved all along the channel within 3%. This result can be obtained in spite of the small size of the channel, thanks to laser anemometer, which is the best means of measurement. The density flux gives an idea of the evolution of local convection coefficient on the wall of the thermoconvective channel. The results are presented as a correlation using Nusselt and Rayleigh numbers. The reference temperature which is chosen is the mean temperature in the inlet channel. Our experimental results are summarized in the correlation

$$Nu\ (x) = 0,397\ \left(Ra\ (x)\ \frac{e}{x}\right)^{0,28}$$

This result is in good agreement with these obtained by different authors (6)

4.2. Flow analysis in the cavity

A) Experimental results

The evolution of the velocity profiles shows the formation of the recirculation zone and an horizontal boundary layer on the floor. We can deduce too, that the jet flow is attenuated when entrering in the cavity. The thermal field points out the existence of a vertical temperature gradient, which decreases along y axis and increases with a Trombe wall temperature. The tridimensional aspect of the flow can be deduced from horizontal and vertical velocity profiles. The aspect of the thermal field is different from that of the dynamic one, it's bidimensionnal.

B) Interpretation of the results

The air flow on each boundary layer developped on the vertical walls is computed in several horizontal sections. From this calculation, we deduce that the air flow has similar values in each of the three boundary layers. From the integration of velocity profiles, Streamlines can be drawn. (Fig. 1). We notice the formation of the recirculation zone under the ceiling and the influence of the warm air jet on the shape of inner flow. On fig 2 which represents the isothermal field, we observe that the isotherms are very close to each other under the ceiling and the upperpart of Northwall that is to say just by the impact point of the warm air jet. We notice too the warm air zone due to the recirculation area, and the vertical temperature gradient. The gradient increases from 5° C for ΔT_1 to 12° C for ΔT_3. Then, we remark the influence of thermocirculation on the thermal field, more particulary by the walls. Nevertheless, the vertical temperature gradient still exists, and so the bidimensionnal aspect in the flow core, which has been noticed by different authors. The convective heat transfer for each wall can be evaluated from temperature distributions. Flux density values show the importance of the heat transfer through the ceiling. The south facing wall has little influence on the thermal balance. The amount of heat wasted in the room has been compared to the one, driven by thermoconvection. The margin of error on the thermal balance between the Trombe wall and the room is lower than 10%. This result seems to be sufficient, if we take into account the low thicknesses of the boundary layers on the walls. The average heat transfer on the walls of the room is correctly represented by the Nusselt Nu. The reference temperature is

$$T_{ref} = \frac{T s + Tp}{2}$$

where Ts is average temperature in the outlet channel and Tp is average temperature on the wall.
Our experimental results for vertical walls (East, North, West) are in good agreement with those obtained by different authors (Fig. 3).

5. NUMERICAL APPROACH

On this part of the study we intend to compare our experimental results with numerical ones obtained with a simplified model (7) which is often used to predict thermal performances of solar buildings. The model is based on an electrical analogy ; radiative heat transfers are taken into account. The result's comparison indicates how important is a choice of a correlation proper to each kind of flow ; for instance under the ceiling, where warm air jet has a predominant influence.

6. CONCLUSION

This study shows it is possible to have a good insight into this kind of flow, and its dynamic processes, thanks to laser anemometer. From all the results, the following ones are noticiable :
- the caracteristic zones in the flow have been pointed out such as :
  . The core of flow showing a vertical gradient of temperature
  . The dominant flow by the side
  . The upper recirculation zone, conditionned by the influence of the heated space on the room, which is not negligible in the thermal balance.
- Good agreement has be found between some of our results and those presented by other authors in the same range of Rayleigh number. The comparison between our experimental results and those found for a simplified model, shows that precise definition of caracteristic zone, knowledge of the main dynamic processes and heat transfer coefficients, for each kind of flow, are of the ut most interest as regards prediction of the thermal balance of the system.

The future developement of this work should enable is to deal with systems more similar to real buildings, and to take into account conditions of similitude to caracterize each system by thermal heat transfer equations.

NOMENCLATURE

e	: thickness of channel	Nu(x)	: Nusselt number
H	: heigh of channel	Nu	: Average Nusselt
g	: gravity	Ra (x)	: Rayleigh number
L	: size of the cavity	Ra_L, Ra_H	: Average Rayleigh

REFERENCES

1 - AKBÅRI et BORGERS : Proce. 4th Nat. Passive solar conf. Kansas City 79.
2 - OSTRACK S., P.R. THORNTOM : Trans. ASME 80. 1958.
3 - FRAIKIN & al : ASME/AICHE Nat. Heat Trans. Conferences Orlando. 1980.
4 - GALDIL A. : Univ. of California BERKELEY Ph. D Thesis. 1980.
5 - MILLAN P. : Thèse de Doc. de Spécialité. Mars 1983. Toulouse

6 - JALURIA Y. : H.M.T. Vol. 5. Pergamon Press. 1980.
7 - BALCOMB J.D., HEDSTROM J.C. and Mc. FARLAND R.D. : LASL - NM 87545.
8 - ELDER J.W.: Journal of fluid mechanics. Vol 11. 1958.
9 - POOTS G. : Quart. J. Mech. Appl. Math. Vol 11. 1958.
10 - JANOT M. et MAZEAS C. : Int. Jour. of heat and Mass transfert.Vol.16. 1973.

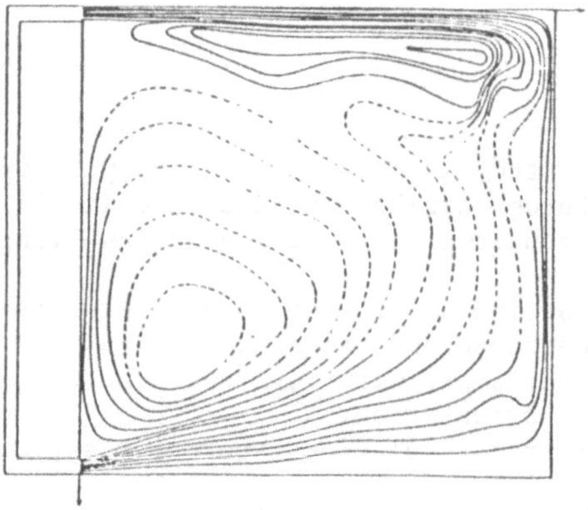

Figure 1 - Streamlines
(plane of symmetry)
$\Delta T = \Delta T_3$

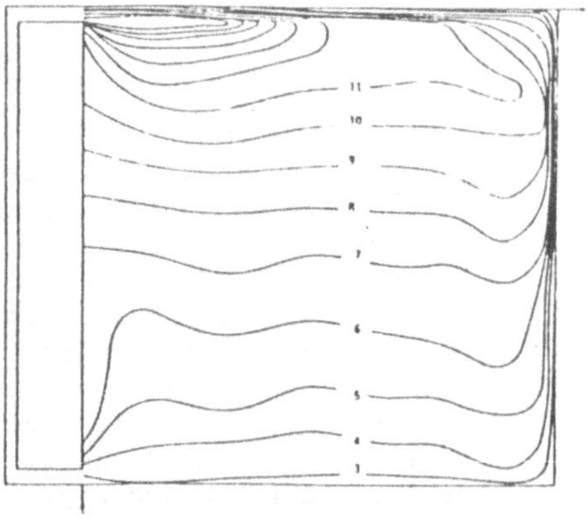

Figure 2 : Isotherms
(plane of symmetry)
$\Delta T = \Delta T_3$

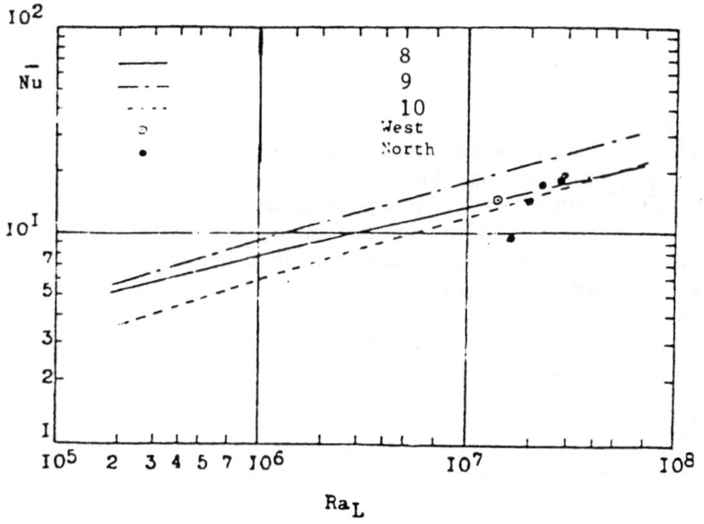

Figure 3 - Correlation between $\overline{Nu}$ and Ra_L

PERFORMANCE OF PASSIVE SOLAR SYSTEMS IN DUTCH LOW-RISE HOUSING

J.J. Habets
Department of Architecture and Urban Design
M.H. de Wit
Department of Physical Aspects of the Built Environment

Faculty of Architecture, Building and Planning
Eindhoven University of Technology
P.O. box 513 5600 MB EINDHOVEN
The Netherlands

Summary

Purpose of the work: The aim of the study is to determine the performance of passive solar systems in Dutch mass-housing under realistic conditions with a view to shading, density, urban tissue and dwelling lay-out.
Method of approach: The following passive solar systems have been taken into consideration: window, greenhouse and trombe wall.
The variables in the investigated models are with regard to the urban lay-out: block, strip-like block and orientation.
Concerning the houses the variables are: insulation, thermal mass, air-changes, V/A ratio and % glazing area.
Significant results: The results show that passive solar systems give only a small contribution to the heat need of a dwelling. Insulation and reduction of ventilation losses are more important to reach low-energy housing.
Urban lay-out and orientation don't have much influence. Prevention of shading is important in winter.
Conclusion: Passive solar systems will only have a modest place in future energy supply of Dutch dwellings.

1. Introduction

The aim of the study is to determine the performance of some passive solar systems in Dutch housing. Up till now most of the computations have been made for ideal conditions. The application of passive solar systems seems to have great influence on the design of houses and urban tissues. A lot of architects have high expectations of the performance of passive solar systems. However application at a large scale will offer less ideal conditions. It is necessary make reliable estimations of the performance of passive solar systems under realistic conditions with a view to urban lay-out, dwelling design etc..

2. Relevant data

The energy consumption for space heating has been calculated for a lot of schemes of low-rise housing projects. The following variables have been taken into consideration:

- Dwellingtypes: Two standard Dutch dwellingtypes: "Z-living"house and "Garden-living"house. Each dwellingtype has the same program, floor area, volume, etc. The lay-outs are showed in figure I.
- Urban lay-out: The SAR tissue method has been used to formulate tissue-

models for different urban lay-outs. These models relate buildings, green elements and urban space in a systematic way. Sections of the models can be seen in figure II. The two models are "block",with normal streets and with narrow streets, and "strip-like block".

- Insulation: Three classes of insulation have been taken into consideration, see figure III. Class I conform to the Dutch standards.
- Room temperatures: Sleeping areas: 16°C, living areas 21°C during the day and in the evening, at night: 16°C.
- Airchange rate: Two classes, n = 0,65 and n = 1.
- Thermal mass: a heavy and a light construction type: concrete and woodframe.
- Passive solar systems: window, greenhouse and trombe wall.
  Glazing percentage: normal and enlarged at the garden side, see figure IV.

3. Results

A representative choice out of the results has been made, see figure V.

4. Concluding remarks

A good insulation and the reduction of heatlosses due to ventilation and infiltration are the most proper ways to reach energysavings in Dutch mass-housing projects.
The application of a greenhouse or a trombe wall only yields a small decrease of the auxiliary heating load.
In all investigated urban lay-outs and orientations low-energy houses can be designed.
A good size and orientation of windows might reduce the energy consumption a little, but don't have to high expectations.
A greenhouse is a nice architectural element but don't overestimate its energy benefits.

FIGURE I: Dwelling types

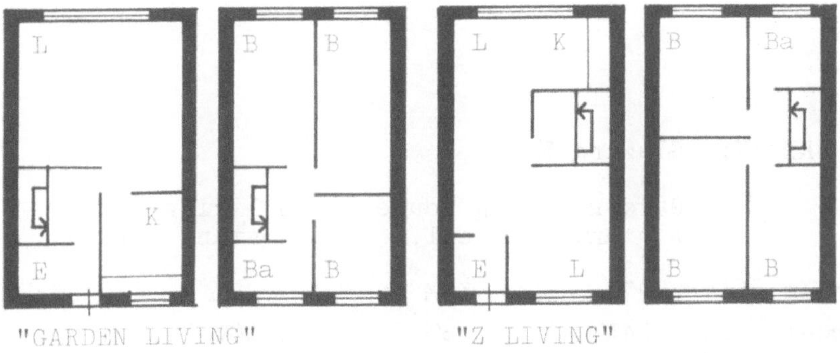

FIGURE II: Urban lay-out

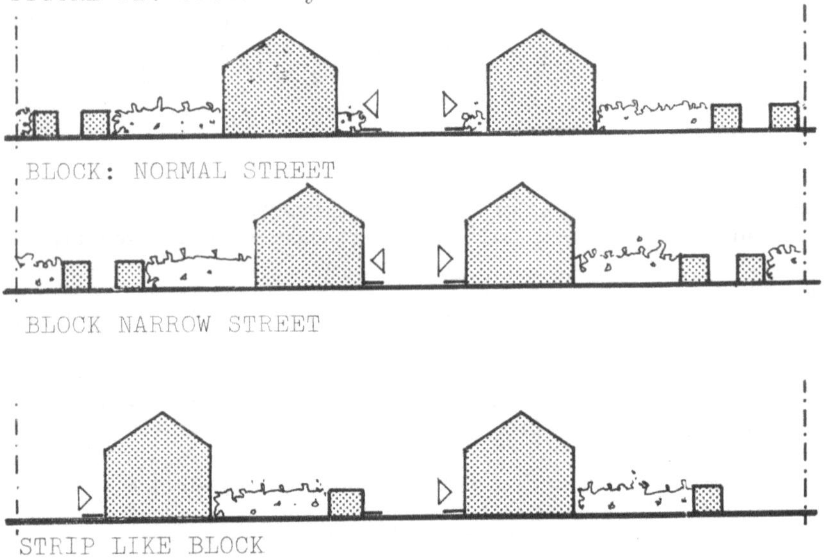

FIGURE III: U-values

	CLASS I	CLASS II	CLASS III
walls	0.68	0.4	0.3
ground floor	0.64	0.64	0.3
roof	0.68	0.4	0.3
windows	5.7/3.0	3.0	3.0/2.0
with shutters		0.7	0.7/0.65

FIGURE IV: Glazing %

	Garden living house A stan.	Garden living house B enl.	Z living house A stan.	Z living house B enl.
garden side	27%	39%	22%	32%
street side	14%	8%	18%	10%

FIGURE V: Selection of results

Z-living block normal streets n = 1		annual energy consumption	solar contribution
insulation I	N	12250	2300
	E	12775	1780
	S	12835	1720
	W	12775	1780
insulation II	N	9532	2145
	NE	9780	1890
	E	10010	1660
	SE	10130	1540
	S	10080	1540
insulation III	N	7760	1920
	E	8240	1440
	S	8240	1440
	W	8240	1440

orientation streetside
glazing normal

Z-living block normal streets insulation II		annual energy consumption	solar contribution
n = 1	N	9532	2145
	NE	9780	1890
	E	10010	1660
	SE	10130	1540
	S	10080	1540
n = 0.65	N	7200	2050
	NE	7550	1700
	E	7640	1610
	SE	7850	1400
	S	7795	1460

Z-living block normal streets insulation II		annual energy consumption	solar contribution
normal windows	N	9530	2145
	NE	9780	1890
	E	10010	1660
	S	10080	1540
enlarged	N	9380	2675
	NE	9585	2475
	E	10075	1980
	S	10415	1640
enlarged with shutters	N	7715	2640
	NE	7920	2450
	E	8410	1950
	S	8750	1620

Z-living block narrow streets enlarged windows insulation II	annual energy consumption	solar contribution
NE	9730	2325
N	9675	2380
Z-living block normal streets enlarged windows insulation II		
NE	9585	2475
N	9380	2675

Z-living block normal streets normal glazing insulation II	annual energy consumption	solar contribution
N	9532	2145
NE	9780	1890
E	10010	1660
SE	10130	1540
with greenhouse 30 m^3		
N	9152	2525
NE	9445	2225
E	9777	1900
SE	9965	1715
with trombe wall 12m^2		
N	8990	2685
NE	9300	2370
E	9680	2000
SE	9890	1790

THE POTENTIAL FOR SOLAR HEATING IN CENTRAL URBAN AREAS: A CASE STUDY

L.J. Matthews
Martin Centre for Architectural and Urban Studies
Cambridge University, 6 Chaucer Road, Cambridge
CB2 2EB England

Summary

Unlike 'idealised' solar sites, cities and towns are characterised by a wide range of building types, uses and forms in close proximity to each other, particularly at their functional centres. Solar assessment in these areas is complicated by high densities, orientation restrictions, and overshadowing. Two general spatial and temporal indicators of solar heating potential are addressed in this paper before a case study computer simulation is described. General indicators draw on the relation between solar attitude, vulnerability to overshadowing, and solar gain through glazing. A further measure is postulated through an analysis of heating regimes and the intensity of solar gain as functions of time. Results of the detailed simulation confirm spatial and temporal indicators, and allow relative solar space heating potential throughout the case study to be assessed at a glance.

1. INTRODUCTION

Throughout Europe, the vast majority of building services energy consumption occurs within urban areas - principally due to their functions as centres of commercial and industrial activity and residential locations. However, surprisingly little is known about the ways in which solar energy may be utilised in such areas, despite the fact that occupied urban buildings have large heating loads which are particularly suited to a low-grade energy source.

Solar heating potential has been well researched and developed for spatially-isolated or suburban contexts, where collected energy can satisfy a considerable proportion of space and water heating requirements. In general terms, the solar contribution depends both on the duration and magnitude of available solar energy, and on the efficacy of the collection, storage, and use of that energy in relation to the heating regime.

An assessment of solar contribution to buildings in existing 'central urban areas' (specifically those characterised by a mixture of uses in close proximity to one another) is further complicated by large numbers of inherent obstructions and a complex range of heating regimes. Steady-state heat loss analysis of central urban buildings also reveals a smaller surface area-to-volume ratio than for typical residential buildings, highlighting the influence of glazing in both heat loss and heat gain relative to that from the rest of the building fabric (1).

This study therefore addresses influences on the availability of solar gain through glazing as indicators of solar heating potential, with specific reference to problems encountered in high density, mixed use development. Two general measures of solar availability are investigated before a more detailed computer simulation of a case study is described.

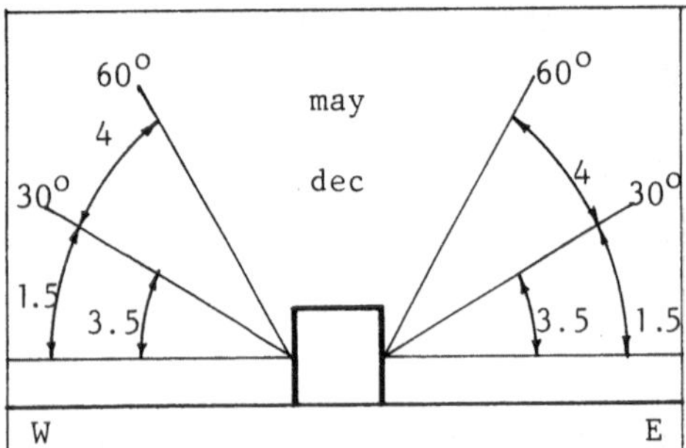

Figure I: Solar Altitudes (W&E)

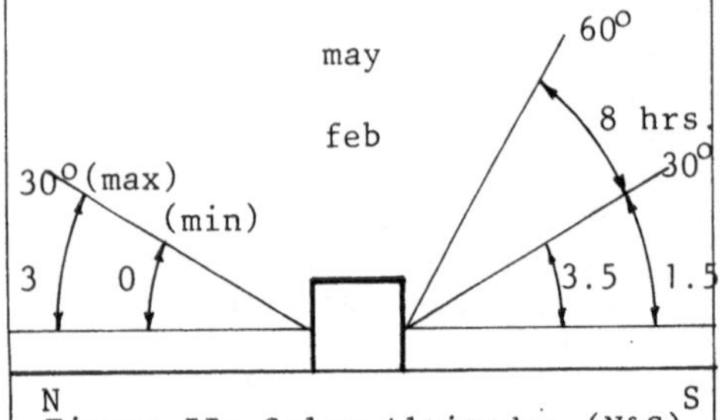

Figure II: Solar Altitudes (N&S)

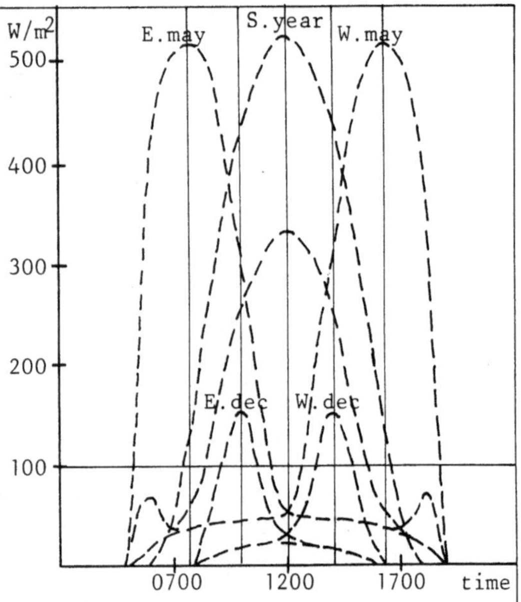

Figure III: Solar Gain for glazing.

2. TEMPORAL AND SPATIAL INDICATORS

The magnitude of clear-sky solar irradiation incident on an unobstructed building facade varies with time according to the position of the sun and the latitude and orientation of the facade. The proportion of incident irradiation passing through glazing then varies with the levels of direct and diffuse components, the angle of incidence of the direct component, and the thermal conductance, absorption and temperature gradient across the glass.

Figures I and II show the suns attitude (isolated from azimuth for clarity) relative to each cardinal orientation of a building at 52° north latitude. The east and west facades receive most of their unobstructed direct solar irradiation at a low vertical angle of incidence, resulting in a high solar gain through glazing (up to 85%) at these times as manifested in Figure III. In fact, measured data from Kew (1967) show that over the heating season aggregated global irrodeation on east and west facades exceeds that of the north and south by almost half as much again. Gain from low solar attitudes is, however, especially vulnerable to overshadowing from nearby buildings in a central urban context.

Although the south-facing facade (Figure II) has a disadvantageous predominant angle of incidence over the heating season, it benefits from a longer duration of direct radiation and a decreased risk of overshadowing at the times of maximum solar intensity than either east or west orientations. The intensity of global radiation from higher solar altitudes is also considerably greater owing to the reduced effects of atmospheric attenuation, producing a maximum at around 1200 hours G.M.T.

Figure III shows the maxima and minima solar gain intensities for each orientation over the heating season. Thermal transmission caused by the inside/outside temperature differential is ignored in this representation because of the wide range of internal air temperatures encountered in mixed use buildings. It may be seen that whereas the absolute duration of incident global radiation is similar for each orientation, the temporal

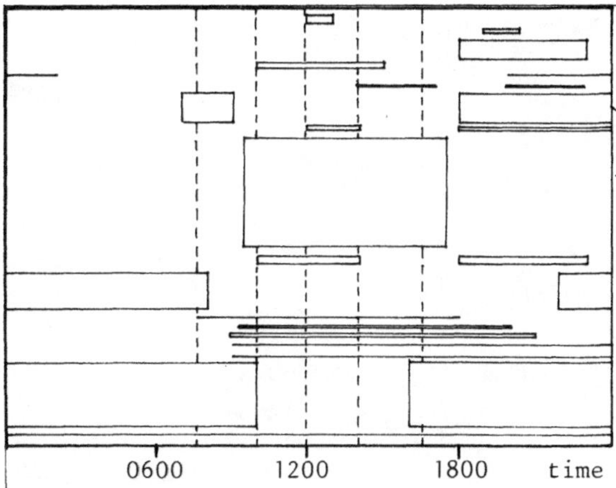

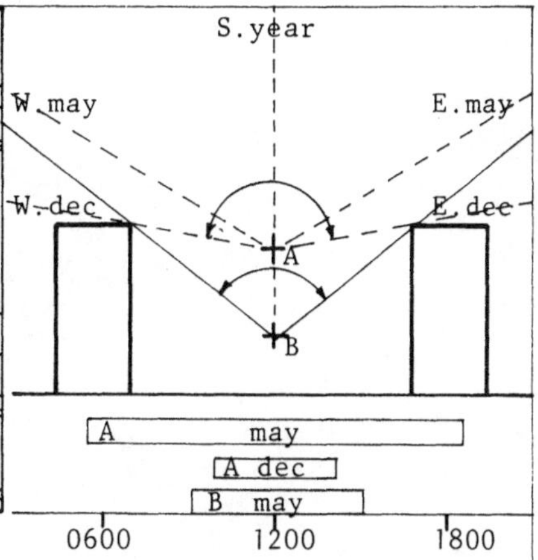

Figure IV: Occupation Schedules

Figure V: Vertical Influence

availability above an arbitrary threshold of 100W/m^2 is considerably reduced. In other words, the higher the irradiation intensity, the more constrained the temporal availability, which in turn may be measured in terms of deviation from the time of maximum gain (denoted by S.year, and the ranges W.dec to W.may and E.dec to E.may).

All of the solar gain over the heating season (Figure III) occurs between 0500 to 1900 and 0800 to 1600 hours when buildings of residential use are largely unoccupied. Those premises which cannot directly use this gain must rely on thermal collection and storage during available hours in order to reduce their space heating load when the premises are actually occupied. Figure IV shows the range of occupancy schedules in a case study group of 30 buildings containing offices, shops, restaurants, pubs and residences in the central urban area of Cambridge. Overlaying the times of maximum solar gain intensity for each facade enables a rough assessment of potential gain from either direct or stored means to be made by measuring the deviation from these times.

The availability of solar radiation is further influenced by locational factors in the vertical dimension. Figure V illustrates the principle of duration of direct irradiation being proportional to height, relative to surrounding biilt form. The longest duration, or acceptable temporal deviations, therefore, occur in elevated positions. Thus, depending on occupancy schedules (Figure IV) uses may be appropriately located according to either exposed orientations, or relative vertical placement.

3. A CASE STUDY

Figure VI shows a case study block of buildings mentioned previously, together with surrounding obstructions viewed from the south-west corner. A simple steady-state model and quarterly fuel bills were used to quantify all incidental heat gains and all heat losses for hourly and heating season totals (1). In terms of solar potential, initial investigations reveal little preferential orientation, and a poor relationship between occupation schedules and spatial locations within the block.

Simulation of solar gains through glazing was made by compiling 3 sub-programs, dealing consecutively with the obstructions to each building, manipulation of resultant data, and finally their graphical representation. The first of these sub-programs was written by Francois Penz of the Martin

Figure VI: The Case Study from S-E

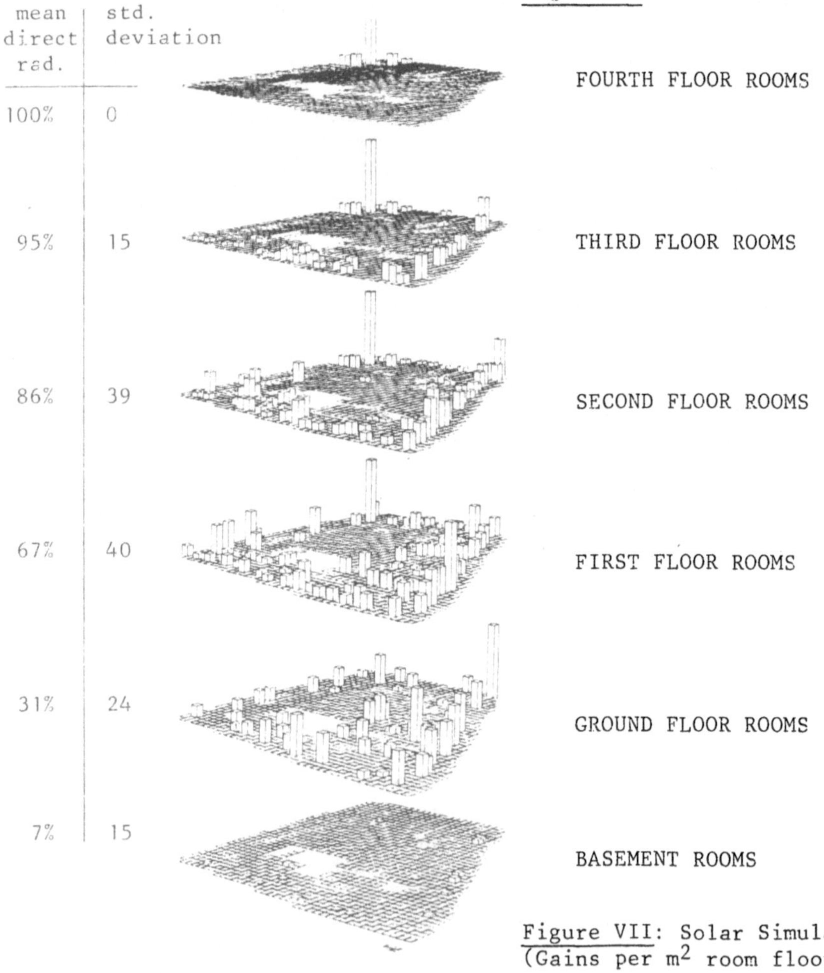

Figure VII: Solar Simulation (Gains per m^2 room floor area)

Centre (2) and is based on the Berlage/Basnett method of determining solar azimuth and altitude at any hour of the day for a specified latitude, and splitting input global radiation into direct and diffuse components (3). Input global data derives from Kew measurements during an average heating season (1967), October to May inclusive. Obstructions to the mide point of each building facade are defined by their altitude and azimuth, enabling obstructed direct radiation to be calculated for those times when the sun is both below the altitude and within the azimuth of the obstructions.

A fortran sub-program was written to aggregate direct irradiation for rooms and to apportion gains to each of the 51 use types and the 30 buildings comprising the case study area. Results are then available for statistical interpretation or coupled to a graphics routine (4) which transforms the array of gridded values into a 3-dimensional histogram representation (Figure VII).

4. RESULTS AND CONCLUSIONS

The results of the solar simulation show that glazing over the entire block of buildings receives only 65.88% of the direct radiation that would otherwise be incident in an unobstructed case. The influence of vertical location can be seen in Figure VII where incident direct irradiation ranges from 7% for basement windows to 100% at fourth floor level. Further locational influences may be gauged from the effect of street orientation; buildings on the south-facing street receiving marginally more than those on the west, east and north-facing streets (with 81, 78 and 64% as much as the south, respectively). On the average, however, west-facing windows receive 58% as much direct radiation as those facing south, east receives 51%, and north only 1.15%.

The result is that while some buildings are well endowed with available solar energy, others are not: the best case receives 85% and the worst 16%. Disparities between premises are more pronounced, ranging from 14.3% to 100% of the unobstructed value, with a mean of .65 and a standard deviation of .354. The worst cases are mostly those premises in basements, or heavily obstructed ground floor locations.

A similar pattern emerged from analyses of other incidental gains - such as those from office and display lighting, people, and appliances - and of thermal losses. Put into perspective, solar gains presently appear to contribute relatively little to the overall heat balance of the case study block, but it is clear that some buildings - and, more particularly, certain premises and rooms - avail themselves to utilisation of ambient solar energy for space heating considerably more than others. Graphical representation of solar availability (Figure V11) enables a ready assessment to be made of the relative solar space heating potential at locations within the urban fabric.

REFERENCES

1. MATTHEWS, L.J. (1983) Steady State Heat-Loss Analyses for Central Urban Buildings. Working Paper No. 5. Martin Centre, Cambridge University.
2. PENZ, F.A. (1983) Obstruction algorithm :- as part of doctoral thesis 'Passive Solar Heating in the UK Existing Housing Stock' Cambridge Univ.
3. BASNETT, P. (1975) Estimation of Solar Radiation falling on vertical surfaces from measurements on a horizontal plane. ERSC Cape., Chester.
4. STIBBS, R.J. (1976) Histogram and Surface Drawing Suite Cambridge University Computing Centre.

AIR SOLAR COLLECTING ROOF FOR A LOW-ENERGY SPORTS-HALL BUILDING

A. SPENA

Dipartimento di Tecniche dell'Edilizia e del Controllo ambientale, Facoltà di Ingegneria, Università degli Studi di Roma "La Sapienza", Via Eudossiana 18, 00184 Roma - Italy

Abstract

Simulation of thermal behaviour of a sports-hall building using either active and passive solar energy devices is presented. An air collecting roof is proposed, integrated by a flat plate solar collector's water system. Results are presented in terms of internal air temperature, as running unknown quantity, trends.

1. INTRODUCTION

Sports-halls appear, among buildings occupied only during the day, the most suitable users of solar energy because of their low internal temperature requirements, and of people contribution to sensible heat gains. In addition, these buildings are characterized by some design freedom, large empty/full volume and gross volume/external surface area ratios; while intermittent occupancy allows the use of low-inertia structures and air heating systems.

Present paper illustrates a feasibility study of a medium-size (about 16.000 m^3) sports-hall equipped with a roof acting as an air heater during winter season, and as a building cooler during summer:[1] in winter, additional heat to air can be supplied by a solar collector's plant operated all the year for sanitary water.

2. DESCRIPTION OF THE SYSTEM

The building is located in Rome, 41.5 lat.N; it has all the collecting areas facing South (figs.1-4). A wide double-glass window area faces North (fig.2), while satisfactory natural lighting is guaranteed also from the South, using (fig.5) shading curtains; the natural light factor η results of 0.12 in the hall, and of about 0.05 in the ancillary rooms, thus minimizing the need for electrical lighting. The roof is constituted by a single, low tilted (6°), pre-stressed concrete structure; external walls are well-insulated, low inertia panels. A ventilation rate of 0.5 volumes per hour is assumed; 500 people are at least expected.

Collecting device is constituted by a deck making air channels[2] connected to manifolds and ducts which deliver air to the ambient or to the environment (figs.6,7) depending on the value of the internal air temperature. Among several available options (fig.8), in present paper an opaque, black painted, thin metal(b) commercial deck to be applied on the roof is studied, which optimizes the pay-back time of the investment. The available roof area is of about 1500 m^2. To reduce noise and pressure drop, a channel cross-section area of about 0.03 m^2/m, divided over no more than 3 channels per m has been adopted. No air heat storage is provided.

The commercial flat-plate, simple-glazed solar collector's system (up to 240 m^2) is constituted by a symmetric double array of collectors, each composed of 3 rows (fig.9). The tilt is 60°, while Δt=7°C for each collector; array outlet temperature is 50 °C. A dayly warm water storage is provided, consisting in two boylers of 5 m^3 each.

3. SYSTEM SIMULATION AND RESULTS

A mathematical model has been built to simulate thermal behaviour of the complete system, with a time-step of 0.5 hours, over a one-year period of time. Climate is simulated in terms of external temperature, beam and diffuse separated available radiations[3,4], and of statistical probability of sky cloudness. A previously developed model[5] has been used to determine solar heat gain through the windows and the walls, together with energy collection by the roof. Air forced convection heat transfer coefficient in the channels is evaluated as[6] (turbulent flow):

$$Nu = 0.023\ Re^{0.8}\ Pr^{0.33}$$

thus obtaining the air temperature at the end of each channel by means of the temperature field determination over the entire roof. Collector's performances are evaluated following the method of[5].

Ventilating requirements, air leakages, thermal inertia of some parts of the building structure are also considered; heat transmission through the walls takes into account walls damping and delay in the simplified form[7] (opaque walls):

$$q(\tau) = HS\left[(\bar{t}_{sa} - t_a(\tau)) + \sigma\,\theta(\tau - \varphi)\right]$$

where q is the power transferred, $\bar{t}_{sa}$ is the mean daily sol-air[3] temperature, θ are the relative hourly departures, τ is time, S is the surface area, H is the overall heat transfer coefficient; σ and φ, namely damping and delay, are defined in[8]. Glass walls have naturally φ=0 and an additional direct heat gain from radiation. Anyway, external walls resulted of an average low damping and delay, thus introducing almost negligible inertia effects of the structure on thermal loads (roof effects are neutralized by the deck).

As a result of the simulation, internal air temperature trends are reported in figs.10,11,12, together with values of the external temperature, for typical winter, mean-seasonal, and summer conditions. A negligible time-shift of the effects with respect to their driving forces may be observed; also negligible results (lower than 3%) the heat load fraction directly transmitted to the hall by the roof structure. During summer, t_a never reaches intolerable values, while solar radiation appears strongly subdued; fig.12 shows the values that t_a could attain in case of not-treated roof. Thermal energy available from solar collector's system is finally reported in fig.13. Histograms of the computed efficiency ε of the roof as solar collector, and of minimum and maximum daily values of t_a are respectively shown in figs.14 and 15.

As far as the economy is concerned, heat collected by the roof can be estimated as follows:

- winter season (6 months):280 10^6 kCal, namely 1000 kCal/m^2day, or about 90 kCal/m^3day; globally, about 35 TEP.
- summer season (4 months):280 10^6 kCal, namely 1500 kCal/m^2day subtracted by the roof.

Winter savings allow an additional cost for the roof deck ranging from 0.12 TEP/m^2 with a life-time of 5 years, up to 0.48 TEP /m^2 assuming 20 years, at constant prices.

Heat collected by the water system may be estimated on the order of 1000 kCal/m^2day over the entire year, thus allowing a specific cost of the solar active system ranging from 0.23 TEP/ m^2 with a life-time of 5 years, up to 0.9 TEP/m^2 assuming 20 years. It must be observed that, being averaged the used values, and in absence of congruent storage systems, the obtained results should be carefully considered for practical purposes.Finally,fan consumption results lower than 4% of the overall removed heat.

4. CONCLUSIONS

The overall solar heat gain of the proposed system appears able to give, for the simulated climatic conditions and for the assumed building characteristics, tolerable values of the ambient air temperature during consistent building occupancy time, without conventional thermal utilities integration. Referring to current costs,collecting roof results a profitable option; while the active solar collectors system, though well integrating the passive system from a merely technical point of view, appears still just over a maximum allowable costs level.

References

1. Givoni.B.,"Man,climate and architecture",Appl.Sci.Publ.,London,1976.
2. VanDresser,P.,"Homegrown sundwellings",The Ligh.Tree,S.Fè,1977.
3. "ASHRAE Handbook 1981 Fundamentals",ASHRAE,Atlanta,1982.
4. "Annuario di statistiche meteorologiche",Vol.XV,ISTAT,Roma,1976.
5. Spena,A.,"Un metodo generale per la determinazione delle prestazioni di collettori solari disposti in schiere parallele",Proc. XXXVIII ATI Congress,Bari,sept.1983.
6. Kreith,F.,"Principles of Heat Transfer",Int.Ed.Publ.,N.Y.,1973.
7. Fontana,D.M.,private communication,oct.1982.
8. MacKey,C.Q.,Wright,L.T.,ASHVE Transactions,Vol.54,1948.

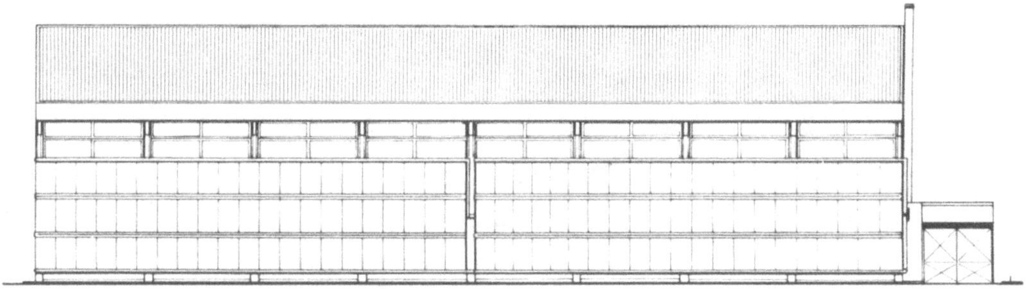

Fig.1.South view.

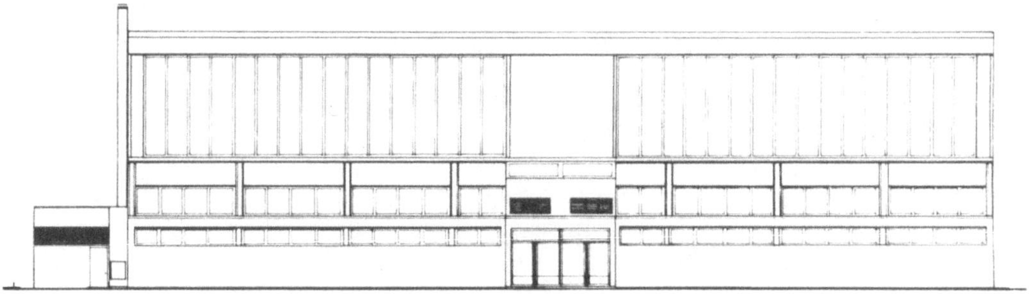

Fig.2.North view.

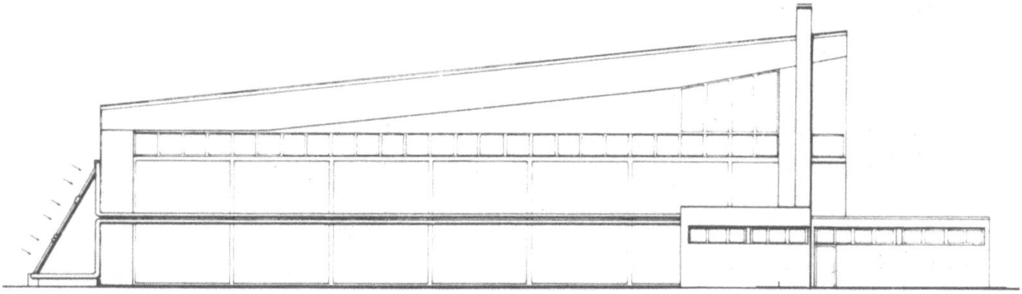

Fig.3.Lateral view.

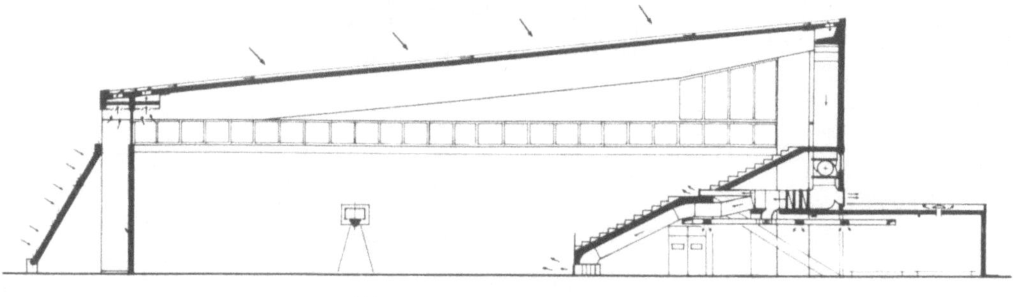

Fig.4.Cross-section.

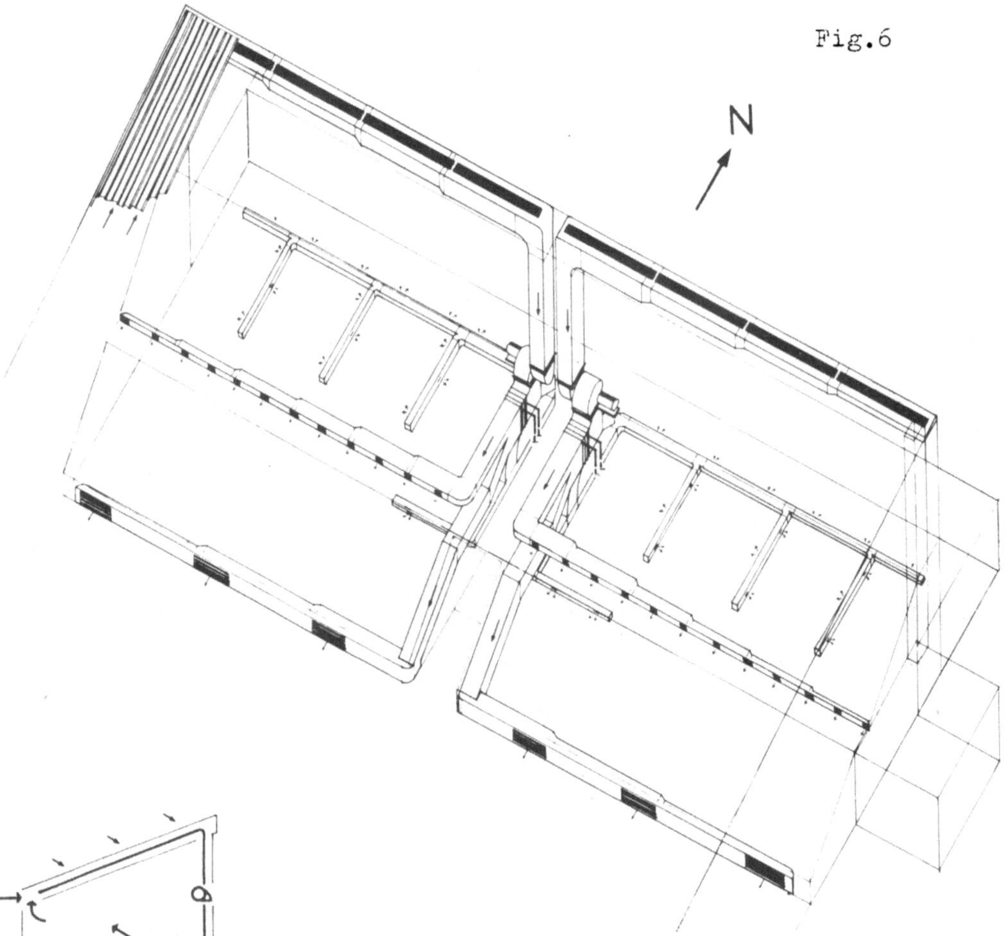

Fig.6

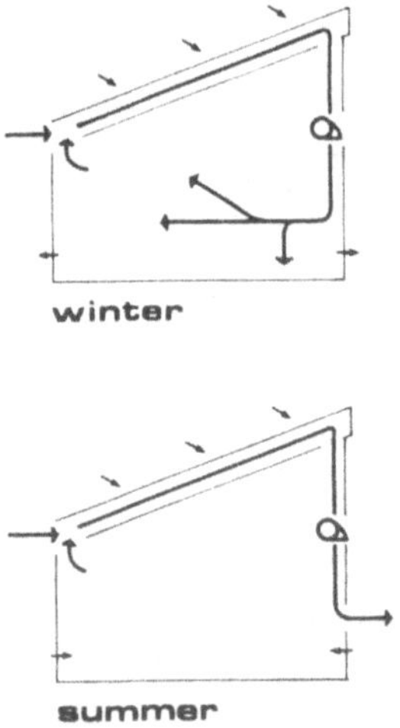

Fig.7

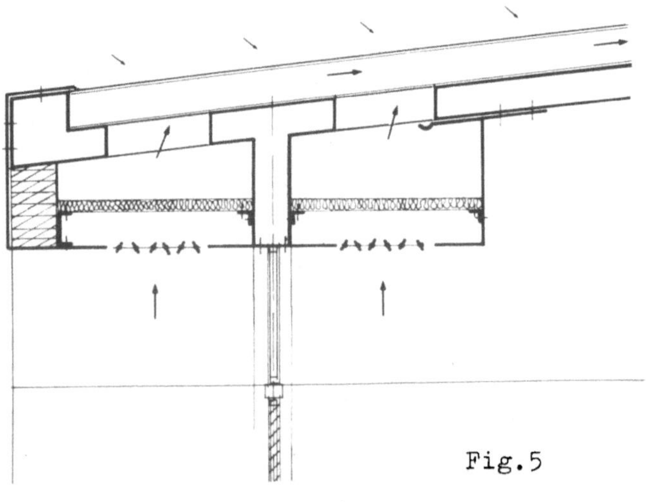

Fig.5

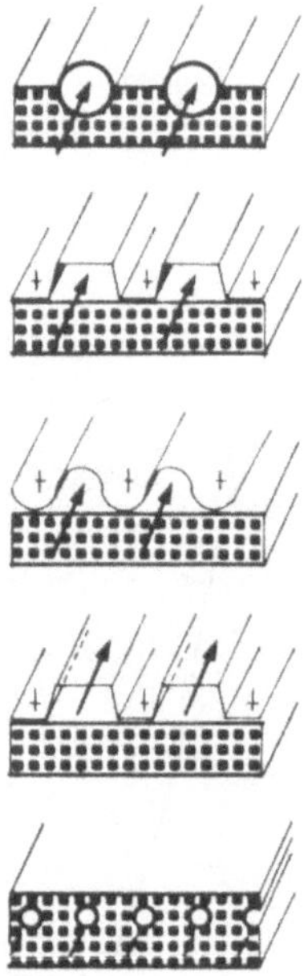

Fig.8

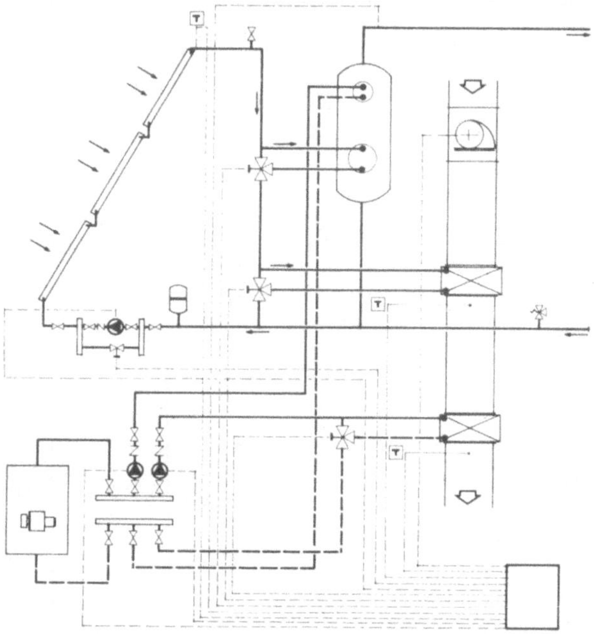

Fig.9

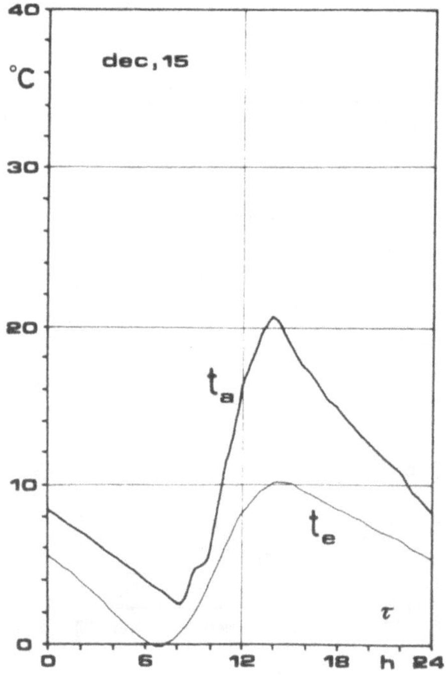

Fig.10

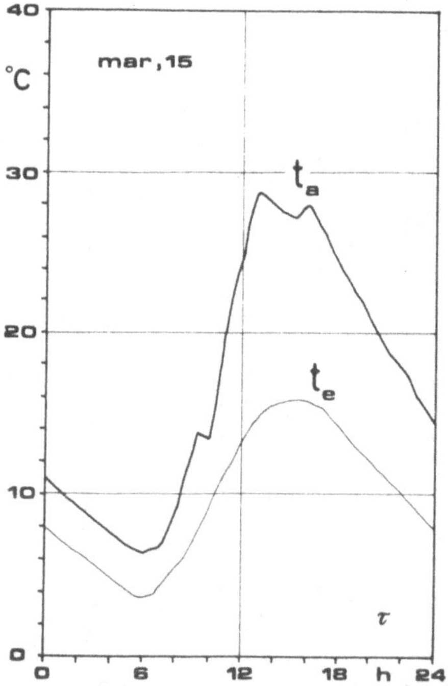

Fig.11

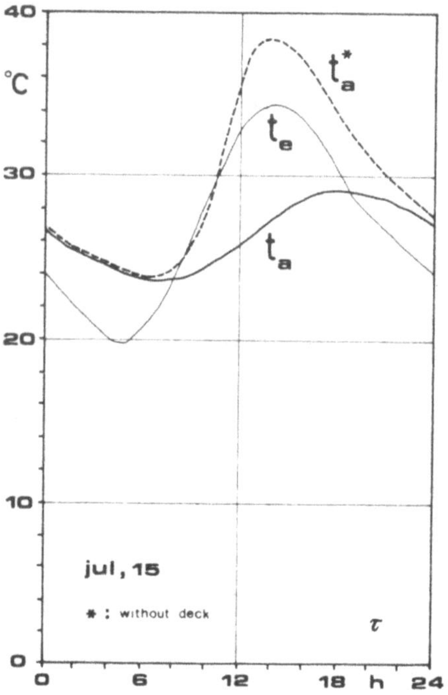

Fig.12

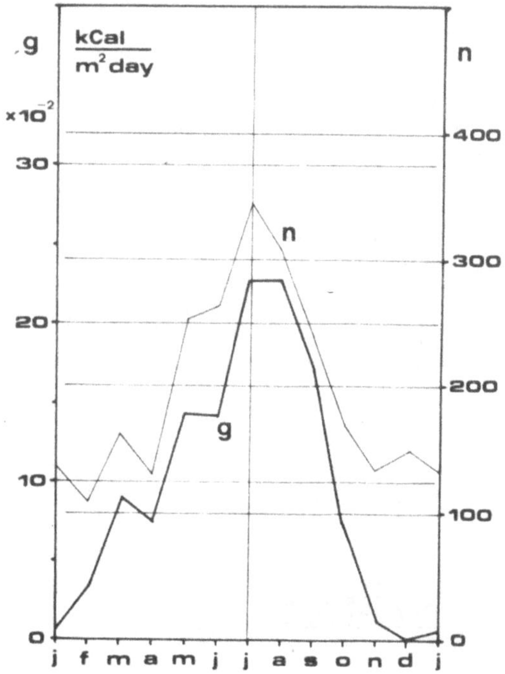

Fig.13.g,solar collectors lo ad;n,actual monthly sunny hs.

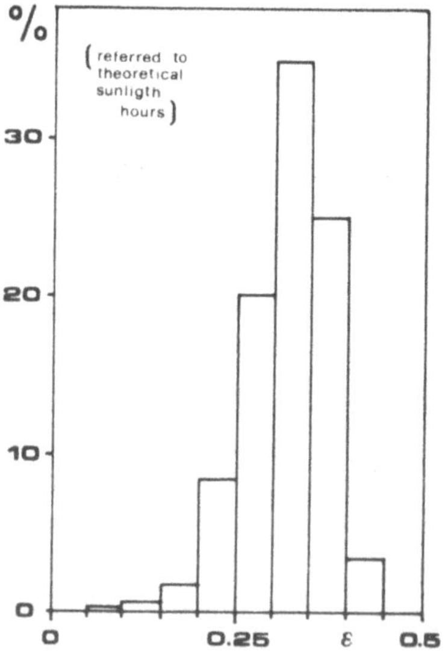

Fig.14.Instantaneous efficiency ε of the roof deck.

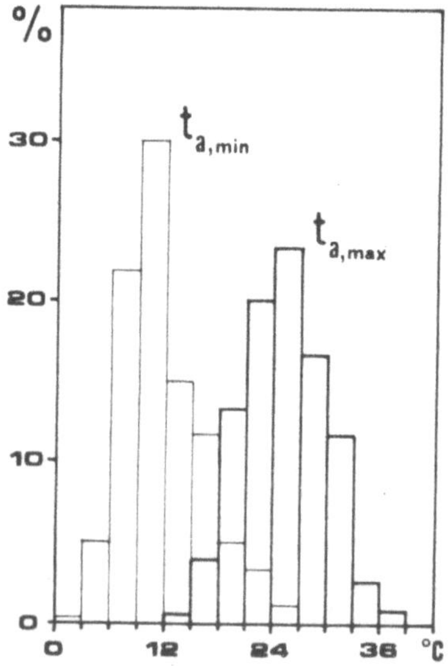

Fig.15

A REVIEW OF THE USE OF SELECTIVE ABSORBERS IN FOIL FORM FOR PASSIVE SOLAR COLLECTORS

P.C.JONES and J.J.MASON
INCO Selective Surfaces Limited,
Wiggin St., BIRMINGHAM B16 OAJ,
England.

Summary

This paper details the theoretical and experimental justification for the use of selective absorbing surfaces in mass walls. Methods of coating walls with foils are described, which are compatible with on-site techniques used in the building industry. Selective surfaces are shown to be a cost effective means of increasing the thermal output of storage walls for space heating.

1. INTRODUCTION

Since the late 1970s, there has been increasing interest in the use of selective absorbing surfaces in both passive solar designs and active systems, such as the flat plate collector. This upsurge of interest has arisen in part from various analytical exercises combined with experimental verification, which have demonstrated the benefits of selective surfaces over conventional neutral absorbers such as black paint. Selective absorbing surfaces combine strong absorptance in the solar spectrum with high reflectivity at longer wavelengths. This reduces the amount of re-radiated energy and thereby improves overall efficiency by retaining a higher fraction of captured solar energy. The increased interest is also due to the availability of low cost, durable coatings in forms suitable for constructing passive walls and for applying to flat plate solar collectors.

One such coating, which was developed by Inco Alloy Products Technical Centre and marketed under the trade name MAXORB*, consists essentially of a nickel oxide coating, which is produced continuously, on a thin nickel foil. The material is supplied in three forms: first, a "dry" foil which can be applied to walls pre-treated with an adhesive; secondly, with the reverse side coated with a temperature stable silicone pressure-sensitive adhesive; and thirdly, with the foil bonded to a copper or aluminium strip.

2. CHARACTERISTICS OF MAXORB FOIL

The absorptance of Maxorb Solar Foil, calculated by integrating the reflectance-wavelength curve for an air mass 2 spectrum, lies between 0.96 and 0.98. The thermal emittance has been determined by a number of methods (see Mason(1)) which give a total hemispherical emittance of 0.08 at 100°C.

A number of studies (see Masters(2) and Mason(3)) have been undertaken to establish the resistance of Maxorb to degradation under conditions prevalent in solar collectors and under conditions which are considered to accelerate the degradation process. The surface optical properties, after ageing in air for 500 days at 200°C, 84 days at 250°C and 10 days at 300°C, show no significant change. The results of outdoor exposure tests after 3 years under stagnating conditions show a small reduction in solar absorptance without any significant increase in thermal emittance. Moisture resistance has been determined using both cyclic and steady state testing and confirms the high degree of stability of the Maxorb surface.

It is interesting to note that in one of the most recent studies of the durability of solar absorber coatings(4) in which coatings were classified as selective ($E \leq 0.2$), partly selective ($0.2 < E < 0.5$) and non-selective ($E > 0.5$), the durability of selective surfaces, which included Maxorb, were as good as non-selective coatings. Only the partly-selective coatings were less stable than the well made selective or non-selective coatings.

3. SELECTIVE SURFACES FOR PASSIVE APPLICATIONS

3.1. Theoretical Studies

Several workers have calculated the effect of design parameters on the performance of Trombe walls. The results of Wieringen's studies (5) for a NW European climate are shown in Table 1. They have been calculated for four types of Trombe Wall, all having a thickness of 10cm and 0.7 $Wm^{-1}\ K^{-1}$ thermal conductivity.

TABLE 1: Thermal Output (KWh/m^2) of Trombe wall, according to Wieringen(5)

Wall Type	Heat Loss Coefficient $U_L\ Wm^{-2}\ K^{-1}$	Thermal Output KWh/m^2
Single glazed, neutral black painted wall	6	-54
Double glazed, neutral black painted wall	3	17
Single glazed selective absorber	2.5	62
Evacuated tube with indium oxide coating	1.5	73

The calculations clearly show that the single glazed selective wall outperforms the double-glazed neutral absorbing wall, because of its better overall efficiency, which results in a reduction of thermal losses at night. The model used by Wieringen has been validated using small test cells (6) one with a neutral black absorbing surface and the other with Maxorb Solar Foil glued to a plastered brick surface.

Using methods developed at Los Alamos, McFarland et al (7) has calculated the effect of selective and neutral surfaces on the output of Trombe and Water Walls for various locations in USA. The results show that, for all conditions studied, the output of single glazed selective surfaced walls exceeds those of double or even triple glazed neutral walls.

Hyde (8) of Los Alamos, has measured the performance of test cells, comparing the effect of selective and non-selective surfaces, single and double glazing and night insulation. The measurements show that selective surfaces increase the performance of Trombe and Water walls. Furthermore, he goes on to show that the performance of Water walls with selective absorbers is about equal to the performance of neutral surfaces with night insulation. On Trombe walls, the selective surface appeared to outperform night insulation under their test conditions. Since movable night insulation is normally an inconvenient, sometimes unattractive and relatively expensive option, its replacement by a selective absorber has many advantages. It is a truly passive solution to the problem, requiring no mechanical energy, control systems or physical effort by the inhabitants.

Gordon (9) has used his analytical model for predicting the thermal performance of passively heated houses to include selective surfaced wall elements. Good agreement between the model and Hyde's results were obtained, therby giving confidence in using the model to predict performance in as yet untested experimental configurations.

3.2. Selective Coated Solar Walls

Maxorb Solar Foil is normally supplied with the rear surface coated with a pressure-sensitive adhesive, either Dow Corning 282 or General Electric 6574, which is protected prior to use with a release liner. This system was developed for the application of Maxorb to metal substrates, but it has been reported that it is possible to successfully apply Maxorb Foil rear coated with these standard adhesives to concrete. However, this is only effective with a freshly prepared smooth surface. Normally, the adhesive pulls away, due to the friable and sometimes uneven nature of the surface. We have examined the problem and looked at various surface pre-treatments; one meeting the requirement of temperature stability with ambient temperature cure, was a silicone resin produced by Dow Corning designated R-4-3117 Conformal coating.

Concrete, having an irregular surface, calls for the application of successive layers of materials having progressively smoother surfaces. Since the usual thickness of the pressure-sensitive adhesive is 30 to 50 micron, the surface roughness must be reduced in order to increase contact area. Cement rendering followed by a finishing plaster is adequate, but the use of a proprietary plastic base filler improves temperature capability.

Brick, having a lower absorptivity to liquids and generally a smoother surface than concrete, responded to single surfacing routes. Direct treatment with Conformal coating was found to be effective. It is advisable to apply the foil to the wall in vertical strips, overlapping each strip by 2 to 5mm. In order to give added security, the foils can be mechanically fixed with wooden battens at the top and bottom. These fixings can also be used for mounting the glazing units.

Another method of preparing a concrete surface, prior to application of a pressure-sensitive adhesive coated foil, is being currently explored. This involves spray coating a silicone weather coating which was initially developed for protecting roof decks. Spray coating a concrete wall is a fast, convenient, low labour cost process and is compatible both in terms of trained labour and working environment with building industry practices. A fairly thick layer of silicone, typically 0.25mm thick, is applied which cures to become tack free within an hour. The pressure-sensitive adhesive coated foil can then be directly applied to the wall. The initial adhesion between the foil and the Weather Coat is relatively weak, allowing strips of foil to be removed if misaligned and repositioned without damage to either component. Tests have shown that the bond strength builds up with time and temperature to levels similar to those developed when the pressure sensitive adhesive coated foil is applied to well prepared metal substrates. The times and temperatures involved in this process are such that the proper development of the adhesive bond will take place in situ after glazing the wall.

An alternative route of applying foil to brick or concrete walls utilizes an adhesive that is conformal, is air/moisture cured and easily applied with non-adhesive backed Maxorb foil, directly to the wall. Installation trials have shown that Maxorb can be applied directly in a "dry" condition to concrete coated with General Electric RTV161 or 162 silicone rubber adhesives. The RTV162 adhesive was trowelled onto the cement rendered surface after dusting off with a dry brush. A stainless steel roller was used to "doctor" the surface and the Maxorb Foil was laid on the wall and rolled down to remove air pockets. Foils could be moved after application, enabling adjacent pieces to be accurately positioned without overlap. Using this technique, the Maxorb Foil has recently been successfully used to coat Trombe walls in new houses erected in Bath (England). The foil coated walls have been in service for nearly 18 months. There is no evidence of foil delamination and the system is reported to be operating satisfactorily.

A similar technique has been used to foil coat one mass wall of a series of three cells, the others being non-selective, one with and one without a blind. These three test cells are sited at Peterborough (England) and are being monitored for Energy Technology Support Unit (ETSU) by workers from the Polytechnic of Central London. It has been reported that the results are encouraging for the selective surfaced wall, the cell air temperature being usually higher than the other cells (10).

The RTV161 and 162 adhesives were designed to meet military standard specifications and have found applications mainly in electrical and electronic applications where neutral cure by-products are required.

Recently, General Electric have introduced a lower cost equivalent, designated CRTV122, for the building industry. This has the same moisture vapour curing system, which releases alcohol vapour, as RTV161 and RTV162.

Initial laboratory tests suggest that CRTV 122 will be as effective as RTV162 in adhering foil to masonry surfaces.

3.3. Double Envelope System

In this concept of passive space heating, air heated in a large air collector, which forms part of the southern face of a building, is circulated by fans to the main thermal mass of the construction. This segregation of the absorption and storage functions should provide a method of reducing Trombe wall losses. Two houses have recently been constructed, designed by Helix Multiprofessional Services, utilising this concept (11). Of the total 45.9m^2 of south-facing glazed area, some 27m^2 comprises the air heater. This consists of low thermal mass aluminium panels, to which Maxorb Solar Foil has been applied using the pressure-sensitive adhesive. Because of the flexibility of the production system, coated panels in excess of 3m in length were produced. Since completion of this installation, a process has been developed to bond, using a high temperature adhesive, Maxorb foil to copper and aluminium strip in lengths of up to several hundred metres and widths at present up to 0.5m. This continuous process offers the possibility of producing large areas of self supporting absorber cost-effectively. Thicknesses of 0.05 to 0.25mm aluminium or copper can be processed in this way.

The integrity of the laminate has been evaluated in both short term breakdown and long term stagnation tests. Both the aluminium and copper laminates have performed similarly in these tests.

The laminate can withstand temperatures of 300°C for a few minutes or 250 to 275°C for one hour before there is evidence of breakdown. To date, the laminate has withstood exposure for 100 days at temperatures of 150 and 80°C and thermal and humidity cycling to MIL-STD-810B, again for 100 days, without any significant degradation of the product. These tests are continuing.

4. CONCLUSIONS

The development of large area durable selective absorbing surfaces in foil and laminate form is seen as an important factor in assisting the viability of passive solar installations.

In the case of mass walls, techniques compatible with building industry practices are being developed for the application of selective foils during construction. Whilst laboratory tests have indicated the long term durability of the systems explored, real experience of foil-clad walls is limited to about 18 months.

A typical built cost of an external cavity wall to current U.K. building regulations is £45/m^2. A mass wall, 190mm thick, internally and externally plastered to accept foil coating, together with single glazing, can be constructed at a virtually identical cost.

Since a single-glazed selective coated wall can develop 60KWh/h per square metre in a N.W. European climate (5), this allows £30/m^2 for applying the coating to the wall, assuming a cost of prime energy of £0.05 per KWh, and a payback period of 10 years. The material costs of the coating system are about £10/m^2, divided equally between adhesive and foil components. Labour costs should be kept below £10/m^2, if the techniques developed to date are used on large area installations. It therefore appears that single glazed selective foil coated walls can be cost effective in energy terms in N.W. European climates at present energy cost levels. Of course, in Mediterranean climates where wall output during the heating season can be twice that of N.European , the cost-effectiveness is more striking. More experimental work, both of wall performance and durability is clearly justified in the light of evidence to date.

REFERENCES

1. J.J.MASON and B.WRIGHT, Proc. 2nd Int. Solar Forum, Hamburg, 1, pp337, 1978.
2. L.W.MASTERS, J.F.SEITER, E.J.EMBREE and W.E.ROBERTS, NBSIR, 81-2232, 1981.
3. J.J.MASON and T.A.BRENDEL, SPIE, Proc. 324, pp345, 1982.
4. U.FREI and J.KELLER, Contribution to IEA Task III. Workshop on Service Life of Solar Collector Components and Materials. Technical University of Denmark, Dec. 6-7, 1983.
5. J.S.WIERINGEN, Applied Energy, 7, pp67, 1980.
6. J.S.WIERINGEN and J.VENIS, Energy and Buildings, 3, pp335, 1981.
7. R.D.McFARLAND and J.D.BALCOMB, 3rd National Passive Solar Conference, San Jose, 3.1, pp54, 1979.
8. J.C.HYDE, 5th National Passive Solar Conference, Amherst, 5.1, pp277, 1980.
9. J.M.GORDON, Solar Energy 29, No.1, pp13-17, 1982.
10. J.G.F.LITTLER, Private Communication.
11. J.KEABLE, Solar World Forum, Proc. ISES Congress, Brighton, England, 3, 1954, 1981.

*MAXORB is a Registered Trade Mark

PASSIVE SOLAR DESIGN IN NON-DOMESTIC BUILDINGS

I.P. Duncan
Building Design Partnership
24 St John Street
Manchester M3 4FB

D. Hawkes
The Martin Centre for
Architectural and Urban Studies
University of Cambridge

SUMMARY

The main purpose of the work reported here was to make a broad assessment of the potential contribution which passive solar design might make to the demand for energy in both new and existing non-domestic buildings in the UK. It was found that the basic definition of a passive solar system requires extending when applied to non-domestic buildings, in order to properly realise the full potential of the passive solar approach to building design in the areas of ventilation and lighting, as well as heating. In the existing non-domestic stock, multi-residential buildings and schools seem to hold particular potential for passive heating and ventilation. The greatest potential in new buildings seems to reside in offices and factory/warehouse buildings, in heating, ventilation and lighting. Further effort in technical, economic and institutional areas is required to meet the complexities of the non-domestic sector.

1. INTRODUCTION

In 1978, 56% of the energy delivered in the UK was used in buildings; non-domestic buildings accounted for slightly over 50% of that. There has been a good deal of discussion of technical means of making energy savings in non-domestic buildings, including the adoption of much higher standards of thermal insulation, and there have also been a number of inventive design proposals for non-domestic buildings which offer energy savings (1,2). There has not, however, been any significant attempt to analyse the nature of non-domestic buildings statistically and to make a broad review of the potential for energy savings in the stock overall.

In 1980, Building Design Partnership and the Martin Centre for Architectural & Urban Studies at Cambridge began a preliminary study of the potential for applying passive solar design to non-domestic buildings in the UK, as part of the Department of Energy's Solar Heating Research & Development Programme. Its main aim was to provide a basis for future work in this field by exploring the major issues involved, indicating the range of possible applications and estimating the potential energy contributions from passive solar design in both the existing building stock and in new building design. The work was seen to be complementary to an earlier study of passive solar housing in the UK (3).

2. NON-DOMESTIC BUILDINGS AND PASSIVE SOLAR DESIGN

Non-domestic buildings vary widely in their age, form, size, construction and usage. This means that any comprehensive analysis must be preceded by the definition of a typology. For the purposes of the present study five principal use categories were identified in which significant amounts of energy are consumed. Data for the delivered and primary energy use of each type for space heating and lighting were collected (Figure 1).

This analysis established a broad understanding of how energy is used in non-domestic buildings. It allows the various categories to be ranked in order of gross consumption and reveals the high proportion of total primary energy which is consumed by artificial lighting in some types of non-domestic buildings. It is clear that passive systems for space heating may be applied effectively to non-domestic buildings, but they should be seen as elements of a more complex system in which they are combined with a careful approach to natural lighting design. This itself depends upon the careful control of artificial lighting, by either automatic or manual means, if its full potential is to be realised. The need for a more integrated approach to daylighting as an element of passive design has been acknowledged and the Building Research Establishment is now actively pursuing this area (4).

It should also be noted that ventilation requirements can make up a large proportion of the heating demand in many non-domestic buildings, and a number of passive design measures are ideally suited to providing pre-heating of such ventilation. This leads to the proposition that the term 'passive design' has a different, more extensive meaning in the non-domestic sector than is customarily taken in considering its application to dwellings :

> 'A passive design is one in which the form, fabric and systems of a building are arranged and integrated in order to increase the benefits from ambient energy for heating, ventilation and lighting and reduce the consumption of conventional fuels'

In a study of this kind it is necessary to take into account the contribution which may be made both in new building designs and in retrofit applications to existing buildings. In order to provide a realistic basis for the analysis of existing buildings it was necessary to develop a more elaborate typology for each of the principal functional types. This was based upon an analysis of the built form. Although generalisations about buildings must always be made with caution, a number of studies have shown that buildings often conform, in most technical respects, to some established type of solution which evolves slowly over time. This is usually the product of prevailing legislation, technology and user requirements. Either by making use of existing typology analyses or by making our own working assumptions, it was possible to build up a quite detailed picture of the characteristics of the non-domestic building stock and, from this, to make an analysis of passive potential.

The application of passive design, as with other energy conserving measures, is by its nature building specific. The diversity of the non-domestic sector creates a significant problem when attempting to assess the national energy saving that might accrue from passive design measures applied to the total UK stock of non-domestic buildings. In this context, the detailed analysis across all of the sectors of the many factors that would affect the passive design of an individual building was found impossible. Not only would this demand a huge input of effort and resources but much of the detailed data and national statistics required are simply not available. The methodology adopted, therefore, had to be one of successive approximations using the information that was available.

3. ANALYSIS OF PASSIVE POTENTIAL

3.1 Space Heating

The first step was to identify which of the varied factors affecting the application of passive design would be the most significant when assessing the national potential for solar energy contributions. These factors for passive space heating are :

(1) Building age
(2) Built form
(3) Building size
(4) Passive system performance

(5) Passive system size
(6) Number of buildings in the UK stock
(7) Fraction of total building stock suitable
(8) Market penetration by passive solar design technologies

Having defined the built form and average size of a typical building, the type & size of passive system that might be applied could be assessed. The number of 'typical' buildings either already existing in the UK stock or assumed to be built over the next 40 years was identified. Informed judgement was then used to assess the fraction of the total stock likely to be suitable for passive design. Finally, a suitable model of market penetration was used to indicate the possible take-up of the technology of passive design in each sub-group of the various non-domestic categories (see Fig 2).

There are a large number of factors that can affect passive system performance and most of them will vary from building to building. It was therefore clearly impossible to assume a specific level of performance. An estimate of the likely range of passive performance was therefore made. A low figure was based upon a mean of the low end of the range of estimates for performance of passive systems in housing, as given by various researchers in this field (3). A high figure was based upon comparison of each non-domestic building type and housing, with regard to the likely effect on passive performance due to differences in such factors as occupancy patterns, level of energy demand, ventilation requirements, types of construction and compatability of different passive systems.

3.2 Lighting

The assessment of potential energy savings which may follow from the more effective use of daylight in passive buildings was based upon a somewhat less detailed analysis. This was due to an even greater lack of information and data than found in the analysis of space heating. The main source of reliable data on this subject was in the area of photo-electric control of lighting in relation to office and school spaces, based upon research by the Building Research Establishment. The analysis took into account the quantitative relationship between daylighting & artificial light levels for a full range of non-domestic buildings, the savings which effective controls may offer, and estimates of technical potential and market penetration in both existing and new buildings.

3.3 Results

The analysis of potential energy savings was made in terms of two indices : technical potential, which is the saving attainable when only technical constraints are considered and dateline contribution, which also takes into account economic, social and institutional constraints. The results for each of the building types considered in this study are shown in Figure 3 for the year 2020. The estimated growth in potential savings for the non-domestic sector as a whole is shown in Figure 4.

In the absence of appropriate performance measures and reference datums for the non-domestic sector, the above analysis was based upon a range of intentionally conservative assumptions. This explains the diversity of the results and suggests that they can only provide a general indication of the potential for passive design.

It should be noted that, in addition to the increased beneficial utilisation of solar energy, passive solar design also tends to improve the energy conserving nature of a building. Due to the complexity of the design process, the range of building types and the type of statistical analysis undertaken here, it was, however, only possible to assess the marginal energy benefits from increasing solar gains and not to properly account for this fuller energy saving potential. If the full scope of passive solar design were emphatically demonstrated across the range of building types, it should be possible to substantially increase both the technical and market potential from that indicated by the above analysis.

4. CONCLUSIONS

Passive solar design - in an extended definition - can make a significant energy contribution to many new and existing non-domestic buildings in the UK and thereby reduce the Nation's total energy bill. Further effort in passive design development and technical evaluation is, however, required to more effectively meet the complexities of the non-domestic sector and so maximise the benefits.

In the existing non-domestic stock, multi-residential buildings and schools seem to hold particular potential for passive heating and ventilation. The greatest potential in new buildings seems to reside in offices and factory/warehouse buildings, in heating, ventilation & lighting. Initial indications are that there is a particular potential for atria and conservatories within a number of building types.

Passive design costs and benefits need further detailed evaluation but there are a number of passive design buildings that already indicate the cost-effectiveness of this approach. Possible barriers to passive design implementation and development - technical, economic, environmental and institutional - require more detailed study and both government and professional bodies might need to take a lead role to help overcome them.

ACKNOWLEDGEMENTS

The work described in this report was performed under contract for the Energy Technology Support Unit (ETSU) on behalf of the UK Department of Energy. The views and judgements expressed in the report are those of the contractor and do not necessarily reflect those of the Unit or the Department. A full report on this work has been submitted to ETSU, Reference No. ETSU S - 1110.

REFERENCES

(1) Kasabov, G. Buildings : the Key to Energy Conservation
RIBA London 1979

(2) Hawkes D. & Owers J. (Eds) The Architecture of Energy
Construction Press/Longmans, London 1982

(3) Energy Conscious Design Passive Solar Housing in the UK
ECD London 1980

(4) Crisp V H C, Littlefaire P J, Cooper I, McKennan G, Daylighting as a Passive Solar Energy Option; ongoing study for ETSU.

1 Magnitude of existing market

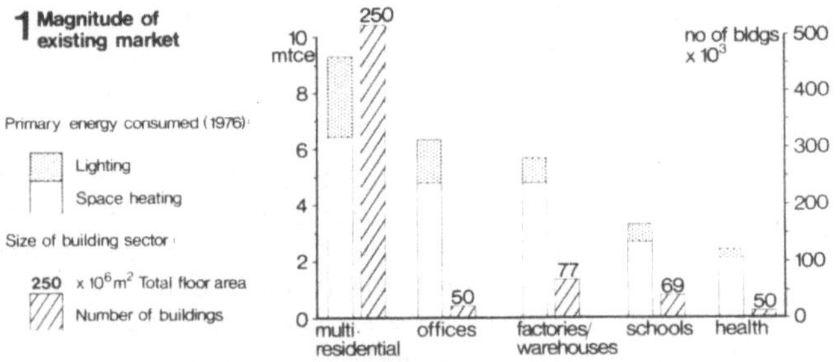

1 mtce = 1 metric ton of coal equivalent
= 2.7×10^7 GJ = 7.5×10^9 kWh

2 Market model

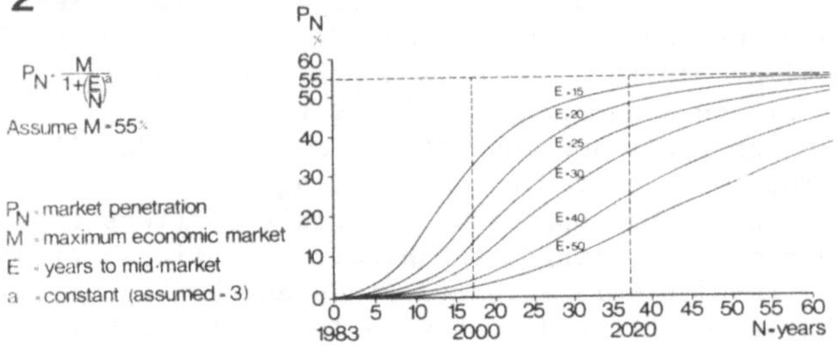

3 Potential energy savings

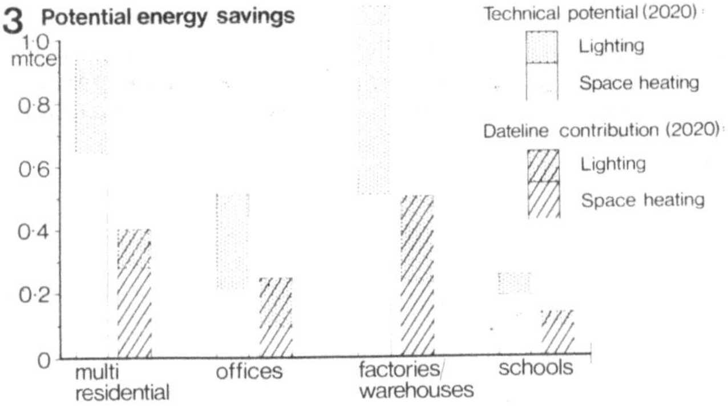

4 Predicted growth in non-domestic energy savings

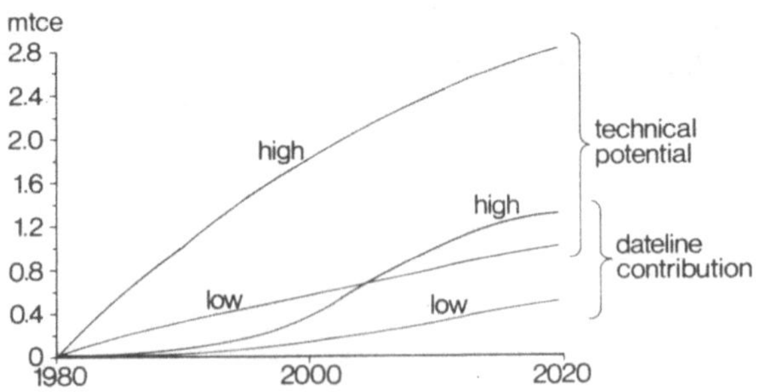

THE USE OF PASSIVE SOLAR GAINS FOR THE PRE-HEATING OF VENTILATION AIR IN HOUSING

Baker N V
Energy Conscious Design
11-15 Emerald Street
London WC1N 3QL

SUMMARY

This paper describes the content and preliminary results of a study funded by the UK Department of Energy and administered by the Energy Technology Support Unit. The study is concerned to increase the performance of passive solar systems, mainly in housing, by direct coupling of ventilation heat load and solar gains.

1. INTRODUCTION

It is well established that the absolute performance of a passive solar system is sensitive to the size of the heating load. It is also clear that, as buildings become better insulated, so the ventilation heat loss (which in the absence of heat recovery, has a lower limiting value), becomes the major part. The study described here is concerned with the deliberate linking of the ventilation load with the passive solar gains.

The ventilation load has two important characteristics which make this approach is very promising. Firstly, unlike heat loss through the fabric, the ventilation load can be very local. If we consider a hypothetical (and ideal) case where all the ventilation air is drawn in through a single apperture, then the heating load will occur at this aperture, i.e. where the cool air enters the heated room. Ideally, we would then design our building so that this position of maximum ventilation load coincides with the position where the solar gains are made. This location could be a sunspace or Trombe wall, or a direct gain room.

The second important characteristic is that the temperature threshold of the ventilation load is set by the external air temperature rather than the internal temperature. To illustrate this point consider the operation of a sunspace in a conventional circulation mode. In this case the gains made in the sunspace are conveyed to the heated interior as warm air, where they offset total fabric and ventilation loss. However, if the temperature of the air in the sunspace is less than the temperature of the heated space, the gains cannot be used. This severely limits the number of hours that the passive element will be producing useful energy although there is solar energy available.

If on the other hand, the infiltration rate of the house has been reduced to such an extent that air has to be deliberately let into the building, and if this air is drawn in via the sunspace, then any temperature increment, however small will be useful. This will greatly increase the 'operational hours' of the system.

It is easy to accept the advantages of ventilation pre-heat in principle - but two important questions arise. Firstly, how possible at a practical level will be the control of natural ventiltion and its coupling with the solar gain, and secondly what energy contribution will this make.

2. MODELLING

The author has already reported encouraging results from a simple modelling exercise (1). Here, the performance of a conservatory house was compared using varying arrangements of lightweight and heavyweight construction and using two modes of operation, ventilation pre-heat and forced circulation with a fan switched by temperature differential. The results showed that the ventilation pre-heat mode was approximately 3 times higher in performance than the circulation mode, and furthermore was relatively insensitive to the disposition of thermal mass.

However, an important assumption had to be made in the modelling. This was that an arbitrary fraction of the ventilation air was drawn from the conservatory. Furthermore, this fraction varied only with occupancy pattern and bore no relation to wind speed, direction and temperature difference, the climatic parameters which influence infiltration rates in reality. The encouraging performance was based upon an assumption that 60% of the ventilation air was drawn through the sunspace.

For the modelling exercise described here, we have combined a simple air flow model, driven by wind pressure and stack effect, with a thermal resitance network model FRED, described by Penz (2). We have sought to simplify the air flow model as much as possible whilst still retaining sufficient detail to demonstrate the pre-heat mechanism, and to relate to measurable climatic parameters such as windspeed, direction and air temperature and measurable building parameters such as building height and air permiabilities.

The basis of the model is a thermal resistance network representing a three zone building, one unheated zone (sunspace) and two heated zones south zone and north zone respectively. Fig (1).

The airflow model is illustrated in Fig (2). MW is the wind induced transverse airflow, MN and MS are stack induced airflows and ME a transverse stack unduced airflow due to assymetry in the airflow permiability distribution. We can also introduce the terms WN, WS, and WE which are similar to the stack terms but are induced by wind suction through the roof.

Four permiabilities are ascribed to the envelope. North wall, south wall, north roof and south roof. It is assumed that no resistance to flow occurs within the building. In practice we have adjusted values of permiabilities to give plausible infiltration rates, thus we must consider these permiabilities to be 'effective permiabilties'.

Many field studies of infiltration have established that at low windspeeds airflow is dominated by stack effect whilst at higher wind speeds it is dominated by wind pressure, and the two effects are not simply additive. Various methods are adopted to describe the transition from one mode to the other, the simplest being a point of transition at a critical windspeed. More sophisticated transitions involve interpolation and we have used a hyperbolic function to describe the transition. This approach must be described as empirical since it is not based upon a model of the actual physical process.

2.12 The Building

The hypothetical building which was modelled corresponds to a 95 m^2 three bedroom terraced house insulated to current UK building regulations with the addition of floor insulation. The house has a lean-to conservatory on the south side. The house is modelled as three zones only, sunspace, south zone and north zone.

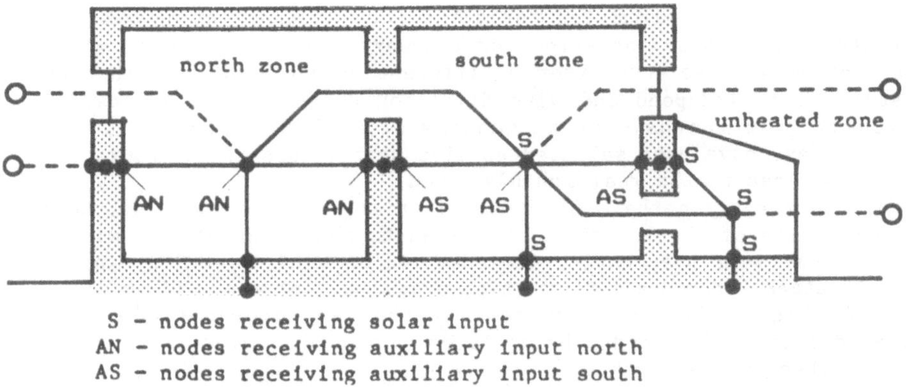

Fig 1 Thermal resistance network.

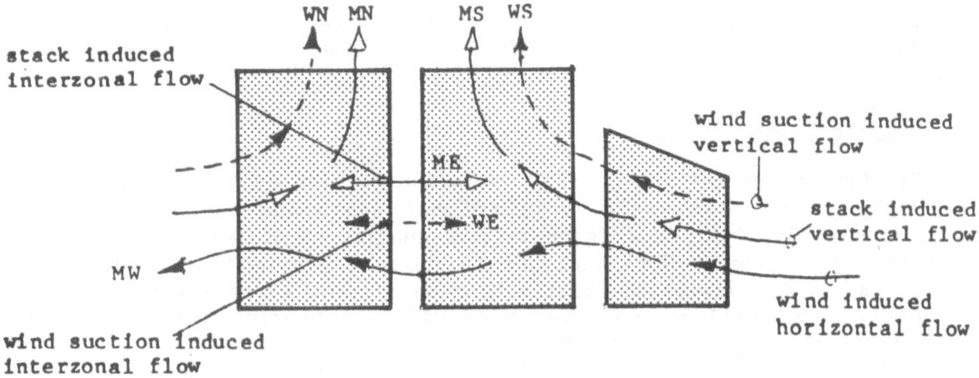

Fig 2 Airflow model.

		day 1	day 2	day 3	day 4	day 5
WEATHER						
mean ambient temp	degC	4.5	3.1	4.9	3.8	4.9
mean windspeed	m/s	2.3	2.9	5.4	5.8	2.3
ave. wind direction	deg	350	270	226	163	350
total solar (vert.south)	kWh/sqm	1.9	1.0	0.4	2.3	0.5
SYMETRIC PERMIABILITIES						
total auxiliary south	kWh	15.4	23.3	35.7	41.5	19.8
total auxiliary north	kWh	26.5	17.7	13.2	12.8	26.3
total auxiliary	kWh	41.9	41.0	48.9	54.3	46.1
total pre-heat	kWh	2.5	4.1	4.4	15.7	1.5
pre-heat fraction	%	5	10	9	29	3
mean infiltration rate	ac/h	0.9	1.1	1.3	1.9	0.9
ASSYMETRIC PERMIABILITIES						
total auxiliary south	kWh	17.4	27.0	49.3	63.9	23.9
total auxiliary north	kWh	22.7	18.4	14.6	13.9	23.0
total auxiliary	kWh	40.1	45.4	63.9	77.8	46.9
total pre-heat	kWh	4.0	5.1	6.0	20.0	2.2
pre-heat fraction	%	10	11	10	26	5
mean infiltration rate	ac/h	0.9	1.2	2.0	3.0	0.9

Fig 3

The building was simulated for a five day sequence using real hourly temperatures and solar radiation for February 1967. The hourly windspeed and direction were selected from a different sequence of days in order to illustrate the windspeed and wind direction dependence. In later use of the model eight months heating seasons of '5-day' months will be used where the days have been selected to give mean values of the four climatic parameters close to the real monthly mean.

For the first run the same permiabilities were ascribed for the north and south parts of the building. The results show that at low windspeeds there is some vent pre-heating irrespective of wind direction, this being due to be stack induced air flow which draws half of the air via the south side. As the windspeed increases so the vent pre-heat contribution becomes direction dependant, being largest when the wind is in the south and zero when in the north.

In the second run, the permiabilities have been made assymetic, but arranged to give approximately the same ventilation rate at low windspeeds. The north roof permiability has been made four times as great as the south, whilst the south wall permiability has been made four times greater than the north wall. This leads to an increase in vent pre-heating performance. In particular it is now possible to draw the majority of the ventilation air from the south at low windspeeds, under stack effect.

2.2 Results of simulation

Fig (3) summarises the hourly simulation results on a daily basis. The shift from stack dominated ventilation on days one and two, to wind dominated on day four, due to prevailing conditions of both windspeed and direction is evident. Note that the adoption of assymetic permiabilities significantly increases the contribution on days of low windspeed.

Daily mean infiltration rates vary widely - hourly values show even greater variation. It is apparent that the greatest vent pre-heat contribution occurs under conditions of over-ventilation; which in the absence of better infiltration control, is a valuable function.

It must be noted that the period modelled represents severe conditions for the UK, with relatively low temperatures, high windspeeds and poor solar radiation. At no time did the conservatory temperature reach above room temperature, and thus the solar contribution in circulating mode would have been zero.

These preliminary modelling results indicate that ventilation pre-heating can be enhanced by the control of premiability distribution. They also show that vent pre-heat can be attained with stack induced ventilation alone. Since wind induced ventilation is much more variable than stack effect, and since the wind often results in over-ventilation in the winter, it suggests a solution whereby the building is sheltered to such an extent that (except in exceptional high wind conditions), the ventilation remains stack effect dominated.

A simplification in the modelling which may be significant is the way that the wind pressure coefficient is dealt with In field measurements, pressure coefficients are shown to be dependent on both wind directions and strength, and on position over the building envelope. In the model, this variation is dealt with empirically and rather crudely with a cosine function of wind direction, and by ascribing the wind suction term causing air flow out through the roof. However, local effects of other buildings, walls, earth mounds and vegetation, causing siginficant variations of pressure coefficient across individual faces of the building could be important and will be considered later in the study.

One possible engineering solution to the variability of windspeed and direction, applicable in a construction where the 'base level' permiability is very low, would be to use wind induced suction at roof level to lower the mean pressure of the building interior. This would have the effect of elevating the neutral pressure zone on the lee side (south) of the building, still allowing the major fraction of ventilation to be drawn in from the south. However, the degree of suction would have to be controlled to take account of windspeed and direction, and temperature. This could be done mechanically or electro mechanically using microprocessor based control logic.

Another possible solution is to dominate air flow for ventilation by mechanical means, thereby, apparently offering total control. A fan mounted in the wall between a conservataory and the heated zones of the house is in effect a way of guaranteeing positive pressure where and when it is required. Or, mechanical extraction could provide a controllable negative pressure which generate infiltration from the appropriate point. However, caution must be taken not to use the mechanical ventilation to ventilate over and above what is already adequate natural ventilation. Furthermore for mechanical ventilation to dominate natural ventilation significantly, would require high levels of envelope air tightness and considerable fan power.

Applications of ventilation pre-heating are not limited to new building. Many existing buildings have poor insulation levels and may be over ventilated. The addition of a sunspace or a conservatory may serve the triple function of protecting a poorly insulated facade, reducing the excessive ventilation, as well as pre-heating the ventilation actually required.

3. CONCLUSIONS

At this early stage in the study we feel confident that the pattern of airflows into a passive solar building for the purpose of ventilation, has a considerable influence upon the solar performance, over and above the influence of a scalar infiltration rate. We have less confidence in how the necessary control of this air flow can be exercised at a practical level. However, we feel that a combination of manipulating external wind conditions, together with careful control of air permiability distribution, could lead to a significant improvement in passive solar performance.

Although explicit studies of the ventilation pre-heat concept are not known to the author, many designers have already conceived their passive solar designs in these terms. Indeed the term 'buffer space' implies this principle. We have described a primary school design, now nearing completion, in which ventilation pre-heating is the main passive solar design influence (3). However, we feel that there is a need for further analytical effort, followed by physical testing and field trials to lead to an optimisation of this particular aspect of passive solar designs.

REFERENCES

1. Baker N. Mass optimisation in conservatories. Proc. Conf. Solar Architecture, Cannes Decemeber 1982. EUR 8563.
2. Penz F. Passive Solar Heating in existing dwellings. Martin Centre report to E.T.S.U. (UK Department of Energy) November 1983.
3. Baker N. The influence of thermal comfort and user control on the design of a passive solar school building. Energy and buildings, 5, 1982. north zone

THE SOPHIA ANTIPOLIS PASSIVE TEST FACILITY

P. BACOT - B. BOURGES - G. KRAUSS - A. NEVEU **
F. NEIRAC - R. REGAS *

ECOLE NATIONALE SUPERIEURE DES MINES DE PARIS
Centre d'Energétique
60 Bd St Michel **
75006 PARIS
Sophia Antipolis *
06565 VALBONNE Cedex
FRANCE

Summary

The Centre d'Energétique de l'Ecole des Mines de Paris has built in 1981 a passive test facility of reasonnable size, including two rooms, in order to test and to validate various simulation models. With a complete equipment (meteo, air and walls temperature, hygrometry etc...), it allows all kind of studies in transient conditions with a very good knowledge of different parameters.

1. INTRODUCTION

La validation d'un modèle mathématique nécessite des données et des résultats concernant le phénomène physique étudié. C'est pourquoi l'Ecole des Mines s'est dotée d'un bâtiment à l'échelle 1, construit sur l'aire d'essai à Sophia Antipolis, près de Nice.

1.1 Description

Cette cellule test est composée de deux pièces indépendantes; une pièce Sud (70 m3) délimitée au Sud et au Nord par deux murs en béton armé de 25 cm d'épaisseur, et une pièce Nord (50 m3) qui fait office de local technique. L'ossature de la cellule est formée par une charpente métallique sur laquelle viennent se greffer les murs et la toiture.

1.2 Aspect thermique de la cellule solaire passive

Le bâtiment est fortement inerte, cette inertie étant concentrée au niveau de la dalle, des murs Sud et central. Le mur Sud possède une isolation mobile, et les fenêtres Sud, inclinées à 30°, sont occultables par des volets isolants. Les fondations sont isolées du sol par des plaques de polystirène de 8 cm d'épaisseur, les murs Est, Ouest et Nord étant isolés de l'extérieur grace à un complexe de type "washperle" (constitué d'un isolant, d'un treillis en fibre de verre et d'un enduit en ciment).

La toiture est composée d'une isolation autoporteuse (type rexocal) recouverte par des tuiles bitumeuses. Quant aux ponts thermiques introduit par l'isolation, ils ont été supprimés pour assurer la continuité de cette isolation.

1.3 Aspect évolutif de la C.S.P

L'architecture de la cellule permet d'envisager plusieurs configurations pouvant se combiner entre elles.

Stockage: Mur et vitrages Sud ne sont pas isolés pour permettre au bâtiment de capter le rayonnement solaire incident.

Nuit: Mur et vitrage Sud sont isolés, pour limiter les déperditions thermiques.

Jour sans soleil: Seuls les vitrages Sud ne sont pas isolés pour permettre à la lumière du jour de pénétrer (rayonnement diffus).

1.4 Caractéristiques thermiques de la CSP

Le calcul du coefficient K moyen et du coefficient G de la CSP a été fait pour apprécier ses performances thermiques:

	Mur Sud isolé Vitrage Sud isolé	Mur Sud isolé Vitrage Sud non isolé	Mur Sud et vitrage Sud non isolés
K moyen	0,48	0,70	0,99
G sans renouvellement d'air	0,55	0,80	1,14
G avec 1 V/h de renouvellement d'air	0,84	1,14	1,48

1.5 Intrumentation

L'acquisition des données de la CSP est entièrement automatique, 80 mesures sont effectuées toutes les 6 minutes.

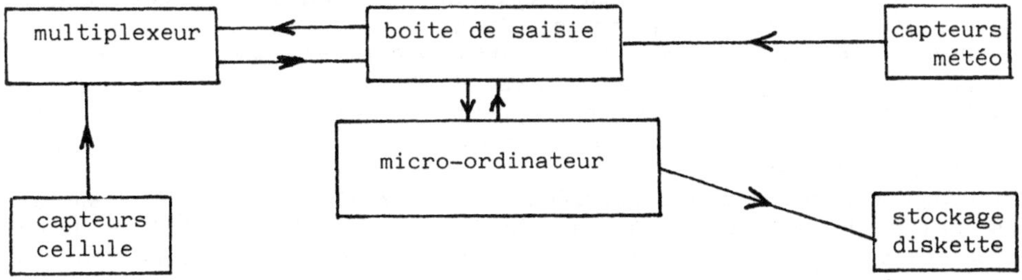

. Les capteurs météo mesurent l'ensoleillement direct, diffus et global reçu sur des plans horizontaux ou inclinés, la température extérieure, le degré d'hygrométrie, la vitesse et la direction du vent.

. Les capteurs cellule sont des thermistances étalonnées, mesurant la température résultante sèche, les températures d'air et de surface en différents points de la cellule.

2. APPLICATIONS ACTUELLES

Les mesures faites depuis la mise en service de cette cellule ont été exploitées:

2.1 Pour le recalage de "Minerve"

Le code de calcul Minerve est conçu pour répondre à différentes utilisations dans le cadre de la thermique du bâtiment. Il vise principalement l'étude du confort thermique et l'analyse des modes de transfert énergétique. A cela s'ajoute le recalage et la mise au point de méthodes simplifiées, Minerve servant de référence.

La conception du code intégre un certain nombre de contraintes:

- simulation d'enveloppes multizones complexes en régime variable pour tout climat.
- possibilité de prendre en compte la variation dans le temps et la non linéarité des fonctions d'échanges, les transferts par chaleur latente et l'hygrométrie, l'introduction de composants particuliers (mur Trombe, mur diode, stockage en lits de cailloux, etc...).
- facilité d'accès qui se traduit par l'emploi d'un langage évolué (Fortran).

Les séquences de mesures enregistrées sur la cellule test ont été utilisées en données d'entrée du programme Minerve. Les réponses réelles du bâtiment et les réponses simulées ont pu être comparées et les résultats satisfaisants obtenus prouvent la validité des hypothèses adoptées dans Minerve. (fig.2)

2.2 Pour l'analyse modale

L'analyse modale peut être considérée comme une approche unifiée de trois méthodes courantes de calcul thermique en régime transitoire: les différences finies, les facteurs de pondération et la méthode analytique de séparation des variables. C'est un formalisme mathématique destiné à caractériser rigoureusement la notion d'inertie thermique.

Une des particularités de la méthode est sa très grande concision: une conséquence importante de ce fait est que, dans la mesure où le nombre de paramètres caractérisant le comportement d'un bâtiment est très restreint, des applications opérationnelles sont envisageables;

- l'identification des paramètres à partir de séquences de mesures
- la commande de système de chauffage (gestion de l'intermittence)
- le diagnostic des bâtiments existants (identification des caractéristiques statiques et transitoires)

Parmi les perspectives d'utilisation de la méthode, on peut citer la caractérisation des ponts thermiques, du comportement des sols, l'identification in situ des paramètres d'inertie pour une régulation auto-adaptative, la mise au point d'une méthode informatisée d'évaluation du confort thermique etc...

Grace aux mesures enregistrées sur la cellule test, le résultats essentiel de la méthode d'analyse modale, à savoir la possibilité de caractérisation d'un bâtiment par un nombre très réduit de paramètres, est désormais opérationnelle. (fig.3)

3. CONCLUSION

La connaissance fine des transferts thermiques dans les enveloppes du bâtiments est donc un point de départ, d'une part à la mise au point de composants solaires passifs et à l'apparition d'une architecture climatique, et d'autre part à l'optimisation des systèmes de production/distribution de chaleur.

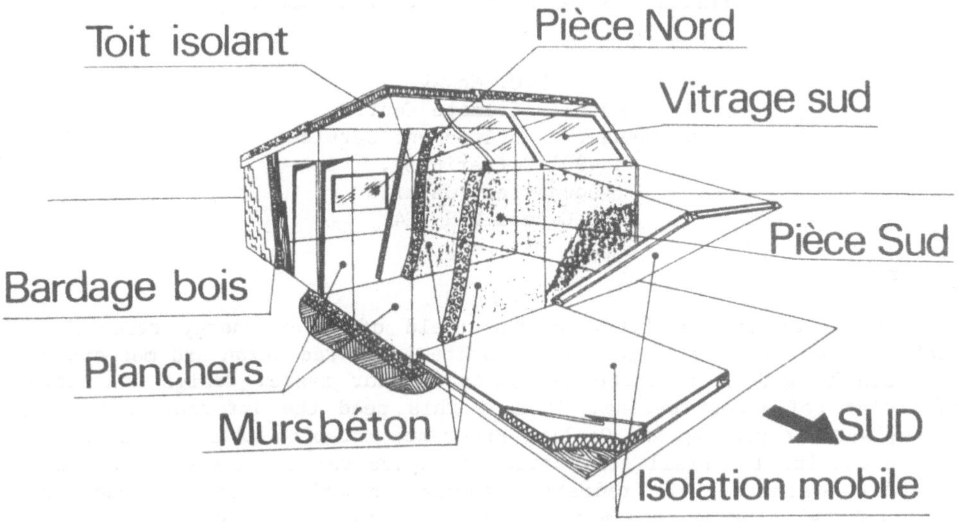

FIGURE 2

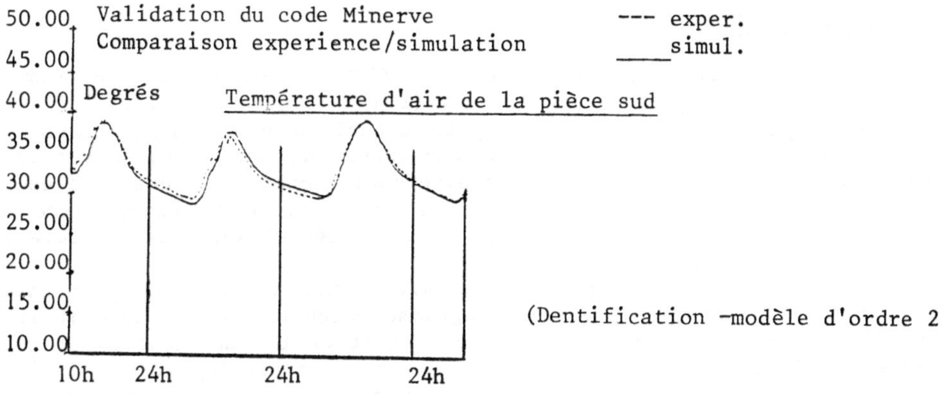

FIGURE 3

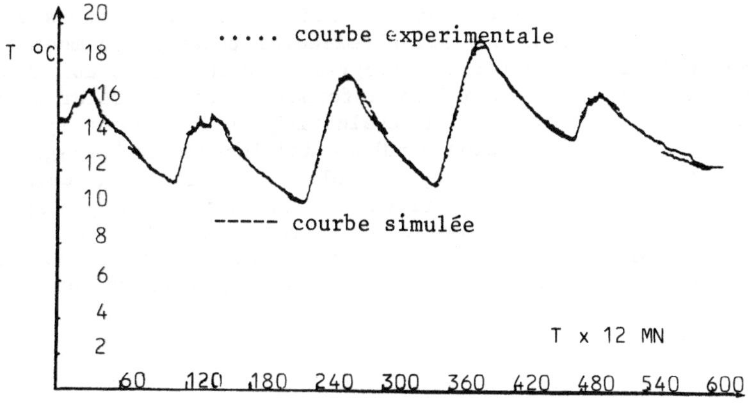

SOLAR RADIATION AND PYRANOMETRY STUDIES FOR SOLAR ENERGY APPLICATIONS: AN OVERVIEW OF IEA TASK IX

D.C. McKay
Canadian Climate Centre
Atmospheric Environment Service
4905 Dufferin St.
Downsview, Ontario
Canada M3H 5T4

Summary

With increased activity in the field of solar energy research and application, there is a need for accurate solar radiation and meteorological data to aid in resource assessment, solar system design evaluation, and solar collector testing. To meet this need the International Energy Agency's Solar Heating and Cooling Programme established a programme, Task IX, to examine the small scale time and space variability of solar radiation, to evaluate solar radiation simulation models, and to demonstrate improvements in irradiance measurement for solar collector testing.

The paper will describe in detail research and development undertaken by the task participants in the above areas, current results and plans for the future.

1. INTRODUCTION

As part of the Organization for Economic Co-operation and Development (OECD) the International Energy Agency (IEA) was established in 1974 with one of its major objectives being to establish co-operation among IEA participating countries to reduce excessive dependence on oil through energy conservation, development of alternative energy sources, and energy research and development. To meet this objective among others, the IEA participants established a number of thrusts, one of which is the development of renewable energy.

One of the programmes initiated within the renewable energy thrust is the Solar Heating and Cooling Programme which was established in 1977. The objective of the programme is to facilitate the application of solar heating and cooling products and systems by contributing to the information and technology base for use by industry, government and researchers.

With this increased activity in the field of solar energy research and application, it became apparent that there was a need for accurate solar radiation and meteorological data to aid in resource assessement, solar system design evaluation, and solar collector testing. To meet this need the IEA Solar Heating and Cooling Programme executive established a number of tasks to provide the required information to the solar energy engineer and designer responsible for evaluating solar energy systems. Task IV, Development of an Insolation Handbook and Instrument Package, was responsible for producing an inventory of solar radiation data which is available from IEA participating countries and an information package on instruments that are used for measuring solar radiation. Task V, Use of Existing Meteorological Information for Solar Radiation, was responsible for enhancing access to, and facilitating the measurement of solar radiation and other meteorological data needed for solar energy thermal applications.

To expand on the work performed within Tasks IV and V, Task IX, Solar Radiation and Pyranometry Studies, was initiated in October, 1982 and consists of eleven participating countries (Table 1). The objective of Task IX is to advance the state-of-the-art in predicting and measuring solar radiation for solar energy research and engineering applications. Task IX is divided into three subtasks, each run by a specific subtask leader:

Subtask (a) Small Scale Time and Space Variability of Solar Radiation (Dr. F. Neuwirth, Austria)
Subtask (b) Validation of Solar Radiation Models (Dr. D.C. McKay, Canada)
Subtask (c) Pyranometry. (Dr. D.I. Wardle, Canada)

The paper will examine each subtask separately.

Table 1: Participating Countries in Task IX

Austria	Japan
Canada	Netherlands
Denmark	Norway
Federal Republic of Germany	Sweden
	Switzerland
Italy	United States

2. SUBTASK (a) SMALL SCALE TIME AND SPACE VARIABILITY OF SOLAR RADIATION

2.1 Introduction

It is not feasible to build networks that monitor solar irradiance at every location where a solar energy system might be installed, yet location specific irradiance data are required in order to design and evaluate solar energy systems properly. Small scale temporal and spatial variations in solar irradiance caused by local topography and climate make it very difficult to interpolate the needed location specific data from the network data. The purpose of subtask (a) is to improve the understanding of small scale time and space variability of solar irradiance, as it affects the ability to interpolate solar irradiance values for locations not represented in a regular solar irradiance monitoring network.

2.2 Method of Approach

A questionnaire soliciting data and information appropriate for examining small scale time and space variability and pertinent literature was distributed to all IEA member countries.

As well, each expert member of Task IX undertook a literature review within their own country.

2.3 Results

From the returned questionnaires and the literature review a report will be written which will contain:

(i) documentation of the results of the questionnaires from individual countries;
(ii) some typical examples of existing small scale variabilities of solar radiation;
(iii) a literature review,

(iv) discussion on the use of satellite data for estimating solar radiation for small scale space variability,
(v) recommendations.

3. SUBTASK (b) VALIDATION OF SOLAR RADIATION MODELS

3.1 Introduction

There is a need to provide spatial and temporal data on the amounts of solar irradiance and on its direct beam and diffuse components at particular locations. In most instances, meteorological networks are poorly prepared to provide such data, particularly on tilted surfaces. Therefore, there is a requirement for numerical modelling procedures which can be used to provide estimates of solar irradiance for locations where measurements are not made or where there are gaps in the measurement record. A number of researchers in various countries have developed and are using models to predict solar irradiance at any location. Some of these models may be considered as general models with universal application. However, in most instances the models have not been verified at an extensive number of locations. The purpose of subtask (b) is to improve the understanding of solar irradiance models used to predict solar irradiance on horizontal and tilted surfaces for solar energy applications and to determine if these various models have a universal application.

3.2 Method of Approach

A request was sent out to researchers in IEA countries requesting solar radiation models which simulate solar radiation on horizontal surfaces and which simulate solar radiation on tilted surfaces. As well, a literature search was undertaken to identify solar radiation simulation models which have appeared in the recent scientific literature. Along with the request for solar radiation models a request was made for data sets which could be used in the validation. The data sets would include meteorological data and measured solar irradiance data. When all the solar radiation simulation models and data sets are received by the co-ordinating country, Canada, a mini-validation test will be undertaken using a sub-set of the data received to ensure that all the models have been implemented correctly on the computer. After this, the final validation run will be performed using the entire data sets, and statistical analyses run on the output.

3.3 Results

Table II and Table III list the models that are presently coded on the computer of the co-ordinating country. Table II lists models which simulate solar radiation on horizontal surfaces. Table III lists models which simulate solar radiation on tilted surfaces. Models as well as validation data are currently being received. The output from the subtask will include a report and a magnetic tape. The magnetic tape will contain all the models tested and those data sets used in the validation which have no restrictions to distribution. The report will contain:

(i) a description and listing of all the models,
(ii) list of the data sets used,
(iii) the analysis proceedure used;
(iv) the statistical analysis;
(v) recommendations.

Table II: Models for Estimating Solar Radiation on Horizontal Surfaces

Atwater and Ball (1978) [1]
Barbaro et al (1979) [2]
Collares-Pereira, and Rabal (1979) [3]
Davies and Hay (1980) [4]
Hoyt (1978) [5]
Orgill and Hollands (1977) [6]
Paulin (1980) [7]
Revfeium (1981) [8]
Rietveld (1978) [9]
Suckling and Hay (1977) [10]

Table III: Models for Estimating Solar Radiation on Tilted Surfaces

Bugler (1977) [11]
Hay and Davies (1980) [12]
Klucher (1978) [13]
Page (1961) [14]
(1978) [15]
Puri et al (1980) [16]
Revfeim (1982) [17]
Temps and Coulson (1977) [18]

4. SUBTASK (c) PYRANOMETRY

4.1 Introduction

Accurate measurement of solar irradiance is a necessity in several areas of solar energy development. In a 1979 IEA Task III report [19] it was shown that the poor quality of irradiance measurements are the main cause of descrepencies in collector testing techniques. A conclusion of this report is that "The participants had difficulties to ascertain the nominal accuracy of $\pm$ 3% for their pyranometers", and subsequent Task III work revealed that uncertainties of up to 10% could sometimes occur.

A major cause of these uncertainties is that calibration factors measured by different institutions are differently and often imprecisely defined and are sometimes poorly effected. Another is that pyranometer signals do not depend only on irradiance, and that corrections for the interfering effects of other variables are needed. These include environmental variables such as wind and temperature, as well as the irradiance distribution and the orientation of the instrument. Moreover, the instruments themselves have certain fundamental limitations. The purpose of subtask (c) is to demonstrate improvements in irradiance measurement for solar collector testing that can be achieved with detailed characterization of pyranometer responsivities.

4.2 Method of Approach

A round robin experiment to test both pyranometers and pyranometer methodologies has been established. There will be several participating laboratories or institutes which will make characterizing measurements both indoors and in the field on various groups of pyranometers at different times. The laboratory of the co-ordinating country, Canada, in addition to making measurements, will procure instrumentation for the experiment, will co-ordinate circulation of the instruments among the other laboratories will maintain control instruments and standards, and will liase with pyranometer manufacturers.

The discrepencies between calibrations done at different participating laboratories will first be analyzed using the data from the round robin. Next, individual instrument response functions will be derived for each of the tested pyranometers so that typical transfer functions for all major pyranometer types can be established.

4.3 Results

Thirty four pyranometers have been procured by the co-ordinating country and have all undergone field characterization at the co-ordinating country's laboratory. The round robin experiments have commenced and sub-groups of pyranometers have been sent to the Federal Republic of Germany, Sweden, Austria, the Netherlands, United States, and New Zealand.

A final report will be published which will include:

(i) documentation on the results of the round robin testing and an indication of improvements in pyranometry,
(ii) summary data collected during the testing,
(iii) analysis techniques used,
(iv) precision information,
(v) detailed methodologies used in the testing,
(vi) recommendations.

5. CONCLUSIONS

With increased activity in solar energy research and development occurring in many countries the need for solar radiation and meteorological data to aid engineers and designers in their particular solar energy applications has become obvious. To accommodate this need the IEA Solar Heating and Cooling Program established a number of tasks to provide the required information. Task IX; Solar Radiation and Pyranometry Studies is one of these tasks which was established to examine the small scale time and space variability of solar radiation; to evaluate solar radiation simulation models; and to demonstrate improvements in irradiance measurements. Through the co-operation of the eleven IEA countries it is opined that the task will advance the knowledge in the above mentioned areas and provide useful information necessary to make solar energy a viable energy alternative.

References

[1] Atwater, M.A. and J. Ball (1978): A numerical solar radiation model based on standard meteorological observations. Solar Energy, 21 163-170.

[2] Barbaro, S., S. Coppoline, E. Leone and E. Sinagra (1979): An atmospheric model for computing direct and diffuse radiation. Solar Energy, 22, 225-228.

[3] Collares-Pereira, M. and A. Rabl (1979): The average distribution of solar radiation correlation between diffuse and hemispherical and between daily and hourly insolation values. Solar Energy, 22, 155-164.

[4] Davies, J.A., and J.E. Hay (1980) Calculation of solar radiation incident on a horizontal surface. Proc. First Canadian Solar Radiation Data Workshop, edited by T.K. Won and J.E. Hay 32-38.

[5] Hoyt, D.V. (1978): A model for the calculation of solar global insolation. Solar Energy, 21, 27-35.

[6] Orgill, J.A. and K.G.T. Hollands (1977): Correlation equation for hourly diffuse radiation on a horizontal surface. Solar Energy, 19, 357-359.

[7] Paulin, G. (1980): Simulation de l'energie solaire au sol. Atmosphere Ocean, 18(4), 286-303.

[8] Revfeim, K.J. (1981): Estimating solar radiation income from "bright" sunshine records, Q.J.R. Met. Soc., 197, 427-435.

[9] Rietveld, H.R. (1978): A new method to estimate the regression co-efficients in the formula relating radiation to sunshine, Agric. Met. 19, 243-252.

[10] Suckling, P. and J.E. Hay (1977): A cloud layer-sunshine model for estimating direct, diffuse and solar radiation. Atmosphere, 15, 194-207.

[11] Bugler, J.W., 1977: The determination of hourly insolation on an inclined plane using a diffuse irradiance model based on hourly measured global horizontal insolation. Solar Energy, 19, 477-491.

[12] Hay, J.E. and J.A. Davies, 1980: Calculation of the solar radiation incident on an inclined surface. Proceedings First Canadian Solar Radiation Data Workshop. J.E. Hay and T.K. Won, eds., Atmospheric Environment Service, Toronto, 59-72.

[13] Klucher, T.M., 1978. Evaluation of models to predict insolation on tilted surfaces. U.S. Department of Energy, Division of Solar Energy, DOE/NASA/1022-78-28, 28 pp.

[14] Page, J.K. 1961 The estimation of monthly mean values of daily total short-wave radiation on vertical and inclined surfaces from sunshine records for latitudes 40N-40S. Proc. U.N. Conference on New Sources of Energy, Paper No. 35/5/98, 378-389.

[15] Page, J.K., 1978. Methods for the estimation of solar energy on vertical and inclined surfaces. Proceedings 5th Course on Solar Energy Conversion, Dept. of Physics, Univ. of Waterloo, Canada, 37-39.

[16] Puri, V.M., R. Jiminez and M. Menzer, 1980. Total and non-isotropic diffuse insolation on tilted surfaces. Solar Energy, 25, 85-90.

[17] Revfeim, K.J.A., 1982: Estimating global radiation on sloping surfaces. N.Z. Journal of Agricultural Research, 25, 281-283.

[18] Temps, R.C. and Coulson, K.L., 1977. Solar radiation incident upon slopes of different orientations. Solar Energy, 19, 179-184.

[19] Talarek, H.D., 1979. Results and Analysis of IEA Round Robin Testing: Document III.A.1 Kernforschungsanlage Jülich GmBh. Jülich, Federal Republic of Germany, 62 pp.

THE DESIGN AND ESTABLISHMENT OF A SOLAR ENERGY RESEARCH CENTRE IN GREECE

C.N.D.Stambolis
Heliotechnic (Hellas) Ltd
3, Alex Soutsou, 106 71, Athens, Greece

Summary

This paper concerns a proposal for the setting-up and the design of a solar energy research Centre in Greece. The Cenetre will be involved in the broad spectrum of renewable energy technology and emphasis will be placed more on information dissemination and development work than on pure research. The current proposals are based on a study which the author undertook on behalf of UNESCO for the establishment of an Energy Centre in Crete. The views and ideas expressed in this paper are a further development of the feasibility study for the Crete Energy Centre. The author believes that the creation of a Solar Energy Research Centre in Greece will come to fill a growing information and education gap in the field of renewable energy sources. The proposed Centre could also help the urgent needs of the local booming solar industry thus helping ensure better product quality and consumer protection. Finally, the Centre's role is seen on a national and regional basis. Furthermore, the Centre could greatly contribute to Greece's growing international activities in renewable energy.

1. Introduction

Since 1961 several proposals have been made (1, 2) for the establishment of a Solar Energy research centre in Greece. Most of these proposals were concerned with the creation of a research establishment which would undertake pure and applied research in several areas of solar technology. Almost none of these proposals considered the educational aspects of such a centre. In 1981 the Greek Government expressed an interest for setting up a regional centre for the promotion of renewable energy sources. The government's interest arose from a proposal from the Crete Energy Centre, a non- profit making learned society, based in the town of Chania, in Crete. As a result of this interest UNESCO commissioned the author to undertake a full feasibility study for the design of the Centre's facilities in Crete (3). This study was completed in March 1982 but had since had little impact in the ensuing discussions and deliberations for the setting up of a Solar Research Centre in Greece. Present government thinking seems to favour the creation of fully fledged research centre which will carry out pure research on several areas of solar technology including collectors, materials and photovoltaics. The government has sought help from the EEC and in June 1983 convened a special two day conference in Athens to discuss the whole matter. As the proposals put forward by the government at that time were ill defined and lacked any background data and justification (4) EEC agreed to a proposal by Greece's Ministry of Research and Technology to appoint a consultant for the preparation of a feasibility study. As far as we have been able to determine this study is still in progress.

In view of the continued interest, both by the Greek government and the EEC, for the setting up of Solar Centre in Greece, we felt that it would be useful to put forward our ideas in this area. The ideas presented in this paper and the resulting proposals in no way do they represent Greek

Government or UNESCO thinking on this matter. In fact our current proposals depart considerably from the results reached in the Unesco funded feasibility study. This paper summarises the work that has been undertaken by the author since 1982 within the context of a broad research project, funded by Heliotechnic (Hellas) Ltd, on the utilisation of renewable energy sources in Greece (5).

2. Proposed Centre Activities

The conceptual diagram of the proposed activities of the Centre is given in Fig. 1. There are five main groups of activities (education, information, demonstration, testing and small scale research) through which the Centre relates to the outside world. The main objective in the creation of the proposed Centre is to actively contribute in the development of renewable energy technology in Greece and thus directly assist: the consumer, (e.g. through the proposed public information programme and demonstration projects) the manufacturing companies (equipment testing), the professionals involved in this field, (engineers, technicians - through the organisation of courses and lecturers), the universities and other institutes of higher education (provide facilities for Diploma and graduate students to carry out small scale research work). It is envisaged that the Centre becomes an information point and service centre for all people working in the broad renewable technology area. The Centre should engourage human interaction and cross-fertilisation of ideas as well as provide a platform for discussion in a national and regional context.

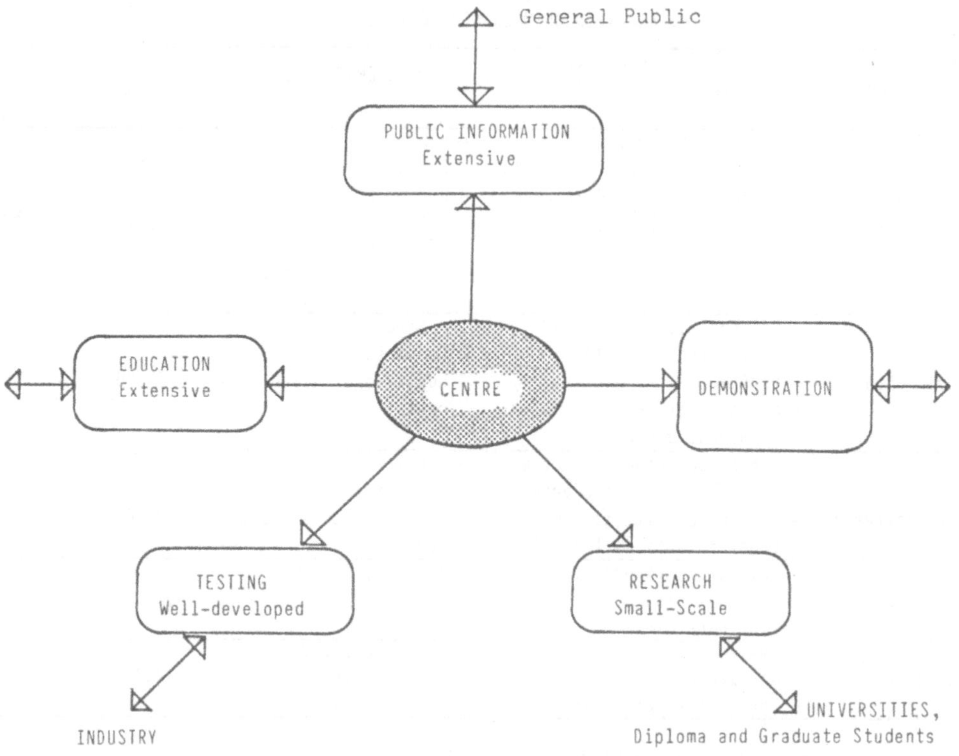

Fig. 1: Conceptual Organisation of Centre Activities

3. Operational Framework

Crucial for the realization of the Centre is the need for permanent staff and housing facilities. It is also important for the Centre to maintain a degree of relative autonomy regarding the organization of its activities and its links to the various national, regional and international bodies involved in energy work. The Centre's legal status should preferably be that of an independent non-profit making company, (in the case of Greece that would result in a Foundation) to be managed by a board of trustees. Representatives from the Ministries of Energy, Research and Technology, the Greek Solar Energy Society (Greek Section of ISES) could be appointed in the board. An annual budget in the region of half million U.S. $ should be provided mainly from Government sources but also from other sources such as UN, the EEC, Greek public companies and industries. This budget could be augmented by outside contributions, mainly from Public Organisations, depending on needs and programme requirements. The Centre should gradually build up its own money-earning activities such as equipment testing, seminars,courses, publications and generally selling of information. In table 1 an analysis is made of the Centre's main activities and proposed budget allocation. During the first phase of the Centre's operation it is anticipated that a total of 20 full time staff will be required. The Centre team should include the following:
Administrative staff (4), Technicians (4), Engineers, Phycisists (4), Information Technology scientists and assistants (4), Communication Experts (3), Keeper (1). Working facilities should also be provided for up to five visiting staff members.

TABLE 1: ANALYSIS OF MAIN CENTRE ACTIVITIES

MAIN ACTIVITY GROOPS	SPECIFIC ACTIVITIES AND DEFINITIONS	INTERACTIONS	ANNUAL BUDGET ALLOCATION * US $	%
Public Information	- Public Information (phone service, leaflets, booklets) - Renewable Energy Data Bank - Reference Library	- Public - Schools - Higher Education Establishments - Data banks - Private and Public Sections companies	150.000	30
Education	- Courses (technician and professional level) - Seminars - School lectures - Permanent exhibition	- Public - Companies - Educational establishments	150.000	30
Demonstration	- Centre buildings to act as a demonstration projects - Centre to initiate demonstration projects at local level	- Public - Companies - Educational establishments	50.000	10
Testing	- Collectors (thermal) - P.V. systems - Wind generators	- Industry	100.000	20
Research	- Small Scale - Applied research - Cooperative projects	- Research Institutes - Universities - Diploma students	50.000	10

* Including administrative overheads

4. Planning of the Centre's Permanent Facilities

Following careful analysis of each of the proposed Centre's activities a total required area of 1,400 m² was arrived at for the Centre's first phase of operation. Figures 2 and 3 show respectively the functional and space organisational diagrams. A second phase of operation assumes an expanded Centre with a total of 2,200 m² floor area. The construction budget for the first phase is currently estimated at 160-150 million Greek Drachs (i.e. 1.5 million U.S.$) including the purchase and installation of all workshop, laboratory and computer equipment. Several possible sites have been considered for the location of the Centre. Given the nature of work to be conducted at the Centre, its size and number of people to be employed, the Centre's design at this stage is site independent. For the Centre to be able to function efficiently and serve the greatest number of people a location near Athens, but not necessarily within the greater Athens area, would be advisable.

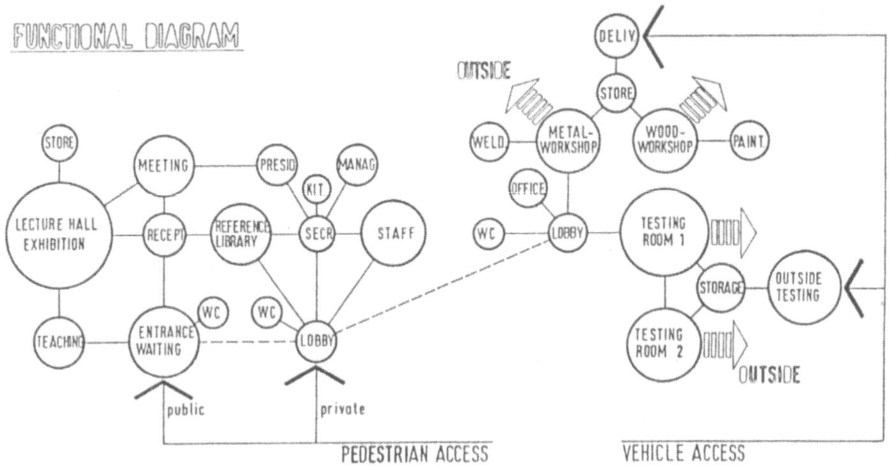

Fig. 2: Interelationship of Centre's functions

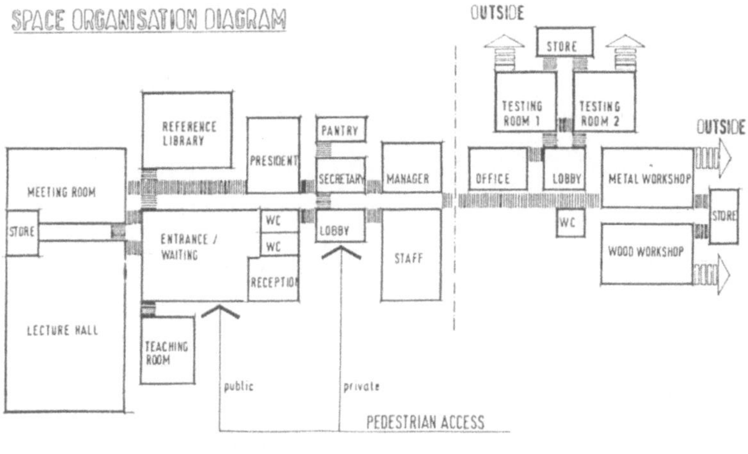

Fig. 3: Centre's Space Organisation diagram

References

(1) A.SPANIDES and A.HADJIKAKIDES. Private communication (several proposals were put forward through COMPLES to the Greek authorities from 1970-75).

(2) C.STAMBOLIS. Design Proposals for a Solar and Aeolian Energy Research Centre in Greece, COMPLES Conference Proceedings, Athens 1975.

(3) HELIOTECHNIC ASSOCIATES INTERNATIONAL. Crete Energy Centre, Feasibility Study, London February 1982. Commissioned by UNESCO, (Paris).

(4) FINANCIAL TIMES. World Solar Markets, July 1983; see also Greece's Weekly for Business and Finance, 3, August, 1983.

(5) SEVERAL SPECIALIZED STUDIES are under preparation within the context of this long term project. See for example paper on "Performance and Cost Evaluation of Solar Passive Buildings in Greece" presented at this conference. Further information on this project can be provided by writing to C.Stambolis, Heliotechnic (Hellas) Ltd, 3 Alex Soutsou, Athens 106 71, Greece.

SOLAR HEATED SWIMMING POOLS IN EUROPE

an EC Demonstration Programme

Ulrich Luboschik IST Energietechnik GmbH, Ritterweg 1
D 7842 Kandern-Wollbach

Willi Kaut Commission of the European Communities, DG XVII
200 rue de la Loi,B 1049 Brussels, Belgium

SUMMARY

In 1975 the Commission decided to support the use of renewable energy sources. Apart from research and development activities a programme of demonstration projects was issued. Technologies, ready to be used which still imply a financial risk are subsidized up to 40% of their total cost. The energetical results of these projects have to be monitored during at least two years, in order to be able to quantify the project' s economical result.

About 50 swimming pools in nine countries of the EC are included in this demonstration programme. The date handling is done in the author's office. The data acquisition varies from hand filled forms (three times per day) to automatic systems. At the end of 1983 data of 32 pools were available.

For each pool a standardized analysis is undertaken which focusses on five aspects:

- Monthly solar radiation on the collector and useful monthly energy
- Tøtal monthly energy used in the pool and resulting solar fraction
- Efficiency curve of the solar system
- Daily input/output diagram
- Seasonal energy balance (Sankey Diagram)

Due to the different levels of the monitoring campaigns, a clear definition of the different energy flows had to be given. Pools with higher levels of monitoring are analysing the influence of wind speed, humidity of the air, pool covers, etc.

The results till 1983 show, that the approach to the monitoring system gives results which are interesting for the pool owners as well as for the manufacturors and installation firms of the solar system. The mean efficiency values of the solar system s vary between 20 and 55%. The results demonstrate that unglazed collectors are able to heat open air and indoor swimming pools during the summer season from England to Southern countries.

The programme will end in 1985.

2. SHORT DISCUSSION OF THE RESULTS

2.1 Data Sheet Treatment

The incoming data sheets were first manually checked on their plausibility. Cross checking helped to identify if the values were physically possible. For example the solar pump's running time was checked and decided if the midday values could be used for the efficiency analysis. The weather data were helpful to cross-check the heat-meter and the pump's running time. Sometimes when the pool became too hot, the solar installation was cut off. These values had to be eliminated for the daily input/output diagram. Quite often these days were only identified when the final graph was plotted.

It was found to be helpful if during an on-site visit a serie of readings of the instantaneous values was taken.

Once the data sheet were checked they were registered as raw-data on a floppy disk. The used PC computer was an Apple IIe.

2.2 Data Analysis on PC Computer

The total amount of data entered into the computer is approx. 270 000 numbers. The output of five figures on a plotter demanded approx. 16 different programmes. The programmes were written in PASCAL.

2.3 Remarks concerning the Plots

In order to enable a direct comparison between different solar installations all figures of different pools have the same scaling e.g. Fig. 2 of Quillan (F) can be directly compared with Fig. 2 of Dragon School (GB).

The averages computed are always mean daily values. If only 20 days of one month were available, the average of these values was taken as the mean daily value of the respective month. The scaling of the plots which was origionally defined during the workshop in Brussels (March 1983) had to be partly abandoned. If the actual values were bigger than the scaling, the histogram is left open.

2.3.1 Fig. 2 - Solar energy and system efficiency

The lower, striped part is the measured solar energy delivered to the swimming pool system.

FIGURE 2 - MONTHLY SOLAR ENERGY

SE-P-NO: 33.5/83 GUIPRY (35)

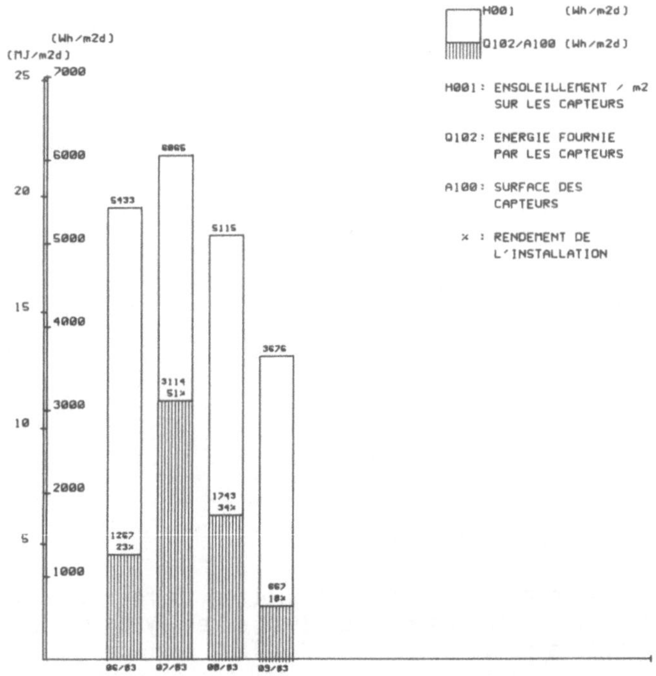

2.3.2 Fig. 3 - Solar fraction

This plot shows the pool management how much of the actually needed energy came from the solar plant and how much had to be generated by the auxiliary heating system. As the mean temperature of the pool is very important for the energy consumption the pool mamangement is able to take some decisions e.g.

- pool temperature was high, but the auxiliary heating was still (unnecessarily) functioning.

- solar fraction is low, perhaps additional solar collectros could make the pool independent of auxiliary heating

In order to indicate different climats the average daily air temperature is plotted as well. It was calculated as

Tmorning + Tnoon + 2x Tevening, the sum devided by 4

$$\text{average } T = \frac{Tm + Tn + 2Te}{4}$$

Figure 3 - MONTHLY SOLAR-FRACTION

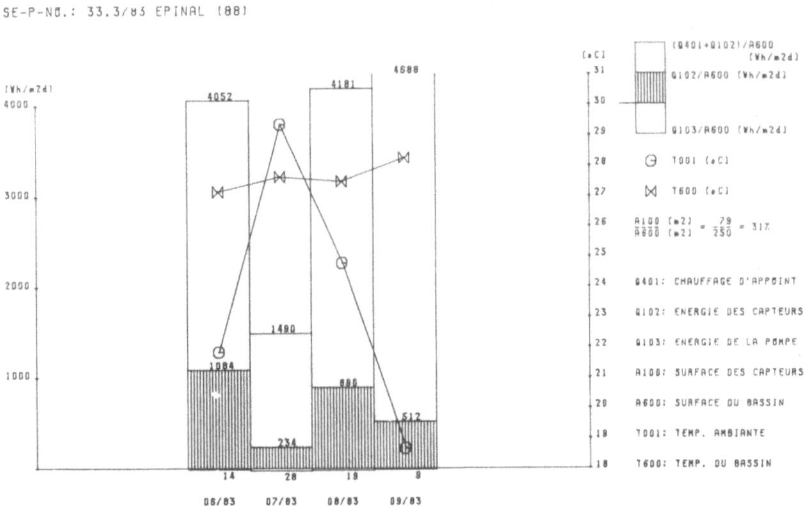

2.3.3 Fig. 4 - Sankey diagram

The seasonal energy flow is plotted in form of a Sankey diagram. The solar energy received directly by the surface area of the pool is not included. The surface reflection, the radiation against the sky, and the absorption of the pool itself are unknown.

The monthly mean values of collector efficiency and solar fraction should correspond to the values of the figures 2 and 3. For the Sankey diagram they are calculated as the difference of the values at the opening and closing day, whereas for Fig. 2 and 3 they are the average of the monthly daily values. Variations are therefore possible if daily readings are missing.

Figure 4 - SANKEY-DIAGRAM

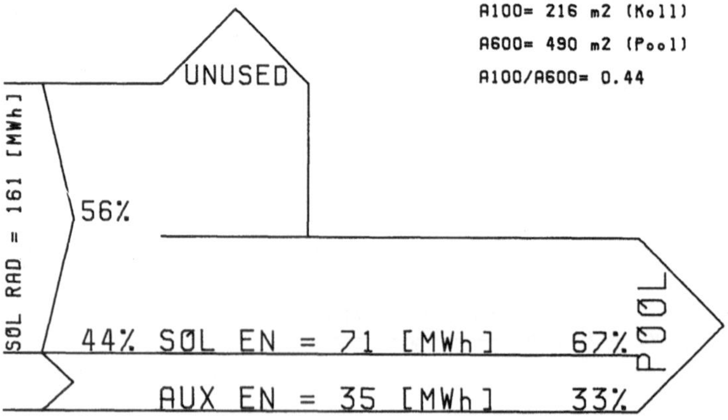

2.3.4 Fig. 5 - Collector array efficiency

This plot is especially interesting to compare the functioning of solar installations in different climats. The histogram traced on the same plot, indicates under which conditions the solar installation was functioning. The maximum of these beams

Figure 5 - COLLECTOR ARRAY EFFICIENCY

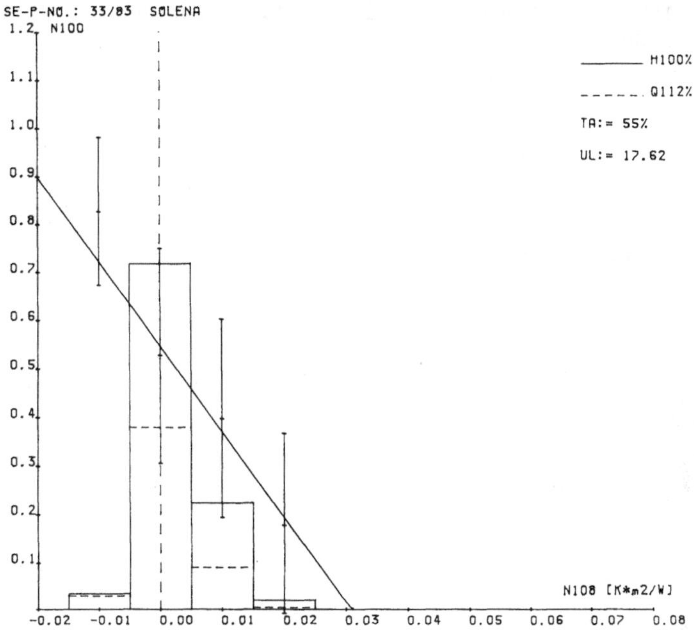

should be as near as possible to zero on the dt/I axis. The vertical lines give the standard deviation of the measured points around the mean value. This graph needs a lot of points during stable weather conditions. If the weather is changing at the moment of the reading, the internal thermal masses of the installation falsify the data completely. It was necessary to make a close checking of the daily data.

For the French swimming pools it was very helpful to have the readings during the on-site visit, especially for some installations the flow rate varies during the day. For a few pools it was not possible to establish this graph. In 2 pools the data were not measured, in 6 pools the equipment failed or there dit not exist enough readings.

2.3.5 Fig. 6 - Daily input/output diagram

In this graph we plotted the result of the daily solar input (absciss) and the output of the collector (ordinate). Typical for all lines is that the absciss is not cut in the zero point but slightly to the right. If the daily radiation is not above a certain threshold the solar installation does not function at all.

From this plot further information can be taken as internal thermal masses and mean efficiency of the installation. The line is calculated analytically, visually one would trace it in a few cases a bit differently. Further analysis will be undertaken.

Figure 6 - DAILY INPUT/OUTPUT

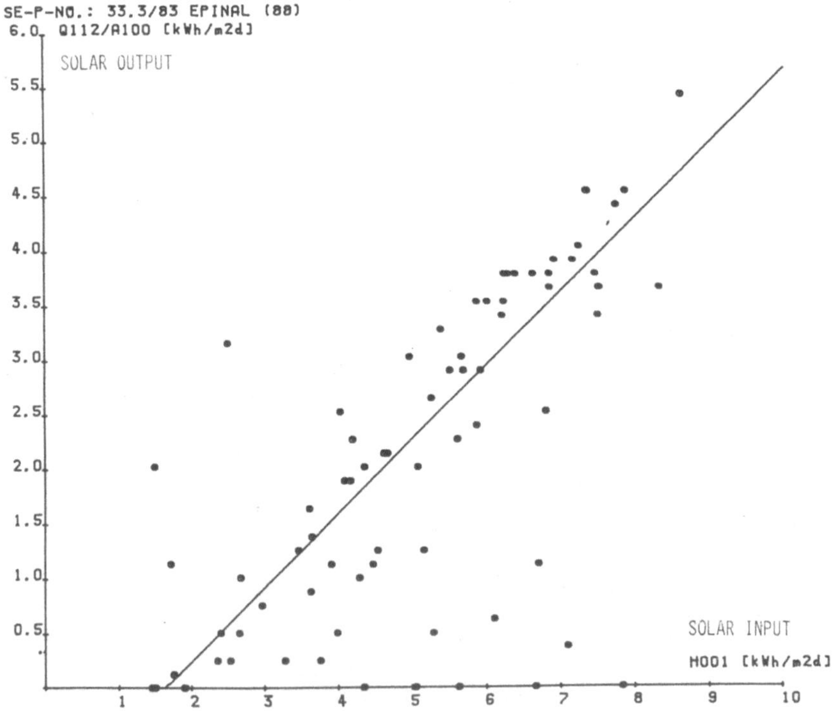

SESSION V - SOLAR COLLECTORS AND SECOND GENERATION DHW AND SH-SYSTEMS

Design of static concentrators with the receiver immersed in a dielectric tube

Computer-aided development of solar domestic hot water systems

Solar domestic hot water systems in Denmark - Design, performance and costs

Low-cost, low-weight and medium-efficiency solar collectors, construction and examples

Investigation of solar-assisted heat-pump systems using the heat recovery glazing concept

A detailed theoretical model for a flat plate natural convection loop-type solar air collector

Optimal control theory applied to dwelling heating system

Performance and reliability of inexpensive single family SDHW package systems

Performance predictions for all glass tubular evacuated absorbers

Analysis of the climatic data necessary - Problems to be solved and application to flat-plate collectors

Experience with collector draining and rigid mineral insulation in continental climate (Capellen - Luxembourg)

The limits of the transient method applied to the solar collectors performance measurement

Etude expérimentale de capteurs solaires à air en région parisienne

A second generation thermosiphon solar water heater

Design of domestic hot water- and solar heating systems with Philips' evacuated collectors

A simplified solar domestic hot water and heating system with evacuated tubular heat pipe collectors

Chauffe-eau solaire domestique à stockage intégré

Spectrally selective transition-metal oxinitride coatings on rough substrates for application in high performance solar collectors

Hydrophile solar collector system

Balancing of solar heating options

On the importance of wind velocity and ambient temperature to the performance of flat-plate collectors

Performance of pipe type solar water heater

Research project "Guidelines for optimal planning and construction of solar energy systems"

Shared solar space and water heating

EIGHT YEARS OF COLLECTOR TESTING IN EUROPE
WHAT HAVE WE LEARNED?

W B GILLETT
Sir William Halcrow & Partners
Burderop Park
Swindon SN4 0QD, UK

and

J E MOON
Solar Energy Unit
University College
Cardiff CF2 1TA, UK

Summary

The Collector Testing Group has been working on the development of test methods for solar collectors under the direction of the Commission of the European Communities DGXII and the Joint Research Centre at Ispra since 1975. As a result, it is now possible to test the thermal performance of solar water heating collectors with confidence both outdoors and in solar simulators. Further work is required before air collectors can be tested with the same level of confidence. Future work on solar water heating systems should concentrate on the development of durability tests for collectors, and on short term performance tests for complete solar water heating systems.

1. INTRODUCTION

Collaborative work on the development of test methods for flat plate liquid heating collectors began in Europe in 1975 with the formation of the CEC Collector Testing Group (CTG). The initial objectives of the Group were to develop thermal performance test methods for use in European weather conditions, and with sufficient accuracy to increase the credibility of the European solar heating industry. Later, the objectives were expanded to encompass the development of durability test methods and the publication of guidelines on solar collector design with a view to encouraging more rapid implementation of solar heating systems.

The work of the Collector Testing Group has involved two concerted action programmes covering the periods 1975-1979 and 1979-1983. The CTG consists of specialists and experts from more than twenty European laboratories; each country in the Community provides at least one participant. The results of the work of the CTG were first published in 1980 as "Recommendations for European Solar Collector Test Methods" (1), and a more comprehensive publication entitled "SOLAR COLLECTORS Test Methods and Design Guidelines" has recently been completed (2).

The work of the CTG has involved a number of concerted activities in which similar tasks have been performed in the different European laboratories. These activities have allowed outdoor test procedures to be subjected to the different climatic zones in Europe, and have provided a baseline of common experiences from which the Group has been able to formulate its recommendations.

Details of the work carried out by the CTG have recently been reported extensively (3) and will therefore be referred to only briefly here. It is important now to identify what has been learned from the work, and to indicate clearly what new R & D activities are needed.

2. COLLECTOR THERMAL PERFORMANCE ANALYSIS AND STANDARDS

The first public document to be produced by the CTG was a Reporting Format for presenting the thermal performance test results for a collector. This format has been adopted not only by European countries but also by many others through their participation in the Collector Testing Task of the International Energy Agency (IEA).

The Format Sheets have been improved and expanded over the years, but the basic performance characteristic for a solar collector is retained either as a first order relationship:

$$\eta = \eta_o - UT^* \qquad \text{- Equation (1)}$$

or as a second order relationship

$$\eta = \eta_o - a_1 T^* - a_2 G (T^*)^2 \qquad \text{- Equation (2)}$$

Where T^* is defined in terms of the mean fluid temperature in the collector (Tm) as

$$T^* = (Tm - Ta)/G \qquad \text{- Equation (3)}$$

The format sheets and test method recommendations produced by the CTG have been used as a basis for the preparation of National Standards in Europe (4, 5, 6), and are also being used by the Solar Energy Committee of the International Standards Organisation ISO TC/180 which is currently drafting a Standard for solar collector testing.

3. STEADY STATE OUTDOOR TESTING OF WATER HEATING COLLECTORS

When the work began in 1975, the first main activity was for each laboratory to perform a steady state outdoor test on a matt black, single glazed collector, and the results varied widely between the laboratories. Improved designs of test loop, better instrumentation and better defined procedures have greatly reduced the variations between laboratories and today the results from different laboratories in Europe can be expected to agree on values for the collector performance parameters η_o to within about $\pm$ 0.02, and U to within about $\pm$ 0.5 W/m^2K. Better agreement is usually possible with low heat loss collectors such as evacuated tubular collectors, but concentrating collectors are more difficult to test.

Few European laboratories however would choose to test collectors outdoors today, even if their test sites were situated in the South of Europe. The variations in outdoor conditions restrict the periods on which steady state testing can be performed to a very few days per year in Northern Europe, and are a significant inconvenience in the South. Current and future work on collector testing seems likely to be carried out more and more in solar simulators. Three Northern European countries have developed outdoor test methods which can be used in variable outdoor conditions, but these have not yet become widely accepted and appear likely to be of most use for testing large collector arrays or collectors which for some reason cannot be tested with confidence in solar simulators.

4. STEADY STATE TESTING OF AIR COLLECTORS

Collaborative work on air collector testing by the CTG began in 1982 when 15 test facilities were constructed with CEC support.

A round robin collector from Belgium was distributed for testing in 1983 and preliminary results have been obtained using an agreed procedure. The results show considerable scatter and this is attributed to three main causes:

- variations in collector air leakage
- poor accuracy of flow rate measurements
- poor accuracy of inlet and outlet air temperature measurements

Improvements in the accuracy of air flow rate and temperature measurements are required in order to permit air leakage effects to be better defined and to determine efficiencies more accurately. At present there is very little experience of air collector testing in Europe and the accuracies in efficiency measurement vary greatly between laboratories.

5. SOLAR SIMULATORS FOR COLLECTOR TESTING

More than 12 solar simulators have been built in Europe for collector testing in recent years, and, with the exceptions of Greece, Ireland and Luxembourg, each country in the EC has at least one simulator.

The largest solar simulator in Europe is at the JRC in Ispra, where a "Workshop on Solar Simulators" was held in February 1982. The proceedings of this workshop (7) contain descriptions of most of the world's solar simulators that are used for collector testing (30 in all).

Special attention needs to be paid to the measurement of simulated solar irradiance and thermal irradiance in solar simulators, and test procedures for use in simulators are published in the new CEC publication (2), and in the British and French Standards (4, 5).

The more recent designs of simulator have included provision for rotating the lamps to permit the measurement of incidence angle modifiers and collector tilt effects as shown in Figure 1.

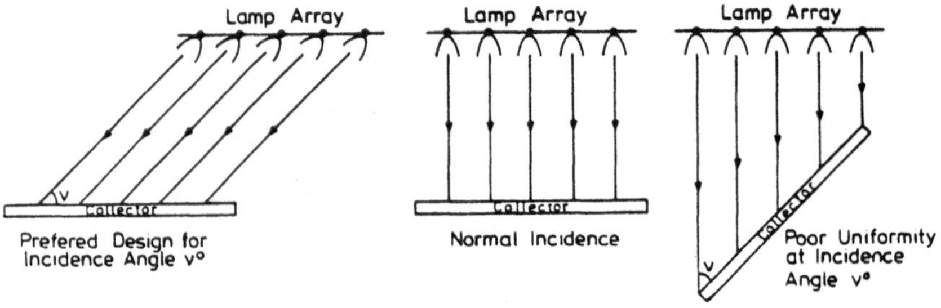

Figure 1 Rotating Lamp Array for Incidence Angle Modifier Measurement

6. COLLECTOR DURABILITY

With the assistance of experts from each of the EC countries, a survey of approximately 2500 m^2 of installed collectors (85 installations), was carried out in 1982, using a standard reporting format. The results from this survey have been used to prepare guidelines on collector design (2), and also to confirm which aspects of collector durability should be tested before a new design is put onto the market.

Many of the design considerations, which are important for solar collectors, are well known in other fields. For example, it is important to make provision for differential thermal expansion, to select materials with appropriate resistances to temperature and to ensure that connected materials are compatible from the point of view of corrosion.

Design guides and handbooks are available to assist with both temperature and corrosion effects, and relatively simple tests can be used to evaluate the temperature resistance of collectors. The corrosion resistance and long term durability of collector materials however, can be very expensive to test, and it is often more cost effective to check designs using data from the materials suppliers and from engineering handbooks. For some collectors, a salt mist test will give a good indication of corrosion resistance, but more experience is required before collector lifetimes can be reliably predicted from salt mist tests.

The following "Qualification Tests" are under development by the CTG for evaluating collector durability:

- High temperature stagnation
- Rain penetration
- External thermal shock
- Absorber fluid over-pressure

Outline procedures for carrying out these tests at relatively low costs are given in the new CEC publication (2).

An aspect of collector design which appears to warrant further study is that of ventilation. Condensation in collectors can cause rapid degradation of the absorber surface and insulation, and a reduction in collection efficiency. Clearly there is a need for both drainage and ventilation, but how to optimise the positioning and sizing of the holes has not been determined.

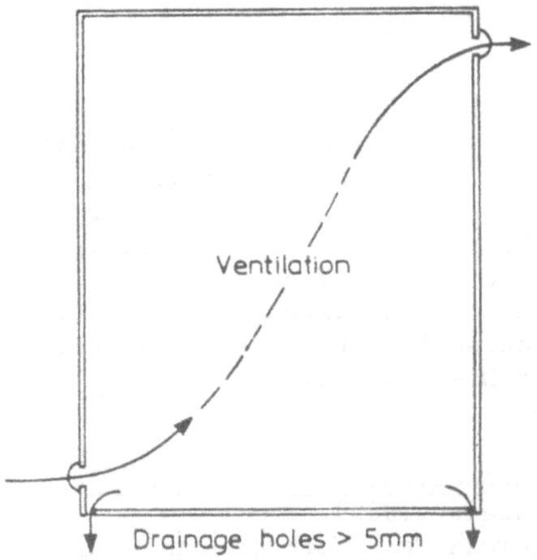

Ventilation Air Flow and Drainage Holes

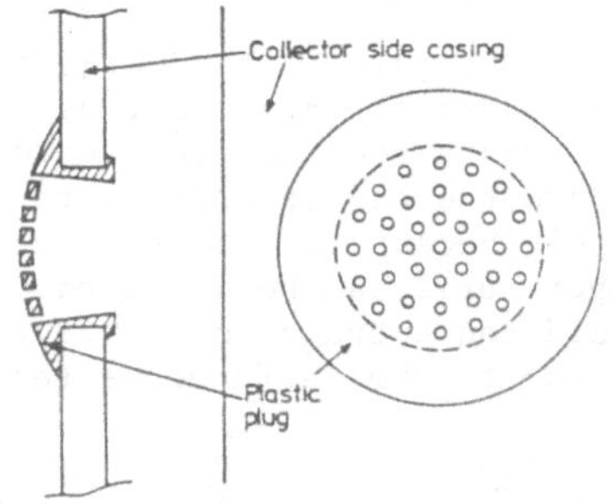

Ventilation Plugs

7. OPTICAL PROPERTY MEASUREMENT

Round robin activities were carried out by a group of European Specialists in 1982 and 1983 to compare measurements of the optical properties of absorber surfaces and collector cover materials. The results indicated that measurements of solar absorptance and solar transmittance can be made with an accuracy of better than $\pm$ 2% by most laboratories but that better agreement could be achieved if instrument calibration surfaces and data analysis procedures were standardised.

The measurements made of emittance and transmittance for the thermal infra-red wavelengths varied markedly between laboratories, and demonstrated a clear need for more well defined procedures and for better calibration surfaces.

These measurements are needed to allow designers to choose the best surfaces and materials, to permit quality control checks during manufacture, and for durability studies.

8. TESTS FOR SOLAR WATER HEATING SYSTEMS

The most widespread application of solar collectors in Europe is to heat domestic hot water (DHW), and test methods for complete solar DHW systems are currently under development. In the USA test methods have been developed to give certification and ratings for solar DHW systems. The European approach is more general, with the aim of developing short term tests which can be used with either simplified or full simulation models to predict long term system performance. Models have already been developed by the European Modelling Group, and it now remains to develop the test methods and to validate them in typical European climatic conditions for different system types, and demand patterns.

The performance of solar DHW systems depends on component characteristics, weather conditions and hot water demand. The test procedures under development include outdoor and simulator tests for whole systems, and the use of in-line heaters to simulate the collector array. Few thermosiphon systems have been tested in Europe, but the market for these seems likely to grow. Thermosiphon system testing will therefore form an important part of future work programmes.

9. CONCLUSIONS

As a result of the work of the Collector Testing Group, the thermal performance of water heating collectors can now be determined with appropriate accuracy both outdoors and in solar simulators. However, the performance of air collectors cannot yet be determined with certainty, largely because of the poor accuracy to which air flow rates and temperatures can be measured. Future work on solar water heating systems should concentrate on the development of durability tests for collectors and short term performance tests for solar DHW systems.

References

1) Derrick, A & Gillett, W B (Eds) Recommendations for European Solar Collector Test Methods CEC DGXII Brussels (1980).
2) Gillett, W B & Moon, J E SOLAR COLLECTORS Test Methods and Design Guidelines (to be published).
3) Gillett, W B, Aranovitch, E & Moon, J E Solar Collector Testing in the European Community Int J. Solar Energy Vol 1 pp 317-341 (1983).
4) DD77: 1982 Draft for Development: Method of Test for the Thermal Performance of Solar Collectors British Standards Institution (1982).
5) AFNOR P50-501 Liquid Circulation Solar Collectors. Determination of Thermal Performance Association Française de Normalisation (1980).
6) DIN 4757 Part 4 Solar Collectors. Determination of efficiency, heat capacity and pressure drop Beuth Verlag, Berlin (1982).
7) Aranovitch, E & Gillett, W B Workshop on Solar Simulators. Proceedings JRC, Ispra (1982).

NRCC AIR COLLECTOR TESTING FACILITY

M.A. BERNIER
National Research Council Canada
Division of Building Research
Ottawa, Canada, K1A 0R6

Summary

An outdoor air collector testing facility has been established at NRCC during 1983 to develop improved thermal performance test methods and equipment and better characterizations of air collector thermal performance. The facility was calibrated using an overall calibration approach. At high flow rates (0.045, 0.080 and 0.160 kg/s), accuracy of the measured rate of collected energy, is within the requirements of ASHRAE Standard 93-77.

1. INTRODUCTION

Experimental evaluation of the thermal performance of air solar collectors presents problems related to the experimental procedures and equipment and to the nature of air collectors. An experimental facility capable of performing a complete ASHRAE Standard 93-77 thermal performance test (1), was established at the Division of Building Research. The facility, to be used to evaluate the thermal performance of air collectors and for the development and verification of new measurement techniques and equipment, consists of an altazimuth frame and an air test module (Figure 1).

2.1 THE ALTAZIMUTH FRAME

The altazimuth frame, a welded steel structure mounted on four wheels, supports the collector under test (up to 2.7 × 2.7 m size) and the air test module. The on-board computer, using the date and time entered by the operator, positions the altazimuth frame so that the face of the collector is perpendicular to the sun's rays (or at a specified angle if an incident angle modifier test is performed) to within ±1°. A wind vane anemometer and a Precision Spectral Pyranometer (PSP) are attached to the altazimuth frame to measure local wind speed and solar radiation in the plane of the collector. The altazimuth frame rotates on a leveled concrete pad.

2.2 THE AIR TEST MODULE

The air test module (Figure 1) measures mass flow rates before and after the collector using orifice plates according to ASME procedures (2). By selecting one of five orifice plates, collectors can be tested at flow rates up to 0.20 kg/s. Flow through the collector is regulated (to within ±0.6% of the desired value) by bleeding air through a computer-controlled bypass (3).

Inlet and outlet temperatures are measured using both silicon transistors and platinum resistance temperature sensors. To obtain the average temperature and decrease the measurement uncertainty, each type of sensor is connected in series and arranged in an array of five elements (Figure 2). Temperature measurement errors due to radiation exchange between the sensors and the duct wall are minimized by insulating the temperature measurement section with 20 mm of armaflex insulation and by partially shielding each sensor with a stainless-steel tube. The silicon sensors have a negative temperature coefficient and the platinum ones have a positive coefficient, so that an error due to a drift in the computer's A/D converter can be detected as an apparent divergence of the inlet temperature readings (4). The ambient temperature and the temperature of the air at both orifice plates are measured using five-element platinum resistance temperature sensors. The air temperature at the collector inlet can be controlled within ±0.2°C by a computer-regulated, multi-element strip heater (9.4 kW).

The on-board computer is accessed via a remote terminal. The operator sends a series of commands (3) to the computer and the test is carried out automatically. Data are stored in the on-board computer until the end of the test, when they are transferred via underground cable to a DEC PDP 11/23 computer for analysis and storage.

3.1 CALIBRATION TECHNIQUE AND INSTRUMENTATION

An overall calibration was undertaken to verify the accuracy of the measured rate of collected energy and compare it with the requirements of ASHRAE Standard 93-77. The rate of energy collected by a solar collector with no leakage is defined as:

$$Q_c = \dot{m} \cdot C_p \cdot \Delta T \tag{1}$$

where Q_c = rate of collected energy (W), $\dot{m}$ = mass flow rate (kg/s), C_p = specific heat (J/(kg•°C)) and ΔT = temperature rise across the solar collector (°C).

In the overall calibration approach, the solar collector is replaced with the Calibration Heat Source (CHS) (5), which transfers an accurately measured quantity of electrical energy (P_{in}) to the circulating fluid (air). The accuracy can then be determined by comparing Q_c, as measured by the test loop, with the known input (P_{in}).

In the CHS used (Figure 3), entering air forms a "thermal guard" around the heater. This reduces net heat losses by reducing the temperature difference across the exterior insulation. The heat losses from the CHS are proportional to the difference between air inlet and ambient temperature. To compensate for these losses a correction factor (1.26 W/°C) must be applied to P_{in} (5).

The power supplied to the CHS is regulated and the power level controlled by a variable transformer. Based on the observed stability of the power input and on the accuracy of the power transducer used for measurements, the accuracy of P_{in} is estimated to be ±0.5%. Heat losses from the CHS introduce an additional uncertainty (estimated at ±0.5%) in the energy transfer to the circulating fluid.

3.2 CALIBRATION RESULTS

Table I gives the results of a typical calibration test run, performed indoors under steady state conditions. The results, for a flow rate of 0.08 kg/s, are shown on Figure 4, where the accuracy of Q_c is plotted against inlet temperature minus ambient temperature ($T_i - T_a$). The accuracy of Q_c is defined as: $\left[\frac{P_{in} - Q_c}{P_{in}}\right] \times 100$.

Heat losses in the pipe sections from the temperature measuring locations to the collector (header heat losses) could account for much of the difference between Q_c and P_{in}. The combined header heat losses were measured by joining the headers end-to-end, and measuring the temperature drop of heated air passing through them. The heat loss coefficient $[(UA)_{HHL}]$ based on the difference between the average fluid temperature $(T_i + T_o)/2$ and the ambient temperature (T_a) was found to be 0.78 (W/°C). For well insulated headers of approximately the same size, the header heat losses (Q_{HHL}) are given by: $Q_{HHL} = (UA)_{HHL} [(T_i + T_o)/2 - T_a]$, where $(UA)_{HHL} = 0.78$ (W/°C) in this case.

The values of Q_c based on the measured T_i and T_o have been corrected for header heat losses and the results superimposed on Figure 4.

The random uncertainty allowed by ASHRAE Standard 93-77 on the measured rate of energy collected was evaluated (Appendix A). No uncertainty was included for header heat losses. All data points lie within the allowable uncertainty intervals, indicated by the vertical bars (Figure 4).

Less extensive calibration tests were performed at four other flow rates (results in Table II). The accuracies of Q_c at 0.160, 0.080, and 0.045 kg/s are acceptable, whereas the accuracies at 0.025 and 0.015 kg/s are outside the uncertainty interval allowed by the Standard. Such problems as incomplete mixing upstream of the temperature sensors, radiation errors, and eccentric location of orifice plates in the pipe, could explain the inaccuracy of Q_c at low flow rates.

4. REFERENCES

1. Methods of Testing to Determine the Thermal Performance of Solar Collectors. American Society for Heating, Refrigerating and Air-Conditioning Engineers, ASHRAE Standard 93-77, 1977.
2. Fluid Meters, Their Theory and Application, American Society of Mechanical Engineers, Fifth Edition, New York, 1959.
3. Outdoor solar collector test equipment at the National Solar Test Facility. Ontario Research Foundation, Report No. PHYS. G.P., 81-15, 1981.
4. NIELSEN, V.H. (1981) Solar Collector Thermal Performance Measurements at the National Solar Test Facility. Solar Energy Society of Canada - Montreal.
5. WRIGHT, J.L. and HOLLANDS, K.G.T. (1980) A Calibration Collector for Solar Air Heater Test Loops, University of Waterloo Research Institute, Final Report for DSS contract 07SU.31155-0-2605, Waterloo.
6. KLINE, S.J. and MCCLINTOCK, F.A. (1953) Describing uncertainties in single sample experiments, Mechanical Engineering, January.
7. Measurement of fluid flow by means of orifice plates, nozzles and venturi tubes inserted in circular cross-section conduits running full. ISO Standard 5167-1980(E), ISO Standards Handbook 15, 1983.

5. APPENDIX A UNCERTAINTY ALLOWED ON Q_c

The random uncertainty allowed by ASHRAE Standard 93-77 on the measured rate of collected energy (Q_c) can be evaluated using the propagation of random measurement uncertainty technique (6). Equation 1 can be expressed in terms of the measured parameters:

$$Q_c = K \cdot C \cdot A \cdot \sqrt{\frac{\Delta P \cdot P_{abs}}{T_{abs}}} \cdot C_p \cdot \Delta T \qquad \text{(A-1)}$$

where K is a constant, C is the discharge coefficient, A is the area of the orifice, ΔP is the pressure drop across the orifice, P_{abs} is the orifice absolute pressure and T_{abs} is the orifice absolute temperature.

The accuracy requirements of each individual measurement, as given in ASHRAE Standard 93-77, are:

$W_A = \pm 0.2\%$ $W_{P_{abs}} = \pm 2\%$ $W_{\Delta T} = \pm 0.1°C$

$W_{\Delta P} = \pm 1\%$ $W_{T_{abs}} = \pm 0.5°C$ $W_{C_p} = 0.5\%$ (assumed).

The standard does not specify any uncertainty on C: it is assumed to be ±2% (7).

The uncertainty in the dependent variable (W_{Q_c}) can be expressed in terms of the uncertainties in each of the independent variables (W_A, $W_{\Delta P}$, etc) (6). Performing an analysis as described in reference 6, we obtain:

$$\frac{W_{Q_c}}{Q_c} = \left[5.54 \times 10^{-4} + \left(\frac{-0.25}{T_{abs}}\right)^2 + \left(\frac{0.1}{\Delta T}\right)^2\right]^{\frac{1}{2}}$$

By substituting the corresponding values of T_{abs} and ΔT, values of $\frac{W_{Q_c}}{Q_c}$ are obtained (bars, Figure 4).

TABLE 1 Typical Calibration Results*

ΔT (°C)	$T_i - T_a$ (°C)	P_{in} Corrected for CHS Heat Losses (W)	Q_c (W)	Accuracy of Q_c (%)	Q_{HHL} (W)	Q_c Corrected for HHL (W)	Accuracy of Q_c Corrected for HHL (%)
11.45	31.7	960.1	924	3.8	28.4	952.4	0.8
11.48	31.6	960.2	926	3.6	28.3	954.3	0.6
11.49	31.5	960.3	927	3.5	28.3	955.3	0.5
11.50	31.4	960.4	928	3.4	28.3	956.3	0.4

*Flow rate ($\dot{m}$) = 0.080 kg/s
P_{in} = 1000 W
Values of $\dot{m}$, ΔT, T_i, T_a were measured at 5-second intervals and averaged over a 5-minute period.

TABLE II Measured and Allowable Accuracy of Q_c ($T_i = T_a$ in all cases)

Upstream Flow Rate (kg/s)	Orifice Plate	P_{in} (W)	Measured[d] ΔT (°C)	Accuracy of Q_c (corrected for HHL) (%)	Allowed random uncertainty[c] (%)
0.160	A	1000	6.0	2.6[a]	2.9
0.080	B	1000	12.3	0.8[a]	2.5
0.045	C	1000	21.5	1.9[a]	2.4
0.025	D	1000	38.2	3.7[a]	2.4
0.015	E	500	29.2	15.1[b]	2.4

[a] Average of 4 measurements
[b] Average of 3 measurements
[c] Method described in Appendix A
[d] Silicon transistors

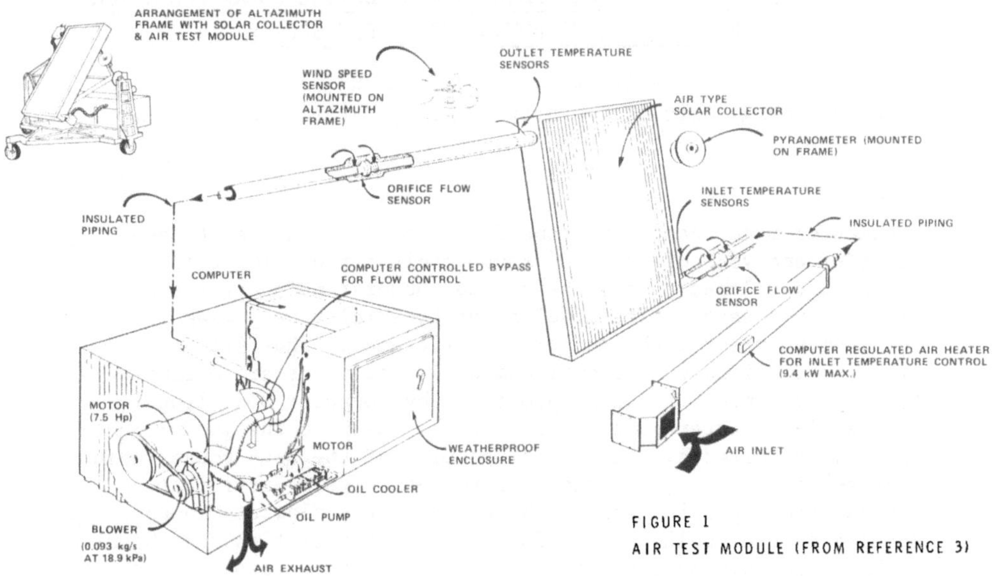

FIGURE 1
AIR TEST MODULE (FROM REFERENCE 3)

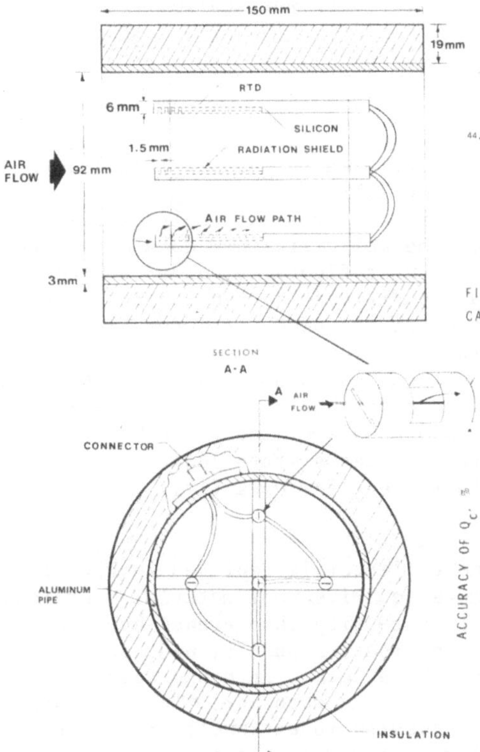

FIGURE 2
TEMPERATURE MEASURING SECTION

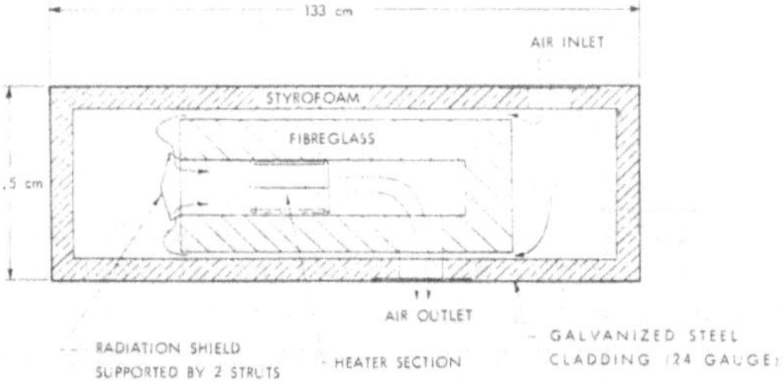

FIGURE 3
CALIBRATION HEAT SOURCE (FROM WRIGHT AND HOLLANDS)

BR 6510-3

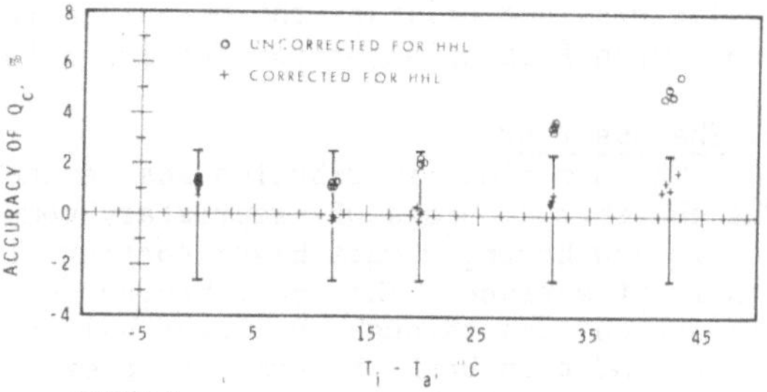

FIGURE 4
ACCURACY OF Q_c VS $T_i - T_a$ (FLOW RATE 0.08 kg/s)

BR 6510-4

COMPUTER MODEL OF THE NOBELS PEELMAN AIR HEATING SOLAR COLLECTOR USED IN THE CEC6 ROUND ROBIN TEST SERIES

A.A. Green and J.E. Moon
Solar Energy Unit, University College, Cardiff, Wales, UK.

Summary

A computer model based on a one-dimensional, steady-state heat transfer analysis of the CEC6 round robin air heating collector has been developed to aid theoretical understanding of its heat transfer processes. The model predicts a relatively weak flow-rate dependence for the collector efficiency. The effects of likely variations in the environmental conditions are estimated to be very similar to those predicted by computer modelling for double-glazed water heating collectors. Despite the good repeatability exhibited by the initial results of the CEC6 round robin test series, there are large differences between participating laboratories, particularly with regard to the flowrate dependence. Model-predicted collector performance characteristics are shown to be reasonably consistent with those CEC6 test results which exhibit the expected insensitivity of the zero loss efficiency to the air mass flowrate. It is concluded that round robin collector efficiency measurements of greater consistency and accuracy are required, to make the intended application of the model to the general normalisation of test data worthwhile.

1. Introduction

The Nobels Peelman air heating collector, used in the sixth collaborative round robin test series of the CEC solar collector testing programme, features a complex air flow distribution which ensures extremely efficient absorber to air heat transfer. Collector thermal efficiency data have been obtained initially at three recommended air mass flowrates, under a wide range of outdoor and indoor environmental conditions, by ten laboratories in the European Communities. A computer model of the CEC6 collector has been developed to aid theoretical understanding of its heat transfer processes, and to estimate their sensitivity to variations in parameters characterising the operating conditions. This specialised model has evolved from earlier thermal performance modelling of single- and double-glazed water heating collectors, and utilises the one-dimensional, steady-state analytical method and Newton-Raphson algorithm developed in the earlier work (1).

2. The absorber

The air flow distribution inside the collector is depicted in Fig. 1 . The absorber consists of proprietary compressed wire wool material, pervious to air and having a matt black coating of near-unity absorptance baked onto its front surface. The space behind the absorber is partitioned, forcing the air to pass through the lower half of the absorber into the space in front, and then back through the upper half of the absorber. The permeation of the wire wool material by the air flow results in extremely efficient absorber to air heat transfer, even at low to moderate flowrates.

Calculations based on empirical convective heat transfer data reported in the scientific literature (2), for insolated air-cooled porous beds

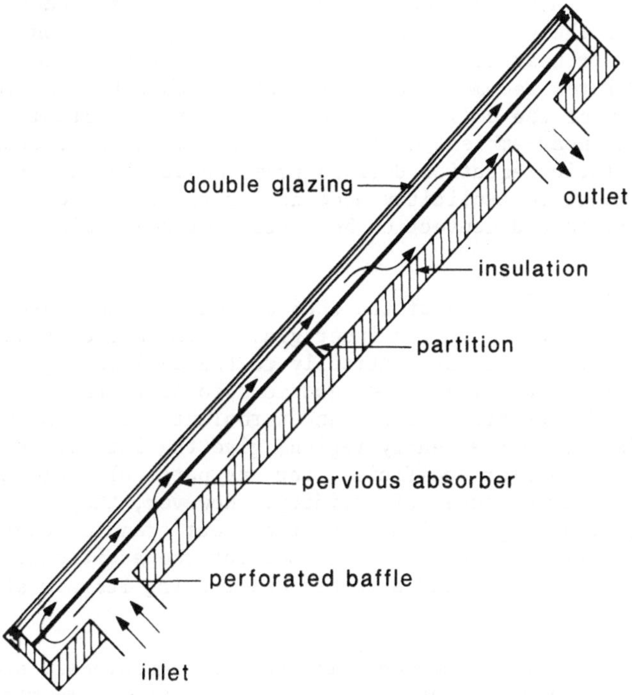

Fig. 1 Sectional view of the CEC6 air heating collector, showing its air flow distribution.

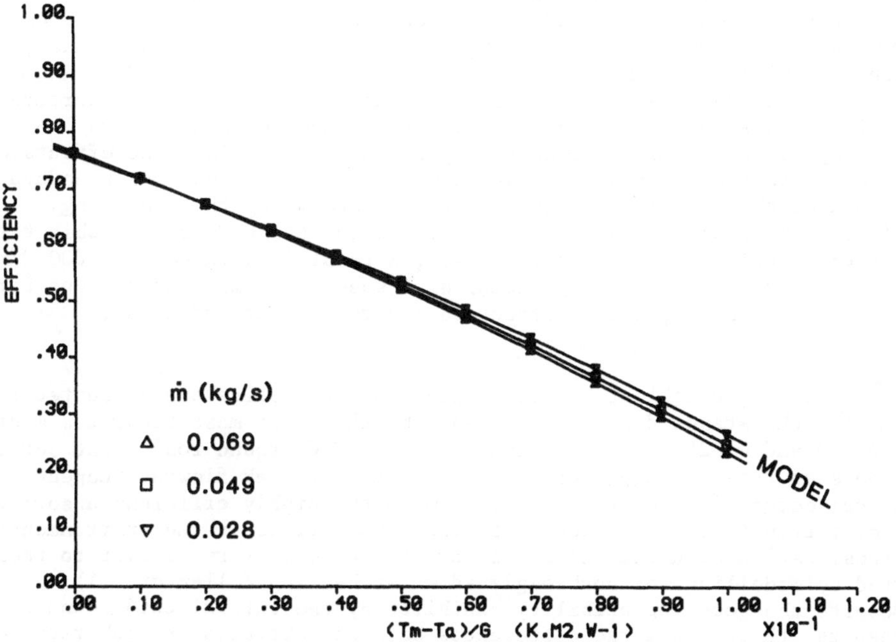

Fig. 2 Model-generated CEC6 collector performance characteristics for the standard conditions and the three recommended air mass flowrates.

consisting of randomly-stacked blackened wire meshes, indicate that the thermal performance of the CEC6 collector should closely correspond to an infinite absorber to air heat transfer coefficient (3) at the moderate to high air mass flowrates recommended for the round robin test series. The thermal efficiency of the collector should therefore exhibit little flowrate dependence at near-ambient heat transfer air mean temperatures T_m. Nevertheless, a correlation equation derived from the published porous bed data is used in the model to explicitly account for the flowrate dependence of the absorber to air forced convection heat transfer coefficient.

3. Flowrate dependent heat losses

In the heat transfer analysis of the CEC6 collector, the convective heat fluxes from the heated air to the inner glass cover and duct walls are the most difficult to represent mathematically to the desired degree of accuracy. Of the applicable correlation equations found in the scientific literature, those in (4) for transitional and turbulent flows, which account for mixing in the classic thermal entry region, give the largest average forced convection heat transfer coefficient for any particular air mass flowrate in their respective ranges of validity. However, they are likely to underestimate actual average forced convection heat transfer coefficients for the CEC6 collector since they do not account for the additional mixing caused by the baffles, partition and absorber obstructing its air stream.

4. Method of solution

The rest of the heat flux components in the one-dimensional, steady-state analysis of the CEC6 collector are represented in the same manner as for double-glazed water heating collectors (1). An algorithm based on the Newton-Raphson numerical method of solution (1) is used to determine, for specified operating conditions, the unique combination of steady-state temperatures within the collector for which the total nett heat fluxes into the duct walls, absorber, and inner and outer glass covers are all zero.

The instantaneous steady-state efficiency of the collector, based on its $1.835\,m^2$ aperture area, is computed for consecutive values of T_m increased in increments of $0.01 \times G$ K from the ambient air temperature T_a to 370 K, where G is the total irradiance (in W/m^2) in the plane of the collector aperture. To facilitate intercomparison, collector performance characteristics are usually computed for a set of standard operating conditions. The effects of variations in test conditions are estimated by varying, about its standard value, each of the appropriate parameters characterising the operating conditions. The standard conditions specified for the CEC6 round robin test series are:- a collector tilt angle of 45°, a total irradiance G of 800 W/m^2, no diffuse irradiance, an average local wind speed of 3 m/s, an ambient air temperature T_a of 288 K, and an effective sky temperature of 278 K.

5. Results

Model-generated collector performance characteristics are presented in Fig. 2, for the standard conditions, and the three air mass flowrates $\dot{m}$ of 0.069, 0.049 and 0.028 kg/s recommended for the CEC6 round robin test series. It can be seen that the model predicts a relatively weak flowrate dependence for the collector efficiency, consistent with the highly efficient absorber to air heat transfer. The effects of likely variations in the environmental conditions, estimated using the model, are found to be very similar to those predicted by modelling for double-glazed water heating collectors (1).

Despite the good repeatability exhibited by the initial CEC6 round robin test results, there are large differences between participating laboratories, particularly with regard to the flowrate dependence. This is probably attributable to flowmeter and temperature sensor calibration difficulties and collector leakage. Empirical efficiency data, for the three recommended air mass flowrates and a wide range of outdoor and indoor environmental conditions, are plotted on the graph in Fig. 3 . No general flowrate dependence is evident, though contradictory results from individual laboratories indicate an appreciable sensitivity of the collector efficiency to flowrate.

Raw efficiency measurements of sufficient experimental accuracy are generally normalised to the standard operating conditions, in order to aid their meaningful intercomparison. The model can be used for this purpose. Each efficiency datum is normalised by adjusting its value by the difference between the model-predicted efficiencies for the standard conditions and for the actual conditions under which the datum was measured.

CEC6 test results from some of the participating laboratories exhibit the expected insensitivity of the zero loss efficiency to the flowrate. Two such sets of efficiency measurements (one obtained outdoors and the other in a solar simulator), normalised to the standard conditions and an air mass flowrate $\dot{m}$ of 0.049 kg/s, are compared in Fig. 4 with the model-generated collector performance characteristic for those conditions. It can be seen that the agreement at low values of $(T_m - T_a)/G$ is excellent. These results indicate that the overall heat loss coefficient is actually somewhat higher than predicted by the model, probably owing to collector leakage and the additional mixing induced by the obstructions in the air stream, which the model does not take into account.

6. Conclusions

The reasonable agreement obtained between predicted and measured efficiencies, despite the inadequacies of both model and test data, tends to confirm our theoretical understanding of the thermal processes within the CEC6 collector. Round robin collector efficiency measurements of greater consistency and accuracy are required to make the intended application of the model to the general normalisation of test data worthwhile.

Much research remains to be done on air heating solar collectors, in refining test equipment and procedures (5), and developing more accurate mathematical representations of forced convection heat transfer processes for complex air flow distributions. This work will be directed toward achieving good reproducibility of test data between laboratories, and more satisfactory agreement with computer modelling results.

References

1. Green, A.A. (January 1979). The influence of environmental parameters on flat plate solar collector performance. UK-ISES Conference Proceedings C18 Meteorology for Solar Energy Applications. Pages 95-107. London.

2. Hamid, Y.H. and Beckman, W.A. (April 1971). Performance of air-cooled radiatively heated screen matrices. Trans.ASME Journal of Engineering for Power. Vol.93. Pages 221-224.

3. Beckman, W.A. (January 1968). Radiation and convection heat transfer in a porous bed. Trans.ASME Journal of Engineering for Power. Vol.90. Pages 51-54.

4. Hollands, K.G.T. and Shewen, E.C. (November 1981). Optimization of flow passage geometry for air-heating, plate-type solar collectors. Trans.ASME Journal of Solar Energy Engineering. Vol.103. Pages 323-330.

5. Green, A.A., Cross, B.M. and Moon, J.E. (April 1984). Thermal performance testing of air heating solar collectors. IEE Conference Publication No. 233 Energy Options: The Role of Alternatives in the World Energy Scene. Pages 43-46. London.

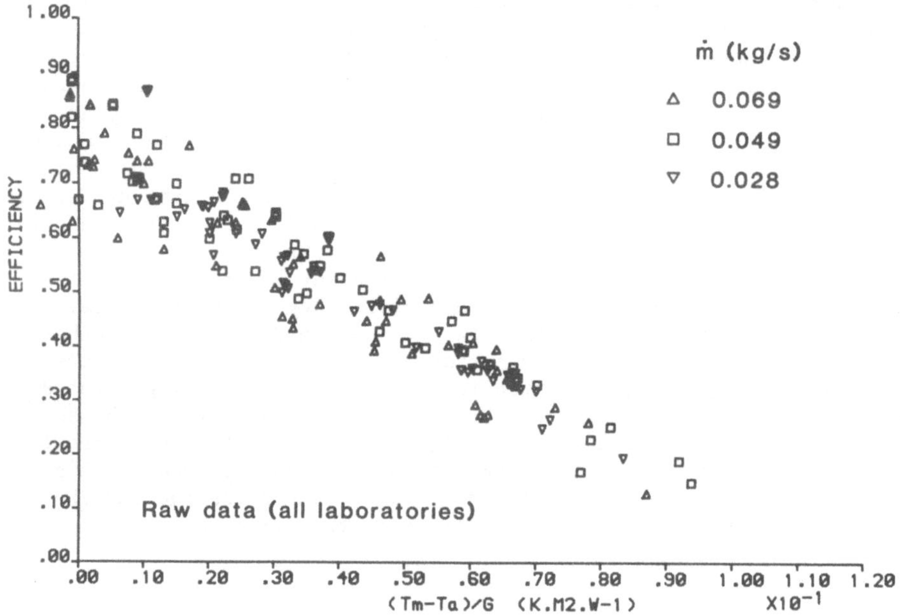

Fig. 3 Initial CEC6 round robin collector efficiency data from nine laboratories, for three nominal air mass flowrates.

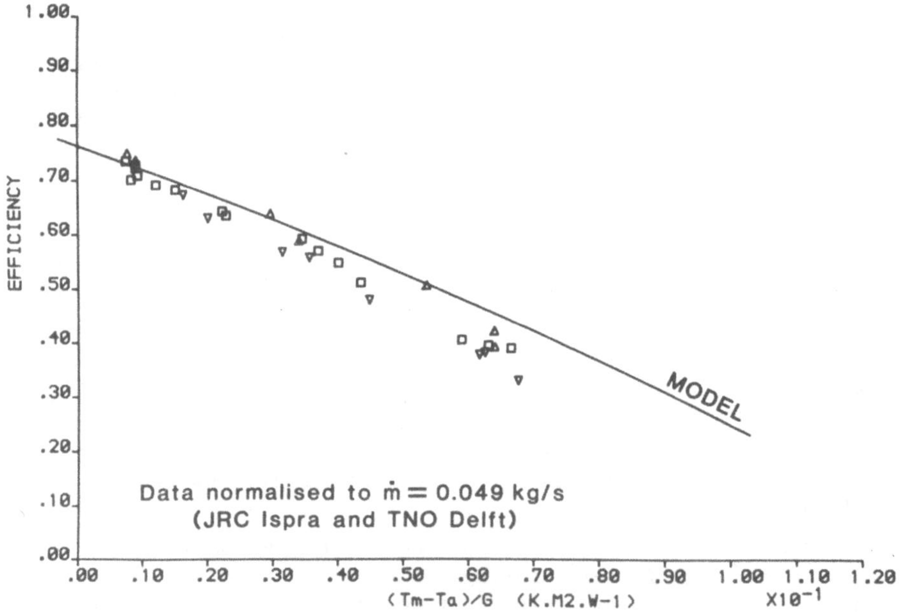

Fig. 4 Two selected CEC6 collector efficiency data sets normalised to the standard conditions and an air flowrate of 0.049 kg/s, compared with the model-generated performance characteristic for those conditions (the symbols indicate the flowrates at which the data were measured).

EXPERIMENTELLE UNTERSUCHUNG AN LUFTGEKÜHLTEN FLACHKOLLEKTOREN
(Experimental Investigation of Solar Air Collectors)

N. Fisch and E. Hahne
Universität Stuttgart
Institut für Thermodynamik und Wärmetechnik

Zusammenfassung

Es wurde der Einfluß des Massenstromes auf die Wirkungsgradkennlinie, die Zeitkonstante und den Leckluftanteil von luftgekühlten Flachkollektoren untersucht. Zur Bestimmung des Wirkungsgrades in Abhängigkeit von Betriebs- und meteorologischen Parametern wurde ein Nomogramm entwickelt. Der Wirkungsgrad wird mit zunehmendem Luftdurchsatz größer, während dagegen die Zeitkonstante abnimmt. Für zwei verschiedene Kollektoren wurden - in einem Massenstrombereich von ca. 35 bis 70 kg/(h·m²) - Zeitkonstanten zwischen 5 und 15 Minuten ermittelt. Es wurde ein einfaches Rechenverfahren entwickelt, um aus den - unter instationären Bedingungen durchgeführten Messungen - mit Hilfe der Zeitkonstante, den stationären Wirkungsgrad zu bestimmen. Die Leckluftanteile werden mit zunehmendem Massenstrom und statischer Druckdifferenz (zwischen Kollektor und Umgebung) größer. An einem Kollektor wurden Leckluftanteile (bezogen auf den "Eintrittsmassenstrom") bis zu 28 % für den "Saugbetrieb" und maximal -22 % für den "Druckbetrieb" ermittelt.

Summary

The effect of flow rate on the efficiency curve, time-constant and air-leakage fraction of air-cooled solar collectors was investigated. A nomogram is introduced giving the collector efficiency in its dependence of operational and meteorological parameters. The efficiency increases with increasing mass flow rate, while the time constant decreases. Time constants between 5 and 15 minutes were determined - for flow rates from about 35 to 70 kg/(h·m²) - for two different collectors. A simple calculation procedure was developed to determine the steady state efficiency from measurements under transient conditions, using the collector time constant. The air-leakage fraction increases with mass flow rate and statical pressure difference between collector and surroundings. Depending on the collector, an air-leakage fraction (based on inlet mass flow rate) up to 28 % was obtained for suction operation and to a maximum of -22 % for pressure operation.

Danksagung: Diese Untersuchung wurde vom BMFT unter dem Zeichen 03 E 8021 gefördert; dafür danken die Autoren.

1. EINLEITUNG

Am Institut für Thermodynamik und Wärmetechnik (ITW) der Universität Stuttgart wurde 1978 mit dem Aufbau einer Außen- und einer Labortestanlage für luftgekühlte Flachkollektoren begonnen (s. Lit. [1] und [2]). An verschiedenen Sonnenkollektoren wurden Messungen entsprechend der ASHRAE- [3] und der BSE-Richtlinie [4] durchgeführt. Die Ergebnisse dieser Untersuchungen sind in [1] und [2] dargestellt. Die Ziele waren einerseits die bestehenden Testrichtlinien anzuwenden, untereinander zu vergleichen sowie gegebenenfalls Veränderungen vorzuschlagen und andererseits das thermische Verhalten dieser Kollektoren ganz allgemein zu untersuchen. In der vorliegenden Arbeit wird der Einfluß des Massenstromes auf den Wirkungsgrad, die Zeitkonstante und den Leckluftanteil gezeigt.

2. VERSUCHSANLAGE

Das Bild 1 zeigt das Anlagenschema des 1982 umgebauten Außenteststandes für Luftkollektoren. Im Vergleich zu [1] wurden zwei wesentliche Änderungen an der Anlage durchgeführt. Durch den Einbau von elektrischen Lufterhitzern - vor jedem Kollektor - können die Lufteintrittstemperaturen jetzt individuell geregelt werden. Mit Hilfe von kalibrierten Durchflußmeßblenden am Ein- und Austritt der Kollektoren, werden deren Leckluftanteile bestimmt (s. Bild 1).

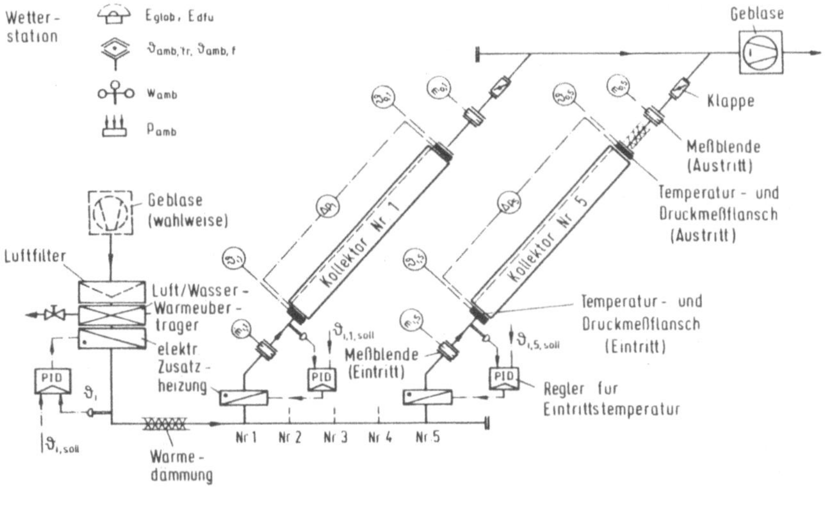

Bild 1: Anlagenschema des Außenteststandes für luftgekühlte Kollektoren

3. ERGEBNISSE DER KOLLEKTORUNTERSUCHUNGEN

3.1 Wirkungsgrad in Abhängigkeit vom Luftdurchsatz

Der thermische Wirkungsgrad eines Sonnenkollektors wird mit zunehmendem Massenstrom größer (s. [1] und [2]); die Temperaturerhöhung des Fluids (ϑ_o - ϑ_i) wird dagegen kleiner. Die

Antriebsleistung eines Gebläses nimmt proportional mit der dritten Potenz des zu fördernden Massenstromes zu. In Solaranlagen mit Luft als Wärmeträger hat der Luftdurchsatz einen bedeutenden Einfluß auf deren solaren Deckungsanteil (s. [5]), Systemwirkungsgrad und Leistungsziffer. Es ist deshalb erforderlich die Abhängigkeit der Wirkungsgradkennlinie vom Massenstrom zu bestimmen. Das Bild 2 zeigt ein Nomogramm zur Ermittlung des Wirkungsgrades in Abhängigkeit vom Massenstrom, der Lufteintrittstemperatur und den meteorologischen Parametern (E_{glob}, ϑ_{amb} und w_{amb}). Der rechte Teil des Nomogramms ent-

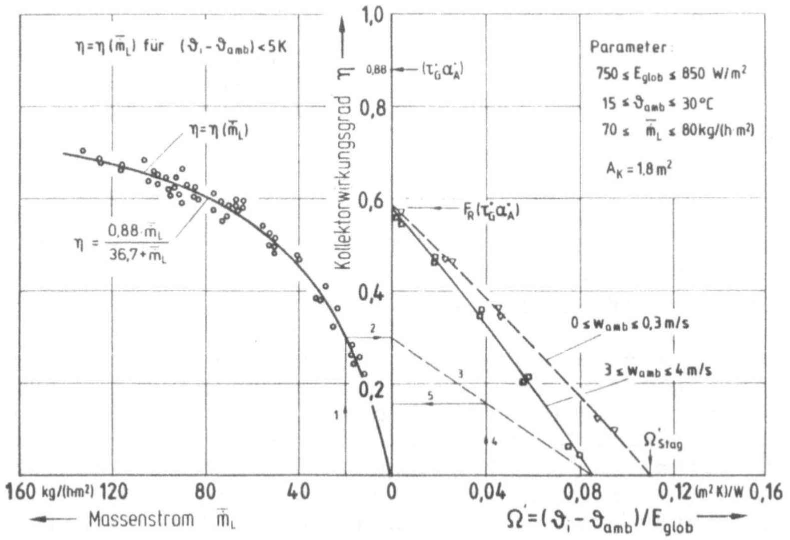

<u>Bild 2:</u> Wirkungsgrad-Nomogramm eines Luftkollektors

spricht der Darstellung des Wirkungsgrades nach ASHRAE [3]. Es wurden zwei Wirkungsgradkennlinien, einerseits für erzwungene Konvektion ($3 < w_{amb} < 4$ m/s) und andererseits für annähernd freie Konvektion ($w_{amb} < 0,3$ m/s), an der Kollektorabdeckung bestimmt. Im linken Teil des Nomogramms (Bild 2) ist der Wirkungsgrad für $\vartheta_i \approx \vartheta_{amb}$ in Abhängigkeit vom Massenstrom dargestellt. Die Meßpunkte wurden durch eine gebrochen rationale Funktion (s. Bild 2) approximiert. In eigenen Untersuchungen [1] und [2] wurde gezeigt, daß sich die Wirkungsgradkennlinien für verschiedene Luftdurchsätze - aufgetragen nach der ASHRAE-Richtlinie [3] - im Stagnationspunkt schneiden. Damit kann der Wirkungsgrad des Kollektors - in Abhängigkeit vom Massenstrom $\bar{m}_L$ und dem Betriebskoeffizienten Ω' - aus dem Nomogramm ermittelt werden. Die Vorgehensweise ist in Bild 2 für $\bar{m}_L = 20$ kg/(h·m²), $\vartheta_i = 30$ °C, $\vartheta_{amb} = 0$ °C und $E_{glob} = 750$ W/m² als Beispiel dargestellt.

3.2 <u>Zeitkonstante in Abhängigkeit vom Luftdurchsatz</u>

Nach der von ASHRAE [3] vorgeschlagenen Methode wurde die Zeitkonstante bei sprungartiger Änderung der Einstrahlung bestimmt. In Bild 3 sind die experimentell ermittelten Zeitkonstanten, für zwei verschiedene Kollektoren, in Abhängigkeit vom Luftdurchsatz dargestellt. Die mit zunehmendem Massenstrom abnehmenden Zeitkonstanten liegen in dem hier untersuchten Bereich [35 kg/(h·m²) < $\bar{\dot{m}}_L$ < 70 kg/(h·m²)] zwischen 5 und 15 Minuten. Betrachtet man den Kollektor als ein "Übertragungsglied erster Ordnung" und unter der Annahme eines linearen Temperaturverlaufes in Strömungsrichtung, so ergibt sich für die Zeitkonstante die folgende Gleichung:

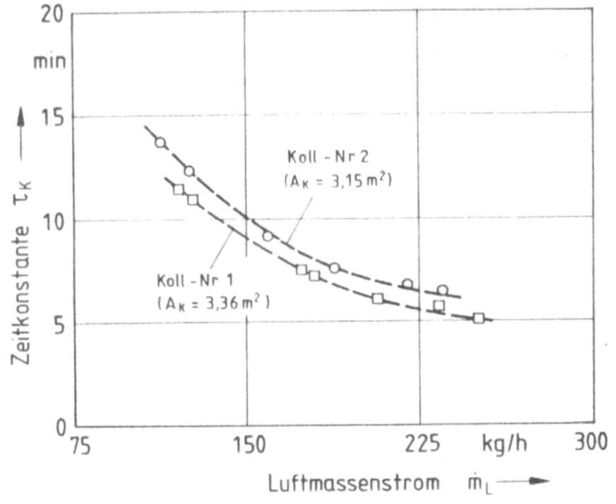

Bild 3: Zeitkonstante in Abhängigkeit vom Luftdurchsatz

$$\tau_K = \frac{C_{eff}/A_K \cdot F' \cdot k_{eff,m}}{1 + \frac{2 \cdot c_{p,L}}{F' \cdot k_{eff,m}} \cdot \bar{\dot{m}}_L} \qquad (1)$$

Die Meßpunkte wurden deshalb durch eine gebrochen rationale Funktion in der Form wie Gl. (1) approximiert (s. Bild 3). Das transiente Verhalten des Kollektors wird durch dessen Zeitkonstante bestimmt. Es wurde ein einfaches analytisches Verfahren entwickelt um aus Messungen - die unter "quasi stationären" Bedingungen durchgeführt wurden - den "stationären" Kollektorwirkungsgrad zu berechnen. Das Bild 4 zeigt für zwei wolkenlose Tage, an denen die Testbedingungen nach ASHRAE [3] eingehalten sind, daß die "momentanen" Wirkungsgrade innerhalb

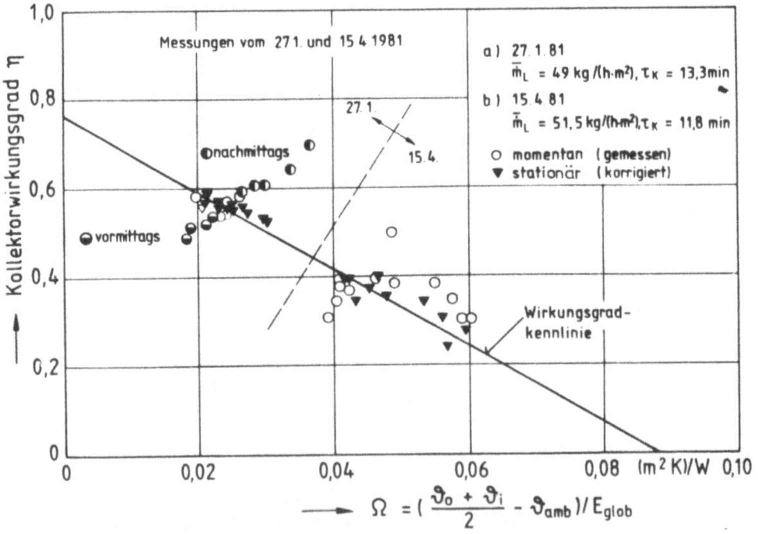

Bild 4: "Momentane" und analytisch korrigierte "stationäre" Kollektorwirkungsgrade

eines relativ großen Bereiches von vormittags zu nachmittags erheblich zunehmen. Dies ist auf die Wärmekapazität des Kollektors zurückzuführen. Mit Hilfe der Zeitkonstante τ_K und dem zeitlichen Verlauf der Luftaustrittstemperatur $\vartheta_o(t)$ des Kollektors wurden die "momentanen" Wirkungsgrade nach Gl. (2) und (3) korrigiert. Das Bild 4 zeigt die sich damit ergebende

$$\vartheta_{o,stat} \approx \vartheta_o(t) + \tau_K \frac{\Delta\vartheta_o(t)}{\Delta t} \qquad (2) \qquad\qquad \eta_{stat} = \frac{\bar{\dot{m}}_{L,i} \cdot c_{p,L} \cdot (\vartheta_{o,stat} - \vartheta_i)}{E_{glob}} \qquad (3)$$

sehr geringe Streuung für die berechneten "stationären" Wirkungsgrade η_{stat}.

3.3 Leckluftanteil in Abhängigkeit vom Luftdurchsatz

Close und Yusoff haben in einer theoretischen Arbeit [6] den Einfluß des Leckluftstromes auf den thermischen Kollektorwirkungsgrad untersucht. An der Außentestanlage (s. Bild 1) des ITW wurden die Leckluftanteile verschiedener Kollektoren in Abhängigkeit vom Massenstrom experimentell bestimmt. Das Bild 5 zeigt den Leckluftanteil - definiert nach Gl. (4) - in Abhängigkeit vom Massenstrom am Eintritt des Kollektors $\dot{m}_{L,i}$. Die Messungen wurden einerseits im "Saugbetrieb" - dabei ist

$$\text{Leckluftanteil} = \frac{\Delta\dot{m}_L}{\dot{m}_{L,i}} = \frac{\dot{m}_{L,o} - \dot{m}_{L,i}}{\dot{m}_{L,i}} \qquad (4)$$

der statische Druck im Kollektor kleiner als der Umgebungsluftdruck ($\dot{m}_{L,o} > \dot{m}_{L,i}$) - und andererseits im "Druckbetrieb" ($\dot{m}_{L,o} < \dot{m}_{L,i}$) durchgeführt. Die Leckluftanteile nehmen mit dem Luftdurchsatz zu, wobei im "Saugbetrieb" stets größere Leckluftströme als im "Druckbetrieb" vorhanden sind.

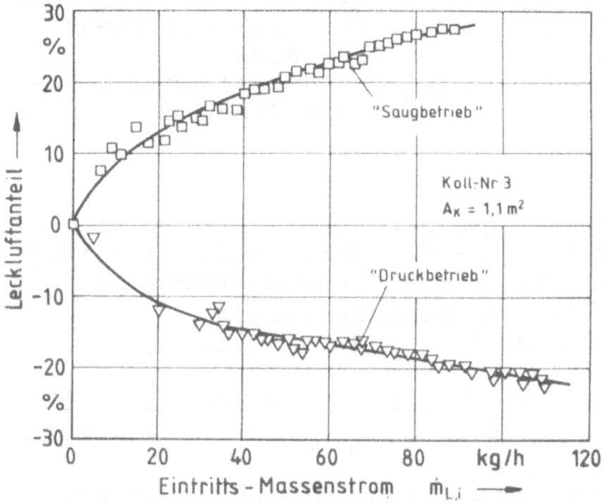

Bild 5: Leckluftanteil in Abhängigkeit vom Massenstrom $\dot{m}_{L,i}$ für "Saug-" und "Druckbetrieb"

In Bild 6 ist der Einfluß der statischen Druckdifferenz am Kollektoreintritt ($\Delta p_{stat,i} = p_{stat,i} - p_{amb}$) auf den Leckluftanteil dargestellt. Es zeigt sich, daß - bei gleichem Massenstrom - die Verluste umso größer sind, je größer die statische Druckdifferenz $\Delta p_{stat,i}$ ist. Beim Vergleich experimentell ermittelter Leckluftanteile und insbesondere der Kollektorwirkungsgradkenn-

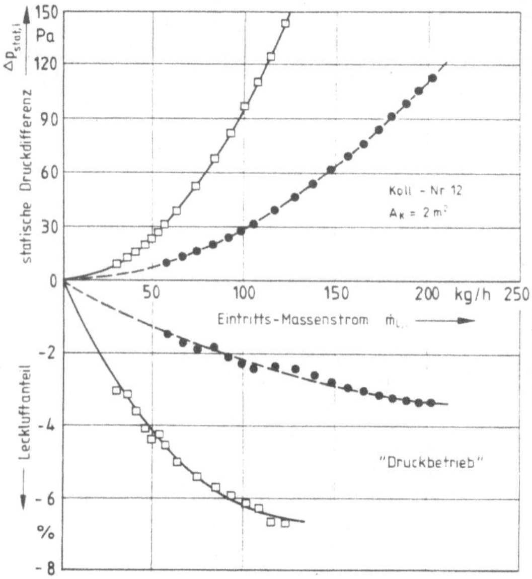

linien sind neben dem Massenstrom die statische Druckdifferenz zwischen Kollektor und Umgebung zu beachten.

Bild 6: Leckluftanteil in Abhängigkeit vom Massenstrom $\dot{m}_{L,i}$ für unterschiedliche Druckdifferenzen.

Nomenklatur

A_K	transp. Kollektorfläche	$p_{stat,i}$	stat. Druck am Eintritt
C_{eff}	eff. Wärmekapazität (Kollek.)	t	Zeit
$c_{p,L}$	spez. Wärmekapazität der Luft	w_{amb}	Windgeschwindigkeit
E_{glob}	Globalstrahlung	η	Kollektorwirkungsgrad
F'	Kollektorwirkungsgradfaktor	ϑ_i	Lufteintrittstemperatur
$k_{eff,m}$	Gesamtwärmedurchgangskoeff.	ϑ_o	Luftaustrittstemperatur
$\dot{m}_{L,i}$	Massenstrom am Koll.-eintritt	ϑ_{amb}	Umgebungstemperatur
$\dot{m}_{L,o}$	Massenstrom am Koll.-austritt	τ_K	Zeitkonstante des Koll.
$\overline{\dot{m}}_L$	Massenstrom bezogen auf die Kollektorfläche	Ω'	Betriebskoeffizient nach ASHRAE
p_{amb}	Umgebungsluftdruck	$(\tau_G^* \alpha_A^*)$	Transmissions-Absorptions Produkt

LITERATURVERZEICHNIS

1. ARAFA, A., FISCH, N., et al. Technische Nutzung Solarer Energie. Forschungsbericht T 83-300, BMFT, S.28-54,(1983)
2. FISCH, N., HAHNE, E. Experimentelle Untersuchungen an luftgekühlten Flachkollektoren. Statusbericht: "Nutzung der Sonnenenergie in der Landwirtschaft", BMFT, S.14-34, (1982)
3. ASHRAE-Standard 93-77. Method of Testing to Determine the Thermal Performance of Solar Collectors, ASHRAE, New York, (1977)
4. BSE-Richtlinien. Wirkungsgradtest von Solarkollektoren. Bundesverband Solarenergie, Essen, (1980)
5. FISCH, N., HAHNE, E. Simulation of solar air-heating systems. Proceedings of the First EC Conference on Solar Heating, Amsterdam, (1984)
6. CLOSE, D.J., YUSOFF, M.B. The Effects of Air Leaks on Solar Air Collector Behaviour. Solar Energy, Vol. 20,(1978)

THE 'ALBEDO' FLAT-PLATE SOLAR COLLECTOR

H. TABOR
(The Scientific Research Foundation, Jerusalem, Israel)

ABSTRACT

In the 'albedo' flat-plate collector, the rear solid insulation is replaced by an air-gap and window. With the collector mounted some distance above a white roof or ground, the absorber is irradiated also on its rear face, by sun and sky radiation reflected from the roof, thereby increasing the income of radiation in the order of 20 - 60%. For a typical DHW installation, the collector area needed is estimated at 74% that of a conventional collector despite the increase in loss factor U - which increase is less than might be expected. The units are long and narrow, assembled in parallel rows, spaced to minimise shadowing of the roof.

First applications are for large (condominium) DHW and industrial or commercial HW installations; heating of buildings. Summer air-conditioning is the next stage.

1. INTRODUCTION

A primary goal in solar R & D is to obtain the maximum energy yield at the minimum cost: cost here includes charges on the capital outlay for the equipment, cost of installation in the field and O and M (operation and maintenance) costs. This has been the motivation behind the development of the albedo collector described in this paper: the quadruple aim has been (i) to reduce the cost per unit area of collector; (ii) to increase the yield per unit area of collector; (iii) to reduce transportation and site erection costs to a minimum and (iv) to keep down O & M costs.

2. THE ALBEDO CONCEPT

In the albedo collector, a window replaces the rear insulation of the conventional flat-plate collector so that the rear side of the absorber plate can receive albedo i. e. solar radiation reflected from the ground. Thus the absorber panel receives radiation on both sides. Some "double-sided" flat-plate collectors have been described, where the rear side is irradiated by mirrors. These mirrors require that their tilt be changed from time to time which adds to the O & M costs.

In the present system, instead of mirrors, the roof on which the collectors are mounted is painted white - or made of light-coloured material. The collectors are mounted some distance above the roof (or ground) in horizontal rows, with some space between the rows: the closer the spacing, the more the ground surface is shaded so that the albedo contribution is reduced. (Fig. 1)

Fig. 2 shows that the collector is the recipient of five different energy inputs i. e. in addition to the beam I_{bt} and sky I_{st} components on the upper (tilted) face, there is sky radiation I_{sr} reaching the rear and two albedo contributions from the ground i. e. I_{af} on the front and I_{ar} on the rear: each of these two is made up of two components, i. e. reflected sky radiation and relected beam radiation. It is possible, for a given geometry of spacing between rows and the height of the rows above the ground, to determine the positions of the light and shaded areas of the ground for any given position of the sun in the sky. The intensities of the light and dark bands are calculable if the beam and sky intensities are known. The radiation view-factor of each band, with respect to a collector, can be computed so that the total albedo 'seen' by the collector can be summated. Allowing for the fact that the ground reflectance will be less than 1.0 - and for the $(\tau\alpha)$ product of each of the energy inputs, it is possible to arrive at the amount of effective albedo gain I_a of the collector: a computer program will indicate the closest spacing between collector rows that can be allowed without serious loss of integrated albedo (space is often limited) as well as the lowest height of the collectors above the ground, leading to shorter legs and lower wind forces. An initial study suggests that if W is the width of the collector, the spacing between rows should be 3W (or more) and the height above the ground 1.5W (or more). The method of calculating the albedo contribution I_a will be described in another paper. Because I_a is a function of the geometry and of the fraction ρ of diffuse (sky) radiation, then, for a given geometry and site, it is found that the albedo fraction a, defined as I_a/I_t can be expressed as:

$$a = p\rho + q(1 - \rho) = q + j\rho \quad \ldots\ldots\ldots\ldots \quad (1)$$

and that, averaged over a day or month, the mean albedo fraction $\bar{a}$ can be very closely represented by:

$$\bar{a} = P\bar{\rho} + Q(1 - \bar{\rho}) = Q + J\bar{\rho} \quad \ldots\ldots\ldots\ldots \quad (2)$$

where $\bar{\rho}$ is the mean diffuse fraction and p, q, j, P, Q, and J are constants for a given geometry, latitude and month. The total instantaneous collector income is $I_t(1 + a)$ where $(1 + a)$ is the albedo factor, represented in Figs. 3 and 4 as a "concentration" factor C, and the daily income is $H_t(1 + \bar{a})$ where H_t is the daily total on the tilted surface.

Table I shows one set of results, with eleven-year average $\bar{\rho}$ values for Israel. It is seen that $(1 + \bar{a})$ varies from 1.33 to 1.58. (For completely overcast conditions, a is a constant (0.66) for all hours and dates since only the diffuse components count and these are purely geometrical functions, if a hemispherical sky is assumed).

Construction. A new concept for double-sided glazing - possible when the absorber is long and narrow - is a single sheet of fibre-glass bent to form a transparent (self-supporting) elliptical envelope. Thus, except for two small end plates, there is no enclosing box. (Durability tests abroad indicate the glazing material to be satisfactory over long periods of time out of doors).

Heat-loss effect. The heat-loss coefficient for a double-sided collector is higher than for a conventional single-sided unit but is not double for two reasons: (i) the rear solid insulation of the single-sided collector - which is replaced by an air-gap - is not itself a perfect insulator; (ii) the rear air-gap for a tilted collector has a lower convection loss component - due to heat flow downwards - than the front-side air-gap. Calculation shows that the U factor for a double-sided collector with a selective absorber ($E \sim 0.1 - 0.15$) will usually be some 40 - 50% higher than for a single-sided collector.

3. PERFORMANCE GAIN

The (instantaneous) yield Y_1 of a single-sided collector may be written:

$$Y = E_o I_t - U\,\Delta T \quad \ldots\ldots\ldots\ldots\ldots\ldots \quad (3)$$

where E_o is the efficiency for zero temperature rise ΔT above ambient.

For the albedo collector

$$Y_2 = E_o I_t(1 + a) - U(1 + u)\,\Delta T \quad \ldots\ldots\ldots \quad (4)$$

where u is the fractional increase in the loss factor U.

Then $Y_1 = Y_2$ when $\Delta T = \frac{E_o I_t}{U} \cdot \frac{a}{u}$ . . . . (5)

For all values of ΔT less than this, $Y_2 > Y_1$ i.e. the double-sided collector gives a greater yield. As an example, if $E_o = 0.75$, $U = 4\ W/m^2 {}^oC$, $u = 0.5$, we obtain:

For albedo contribution a =	0.2	0.3	0.4	0.5	
Cross-over ΔT^oC (above ambient) =	67.5	101	135	169	For I_t = 900 W/m^2
	45	67.5	90	112	= 600 "
	22.5	34	45	56	= 300 "

We can write equation (4) to give a pseudo efficiency - i.e. yield/input on the tilted surface I_t - which we will call "efficacy"* E_f i.e.

$$E_f = Y_2/I_t = E_o(1+a) - (1+u)\ U.\ \Delta T/I_t \quad (6)$$

$$E_f/E_o = (1+a) - (1+u)Z, \text{ where } Z \equiv \frac{U}{E_o} \cdot \frac{\Delta T}{I_t} \quad (6a)$$

Without the albedo effect, the efficacy is that of a flat plate collector with the slope increased by the factor $(1+u)$: as the albedo contribution is added, the efficiency curve moves upwards an amount $E_o a$, maintaining the same slope. This is shown in Fig. 3. **

The gain factor G of an albedo collector over the conventional collector - whose efficiency is $E_o(1-Z)$ - is:

$$G = \frac{(1+a) - (1+u)Z}{1-Z} = 1 + \frac{a - uZ}{1-Z} \quad (7)$$

Fig. 4 shows G v. Z for u = 0.5; Values of $\Delta T/I_t$ are also plotted for the case $U = 4\ W/m^2$, ^oK; $E_o = 0.75$.

Figs. 3 and 4 give instantaneous efficiency and gain values. Average values over a day (or longer period) can be determined if the march of radiation during the day is known. Daily yield Y_{2D} is:

$$\int E_f \gamma I_t dt = \int (\gamma I_t E_o(1+a) - (1+u)U.\ \Delta T)dt = E_o \bar{H}_t \bar{\gamma}(1+\bar{a}) - (1+u)U.\ \Delta T.\ h_a \ldots \quad (8)$$

where γ is the incident angle modifier $(\tau\alpha)/(\tau\alpha)_n$; $\bar{H}_t$ = total daily radiation in Joules per m^2 on the tilted surface, the bars on γ and a indicate daily mean values, and h_a is the seconds per day that the collector operates.

For the conventional collector $Y_{1D} = E_o H_t \bar{\gamma} - U.\ \Delta T.\ h_c$ . . . . (9)

whence: $G = Y_{2D}/Y_{1D}$

Table I shows, for the sample geometry chosen, the mean values of $\bar{a}$ and $\bar{\gamma}$: $\bar{a}$ goes from 0.2 - 0.3 in winter to 0.5 - 0.6 in summer. Fig. 4 suggests that the albedo collector may be attractive for summer air-conditioning applications. (In cloudy weather $\bar{a} > 0.6$ and the gain is over 60% though the absolute yields are, of course, low).

4) APPLICATION

(a) Application to f-chart analysis. The integrated factors in equations (8) and (9) suggest that these could be applied to f-chart analysis thereby estimating the benefit of the albedo***. For the Y axis of the f-chart i.e. $AF'_R(\overline{\tau\alpha})H_tN/L$, we substitute $AF_r'\bar{\gamma}(1+\bar{a})H_tN/L$: in the X axis we substitute $(1+u)U$ for $F_R U_L$.

* Efficacy instead of efficiency because it can exceed unity!

** The characteristic is shown curved because U is a weak function of ΔT. See Ref. 2.

*** A basic check on whether the f-chart analysis (3) derived from simulations on conventional collectors has not yet been conducted to ascertain whether these will apply rigidly to the albedo collector.

We considered the example of a system to provide 100 litres of domestic hot water per day per family for the sample geometry and the local (Israel) insolation, the water to be heated 40°C above the temperature of the cold (mains) water supply. For the conventional collector, the solar fraction for the best month was chosen as 80% - which led to an area of 1.604 m^2 per family: for the albedo collector, the best month solar fraction was set at 85% (because of the larger swing between winter and summer) and the area needed was 1.177.

For the whole year, the values of f obtained were, in %:

	I	II	III	IV	V	VI	VII	VIII	IX	X	XI	XII	Annual Average
Conventional collector	52	61	65	72	75	75	76	80	78	70	61	49	67.8%
Albedo collector	47	59	67	75	80	85	84	78	77	69	58	44	68.7%

This example shows that, on yearly average performance, the albedo collector required only 73.4% the area of the conventional collector for approximately the same yearly solar fraction.

If the f-chart analysis is applied to a heating and cooling system, the benefit of the albedo collector will be more pronounced because of the high albedo fractions in the summer months.

(b) Installation. The units (current models are 40 cm wide x 2 m long) are light weight (12 kg per unit) and easily installed by one man. Fluid connections at opposite ends give in-line assembly which greatly reduces installation costs. The collector is intended primarily for flat roofs and the major applications are for large installations i.e. factories, hotels, condominiums, etc.

5) ECONOMICS

A full economic analysis will be made at a later date, but the major goals of reduced cost per unit of output indicated in the introduction have been achieved i.e. a higher yield per m^2, a lower cost per m^2 and lower installation costs. O and M costs have not yet been determined in comparison with conventional flat-plate collector installations (the roof may need to be repainted every few years) but it is reasonable to believe the O and M costs will be less than for any system employing mirrors.

REFERENCES

(1) Farber, E.A. and Morrison, C.A. "Clear Day Design Values", Mech. Eng. Dept., U. of Florida, Gainsville, Fla. 32611 or Chapter IV in "Applications of Solar Energy for Heating & Cooling of Buildings", Ed. R. C. Jordan, ASHRAE Publication GRP 170 (1977).

(2) Tabor, H. "Letter to the Editor" (re SEIA testing), SOLAR ENERGY, 24, pp. 113-115 (1980).

(3) Beckman, W.A., Klein, S.A. and Duffie, J.A. "Solar Heating Design", Wiley (1977).

ACKNOWLEDGEMENTS

SRF gratefully acknowledges financial support received from the Wolfson Foundation, London.

TABLE I. ALBEDO FACTORS FOR SPECIFIC EXAMPLE

Latitude 32°N; tilt 42°; height = 1.5 W; spacing between rows = 3 W; ground reflectivity 0.8

MONTH	I	II	III	IV	V	VI	VII	VIII	IX	X	XI	XII	(4)
Q	.233	.263	.315	.410	.553	.639	.553	.410	.315	.263	.233	.226	
J	.272	.284	.284	.235	.099	.022	.099	.235	.284	.284	.272	.270	
$\bar{\rho}$ (2)	.391	.346	.353	.342	.273	.228	.232	.231	.233	.284	.319	.392	1.00
$(1+\bar{a})$	1.339	1.361	1.415	1.490	1.580	1.644	1.576	1.464	1.381	1.344	1.320	1.332	1.66
$\bar{\gamma}$ (3)	.952	.940	.934	.924	.904	.894	.904	.924	.934	.940	.952	.969	.894
$\bar{\gamma}(1+\bar{a})$ (1)	1.27	1.28	1.32	1.38	1.43	1.47	1.42	1.35	1.29	1.26	1.26	1.29	1.48

NOTES :

(1) This factor is used in place of $(\tau\alpha)/(\tau\alpha)_n\, H_T$ in the expression for H_T in f-chart calculations. This is for the first row in a multi-row array : subsequent rows show a factor about 3% lower.

(2) This is the long term (11 year) mean monthly average diffuse fraction (daily horizontal diffuse/daily horizontal total) for Beth Dagan.

(3) For the values of $\bar{\rho}$ given : it varies very slightly with the fraction of diffuse radiation.

(4) For fully overcast conditions.

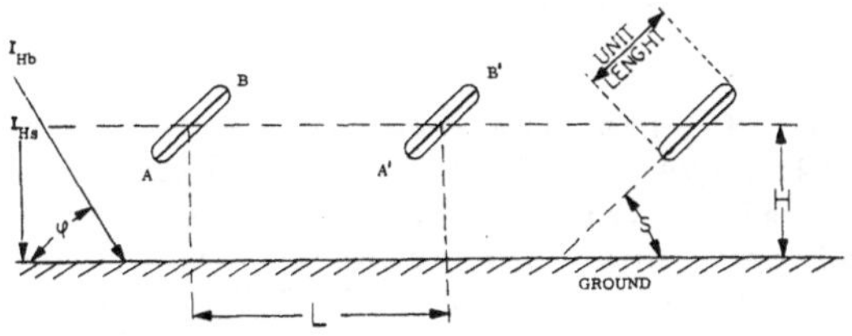

FIG. 1.

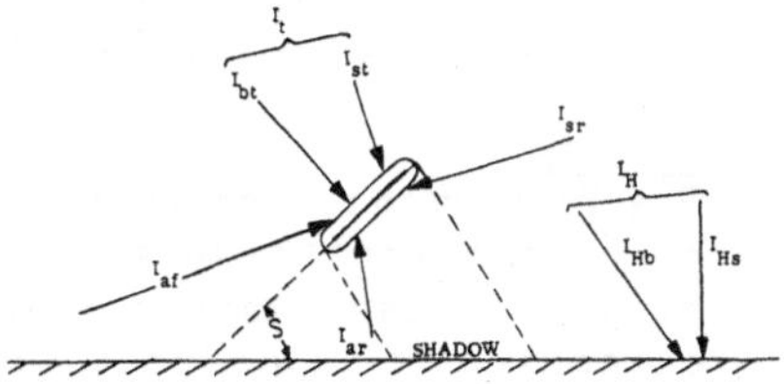

FIG. 2.

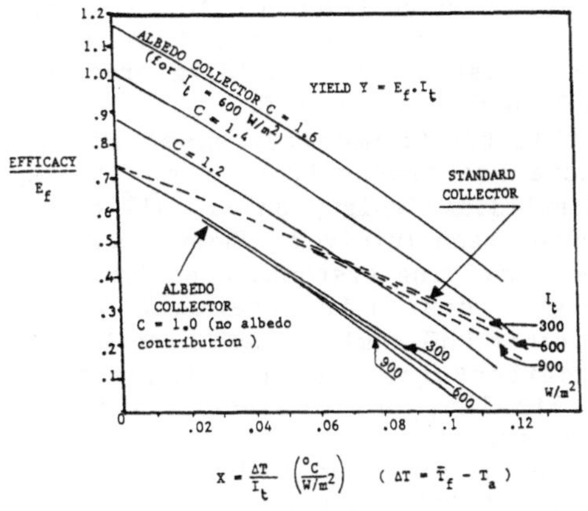

FIG. 3.

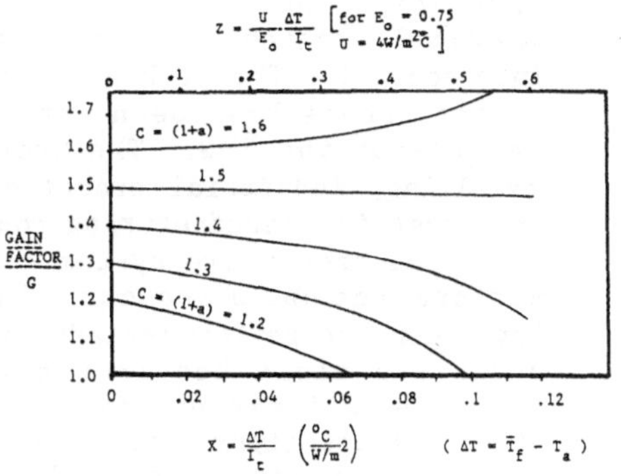

FIG. 4.

SYSTEM PERFORMANCE OPTIMISATION STUDIES, FOR 2ND GENERATION DHW AND SH SYSTEMS

R FERRARO and R GODOY
(on behalf of the CEC Performance Monitoring Group)
Energy Conscious Design
11-15 Emerald Street
London WC1N 3QL
England
(Tel: 01-405 3121)

MEMBERS OF THE PERFORMANCE MONITORING GROUP (1982/83)

Co-ordinator	- Energy Conscious Design, London, UK
A Debosscher	- Katholieke Universiteit Leuven, Belgium
P E Kristensen	- Thermal Insulation Laboratory, Lyngby, Denmark
G Kuhn	- CNRS - CRTBT, Grenoble, France
U Luboschik	- IST Energietechnik GmbH, Germany
J Owen Lewis	- University College Dublin, Ireland
F Cecchi Paone	- Fidimi Consulting Spa, Rome, Italy
D Brethouwer	- Institute of Applied Physics (TNO), Netherlands

Persons responsible at the Commission: T C Steemers and W Palz
Scientific adviser to the Commission: C den Ouden

Summary

The Performance Monitoring Group (PMG) has been receiving performance data from monitored solar heating systems designed 4-8 years ago. Most of these systems are generally over-complicated, oversized, economically unattractive and have poor performance. Based on experience and evidence from modelling exercises, the PMG believed that it was possible to design solar heating systems with higher outputs per unit collector area and/or lower capital costs. Such systems will provide better cost-benefit balances and a greater potential for market penetration.

The purpose of the work has been to develop a series of designs for better optimised solar heating systems, based on the experience gained in earlier phases of work. The designs have been developed using simulation models for performance predictions; they have been rigorously costed, and take current industrial and building practice into account. The work has been 100% funded by the CEC.

Design Groups have been set up within the Community, based around members of the PMG. The Groups have experience in system design, modelling, industrial practice and building design, and in all cases an industrial manufacturing company has been involved. Each Group has developed one or two system designs, involving systematic analysis of performance and a structured process of design improvement. There have been two review periods, where other PMG members have made inputs under the topic headings of: architectural aspects, industrial practice and system modelling. Finally, performance and cost-effectiveness figures have been compared.

The studies have resulted in designs, produced under realistic practical constraints, which show improvements in performance-to-cost ratios by a factor of 1.5 to 2.5, compared to most projects studied previously by the Performance Monitoring Group in Phases 1 and 2 of its work (1979-81).

1.0 INTRODUCTION

The main aim of the CEC Performance Monitoring Group (PMG) has been to draw conclusions about the likely benefits to be gained from installing solar heating systems in different parts of the Community. (Please refer to the related paper in Session 1B) (1,2,3,4). The work reported in this paper results from the third period of work carried out by the PMG from 1st January 1982 to 30th November 1983, and the paper consists of edited extracts from the relevant Final Reports for this period (5,6).

In the past, the PMG received performance data from monitored solar heating systems designed 4-8 years ago. These systems were generally over-complicated, oversized, and economically unattractive. Based on experience and evidence from modelling exercises, the PMG believed that it was possible to design solar heating systems with higher outputs per unit collector area and/or lower capital costs. Such systems provide better performance-to-cost ratios and a greater potential for market penetration.

1.1 Aims and Objectives

The main objective of this third phase was to produce a series of solar system case study designs, based on a set of realistic practical constraints, industrial expertise, simulation modelling and existing experience. The case studies were to be designed to achieve the highest possible performance-to-cost ratios and are presented in such a way that they could act as the basis for detailed proposals for future monitored field trials and pilot projects, from which measured data could be obtained to confirm the predicted improvements in performance.

The work should therefore be seen in its overall context, where one progresses from the collection of measured data by the PMG in Phase 2, via the Case Study designs, to Future Projects. As will be seen from this paper, the results of the work are promising, both in terms of potential performance improvements and cost reduction.

The Studies have been carried out by making local extensions to the existing PMG structure, in the form of Design Groups based around individual members of the main Group. These Design Groups have been co-ordinated by the PMG and in each case include local expertise in system design and modelling, architectural and building practice, manufacturing expertise, and costing experience.

At the same time, other PMG members undertook Review Tasks in relation to all the Design Groups' work, thereby assisting in the design process, and in order to compare the designs produced in different contexts. These reviews cover Architectural Aspects, Engineering Design/Cost, and Comparative Modelling. It is of special interest that a successful working relationship was established with the European Modelling Group. Performance modelling of most of the designs has been carried out on their model EMGP2, firstly at intermediate design stage, and finally after completion.

2.0 COSTINGS AND PERFORMANCE-TO-COST RATIOS

Cost benefit calculations have been made by the PMG on various monitored solar heating systems in the past, and on all the systems presented in this study. Data collected by the PMG in Phase 2 showed that the best active space heating systems had a 'merit figure' of 2MJ (0.56 kWh) saving in energy per year for each ECU spent on the system. The best domestic water heating systems had a merit figure of 3.1 MJ p.a./ECU, (0.86 kWh p.a./ECU). Reliable figures for passive systems were not available at that time.

It is in relation to these figures that the performance-to-cost ratios of the improved designs are to be judged, although it is difficult to draw generalised conclusions on cost aspects which are applicable to all parts of the European Community. Fuel prices do vary nationally, as do VAT, tax relief systems and incentive schemes. However, all systems developed by the Design Groups have been costed in the market place of the country of origin, and 'merit figures' produced as previously described. Cost figures have been compiled for one-off systems and 100-off where possible, and the effects of local VAT, tax relief and incentives presented separately.

2.1 System Optimisation Methods

Considerable work was done by the PMG on methods for system optimisation, and these were reported in the final reports of Phase 2 (1). In principle, the method in the Performance Optimisation Studies has been taken from that work and consists basically of the following four steps:

a) The solar outputs for different combinations of collector areas and storage volume are calculated using a simulation model.

b) The various systems are costed, including extra-over building costs where they arise, and an allowance for additional floor space required, if any.

c) A diagram is constructed showing useful solar output and total costs at given storage volumes, as a function of the collector area.

d) The combination of collector area and storage volume giving the highest ratio of energy savings to installation cost (which represents the optimum) is then read from the diagram.

3.0 PERFORMANCE OPTIMISATION STUDIES

Due to lack of space, it is not possible to describe each system individually, and detailed information is contained in ref (5). Summary information is contained below, with a summary of results in section 5.0.

3.1 Design of a Solar Water Heating System for Northern Europe (Design Group 1)

This solar water heater comprises a single $5m^2$ collector and an 'inverted' solar and auxiliary storage tank. The storage is integrated into a unit which also includes the ancillary equipment and which, being partly in the loft and partly suspended from the ceiling, does not occupy floor space. Excess solar heat can be used for space heating by connecting the storage to the space heating system. The system was designed for integration into a typical Danish house.

3.2 Design of a Solar Water Heating System for Northern/Central Europe (Design Group 2)

The design of this solar water heater uses a novel type of drain-down system and a low cost storage tank which is located in the central service core (in a typical Dutch house). This arrangement simplifies the design, reduces the total cost of the system and does not take up valuable floor space. Auxiliary heating is by way of a new type of multi-point heater. A detailed simulation and costing exercise has been carried out to obtain the best performance-to-cost ratio.

3.3 Design of a Solar Water Heating System for Southern Europe (Design Group 3)

This case study consists of $100m^2$ of solar collectors serving a block of flats. The collectors are mounted in the roof, supported by a light metallic structure. A technical shelter, also on the roof, contains the heat storage tank and all ancillary equipment.

The heat distribution system is of mono-tube type, which has a better predicted performance-to-cost ratio than conventional loop distribution systems.

3.4 Design of a Solar Water and Space Heating System for Northern Europe (Design Group 1)

This design is based on a low-energy single storey detached house currently available on the Danish market. The solar collector circuit is of indirect type and the solar heating system uses a 'dedicated' heat exchanger in the ventilation system, and the auxiliary operates on radiators. The size of the heat emitters, storage tanks and collector array have been optimised using predicted performance figures and estimated current system costs.

3.5 Design of a Solar Water and Space Heating System for Northern/Central Europe (Design Group 2)

This design is based on a three bedroom terraced house with two storeys and attic, thermally insulated above local regulations found in the Netherlands, from where the house type is derived. The solar collectors have been dimensioned to ensure that they can be integrated into the roof using the well proven techniques applied to dormer windows. The solar collector is of a novel drain-down design and uses submerged pumps. The heat emitters are fan coil units with separate heat exchangers for solar and auxiliary, operating on the counter-flow principle. The system was sized using predicted performance data, current cost of the components and an optimisation technique which is illustrated in ref (5).

3.6 Design of a Solar Space Heating System for Southern Europe (Design Group 3)

This case study is based on a low-cost single family house, typically found in southern France. The solar heating system has been simplified by using the solid floor of the house as the heat emitter and heat store of the solar heating system. This results in a long response time which is well adapted to climatic conditions in the Mediterranean climate.

3.7 Design of a Passive Solar Space Heating System for Central Europe (Design Group 4)

This is an investigative design of a sun-space passive house, suitable for the UK market. The purpose of the exercise was to establish whether current experience with passive designs is sufficient to develop optimised systems. The study contains a design checklist for direct gain systems and conservatories. Four single family two-storey houses with integrated conservatories were designed to illustrate some of the issues discussed in the checklists. Two of the designs were costed in detail and their performance modelled using a thermal network simulation model (FRED).

4.0 PERFORMANCE OPTIMISATION STUDIES: REVIEWS

All designs were reviewed twice by members of the Group with particular reference to various aspects of the designs: Architecture and Design; Engineering and Cost; Performance Simulation. The first review resulted in a direct input to the design process in each case.

4.1 Architecture and Construction Review

This review studied the house designs and construction details of the design proposals, the thermal characteristics of the buildings, the integration of the solar components into the house, and the site planning aspects. Possible planning alternatives were also suggested.

4.2 Engineering Review

This analysis considered the technical feasibility of the proposals and the different national design requirements. The study also considered the availability, reliability and cost of the main components.

4.3 Modelling and Performance Review

The performance of all the active solar heating systems proposed by the Design Groups was calculated independantly using the simulation model being developed and validated by the CEC Modelling Group (EMGP2). This method was used to compare the different designs under similar criteria and to increase the degree of confidence achieved in the predicted performance of the systems. It was also possible to study the detailed performance of components which were critical to the performance of the system. (However, neither model validation nor intercomparison between models part of the brief.)

5.0 PERFORMANCE OPTIMISATION STUDIES: RESULTS

A summary of predicted annual performance figures in MJ p.a./m^2 (and kWh p.a./m^2) is shown in Fig. 1 and a summary of corresponding performance-to-cost ratios in Fig. 2, in MJ p.a./ECU (and kWh p.a./ECU).

The range in performance is due to the effect of increased hot water loads (from approximately 100 to 180 l/day). Where costs are given as a range, they reflect savings compared to a single system incurred when building 30 to 100 similar systems, and tax incentives, where applicable.

DESIGN GROUP	DHW SYSTEM	ACTIVE SYSTEM	PASSIVE SYSTEM
Northern Europe:	1,228 - 1,445	918 - 1,026	-
	(341 - 401)	(225 - 285)	
Northern/Central Europe:	1,022 - 1,393	767 - 968	-
	(284 - 387)	(213 - 269)	
Southern Europe:	3,006	1,636	-
	(835)	(454)	
Central Europe:	-	-	360 - 472
(Passive Study only)			(100 - 131)

All units in MJ p.a./m^2 (kWh p.a./m^2)

Fig 1. Performance Optimisation Studies: Predicted annual performance

DESIGN GROUP:	DHW SYSTEM	ACTIVE SYSTEM	PASSIVE SYSTEM
Northern Europe:	3.5 - 5.4	2.4 - 3.0	-
	(0.9 - 1.5)	(0.7 - 0.8)	
Northern/Central Europe:	3.0 - 4.1	2.4 - 3.4	-
	(0.8 - 1.1)	(0.7 - 0.9)	
Southern Europe:	6.4 - 10	3.4 - 3.9	-
	(1.8 - 2.8)	(0.9 - 1.1)	
Central Europe (Passive Study only):			2.1 - 6.7
			(0.6 - 1.9)

All units in MJ p.a./ECU (and kWh p.a./ECU)

Fig. 2. Performance Optimisation Studies: Annual energy savings per invested ECU (Performance-to-cost Ratios)

To put these figures into context with respect to measured performance data from monitored field trials (from the PMG data collection exercises) and with earlier target predictions from modelling exercises, made at the end of Phase 2 of the PMG work, Fig 3 has been included. Some of this information is represented graphically in Fig. 4, clearly indicating the improvements achieved.

SYSTEM TYPE	PHASE 2 [3] Measured	Target Predictions	PHASE 3 [3] min	PHASE 3 [3] av.	PHASE 3 [3] max	Performance Optimisation St
DHW	820	1080 - 1440	387	1,070	2,234	1,022 - 3,006
	(225)	(300 - 400)	(108)	(297)	(620)	(284) (835)
A-SH	560	1000	126	571	1,288	767 - 1,636
	(154)	(275)	(35)	(159)	(358)	(213) (454)
P-SH [1]: DG	na	70 - 435	494	835	1,096	na
		(20 - 120)	(137)	(232)	(304)	
P-SH [1]: TW	290 [2]	na	na	496 [2]	na	na
	(81)			(138)		
P-SH [1] SS	na	220 - 545	na	279 [2]	na	360 - 472
		(60 - 150)		(78)		(100) (131)

[1] measured useful solar contribution [2] one project only
[3] some projects are in-common

DHW: domestic hot water system
A-SH: active space heating system
P-SH: passive space heating system

DG: direct gain
TW: Trombe Wall
SS: Sun space (conservatory)

Fig 3 : Comparison of System Performance for Phases 2 and 3, Design Targets from Phase 2 and Design Predictions (Phase 3). Figures in MJ p.a./m^2 (and kWh p.a./m^2).

6.0 CONCLUSIONS

Detailed conclusions concerning each system design cannot be included here, but the overall conclusions are as follows:

6.1 Performance

- The results from the Performance Optimisation Studies confirm the performance predictions made at the end of Phase 2, and indicate that significantly improved performance is achievable in better optimised solar heating systems, compared to most existing solar heating projects whose performance has been studied in the PMG data collection exercises.

- There are already some new projects with better designed passive or active systems that tend to confirm that the performance figures predicted in the Performance Optimisation Studies are achieveable.

6.2 Performance-to-Cost Ratios

- In overall terms, the Performance Optimisation Studies have resulted in designs which show improvements in cost effectiveness by a factor of 1.5 to 2.5 with respect to most projects studied in the PMG data collection exercises, in Phases 2 and 3.

- The study has shown that:

a) Some types of solar heating system could be cost effective now, e.g. simple passive systems (throughout Europe), and solar water heaters (in southern Europe).

b) Some types of solar heating system could become cost effective in the near future, e.g. advanced forms of passive and solar water heaters (throughout Europe).

c) Other types of solar heating system require further optimisation and development to improve cost effectiveness, e.g. active space heating systems.

Fig. 4. Performance and cost Diagram

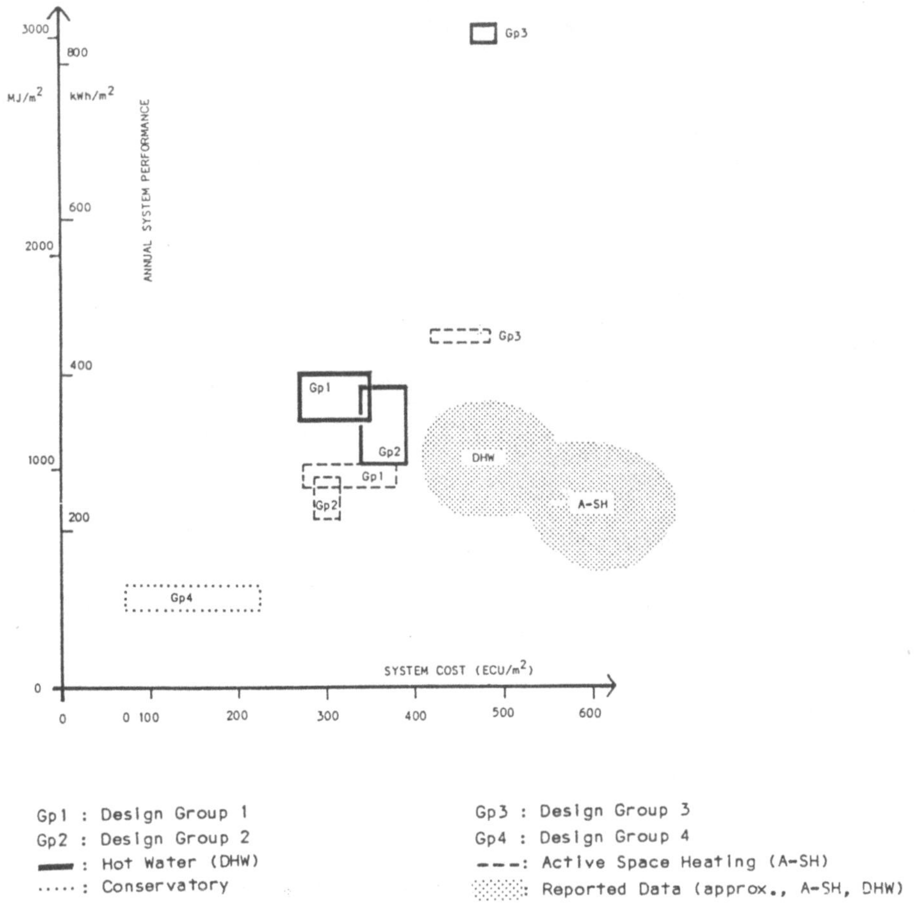

References

1. Peformance Monitoring of Solar Heating Systems in Dwellings: Executive Summary and Recommendations. Commission of the European Communities. EUR 8002. 1982.
2. Solar Water Heating: An analysis of design and performance data from 28 systems. Commission of the European Communities. EUR 8003. 1982.
3. Solar Space Heating: An analysis of design and performance data from 33 systems. Commission of the European Communities. EUR 8004. 1982.
4. Solar Heating: Performance of Recent Systems. Commission of the European Communities. EUR 8947. 1983.
5. Solar Heating: Performance and Cost Improvements by Design. Commission of the European Communities. EUR 8948. 1983.
6. Solar Heating: Performance Monitoring 82/83 - Executive Summary and Recommendations. Commission of the European Communities. EUR 8949. 1983.

EXAMPLES OF PRACTICAL APPLICATIONS OF COLLECTORS FOR 2ND GENERATION SOLAR ENERGY SYSTEMS

M. Baardman
EBS, Energie Besparende Systemen B.V.
GRAVE, The Netherlands

SUMMARY

Recent developments of various collector types to be manufactured with the same standard absorber are shown. The main reason for our approach was to achieve a significant cost reduction of solar energy systems by using a uniform way to manufacture collectors for various applications. Another important factor resulting in cost reduction is the development of well-designed solar energy systems of which all the components are carefully matched. Many of the systems we market today have been optimised in cooperation with research institutes and we contributed to 'concerted actions' of the CEC [1].

For many applications an overall system approach, starting from the most appropriate collector design, to the last detailed component has been worked out. Solutions as realised recently for various applications are discussed in this paper, such as: - energy roofs, - uncovered collectors for swimming pool heating, - covered collectors for swimming pool heating and shower facilities, - covered collectors for roof integration used for solar DHW-systems and combined SH and DHW-systems and special collectors for agricultural and industrial processes including solar drying.

INTRODUCTION

In many international conferences on solar heating the importance of cost reduction for solar energy systems was stressed by many researchers. Only very few representatives of the solar industry showed practical examples of ways to realise such a cost reduction and still maintain durable and reliable total solar installations. In this paper I will show what our firm EBS has been able to realize so far.

In all the development work we have carried out over the last 8 years we have always kept in mind the following starting points:

- cost reduction of solar energy systems can only be achieved if a uniform way can be found to manufacture low cost collectors for various applications in several countries;
- another important factor resulting in cost reduction is the development of well-designed systems of which the components are carefully matched.
- all our development and design work had to result in manufacturing procedures needing only a rather small investment in manufacturing equipment.

With these starting points we have succeeded in developing several collector types all of them having the 'sunstrip' as basic absorber material. Each of these collectors is suitable for various applications.

SECOND GENERATION COLLECTORS FOR LOW TEMPERATURE APPLICATIONS, TYPES LP AND LP-E

Both solar collector types are uncovered 'sunstrips' of which both ends of the fin have been bent in such a way (see figures 1a and 1b) that a flexible collector module can be made (see fig. 2), which can easily be mounted on horizontal roofs and vertical façades.

Applications for these types of collectors are:

- energy roofs, collectors in combination with a heat pump (operating temperatures 0 °C - 20 °C, type LP-E without insulating material at the back);
- open-air swimming pool heating (uncovered with backside insulation, type LP for operating temperatures up to about 30 °C).

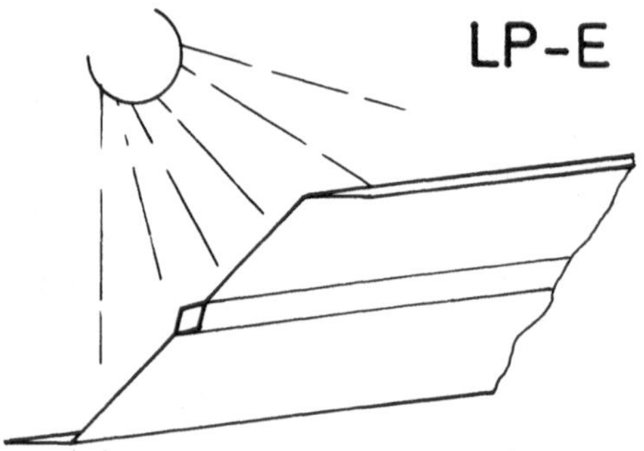

Figure 1a: Sunstrip with bent fins.

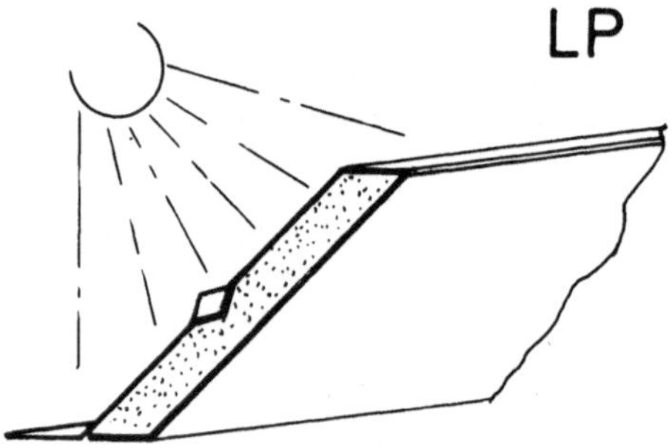

Figure 1b: Sunstrip with bent fins, and back insulation.

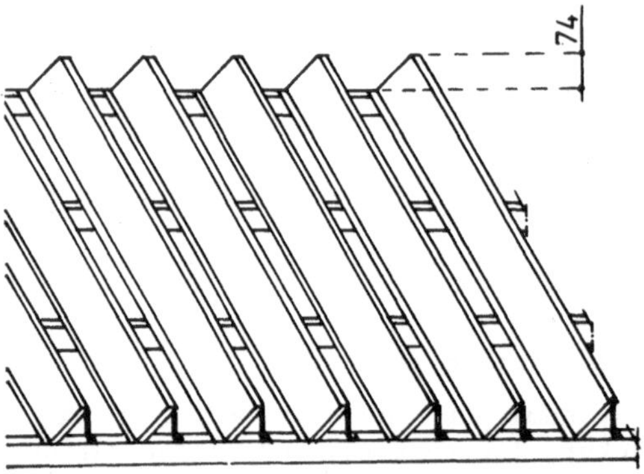

Figure 2: Collector module for LP and LP-E Collectors.

In figure 3 an example of the LP collector mounted onto a pergola for heating a 140 m^2 swimming pool is shown. Figure 4 shows one of the largest solar heated swimming pools in the Netherlands, 800 m^2 of LP collectors are installed partly on a flat roof and partly on pergolas.

Figure 3: LP-collector mounted on a pergola.

Figure 4: 800 m^2 of LP-collectors mounted partly on a flat roof and on a pergola.

SECOND GENERATION COLLECTORS FOR APPLICATIONS UP TO ABOUT 60 °C, TYPE LP-A

This type of collector combines the advantages of the other two types e.g. - flexibility in mounting on horizontal roofs and vertical façades, - simple manufacturing possibility. It consists of a spectral selective sunstrip absorber mounted in a specially designed casing, suitable for installing the back-insulating material and a transparent cover asking very little manpower. Similar collector modules as used for the types LP and LP-E can be manufactured, which are easy to install on horizontal roofs and vertical façades (see figure 5).
In figure 6 a prototype of this LP-A collector is shown. Applications for this collector type are: - shower facilities in swimming pools, - several agricultural and industrial solar plants asking hot water up to about 60 °C.

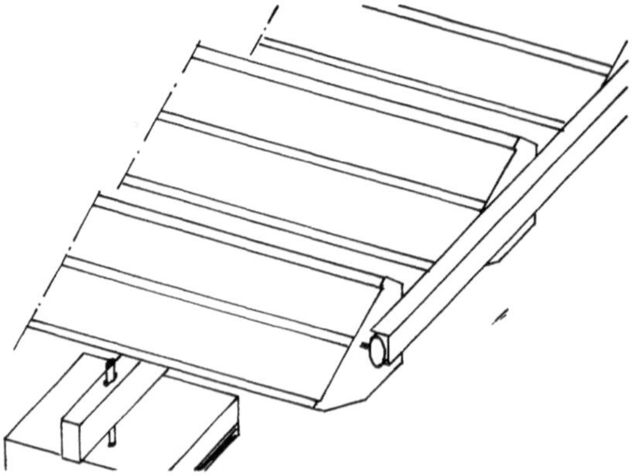

Figure 5: Collector module with LP-A collectors.

Figure 6: Prototype of LP-A collector module.

SPECIAL COLLECTOR FOR DOMESTIC SOLAR WATER HEATERS

Sloped roofs, usually covered with tiles, demand special collector types of which architectural integration details have to be carefully adapted to the local building tradition. For groups of terraced houses we developed a solar DHW-system, using a special collector for this purpose, and a special preheat storage tank. More than 1000 of these solar water heaters are installed in the Netherlands and for an average 4-person household 50% of the heating bill for hot tapwater is covered by solar energy. The installed price per system has been cut down to less than 900 US$. This could be achieved by careful detailing not only of the solar energy system but also of the integration details for installing the whole system. In figures 7 and 8 the integrated roof collector and the installed preheat tank on the attic space behind the collector are shown. More details can be found in [2].

SYSTEMS FOR VARIOUS OTHER APPLICATIONS

Combined systems for solar heating and domestic hot water supply ask for larger integrated roof collectors. The same holds for special solar energy systems in the agricultural sector.

Figure 7: Integrated roof collector for a DHW-system.

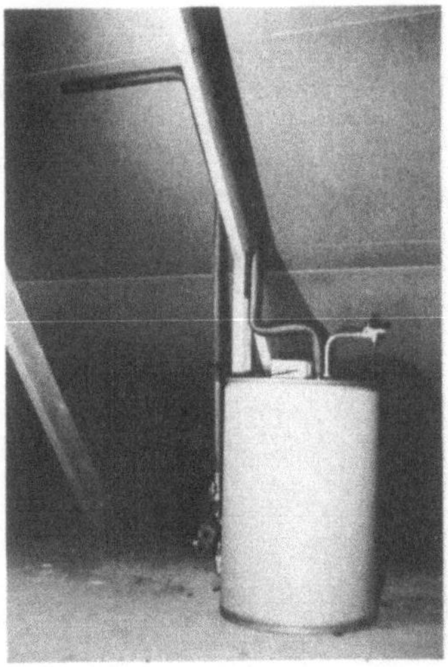

Figure 8: Solar preheat tank of a DHW-system.

Several of such systems have been designed and built as can be seen in figures 9 and 10.

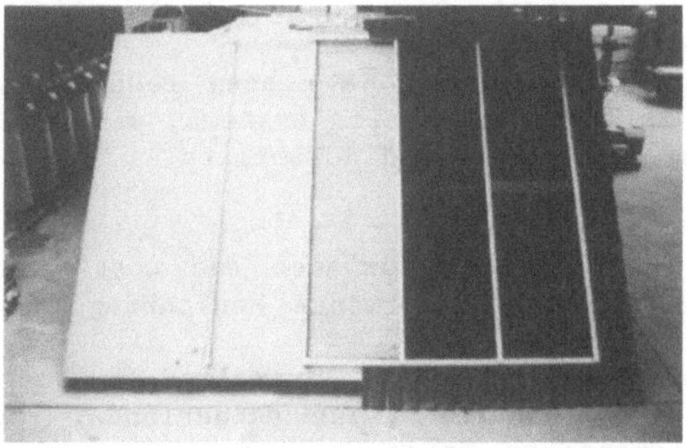

Figure 9: Prototype of integrated collector roof.

Figure 10: Integrated collector roof for a combined SH and DHW-system.

We also designed and built several special systems for other - usually sunnier climates. In figure 11 a recently built special system for a tobacco drying process is shown.

Figure 11: Special system for a tobacco drying process.

CONCLUDING REMARKS

- A good collector for a certain application is a good starting point but much more important is a good solar installation detailed in every respect, of which integration of the collectors in the built environment is almost a must to achieve cost reduction of the installation installed.
- The total cost of our 1984 systems have been reduced by 40 t0 60% compared to the cost of first generation systems, without compromising with respect to system performance and durability.

REFERENCES

[1] Ferraro, R. and R. Godoy, Performance and cost improvements by design, CEC report of the Performance Monitoring Group. Document EUR 8948

[2] Baardman, M. and C. den Ouden, Examples of solar collector integration into architecture in The Netherlands. Proceedings of the First EC Conference on Solar Collectors in Architecture, pp. 77-86, D. Reidel Publishing Company.

PERFORMANCE TESTS ON COMMERCIAL ONE-FAMILY DHW SOLAR SYSTEMS

M. Antinucci, M. Cocilovo, L. Grassi
PHOEBUS S.p.A.
Via G. Leopardi, 148 - 95127 Catania
F. Parrini
ENEL-CRTN
Via Rubattino, 54 - 20124 Milano
ITALY

Summary

In order to contribute to the R&D efforts toward the optimization of solar DHWS, a test facility has been set up by PHOEBUS, Catania. Six commercial single family systems can be simultaneously operated for a long time. After a description of the test facility and its operation mode, data about the performance of a reference DHWS during several months are reported. A detailed thermal analysis takes into account the separate contribution to the losses by the various system component. Major defects are identified and some suggestions for improvement are indicated. Six commercial DHWS are presently operated; preliminary service results of two thermosyphon systems are reported.

1. INTRODUCTION

System efficiency in solar DHWS is always much lower than collector efficiency. The investigation on causes for such lower values is of great importance in order to improve system design for 2nd generation DHWS.
A test facility has been set up to operate six single family systems in order to identify the penalty factors or the system efficiency and to suggest useful modifications.

2. METHOD OF APPROACH

A pre-programmed load,consisting on a draw-off profile, is applied to the system. A single day thermal balance is elaborated, taking into account the heat content variations of the storage volume, as measured by six temperature probes dipped into it. An alternative approach would have been the evaluation of long-term thermal balances, thus neglecting the effects of storage heat content variations. This mode is more satisfactory for comparative tests among different systems, but is less helpful when the effect of different modes of operation or the response to meteorological conditions are investigated.

Unfortunately we were not able to obtain continuous long term records of system operation due to faults of the acquisition and control system. Calculation of system efficiency η_s and solar fraction f are performed referring to a method previously published by J. Anhalt et al.(1). System efficiency is given by $\eta_s = \frac{Eu}{Aa.I}$, where Aa is collector area, I is total daily irradiation on the collector plane and Eu, useful solar energy is:

$$Eu = E_1 - \eta_{sto}(Ea + Ep)$$

where E_1 is the sum of extracted heat and storage thermal content variation,

Ea and Ep are the auxiliary heater and pump consumptions. The storage efficiency η_{sto} is defined as:

$$\eta_{sto} = \frac{E_1}{Es + Ea}$$

where Es is the solar energy input to the storage.

A reference system (system no.1) has been operated for six months under different operating conditions (different thermal load and auxiliary heating set temperature) in order to verify the effect of these conditions on system efficiency.

A temperature set value of 50°C for the auxiliary heater was then chosen, while the following standard load profile was selected.

Hour	8.30	11.00	14.00	17.00	20.00	Total
heat (Wh)	730	970	1000	700	1300	4700
ref.volume (l)	21	28	29	20	37	135 (Tout=45°C)

On such conditions other two systems have been operated for a couple of months, together with the reference one, while presently six systems are under test. Main system characteristics are as follows:

System no.	Coll.area (m^2)	Storage vol.(l)	Flow	Auxil poxer(w)	Pipe (m)
1 (Ref)	3,0	180	forced	1600	20
2	3,6	200	natural	1600	3
3	3,0	220	natural	1200	2

Pipe insulation: expanded foam, 15 mm thick, for all systems.

3. RESULTS

Figure 1 reports the results obtained on system no. 1 during 7 periods composed of uninterrupted test days, characterized by a minimum daily irradiation of 2500 Wh/m^2. System efficiency (a) and storage efficiency (b) are defined in sect. 2. Collector field efficiency (c) is defined as the solar energy input to the storage divided by solar energy incident on the collectors. For better understanding, daily efficiencies of collectors, pipes and on/off operation were separated.

Collector efficiency (d) is defined as the collectors heat output divided by solar irradiation integrated only when the pump is on. Pipes efficiency (e) is the ratio between heat input to the storage and collectors heat output. Control operation efficiency (f) is the ratio of the solar irradiation integrated only when the pump is on to the total irradiation. Each of the seven periods is characterized by different conditions:

- thermal load: 7100 Wh/d for period 1, 4700 Wh/d for period 2 to 7;
- thermostat set value: 60°C for period 6 and 7, 45°C for period 1 to 5;
- summer conditions: period 1, 2, 3;
- fall conditions: period 4, 5;
- winter conditions: period 6, 7.

The anomalous values of control operation efficiency in periods 2, 3 are due to defects of the insulation of the collector sensor cable. System efficiency (a) which ranges from 20 to 30% in periods 1 and 4, is only 10-15% in periods 6, 7. This can be attributed partially to lower collector field efficiency (c), and significantly to the storage efficiency (b), that falls down to 50% in period 7.

Pipes efficiency (e) which exceeds 90% in period 1, is less than 80% in period 4. This can be explained by high collector temperatures due to large irradiations and low outside air temperatures. Intermediate values are recorded in periods 6, 7.

Control operation efficiency (g) is disappointingly low, ranging between 60% in period 7 to about 80% in period 1.

To better evaluate the reason of the low value of this "control operation efficiency" the calculation of the performance of an ideal controller was done and its control operation efficiency compared with the actual one. Table II reports three types of evaluation of the ideal control operating efficiency: the first ideal control switches on the pump whenever the energy output of the collectorscalculated from its instantaneous efficiency curve, using storage temperature and meteorological data every five minutes, exceeds the energy absorbed by the pump. The second calculation includes the heat exchanger penalty factor, and the third one includes also a 20% pipe losses.

Table II shows that a difference of 10-20% between experimental values and calculations, even applying the third type of calculations, is still present, and this difference increases in winter time, when sunshine duration is lower, even at similar total irradiation values.

Two thermosyphon flow solar DHWS have been submitted to test together with the reference one. Fig. 2 reports the integrals of solar irradiation (I), thermal load (L) and useful energy (E) and the average values of system efficiency and solar fraction for several sets of test data.

4. CONCLUSION

From the analysis performed on system no. 1 a set of considerations can be drawn:

a) storage losses have to be drastically reduced, especially in the upper part of the cylinder, in order to contain storage efficiency over 80% also in winter conditions;
b) pipe losses are still too high, especially in intermediate seasons;
c) a large fraction of the incident solar energy is lost because of the operation of the electronic controller. The analysis suggests that this section of system performance is worth of R&D for improvement;
d) collector surface and tank volume are not optimized.

The small amount of data obtained from system nos 2 and 3 prevents from definite evaluation of their performance. The data acquisition is at present continuing on these and other four systems, and our aim is to collect a more continuous and lenghtened set of data for a better analysis.

REFERENCES

1. ANHALT J. et al (1982). Study of a Solar DHW System, to be published in annual report 1982. Inst. of Kernphysik, KFA, Jülich, GmbH.

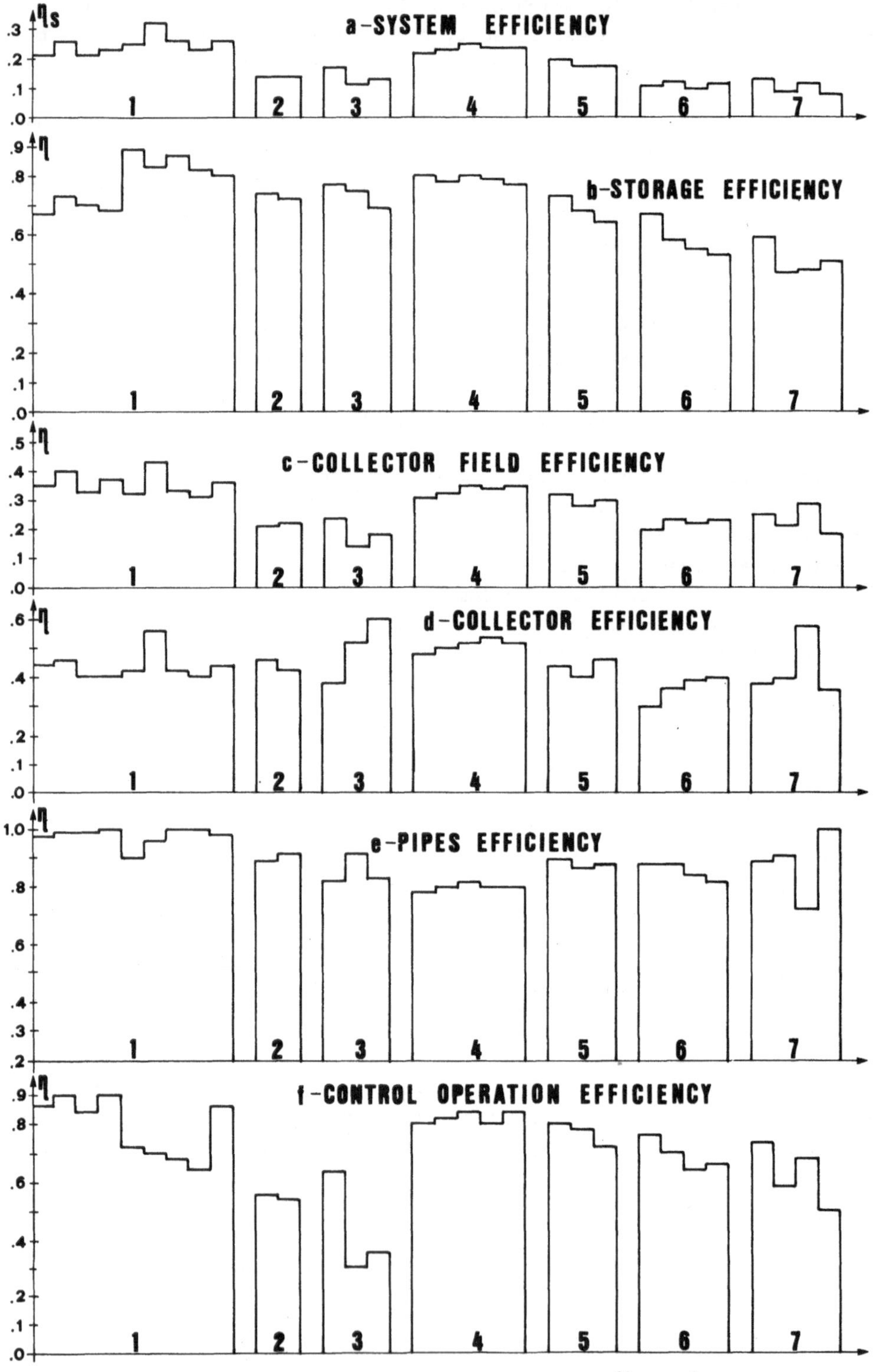
a-SYSTEM EFFICIENCY
b-STORAGE EFFICIENCY
c-COLLECTOR FIELD EFFICIENCY
d-COLLECTOR EFFICIENCY
e-PIPES EFFICIENCY
f-CONTROL OPERATION EFFICIENCY
1
2
3
4
5
6
7

figure 1

TABLE I - Calculated and measured value of control operation efficiency

DATE	Irradiation (Wh/m^2.d)	η_{th}^1	η_{th}^2	η_{th}^3	$\eta_{exp.}$	period no.
Oct. 7	5761	.94	.93	.91	.81	4
Oct. 8	5738	.94	.94	.91	.82	4
Oct. 9	5458	.93	.93	.92	.83	4
Oct.10	5555	.92	.92	.90	.79	4
Oct.11	5550	.94	.93	.92	.83	4
Nov.10	4658	.89	.89	.88	.80	5
Nov.12	4064	.88	.87	.86	.73	5
Dec.23	4320	.90	.90	.87	.77	6
Dec.24	4676	.89	.88	.87	.70	6
Dec.25	4921	.90	.89	.87	.66	6
Dec.26	5150	.91	.90	.89	.67	6
Jan.27	3277	.83	.83	.81	.59	7
Jan.28	4088	.88	.88	.85	.69	7

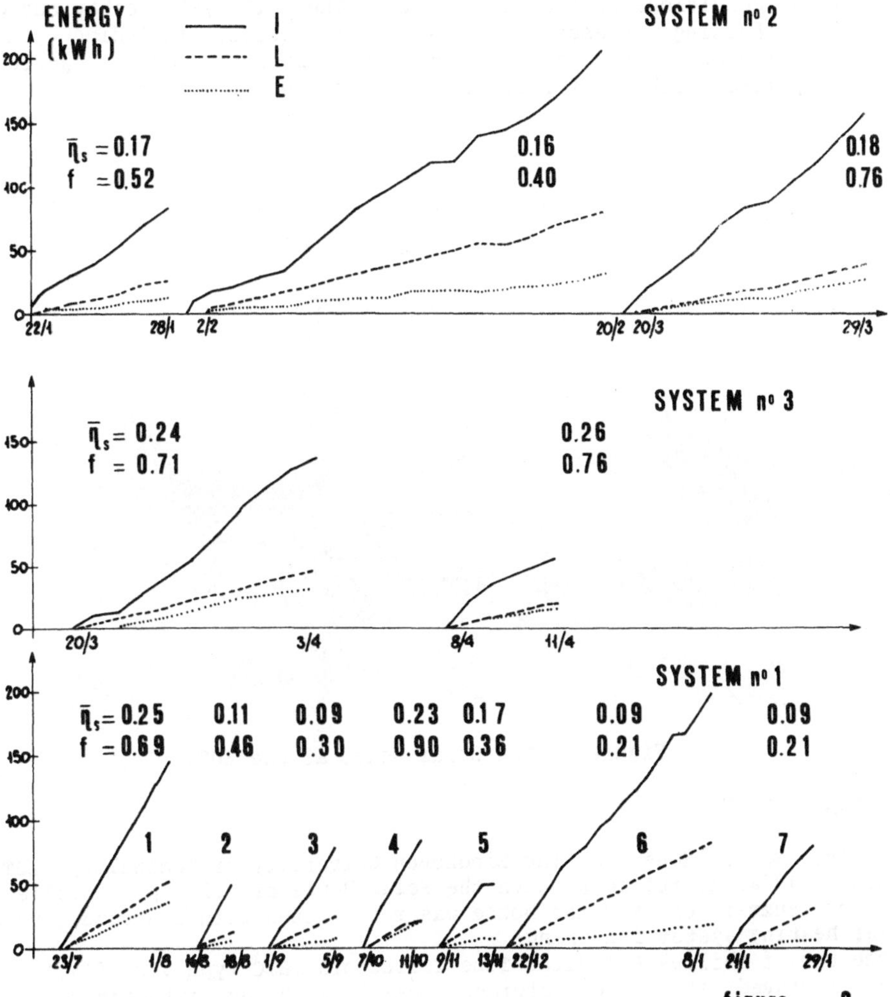

figure 2

PERFORMANCE OF THE SECOND GENERATION SOLAR HEATING SYSTEM IN THE SOLAR HOUSE OF THE EINDHOVEN UNIVERSITY OF TECHNOLOGY

R.W.G. Bisschops, C.W.J. van Koppen and W.B. Veltkamp
Eindhoven University of Technology
Department of Mechanical Engineering
Postbox 513, 5600 MB Eindhoven, The Netherlands

Summary

Summer 1981 a new solar heating system has been installed in the Solar House at the E.U.T. The principal features of the system are Philips VTR 261 evacuated tube collectors, integration of the auxiliary heater with the (stratified water) storage and application of the new, balanced flow control strategy. The system is continuously monitored according to IEA requirements, with the objective of establishing the performance of the 2nd generation system, and components under real load conditions in the Dutch clmate. The full system behaviour during the heating season '83-'84 is discussed, including practical experiences with the various new components and a performance comparison with the former flat-plate collector system.

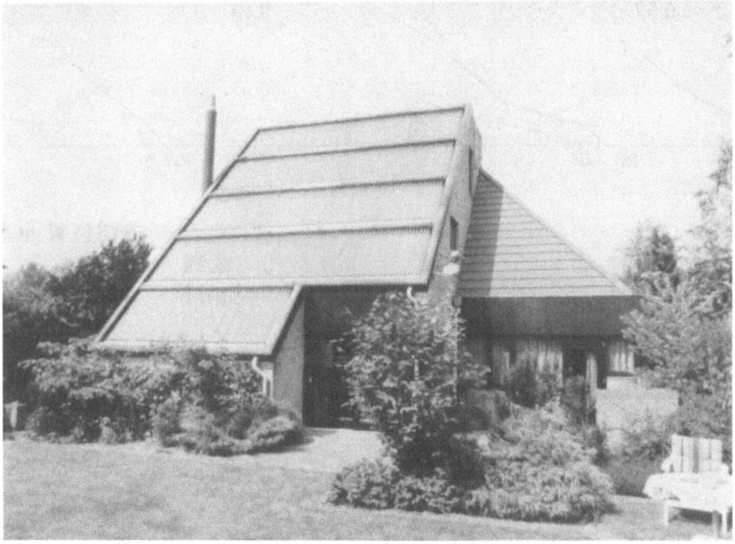

Figure 1. The Solar House at the EUT.

1. INTRODUCTION

The Solar House at the Eindhoven University of Technology (EUT) was built in 1976. At its completion the Solar House had flat plate collectors. In the summer of 1982 the house was retrofitted with a second generation solar heating system featuring:

- the use of Philips Evacuated Tube Collectors (ETC type VTR 261),
- an improved stratified storage vessel with an integrated auxiliary burner, and

- a novel control strategy.

The heat demand is 23000 kWh/yr for space heating and 3700 kWh/yr for hot water. The objectives of the monitoring period of two and a half years are:

- to gain performance and operational data for the ETC's under real load,
- to establish a well validated basis for the comparison of conventional and ETC system performance, and
- proving the novel control strategy, which implies roughly equal daily throughputs for the collector loop and the load circuits, and
- a proof of concept test for the integrated heater and storage design.

The project has been incorporated in Phase II of the Dutch National Solar R&D Program, and forms part of the ETC research of the IEA (Solar Task Group VI).

2. SYSTEM HARDWARE

The collector array consists of 23 modules of 16 ETC's each. All modules are connected in series to the array line, zigzagging upward on the roof. The aperture area of the array is 47.15 m^2 and the gross area is 57.66 m^2. System fluid is water without additives. Behind the tubes is a flat Tedlar coated aluminium reflector. Orientation is 7° West from South and tilt is 48°.

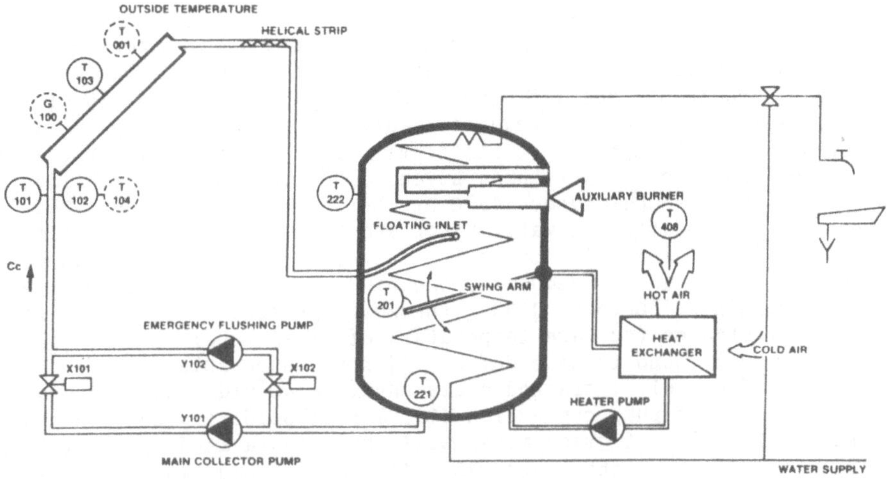

Figure 2. Simplified installation scheme

The mass of the storage fluid is 3700 kg and the heatloss factor of the vessel is 6.2 WK^{-1}. A nitrogen pocket in the top of the storage acts as expansion volume. The stratification is preserved by: using a floating inlet, a swing arm outlet and by limiting all entry velocities to 0.2 ms^{-1}. The efficiency of the burner amounts to 87% (ref. to upper caloric value). To avoid condensation in the flame tube the minimum temperature in the top part of the storage is 60^oC and to prevent scaling inside the hot water coil the maximum temperature for the toppart of the storage is 80^oC. The theoretical heat transfer capacity of the air heater is 1500 WK^{-1}. The garage is heated by waste ventilating air.

3. CONTROL STRATEGY

Collector. In the normal collector mode the flow is controlled by and proportional to the difference $\Delta T = T103 - T221$ (see figure 2 and list of symbols). Collecting starts at $\Delta T > 10$ K and stops at $\Delta T < 6$ K. The flow runs from 100 kg hr^{-1} at $\Delta T = 6$ K to 1100 kg hr^{-1} at 17 K. Overheat protection (T222 > 80 ^{o}C) is effected by stopping the collectorpump and letting the array line boil empty. If T 101 < 2 ^{o}C the array line is flushed for 10 minutes to prevent freezing.

Storage. During the storing of energy all collector modes are possible. The auxiliary burner ensures a minimum of energy in the (top part of the) storage vessel. After starting the auxiliary burner operates untill the temperature in the top-part of the vessel is incremented by 10 K.

Comfort. A thermostat in the living room controls the air heater pump and, simultaneously, the heater fan in on-off-way.

Note: The control just described has been simplified, as compared to the original one [1], on the basis of practical experiences.

4. EXPERIENCES ON COMPONENT BEHAVIOUR

Earlier experiences were reported in [1]. Additional favourable experience was gained concerning the effect of the helical strip in the highest point of the array line. The strip creates a centrifugal pressure field forcing any gas bubbles to the centre of the tube. Experiments proved this method of degassing to be so effective that all other other provisions (ventpot, steallead, see[1]) could be abandoned. The design of the integrated auxiliary burner requires further investigation, because its operation goes with extreme stratification in the top part of the storage, e.g. water surface temperatures beyond 105 ^{o}C (hot plume from flame tube) and water bulk temperatures 45 ^{o}C at the level of the burner. The latter leading to condensation of water vapor and to corrosion in the flame tube. Baffles will be installed on both sides of the flame tube to augment the natural convection and consequently the mixing in the affected region.
In the beginning of October 1983 a thermovision (infra red) check was made on the performance (surface temperature of glass tube) of the collectors. Ten tubes were found to have defects (1 non-condensable gases in heatpipe (faulty manufacture), 1 cracked glass envelope at mid length (vandalism?), 1 broken off end of pump connection (?), 7 poor soldering between heat pipe and metal ferrule of glass metal seal; such aggravated by bending of the flexible condensor necks over 10 ^{o} (!) for the mounting of modified end supports, in May 1983). From detailed records of storage temperatures the floating inlet was found to operate well. The Tedlar coated reflector behind the collector has kept its shining appearance.

5. PERFORMANCE OF ETC'S

The performance of the ETC's meets the good expectations for the Dutch climate. Fig. 3 shows the I/O-characteristic for March 1984. Table I shows Q112 (see list of symbols) as a function of H001. The values in this table are obtained with a multivariable regression method according to the prescriptions of Task VI of the IEA.
Q112 is written:

$$Q112 = C_1 \; H001 + C_2$$

The results indicate a relatively high value of C_1 (about 0.67 pointing to a good optical efficiency) and a low $|C_2|$ (about 1, implying small heat losses).

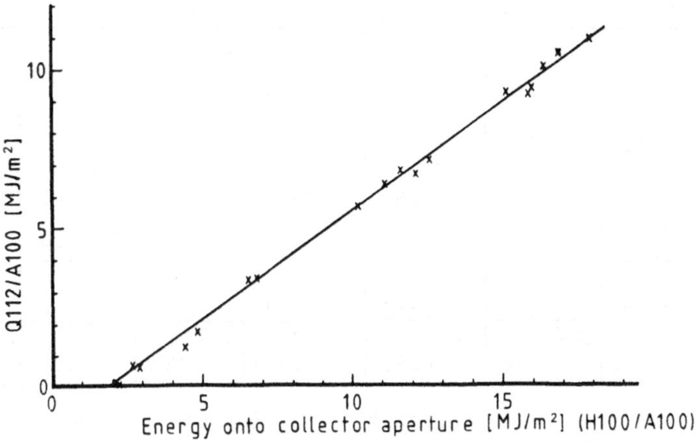

Fig. 3. I/O-characteristic March 1984

As yet no generally accepted method is available to characterise ETC's; the Hottel-Whillier equation has been found not to apply [1]. As far as the deriviation goes the I/O curves may be interpreted to indicate an optical efficiency in the order of 0.70 and a heat loss coefficient below 1 $Wm^{-2}K^{-1}$.

Table I. Daily input/output.

Month	C_1	C_2	x intercept	nr of points	Correlation coefficient
Nov. '83	0.66	-0.44	0.67	23	0.98
Dec. '83	0.61	-0.43	0.70	31	0.99
Jan. '84	0.57	-0.39	0.69	28	0.98
Feb. '84	0.63	-0.66	1.05	28	0.99
March '84	0.68	-1.26	1.86	25	1.00

6. COMPARISON OF THE FLAT PLATE COLLECTOR (FPC) SYSTEM AND THE NEW ETC SYSTEM.

In the former solar heating system the collector consisted of aluminium finned tube absorber plates, coated with black chrome. The measured optical efficiency and heat loss coefficient were 0.77 and 5.5 $Wm^{-2}K^{-1}$ respectively. The net aperture area amounted to 51 m^2 and the collectorflow was constant (860 kg/hr). The volume of the storage was 4.1 m^3, the heat capacity 17 MJk^{-1}, the theoretical loss coefficient 6.62 WK^{-1} and the theoretical loss coefficient 6.62 WK^{-1}. The auxiliary burner was a natural gas boiler, make Vaillant.
Since the new data aquisition system is fully operational and the measured data was found to be reliable from October 18.1983, data about almost the entire heating season 1983-1984 is available. This permits comparison with the performance of the former system in the heating season '79-'80, documented by Kemp in [2], see table II (values in kWh, $\overline{T001}$ in ^oC).

Table II. Comparison of heating seasons '79-'80 [2] (FPC) and '83-'84 (ETC)

Month	H001	Q112	Q200	Q210	Q220	Q201	Q256	$\bar{T}001$
Nov. '79	2443	795	795	2767	397	2927	-	5,8
'83	2863	1578	1465	2983	154	2673	64	6,1
Dec. '79	883	188	188	3555	388	4154	-	5,6
'83	1882	936	898	3526	161	3629	61	3,8
Jan. '80	1707	403	403	5123	325	5390	-	0,5
'84	1733	732	681	4006	217	4482	32	3,4
Feb. '80	2481	822	822	2910	278	2945	-	5,1
'84	2606	1376	1297	3842	232	3675	34	2,0
Mrch '80	2952	841	841	2806	283	2963	-	5,4
'84	4404	2472	2352	3044	243	1926	31	4,0
'79-'80	10466	3049	3049	17161	1671	18379	-	4,5
'83-'84	13488	7094	6693	17401	1007	16385	222	3,9

Table III shows some characteristic values for the old and the new system. Conclusions are:

1) The quotient Q112/H001, i.e. the solar efficiency of the ETC system in the heating season '83-'84 is larger than the solar efficiency of the FPC system in the season '79-'80 by a factor of about 1,8 . Most of this difference is due to the higher quality of the ETC's (see months of similar sunshine, Nov. Jan. and Febr. in table 2) and only a minor part to higher solar radiation (Dec. and March in Table 2).
2) The utilisation efficiency Qtt/Qtf is some 10% lower for the new system as compared to the old one. In particular the storage and piping losses of the new system will have to be studied. It is remarked that these losses are not energy losses in a strict sense, because the storage and most of the piping is located indoors.

Table III. Characteristic values for the old FPC and the new ETC system.

System	Heating season	Q112/H001	Qtt kWh	Ns	Qtf kWh	Qtt/Qtf
FPC	'79-'80	0,29	21428	0,14	18832	0,88
ETC	'83-'84	0,53	23078	0,29	18630	0,81

Symbols (energy in kWh, temperature in ^{o}C):

H001	Total solar incident	Q256	Storage losses while flushing
Ns	Fraction solar to storage	Qtt	Q200 + Q201
Q112	Solar energy collected	Qtf	Q210 + Q220 + Q256
Q200	Solar energy to storage	T001	Ambient temperature
Q201	Energy from auxiliary burner	T100	Average collector temperature
Q210	Energy to space heating	T103	Temperature 2/3 collector
Q220	Energy to domestic hot water	T221	Bottom temperature storage

REFERENCES

1. R.W.G. BISSCHOPS, C.W.J. VAN KOPPEN AND W.B. VELTKAMP. Experimental performance of the second generation solar heating system in the solar house of the EUT, Solar World Forum, Perth 1983, Pergamon Press.
2. J.W.C. KEMP, Analysis of the measuring data from the Solar House of the EUT (in Dutch), EUT Report WPS3-81.01.R383,

EVACUATED COLLECTORS IN COOPERATIVE INTERNATIONAL ENERGY AGENCY RESEARCH AND DEVELOPMENT

W.S. Duff
Solar Energy Applications Laboratory
Colorado State University
Fort Collins, Co. 80523 USA

K. Vanoli
IST Energietechnik GmbH
Ritterweg 1, D 7842
Kandern-Wollbach

Summary

The IEA Task on evacuated collector systems has been active since more than four years. The task includes research and development projects for solar cooling, heating, and domestic hot water in buildings; industrial process heat; district heating; and a test bed with simulated load. The countries participating are Australia, Canada, the CEC, the Federal Republic of Germany, Japan, the Netherlands, Sweden, Switzerland, the United Kingdom and the United States of America.

Nearly all available types of evacuated collectors and most relevant applications are included in the task. All installations are extensively monitored and a common reporting format is prescribed. The detailed record of performance produced provides the information needed to assess the relative advantages of these collectors in systems as well as a basis for extrapolation of results to other climates and applications.

This paper will describe the task installations and collectors. Performance of both collectors and systems will be presented. System operating experience will be discussed. Reliability, durability, and maintainability will also be presented.

DESCRIPTIONS OF EVACUATED COLLECTORS USED IN TASK VI INSTALLATIONS

There are two main categories of evacuated tubes: 1) the metal absorber plate in a glass tube and 2) the Dewar flask with a selective surface on the outer surface of the inner glass tube. In the first type heat can be removed either directly by circulating a fluid through a tube bonded to the metal absorber plate or indirectly by a heat pipe bonded to the metal absorber plate. For the second type heat removal can be by a fin or tube in contact with the inner surface of the inner glass tube or by direct liquid contact with the inner surface of the inner glass tube. The collectors previously used, in use or planned for Task VI installations are given in Table 1.

TABLE 1. Collectors Used in IEA SHAC Program Task VI Installations

Collector	Type	Heat Removal
Corning	Metal in glass	Direct
General Electric	Dewar	Fin
Owens Illinois	Dewar	Direct liquid contact
Philips VTR series	Metal in glass	Heat pipe
Philips MKIV	(non-conventional)	
Sanyo	Metal in glass	Direct
Sanyo	Metal in glass	Heat pipe
Solartec	Dewar	Direct liquid contact
Sunmaster	Dewar	Direct liquid contact
Sydney University	Dewar	Fin

DESCRIPTION OF TASK INSTALLATIONS

The descriptions of the IEA SHAC Program Task VI installations and collectors are provided in Table 2.

Table 2. Performance of Solar Heating, Cooling and Hot Water Systems in the IEA SHAC Program Task VI

Installation Location	Applications	Evacuated Collectors Used/Planned	Description
Australia Sydney	Heated, cooled University offices	Sydney Univ. Collectors Flat Plate Collectors	A domestic sized solar assisted heating and cooling system is installed on the roof of the Mechanical Engineering Dept. It provides heating and cooling for 4 of the offices. An initial system of 34.4 m^2 of flat plate collectors has been replaced by 41.4 m^2 of Sydney University ETC's. There is 3,200 l of hot water storage.
Canada Edmonton	Industrial Process Heat	Solartech (O-I tubes)	The solar system was installed in a bottling plant building. It was designed to assist in maintaining a caustic soda solution used for the washing of re-usable empty soft drink bottles at a temperature of 75°C.
CEC Joint Research Centre, Ispra	Heated, cooled laboratories	Sanyo with and without reflectors Philips (Neopentane working fluid) Flat plate Collectors	A 4kW solar assisted chiller provides cooling for the highly insulated Joint Research Centre Solar Laboratory. The floor area of 160 m^2. Energy from 144 m^2 of solar collectors, including various types of flat plate and evacuated tube collectors is available.
Japan Osaka	Single family heating, cooling and hot water	Sanyo General Electric TC-100 Sanyo Heat Pipe	The solar house, built in 1977, is a two story, reinforced concrete building with a conditioned living area of 118.52 m^2. It was designed as a single family residence utilizing solar energy for space heating, space cooling and domestic hot water.
Netherlands Eindhoven	Single family heating and hot water production	Philips MKI Heat Pipe (Isobutane working fluid)	The solar house, built in 1976, has a conditioned living area of 220 m^2. It was designed as a single family residence and is currently occupied by a family of three. Solar energy is utilized for space heating and domestic hot water supplies.
Sweden Knivsta	District Heating	Philips MKI Heat Pipe (Isobutane working fluid) General Electric TC-100 Owens-Illinois	Three systems are installed in the district heating plant building. The heat is transferred directly to the return line of the district heating system. Since the heating load is always larger than the energy production from the collectors, there is no storage. The district heating plant supplies space heating and domestic hot water to surrounding buildings.

Sweden Södertörn	District Heating	Philips (Isobutane working fluid) General Electric TC-100 Flat Plate Collectors	A total of more than 800 m^2 of both flat plate and evacuated tube collectors are used to augment the conventional heating source for the district heating system. Each bank of collectors is separately connected to the main district return pipe. An anti-freeze solution is circulated in the collectors loops.
Switzerland Geneva	District Heating	Corning Sanyo Swiss Evacuated Flat Plate Collectors	Two solar systems are connected directly without intermediate storage, to the district heating network. Since the minimum heat demand is always much higher than the heat produced by solar, the district heating network supplies space heating and domestic hot water to surrounding buildings.
Switzerland Hallau	District Heating	Corning Concentrating Collectors	The solar energy system is installed on the roof of a factory that produces fruit juice and wine. An area of 350 m^2 of evacuated collectors is installed. The energy collected is used for space heating and process heat. Heated water is stored in two tanks of 10 m^3 and 23 m^3.
United Kingdom Bracknell	Simulated Load Domestic hot water emphasis	Philips MKI Heat Pipe (Isobutane working fluid)	The solar energy system is installed in Building Services Research and Information Association building and is connected to space heating and domestic hot water storages in the laboratory. Heat is removed from the storages by systems which simulate the space heating and domestic hot water installations of a single-family dwelling.
United States Fort Collins	Single family heating, cooling and hot water production	Corning Philips MKIV Philips MKI Heat pipe (Isobutane working fluid) Philips MKI Heat pipe (Water working fluid and longer tubes) Owens-Illinois Sunmaster Flat Plate Collectors	Solar House I, completed in 1974, is a wood frame, two story, three bedroom, residential building utilized for offices. The conditioned living area is 249 m^2. Solar energy is used for space heating, space cooling and hot water.
Germany Freiburg	Multi family heating and hot water production	Corning Philips MKIV Philips MKI Heat pipe (Isobutane working fluid)	The solar house is a three story apartment building with 656 m^2 of conditioned living area. Approximately 25 persons are permanently living in the 12 apartments. Solar energy is used for space heating and hot water.

SYSTEM OPERATING EXPERIENCE

Some excerpts of the most significant conclusions from the first task interim report are given in Table 3.

TABLE 3. Excerpts of Conclusions From the First IEA SHAC Program Task VI Interim Report

1. Although excellent performance results were achieved for all task experiments relative to conventional systems, not all evacuated collectors or systems performed the same. There was a substantial range of results among the different experiments, collectors and climates.
2. The highest seasonal system efficiencies recorded by Task VI installations were the Corning collector at Colorado State University (CSU) Solar House I from 1975 through 1978. Values were 44% for heating and 47% for cooling.
3. For the evacuated collectors, the ratio of monthly energy produced to the energy falling on the collector aperture ranged up to two and one-half times that of the flat plate collector.

4. For the same daily average insolation levels and average ambient to collector fluid temperature differences, there are significant differences between summer and winter operation in the daily energy outputs of evacuated collectors. All causes of these differences are not yet adequately modelled.
5. The Corning collector operated at the CSU Solar House from 1975 to 1978, and at the Solarhaus Freiburg from 1978 to the present. During that time one tube was broken during installation and less than 2% lost their vacuum. Of those, most lost vacuum during shipment or installation.
6. During the eight years of operation of the Corning Collector there has been no measurable degradation of collector performance.
7. Fewer than 1% of the Philips VTR141 tubes removed from CSU Solar House I after two-years operation had lost vacuum. Several tubes, less than 2%, were broken in shipping, unpacking or installation.
8. No collector maintenance has been required at any of the installations. System maintenance needs have been minor and, for the most part, have been for the conventional parts of the system.

CURRENT AND FUTURE TASK VI ACTIVITIES

Task VI is currently preparing a second interim report with a projected publication date of late 1984. In addition to further results beyond the first interim report, including results from new installations, there will be extrapolation of performance to other applications and climates and new approaches to analyzing the cost and performance of systems.

In the final task report, due in early 1986, description of proper design of evacuated collector systems will be a major emphasis.

RESULTS OF DURABILITY TESTS ON VARIOUS TYPES OF COLLECTORS.

G. RIESCH and P. WEISGERBER
Joint Research Centre, Ispra
Commission of the European Communities.

Summary.
Tests proposed in the literature for evaluation of the durability of solar collectors have been examined experimentally.
The following tests are retained as being the most significant ones:
In the field of qualification tests: Thermal shocks, hail test, overpressure and leak test (of absorber).
In the field of efficiency degradation: Dry outdoor and simulated indoor exposure, rain tightness test.

Other tests, as for instance exposure to simulated loads (pressure, suction or longitudinal loads), thermal cycling, ultraviolet irradiation, humidity treatment, ozone treatment and hail tests on flat plate collectors have been found to be less significant.
Results of the tests and conclusions are reported in the paper.
Standard corrosion test procedures showed effects under certain conditions leaving doubts however for their correlation to real service conditions. A series of weathering experiments under simulated service conditions in different typical climate zones was started in order to get confirmed data.

1. Introduction.

The durability of various types of thermal collectors has been studied experimentally since 1978 at the Joint Research Centre of the Commission of the European Communities. This work was done in the frame of the "EC Collector Testing Group" and has already been used to define a provisional, first set of minimum requirement tests.
Tests on complete collectors, almost all with non-selective, black paint absorbers, were executed in the following fields:
1. Tests on destructive failures (qualification tests) .
2. Tests on slow efficiency degradation.
3. Tests on atmospheric corrosion (case, outside of absorber).

1. Destructive tests.

1.1 Thermal shocks (AT-6)

Two types of thermal shocks are studied:
- outer thermal shocks by spraying cold water onto the hot collector
- inner thermal shock by filling the hot collector with cold water.

In both cases the collector is heated up by irradiation from the solar simulator AT-6.
Outer thermal shocks did not cause any damage to six different collectors treated, some temporary condensation in one case excepted.

In the case of the thermal shock by filling with water the water evaporates in the first moment and causes hence not only a thermal shock but also a slight, transient overpressure on the absorber.
In one case of six collectors tested so far by this method the absorber was visibly lifted off from the thermal insulation for several centimeters, but this bowing was partially reversible. There remained however a permanent bowing of the absorber plate of about 1 cm in the upper part of the collector.

1.2 Hail test. (AT-4).

It has been found experimentally with the hail gun (AT-4), that hail at our standard test condition of 2.5 J of impact energy (ice balls of 27mm of diameter (≈9.3 g) and with 23.1 m/s of speed) is unproblematic for thermal collectors. Even ten times higher impact energies that occured in September 1982 at our site in reality (estimated to have been 26 Joule) did not damage any flat plate collector.
Hail can hence be considered to be unproblematic for flat plate collectors of modern design.
For evacuated tube collectors however hail resistance can be an important criterion: There are samples of one manufacturer that broke already at impact energies of 1 to 2 Joule, there are samples of another manufacturer, that broke at 3 to 4 Joule and there are others that resist still at 12 Joule and more of impact energy.
It is difficult to fix the value of the impact energy to which a solar device must resist, as this depends on the meteorological conditions of a given site. A rather reasonable value for european weather conditions seems to be about 2.5 Joule (our standard impact energy). Devices that break at lower impact energies are probably not apt for longer life.
For details on hail tests see (1).

1.3 Static pressure tests of absorbers (AT-5).

Resistance of absorbers to overpressure.

This test is executed at an overpressure of 1.5 times the nominal operating pressure by applying a hydraulic hand operated pressure source. From 16 tested collector types two (roll bonded aluminium absorbers) were severely deformed by the applied pressure. No bursts of absorbers were encountered.
Leak tightness test.
Several types of collectors and bare absorbers were tested for leakage with pressurized air at nominal operating pressure. Very small leaks corresponding to a pressure drop of 0.02 bar/minute represent the detection limit of the method. Smaller leaks can be detected with a helium leak detector now available.
These two tests can easily be performed and do not require expensive apparatus. It can be recommended to execute these simple tests on all collectors with roll-bonded absorbers before they are mounted on the roof (or even better before they leave the factory).

1.4 Resistance to life loads (AT-3).

For collectors of current design with glass covers the resistance to wind pressure loads (up to 2400 Pa) or to snow loads (up to an additional 3000 Pa) does not seem to be a problem. In fact the glasses are normally 3 to 4 mm thick and are hence sufficiently resistant.

Also for collectors with plastic glazing of the current design the resistance to mechanical deformation by wind and snow pressure seems to be quite sufficient, at least at room temperatures.

2. Tests on slow efficiency degradation.

2.1 Rain water tightness.

Penetration of rain water into collectors is an often occuring event. It seems to be obvious, that rain water entered into a collector can damage the collector in the long run (corrosion inside of the box, soaking of the thermal insulation) and that it also can reduce its efficiency (condensate on the inside of the glazing.)
Two types of apparatus are used to study this effect: a grid of orientable spray nozzles inside of a pressurizing chamber, that allows to simulate "breathing" of the collector as under the influence of gusty winds,
and a simple shower device, applied on the collector in normal operation position.
The water penetration into the collector is determined by weighing the collector before and after the shower treatment on an accurate balance (precision +-1 g relative).
Ten collectors were exposed to the shower during 2 hours on the rack facility. The measured weight increase varied from 7 g to 281 g showing that no one of the tested collectors was completely water tight and that two were really very leaky. The water penetration in the cyclic underpressure device was about the same as under the showers.
It has to be said, that the aspect of water tightness of collectors needs further investigations.
In any case it seems to be preferable that all collectors, which, as it turns out, cannot easily be made water tight should be ventilated or at least be provided with suitably located drain holes on the bottom.

2.2 Exposure to ultraviolet irradiation (AT-2).

For the irradiations three collectors with materials susceptible to uv-damage have been selected. Irradiations to ultraviolet doses equivalent to several years of outdoor exposure were applied at temperatures around or somewhat less than the normal operating conditions of the collectors. For details see (2).
No significant damage could be observed after the exposure on any one of the three collectors. Based on this rather limited statistics one might conclude that thermal collectors of the current design are not sensible to uv-irradiation.
The AT-2 facility is hence now used only for screening uncovered plastic collectors.
UV irradiations should be done, if any, in rather severe conditions i.e. at high temperature, with high humidity of the surrounding atmosphere and with intermittent water spray on the samples.

2.3 Outside exposure to sun and weather in dry conditions. (with analysis of the Dry Exposure Temperature).

The efficiency of a flat plate thermal solar collector can decrease with its lifetime due to reduction of the transmission of the coverplate, the reduction of the absorption of the absorber and the increase of the heat losses.
A well established method to test the durability consists in the dry

exposure of the collector during several months. During this dry exposure the collector is subject to daily thermal cycles, to general weathering (rain, wind, freeze) and to maximum temperatures, that are higher than the ones reached in normal operation. It is useful to have during the long exposure, which lasts at least 2 months, already some quantitative measure of an eventual efficiency degradation. (as frequent efficiency measurements are not easily to be made).
This information can be obtained by monitoring with a data logger during the exposure: the temperature of the collector absorber, the ambient air temperature, the global irradiation and the wind speed.
The data are then analysed in different ways: either the extrapolated maximum stagnation temperatures are determined or heat balances on absorbed and lost heat of the collector are made, thus defining a collector "quality".
Long term changes of these values would indicate a degradation of the collector.
For details see (3).
Two collectors exposed in dry conditions on the roof have been monitored since January 1982 in this way. The analysis did reveal no loss of efficiency of the two collectors. This result was confirmed by measurements of the instantaneaous efficiency in the solar simulator LS-1 at the end of 1982 and at the beginning of 1984.

2.4 Exposure to simulated sun (AT-6 facility).

The outdoor exposured as described under point 2.3 can be substituted by an indoor exposure. If one supposes, that the major effects for the degradation of the efficiency are caused by the high temperatures reached (which is possible) then by increasing the temperatures one would expect an accelerated degradation. The high temperatures of the whole collector are obtained by irradiation under the solar simulator AT-6. In order to check whether during the long term irradiation, which is of the order of 200 hours or more, the efficiency of the collector has decreased a continuous measurement of the same data as in the outdoor exposure is made, which permit analysis.
Five collectors were treated by this method so far but no efficiency decrease has been found (in agreement with outdoor exposures).
The sensibility of the method has been checked however by artificially decreasing the transmittance of a collector (putting an extra glass pane between the light source and the collector).

3. Tests on atmospheric degradation and corrosion.

3.1 Climate chamber tests in atmospheres containing sulfur dioxyde.

No visible effects were observed after 100 h exposures in atmospheres according to IEC standard 68-2-42, i.e. 25°C, relative humidity 75%, SO2 concentration 25 vpm and variations up to 40°C, r.h. 90%, SO2 concentrations 100vpm.
Considerable corrosion damages could be produced however on a collector in an atmosphere at 40°C, r.h. 90%, and SO2 concentration 50 vpm cycling the collector temperature below the dew point by intermittent feeding of cold water (12°C) into the absorber. Due to the ventilation apertures which the collector is provided with, also staining of the absorber surface occured.

3.2 Weathering under service conditions in typical atmospheres.

Efforts were put on bringing out evidence of corrosion damages on collectors in real service. Some damages were reported by a contractor, missing however information about service conditions and atmospheric constituents.

In order to obtain confirmed data it was concluded to start weathering experiments. A simple temperature control rig was designed, enabling to keep different collectors with solar irradiation under equal simulated service conditions, and approved in a one year prototype experiment. The choice of five different commercial collector models was made basing on various construction materials. A series of collectors of each model was stored in order to equip weathering experiments in some typical climate zones, to carry out climate chamber tests varying some conditions, and for reference.

Two test fields, each equipped with 15 collectors, 3 of each model, were taken into operation:

CTF-1 in clean air and as reference experiment at the JRC Ispra, operating since April 83.

CTF-2 in an industrial zone 40 km from Ispra, operating since December 83.

A third one, CTF-3, in a coastal zone to study marine atmosphere is in planning.

In the operating test fields the following data are continuously recorded: Collector inlet and outlet temperatures, solar irradiation, ambient temperature, relative humidity, and for CTF-2 also the concentration of SO2. The corrosion resistance of the collectors is followed by regular visual inspections.

At the end of the experiments, after several years, each collector will be checked for eventual thermal efficiency losses and dismantled for thorough inspection of every part. Areas attacked by corrosion will be characterized and evaluated quantitatively by means of the known standard procedures.

Summary.

It has been shown, that some of the tests proposed in the literature on the durability of collectors are not significant. The tests described in this paper seem to be the essential ones permitting a reasonable and rather complete evaluation of the durability caracteristics of collectors with black painted absorbers. Experience with selective surface absorbers will be collected in the future.

These results have been communicated to international Standards writing organisations, such as the UEAtc.

References.

(1) Solar Collector Resistance to Simulated Hail in
Solar Energy Programme Progress Report, July-Dec.1982, p.19.

(2) Ultraviolet irradiation of Collectors with the AT-2 Facility of ESTI in
Solar Energy Programme Progress Report, Jan.-June 1982, p.19.

(3) Method for the Evaluation of the Efficiency Degradation of Flat Plate Collectors in
Solar Energy Programme Progress Report, July-Dec.1982, p.26.

AGEING RESISTANCE OF SOLAR COLLECTORS

J-L CHEVALIER, P. COMTE, J.ORTS, Y. PERRAD
Centre Scientifique et Technique du Bâtiment
BP 21
06562 VALBONNE Cédex
France

Summary

We present in this paper the successive steps of a study on durability of solar collectors. The work initiated on the material point of view led to define specific durability tests for the components working in severe conditions. But, the necessity of a different approach considering the whole collector appeared rapidly. The results of a program of durability inspections on real installation and the knowledge we obtained during the french qualification procedure on solar collectors were used to assess a set of qualification tests and ageing process on the whole collector. This permits to apreciate the aptitude of the collector to satisfy its functionnal requirements along its service life.

1. THE PURPOSE

The ageing of the solar collectors is an essential data for the cost effectiveness of the solar energy uses in building. The first idea is that failures of the collectors are due to material problems. But it appears rapidly that many degradations are related to fitting, assembly of the components. We shall present now the result of five years working in this field : first, we will insist on the approach by describing the various steps of our study. Then, as a provisionnal conclusion it will be proposed a complete procedure to apreciate in a short time the durability and the reliability of any new solar collector appearing on the market.

2. THE PROCEEDING OF THE WORK

2.1 A durability survey

To perform a serious durability survey of a material or a product, the following steps are necessary :

- to determine fonctionnal requirements and to give limits to fonctionnal characteristics,
- to identify the material or the product and the environmental factors in the foreseen application,
- to find degradation index, and to define an artificial ageing test,
- to compare the degradations in natural and artificial weathering,
- to propose a correlation and eventually an estimation of the service life.

The ageing test so defined is a set of operations including :

- a qualification test on the material or product,
- an ageing process,
- the same qualification test performed after ageing.

We shall introduce later on specific ageing test either components materials, or on the whole collector.

2.2 The "material" approach

Using the experience of the laboratory in the field of building materials durability (1) (2) (3), we first try to investigate the working conditions of the materials used as solar collector components. These materials are generally well know in the building applications but they now work in different external conditions, and this could change their ageing behaviour. This approach has been practiced by others laboratories (4). The working temperature is a major factor and we realized in Grenoble a plateform to follow the temperature of each component of ten different commercial collectors during one year in conditions which we can call "partial stagnation". The results appear on fig. 1 where we give on histogrammes the temperatures reached by absorbers, covers, insulation and sealants. This work revealed that two components caused their material to work in very severe temperature conditions :

- The absorber coating, submitted too to cycling temperatures.
- The cover when plastic made.

Sealant reaches the same temperature it can obtain in building uses, and insulation is not submitted to high temperature if there is no contact between it and the atmosphere.

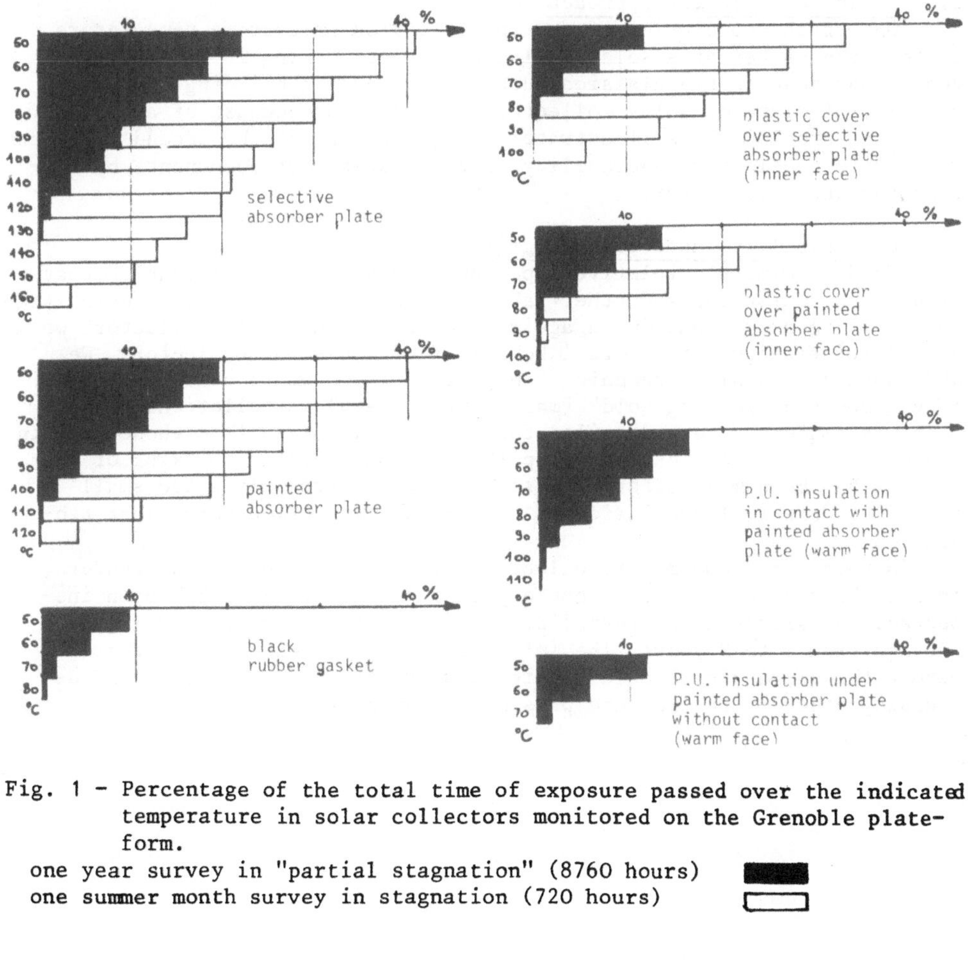

Fig. 1 - Percentage of the total time of exposure passed over the indicated temperature in solar collectors monitored on the Grenoble plateform.

one year survey in "partial stagnation" (8760 hours)

one summer month survey in stagnation (720 hours)

2.3 Specific ageing tests on absorber coatings and on plastic covers

For these two components, the procedure presented for a durability survey has been applied.

For plastic covers, a combined ultra-violet and temperature exposure with intermittent water spraying, using mini-collectors, has been retained as ageing process. The qualification test before and after are solar transmittance measurement and if necessary mechanical properties. This test has been experimented on commercial cover materials (5) and successful correlation has been made with results in natural weathering for 4 years. A very similar test is used in the United States (6) to appreciate cover material durability.

For absorber coatings, we had to reproduce high temperatures but also temperature cycling. The test includes 500 temperature cycles between ambiant temperature and the expected conventional stagnation temperature in the foreseen application, and optical measurements (solar absorptance and thermal emittance) as qualification tests. The correlation of this accelerated test with natural degradation was difficult, because it appears very hard to find actual degradation of this type, in agreement with others laboratories (7). Nevertheless, this test has to be kept on to prevent the use of less durable materials in the future.

2.4 The whole collector approach

One of the conclusion of our investigation on extreme temperature is that the gasket of a solar collector is not working in more severe conditions than in the classical uses of rubbers in building. But, in practice, failure of solar collectors often occurs because of a sealant or gasket problem. This illustrates the usefulness of the second approach: we have to consider the durability of the assembly of components by testing the whole collector.

2.5 The durability inspection program

To determine the relative importance of all the qualification tests proposed by ourselves and other laboratories (8), and to select which factors have to be present in an ageing process on the whole collector, we needed informations of the real behaviour of the solar collectors. We underwork a program of technical inspections related to collector durability aspects on working "old" (more that 2 years) installations. This work was finded by very detailed format sheet inspired from those used in the C.E.C. Collector Testing Group (9). An intermediate checking of the results has been made after 35 visits, which is probably a too small number to draw real conclusions (10). But we can notice very clear tendancies.

A total of 3 000 m^2 of collectors, from 9 different manufacturers, in all the regions of France but especially in south East has been inspected. The sizes of the installations varies from 2 m^2 to 500 m^2, and the age from 2 to 9 years. The informations collected in this program permit the summary presented on fig. 2 where the numbers are in the decreasing order of the failure presence frequency.

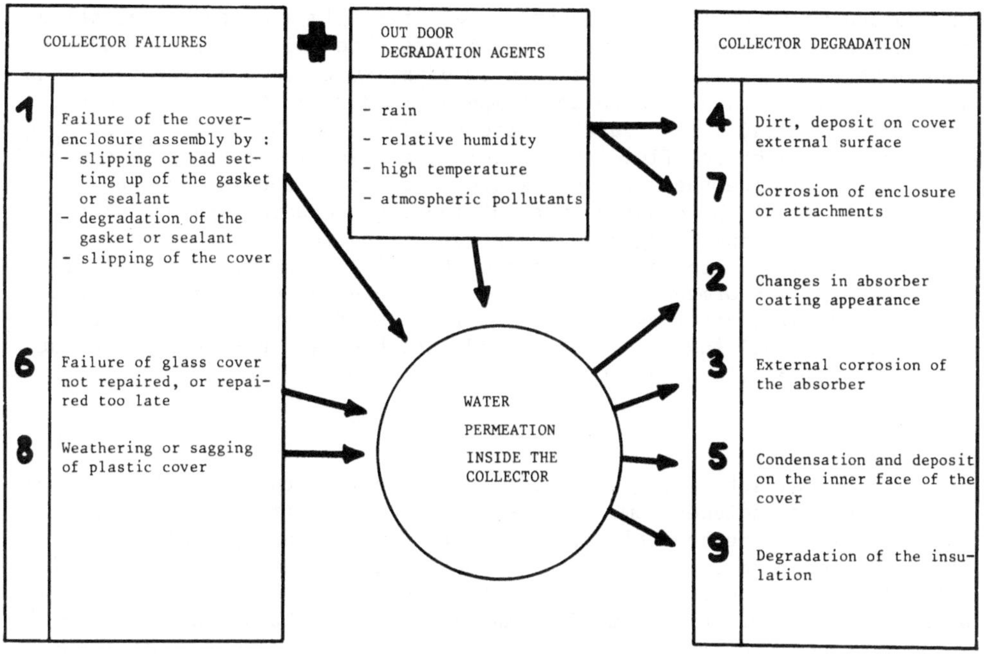

Fig. 2 : Degradation modes of the solar collectors checked from durability inspection of 35 real installations.

2.6 <u>An ageing test on the whole collector</u>

The first practice in this field has been to perform a natural ageing process of one year in stagnation condition, preceded by thermal performance measurements and followed by the same qualification test if the visual observation and disassembly after ageing let fear a decrease of thermal performance.

This test has been performed on about 40 different collectors as a part of the procedure for the french "technical advice". When considering this sum of test, it appears that the more frequent problems are related to plastic covers, humidity penetration in the collector, fitting of gaskets and sealants, corrosion by screws on the metallic enclosures, and sometimes degradation of selective coatings. Another conclusion is that most of these failures are revealed in the first two months of the exposure, if we except UV effect on plastic covers and corrosion effects in the air plate.

The new investigations have been from this moment oriented to a shortening of the ageing process and an improving of the qualification test set.

The new proposed ageing process on the whole collector is a compromise between a shortened natural exposure in stagnation conditions, and an artificial process based on indoor thermal simulation, experimented and practiced by another french laboratory (11).

We propose to expose the collector to "artificially assisted" stagnation conditions, and natural atmospherical agents. The collector, exposed outdoor more than 30 days and less than 60 days, must approach for less than 10 % its conventionnal stagnation temperature during at least 30 hours, by mean of natural solar radiation eventually helped by a loop permitting the circulation of hot fluid in the collector. The interest of this process is to minimize the differences due to climatic conditions between various places of the country, without loosing the advantage of natural exposure to atmospheric agents and daily cycling temperature.

The qualification tests associated to this ageing process still include thermal performance measurement before ageing repeated after if necessary, but also after ageing hail resistance of cover, rain penetration, over pressure, bulling out of cover and attachments. These tests are not described in details here (10).

3. CONCLUSION

The actual set of tests proposed to appreciate the durability of solar collectors includes some procedures to test material behaviour when they work in severe conditions, and others to test the aptitude of the whole collector to satisfy its functionnal requirements along its service life. After five years of work in this field, we can say now that our experience in durability of building materials in one hand, in testing and inspecting commercial collectors or real installations in the other hand has been improved, and we think that our set of tests is a "ran in" tool to evaluate a new collector appearing on the market. We really hope that this will occur, so we shall have opportunities to perform and eventually simplify the set of tests to reduce its duration and cost.

BIBLIOGRAPHY

(1) "Short term wheathering of polymeric materials" R. COPE, G. REVIRAND- C.S.T.B. - Durability of Build Materials (1 - p 225-240) (1982)
(2) "La durabilité des matières plastiques dans leurs emplois en captation solaire" P. EURIN et al - GPCP - Journées d'études PARIS (1979)
(3) "Etude du vieillissement des joints en polychloroprène et en EPDM" J-L CHEVALIER et al - Cahiers du CSTB n° 1616-Livraison N° 205 (1979)
(4) "Use of plastics in solar energy applications" A. BLAGA - N.R.C.C. - Solar Energy - Vol. 21 p 331-338 (1978)
(5) "Durabilité des matériaux plastiques transparents utilisés en couverture de capteurs solaires plans" J-L CHEVALIER, A. BALME, H. SALLEE Conférence internationale sur les Matériaux pour la Convertion Photothermique de l'Energie Solaire - AJACCIO (1980)
(6) "Standards for cover plates for flat plate solar collectors" E. CLARK et al - N.B.S. - Technical Note N° 1132 (1980)
(7) "Surface constancy of absorbers" P. ANDERSEN - IEA Paper réf 295 3675/81 (1983)
(8) "Qualification and durability of thermal solar collectors" G. RIESCH- Document CEC EUR. 8786-EN (1983)
(9) "Collector test method recommandations" Draft document from CEC Collector Testing Group to be published (1984)
(10) "Compte-rendu d'Activité 1983" à paraître - Cahiers du C.S.T.B. (1984)
(11) "Contribution à l'établissement de critères de qualité et de fiabilité des capteurs solaires" Rapport CETIAT-SOL 4 (1983).

FOUR YEARS EXPERIENCE IN COLLECTOR TESTING IN THE FRAME OF THE ENEL PROMOTION PROGRAMME FOR DHW SYSTEMS

F. Parrini
ENEL-CRTN
Via Rubattino, 54 - Milano
M. Antinucci - M. Cocilovo
PHOEBUS S.p.A.
Via G. Leopardi 148 - 95127 Catania
ITALY

Summary

In the frame of the ENEL promotion campaign for 100.000 m^2 of solar DHW in Italy, PHOEBUS was commissioned by ENEL to work out a testing procedure for qualification of solar collectors and to perform such tests on commercial collectors admitted to the campaign. Scope of such short-term test programme is to detect defects which could lead to problems in the early lifetime of the collectors. Significant results of characterization tests performed on 37 commercial types of collectors are reported.

Experimental results of a one year dry stagnation exposure carried out on ten out of the 37 collectors previously characterized are also reported. No critical defects arose that could not be detected by the short term PHOEBUS-ENEL test programme.

1. INTRODUCTION

ENEL decided in 1980 to study and promote a campaign for the diffusion in Italy of solar DHWS for a total amount of 100.000 m^2 solar collectors. Incentive to the users was foreseen as a low interest credit programme up to 70% of the DHWS cost. Since at that time a national standard for complete characterization of solar collector was missing, ENEL, in order to protect the users, commissioned PHOEBUS to work out an internal test specification and to perform the characterization of commercial collectors of manufacturers applying to participate to the promotion campaign. The promotion campaign started in 1983 and is still going on.

The experience gained in four year use of the spec. (PHOEBUS 80/PBD/001) has been submitted to the relevant sub-committee of Comitato Termotecnico Italiano, who is presently developing an official national standard.

2. PHOEBUS 80/PBD/001 SHORT TIME TEST PROGRAMME

From the practical point of view, such testing programmes on one side must be indicative as possible about the reliability of solar collectors especially in their early lifetime, on the other hand must be performed in a reasonably short time.

Test of course can be made only on collectors manufactured with production uniformity and with efficient quality control organization. Such conditions were verified by an inspection and sampling programme.

The approach of 80/PBD/001 is similar to NBSIR 78/1305A (1). The only basic difference is the addition of a 500 hours salt mist exposure, that was introduced after many cases of unsatisfactory service for corrosion problems.

The main test-programme (first efficiency-dry stagnation-second efficiency) was performed on three collectors, in order to check production uniformity. Moreover, since European insolation is unfortunately lower than American, the duration of the dry stagnation test was decreased from 30 to 20 days, leaving unchanged the minimum value of 4.7 kWh/m^2 day. Four hours exceeding 950 W/m^2 solar irradiance are also quite rare in Europe, therefore this condition was not required.

In order to assess the consistency of these reductions, a one year dry stagnation exposure was performed on ten collector models, three samples each. Such collectors had already passed the test programme, were bought by local distributors and selected in order to represent several typical materials and technologies.

3. RESULTS

Only forty-six per cent of the collector models submitted to the test programme resulted satisfactory, while the remainders had to be modified. The most critical defects evidenced were 35, verified in 20 collectors. Sixty per cent of the total defect number was evidenced during the stagnation-thermal shocks - rain simulation cycle. Nine per cent was evidenced in salt mist test of the whole collector, and thirty per cent in salt mist on the absorber alone.

Figures 1 and 2 summarize the distribution of defects in the overmentioned tests. Twenty defects, considered of minor importance, are also reported, as "oxidation of localized parts" and "galvanic couples" in fig.2. Table I lists the defect types, their cause, and the modifications, suggested by the manufacturer and then verified.

The comparison among efficiency curves obtained contemporaneously on three samples of each collector model did not evidence any case of efficiency difference exceeding +/- 10%.
The efficiency test was repeated at the end of the stagnation exposure cycle on two of the three exposed samples: no significant changes of the efficiency curve were found in any collector model.

The long term stagnation exposure on ten collector models has completed its first year and is still going on for a second year. Performance characteristics were checked before and after the first year of exposure. No efficiency degradation was observed on the exposed models up to now.
The performance characteristics were verified repeating the efficiency curve measurement before and after the exposure, but also were monitored by a simpler and faster method. This method is based on the periodical measurement of the peak stagnation temperature (Tp). The solar irradiance (I) and ambient temperature (Ta) are also measured in a half-an-hour period about solar noon, in sunny days, and the quantity I/(Tp-Ta) calculated. The same measurement and calculation is contemporaneously performed on a reference collector, not submitted to long term stagnation exposure, and the values of tested and reference sample compared, as shown in Table II. As can be

seen, in spite of the large variation of the absolute value of the ratio I/(Tp-Ta) in the periodical measurements, repeated every three-four months under the actual outdoor conditions, the difference between exposed and reference sample remains fairly constant. This is consistent with the slight variation of the efficiency curve parameters, also reported.

Minor defects, caused by the stagnation exposure, were evidenced as condensation of small oily droplets on the glass inner surface and gas bubbles between the insulating foam and the protective sheet. The time of their occurrence is graphycally reported in fig.3. The abscissa reports the progressive number of "useful" stagnation days, as defined in the requirements for stagnation test in the 80/PBD/001 test programme. Our one year exposure corresponded to about 180 such days. Disassembly of one collector per each model at the end of the first year of exposure showed oxidation of poliurethane foams, turning their colour from yellow to brown for a depth of 10-15 mm, even if protected by reflecting sheets or glass-wool layer. Nevertheless, there is no evidence of any consequence of these material degradations on thermal performance at the end of the first year.

4. CONCLUSION

The test programme described in the spec.80/PBD/001 has proved to be efficient in detecting design defects, wrong material use or assembly unaccuracies that could lead to service problems especially in the early stages of collectors lifetime.

The long term stagnation exposure, which has terminated its first year, has not evidenced critical defects or efficiency degradation in ten collector models which had satisfied the 80/PBD/001 spec. Two consequences can be deduced from the results obtained up to now: a) the slight modification introduced by us to the NBSIR 78/1305A cycle are not of critical importance for the efficiency of the test procedure; b) a long term exposure test is a useful tool for testing solar collector reliability, but the most critical defects can be more rapidly detected by suitably selected short term programmes like 80/PBD/001.

Further work is necessary on this subject, nevertheless the correspondance between the results of inspection reports on existing installations collected by CEC-CTG (2) and our test experience encourages along this way.

ACKNOWLEDGEMENTS

The assistance of Dr.F.Aleo of CONPHOEBUS S.p.A. is gratefully acknowledged

REFERENCES

1. NBSIR 78-1305A (1978). Provisional Flat Plate Solar Collector Testing Procedures: First Revision, June 1978
2. GILLET W.B. and MOON J.E. (1983) Solar Collector Design Guide, First Draft.
3. ASTM E 823-81 (1981). Standard Practice for Nonoperational Exposure and Inspection of a Solar Collector.
4. LENEL U.R. and MUDD P.R. (1984). A Review of Materials for Solar Heating System for Domestic Hot Water. Solar Energy. Vol.32, No.1, pag.109

TABLE I - List of the principal defects, causes and modification in rain simulation, stagnation exposure and salt mist tests

DEFECT TYPE	CAUSES (x) = no. of cases	MODIFICATIONS
1. Condenses (not induced by water intake)	a) moisture: insufficient ventilation (4)	.modification of the ventilation system
	b) grease and other oily substances: insufficient degreasing of metallic parts before painting	.better cleaning and degreasing of metallic parts;
2. Breakage of covering glass during external thermal shock or during stagnation	a) tension in the glass due to a rigid fixing (1)	.new fixing system allowing for thermal expansion
	b) glass high mechanical stresses (1)	.use of tempered glass;
3. Water untightness during rain simulation or external thermal shock	a) temperature induced degradation of sealing thermoplastic materials (2)	.use of silicone based thermosetting sealing compound;
	b) design defect of the i/o connection tubes gaskets (4)	.new design of the gasket and choice of suitable material;
	c) assembly defects as discontinuities and shortness of sealing product, unsealed joints at case corners (2)	.greater care in application, new design, introduction of a quality control station;
4-5. Sealing profile or reflecting sheet displacement	a) temperature-cycling effect on sticked parts (5)	.new design of sealing system or use of suitable adhesive;
6. Change of colour of the selective coating (BlCu)	a) thermally induced degradation of the selective coating (1)	.improvement of the electroplating process;
Salt mist:		
3-4. Oxidation and blisters on the outside case surfaces	a) unsuitable corrosion protection (3)	.substitution with suitable materials and coatings;
5. Paint detachment and blisters on the absorber surface	a) poor quality or poor adhesion of the painting products (7)	.new painting cycle, better degreasing, introduction of primers;
6. Absorber selective coating (BlCr and BlNi) pitting	a) insufficient smoothness of the surface before coating (manufacturer's analysis) (2)	.smooth surfaces, increase film thickness (N.B.: proposal of the manufacturer)
	b) insufficient protection (1)	
1-2. Oxidations or galvanic couples	a) unsuitable materials and coatings	.material change or collars.

TABLE II - Long term stagnation exposure effects on collector thermal performance

Collector type no.		Initial efficiency o	F'UL (W/m^2°C)	Peak stagnat. measures: I/(Tp-Ta) (W/m^2°C) 1st	2nd	3rd	4th	Final efficiency o	F'UL (W/m^2°C)
1	exposed	.85	8.20	8.56	9.62	9.45	-	.86	7.81
	referen.	.84	7.45	8.70	9.51	9.93	-	.85	7.90
2	exposed	.85	8.00	9.56	8.55	10.31	-	.86	7.63
	referen.	.85	8.00	10.11	8.61	10.17	-	.84	7.53
3	exposed	.82	7.69	11.1	11.1	12.3	9.7	.82	7.82
	referen.	.81	7.41	11.0	11.0	12.1	9.7	.82	7.42
4	exposed	.80	7.27	7.46	8.02	9.42	-	.81	7.59
	referen.	.81	7.41	7.53	7.95	9.50	-	.82	7.42
5	exposed	.79	7.31	8.46	11.27	11.91	8.94	.77	7.24
	referen.	.77	7.00	8.22	11.22	12.07	8.89	.76	7.06
6	exposed	.76	7.62	11.75	11.96	8.69	-	.73	7.26
	referen.	.73	7.26	11.64	12.08	8.16	-	.72	6.61
7	exposed	.85	9.05	11.92	8.91	-	-	.88	8.78
	referen.	.82	9.13	11.67	8.87	-	-	-	-
8	exposed	.73	6.96	11.02	8.29	-	-	-	-
	referen.	.74	7.33	11.16	8.65	-	-	-	-
9	exposed	.78	8.12	9.04	-	-	-	.83	8.93
	referen.	.79	8.08	8.99	-	-	-	.83	8.88
10	exposed	.75	7.00	7.96	-	-	-	.73	6.74
	referen.	.76	6.82	7.75	-	-	-	.76	7.02

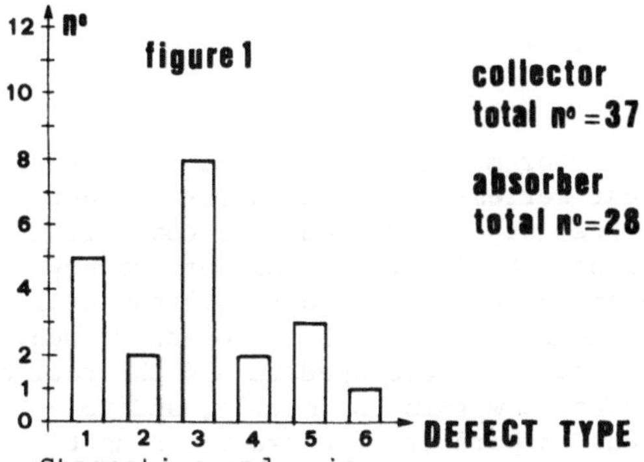

Stagnation and rain

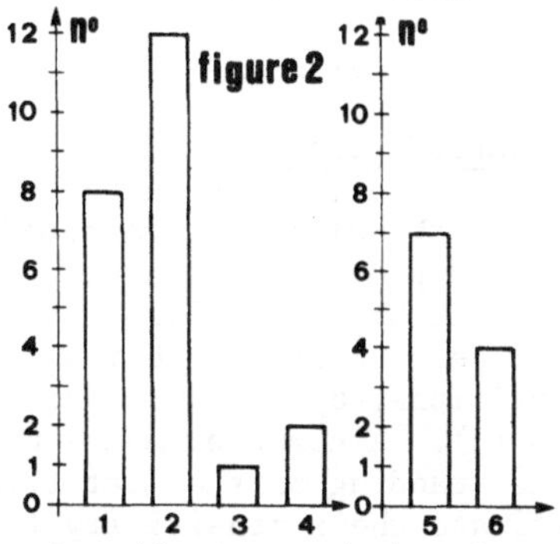

Salt mist:collector and absorber

Defect type numbers correspond to the numbers in Table I

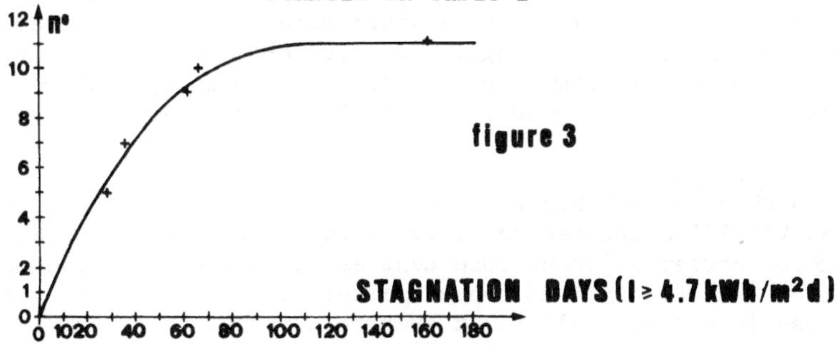

COMPUTER MODEL OF THE PHILIPS EVACUATED TUBULAR SOLAR COLLECTOR USED IN THE CEC5 ROUND ROBIN TEST SERIES

A.A. Green
Solar Energy Unit, University College, Cardiff, Wales, UK.

Summary

The fifth round robin test series of the CEC solar collector testing programme used a Philips evacuated tubular collector. A computer model based on a one-dimensional, steady-state heat transfer analysis of the collector module was developed to aid theoretical understanding of its heat transfer processes and to normalise test data to standard operating conditions. Model-predicted collector performance characteristics are in excellent agreement with the CEC5 round robin test results, for both kinds of diffuse back reflector used in the test series. The effects of variations in local wind speed, effective sky temperature, percent diffuse irradiance and collector tilt angle (for angles in the range $25^{\circ} - 65^{\circ}$) are estimated to be small. Because the absorber fins and heat pipe evaporators in the collector module are well isolated from the environment, their heat transfer processes are only weakly influenced by variations in environmental conditions. It is concluded that the observed scatter in the round robin test results, which was found to be non-reducible by normalisation, is therefore probably attributable to instrument calibration errors, and to heat losses through cracked glass envelopes and gaps in the header insulation.

1. Introduction

The Philips evacuated tubular collector module, type VTR 151, used in the fifth collaborative round robin test series of the CEC solar collector testing programme, featured isobutane heat pipes for good heat conduction from the absorber fins to the header pipe. Collector thermal efficiency data were obtained under a wide range of outdoor and indoor environmental conditions, by eighteen participating laboratories in the European Communities (1). A computer model of the collector was developed to aid theoretical understanding of the heat transfer processes within the module, and to estimate their sensitivity to variations in parameters characterising the operating conditions. The model was also developed for implementation in normalising the collector thermal efficiency data to standard operating conditions, in order to aid meaningful intercomparison of the round robin test series results. This specialised model was evolved from earlier thermal performance modelling of single- and double-glazed water heating collectors, and utilised the one-dimensional, steady-state analytical method and Newton-Raphson algorithm developed in the earlier work (2).

2. The collector

The CEC5 round robin collector consisted of a module of fifteen identical Philips VTR 151 evacuated solar collector tubes mounted at an average spacing of $\simeq 75$ mm between adjacent tube centres, with the heat pipe condensors clamped to a 15 mm o.d. (13 mm i.d.) copper header pipe. The absorber fins and heat pipes lay in a common plane, an average distance of $\simeq 50$ mm in front of a removable planar diffuse back reflector. A pair of interchangeable back reflectors was supplied with each module, having diffuse solar reflectances of ≈ 0.72 (white reflector) and ≈ 0.05 (black reflector). Performance data and a description of the major features and mode of operation of the VTR 151 solar collector tube have been published in the scientific literature (3).

3. Heat pipe thermal conduction

The following correlation equation for the thermal resistance R (in K/W) between the absorber fins of a VTR 151 collector tube and the water in the header pipe was derived from data measured by Philips (4), by computing the least square regression line of $\log_e(R - 0.79)$ on the water mass flowrate $\dot{m}$ (in kg/s):

$$R = 0.79 + 1.224e^{-65.23\dot{m}} \qquad (1)$$

The empirical data, obtained by Philips for water mass flowrates down to 0.033 kg/s, are plotted on the graph in Fig. 1, together with equation 1. Many of the CEC5 round robin tests were performed using water mass flowrates near an initially-recommended value of 0.0167 kg/s (60 kg/hr). Application of the model to normalising the results of those tests to the standard water mass flowrate $\dot{m}$ of 0.033 kg/s therefore involved extrapolation of equation 1 significantly outside the range of the correlated empirical data, as can be seen from Fig. 1.

Equation 1 is invalid for high condensor temperatures since, at temperatures above about 375 K, the condensors become filled with isobutane and parts of the evaporators become superheated, resulting in a sharp drop in the evaporator to condensor conductance (3).

4. Heat losses

The heat flux components and temperature nodes of the one-dimensional, steady-state heat transfer analysis of the CEC5 collector module are summarised in Fig. 2. The negligible conductive and convective heat transfer across the < 100 Pa vacuum between the absorber fins and their cylindrical glass envelopes was disregarded in the analysis. Heat transfers to and from the glass envelopes were analysed separately for the hemicylinders above and below the plane of the absorber fins, to enable the significantly different radiative sink temperatures of their oppositely-directed fields of view to be taken into account. The analysis also accounted for the significant

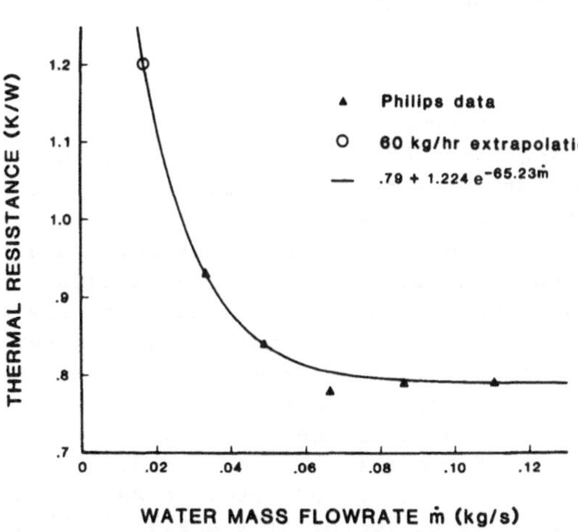

Fig. 1 The $R - \dot{m}$ relationship for a CEC5 collector tube, showing equation 1 and the data measured by Philips.

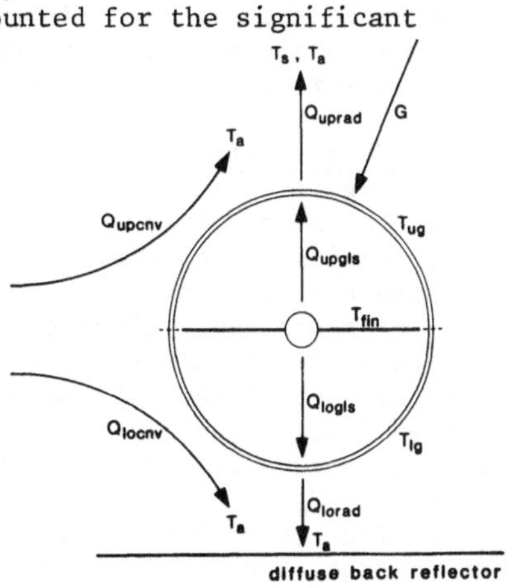

Fig. 2 CEC5 collector tube temperature nodes and heat flux components.

variation of the hemispherical emittance of the black cobalt absorber fin surfaces with temperature, reported by Philips (4).

In order to account for transitions between laminar and turbulent air flow around the glass envelopes, occurring at low local wind speeds, geometric means of the appropriate free and forced convective heat transfer coefficients were used (2). The nett radiative heat exchanges between the outer surfaces of the upper and lower glass hemicylinders and their respective fields of view were approximated to those from a plane glass surface, of area equal to the total projected area of the glass hemicylinders, in the plane of the absorber fins. A detailed analysis of the radiative exchanges for the convex glass surfaces would have been an unnecessary complication, since wide variations in the longwave radiative environment generally have relatively little influence on the thermal performance of evacuated tubular collectors.

5. Absorption of solar radiation

For the specialised application of the model to normalising the round robin test results to standard conditions, the assumption was made that, in the tests, the total irradiance G in the plane of the absorber fins could be adequately represented as consisting of a beam component of direct radiation incident normal to the plane of the absorber fins, together with a hemispherically isotropic diffuse component. Spectral near-normal reflectance and normal transmittance measurements were made respectively on samples of the absorber fins and glass envelopes, using an integrating sphere spectrophotometer. Weighted mean absorptance and transmittance values appropriate to the Air Mass 2 solar spectrum were calculated from the spectral measurements. The mean transmittance of the cylindrical glass envelopes for beam radiation incident normal to the plane of the absorber fins was inferred from the normal transmittance by utilising published results of a detailed optical analysis of similar cylindrical glass envelopes (5). The transmittance for isotropic diffuse radiation was approximated to the angular transmittance for beam radiation incident at an angle of 60° to the surface normal (6).

Solar radiation passing through the gaps between adjacent absorber fins was assumed to fall at near-normal incidence on the illuminated strips of back reflector directly behind the gaps, and to then be diffusely reflected with a Lambert cosine distribution. The absorption of beam and diffuse solar radiation by the glass envelopes, and the difference in absorptance between the black cobalt absorber fin surfaces and the oxidised copper evaporator surfaces, were taken into account.

6. Method of solution

An algorithm based on the Newton-Raphson numerical method of solution (2) is used in the model to determine, for specified operating conditions, the unique combination of steady-state temperatures within the collector module for which the total nett heat fluxes into the absorber fins and the upper and lower glass hemicylinders are all zero. The instantaneous steady-state efficiency of the collector module, based on a $1.152\,m^2$ aperture area equal to the product of the 1.2 m width of the module and the 0.96 m length of the absorber fins, is computed for consecutive values of the heat transfer water mean temperature T_m increased in increments of $0.01\times G$ K from the ambient air temperature T_a to 370 K, where G is the total irradiance (in W/m^2).

To facilitate intercomparison, collector performance characteristics were computed for a set of standard operating conditions. The standard conditions specified for the CEC5 round robin test series were:- a water mass flowrate $\dot{m}$ of 0.033 kg/s in the header pipe, a collector tilt angle of 45°, a total irradiance G of 800 W/m^2, no diffuse irradiance, an average local wind speed of 3 m/s, an ambient air temperature T_a of 288 K, and an effective sky temperature T_s of 278 K. The effects of variations in test conditions were estimated by varying, about its standard value, each of the appropriate parameters characterising the operating conditions. The model was also used to normalise the round robin test series results to the standard conditions. Each efficiency datum was normalised by adjusting its value by the difference between the model-predicted efficiencies for the standard conditions and for the actual conditions under which the datum was measured (1).

7. Results

The normalised round robin collector efficiency data from all eighteen participating laboratories, for the CEC5 collector module with the white diffuse back reflector, are plotted on the graph in Fig. 3 together with the model-generated collector performance characteristic for the standard conditions and white reflector. Excellent agreement was also obtained between the model-predicted performance characteristic and the round robin efficiency data for the CEC5 collector module with the black diffuse back reflector (1). The effects of likely variations in local wind speed, effective sky temperature, percent diffuse irradiance and collector tilt angle (for angles in the range $25^o - 65^o$), estimated using the model, were found to be very small.

Collector efficiency data measured by Philips at water mass flowrates of 0.06 and 0.016 kg/s, before and after being normalised using the model, are plotted on the graph in Fig. 4 together with the model-generated collector performance characteristic for the standard conditions and white reflector. The generally good agreement between model predictions and empirical results, exemplified here, corroborates use of the extrapolation of equation 1 significantly below the flowrate range of the thermal resistance measurements from which it was derived.

8. Conclusions

The excellent agreement obtained between predicted and measured efficiencies tends to confirm our theoretical understanding of the thermal processes within the CEC5 collector module. Because the absorber fins and heat pipe evaporators were quite well isolated from the environment, their heat transfer processes should have been only weakly influenced by variations in environmental conditions. The observed scatter in the round robin test results, which was found to be non-reducible by normalisation, is therefore probably attributable to instrument calibration errors, and to heat losses through cracked glass envelopes and gaps in the header insulation.

References

1. Moon, J.E., et al. (1983). Results and analysis of a round robin test series using an evacuated tubular solar collector. EUR 8757. Commission of the European Communities. Luxembourg.

2. Green, A.A. (January 1979). The influence of environmental parameters on flat plate solar collector performance. UK-ISES Conference Proceedings C18 Meteorology for Solar Energy Applications. Pages 95-107. London.

3. de Grijs, J.C., Bloem, H. and de Vaan, R.L.C. (August 1981). Evacuated tubular collector with two-phase heat transfer into the system. ISES Conference Proceedings: Solar World Forum, Brighton. Pergamon. Oxford.

4. de Vaan, R.L.C., et al. (1982). Private communications. Nederlandse Philips Bedrijven B.V.

5. Felske, J.D. (June 1979). Analysis of an evacuated cylindrical solar collector. Solar Energy. Vol.22. Pages 567-570.

6. Duffie, J.A. and Beckman, W.A. (1980). Solar engineering of thermal processes. Wiley. New York.

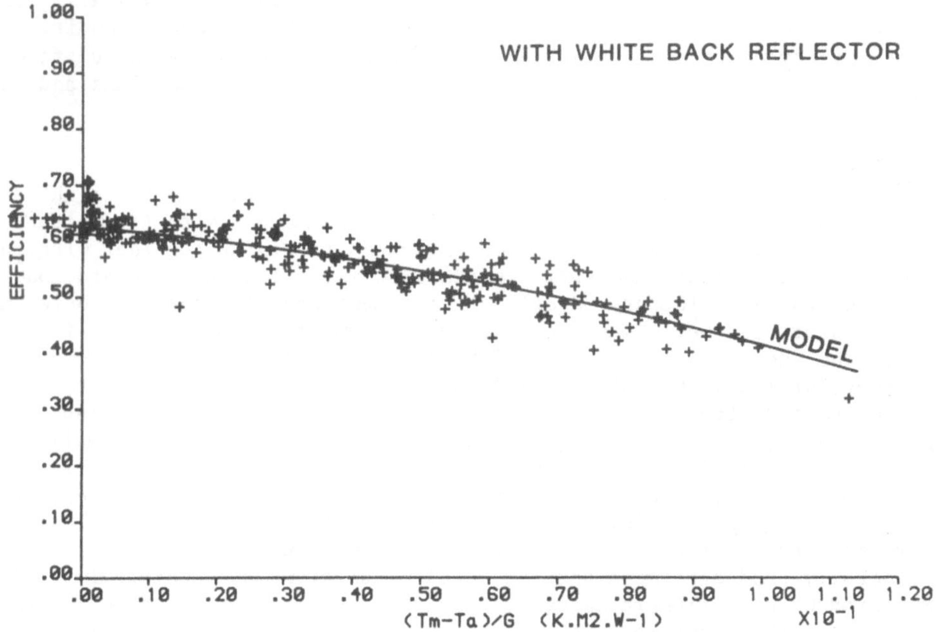

Fig. 3 Normalised CEC5 round robin collector efficiency data from eighteen laboratories, compared with the model-generated performance characteristic for the standard conditions.

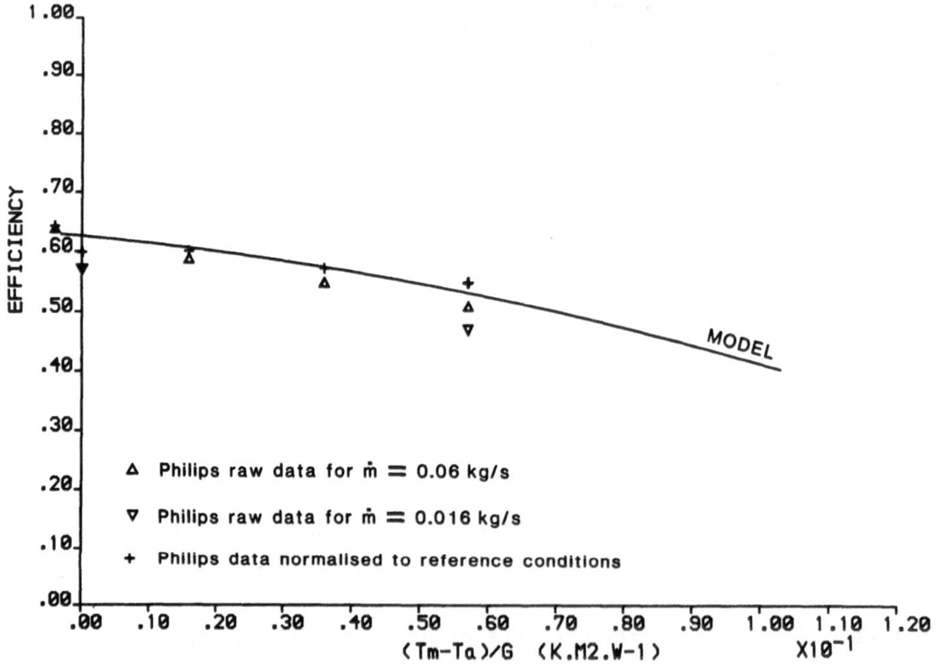

Fig. 4 CEC5 collector efficiency data measured by Philips at two flowrates, compared with the model-generated performance characteristic for the standard conditions and showing the effect of data normalisation.

DESIGN AND CHARACTERISTICS OF A FLAT PLATE SLIT HONEYCOMB SOLAR COLLECTOR

S.J.M. Linthorst and C.J. Hoogendoorn
Dept. of Applied Physics
Delft University of Technology
Stichting FOM, Utrecht

Summary.

A slit honeycomb has been designed for application as convection suppressing device in a commercial flat plate solar collector. Based on numerical and experimental results optimum working conditions has been found for a slit honeycomb of $A_x = 0.10$ with a height of 40 mm in combination with a gap of 3 mm between the honeycomb and the absorber. Numerical results show the interaction between the different heat transfer modes and good agreement have been found between measured and predicted heat losses. Measurements show that the solar transmittance of the slit honeycomb amounts 0.99 and 0.97 for irradiation angles of 90 and 70 degrees. Application of the slit honeycomb in the collector results in a change of F_R $(\alpha\tau)$ from 0.763 to 0.751 and a decrease of F_R U_0 from 4.35 W/m^{2}K to 2.97 W/m^{2}K. The overall effect results in a significant higher instantaneous efficiency at large temperature differences. We aim that further development will lead to a slit honeycomb collector with an efficiency ($F_R(\alpha\tau) \approx 0.83$, $F_RU_0 \approx 2.5$ W/m^{2}K) which approaches that of evacuated tube collector.

1. INTRODUCTION.

The main heat loss of spectral selective flat plate solar collectors is the heat loss by natural convection between the absorber and the cover plate. Reduction of this heat loss can be obtained by application of convection suppressing devices (abbrev. CSD's) which suppress the development of natural convection and reduce the natural convective heat loss. Since the CSD should be implemented in the enclosure between the absorber and the cover plate of the collector the CSD-material used should have high solar transmissivity, good uv-stability, good thermal stability and a small thickness to minimize the conductive heat loss through the material. Several investigations concerning the application of CSD's in flat plate solar collectors have been performed. Hollands et.al. [1],[2] and [3] have been interested mainly in suppression of the natural convective and total heat loss by the CSD's. The overall influence of the CSD's on the efficiency of the solar collectors have been investigated by for instance Baehr [4], Marshall [5] and Buchberg [6]. More recently Symons investigated the solar transmissivity of CSD's [7] and also the effect of the CSD's on the efficiency of the solar collector and the natural convection [8]. The CSD's used for these investigations have been made of different materials (such as glass, lexan, mylar and FEP teflon) and different shapes (square cells, slits, v-slits, cylindrical, sinusoidal). All experiments showed that effective suppression of natural convection could only be obtained if the dimension of the convective cell was reduced to an aspect ratio $A_x \approx 0.1$ (the ratio of the height (h) of the cell along the absorber and the distance (D) between the cover and the absorber). Moreover the efficiency results showed that for non-spectral selective collectors large efficiency improvement could be obtained, due to both suppression of natural convection and reduction of reradiation heat losses. For spectral selective collectors the improvement of collector efficiency has been less than expected. The purpose of our work is to design a CSD which can be easily implemented in flat plate solar collectors and which will be competitive with vacuum tube collectors.

2. METHOD OF APPROACH.

Experiments have been performed with a heat transfer apparatus to determine the natural convective and total heat transfer in a model of a flat plate solar collector. In our investigation slit honeycombs have been considered only. The slit honeycombs used for the experiments have been constructed by stretching Hostaphan BN 50 foil (thickness $h_w = 50\ 10^{-6}$ m, $\lambda = 0.13$ W/mK) cross slope over the hot plate. The cold wall of the enclosure has been painted black ($\varepsilon_c \approx 0.95$), the hot aluminum wall had an emissivity of $\varepsilon_h \approx 0.25$, thus presenting a spectral selective solar collector. The natural convective heat transfer has been measured for slit honeycombs of varying aspect ratio A_x. Results in the form of Nu - Racos(ϕ) dependence have been given in fig. 1.

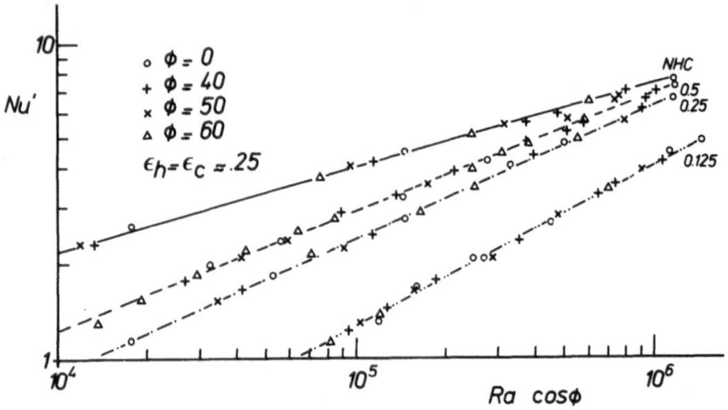

Figure 1: Measured natural convective heat transfer for slit honeycombs of different aspect ratio A_x.

The increased suppression of natural convection for decreasing A_x can be seen quite clearly in the figure. Effective suppression of natural convection at the Rayleigh range of interest ($10^5 < Ra < 5\ 10^5$) will be obtained with slit honeycombs of aspect ratio 0.1. In fig. 2(a) have been given the total heat transfer results (radiative and convective heat transfer) for stratified air situations (ϕ = 180 degrees, hot wall above cold wall) dependent on the distance D between the hot and cold wall. As expected the situation without slit honeycomb gives the smallest heat transfer for stratified air conditions (ϕ = 180 degrees). Larger total heat transfer has been found for situations with a slit honeycomb applied. The increase of heat transfer is due only for a small amount to the extra conduction heat losses through the honeycomb material and can only be explained by radiation-conduction interaction. On the other hand in inclined situations (hot wall under the cold wall) the total heat transfer is smallest for $A_x = 0.10$ with a significant minimum value for $D \approx 40\ 10^{-3}$m (see fig. 2(b)). This is due to convection suppression by the honeycomb. To comprehend the coupled radiative-conductive total heat transfer process a numerical program has been developed to calculate the coupled total heat transfer in slit honeycombs. With this program the experimental stratified air conditions measured for aspect ratio 0.1 have been compared with the numerical derived results. Calculations and measurements agreed within the experimental accuracy.

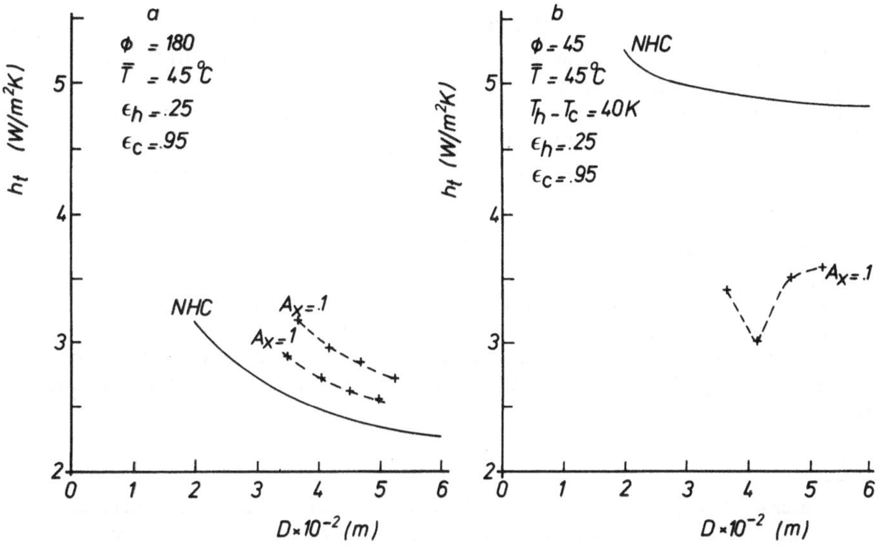

Figure 2: Measured total heat transfer in slit honeycombs.

Interaction of radiation and conduction occur at the side walls. Due to the small thickness of the honeycomb material the temperature distribution of the side walls will be determined for a part by radiative heat exchange. As a result of the coupled heat transfer process large temperature gradients exist in the side walls and the air near the hot wall. To diminish the interaction between the heat transfer mechnisms we introduced a gap between the slit honeycomb and the hot wall. By varying the gap width (s) and slit honeycomb width (d) (plate distance D = d + s) we found a d,s combination where minimum heat transfer occurs (ΔT < 70 K). The dimensions of this optimum slit honeycomb are d = 40 mm, s = 3 mm (see fig. 3). When convection is no longer suppressed the total heat transfer increases almost linearly with the temperature difference. An optimum between minimum coupled heat transfer and maximum suppression of natural convection has been found for s = 3 mm (= 7% of the plate distance D).

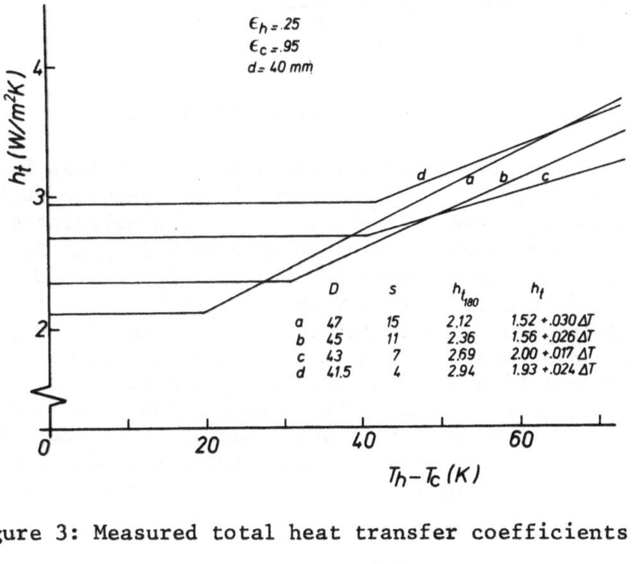

Figure 3: Measured total heat transfer coefficients (ϕ = 45).

3. DESIGN SLIT HONEYCOMB SOLAR COLLECTOR

For the design of the slit honeycomb solar collector attention has been given to the following aspects:

(1) To reduce the natural convective heat losses a slit honeycomb has been chosen of $A_x = 0.1$ made of Hostaphan BN 50 ($h_w = 50\ 10^{-6}$ m).

(2) To diminish the radiative-conductive interaction we applied a gap between absorber and slit honeycomb. Lowest heat transfer of the slit honeycomb has been found for d = 40 mm, s = 3 mm.

(3) The commercial EBS sunstrip collector has been chosen to applicate the slit honeycomb, where a high absorptance spectral selective layer ($\alpha \approx 0.94$; $\varepsilon \approx 0.10$) has been used in combination with a high transparent glass cover (tempered low-iron glass, $\tau_c \approx 0.9$).

(4) Special attention has been given to unnecessary heat losses through conduction and improved insulation has been applied at the back and sides of the collector.

4. RESULTS.

With the numerical program the total heat transfer has been calculated as function of the emissivity ε_{hc} of the slit honeycomb material for the prototype ($\varepsilon_h = 0.10$, $\varepsilon_c = 0.84$, s = 3 mm, d = 40 mm). In fig. 4 the results have been given.

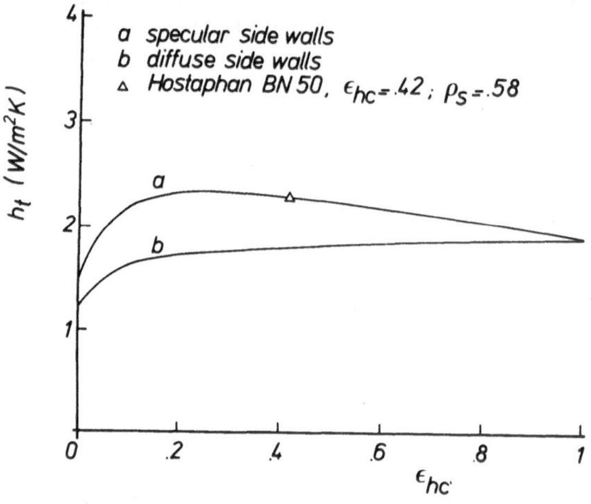

Figure 4: Calculated total heat transfer coefficients.

Two extreme cases have been given; curve 1 shows the h_t - values for diffuse reflective side walls ($\rho_d = 1 - \varepsilon_{hc}$), curve 2 for specular reflective side walls $\rho'_s = 1 - \varepsilon_{hc}$). The triangle indicates the calculated h_t - value for the Hostaphan honeycomb. This value amounts 2.22 W/m^2 K, which is significantly lower than for the non-CSD situation ($h_t \approx 4.5$ W/m^2 K) but larger than the theoretically lowest value ($\varepsilon_{hc} = 0$, $\rho'_s = 1$ gives $h_t = 1.38$ W/m^2 K). In fig. 5 the results of the efficiency measurements have been given. The measurements have been performed indoors with a solar simulator. As shown in the figure the collector applied with the slit honeycomb has only a slightly smaller η_0 but a considerably smaller heat loss factor (U_0) which results in significantly higher efficiency at temperature differences of interest. At very large temperature differences convection is no longer suppressed and the efficiency decreases rapidly.

This results in a not extremely high stagnation temperature, which is advantageous considering the materials used. We expect that further development will lead to a $F_R\ (\alpha\tau) \approx 0.83$ and a $F_RU_0 \approx 2.5\ W/m^2K$, which should result in an efficiency curve as given by curve c in fig. 5.

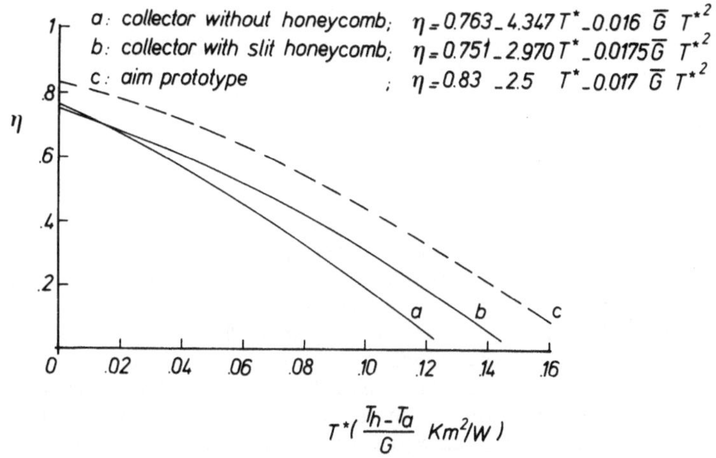

Figure 5: Efficiency measurements of the slit honeycomb collector.

5. CONCLUSIONS.

Application of a slit honeycomb as natural convection suppressing device in a flat plate solar collector increases the collector efficiency significantly. The increase of efficiency is for spectral selective collectors less than may be expected on a independent mode heat transfer analysis. This is due to interaction between radiative and conductive heat transfer. To diminish the interaction between the heat transfer modes an optimum gap of $\approx 7\%$ of the absorber to cover spacing should be applied between the slit honeycomb and the absorber. Considering the small top heat losses of a slit honeycomb solar collector special attention should be given to heat losses through conduction. Further to obtain a large η_0 an as high as possible absorptivity of the absorber, transmissivity of the glass cover and heat removal factor is desireable. This should result in flat plate solar collectors which are competitive with vacuum tube collectors.

REFERENCES

1,2,3. Hollands, K.G.T. et.al.(1976,1978,1979). Methods for reducing heat losses from flat plate solar collectors. University of Waterloo Research Institute.

4. Baehr, A; Piwecki, H; Rigolini, L. (1978). Measurements on a flat thermal solar energy collector with a cellular honeycomb after Francia. International Symposium Workshop on Solar energy, Cairo, Egypt.

5. Marshall, K.N; Wedel, R.K; Dammann, R.E. (1976). Development of plastic honeycomb flat-plate solar collectors. Lockheed Missiles & Space Company Inc., Palo Alto California, USA.

6. Buchberg, H; Edwards, D.K; Mackenzie, J.D. (1977). Transparent glass honeycombs for energy loss control, School of Engineering and Applied Science, University of California.

7. Symons, J.G. (1982). The solar transmittance of some convection suppression devices for solar energy applications: An experimental study, J. of Solar Energy Engineering, vol. 104, pp. 251-256.

8. Symons, J.G;Peck,M.K. (1983). An overview of the CSIRO project on advanced flat plate solar collectors, ISES Solar World Congress, Perth, Australia.

DESIGN OF DIRECT SOLAR FLOORS AND PRACTICAL RESULTS

By : Michel CHATEAUMINOIS, Daniel ROUX, and Daniel MANDINEAU
Département Thermique
ECOLE SUPERIEURE D'INGENIEURS DE MARSEILLE
28, rue des Electriciens - 13012 MARSEILLE FRANCE
Tél. (91) 49.91.40

Summary

Classical solar space heating systems by flat collectors exhibit high costs and poor rehiability. A new concept, "DIRECT SOLAR FLOORS" (in abreviate "D.S.F.") offers a drastic improvement from this condition : it consists merely of a direct coupling of collectors to a classical heating floor, without any external storage. The floor, which ensures simultaneously heat storage, regulation and space heating has to be thicker than usual. We expose the main features of this system, our theoretical modelling and the conclusions for behaviour and rules of sizing for DSF. A practical example is described and measured performances are discussed. D.S.F. appears as a very powerfull space heating system of low cost, high thermal yields and improved reliability.

1. DESCRIPTION

Our experience of various heating systems using solar flat collectors in real conditions led us to point some frequent sources of troubles or poor thermal yields. From these basis, we started to study a new concept rather simple, but which proved soon to be very interesting : direct coupling of collectors to an heating grid embedded in the concrete floor, without any water tank or intermediate exchanger. By doing this, our aim was to realise a gain on three points :

- less total cost of the system, by reducing components number, and saving the important floor occupation of the classical water storage,
- better thermal yields, by the use of lower fluids temperatures at each stage of the process,
- and, perhaps most important of all, improved reliability, due to an extreme simplicity of the circuitry and regulators.

Our previous theoretical studies shown that these objectives could be reached, complying with a few rules, which define what we call the typical Direct Solar Floor (D.S.F.). Immediately, we have to specify two main kinds of DSF :

- first, what we call "on the ground thick slab", a concrete floor of about thirty centimeters thickness, directly built on the ground, and better fit for individual houses, or single levelled buildings.
- and then, what we call "level floors", used for collective dwellings, with less thickness, but thermal emission on both faces, and allowing a reduction of weight constraints in building structure.

The technology of the floor in itself doesn't differ from the usual realizations of heating floors. For the "ground thick slab", one can found :

- compacted and planed natural soil,
- polyethylene foil,

- thermal insulation ensured by polymeric foam (a 4 cm thickness should be a minimum),
- the heating grid, made of polymeric piping or any other classical type of heating pipe, but with a sligtly tighten spacing between tubes. This grid may be fixed on the iron frame work, either directly against thermal insulation, or upon a first thin concrete slab,
- thick concrete slab itself, of about thirty centimeters, including eventually the thickness of a tiled or other heavy masonry coating.

The "level floor" is fully identical to classical heating floor, a somewhat greater thickness excepted (about 20 cm).

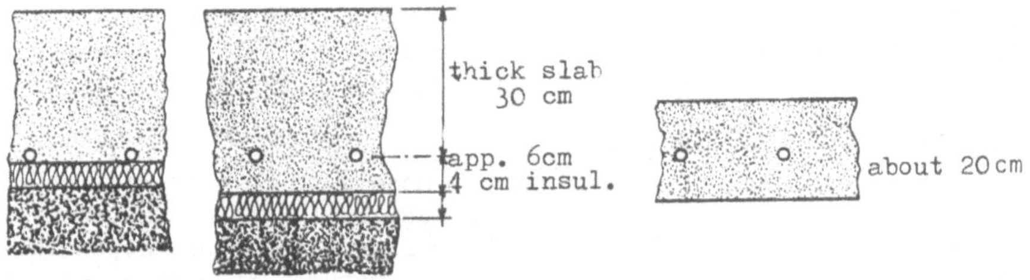

Figure 1 : On the ground thick slabs and level floors.

The only regulation between flat collectors and grid floor is a pair of differential temperature sensors (between outlets of collectors and return from grid), driving the pump of the loop when solar power is sufficient in respect to instant temperature of the slab. Additional Domestic Hot Water supply is shown, providing substantial energy saving during summer, and also some protection against over heating of collectors. The drawing shows a disposition which proved very efficient, with DHW loop connected in parallel with heating floor, only by natural thermosyphon.

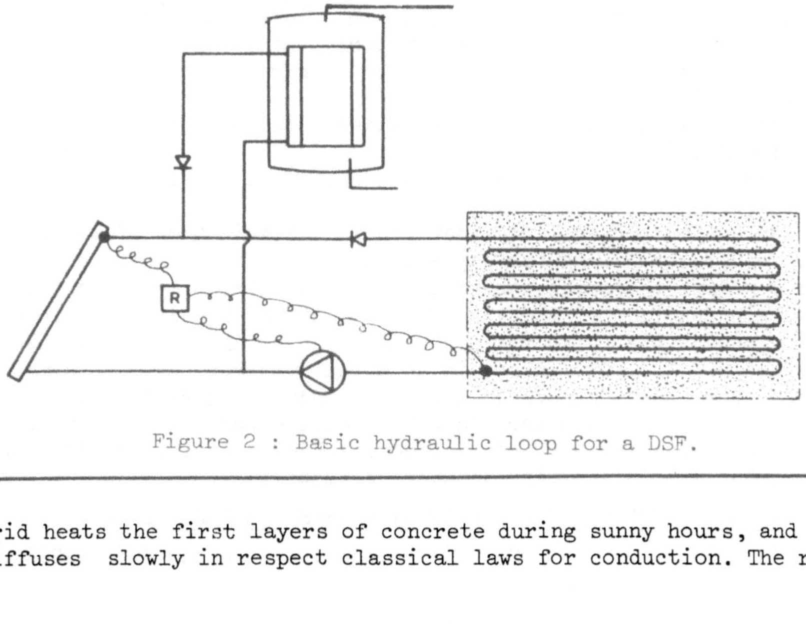

Figure 2 : Basic hydraulic loop for a DSF.

Grid heats the first layers of concrete during sunny hours, and then, heat diffuses slowly in respect classical laws for conduction. The result is an heat flux from the surface of the floor to the inside of the house, which presents certain particularities. It appears first a phase difference between internal flux emitted and solar energy received. Like than for a well known "TROMBE wall" this phase difference is about 7 to 8 hours for a concrete thickness of 30 cm. But the most important is the very low residual amplitude of internal flux emitted over several days. This amplitude is as small as 6 % of mean value for the 30 cm slab.

So, the DSF appears as a quite constant flux emittor, which can provide a basic heating, persistent even after several days of bad weather. In fact, in mathematical model, if we "shut off" the sun after a long period of maximum "sunny weather", it may take more than 3 or 4 days for the flux fall down until half value of the maximum rate.

On the other hand, it is evident that auxiliary heating cannot be distributed by the same circuit, and has to be achieved by a fully separated system, able with quick response of peak heating power.

2. MODELIZATION AND MAIN RESULTS

We have developped a few mathematical models to characterize the thermal behaviour of DSF. All of them are based on the following assumptions :

- flat collectors are figured by their linear equation, depending upon instant solar power (E, W/m^2) ; and mean temperature difference between thermal fluid and external temperature (θ, °C) as : $p = \nu E - K\theta$
- internal ambient temperature is constant and equal to a fixed level. Additional heat is furnished by a perfect auxiliary system, instantly when needed. If diffused heat from the floor exceeds the need, we assume it is evacuated, by unspecified conveniences, so there is not any overheating,
- solar power is figured by mathematical arbitrary functions (half sine from t), sized to represent weather sequencies,
- external temperature is also given at each step of time by arbitrary functions with combined variations in mean level, amplitude and phase difference with solar power.

These models are treated by fully conventionnal methods of numerical resolution of the heat equation, step by step in the concrete mass, with external conditions according to the above assumptions. Investigations over a large range of shapes, size of components and weather sequencies led us to the conclusion that thermal yields from the floor can be expressed in function of a very reduced number of parameters, which are :

- mean value of solar energy received during the choosen sequencie (E_R expressed in $kWh/m^2.day$),
- temperature difference between fixed internal level and mean external temperature (Δt, in °C),
- ratio between areas of Solar collectors and heating floor (R, no dimension).

We can notice particularly that precise distribution of solar power or external temperature along the day showed no significant effect. Such a result allowed us to express thermal yields calculated by the detailed program, only in function of the variables E_R, Δt and R, for a few interesting configurations of DSF, with fixed characteristics and size, and appearing as "standard models", Of course, in an actual project, one can draw somewhat aside from these "standards", but the induced thermal variation stay rather small, and can be described by corrective terms of low importance.

From this work, we have developped a calculation method, now published in France, but usable in every European climate (1).

3. EXPERIMENTAL RESULTS : EXAMPLE OF EYGUIERES'S PRIMARY SCHOOL

This school is built in South of France, not so far from MARSEILLES. It is a single level building with South side organized to catch a maximum of passive solar incomes. Rear spaces are equiped with a "thick slab on the ground" DSF for 412 m^2 total area, supplied by 60 m^2 of solar collectors (common black type). DSF is built directly upon natural soil compacted with a polyethylene film and 3 cm polystyrene foam insulation.

Experimental measurement were done on this installation by an independant laboratory (ECOLE DES MINES D'ALES). Main physical values measured were :

- heat counter for the energy income into the slab,
- solar energy received and external temperatures on the location,
- mean temperature in the rooms heated by the DSF.

The table below gives a summary of these results in comparison with our calculation method.

Month	Sol. en. kWh/m^2.d	Ext. Temp. mean	Int. Temp. mean	Incomes(A) kWh/month	Emission (B) kWh/month
October	4,65	14,3	20,5	2414	2281
November ...	4,40	9,8	18,8	2615	2970
December ...	2,46	7,3	17,5	1511	1307
January	3,38	9,6	19	2896	2327
February ...	3,25	8,7	18,9	2185	1822
Marsh.......	4,74	9,5	20,4	4606	3190
April	5,24	13,2	22,2	4075	3513
			Total	20300	17410

The column (A) "incomes, is the energy supplied by collectors to heating grid, including thermal losses through bottom insulation, whereas column (B), "emission", represents calculated useful energy emitted by upper surface of the floor. Losses are very uneasy to estimate, but 15 % difference between two columns looks a very realistic value, so there is a very good agreement between calculated and experimental values. Annual specific productivity appears as high as 290 kWh/m^2 for heating only in this example, but even more high results can be reached in mountain climates, up to 650 kWh/m^2, including hot water supply.

oOo

(1) CALCUL DES PLANCHERS SOLAIRES DIRECTS, Collection de l'ESIM, EDISUD La Calade - 13090 AIX-EN-PROVENCE. (FRANCE).

ANNEX 1 : A theorical example of thermal emission from a 40 cm thickness DSF.
Notice the very long decreasing time for emission : after three days without sun, it remains over one half of the maximum main value reached.
Notice also the very low residual oscillation.

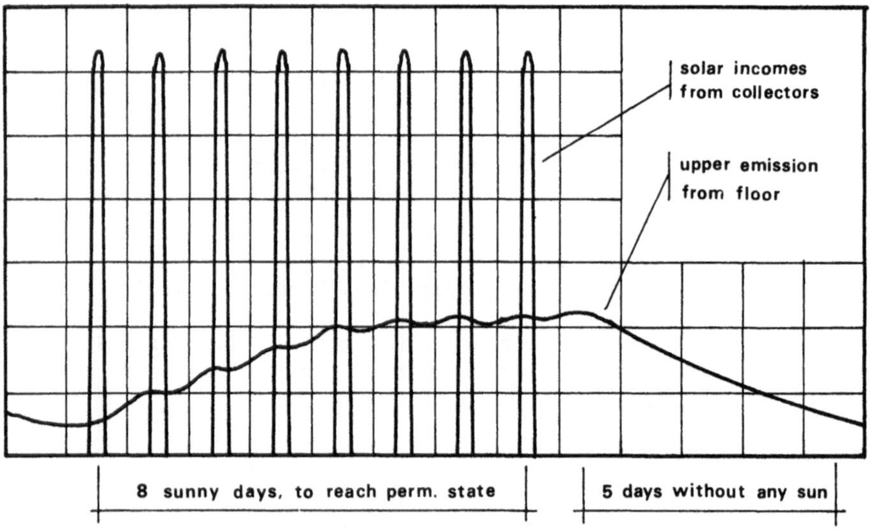

ANNEX 2 : An example of performance diagram (from our book "CALCUL DES PLANCHERS SOLAIRES DIRECTS").

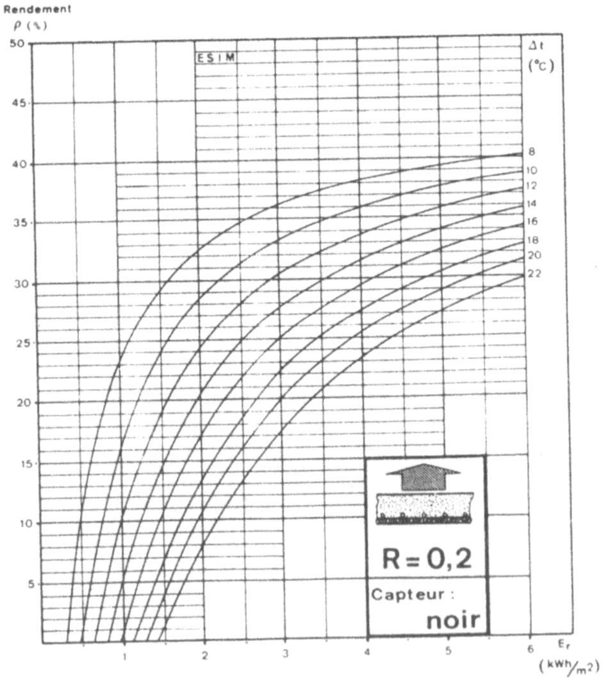

Notice the high thermal yield : about 30 - 35 % for medium use conditions, and up to 40 % for season's end condition.

These curves may be described by analytical formulas which give very compact subroutines for calculation programs.

A CONTRIBUTION TO THE LAYING DOWN OF DURABILITY AND RELIABILITY TESTS FOR SOLAR COLLECTORS

R.DIETZ and J.VINCENT
CETIAT - VILLEURBANNE - FRANCE

Summary

The purpose of this work is to provide some results for the laying down of a standardized test method to evaluate durability and reliability of solar collectors.
Ten solar collectors have been choosen as representative of the effective french production. They were submitted to qualification tests such as moist air penetration, pressure (inside absorbers), freezing, thermal performance. They were also submitted to an artificial accelerated ageing by travelling fluid at conventional stagnation temperature during 80 hours. The main results are that the qualification tests are easily performed except moist air penetration ; ageing procedure is encouraging and can be considered in standardized test methods.

1. INTRODUCTION

A set of ten solar collectors building a representative sample of french industrial production has been submitted to two kind of tests :
- Qualification tests of collectors to caracterize their behaviour to moisture, pressure, freezing, external thermal shocks.
- Artificial accelerated ageing process with a view to estimate the maintenance of collectors' thermal performance.

2. QUALIFICATION TESTS

2.1 Moisture penetration

Collectors are kept for 3 to 4 hours in an enclosure whose ambiant conditions are tropical (40 °C, 80 % R.H.), then suddenly cooled to display through condensation moisture penetration. This test did not show convincing results, it simply indicates there is no simple test to estimate moisture penetration. Moisture penetration has not such important consequences it could justify an uneasy and costly test.

2.2 Pressure test

Collectors' absorbers are submitted to test pressure of 1.5 time the nominal pressure. Every collector has been tested successfully. The simplicity and low cost of this test are of great interest.

2.3 Freezing test

The collectors are filled with water and closed then submitted to three cycles of freeze and thaw being alternatively placed and withdraw from an enclosure whose ambiant temperature is - 10 °C. The tests results showed that only one cycle was unsufficient.
When technical directions for use indicate that the collector must not be used without antifreeze then it is not submitted to freezing test.

Three out of seven collectors undergo this test successfully. Such a test is part of french standard (NF P 50502) with lower temperatures. The limitation for these tests to be realised lay in the required temperature. The lower is this temperature the higher are the investments for the laboratories.

2.4 External thermal shock

It was supposed that this test was destructive so it was carried out after thermal performance measurements following artificial ageing. The collectors are travelled by a fluid at conventional stagnation temperature during 15 minutes at a flowrate of 3 $l/min/m^2$.
In reality this test revealed itself as a rain penetration test coming after artificial ageing the thermal shock test did not produce further damage.

3. ACCELERATED ARTIFICIAL AGEING

Thermal performance tests of ten collectors have been carried out during this work before and after the ageing process here under described. On a test bench specially designed for this purpose the collectors are brought to their conventional stagnation temperature during ten hours. then they are cooled and investigated. After it temperature is raised again to the same stagnation temperature and maintained during 20 hours. After cooling and investigation the test goes on for 50 hours. At the end of it the state of the collector is written down and the thermal performance test is carried out. This ageing process seems to be significant in spite of little effect on thermal performances, effects on outgasing of insulation products and on maintenainance of prime collector thightness are noticeable. It is remarkable that seven out of ten had less than 10 % of decrease in efficiency and no damage.

4. CONCLUSIONS

The set of tests described is a part of a wider set of tests to be carried out on solar collectors. An ageing test is made of qualification tests coming before and after an ageing process. Generally speaking the topics to be examined are thermal performance mechanical performance and optical properties. The durability is estimated through the change in performance or properties after specific ageing process. The accelerated artificial ageing process used in this work can be combined with a natural one so that the lenght of time of the ageing process could be kept into reasonable limits independently of weather conditions.

LEISTUNGSVERGLEICH VON ZWEI SONNENHEIZUNGSANLAGEN ZUR WASSERERWÄRMUNG

A. Höß
Technischer Überwachungs-Verein Bayern e.V., Zentralabteilung Wärmetechnik

Zusammenfassung

Mit beiden Sonnenheizungsanlagen zur Wassererwärmung konnte nach entsprechenden Änderungen eine hohe Leistungsfähigkeit erreicht werden. Die Pumpenanlage und die Schwerkraftanlage erreichten annähernd gleich gute Werte. Aufgrund des angewendeten Prinzipes wurde zwar bei der Schwerkraftanlage von den Kollektoren eine geringere Energiemenge geerntet als bei der Pumpenanlage, dies wurde jedoch z.T. wieder ausgeglichen durch den geringeren Wärmeverlust des Speichers.

Für die Anwendung ist von Bedeutung, daß bei einer Pumpenanlage die Fühler der Regeleinheit zweckmäßig angeordnet sind und die Regeleinheit richtig eingestellt ist.

Die Pumpenanlage zeichnet sich durch
- große Freizügigkeit bei der Aufstellung des Speichers und durch
- große Freizügigkeit bei der Anordnung der Leitungen des Solarkreises aus.

Für die Anwendung der Schwerkraftanlage ist ein ausreichend hoher Durchfluß im Solarkreis wichtig. Der Startwert für den Solarkreis sollte allerdings einstellbar sein, damit den Bedürfnissen der Nutzer und den meteorologischen Gegebenheiten besser Rechnung getragen werden kann.

Die Schwerkraftanlage zeichnet sich durch
- die Verhinderung einer Rückströmung bei niedrigen Temperaturen in den Kollektoren,
- eine geringe notwendige Höhendifferenz zwischen den Kollektoren und dem Speicher und
- das Beladen des Speichers von oben nach unten aus, wodurch ausreichend warmes Wasser bereits unmittelbar nach dem Start der Anlage zur Verfügung steht.

Summary

The efficiency of two domestic hot water systems had to be tested under the same environmental influences and capacities.

One system worked with forced circulation (circulation pump system) and the other with gravity circulation (thermosyphon system). In the thermosyphon system the liquid circulates by means of differences in its density. In this case a circulation through the collectors and the storage tank will start only if the liquid in the collectors warms up to a given, a adjustable temperature which is equivalent to an extension of volume. By adjusting the overflow level, this temperature difference is fixed.

With both systems a high efficiency was possible. The thermosyphon system got less energy from the collectors but wasted less heat on the storage tank.

Essential for a high efficiency of the pump system was the regulation system.

Essential für a high efficiency of the thermosyphon system was the flow rate through collector and storage tank.

The advantage of the thermosiphon system is

- the high water temperature after the start of the system
- the prevention of a backward flow at low temperatures in the collectors and
- the possibility of little difference in the level between the collectors and the storage tank.

*****

1. Zweck

Es war die Leistungsfähigkeit von zwei Sonnenheizungsanlagen zur Wassererwärmung unter gleichen Umweltbedingungen und Lastbedingungen zu ermitteln. Eine Anlage wurde mit Zwangsumwälzung im Solarkreis, im folgenden Pumpenanlage genannt, Bild 1 und eine Anlage mit Naturumwälzung im Solarkreis, im folgenden Schwerkraftanlage genannt, Bild 2 ausgeführt.

2. Beschreibung der Anlagen

Beide Anlagen waren als Zweikreisanlagen aufgebaut. Im Solarkreis wurde ein Gemisch aus 50% Glykol : 50% Wasser verwendet. Beide Speicher wurden über von oben nach unten durchströmte Wärmetauscher beladen. Kaltes Wasser wurde den Speichern von unten zugeführt. Warmes Wasser wurde an den Speichern ebenfalls unten entnommen.

Die beiden Anlagen wurden auf einem Modelldach mit einer Neigung von 45°, dessen First in Ost/West-Richtung verläuft, errichtet und unter vollkommen gleichen meteorologischen Bedingungen betrieben. Aus jeder Anlage wurde nach folgendem Programm Wasser entnommen.

11.00 Uhr 20 l Wasser
13.00 Uhr 30 l Wasser
19.00 Uhr 50 l Wasser
20.00 Uhr 100 l Wasser (Leerzapfen)

Jede Anlage bestand aus

2.1 Kollektor Jeweils zwei Module, direkt durchflossene Vakuumröhren mit 2 X 6 Glasröhren, Außendurchmesser 100 mm
Absorberfläche 2 X 1,13 m^2 = 2,26 m^2
Abmessungen eines Modules 710 mm X 2620 mm X 152 mm
Gewicht eines Modules 41 kg

2.2 Speicher — Jeweils stehend angeordnet, Inhalt 145 l, Werkstoff Nr. 1.4301, Betriebsdruck max. 6 bar und 60 mm Weichschaum Wärmedämmung

2.3 Steuerung — Pumpenanlage
Umwälzpumpe durch Regler über Differenztemperatur zwischen Kollektor und Speicher gesteuert, Pumpenleistung 25 W
Schwerkraftanlage
Durch Volumenausdehnung des Mediums im Kollektor bei Erwärmung wird der Solarkreis über einen Überlauf durch Naturumwälzung in Gang gesetzt.

3. Beschreibung der Meßtechnik

Zur Ermittlung der Leistungsfähigkeit der beiden Anlagen wurde jeweils gemessen:

3.1 Allgemeine Daten
Außenlufttemperatur
Umgebungstemperatur der Speicher
globale Bestrahlungsstärke in der 45° geneigten Ebene
globale Bestrahlungsstärke in der horizontalen Ebene
Sonnenscheinstunden
Niederschlagsmenge
Windgeschwindigkeit

3.2 Solarkreis
Kollektortemperatur
Volumenstrom

Vorlauf- und Rücklauftemperatur vor den Speichern

3.3 Wasserkreis
Kaltwassertemperatur
Warmwassertemperatur
Speichertemperatur an 5 Orten über ihrer Höhe
gezapfte Wassermenge und entsprechende Wasserwärmemenge

4. Ergebnisse

4.1 Kollektor
Der Wirkungsgrad eines Kollektormodules wurde in Bild 3 dargestellt.
Die höchste mögliche Absorbertemperatur über der Außenlufttemperatur der einzelnen Röhren der Module wurde für eine globale Bestrahlungsstärke von 1000 W/m^2 zwischen 199,4 K und 224,0 K gemessen. Die Stillstandtemperatur errechnete sich bezogen auf eine globale Bestrahlungsstärke von 1000 W/m^2 und eine Außenlufttemperatur von 32 °C im Mittel zu 246,7 °C.

4.2 Speicher

Der Wärmeverlust der Speicher wurde für einen Temperaturbereich des Wassers von 50 bis 60 °C bei einer Raumlufttemperatur von rd 18 °C
für die Pumpenanlage zu 2,05 W/K und
für die Schwerkraftanlage zu 1,60 W/K
ermittelt. Der Unterschied ist auf die Wärmeableitung und die Unterbrechung der Wärmedämmung des Speichers durch das Vorlaufrohr des Solarkreises zurückzuführen, das bei der Pumpenanlage von oben in den Speicher eingeführt war.

4.3 Startwert für den Solarkreis

Die Pumpenanlage startete bei einer Kollektortemperatur von rd 10 K über der Speicherwassertemperatur, während die Schwerkraftanlage erst bei einer Kollektortemperatur von etwa 70 °C in Gang kam.

4.4 Nutzwärmemenge

In Bild 4 ist die Nutzwärmemenge für die Pumpenanlage dargestellt. Aus der Anlage ist zu ersehen, daß eine Erhöhung der Nutzwärmemenge durch eine geeignete Regeleinrichtung möglich ist. Sie wurde durch Anordnung des Kollektorfühlers direkt im Kollektor anstelle an der Vorlaufleitung erzielt. Die Änderung hatte einen kleineren Startwert für den Solarkreis zur Folge. Im Solarkreis wurde etwa 150 l/h umgewälzt.
In Bild 5 ist die Nutzwärmemenge für die Schwerkraftanlage dargestellt. Aus der Anlage ist zu ersehen, daß eine wesentliche Erhöhung der Nutzwärmemenge durch eine Vergrößerung der Umwälzmenge im Solarkreis erzielt wurde. Ursprünglich wurden zwischen 5 und 25 l/h umgewälzt, die dann auf maximal 150 l/h erhöht wurden.

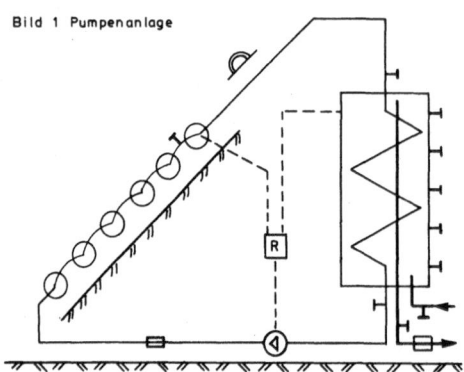

Bild 1 Pumpenanlage

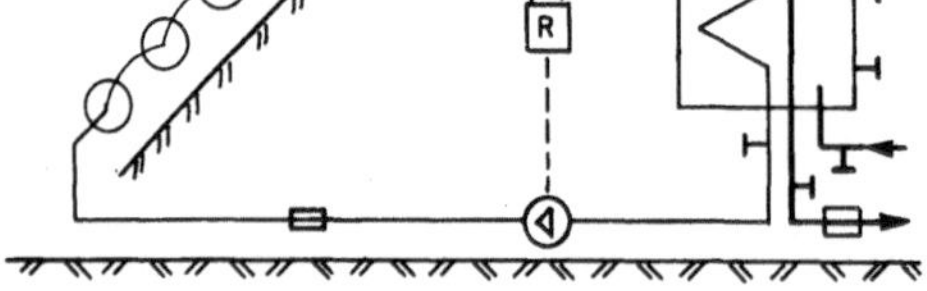

Bild 2 Schwerkraftanlage

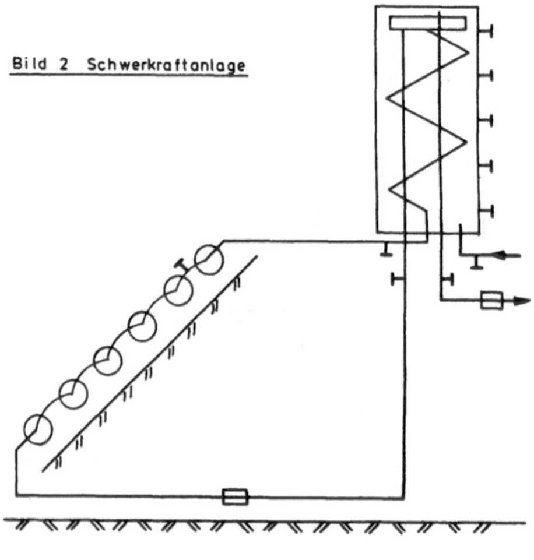

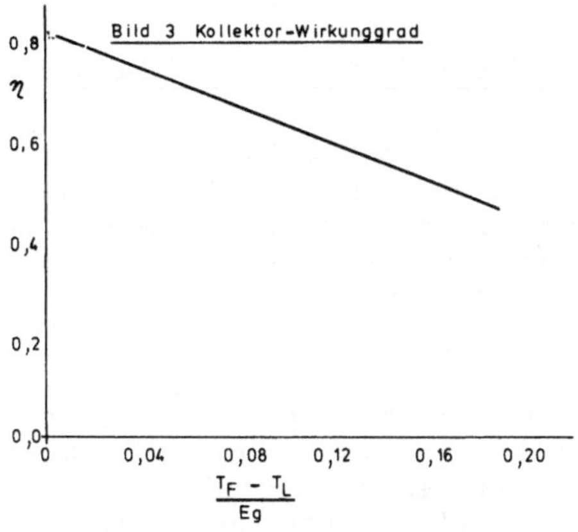

Bild 3 Kollektor-Wirkunggrad
0,8
η
0,6
0,4
0,2
0,0
0
0,04
0,08
0,12
0,16
0,20
$\frac{T_F - T_L}{Eg}$

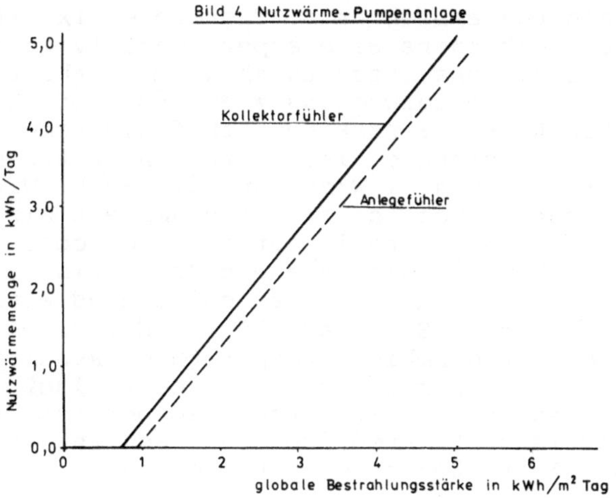

Bild 4 Nutzwärme-Pumpenanlage
Kollektorfühler
Anlegefühler
Nutzwärmemenge in kWh/Tag
5,0
4,0
3,0
2,0
1,0
0,0
0
1
2
3
4
5
6
globale Bestrahlungsstärke in kWh/m² Tag

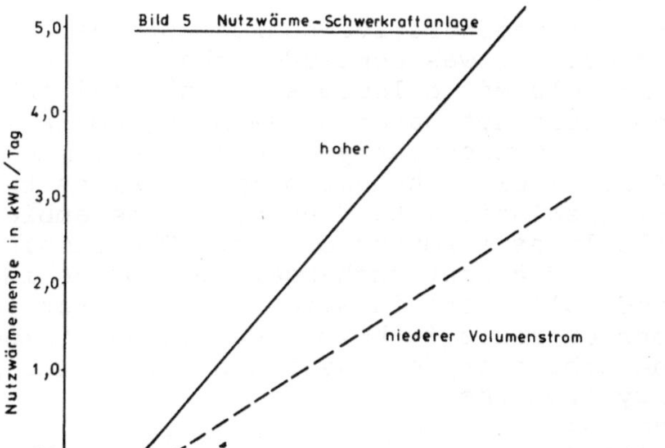

Bild 5 Nutzwärme-Schwerkraftanlage
hoher
niederer Volumenstrom
Nutzwärmemenge in kWh/Tag
5,0
4,0
3,0
2,0
1,0
0,0
0
1
2
3
4
5
6
globale Bestrahlungsstärke in kWh/m² Tag

PERFORMANCE OF UNGLAZED FLAT PLATE HEAT EXCHANGERS AT LOW TEMPERATURES

W. Schwaigerer, E. Hahne, J. Sohns
Institut für Thermodynamik und Wärmetechnik
Universität Stuttgart
Pfaffenwaldring 6, D-7ooo Stuttgart 8o

SUMMARY

The heat source of a heat pump for space heating or domestic hot water production is in general ambient air. It was proposed to use unglazed flat plate heat exchangers, so called "solar absorbers", in the heat pump cycle to gain heat from the outdoor air and radiation from the sun. It is shown to what extend "solar absorbers" get energy from the sun and the ambient air. Therefore six different flat plate heat exchangers were experimentally investigated under outdoor-conditions. Results show, that the flat plate heat exchangers absorb between 53 % and 88 % of the solar irradiance, when they are working with fluid temperatures near at ambient temperature. The overall heat transfer coefficients are between 11 W/m² K and 32 W/m² K. With heat transfer by forced convection from the ambient air (wind speed about 4 m/s) the overall heat transfer coefficient reaches up to 4o W/m² K. Using these characteristic values, the yearly obtained useful heat was calculated with weather conditions, measured in Stuttgart 1981 and 1982. All absorbers operated 5 K below the ambient temperature. Results show, that an unshaded absorber gains about 1oo to 13o% more than a shaded absorber, in case the absorber ist back isolated. If the absorber is ventilated from the back the unshaded ones gain about 6o to 8o% more than the shaded absorbers.

1. Introduction

In search of regenerative sources of energy for domestic heating techniques it was concluded that the application of conventional glazed collectors is only suitable for heating of domestic hot water or swimming-pools in a Middle-European climate. Consequently an effort was made to raise the temperature level with heat pumps. Besides the energy of the global insolation, the energy of the ambient air is also available as a source of heat. The combination of a collector and a heat exchanger led to the idea to apply unglazed collectors in which the temperature of the heat transfer fluid is below the ambient temperature. Such solar absorbers or "energy roofs" offer the possibility to gain energy from the

1.) ambient air
2.) insolation
3.) wind
4.) humidity
5.) and precipitation (1).

Numerous types of structures were developped as energy

absorbers, such as energy facades, energy blocks and energy fences, which mainly perform as heat exchangers (2). The first experiments on systems with heat pumps and energy absorbers showed, that the gain from humidity, rain, wet snow and the increase of the surface through freezing (built up of frost) is negligible (3). Even the improvements of the heat transfer coefficient by wind and the heat gain from insolation were doubted (4, 5, 6). Therefore it would not make any difference whether the absorber is shaded or not. The purpose of this study was to investigate the influence of these parameters on the heat gain of several absorbers and to calculate the useful heat for a year.

2. Tested absorbers

Several flat plate absorbers (German products) were experimentally investigated. Figures 1, 2, and 3 show the objects as they were mounted on the test rig at the institute. Table I sums up the most important technical data. The absorber with the glass cells in fig. 1 is excluded from the following investigation, since it is considered more as a collector. Absorbers 1, 2, and 4 have back insulation, absorbers 3, 5, and 6 are ventilated from the back. Absorber 5 and 6 were mounted on an aluminium trapezoid profiled surface. None of the absorbers has selective coating.

Figure 1, 2, 3:

Tested flat plate absorbers

No	material	area [m²]	coating	conversion factor	heat-transfer-coefficient [W/(m²·K)]
1	copper	1.822	copper oxide	0.533	10.868
2	steel	2.174	black paint	0.534	13.072
3	polypropylene	1.034	black	0.716	21.610
4	ethylene-copolymeride-bitumen	2.0	black	0.820	14.795
5	aluminium	1.978	brown eloxadized	0.883	30.244
6	steel	1.718	brown paint	0.829	31.761

Table I:

Technical data of the absorbers

3. Test conditions

The absorbers were mounted at a tilt angle of 48.5° on the test rig at the ITW. The heat transfer fluid is water, the flow direction is from bottom to top, except for absorber 4, where, according to the instructions of the producer, the fluid flows in the opposite direction. Figure 4 shows the test loop scheme of the set up. A chiller is used for the inlet temperature, which is below the ambient temperature. The inlet temperature and the water flow rate is changed systematically. Measurement without insolation is carried out at night. The wind speed was simulated with a fan (see figure 3).

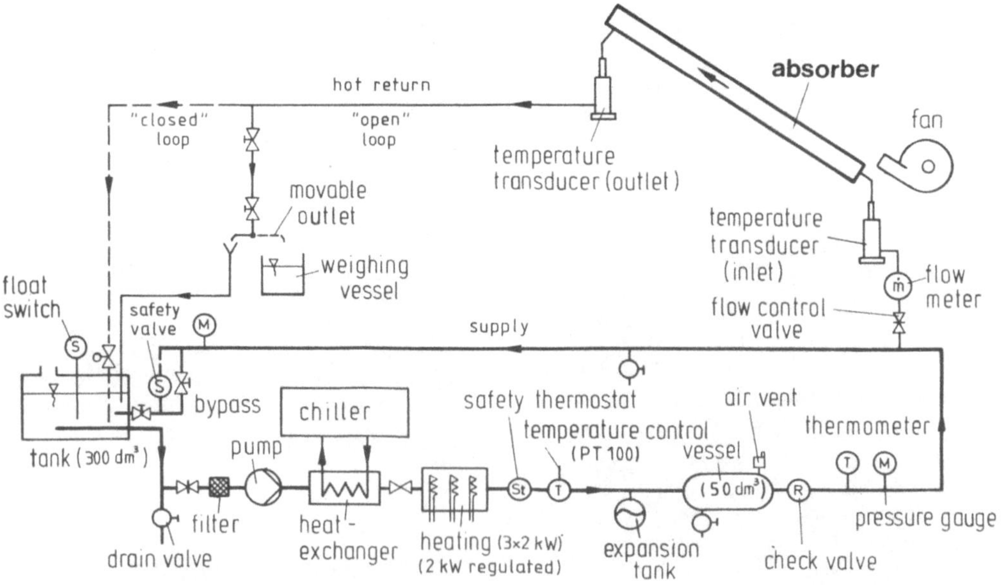

Fig. 4: Scheme of absorber test loop

4. Results

4.1 Experiments

The heat loss or gain of a absorber is plotted versus the temperature difference between the mean fluid and ambient temperature and is classified to five insolation groups. The calculation of the straight lines is done according to the least square fit method. Figure 5 shows the results for absorber 5. The lines for forced convection are remarkably steeper than the lines for natural convection. The lines for the various insolations are almost parallel, except for those with high insolation. The heat gains of the absorbers are defined with sufficient accuracy by the conversion factor, which is the ratio of the useful energy per square meter to the insolation at ambient temperature, and the heat transfer coefficient k_{eff}, which is dependent on the wind.

This characteristic values are given in the last two columns of table I. k_{eff} is caluclated from the following equation

$$k_{eff} = a - b \cdot w_{amb}^{0.8}$$

for a wind speed of 2 m/s, according to (7). In case of forced convection and at a mean fluid temperature of 5 K below the ambient temperature the absorber delivers about 980 W/m² useful energy, if the insolation is 800 W/m². The absorber delivers only 240 W/m² without insolation, which means that the absorber obtains 25 % of the energy of the ambient air. The influence of the mass flow rate on the absorber performance is negligible. Further no improvement of the heat gain was confirmed through rain or condensation of humidity. Additionally the time constants for sudden changes of insolation are measured for absorber 5 and 6. The values are between 1.5 and 2.5 minutes.

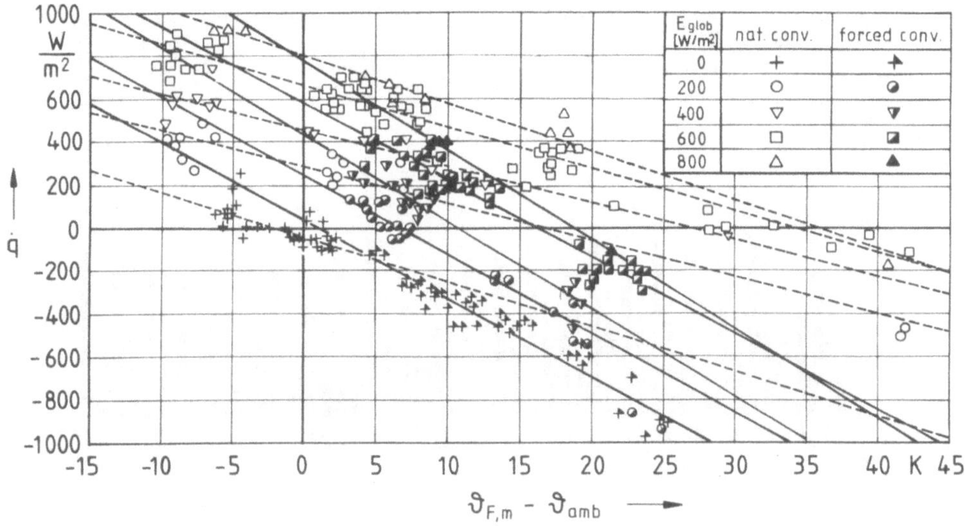

Figure 5: Heat gain of absorber 5

4.2 Calculations

If the absorbers are working the whole day or for a complete year, the influence of the insolation on the useful energy of the absorbers decreases. Therefore the yearly heat gain of the absorbers investigated was computed with weather-data, measured at Stuttgart 1981 and 1982. The difference of the mean fluid temperature $\vartheta_{F,m}$ and the ambient temperature ϑ_{amb} is kept constant at -5K, but the mean fluid temperature may not exceed 18°C.

Fig. 6 shows the results. The bars at the top describe the monthly insolation, the smaller bars in the middle describe the heat gain of the absorbers, if they are shaded, and the bars at the bottom show the heat gain of the unshaded absorbers. The heat exchangers, which are ventilated from the back (small bars 3, 5, and 6), are better than the isolated absorbers. Over one year the heat gains of the absorbers are nearly the same. If the absorbers are working in the sun, you can see, how they use the supply of sun. In summer the improvement of the heat gain is remarkable. In average the absorbers isolated backwards deliver about 1oo to 13o % more than the shaded absorbers. In case of back ventilation the absorbers obtain 6o to 8o % more.

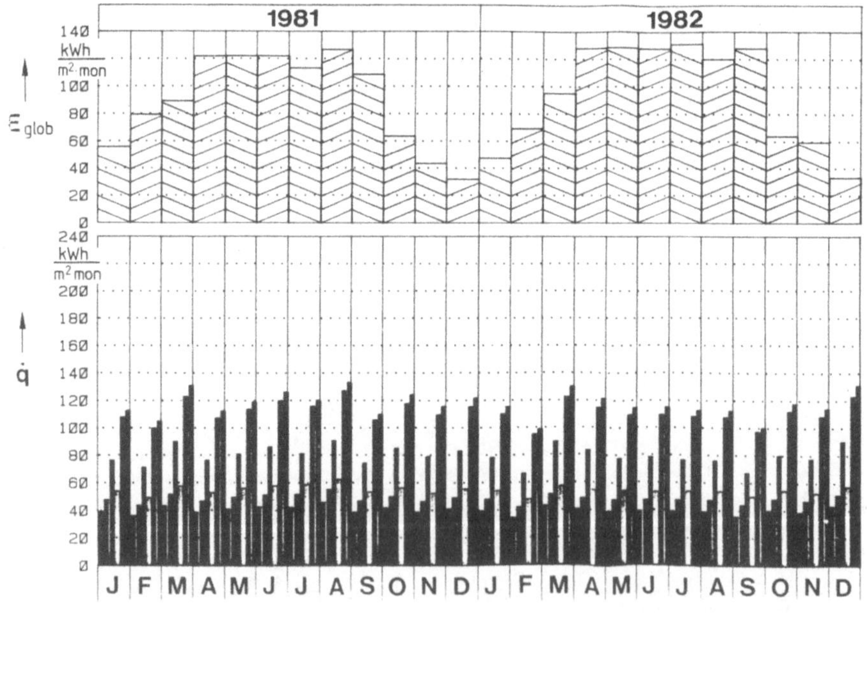

Fig 6. Yearly heat gain of the absorbers (centre : shaded, below : unshaded)

5. Conclusions

The results show, that unshaded absorbers have more heat gain than the shaded absorbers. This is also valid for the whole year calculations. These advantages may be destroyed by the controlling of the system. Especially the time delay between the energy supply in summer and the energy demand in winter prevents the use of energy delivered by the absorbers (8). This is again the old problem of insufficient storage techniques.

6. Nomenclature

a, b	constants	-
E_{glob}	global insolation	W/m^2
k_{eff}	effective heat transfer coefficient	W/m^2
$\dot{q}$	heat gain per square meter	W/m^2
w_{amb}	wind speed	m/s
ϑ_{amb}	ambient temperature	°C
$\vartheta_{F,m}$	mean fluid temperature	°C

7. References

1. Schimmelpfennig, K.: Flächenwärmetauscher. Verlag C.F. Müller, Karlsruhe, 1982
2. Kaufmann, K.-D.: Energiedächer. IDEA-Verlag, 1981
3. Böbel, A.: Nutzung der Wärme aus der Umwelt durch Solarsysteme, Energiewirtschaftliche Tagesfragen, Heft 9/lo, 1978
4. Kersten, R.: A comparison of solar heating systems and heat pump systems. 3. Internationales Sonnenforum, Hamburg 198o
5. Müller, H., and Gärntner, K.: Planung und Ausführungen von Wärmepumpenanlagen mit Energieabsorbern. BSE-Fachtagung Köln 1983
6. Dietrich, B.: Betriebsergebnisse von Energieabsorberanlagen. BSE-Fachtagung, Köln 1983
7. Bundesverband Solarenergie e.V.: Gebrauchstauglichkeit von Energieabsorbern. Pomp & Sabkowiak, Essen 1982
8. Posorski, R.: Energieabsorber-Wärmeübergänge und Systemverhalten. BSE-Fachtagung, Köln 1983 .

8. Acknowledgment

The investigations have been financed by the Federal Ministry of Technology under 03 E 8021 A. The authors gratefully acknowledge this support.

OPTIMISATION OF SOLAR COLLECTOR AREA FOR SOLAR THERMAL SYSTEMS

N.K. BANSAL and AMAN DANG
Centre of Energy Studies,
Indian Institute of Technology,
New Delhi, India.

Summary

Invariably solar energy systems are provided with an auxiliary energy source to meet the energy requirements of a system operating at a constant temperature. A technoeconomic analysis has been developed in this paper to calculate the area of solar energy collector, which yields minimum cost for the derived useful energy. The analysis has been illustrated by specific examples of a solar drying system and a solar water heating system.

1. INTRODUCTION

All solar energy systems invariably have a high initial investment cost followed by a low operating cost. To be economically viable, the life time of these systems must be large enough to enable the investor to get adequate returns on the investment. Majority of solar energy systems are provided with an auxiliary energy source for its round the year utility and to enable it to provide the required energy demand at the desired temperature. Optimisation of solar energy collector area can not usually be made with respect to the highest energy demand and/or to the availability of the minimum solar radiation. Ultimate quantity of interest is naturally the price of the delivered useful energy. Factors, which should be taken into account, to calculate this cost are the price of the system, maintenance, rate of interest, salvage value and the expected life time of the system. It is often uneconomical to design the solar energy collector area for the peak energy demands (1). The purpose of this paper is to develop an analysis for finding the fraction of solar energy that should be used in hybrid solar thermal systems to keep the ultimate cost per unit useful energy to a minimum level. Two typical examples, namely solar drying and solar water heating, have been considered to illustrate the method of analysis.

2. ANALYSIS

Derivation of accurate system's cost must start with an initial cost estimate for a system near its final size. Fixed costs for pumps, controls, piping and electrical work are estimated first and then purchased costs for collectors, heat exchangers and storage tanks are added. Installation cost is included and distributed among the components. The total installed cost is then divided by the total installed collector area to produce the unit system cost.

Let Q_0 be the peak energy demand, a fraction x of which is met by solar energy and the rest by the auxiliary source. The annual cost of the system can then be written as

$$C = C_1' x + C_2' (Q_0 - x) + C_R \quad (1)$$

where C_1' is the annual fixed cost per unit usefulenergy of the solar energy system, C_2 is the annual fixed cost per unit useful energy of the auxiliary source and C_R is the total maintenance cost. The annual fixed cost of a system is calculated by multiplying the initial investment with the annuity factor viz.

$$C_a = I q^L (q-1) / (q^L - 1) \quad (2)$$

where q = (1 + z/100); z being the interest rate and L the life time of the system in years. The ratio of C_a to the annual useful energy derivable from the system yields the annual cost per unit useful energy i.e. C_1' and C_2' in Eq. (1). Eq. (1) in practice has to be written as

$$C = C_1' x + C_2' (Q_0 - x) + n C_2'' \int (Q(t) - x)\, dt + M_s + M_a \quad (3)$$

where Q(t) is the energy requirement at a particular time, M_s , M_a being the annual maintenance cost of the solar system and the auxiliary system respectively.

1.1 Solar Drying

In a typical drying situation, the energy requirement goes on decreasing with time and Q(t) can approximately be written as

$$Q(t) = Q_0 \exp(-\alpha t) \quad (4)$$

Eq. (3) in this case can be written as

$$C = C_1' x + C_2' (Q_0 - x) + \frac{n C_2''}{365} \int_0^{t_1} (Q_0 e^{-\alpha t} - x)\, dt + M_s + M_a \quad (5)$$

where t_1 corresponds to the particular drying period and n is the number of such drying periods (harvesting days). It also follows that at drying situation, at time t_1 , $x = Q_0 \exp(-\alpha t_1)$ or,

$$t_1 = \frac{1}{\alpha} \ln (Q_0 / x) \quad (6)$$

Carrying out the integration in Eq. (5) and substituting the value of t_1 from Eq. (6)

$$C = C_1' x + C_2' (Q_0 - x) + n C_2'' \left[\left(\frac{Q_0 - x}{\alpha}\right) - \frac{x}{\alpha} \ln \frac{Q_0}{x} \right] \quad (7)$$

For C to be minimum $\partial C / \partial x = 0$, a condition which yields

$$x / Q_0 = \cdot \exp \left[365 (C_1' - C_2') \alpha / n C_2'' \right] \quad (8)$$

1.2 Hot Water Heating

For the case of hot water heating, Q(t) can usually be taken to be constant and put equal to Q_0, hence

$$C = C_1' x + C_2' \ (Q_0 - x) + n C_2'' \ (Q_0 - x) + M_s + M_e \qquad (9)$$

The condition $\partial c / \partial x = 0$, in this case, yields

$$C_1' = C_2' + n C_2'' \qquad (10)$$

This criterion has to be satisfied to keep the cost of useful energy to a minimum value

3. RESULTS AND DISCUSSION

To illustrate the applicability of the developed analysis, a typical case of tobacco drying is considered (2). Total curing time for one loading is 5 days and the initial moisture content in the tobacco leaves is 2400 Kg. Moisture left on the fifth day is 50 Kg, thus yielding α =0.77. Initial energy requirements are 800 Kwh. The cost of energy per Kwh as a function of solar fraction is shown in Fig.1 for different collector prices. It is evident that beyond a certain fraction of solar energy, the cost of useful energy goes on increasing. Looking at the results of the analysis it is advisable to install collector area corresponding to the solar fraction for which the cost of energy is minimum. As the price of the solar energy collector goes down, the contribution of solar energy increases and very cheap collector is able to provide the total energy requirements at a

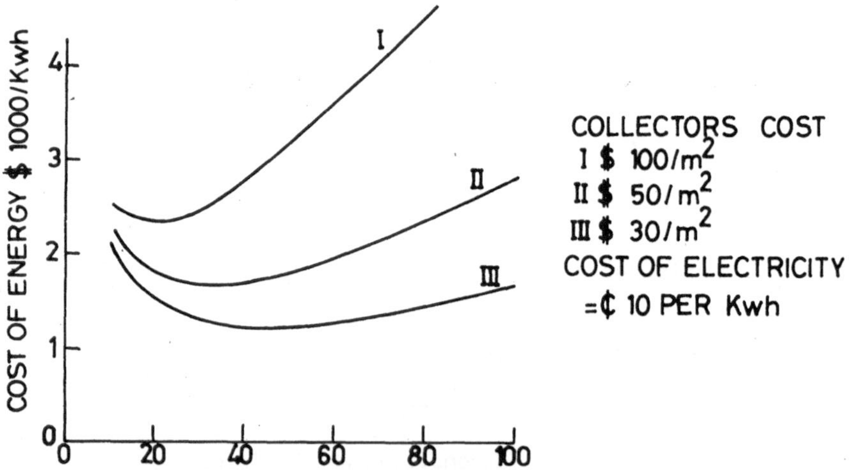

FIG. 1-COST OF ENERGY IN $/Kwh FOR DIFFERENT PERCENTAGE OF SOLAR CONTRIBUTION AS A FUNCTION OF COLLECTOR'S COST.

competitive price. Solar plastic collectors in this context can play a major role in popularising the use of solar energy for typical applications (3, 4).

For hot water heating system, we take up the case of 10,000 litres capacity system. To heat this water quantity from an initial temperature of 10°C to 50°C, 465 Kwh energy is required per day. The annual energy cost variation as a function of solar fraction is shown in Fig.2. Annual average solar energy incident on the collector has been taken to be around 1500 Kwh/m^2, a typical value for climates like southern region of the Federal Republic of Germany. Again the cost is seen to be minimum for a certain percentage of solar contribution.

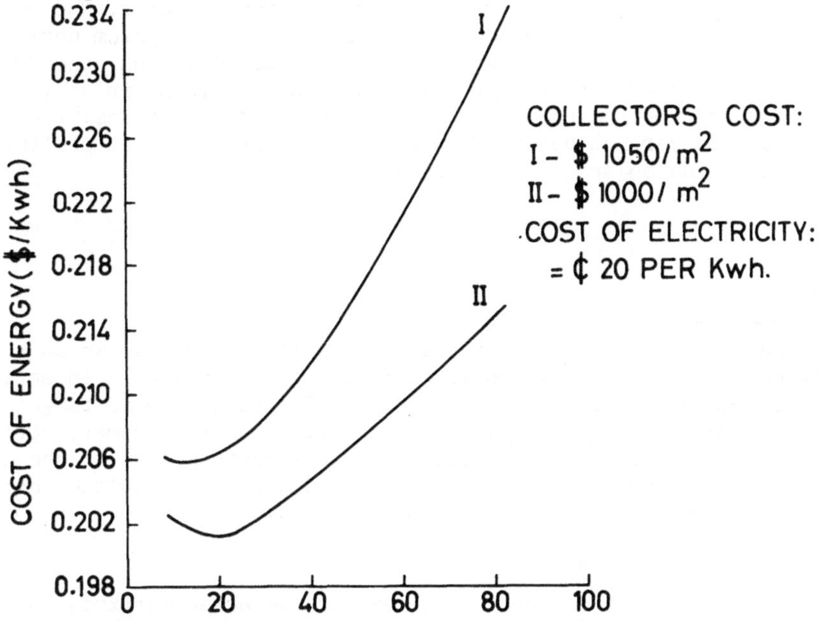

FIG. 2-PERCENTAGE OF SOLAR CONTRIBUTION.

REFERENCES

1. SODHA, M.S., AMAN DANG and BANSAL, N.K. (1984), A simple cost analysis for hybrid systems, Solar Energy, 32(4), 561.

2. BALAGOPAL, K., MURTHY, C.R.K., RAO, M.R. and THOMAS, A. (1981), System design for tobacco curing by utilisation of solar energy, Symposium on Solar Science and Technology, Bangkok, 25 Nov. - 4 Dec., 1980, Proceedings 2, 223.

3. BANSAL, N.K. and BOETTCHER, A. Efficiency and Economic Qualifications of solar flat plate collectors, 4th Internationales Sonnenforum, Berlin 2-6 Oct., F.R.G.

4. BANSAL, N.K., BOETTCHER, A and UHLEMANN, R. Thermal potential of solar air and water heaters made from different plastic materials, 4th Internationales Sonnenforum, Berlin 2-6 Oct., F.R.G.

INVESTIGATION OF INDOOR- AND OUTDOOR PERFORMANCE MEASUREMENTS ON SOLAR DOMESTIC HOT WATER SYSTEMS

D. Gilliaert, H. Hettinger, P. Rau, C. Roumengous, P. Tebaldi
Commission of the European Communities
Joint Research Centre - Ispra Establishment
21020 Ispra (Va) - Italy

Summary

For the characterisation of the performance of solar domestic hot water systems, two different methods are investigated. One method is to perform the tests outdoor and to determine the system performance over a period covering different weather conditions. The other method is to perform the tests indoor, using a solar simulator in a climatic chamber under standard meteorological and load conditions. The investigation is carried out on four different systems. First preliminary results, which are very informative, are given.

INTRODUCTION

Based on a previous assessment on solar collectors, which resulted in a recommended test-procedure for efficiency measurements, an action has been set up at the JRC for SDHW systems. The first aim is to get insight in the behaviour of the systems and to determine the primary performance factors. This should result, later on, in a contribution to the development of recommended test procedures for SDHWs, which could be enhanced by an international cooperation on this subject.

TEST FACILITIES AT THE JRC

Four different commercial systems are selected for this investigation:
- a pumped direct system;
- three pumped indirect systems; heat exchanger in the tank, outside the tank and a wrap-around heat exchanger.

All systems have a 200 l storage tank, 3 m^2 of collector surface and have an electrical heater in the upper half of the tank. One copy of each system is mounted outdoor, after measurements carried out on the components. A second copy is installed indoor, to investigate the indoor test methods.

INVESTIGATION PROCEDURE

The investigation intends, at a first stage, to get insight in the behaviour of a SDHW system, to determine the main energy flows through the system and their sensibility to the ambient conditions. The systems mounted outdoor have the collector oriented to the south and inclined at 45 degrees. The storage tank and the control unit are installed in an environment at a constant ambient temperature of 10°C. The system is subjected to a specified diurnal draw-off cycle. This consumption is specified in units of heat. This approach eases a comparison between systems, as the thermostat setpoint of the auxiliary energy supply becomes less important.

The system is instrumented in a way that all energy flows can be considered separately. The storage tank is considered as a central unit to which all energy flows are measured as an in- or output. Following energy contributions are measured (see also Fig.1):

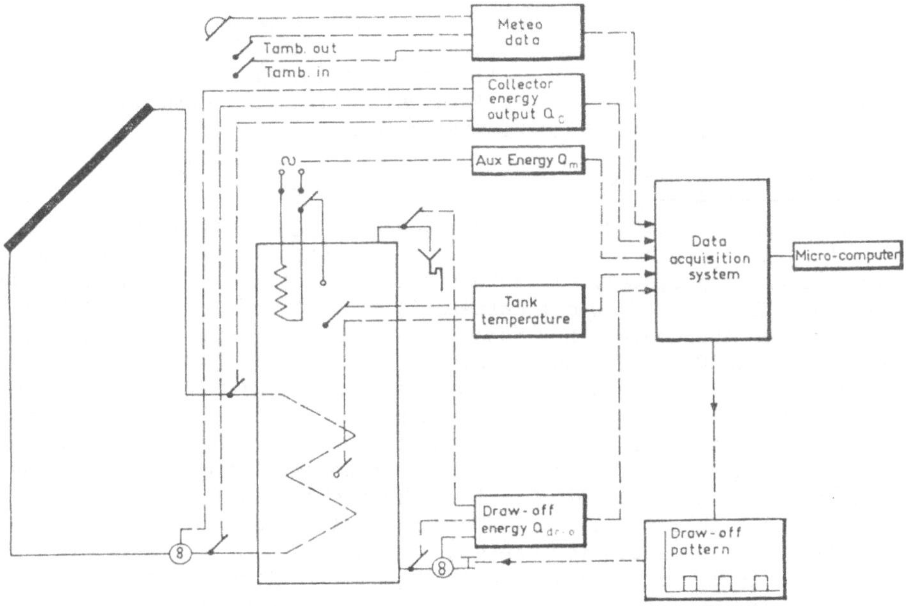

Fig. 1 : Investigation procedure SDHW systems

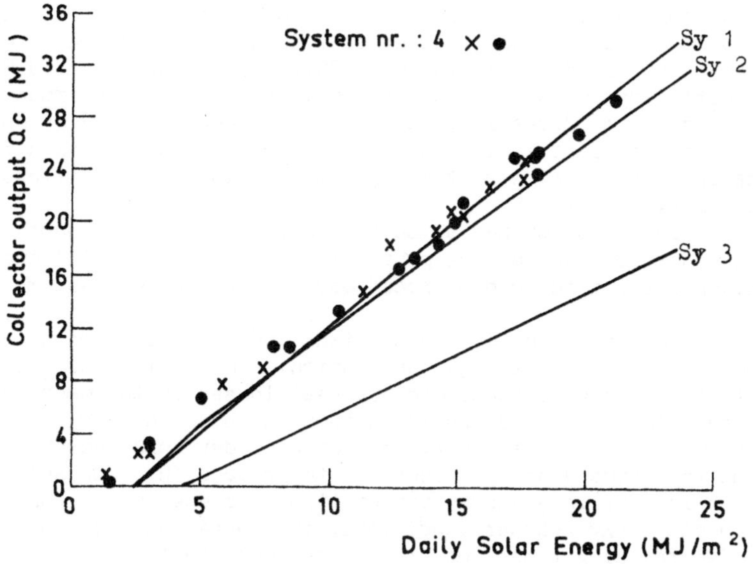

Fig. 2

- collector input energy to the tank (T1p, T2p and flow mp);
- auxiliary heating energy (here electricity Qaux);
- draw-off system energy (T1d, T2d and flow md).

These energies are calculated, based on the measurements of two absolute temperatures and the appropriate flowrate. The data are taken at a 30 sec interval by a microcomputer controlled data acquisition system and are integrated and stored on a 30 minute interval. All temperature sensors and flowmeters are calibrated before installation.

EXPERIMENTS - SYSTEM TESTS

The systems mounted outdoor have been working for more than three months, the first month (December 1983) as a preheating system and two months as a heating system. The setpoint temperature for the auxiliary heater is about 50°C. The draw-off pattern is choosen arbitrarily, as there is no consensus on this point, but reasonably there are daily three draw-offs of 10500 kJ at 9h30, 13h30 and 18h30.

The performance of a SDHW system is primarily influenced by the performance of the solar subsystem (collector energy input to the tank). This performance is also determined by system dependent factors such as the collector characteristics, the solar circuit control strategy, the storage tank stratification and eventually the heat exchanger. The solar performance is set out as a function of the daily solar radiation (in the collector plane). The results for one system are represented in Fig.2.

The results are remarkable and informative. Although one can distinguish important absolute differences between the different systems, all figures have a similar form. There is a threshold value of the solar daily radiation to start the system. Above this value increases the collected heat linearly with the daily solar radiation. There are two ambient parameters included, the outdoor ambient temperature and the length of the solar day. A higher ambient temperature increases the collector performance, as there are less collector heat losses. A longer solar day results, for the same daily radiation, in a longer period that the solar system is active (more heat losses). This explains the decrease in collector performance. The average solar performance of the four systems is set out in Fig.2 and is compared with the weather data in Europe, for the meteo-stations of Ispra (45 N), Brussels (Ukkel) (50.8 N), and Hamburg (53 N). There is a large difference between Brussels and Ispra, as a large amount of the yearly solar energy at Brussels is available between 10-15 MJ/m^2.day, while at Ispra the greater part is available at 20-25 MJ/m^2.day. This affirms the importance of the choice of the right meteorological conditions for short time tests.

The solar contribution has to be considered as one component in the whole energy contribution for the production of warm water. All energy values are given for a clear and overcast day for the tested systems in Table 1.

The collector energy output to the tank does not correspond to the reduction of the auxiliary energy consumption. A part of this collector output is lost in the tank as supplementary heat losses in the lower part of the tank. The solar circuit circulation pump has also an energy consumption which has to be taken into account. One can consider this energy or take into account the primary energy necessary to produce this mechanical energy. The reduction in auxiliary energy consumption is presented in Table 2.

Even in clear sky weather conditions, the energy savings vary from 26 to 58%. Generalising these results, it can be said that even for a good system the tank heat losses (7000 - 8000 kJ/day) are 20 to 25% of the useful draw-off energy and that on a clear day 15 - 25% of the collector output energy is lost in supplementary tank heat losses and pump energy consumption.

TABLE 1 - Energy flows in SDHW systems for a clear and an overcast day

	Overcast day (25/26/27-2)			Clear day (9/10-2) 19150 kJ/m^2.day Tout:9.4 C(9-17h)			
	Aux.	Dr.$^{-o}$	THL	Col.	Aux.	Dr.$^{-o}$	THL
	units: kJ/day			units: kJ/day			
SY 1	40450	31650	8800	23850	20700	31850	12700
SY 2	38850	31850	7000	22000	19250	31750	9500
SY 3	38050	31600	6450	14500	25800	31900	8400
SY 4	39700	31400	8300	27200	14860	31700	10000

Aux: electrical auxiliary energy THL: tank heat losses T:10°C
Dr.$^{-o}$: draw-off energy Col: collector output energy

TABLE 2 - Auxiliary energy savings for the four SDHW systems

	Overcast day Aux. en.	Clear day$^+$ Aux. en.	Pump energy	Energy saving	in %
	kJ/day	kJ/day	kJ/day	kJ/day	
SY 1	40450	20700	1150 (3450)*	18600 (16300)	46 (40)
SY 2	38850	19250	1500 (4500)*	18100 (15100)	46.5(39)
SY 3	38050	25800	2400 (7200)*	9850 (5050)	26 (13)
SY 4	39700	14860	1500 (4500)*	23340 (20340)	58 (51)

+ Daily solar radiation: 19150 kJ/m^2.day Tout: 9.4^{o}C
* Primary energy to produce the mechanical energy (* 3)

INDOOR MEASUREMENTS

The indoor test conditions have to be chosen in such a manner that they will not favour a certain system nor produce optimistic results which cannot be met in normal operation. The test method consists in measuring the daily performance of a SDHW system under prescribed meteorological and load conditions until the steady-state periodic one-day performance is achieved. The advantage of such an indoor standard method is that both consumer and manufacturer can make comparisons of different systems under identical experimental conditions. As a starting point for choosing the test conditions, the ASHRAE 95-81 standard was considered. The systems to be tested are installed according to the prescriptions given by the manufacturer in the Solar Simulator LS-1 of the JRC Ispra. The standard test conditions are:

- ambient temperature $T_a = 20 \pm 1^oC$
- wind speed $u_s = 4 \pm 1$ m sec^{-1}
- tank inlet temp. $T_{in} = 15 \pm 1^oC$

The simulated solar radiation during a test day is 16800 kJ/m^2 to represent a summer day and 10400 kJ/m^2 to represent a winter day.

The standard daily test load Q_{DL} = 43200 hJ is taken in three draws at 9h, 13h and 17h. Each draw is stopped when 14400 kJ are taken from the tank or if the delivery water temperature falls below 25°C (solar only) or 35°C (solar + supplemental heating).

The test day is repeated until convergence or until the fifth day, which ever occurs first. The indoor experiments started in March 1984, therefore the result for only one system can be reported.

System 4; collector aperture surface: 3.05 m^2; units kJ

	Solar only summer day	Solar only winter day	Solar + suppl. summer day	Solar + suppl. winter day
Daily solar radiation	51240	31720	51240	31720
Solar energy to tank	28790	19620	26937	17287
Draw-off energy	29310	19910	38001	37391
Auxiliary energy	-	-	13608	23598
Parasitic energy (pump, ...)	2300	1730	2300	1730
Standard test load	43200	43200	43200	43200

CONCLUSIONS

Although a solar domestic hot water system is subjected to a large variety of weather conditions, one can establish a univocal dependence on these parameters. Each attempt to characterise a SDHW in a short time (cfr. indoor test) has to use "representative" days for the solar irradiation and the ambient temperature.

Outdoor testing enables at least to compare systems. The collector output to the storage tank as a function of the daily solar radiation seems to be a valid basis, but completed with data of the supplementary tank heat losses and the pump energy consumption.

The indoor test results, available for one system, reflect the outdoor behaviour, if they are corrected for the ambient conditions.

The overall precision of the results depend on the number of measurements taken into account, on the frequency of measuring and the precision of the instruments. This means in practice, that all energy flows have to be measured.

A micro-computer controlled data acquisition system is prefered over heatmeters because these latter ones are not adapted for this application due to too small temperature differences in the collector circuit.

DESIGN OF STATIC CONCENTRATORS WITH THE RECEIVER IMMERSED IN A DIELECTRIC TUBE

J.C. Miñano
Instituto Energía Solar
E.T.S.I. Telecomunicación. Ciudad Universitaria
Madrid-3. Spain

Summary

The immersion of the receiver of a linear static concentrator in a dielectric tube is proposed. The use of such medium increases n times the power reaching the receiver (n being the index of refraction of the dielectric). It has been commonly avoided because of the excesive volume of dielectric needed in conventional static concentrators (CPC type). The proposed structure reduces up to the 98% of this volume in the case of a tubular receiver. Another concentrator with a bifacial flat receiver is suggested for photovoltaics.

1. INTRODUCTION

Static concentrators are an intermediate step between sun tracking concentrators and flat static plates. Their interest lie in the fact that they do not require any tracking or tilt adjustement for obtaining higher irradiances on the receiver than the ones obtained by conventional flat plates.

Non-imaging optics permits the design of simple optical elements which approach the maximum physically allowable concentration ratio for a given range of incidence angles (1). For instance, this maximum concentration ratio states, in terms of energy and for the yearly averaged local conditions of Madrid that, if a concentrator collects 85% of the energy reaching its entry aperture, the irradiance on the receiver can be increased (by means of a concentrator) by a factor of 1.9. Lower fractions of the collected energy lead to higher possible irradiances on the receiver (2).

Static concentrators for thermal and photovoltaic applications designed according to nonimaging optic techniques have been developed succesfully (3)(4)(5). In particular, the Integrated Stationary Evacuated Concentrating (ISEC) collector tube which is a static concentrator for thermal applications, can operate at efficiencies comparable to the ones of tracking collectors for temperatures down to 200°C (3).

From the beginning of the developement of nonimaging concentrators it is well known that filling the concentrator with a transparent dielectric medium increases the concentration without varying its "angular field of view", ie. the irradiance on the receiver increases, without varying the fraction of energy collected from that reaching the entry aperture (1)(2). The increase can be a factor of n^2, being n the index of refraction of the dielectric medium. Nonimaging concentrators using dielectric medium are well known (1) but they use a large amount of dielectric material which rises the cost of the concentrator. The use of such medium is only justified in thermal applications of concentration if its cost is small. In photovoltaics, where the cost of the cell allows for using costly concentrators, the reduction of dielectric volume is also desired.

Another factor is involved in the use of dielectrics in thermal applications: the Stephan-Boltzmann constant is proportional to n^2, and consequently, the radiative losses of the receiver increase by a factor of n^2 due to the use of the dielectric (6).

2. DESIGN PROCEDURE

The design procedure is fully described in ref. (7) and its is based in a general method of design of two dimensional nonimaging concentrators given by Winston and Welford (8). The use of differential equations in reference (7) has been avoided by a proper application of Fermat's principle.

The concentrators obtained are optimal and ideal with respect to the rays contained in a plane transversal to the axis of translational symmetry, i.e.: a) any ray hitting the entry aperture within the acceptance angle reaches the receiver (the concentrator is ideal) and b) any reverse ray leaving the receiver traverses the entry aperture within the acceptance angle (the concentrator is optimal). The concentrator, in general, is neither ideal, nor optimal with respect the remaining rays.

3. RESULTS

The mirror profile has in general a part which is inside the dielectric tube and another part which is outside this tube. Two concentrators appear specially interesting: the first one is designed for a bifacial flat receiver and has all the mirror profile inside the dielectric tube (see Fig. 1). Its interest is for photovoltaic applications where bifacial cells are the

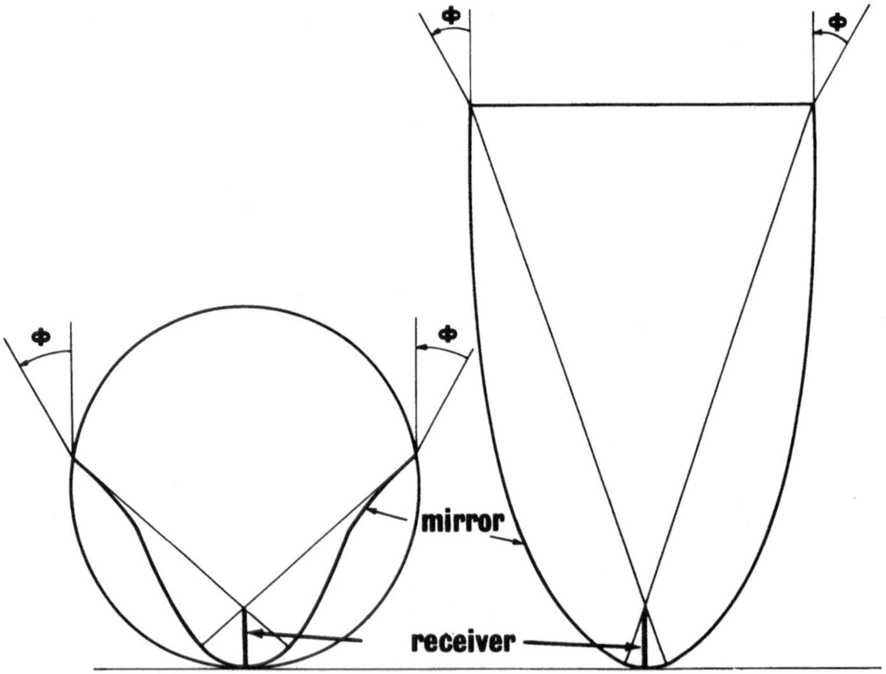

Fig. 1. Concentrator with bifacial flat receiver designed to be inside the dielectric tube (left) compared to an equivalent CPC (right). The angular acceptance (2 ϕ) is 60° and the index of refraction of the dielectric 1.5.

most appropriate for static concentration (9)(10). The reduction of dielectric volume and size of the concentrator with respect to a similar conventional CPC is shown in Fig. 1. The receiver of the second concentrator is tubular and all the mirror is outside the dielectric tube. The interest of this concentrator is in thermal applications. The increase of concentration with respect to a similar concentrator which does not use dielectric is n (see Fig. 2). The reduction of dielectric volume with respect a similar conventional CPC can be verified looking at Fig. 3 (in the cases shown in Figures 2 and 3 this reduction is the 98%). The radius of the dielectric tube is n times that of the receiver.

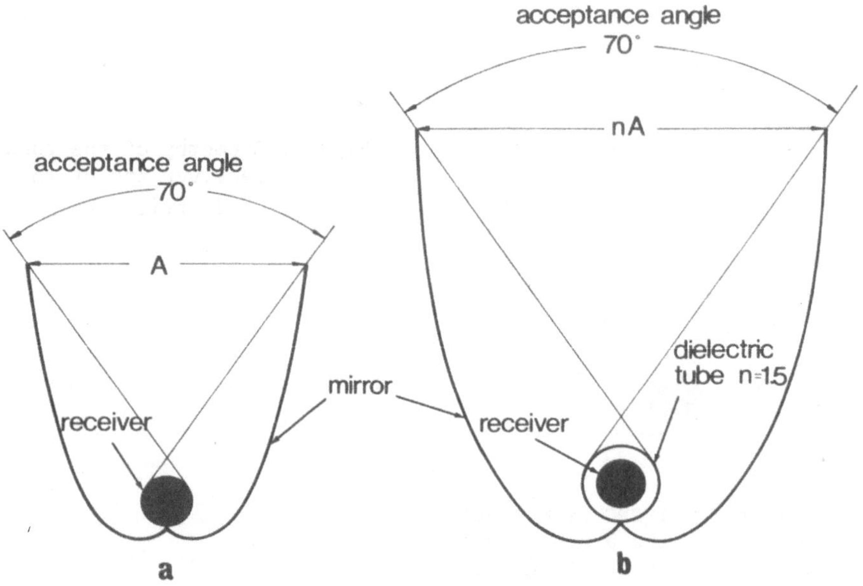

Fig. 2. Two concentrators with tubular receiver.The one at the right uses a dielectric tube surrounding the collector allowing it to have a greater concentration than the one at the left.

To truncate these concentrators by eliminating the higher parts of the mirror is a practical tool commonly used. This does not vary appreciably the optical characteristics of the concentrator(11,12). If we truncate the concentrators of Fig. 2 at the same height, the one of Fig. 2b has a greater concentration.

Let us compare two linear concentrators: one having the concentrator of Fig. 2a as cross-section and the other one having Fig. 2b as cross section. The energy reaching the receiver of the concentrator of Fig. 2b is n times the energy reaching the receiver of Fig. 2a. Nevertheless, the receiver of Fig. 2b radiates n^2 times more than the one of Fig. 2a at the same temperature, due to the aforementioned dependence of the Stephan-Boltzmann constant on n^2. (For the sake of simplicity, we assume that the index refraction of the dielectric is constant in the infrared and visible regions of the spectrum). When we consider all the rays in the concentrator, and not only the ones contained in a transverse plane, we find that an important amount of the "reverse" rays leaving the receiver suffer a total internal reflection at the dielectric tube surface and come back to the receiver (Fig. 4). If we assume that the receiver radiates isotropically only 1/n of

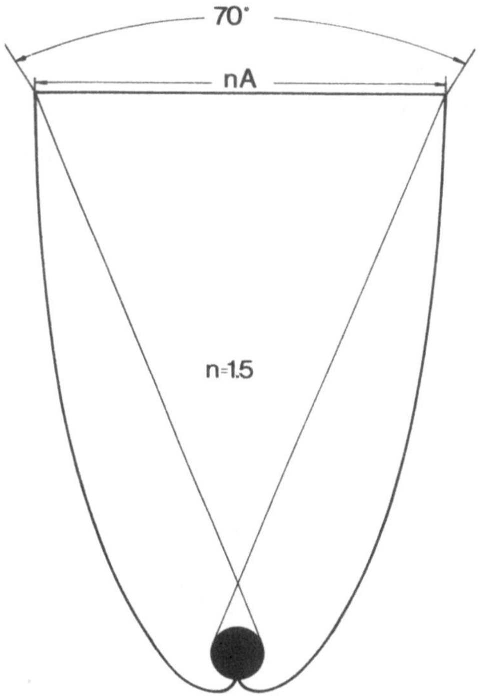

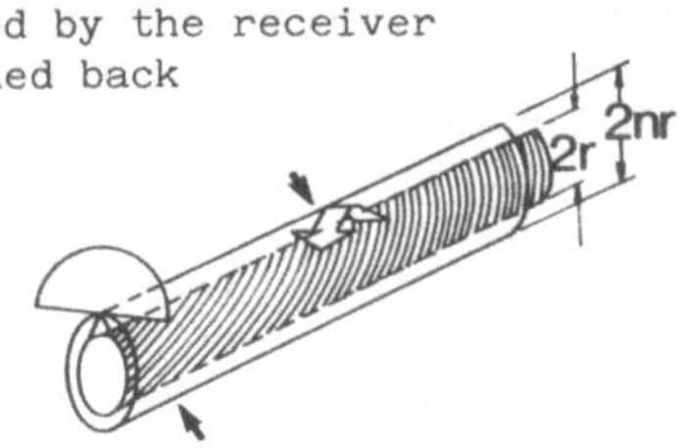

Fig. 4. Geometry of the tubular receiver and of the dielectric tube.

Fig. 3. Conventional CPC with tubular receiver using a dielectric of index of refraction 1.5. The concentration is the same as the concentrator of Fig. 2b but it uses much more dielectric medium.

the power radiated by the receiver can leave the dielectric tube (7). Then, the concentrator of Fig. 2b radiates n times more than the one of Fig. 2a, and so, the difference between the power reaching the receiver and the power radiated by it is n times greater for the case of Fig. 2b. This means, for $n = 1.5$, that the power on the receiver of Fig. 2b is 50% more than that on the receiver of Fig. 2a.

The interest of the concentrator of Fig. 2b lies in the small amount of dielectric volume used. A series of concentrators using increasingly volume of dielectric can be designed ranging from the one of Fig. 2b to the one of Fig. 3. The increase of dielectric does not vary the optical characteristic with respect to the transverse rays, but increases the angular acceptance for rays not contained in a traverse plane.

The receiver surrounded by the dielectric tube can be considered by itself as a concentrator, which constitutes an enhacement with respect to the receiver alone. Another feature of such structure (dielectric tube + receiver) is that if the dielectric is opaque to infrared radiation and has a low thermal conductivity it would make the receiver to behave as if it has a selective surface.

CONCLUSION

The utilization of a dielectric medium in static concentrators for thermal applications (which has been generally avoided because of its high cost) is suggested using a structure such as the one shown in Fig. 2b. This concentrator has a small amount of dielectric, and achieves most of the advantages that its use allows (fundamentally to increase n times the power on the receiver).

REFERENCES

(1) W.T. Welford, R. Winston. The Optics of Nonimaging Concentrators. Academic. New York (1978).

(2) J.C. Miñano, A. Luque. Limit of concentration under non-homogeneous extended light sources. Applied Optics 22 (1983) 2751.

(3) J.J. O'Gallagher, K. Snail, R. Winston, C. Peek, J.D. Garrison. A new evacuated CPC collector tube. Solar Energy 29 (1982), 575.

(4) J.D. Garrison, R. Winston, J.O'Gallagher, G. Ford. "Two-fixed evacuated, glass, solar collectors using nonimaging concentration". International Conference on Nonimaging Concentrators. Mitchell C. Ruda, Editor. Proc. SPIE 441, p 97 (1984).

(5) A. Luque, J. Eguren, J.M. Ruiz, A. Cuevas, J. del Alamo, J.M. Gómez, M. Acuña, G. Sala. New concepts for static concentration of direct and diffuse radiation. Third E.C. Photovoltaic Solar Energy Conference. Cannes (France) 1980. W. Palz, Editor. p 396.

(6) A. Rabl. Comparison of solar concentrators. Solar Energy 18 (1976), 93.

(7) J.C. Miñano, J.M. Ruiz, A. Luque. Design of optimal and ideal 2-D concentrators with the collector immersed in a dielectric tube. Applied Optics 22 (1983), 3960.

(8) R. Winston, W.T. Welford. Two dimensional concentrators for inhomogeneous media. J. Opt. Soc. Am. 68 (1978), 289.

(9) A. Luque. Theoretical bases of photovoltaic concentrators for extended light sources. Solar Cells, 3 (1981), 355.

(10) A. Luque. "Bifacial solar cells" in "Silicon Processing for Photovoltaics" edited by Khattak/Ravi (in press).

(11) A. Rabl. Optical and thermal properties of Compound Parabolic Concentrators. Solar Energy 18 (1976), 497.

(12) J.C. Miñano. Directional intercept factor of truncated CPC's. Applied Optics 22 (1983) 2747.

COMPUTER-AIDED DEVELOPMENT OF SOLAR DOMESTIC HOT WATER SYSTEMS

S. MELSON and K. ELLEHAUGE
Thermal Insulation Laboratory, Technical University of Denmark.

Summary

The results from the methodical work with small solar domestic hot water systems which has been carried out at the Thermal Insulation Laboratory are presented.

The poster with the title "Solar Domestic Hot Water Systems in Denmark" deals with the design of the systems and the performance and cost. This poster deals with the methodical work og performance monitoring and computational treatment of results which has been carried out to develop and evaluate the systems.

The work has been carried out as a systematic three-step process of:

- systems design and construction
- performance monitoring
- computational treatment of measurements.

This process has in 1980 led to construction of the best performing DHW-system reported at the Performance Monitoring Group at the CEC Solar Energy R&D Programme (the "Lyngby-system").

The systems have been carefully monitored and data have consisted of hourly averages for energy flow, temperature, solar irradiation etc. The computations are carried out as computer-simulations of the systems using models developed at the laboratory. The models are validated using measured data. After validation series of parametric variations provided information of the effect of changing components, DHW-consumption, solar irradiation and other system-parameters.

The results of this work during the last 3 years as well as its influence on system designs and monitored performance of consecutive systems are presented in this paper.

1. INTRODUCTION

The economic ressurces for the Danish solar R&D programme have been decreasing since the disappointing results of the first-generation solar DHW-systems. For this reason there has been constructed only a few second-generation systems at the Thermal Insulation Laboratory. The emphasis has been laid on very careful monitoring - often more than one year - as well as computational treatment of the measurements.

The detailed monitoring is essential for the following detailed computations. Though most of the model simulations have been carried out by local developed models, not by far as powerful as the EMGP-2 model, the computations have provided a very detailed knowledge of designs as well as components. This knowledge has been used in the design phase of every new project. The result has been a series of very well performing small solar DHW-systems.

The design-monitoring-computational cycle is shown in the diagramme below

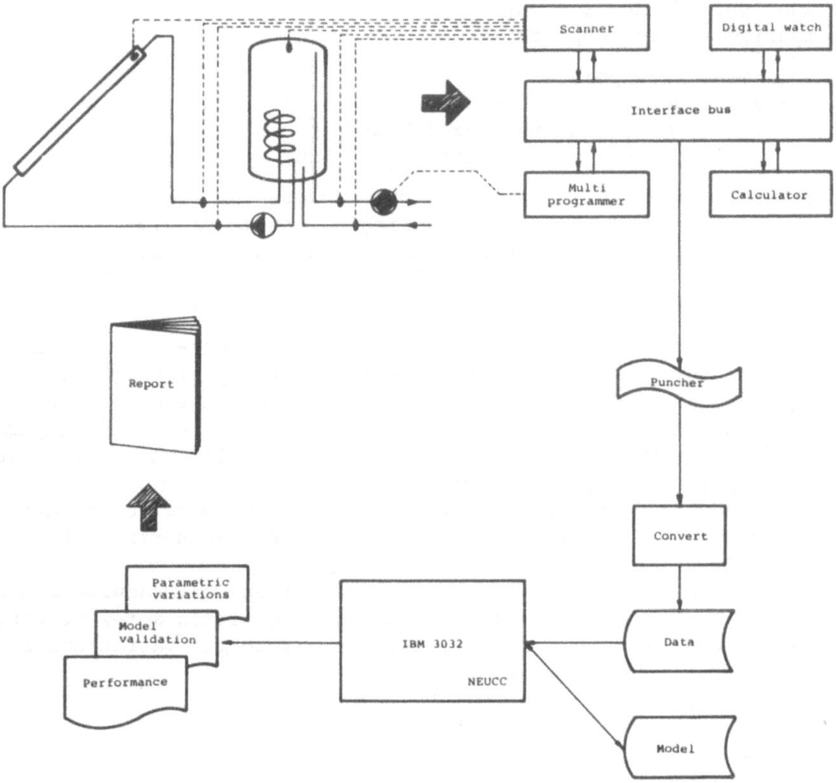

Figure I. The design-monitoring-computational cycle.

2. THE MONITORING SYSTEM

The monitoring system has for all recent systems been based on Hewlett-Packhard-equipments, as follows:

HP 9815A	Controller/calculator
HP-IB-59309A	Digital watch
HP-IB-98135A	Interface Bus
HP 5900 A	Multiprogrammer Interface
HP 6940 B	Multiprogrammer for pulsecounting (flow-measurements)
HP 3455A	Digital voltage-meter (temperature-,insolation- and wind-speed measurements)
HP 3495A	Scanner
FACIT 4070	Papertape Puncher

This measuring system has proved to be very stable and reliable. At the

"Gl. Holte" system for instance, it has produced valid measurements 362 full days during 1982, 24 hours a day.

The measuring system supports

- up to 64 measuring points
- up to 5 scannings each minute
- integration of measurements over 1 hour
- punching measurements on papertape every hour

The measurements are very accurate and furthermore the Laboratory has got detailed experience in setting up measuring points to provide useful, representative, knowledge of the proporties of the DHW-system.

3. COMPUTATIONS

The measurements are converted from papertape to direct-access-files at NEUCC, the computer Centre of the Technical University of Denmark.

Systems evaluation includes annual and monthly computation of tapped energy, heat losses from different components, average- and extreme temperatures and others. Furthermore: calculation of the overall systems efficiency, the DHW-comsumption, the solar fraction and so on.

The performance is compared to results predicted by model-computations. Discrepancies has lead to changes of design and/or components or to adjustment of the model.

The model validation is based on hourly comparisons of measured and predicted values. The models predictions are accurate within a few percent. Models have proved to be extremely useful for parametric variations as shown in a few examples following here:

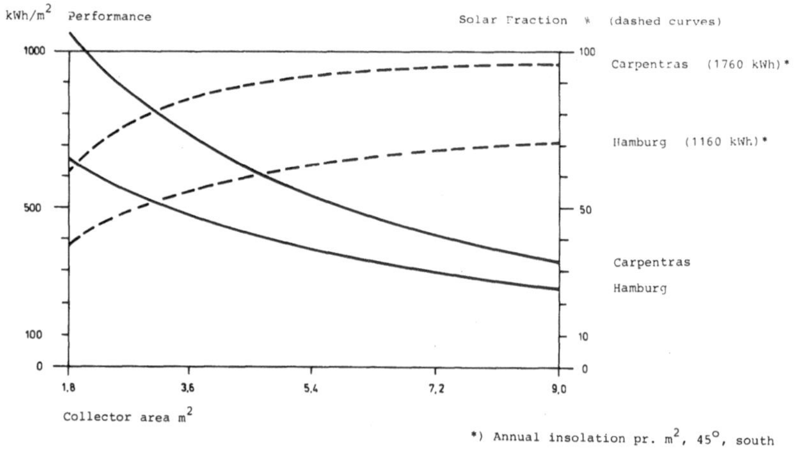

Fig. II. Performance and Solar Fraction of a Standard-design DHW system at various locations.

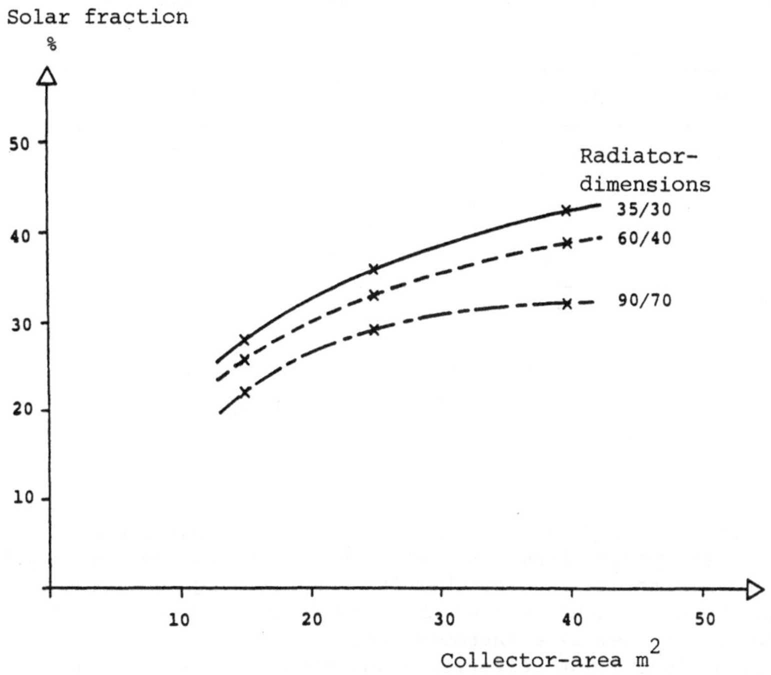

Figure III. Solar-fraction of heat for DHW depending on collector-area and radiatordimensions.

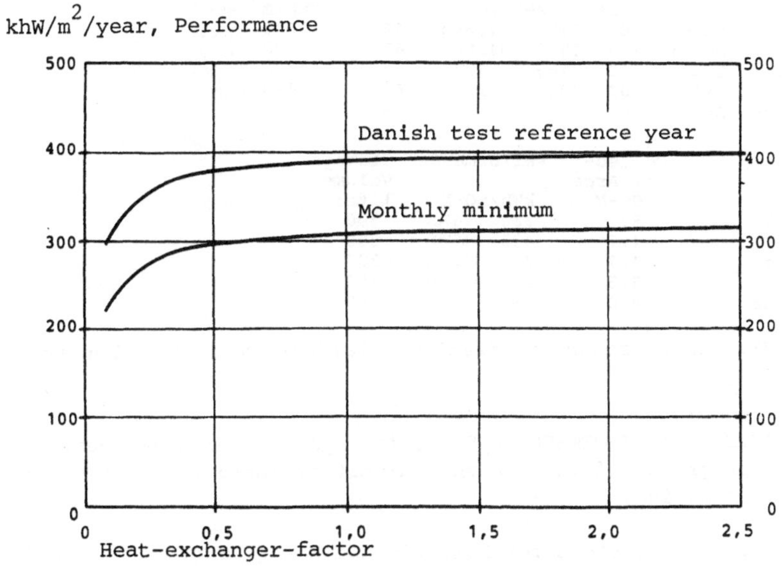

Figure IV. Performance depending on the effective area of the heat-exchanger expressed by a factor relating to a Danish standard DHW-system.

Storage volume	Storage Height	Performance	Solar fraction	Performance
ℓ	cm	kWh/year	%	kWh/m^2/year
150	123	1935	64,5	352
200	135	1950	65,0	355
250	135	1965	65,5	357
300	135	1980	66,0	360
150	175	1950	65,0	355

Hot Water Demand: 3000 kWh/year
Heat-loss coefficient storage: 2,0 W/$^\circ$C

Figure V. Performance depending on storage volume and -geometry.

4. SYSTEMS

Figure VI lists a few parameters for five small second-generation DHW-systems. The last system "Lundtofte" was constructed in the summer 1983 and has been monitored since october 1983. Monitoring continues until autumn 84. The performance during the cold season has proved satisfactory. The "Lundtofte"-system is a thermosyphon.

More informations about performance and cost is presented in the paper "Solar Domestic Hot Water Systems in Denmark. Design, Performance and Cost."

Systems		Hot Water Demand l/day	Hot Water Demand kWh/year		Solar Fraction %	Solar Irradiation kWh/m^2/year		Systems Efficiency %
Lyngby	Pilot	185	3100	(11200)	66	1000	(3600)	38
BV300	Pilot	200	3085	(11100)	67	1190	(4300)	37
GL-Holte	Demo	140	2185	(7900)	69	1100	(4000)	29
VV150	Pilot	200	3000	(1080)	66	1190	(4300)	30
Lundtofte	Demo	-	-	-	-	-		-

	Collector Area SQ-M	Performance kWh/SQ-M		Storage Volume liter
Lyngby	5.4	380	(1370)	300
BV300	4.7	435	(1570)	300
GL-Holte	4.7	320	(1150)	300
VV150	5.5	355	(1280)	150
Lundtofte	4.0	-	-	150

Figure VI. Small second-generation solar DHW-systems in Denmark.

REFERENCES

MIKKELSEN, SV.E. and JØRGENSEN, L.S.: Solar systems for Space Heating. The solar energy programme of the Danish Ministry of Energy, report no. 15, august 1981. In Danish.

ELLEHAUGE, K.: Solar Systems for Domestic Hot Water. Thermal Insulation Laboratory, Technical University of Denmark, report no. 123, november 1982. In English.

MELSON, S.: Solar Domestic Hot Water System in Gl-Holte. The solar energy programme of the Danish Ministry of Energy, report no. 24, january 1984. In Danish.

SOLAR DOMESTIC HOT WATER SYSTEMS IN DENMARK. DESIGN, PERFORMANCE AND COSTS.

K. ELLEHAUGE
Technical University of Denmark
Thermal Insulation Laboratory

Summary

In Denmark the last years of work with solar systems for domestic hot water has shown some good results. The work has consisted in monitoring of experimental projects and demonstration projects. The monitored data have been analysed with computer simulation programs. The designs, the performance and the costs of the systems are presented.

1. BACKGROUND

At the Thermal Insulation Laboratory projects with solar energy systems have been carried out since the middle of the 1970's. Some of these projects have been financed by the Danish Ministry of Energy. Others have been financed by the CEC.

This paper deals with the part of the work, which has been development of solar systems for domestic hot water. In this work the Thermal Insulation Laboratory has had sucessful results, which have been of interest outside Denmark. In Denmark the solar industry has had advantage of the work carried out, often in collaboration with the industry.

The basis of the work was made with some demonstration projects which were carefully monitored in the late 1970's. The performance of these systems were poor, but due to the detailed monitoring programme and a subsequent through analysis of the monitoring data it was however, possible to draw detailed conclusions as to why the systems did not perform as expected.

2. DESIGN

On the basis of this it was decided to construct some new and improved "2. generation" systems. The work with these systems has been carried out since 1980 and has shown some very good results. The systems have been simple and have had thermal performances of 320-400 kWh/m^2 solar collecter per year with solar fractions of 60-70%. All the systems have been designed for delivering of hot water for a single family dwelling i.e. 150-200 litres hot water per day.

The first system, the BV300 system, was reported in the publication from the Performance Monitoring Group, called "Solar Water Heating, the analysis of design and performance data from 28 systems", Dec. 1981 (the "Lyngby" -system). In this report the system had the best combination of annual performance and solar fraction. The BV300 system consist of a 5,4 m^2 solar collector and at storage unit of 300 litres.

As part of the BV300 project a detailed computer simulation model of the system has been validated using the measured weather data and it has been possible to predict the monthly and yearly performance of the system very accurately. Using the validated model it has been possible to make parameter variations to find the optimal design of the system and of new systems. This working methodology is described more detailed in the paper entitled "Computer-aided development of solar domestic hot water systems".

The BV300 experimentel was followed by a demonstration project on a single family dwelling in Gl. Holte, a suburb to Copenhagen. The Gl. Holte system had 4,8 m^2 solar collecter and a storage unit of 300 litres. It was monitored all through 1982, and the performance was as expected, i.e. 320 kWh/m^2 per year with a solar faction of 69%. The performance was not as high as for the BV300 system, because of a smaller hot water consumption.

At the laboratory the BV300 system was followed by 3 experimental systems one of these was constructed with selfcirculation in the collecter circuit. After these experiments a demonstration project was started in the summer of 1983. The system had selfcirculation in the collecter circuit. It had a collector area of 4 m^2 and a storage of 150 litre. The exchanging of heat from the collector occurs through the mantle of the storage. The system is being monitored for a year and the preliminary results indicate that the system will perform as good as the other systems, i.e. an annual performence of 350-400 kWh/m^2 and a solar fraction of 60-70%.

Through the work with the systems it was found that the following criteria were important for a good result.

- System configuration is simple and reliable.
- Collector size is dimensioned for the actual hot water consuption.
- Solar collector has a high efficieney.
- Only small heat losses from storage tank and piping. All cold bridges to the storage are in the bottom.
- No selfcirculation of heat from the storage into the connected piping.
- Good temperature stratification in the storage tank.
- On efficient control system.

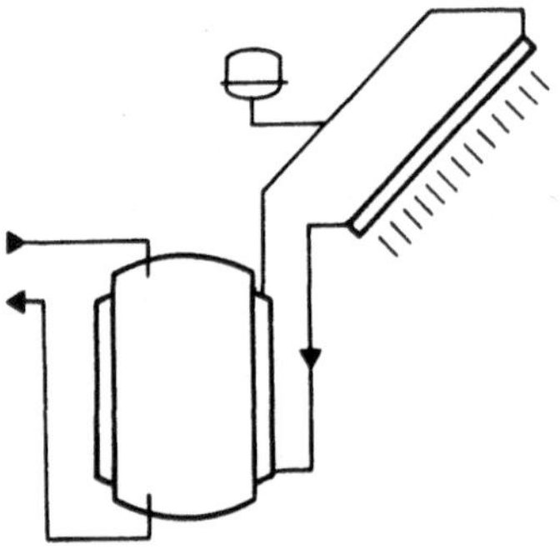

Pump system

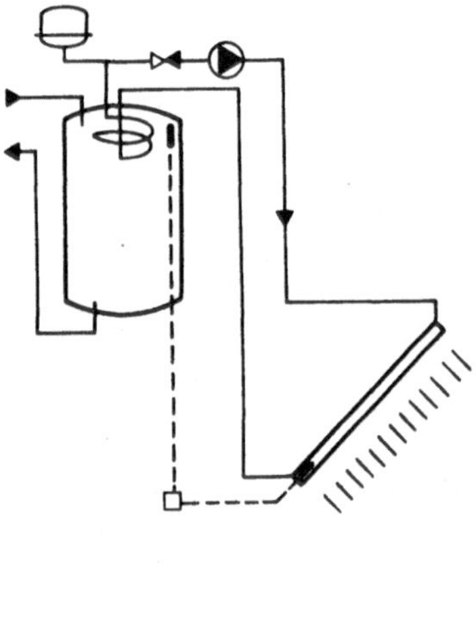

Selfcirculating

3. PERFORMANCE

It is found that if the above mentioned criteria are followed and if a collector is used with an effectivety as shown in the diagram then the performance is mainly found from the hot water consumption and the collector area. The size of the storage is not important if it is held within reasonable limits.

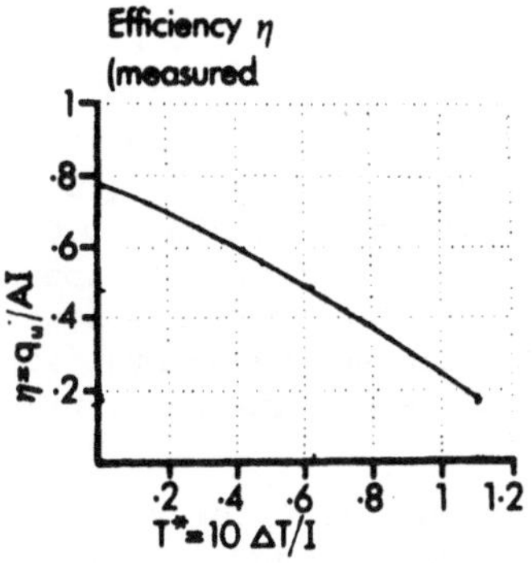

The performances can then be found from the following diagrams.

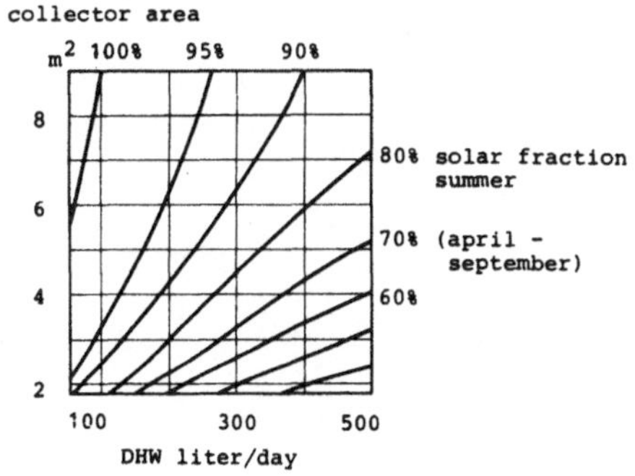

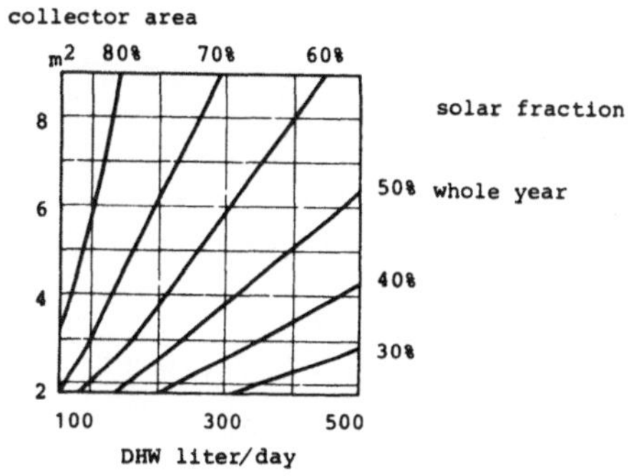

It is often advantageous to dimension the systems to deliver 90-95% of the hot water consumption during the summer months. In this way the auxiliary energy source may be turned off in the period where there is no heating demand for room heating. In Denmark this period is often 4½ months in the summer.

Example

Daily hot water consumption 200 litre/day heated from $8^{o}C$ to $45^{o}C$, solar collecter area $5m^2$, solar fraction in summer months 93%, yearly solar fraction 66%. Yearly performance: (45-8)•200•365•1,163•0,66 = 2050 kWh/m^2 or 410 kWh/m^2 solar collecter.

In a house with an existing oil burner the energy savings will often be twice the performance of the solar system due to saved no load losses from the oil burner in summer.

For example:

Delivered from solar system	2000 kWh
Saved no load loss from oil burner	2000 kWh
Energy savings	4000 kWh
or 450 litre oil	

4. COSTS

The cost of 450 litre oil is in Denmark 1500 D.Crs. or 140 ECU. The first demonstration project in Gl. Holte had a total price of 26000 D.Crs. or 3200 ECU while the latest demonstration project with selfcirculation had a price of 21000 D.Crs. or 2600 ECU, but some manufacturers are now selling similar systems at even lower prices. The Danish govermental support of solar systems are now 30%, so the simple pay-back time for a private person will often be less than 10 years.

There is no doubt that this price could be brought to a lower level, if the manufacturers could bring about a higher production and when installed in new houses, but it also seems that there still is work to be done with respect to using cheaper materials in the systems and with respect to rationalizing the production and installation.

LOW-COST, LOW-WEIGHT AND MEDIUM-EFFICIENCY SOLAR COLLECTORS, CONSTRUCTION AND EXAMPLES

Svend Erik Mikkelsen
Thermal Insulation Laboratory
Technical University of Denmark

Summary

In the Danish Solar Energy Programme 22 solar collectors from 1980 were tested for reliability and durability. The result was that more than half of them was not acceptable. The most important was that they were not rainproof and had outgassing on the cover, mostly from insulation foam. Most of them was complicated heavy construction with glass as cover. On this background we tried to show new constructions which should be rainproof and without the other problems we identified. They should also be light-weight and low-cost collectors. Some of the designs have been produced and tested and have shown good results. The conclusion is that it is possible to construct efficient collectors with a weight about 10 kg/m^2, that these collectors seem to have a good durability and the production price can be low.

1. TEST OF RELIABILITY AND DURABILITY OF SOLAR COLLECTORS

The background for this project was a project within the Danish Solar Energy Programme where 22 solar collectors were tested on the laboratory's equipment for indoor test of the reliability and durability of solar collectors. The collectors represented the Danish market in 1980.

The following tests were made.

Air-leakage
Stagnation test
Temperature shock of absorber and cover
Thermal Cycling
Rain-leakage with and without simulated wind load on cover.

After the indoor test the collectors were placed outdoors unconnected in stagnation for 1½ year, in order to carry out a kind of accellerated test and to verify the indoor test.

The result showed that the indoor test was usable and especially the stagnation test and the rain-leakage test gave good information on the reliability and durability of solar collectors.

40% of the collectors passed the indoor test without remarks. The same collectors did well in the outdoor test.

Another 40% of the collectors showed serious problems in the indoor test and for these coolectors the expected operation time with a satisfactory output is not more than a few years. Many of them showed widespread corrosion after one year outdoors. The last 20% of the collectors showed problems which could be solved easily.

The most serious problems were outgassing on the cover from plastfoam (polyurethane) and rain-leakage.

The outgassing results in a lower transmission and after some time in a degradation of the insulation material. For some collectors the outgassing became so severe that the absorberplate could not be seen. This happens when dust efter some time is combined with the outgassing.

In the climate of Nothern Europe it is very important that collectors are raintight and that there is suitable ventilation in the collector so that moisture accumulating during nighttime can be ventilated out of the collector during daytime. Otherwise the absorber plate will corrode, as it was seen with many collectors in this test.

The tested collectors belong to the 1st generation of collectors and the test pointed out some serious problems. On the other hand the test also proved that some collectors were satisfactory.

2. CONSTRUCTION OF SOLAR COLLECTORS

After these disappointing results we tried to construct collectors which first of all dit not show the problems we found in the tests.

Only liquid-based flat-plate collectors were considered and for several reasons it is important that the principle is simple, and that the collector can be produced in small factories.

Glass is almost always used as cover for solar collectors because it is cheap and common. But the use of glass involves several problems regarding the construction. It is heavy and fragile, and therefore a rigid frame is needed, which makes the solar collector even heavier. A usual weight is from 20 to 35 kg per sq. m of collector. More important is that the connection between the glass and the frame is difficult to make rainproof for 15 years or more, and that it is almost impossible to dismantle the glass for repear or replacement on top of the building.

Therefore plastic films have been used as cover. These may have a shorter lifetime than the glass, it is therefore important that the film can be replaced easily. Today plastic films are often more expensive than glass, but new and cheaper films can be expected in the future.

When the collector weight goes down (per sq. m), the size of the collectors can be increased without making it impossible to handle the collector. This will make installation easier.

3. EXAMPLES

In the following examples only the important details are shown. One problem is to keep the plast film tight. In some cases the film will keep tight if it is mounted tight while the box is slightly deformed. Another solution is the use of elastic mounting like example F and G.

In example G the collector is roofing as well. It will be useful if the piping in this example leads to the rear of the collector.

The good durability and reliability of these collectors comes from the combination of the rainproof construction and ventilation through the insulation, which ensures a good dry climate in the collector. This is from the tests referred to essentiel for the durability.

The efficiency of these collectors is the same or slightly better than for a conventional collector with selective surface and single glazing

4. REFERENCES

1. PEDERSEN, P.V. and MIKKELSEN, S.E. (1983). Reliability and Durability of Solar Collectors - evaluation of 22 solar collectors based on tests. Thermal Insulation Laboratory, Lyngby (in Danish).
2. PEDERSEN, P.V. (1982) Reliability and Durability of Solar Collectors in Denmark. Paper from Solar World Forum, ISES, Brighton.

A

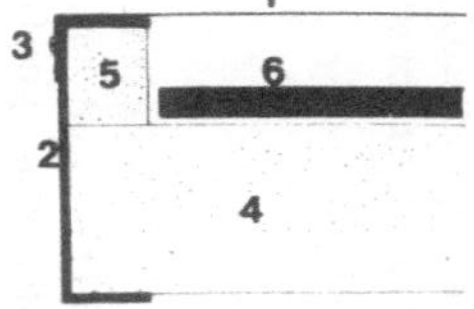

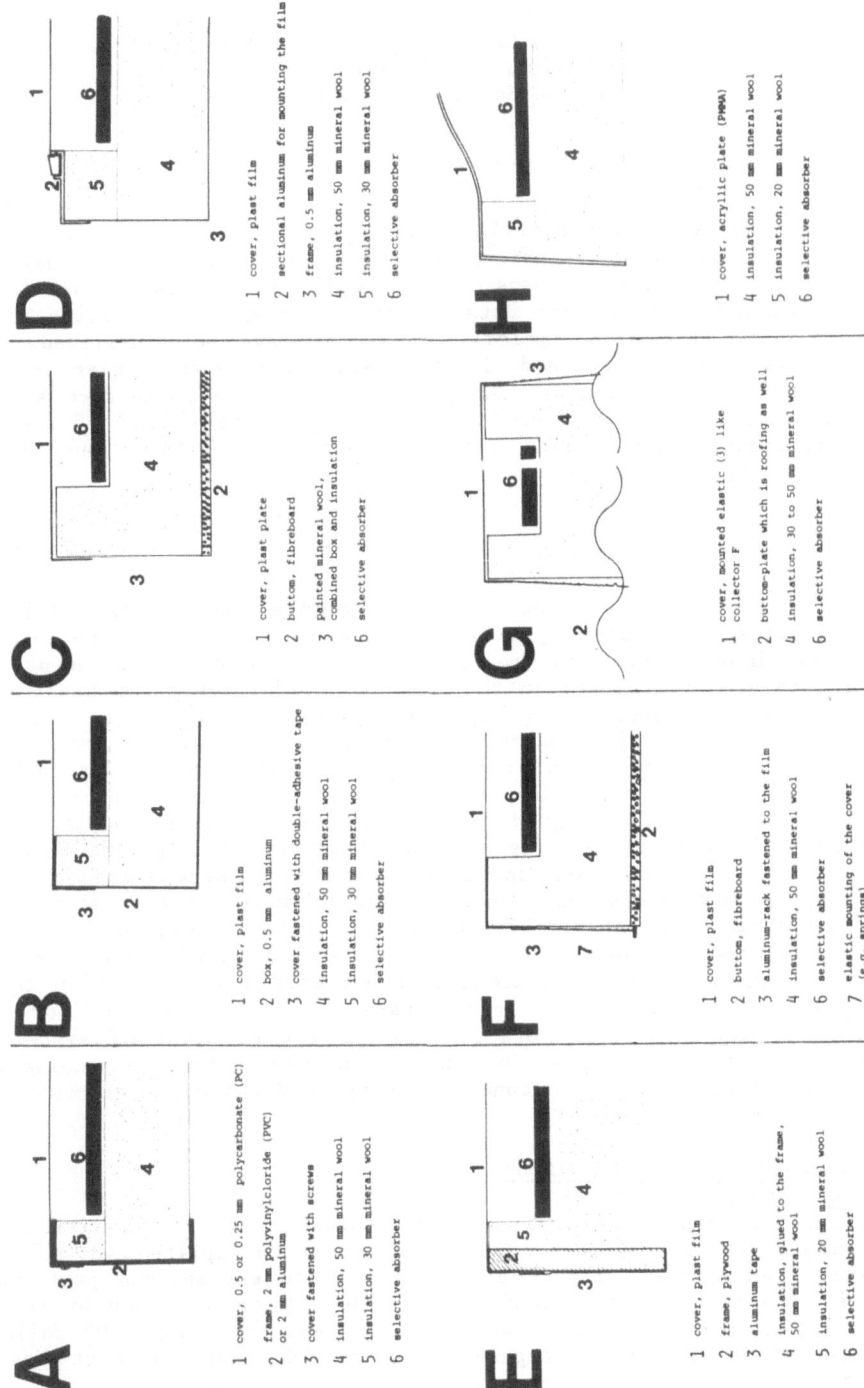

1 cover, 0.5 or 0.25 mm polycarbonate (PC)

2 frame, 2 mm polyvinylcloride (PVC) or 2 mm aluminum

3 cover fastened with screws

4 insulation, 50 mm mineral wool

5 insulation, 30 mm mineral wool

6 selective absorber

B

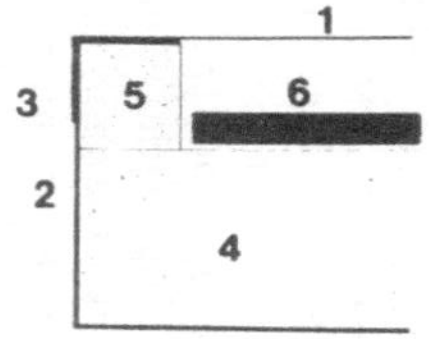

1 cover, plast film

2 box, 0.5 mm aluminum

3 cover fastened with double-adhesive tape

4 insulation, 50 mm mineral wool

5 insulation, 30 mm mineral wool

6 selective absorber

C

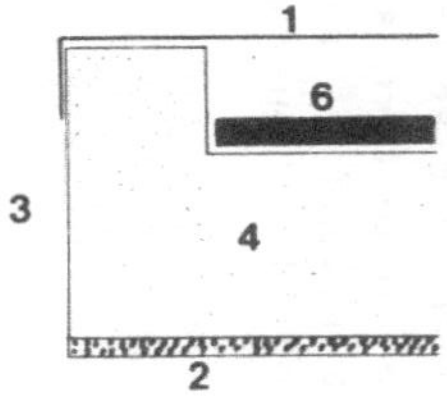

1 cover, plast plate

2 buttom, fibreboard

3 painted mineral wool, combined box and insulation

6 selective absorber

D

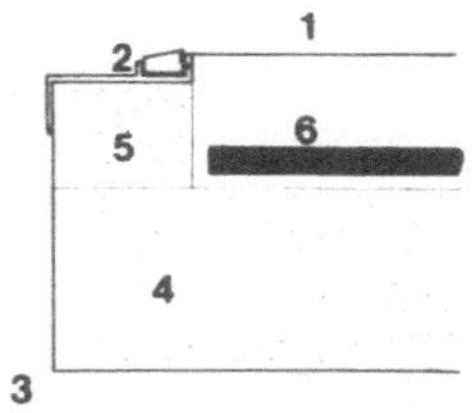

1 cover, plast film

2 sectional aluminum for mounting the film

3 frame, 0.5 mm aluminum

4 insulation, 50 mm mineral wool

5 insulation, 30 mm mineral wool

6 selective absorber

E

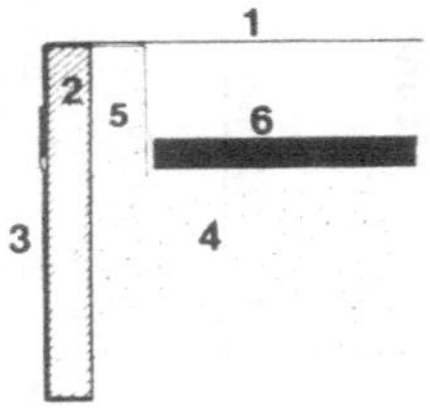

1 cover, plast film

2 frame, plywood

3 aluminum tape

4 insulation, glued to the frame, 50 mm mineral wool

5 insulation, 20 mm mineral wool

6 selective absorber

F

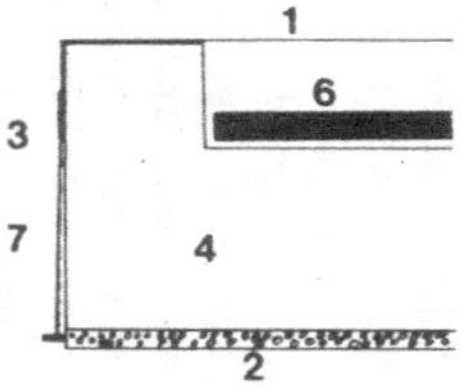

1 cover, plast film

2 buttom, fibreboard

3 aluminum-rack fastened to the film

4 insulation, 50 mm mineral wool

6 selective absorber

7 elastic mounting of the cover (e.g. springs)

G

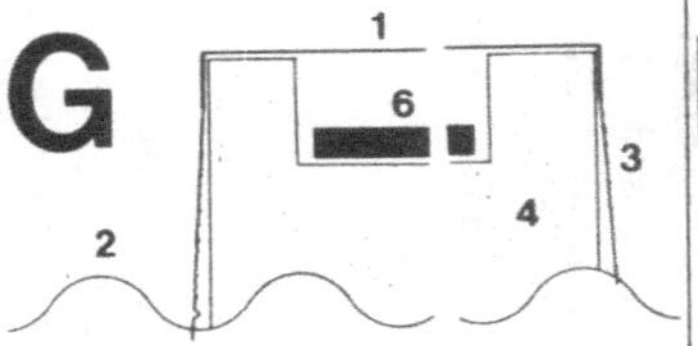

1 cover, mounted elastic (3) like collector F

2 buttom-plate which is roofing as well

4 insulation, 30 to 50 mm mineral wool

6 selective absorber

H

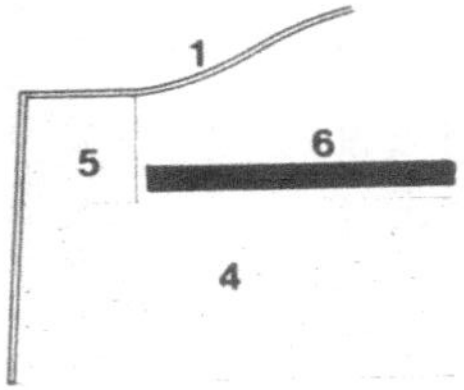

1 cover, acryllic plate (PMMA)

4 insulation, 50 mm mineral wool

5 insulation, 20 mm mineral wool

6 selective absorber

INVESTIGATION OF SOLAR-ASSISTED HEAT-PUMP SYSTEMS USING THE HEAT RECOVERY GLAZING CONCEPT.

P SIKORA*, G MORAN & J BANNARD.
The National Institute for Higher Education,
Limerick, Ireland.
*HRG Ltd., 40 South Mall, Cork, Ireland.

Summary

Several modes of two-pass air collector are under study with a view to optimising their dimensions referred to typical applications. The collector is used as a means of raising the source temperature into a heat pump. Manufacturers data for the performance of rolling piston compressors servicing typical heat pumps have been utilised. The load has been modelled and climatic data obtained from Irish stations have been used to compare the performance of the collectors as heat pump sources. It is found that very attractive values of seasonal coefficient-of-performance may be obtained for the purpose of domestic space heating.

1. INTRODUCTION

In the past heat pumps have been regarded as unreliable and barely cost-effective. In recent years there has been a renaissance brought about by improvements in reliability, but major improvements in performance are anticipated in the near future following the development of rotary compressors. This paper describes work which is aimed at showing that the major advantages occurring at higher evaporating temperatures may be utilised if air collectors are used to increase the source temperature. A number of different modes of a two-pass air collector have been designed which can be directly coupled to a building. The collector is termed Heat Recovery Glazing (HRG) and may both collect solar radiation and also recover some heat losses from the building. It may be envisaged as a unitary collector or as a part of the building fabric. It is suitable as both an original building element or as a retrofitted component. The collector may be regarded essentially as a passive component or as an element of a hybrid system in which it may operate in the closed loop (refrigerated air recirculation) or open loop. This paper gives comparative results for a number of systems of collector, heat-pump and load for some of the many different combinations in the open loop, and it is found that very good seasonal performance is obtained from the systems.

2. THE COMPONENTS OF THE SYSTEM.

2.1 The Heat-pump

Figure 1 shows a plot of theoretical Carnot Coefficient of Performance for an air-to-air heat pump with 9°C splits and 43°C air delivery. Also shown is the hatched zone which encompasses performance data for a range of reciprocating piston heat pumps compiled by Andrews et al [1]. It is found that with such conventional pumps COP fails to increase above a source temperature of $\sim$15°C, and for most designs excessive operating pressure limits use to little more than this temperature. On the other hand, COP values have been calculated using the manufacturers performance data for the rolling piston compressors, with 1kW and 4kW (electrical) capacity, also with $9^{o}C$ splits and with 12% peripheral power allowance. It is clear that not only is the performance much closer to the theoretical but also no temperature limit exists within the required range and that the performance is markedly better than reciprocating-piston-machines at the higher source temperatures. Similar advantages have been measured for rolling-piston rotary compressors by Ecker [2].

In addition to the obvious performance advantage shown in Figure 1, the rotary compressor is also expected to provide better reliability in the heat pump application. This is because of the less constricted gas path through the compressor, the absence of inlet valves, and the superior volummetric efficiency of the design. These factors allow the rotary machine to function efficiently at the very high mass flow rates obtained at elevated evaporating temperatures, and hence to follow the trend of the Carnot curve.

2.2 The Collector.

Measures for reducing the glazing losses of flat plat collectors are both expensive and self-defeating in that they reduce the solar transmission. A promising idea was introduced [3] which offers an alternative method of reducing the heat loss from glazing without incurring any penalty in reduced solar transmission. Called the Thermal Trap principle the collector fluid is passed between thick slab outer glazing layers, cooling them, and flowing inwards towards the collector surface where the energy is trapped by means of the radiation-selective properties and low thermal conductivity of the glazings. The HRG collector design uses the principle to reduce losses to the ambient by passing the cold air stream between the layers of wide-gap double glazing and thence along the collector surface. Furthermore, by coupling to a heated space / thermal store it aims to recover internal or stored energy. This latter feature will produce better performance during periods of solar gain and will enable the collector to continue operating during periods of zero or low solar gain when heating needs are greatest.

Recent work on collectors using the Thermal Trap principle [4,5] have shown experimentally and theoretically their advantages over conventional collector types. For air collectors, where the number of passes is limited, the 2-pass collector is shown to be superior in collection efficiency to the single pass mode by 15 - 20%. A mathematical model for these collectors has been reported [5]. It makes the assumption that temperature variations of the glazing slabs along the direction of fluid flow can be characterised by an averaged nodal temperature and it pays particular attention to the effects of slab thickness on performance. Whilst these results are encouraging the HRG collector differs from the Thermal Trap types studied in two important respects, (a) the trapping glazing layers used are of thin highly conducting glass or polycarbonate and therefore temperature variations along the flow direction are significant, and (b) back losses are significant. A new analysis was therefore undertaken which takes account of the back losses and the temperature variations along the flow direction.

Studies are underway to develop an analytical model of the HRG type capable of being easily connected to heat pump, evaporator and building load model. In tandem with this effort a numerical model has been developed of the HRG operation in a number of different modes and two of these modes are shown in Figure 2. These are the Trombe wall and the conservatory mode, both of which may be ground-coupled in order to profit both from the short-term and seasonal storage. The other modes are the ground-coupled collector, and the HRG window. This latter allows solar gain to heat the building directly. Results for each of these modes, all operating in the open loop, using the numerical model, are shown in Figure 3. Here the heat collected is shown with respect to time, averaged about solar noon, for an average March day (at Valentia, Ireland) compared with the incident direct radiation. An identical 2-pass HRG collector, uncoupled but using back insulation is included for purposes of comparison.

The advantage of building/storage coupling is immediately apparent. All of the modes looked at, with the exception of the window, perform better than the ordinary collector during the hours of sunshine. However the direct gain to the building through the window has not been considered here. But it is during the hours of zero or low solar gain that these advantages are most apparent. In all the cases studied the collectors continue to collect/recover heat when the ordinary collector would be inoperative. The HRG window shows the highest night-time performance.

2.3 The Load.

The system load is based on a typical detached family house having an overall loss coefficient of $0.6W/m^2K$ and air change rate of 0.5/hr. The south-facing glass is taken to be 10% of the floor area. Heating load is calculated according to the standard procedures, and U-values are adjusted for the effect of window-collector, Trombe-wall or conservatory as these are directly coupled to the house.

3. SYSTEM PERFORMANCE.

The system of collector, heat-pump and heating load offers an almost limitless number of permutations. Most households will have a secondary form of heat, and for this reason the data presented in Figure 4 is particularly encouraging. This is because in this model the total heating load has been assumed to be met by the HRG system and no allowance has been made for other forms of heating, e.g. open fires, cookers, etc., the contribution from which may be considerable. Here three of the HRG modes are compared with the rotary compressor heat-pump operating without a collector. The figure shows clearly that advantages are to be gained in terms of the overall performance of the heat-pump when coupled to a collector. The data for all the systems shown are obtained in the open-loop (non-recirculating), and COP values calculated using the data in Figure 1. Shown are the contributions of the passive solar gain, and also the driving energy and peripherals of the heat-pump. The data is for a representative day for each of the four seasons. The house is heated to 18^oC from 0800 hrs to 2300 hrs daily and average outside temperatures are 7^oC in January (winter) and 15^oC in July.

The results presented here show the air-collector/rotary-compressor system in an encouraging light, although space does not permit a full analysis. It is clear that with seasonal values of COP of over 4, a fresh look is needed at heat-pump systems of the type described here.

REFERENCES.

1. Andrews, J., Kush, E.A. & Metz, P.D. Brookhaven Nat. Lab. Report 50819 UC-59C. (1978).
2. Ecker, A.L. ASHRAE Tans 86 Part 1., Paper 2578 (1980).
3. San Martin, R.L. & Fjeld, G.J. Solar Energy 17 (1975) 345.
4 Samuel, T.D.M.A., & Wijeysundera, N.E. Solar Energy 26 (1981) 65.
5. Satcunanathan, S & Deonarine, S. Solar Energy 15 (1973) 41.

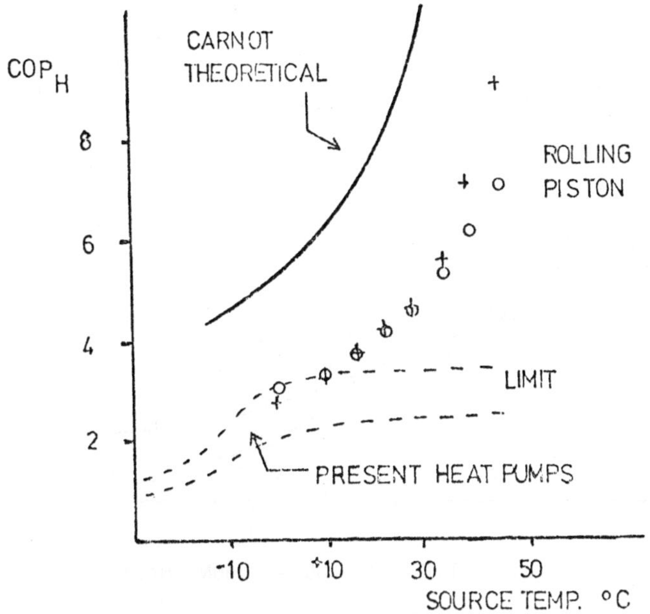

FIGURE 1. Coefficient of Performance of Heat Pumps.
oooo Rotary 1kW$_e$ compressor
++++ Rotary 4kW$_e$ compressor.

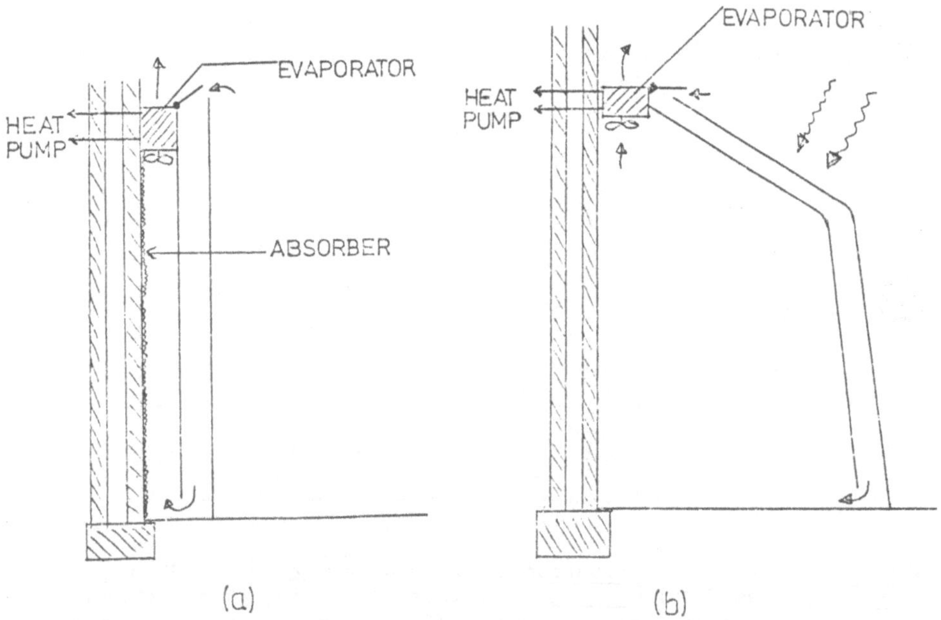

FIGURE 2. Two of the modes of heat-recovery glazing.
(a) Trombe wall (b) Conservatory.

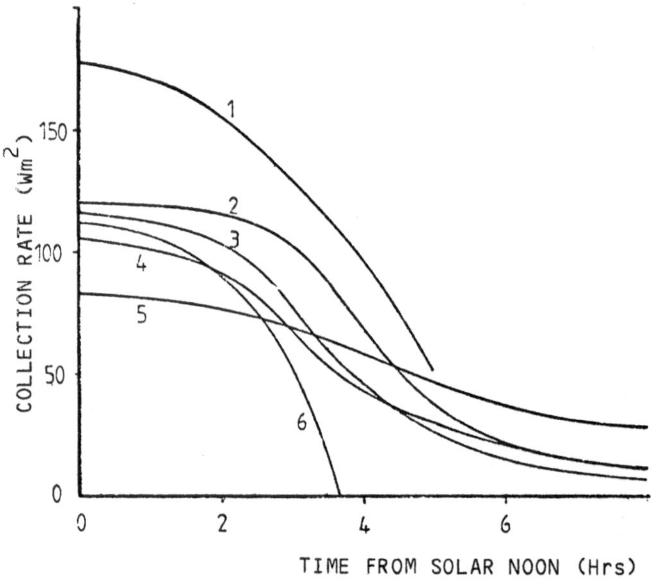

FIGURE 3. Performance of HRG collectors compared with the solar radiation about the solar noon. (1) Solar radiation; (2) HRG Trombe-wall (3) HRG Ground-coupled collector; (4) Conservatory; (5) HRG Window (6) Uncoupled 2-pass collector.

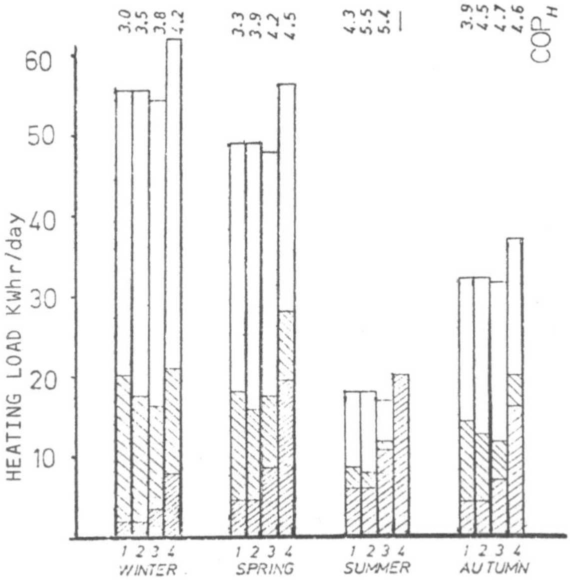

FIGURE 4. Seasonal distribution of passive [hatched box], Heat pump [hatched box], and Total [open box] heating loads for four systems: (1) Air-source heat-pump; (2) Ground-coupled collector; (3) Trombe wall collector; (4) Window collector

A DETAILED THEORETICAL MODEL FOR A FLAT PLATE NATURAL CONVECTION LOOP-TYPE SOLAR AIR COLLECTOR

A. D. Sferides and T. P. Tsingas
Public Power Corporation
P.O. Box 10757
541 10 THESSALONIKI
GREECE

Summary

The work to be described has served as the preliminary stage in a project toward the optimization design of a natural convection loop-type solar air collector. The final stage of this project would be the monitoring of some real collector models. To discriminate among the variety of possible solutions, so that a limited number of collectors could be constructed, effort has been made to make the theoretical model as accurate as possible. Numerical processing is based on careful thermal balancing in the absorbing surface(body) and in the collector glazing. This procedure has led to a lengthy iterative calculation incorporated in a computer program. The result, despite its expected shortcomings, has been a valuable tool for the investigation of the exact role of the various parameters in the design of natural convection flat-plate solar air collectors. Further the model has completely cleared out thermal situations which usually are (often erroneously) taken as granted.

1. INTRODUCTION

The analysis of flat-plate solar collectors most usually follows the well-known Hottel-Whillier-Bliss (HWB) derivation. It is accepted that the HBW model is the result of many simlifying assumptions, which are neccessary in order to give a general picture of what is happening inside the collector, without resort to complicated theoretical considerations and computer assistance. Arithmetical results obtained with the HBW derivation are well within usual engineering computational errors. Though the existence of other simplifying views of the collector behaviour and performance cannot be overseen, the observed differences to the HBW model are rather of minor importance.

After all those simplifying assumptions, the fact that small design modifications do not influence the numerical results given by simlified models, is of natural consequence. Yet design optimization is looking forward to as many such modifications as possible, so that a general considerable improvement may result.

So, an exact model appears to be valuable in optimization research. More broadly, it assists in the better understanding of the extremely interesting thermal processes which happen in a collector and brings out new items to be incorporated in a simplified calculation too.

2. DEVELOPMENT OF THE MODEL

2.1 Description of the model

The model has been described in an earlier brief presentation (1); yet because of the details which have since that time come into consideration, a brief description will also be given here.

The colector is a flat plate one, to be placed on outer wall(s) of buildings for the heating of the air of corresponding rooms by natural circulation. In this stage - to serve the before mentioned purpose - the collector has been considered vertical, though provision for inclination will be taken in a later work. As for the path of the heating fluid, it is considered a front stream collector, and this assumption is in accordance with the aim to keep the construction cost low. In all other aspects, it is a classical air collector with good back and edge thermal insulation, one-plate glazing (again for cost reasons), and provision for the possibility of using non-metal absorbing surface(body).

2.2 Assumptions used

The basic assumption, the one making the vital difference among all simplified models, lays in considering variable temperature distribution on the surface of the absorbing plate and on both surfaces of the glazing.

It is for this extremely complicating assumption why another, natural consideration has inavoidably been omitted : To keep empirical or semiempirical relations out of the "clean" theoretical considerations, losses due to external wind speed have been omitted.

To comply with the basic assumption of temperature gradient in the direction of flow, conduction in this same direction in the absorber and in the glazing body has been considered.

All calculations have been performed for the steady state of the collector operation.

3. THEORETICAL CONSIDERATIONS

For the calculation of the heat quantities exchanged by convection and radiation between the collector surfaces and the air, the general principle of superposition has been used (2,3). The temperature profiles have been analysed to a number of small temperature step variations. Then the thermal effect of each step has been computed separately and all the effects added.

Computation of the effect of each step must take place according to the position of the step either in the laminar or in the turbulent flow region. For the problem of defining the transition distance, an algorithm has been developed by means of the relation given by Raithby and Hollands (4)

$$\Delta_L = 4/3 \cdot \Delta_T$$

with Δ_L and Δ_T defining the laminar and turbulent equivalent conduction thicknesses.

For the calculation of the heat transfer in the laminar region use has been made of relations resulting from the exact solution of the similarity equations for the laminar flow.

For the computation of the turbulent heat transfer, the relations resulting from the approximation of the temperature and velocity profiles through power functions, suggested by Eckert and Jackson have been used. But, instead of the 1/7 exponent profile suggested by Eckert and Jackson (5), the exponent 1/10 as suggested by Bayley (6) has been used in the calculations.

4. LAYOUT OF THE PROCEDURE

As mentioned in the Introduction, the computational procedure is essentially an iterative lengthy calculation. It is composed of a main loop which contains two minor loops. Each sub-loop calculates the temperature profiles of the absorber surface and correspondingly of the glazing surfaces.

A block diagram with the general steps of the procedure is given in Fig. 1.

5. NUMERICAL RESULTS

In Fig. 2,3 and 4, the computed temperature profiles are given for a collector as follows :

-Height 2.00m
-Width 1.00m
-Absorbing plate to glazing distance .20m
-Glazing material : sheet commercial glass with
  Transmittance for solar radiation 0.80
  Transmittance for infrared spectrum 0.03
  Thermal conductivity 1.05W/m °C
  Emittance at 20°C 0.94
  Reflectance 0.15
-Absorbing plate : a non-metal material painted dark green with
  Emittance at 100°C 0.95
  Reflectance for solar radiation 0.05
  Thermal conductivity 0.28W/m°C
-Insulation : polyurethane foam with
  Thermal conductivity 0.024W/m °C
-Temperature of room to be heated 18°C
-Ambient (outdoor) temperature 6°C
-Insolation (total value computed for the inclination of the

collector) 1000W/m^2

As is clearly shown in the Figures, temperature profiles are markedly non-uniform. On those, the influence of the transition from laminar to turbulent flow is clear, as is the form of a certain profile on the others.

It is worth while to note the role of the inner glazing surface : It may function either as a heat source or as a heat sink for the air in the collector, depending on the high or low value of the insolation.

6. CONCLUSIONS

The effort to build up an accurate theoretical model for a flat plate natural convection loop type solar air heater has resulted to the device of an iterative computational scheme. By means of this, accurate calculation of exchanged heat quantities and temperature profiles on all surfaces involved is performed. The computational procedure and the results, besides being valuable tools for optimization design, are aids to the better understanding of the thermal processes in the collector.

Validation of the model through experimental methods is considered necessary. And, as no theoretical model can replace reality, it is expected that conclusions from real model operation will result in improving - by feed back - the theoretical one.

REFERENCES

1. A.D. SFERIDES, M. PAPADOPOULOS (1983) : Design and development of an easy made convective type solar air colector - A presentation at the 1st CEC Contractors meeting at Meersburg, Germany.
2. ECKERT E.R.G., DRAKE R.M. : Heat and Mass Transfer - McGraw-Hill-Kogakusha, 1959
3. H. SCHLICHTING : Der Waermeuebergang an einer laengsangestroemten ebenen Platte mit veraenderlicher Wandtemperatur - Forch. d. Ingrws., 17, 1, 1951
4. G.D. RAITHBY, K.G.T. HOLLANDS : A general method of obtaining approximate solutions to laminar and turbulent free convection problems - from "ADVANCES IN HEAT TRANSFER", Academic Press, 1975
5. ECKERT E.R.G., JACKSON T.W. : Analysis of turbulent free convection boundary layer on a flat plate - NACA TN 2207, 1950
6. BAYLEY F.J. : An analysis of turbulent free convection heat transfer - Proc. Ins. Mech. Eng., 169, 20, 1955

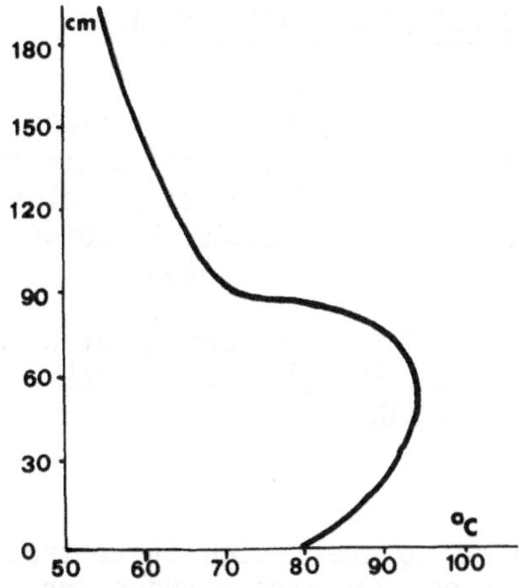
cm
180
150
120
90
60
30
0
°C
50
60
70
80
90
100

Fig. 2 Temperature profile for absorber surface.

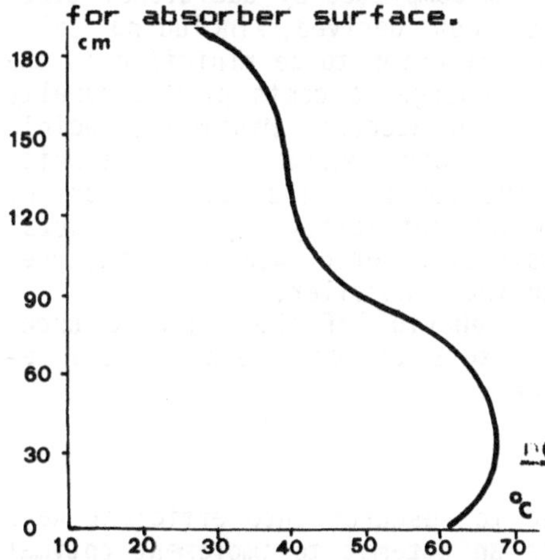
cm
180
150
120
90
60
30
0
°C
10
20
30
40
50
60
70

Fig. 3 Temperature profile for glazing inside surface

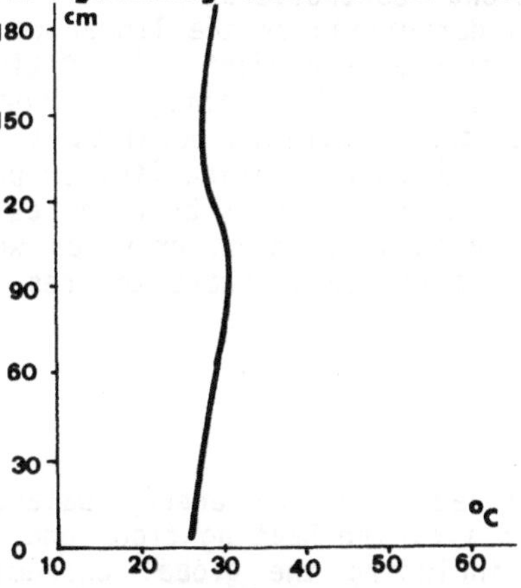
cm
180
150
120
90
60
30
0
°C
10
20
30
40
50
60

Fig 4. Temperature profile for glazing external surface

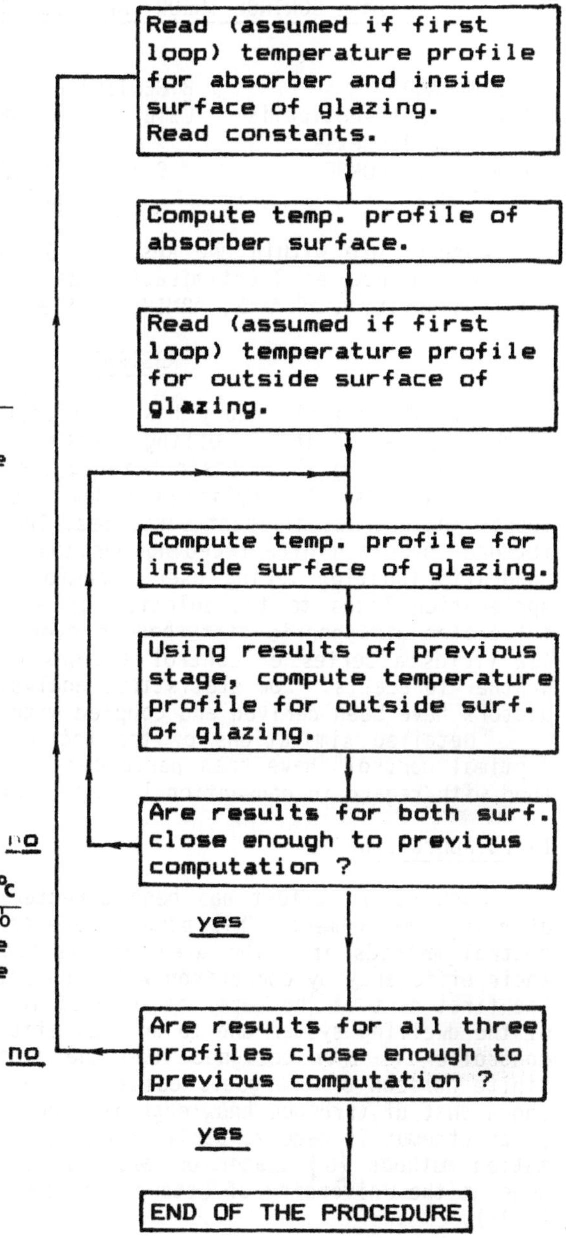
Read (assumed if first loop) temperature profile for absorber and inside surface of glazing. Read constants.
Compute temp. profile of absorber surface.
Read (assumed if first loop) temperature profile for outside surface of glazing.
Compute temp. profile for inside surface of glazing.
Using results of previous stage, compute temperature profile for outside surf. of glazing.
Are results for both surf. close enough to previous computation ?
no
yes
Are results for all three profiles close enough to previous computation ?
no
yes
END OF THE PROCEDURE

Fig. 1 A block diagram of the computatioal procedure.

OPTIMAL CONTROL THEORY APPLIED TO DWELLING HEATING SYSTEM

J.M.CARDI
ARMINES Sophia-Antipolis
rue Claude Daunesse
06565 VALBONNE
FRANCE

B.GENTILI ; J.MARTIN
CSTB Sophia-Antipolis
BP 21
06562 VALBONNE
FRANCE

P.MORAND ; P.PARENT
AFME Sophia-Antipolis
Route des Lucioles
06565 VALBONNE
FRANCE

Authors are within the Research Group IRCOSE (Institut de Recherche sur la Commande et l'Optimisation de Systèmes Energétiques) which is a joint association of AFME, ARMINES, CSTB, INRIA, GDF.

ABSTRACT

Optimal Control Theory has been applied to the design of controller to be implemented in a dwelling heating system. The system under consideration includes a floor basic heating system completed by additional electric convectors. A simplified model has been derived, including solar inputs and user significant variables. The criterion to be minimized is the integral over a finite receding horizon of energetic costs plus a penalty quadratic function taking user's comfort into account. Minimum principle application leads to the solution of a two point boundary problem which integrates atmospheric disturbances over the horizon. The derived control law yields a series of control actions which optimality depends on local weather forecasts. From statistical analysis of a set of weather data, predictors have been derived and coupled with the controller.

Detailed simulations of the thermal behavior of the building under "optimal control" have been performed to assess the efficiency of the method with regard to conventional controllers.

I- INTRODUCTION

Much recent effort has been directed to research into efficient ways of energy management. This paper presents an attempt to implement optimal control methods into the area of dwelling heating system and to assess their efficiency by comparison with conventional controllers.
The first part of the paper is devoted to a definition of the linear model of the dwelling system and to an evaluation of user's comfort [2]. Optimal controller is then designed from the minimization of an integral over a finite horizon of energetic costs plus a quadratic comfort penalty. It is shown that disturbance knowledge is necessary to ensure optimality. In part 3, an attempt is made to derive good local weather forecasts by using estimation methods [5]. Later on detailed simulations based on previous work done at the University of Ottawa will be performed to validate the results (1) (6).

2- DWELLING THERMAL MODEL

2.1- Thermal balance of the dwelling

A simplified model has been derived from the energy balances within the main components of the dwelling ; i.e. the heating floor and the shell which leads to a 2nd order system including the global characteristics of dwelling. Applying thermal laws within the shell and floor, the equation of the system may be written as follows :

$C_s\dot{T}_s = H(T_f - T_s) + B1E_s + GV(T_{ex} - T_s) + P1$ (1) within the air + shell

$C_f\dot{T}_f = H(T_s - T_f) + B2E_s + P2$ (2) within the floor.

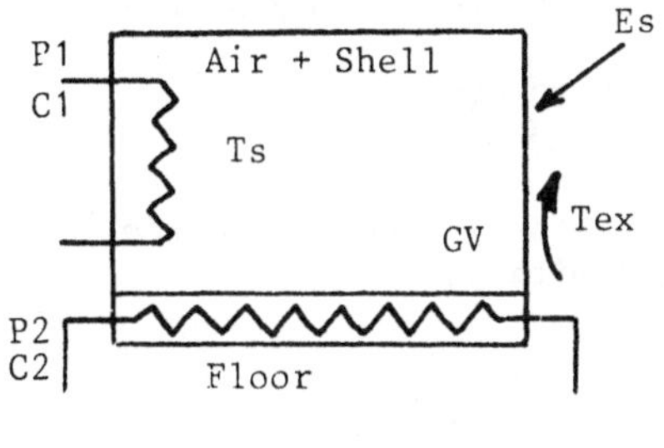

Fig.1 - System Description

Ts and Tf are respectively average temperatures of shell and floor. Cs and Cf are their corresponding thermal inertia. H and GV are global exchange coefficients, Tex and Es are external temperature and solar inputs. P1 and P2 are powers supplied to housing with respective time varying costs C1 and C2.

These equations may be written in a state space vector form :

$$\dot{X} = AX + BU + EW \qquad (3)$$

where X is the state vector $(Ts, Tf)^T$, U the control vector $(P1, P2)^T$ and W the disturbance vector $(Es, Tex)^T$.

The coefficients of matrices A, B and E are derived from actual systems by using least squares identification methods.

2.2- Definition of comfort cost :

Previous work on comfort(2) has shown that users are sensitive to air and radiant temperatures among other parameters (dampness, airflow rate). It has been pointed out some comfort equivalences within the subspace air-radiant temperatures represented by a cluster of straight lines.

Introduction of this comfort representation as a constraint within the optimization problem was leading to a solution which implementation did not appear to be very nice. It turned out more useful to derive a quadratic penalty function to assess user's comfort. FIG.2 shows how an elliptical domain is associated with the penalty function. The main axis of ellipses keeps on the direction of the cluster, the axis ratio is assumed to be 10 and the 3rd parameter (ρ) is just a scaling factor, associating a cost to user's comfort.

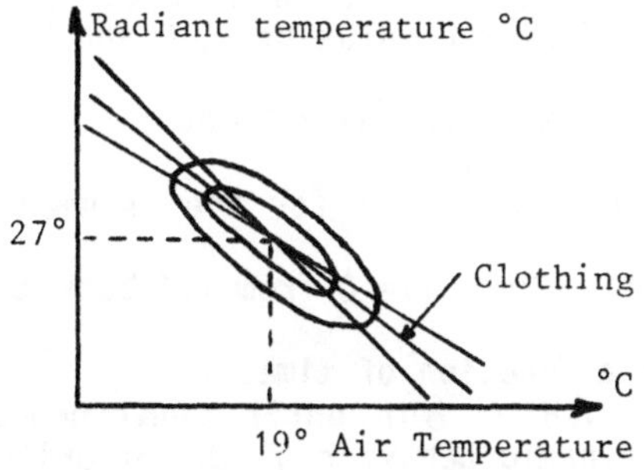

Fig.2 : Comfort equivalence lines parameter : clothing

In standard conditions, we can derive the following penalty function

$$\frac{1}{2}\rho\left[X - X_c\right]^T \begin{bmatrix} 1 & .41 \\ .41 & .18 \end{bmatrix} \left[X - X_c\right] \quad \text{with } X_c = \begin{bmatrix} 19 \\ 24 \end{bmatrix} \text{°C}$$

3- APPLICATION OF MINIMUM PRINCIPLE

3.1- Performance Criterion

The purpose of the control system is mainly to minimize energetic costs over a finite horizon corresponding to system dynamics.
With the additional cost due to comfort, the criterion may be written :

$$J = \int_{t}^{t+T} \left\{ \tfrac{1}{2} \left[X - Xc\right]^{T} Q \left[X - Xc\right] + C^{T} U \right\} d\tau \qquad (4)$$

with $0 \leqslant Uj \leqslant Uj\max$ j is the index for J^{th} energy source.
C is the energetic costs vector associated with U.

3.2- Solution of the problem

The Hamiltonian of the system is defined as :

$$H = \tfrac{1}{2} \left[X - Xc\right]^{T} Q \left[X - Xc\right] + C^{T} U + P^{T} \left[AX + BU + EW\right] \qquad (5)$$

He turns out to be linear with respect to the command U :

$$H = \tfrac{1}{2} \left[X - Xc\right]^{T} Q \left[X - Xc\right] + P^{T} EW + P^{T} AX + \left[C^{T} + P^{T} B\right] U \qquad (6)$$

Necessary conditions for optimality lead to the following equations :

$$\dot{P} = - \delta H / \delta X - A^{T} P - Q \left[X - Xc\right] \qquad (7)$$

$$\dot{X} = \delta H / \delta P = AX + BU + EW \qquad (8)$$

$$H (X^*, P^*, U^*) \leqslant H (X^*, P^*, U) \qquad (9)$$

Since H is linear with U (9) gives the following optimal contral law:

$$\left.\begin{array}{llll} Uj^* = Uj\max & \text{if} & \left[C^{T} + P^{T} B\right] j < 0 \\ Uj^* = 0 & \text{if} & \left[C^{T} + P^{T} B\right] j > 0 \end{array}\right\} \qquad (10)$$

Singular control occurs if $\left[C^{T} + P^{T} B\right] j = 0$ in that cas Uj* can take any value form 0 to Ujmax resulting from the value of X, P and W.

To solve this problem a first order gradient algorithm is used :

- Assume a suitable u(t) as a starting point.
- integrate (8) with initial condition X(t) = X for the assumed u(t) from t to T + t.
- Integrate (7) with initial condition P(t+ T)= 0 from t+T back to t.
- Evaluate $\delta H / \delta U = C + B^{T} P$ as a function of time.
- Replace U(t) by u(t) + K*Sat(δH / δU) until convergence is achieved. K is chosen from the dynamic behavior of the algorithm in order to avoid either slow convergence rate or unstable swings.

Equation (8) shows the need for disturbance prediction over the horizon, which is attempted in the following.

4- DISTURBANCES MODELING

Atmospheric disturbances such as external temperature and solar irradiation are of importance in the determination of optimal control laws. From a statistical analysis of a large set of weather data, preliminary studies have shown that external temperature presented some reproducible pattern and could be predicted within 6-8 hours ahead of time.

In the case of solar radiation, previous studies (4) yielded autoregressive filters of order 1 for one hour ahead predictor. Due to the shape of the autocorrelation function, it seems useless to derive predictors which will go further ahead in time.

-External temperature predictor

Just by plotting hourly temperature evolutions over several years, it can be shown that some periodic structures are imbedded within the stochastic process : basically yearly and daily variations.

Since interest is for several hours predictors, average and seasonal variations of temperature have been removed from the set of data :

$$\bar{T} = 10.6 - 2.7 \operatorname{Sin} \frac{2\pi t}{8640} - 7.6 \operatorname{Cos} \frac{2\pi t}{8640} \quad \text{(t in hours)}$$

From the modified set of data, autocorrelation has been derived. A 24 hour periodic structure can be shown coupled with a forgetting factor :

$$R(t) = .73 \operatorname{Exp} -\left(\frac{t}{67.5}\right) + .27 \operatorname{Cos} \frac{2\pi t}{24} \quad \text{(t in hours)}$$

Using a least squares minimization technique on prediction error (5) temperature predictor filters can be computed along with their associated error. In fact this error shows how much the natural variance of the temperature is reduced.

6 - hours ahead predictor can be written as follows :

$$\hat{T}(t+6) = .79\ T(t) - .25\ T(t-6) + .19\ T(t-12) + .63\ T(t-18) - .5\ T(t-24)$$

with a 70 % confidence interval equal to $\pm$ 1.9 deg.C.

5- OPTIMAL CONTROL VALIDATION

5.1- Precise modeling of dwelling

Heat transfers within a room are complex, therefore models describing thermal evolution of buildings are usually huge pieces of code. In order to assess the efficiency of an optimal control law, the model should be of a sufficiently reduced order to run fastly.

Different approaches for modeling thermal phenomena in dwelling have been attempted. Due to its reduced number of nodes describing the components and its ability to take coupling effects into account, a transfer function method has been retained (1).

The basic cell is described with seven nodes - (front, ceiling, heating floor, internal partition walls, glazing, partition walls with buffer space, indoors air) coupled with external perturbations (outdoors temperature, buffer space temperature, solar radiation).

5.2- Closed loop simulation

The first part of this chapter is reserved to the analysis of the system under optimal control for typified weather perturbations. From this analysis, much insight has been gained of optimal controller response to system perturbation, and variation of energy costs. Fig.3 shows a two day typical response of a high inertia building. From top to bottom, the first set of curves represents the time varying cost along with the decision function for each command.

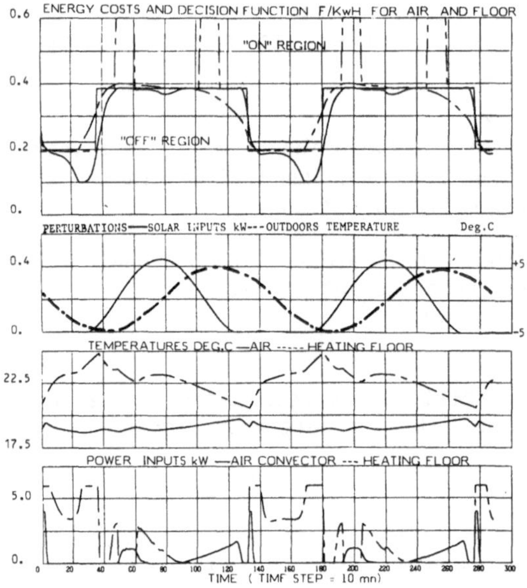

Fig. 3. Optimal control on dwelling BI energy system

Whenever the decision function is above (below) the cost line, power is switched on (off). When it is tangent to cost line power is taking intermediate value to compensate for heat losses. The two second curves are representing typical weather perturbations. The third set of curves shows temperatures within the floor and air, and the last set, the powers introduced within the floor and the air. The main result, which was an expected one, is the storage of energy within the floor whenever the associated cost is low and thermal needs are expected. When the cost is high, stored-heat is released from the floor and the air convectors are keeping air temperature within comfort range

6 - CONCLUSION

This paper presents new analysis and preliminary results relating to the optimization of energy management,taking user's comfort into account. Interest has been shown when dealing with high thermal inertia systems with time varying energy costs. Savings up to 25% can be expected from optimal controllers over more conventional types.

Subsequent research will go on with different classes of dwelling and further with commercial buildings. Some attempts will also be made to the optimum design of the building components (floor inertia, insulation, nominal power) when using optimal controllers.

Later on, optimal control laws will be applied to actual heating systems and downloaded onto microprocessor module for industrial application.

REFERENCES

[1] STEPHENSON, MITALAS "Cooling Load Calculations by Thermal Response Factor Method". Division of Building Research University OTTAWA 67

[2] J.C DEVAL, X.BERGER "Ambiances de Confort", CNRS - Laboratoire d'Ecothermique Solaire- 06562 Valbonne- FRANCE.

[3] A.E BRYSON Jr., and C.Y HO "Applied Optimal Control" Blaisdell Publishing CO, 1969.

[4] T.N. GOH and K.J. TAN "Stochastic Modeling and Forecasting of Radiation Data".Solar Energy,Vol.19.pp.755-757 Pergamon Press 77

[5] A.PAPOULIS "Maximum Entropy and Spectral Estimation : a review". IEEE, Transactions on Acoustics, Speech and Signal Processing. Vol.ASSP-29, n° 6, Dec. 1981.

PERFORMANCE AND RELIABILITY OF UNEXPENSIVE SINGLE FAMILY SDHW PACKAGE SYSTEMS

D. BIENFAIT - A. FILLOUX - J.M. CARDI and M. GSCHWIND
CENTRE SCIENTIFIQUE ET TECHNIQUE DU BATIMENT (CSTB)
ECOLE NATIONALE SUPERIEURE DES MINES DE PARIS (ENSMP)

Summary

Unexpensive package devices, most of them with thermosyphonic circulation, yields a promising trend for single family SDHW systems. New relevant test methods intended to assess the quality, reliability and thermal performances of such systemes have been defined :

1. DIFFERENT KIND OF SOLAR DOMESTIC HOT WATER SYSTEMS

Single family solar domestic hot water systems may be classified as follow :

- close-couped thermosyphon SDHW
- thermosyphon SDHW with tank distant from collectors
- bread-box kind SDHW systems
- usual systems with pump and differential controller

(more detailed classification might be further established according to the back-up energy or heat exchanger presence).

All these systems (but the bread-box SDHW systems) are now rather widely marketed in France ; during march 1984, tests supported by the governmental agency, AFME, have been carried on in Sophia-Antipolis (south of France) on thirty different SDHW systems.

Fig 1 : 30 SDHW test array in Sophia-Antipolis

2. THERMAL PERFORMANCE TEST (CSTB and ENSMP)

Scope

Efficiency assessment of Solar Domestic Hot Water systems (SDHW) compared to solar collectors is a difficult task for two reasons :

1 - Because of the thermal mass of the system, it requires a several hours long test, therefore it is hard in outdoor conditions to achieve any perfect repeatability of a test method. On the other hand the use of a solar simulator is not always possible.

2 - The efficiency of SDHW systems is highly dependent on the value and the pattern of both solar irradiations and hot water draws. Therefore the comparison between the efficiency of two SDHW systems may be different, according to the choice of solar irradiation and hot water draws pattern.

Given these points, CSTB and ENSMP chose to develop a test method that enables an easy and consistent thermal performance rating.
The goal is to derive from test results the value of the following parameters :

η_o	(n.d.)	Solar collector efficiency when fluide temperature is equal to ambient temperature
a_1	($W/m^2 . °K$)	Solar collector heat loss coefficient
K_b	(W/K)	Tank heat loss coefficient
M	(kg)	Storage thermal mass
$A . \sqrt{p}$	(m^2)	Product of solar collector area, A and solar loop efficiency, $\sqrt{p}$

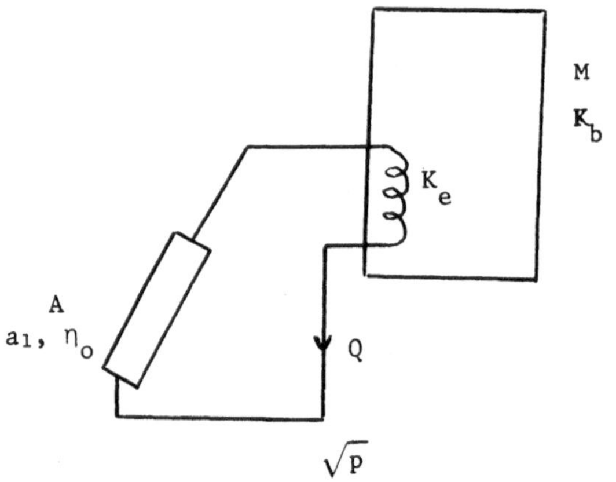

fig 2 : Parameters involved in the rating of SDHW thermal performance

Solar loop efficiency, $\sqrt{p}$

The solar loop efficiency is intended to take into account the drop in efficiency due to the heat exchanger and the energy flow rate value of the solar loop

K_e (W/K) : heat exchanger transfer coefficient

Q (W/K) : solar loop energy flow-rate

The expression of $\sqrt{p}$ can be theoretically derived from a_1, A, K_e and Q according to the following relation :

$$\sqrt{p} = \frac{\dfrac{Q}{a_1.A}}{\dfrac{1}{\exp(\frac{Ke}{Q}) - 1} + \dfrac{1}{1-\exp(- \frac{a_1.A}{Q})}}$$

Which, when there is no heat exchanger, or when flowrate walue turns out to be infinite leads to the following well-known formulas :

$$K_e = \infty \ : \qquad \sqrt{p} = \frac{Q}{a_1.A} \cdot \left(1 - \exp - (\frac{a_1.A}{Q})\right)$$

$$Q = \infty \ : \qquad \sqrt{p} = \frac{K_e}{K_e + a_1.A}$$

The values of Q, and, to a certain extent, K_e, may not always be thoroughly measured ; for instance there is no simple way to measure the flowratevalue of thermosyphon without severe alteration of its value. It can then be seen that a direct determination of $\sqrt{p}$ could eliminate the difficulties encountered in the measurement of Q and K_e.

This point constitutes the major feature of the test method.

Test method :

In a first step, the values of the parameters η_0, a_1, K_b and A are determined through a set of indoor thermal tests.

In a second step the SDHW is submitted in outdoor conditions to the following operations :

- circulation of cold water through the system in order to achieve uniformity of its temperature
- exposure of the SDHW to a solar irradiance of 800 W/m² on average, for a 4 hours period, during which ambient temperature and solar irradiance are recorded (in fact, the period duration is adjusted in order to take into account the actual value of solar irradiation)
- tapping of the SDHW and record of the inlet and outlet temperature denoted as T_i and T_o.

The temperature records during the tapping are depicted in the figure3 for 3 of the 30 SDHW system that have been tested in Sophia-Antipolis

Data-processing of these records enable to calculate the mass M of storage involved by solar heating as well as the amount of energy, E, stored in the tank :

$$M = \frac{\int (T_o - T_i).\, dM}{T_m - T_i}$$

$$E = 4.180. \int (T_o - T_i).\, dM$$

Where T_m holds for the maximum value of T_o during the tapping.

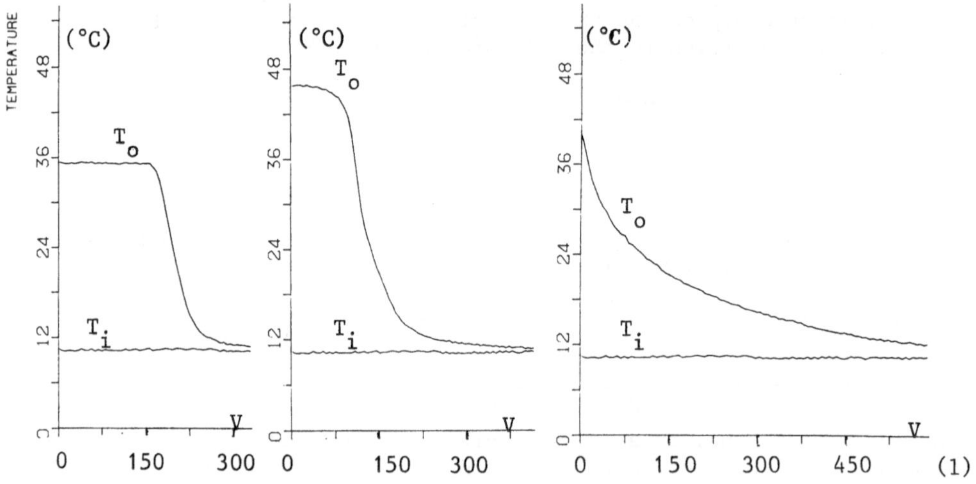

Typical records of tank outlet temperature, T_o and tank inlet temperature T_i versus the volume V of water during the withdrawal of 3 distinct SDHW systems.

The histogramm below depicts the value of $\sqrt{p}$ that have thus been determined on the 30 SDHW.

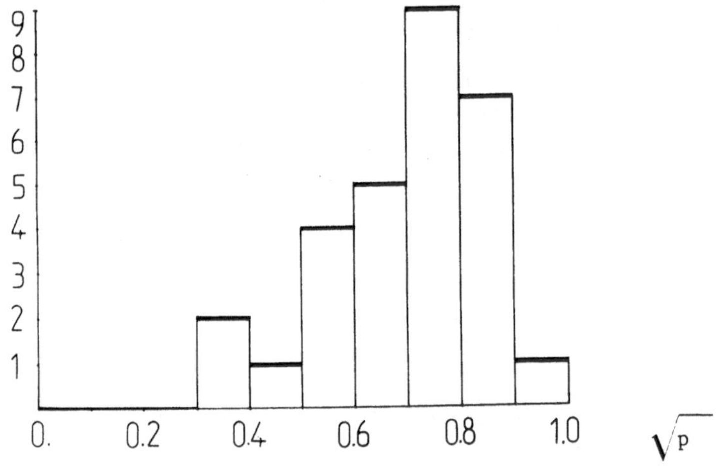

Histogramm of the 30 $\sqrt{p}$ values

The poor values of $\sqrt{p}$ (i.e : $\sqrt{p} < 0,6$) registered for some SDHW may be accounted either by a too low value of the thermosyphon flowrate or by an initial slight overestimation of the solar collector efficiency.

CONCLUSION

The termal test method that have been developed and experimented by CSTB and ENSMP enables the thermal rating of SDHW system by a set of 5 parameters which make it possible, using an appropriate computation method to determine the annual thermal performance for whatever geographical location or hot water usage is considered.

It should also be pointed out that since it is based on actual operation of a SDHW, the $\sqrt{p}$ parameter of the test method accounts in a certain way for all factors such as solar loop thermal inerty, pipes heat losses, differential controller deadband, ...) that affect the overall thermal performance though they are not taken into consideration in the theoretical analysis.

3. RELIABILITY OF SDHW SYSTEMS (CSTB)

A survey of SDHW systems installed throughout France have been done in order to point out the most frequent causes of failures or poor operation : Pump driven SDHW system.

The defaults registered are often related to the flowrate control : electronic failure of controller (often as a result of thunder falling on electricity supply lines), bad adjustment of differential controllers, leakage or absence of check valve, inappropriate placement of temperature probe, ... In order to avoid these hazards, some manufacturers are now developing simplified and safer control devices. Defaults on solar colector themselves are now less and less frequent ; they may be relevant to glass breakage (hail stone or thermal shock) or corrosion (for instance on poor quality soldering).

Close-coupled thermosyphon SDHW systems

Defaults are mainly the following :

- High sensitivity to freezing hazards : since the tank lies in the open air, tank and water supply pipes have to be protected against frost. Different kind of protections (automatic emptying, electric resistors, ...) are proposed by the SDHW manufacturers. However, at the time being, due to the lack of experience, none has proved to be perfectly safe and further investigations seem necessary ; a freezing chamber suited for SDHW have been constructed for that purpose in CSTB.
- The tank has to be located above the collector, othervise, unless appropriate provisions are taken, this leads, as a rule, to severe heat losses because of reverse thermosyphon during night time.
- In most of closed coupled SDHW systems, the tank axis is horizontal, which makes it difficult to yield any sufficient temperature stratification ; it would then be rarely advisable to use back-up energy, even in the tank upper part.

PERFORMANCE PREDICTIONS FOR ALL GLASS TUBULAR EVACUATED ABSORBERS

P. BONIFAY - M. CHANTANT
CEA - Centre d'Etudes Nucléaires de Cadarache
Service d'Etudes Energétiques
B.P. 1 - 13115 SAINT PAUL LEZ DURANCE
FRANCE

Summary

A solar collector using all glass evacuated tubes associated with a back diffusing plane was built and tested.
The tubes were fabricated by CEA-CEN Grenoble.
During the tests, we studied the influence of the geometrical parameters of such a collector ; two heat extraction systems were compared.
A computer model was developped. The comparison between the calculations and the experiments is presented.

1. INTRODUCTION

CEA-CEN Grenoble developped and fabricated high performance evacuated absorber tubes, with selective coating (α = 0,91 $\varepsilon \simeq$ 0,11). Our purpose was to test a collector, composed of such tubes and with a white diffusing plane behind them, and to optimize the performances of it.

1.1 Description

The collector was composed of 8 evacuated tubes
OD = 48 mm
d = outer absorber diameter = 38 mm
L (tube) = 1.400 mm

The measured reflectivity of the diffusing plane was ρ = 0,63. During the tests, the constitutive geometrical parameters of the collector (e : distance between the tubes, h : distance between the tubes and the diffusing plane) varied

$$2 < \frac{e}{D} < 4$$

$$1 < \frac{h}{D} < 4$$

- The inlet fluid temperature varied from 30 °C to 140 °C. The optical efficiency of such of collector is varying essentialy with θ : (angle between the projection of the sun in the transverse plane and the longitudinal plane of the collector).

θ varied in the tests from - 30 to 30 (θ = O at 12 h TSV).

The efficiency is calculated relative to the back diffusing plane surface delimited by the external tubes.

The results of the experiments are presented in Fig. 1 for θ = 0°

The daily efficiency is also recorded during the tests.

- A computer model of this type of collector was jointly developped. The optical analysis deals with the direct and reflected part of each component of the radiation (direct and diffuse). In the model the back plane is considered perfectly diffusing.

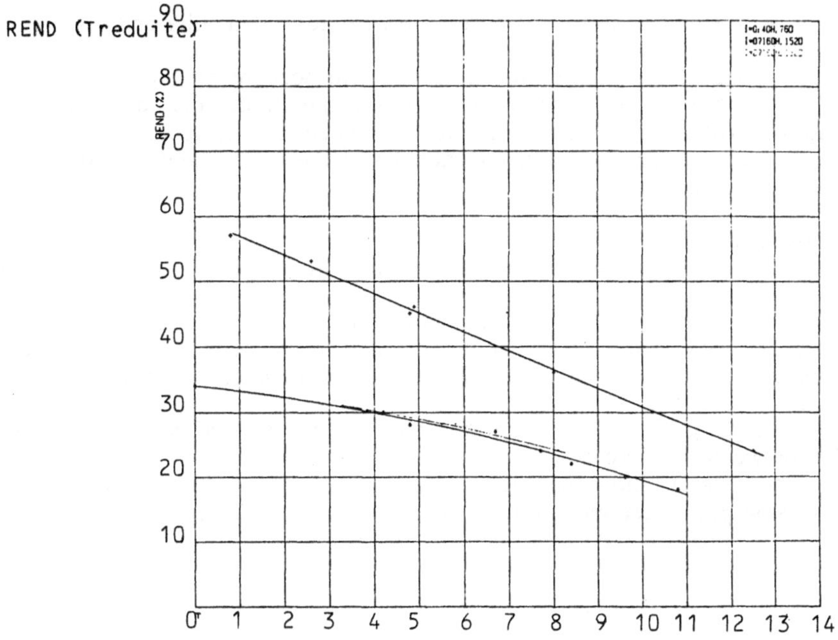

$$100 \times \left(T_{red} = \frac{T_m - T_a}{\emptyset} \right) \quad \frac{m^2\ °C}{W}$$

Fig. 1 : Experimental results (Efficiency - $Tr = \frac{T_{mf} - T_a}{\emptyset}$) $\left(\frac{m^2\ °C}{W} \right)$

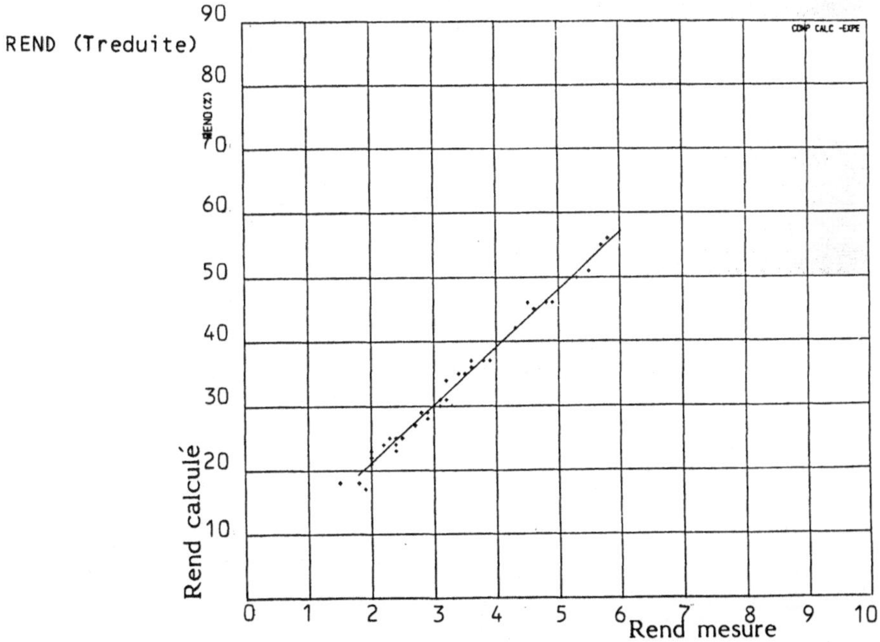

Fig. 2 : Comparison between calculations and experimental results

The thermal analysis computes the heat loss of a tube and of the connections.

The heat extraction system is taken into account by the heat transfer coefficient between the absorber surface and the fluid. This coefficient was previously evaluated using a finite element model of the tube.

The comparison between the model and the experimental results is presented in Fig. 2.

1.2 Discussion

- The h parameter (distance of the tube to the back plane) has no marked influence.
- The overall heat loss coefficient of the collector is depending of the number of tubes used. For elevated temperature (90-100 ° C), it is interesting to separate the tubes ; the incoming energy per tube beeing increased and the heat loss decreased
- with a good quality reflecting plane ($\rho \simeq 0,85$) the efficiency gain of the collector would be 5 to 8 %.
- the results show the need of using an optical reflecting system in place of diffusing plane, for elevated temperatures (90 to 140 °C).

ANALYSIS OF THE CLIMATIC DATA NECESSARY-PROBLEMS TO BE SOLVED AND APPLICATION TO FLAT-PLATE COLLECTORS

J. ADNOT, G. WATREMEZ
Centre d'Energétique
Ecole des Mines de Paris
60, bld St Michel
F 75272 - PARIS CEDEX 06

Financial support :
Commission of the European Communities and Agence Française pour la Maîtrise de l'Energie.

Summary :
The analysis of the climatic data necessary for an accurate computation of the Energy output of flat-plate collectors leads to the following conclusions :

- averaging of meteorological quantities creates a loss of information from 0 to 10 % (under-estimation of the output) ;
- inertial losses in collectors may very from 0 to 20 % of available Energy ;
- Simplified Inertial Models (like in TRNSYS, EMGP2, ...) are preferable to the non-inertial model for simulation purposes although this last one remains acceptable ;
- The "Cumulative Frequency Curves" are a suitable form of reduction of data and - due to some compensations - their simplest form (no inertia-correction, no special outside temperature) is more accurate.

Introduction :

The use of climatic data for the computation of building envelopes and solar equipments arises from our point of view three types of problems :

1 - Problems of VALIDITY (accuracy of measurements and reconstitutions)

2 - Problems of SENSITIVITY (analysis of the data needed : which variables have an influence on the computed performance, and with which time step ?)

3 - Problems of DATA REDUCTION (statistical treatments making the data more compact and so more user-friendly)

The present work is related with the second question and only in the case of flat-plate collectors :

- Non Selective between 10 and 50°C
- Selective between 10 and 50°C
- High Performance between 50 and 150°C

1 - PRELIMINARY DEFINITION

We know that different models may be used for Solar Collectors and that some meteorological phenomena are not correctly depicted by the usual (hourly) measurements, like in the following examples :

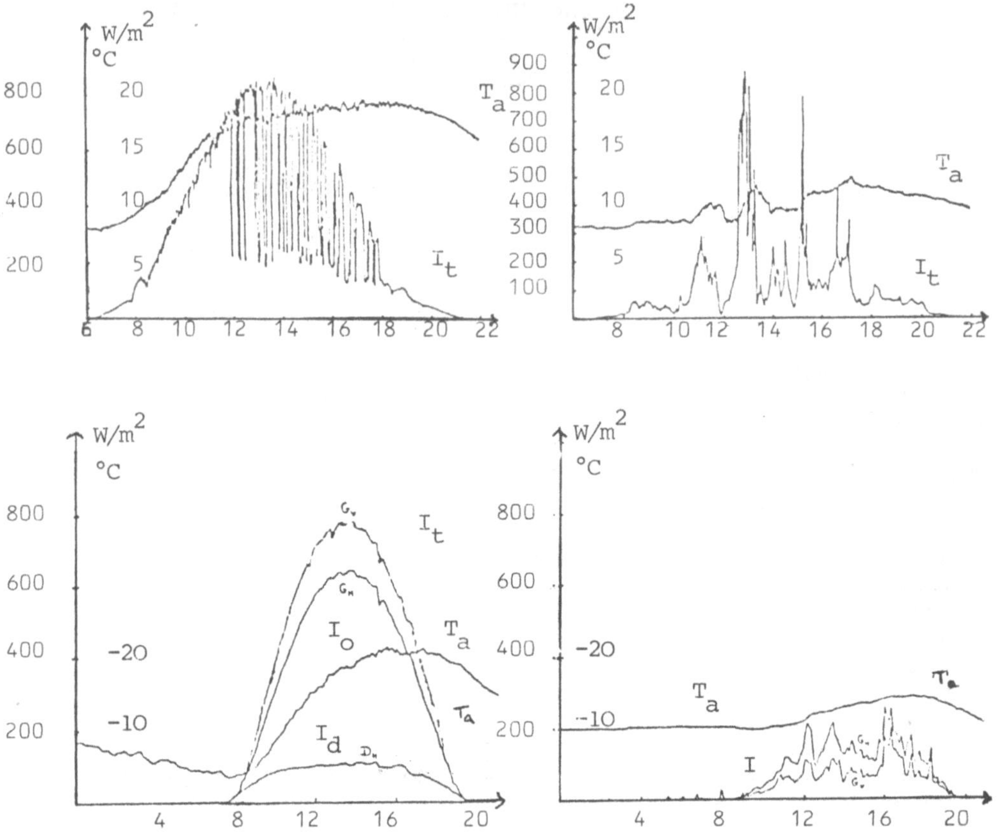

The study allowed for the comparison of three models (Reference MINERSOL : R ; Inertial : I ; Non-Inertial : H) and Data Basis more or less Reduced (measurements every minute or each 6, 30, 60 minutes ; Cumulative Frequencies Curves ; ...) for two locations of the Parisian Region (Le landy and Trappes).
The model and its Data cannot be chosen separately and there are optimal couples in the following table :

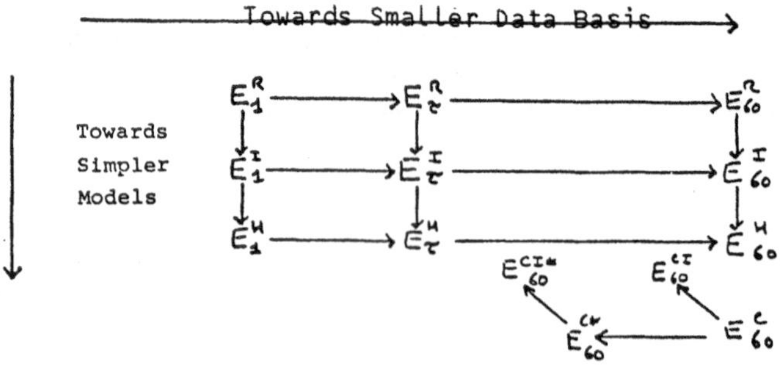

2 - STUDY OF THE PENOMENA

When one moves towards small data bases he looses information ; when one moves towards a simpler model he treats uncompletely or forgets the inertia phenomena (inertial loss). Losses of information and Inertial losses get a large importance for low utilisabilities, as for instance :

- loss of Information on the reference model :

$$E^R_6 - E^R_{60} = E^R_6 \times 0.175 \exp(-4.8\ \phi_R^{60})$$

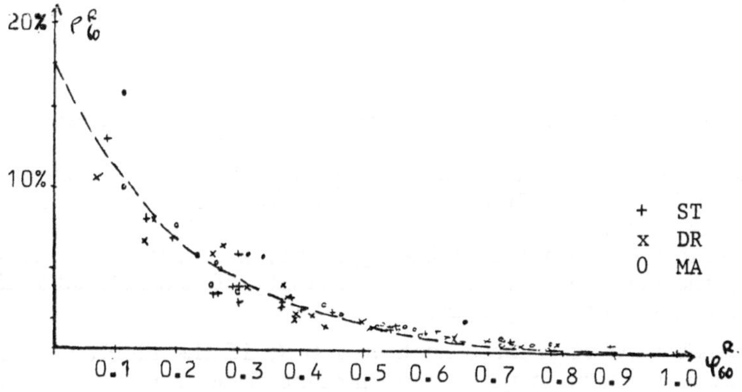

- inertial loss (actual, with the most accurate data) :

$$E^H_6 - E^R_6 = E^H_6 \times \tau_c^{1.29} \times \exp(-5\ .\phi^H_6)$$

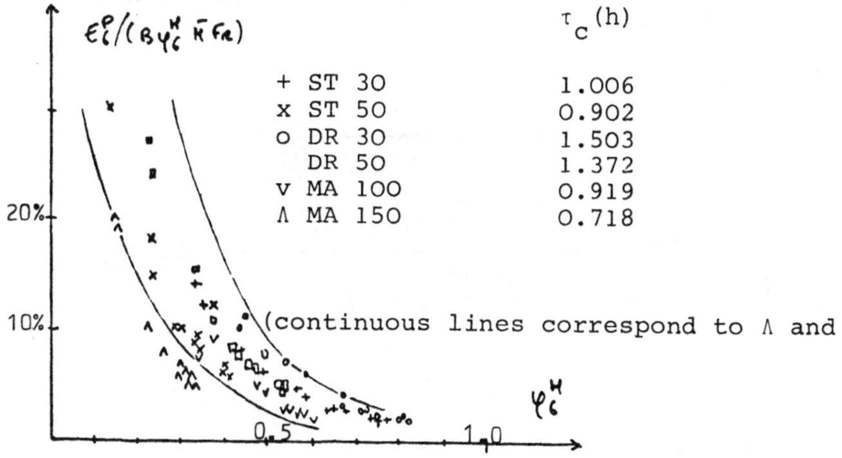

3 - PRACTICAL CONCLUSIONS

The Simplified Inertial Models in EMGP2, SYSTHERM, TRNSYS, etc ...) used with hourly time steps (this is always the case) give results very close to those of MINERSOL : error, all cases together, of 4.5‰, maximum error 12 ‰. They are statistically preferable to the non inertial model although this last one remains acceptable and gives smaller computing times:

	INERTIAL SIMULATION	NON-INERTIAL SIMULATION
Confidence Interval at 95%	(-2.7%, +0.5%)	(-1.1%, +5.7%)

The reduction of data "Cumulative Frequencies Curves" has an acceptable accuracy. It can be used under its simplest form (no inertial correction, outside temperature averaged over 24h) or, if we want to let the designer feel the influence of collectors inertia, under a more complete but less accurate form (inertial correction, outside day-time temperature) :

	CFC-SIMPLE FORM	CFC-COMPLETE FORM
Confidence Interval at 95%	(-6.6%, +4.6%)	(-9.0%, +5.8%) (or even more)

The confidence Interval was enlarged by a factor about 3 in the course of the reduction of Data.

4 - COMPLEMENTS

If one makes an analysis of typical cases of use of the models, he may conclude that :

- for Space Heating, the advantage in using for simulation purposes the inertial model and for Simplified Methods the simplest CFC is larger than on the average ;
- for Domestic Water Heating on the opposite, all procedures are equivalent and only the consideration of the simplicity of use will make the decision (no discrepancy larger than ± 1 %) ;
- in the generation of Process Heat at 100 K, if the simulation may be done with or without inertia, the simplest form of the CFC is to be used (to avoid a bias of about -4 %).

In the climatic conditions of the Parisian Region a non inertial estimator without bias would directly obtained if the integration time-step of the Meteorological Office is always about twice the caracteristic time of collectors. The extrapolation to other climates seems promising but has not been yet exactly performed.

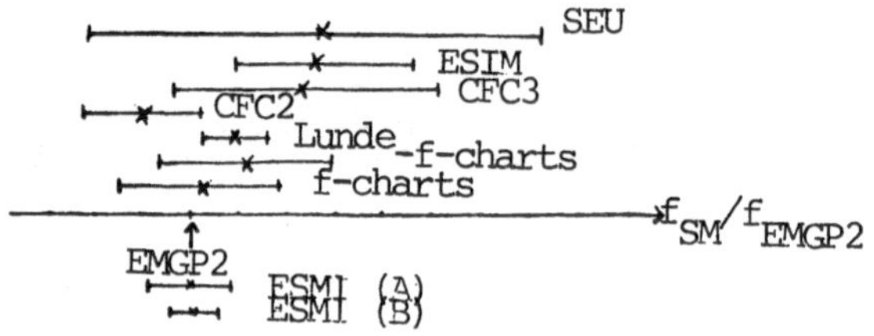

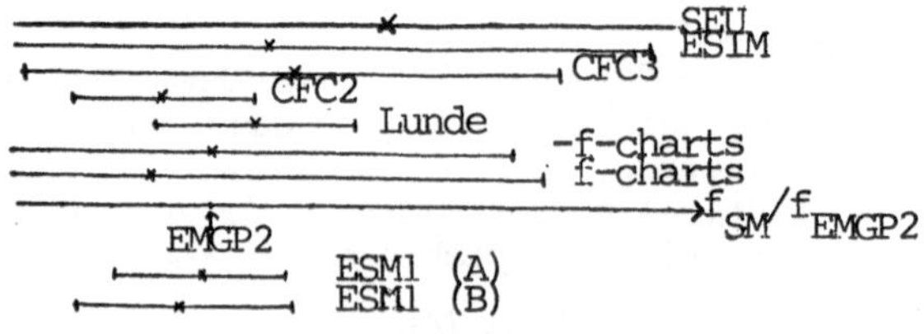

EXPERIENCE WITH COLLECTOR DRAINING AND RIGID MINERAL INSULATION IN CONTINENTAL CLIMATE (CAPELLEN-LUXEMBOURG)

J. REMOVILLE
Commission of the European Communities, detached
at GRADEL S.A.

Summary

Besides checking the economical interest of solar domestic hot water systems in the local climate and water conditions of Luxembourg (performance, amortization time), and comparing the solar collector concepts of its industry to foreign manufacturing, the solar research programme of Luxembourg aimed at getting experience with the type of reliability, durability and maintenance requirements which could be expected from commercially available items of solar systems.
The paper emphasizes the aim of the Luxembourg solar research programme, linked mainly to self-supporting mineral insulation for collectors and to open circuits with automatic drainage avoiding possibly poisonous anti-freeze. It describes briefly two original types of collectors, compares two closed circuits to an open one during 4 years of operation in Capellen as far as reliability, maintenance requirements and overall performance are concerned. In conclusion, it is pointed out that DHW heating with present oil prices are not yet competitive in Luxembourg climate, explaining thus the present stop in solar thermal research and development experienced in Luxembourg after four years of participation into the CEC collector performance testing programme.

1. INTRODUCTION

The main goals settled in 1978 by Luxembourg to its solar research were
. checking the percentage of solar gain which could be expected in its local climate
. comparing its own collector designs to foreign manufacturing
. getting an idea on the quality requirements of commercial solar components and on the corresponding reliability, durability and maintenance requirements of solar facilities.

The present paper focuses mainly on the following aspects of that programme: the self-supporting mineral insultation, an original design and mounting approach, the automatic draining of collectors, a feature most related to the local continental climate.

To achieve these general goals, progressive steps were considered: a prototype designing and testing phase, sponsored at 75% by the Government (two collectors), an instrumented demonstration facility (three domestic hot water circuits) built from governmental credits for the "Service D'Electricité de l'Etat" at Capellen, and an exploitation phase, involving natural ageing of the prototypes in dry conditions, and monthly operational control of the Capellen circuits. Gradel was further involved in the CEC more theoretical activities of flat plate collector performance measurement (1).

2. MINERAL SELF-SUPPORTING THERMAL INSULATION

The interest of mineral with respect to organic insulation materials came out from preliminary collector studies. They have no such drawbacks as thermal instability (outgassing, decomposition, loss of thickness), sensitivity to pests, fungus, humidity and limited mechanical resistance. Their actual disadvantage, their weight, must be turned out into a plus point by letting the insulation panel participate to the bearing structure of a building, and be integrated to it. The raw material, the sand, is abundant,well distributed over the earth crust and inexpensive.

Two prototypes were built from commercially available light building materials (Table I.).The first collector involves prefabricated thick panels of hydrogen expanded concrete, reinforced, 2 m long, spanned between two metallic supports, side by side, as a base for conventional copper absorbers. A single peripheral band of the same material cemented around the absorbers provides for the lateral insulation and receives the profiled glasses giving an esthetic aspect to the assembly. In a second type of collector, thinner plates of fiber cement, mounted together as wooden panels (glued and stapled), form a casing: the bottom insulation is completed by glass foam loose panels. A single glass pane covers the conventional absorber, and is maintained by aluminum profiles screwed in the casing sides.

The dry stagnation tests showed the interest, for these porous materials directly exposed to atmospheric agents, to prevent the occurence of stagnating water. The coating, which must be permeable to water vapour (escape of condensed water), is not always homogeneous on the porous structure. Water trapped behind the coating loosens first this latter from the support when frost occurs. Further the water is soaked in the open porosity and destroys the material in successive layers (tight materials), or in its integrity (porous materials). This was the case for the most porous fiber cement used in the second prototype which disintegrated in powder form after one winter. Besides, too large a span induced mechanical stresses (coating and glue crippled), the aluminum cover thermal stresses (profiles torn out) and the tars of the glass foam deposited behind the glass pane. The thicker porous concrete reacted much better in that respect: from the design stage on, outside water was kept away (protective aluminum cover on top, glass profiles extended downwards and sidewards, glass profiles supports sealed to material;) improvements were brought later on against condensing water (gutters under glass profiles and bottom drains).

As far as the thermal properties are concerned, they were generally poor, particularly for the porous cement prototype which exhibits a quick dropping performance curve and a very high thermal capacity, preventing it to exploit solar incidences of short duration.

The successful mineral based collector compares favourably with the foreign manufactured ones, an all organic type (P.V. foam encased in glass fiber reinforced P. Ester) and a more metallic one (rock fiber in aluminum casing), as far as durability is concerned, unfavourably as far as thermal properties are concerned. Four of this type are mounted in one of the demonstration circuits.

3. OPEN CIRCUIT WITH ANTI-FROST DRAINING

The interest shown for open circuits is related to the consumer's safety (no poisoning risks) and to the possible added performance increase (no heat exchanger). The draining of collectors, one of the means to avoid freezing, became an important goal to master, in the local cold weather conditions.

Off the three domestic hot water circuits put into operation in april 1980 (2) one has been provided with all the risky novelties. It supplies water to the change rooms of a garage, unpermanently occupied, so that the disturbance to the consumers in case of an incident is minimized. The novelties encompass the mineral collectors of Luxembourg design (3), the open circuit running hot hard water and the automatic draining. Two family houses benefit from the comparative domestic hot water systems, closed loops with anti-freeze, with respectively fuel and electric back up supplies. The three of them are instrumented with continuous registering of solar energy collected, make-up energy consumed, total energy delivered.

The experience with the open circuit and its drainage, puts forward that the designers fears related to the malfunctionning of frost exposed components (sensors, air purges...) were unjustified, and that the effects of solar uncontrolled temperature rise on components as draining valves was largely underestimated, despise the water softener incorporated in the circuit.

A first period of one and a half year encompassed the debugging of mounting, acceptance, recording and interpreting faults, mainly attribuable to a deficient organisation diluting the responsibilities between designer, plumber and user. Consequently several incidents and two major accidents were reported: a first freezing in jan. 1981, affecting three collectors out of four (draining valve plugged because of incorrect mounting and connecting, and having escaped notice during acceptance tests, added to an overdimensionning of collector area with respect to the over estimated water consumption), a second in Jan. 1982 affected the lower manifold (unsufficient slope undetected at repair acceptance).

The routine period, under the sole responsibility of the designer, emphasized the water hardness problem, any incident increasing the temperature or the water circulation rate leading to valve plugging. Two casual short circuits in the differential regulator, one in March 1982 closing the valves, the other in October 1982 we experienced. The second feeding constantly the pump and circulating the boiler water in the collectors, led to valve plugging within weeks. A monthly control revealed a last plugging (Aug. 1983). Safety minded draining design (drainage occurs in case of electric breakdown), involves a valve constantly fed and an unavoidable local heating. Incidently, the standard exchange of the first regulator with one settled at a draining temperature of 8°C instead of 2°C, certainly helps to reduce the valves plugging frequency.

Comparing the overall circuits performances does not exhibit very reliable results: such factors as water consumption of the user, differences in energy data recording (fuel or electricity) make the circuits too different to judge for their respective merits. The solar contributions remain rather deceiving. (Table II.)

That of the mineral insulated collector circuit suffered from the low water consumption compared with those of the permanently occupied houses.

4. CONCLUSION

Although the indulgent approach towards the user impairs a sound comparison of the prototype circuit with the commercial ones, the mineral insulation option tested is satisfactory at least as far as durability is concerned. The draining concept must be carefully studied and instrumented, mainly when hard waters are encountered. The solar contribution was deceiving in general and the water consumption assumptions overestimated.

The investments are not significantly different from one circuit to the other, and in the order of 220.000 FB for 8 m^2 each. The annual economies, although growing with the energy prices, do not reach yet the amount of annual interest the bank sector can give for the same money. This has been the major reason why Luxembourg does not pursue anymore any research in the field of solar thermal energy, independantly from the significant maintenance costs and the survey time devoted to these facilities, essentially attribuable to the open circuit with drainage. Indicatively the repair/survey costs amounted to 50.000 FB/year for the three circuits, two years long, and sunk to 25.000 FB/year (mostly survey costs).

REFERENCES

1. REMOVILLE J. "The limits of the transient method applied to the solar collectors performance measurement" same Proceedings
2. REMOVILLE J. "Solar hot water systems at Capellen" (1980) Gradel report nr 813.
3. REMOVILLE J. "Element de paroi solaire pour bâtiment" (1980) Lux. Patent.

Glassfoam Material	Use/ thickness	Density kg/dm3	Thermal conductivity KCal/mh°C	Fixation means	Sealant	Coating
1st collector						
H2 expanded light cement SIPOREX	bottom 150mm side 125mm	0,55	0,09	SIPCOL cement glue	SILPRUF silicone based	FESTOL 450
2nd collector						
Heavy PROMABEST"H"	bottom 15mm	0,75	0,118	PROMAT PSG 32 Heat resisting glue + stapples	SILPRUF	FIXA-CRIL SAPTO-LITE
Light PROMABEST"L"	side 50mm	0,43	0,071			
FOAM-GLASS T2	insulation 50mm	0,14	0,044	loose		

Table I. Characteristics of commercial insulating materials used for Luxembourg collector prototypes.

	Global 1980-1983	Summer 1980
House I. with exchanger oil back-up energy	33%	45%
House II. with exchanger electrical back-up	51%	66,7%
Garage open circuit electrical back-up	33%	38%

Table II. Compared solar contributions to the DHW Systems of Capellen.

Fig. 1. - Metallic (left), mineral (middle) and organic (right) based collectors of the 3 DHW solar circuits of Capellen.

Fig. 2. - Self-supporting - Porous concrete insulation and glass profiled cover of the mineral based Luxembourg designed collectors.

THE LIMITS OF THE TRANSIENT METHOD APPLIED TO THE SOLAR COLLECTORS PERFORMANCE MEASUREMENT

J. REMOVILLE
Commission of the European Communities, detached at GRADEL S.A.

Summary

The paper reviews shortly the three methods basically used within the 20 laboratories involved in the CEC collector performance testing programme, for measuring the thermal performance of flat plate collectors. It describes the principle of the transient method and the measuring facilities used in Steinfort. It gives the standard divergence in results obtain using this method compared to the others for testing collectors CEC III and IV (flat plate type). It sorts out the paradoxal results obtained with the collector CEC V, of tubular evacuated heat pipe type, mainly at low flow rates. It concludes that the transient method is not applicable to heat pipe type of collectors, principally because it does not allow the gaz and liquid phases in the heat pipes to stabilize. Managing for this stabilization by increasing the flow or correcting for the latent heat take the interest of the method away, which was reduced investment in equipment and measuring time.

1. INTRODUCTION

In 1978, Gradel joined the European solar collector testing group, the aim of which was comparing the methods, the equipments and establish recommendations (1) for measuring the performance of thermal collectors in the european climate. The concerned laboratories had already commissionned their testing facilities, tested a first collector and were busy with the measurement of a second. Therefore, the CEC awarded contract involved but a partial dotation, which added to the Luxembourg own limited financial means (government and industry), incited it to design and run the cheapest possible facility, together in investment and man-power needs. It is commented hereafter on the transient method which appeared to fulfill these economical requirements. In the fall of 1982, Luxembourg stopped its research programme on thermal solar energy.

2. THE TRANSIENT METHOD

Among the three main methods followed in Europe for measuring solar collector performances, the simulator one involves large investments (lamp array, computer...), the outdoor quasi-steady state method requires inlet temperature regulation and good sunshine conditions seldom in Europe, whereas the combined in-door, out-door method relies on two loops and frequent collector transfers.

The transient method first described by the JRC Ispra (2) claimed further that it could reduce the time necessary to test one collector, to probably two sunny days. It allows the sunshine to progressively heat the reservoir water, and the inlet temperature to rise during the measurement (at a limited rate, playing with the reservoir volume). The corresponding expression of the performance η is recalled hereafter

$$\eta = \frac{1}{IA} \left[\dot{m}f\ (Te - Ti) + (mc) \frac{\Delta Tm}{dt} \right]$$

with the solar irradiance I, the collector area A, the mass flow $\dot{m}f$, the inlet, exit and collector mean temperatures Ti, Te and Tm, the collector heat capacity (mc) and the time t.

The equation introduces a correcting term, taking into account the energy stored in the collector by the rate of temperature increase. Two new dimensions have to be determined, the collector heat capacity (measured) and the rate of temperature increase (determined by calculation or graphically).

The measuring procedure, with the scarcity of even half-days on continuous sunshine in Luxembourg, has been a compromise between Ispra recommendations and the French standard (3). The sun was allowed to increase the water temperature during one hour measurement. Another temperature level was reached with the help of the reservoir electric heater. Further one hour measurement followed after the temperature stabilization. The temperature level steps were generally 15 to 20°. The rate of temperature solar increase was limited to 10°C/h, reducing the influence of the correction term.

A rough transient method (4) was used for determining the heat capacity, using the fact that the heat losses are equal for equal collector mean temperatures: two heat losses experiments (heating the reservoir water by a constant rate) were pursued for two different temperature increase rates (varying the reservoir water volume). The measured data were also directly useful for the heat losses curve.

3. LOOP SHORT DESCRIPTION (Fig. 1.)

The used loop is very simple, with following peculiarities: the water reservoir is open to air, necessitating a pump rather than a circulator for flowing the water. Collector hot water falls from the reservoir top and helps a quicker mixing of the stored water, mainly during or after the electric heating periods. A by-pass valve allows to adjust the flow. Temperature measurements used first thermometers read every 6 minutes, then differential thermo-elements in series continuous recorded, finally a multi-point thermorecorder (every 2 min.). Flow measurements used rotameters. The differential temperature measurement did not work properly. The pump made it difficult to get a reliable flow adjustment.

4. RESULTS

The method was successfully applied to the round robin collectors III and IV of the CEC performance tests programme. Generally the method is less reliable and less precise than the main ones, even though the dispersion experienced in the measurement results of Luxembourg improved for the second collector. Whereas Luxembourg scattering was significant compared to that of the other laboratories (20% instead of 7%), with respect to the mean values obtained for collector III, the dispersion reduced to respectively 7% and 5% for collector IV. This can be attributed to both experience gain and instrumentation improvement.

Applied to collector V, a tubular evacuated heat pipe type of collector, the method yielded erratic results at low flow, namely increased performance with increasing mean collector temperature. Repetitive tests at same flow conditions gave similar results: the apparent heat gain is expressed by a positive slope of the same value for all the performance curves obtained, whereas the ηo values of the curves are very much dispersed from one day to the other. Other reported anomalies are: outlet-inlet temperature differences sharply fluctuating (up to $\pm$ several degree within 6 minutes), not directly decreasing with decreasing incident energy, the answer time being shifted not negligibly (Fig. 2). The thermal loss curves exhibit a similar erratic decrease with increasing collector mean temperature, and the corresponding temperature difference variation in quite peculiar (sharp increase first, sharp decrease, stabilization followed by a second increase).

Increasing the coolant flow (from 601, to 100-150 l/h) reestablishes logical reactions (η curves of more classical aspect, heat losses increasing with Tm-Ta), whereas the temperature sudden fluctuations are still observed, with periodical sharp drops .

5. INTERPRETATION

As the heat pipe filling conditions (6)(working fluid level at ambient temperature and pressure), are only known from the manufacturer, further research related to the thermal properties of isobutane (critical points coordinates, latent heat of condensation...)or related to the phases displacement with respect to the critical point, was not undertaken. Nevertheless, it is assumed that two phases are present in the heat pipe, at least at normal temperature.

Analysing the collector behaviour, one can attribute some of its anomalies to the heat pipe thermal irreversibility and to its geometrical manufacturing (unequal mass distribution, thus unequal thermal capacity distribution).

In steady state conditions, all heat transfers in the collector are equal: solar input, heat given to the working fluid, heat required by the evaporation, heat restituted by condensation, heat transmitted to the coolant. In transient conditions, as the heat transfer rates involved in the above mentionned successive operations are not equivalent, and as several may even change in the course of the measurement (if one phase is left instead of two, the evaporation process becomes a mere conduction process), discontinuous values of coolant inlet-outlet temperature differences are experienced: as high as 1° and even $2^{o}C$ variations of the mean ΔT value are recorded at low flow (Fig. 2), in the chosen measuring intervals of 6 min, with practically invariable solar radiances. The main term in the performance calculation is thus already queried, forgetting even the correction term, where the mean temperature derived, $\frac{dTm}{dt}$, varies of sign as of orders of magnitude.

A short survey of coolant temperature increase versus time (Fig. 2) shows the following: first, a sharp ΔT rise where solar heat transfer in the heat pipe (2 phases) is optimum. Á sharp ΔT decrease follows, probably imputable to liquid disappearance in the heat pipe, the energy stored into the copper heads participating alone to the coolant temperature increase. A more steady portion of heat transfer is established, a new unstable equilibrium between the hotter heat pipe wall, and the copper sink closer to coolant temperature: successive appearances-disappearances of liquid may even be enhanced in suddenness, by phase change delays (flash distillation). Occurence frequency recorded varies between 10 minutes and an half-hour.

Reversedly, the behaviour of temperature decrease versus time during heat losses measurement (Fig. 3) expresses the heat pipe irreversibility: optimum transfer as long as the copper heat sink has not reached coolant temperature, then difficult transfer along the heat pipe walls. The same steadier portion follows, as before, with intermittent vaporisations inside the heat pipe.

Increasing the flow under solar irradiation, improves the heat removal rate from the copper heat sink (as the heat pipe becomes emptied from liquid): it prevents the copper to store too much heat. The time delay between solar input and ΔT reaction depends on the coolant flow. At low flow (Fig. 2), this delay is seemingly half an hour or one hour. As the method

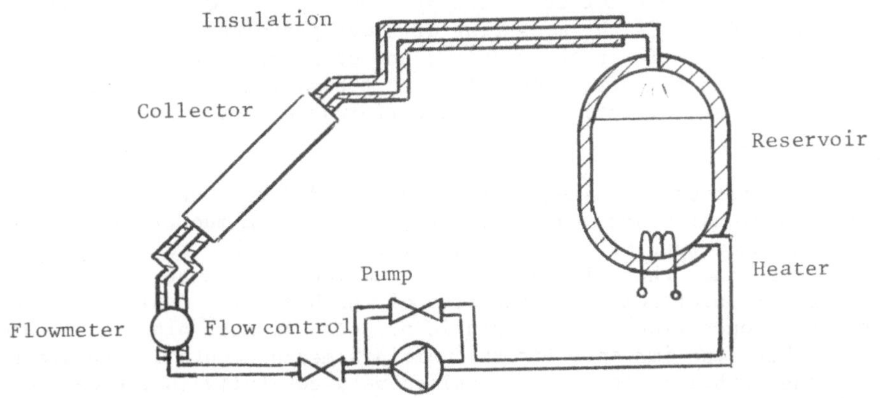

Fig. 1. Schematic circuit of the transient method used.

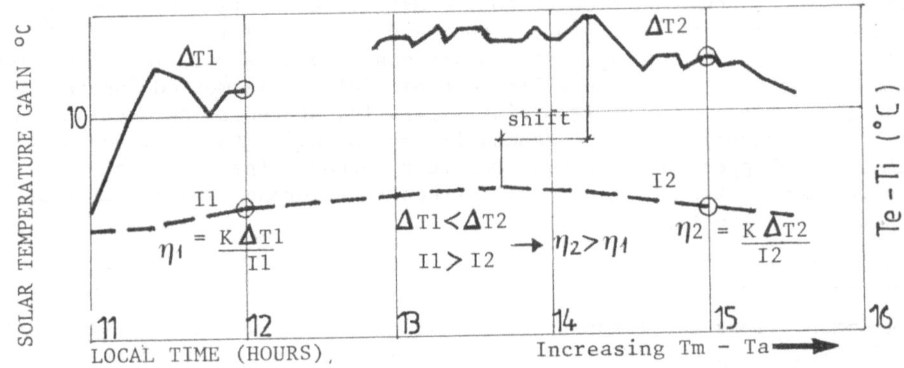

Fig. 2. Fluctuations of solar temperature gain and aspect of solar incident energy versus time at low flow, showing time shift.

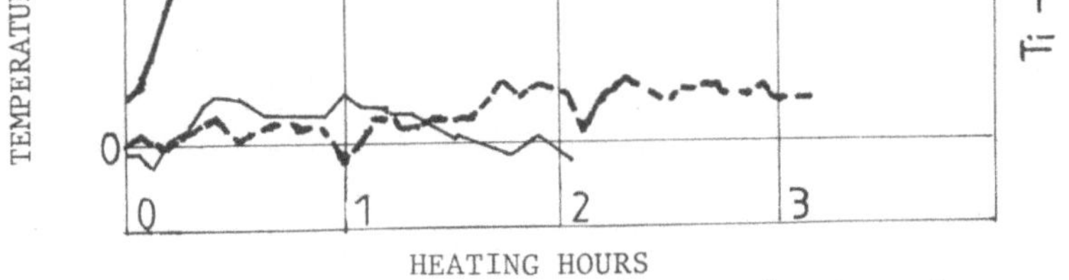

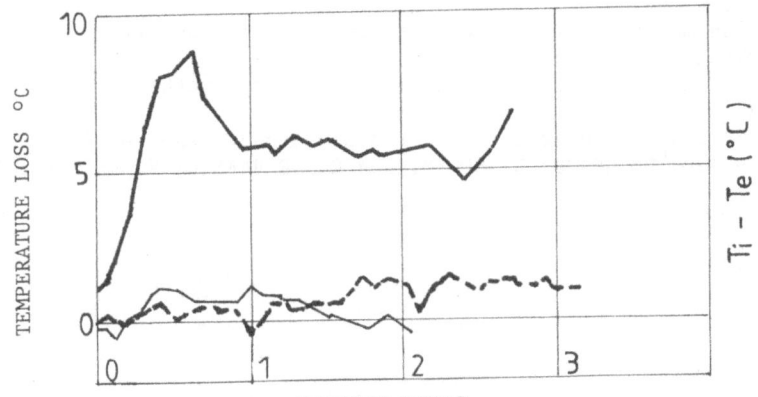

Fig. 3. Temperature difference fluctuations during heat loss measurements at low flow.

involves a one day duration for the efficiency measurement, the high solar inputs of the morning have not achieved high Δts yet, whereas the maximum ΔT of the afternoon are recorded as the pyranometer reading has already considerably decreased. Hence the measuring of lower performances (proportional to $\frac{\Delta T}{I}$) at low Tm - Ta in the morning, and of higher performances at high Tm-Ta (in the afternoon). At high flows, the delay falls to 10 minutes at the utmost, affecting less the performance curves aspect.

6. CONCLUSION

The above described transient method has yielded reasonably good results with traditional flat plate collectors. It has been much less reliable for heatpipe tubular type of collectors, mainly at low flow rates. This is imputable to the inner thermal inerty of the collector concept, probably linked to the irregular heat transfer rate of the heat pipe (shifting from two to one phase operation). The intrinsic property of heat pipes make them heat irreversible, and improper for thermal loss measurements in transient conditions: the method is thus to be applied very carefully in the case of this last type of collectors, controlling in particular whether both phases of the fluid are constantly present, guaranteeing a single heat transfer process in the course of the experiment.

Improvements could be brought to the method: enhancing the flow is not sufficient; reducing the temperature increase rate (from 12°C/hour to 3°)in order to get closer to steady state conditions, involves more measuring time, requires more sunny hours, cancelling out partially this method inherent advantage; reducing the data recording intervals, or recording continuously, although more expensive is the reasonable way, allowing to apply computer programmes developped for advanced transient methods. The transient approach remains the most interesting one for collector performance measurement in the european climate (5).

7. REFERENCES

(1) Recommendations for european solar collector test methods by A.DERRICK W.B. GILLETT, Jan. 1980 - Edited by University College Cardiff.

(2) Determination of the instantaneous efficiency of a solar collector by a transient method by E. ARANOVITCH, M. LEDET, C. ROUMENGOUS Dec. 1976 (J.R.C. Ispra)

(3) French Norm P.50.501. Capteurs solaires. Mesure des performances thermiques Dec. 1977

(4) Simplified measurement of the heat capacity of a solar collector, by J. REMOVILLE - Luxatom 1979

(5) The development of CEC transient collector test methods, by N.B.GILLETT May 1982

(6) Evacuated tubular collector with two phase heat transfer into the system - J.C. DE GRIJS, H. BLOEM, and R. DE VAAN, N.V. Philipps 1980

ETUDE EXPERIMENTALE DE CAPTEURS SOLAIRES A AIR EN REGION PARISIENNE

J.L. DUFRESNE, P. EHINGER, J.C. FAYOLLE, A. LAHELLEC, A. LAKHSASSI.
Equipe R.A.M.S.E.S. (CNRS-PIRSEM) - Centre Universitaire, Bât. 208
91405 ORSAY - FRANCE

Summary

During the year 1982, we have tested, in external climatic conditions, two air cooled solar collectors arranged to have a very good air/plate absorber thermal transfer (ripple plate and bottom) and specially designed as a wall component. In some limits of quasi-stationary climatic conditions, the classical model of DUFFIE, BECKMAN and KLEIN fits well the data. For unstationary periods - most of cases - we propose a model taking into account the inertia of the bottom of the collector. A good fit is obtained without any cut on climatic conditions. The global heat transfer coefficient, h, is determined as a function of R_e.

1. L'EXPERIMENTATION

De mi-79 à mi-81, nous avons effectué le suivi thermique d'une maison solaire dans la région parisienne, équipée de capteurs solaires à air et d'un système de stockage sur lit de roches (Réf. 1). Nous avons noté un faible rendement durant les périodes d'alternance soleil/nuages fréquents à Paris au printemps et en automne. C'est pourquoi nous avons testé en 1982 et 1983, des capteurs à air conçus pour être mieux adaptés à de telles conditions.

Sur notre banc d'essais de capteurs, réalisé à ORSAY en 1981 (Réf. 2), nous avons testé en atmosphère naturelle, deux capteurs conçus par notre équipe pour assurer un très bon transfert thermique air/absorbeur (Figure 1) Il s'agissait, en tenant compte des propriétés mécaniques et thermiques d'un matériau isolant de conception récente, le PHENOPERL (Réf. 3), de concevoir un capteur à air qui soit également un élément de mur. L'absorbeur, à faible inertie, est une feuille d'aluminium de 1, 7 kg/m^2 en forme d'onduline. La couverture est une vitre de 6 mm d'épaisseur.

Les conditions expérimentales ont été les suivantes :

- Débit de l'air caloporteur	:	30 à 250 m^3/heure (soit 21 à 173 m^3/h/m^2 capteur)
- Perte de charge dans le capteur	:	0 à 150 Pascal/m^2
- Diamètre aéraulique	:	0,0138 m2
- Distance absorbeur-fond	:	2 cm
- Surface d'échange thermique	:	1,45 m^2
- Nombre de Reynolds	:	1600 à 13000
- Flux solaire	:	300 à 800 watts/m^2
- Vent	:	0 à 2 m/s/

Environ 50 points de mesure (débit, température, pertes de charge, conditions météorologiques) sont enregistrés toutes les 6 minutes.

L'étude systématique des conditions météorologiques nous a permis de tracer le profil de chaque journée d'expérience. Nous avons pu en sélectionner un bon nombre pour lesquelles le flux solaire, en milieu de journée, ne présente que de faibles variations autour d'une valeur moyenne.

Cela nous a permis de définir des conditions de quasi-stationnarité :

Flux ϕ = Constante $\pm$ 20 watts/m^2

pendant Δt = 20 minutes minimum.

Une étude statistique détaillée des fluctuations d'ensoleillement a confirmé le caractère draconien de ce critère (Réf. 4).

2. RESULTATS EN REGIME QUASI-STATIONNAIRE

Pour caractériser le fonctionnement des capteurs, nous avons cherché à comparer nos résultats expérimentaux avec les résultats simulés du modèle classique de DUFFIE, BECKMAN et KLEIN (Réf. 5 et 6), sur l'ensemble des pointés de chaque journée expérimentale.

Dans une première étape, nous avons testé sans précaution préalable, le modèle utilisant d'une façon classique le coefficient d'échange d'un absorbeur plat donné par la relation de KAYS (Réf. 5) :

$$h = 0{,}0106\ Re^{0.8}$$

Les résultats fournis par ce modèle sont en mauvais accord avec les résultats expérimentaux (Figure 5.A). Pour améliorer le modèle, deux paramètres sont accessibles à un meilleur ajustement : le coefficient d'échange, h, et l'absorptivité, α, de l'absorbeur.

Pour tenir compte de la forme onduline de l'absorbeur, nous avons donc mesuré, pour chaque journée expérimentale, la valeur du coefficient d'échange h (exp.). La loi d'échange établie est :

$$h(exp.) = 0.225\ Re^{0.59} \quad \text{(Figure 2)}$$

Pour notre configuration, nous avons donc trouvé un h supérieur à celui de la relation de KAYS. En particulier, à 100 m^3/heure :

$$h(exp.) \simeq 3\ h\ (KAYS).$$

En ce qui concerne l'absorptivité, α, de l'absorbeur, valeur connue sans grande précision, nous avons cherché la meilleure valeur qui, dans l'ajustement du modèle, minimise la différence des intégrales des puissances calculée et mesurée. Pour toutes les journées de mesure, on a trouvé une dispersion faible autour de α = 0.87, valeur finalement adoptée.

Les résultats de l'ajustement du couple h(exp.), α sont montrés sur les courbes de la Figure 5-B, sur lesquelles sont portés le flux ϕ et les puissances utiles calculée et expérimentale. On constate un assez bon accord expérience/modèle, avec toutefois la présence d'inertie dans le capteur.

L'étape ultérieure est donc l'étude du capteur en régime dynamique.

La Figure 3 présente le rendement du capteur en régime quasi-stationnaire.

3. RESULTATS HORS REGIME STATIONNAIRE

Nous avons élaboré un modèle thermique qui prenne en compte l'inertie du capteur, c'est-à-dire celle qui provient du fond et des bords, la contrainte que nous imposons au formalisme étant de conserver l'aspect systématique global du modèle. Le circuit analogique représentant ce modèle est décrit dans la Figure 4 :

- Q_A est le flux disponible à la surface de l'absorbeur (= P.utile du modèle sans inertie),
- R caractérise les échanges fluide-absorbeur et fluide-fond que nous supposons semblables, et se calcule en fonction de h, R' et R_S,
- La résistance R' traduit les transferts convectifs absorbeur-fond,
- R_S représente les échanges radiatifs linéarisés absorbeur-fond,
- r et C caractérisent l'inertie thermique,
- T_f T_a T_{si} T_i sont respectivement les températures du fluide, de l'absorbeur

de la surface d'isolant et de l'isolant.

Les ordres de grandeurs attendus sont les suivants :

- C estimée comprise entre 10 et 20 Wh/°K (masse d'eau de la peau du fond et des bords et d'une partie de la mousse isolante)
- R_s estimée à 0.65 m^2 K/W
- r de l'ordre de 0.02 m^2K/W (peau)
- R_{ext} infinie (notre modèle d'inertie conserve l'énergie disponible)
- R' = 1.2 R.

Afin d'assurer un formalisme globalisant pour notre modèle, nous avons opté pour une méthode de représentation des fonctions de transfert par facteurs de pondération (Réf. 7, 8 et 4). L'introduction dans ce modèle, des valeurs attendues des paramètres a donné les résultats illustrés dans la Figure 5-C. Les effets observés d'amortissement et de déphasage d'une part, de leur dépendance du débit d'autre part, sont qualitativement bien pris en compte par le modèle. On constate sur une série de mesures, une diminution significative du χ^2. Ceci montre que le modèle permet une meilleure approche de l'aspect "élément actif" du capteur dans un système solaire en projet.

Ce modèle nous a permis de préciser les conditions de quasi-stationnarité pour la détermination du coefficient d'échange, h. On constate aussi que la constante de temps du capteur étudié varie de 0,5 heure, aux forts débits, à 1,5 heure, aux débits faibles.

REFERENCES

1. CLEMENT, D. (1983) Etude expérimentale des systèmes de chauffage solaire pour habitat individuel avec capteurs solaires à air et stockage thermique dans un réservoir de cailloux. Thèse de 3ème cycle ; Université PARIS VII, Octobre 1983
2. FAYOLLE, J.C., LAKHSASSI, A. and PICARD, D. (1983). Etude expérimentale de capteurs solaires à air en climat océanique. Congrès J.I.T.H. 83, Université de MONASTIR (Tunisie), Avril 1983
3. Brevet Français n° 77.08.250 du 18/3/77 de la SICA
4. LAKHSASSI, A. (1983). Etude expérimentale de capteurs solaires à air en climat océanique. Thèse de 3ème cycle ; Université PARIS VII, octobre 1983
5. DUFFIE, J.A. and BECKMAN, A. (1974). Solar Energy Thermal Processes, publié par WILEY-Interscience (New-York)
6. KLEIN, S.A. (1975). Calculation of flat plate collectors loss coefficients. Solar Energy n° 17
7. NESSI, A. and NISOLLE, L. (1975). Régime variable de fonctionnement dans les installations de chauffage central. Publié par les Editions DUNOD (PARIS)
8. MITALAS, G.P. and STEPHENSON (1967). Room thermal response factors. ASHRAE transactions III 2.1.2.10 Part 1.

++

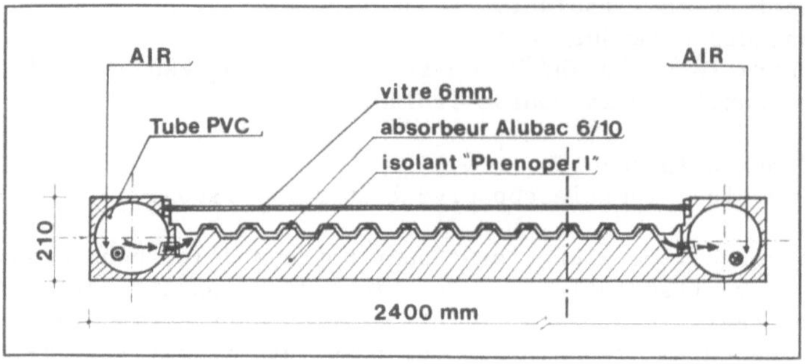

FIGURE 1

Coupe longitudinale du capteur étudié. deux capteurs identiques - C1 et C2 - ont été testés en parallèle. Le caisson est en matériau isolant PHENOPERL (cf. Ref. 3)

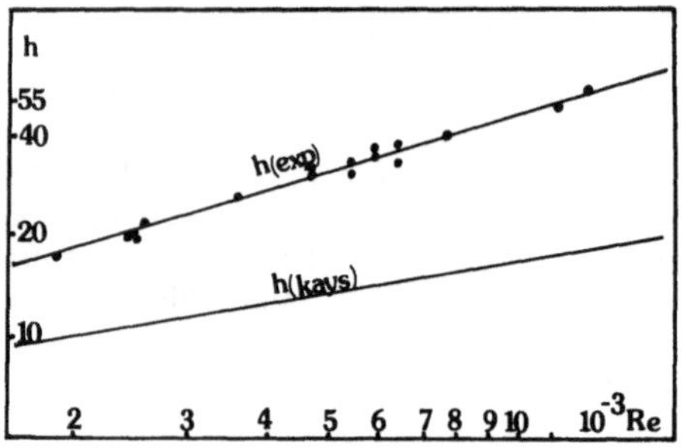

FIGURE 2

Variation du coefficient de transfert thermique air/absorbeur, h , en fonction du nombre de REYNOLDS. Relation de KAYS et résultat expérimental mesuré pour l'absorbeur de la Figure -1-

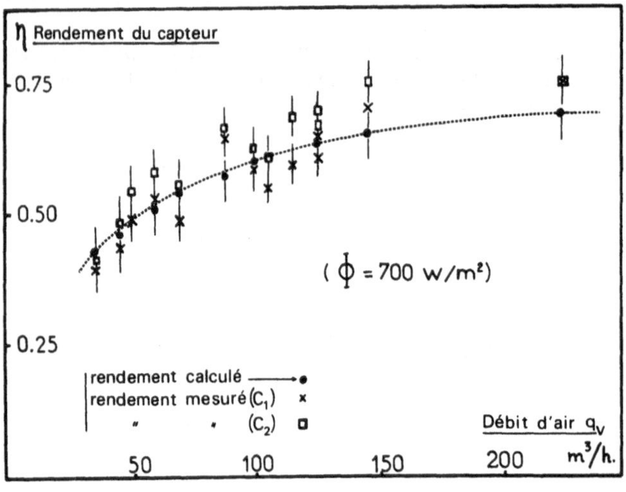

FIGURE 3

Rendement mesuré des capteurs C1 et C2 de la Figure 1 en régime quasi-stationnaire. Comparaison avec les valeurs calculées par le modèle classique; les barres d'erreur sur les valeurs mesurées ou calculées sont indiquées.

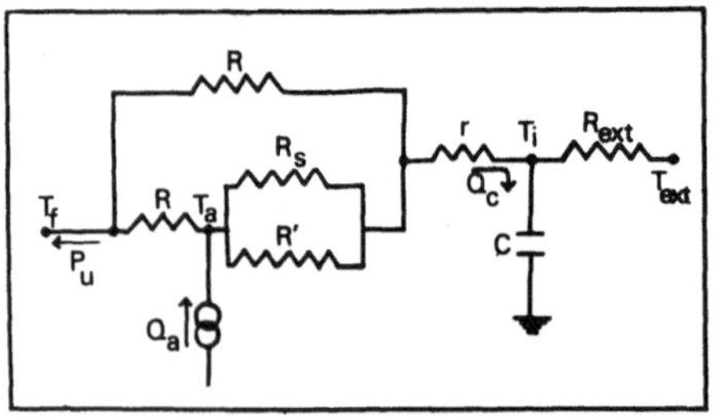

FIGURE 4

Circuit électrique analogue du modèle thermique utilisé pour rendre compte des données recueillies hors régime stationnaire.

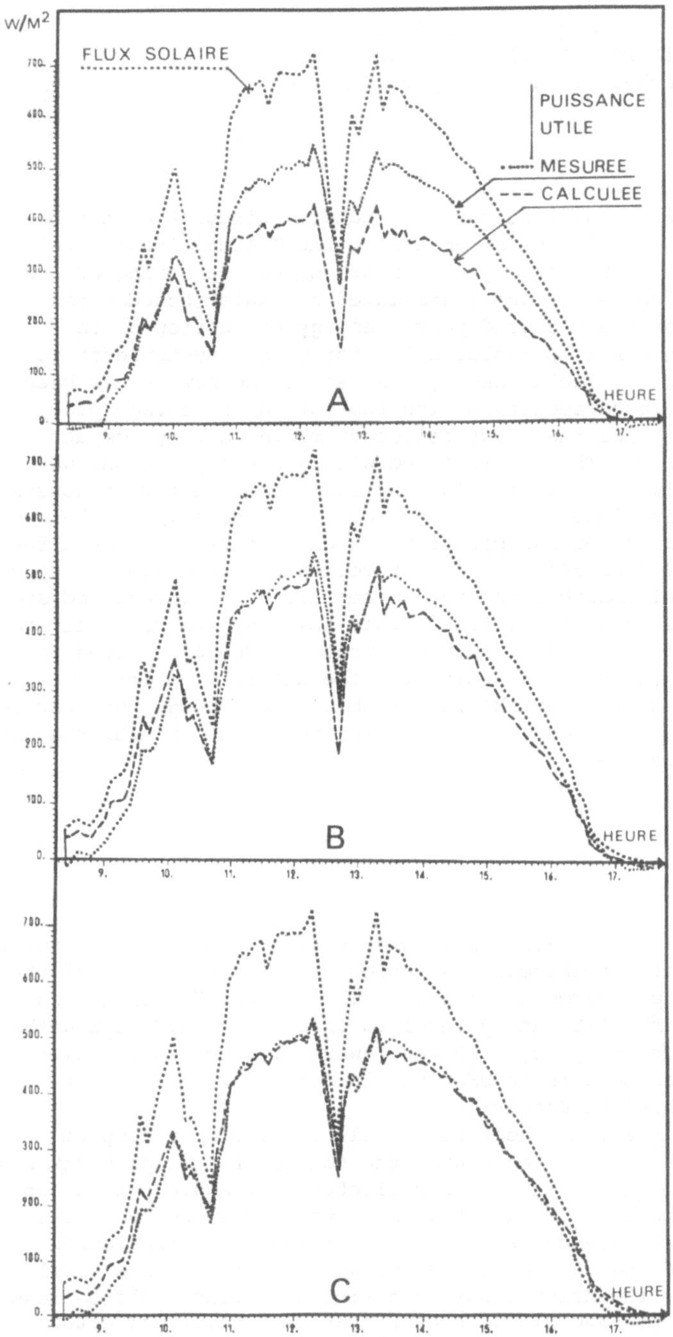

FIGURE 5

Comparaison des puissances utiles mesurées et calculées par m² de capteur de la Figure 1.
[jour 54 - 1982]
débit = 170 m³/h.m²

Cas - A -
modèle sans inertie avec h(KAYS)

Cas - B -
modèle sans inertie avec h(exp.). Valeur du χ^2 d'ajustement mesures / calculs
χ^2 = 2.2 / point

Cas - C -
modèle avec inertie et h(exp.)
χ^2 = 0.9 / point

A 2ND GENERATION THERMOSIPHON SOLAR WATER HEATER

Peder Vejsig Pedersen
Thermal Insulation Laboratory
Technical University of Denmark

Summary

An experimental solar water heater with natural circulation has been in operation with detailed measurements at the Thermal Insulation Laboratory, Technical University of Denmark since the spring of 1981. With 4.5 m^2 solar collector and a simulated hot water concumption of 4x50 litres per day the measured yearly energy production of the system was 439 kWh per m^2 of solar collector with a system efficiency of 40% of the insulated solar energy. As something new it has been possible to have a good operation with the hot water storage placed at the same level as the solar collector making a very compact unit design possible. This is achieved with a new type of one way valve with low flow resistance. In a building project for 55 houses in Denmark which is finished by the end of 1984, the compact thermosiphon solar water heater has been integrated as part of a south facing wall with a tilt of 59°. The solar collector has a 3.5 m^2 selective stainless steel absorber and the 16 cm thick wall integrated storage tank has a volume of 150 litres. This solar system will only represent an extra investment of 10.000 D.Crs. or 1000 $ compared to a normal electric water heater which was the alternative for the houses. The savings in Denmark will be about 1400 D.Crs. per year, so a decrease of the total costs of living in the houses is achieved already from the first year.

1. INTRODUCTION

In the autumn of 1981 a small solar water heating system with natural circulation was built at the Thermal Insulation Laboratory as part of the solar heating research programme in Denmark. The project was initiated because researchers of the laboratory involved in co-operative work within the International Energy Agency, IEA were aware of good results with thermosiphon solar water heaters in USA even in regions with frequent frost periods like we have in Denmark.

At the same time it was the hope to be able to develop a compact system which could work properly even when the hot water storage was placed at nearly the same level as the solar collector. To achieve this, two basic problems had to be solved. First we had to find a one way valve with a very low flow resistance, which could prevent back circulation during night and second, we had to design the whole system with a small flow resistance because the driving forces during operation would be less than when you have the storage placed higher than the solar collector. It was our hope to develop a so-called "single pass" system with a rather high temperature rise over the collector, which has been predicted to have a good efficiency by the well known solar scientist Dr. Harry Tabor and others.

2. RESULTS WITH THE THERMOSIPHON SYSTEM

The solar collector was made as a roof-integrated solar collector with 45° inclination and solar collector with 4.5 m^2 selective absorbers which had a low flow resistance. For the hot water storage a cheap and well insulated 200 litre steel tank with a wrap around heat exchanger was chosen. The solar collector was connected to the wrap around heat exchanger with flexible tubes so the storage could be tested in different hights. Polypropylen glucol and water was used as heat transfer fluid to prevent freezing. The system has functioned with hourly measurements of solar insolation, collector flow, temperature in the storage and energy to and from the storage since Feberuary 1982. There is a simulated hot water consumption of 4 times 50 litres per day with the help of a control clock and a magnetic valve. The system has been tested with the storage placed in different heights compared to the collector. At the lowest, the bottom of the wrap around heat exchanger was placed 550 mm lower than the top of the collector absorber, that means that 2/3 of the storage was placed at the same level as the solar collector. To prevent back circulation during night a one way valve with nearly no flow resistance was placed in the collector circuit.

It was found that the energy production of the system was not changed when the storage was placed in the lowest position compared to when it was placed higher than the collector. But the flow in the collector circuit was halved from a maximum of 2 litres per minute (0.44 l per minute per m^2 collector) to 1 litre per minute (0.22 litre per minute per m^2 collector). The system efficiency per day and per month was nearly constant, varying between 39% and 43% in the seven summer months. The measured yearly production of the solar system is 7115 MJ or 439 kWh per m^2 of solar collector which is 40% of the solar insolation. This is higher than any other comparable system in Europe with reported detailed measurements. There is a nearly constant temperature stratification in the storage on good solar days of 40-50°C. The constant cold bottom of the storage ensures that the fluid entering the collector has a low temperature and it takes only a short time of operation to get a usable high temperature in the top of the hot water storage because of the high temperature rise over the solar collector of 25-40°C.

It can be concluded from the experiement that the compact thermosiphon system has worked very well in all senses and it certainly was possible to achieve a good operation and yield compared to more complicated pumped systems. This simple and compact solar water heating system design has benefitted successfully from the "single pass" function. The reason is of course that the task of the collector is basicly the same, to increase the temperature level of the storage tank from one temperature to another.

3. DEVELOPMENT OF A 2ND GENERATION COMPACT SOLAR WATER HEATER

Based on the good results with the compact thermosiphon system experiment and experience with 1st and 2nd generation active solar domestic hot water heating systems in Denmark, development of a 2nd generation thermosiphon solar water heater was initiated in co-operation with industrial designers and a Danish solar collector manufacturer. The result is a very compact solar water heater unit which for example can be used either as an outdoor unit where a south oriented solar collector with a vacuum formed completely raintight acrylic lid as collector cover is leaning on to a thin insulated storage unit with the same width as the collector, which again can be placed in front of a south facing wall of a house ensuring

MEASURED ENERGY PRODUCTION OF THE EXPERIMENTAL COMPACT THERMOSIPHON SOLAR HOT WATER HEATER AT THE THERMAL INSULATION LABORATORY IS SHOWN IN FIG. 1, AND THE WALL INTEGRATED COMPACT HOT WATER UNIT FOR 55 SOLAR LOW ENERGY HOUSES IN DENMARK IS SHOWN IN FIG. 2.

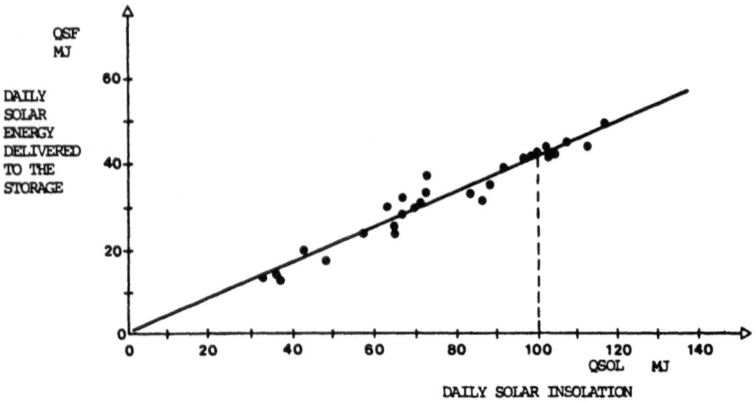

Fig. 1 The daily measured energy production from 21/7-22/8-1982 compared to the daily solar insolation. There is a rather constant system efficiency with a mean of 42%.

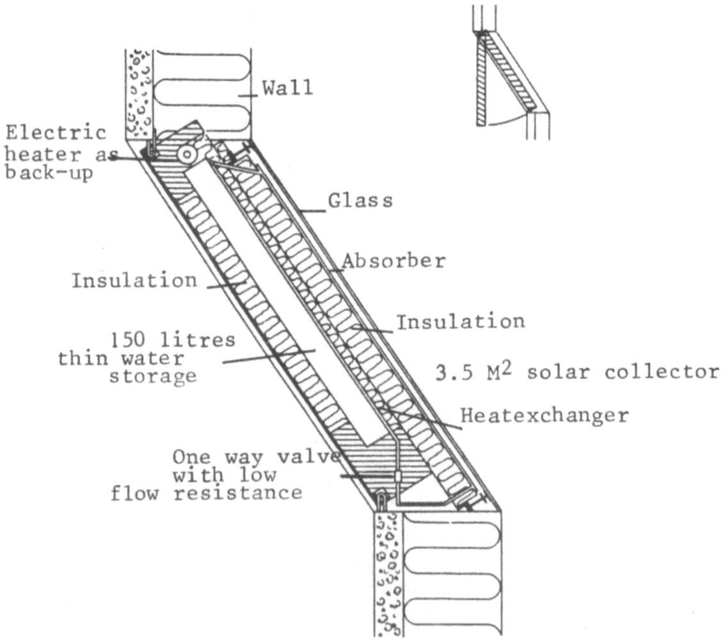

Fig. 2 Wall integrated compact solar hot water unit with natural circulation

non freezing water connections to the unit. A thin 16 cm hot water tank of stainless steel with a stainless steel heat exchanger has been specially developed for the storage unit. With a volume of 150 litres it has a weight of only 50 kg and allows pressure up to 13 bar. An electric heater is built into the storage tank as back up energy, making the solar unit an alternative to a normal electric water heater. The solar unit can also be built into the house for example in the roof. Here it will be possible to place the hot water storage just behind the collector with the same tilt as this, leading to an even more compact solar system.

In a building project for 55 houses in Denmark which is finished by the end of 1984, the compact thermosiphon solar water heater has been integrated as part of a south facing wall with a tilt of 59°. The solar collector has a 3.5 m^2 selective stainless steel absorber and the 16 cm thick wall integrated storage tank has a volume of 150 litres. This solar system will only represent an extra investment of 10.000 D.Crs. or 1000 $ compared to a normal electric water heater which was the alternative for the houses. The savings in Denmark will be about 1400 D.Crs. per year, so a decrease of the total costs of living in the houses is achieved already from the first year.

4. REFERENCES

1. FANNEY, A.H. and LIU, S.T. (1980). Comparison of experimental and computer predicted performance for six solar domestic hot water systems. Ashrae symposium on solar hot water systems, Los Angeles, Ca.
2. GORDON, J.M. and ZARMI, Y. (1981). Thermosiphon systems: single pass VS muilt pass. Technical Note, Solar Energy Volume 27 number 5.
3. TABOR, H. (1969). A note on the thermosiphon solar hot water heater. COMPLES 17,33.
4. MERTOL, A, PLACE, W, WEBSTER, T. and GREIF, R. (1980). Thermosiphon water heaters with heat exchangers . Proc. Am. Section of ISES, Phoenix, Arizona.

DESIGN OF DOMESTIC HOT WATER- AND SOLAR HEATING SYSTEMS WITH PHILIPS' EVACUATED COLLECTORS

R.L.C. DE VAAN

Nederlandse Philips Bedrijven B.V.,
Development Department Solar Collectors
Eindhoven, The Netherlands

Summary

Philips have been active in the field of solar energy for more than a decade, and since 1979 they have concentrated their efforts on evacuated collector tubes. Some of the underlying reasons for this approach are quoted here.
A survey is given of the generic types of domestic hot water systems and their relative prices on the U.S.A. and Japanese markets in 1983, together with an indication of the expected prices for evacuated tubular collector systems. Also details are outlined of the more important field tests carried out on Philips' evacuated tubular collector systems, including data on reliability.
Finally, a few brief comments are made about simplifying systems design when using Philips evacuated collector tubes (E.T.C.'s).

1. INTRODUCTION

Philips' activities in the field of solar energy started in 1974 under the name of Philips Energy Systems, and since 1979 have been incorporated in the Philips Lighting Division. During the period from 1974 to 1979 a number of field tests were realised, not only in solar energy, but also in energy saving and other energy options. Some of the important projects during this period included the solar house in Aachen (Federal Republic Germany) in 1975 (ref. 1), the project Cobbeek, Veldhoven (The Netherlands) in 1978 (ref. 2), projects in Givenchy (France) in 1977, Nuenen (The Netherlands) in 1978, and Comerio (Italy) in 1979.

All these projects, with the exception of the solar house in Aachen, were equipped with different types of Philips' flat-plate collectors. All the projects carried out since 1979, however, have been equipped with evacuated tubular collector tubes (E.T.C.'s). Table V shows a survey of these projects, together with data on reliability. The aim of these activities was to evaluate and improve the performance of components developed by Philips for solar energy systems, development and evaluation of computer models suitable for system simulation and market research. As a result of these activities Philips have concentrated their efforts since 1979 on the development and production of collector tubes with heat pipe, which are available on the market as types VTR 261 and VTR 361.

The principal reasons for the E.T.C. approach were:

a) Solar energy has to compete with many other energy sources and energy-saving options (with possible exceptions in developing countries). Only a slow penetration of the market can be foreseen, which is in turn based on the expected price for solar collector systems now and in the future. To achieve mass production in such a situation, one needs a component suitable for most applications. For temperatures up to 140°C such a component is the Philips' E.T.C. (ref. 3 and 4).
b) Internal studies led to the conclusion that, with medium-scale mass production (10^6 tubes a year), the expected factory price for E.T.C.'s would be about equal to the factory price of flat-plate collectors for a similar active area.
c) Reliability, durability and system availability for conventional systems are extremely high. Customers expect the same of a solar collector system. All field tests and other tests show undoubtedly that Philips' e.t.c.'s can fulfil these requirements.
d) Evacuated tubular collectors with heat pipe cannot only be used in large(r) systems, but are also extremely well suited to small scale applications (even in single tube systems, e.g. solar cookers). These are important advantages in marketing such a component.
e) From system studies, computer simulations and field tests it appears that E.T.C.-based systems offer the highest gains, the best reliability, the lowest maintenance cost and, if mass production is realised, also the lowest system cost. Therefore Philips believe that the evacuated tubular collector is the key component for a successful operation in the solar energy market.

In the following paragraphs these points are explained further, together with results of Philips' marketing research and field tests.

2. COLLECTOR SYSTEMS AND THE MARKET

Designing and using solar collector systems may be interesting activities from a scientific point of view, but the aim of industry is of course to develop products which can compete with other energy sources. This means that it is necessary to have sufficient market information for creating a successful design.

In the product design is is also necessary to take into account at as early a phase as possible the limitations set by politics, economy and climates. Information on the level of penetration of solar systems can, for example, be found in ref. 5. Based on these data one would expect a penetration level of approx. 10% for the market if a pay-back period of five to seven years could be realised. From the available information on the solar markets in Japan and the U.S.A., this seems indeed to be the case in sunny, non-freezing regions.

Table I shows a survey of generic types of domestic hot water systems available on the market with their relative price range (ref. 6 and 7). The number of systems sold on the Japanese and U.S.A. markets in 1983 was well over 800,000; this number is already so high that a sharp decrease in the price of solar systems is not expected in the near future.

Table 1

Survey of DHW Systems (generic types) and price range in Japan/U.S.A. (1983)					
System type	Configuration*	Ease of installation	Active area (m^2)	Price range (in U.S.$, not installed)	
				U.S.A.	Japan
Thermosiphon	1,2	+ +	2	700,–/1400,–	700,–/1000,–
Open loop	1,2,3,4	+	3	1000,–/1800,–	—
Closed loop	1,2,3,4,5,7,8	0	3	800,–/2300,–	1300,–/2500,–
Drain down	1,2,3,4,6,7	–	3	1000,–/2700,–	—
Drain back	1,2,3,4,7,9	– –	3	1500,–/2000,–	1100,–/2000,–

*Configuration: 1 = collector(s); 2 = storage; 3 = pump(s); 4 = on/off control; 5 = heat exchanger; 6 = drain valves; 7 = expansion vessel; 8 = special heat transfer fluid; 9 = special component(s)

If medium scale mass production could be realised (10^6 tubes/yr) the expected price for an ETC-based thermosiphon system is: U.S. $ 800,– ,active area $2m^2$; U.S. $ 1300,– ,active area $4m^2$

In table I is also given the expected price for an E.T.C.-based system in the case of mass production. As this price is equal to that for flat-plate based systems, the price performance ratio for E.T.C.-based systems will be much better. The problem of course is how to achieve the phase of mass production, not because mass production itself is a problem, but the argument lies in the fact that the strategy of energy and economy is strongly politically influenced.

3. SYSTEM RELIABILITY AND PERFORMANCE

Two most important and often neglected aspects of solar collector systems are reliability and related maintenance cost. Far too many systems sold in the past have shown very poor performance (ref. 8, 9, 10, 11). The major reason for this fact is that most systems incorporate too many cheap, unreliable components. It is no solution, however, if more expensive and hopefully more reliable components were used, resulting in unacceptable pay-back periods. If one realises that most of the components in table II have already been mass-produced for many years, there seems to be little chance that a complete set of more reliable, cheap components will be offered, compared with those currently available on the market.

Table 2

M.T.B.F. values for solar collector system components			
Component	M.T.B.F. (yr)	Component	M.T.B.F. (yr)
Collector panel	7 - 10	Powered valve	2 - 20
Storage/expansion tank	5 - 10	Check valve	10 - 20
Hose	3 - 5	Press. relief valve	10 - 20
Soldered joint	> 25	Air vent	0.5 - 8
Pump	2 - 40	Control system	2 - 5
Fan	10 - 40	Heat exchanger	8 - 50

This leads to one clear conclusion: one should produce a solar collector system with a minimum of components. This means either a thermosiphon system or an open-loop system, (see table I), or simplification of the other systems. At this point the question may arise whether this also means that one should not use E.T.C.'s because - with their rather small active area per tube - a collector system requires an array with many tubes. The answer of course depends on the reliability of an individual tube and on the effect that failure of this tube would have on the system. On this last point the Philips' E.T.C. incorporates a heat pipe connected to the system pipe by a clamping block, thus providing heat transfer from the absorber to the system. This infers that, as a result of the thermal diode effect of the heat pipe (ref. 12) and the fact that the only connection between the system and the collector is a piece of metal, failure will only result in a somewhat lower contribution. On the question of the reliability of tubes, an analysis of the field tests with the highest failure rate is given in table III.

From this analysis a mean time between failure of 150 years is found. The mean time between failure found from all data until now is more than 400 years. (Shown in table V.) The reason for this is a change in the sealing process since installation of the Valencia project. From the data currently available it seems fairly certain that the useful life of Philips' E.T.C.'s is more than ten years, taking into account some long stagnation tests and comparison with information on some related lamp types. Whether this period of 10 years could exceed 20 years - as suggested from calculations - depends on the rate of deterioration, details of which are still unknown.

Table 3

Valencia collector array reliability
1 year of operation (1981-10-01/1982-10-01)
System configuration: 34 groups of 57 tubes
1 groups of 38 tubes

Failures per group	Number of groups	% of total	% calculated with M.T.B.F. of 50 yr	100 yr	150 yr	200 yr
0 failures	24	68	32	56	68	75
< 2 failures	33	94	69	89	94	97
< 3 failures	35	100	90	98	99	100

The only comment that can be made at this point is that 430 tubes in an outdoor stagnation test for the last three years have not shown significant signs of degradation. This also applies to a small array with 15 tubes in an outdoor stagnation test which has been running for more than five years.

4. THERMOSIPHON AND OPEN-LOOP SYSTEMS

From the foregoing paragraphs one can conclude that thermosiphon and open-loop systems can be considered the preferred types. They combine low cost, good reliability and low maintenance cost. However, a severe limitation in the use of flat-plate based systems is their sensivity to freezing. Normal use of these systems is therefore restricted to non-freezing and occasionally freezing climates. In this situation in particular, E.T.C.-based systems may offer a breakthrough. Freeze protection can be realised by using simple recirculation. This is not recommended in flat-plate based systems because night losses are similar in magnitude or even exceed daily collector gains.

With E.T.C.-based systems the matter is totally different, especially with those that incorporate a heat pipe. A small extra collector area - two to six tubes, depending on the climate - can fully compensate for the night losses. This is shown in table IV, where calculated results for such systems are given.

For pumped systems it is possible to reduce recirculation losses by means of an on/off or high/low control on the pump. One finds from calculations that this may reduce losses by approx. 30%. This does, however, still result in unacceptable values for flat-plate based systems. In the case of systems based on the VTR 261 and VTR 361 types one should consider the possibility of having the pump running continuously. The pressure drop over well designed systems is so low that a 10W pump is more than sufficient. This means that, in a worst case situation, extra pump losses amount to about 140 MJ a year. This is adequately compensated by an extra tube, and it is to be expected that the reliability and durability of the tube is better than that of the control system. Of course the same also applies to the other systems, although one should consider in that case to use a simple timer as an on/off control as well.

Table 4

Recirculation losses for a DHW system with $4.5 m^2$ active area; $\Delta T = 40K$				
Collector type	System loss* Coefficient (W/K)	'Night' losses (18h) (MJ)	Daily gains** (MJ)	Result
Flat plate, non sel.	37	96	1.3	- -
Flat plate, sel.	19	48	2.7	- -
Philips VTR 261/361	2.4	6.2	6.6	+

* System loss coefficient = active area . U_1 + header loss coeff. + system pipe coefficient

** Calculated for an average $\Delta T = 30K$ and total insolation $Q_d = 3 MJ/m^2$, day

Table 5

Survey of several field tests with Philips ETC's						
Country	Location	Number of tubes	Tube type VTR	Application	System start up	Failures until 84-04-01
Canada	Ottawa	990	141	DHW	N.R.	N.R.
	Toronto	N.R.	141	N.R.*	N.R.	N.R.
Germany	Aachen	72	261	DHW*	N.R.	∅
	Freiburg	210	261	DHW	1982-06-01	1
Italy	Ispra	84	361	Cooling	N.R.	N.R.
Netherlands	Eindhoven	15	141	Stagn. Test	1979-01-01	∅
	,,	120	141	DHW	1979-06-21	∅
	,,	420	141	Cooling	1980-06-01	2
	,,	120	141	DHW	1981-03-15	∅
	,,	98	141	DHW	1982-12-01	∅
	,,	72	361	Test	1982-01-01	∅
	,,	60	141	DHW	1982-04-01	1
	,,	368	261	DHW + RH	1982-07-01	1 ∅
Spain	Valencia	1976	141	DHW	1981-08-24	36
Sweden	Södertörn	1672	141	District heating	1982-04-15	17
	Knivsta	342	141	District heating	1981-06-01	∅
	Studsvik	76	261	Test (performance)	N.R.	N.R.
	Studsvik	56	361	Test (performance)	N.R.	N.R.
U.S.A. (Colorado)	Fort Collins	336	361	Cooling + RH	1982-05-15	14
U.K.	Bracknell	300	141	Room heating	1981-09-01	∅

* N.R.: Not reported

REFERENCES

1. HOERSTER, H. (1980). Wege zum Energiesparenden Wohnhaus. Philips Fachbücher.
2. LANGE, F. DE (1981). The Veldhoven Project. Philips Gloeilampenfabrieken B.V. 'Evoluon', Eindhoven, The Netherlands.
3. The Evacuated Tubular Collector VTR 261. (1983). Philips product brochure.
4. GRIJS, J. DE and JETTEN, R. (1983). Evacuated Tubular Collectors with Heat Pipes simplify Module and System Design. Paper for ISES Conference, Perth, Australia, to be published.
5. LITTLE, A.D. INC. (1981). solar Heating Market. Final Report.
6. SOLAR VISION INC. (1983). solar Products Specification Guide. Harrisville (N.H.) U.S.A.
7. U.S.A. and Japanese product brochures. (1983).
8. URBANEK, A. (1983). Beim teuersten Solartest fehlten viele guten Anlagen. Sonnenergie & Waermepumpen. Vol.8,6.
9. AURIS, R. and DRAVING, W. (1981). Solar Domestic Heating Performance Test Programme. Solar Engineering 1981.
10. MAVEC, J., WATTE, E. and MOLOSEWICZ, R. (1981). Evaluation of Reliability of Operational Solar Energy System. Solar Engineering 1981.
11. WANG, P. and MOLOSEWICZ, R. (1981). Reliability Assessment of Solar Domestic Hot Water Systems. Solar Engineering 1981.
12. GRIJS, J. DE, BLOEM, H. and VAAN, R. DE (1981). Evacuated Tubular Collector with Two-Phase Heat Transfer into the System. Proceedings of the Intern. Solar Energy Soc. Congress, Brighton, U.K.

N.B. Solar Engineering 1981: Proceedings of the ASME Solar Energy Division, Third Annual Conference, April/May 1981.

A SIMPLIFIED SOLAR DOMESTIC HOT WATER AND HEATING SYSTEM WITH EVACUATED TUBULAR HEAT PIPE COLLECTORS

R. Kersten, B. Vitt
Philips GmbH Forschungslaboratorium Aachen
P.O.Box 1980, 5100 Aachen,FRG

Summary

This paper presents experimental results and system analysis investigations on a simple solar system concept, which is characterized by a little effort if extending the solar hot water system to heating. Measured data on a test installation comprising 12 m² of Philips heat pipe evacuated tubular collector (ETC), connected to a single pressurized storage device for both hot water and heating supply, have been used for the validation of a computer model. Based on this, monthly and annual system performance for a typical central European domestic application has been analyzed. The results show that the useful solar energy of a hot water system, overdimensioned with respect to the collector area, is considerably improved by this simple coupling to the heating circuit. In a well designed system of this type the yearly solar energy used can be as high as 400 kWh per m² collector aperture area if high efficiency ETCs are applied. Simulation results are presented concerning the system performance with respect to flat plate collectors (FPCs), different heating loads and varying collector and storage dimensions.

1. INTRODUCTION

The high system efficiency of solar hot water systems comprising evacuated tubular collectors (ETCs) is well known /1-4/. In the central European climate the extension of such a system to heating normally results in a reduced economy. General reasons for this are a worse solar to load ratio for heating with a reduced system efficiency and a worse cost to benefit ratio for the additional installation of collector modules, a second storage unit, piping, valves and controls. The cost to benefit ratio for the extension to heating can be improved by the following simple system concept with Philips heat pipe ETCs /5,6/, being applied in a direct flow solar system. The system design makes use of an efficient natural stratification /7/ within a single tank for both hot water supply and heating. This type of system has been installed and tested at the Philips Experimental House in Aachen (Fig. 1). Its thermal performance has been analyzed with the help of computer models.

2. SYSTEM DESCRIPTION

2.1 Solar circuit

The collector (aperture area = 12.4 m²) consists of six modules (Philips - Stiebel Eltron), each module containing 12 Philips evacuated collector tubes VTR 261 /5/ with an aluminium "ripple" reflector. The header tube is directly coupled to the 650 l domestic hot water storage

3. SYSTEM PERFORMANCE

3.1 Validation of the simulation model

The measured data of selected periods in winter 1983/84 have been used for the validation of the computer simulation model, which is applied to calculate the seasonal performance of this type of solar system. As an example, a comparison is shown between measured and calculated energy flows for a clear (Fig. 3a) and a cloudy (Fig. 3b) winter day. Even on an hourly basis, simulated and measured data agree well. It should be noted that a linear collector model with constant (effective) values for η_0 and $F'U_L$ was used, neglecting incident angle modifier and non-linearities of both the tubes heat loss coefficient and the heat pipe transport characteristics. This collector model is sufficiently accurate though it underestimates the solar gain at very low radiation levels (Fig. 3b).

Hourly comparison of measured and simulated heat flows on a clear (a) and a cloudy (b) winter day in Aachen 1984 :

Fig. 3a

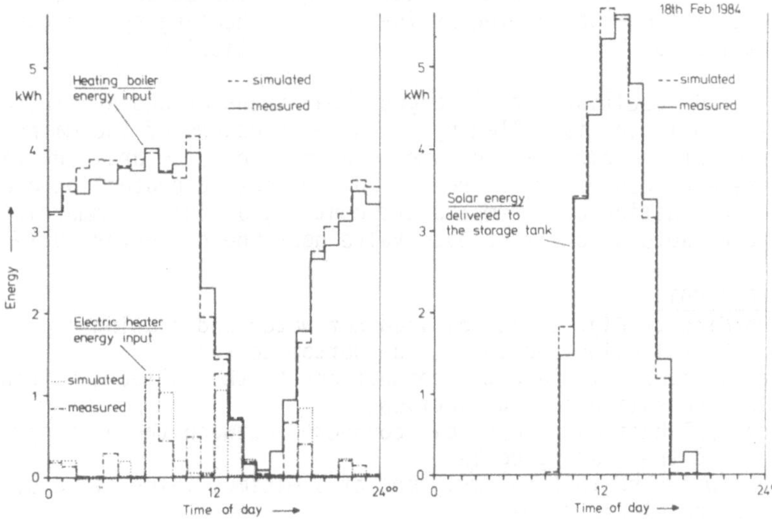

Fig. 3b

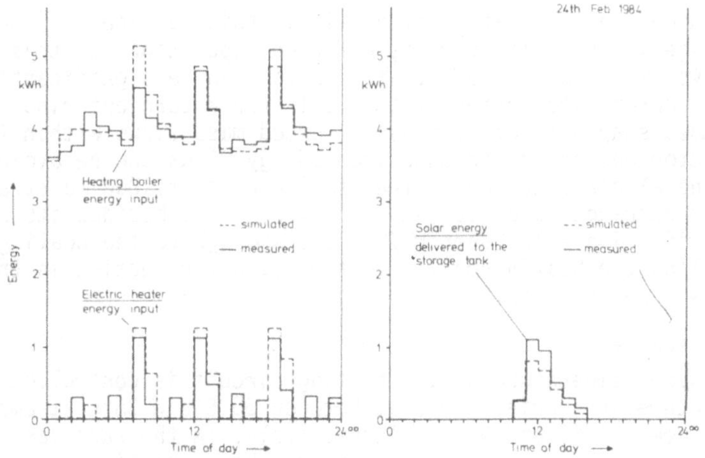

Fig. 1 The Philips Experimental House with its on-roof-collector installation

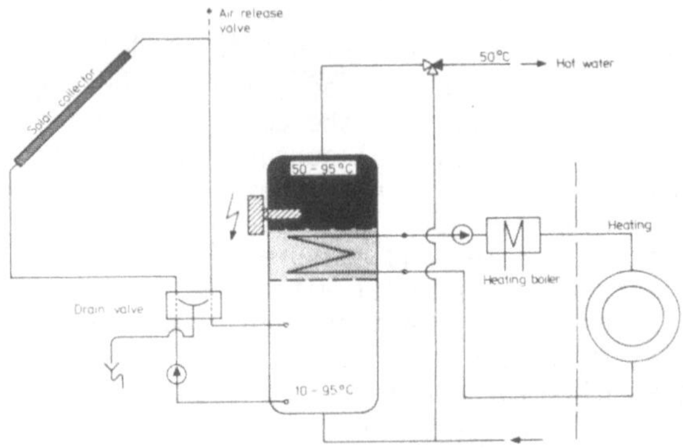

Fig. 2 The solar hot water and heating system (schematically)

tank (Fig. 2) resulting both in a good heat transfer and hydraulic performance. The collector is filled by the water pressure of the mains. The on/off control of the circulating pump is performed by an ordinary temperature difference control. The freeze and overheating protection operates with the combination of a temperature control, a single commercial drain valve and an automatic air release valve near the collector outlet.

2.2 Storage unit

According to Fig. 2 the combined hot water and heating storage consists of the following sections (from bottom to top):

-- the cold water inlet and the connections to the collector circuit,
-- a solar preheating storage section,
-- the central heat exchanger coil connected between the heating circuit return and the heating boiler,
-- the hot water comfort volume, kept at a minimum temperature of 50°C by means of an electric heater.

Maintaining a good stratification within this combined hot water and heating storage is essential for its efficient operation. In this direct flow solar system the degree of stratification can be significantly improved by an appropriate design of the collector return entering the storage: a trumpet shaped inlet suppresses forced mass flows within the tank during collector operation. So undesired energy flows can be avoided, e.g. those from the electric heater to the heating circuit. Due to stratification the heat exchanger operates in both directions: besides its normal function transferring solar energy from the storage to the heating circuit it may also link the heating boiler to the hot water section in the case of cold and less sunny periods.

2.3 Heating circuit

The forward temperature of the heating circuit is controlled by the ambient temperature and varies linearly between 55°C and 35°C at ambient temperatures between -15°C and +15°C, respectively. In the rare case of solar heating surplus with forward temperatures too high, heating control is performed by a mass flow reduction due to the single room thermostatic valves.

For the following simulation the assumed performance data of the Philips ETC with heat pipe (η_0 = 0.70, $F'U_L$ = 1.0 W/m²K) refer to an improved version with aluminium reflector. In the case of flat plate collectors (FPCs) the corresponding parameters are η_0 = 0.75 and $F'U_L$ = 6 W/m²K for a non-selective and 4 W/m²K for a selective version, respectively.

In accordance with our test installation additional assumptions on the system components are a collector flow rate of 35 l/m²·h, loss factors of 5 W/K for the complete collector circuit piping and 4 W/K for the storage tank (volume 650 l). Whereas the Philips ETC operates in a direct flow system, in the case of the FPC the heat transfer factor of the heat exchanger is 30 W/m²K.

The annual hot water energy consumption assumed is 3000 kWh throughout. The heating energy demand of a typical single family house was taken into account as a reference, comprising a good thermal standard (yearly heating demand Q_H = 15200 kWh) and a moderate use of passive solar energy.

3.2 Seasonal performance

The monthly distribution of the total energy demand varies according to Fig. 4.

The corresponding distribution of the relevant solar energy flows are shown in Fig. 5a assuming a 12 m² installation of Philips ETCs.

Fig. 4
Monthly distribution of the energy demand for a one family house, calculated with weather data of Aachen 1976.

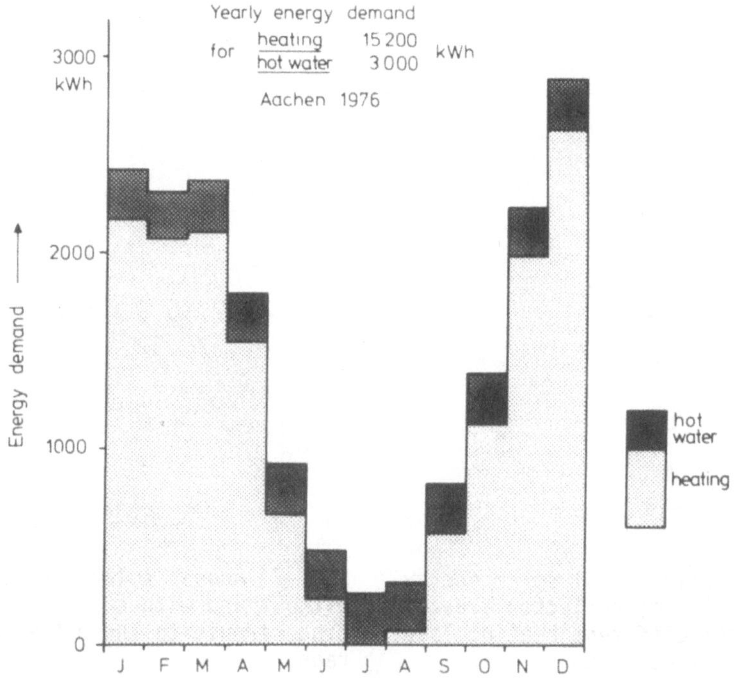

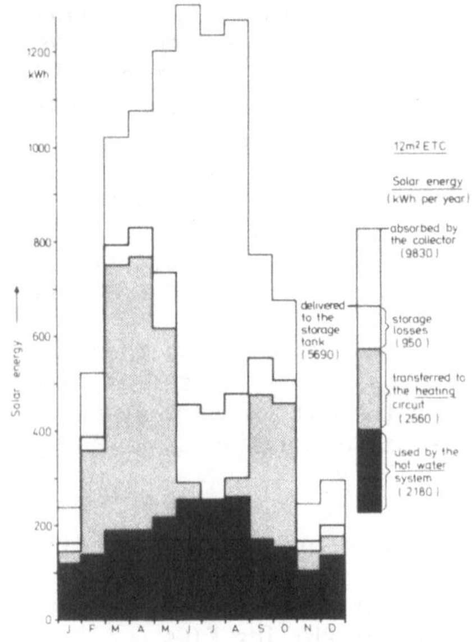

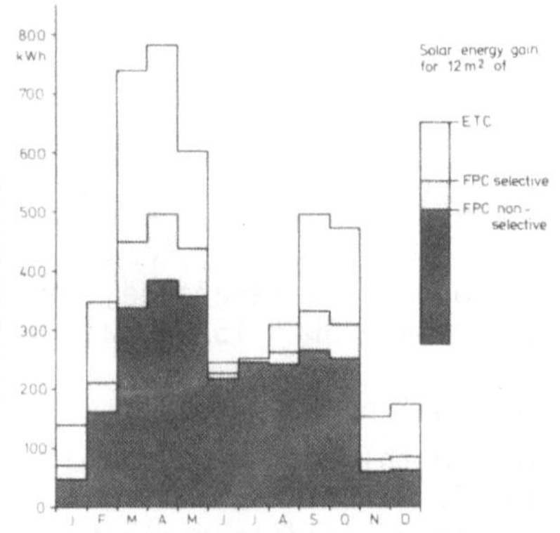

Fig. 5a Monthly solar energy flows. Climate and load structure refer to Fig. 4 (annual insolation: 1170 kWh/m²).

Fig. 5 b Comparison of useful solar system energies for different types of collectors.

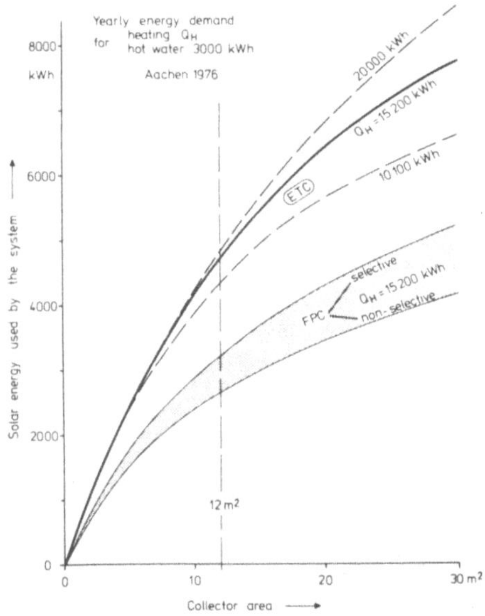

Fig. 6 Yearly useful solar energy as a function of collector area. Parameters: different heating loads and types of collectors

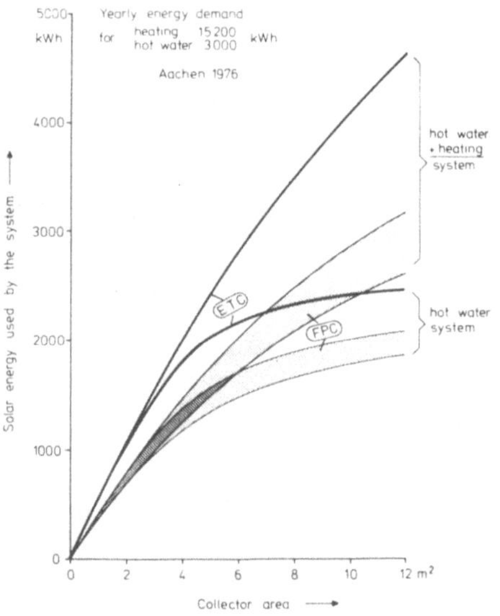

Fig. 7 Annual solar energy gain without and with coupling the heating circuit to the hot water storage.

The solar energy absorbed by the collector is dissipated in different ways:

-- the solar contribution to the hot water system which is comparable to that of a pure solar hot water system,
-- the solar contribution to the heating circuit which is significant in spring and autumn,
-- the solar induced storage losses, especially during summer when the overflow situation accurs, and
-- thermal losses of the collector circuit, including overflow (white area).

The superior performance of the ETC is shown in Fig. 5b in comparison to FPC-installations in the same solar system design.

3.3 Yearly performance

The solar system performance is mainly influenced by the type and area of the collector used. The result of a sensitivity analysis is given in Fig. 6. Obviously an efficient solar hot water and heating system with ETCs can use approximately 400 kWh/m^2 for an optimized design (10-14 m^2) in Middle Europe, and, at the same time is capable of reaching a solar fraction of 25-35 %. In the case of an improved FPC the yearly solar gain is limited to 250 kWh/m^2.

It is worth to note the advantage of this extended solar system in comparison to the pure solar hot water system. The different performance is shown in Fig. 7 which demonstrates that the well known saturation effect of solar hot water systems can be avoided. Obviously, the improvement of the performance is higher if ETCs are used.

Finally, a sensitivity analysis (Fig. 8) with respect to the preheating volume of the storage shows that enlarged volumes can hardly be cost-effective. Consequently, easily manageable storage volumes of about 500 l are sufficient for the system concept presented here, provided the storage with its components is well designed to assure a proper stratification.

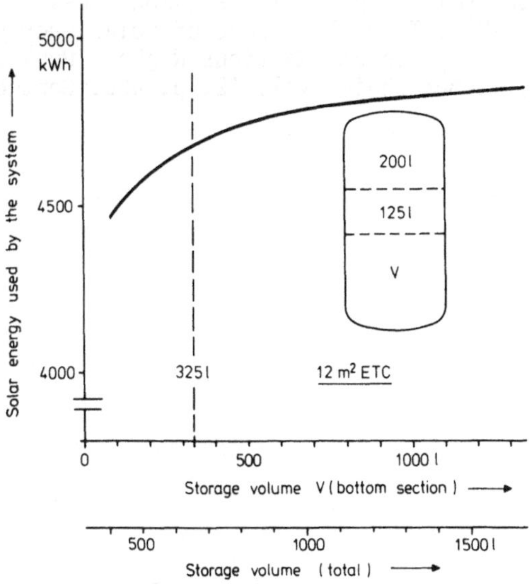

Fig. 8 Useful solar energy versus volume of the preheating section.

4. CONCLUSIONS

For middle European climatic conditions a solar hot water system can easily be extended to heating. Apart from an increased collector area only little effort is required: A slightly enlarged hot water storage with an additional heat exchanger connected to the heating return.

The use of a carefully designed direct flow solar system with ETCs is advantageous in such a system.

For ETCs the annual solar energy used per unit area of collector (400 kWh/m^2) is slightly reduced in comparison with a pure solar hot water system (500 kWh/m^2).

Presently, the economy of solar hot water systems is nearly competitive with conventional systems /8/. The proposed extension may be attractive as a next step because of its simple design.

ACKNOWLEDGEMENTS

We thank E. Kuhl and W. Braun for performing the experiments.

REFERENCES

1. Turrent, D. et.al. (1983). Solar thermal energy in Europe. D. Reidel Publishing Company.
2. Jäger, F. (1981). Solar energy applications in houses. Pergamon Press
3. Hörster, H. Ed. (1980). Wege zum energiesparenden Wohnhaus. Philips Fachbuchverlag Hamburg
4. Schreitmüller, K.R. and Vanoli, K. (1983). Proceedings of the Solar World Congress, Perth, Vol. 2, p. 816. Pergamon Press
5. de Grijs, J.C. and de Vaan, R.L.C. (1983). Proceedings of the Solar World Congress, Perth, Vol. 2, p. 1061. Pergamon Press
6. de Grijs, J.C. and Jetten, R.H.J. (1983). Proceedings of the Solar World Congress, Perth, Vol. 2, p. 1071. Pergamon Press
7. den Ouden, C., Ed. (1980). Thermal storage of solar energy. Proceedings of an international TNO-Symposium, Martinus Nijhoff Publishers
8. Bergmann, G. (1983). Wärmetechnik, Vol. 12, p. 426. Gentner-Verlag Stuttgart

CHAUFFE-EAU SOLAIRE DOMESTIQUE A STOCKAGE INTEGRE

J. BOUGARD*, C. BOUSSEMAERE*, G. DESCY**, M. VAN GYSEL**
* Faculté Polytechnique de Mons - Service de Thermodynamique
** Industrie-Développement-Energie S.A., Rochefort.
Belgique.

Résumé

L'intégration d'un absorbeur solaire, d'un réservoir de stockage de la chaleur et d'un échangeur de chaleur dans une structure unique conduit à un chauffe-eau solaire présentant diverses particularités avantageuses : ensemble autoportant facile à installer, stockage sous forme d'eau pure non sanitaire, sous pression atmosphérique, résistance passive au gel, absence de pompe et de régulation, coût réduit.
Un tel prototype a fait l'objet d'essais stationnaires et dynamiques.
Les caractéristiques principales sont : η_o = 0,74; U_m = 3,5; constante de temps = 12 heures (prototype I), ajustable.
Le collecteur se comporte comme un intégrateur, avec un rendement de conversion diurne variant de 42 à 53 % selon la saison et par ciel clair. Le rendement de stockage varie de 39 à 50 % mais sera considérablement augmenté par l'utilisation d'un interrupteur thermique breveté.
La protection contre le gel est passive et ajustable de 5 à 20 jours sous une température ambiante de -10°C.
Une opération de démonstration de 1000 m^2 est en cours, à l'initiative du Ministère de l'Energie de la Région Wallonne.

Summary

A domestic hot water heater integrating a solar absorber, a heat storage vessel and a heat exchanger in a single casing possess a variety of interesting characteristics : self supporting system, easy to install, storage using normal pure water under atmospheric pressure, passive anti-freezing protection, absence of pump and regulation systems, reduced investment and installation costs.
A prototype has been tested under stationary and transient conditions.
Its main characteristics are : η = 0,74; U_m = 3,5; adjustable time constant [12 hours for prototype nr I].
This solar heater behaves as an integrator with a diurnal solar conversion efficiency in the range 42 to 53 % under clear sky conditions, with seasonal variations.
Storage efficiency varies from 39 to 50 % but will be drastically increased by using a patented thermal switch.
Anti-freezing protection is passive and adjustable from 5 to 20 days for -10°C ambient temperature.
A 1000 m^2 site demonstration plant sponsored by the Regional Ministry of Energy of Wallonia is scheduled for 1984.

1. INTRODUCTION

Le coût d'achat et d'installation d'un chauffe-eau solaire classique unifamilial est généralement élevé, suite à diverses contraintes conceptuelles, techniques et architecturales :
- liaisons hydrauliques de plusieurs mètres entre capteurs et stockage,
- réservoir de stockage d'eau conforme aux réglementations sanitaires,
- alimentation électrique de la pompe et de la régulation,
- capteurs et circuit solaire résistant au gel,

etc.

Les collecteurs solaires et le stockage d'eau chaude sanitaire peuvent être regroupés et la pompe de circulation supprimée dans les chauffe-eau solaires à circulation naturelle. Ceci introduit cependant des contraintes architecturales et esthétiques supplémentaires liées à la nécessité d'avoir le réservoir de stockage en position la plus élevée.

Le chauffe-eau décrit plus loin réduit les principales contraintes grâce aux particularités suivantes :
- intégration des systèmes de captation et de stockage,
- ensemble autoportant s'installant comme un collecteur seul,
- stockage de chaleur sous forme d'eau pure non sanitaire, sous pression atmosphérique,
- résistance passive au gel,
- absence de pompe et de régulation.

2. CAPTEUR A STOCKAGE INTEGRE

La figure 1 montre la conception générale du capteur solaire à stockage intégré (licence ADL - Brevets IDE).

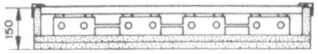

Le module de base a une ouverture optique de 1,88 m^2. La conversion thermique de l'énergie solaire est réalisée à l'aide d'un absorbeur en cuivre, de 0,4 mm d'épaisseur, recouvert d'une couche sélective de chrome noir.

Le réservoir de stockage est constitué de quatre bacs verticaux en cuivre ou d'un seul, horizontal, dans le prototype n°2, d'une capacité totale de l'ordre de 30-40 litres/m^2, ajustable en fonction du climat et de l'utilisation, contenant de l'eau pure, sous pression atmosphérique. La liaison thermique entre l'absorbeur et le stockage est réalisée grâce à un tube carré en laiton, soudé sur les bacs et riveté sur l'arbsorbeur.

La protection contre le gel est assurée par inertie thermique et déformation élastique tandis que la surchauffe est empêchée par un condenseur extérieur.

Le circuit d'eau sanitaire est constitué d'un seul tuyau en cuivre, noyé dans le réservoir de stockage et d'une surface d'échange de 0,93 m^2 par module.

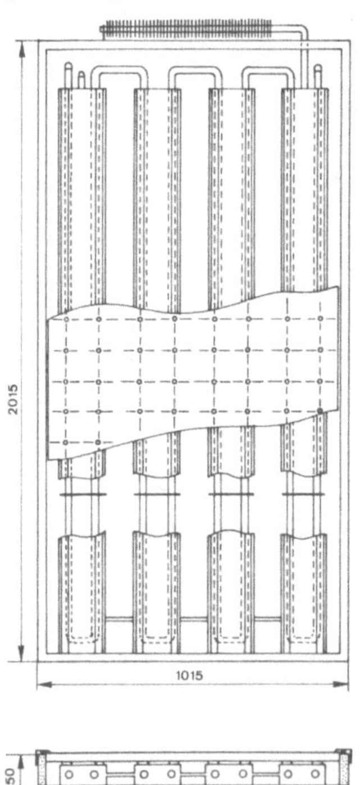

Fig. 1. Schéma du collecteur à stockage intégré

Tableau I. Caractéristiques : prototype n° 1		
	mesuré	calculé
η_o	-	0,74 ± 0,02
U_m (W/m^2.K)	3,5	3,56 ± 0,2
τ (h)	12,5	12,6
C (kJ/m^2.K)	-	161

Vu la très grande constante de temps τ du collecteur, plus de 12 heures, les méthodes classiques d'essais extérieurs ne conviennent pas. Le coefficient global de pertes a été mesuré directement tandis que les valeurs de η_o et de la capacité thermique apparente C ont été obtenues par calcul, confirmées par des essais transitoires sous soleil artificiel (fig. 2).

Au cours de la phase de fourniture d'eau chaude sanitaire, par extraction de chaleur du stockage, le coefficient global de transfert de chaleur obéit à la loi $h = 109.\Delta T^{0,25}$ pour des tuyaux verticaux (prototype n°1) et $h = 273.\Delta T^{0,25}$ pour des tuyaux horizontaux (prototype n°2).

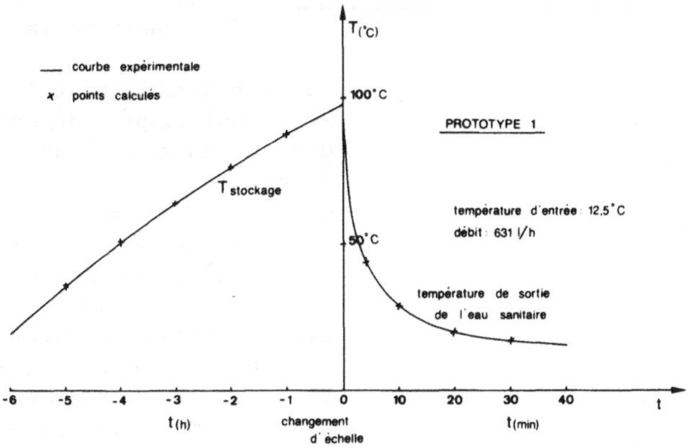

Figure 2. Essais de charge et de soutirage.

3. CHAUFFE-EAU A STOCKAGE INTEGRE

Considérons un chauffe-eau solaire constitué simplement de deux capteurs à stockage intégré, à serpentins horizontaux (prototype n°2), placés en série.

Tableau II. Performances du chauffe-eau solaire Latitude : 45° - 21 mars			
H(kWh/m^2.j)	η_{diurne} (%)	η_{24h} (%)	$\eta_{stockage}$ (%)
1	46,9	18,3	39
2	47,2	18,3	39
3	47,1	18,5	39
4	47,1	18,4	39
5	47,1	18,4	39
6	47,1	18,4	39
7	47,1	18,4	39

En supposant le système à la température ambiante au lever du soleil et une distribution sinusoïdale de l'exposition énergétique du capteur au cours de la journée, on peut calculer les performances du système en fonction du jour de l'année, de la latitude et du mode de soutirage.

Le tableau II donne, à titre d'exemple, les rendements moyens de conversion sur la période d'insolation et sur 24 heures,sans soutirage, ainsi que le rendement de stockage nocturne.

On constate que les performances sont indépendantes de l'exposition énergétique journalière H. Les collecteurs se comportent donc comme des intégrateurs parfaits.

Tableau III. Performances de captation, sans soutirage.
Latitude : 45°

Date	η_{diurne} (%)	η_{24h} (%)	$\eta_{stockage}$ (%)
21 décembre	53	16	30
21 mars	47	18	39
21 juin	42	21	50
21 septembre	47	18	39

Les durées du jour et de la nuit, variables avec la saison, influencent naturellement les performances, ainsi que le montre le tableau III.
Le rendement de stockage est naturellement défavorisé par la longueur de la nuit.

Les performances utiles, avec soutirages, dépendent de la distribution de ceux-ci. La figure 3 montre la variation de la fraction solaire dans le cas d'un soutirage unique, à 19 heures, avec un débit de 10l/min en fonction de la quantité soutirée et pour un accroissement de température de l'eau chaude sanitaire demandé égal à 30°C. Le chauffe-eau solaire fonctionne en préchauffage et l'appoint est fourni en série.

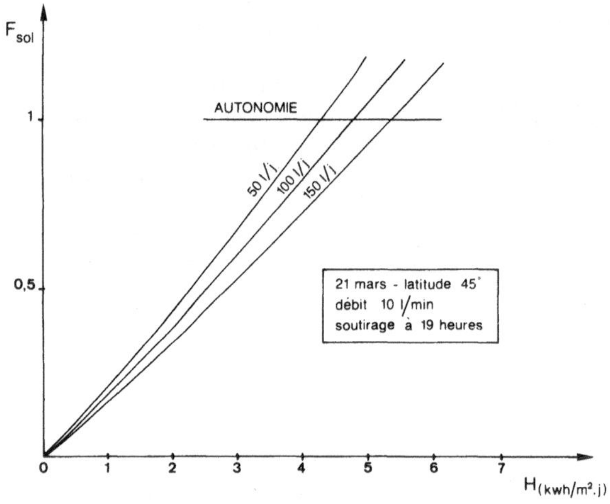

Fig. 3. Fraction solaire en fonction de l'exposition énergétique journalière et de la consommation.

4. PROTECTION CONTRE LE GEL

L'inertie thermique du stockage assure une protection suffisante contre les gelées nocturnes. Lors des périodes de températures négatives permanentes, une fraction de l'eau du stockage peut geler, sans dommage pour le système, empêchant le gel du circuit sanitaire.

La durée de la protection est donnée dans le tableau IV.

On voit que la protection est suffisante, d'autant plus que les basses températures ambiantes s'accompagnent généralement de ciel clair donc de soleil pendant le jour.

Tableau IV. Durée de protection passive antigel (en jours) t_a = -10°C				
	H(kWh/m^2.j)			
C (l/m^2)	0	0,2	0,4	0,6
38	4,2	5,1	6,6	9
80	8,8	10,7	13,9	19

5. AMELIORATION

Un interrupteur thermique, breveté, utilisant les propriétés des alliages à mémoire de forme, est en cours d'étude. Il permettra de déconnecter thermiquement l'absorbeur du stockage lorsque l'insolation est insuffisante, en particulier pendant la nuit.

Le coefficient global de pertes de stockage passerait de 3,5 à 2 W/m^{2}K améliorant ainsi sensiblement le rendement de stockage. La figure 4 illustre clairement cet effet.

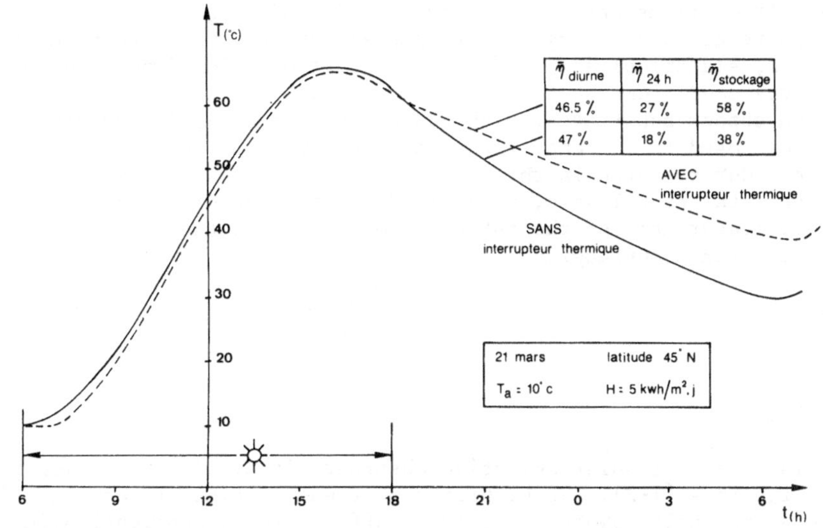

Fig. 4. Influence d'un interrupteur thermique.

6. DEMONSTRATION

Une opération de démonstration au cours de laquelle 1000 mètres carrés de tels chauffe-eau solaires seront installés en Wallonie sur des habitations sociales existantes est en cours de réalisation, à l'initiative de la Région Wallonne (Ministère de l'Energie).

L'organisation de l'opération est assurée par la Société Nationale du Logement et le suivi scientifique in situ par la Faculté Polytechnique de Mons.

7. CONCLUSIONS

L'intégration d'un stockage de la chaleur à un capteur solaire permet de concevoir un chauffe-eau solaire de faible coût, évitant les nombreuses contraintes d'installations affectant les systèmes solaires classiques, sans consommation d'énergie pour la circulation des fluides ni la régulation.

SPECTRALLY SELECTIVE TRANSITION-METAL OXINITRIDE COATINGS ON ROUGH SUBSTRATES FOR APPLICATION IN HIGH PERFORMANCE SOLAR COLLECTORS

E. Vogelzang, A.A.M.T. van Heereveld, M. Sikkens and J.C. Francken
Department of Applied Physics, University of Groningen
Nijenborgh 18, 9747 AG Groningen, The Netherlands

Summary

For application in high performance solar collectors, a spectrally selective absorber with a high solar absorptance (≈ 0.95) and a low thermal emittance ($\lesssim$ 0.05) is required. Since these collectors can reach high stagnation temperatures (around 300-500°C), the thermal stability of the selective absorber is of particular importance. The purpose of our work is the development of transition-metal (oxi)nitride/reflector tandems that can meet these requirements. Amongst the transition metals, Cr and Ti are promising due to their relatively low cost and the high thermal stability of their nitrides. Thin chromium and titanium oxinitride films are prepared by reactive DC bias sputtering in an atmosphere containing argon, nitrogen and oxygen. The structure, composition, electrical and optical properties of films deposited under various conditions have been investigated.
Rough copper substrates, with roughness dimensions in the order of 0.5 μm, are used to increase the absorptance without increasing the emittance. Our research in this field concentrates on the production of rough metal base layers, a detailed description of the surface structure, measurement of the optical properties and the theory of statistically rough surfaces.

1. INTRODUCTION

A wide range of solar selective absorbers is available for low temperature applications [1-2]. The stability of these coatings is usually limited to temperatures below 300°C. This is insufficient for application in high performance collectors, operating above 100°C, due to the high stagnation temperatures that can be reached (around 300-500°C). Furthermore, in the production process of evacuated collectors, a short bake-out at 400-500°C is usually necessary. It will be clear that more stable coatings are needed in this field of application. The relatively high operating temperature requires a low thermal emittance, in particular in applications where a low concentration ratio or no concentration at all is used.
A few candidate materials have been investigated [3-5], but the required low emittance (≈ 0.05 at 150°C) limits the solar absorptance to values in the order of 0.75-0.85. Grading the film composition can overcome this problem but such graded films tend to become more homogeneous again at high temperatures [6].
The purpose of our work is the development of a selective absorber with high absorptance (≈ 0.95), low emittance (≈ 0.05 at 150°C) and high thermal stability (up to 500°C in vacuum). Transition metal nitrides are promising materials in this respect. Amongst these, chromium and titanium nitrides are good candidates due to their relatively low cost and high stability. The rather moderate absorptance can be increased by using a rough substrate

instead of a smooth one.

2. METHOD OF APPROACH

2.1. Nitride films

Chromium and titanium oxinitride films are deposited by reactive DC bias sputtering in an atmosphere containing argon (7 Pa), nitrogen (max. 10 Pa) and dried air (1.3 10^{-3} Pa). The purpose of the latter is to incorporate some oxygen into the films. The base pressure of the system is about 10^{-4} Pa.
A cathode voltage of 1.8 - 2.2 kV and a current density of 5 A/m^2 are used. A negative bias voltage of 0 -190 V is applied to the substrate to control the oxygen content of the films. The deposition rate of the system is 0.3 - 3.7 Å/s, depending on the conditions.
Films, deposited under various conditions, are analyzed using diffraction methods, Rutherford backscattering combined with nuclear reaction analysis (RBS/NRA), X-Ray Photo-electron Spectroscopy (XPS), resistivity measurement and electron microscopy. The optical properties are investigated by means of reflectance and transmittance measurements in the spectral region 0.4 - 25 μm, together with solar absorptance and thermal emittance measurements.

2.2. Rough base layers

Rough copper base layers are deposited electrochemically from a standard copper plating bath containing small particles [7]. These particles are incorporated in the growing film, creating a rough surface. The roughness can be influenced by varying the layer thickness, current density, size and type of particles and additives. This work is carried out at the Metal Institute TNO, Apeldoorn.

3. EXPERIMENTAL RESULTS

3.1. Properties of TiN_xO_y films

Thin TiN_x have been sputtered in an Ar/N_2 atmosphere without a substrate bias and with a bias of 100 V.
Fig. 1 shows the resistivity of the films as a function of the nitrogen pressure during deposition. XPS measurements indicate an oxygen content of about 40 at.% in the zero-bias films [8], which accounts for the high resistivity of these films as compared to the bias sputtered films. Therefore, the zero-bias films actually are TiN_xO_y films, whereas the low resistivity indicates that the films sputtered at a 100 V bias are TiN_x films.
From diffraction experiments it has been found that the resistivity maximum at 0.02 Pa corresponds to a phase transition from the h.c.p. Ti to the f.c.c. TiN structure. This has been observed before [9]. Unfortunately, even at the resistivity maximum, these films are too metallic to be of practical use as selective absorbers.
The TiN_xO_y films sputtered at zero bias have a dark green or grey appearance, in contrast to the golden appearance of the TiN_x films sputtered at a bias of 100 V, and seem more suitable as a selective absorber. However, since the oxygen in these films originates from the residual gases in the sputtering system and from desorbed species from the walls, the films don't reproduce well. For this reason, TiN_xO_y films have also been deposit-

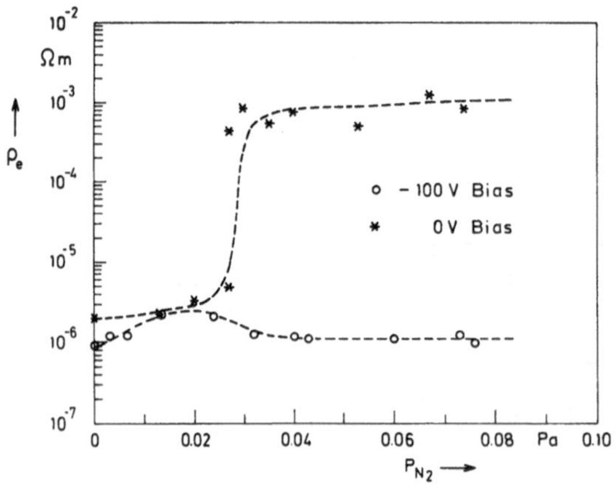

Fig.1: Resistivity $TiN_x(O_y)$ films vs the nitrogen pressure.

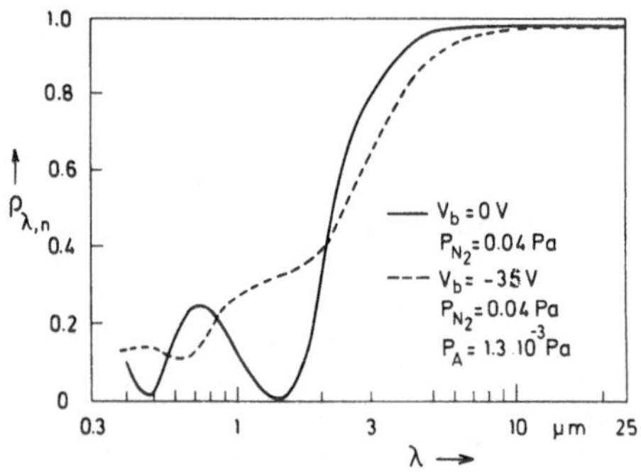

Fig.2: Reflectance of TiN_xO_y films of thickness 0.1 μm on polished copper substrates.

ed in the presence of air at low bias voltages. By varying the bias voltage and the air pressure P_A, the oxygen content can be controlled.
Fig. 2 shows the reflectance of two TiN_xO_y films with thickness of about 0.1 μm on polished bulk copper. The samples deposited at zero bias typically show a solar absorptance α_s of about 0.85 and a normal emittance ε_n of 0.05. The samples deposited at a bias voltage of 35 V show values of α_s ranging from 0.80 - 0.85.
The thermal stability of TiN_xO_y/copper tandems has been tested by heating to 400°C in vacuum at a base pressure of 10^{-2} Pa. After a small initial change during the first 2 hours, the optical properties appear to be stable.

3.2. Properties of CrN_xO_y films

As with TiN_xO_y films, the resistivity of $CrN_x(O_y)$ films sputtered in an Ar/N_2 atmosphere decreases drastically when a substrate bias is applied during the deposition (see figure 3). Diffraction experiments show a Cr_2O_3 structure in zero-bias CrN_xO_y films deposited at low partial nitrogen pressures, together with the b.c.c. Cr structure. At higher nitrogen pressures the f.c.c. CrN phase has been found.
The diffraction pattern of bias sputtered Cr films only shows the b.c.c. Cr structure at $P_{N_2} = 0$. At $P_{N_2} = 0.13$ Pa a mixture of the b.c.c. Cr phase and the h.c.p. Cr_2N phase has been found whereas at higher nitrogen pressures the CrN_x films consist of the f.c.c. CrN phase.
CrN_xO_y films sputtered at zero bias have a dark grey colour. Bias sputtered CrN_x films have a silver-metallic appearance; only at nitrogen pressures above 1 Pa films with a grey colour have been obtained.
Figure 4 shows the reflectance of a CrN_x and a CrN_xO_y film on a polished copper substrate. The high reflectance of CrN_x in the solar spectral region, limits the α_s of the CrN_x/copper tandem to only 0.60. Zero-bias CrN_xO_y films show, however, a much higher solar absorptance: $\alpha_s = 0.87$, $\varepsilon_n = 0.05$.
The thermal stability of CrN_x films is excellent. After heat treatment for 2 hours at 400°C in vacuum and 15 hours at 300°C in air, no changes in the reflectance of the samples could be detected.

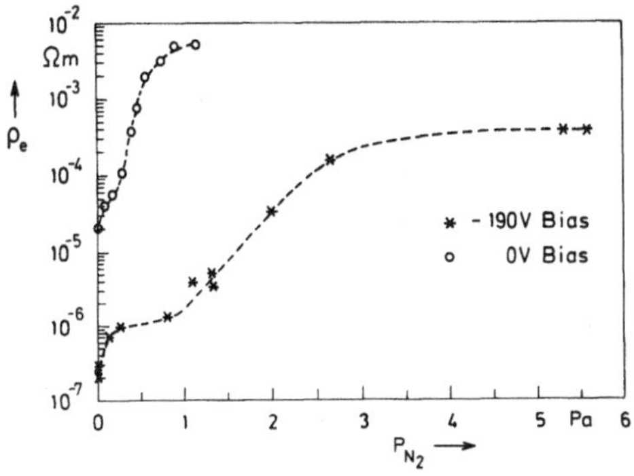

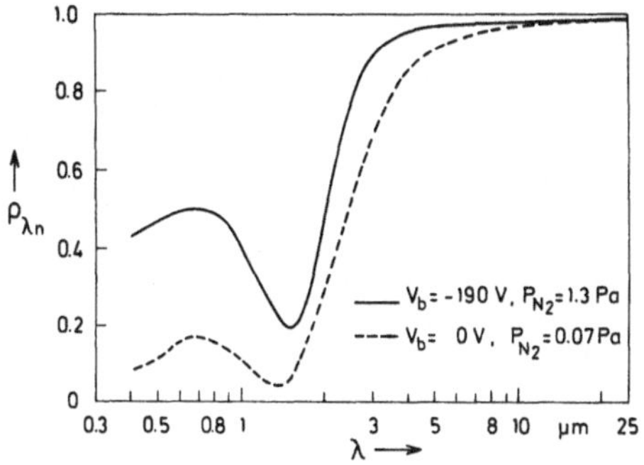

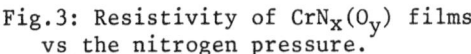

Fig.3: Resistivity of $CrN_x(O_y)$ films vs the nitrogen pressure.

Fig.4: Reflectance of a CrN_x/copper and a CrN_xO_y/copper tandem. The thickness of the films is 0.1 μm.

3.3. Properties of rough copper layers

To enhace the moderate solar absorptance of the (oxi)nitride films, we have chosen the method of surface roughening. The roughness can be introduced conveniently by using a rough substrate onto which the film is sputter-deposited. The film is thin enough to follow the substrate contours. To avoid an increase of the emittance, the roughness dimensions should be chosen carefully. As a rule-of-thumb, the characteristic dimensions should be about 0.5 μm, the order of magnitude of solar wavelengths.
Fig. 5 shows a SEM micrograph of a typical copper layer, deposited with the technique described in 2.2. SEM stereo micrograph pairs are analyzed using a calibrated stereo viewer, based on point-by-point parallax measurement. Measurements of the surface profile are analyzed statistically, yielding a surface height distribution with characteristic parameter σ, the r.m.s. height, and a surface correlation function with characteristic parameter T, the correlation length. Both σ and T should be about 0.5 μm according to our rule-of-thumb. These characteristic dimensions have indeed been measured in rough substrates with suitable selective properties.
Several methods are used to determine the optical properties of rough surfaces. A modified carbon arc and optical bandpass filters are used as light source for spectral measurements of both the hemispherically integrated reflectance and the bidirectional reflectance (BRDF) at wavelengths from 0.4 to 20 μm. The hemispherical reflectance for near-normal incidence of copper surfaces with varying r.m.s. roughness is shown in fig.6. BRDF-measurements supply more detailed information to be used in conjunction with theoretical models. For rough surfaces with $\sigma \approx 0.5$ μm, we have found that the reflected radiation is predominantly diffuse at solar wavelengths and specular at thermal wavelengths, the diffuse-to-specular transition occurring close to the reflectance edge.
The solar absorptance and thermal emittance of samples can either be calculated from the reflectance or measured directly with calorimetric techniques.
Preliminary results on a sputtered CrN_xO_y film on a rough copper substrate of approximately correct dimensions showed an increase in absorptance from $\alpha_s = 0.74$ to $\alpha_s = 0.90$ together with a slight increase in the normal emittance from $\varepsilon_n(100^oC) = 0.035$ to $\varepsilon_n(100^oC) = 0.042$. Further tuning of roughness and film properties is expected to give even better results. To enable

Fig.5: SEM micrograph of a typical rough copper surface.

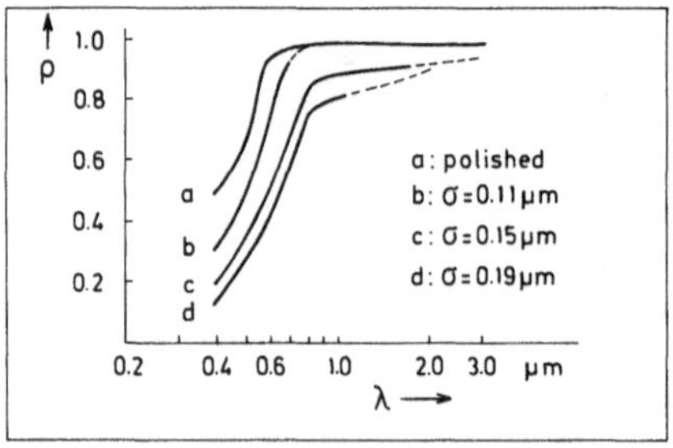

Fig.6: Hemispherical reflectance of rough copper surfaces with various r.m.s. heights σ.

such an optimization, a theoretical model of the optical properties of rough surfaces is being developed.
After 1 hour at 500°C in vacuum, no changes in the roughness structure have been observed.

4. CONCLUSIONS

Chromium and titanium oxinitride films on rough copper substrates appear to be excellent candidates for application as a selective absorber in collectors operating above 100°C. So far, an α_s of 0.90 and an ε_h(150°C) of 0.066 have been obtained, which comes close to the "ideal" values of 0.95 and 0.05, respectively. With a better understanding of the optical properties of rough surfaces and the role of oxygen in the films, these values seem to be within reach.
The thermal stability appears to be adequate, but more experiments will be necessary to estimate the durability of the tandem under operating conditions.

5. REFERENCES

1. R.H.Hahn and B.O.Seraphin, Phys. of Thin Films 10 (1978) 1-69.
2. SERI Report, subcontract no. AY-2-02067 (1983) Ch.5.
3. J.A.Thornton and J.L.Lamb, Thin Solid Films 83 (1981) 377-385.
4. R.B.Pettit, R.R.Sowell and I.J.Hall, Sol. En. Mater. 7 (1982) 153-170.
5. Y.Zhiqiang et al.,Proc. ISES, Perth (1983) 1091-1095.
6. S.Craig and G.L.Harding, Thin Solid Films 101 (1983) 97-113.
7. M.Sikkens et al., Thin Solid Films 108 (1983) 229-238.
8. E.Vogelzang, M.Sikkens and A.A.M.T.van Heereveld, Proc. ISES, Perth, (1983) 996-1000.
9. W.Posadowski et al., Thin Solid Films 62 (1979) 347.

HYDROPHILE SOLAR COLLECTOR SYSTEM

V. KORSGAARD
Thermal Insulation Laboratory
Technical University of Denmark

Summary

In the paper a new principle of a solar collector system is described in which the heat is transferred from the absorber surface to the heat storage through an insulating layer of mineral wool by diffusion and condensation. The advantages of the system should be a simpler and cheaper collector system with lower stagnation temperature and less risk for freezing during cold nights without using drain back or antifreeze solutions. A computer model has been set up, and theoretical calculations have been compared with measurements on a small prototype in the laboratory with simulated solar radiation. Good agreement was found. At high solar intensities and storage temperatures the efficiency is approx. the same as for a traditional collector with selective surface.

1. INTRODUCTION

The hydrophile solar collector system is different from the traditional solar collector systems in which the absorbed solar heat is transferred from the absorber to the storage by means of a circulating fluid or gas. In the hydrophile collector the heat is transferred from the absorber to the heat storage by evaporation and condensation of a fluid, preferably water. The principle is shown in fig. 1.

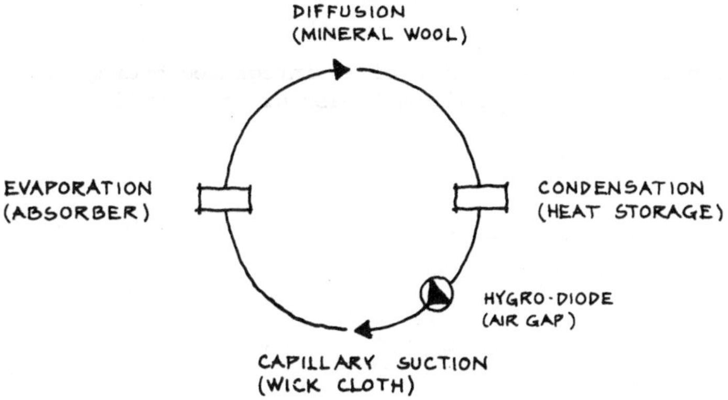

Fig. 1. Principle of heat and mass transfer of the hydrophile solar collector.

2. DESCRIPTION

In the hydrophile solar collector the absorber surface consists of a wetted fabric with good capillary suction qualities (wick cloth) in close contact with a mineral wool batt placed in front of the heat storage with a narrow air gap between the batt and the storage surface. The solar heat absorbed by the wick cloth will evaporate water, and the water vapour will diffuse through the porous batt and condense on the colder front surface of the heat storage. The condensate will by gravity run down the surface of the storage to the bottom of the collector envelope where it will be sucked in by the wick cloth. In this way a closed loop is established in the same way as in a heat pipe. The large area of evaporation and condensation and the short distance between the two will compensate for the slow rate of diffusion through the airfilled mineral wool batt. In the periods where the temperature of the storage is higher than that of the absorber surface, the wetted storage surface will soon be dry, and the heat transfer by evaporation and diffusion the opposite way will stop. The efficiency will increase substantially with the wick cloth covered by a selective foil.

In fig. 2 a simplified cross section of an integrated hydrophile solar collector and a thermal mass or water wall is shown. To obtain a high efficiency it is important that the sealed collector box is placed in good thermal contact with the thermal mass or water wall.

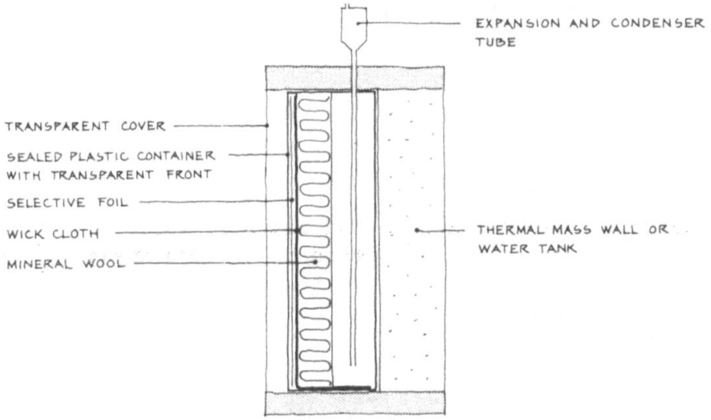

Fig. 2. Simplified cross section of an integrated hydrophile solar collector and thermal mass or water wall.

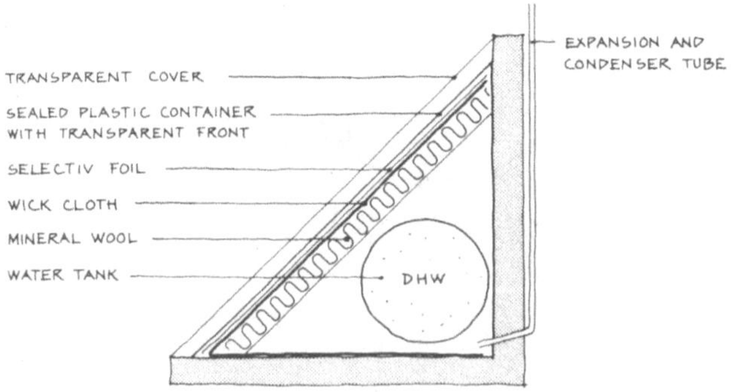

Fig. 3. Integrated hydrophile solar collector and domestic hot water tank.

In fig. 3 is shown how the hydrophile solar collector can be integrated with a domestic hot water tank. The condensation will take place directly on the tank surface and preferably on the coldest areas which should increase the overall efficiency.

To prevent overpressure in the collector an open expansion tube is fitted to the box. The tube should be placed as cold as possible but frost-free. In the tube the vapour in the expanding air will condense and drain back to the wick cloth. Only a small amount of water will thus escape from the collector, so that adding water should be restricted to once a year.

3. COMPUTER MODEL

In setting up a computer model for the heat and mass transfer between the absorber and storage through the mineral wool layer, the following two equations are used:

$$\frac{\partial u}{\partial \tau} = D \frac{\partial^2 w}{\partial x^2}$$

$$\gamma c_p \frac{\partial \theta}{\partial \tau} = \lambda \frac{\partial^2 \theta}{\partial x^2} + r \frac{\partial w}{\partial \tau}$$

where:
- u moisture content in the mineral wool
- D diffusion coefficient for the mineral wool
- τ time
- x space coordinate
- w vapour concentration
- γ mass density of mineral wool
- c_p specific heat of mineral wool
- λ heat conductivity of mineral wool
- θ temperature
- r heat of evaporation

The computer program calculates the time dependency of temperature and moisture content by means of the finite difference method.

The results of the calculations carried out for the test model described in the following section are shown in the table.

In fig. 4 the characteristic, corresponding temperature and moisture content in the 6 cm thick mineral wool batt (see fig. 8) is shown during a heating up period with constant solar intensity.

The time dependent moisture content of the mineral wool during a heating up and cooling down cycle is shown in fig. 5. In fig. 6 the heat loss through the moist mineral wool batt is shown for a cooling down period and compared with the heat loss if the mineral wool had been dry. It will be noticed that the insulation will dry out within 12-13 hours, which means there is no risk for accumulation of moisture over longer periods of time.

4. EFFECTIVITY CURVES

The effectivity of the hydrophile solar collector system has been calculated as a function of the difference between the storage and the ambient temperature at different solar intensities. The results are shown in fig. 7 for an ordinary black absorber and a selective one. The curve for an ordinary collector with a selective absorber is also shown.

At low storage temperatures the effectivity of an ordinary selective absorber is 15-20% higher than for the corresponding hydrophile collector.

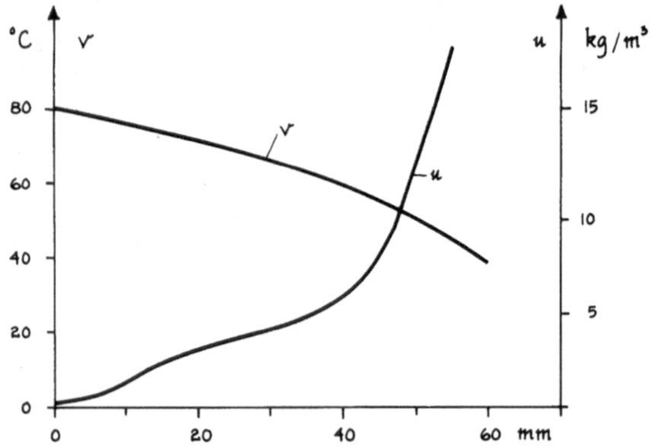

Fig. 4. Characteristic corresponding temperature and moisture content distribution in the 6 cm thick mineral wool batt during a heating up period with constant solar intensity.

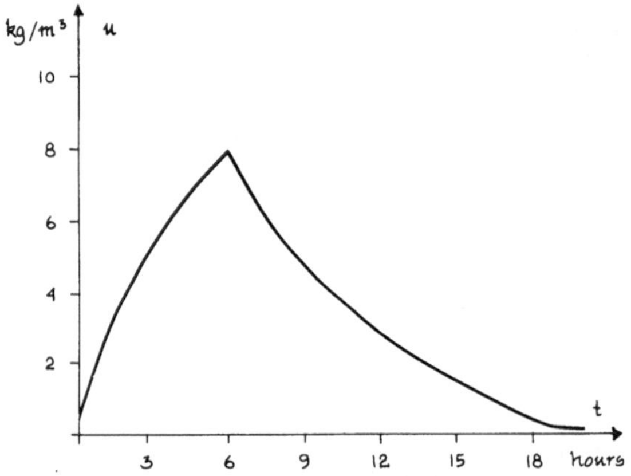

Fig. 5. The time dependent moisture content of the mineral wool batt during a heating up and cooling down cycle.

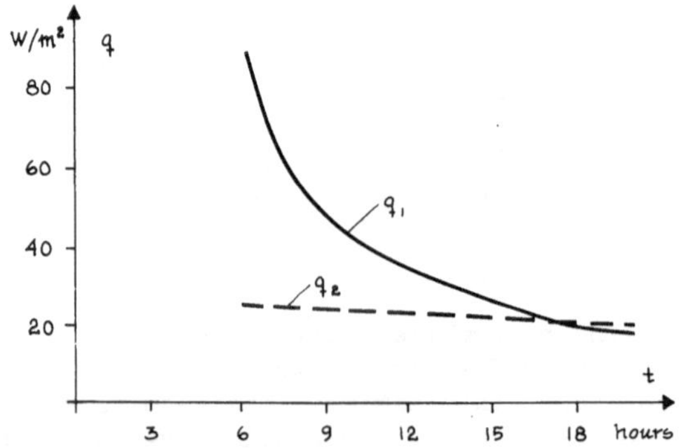

Fig. 6. The heat loss through the moist mineral wool batt during a cooling down period compared with the heat loss if the batt had been dry.

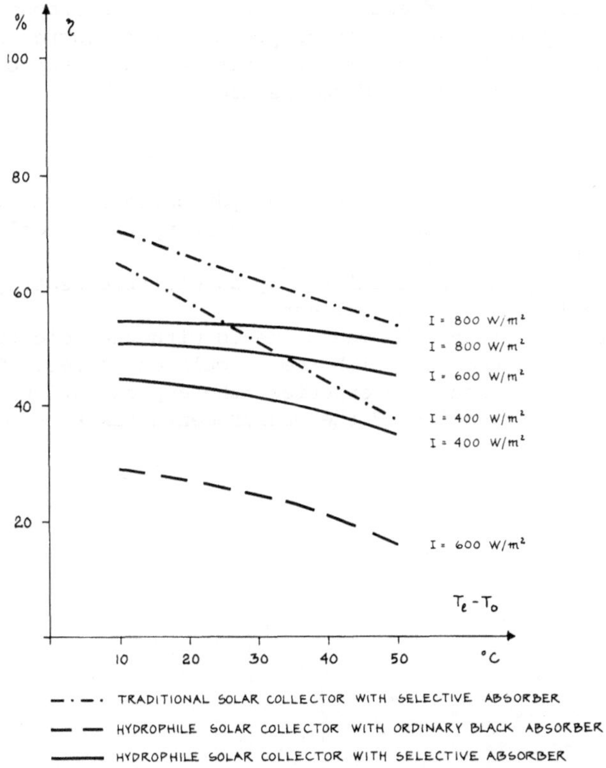

Fig. 7. Efficiency curves for traditional and hydrophile solar collectors

		Measured			Calculated		
H	T_l	T_a	Q_u	η	T_a	Q_u	η
W/m^2	^oC	^oC	W/m^2	%	^oC	W/m^2	%
193	22.0	44.0	66	34	45.5	61	32
346	38.0	60.5	139	40	63.3	129	37
487	26.0	67.0	235	48	70.6	221	45
333	22.5	57.5	139	42	59.5	132	40
482	47.5	70.0	220	46	74.2	204	42
634	30.5	73.5	335	53	79.9	310	49
876	38.0	82.5	505	58	90.5	472	54
516	24.0	68.5	255	49	72.5	239	46

Table: Measured and calculated values of effectivity of the hydrophile solar collector with a selective absorber.

H simulated solar intensity
T_l temperature of thermal storage
T_a temperature of selective absorber
Q_u stored heat
η effectivity

At storage temperatures of 50-60^oC and a solar intensity of 800 W/m^2 the effectivity is almost the same. For a hydrophile collector with an ordinary black absorber the effectivity is essentially lower in the whole temperature range.

The low slope of the effectivity curve for the hydrophile collector is due to the steep increase in vapour pressure with increasing absorber temperature. At 80°C the slope of the saturated water vapour pressure curve is 15 times larger compared to the slope at 20°C.

5. TEST MODEL

To demonstrate the principle of the hydrophile solar collector system and to verify the calculations a small laboratory test model was constructed as shown in fig. 8.

The solar radiation was simulated by an electric heating foil glued to the transparent front of the collector.

The results of some of the measurements with the selective absorber is shown in the table together with the calculated values. The agreement seems to be as good as could be expected, the approximation of the calculations and the uncertainties of the measurements taken into consideration.

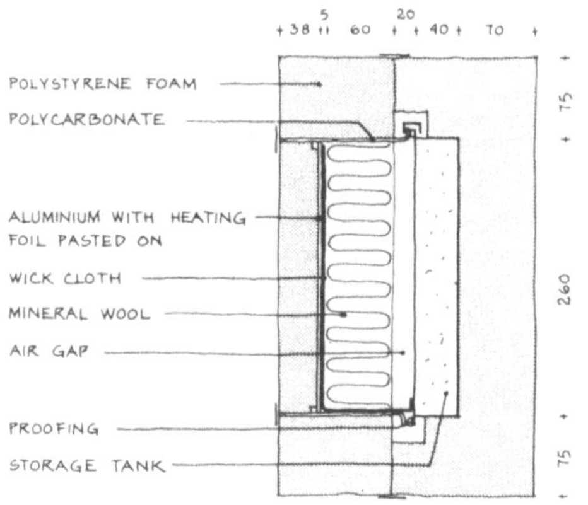

Fig. 8. Cross section through laboratory test model of the hydrophile solar collector.

6. CONCLUSION

Test results as well as calculations have shown that the hydrophile solar collector system is able to utilize the solar radiation with the same efficiency at higher temperatures and intensities as traditional collector systems. However, before the system can be introduced on the market further testing and investigations must be carried out.

The system is patent pending.

Acknowledgement

The calculations and the testings have been carried out by M.Sc.Eng. Casper Paludan-Müller.

BALANCING OF SOLAR HEATING OPTIONS

W.B. Veltkamp and C.W.J. van Koppen
Eindhoven University of Technology

Summary

In the field of energy conservation many options are presently competing.This study aims at providing more rational criteria for selection between these options.The options considered are; insulation of the walls, regeneration of the heat in the waste air, double glazing, attached sunspace at the south facade, solar water heating, solar space heating, night shutters and increasing the internal mass.For each option the investments are defined and subsequently each option is internally optimized, as far as applicable, and a comprehensive computer program is used which enables selecting the optimal combination of the options. Such is considered to be the combination leading to the minimum auxiliary energy need for a given investment. The decission regarding the latter is left to the user, but some aids to this end are provided. For the Dutch conditions the sequence of preference, appears to be: wall insulation, heat regeneration, double glazing, solar water heating and an attached sunspace. At the costs considered the other options do not enter into the optimum combination for an investment of up to ƒ 10,000.- .

1. INTRODUCTION

An optimum design of a house and its heating system can be anything, depending on where ,when and for whom it is assessed. As an optimization study should have a wider applicability than for one person, one location and one date, we decided to split the problem in two parts, an objective and a subjective, individual one. In the first part we asses for a range of investments the system configurations and sizes with the lowest remaining auxiliary energy consumption. In the second part it is left to the user to fix the investments he is prepared to spend on energy savings, but the consequences of his decision in terms of the marginal energy saving/investment ratio are given. The first part still depends on prices, longevity and similar features, but avoids the individual and unpredictable parameters like energy prices, tax incentives, loan and other interest rates, security of energy supply etc. The user can choose his own economical optimization criterion. In this study we try to make a synthesis of all major energy conserving options in the built environment, with the objective to minimize the use of primary energy for a given comfort level. As regards the mathematical aspects of the optimization a well known search algorithm is used to find a minimum of the non-linear auxiliary energy function under the constraint of the non-linear costfunction and other constraints (e.g. a minimum window area).

2. OPTIMIZATION METHOD

Opinions about optimization of a solar energy system differ widely. In fact an optimum system is different not only for every individual, but also for every location and every point of time. As a consequence it is impossible to calculate the optimum for every individual case, without creating a huddle of numbers from which a decision is difficult to derive.

2.1 Bisection of the optimization problem

To resolve this problem we try to eliminate the part that is determined by the individual and to keep the part that is common to nearly everyone. This is possible because the costs, or better, the ratios of the costs of the parts of the system are roughly the same in most situations. Starting from the costs of all energy conserving options (as a function of their size) we subsequently try to find the optimal distribution of a certain investment among them, i.e. the distribution for which the associated auxiliary energy is minimum. For a range of investments we thus obtain the auxiliary energy and the sizes of the energy conserving options. On the basis of these data the user is then free to decide how much he wants to invest for the energy saved, using the economical optimization criterion of his own choice.

2.2 Technical optimization

In previous work [1] we performed a technical optimization of a system with a solar energy installation. In these studies we investigated the optimum values of the technical parameters, which incur minor or no costs. The most important results were that a solar heating system performs best at relatively low flow rates through collector and heating system in combination with a thermally stratified storage. The just mentioned optimization of the flows has been implemented in the fast program SISOEN, which is used as object function in this study.

2.3 Elimination of variables

In a complete optimization many parameters have to be considered. From a more practical point of view, however, several variables can be eliminated.

- the orientation is not a free variable and the optimum is well known
- the difference between the heater in series or parallel is negligible, if good controlled
- an air heating system appears to the better solution for a solar heating system
- as water is cheap and easy to handle we only implemented water storage and collector
- the price/performance ratio of an evacuated collector outperforms other collectors

3. THE RESTRICTIONS

In this study we have only considered a typical Dutch serial house;

- volume of the house	245	m^3
- facade, roof and floor area	136	m^2
- internal wall area	290	m^2
- internal heat production	330	W
- heat leaks in the envelope	19	WK^{-1}
- ventilation rate	.5	h^{-1}
- infiltration rate	.2	h^{-1}
- room settemperature	18	^0C
- daily hot water load	16	MJ
- windowarea SW	5.4	m^2
- windowarea NE	3.7	m^2

The orientation is SW and the collectortilt is 45^0 . The air temperature is controlled such that the mean of the radiation temperature of the surrounding walls and windows and the air temperature remains at 18^0C.

3.1 The energy conserving options

The options and their properties are listed below;

- insulation of the cavity walls		.036 $Wm^{-1}K^{-1}$
- waste air heat regenerator counterflow		
- double pane window	UL= 3 $Wm^{-2}K^{-1}$	tr=.69
- attached sunspace single pane	10.5	m^2
- evacuated tubular collectors	1.69 $Wm^{-2}K^{-1}$	ta=.75
- storage stratified water		
- insulation storage		.036 $Wm^{-1}K^{-1}$
- HSW coil running across storage		

- water–air heater in counterflow
- internal walls, brick a=1.3 m^2s^{-1}
- insulation night shutters .036 $Wm^{-1}K^{-1}$

3.2 The costs

The assumptions on the costs are listed below (1 Dfl ≈ .33 US $):

- collector array initial	500.-	Dfl
- collector variable	300.-	$Dflm^{-2}$
- storage initial	250.-	Dfl
- storage per unit of area	25.-	$Dflm^{-2}$
- storage insulation variable	300.-	$Dflm^{-3}$
- storage housing	200.-	$Dflm^{-3}$
- hot water coil initial	100.-	Dfl
- hot water coil variable	.3	$DflW^{-1}K$
- air heater variable	.4	$DflW^{-1}K$
- regenerator initial	600.-	Dfl
- regenerator variable	.8	$DflW^{-1}K$
- insulation walls initial	3.9	$Dflm^{-2}$
- insulation walls variable (<.08 m)	64.-	$Dflm^{-3}$
- insulation walls (>.08m) extra	194.-	$Dflm^{-3}$
- single glazing (incl walls)	150.-	$Dflm^{-2}$
- double glazing (incl walls)	210.-	$Dflm^{-2}$
- attached sunspace	225.-	$Dflm^{-2}$
- night shutters initial	250.-	$Dflm^{-2}$
- insulation night shutters	300.-	$Dflm^{-3}$
- added mass to interior walls	400.-	$Dflm^{-3}$
- connection heating system	100.-	Dfl

4. RESULTS AND DISCUSSION

For the following combinations of energy conserving options the best ditribution between the options of the investments of upto ƒ9000.- is investigated;
- insulation of the walls, single glazing
- insulation of the walls, double glazing
- insulation of the walls, single glazing, regeneration
- insulation of the walls, double glazing, regeneration
- insulation of the walls, double glazing, greenhouse, regeneration
- insulation of the walls, double glazing, regeneration, solar hot water
- insulation of the walls, double glazing, greenhouse, regeneration, solar hot water
- insulation of the walls, double glazing, regeneration, solar heating, solar hot water
- insulation of the walls, double glazing, regeneration, solar heating, hot water, greenhouse

A summary of the results is presented in fig. 1. Each separate curve represents the auxiliary energy for the combination of options considered. The best choice of combinations is represented by a fat curve(i.e. the curve parts with the lowest remaining auxiliary energy). The sequence in which the options enter the optimum combination and their investment is;

-1. insulation at ƒ 400.-

-2. regeneration of heat from the waste air at ƒ 1200.-

-3. double glazing at ƒ 1950.-

-4. solar water heating at ƒ 3800.-

-5. attached sunspace at the south facade at ƒ 6500.-

The other options do not enter into the optimum combination for investments of upto ƒ9000.-.Solar heating, though, is only marginally less favourable than solar water heating for investments of above ƒ5000.-. The lower part of the fig. is enlarged in fig. 2 to show more clearly the active and passive options. The differentials of the energy saved with the investment (a well known investment criterion) at which the options enter the optimum combination are;

- insulation	105	MJƒ-1
- regeneration	50	MJƒ-1
- double glazing	14	MJƒ-1
- solar water heating	3.5	MJƒ-1
- greenhouse	2.5	MJƒ-1

The optimum sizes of the options as a function of the investment are shown in the following figures. The curves are discontinuous when a new option enters the optimum combination, as at these points a redistribution of the investments occurs. The results hold only for the parameters and costs mentioned under Dutch conditions, but the sequence of options will not differ markedly in most situations.

4.1 Design aids

The results of this study can be extended to a simple graphical design aid. The method presented is currently being implemented on a micro computer in PASCAL. With this method the optimization can be personalized. Some extensions, like the effects of infiltration, heat leaks and seasonal storage, deserve attention.

5. REFERENCES

1. Veltkamp, W.B. (1982). Optimisation of the flows in a solar energy system. Report, Eindhoven University of Technology, WPS3-82.09.R335.

2. Veltkamp, W.B. and C.W.J.van Koppen (1982). Optimization of the energy management of low-energy houses with a solar heating and hot water system. Report, Eindhoven University of Technology, WPS-82.09.R334.

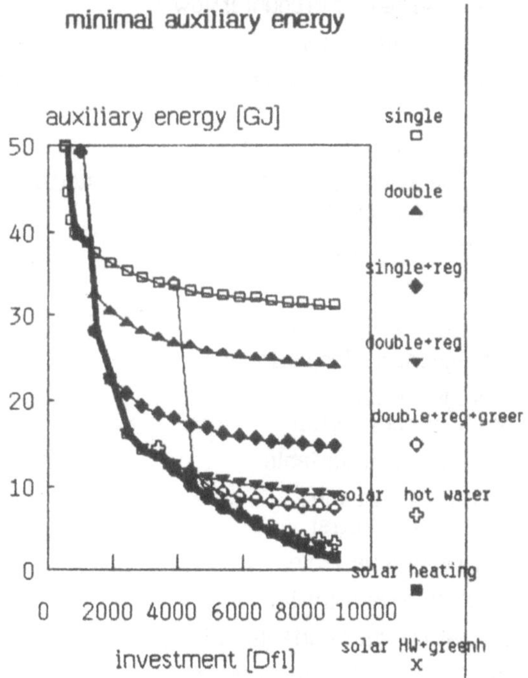

Figure 1

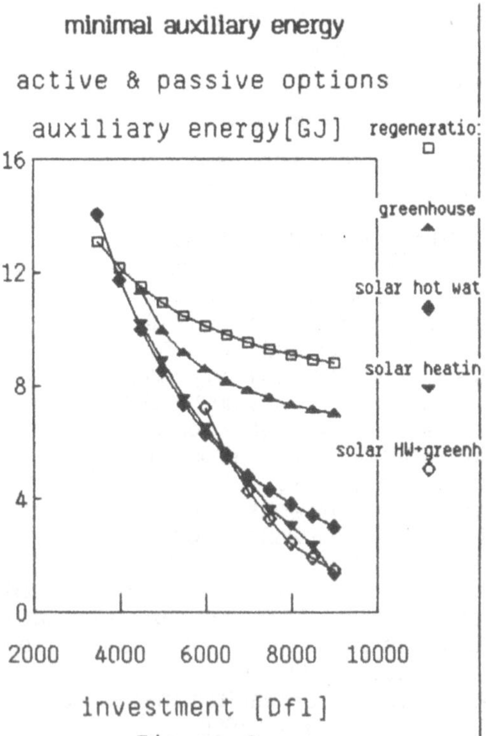

Figure 2

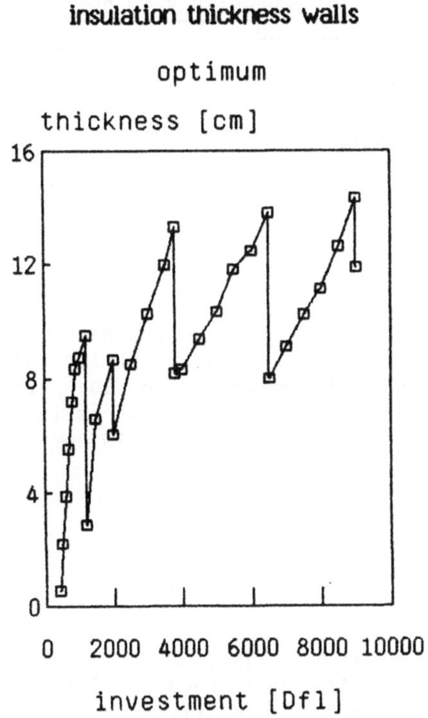

Figure 3

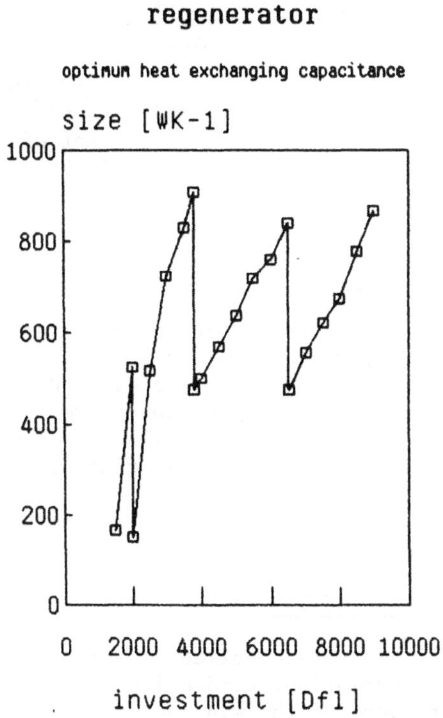

Figure 4

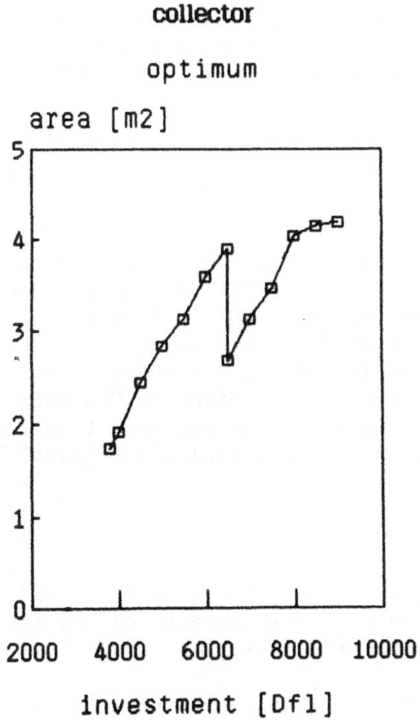

Figure 5

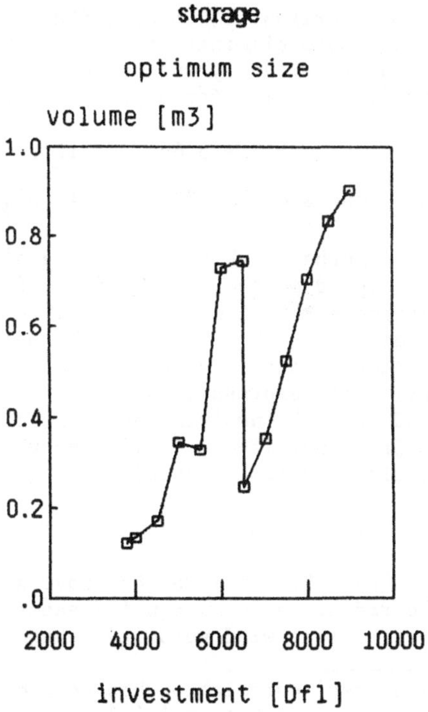

Figure 6

ON THE IMPORTANCE OF WIND VELOCITY AND AMBIENT TEMPERATURE TO THE PERFORMANCE OF FLAT-PLATE COLLECTORS

A. Honarbakhsh
Chalmers University of Technology
Department of Energy Conversion
412 96 Gothenburg
Sweden

Summary

The different modes of convective heat transfer from the outmost cover plate of a flat-plate collector are investigated. The effect of the daily variation of wind speed and ambient temperature on the performance of the flat-plate collector are studied. This effect is large if a non-glazed collector is considered but it is less than 10 percent if the collector is glazed.

1. INTRODUCTION

Flat-plate collectors are subjected to various kinds of heat losses, i.e. radiation, conduction, and convection. The last-mentioned is highly dependent on the wind velocity and on the wind regime. The first comprehensive work attempting to evaluate convective losses from a flat plate seems to be that of Jürges [1]. His experiment is based on heat transfer from a horizontal, square, flat plate with dimensions 0.5 x 0.5 m. Although he tried to eliminate the radiation component in his experiment, his results are slightly surface dependent. For instance, Jürges' correlation,1) eqn (1) is valid for a gray surface.

$$h_w = 5.3 + 3.6v \qquad \text{if } v < 5 \text{ m/s}$$
$$h_w = 6.47v^{0.78} \qquad \text{if } v \geqslant 5 \text{ m/s} \tag{1}$$

Another weak point of Jürges' correlation is that the correlation does not depend on the length of the flat plate. This correlation has been used for the calculation of convective heat transfer in fields such as building science, and the design of thermal converter devices, including solar collectors and solar distillation plants. Rowley et al. [3] found an experimental expression similar to that of Jürges. They defined a "surface conductance" involving both the convection and the radiation heat transfer coefficient. If one eliminates the radiation term from Rowley's original result, eqn (2) is obtained:

$$h_w = 2.8 + 3v \qquad 0 < v < 7 \text{ m/s} \tag{2}$$

Eqn (2) confirms the above statement, that eqn (1) actually includes the radiation term. Eqn (2) seems to be more realistic because of the fact that at zero wind velocity, the natural convective term, i.e. the value

1) Jürges' correlation has been called McAdams' relation in the literature because of McAdams' [2] presentation in Heat Transfer.

2.8, is closer to the result obtained from the correlation of free convection on a horizontal plate reported by Parker et al. [4].

Sparrow et al. [5] have studied the effect of wind on rectangular plates at various slopes in wind tunnel experiments. Eqn (3) is the conclusion of their results:

$$h_w = 0.86\ k/L_c \cdot Re_{L_c}^{1/2} Pr^{1/3}, \quad \text{where} \quad 2x10^4 < Re_{L_c} < 9x10^4 \qquad (3)$$

where L_c is the characteristic length of the flat plate, four times the heat transfer area divided by the length of the periphery. A comparison between Polhausen's [6] and Sparrow's correlations confirms the validity of eqn (3) up to a Reynold's number of 10^6.

Sparrow's correlation is based on the assumption that the heat loss from the collector to the surrounding is due to forced convection. Jürges' correlation, on the other hand, assumes mixed convection. The standard method for characterizing the mode of convection from the collector to the surrounding is by comparison of the associated inertial and buoyancy forces. Whether inertial forces or buoyancy forces are dominant depends on the wind velocity and the temperature difference between the surrounding and the outmost collector plate.

Fig. 1 presents the temperature dependence of the ratio of the Grashof number to the Reynolds' number, i.e. the buoyancy forces to the inertial forces. In the upper part of the figure, the range of the temperature difference, ΔT, for collectors with different numbers of covers is shown. The common range of the wind velocity is about 1-5 m/s. Thus the shaded area in the figure represents the form of convection at the collector's outmost surface in the most common operating conditions. As the figure shows, the convection losses occur in all modes of heat transfer, i.e. natural, mixed or forced convection, respectively going downwards in the figure. For an uncovered collector around noon when the temperature difference is maximum, the convection losses should be considered to be in the form of natural convection. Generally speaking, the mixed convection mode is the most realistic case for the flat-plate collector. Unfortunately, to the author's knowledge, no investigations on mixed convection heat transfer applicable to the flat-plate collector have been published. Considering the above discussion, Sparrow's correlation is not quite applicable to the flat-plate collector, where natural convection prevails and the Reynolds' number is small.

Although it is necessary to study the mode of convection heat transfer from the collector over a long term period, this is not within the scope of the present work. The purpose of the present work is to show the effect of the above correlation on the prediction of the performance of flat-plate collectors and also to evaluate the influence of diurnal variations of wind velocity and outside temperature on the collector performance. To the author's knowledge, this kind of study has not been presented before.

2. THEORY

The thermal performance of a flat-plate collector is expressed by the Hottel-Whillier-Bliss (HWB) model, see e.g. [7]. Under quasi-steady-state conditions, eqn (4) describes the model:

$$q_u = F' \cdot [H(t) \cdot f(\tau\alpha) - U_L(v,t) \cdot (T_p - T_a(t))] \qquad (4)$$

Fig. 1

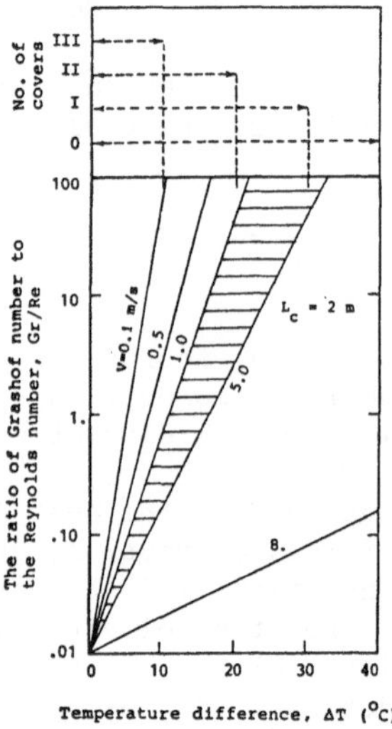

Fig. 2

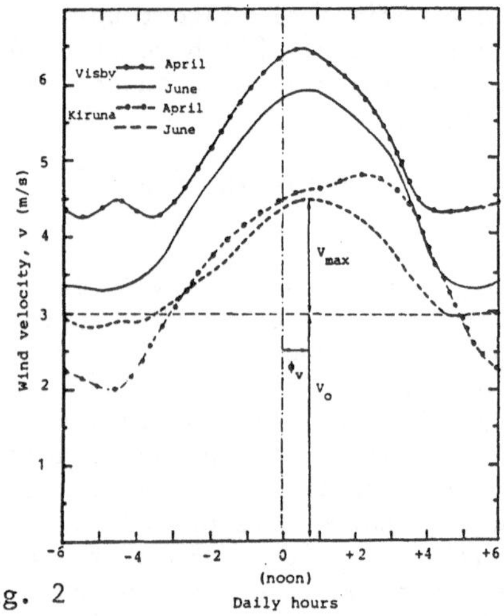

Fig. 1. Variation of Gr/Re with temperature difference between the outmost cover of the flat-plate collector and the surrounding, Δt, at different wind velocities. In the upper part of the figure the practical range of Δt for collectors with different number of covers is shown.

Fig. 2. Monthly average hourly wind velocity at Visby and Kiruna Latitude, (57° 30') and (67° 45'), respectively, in April and June, Eriksson (8). V_{max} and V_0 are parameters of eqn (8).

The flat-plate collector is connected to a storage unit. The energy equation for the storage-collector is expressed as follows:

$$M \cdot c_p \frac{dT_s}{dt} = q_u - q_{1s} \tag{5}$$

In eqn (5), it is assumed that the storage is well-mixed, and the whole system is at the same temperature, T_S. If we also assume that storage losses are negligible, using eqns (4) and (5) we have

$$\frac{dT_s}{dt} = \frac{1}{Mc_p} [F' \cdot f(\tau\alpha) \cdot H(t) - F' \cdot U_L(v,t) \cdot (T_s - T_a(t))] \tag{6}$$

2.1 <u>The solar radiation model</u>

If we assume that the solar flux at ground level, incident on a tilted surface, varies in a quasi-steady form and can be expressed in a cosine form, we have

$$H_T(t) = - H_{max} \cos\left(\frac{\pi t}{12}\right) \tag{7}$$

2.2 The wind velocity model

The wind velocity model is also assumed to be in the quasi-steady-state regime and is formulated by eqn (8)

$$v = V_o - V_{max} \cos\left(\frac{\pi t}{12} + \phi_v\right) \tag{8}$$

Fig. 2 shows the variation of the monthly average hourly wind velocity for the Swedish towns of Visby and Kiruna at latitudes 57°30', and 67°45', respectively. Some values of the parameters of eqn (8) are demonstrated in the figure.

Fig. 2 represents a typical coastal climate with a mesothermal forest climate, precipitation all through the year and constantly moist. However, for other classes of climates the monthly variation of the wind velocity may be quite different. Unfortunately, meteorological observations of this kind are rare.

2.3 The ambient temperature model

In the same manner as the solar radiation model and the wind model, the ambient temperature is modeled by a cosine function. Eqn (9) shows this model

$$T_a = T_m + \frac{\Delta T_{max}}{2} \cos\left(\frac{\pi t}{12} + \phi_a\right) \tag{9}$$

Fig. 3 shows the monthly average hourly ambient temperature in Visby and Kiruna in April and June. The parameters of eqn (9) are presented in this figure.

Since the number of parameters in the above models are too many for a meaningful parameter study, a non-dimensional procedure is used. Eqns (4), and (7)-(9) are inserted in eqn (6), in which the following dimensionless parameters are introduced:

$$T = \frac{T_s - T_m}{T_i - T_m}$$

$$\tau = \frac{t}{P} \tag{10}$$

$$\phi = \frac{\phi}{P}$$

After some manipulation we get

$$\frac{dT}{d\tau} = - \Pi_1 \cos 2\pi\tau - \left(\Pi_2 - \Pi_3 \cos\,(2\pi\tau + \phi_v)\right)T +$$

$$+ (\Pi_4 - \Pi_5) \cos\,(2\pi\tau + \phi_a) \tag{11}$$

The dimensionless group Π_1 represents the absorption function. Π_2 stands for losses due to a constant daily wind velocity. Π_3 indicates the effect of daily variation of wind velocity on the performance of the flat-plate collector. Finally, Π_4 and Π_5 show the influence of daily variation of ambient temperature.

The absorption function, Π_1, may be demonstrated with the help of Fourier's and Kirpichev's numbers, [9], i.e.

$$\Pi_1 = F \cdot Ki_{max} \cdot Fo$$

where

$$F = F' \cdot f(\tau\alpha)$$

$$Ki_{max} = \frac{H_{max} \cdot L_c}{k(T_i - T_m)}$$

$$Fo = \frac{k \cdot P \cdot L_c}{M_s c_p}$$

In the above relationship, F represents the design parameters of the collector and it is independent of the operating conditions. M_s is the whole storage mass.

All other numbers, i.e. Π_2-Π_5, represent the losses from the collector. They may be expressed as follows:

$$\Pi_{2,3} = F' \cdot Fo \cdot Nu$$

$$\Pi_{4,5} = F' \cdot Fo \cdot Nu \cdot \frac{T_{max}}{T_i - T_m}$$

where

$$Nu = \frac{U_n \cdot L_c}{k}$$

n is 2, 3, 4 or 5, and U_n is the appropriate heat loss coefficient in the Π-numbers.

The thermal performance of the collector is expressed by the daily averaged efficiency, $\bar{\eta}_D$, which is defined by the following relationship:

$$\bar{\eta}_D = \frac{\int_{t_i}^{t_{st}} q_u \, dt}{\int_{t_i}^{t_{st}} H_T \, dt}$$

where t_i is the initial time and t_{st} is the time of occurance of the stagnation temperature. The initial time is the time at which the collector temperature is equal to the initial temperature of the storage tank.

3. RESULTS

3.1 Variable wind velocity

Fig. 4 presents the variation of $\bar{\eta}_D$ with wind velocity, v, for an uncovered collector. The influence of eqns (1) and (3) is shown in the figure. The figure indicates the fact that the difference between Jürges' and Sparrow's correlations is more than 20% of the daily efficiency and this increases when the wind velocity increases.

Fig. 5 shows the same relationship as Fig. 4 but for collectors with one cover. The trend in this figure is the same as of Fig. 4 but with a weaker effect.

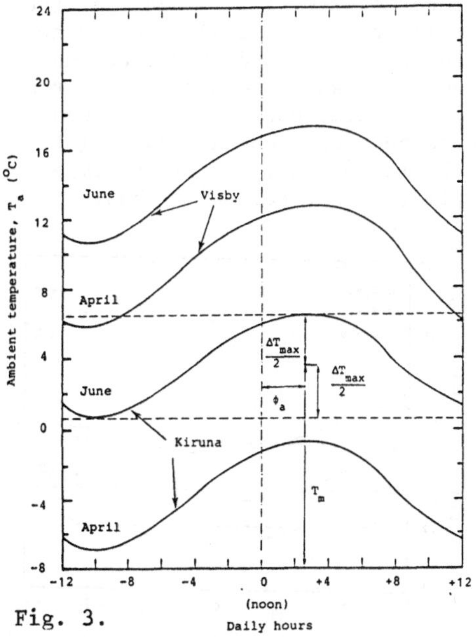

Fig. 3.

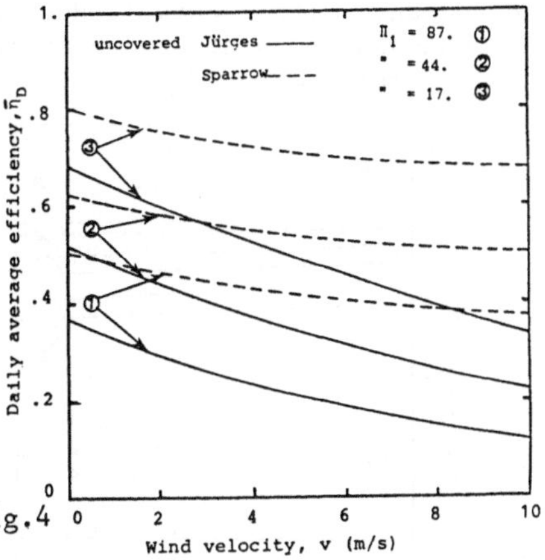

Fig.4

Fig. 3. Monthly average hourly ambient temperature in Visby and Kiruna, in April and June, Eriksson (8).

Fig. 4. Variation of daily average efficiency, $\bar{\eta}_D$, of an uncovered flat-plate collector with wind velocity for different storage volumes and two different correlations, i.e. Jürges (solid line) and Sparrow (dashed line). Nos. 1, 2, and 3 stand for various values of absorption factor, Π_1 , i.e. 87., 44., and 17..

Figs. 6(a) and 6(b) indicate the variation of $\bar{\eta}_D$ with delay ϕ for Π_3 = 10., and 3.. The dashed lines show the influence of the daily average wind speed on the collector efficiency, i.e. the standard method of considering the influence of wind velocity. Differently formulated, the distance of the dashed line from the solid one shows the error introduced in estimating the performance of uncovered flat-plate collectors using the average wind speed. Figs. 6(a) and 6(b) also show that the error decreases as Π_3 decreases, which could be excepted.

In order to study the effect of thermal capacity of a system on the performance of the flat-plate collector the concept of "system coefficient", C, is introduced. The system coefficient is defined as a coefficient by which the size of the thermal capacity of a system is compared to an arbitrary reference.

3.2 Variable wind velocity and ambient temperature

Fig. 7 shows the influence of diurnal variation of ambient temperature on the daily average efficiency of the flat-plate collector at different phase differences of the ambient temperature, ϕ_a = 0, 2/24, 4/24. For the sake of simplicity, in this figure, curve ① from Fig. 6(a), and the curve of constant wind velocity from that figure are drawn as dash-dotted and dashed lines, respectively. From this figure it can be seen that a diurnal ambient temperature with a maximum at around 2 a.m. does not cause much deviation from that of the daily average temperature (compare curve ② and the dash-dotted curve). However, if the maximum daily ambient temperature changes to an afternoon or morning time, this result will change (e.g., compare curve ③ and the dash-dotted curve).

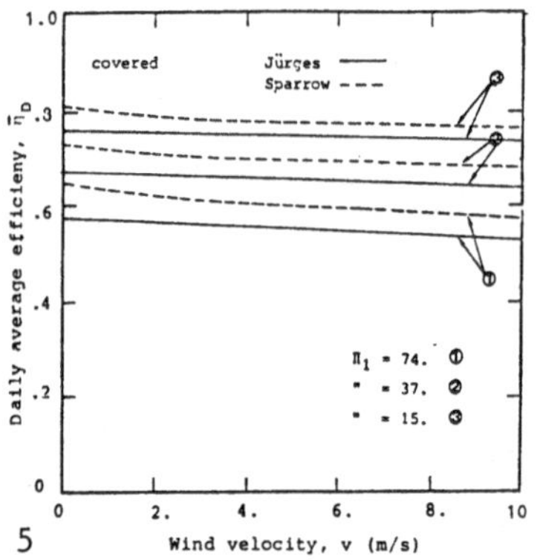

Fig. 5

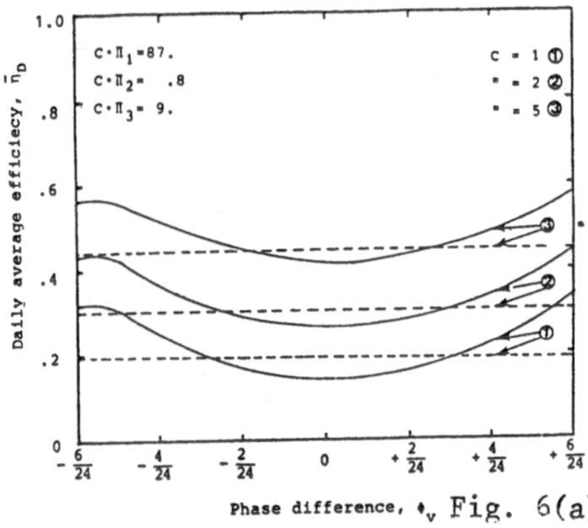

Fig. 6(a)

Fig. 5. Variation of daily average efficiency, $\bar{\eta}_D$, for an one-glass covered flat-plate collector, with wind velocity for different storage volumes and two different correlations, i.e. Jürges (solid line) and (Sparrow) dashed line. Nos. 1, 2, and 3 stand for various values of absorption factor, Π_1, i.e. 74., 37., and 15..

Fig. 6(a) Variation of daily average efficiency, $\bar{\eta}_D$, of an uncovered flat-plate collector with the phase difference between the variation of wind and solar radiation, ϕ_V. Nos ① , ② , and ③ in circles denote values of the system coefficient, 1, 2, and 5. The dashed lines stand for the effect of average wind velocity corresponding to different system coefficients. The outlet temperature is assumed to be constant and the Jürges' correlation, eqn (1) is used. Other parameters are shown in the figure.

Fig. 6(b)

Fig. 7

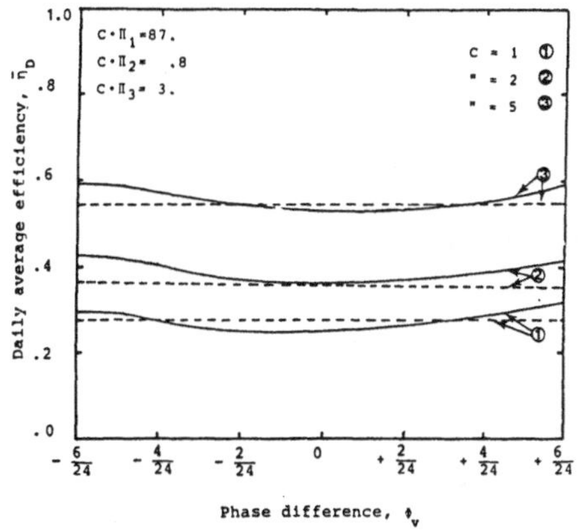

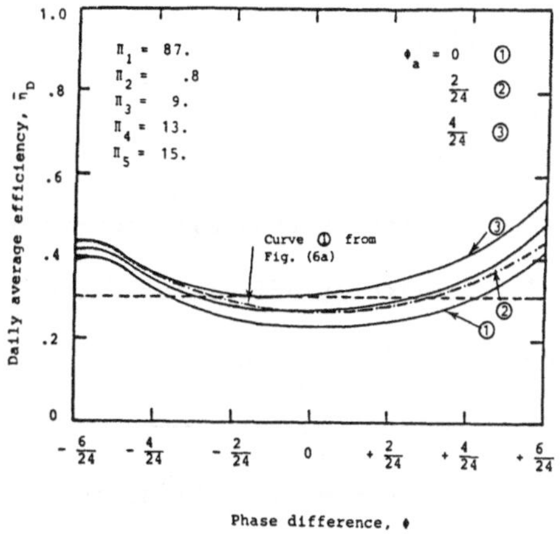

Fig. 6(b) Variation of daily average efficiency, $\bar{\eta}_p$, of an uncovered flat-plate collector with the phase difference between the variation of wind and solar radiation, Φ_v. Nos. ① , ② , and ③ in circles denote values of the system coefficient, 1, 2, and 5. The dashed lines stand for the effect of average wind velocity corresponding to different system coefficients. The outlet temperature is constant and the Jürges' correlation is used. Other parameters are shown in the figure. The difference between this figure and the former one (6.a) is that different values of Π_3 have been used.

Fig. 7 Variation of daily average efficiency, $\bar{\eta}_p$, of an uncovered flat-plate collector with phase difference between variation of solar radiation and wind velocity, Φ_v, for different phase difference of the ambient temperature, Φ_a. Dimensionless parameters of the system are noted in the figure.

Comparing curve ③ with that of constant wind speed and ambient temperature (dashed line) indicates that using the latter for design purposes will be quite satisfactory if the maximum of the wind variation is around noon and that of the ambient temperature is at around 4 a.m. However, in most cases this form of combination of wind and ambient temperature does not exist.

4. CONCLUSION

The conditions under which different modes of convection heat transfer occur from the outmost cover plate of the flat-plate collector to the surrounding are studied. The difference between correlations expressing these losses are studied. The differences are large if an uncovered collector is considered, but they are less than 10% of the daily efficiency if the collector is covered.

The effect of a diurnal variation of wind speed and ambient temperature on the performance of the collector has also been evaluated. This effect has been compared with that of daily average wind speed and ambient temperature. If the maximum values of wind velocity and ambient temperature occur at around 2 p.m., then the assumption of average daily values is reasonable. Otherwise, an efficiency corresponding to average daily values may cause considerable errors.

Acknowledgements - I wish to thank Dr Per Albråten and Mr. Tor Broström for their advice in the preparation of this article.

NOMENCLATURE

c_p thermal capacity, J/kg°C

C system coefficient

f(τα) effective transmittance-absorptance product

F' collector efficiency factor

Fo Fourier's number

Gr Grashof number, $\frac{g\beta\Delta T L_c^3}{\nu^2}$

HWB Hottel-Whillier-Bliss model

H solar flux, W/m^2

h thermal loss coefficient, $W/m^2 {}^{\circ}C$

k thermal conductivity, W/m°C

Ki_{max} Kirpichev's number, maximum value

L_c characteristic length

M storage mass

Nu Nusselt's number

n index number

Pr Prandtl number, $\frac{\mu c_p}{k}$

p period, 24 hours

q power, W/m^2

Re Reynolds' number, $\frac{VL_c}{\nu}$

t time, second

T temperature, °C

U thermal loss coefficient, $W/m^2 {}^{\circ}C$

v wind velocity, m/s

V see Fig. 1

α absorptance

τ transmittance

$$\Pi_1 = \frac{F'f(\tau\alpha)H_{max}P}{Mc_p(T_i-T_m)}$$, dimensionless

$$\Pi_2 = \frac{F'U_LP}{Mc_p}$$, dimensionless

$$\Pi_3 = \frac{F'UP}{Mc_p}$$, dimensionless

$$\Pi_4 = \frac{F'U_L\Delta T_{max}P}{2\ Mc_p(T_i-T_m)}$$, dimensionless

$$\Pi_5 = \frac{F'U\Delta T_{max}P}{2\ Mc_p(T_i-T_m)}$$, dimensionless

ϕ delay, see eqn (8)

Φ dimensionless delay, see eqn (10)

ΔT temperature difference

$\bar{\eta}_D$ daily average efficiency

Subscripts

a ambient temperature
i initial time
ls storage loss
L losses
p plate
m average value
T tilted surface
st stagnation temperature
s system
u useful
w wind

REFERENCES

1. JÜRGES, W. (1924). Der Wärmeübergang an einer ebenen Wand, Beihefte zum Gesundheits-Ingenieur, Reihe 1, Beiheft 19, München.
2. MCADAMS, W.H. (1954). Heat Transmission, McGraw-Hill, New York.
3. ROWLEY, F.B., ALGREN, A.B., and BLACKSHAW, J.L. (1930). Surface Conductance as Affected by Air Velocity, Temperature and Character of Surface, ASHVE Trans. 36, pp. 501-508.
4. PARKER, J., BOGGS, J.H., and BLICK, E.F. (1974). Introduction to Fluid

Mechanics and Heat Transfer, Addison Wesley, London.
5. SPARROW, E.M., RAMSEY, J.A.N., and MASS, E.A. (1979). Effect of Finite Width on Heat Transfer and Fluid Flow about an Inclined Rectangular Plate, Journal of Heat Transfer, 101, p. 2.
6. POLHAUSEN, E., and ANGEW, Z. (1921). Math. u. Mech. 1, pp. 115-121.
7. KREITH, F., and KREIDER, J.F. (1978). Principles of Solar Engineering, McGraw-Hill, Washington.
8. ERIKSSON, B. (1977). The Daily and Annual Variation of Temperature, Humidity and Wind Velocity at some Places in Sweden, The Swedish Meteorological and Hydrological Institute (SMHI) Report, Meteorologi och Klimatologi, No. RMK 8. (In Swedish).
9. LUIKOV, A.V. (1968). Analytical Heat Diffusion Theory, Academic Press, London.

PERFORMANCE OF PIPE TYPE SOLAR WATER HEATER

N. I. NGOKA

University of Ife,
P. O. Box 1035
Ile-Ife, NIGERIA

Summary

This paper gives relevant information for the design, construction and performance of pipe type solar water heater which is considered suitable for domestic application in a developing country such as Nigeria. It consists of series of pipes positioned inside circular blackened collectors enclosed in a glass shielded wooden box.

A 10cm thick fibreglass insulation is provided at the bottom of the collector system. A 59 litre hot water storage tank is connected by feed and return pipes to the tilted collector. The heated water rises from the collector into the tank and the cold water decends from the tank back to the collector by thermosiphon circulation. The tank is insulated with sawdust obtained locally from a carpenters shop.

Experiments were performed on the prototype of the solar water heater under varying conditions in order to assess its performance. This involved the alteration of the angle of tilt of the collector, and monitoring of water temperatures during various periods of the day, from the 5th to 11th of May 1983.

It was found that 18° was an approximate angle of tilt for a collector of this nature located at Ile-Ife, Nigeria. Adequate hot water would be provided by the system with maximum temperatures being attained at about 1500 hours. The unit efficiency varies form 52% to 85% depending on the incoming solar radiation and the hot water temperature.

1. INTRODUCTION

It is generally accepted that solar water heating is the most common application of solar engineering available in the world, especially in the developed countries. In these countries solar water heaters are now in commercial production and are gainnning widespread acceptance due to the present need to reduce energy costs. In some countries such as the United States of America, and the United Kingdom, tax concessions are granted to home owners who incorporate solar water heating in their houses. As a result of this, there is a growth in market potentials for this garget. Several companies have been set up quickly to capitalise on solar water heating installations. Some of these companies give misleading performance data which results in poor performance of the system chosen by the user.

In most developing countries such as Nigeria, commercialisation of solar water heating is still at its infancy. Although the technology of solar water heating has been well known for several years (1,2,3,4,5,6) researchers in Nigeria (7,8,9,10) are still involved in the development of prototypes suitable for application to local conditions. The apparent increase in the domestic use of energy (11) in Nigeria has resulted in the unsteady supply and distribution of electricity in the country. In the urban areas, hot water is essential for bathing, washing, especially among the middle and upper class. As a result of this, electric water heaters are usually installed in houses and flats designed and constructed for this category of people. Most of these cannot be used when needed due to unreliable supply of electricity.

The thermosiphon circulation system proposed in this paper could supply sufficient domestic hot water needs for urban dwellers. It also ensures instant supply of hot water when needed.

2.0 SYSTEM DESIGN

As the principal objective of this project is the provision of a cheap and easy to construct solar water heater, efforts were made to introduce locally available materials and simple technology which could be conveniently understood by an average artisan. This system which was designed by the author was fabricated at the Agric Engineering Workshop by a semi-skilled welder.

Figures 1 to 4 illustrate the details of the construction techniques adopted for this solar water heater. The heat collectors consist of six tubes, each of which are 75 cm long with a diameter of 13.0 cm. These tubes were painted black for effective solar absorption. Each of these collector tubes has a 13 mm diameter water pipe passing through it. These are connected to 25 mm diameter heater pipes which enter the tank at two levels.

Both the plumbing system and the tubular solar collectors are placed in the metal cabiner as shown in Fig.3. A 10 cm thick fibreglass insulation is provided at the bottom of the arrangement, while a 4 mm thick plain glass covers the top of the metal casing. The whole arrangement is mounted on an adjustable base which allows for variation in the tilt of the collector.

A run-off tap is fitted very close to the bottom of the 59 litre tank for hot water collection. Cold water was usually poured into the tank when the top lid is removed. The tank is insulated with 25 mm saw-dust cover.

The use of just one fixed angle of tilt was considered suitable for this experiment because the more efffective sun tracking system would be rather complicated and expensive.

It was therefore essential to find out the most suitable angle of tilt.

It has been shown (12) that collectors should be oriented towards the equator and inclined at an angle equal to the latitude of its location plus 10° (in Northern Hemisphere) or minus 10° (in Southern Hemisphere). The solar water heater is to be installed at Ile-Ife which is located on longitude 4.6° E and latitude 7.5° N. By adding 10° to 7.5° we get the suggested (12) inclination of 17.5°. For simplicity of construction, the system designed has the capability for adjustments to 16°, 18°, 20°, 22°, and 24°, tilts.

3.0 HEATER PERFORMANCE

The principal objective of the experiments performed with this systems is to test whether it is capable of sypplying adequate hot water when needed, and also to assess its efficiency.

3.1 Installation

The solar water heater (Fig. 1.) was installed in an open space in front of the Faculty of Environmental Design Building. The collector was oriented due south, its angle of tilt adjusted accordingly during the experiment. The tank was mounted due North.

3.2 Experimental Work

In order to assess the performance of the solar water heater constructed, the following measurements were made :

i. Daily global radiation measurement were made with a Bimmettalic Actinograph.

ii. An electronic probe thermometer was used to measure the following at hourly intervals from 8a.m. to 8p.m daily :

- temperative of water in tank (T_T)
- temperative of water in collector pipes. T_P
- drain off water temperature (T_m)
- ambient temperature (T_a)

3.3 Computation of Efficiency

The Hottel-Whillier - Bliss equation (13-16) was used to evaluate the useful heat collected (Q) per unit area. The factors taken into consideration for this equation are :

i. Incident solar radiation normal to the collector plate, G_c

ii. The mean temperature of the heat removal fluid (water) in the collector, T_m

iii. Ambient temperature, T_a

This equation is expressed as follows :

$$Q = F\left\{(\tau\alpha)\ G_c - U\ (T_m - T_a)\right\} \quad \ldots\ldots\ldots\ldots\ (1)$$

Where :

F = factor related to effectiveness of heat transfer from the collector surface to the water.

T = transmittance through the covers.

α = collector plate absorptance.

U = coefficient of heat loss.

η = overall efficiency of collector.

The thermal performance and the overall efficiency of the collector is given by :

$$\eta = \frac{Q}{G_c}$$

and expressed as follows :

$$\eta = \frac{Q}{G_c} = F\ (T\alpha) - \frac{U}{G_c}\ (T_m - T_a) \quad \ldots\ldots\ldots\ldots\ldots (2)$$

Based on suggestions by Duffie (17), the following values were assumed for the components used in this investigation :

F = 0.90 $\quad$ a = 0.89

T = 0.78 $\quad$ U = 3.5

The efficiency of the water heater was calculated based on the obove relationship and assumptions.

4.0 RESULTS

The following represent analysis of some of the data considered relevant from this experiment.

Fig. 5 shows curves of the results obtained by altering the angle of tilt of the collector. It is evident from this that angle of 18° was most ideal for solar radiation absorption of the collector for Ile-Ife.

Fig. 6 shows the influence of solar radiation intensity on heat gain in the water heater. The system gained more heat during a bright sunny day than on a cloudy day.

Fig. 7 shows the results obtained during four days of the experiment when the water in the tank was not replaced with fresh one. The initial water temperature for both the tank and at drain off was same at the beginning of the experiment.

This was approximately 2.5° C lower than the ambient temperature. On the first day the drain off temperature rose steadily from 2.5° C at 8.a.m. to about 70° C later in the day. Although there were heat exchanges with the environment, the system maintained a temperature above 40° C from the second day. The temperature in the heater was maximum at about 1500 hours.

Fig.8 shows the temperature profile when 5 litres of water was drained off every hours for daytime use. The maximum water temperatures was also obtained at about 1500 hours.

Fig. 9 shows that the hourly efficiency for the unit varies during the day. For the seven days during which efficiency was assessed, it has been found that the efficiency ranged from 52 to about 85 percent. Maximum efficiencies were attained during periods of high insolation.

5.0 CONCLUSIONS

From this experiment it has been found that a pipe type solar water heater is feasible and could serve as a means of domestic water supply in the urban areas of Nigeria.

The system designed is simple and could be fabricated by an average skilled welder. Local materials could be obtained readily for the construction of this unit.

With adequate insulation of the tank and the pipe cabinet, it is possible to obtain efficiencies of up to 85 percent.

The success achieved with this prototype, could lead to widespread application of the system in domestic buildings in the developing countries. EEC could support further work in this area for the benefit of the less developed countries (LCD).

REFERENCE

1. Brooks, F.A., Solar Energy and its uses for heating water in California, Bull. Calif. Agric, Exp. Sta., N° 602, 1936

2. Hottel, H.C. and Woettz, B.B., The performance of flat plate solar heat collectors, Trans. ASME, 64, 91-104, 1942

3. Heywood, H., Solar Energy for water and space heating, J. Inst.Fuel 27, 334-347, July 1954

4. Morse, R.N., Solar Water Heater, Proc. World Symposium on Applied Solar Energy, Stanford Research Inst., University of Arizona, Phoenix, Arizona, 191-202, 1956

5. Chinnery, D.N.W., Solar Water Heating in South Africa, National Building Research Institute, Bulletin 44, CSIR Research Report 248, Pretoria, South Africa, 1967

6. Tabor, H., Solar Energy Collector Design, Bull, Res. Coun., 5C, N° 1, Israel, 1955

7. Doyle, M.D.C. and Shishodia, K.S., Design Considerations for flat plate solar water heaters, paper presented at the Solar Energy Society of Nigeria Conference, Bida, May 1983

8. Ngoka, N.I., Design of a low cost solar water heater, paper presented at the Solar Energy Society of Nigeria Conference, Bida, May 1983

9. Ofi. O., The design, construction and evaluation of a solar water heater, paper presented at the Solar Energy Society of Nigeria, Conference, Bida, May 1983

10. Suleiman A.T., and Ahiome, G.E., The Development of a simple solar water heater, paper presented at the Solar Energy Society of Nigeria Conference, Bida, May 1983

11. Ngoka, N.I., Appraisal of the Energy Policy in Nigeria, proceedings of the Third Miami International Conference on Alternative Energy Sources, Vol.3, 437-454, Hemisphere Publication Publishing Corp., New York 1980

12. Veziroglu, T.N., Solar Energy International Progress, Vol. 2, Pergamon Press, New York 1978

13. Whillier, A., Solar Energy Collection and its utilization for house heating, ScD. Thesis, MIT, 1983

14. Hottel, H.C. and Whillier, A., Evaluation of flat plate collector performance, proceedings of conference on the use of Solar Energy, 2 (1), 74, University of Arizona Press, 1958

15. Duffie, J.A., and Beckman, W.A., Solar Energy Thermal Process, John Willey and Sons, New York, 1974

16. Smith,C.T., and Weiss, T.A., Design applications of the Hottel-Whiller-Bliss equation, ISES Congress. Los Angelius, Book of Extended, Abstracts, paper 34/6, 1975

17. Duffie ,J.A., et all, Report of working group on materials and components for flats plate collectors, Proc.. of workshop on solar collectors for Heating and Cooling of Buildings, NSF-RANN-75-019, May 1975

FIGURE 1 Illustration of the assembled solar water heater

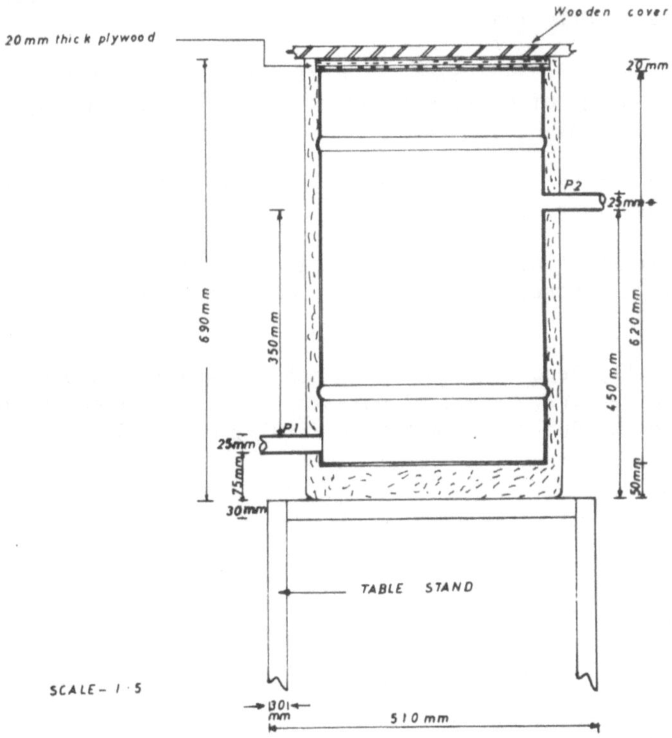

FIGURE 2 Section of water tank

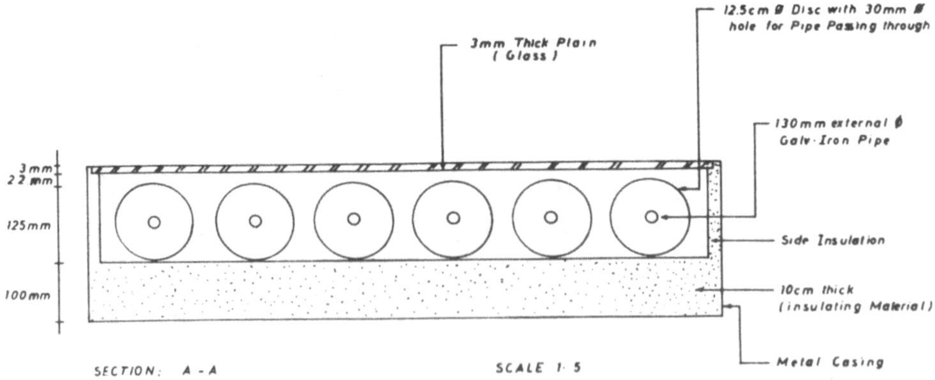

FIGURE 3 Cross-section of collector arrangement

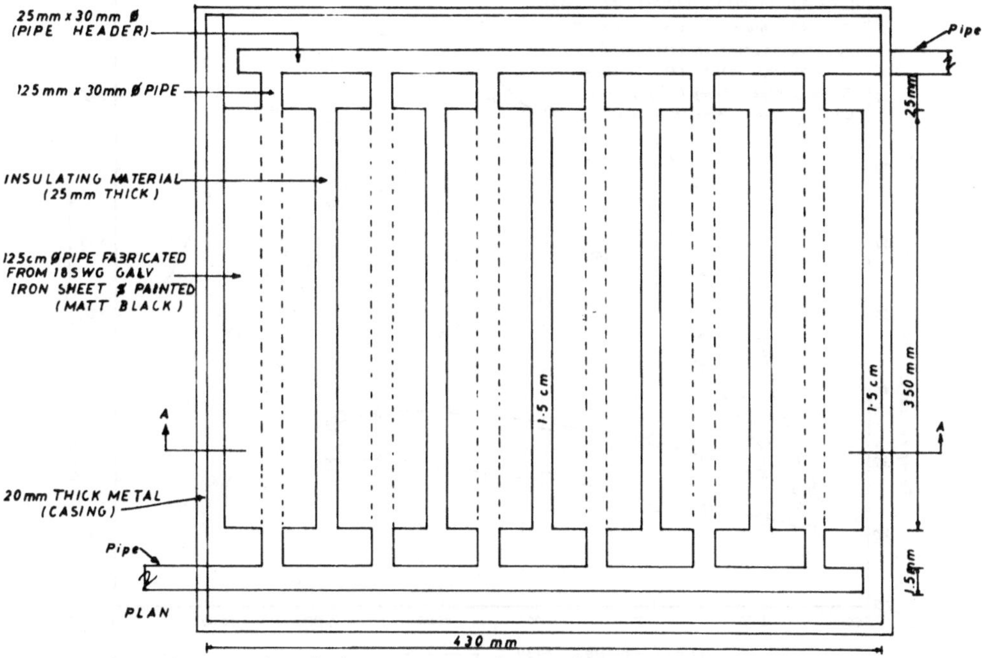

FIGURE 4 Sectional layout or solar collector

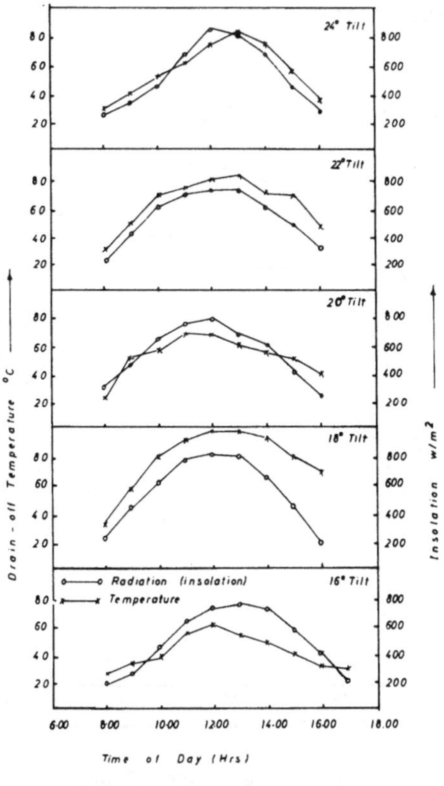

FIGURE 5 Effect of angle of tilt on solar radiation absorption

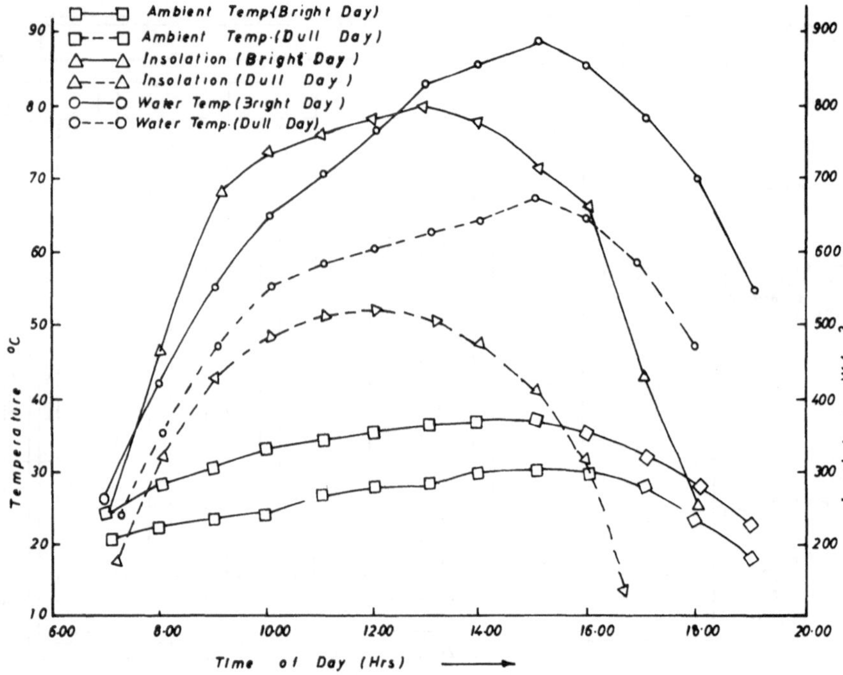

FIGURE 6 The influence of solar radiation on heat gain in the water heater

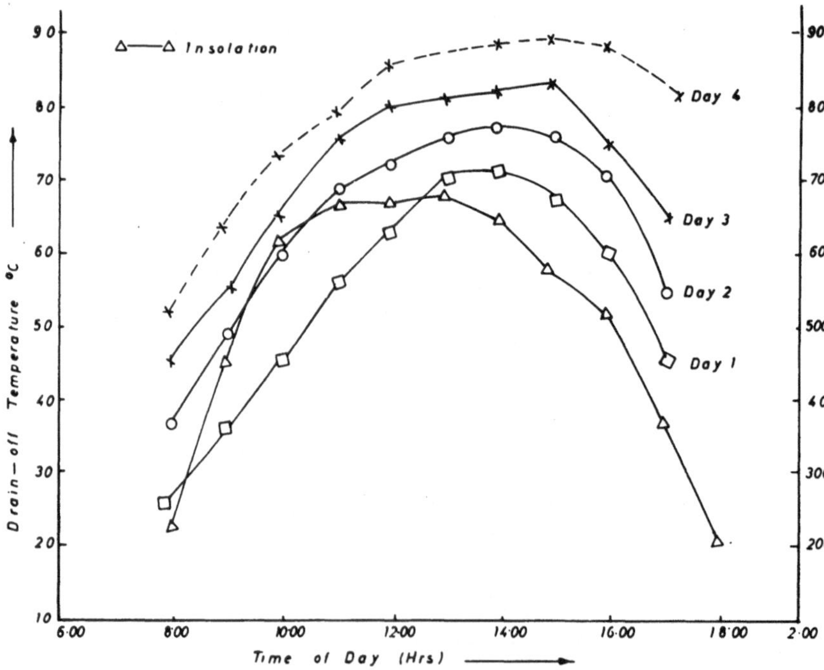

FIGURE 7 Drain-off temperature of water for four days

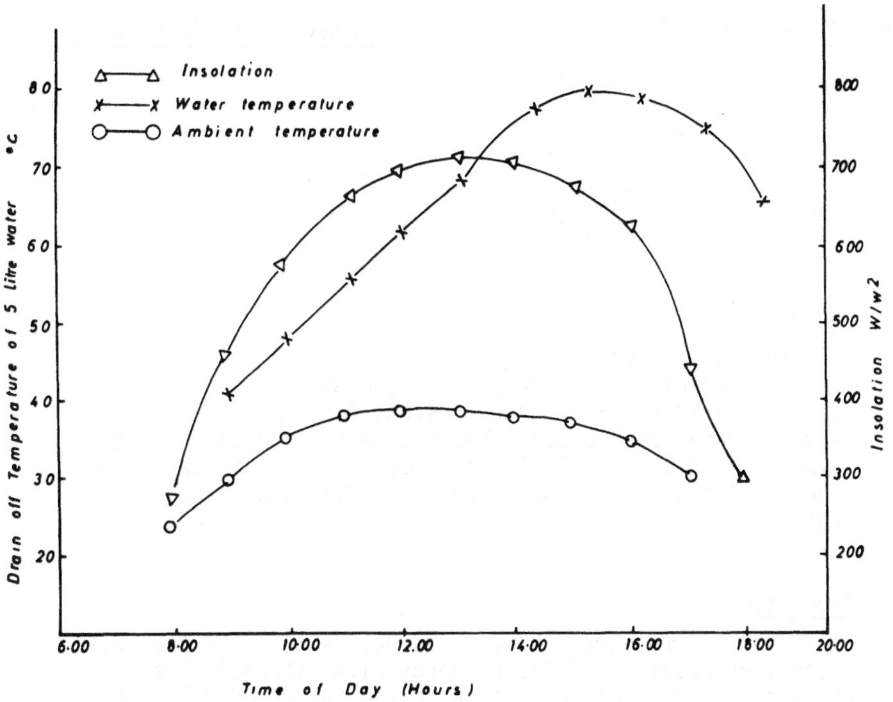

FIGURE 8 Temperature profiles for hourly drain-off of five litres of water

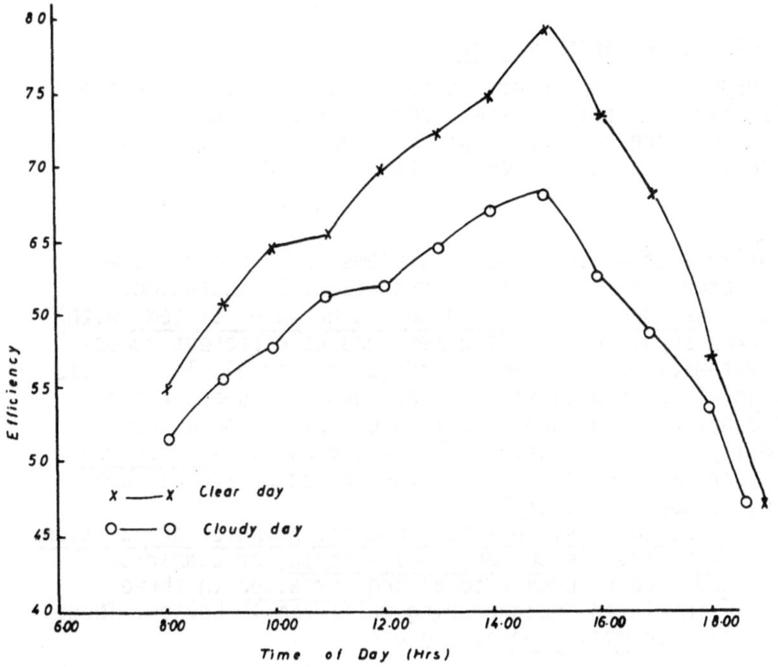

FIGURE 9 Variation of efficiency during the day

RESEARCH PROJECT: "GUIDELINES FOR OPTIMAL PLANNING AND CONSTRUCTION OF SOLAR ENERGY SYSTEMS"

B. Schläpfer
c/o Ernst Schweizer AG
Metallbau
8908 Hedingen
Switzerland

Summary

Within the framework of the IEA, the Swiss Professional Association of Solar-Energy Firms (SOFAS) will design and install various solar energy systems for domestic hot-waterpreparation and space heating. After monitoring the systems, a computer simulation programme will be validatet. The results will be generalized, guidelines will be elaboratet.

1. Introduction

A two year's research project has been started at the Swiss Professional Association of Solar Energy Firms (SOFAS). The project is executed in cooperation with the Swiss Federal Institute for Reactor Research (EIR) and with the Engineering School of Burgdorf (ISB). The major part of the work is done within the framework of the solar heating and cooling programme of the IEA. The project is co-financed by the Swiss National Energy Research Foundation NEFF.

2. Purpose of the Research Project

The purpose of the project is to improve and validate the cost/benefit ratio of solar systems for domestic hotwater preparation and space heating. Engineer's and installer's guidelines for designing solar systems shall be worked out.

3. Method

Five simple domestic hotwater systems and three combined domestic hotwater and space heating systems will be studied.

We have designed fife solar domestic hotwater systems with electric backup supply where the dimensions of different components are systematically varied (see figures 1 to 5). The systems are installed at one single site to get identical weather conditions. The hot water demand is simulated by using a known realistic daily consumer profile. The systems will be monitored during one year (meteo, heat gain of the collector arrays, heat consumption, backup, auxiliary).

Two combined solar systems for domestic hotwater preparation and space heating, and one conventional systeme for comparison will be designed (see figures 6 to 8) and installed in three identical well insulated, neighbouring multi-family houses. These systems will also be monitored during one year.

The computer simulation programme, developped at the engineering school of Burgdorf, will be validated for most of the systems mentioned, by comparison of calculated and measured values of heat gain of the collector arrays, heat losses of the system and backup consumption.

Generalisation of the achieved results of the studied systems will be possible by means of variation of parameters in computer simulation.

Optimization of the cost/benefit ratio will lead to guidelines for optimal planning an construction of solar systems.

4. Systems

Fig. 1. DHW-system with one storage tank

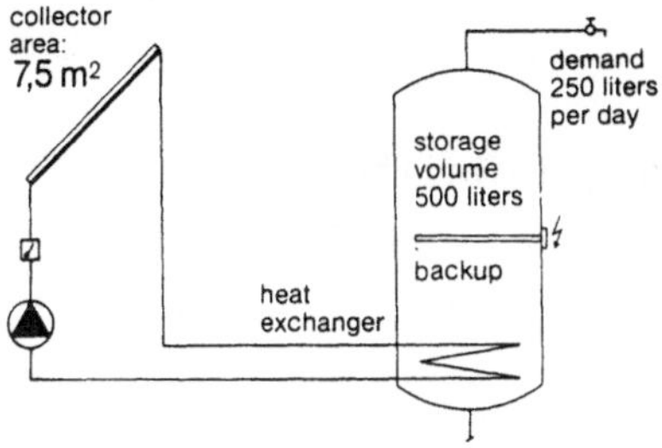

Fig. 2. Large system with one storage tank

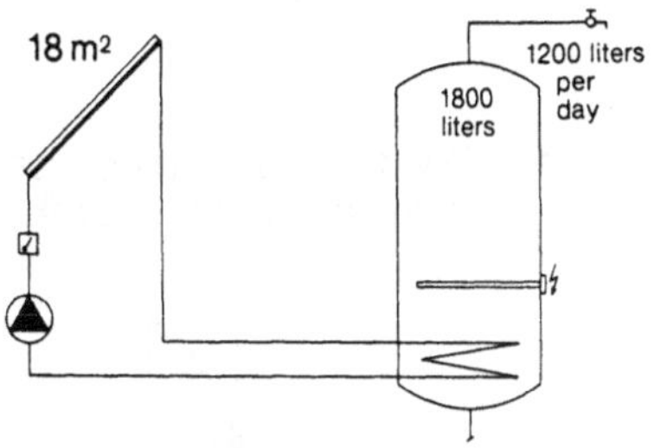

Fig. 3. System with two storage tanks

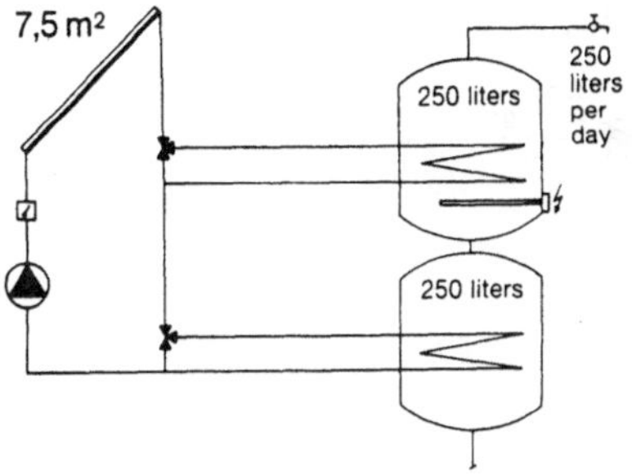

Fig. 4. Thermosyphon system

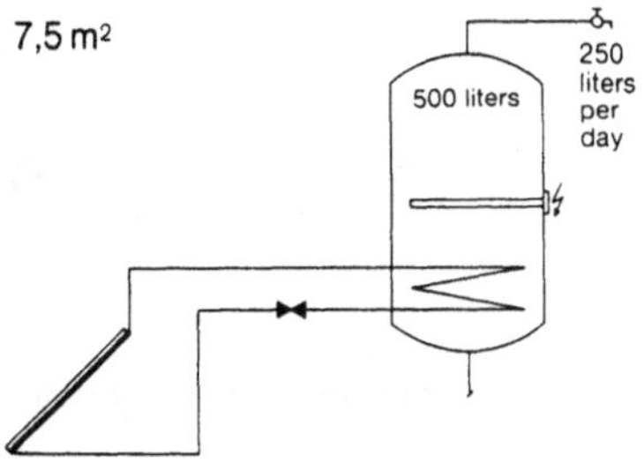

Fig. 5. Variable DHW-system to compare different dimensions, different strategies, etc.

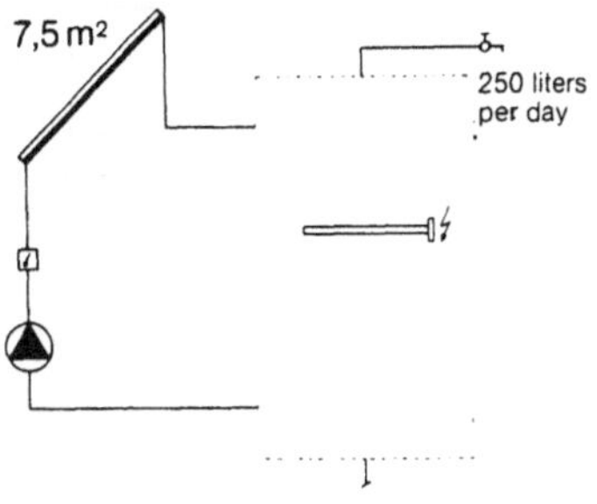

Fig. 6. System with combined storage tank for domestic hotwater and space heating

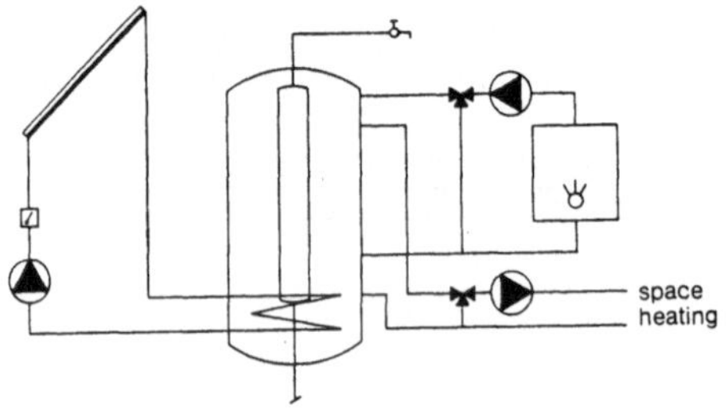

Fig. 7. System for direct space heating (possibility to bypass storage tank)

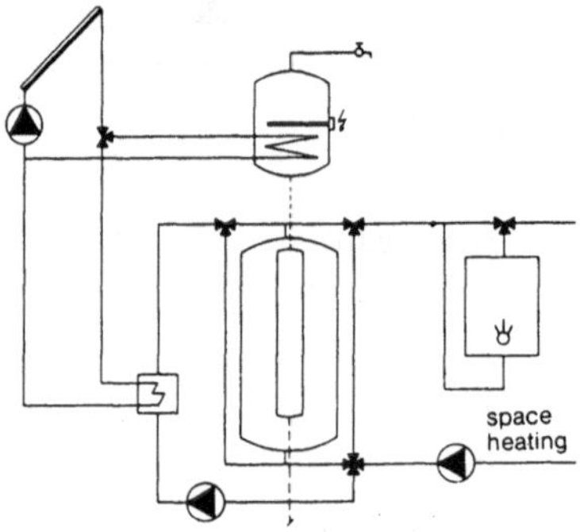

Fig. 8. Conventional system

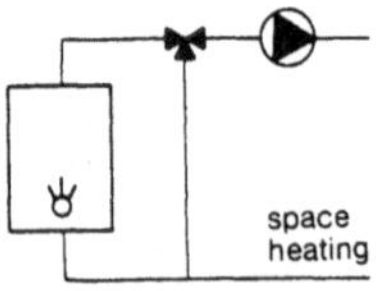

SHARED SOLAR SPACE AND WATER HEATING

J B THRING and D BOYD
Built Environment Research Group
Polytechnic of Central London

Summary

Cost effectiveness analyses of alternative configurations of solar space and water heating systems applied to houses are summarised. A system serving a group of 15 houses is being built in north London and is outlined. Seven houses are roofed with 140m² of flat plate collectors. These will provide about 40% of the heat requirements of eight flats and 27% of the water heating in the seven houses with collectors, via a seven cubic metre store. The Commission for the European Communities have granted £106,000 towards the project. The London Borough of Islington and the Polytechnic of Central London (PCL) are contributing the remaining 60% with some support from the London Energy and Employment Network. The system was designed by the Built Environment Research Group (BERG) and integrated into the basic layout and design of the dwellings in association with David Ford and Partners, the Architects of the scheme for Islington Council.

ORIGINS AND OBJECTIVES

The system is essentially a Mk 11 version of the PCL Solar House at Milton Keynes, which has performed with increasing success for the last 10 years and which provides over half the space and water heating requirements of that house. The main issue has been cost-effectiveness. At £5-6,000 in 1974, the system was slightly more costly than justifiable within a 25 year economic life-span because it was one of the first of its kind in the UK, with many experimental features.

A study by BERG for the UK Department of Energy's ETSU (Thring, Boyd et al, 1982) has since shown how various cost and performance benefits can be derived from design and control improvements. The most significant of these are the economies of scale from sharing a single store and control system amongst ten or more dwellings. It was also found that the extra heat capacity of larger stores for a given consumption is not significant enough to warrant the lower costs per cubic metre under capital cost constraints. Pitching collectors near the optimum angle of 55°C in London, gives an extra 9% delivered energy but it was not possible in the present scheme to increase the pitch from the conventional 30° without incurring on-costs. Dropping the emitter temperature in heated spaces from the normal 50°C to 24°C allowed an extra 1,250 kWh to be extracted from the system with no extra cost or discomfort, and this has been tested in the Milton Keynes house satisfactorily.

The analysis also shows the cost-effectiveness of other variables such as the benefits of certain selective-coated plates over evacuated tubes; of store insulation thickness; of emitter effectiveness and of the quantity of energy consumed.

The main object of the demonstration has been to incorporate most of these cost effective design improvements and to monitor them in unison. It has not been possible, however, to design the housing with solar heating incorporated from the outset. The scheme is therefore a 'new retrofit'. Four earlier attempts have been made with housing architects to incorporate the system, always resulting in an adaptation of existing drawings. They suffered from financial rather than purely technical constraints.

OUTLINE OF THE SCHEME BEING BUILT

The overall layout of the final scheme is illustrated in Fig 1 with the solar panels in black. The store is under the south-west corner flats in space originally for individual refuse bins, now amalgamated into two containers. The flats with space and water heating are outlined in a broken line. The houses receiving summer hot water only are beneath the collectors.

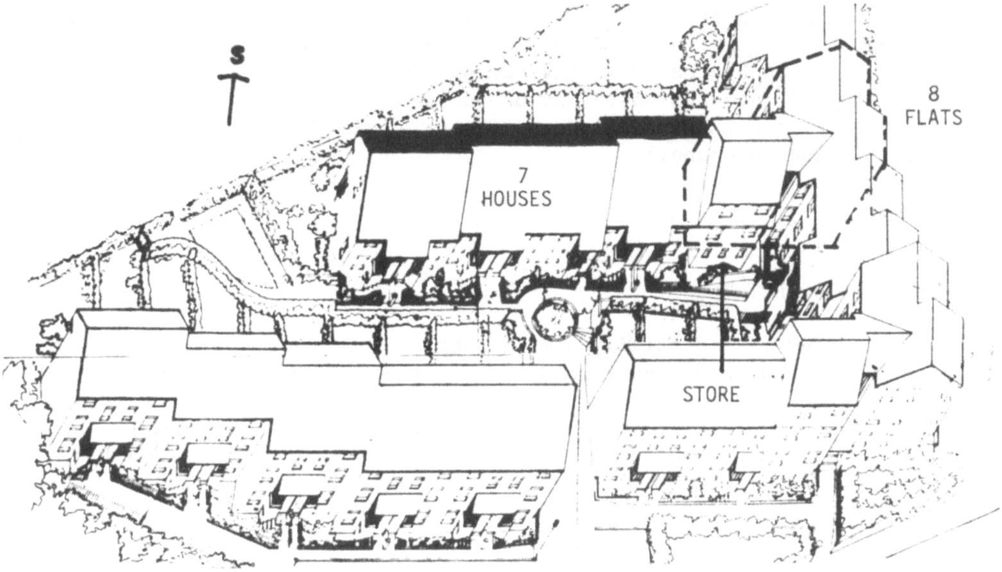

Fig 1. THE HIGHLANDS, ISLINGTON: LAYOUT, COLLECTORS (black) AND FLATS (dotted)

The 140m² of selective flat plate collectors will be built in situ on the roof (Fig 2) connected in parallel to the main flow and return pipes (on the ridge and under the eaves) and covered with 6 mm toughened glass in aluminium bars. Water with 20% monopropylene glycol and Fernox corrosion inhibitor, will be pumped from the collectors to finned copper heat exchangers in the 7m³ steel storage tank. From the store, water will be circulated to the 8 flats and 7 houses. Space heating is via a fanned warm air convector serving bedroom, living room and hall (Fig 3). Water heating is via a twin coil copper cylinder (Fig 4) with 80 mm insulation. Supplementary heat comes from a gas boiler in each flat as originally designed, via the same cylinder and radiators.

Fig. 2 THE SCHEME UNDER CONSTRUCTION: COLLECTOR ROOF ON HOUSES (with summer water heat)

Two levels of heat exchanger in the store (Fig 5) allow enhanced stratification, aided by three levels of input for return water from the flats and houses. Electronic controlled motorized valves input water according to the temperature differences between it and water at various levels in the store. Controls in each dwelling determine heat flows to the space and water heating units according to room and cylinder thermostats and the solar heated flow temperature. It is computed that about 20 MWh will be provided for space and about 15 MWh a year for water heating.

The system will be monitored with electronic heat meters, on the space and dhw sides of the gas and solar heating in the flats and on the dhw only in the houses (Fig 4). These will be read monthly at a central display panel in the main store. Monitoring will continue for at least one year. Sensors have also been included for more detailed monitoring and analysis.

REFERENCE

1. THRING J, BOYD D, HUGHES D et al (1982). Design and Cost effectiveness of Active Solar Space and Water Heating. BERG, PCL, for Energy Technology Support Unit.

Fig 3. FAN CONVECTOR FOR FLATS

Fig 4. DHW CYLINDER IN FLAT

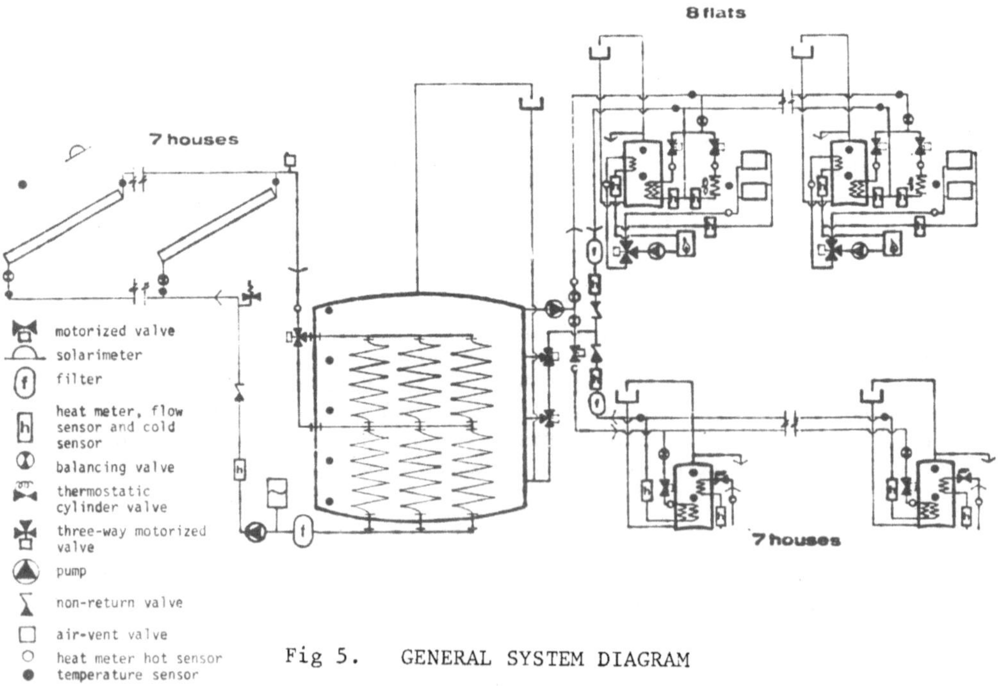

Fig 5. GENERAL SYSTEM DIAGRAM

SESSION VI - SOLAR ENERGY APPLICATIONS IN AGRICULTURE AND INDUSTRY

Solar energy storage for zootechnic applications

Heat transfer in a solar in-door cooker

THE USE OF AIR COLLECTORS IN DRYING PROCESSES OF AGRICULTURAL PRODUCTS

W.L. DUTRE and W. D'HOSE
Laboratorium voor Warmteoverdracht en Reaktorkunde
Katholieke Universiteit Leuven
Celestijnenlaan 300 A
B - 3030 Heverlee, Belgium

Summary

A simulation model for a drying silo is presented and is applied to the drying proces of maize in a small size drying silo with batch loading. The influence of air flow rate and air inlet temperature on the drying time and energy demand is investigated. Energy savings resulting from the use of a recuperation heat exchanger or from partial recirculation of exhaust air are calculated for different operating conditions. The additionally obtained energy savings from solar collectors, when used in combination with a recuperation heat exchanger and partial air recirculation are investigated for a continuously operated maize dryer.

1. SIMULATION MODEL FOR A MAIZE DRYER

1.1. General description of the simulation model

Drying involves simultaneous heat and mass transfer, heat being transferred from the drying air to the product to be dried and moisture being transferred from the product to the air. The proces is governed by the coupled differential equations describing the heat and moisture diffusion in the considered product and the transfer of heat and moisture at its surface to the drying air. Because the initial moisture content of maize is relatively low, drying occurs at gradually increasing temperature and decreasing moisture diffusion rate from the beginning of the proces.

The transient simulation model for a maize dryer with batch loading as used here, is a one dimensional model in which the drying silo is represented by a number of layers, each layer being described by a humidity balance equation, the humid air enthalpy equation, the energy equation for the product and the moisture diffusion equation. The moisture diffusion is described by a set of empirical correlations applicable to thin layers of maize, rather than the physical differential equation for moisture diffusion. It follows that the thickness of the layers considered in the simulation, should be sufficiently small for the empirical correlations to be valid. A layer thickness of 12.5 cm has been used.

The empirical relations describing the moisture transfer proces are based on the correlation for the equilibrium moisture content (E.M.C) of maize, as derived by De Boer [1] from experimental measurements. This correlation describes the EMC in terms of the temperature and relative humidity of the air. This EMC-correlation is combined with an empirical equation describing the time dependence of the moisture migration proces, derived by Troeger and Hukill [2]. This correlation can be applied to the transient desorption proces, in the temperature range from 32 °C to 70 °C. In layers of the maize silo in which the air humidity is high, rewetting of the maize occurs when the local EMC-value exceeds the local maize moisture content. The transient behaviour of this rewetting proces is empirically described by the Del Guidice equation [3]. This phenomenon mainly occurs in the upper layers of the dryer.

1.2. Physical properties of maize

The physical properties of moist maize depend on its moisture content. The following dependence is accounted for in the simulation model.

- specific heat : $c = 1396 + 3.39\ M_{wb}$ (J/kgK)
- thermal conductivity : $k = 0.0902 + 0.00072\ M_{wb}$ (W/mK)
- density : $\rho = 750$ kg/m^3 for $M_{wb} \leqslant 150$ g/kg

  $= 815 - 0.37\ M_{wb}$ for $M_{wb} > 150$ g/kg
- heat of evaporation and desorption of water in maize :

$$h_{fg} = 3640 - 4.164\ M_{db} \quad \text{(kJ/kg)}$$

where M_{wb} and M_{db} respectively represent the wet base and dry base moisture content of the maize.

The volumetric air to maize heat transfer coefficient is large and equals approximately 7000 W/m^{3}K, which is nearly equivalent to an infinite NTU-situation for the heat transfer proces.

1.3. Constraints imposed to the drying proces

In order to preserve the nutritive value of maize, the drying temperature should not exceed 60 °C. Lower drying temperatures are required when the germinal power of the final product is important. Depending on the use of the dried maize, the drying temperature being required, therefore varies from 40 °C to 60 °C. The required final moisture content of the corn depends on the conservation time being requested and varies from 6 % to 15 % db. In the simulations, an average value of 10 % db final moisture content has been imposed, starting from an initial value of 250 g/kg db. The drying proces should proceed at a sufficiently high rate in order to reduce the risk of quality reduction of the product, which may occur mainly during the early stages of a to slowly proceeding drying proces. The total drying time should therefore be reduced by using the highest allowed drying temperature and a sufficiently high air flow rate, accounting for the constraints imposed by the intended use of the maize. During the warm-up transient of the silo, condensation of air humidity may occur, causing a significant and rapidly progressing deterioration of the product. The risk for water vapor condensation mainly occurs in the upper layers of the silo at high air inlet temperatures and low flow rates. The air inlet conditions and flow rate should be such that condensation does not occur during more than five minutes.

2. SIMULATION RESULTS FOR A CONVENTIONAL MAIZE DRYER WITH BATCH LOADING

A parametric study has been performed for a small size silo for maize drying, based on the simulation model and accounting for the constraints as described above. The drying model has been implemented in a modified version of the transient solar system simulation program EMGP2 [4]. A 1 m^3 silo is considered, modelled as a stack of 8 layers of 12.5 cm each. The initial moisture content of the maize is 250 g/kg db and the requested final moisture content is 100 g/kg db. Since no air inlet humidity regulation is applied, the total drying time and the corresponding energy demand only depend on the requested air inlet temperature and flow rate. The solar irradiance in the horizontal plane, the outdoor temperature and the wet bulb temperature of the climatological reference period used in the simulation, are shown in the figures 1 and 2. The drying cycle is assumed to start the first day of this reference period at 8 a.m. and is considered to be uninterrupted. The energy removal from the silo to cool the dried product after completion of the drying cycle, is considered as non recoverable.

The total drying time of the first maize load in the considered reference period is represented in figure 3 as a function of the air flow rate, for different values of the requested inlet temperature. In order to avoid condensation in the upper layers, the flow rate should be at least 1000 m^3/hour. The gross energy demand of one drying cycle is represented in figure 4. It follows that the dryer should operate at the maximum allowed inlet temperature and the lowest possible flow rate for which condensation in the upper layers does not yet occur.

In all cases, the energy consumption can be reduced by means of a recuperation heat exchanger. In figure 5, the influence of the heat exchanger effectiveness on the energy consumption of one drying cycle is represented. The resulting energy savings are most sensitive to the heat exchanger efficiency at low silo inlet temperatures because the drying time is then large as compared to the silo warm-up time.

A similar reduction of the energy consumption can be achieved by means of partial recirculation of exhaust air. The maximum allowed degree of recirculation is however limited by the risk for internal condendation. In figure 6, the influence of the percentage of exhause air recirculation on the energy consumption of one drying cycle is represented. The dotted part of some of the curves, corresponds to operating conditions for which condensation inside the silo occurs for more than five minutes and which are therefore not allowed. The maximum allowed degree of recirculation increases with decreasing silo inlet temperature and increasing flow rate.

In order to increase the safety margin with respect to the condensation problem, the percentage recirculation being applied should remain well below its theoretically maximum allowed value. Applying partial recirculation together with a heat exchanger allows however to achieve high energy savings with a moderate recirculation percentage.

3. SIMULATION RESULTS FOR A SOLAR ENERGY ASSISTED MAIZE DRYER

A drying installation as schematically represented in figure 7 is considered. The system consists of a combination of a recuperation heat exchanger with an efficiency of 50 %, partial recirculation of at most 50 % of the exhaust air whenever possible or reduced otherwise, and solar air collectors. The drying proces is not allowed to be interrupted during the night. Interrupting the proces when no solar energy is available increases the solar energy contribution but also increases the total drying time and requires the silo to be cooled down whenever the drying proces is stopped for several hours in order to preserve the quality of the product. Simulations have been performed for different flow rates, collector surface areas, and requested silo inlet temperatures. Some typical results are shown in the figures 8 and 9, in which the different contributions to the gross energy demand of one drying cycle are represented as a function of the requested silo inlet temperature.

4. CONCLUSIONS

Significant energy savings in silo maize dryers with batch loading can be achieved by means of partial recirculation of exhaust air and heat recovery by means of a heat exchanger. Depending on the operating conditions of the dryer, these energy savings range from 20 % to over 50 % of the gross energy demand when changing the operating conditions from a low flow rate - high temperature (1000 m^3/hr - 55 °C) to a high flow rate - low temperature proces (2000 m^3/hr - 40 °C). Combining the system with solar air collectors approximately yields 10 % additional energy savings, for air collector arrays of 1 m^2 per 50 m^3/hr of the drying air flow rate. Further research is being devoted to discontinuous operation, different system configurations and large industrial dryers in view of minimizing the auxiliary energy demand per ton.

ACKNOWLEDGEMENT

The research reported in this paper is sponsored by the Belgian government as part of the thirth phase of the Belgian National Solar Energy Research Program (1981-1985).

REFERENCES

[1] BROOKER, D., BAKKER-ARKEMA, F., HALL, C. (1974). Drying Cerial Grains. Avi Publishing Company.
[2] TROEGER, J.M., HUKILL, W.V. (1971). Transactions of ASAE, p. 1153.
[3] THOMPSON, T.R., PEART, R.M., FOSTER, G.H. (1968). Transactions of ASAE, p. 582.
[4] DUTRE, W.L. EMGP2, a transient simulation program for solar systems (to be published).

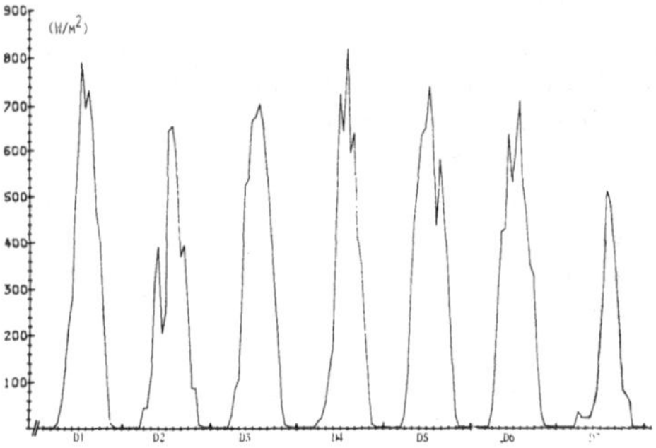

Fig 1. Total Solar Irradiance in the Horizontal Plane

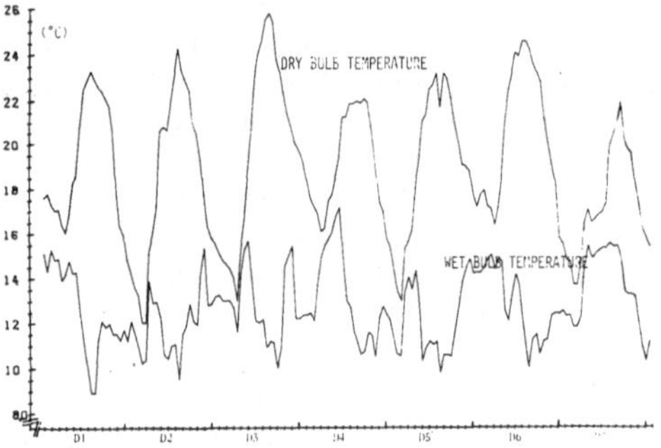

Fig 2. Ambient Temperatures

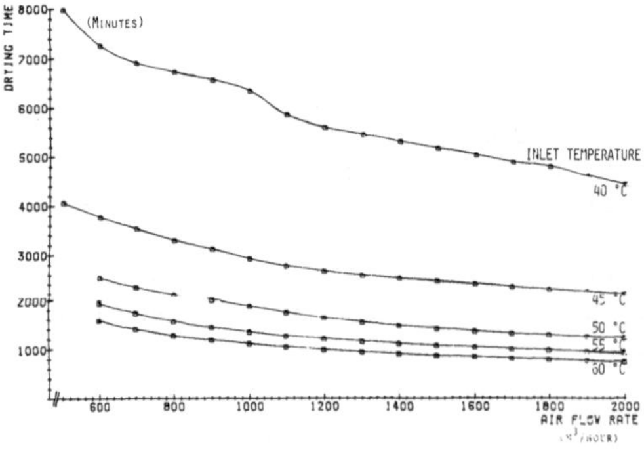

Fig 3. Drying Time

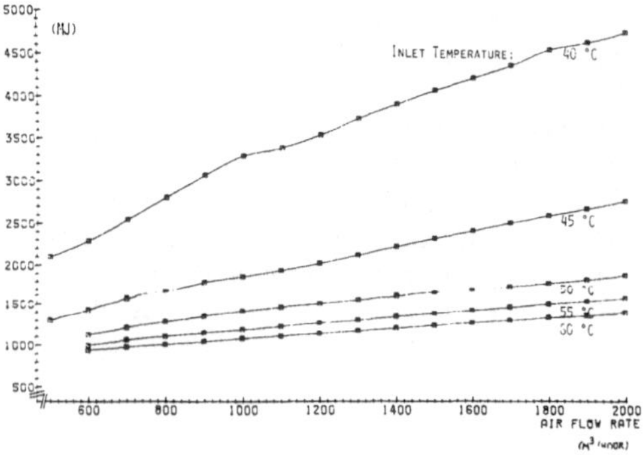

Fig.4. Gross Energy Demand of One Drying Cycle

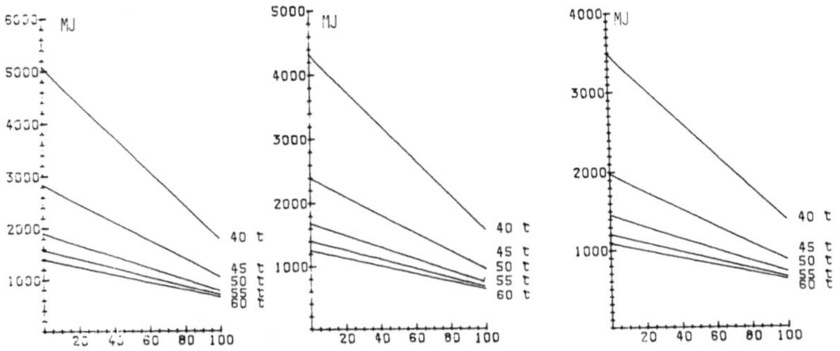

<u>Fig. 5.</u> Influence of heat exchanger efficiency on the energy consumption for three different flow rates and different inlet temperatures

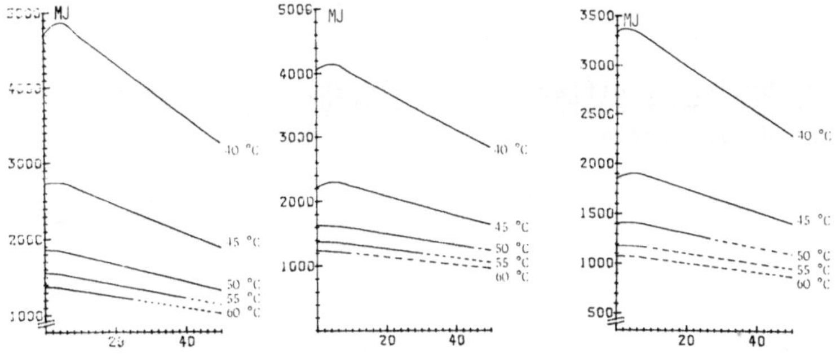

Fig. 6. Influence of partial air recirculation on the energy consumption for three different flow rates and different inlet temperatures

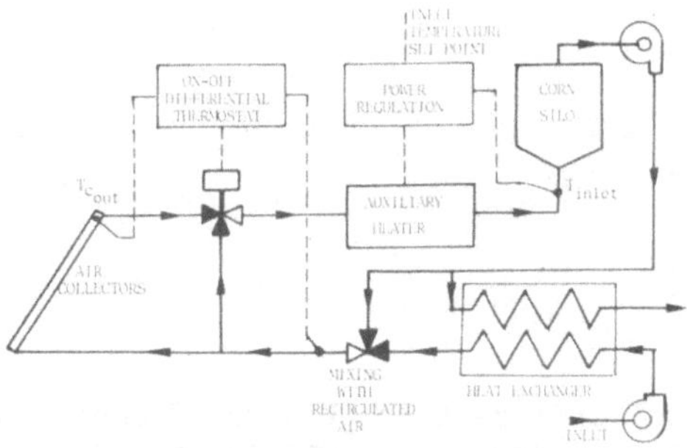

Fig. 7. Drying system with a heat exchanger recirculation-solar air collectors combination

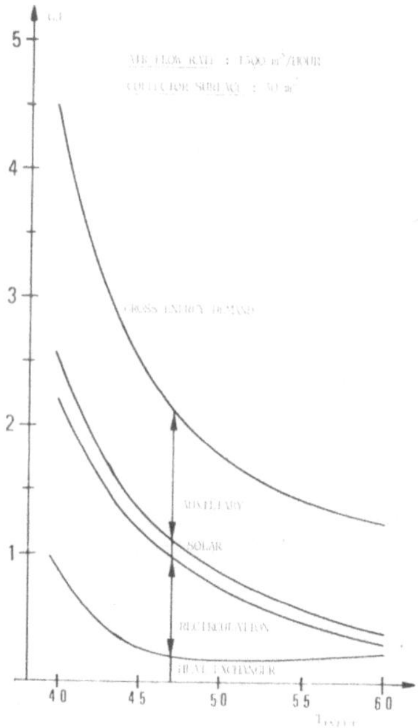

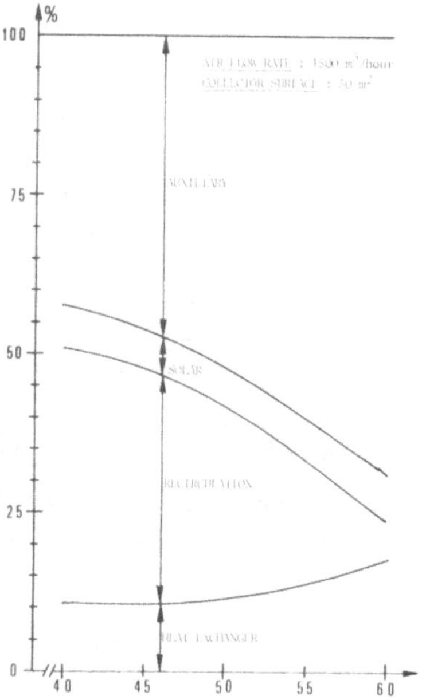

Fig 8A. Contribution of different components to the gross energy demand of one drying cycle

Fig 8B. % contribution to the gross energy demand

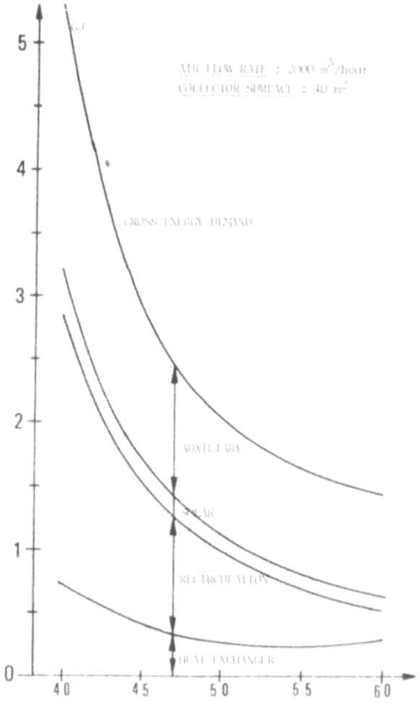

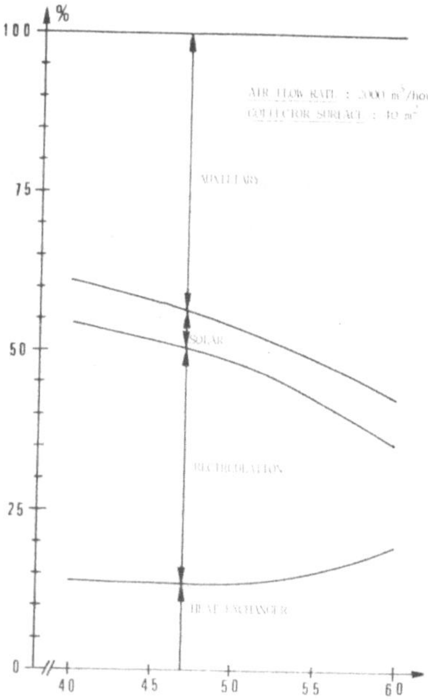

Fig 9A. Contribution of different components to the gross energy demand of one drying cycle

Fig 9B. % contribution to the gross energy demand

A COMPARISON OF FIVE SOLAR SYSTEMS FOR HEATING GREENHOUSES

J. Reichert
Fraunhofer-Institut für Systemtechnik und Innovationsforschung
Karlsruhe, Federal Republic of Germany

Summary

Five greenhouses equipped with different solar heating systems have been built in Bonn (FRG). Performance of the systems has been monitored for about two years to establish energy balances for the houses. During the same period simulation models have been developed and validated with the measurements. The systems considered are: groundwater heat pump, rock bed storage and the greenhouse itself acting as collector, concentrating collectors with water storage, venetian blind air collectors with rock bed storage, flat plate collectors consisting of rotating lamellas with stratified water storage and heat pump. All systems are integrated within the greenhouse, no outside space is required.

Three of the systems delivered good results: the groundwater heat pump (1) with a COP of 4.2, the "passive" house (2) with rock bed storage reaching solar fractions of 15 - 20 %, and the flat plate collector system (5) with solar fractions of 25 - 30 %. The concentrating collector (3) delivered only 3 % solar energy. The air collector (4) turned out to be unsuitable for the strong temperature changes, also problems arose with the system's control. Economic evaluation was difficult because no realistic prices were available for some of the components. Acceptable payback periods will definitely not be achieved for the water systems studied. The payback periods of the studied air systems are about 20 years and have been calculated on the basis of estimated costs for the rock bed storage and climatic data of Hamburg and Bonn. The groundwater heat pump reaching a 9 year bayback period is the most favourable system.

1. INTRODUCTION

Horticultures in Germany operating heated greenhouses have strongly been affected by high energy prices since 1973. Therefore, possibilities for saving energy and substituting fossil fuels by solar systems have been studied.

In 1978 the German Ministry of Research and Development promoted a project for testing solar systems in greenhouses. The Landwirtschaftskammer Rheinland built 5 pilot greenhouses in the vicinity of Bonn and made the investigations on plant growth. The Forschungsstelle für Energiewirtschaft has monitored the performance of the systems and the Fraunhofer-Institut (ISI) carried out systems analyses from the technical and economical aspect.

Constructionwise all 5 greenhouses are nearly identical: an area of 161 m^2, double glazing, increased airtightness, and low-temperature convectors for heating. The solar systems have been integrated in the houses so that the collectors required no additional space which most of the horticultures in Germany do not have anyway. Thus, only the excessive energy in the greenhouses was available for solar heating. Even in well insulated greenhouses excessive energy amounts to only about 30 % of the yearly heating load, i. e. the solar fraction expected was below this value.

In all greenhouses the same suite of plants was grown: living room plants requiring little ligth but relatively high temperatures. Temperatures were set at 20 °C during the day and 18 °C at night throughout the whole year.

2. DESCRIPTION OF SOLAR SYSTEMS

Greenhouse 1 has no solar system and serves only for comparison purposes. Greenhouse 2 has a rock bed storage and acts as a collector itself. If the temperature in the house rises above 24 °C, the air is blown through the rock bed storage by a fan. The cooled air is returned to the greenhouse via air ducts installed under the tables. For heating purposes the airflow in the storage is reversed by electrically driven valves. A second, separately operating system in greenhouse 2 is an electrically driven groundwater heat pump with 14 kW input. Greenhouse 3 has concentrating collectors which track the daily motion of the sun. They serve three groups of water storage tanks which are switched as a cascade. The total volume is 14 m^3. These tanks deliver warm water for the heating system directly. Greenhouse 4 is equipped with rock bed storage, as is greenhouse 2, except the side facing south being equipped with venetian blinds functioning as an air collector.

Greenhouse 5 again has a liquid-based system with flat plate collectorsdesigned as rotating lamellas and mounted on the south side inside the roof. They feed a 19 m^3 water tank from which heat is withdrawn for the heating system either directly or via a small heat pump with 5 kW input.

3. RESULTS

Plant growth in greenhouses 4 and 5 was delayed, in greenhouses 2 and 3, in part, better than in greenhouse 1. No other deficiencies in plant growth such as mycoses were noticed. Because the ventilation flaps in these green-houses are opened less frequently than in conventional greenhouses, a short-age of CO_2 could possibly occur with fast growing plants.

The following figures are based on the year 1981, the global radiation in Bonn being, however, 10 % lower than the average of previous years. The performance of two systems was not satisfactory. The concentrating collector in greenhouse 3 supplied only 3.3 % solar energy. This proved that concentrating collectors are not suitable in regions where the diffuse radiation amounts to more than 50 % of the global radiation. Thermal behaviour of the air collector in greenhouse 4 was very good, outlet temperatures of over 55 °C were reached. During operation, however, it did not withstand thermal and mechanical strains. Very soon the venetian blinds could not be opened anymore so that insufficient radiation fell in the house which, in turn, resulted in a noticable delay in plant growth. Solar fraction of the system was still nearly 16 %, but the electricity for the fan amounted to 45 % of the energy saved.

Greenhouse 5 had a solar fraction of 25 %, the highest of all houses. Because of the high investment cost this system, too, is economically not acceptable.

Economically interesting are two system: the groundwater heat pump and the system of greenhouse 2 (rock bed storage and the greenhouse itself acting as a collector). Heat pumps are no novelty, we would only like to mention the COP of 4.2 and the payback period of 9 years, provided wells exist which is the case in many horticultures.

Regarding greenhouse 2, I would like to report on the results in more detail. The tasks of the Fraunhofer-Institut included the development of simulation models for optimisation of the systems. The first step was the validation of the models on the basis of the data measured hourly.

Fig. 1 is a two day's plot of the hourly measured (solid line) and calculated (dotted line) values of the inside temperature in greenhouse 1. In most cases deviations are under 1 ^{o}C, only at bad and changing weathers and with open ventilation flaps they become greater.

The calculated heating load plotted on Fig. 2 is also pretty close to the measured value, of course not following the oscillations of the heating controller.

The validated models have then been applied to greenhouse 2. With measures such as improved insulation of the base of the house, elimination of existing heat bridges, installation of shades with higher absorption quality on the surface facing outside (in order to prevent radiation to be reflected out of the house), the computed solar fraction increased fom 14,6 to 19,2 %.

Fig. 3 shows the improved monthly solar fraction and the net solar fraction (hatched part), the difference of which being the energy consumption of the fan.

Refinement of control plays an important role in improving the solar fraction. It should, for instance, be prevented to overload the storage in the summer time when there is little heat demand in order to save auxiliary energy for the fan. Analogous to the COP of a heat pump we defined for this system the COP as the ratio between energy output and used auxiliary energy for the fan.

Fig. 4 shows the COP by months whereby the blank part of the bars illustrates the improvements gained through a refined control. The yearly COP is increased from 3.4 to 6.1.

Our calculations also proved that the rock bed storage was overdimensioned. One fourth of the actual size of 100 m^3 would have been sufficient, this means in general a storage volume of 0.2 m^3 per m^2 floor space of the greenhouse. Power input of the fan then has to be about 10 - 15 W per m^2 . Based on rough estimates of the investment cost for the rock bed storage, an economic evaluation leads to a payback period of about 20 years. Cheaper solutions for the rock bed storage have to be found and seem feasible.

The project and the calculations are being continued to further improve the systems. The two liquid-based systems have been abandoned because there are no prospectives for an economic operation. Further efforts concentrate on the above house and house 4 where the venetian blind collector has been replaced by roll-up foils which also serve as a shade. We consider these air-based systems with rock bed storage attractive, also for our climatic conditions, if costs can be lowered. More attractive yet they seem to be in southern regions, especially in developing countries.

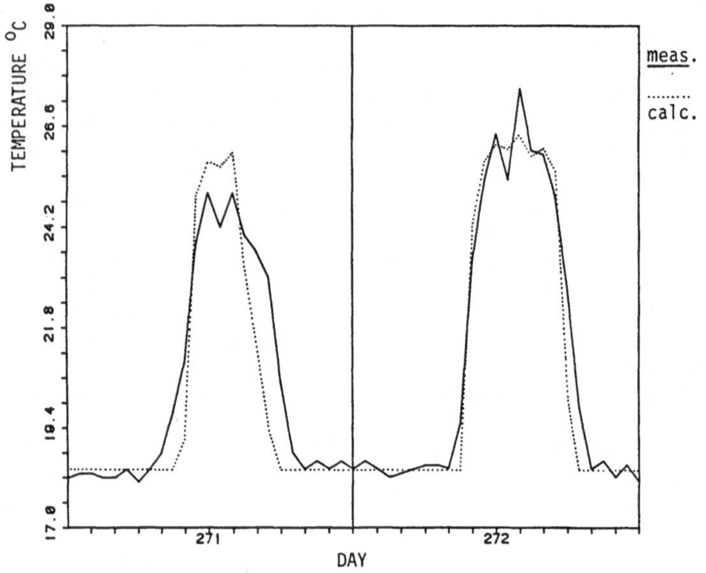

Fig. 1. Comparison of inside temperature

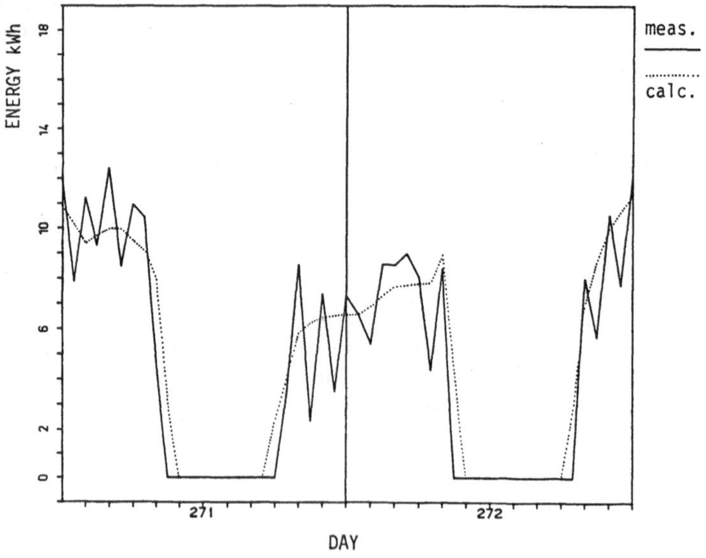

Fig. 2. Comparison of heat demand

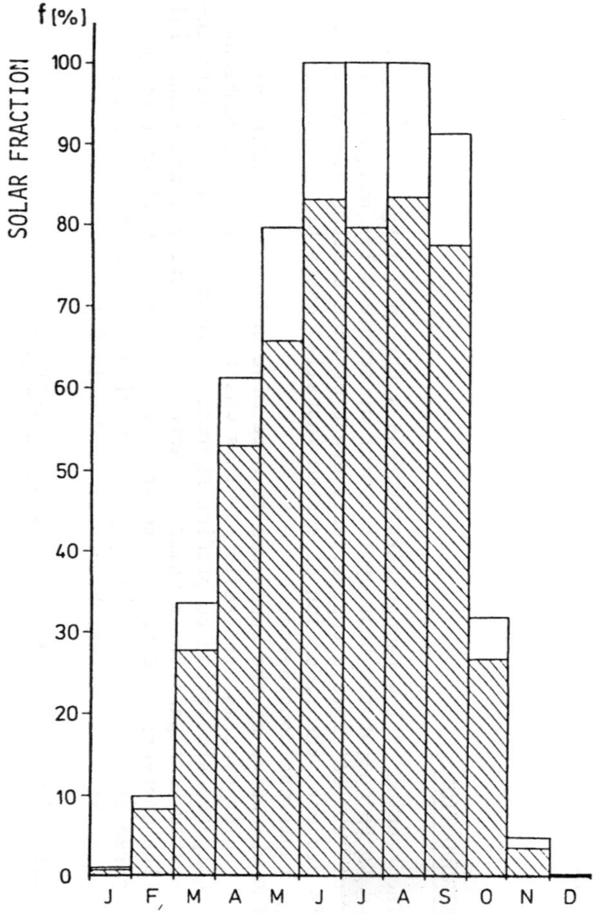

Fig. 3 : Solar fraction of improved run

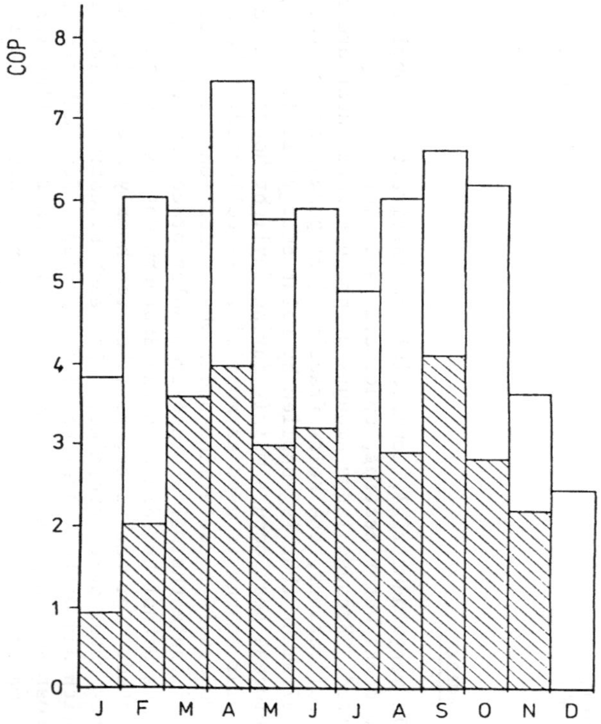

Fig. 4 : Improvement of COP

ANALYSIS AND DESIGN OF SOLAR PROCESS WATER HEATING SYSTEMS WITHOUT STORAGE FOR INDUSTRIAL AND COMMERCIAL APPLICATIONS

P.F. MONAGHAN,
Department of Mechanical Engineering,
University College,
Galway,
IRELAND.

P.J. GOLEM,
Chorley & Bisset Ltd.,
521 Colborne St.,
London, Ontario,
CANADA.

Summary

In many industrial, commercial and institutional applications of solar process water heating, space available for locating collectors is limited. Because large amounts of process water heat are needed for these applications, the solar fraction (the percentage of total energy requirement that is supplied by solar energy) is low. The demand for process heat occurs mainly during daylight hours. As a result, most solar energy is used as soon as it is collected. The objective of this project is to determine under what conditions it is cost-effective to eliminate thermal storage from solar process water heating systems. Five actual buildings in the Southern Ontario area of Canada were selected. For each site, profiles for hourly and daily usage of process hot water were determined. For each situation, solar heating systems with and without storage were designed. For each system, construction cost estimates and estimates of solar energy collection were made. In a majority of the practical situations evaluated, it is cost-effective to design solar process water heating systems without storage. Generalised guidelines on the use of "zero-storage" systems are also presented.

1. INTRODUCTION

The heating of low temperature process water on a large scale is considered to be one of the best applications of active solar heating. A number of efforts to improve cost-effectiveness of such systems have been made. In this project, the reduction of system installed costs by elimination of storage tanks was considered.

The objective of the study was to determine under what conditions 'zero-storage' solar systems are feasible. This was investigated by a combination of computer simulation to predict thermal performance and cost estimation of systems designed for a number of actual institutional and industrial sites. The cost-effectiveness of solar systems with and without storage was compared at each site. This paper summarises the results of a report submitted to National Research Council of Canada (1).

2. SYSTEM DESIGN CONCEPTS AND COMPUTER SIMULATION

Because the systems were designed for a cold northerly climate, a loop with propylene glycol antifreeze fluid was used. The storage and zero-storage systems used as a basis for this study are shown conceptually in Figure 1.

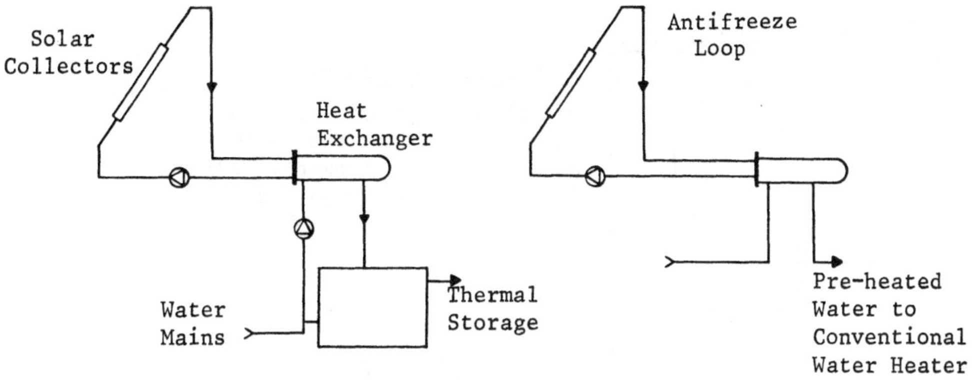

FIGURE 1: Concept Designs of Storage and Zero-Storage Systems.

Because transient response of the collector/heat exchanger loop is so important in zero-storage systems, a transient simulation was developed. The process water heating version of the WATSUN (2) simulation program was modified to accept weather data at five minute intervals, and calculate collector performance on a minute-to-minute basis. Corresponding solar radiation data was specially acquired from Fisheries and Environment Canada. A description of these transient models is not included here. Details are given in the original report (1) and in a follow-up paper (3).

3. RESULTS

Five actual buildings in the South-Western part of Ontario, Canada were selected: i) Milk processing plant; ii) Food processing plant; iii) Commercial laundry; iv) A 20-storey apartment building; v) Hospital. For each site, the actual daily and weekly demand profiles for usage of process hot water were determined. In addition, the maximum installed collector area was estimated. This data is summarised in Figure 2. For each site concept designs of solar heating systems with and without storage were developed. At each site, both systems were based on i) the (maximum) collector area and tilt angles (ii) a single glazed, selectively coated collector iii) heat exchangers sized according to De Winter's recommendations (4) and iv) storage sized at 75 L of water per m² of collector area(5)

Computer simulation results (annual solar energy collected) and installation cost estimates are both presented in Table I, on the basis of 1m² of collector area. The cost-effectiveness measure used was annual solar energy collected (MJ/yr) divided by the installed cost in 1980 Canadian dollars ($).

4. DISCUSSION AND GENERALISATION

A study of the effects of removing storage in each of the five applications indicates the following:

1. Between 15.6% and 25.5% of installation cost could be saved. The average savings were 19.9%.
2. The collected solar energy changed as follows: hospital (+1%); apartment building (-7.4%); laundry (-18.2%); milk processing plant (-20.0%) and food processing plant (-25.5%). The

HOT WATER DEMAND PROFILES

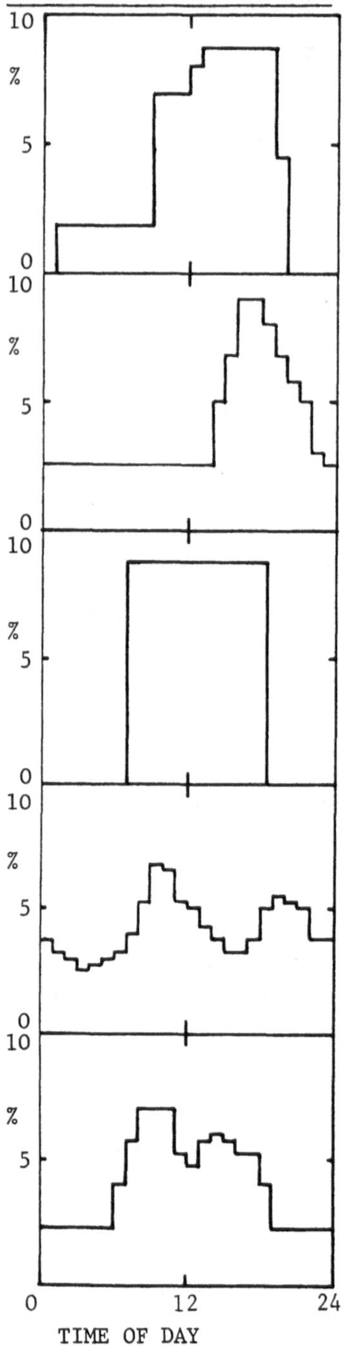

MILK PROCESSING PLANT

Operating Days per Week: 5
Daily Water Quantity: 32400 L/day
Annual Heat Requirement: 2315 GJ
Collector Area (Max): 72 m^2

FOOD PROCESSING PLANT

Operating Days per Week: 5
Daily Water Quantity: 53400 L/day (3 days)
89300 L/day (2 days)
Annual Heat Requirement: 6700 GJ
Collector Area (Max): 144 m^2

COMMERCIAL LAUNDRY

Operating Days per Week: 5 1/3
Daily Water Quantity: 666900 L/day (5 days)
228000 L/day (1 day)
Annual Heat Requirement: 23000 GJ
Collector Area (Max): 144 m^2

20-STOREY APARTMENT BUILDING

Operating Days per Week: 7
Daily Water Quantity: 29800 L/day
Annual Heat Requirement: 2345 GJ
Collector Area (Max): 144 m^2

HOSPITAL

Operating Days per Week: 7
Daily Water quantity: 62600 L/day (5 days)
41300 L/day (1 day)
30000 L/day (1 day)
Annual Heat Requirement: 4310 GJ
Collector Area (Max): 144 m^2

% - Percentage of daily water requirement occurring in a particular hour.

FIGURE 2: A Summary of Process Water Energy Demands and Maximum Possible Collector Area of Each Site.

MILK PROCESSING PLANT	Storage System	Zero Storage System
Solar Fraction	9.50 %	7.60 %
Collected Energy	3056.00 MJ/yr m^2	2444.00 MJ/yr m^2
Installed Cost	700.00 $/$m^2$	581.00 $/$m^2$
Cost Effectiveness	4.36 MJ/yr $	4.21 MJ/yr $
FOOD PROCESSING PLANT	**Storage System**	**Zero Storage System**
Solar Fraction	6.30 %	4.70 %
Collected Energy	2917.00 MJ/yr m^2	2174.00 MJ/yr m^2
Installed Cost	685.00 $/$m^2$	510.00 $/$m^2$
Cost Effectiveness	4.26 MJ/yr $	4.26 MJ/yr $
COMMERCIAL LAUNDRY	**Storage System**	**Zero Storage System**
Solar Fraction	2.00 %	1.60 %
Collected Energy	3125.00 MJ/yr m^2	2556.00 MJ/yr m^2
Installed Cost	695.00 $/$m^2$	558.00 $/$m^2$
Cost Effectiveness	4.50 MJ/yr $	4.58 MJ/yr $
20-STOREY APARTMENT BLDG.	**Storage System**	**Zero Storage System**
Solar Fraction	16.70 %	15.50 %
Collected Energy	2722.00 MJ/yr m^2	2521.00 MJ/yr m^2
Installed Cost	686.00 $/$m^2$	546.00 $/$m^2$
Cost Effectiveness	3.97 MJ/yr $	4.62 MJ/yr $
HOSPITAL	**Storage System**	**Zero Storage System**
Solar Fraction	10.00 %	10.10 %
Collected Energy	2993.00 MJ/yr m^2	3021.00 MJ/yr m^2
Installed Cost	651.00 $/$m^2$	549.00 $/$m^2$
Cost Effectiveness	4.60 MJ/yr $	5.50 MJ/yr $

TABLE I RESULTS OF ANALYSIS OF THERMAL PERFORMANCE and COST

hospital and apartment building had seven-day usage profiles. The food processing plant used energy on five days only and had a skewed daily profile. A zero-storage system can thermally out-perform an equivalent storage system because collector efficiency can be increased due to its continual receipt of the coldest possible inlet fluid.

3. Cost effectiveness of the hospital and apartment building systems improved substantially (by 20.0% and 16.3%, respectively). For the other systems, cost-effectiveness was not significantly affected (between +1.7% and -3.4% change).
4. All systems considered had a low solar fraction (1.6% to 16.7%)

To investigate the sensitivity of storage systems to solar fraction, "typical" five-day and seven-day process water usage profiles were devised. Computer simulation results of the effect of collector area on the solar fractions of equivalent storage and zero-storage systems are presented in Figure 3. This figure shows that, for the seven-day (five-day) profile, it is cost effective to eliminate storage if the solar fraction that would be achieved by a storage system is less than 40% (60%). This statement is based on the assumption that a zero storage system costs 20% less to install than a storage system.

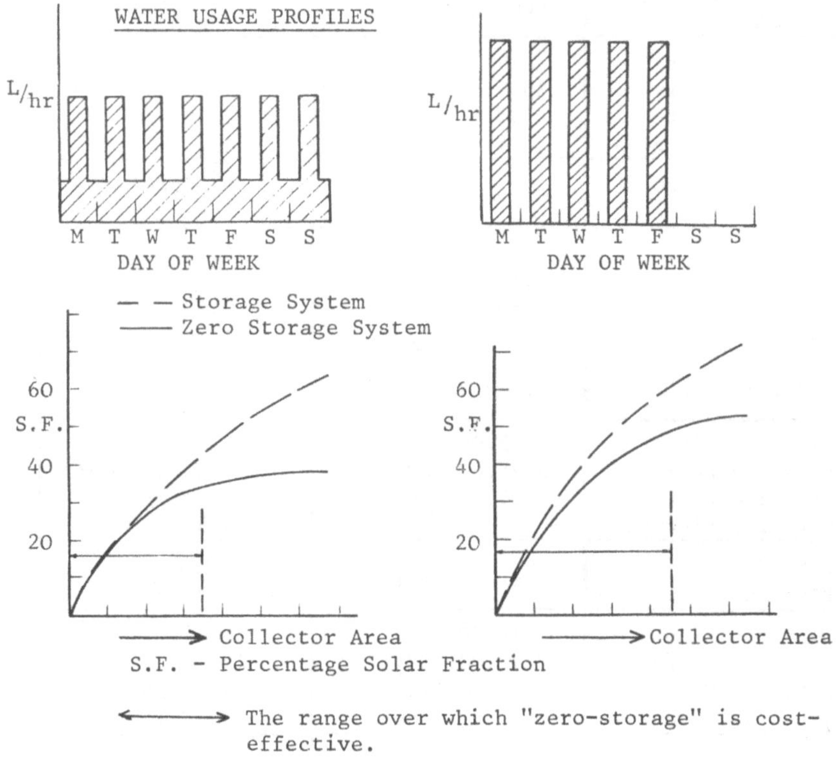

FIGURE 3: Comparison of Thermal Performance of Storage and Zero-Storage Systems for "Typical" 7-day and 5-day Process Water Usage Profiles.

REFERENCES

1. Chorley & Bisset Ltd., "Feasibility of Zero-Storage Solar Process Water Heating Systems". Report Stor-9, Solar Energy Program Publications, National Research Council of Canada, (1980).

2. Hollands, K. G. T., and Orgill, J.F., "The Potential for Solar Heating in Canada", Waterloo Research Institute Report No. 77-01, University of Waterloo (1977).

3. Golem, P.J., "Dynamic Modelling and Simulation of Solar Loops using Methods of Transfer Functions" to be presented at National Conference of Solar Energy Society of Canada, (1984).

4. de Winter, F., "Heat Exchanger Penalties in Double Loop Solar Water Heating Systems", Solar Energy, 17, n6, 335 (1975).

5. Beckman, W.A., Klein, S.A., and Duffie, J.A., "Solar Heating Design by the f-Chart Method, John Wiley & Sons, New York (1974), p.4.

THE PERFORMANCE STUDY OF AN INDUSTRIAL SOLAR HOT WATER SYSTEM

M.N.A. Hawlader
K.C. Ng
T.T. Chandratilleke
National University of Singapore
Singapore 0511

Kelvin Koay H.L.
Manager
Facilities and Engineering Projects
CIAS Airport Services
Singapore 9181

Summary

In this paper, the performance of the solar hot water system at Changi International Airport Services, Singapore is presented. This system uses about 630 m^2 of a solar collector with a storage capacity of 40 m^3 in four tanks to provide a daily average hot water consumption of 46.7 m^3 at a temperature of 70°C. The installation has been fully instrumented and controlled by an automatic controller. A computer program has been developed to evaluate the performance of the system. Observed performance of the solar system, which can provide a solar fraction, 0.60 of the load, agrees well with the predicted values.

1. INTRODUCTION

The use of solar energy for relatively low temperature applications, such as industrial hot water supply and process heating, has been receiving considerable attention in the recent years. These applications are attractive because they, normally, require hot water in the temperature range of 60 - 80°C only. Besides, solar energy applications give an alternative energy source which is clean, renewable and economical in operation.

2. THE SYSTEM

The solar hot water system, described in the paper, is an industrial system located at the Changi International Airport Services (CIAS), Singapore, where hot water is needed for the daily washing and cooking purposes. Fig. 1 shows schematically the solar hot water systems at CIAS and it comprises primary and secondary water circuits with a heat exchanger connected in between. Circulation of water in the primary circuit is effected by pumps which draw water from the heat exchanger to 321 'blue-panel' solar collectors (in parallel feed) whilst the hot water is returned to the heat exchanger where the available heat is then transferred to the secondary circuit. The heat is absorbed by the water of the secondary circuit which is stored in four storage tanks, each having a capacity of 10 m^3, before being fed to the kitchen. A gas-fired anxillary calorifier is used to ensure the hot water supplied to the kitchen to be always at the preset temperature in the event of under supply of solar insolation.

During normal operation, the hot water from the heat exchanger enters each of the four storage tanks through a temperature-controlled valve, situated at the top of the tank. As the tanks are connected in series, when hot water is withdrawn, it leaves from the storage tank no. 1 whilst an equal amount of make-up water enter the system (from the mains) through storage tank no. 4. When the storage tanks attain the preset temperature limit and if the exit temperature of the collector exceeds 85°C, the air cooler unit becomes operational to dissipate excess to the atmosphere.

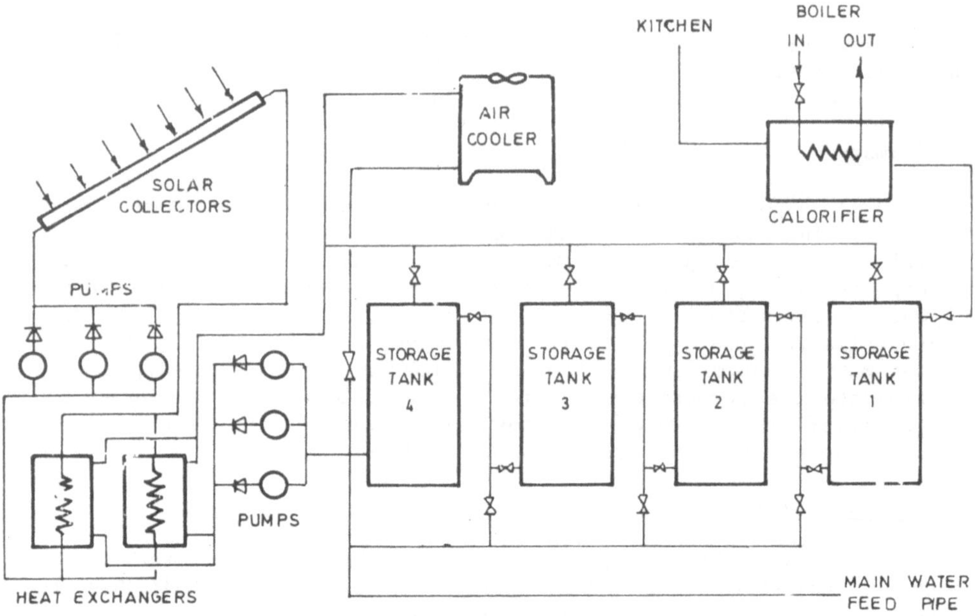

Figure 1. Diagram of the CIAS solar hot water system

Table 1. Monthly average of the daily total global irradiation on a horizontal surface

Months	Monthly average of daily totals, MJ/m^2											
Year	J	F	M	A	M	J	J	A	S	O	N	D
1977	17.7	13.4	18.5	17.2	15.4	13.9	14.7	13.7	14.4	16.1	12.3	14.4
1978	14.7	18.0	16.1	16.5	13.9	14.0	14.7	14.9	15.1	13.7	12.0	14.3
1979	17.4	18.0	15.9	13.3	14.0	12.9	13.9	16.0	15.4	13.9	11.4	15.4

3. SIMULATION METHODOLOGY

A computer program was developed to simulate the solar hot water system at CIAS. This program was used to evaluated the performance over a period of several years using the meteorological data of Singapore (01° 22'N, 103° 55'N). Table 1 shows the average values of the typical monthly solar insolation in Singapore from 1977 to 1979. However, the hourly variation of the direct and the diffuse components of the global radiation was computed from the meteorological office's data, using the method outlined by Hawlader (1). The Performance of the System was evaluated with a demand load pattern which was obtained from the data collected over a period of two days (November 10th and 11th, 1983). The total hot water consumption per day was 52 m^3 as compared with the expected daily variation of 45 to 55 m^3. Table 2 shows the values of the systems parameters used in the computer program. As for the different components of the system, they are represented by equations described in the literature (2,3).

Table 2. System parameters

Collector Area	: 629.16 m^2
Tilt	: 10°
Orientation	: South/North
Collector heat loss Coefficient	: Calculated as function wind speed
Panel Absorptance	: 0.92
Glazing	: Single
Transmittance at normal incidence	: 0.84
at 60° incident angle	: 0.756
Collector efficiency factor	: 0.92
Ground Reflectance	: 0.05
Collector heat capacity panel and water content	: 10,000 J/m^2 K
Collector flow rate	: 20 m^3/hr
Pipe heat loss coefficient	: 0.335 W/m^2
Pipe heat capacity pipe and water content	: 5000 J/m^2 K
Heat Exchanger Conductance-Area Product	: 105235.2 W/K
Store volume 4 stores, 10 m^3 each	: 40 m^3
Heat loss coefficient for stores	: 0.4 W/m^2 K
Supply water temperature (constant throughout the year)	: 29°C
Load demand	: 52 m^3
Considered constant, although variable	
Demand temperature	: 70°C

4. RESULTS AND DISCUSSION

Table 3 shows the measured and the predicted performance of the system over a period of two days, 10th and 11th of November, 1983. As can be seen, the observed and predicted results show good agreement. For a given demand load pattern, the yearly performance of the system was also investigated with the temperatures of the make-up water and the hot water set at 29° and 70°C respectively. The predicted and the measured results are shown in tables 4 and 5 for the year of 1979 and 1983 respectively. As observed in table 5, the daily average consumption of hot water was 46.7 m^3 whilst the total solar fraction supplied by the system was 58%. Assuming the yearly variation in solar insolation is negligible, the predicted annual average solar fraction is 61% which is only 3% higher than the measured value of 1983. Also the months of November and December have their solar fraction well below 50% and this is because of the poorer weather conditions during this time of the year.

Table 3. Daily Performance of CIAS Solar Water Heating System

Day	Nov. 10, 83		Nov. 11, 83	
Variably	Measured	Predicted	Measured	Predicted
Collector Input (GJ)	9.889	9.889	7.840	7.840
Collector Output (GJ)	5.760	5.259	3.960	3.409
Collector Efficiency (%)	58.2	53.2	50.5	43.5
Storage Output (GJ)	2.520	2.804	2.520	2.138
Demand Load (GJ)	4.734	4.734	4.342	4.342
Solar Fraction	0.53	0.59	0.58	0.49

CONCLUSIONS

The simulation study agrees well with the observed performance of the CIAS solar hot water system. The existing solar installation can provide about 60% of the total energy required for water heating at the catering facility. This indicates a considerable saving on fuel bills although the initial capital investment is high.

ACKNOWLEDGEMENT

The authors would like to thank Miss Chan Chew Lan and Mr Chan Chun Yee for their assistance in the computation work.

REFERENCES

1. Hawlader, M.N.A. (1983). Diffuse, global and extraterrestrial solar radiation for Singapore. International Journal of Ambient Energy, Vol. 4, No.4, December.

2. Duffie, J.A. and Beckman, W.A. (1980). Solar Engineering Thermal Process. John Wiley and Sons, New York.

3. Kreith, F. and Kreider, J.N. (1978). Principles of Solar Engineering. McGraw-Hill Book Co., New York.

Table 4. Performance study of solar hot water system at CIAS for the year 1979.

Month	Collector Input (GJ)	Collector Output (GJ)	Collection Efficiency	Store Input (GJ)	Store Losses (GJ)	Demand Load (GJ)	Solar Fraction	Annual Avg Solar Fraction
Jan	350.1	181.7	51.9	181.0	4.40	239.8	0.75	
Feb	322.2	169.4	52.6	168.8	4.06	216.6	0.77	
Mar	304.8	162.8	53.4	162.2	3.95	239.8	0.68	
Apr	239.3	128.0	53.5	127.5	3.16	232.1	0.54	
May	253.6	135.9	53.6	135.5	3.37	239.8	0.57	
Jun	224.7	119.1	53.0	118.6	2.98	232.1	0.51	0.61
Jul	252.0	133.5	52.9	133.1	3.32	239.8	0.55	
Aug	293.2	154.8	52.8	154.3	3.81	239.8	0.64	
Sep	281.1	150.7	53.6	150.1	3.66	232.1	0.64	
Oct	266.1	142.8	53.6	142.2	3.50	239.8	0.60	
Nov	215.6	115.3	53.4	114.8	2.83	232.1	0.49	
Dec	308.2	161.4	52.3	160.7	3.88	239.8	0.66	

Table 5. Measured Solar Fraction of the Energy Required for the CIAS System for the Year 1983

Month	Solar Heat gain (GJ)	Load		Solar Fraction	Annual Average Solar Fraction
		Hot Water Consumed (M^3)	Energy Req'd (GJ)*		
Jan	119.88	1355	232.4	0.52	
Feb	159.12	1198	205.5	0.77	
Mar	159.84	1254	215.1	0.74	
Apr	149.40	1319	226.3	0.66	
May	145.08	1153	197.8	0.73	0.58
Jun	141.84	1486	254.9	0.55	
Jul	135.36	1605	275.3	0.49	
Aug	163.44	1673	287.0	0.57	
Sep	136.80	1489	255.4	0.53	
Oct	143.28	1393	238.9	0.60	
Nov	128.32	1635	280.5	0.45	
Dec	108.72	1512	254.4	0.43	

*Based on water supply temperature of 29°C and load requirement at 70°C

TEST AND MODELIZATION OF A GREENHOUSE USING LOW TEMPERATURE HEATING

O. JOLLIET (1), M. BOURGEOIS (2), L. DANLOY (2), J.-B. GAY (1).

(1) Federal Institute of Technology (EPFL), GRES-LESO, CH-1015 LAUSANNE
(2) European Organisation ofr Nuclear Research (CERN), CH-1211 GENEVE

in cooperation with
Station Fédérale de Recherches Agronomiques (RAC), CH-1964 CONTHEY
Centre Horticole de Lullier (Genève), CH-1254 LULLIER

Summary

An analyses of measured data taken from a heated greenhouse in Geneva discloses the following results :

- The installed thermal screens allow a reduction of the energy consumption of 44 % during the night.
- A first simple static model gives a good estimation of the night requirements in agreement with the measurements.
- Due to the low fraction of solar energy stored in the ground, the nightly heat demand is linearly linked with the outside temperature.
- On the contrary, during daytime, the solar contribution is important and highly variable, for this reason the above representation is meaningless.
- However an appropriated representation can be found by plotting the dayly energy consumption against the ratio of the average solar power over the temperature difference between the inside and the outside. This description is the ground work of a simplified method for evaluating the monthly mean consumption of a greenhouse.
- The measured useable solar energy depends strongly upon the month. On average, the useable solar energy amounts to about 23 % of the incident solar energy.
- Solar gains cover approximately 22 % of the total heat losses.

Due to protective screens, the influence of the wind is relatively small during the night and slightly stronger during the day.

1. INTRODUCTION

An experiment began in 1982 at CERN in Geneva with three main objectives :

- to determine whether heating greenhouses using low temperature industrial thermal rejects is feasible and worthwhile from both an agronomic and economic point of view
- to test different systems that reduce heating costs in greenhouses by maximizing the solar energy contribution and by minimizing the heat losses using thermal screens
- to develop a mathematical model of the greenhouse in order to check and to optimalise its thermal behaviour.

1.1 Experimental Set-up

The test greenhouse has a total area of 663 m^2 and is divided into 5 thermal zones, each zone being equiped with different low temperature heating systems. All systems combine a water-air convector with either pipes layed on the ground, underground heat exchangers, or heated floors.

For the 3 central zones, which are thermally equivalent, the crops are grown in soil.For the two side zones, the planting is done in rock wool (Grodan).

The whole experiment is controlled by an automatic data acquisition system which records for each zone : temperatures, relative humidity, energy flows and local meteorological parameters.

1.2 Period of interest

The second agronometric experiment took place between September 20th and December 31st 1983. During this period 18,000 Chrysanthemums were brought to maturation.

Most of the thermal findings presented below result from this period.

2. ENERGY HEAT BALANCE

The energy heat balance of the greenhouse may be written as follows :

$$P_s \cdot \Delta\Theta \cdot \Delta t \cdot A = (P_{in} + \eta\, F_{inc})\, \Delta t\, A \qquad [MJ]$$

(heat losses) = (heat gains)

where P_s = specific heat losses of the zone $[W/m^2_A K]$

$\Delta\Theta$ = average indoor and outdoor temperature difference $[K]$

Δt = duration $[Ms]$ of the considered period

A = cultivatable soil surface $[m^2]$

The non-solar heat gains may be divided into :

$$P_{in} = P_{aux} + \Delta P_{stock} \qquad [W/m^2_A]$$

with : $P_{aux} = \dfrac{E_{aux}}{A.\Delta t}$ = average power delivered by the auxiliary heating systems

$\Delta P_{stock} = \dfrac{\Delta E_{stock}}{A.\Delta t}$ = measured variation of the energy stored in the ground

The solar gain depends upon :

F_{inc} = average solar radiation $[W/m^2]$

η = useable solar fraction

Dividing by the soil surface area (A) and the time duration (Δt), one gets:

$$P_s \cdot \Delta\Theta = P_{in} + \eta F_{inc}$$

As the greenhouse is equiped with thermal screens, the heat losses during the day and the night vary considerably. For this reason the heat balances have to be considered separately for the day and the night periods.

Day- and nighttime are defined by the light level. Night period begins when F_{inc} is lower than 2 W/m^2.

2.1 Energy balance during the night

For nighttime periods, as far as the variation of the stored energy in the ground is measured, the solar gains vanish and the heat balance equation is reduced to :

$$P_{in} = P_s \cdot \Delta\Theta$$

This means that if one plots P_{in} as a function of $\Delta\Theta$, one expects to get a straight line with a slope proportional to the specific heat losses of the considered thermal zone. Figure 1 shows the experimental results : each point is an averaged value over one night.

As expected, P_{in} depends linearly on $\Delta\Theta$. The observed dispertion is due to perturbations (wind effect, sky temperature ...) that will be discussed later.

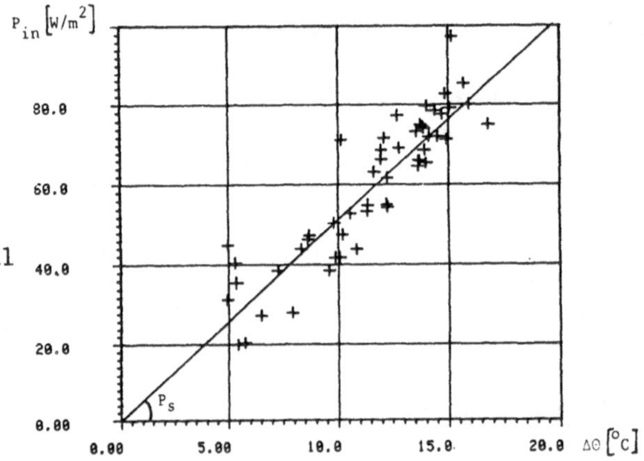

Fig. 1. Heat consumption of the greenhouse as function of $\Delta\Theta$ for the night period.

For night periods this representation is valid, and may be used in order to evaluate the relative effect of the thermal screens, or of the heating systems themself, on the energy requirements.

2.2 Energy balance during the day

For daytime periods, solar gains have to be considered :

$$P_{in} = P_s \cdot \Delta\Theta - \eta F_{inc}$$

For this reason the above representation is no longer valid, as it is shown in figure 2.

Another representation has to be found in order to take into account the solar gains. Such a representation exists if one divides all terms of the above equation by $\Delta\Theta$, giving :

$$\frac{P_{in}}{\Delta\Theta} = P_s - \eta \frac{F_{inc}}{\Delta\Theta}$$

$P_{in}/\Delta\Theta$ may be then plotted as a function of $M = \frac{F_{inc}}{\Delta\Theta}$

This new variable M may be called "meteorological index" ; it gives the ratio of the averaged solar intensity over the indoor - outdoor temperature difference. The net heat demand of the greenhouse is directly linked to this variable, as shown by figure 3.

For zero solar gains, a limit which is never reached experimentally, during the day, one gets: $P_{in}/\Delta\Theta = P_s$. It means that one can estimate the daily heat losses of the greenhouse by extrapolating its thermal characteristics to $M = 0$.

Some additional information obtainable from the above representation is that the useable solar fraction (η) is directly proportional to the slope of the straight line. In our case, η = 50 %.

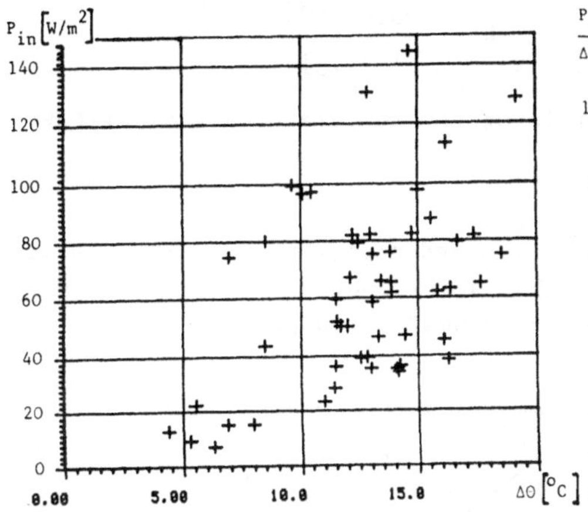

Fig.2. Heat consumption of the greenhouse as function of ΔΘ during day period.

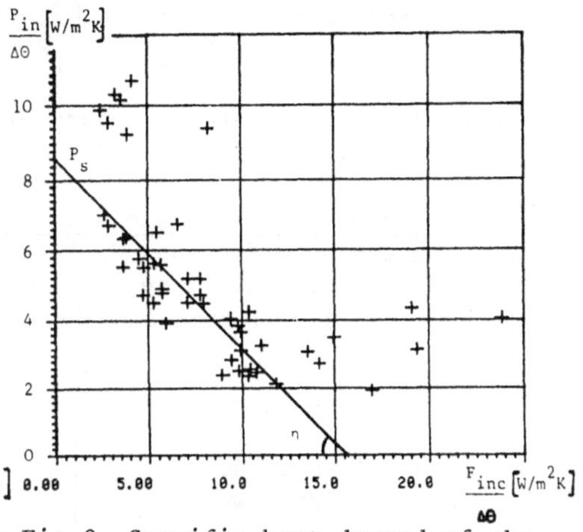

Fig.3. Specific heat demand of the greenhouse as a function of the meteorological index (each point is an average value over one day period.

Rather strong deviations are observed, specifically at small and large M's, and warrant discussion.

a) $M < 5$ [W/m²K]
These days are rather cloudy and cold; some of them correspond also to windy periods where the energy demand is stronger. Following reduction of the wind effect, measurements on these days are again linearly characteristic.

b) $M > 12$ [W/m²K]
Such days are characterised by strong solar gains and low thermal demand For this reason a large fraction of the gains has to be rejected and the useable solar fraction decreases (η). In the morning, however, heat may be still required.

2.3 Influence of the wind

This influence may be observed if one considers the plot of the specific heat losses versus windspeed. An increase of the windspeed of 1 m/s corresponds to an increase of 0.4 W/m²K for the nighttime and 0.9 W/m²K for the daytime of the specific heat loss.

Due to the additional effect of the thermal screen on the greenhouse tightness, the influence of the wind is reduced during the night.

3. RESULTS

3.1 Effect of the screens

In order to measure the economy due to the two thermic screens, they were left rolled up during three nights (7th till 9th January 1984). The measured specific loss without screens was then compared to the specific loss with screens using similar meteorological conditions :

$P_s = 8{,}7 \mp 1$ [W/m²K] without screens

$P_s = 4{,}9 \mp 0{,}5$ [W/m²K] with screens

savings during nighttime : 44 %

Meteo conditions : $\Theta_{out} = 1{,}4$ [°C] $\Theta_{sky} = -5{,}4$ [°C]

$\Theta_{in} = 15{,}5$ [°C] $V_{wind} = 1$ [m/s]

Note that for higher windspeeds or lower sky temperatures the savings would be greater.

3.2 Monthly balance and useable solar energy

Knowing the specific loss P_s for daytime and nighttime, we are able to calculate the total heat losses. The useable solar energy is then the difference between these losses and the measured heat consumption.

fig. 4. Evolution of the incident solar power, of the total heat losses and of the useable solar power between september and december 83

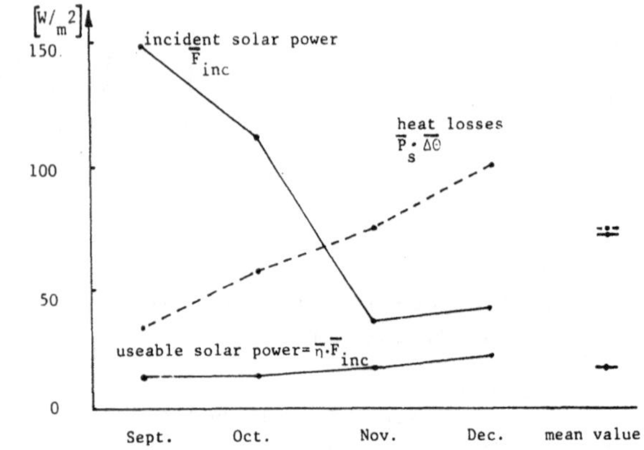

	Sept.	Oct.	Nov.	Dec.	mean value
Useable solar fraction $\overline{\eta}$	8 %	12 %	46 %	53 %	23 %
$M = \frac{\overline{F}_{inc}}{\overline{\Delta\Theta}}$ for daytime [W/m²K]	64.5	29.1	7.0	7.0	19.5
Fraction of heat losses covered by solar energy	33 %	22 %	21 %	22 %	22 %

Useable solar fraction depends strongly upon the month. As shown for daytime measurement in figure 3 (§ 2.3), the monthly mean fraction decreases with the meteorological index M. For November and December, as M = 7 dependance is linearly and η amounts to about 50 %. On the contrary for September and October. M is considerably greater than 12 and η is strongly reduced. These results show that the representation using the meteorological index (§ 2.3) is the groundwork of a simplified method for evaluating the monthly mean consumption of a greenhouse. The solar energy stored in the ground is low and it amounts to 5 % of the incident solar energy. One can notice that solar gains cover during the whole period approximately 22 % of the heat losses.

CONCLUSION

Integral heating of horticultural greenhouses with low temperature water (below 30°C) is possible, as far as appropriate heating systems are used.

- Thermal screens reduce strongly the heat demand and are economically interesting.
- The solar radiation covers a large fraction of the daily heat requirements Detailed results will be published soon in a complete report, which can be obtained from the authors.

THE PERFORMANCE MEASURING DATA OF AN AIR COLLECTOR SYSTEM FOR DRYING HERBS

M. Reuss, K. Meuren, S. Vogt
Bayerische Landesanstalt fuer Landtechnik (BLL)
Voettingerstr. 36
D-8050 Freising-Weihenstephan
Federal Republic of Germany

SUMMARY

A solar air collector system for drying herbs was installed on a farm in Ismaning near Munich. The collector is of a roof integrated type and has a total aperture of 530 m². Fresh air is preheated in the collector and afterwards heated up to required temperature by a conventional oil fired air heater of 650 kW. Furthermore a heat recovery system is installed in the exhaust chimneys of the band drier. Initial measurements in autumn and winter 1983 gave daily collector efficiency values up to 40 %. That means primary energy savings of 340 GJ/a within 600 hours of operation. The energy payback period for the air collector is one year. An economical analysis yields a monetary pay back period of 11.5 years including 25 % subsidies on the total investment. The monetary and energy payback period could be reduced significantly, if the period of operation per year is increased. That means perhaps crowing more herbs, drying grain crop or drying for other farmers.

1. INTRODUCTION

One of the most important applications of solar energy in agriculture is the field of drying. Even in the middle european climate of Bavaria a solar assisted drying plant could work on an economical basis. In 1983 a project of monitoring a solar air collector system for drying herbs financed by the CEC (Projekt H) and the german BMFT (Ministry of Research and Technology) was started. All energy flows in the total plant should be registered, the solar energy input to the conventional parts of the installation, the fuel input to the air heater and the electricity input to operate the fans.

The results of these measurements will be used to improve the plant and to make a detailed economical analysis of the system. The experience will also be introduced to the Planning of further solar assisted drying plants.

2. DESCRIPTION OF THE PLANT

The total installation consists of a solar air collector of a roof integrated type, an oil fired air heater of 650 kW heating power (backup-system), a heat recovery and the 5-band-drier.

The solar collector has an inclination angle of 27° and is facing to south with a total aperture of 530 m². Picture 1 shows the structure of the collector. The backside of the roof is covered with panelled wood fibre boards. A black polypropylen filter mat (19 mm) serves as absorber and is installed between the spares. The collector is covered with gel coates glasfibre reinforced polyester sheets. The air mass flow rate is about 30.000 kg/h, that means an air velocity in the collector of 1.8 m/s.

Ambient air is preheated in the collector. If the required temperature is not reached, the oil fired air heater and the heat recovery system are used to raise the temperature level. The oil burner can operate in two states in correspondence to the needed energy. Heat exchangers of the heat recovery system are installed in 2 of the 3 exhausts of the drier and in the duct next to the conventional air heater.

This heat recovery system can only be used during night time and if the inlet temperature in the dryer is above 60°C. The drying system is of a 5-band-type for continuous drying of fresh materials with high moisture content. The capacity is 60 - 100 kg dry material per hour. The main products to be dried are parsley and peppermint.

3. MEASURING EQUIRMENT

Picture 4 shows an overview of the plant with the places, where the different sensors for measuring temperatures (T), air-humidities (φ), mass-flows ($\dot{m}$) and the electric power for the fans ($\dot{Q}_{el}$) are installed. Furthermore meteorological parameters like global radiation in the collector plane, wind-speed and -direction, ambient temperature and humidity are registered.

For temperature measurement Pt 100 sensors are used. Each 10 of the 20 installed Pt 100 are connected in series with a constant current source of 0.666 mA. The linearisation of the Pt 100 resistors is made by software.

It was not possible to install standard measuring tubes and orifice plates for the large diameters of the air ducts. The air velocity distribution over the cross section is not uniform. So it was necessary to use a continuous method of scanning the air velocity in the cross section. A small anemometer was mounted on a device, which is able to move in 2 directions and to scan the air velocity on a zig-zag-line (see picture 2 and 3). The signal of the anemometer is integrated over the time of 60 sec., the time to scan the whole cross section. From this value a mean air velocity in the air duct is calculated.

The data acquisition is based on a DIGITAL LSI 11/2 minicomputer. The temperature sensors, the humidity sensors, the anemometers for air speed measurement in the ducts and the meteorological sensors are scanned continuously with a rate of 1 cycle (32 channels) per second. The averaged or integrated data of 5 minutes are stored on a magnetic tape for further evaluation.

4. RESULTS OF MEASUREMENTS

Due to the late start of the project a detailed data acquisition could not begin before the end of summer 1983, the end of the drying season, and so no results of the drier and the heat recovery system are available. During autumn and winter measurements on the performance of the collector have been done. Extrapolations from these data, the working time of the plant, the output of dry products and the total fuel consumption an initial analysis could be made. The needed input of primary energy without solar is about 900 GJ/a, this is related to a useful energy of 765 GJ/a. The performance of the conventional air heating system is 0.85. The reason of this good COP is the rather low temperature of operation. During the 600 houres of operation per year the collector delivers more than 245 GJ/a useful energy, giving a solar fraction of 0.32. Per 1 m² of collector 15 l of fuel oil or 17.7 l of crude oil could be saved. Mean daily collector efficiencies up to 40 % are related to the solar energy during the working time.

5. ENERGETIC AND ECONOMIC ANALYSIS OF THE SOLAR SYSTEM

With data of specific energy input of materials /1/ the system was analysed (see also /2/). For the construction of the simple air collector a net energy investment of 500 MJ/m², for the short air ducts of galvanized steel 95 MJ/m² are required. The substituted conventional roof needs an investment of 55 MJ/m², which has to be subtracted. Because of the short operation time of only 600 h/a and the high power demand of 16 kW for the fans the energy payback period is about 1 year. In this calculation the thermal performance of the conventional air heater (0.85) and the production efficiency of fuel oil from crude oil (0.85) is taken in consideration.

The economical analysis was made with a dynamic amortisation method. These calculations take a 10 % energy price escalation rate (recommended by IEA) and 8 % annual rate of discount and 25 % subsidies in consideration. This yields a amortisation period of 11.5 years.

The table gives a list of all relevant data of the system.

TABLE: Relevant date of the system

conventional system	primary energy	900 GJ/a
	useful energy	765/GJ/a
	coefficient of performance	0.85
solar system	useful energy	245 GJ/a
	substituted primary energy	340 GJ/a
	efficiency	0.40
	solar fraction	0.32
	coefficient of performance	17 kWh/kWhe
	energetic amortisation	1.0 a
	monetary amortisation	11.5 a

6. CONCLUSIONS

The collector field was designed that way, that the maximum temperature for drying peppermint (60°C) was not exceeded. It could be proved true, that peppermint could be dried almost without fuel oil.

The preliminary measurements on the collector field gave positive and encouraging results like an energy payback period of 1 year. The monetary payback period of 11.5 years could be reduced significantly if the time of operation could be increased. Possibilities are growing more herbs, drying grain crop and drying for other farmers.

Measurements during the full drying period in 1984 will give more detailed results on the drier himself and the heat recovery system. These results will be incorporated in recommendations for the farmer how to run the plant and in planning further solar assisted drying plants.

REFERENCES

/1/ H.J. Wagner
Energieaufwand zum Bau und Betrieb ausgewählter Energieversorgungstechnologien. KFA Jülich JÜL-1561 Dez. 1978

/2/ W. Schölkopf, M. Reuss
Analysis and Comparison of Air and Water Heating Solar Systems. Proceedings of the Congress on Energy Economics and Management in Industry, in Albufeira, Portugal. Pergamon Press April 1984

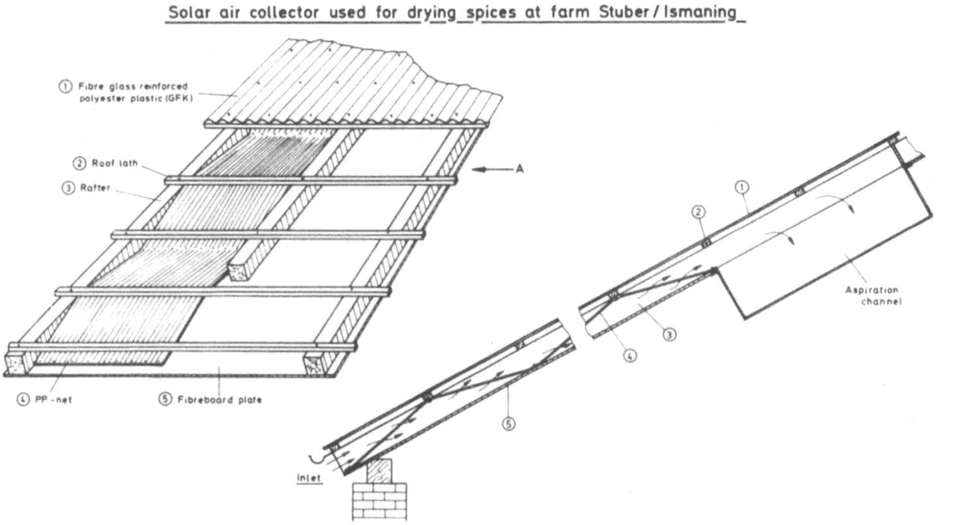

Picture 1: Structure of the solar air collector

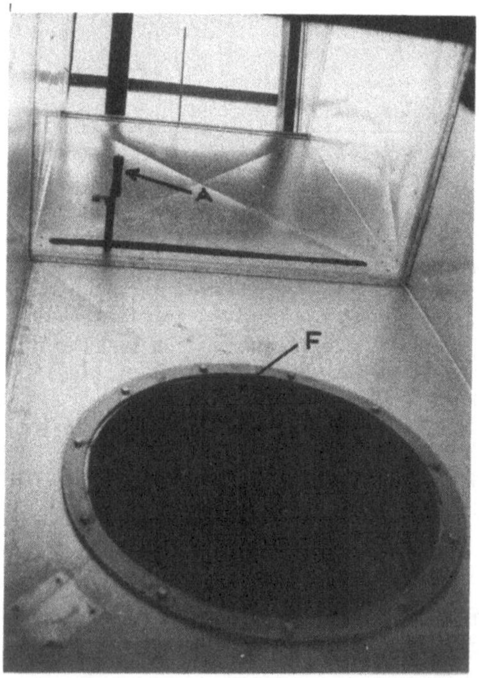

Picture 2: Moveable anemometer (A) installed in the air duct to measure air velocity, (F) outlet to the fan

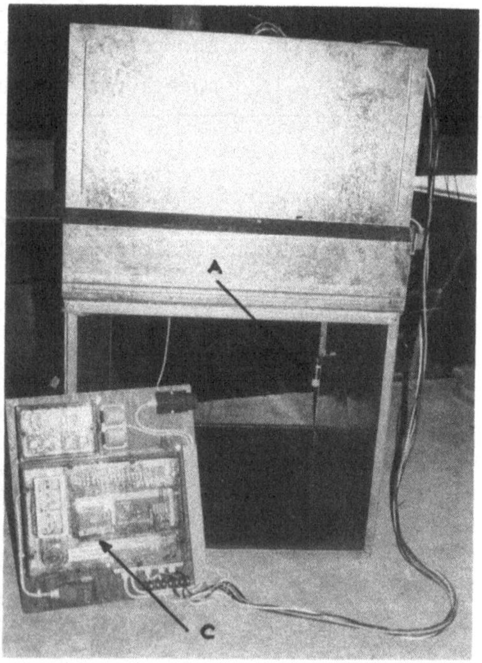

Picture 3: Anemoneter (A) in the air duct and control unit (C)

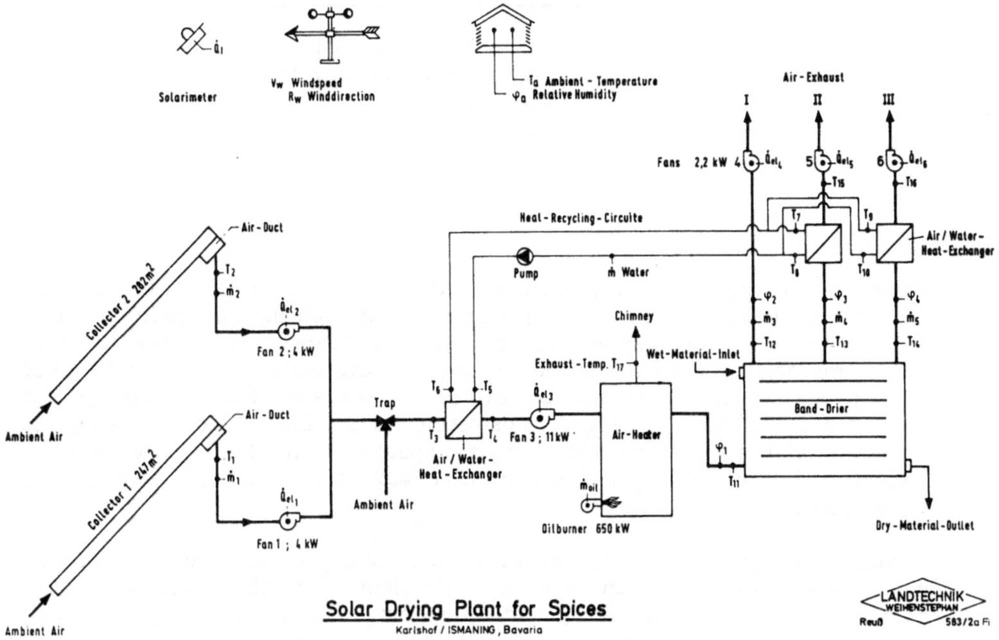

Picture 4: Scheme of the solar assisted drying plant

Picture 5: Photograph of the solar roof collector

Acknowledgements: This research is sponsored by the Comission of the European Communities and the Bundesministerium für Forschung und Technologie der Bundesrepublik Deutschland.

EXPERIENCE WITH ADVANCED SOLAR COLLECTORS FOR WATER HEATING: TECHNICAL AND ECONOMIC ASPECTS

K HAYWARD AND G J WILLIAMS
GEC Mechanical Engineering Laboratory
Whetstone
Leicester

Summary

Under a contract from the European Commission - Energy Conservation Demonstration Programme, a solar water heating scheme has been installed for the Whetstone site canteen of GEC Power Engineering Ltd. The contract covers the design, installation and performance evaluation of the scheme over a two year period with the objectives of assessing the scheme against predicted performance levels, demonstrating its reliability, and evaluating its economic viability in the UK. It is expected that the scheme, with a collector area of 50 m^2, will supply some 40% of the summer domestic hot water demand of 190 GJ.

Although the scheme is not yet fully operational, valuable experience has nevertheless already been gained. Problems directly associated with retrofitting the collector array and interfacing with the existing hot water system have been identified. Initial performance results are however encouraging and it should be possible to make an economic case for the application of the technology in the construction of new buildings.

1.0 INTRODUCTION

Flat plate solar collectors for active water heating have been developed over a considerable period and it is now possible to achieve an acceptable cost effective thermal performance for low temperature applications. Advanced solar collectors, and in particular those that use some form of evacuated glass tube about a high quality absorber surface, offer many advantages over conventional collectors. Firstly, due to their low thermal loss they can generate high fluid temperatures with reasonable efficiency, and secondly, they offer the potential through mass production for significant cost reductions.

In order to encourage the exploitation of these collectors, a demonstration scheme has been installed (with financial support from the European Commission) to supply domestic hot water (DHW) for an industrial canteen located at the GEC Power Engineering Ltd, Whetstone site in the UK. The site has numerous buildings comprising manufacturing facilities and offices where both the hot water and space heating are provided by a high pressure hot water (HPHW) system from a central boiler house. However, the location of the boiler is distant from the canteen and in the summer months, when the space heating is switched off, the HPHW system moves from its optimum operating condition. For this particular case, a system which utilises solar energy at the point of use during the summer months can achieve a significant reduction in energy consumption.

A review of the aspects that affected the overall system design together with operational experience during commissioning is presented in the following sections. The various economic appraisal techniques applicable to energy saving schemes are also discussed.

2.0 TECHNICAL ASPECTS

Fig. 1. 50m² Collector Array

From a preliminary feasibility study of the scheme, it was proposed to construct a system of 150 m² collector area. However, as a result of escalation in the expected costs due to unforeseen installation difficulties, a smaller system of 50 m² has been constructed on the roof of the building adjacent to the canteen. This is illustrated in Figure 1. The major aims of the scheme are, nevertheless, still met but it will not be possible to supply as large a proportion of the canteen DHW demand as originally considered, and other methods (off-peak electricity) are being considered to allow the site boilers to be shut down over the summer months.

2.1 Solar collectors and support frame

Initially, a number of commercially available evacuated tube solar collectors were assessed for suitability within the scheme. One particular collector, manufactured in the UK, was eventually considered preferable in that it offered a number of advantages over the others. For instance, the collector used a heat pipe to transfer the heat generated at the absorber to the system heat transfer fluid and, by selecting an appropriate heat pipe specification, it was possible to limit the temperature rise across the collector system to safe conditions ($< 100^{\circ}$C). This is of course an important consideration for any scheme using an advanced type of solar collector where, depending upon circumstances, high temperatures can readily be generated. In addition, the condenser end of each heat pipe was inserted inside the manifold to improve heat transfer.

Before installation, both planning and building regulations approval for the scheme were sought as some minor structural alterations to the building were necessary in order to support the collector array. The collector support system, designed in accordance with expected wind loading conditions obtained from the relevant British Standard Code of Practice (1), consists of a lightweight interlocking framework attached to rolled steel joists positioned across the exterior supporting walls of the building. This arrangement allows installation of the individual collector modules in three parallel banks at a 45° inclination, to form an array with an orientation of a few degrees west of due south.

2.2 Hot water demand and store size

As the starting point in the design procedure, the accepted figure of 50 litres pre-heat storage capacity per m² of collector array gave a pre-heat tank capacity of 2500 litres. To assess how this related to the hot water demand of the canteen, studies were undertaken to monitor both the water usage and temperature during typical periods of use. Readings were taken, in the first instance, at every hour throughout the day for a period of weeks and subsequently once a day until adequate data had been obtained. The daily readings showed that the demand of some 5000 litres/day did not vary significantly. An example of a typical hourly demand pattern during a day is shown in Figure 2.

At present the hot water demand is for five days only, with the canteen being shut during the weekend. To accommodate this situation where the tank would be expected to reach temperatures approaching 90°C under certain conditions, either a larger capacity tank or some form of fluid mixing at the delivery had to be selected. Of these options, the mixing valve arrangement was preferred; the larger capacity tank would have required improved load bearing supports and, consequently, increased installation costs.

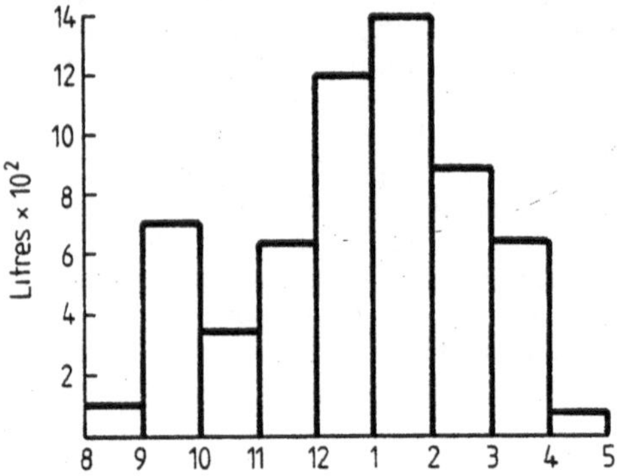

Fig. 2. Typical hourly demand

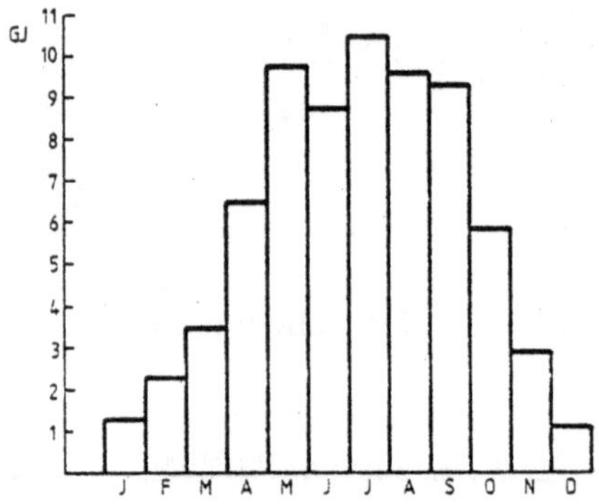

Fig. 3. Predicted energy delivery through the year

Prediction of the system performance with the chosen tank size was considered valuable. The SEU design method (2) was selected as the most appropriate method which would give not only the performance on a month by month basis but also allow a means to verify the choice of store size. In fact the method gave a clear indication that increasing the size of storage had a relatively small effect on the fraction of the heating load carried by the fixed area system. Figure 3 presents the predicted individual monthly energy supplied by the 50 m² collector array and 2500 litre storage tank.

2.3 Controls, tank and pipework

In a temperate country such as the UK, some account has to be taken of the risks due to freezing conditions, and both the collectors and external pipework must be capable of either being drained or used with an anti-freeze fluid. In the case of the latter, an additional heat exchanger is necessary for the storage tank and this imposes a slight efficiency penalty. In view of this, the decision was taken to allow the collector array to drain automatically under freezing conditions. This is achieved by allowing the system fluid to return under gravity to the preheat tank when the system is vented on the action of the frost sensor. By this method, no pre-heat water held in the collector manifold is wasted.

Careful attention was given to the construction of the pre-heat storage tank as this could, as previously mentioned, reach temperatures in the region of 90°C during a weekend. Conventional galvanised mild steel tanks used with copper pipework are not suitable for the storage of water at high temperatures and a tank had therefore to be constructed from a compatible material. Ultimately a polypropylene based material (CELMAR) was selected and a tank fabricated to the required dimensions.

Conventional copper pipework, insulated to a high standard, has been used to transfer fluid between the pre-heat store and the collector array. To ensure that circulation takes place only when the system is able to gain energy, the pump is controlled via a differential control unit which senses the temperature difference between the collector manifold and pre-heat store.

2.4 Installation and operational experience

Installation of the scheme progressed broadly as planned over a twelve week period. Two areas of the installation did however cause some problems. In the case of the collector modules, a total of 500 absorber tubes had to be installed and this proved extremely time consuming due to the nature of the tube to manifold connection arrangement. Also, due to their inherent vulnerability to damage, a small number of the tubes cracked at the neck end and these had to be replaced.

The other area of difficulty related to the system integration with the existing hot water delivery pipework. Here the site maintenance authorities insisted on a dual operation to allow the canteen system to operate in its original mode, if in the future the solar scheme became unreliable. Naturally a complex system of valves had to be included to accommodate this feature.

Commissioning of the system has been recently completed with only minor problems encountered. A computer-based performance monitoring system, illustrated in figure 4, has been installed in the pre-heat tank room to monitor the scheme as it operates under various conditions throughout the year. Various measuring transducers coupled to a data logger are located with reference to the energy transfer processes that occur and these will provide sufficient information to check not only the performance of the scheme against that predicted, but also to evaluate the energy saving.

3.0 ECONOMIC ASPECTS

3.1 Review

By definition a demonstration scheme introduces a new technology which is close to commercial exploitation and where satisfactory operation should confirm future prospects. Preliminary data from the Whetstone scheme indicates that performance is well up to expectations and predictions of energy savings can therefore be quoted with confidence.

By nature solar schemes involve an initial capital investment which may be regarded as yielding a fixed yearly energy saving. Economic analysis is necessary to judge the worth of such schemes, involving, as they do, effects over a number of years. This longer term aspect distinguishes energy saving schemes from the more usual short term business decisions.

Economic assessment criteria take many forms but those most commonly applied to date have been:

i) Simple payback

ii) Discounted cash flow - Net present value

Simple payback involves only calculating the time required to recover the initial capital outlay. Payback is an easily understood concept and is very useful in identifying the immediately feasible short term opportunities. Because of this it does tend to militate against the longer term project. There are disadvantages however in that it takes no account of cash flows as they vary, ignores income after payback and residual asset values.

Discounted cash flow allows for the fact that a unit of cash received in the future is worth less than the present value due to the amount of interest that could have been added. By assuming a rate of interest then all future cost savings can be discounted to a present value. Net present value (NPV) involves summing all the discounted cash savings over the project life and subtracting the initial capital outlay to give the NPV (3). A positive NPV indicates an acceptable project. In a comparison situation, the scheme having the highest positive NPV will of course, be selected. Discounting does not, of itself, allow for fuel cost inflation which can tend to cancel out, or indeed overcome, the interest factor. Clearly an NPV analysis involves a series of assumptions which are, of themselves, open to argument. For this reason the present system is discussed in terms of simple payback but the noted limitations of this approach must be remembered.

As the scheme was to be fitted to an existing building various aspects had to be addressed in order to incorporate the component parts of the system. These naturally introduced costs for which there was little opportunity for quantity discounts. For example, alterations to the building structural load supports and integration with the existing hot water distribution system were required. The Whetstone scheme constructed on this basis cannot therefore be considered as a commercial proposition. Taking site energy costs including transmission losses, and calculating on a single payback basis against the annual energy saving, the investment costs indicate a payback period of around 20 years, which is typical of a small scale, one-off retrofit system.

Fig. 4. Data logger in pre-heat tank room

3.2 Future prospects

Retro-fit on a commercial basis should be linked to the concept of a number of low cost modular packages which would give typical payback periods in the 15-20 year timescale, which is not unreasonable when compared with accepted building cost depreciation periods of around 30 years.

The incorporation of advanced solar heating systems into new buildings presents a much more exciting prospect. In this instance, with imaginative design, there is little reason for costs to be more than those for the collector array. For example storage tanks are already required and it is a simple matter to specify tanks suitable for storage of pre-heated water, with minimal extra cost. In addition, support systems for the collector array can be easily included in the building structure. With all fixed costs for the building already provided for, this should bring payback periods into the 7-12 year bracket which probably represents the present practical limit.

It must be remembered that the above comments apply to the UK. Clearly, countries with significantly higher totals of solar energy or more expensive fuel costs will find solar water heating installations more attractive and, where advanced collector schemes are specified, there is an added advantage of providing a closer match to demand temperatures over an extended heating period.

4.0 CONCLUDING REMARKS

Generally the scheme has been successfully installed to plan with only a few minor difficulties. Furthermore, the initial performance during the commissioning period obtained from the monitoring system is encouraging.

A number of aspects gained from experience with the scheme are worthy of comment in relation to advanced solar collectors and their applications. These are indicated as follows:

- Investigate the possiblities of design or improved production techniques for lower cost collectors whilst still retaining the performance advantage.
- Make associated design improvements to give ease of tube to manifold assembly.
- Explore methods to minimise the use of the absorber tubes through improved reflector designs and materials.
- Avoid individual retro-fit schemes by constructing systems on green field sites on a medium/large scale and adjacent to points of use.
- Concentrate attention on areas that indicate a greater chance of commercial viability which, for example, for the UK means the export market.

To conclude it is considered that the knowledge and experience gained through the introduction of the Whetstone scheme will contribute to many of the above areas.

5.0 REFERENCES

1. British Standards Institution "CP3 Chapter IV, Part II', London 1972.
2. Kenna, J P. "The SEU Design Method for Solar Water Heating Systems". HELIOS 15. University College, Cardiff 1982.
3. Matz, A et al "Cost Accounting, Planning and Control", South Western Publishing Co. 1972.

CONTRIBUTION TO THE SIMULTANEOUS PRODUCTION OF FRESH WATER AND SALT

D.A. TSITSIS
Consulting Engineer
Frank E. Basil, Inc., Consulting Engineers

Summary

A proposed system for the simultaneous production of fresh water and salt by distillation is presented in this paper. The system is identified as the Solar Co-production of Water and Salt (SCWS) and is a modification of Greek patent No. 66725 the "Floating Self Contained Solar Distillator". The proposed system utilizes both solar and wind energy; atmospheric air is the working medium for the process. Air heated by solar energy flows through a glass evaporator and the mixture is condensed as it travels through a separate, air-cooled, cyclonic condenser. The distillate is collected in the lower part of the condenser and the dried air is exhausted through the funnel of the cyclone. The action of the gravity draft in the system is further enhanced by a wind inductor. The SCWS is in the conceptual design phase; the author suggests that it may prove to be an efficient system for the simultaneous production of salt and distilled water.

1. INTRODUCTION

Sodium chloride (NaCl), common salt, is high tonnage material. In addition to its familiar use in the diet of man and animal, large quantities are used in agriculture and as a basic chemical raw material in the production of soda ash, calcium chloride, caustic soda, sodium sulfate, sodium bisulfate, HCl, sodium cyanide, sodium hypochlorite and chlorine. Sea water contains about 35000 ppm of dissolved solids most of which is NaCl. Salt may be removed from sea water by several methods. The most common method is by the distillation of sea water. In Greece and other countries with access to the sea, the prevalent method is by solar desalination, for which extensive, open, shallow ponds are filled periodically with sea water. Under the action of the sun evaporation occurs leaving salt on the basin beds. This process is limited to areas of maximum sunshine; a large land area is required; rains affect production; the evaporated water cannot be reclaimed for agriculture. The advantages of this conventional method is simplicity and minimal fuel costs. Considering an average concentration of salt, 35000 ppm, and a specific gravity (at 20°C) of 1024,3 kg/m^3 for sea water, it is concluded that 35,85 kg. of salt and 988,45 kg. of distilled water are available from one m^3 of sea water or 3,5% of salt and 96,5% of distilled water on a volume basis.

2. DESCRIPTION OF THE SYSTEM

The proposed system consists of one or more modules each comprising:
-An evaporator basin covered by a glass cone or a pyramid,
-A separate air-cooled cyclone type condenser, connected to the top of the evaporator by a duct.

The principle of operation is schematically shown in Fig. 1 "The proposed module of distillation". The atmospheric air is heated by solar energy and as it courses through the glazed evaporator it picks up water vapor released by solar energy. The vapor-laden air is then drawn, due to gravity draft, through the cyclonic air-cooled condenser where the vapor is collected as distilled water in the lower part of the cyclone. From here it is piped to a ground reservoir. The gravity draft effect is further enhanced by a pivoted wind inductor and/or by a wind driven turbine. The accumulated salt in each evaporating basin can be removed manually or taken off by a mobile vacuum unit or a central vacuum system (salt slurry or dry state). The basic plant is supplemented by: an elevated concrete sea water storage tank for filling the evaporator basins; an on-ground concrete distilled water reservoir which collects the distillate by gravity; a wind driven electric generator and a solar photovoltaic generator to charge electric storage batteries for the two sea water pumps (one is stand-by); and other plant auxiliaries, as shown in Fig. 2 "Schematic Diagram of SCWS Plant".

3. METHOD OF APPROACH

Simultaneous heat and mass transfer theory, psychrometric analysis and empirical equations for swimming pool heating, are the considerations for the conceptual design of the SCWS. The theoretical simulation of the process is shown on Fig. 3&4. No analysis is presented in this paper due to space limitations. A computer simulation of a year's operation will be prepared.

4. SIGNIFICANT RESULTS

Preliminary predictions for the SCWS: production, under climatological conditions around Athens-Greece, is roughly estimated to be many times higher than the conventional solar distillation per m^2 of evaporator surface area per day.

5. CONCLUSIONS

The proposed SCWS has several benefits compared to the conventional open solar distillation system for salt production:

The SCWS is a self-contained plant.

No large land area is required to accomodate conventional open evaporating ponds.

The SCWS can be constructed easily in many sizes. The cost of salt and water is expected to be lower, as two products will be produced simultaneously, water and salt.

Salt production per m^2 of evaporator is many times that of the conventional open evaporating ponds.

The production of salt is not affected drastically by rain as is the open system.

The SCWS can be used also as a second stage distillation system for brine disposal of existing or of new water desalination plants.

The proposed system can be modified and used for desalination of brackish or contaminated water or sewage effluent to yield water for irrigation.

The production of salt and water can be increased considerably by using a forced draft system instead of the gravity system described. This improvement requires extra power for the induced draft fans at the premium of an increase in investment and operating costs.

Further analysis is required to determine the most economical size of the distillation module.

An experimental model of the distillation module is necessary to determine actual design parameters and to optimize results.

REFERENCES

1. D.A. TSITSIS 1982-Floating Self-Contained Solar Distillator (In Greek) 1st National Conference on Soft Energy. Aristotle University of Thessaloniki, Greece.
2. ASHRAE Handbook 1981 - Fundamentals Volume

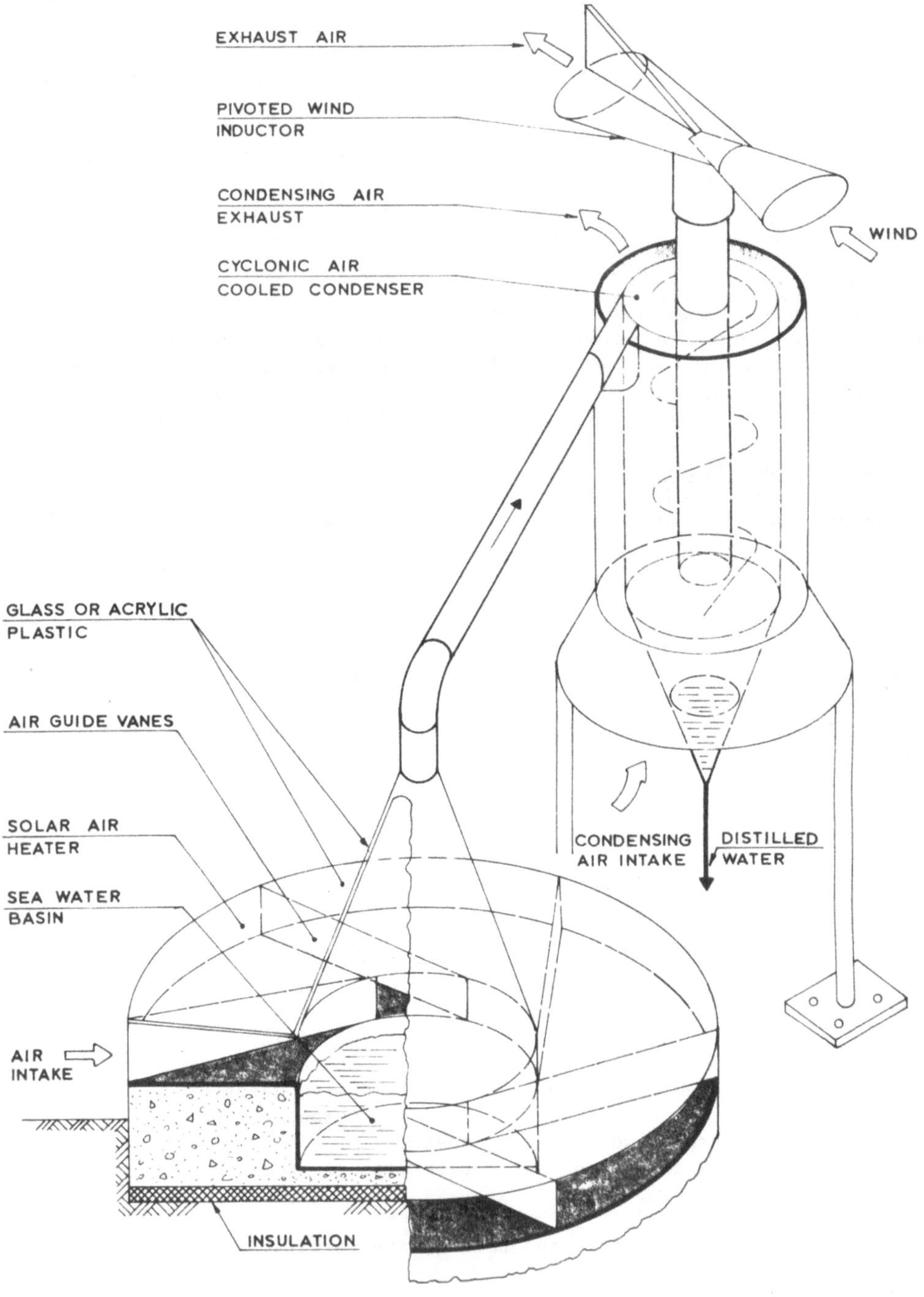

FIG. 1 THE PROPOSED MODULE OF DISTILLATION

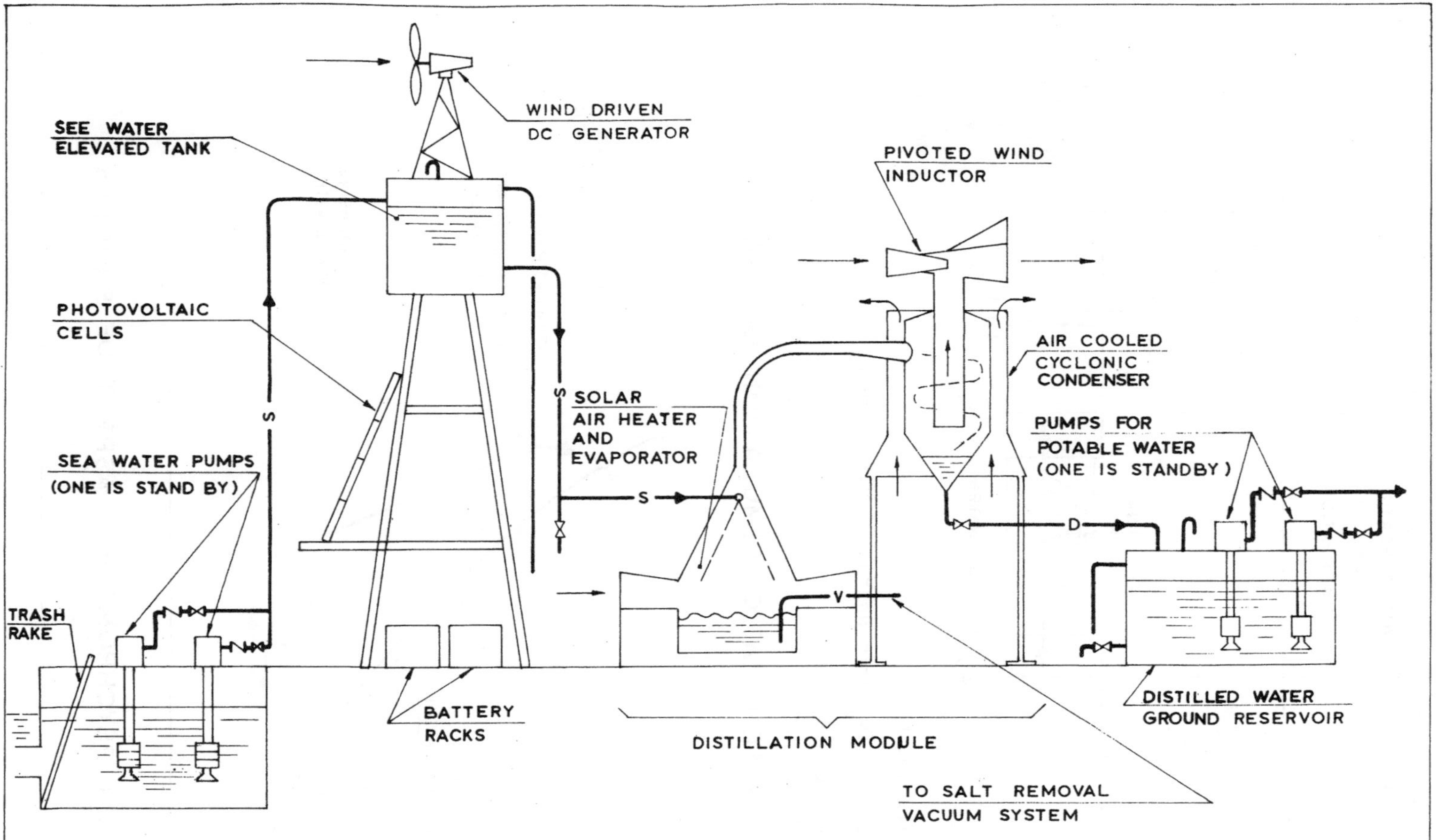
WIND DRIVEN
DC GENERATOR
SEE WATER
ELEVATED TANK
PIVOTED WIND
INDUCTOR
PHOTOVOLTAIC
CELLS
AIR COOLED
CYCLONIC
CONDENSER
S
SOLAR
AIR HEATER
AND
EVAPORATOR
PUMPS FOR
POTABLE WATER
(ONE IS STANDBY)
SEA WATER PUMPS
(ONE IS STAND BY)
D
V
TRASH
RAKE
BATTERY
RACKS
DISTILLED WATER
GROUND RESERVOIR
DISTILLATION MODULE
TO SALT REMOVAL
VACUUM SYSTEM

FIG. 2 SCHEMATIC DIAGRAM OF SCWS PLANT

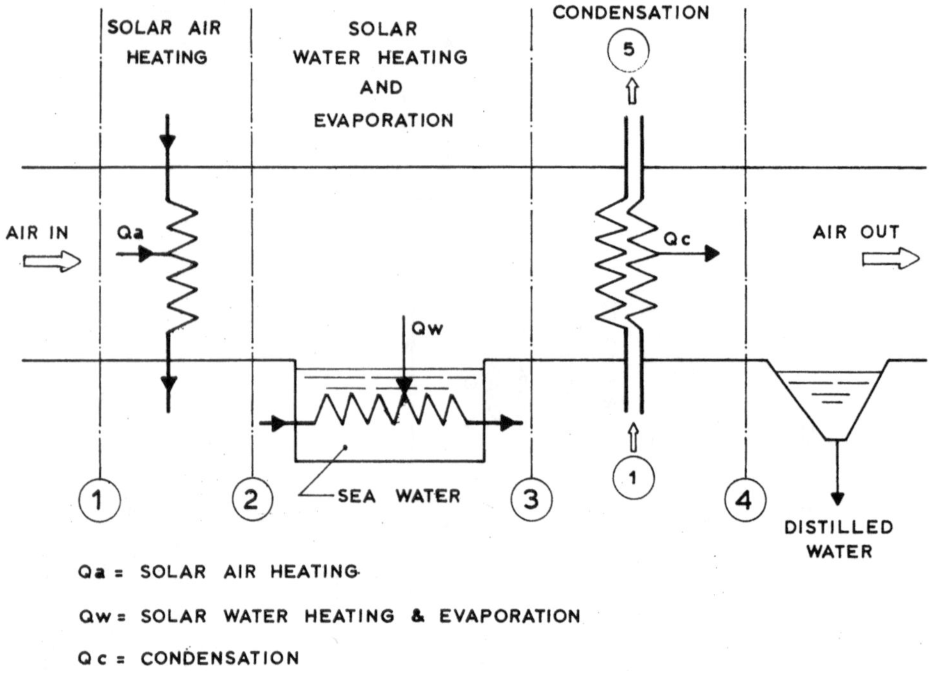

FIG. 3 SCHEMATIC SIMULATION OF PROCESS

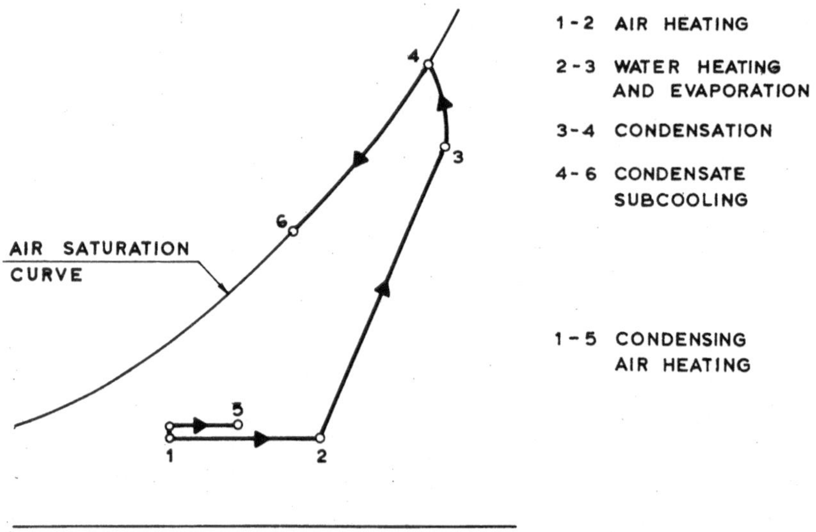

FIG. 4 PSYCHROMETRIC ASPECT OF PROCESS

DRYING OF GRAPES IN CABINET SOLAR DRYER. PRELIMINARY RESULTS

M. Marousis[(1)], M. Tsamparlis[(1)] and G. Raouzeos[(2)]
(1) University of Athens, Department of Physics, Division of Mechanics, Athens 621, Greece.
(2) National Technical University, Laboratory of Unit Operations, GR-106-82 Athens, Greece.

Summary

A pilot scale cabinet solar tray dryer was constructed of 500 Kg capacity of fresh product. An array of air-cooled flat plate solar collectors, of area 5.2 m^2, was installed to heat the drying air stream. Currants were dried in the first drying experiments during autumn '83. Parallel experiments were performed with a similar solar dryer, of 10 Kg capacity, to investigate the effect of the flow rate of the drying agent on the drying time of currants. Currants dried at mean operating temperatures 30° to 40° C and low to moderate flow rates. The process completed in 20 and 13 days of intermittent operation, 180 and 117 hr of real drying time, respectively for the two experiments. No auxiliary heating was used. The drying experiments led to the following conclusions: To reduce the drying time the flow rate and distribution in the drying chamber must improve, and the operating temperatures to increase. Drying time will be more reduced dipping currants before drying. The dried product will be more uniform when thinner layers of currants are dried. The ratio of the total area of the collectors per kilogram of evaporated moisture must increase. The thermal performance of the dryer and the collectors was satisfactory.

1. INTRODUCTION

Currants and seedless sultana grapes are, for Greece, two commercially and economically important agricultural products either fresh or dried. The average annual production is 150 000 tons of which some 57% are exported as raisins with an approximate turnover of $7 \cdot 10^9$ drachmas. They are traditionally open sun-dried either by laying the grapes on the ground, or hanging them in rows above the ground. A small amount of grapes, 10% of the total dried production, is dried in mechanical dryers consuming conventional fuels. Research concerning the drying time and quality of the open sun-dried sultanas had been performed, in the late sixties, from the Greek Ministry of Agriculture at the Institute of Technology of Plants and Products (1). Similar researches are also reported in the literature but nothing is reported concerning the drying characteristics of currants (2,3,4).

Solar drying in Greece appeared only in 1980 as a collaborative attempt of the Agricultural Research Center, Chania-Crete, and the Institute for Agrartechnik of the Hohenheim University of Stuttgart. A cabinet solar dryer of simple design was constructed for drying sultanas. The conclusion was that drying could be completed in 6-7 days only when a systematic experimental study of the operation of such a dryer will have been performed. Currently in Greece exist only three well designed experimental solar drying installations for drying cereals and fruits. They are systematically tested since 1982 but so far no results have been published.

The purpose of the research work conducted at the Laboratory of Mechanics, University of Athens, is to build a solar drying plant for drying fruits, improving thus the yield of the production and the quality of the dried product, which carries economical value for the agriculturer and the country. The present paper describes the experimental cabinet solar dryer that was designed and constructed at this Laboratory, commends upon the performance of the dryer and the solar heating system, reports the preliminary experimental results of drying currants under solar drying conditions, and discusses the design and opperational errors of this attempt.

2. THE SOLAR INSTALLATION

The designed and constructed solar dryer is of pilot scale (fig. 1).It consists of a cabinet tray dryer, an array of four solar collectors, an auxiliary electrical heater, two centrifugal fans and the necessary ducts.

The overall dimensions of the dryer are 1.06mx1.28mx1.61m, its frame is built from steel and its sides from expanded polyurethane foam, 5cm thick sandwitched in aluminium foils 0.1cm thick, providing thus good insulation. It carries five trays with maximum drying capacity 500 kg of fresh product. Air is blown into the drying chamber, at the lower plenum, by a centrifugal fan (0.25 KW) and is simultaneously withdrawn from the top with a similar fan. There is no perforated plate at the lower plenum of the drying chamber for achieving uniform distribution of the drying air stream, nor any kind of baffles.

The array of collectors consists of flat plate solar air heaters of very simple design and cheep. The overall dimensions of each one are 0.93mx x1.47m and the total area of the array is 5.2 m^2. The case and the absorber plate are made of aluminium and a glass cover, 3mm thick, covers the plate. The absorbing plate is painted black with a simple non-selective paint. The air stream flows through a channel of trapezoidal cross-section which is formed between the absorbing plate and the back side of the case. The side and back insulation of the collectors is of expanded polyurethane foam.

The auxiliary elecrtical heater consists of resistances with 2 KW total power.

The installation operates in a batch mode. The drying air stream passes once through the drying chamber but partial recirculation is also possible. The four air collectors are connected in the mixed mode. Drying begins only when the air stream has been warmed up to 30 °C and stops whenever the air stream temperature falls lower than this marginal value. No auxiliary heating is used resulting thus to an intermittent operation. This on/off procedure is done manually due to the lack of an automatic controller.

The first experiments were performed to test the thermal performance of the solar collectors. This was realized in a prototype open air-loop(5). The experiments were performed outdoors during August '83.

The first drying experiment were realized during autumn '83. The material used was currants from Korinth area (Peloponnesos). The fresh product was not dipped before drying and approximately 100 kg of it was layed on each tray forming thus a layer of 15-17 cm deep. The trays were loaded in the chamber and their position was not changed until drying was completed.

During the operation the following parameters were recorded: The ambient temperature and humidity, the total solar radiation incident on the surface of the collectors, the temperature of the drying air stream before and after the array of the collectors, the dry and wet bulb temperatures of the drying air stream before and after the dryer, the temperature profiles along and transverse the dryer, and the volumetric flow rate of the drying air stream. Three sample containers were placed on each tray in order to

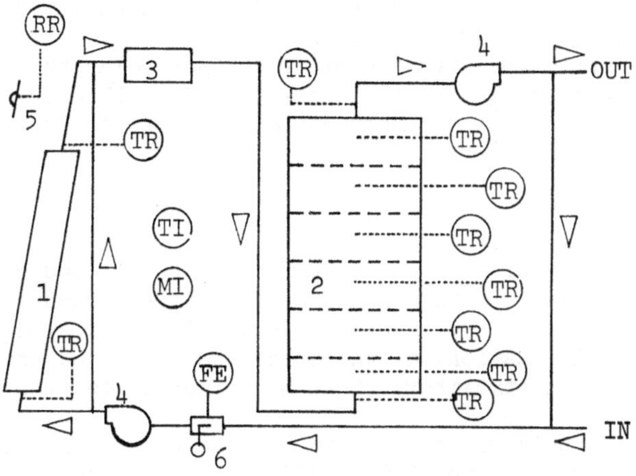

Fig. 1:The experimental installation.

1. Air collectors
2. Cabinet tray dryer
3. Auxiliary heater
4. Centrifugal fans
5. Pyranometer
6. Pitot static tube

TR,TI:Temperature recorders and indicators
MI :Moisture indicator
RR :Radiation recorder
FE :Flow element

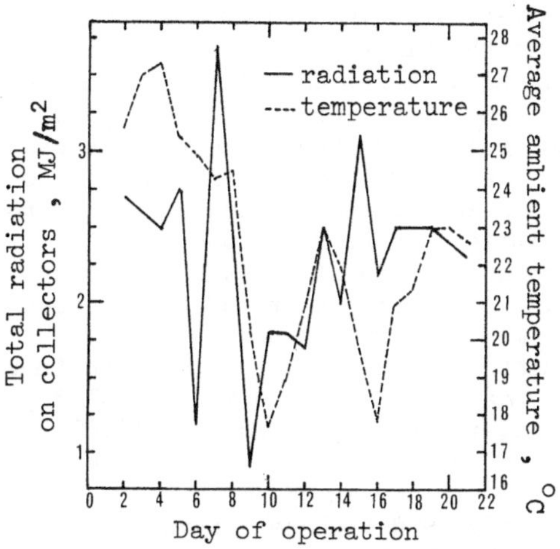

Fig. 2:Weather conditions during drying operation.

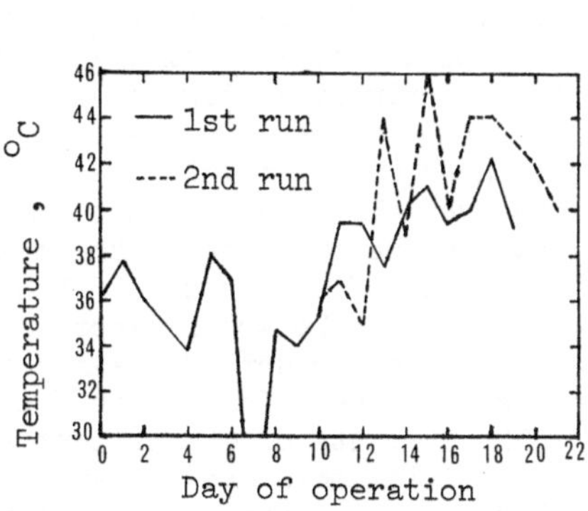

Fig. 3:Operating mean temperature in cabinet dryer.

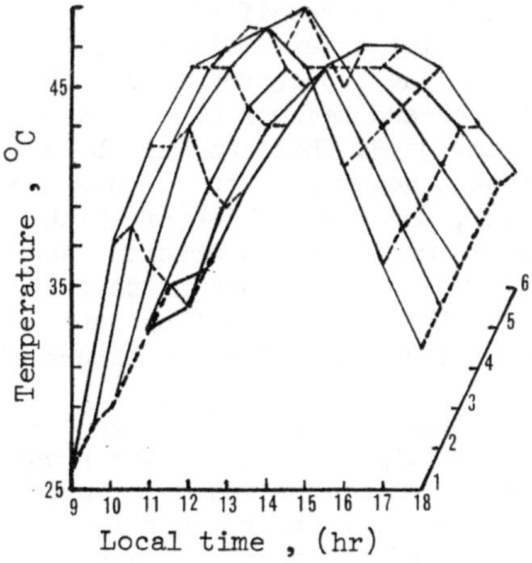

Fig. 4:Temperature profile.

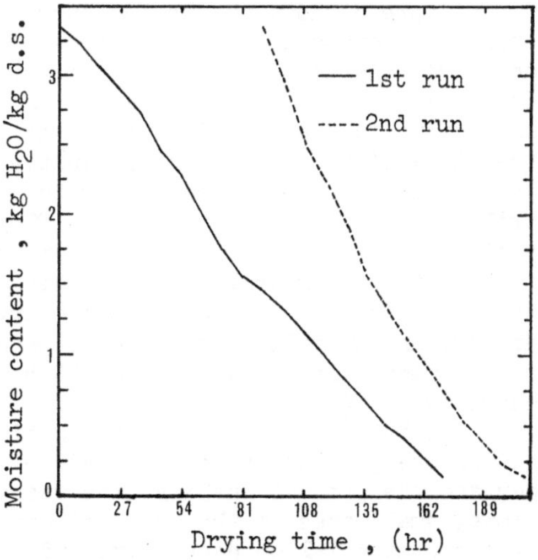

Fig. 5:Drying curve of currants.

monitor the loss of moisture from the processed product. This was realized by weighing the containers on definite time intervals.

Parallel experiments were performed with a similar cabinet solar tray dryer, of 10 kg capacity, connected with a flat plate solar air collector of 2 m^2, to investigate the effect of the volumetric flow rate of the drying agent on the total drying time of currants.

3. RESULTS AND DISCUSSION

Drying experiments on the pilot dryer started on September 10, and on the experimental one on the 20th of the same month, in '83. The recorded weather conditions, at the site of the dryers, are shown in figure 2. The characteristic was the frequent fluctuations of the solar radiation and the rather high relative humidity of the atmosphere (60%) both unusual for the season.

The mean daily operating dry bulb temperature in the chamber of the pilot and experimental tray dryers ranged from 30° to 42°C and 35° to 46° C respectively. Figure 3 shows the mean temperatures in both dryers for every day of operation. The solid line refers to the pilot dryer and the dotted to the experimental. Figure 4 is a tridimensional representation of the recorded temperatures along - dotted lines - and transverse - solid lines - the chamber of the pilot dryer on September 12, '83. From this plot it can be observed that the dry bulb temperature falls as it moves from bottom to top due to the increase of the moisture content of the stream and the thermal capacity of the product. This has tripple effect. First the wet bulb temperature of the stream increases, second the rate of moisture evaporation from the currants decreases and third, the temperature of the product increases. These effects have been experimentally proved too.

The currants of the present experiments had a Brix value 18° when fresh and an initial moisture content 77% (wet basis, 3.35 kg H_2O/kg dry solids, dry basis). The final moisture content was, for both experiments, 13.8% wb. (0.15 kg H_2O/kg d.s.). Drying took place under the conditions described above and with a volumetric flow rate of the air stream 45 and 120 m^3/h for the pilot and experimental dryer respectively. Under these drtying conditions currants in the pilot dryer dried in 180 hours while those in the experimental dryer in 117 hours.

Figure 5 shows the drying curves of currants. This is a plot of the change of moisture content (expressed on dry basis) versus the drying time, and it is characteristic of the material, dried under specified conditions. The solid curve refers to the currants of the pilot dryer while the dotted one refers to the experimental dryer. Observing the two curves we see that the slope of the dotted one is initially steeper and becomes almost parallel to the solid one when drying is nearly completed. This can only be attributed to the higher volumetric flow rate and, generally, the higher operating temperatures, in the experimental dryer, the combination of which results to higher driving forces with regard to moisture evaporation and transfer, from the currants to the drying agent, and consequently to faster drying. In the pilot dryer the temperature and the volumetric flow rate were low requiring thus ten days (90 hours) to reduce the initial moisture of currants by 50%.

Generally the observed drying times were long and elude from the target of the present project to reduce the drying time in favor of the producer and the product quality. The reasons were: Unfavorable weather conditions, small area of the array of air collectors, small fans, deep layers of fresh material on each tray, and no treatment of the fresh currants before drying.

To improve the weather conditions the drying experiments should begin as soon as the harvesting period starts (middle of August). The small surface area of the array and the installation of the small fans was decided upon the available funds at that time. As design error could be considered the absence of some system to distribute equally the air stream in the drying chamber. It was experimentally observed that the product on a tray dried with different rates according to the part it occupied. So the final dried product was of non-uniform final moisture content. The deep layers of material on each tray is an operational error because exceeds, almost four times, the upper limits of loading a tray. This can be improved changing the number of trays in the drying chamber. The treatment of the currants, before drying, is necessary in order to remove the bloom, and check the skin, so that, the additional resistance exerted from the skin is minimized and consequently the rate of moisture evaporation is increased, reducing thus the total drying time. In the literature nothing is reported regarding the pre-treatment of currants. When all the problems and errors are removed currants are expected to dry in maximum 6 days of intermittent operation.

The thermal performance of the dryer, concerning the heat losses and losses due to leakages, was tested when the dryer was empty of product and found to be good. The thermal performance of the solar collectors is described by the efficiency equations as they were deduced from the outdoor experiments. The expressions for the pilot and experimental installation respectively are: $n=0.8-12.5T^*$ and $n=0.9-9.0T^*$

The total investment cost of the installation was 170 000 drachmas and the operating cost, that was actually the cost of the elctricity spent for driving the fans, for 180 hours operation 1 000 drachmas, which is insignificant.

4. CONCLUSIONS

The operating temperature and volumetric flow rate of the drying agent should be as high as possible so that high driving forces are effected and short drying times. The ratio of the total area of the solar collectors, for heating the drying agent, per kilogram of evaporated moisture must increase. The optimum it will only be determined technoeconomically. Dipping of currants must be researched and will eventually reduce the drying time. Currants should dry in thinner layers yielding a more uniform dried product. The thermal performance of the installation is satisfactory, and its cost small.

5. REFERENCES

1. Exarchos, C. and Moisidis, A. (1965). The effect of some parameters on the drying rate of sultanine and on the quality of the final dried product. Bulletin of the Institute of Technology of Plants and Products, Vol. 3 .
2. Radler, F. (1964). The prevention of browning during drying by the cold dipping treatment of sultana grapes. J.Sci.Fd.Agric., Vol. 15, Dec.
3. Grncarevic, M. and Hawker, J.S. (1971). Browning of sultana grapes during drying. J.Sci.Fd.Agric., Vol. 22, May.
4. Ponting, J.D. et al. (1970). Temperature and dipping treatment effect on drying rates and drying times of grapes, prunes and other waxy fruits. Fd.Techn., Vol. 24, December.
5. Testing of Air Heating Solar Collectors. Technical Report of the C.E.C. 1983 Round Robin Series Initial Results, January 1984.

A COMBINED SYSTEM FOR HEAT RECOVERY AND SOLAR ENERGY UTILIZATION IN AUTOMOBILE LACQUERING SHOPS

Dr. F.K. Boese, A. Reich, W. Glinsky, F. Schwick
INTERATOM GmbH, 5060 Bergisch Gladbach 1, West Germany

A combined system for heat recovery and solar energy utilization in automobile lacquering shops

General objective

The economic utilization of solar energy by direct conversion into warm air for drying, combined with a heat energy recovery system is to be tested in this project in two automobile lacquering shops.

The project objective is to considerably reduce the energy consumption of lacquering and drying systems.

Spray painting and drying installations for automobiles are among the largest energy consumers in industry. For health reasons, the acrylic lacquers used in the automobile industry require large amounts of fresh air and inevitably demand large quantities of heat energy. Approximately 20000 automobile lacquering installations each having one spraying and one drying chamber are in operation in the Federal Republic of Germany.

The concept presented here is a possibility for partially supplying the heat energy required by an automobile lacquering shop by way of air-cooled solar collectors, and of utilizing the waste heat energy of the spraying chamber.

The air-cooled collectors which have been planned and constructed in the past can be assigned to the sectors "space heating" and "agricultural" drying of products such as tea, tobacco, green fodder and coffee.

The combination of air-cooled collector systems and a heat energy recovery system has not been used in industry to date.

The available data indicate that there are great potentials for the economic use of air-cooled collector systems combined with a heat energy recovery system in wide sectors of industry.

It is planned to install the system in two automobile lacquering shops in the Cologne area. Technical data of the automobile lacquering shops are shown in Tab. 1.

Functional description

The air-cooled collector fields are mounted on the workshop roofs and connected in series to the existing air intake opening for the spraying chambers.

The size of the planned air-cooled collector fields is dependent on the existing structural conditions and the technical possibilities.

The air-cooled collector fields are dimensioned as follows:

Plant I - approximately 200 m^2
Plant II - approximately 150 m^2

The operation of the air-cooled collectors is quite simple. Fresh outside air is introduced into the hollow collector interior from below. The air then flows through the cavity and is thereby warmed by the sun. It then passes through openings, is collected in channels and conveyed to the user (spraying or drying chamber).

The volume of fresh air required for the spraying chambers is approximately 24000 m^3/h per spraying station. It is planned to draw in some of the intake air for the spraying chamber through the air-cooled collector field.

If the collector field does not supply the required operating temperature of 22 °C, the air will be automatically post-warmed by means of the existing heating facilities. If the amount of heat produced by the collector field is higher than the consumption of the spraying chambers, which can occur at high levels of solar radiation and in summer, the heated air is fed to the intake opening for the fresh air admixture of the drying chambers by means of an additional channel. Operating temperatures of 60 - 70 °C are required here. The collector field can be decoupled as necessary (failures, weekends). Three different modes of operation are planned.

Operating mode 1

If the temperature measured in one of the intake air channels to the spraying chamber is below the specified temperature, a partial flow (approximately 14000 m^3/h) of the volume of fresh air required for the two spraying chambers is preheated in the two collector fields and conveyed to the mixing boxes I and II respectively. This preheated outside air (in each case 7000 m^3/h) is mixed with the fresh air which is sucked in at the outside temperature (in each case 17000 m^3/h). The total air flow is passed through the heat energy recovery system and subsequently heated to 22 °C in the gas heater.

Operating mode 1 is given priority as the collector system supplies most heat in this case.

Operating mode 2

If the temperature measured in one of the intake air channels to the spraying chambers excedds the set specified temperature, the temperature control system changes the positions of the flaps and hence ensures that the fresh air which has been preheated in the collectors is distributed, as a result of which the spraying and the drying chambers can be supplied. Thereby one collector field is available for the two spraying chambers and one collector field for the drying chamber. The fresh air (approximately 7000 m^3/h) preheated in collector field I is conveyed to the mixing boxes I and II respectively. Here it is mixed with the fresh air sucked in at outside temperature (in each case 20500 m^3/h) and conveyed to the spraying chambers via the relevant heat exchanger (to recover the heat energy).

Operating mode 3

If the heating demand of the spraying chambers is covered, i. e. if the specified temperature is reached in one of the intake air channels, the fresh air (approximately 2400 m^3/h) preheated in the two collector fields is conveyed to the drying chamber alone by readjusting the flaps.

Heat energy recovery system

The proposed heat energy recovery system consists of a number of sets of "heat pipes" connected in series, which use the heat contained in the exhaust air to preheat the cold fresh air in addition to the air-cooled collector field.

In spite of the high efficiency of the installed filter mats in separating paint mist, the recovery of heat energy from the exhaust air of lacquering shops is difficult because of the danger of clogging the heat exchanger with paint particles.

The recuperative heat energy recovery system comprises an air-air heat exchanger which operates in "counter-current".

The heat exchanger is installed in an inclined position between the exhaust and the intake channel of the spraying chamber. The lower section acts as the evaporator (warm exhaust air) and the upper section as the condenser (cold intake air).

Since the heat pipes have no external connections and as a modular construction method has been used for the assembly, individual pipe sets can be removed for cleaning purposes even during operation. This means a considerable reduction in maintenance costs when compared to other systems. The heat energy recovery system is dismantled in the summer months.

Status of the project

It is planned to erect the energy saving facilities in atumn 1984, therefore there are no practical results as yet. We expect to demonstrate the facility described above successfully even though working under rough conditions. The construction, control and maintenance of the system will be very simple. This means that the system can be operated by the spray painters.

Conclusion

On completion of the successful performance of this demonstration experiment, there will be an energy saving application which can be easily adapted to the many lacquering shops within the EC.

Table 1: Technical data of the automobile lacquering shops

		I	II
Spraying chambers	number	2	1
Spraying temperature	°C	22 - 24°	22 - 24°
Total input air volume of spraying chambers	m^3/h	2 x 24000	24000
Motor rating of the input air ventilator	kW	2 x 7.5	7.5
Total exhaust air volume of the spraying chambers	m^3/h	2 x 23500	23500
Motor rating of the exhaust air ventilator	kW	2 x 7.5	7.5
Heating capacity of the spraying chamber burner	kW	2 x 232	232
Drying chambers/drying spaces	number	1/3	1
Drying temperature	°C	60 - 70C	60 - 70°
Recirculated air volume of the drying chamber	m^3/h	24000	13000
Motor rating of the recirculating ventilator	kW	7.5	4.0
Fresh air addition during the drying process	m^3/h	2400	1300
Heating capacity of the drying chamber burner	kW	232	145

A Combined System for Heat Recovery and Solar Energy Utilization in Automobile Lacquering Shops

F. K. Boese, W. Glinsky, A. Reich, F. Schwick
INTERATOM GmbH
Bergisch Gladbach, West Germany

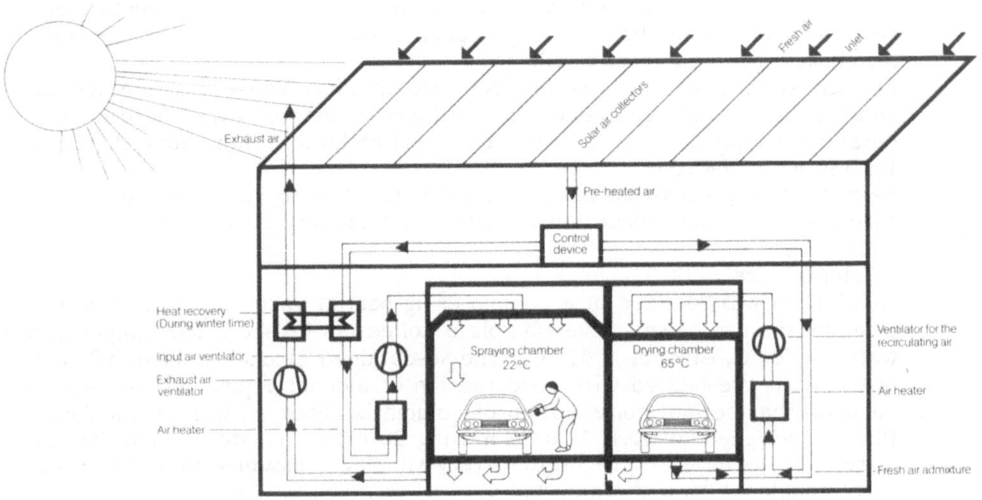

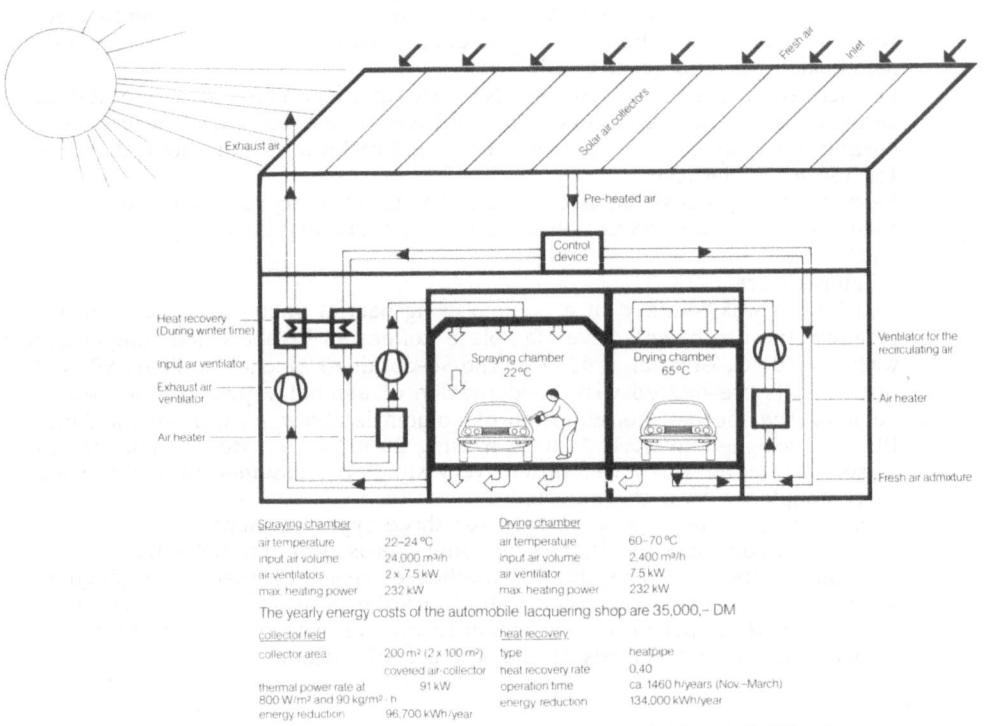

Spraying chamber		Drying chamber	
air temperature	22–24 °C	air temperature	60–70 °C
input air volume	24,000 m^3/h	input air volume	2,400 m^3/h
air ventilators	2 x 7.5 kW	air ventilator	7.5 kW
max. heating power	232 kW	max. heating power	232 kW

The yearly energy costs of the automobile lacquering shop are 35,000,– DM

collector field		heat recovery	
collector area	200 m^2 (2 x 100 m^2)	type	heatpipe
	covered air-collector	heat recovery rate	0.40
thermal power rate at 800 W/m^2 and 90 $kg/m^2 \cdot h$	91 kW	operation time	ca. 1460 h/years (Nov.–March)
energy reduction	96,700 kWh/year	energy reduction	134,000 kWh/year

Two different types of air-cooled collectors are to be mounted on the roofs of the two automobile lacquering shops to warm the intake air for the spraying chamber, and in part also for the drying chamber. The WRG uses the heat energy present in the exhaust from the spraying chamber for additional heating of the cold intake air.

SOLAR DISTILLATION FOR ETHANOL PRODUCTION

B. VERDIER
CEA - Centre d'Etudes Nucléaires de Cadarache
Service d'Etudes Energétiques
B.P. 1 - 13115 SAINT PAUL LEZ DURANCE
FRANCE

Summary

We studied ethanol production from fermented lactoserum. The raw fermented liquor is introduced in a distillation plant able to produce 5 to 10 l/h of azeotropic ethanol. Selective flat plate collectors deliver the required energy to the column. An economic analysis has been performed to study the cost of the alcohol produced from different sources (Jerusalem artichokes, beet, lactoserum...).
Fermentation was performed by ENSA Montpellier. Yeast was selected to ferment lactose and give alcoholic solutions having 10° G.L. The processus duration is typically 7 to 8 hours for 40 g/l of lactose and about 20 hours for 180 g/l of lactose.
Fermentation process has been realized with 85 % to 90 % yield of theoretical ethanol production in following operatory conditions :
- pH = 4.2
- Temperature = 32 °C
Distillation was realized in a column using partial vacuum (250 mm Hg) to allow the use of selective flat plate collectors (evaporation temperatures: water ≃ 70 °C, ethanol ≃ 52 °C). The so-obtained alcohol is about 90° G.L.
The must is pre-heated with condensation of alcohol vapour at the head of column. The thermal consumption of column is about 2 kW/l of alcohol.
Plant is designed to work 24 h/24 h using 20 m³ water storage and 230 m² of solar collectors. The regulation, very simple, is assumed by a three-ways valve coupled with a thermocouple.
Economic assessment is performed for three types of plants :
lactoserum/solar energy, lactoserum/fuel, Jerusalem artichoke/top of Jerusalem artichoke. The analysis method of risk used takes into account the uncertainties relative to economic hypothesis using probabilities of realisation which permits us to evaluate the stability of final price of alcohol and pay-out periods of the whole installation.

1. INTRODUCTION

The installation was realised in Cadarache in South of France (44° N). The basis idea is to valorize lactoserum which is a refuse of milk industry. The concentration of lactoserum for animal alimentation is expensive and the final product is of low interest because of its weak food-value. It is a better way to make ethanol from direct fermentation of lactose.

The production of lactoserum is maximum from March to October according to the maximum solar radiation.

2. FERMENTATION

It was made in a tank of 3 m³ with strains Candida Pseudotropicalis. We obtained alcoholic solutions having 10° G.L.

The alcoholic yield corresponds with the production of 1 degree of alcohol for the consumption of 17 g of sugar/liter. Processus duration is about 7 to 8 hours. So, it is possible to use the fermentation tank three times per day.

The optimal temperature is about 32 ± 3 °C, but fermentation is available in great limits of temperature.

3. DISTILLATION

The objective is to reach an alcoholic degree of solution about 95° G.L. (azeotropic solution).

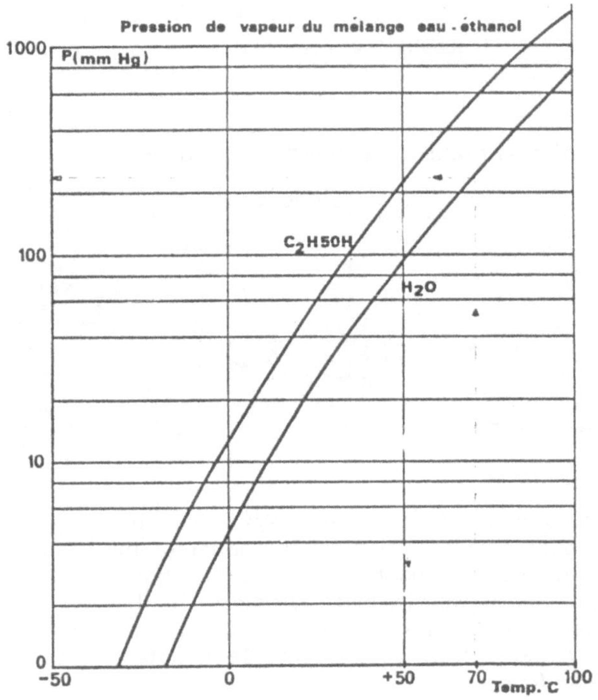

The distillation column is working with low pressure. The operating temperature is about 70 °C. We use 18 plates for stripping section and Raschig rings for rectifying section. The must is preheated by condensation of alcohol vapour at the top of the column. The exchange area for condensing alcohol has been increased with regard to a column working under atmospheric pressure.

Vacuum is made from an ejector

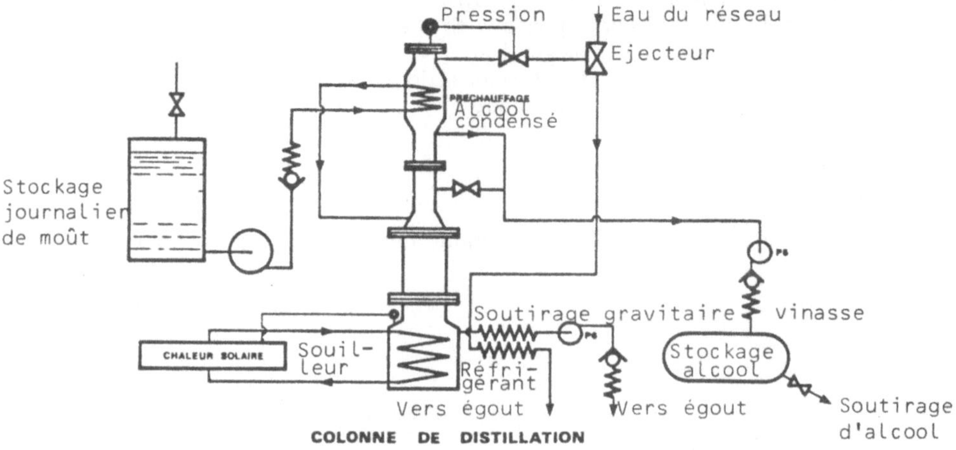

4. SOLAR PLANT

The installation is realised as shown in the figure.

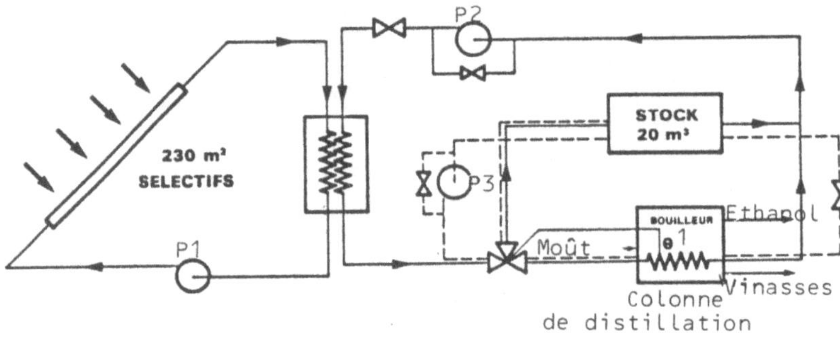

The control is assumed by a three-ways valve which regulates the flow-rate from temperature conditions in the boiler of the column.

Energy consumption is about 1.8 kW/l of alcohol.

The primary loop is working with gilotherm fluid.

The storage is an underground storage of water well insulated.

As soon as collectors are able to provide heat to the system, storage is decoupled from load (distiller).

Plant is designed to work 24 h on 24 during summertime.

5. ECONOMIC ANALYSIS

We particularly studied three types of plant : lactoserum + solar energy, lactoserum + fuel, Jerusalem artichoke/top of Jerusalem artichoke and we compared the obtained costs for these solutions.

Depending of basis economic and technic hypothesis, results took into account an analysis method of risk using probabilities of realisation. Economic study is to be terminated for final publication.

CONCLUSION

This study is an example of simultaneous use of biomass and solar energy. It shows that some light modifications to an usual system (low working pressure, heat exchanger area, regulation and control) permit to have a plant which produces ethanol using solar energy at a cost which is not prohibitive.

SOLAR ENERGY SYSTEM IN ORDER TO CONTROL THE MICRO-CLIMATE FROM THE HORTICULTURE IN GREENHOUSES

J. VAN LOEIJ
Free University of Brussels
Department of Solar Energy
Pleinlaan, 2 - B-1050 Brussels (Belgium)

Summary

In the field of greenhouse-cultures, the energy crisis struck hard. This resulted in the fact that almost 15 % of the enterprises were closed. Because greenhouses use direct solar energy, investigation of the ways to use this energy-source in a rational way is evident.
The aim of the system is to use the warmth, supplied by the sun and the greenhouse-effect as a substitute of the classic energy. The system has proved that from March until September under normal climate-circumstances near aut onomy can be reached and during the other months, the system works as a supply to the classic heating system or to hold the greenhouse frost-proof.
The expense for materials amounted to ± 40.000 Belgian Francs. The work to build the system was done by three men during nine days. The functioning of the system has been followed electronically by a data-logger and a microcomputer. The results are 80 % energy savings for grapefruit cultures and 70 % annually.
Moreover it is possible to reach other fields with the same system : dwellings horticulture, cattle-breeding inside whereas the solar energy system can be built around or near the construction.

1. INTRODUCTION

The recent evolution in the practical use of solar energy has been a motive to important realisations in most countries of the Western world and in the development countries. It is a fact that solar energy applications are linked to the specific climate of the country. A solar application device was built by an orchid specialist in Belgium, Mr. Jan VAN HAUTE. The system works very well and the Free University of Brussels (V.U.B.) became aware of the importance and the possible extension of this application of solar energy use in greenhouses towards all domains where rational and alternative energy consumption can be considered. A workgroup was set up and a solar energy system was worked out and studied by the workgroup in order to control the micro climate of the horticulture in greenhouses.

Taking in regard that in the field of horticulture, and especially in greenhouse-culture, the energy crisis struck hard, and secondly that the energy prices and the selling of the cultures are the main reasons for the regression and the disappearing of quite a lot smaller and bigger enterprises in the glass horticulture ; also the fact that grants are even paid to demolish greenhouses, did make that an intervention concerning rational energy consumption and the use of alternative energy in a greenhouse, based on the well known greenhouse-effect, has become a must.

Moreover it is quite unlogical that during daytime, greenhouses have to be opened to loose the excess of warmth and that during the night, the same greenhouse has to be warmed up by fossile energy to fulfil the energy request.

Consequently it is more efficient to keep the excess of day-warmth in a storage and use it at night in order to maintain the right temperature without any classical heating system.

A system answering to this problem was built out in the horticulture centre of Overijse (near Brussels), where grapefruit is cultivated.

2. DESCRIPTION OF THE SYSTEM (figure 1)

2.1. The components

The components of the system are :

- a flat plate aircollector, composed of an aluminium plate, painted in black, and placed on an insulation plate, wrapped in a plastic foil. This collector is placed in the ridge of the greenhouse
- a pebble bed for energy storage is placed in the soil of the greenhouse. The bed is composed by two horizontal stone levels insulated from the adjacent earch
- an electric ventilator (1000 Watt) is sucking the air out of the collector and is pushing it into the pebble bed
- an air-distribution system which regulates the air-flow. An inversor takes care of the direction-flow of the air in order to warm up the pebble bed or to take the heat out of the bed.

2.2. Used materials

The materials are extremely simple : recuperated off-set aluminium plates, insulation in polystyrene, PVC-tubes, stones, bricks, black board paints and an electric ventilator (1000 Watt). This composition is not only simple but also low-work intensive.

2.3. Operation of the system

The air, heated up in the collector by the sunshine, is sucked by the ventilator and is brought in the upper layer of the stone bed via the distribution system. When the warm air moves, the warmthfront is moving further on gradually. After the traject in the stoneloop, the air is coming back in the greenhouse and is cold. This fresh air avoids in anyway the overheating of the greenhouse so that vents (energy losses) have to be opened less frequently. So waste of energy is limited in a simple way. If there is a heat demand for the greenhouse (at night), the air circulation is inverted by the inversor. Now the air is brought in the under layer of the stone bed and is gradually warmed up so that the energy capacity of the pebble bed is slowly affected. Switches and thermostates regulate the operation of the ventilator.

Moreover the system can even work to cool down a greenhouse, if necessary. In those circumstances cold air is held at night (in the pebble bed) and released during the day. This application could be used in countries with a warm climate.

2.4. Cost of the system

The cost of the investments is very low, the expense for materials amounted to ± 40.000 Belgian Francs. The work to build the whole system was done by three men during nine days.

3. MEASUREMENT CAMPAIGN

The operation of the system has been followed electronically. The instrumentsused are Pt 100 thermocouples, a datalogger and a microcomputer to assure the temperature data-acquisition. The measurements are done by scanning the thermosondes every quartor of an hour without interruption since 18 May 1983. Measurements are also registrated in a reference greenhouse, (without the solar system) in order to compare the results.

4. RESULTS ON ENERGETIC BALANCE

Table 1 and 2 give the result of the energy respectively on energy and oil consumption. The savings are important and reach annually in normal Belgian climate conditions. The minimum temperature requests were : 15°C for the grapefruit culture (May to October) and 5°C for vegetables culture (October to March).

Oil consumption on the boiler of the classic system is indicated in table 2 and this data leads to the estimation of 15 months pay back time in normal conditions. It is undoubtedly sure that the savings obtained with this solar system contributes to cheaper selling prices of the culture products.

5. CONCLUSIONS

The investigation on the grape-greenhouses, heated with solar energy, leads to the realisation of an industrial and commercial product and to the creation of know-how for the use of the rebuilding of the existing greenhouses in alternative, free energy using greenhouses. This example of solar energy device can be applied in Western countries as well as in the developping countries.

The system can be used for heating as well as for cooling and regulation.

Moreover it is possible to reach other fields with the same system : horticulture, cattle-breeding, pre-heating of dwellings and so on.

The research work is going on in this field at the Free University of Brussels which is working towards following finalities :

- methodology for rational use of energy in greenhouses
- simulation programs for the application of the system in several fields
- know-how in order to establish a rational use in greenhouses under several climate conditions.

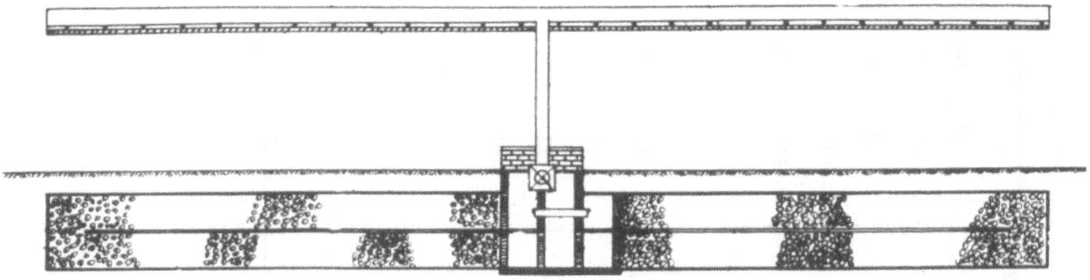

	PERIOD	MEAN SOLAR (W/m²) 24 h			MEAN TEMPERATURE (°C)			ENERGY CONSUMPTION (kWh)						ENERGY SAVINGS (%)			
								Measure			Normal			Measure		Normal	
		Normal	Measure	≠ %	Normal	Measure	≠ %	Solar	Ref	Gains	Solar	Ref	Gains	Month	Period	Month	Period
	March	91	70	77	5.9	6.4	108	-	-	-	-	-		-	-	-	-
	April	138	117	85	9.2	9.3	101	-	-	-	-	-		-	-	-	-
15°	May	185	115	62	13.2	9.4	71	1169	1822	653	40	1093	1053	36		96	
15°	June	206	176	85	16.2	14.5	89	62	216	154	10	191	181	72		95	
15°	July	189	223	118	17.6	19.7	112	9	327	318	94	363	269	97		74	
15°	Aug.	165	158	96	17.5	18.2	104	0	1078	1078	0	1121	1121	100	61	100	81
15°	Sept.	126	112	89	15.2	14.9	98	412	814	402	345	797	452	50		57	
15°	Oct.	73	99	136	11.2	14.7	131	126	343	217	266	425	159	63 (47)		37 (41)	
5°	Nov.	32	34	106	6.3	7.6	121										
5°	Dec.	20	21	105	3.5	5.1	146	738	1070	332	791	1156	365		31		32
5°	Jan.	27	21	78	3.2	4.0	125										
5°	Feb.	52	40	77	3.9	4.0	103										
5°	March	91	83	91	5.9	4.5	76										
	April	-	-		-	-	-	-	-	-	-	-	-	-	-	-	-
								2515	5670	3155	1546	5146	3600		56		70

TABLE I : results (1983 - 10.5 months)

PERIOD	Measurements						Normal					
	Ref		Solar		Gain		Ref		Solar		Gain	
	BF (ℓ)	ℓ/m^2	BF (ℓ)	ℓ/m^2	BF (ℓ)	ℓ/m^2	BF (ℓ)	ℓ/m^2	BF (ℓ)	ℓ/m^2	BF (ℓ)	ℓ/m^2
May → 1/2 Oct. (15°)	35000 (2200)	13.7	13500 (850)	5.3	21500 (1350)	8.4	30500 (1900)	11.9	6000 (375)	2.3	24500 (1525)	9.6
1/2 Oct.→1/2 March (5°)	8000 (500)	3.1	5500 (340)	2.1	2500 (160)	1.0	9000 (560)	3.5	6000 (375)	2.3	3000 (185)	1.2
May → 1/2 March (15°/5°)	43000 (2700)	16.8	19000 (1200)	7.4	24000 (1500)	9.4	39500 (2500)	15.4	12000 (750)	4.7	27500 (1750)	10.7
Annual (15°/5°)	49000 (3050)	19.1	22000 (1375)	8.6	27000 (1675)	10.5	45000 (2800)	17.5	13500 (850)	5.3	31500 (1950)	12.2
pay back time (months)					(18)						(15)	

TABLE II : Oil consumption

EXPERIMENTAL RESULTS OF AIR SOLAR COLLECTOR STORAGE IN SOIL HEATING SYSTEM FOR AN INDUSTRIAL BUILDING

M. RASSAFI, M. MARTIN, R. TORGUET, M. LE RAY

Laboratoire d'Hydrodynamique, d'Aérodynamique et d'Energétique
Université de Valenciennes - Le Mont-Houy - 59326 VALENCIENNES (France)

SUMMARY

An industrial-type building is heated by an air solar collector and a system of air pipes buried in the soil below the building in such a way that the whole energetic chain works approximately half part as an inter seasonnal system and half part as a direct heating system. Some details on the main features of the heating system integrated in the building are given. The mean thermal behaviour during the heating season and, also, the more precise behaviour revealed by outside and inside temperature histograms are briefly analysed.

We recall first some main features of the climate in the city of VALENCIENNES in Northern France. The temperatures are comprised between 2,2° C in January and 17,4° C in August with a still rather low value of 9,1° C in April, and already a moderate value of 10,6° C in October.

The percentage of insolation is comprised between 42 and 45% during six months from April to September, but is only equal to 35% in October and March and falls to 22% in November and January and to 18% in December. Consequently, due to the relatively Northern latitude (50,3°) and to these rather low percentages, the energy received by a south faced roof with a 30 degrees inclination above the horizontal plane has a maximum value of 4,77 Kwh/m2/day in June, but only 0,99 Kwh/m2/day in December with intermediary values of, for instance, 1,80 Kwh/m2/day in February, 3,24 Kwh/m2/day in March and 2,63 Kwh/m2/day in October.

The building which is summarized on Fig.1 has a surface of a 490 m2 a volume of 1830 m3 and is composed of two main parts :

- A great hall with a volume of about 1400 m3,

- Several workshops and auxilliary small rooms with so-

......

me measuring apparatus are located at the southern part of the building below an air solar collector roof composed of an ondulated glass sheet, a first mationless air sheet, a black painted steel absorber sheet, a second air sheet whose air can be circulated by three fans till to six 15cm diameter pipes buried in the soil below the building at a 2,10 meters depth and a seven centimeter thickness layer of insulator. Two of the buried pipes have a 25 meter length and the four others a 38 meter length. The air solar collector roof is south facing with an 30 degrees inclination and an effective 135 m2 insolated surface.

The rate of air flow in the system has been chosen in order to obtain a compromise between the realization of extremely low electric installated power and annual electric consumption and a sufficient high efficiency of the air collectors : using a 3300 m3/hour total air flow corresponding to a specific flow rate of 24 m3/h/m2, one measures an efficiency comprised between 42% and 45%.

Part of the heating fonction is assumed by a very low diffusion through the soil of the thermal waves from the pipes to the floor of the hall.

This part can be considered as a kind of "basis heating" whose effect is relatively moderate, but largely sufficient to have assumed systematically from the beginning of the operation of our system, at the beginning of February 1982 till to April 1984 an inside temperature level systematically above 3 degrees and almost systematically above 4 degrees, even in the case of zero inside energy furnitures from machines, lighting, during the nights and also during some days, and despite of the fact that till this year, the insulation and inertia of the main part of our walls are extremely weak.

The other part of the heating function is assumed by tepid air flowing in the hall after having circulated in open circuit in the solar collector roof and through the buried pipes, or in close circuit with the air of the hall circulating along a loop including the hall air volume and the buried pipes.

This later operation mode can be used, if necessary during some particularly cold nights and during days or part of days with very low incident solar energy levels.

One must remark that, from the end of October till to the mid of April, the sunny days are most generally associated with very cold or cold nights and morgens, and very often from the end of November till to the mid of March with entirely cold or very cold night and full days.

So, direct solar heating is particularly efficient during all the winter period, especially in the cases such that of Agricultural greenhouses (1, 2,3) or industrial building (1,4) where rather moderate levels of inside temperature are required.

We must notice also that with our system, direct heating with air blowing directly in the hall after being passed through the air solar collectors, without a flow along the buried pipes, can be also experienced.

For instance, Mid March we can obtain in the beginning of solar afternoon a temperature of the blowed air after flowing out of the solar collectors of the order of 60 degrees with an outside temperature of the order of 10 degrees.

When this air flows through the buried pipes,its temperature at the ends of pipes is of the order of 40-45 degrees for the 25 meters pipes, and of 35 degrees for the 38 meters pipes.

Some main features of the thermal behaviour of the system can be summarized as follows :

First, as it appears on Figure 2, a mean temperature difference of 5 degrees is maintained between the outside and the inside quite indepedently of the month during the period October 1982 - April 1983. With an effective value of G of an order of 0,8 watts/m3/degree, this result corresponds to about 35000 Kwh furnished by the heating system.

The corresponding electric consumption for an one-year-period going from May 1982 to April 1983 has been equal to 3500 Kwh, i.e one tenth of the heat furnished to the building during the seven colder months.

It is equally important to notice that the total power of the fans in the conditions where they are used, is only 1,5 Kw and that this extremely small power is essentially used during the sunny hours, which arise mainly in the North of France at the end of Winter, in Spring, Summer and the beginning of Autumn.

The histograms of minimum and maximum temperatures (Figure 3 et 4) for the period October 1982 - April 1983 reveal more precisely some interesting characteristics of the behaviour of such a system in an industrial building.

During a third of the total number of days in the period, at the inside, T.Min is lower than 8 degrees, but only 5 days (a little more than 2% of the period) have T.Min lower than 6 degrees, all days having T.Min higher than 4 degrees. A little more than 20% of the days have a minimum inside temperature between 8 and 10 degrees.

45% of the days corresponds to T.Min higher than 10°C.

During about a little more than 20% of the days T.Max is lower than 10 degrees.

During a little less than 30% of the total length of the period T.Max is comprised between 10 and 14 degrees.

Finally, during quite exactly 50% of the cold period T.Max is higher than 14 degrees, the maximum value rising to 26 degrees.

In conclusion, we emphasize upon the fact that the results summarized here must be rather strongly ameliorated in a near future : At present time, the insulation of most part of the walls is very weak (2 cm of plastic between two sheets of steel), but must be strongly increased, due to the fact that a very noisy wind tunnel will be operating at Summer 1984, which requires an efficient acoustical insulation.

REFERENCES

1 Portales B., Martin M., Torguet R., Le Ray M., Stockage longue durée dans le sol pour le chauffage intégral des bâtiments, Proc. Int. Solar Congress Toulouse, France 22-24 oct. 1981.
2 Portales B., Martin M., Le Ray M. Stockage de l'énergie solaire dans le sol pour le chauffage des serres. Proc. Int. Workshop on the Solar Greenhouse, Perpignan, France 2-7 May 1982.
3 Portalès B., Martin M., Le Ray M., Torguet R. Expérimentation of the heating of greenhouses with solar energy stored in subsoil in the north of France, Energex 82, Regina, Canada.
4 Rassafi M., Martin M., Torguet R., Le Ray M. Quasi passive heating of industrial building with a storage below this one, 3d World Congress on Solar Energy, Perth, Australia, August 1983.

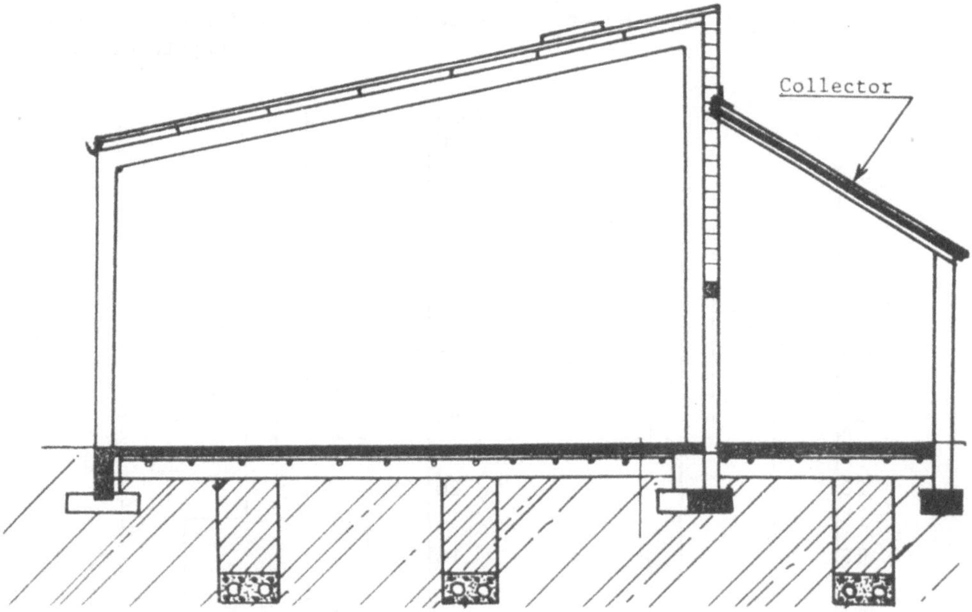

Fig. 1 View of the building.

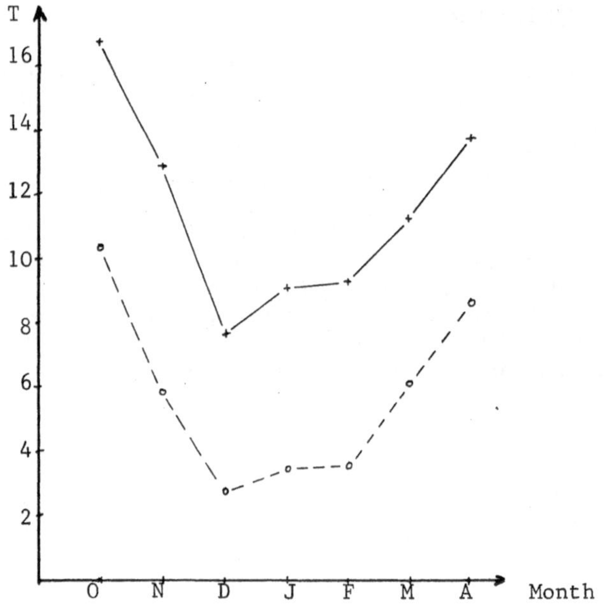

Fig.2 : Mean monthly outside (---o---) and inside (——+——) températures.

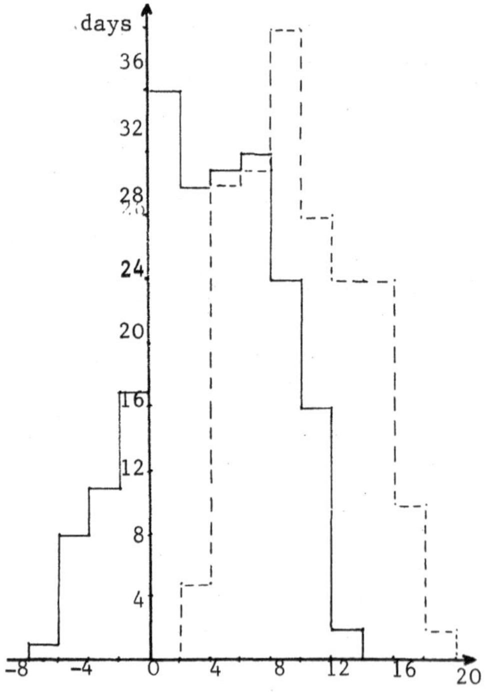

Fig.3 : Histograms of outside (——) and inside (- - - -) minimum temperatures

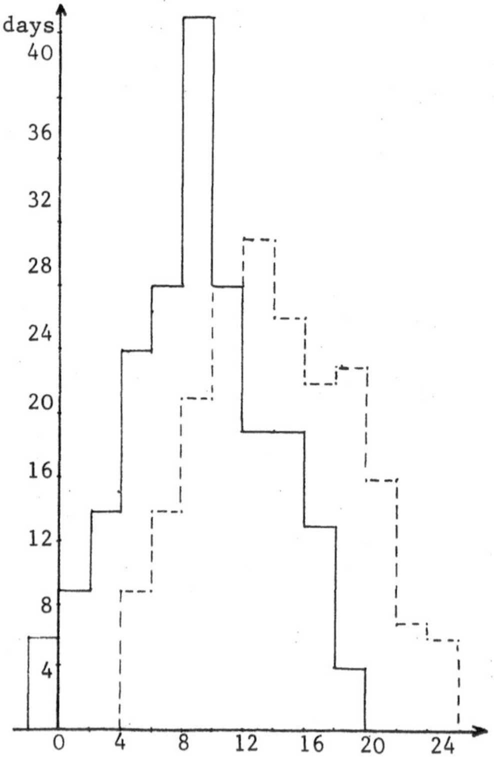

Fig.4 : Histograms of outside (——) and inside (- - -) maximum températures

LOW TEMPERATURE HEAT STORAGE FOR GREENHOUSES

A. Jaffrin* and M. della Faille**

* Laboratoire C.N.R.S. Ecothermique Solaire ;
B. P. 21 - 06 562 VALBONNE CEDEX F -

** SOLVAY and Cie, 33 Rue du Prince Albert,
1050 BRUXELLES B

Abstract : Natural excess heat generated during sunny days in closed greenhouses can effectively be used for night requirements provided one finds a suitable storage medium and an efficient exchanger geometry. A phase change material derived from calcium chloride with a fusion temperature range of 20° to 24°C was developped by C.N.R.S. and SOLVAY with the help of a specific thickener. Various container geometries were chosen for comparative tests. Full scale experiments were conducted in agricultural greenhouses, both operating with an air-loop for solar heat recovery and an air-exchanger structure for the PCM storage. Thermal performance as well as economic prospects are analysed in two different situations : a 500 m² deep-earth rose plantation and a 200 m² green plants nursery on tables. Water condensation on the storage increases heat transfers but can be a threat for the plants at night. Different solutions were tested : either letting vertical containers get rid of the condensation or treating the air by a dehumidifying heat pump. Preliminary measurements show that the fusion temperature range may still be slightly too high, but that a combination of solar heat recovery backed up by a CO_2 generator burner and a small heat pump can make a traditional greenhouse self sufficient in our southern latitudes.

1. Solar heat recovery in agricultural greenhouses.

Basic concepts in plant growth show that vegetal production (in dried mass) is a linear function of the sum of degree-days accumulated by the plant, with a lower limit for the temperature (zero growth temperature) and an upper limit beyond which no benefit can be derived. A solar greenhouse is a greenhouse in which part of the solar heat generated during the day is extracted and stored to be used later at night. There are therefore two main possible goals for solar greenhouses : i) keeping night temperature above the zero-growth limit in unheated units ; ii) taking advantage of excess heat during the day to provide back-up heat at night with the help of classical heating systems. Intermediate goals, like extracting heat at a moderate temperature level, may prove useless in terms of plant growth, but can generate benefits from the reduction of daytime heat losses.

Air and water solar systems are in competition : water systems (1) make use of semi-transparent collectors and water storage and involve an expensive technology. Air systems simply use the plants and the ground as heat absorber and are easy to retrofit. They however imply the use of adequate storage systems built as air heat-exchangers. Rock beds offer an thermally efficient answer to that problem, but the required volume is often prohibitive. Phase change materials (PCM) have long been known to be a potential solution, but until now there were few PCM candidats in the right temperature range.

2. Phase change materials for greenhouse.

Frequently quoted materials like polyethylene glycol, potassium fluoride hydrate or manganese nitrate hydrate are too expensive or improper for the time being. A previous experiment (2) was run with calcium chloride hexahydrate and an excess of water, to decrease the fusion temperature below 25°C. But the remaining heat capacity was weak (100 kJ/kg on a 10°C range). Following a suggestion by Meisingset and Grønvold (3), a lower fusion temperature for $CaCl_2$ hydrates was obtained with the help of other chlorides (KCl and NH_4Cl). A compound using the same thickener as Chliarolithe (4) was developped by C.N.R.S. and SOLVAY for experimental applications. The result is shown in terms of fusion curves (Fig. 1) : a heat capacity of 160 kJ/kg is obtained within a 4°C temperature range starting at 20°C. This is a sharp improvement compared with the previous PCM, but as a draw-back, little heat capacity is involved at lower temperatures. Another consequence is that a larger heat content per unit volume implies a larger surface of exchange per unit volume of encapsulated PCM.

Various container geometries were therefore selected for comparative tests in real conditions : some of them in underground storage volumes in case of deep earth plantations, an other one in arrays below the tables of a green plants nursery.

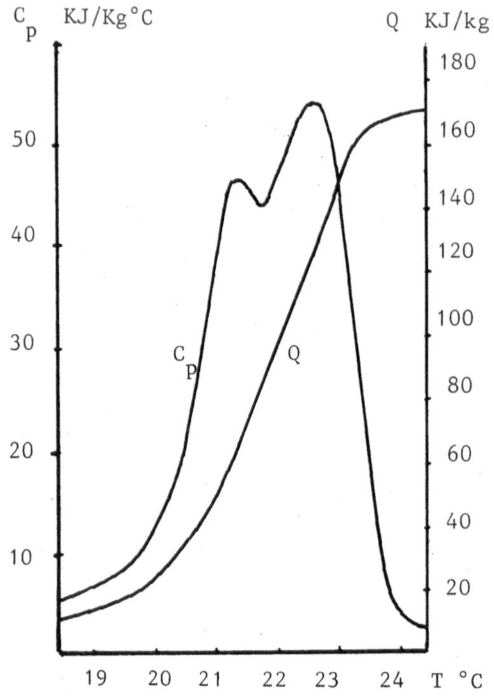

Fig.1 : Specific heat and total heat of fusion of the 20-24° compound .

3. La Baronne solar greenhouse.

The French experimental facility near Nice belongs to A. M. Chamber of Agriculture CREAT and is instrumented by CNRS. A four year experience has already provided precious informations about the micro-climate created by a closed air loop through a PCM underground storage. The original container geometry was maintained for most of the storage (10 T out of 11 T for 500 m²) : flat 1,5 dm³ complex plastic bags laid down on successive shelves in ventilated hemi-cylindrical ducts, as shown in Fig. 2 . Such bags characterized by a surface to volume ratio of 110 m2/m3 and a cost of 700 FF/m3 , are easy to make but offer little mechanical resistance : transport is a problem and the receiving structure is complicated and introduces thermal resistance on the back side . Upper surfaces cannot evacuate water condensates which are then responsible for air saturation at night.

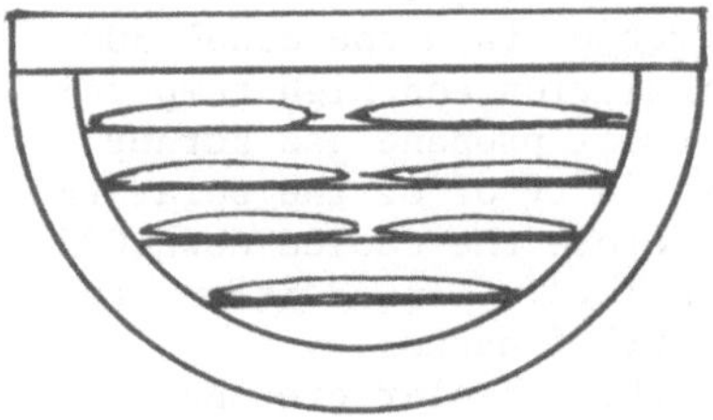

Fig.2 : Latent heat storage using plastic bags on shelves .

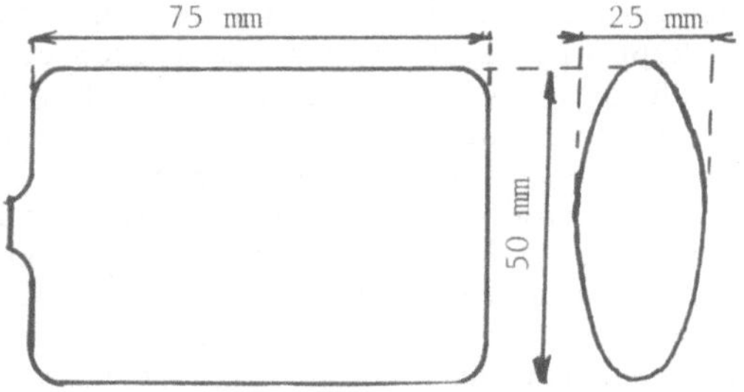

Fig.3 : Geometry of a latent heat pebble.

Still, for simplicity reasons, this configuration already tested for several was maintained for the major part of the storage, and was equiped with the new 20°-24°C compound . However, to prepare more practical applications, another type of container was devised : it makes use of small and rigid polyethylene bottles, with a flat but still convex shape (Fig.3)and a capacity of 60 cc . This choice combines a favorable surface to volume ratio (160 m2/m3) and a good mechanical resistance, so that it is possible to handle without caution the small containers and stack them like rock pebbles in any simple receiving volume . This way leads to the concept of latent heat pebble beds . Air heat exchange takes place on all sides, water condensates can be eliminated by simple trickling . At the present stage , the only limitation concerns the cost of the container and of the filling operation . Although the actual cost in this experiment was 8 times the cost of the bags, reasonable estimates should put it around 3000 FF/m3 .13000 PCM pebbles were made and placed in one of the 8 storage tunnels : they represent a 47 kWh storage capacity under an apparent volume of 1,5 m3 and offering an exchange surface of 125 m2 .

The closed air loop used for solar heat recovery and night heating is sketched in Fig.4.Eight400 W fans, situated along the north wall, extract warm and humid air at a 10.000 m3/h rate, under sunny conditions . Air looses sen-sible and latent heat on its way through the storage and escapes on the south side for a new cycle, by pushing horizontal membranes . When the sun goes down, these fans stop operating . At night, monitored by a thermostat, a single 600 W fan situated on the south side starts extracting air from the storage , the membranes being closed by depression , at a reduced 2000 m3/h rate. and through the exchangers of a heat pump . When the relative humidity becomes excessive, which happens soon, the heat pump comes into play and extracts water from the air : air from the storage is thus dehumidified and over-heated before being sent back to the greenhouse . the benefits derived from this heating mode are twofold : i) sanitary problems arising from an excess of humidity at night are reduced (the heat pump can condense 5 to 10 Kg of water per hour with the action of a 3 KW compressor); ii) the thermal balance of the greenhouse is improved by the reduction of water condensation on the cover and by the direct contribution from the compressor (60 kWh per night) .

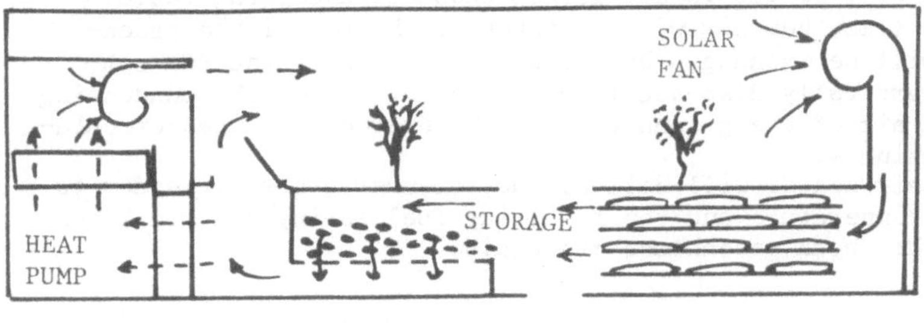

Fig.4 :Day time (full arrows) and night time (dotted arrows)operation modes for the solar air system .

Because photosynthesis consumes carbon dioxyde, the solar greenhouse should not be kept air-tight unless one introduces artificial CO_2; in that case it is then highly advisable to keep a higher concentration than the usual 300 ppm level, to boost photosynthesis under strong radiation and high temperature conditions . This has been done here with a small propane gaz burner (adjustable from 5 to 10 Kcal/h), placed in the vicinity of of the solar fan intakes : hot gazes are thus sent to the storage before the cooled down CO_2 gets distributed to the plants at a 500 tà 900 ppm level . In addition, this simple burner can be used as back-up heater at night if necessary .

The present experiment in La Baronne has shown that a solar greenhouse can be made self-sufficient in our southern latitude with the only help of the small heat pump and the CO_2 generator, in spite of the inevitable defects that are encountered in a first trial of this nature : too large head losses through the new storage network, responsible for a reduced air flow rate, various air leaks on the north wall and along the storage tunnels, and a heat pump that had to be adapted to non standart climate conditions . The data recorded from january to the end of march 1984, under relatively cold weather conditions are summarized in Table I .In january and february, some back-up heat was necessary, but in a weak amount,compatible with the use of the CO_2 generator only; the greenhouse was kept 10°C higher than the outside with the following daily contributions:

Stored solar energy : 250 kWh
(sensible heat:45%;condensation:55%)
Electricity: fans 25 kWh (day)
Heat pump 70 kWh (night)
Gaz (propane): CO_2 40 kWh (day)
Back-up 100 kWh (night)
Total : 485 kWh per day for 500 m2 .

In march, a mean night temperature of 11, °C was kept at night without back-up heat .

These figures will be improved by future arrangements in the air system .

	January	February	March	
T_{ext}	2,5	1,7	4,0 °C	NIGHT
T_{in}	12,0	11,3	11,6 °C	NIGHT
$T_{stor.\,in}$	22°80%HR	24°80%HR	25°70%HR	DAY
$T_{stor.\,out}$	17°90%	18°90%	18°90%	DAY

Table I

Average temperature conditions observed in 1984 in La Baronne

4. <u>Solar greenhouse at Rosignano.</u>

Two 200 m^2 solar greenhouses have just been built in the SOLVAY's works at Rosignano, near Livorno in Italy.

These greenhouses are designed for ornamental plants (Philodendrum sp.,...) nursing on tables (four tables in each greenhouse) and are normally heated by conventional fuel burners when the temperature falls below 16 °C.

One of the greenhouses will be used as a check nursery.

The other will be fitted out with a phase change material heat storage : 4000 brick shaped polyethylene containers (thickness : 3 cm, capacity : 600 cm^3 or 1 kg PCM) will be put in air forced channels under the tables (Fig. 5). During day time, hot air will be taken at the top of the greenhouse and the heat will be exchanged by forced convection on the PCM containers, which are vertically disposed in order to get rid of the condensing water. At night, the air of the greenhouse will be heated by forced circulation on the PCM containers.

The conventional air heater will deliver the balance of heat in order to maintain the preset temperature and the resulting fuel saving will be known by comparison with the check greenhouse run beside.

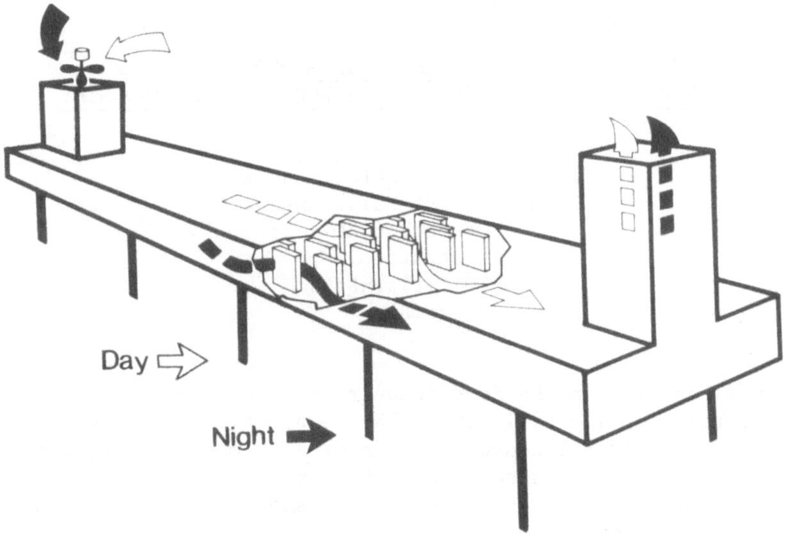

Fig. 5 : PCM containers in array
in air forced channels under the tables.

The equipment of the experimental solar greenhouse is now in construction and the trials will begin in the next winter season.

REFERENCES :

1 - CEA - INRA Solar greenhouses, described for exemple in the Proceedings of the International Solar Congres "Utilisation de la chaleur solaire dans l'industrie et l'agriculture" Nice, Sept 79; A. Fourcy et al.

2 - Latent heat applied to horticulture; A.Jaffrin & P. Cadier; Solar Energy Vol.28,N°4 , 313.

3 - K.K. Meisingset & F. Grøvold, Univ. of Oslo, Communication to the Conf. "Solar energy benefits evaluated technics and results"; Bermingham,Sept.82.

4 - European Patent N° 19573 from ANVAR .

TWO PHASE FLOW GUIDANCE IN A STEAM PRODUCING SOLAR COLLECTOR FOR PROCESS HEAT SUPPLY FOR THE FOOD INDUSTRY

V. Heinzel
Institut für Reaktortechnik
Universität Karlsruhe
Postfach 6380
Kaiserstr. 12
D-7500 Karlsruhe 1
Germany

Summary

Steam is applied in the food industry for heating purpose in the temperature range of 120 till 150°C. The evaporation directly within the collector and natural convection render a primary circuit with a pump and a heat exchanger unnecessary. Flat plate collectors as well as single tube collectors were considered in respect of mastering the two-phase flow natural convection.

The aim at a large height difference between heat source and cold leg in a circuit with one-phase flow leads in case of two phase flow in the riser to geyser like pulsation. In case of a flat plate collector this pulsation can be suppressed, if an orifice in the steam exit causes a slight balancing overpressure or the steam separating drum is installed close to the upper header. If the upper header acts already as an internal steam drum, no pulsations occur.

In case of a single tube collector even with horizontally arranged boiler tubes continuous boiling is achieved, if a properly designed annex takes care on the steam separation and the recirculation of the remaining water.

1. Introduction

Mainly in the food industry production branches consume steam for heating purpose in the temperature range of 120 to 150°C. Condensing steam allows very high heat flow rates. Steam is not poisonous, so that leakages do not harm the food. Often devices for food sterilization are in operation for a few hours daily only. They match to some extent the course of the sun and do not require day storages.

Some solar plant are already installed or suggested for the aforesaid temperature range. They use a closed circuit with e.g. an oil transferring the heat from the solar collector to an heat exchanger. The fluid in the collector circuit is circulated by a pump. On the secondary side of the heat exchanger water is evaporated and the steam guided to the consumer.

In case of the above mentioned applications is seem more convenient, to produce the steam directly within the collector and to eliminate the heat exchanger. In addition natural convection could render the circulation pump abundant. However, the natural convection with two phase flow has to be mastered.

Therefore, the demonstration of stable convection was aimed at for a flat plate like collector as well as for a single tube receiver. In respect of a equator near site or for a east-west orientation of a collector with a booster horizontal boiler tube arrangements had to be considered, too.

2. Flat plate collector

The most known instability of the two-phase flow is the steam blockage, following a rapid pressure loss increase owing to steam production in a channel. As counteraction the friction pressure loss has to be increased. In case of a one-phase flow natural convection the flow rate or the surmontable pressure loss increase with growing height difference of the heat source and the cooler.

A flat plate collector was assumed with several parallel heater or boiler tubes. The boiler tubes are provided with water from a lower header. An upper header collects the steam-water mixture. Through a riser tube the two-phase flow is guided to a steam drum, where the steam separation takes place. The remaining water flows down through a downcomer to the lower header.

Buoyancy and pressure loss for the two-phase flow were calculated. It could be shown, that with a height difference of 0.5 m between upper header and steam drum no steam blockages would occur. For reasons of confirmation a glass modell with 4 parallel boiler tubes provided with electric heaters was build up (Fig. 1). Indeed steam blockages did not appear, but geyser like pulsations with regular periods were found. The periods till 50 sec were observed. They decreased with deminishing water level in the steam drum and boiler tube diameter. The steam pulses started with a rapid increase of the velocity (Fig. 2). The accelerations shaked the experimental device. At the end of the pulses all tubes are streamed through with the colder water of the drum. Then a reheating phase followed and the velocity went down to zero.

At the end of the reheating phase in the upper header or the riser a first steam bubble occurs. The bubble reduces the pressure of the water column and new bubble is formed below. Within less than 1 sec the riser is emptied. The large buoyancy caused then the rapid accelerations. A computer program according to this theory brought corresponding results.

Therefore, an orifice was installed in the steam outlet of the drum. Its purpose was to build up an addition pressure increase due to the higher steam release during the pulse and to balance therewith the pressure loss due to the lacking water in the riser. At fully suppressed puslation the orifice caused an overpressure of 500 Pa.

An other solution was to lower the drum. Only for small boiler tubes 5 mm i.D. the flow became constant if the riser has an inclination of about 45° and the height difference between header and water level in the drum is below 150 mm.

Still superior was an enlarged header 60 mm i.D., which acted also as steam drum. If the boiler tubes were inclined with about 20° over a few centimeters befor entering the header, the receiver can be installed horizontally still operating with continuous boiling.

A collector with 12 parallel tubes and an internal header was manufactured and tested. The receiver had an area of 2 m². Two phase mirrors increased the aperture to 4 m². The collector released at noon time during several days steam at a continuous rate corresponding a power of 1.8 kW (Fig. 3).

3. Single tube collector

Single tube collectors inclined to the horizontal can be operated under boiling condition. If the tube is filled partially the upward flowing steam bubbles stir the water and take care, that the whole tube is wetted. A final steam separation takes place at the upper end of the tube.

In case the receiver tubes have to be installed horizontally, the steam water separation must be achieved in an annex. Either the tubes are connected to a common header then the experiences of the flat plate collector can be applied or each receiver tube has an individual annex. In both cases the receiver tube diameter has to be small enough, that at rated power a surge flow pattern will be reached.

The annex has then the purpose to separate the steam and to guide the remaining water and the returning condensate through an overflow and a recirculating tube back to the other end of the receiver. Overflow and condensate may not enter the receiver tube at the side of the steam exit. Otherwise the steam would condense in the cold water and cause cavitation. The recirculating tube can be installed coaxially in the receiver tube. Therewith addition heat loss area is avoided (Fig. 4). Furthermore such a design can be applied to vacuum collectors.

A collector with an annex shown in Fig. 4 was tested out-door. The receiver with 13 mm i.D. was covered by a selective foil (Maxorb) and housed in a glas tube. The later was vented through a silica dryer to the atmosphere. A parabolic mirror with an aperture of 0.46 m^2 concentrated the light. The collector produced steam continuously. The steam output rates reached a power of 175 W at noon time. The annex and part of the collector are shown in Fig. 5.

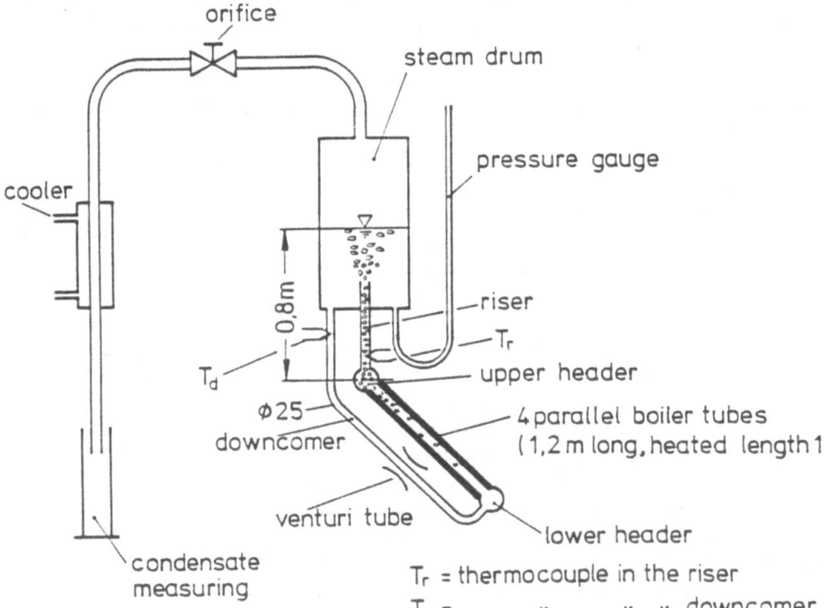

fig.1: modell with 4 parallel electricly heated boiler tubes simulating a flat plate collector

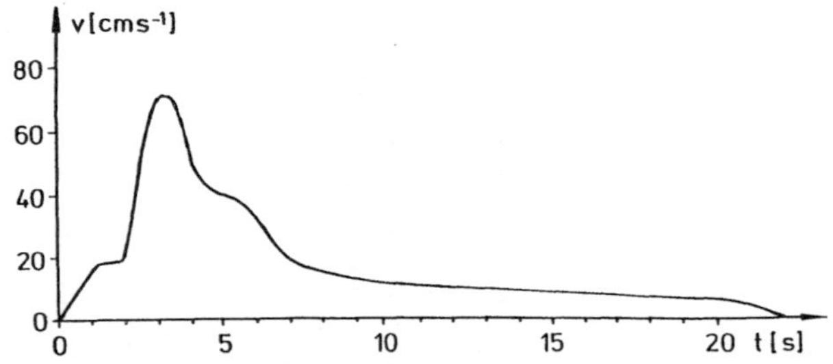

fig. 2 : velocity in the downcomer during a puls (heater power 520 W)

fig. 3 : flate plate collector with two plane mirrors having an aperture of $4 m^2$
steam release corresponding to 1,8 kW

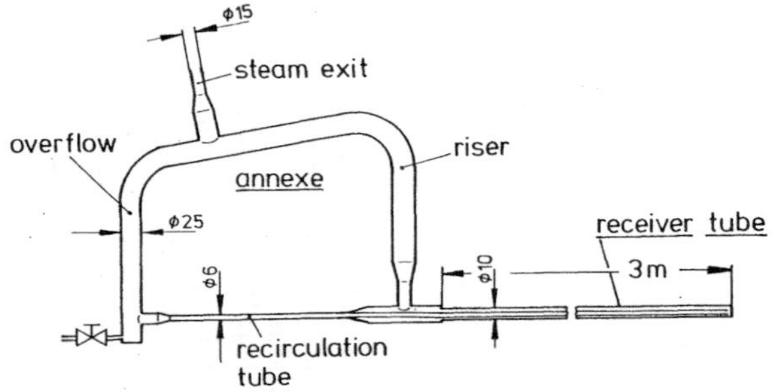

fig. 4 : single tube collector with an annexe for steam separation and recirculation

fig. 5 : annexe of the tested single tube collector

ASSESSMENT STUDY OF THE POTENTIAL USE OF SOLAR ENERGY IN AGRICULTURE

Prof. G. SCHEPENS
D. MAHY
Faculté des Sciences Economiques
Facultés Universitaires N-D. de la Paix
Rempart de la Vierge, 8
B-5000 NAMUR
BELGIUM

SUMMARY

Action 1 of project H of the E.C. Solar Energy R & D Programme aims at the Assessment Study of the applications of Solar Energy in agriculture and most promising related industries.
Its objective is to determine as precisely as possible the opportunities for the application of solar energy in each of the E.C. regions, which identification is based on climatic, agricultural and energy characteristics.
This assessment studies the important energy consuming sectors of agriculture : the greenhouses, the drying sector and the livestock sector; and evaluates the most promising solar development routes against current practice.
Opportunities, obstacles and ways to overcome them are investigated in order to give some prospects of the solar technologies penetration in the agricultural sector and emphasize the key areas where european R & D should focus on.
In addition to Action 1, Actions 2 and 3 inside the Project H are related to field experiments and synthetized separately.

Project running and organizational aspects

An expert group is working on the subject since the start of Project H.

From an organizational point of view, there is a participant to the research in each European country : the work of these experts is coordinated and synthetized by the Belgian team.
Each participant gives the coordination team the raw materials necessary to set up the synthesis at an European level and writes a short assessment study for his own country.

Here follows the composition of the working group :

Coordinators

Prof. G. SCHEPENS (see address at the top of the page)

Subcontractant for technical coordination

Prof. V. GOEDSEELS   Katholieke Universiteit Leuven   BELGIUM

Members of the working group

K. MEUREN	Bayerische Landesanstalt für Landtechnik	FEDERAL REPUBLIC OF GERMANY
M.G. AMSEN	Institut fur Vaeksthus Kulturer	DENMARK
D.W. ROBINSON	Kinsealy Research Centre	IRELAND
J.F. MOLLE	CEMAGREF	FRANCE
Sp. KYRITSIS	Institut Agronomique d'Athènes	GREECE
A. SCARPINI	RENAGRI	ITALY
PH. LEIJENDECKERS	RTB Van Heugten	THE NETHERLANDS
J.L. CARPENTER	Seale Hayne College	UNITED KINGDOM
G. SCHEPENS	Facultés Universitaires Notre-Dame de la Paix	BELGIUM

The European synthesis and the national assessment studies will be published shortly in the book series launched by the Commission of the European Community's Directorate General (XII) for Research, Science and Education, in cooperation with D. Reidel Publishing Company. It will be the Volume 1 of Series H. Volume 2 is now also under preparation and is concerned with the analysis of field experiments covered by Project H, i.e. operational data of solar systems in European Agriculture.
One will refer to these two publications for any information not content in this abstract.

1. CLIMATE, AGRICULTURE AND ENERGY CONTEXT

According to the climatic conditions (solar radiation for one, but also temperature and rainfalls), and the agricultural structure - specially its direct thermal energy consumption - of the different European regions, four main homogeneous areas have been identified, for which separate conclusions and recommendations can generally be drawn, regarding solar energy applications and their future prospects.

The major agricultural activities concerned are protected cultivation, animal rearing, crop drying and specific activities as fish farming or anaerobic digestion of agricultural waste.

Here follows a table showing the direct thermal energy consumption per country of the E.C. and per activity sector (in 10^3 t.o.e.).

Direct thermal energy consumption in E.C. agriculture (in 10^3 t.o.e.)

	Greenhouse heating	Livestock heating	Crop drying
B	381	91.9	7.4
FRG	890	186	136
Dk	265	18.1	10.3
F	750	399	802.7
UK	275	63.8	92
GR	26	16.3	12.6
I	140	350	140
IRL	62	50.1	45.4
L	-	2.2	0.7
NL	2 965	184.6	8.4
TOTAL	5 754	1 362	1 255.5
%	68.7	16.3	15

Of all thermal energy consuming sectors in E.C. agriculture, the greenhouse sector has the highest consumption (representing about 70 % of the direct thermal energy of E.C. agriculture). The share of the greenhouse sector in this energy consumption is generally higher (more than 3/4) in the northern countries than in the southern countries (where it reaches less than 1/3). The Netherlands is by far the heaviest energy consumer in the E.C. greenhouse heating, and by this fact in the E.C. agriculture.

Livestock rearing and crop drying represents each about 15 % of the direct thermal energy of E.C. agriculture, with a relatively high percentage for crop drying in France and for livestock rearing in France and Italy.

2. SOLAR ENERGY USE IN AGRICULTURE

Passive and active solar devices can be used in order to cover a part of these energy needs, or to permit new heating or cooling activities suitable for giving some products a better quality, and some animals a better comfort.

Solar greenhouses of passive or bioclimatic type, with separate collectors or with circulation on the roof have been reported as interesting for southern regions. On the other hand, energy conservation measures and full optimization of the cultivation as a whole are the main ways to save energy in the North.

Barn drying with low temperature air is one of the most promising application fields especially for the central region of Europe. Very simple low cost self-made air collectors are particularly suitable for this application.

Solar fruit drying in small family farms of the southern regions allows the improvement of both the yield and the quality of the crops, as well as their storage.

In the livestock sector, it is in the field of hot water production for calf rearing that solar energy shows the best prospects. Low cost solar air collectors could also help to condition the livestock housing in southern countries of the E.C.

THE PERFORMANCE MEASURING DATA OF A DHW AND STABLE HEATING SYSTEM

M. Reuss, K. Meuren, S. Vogt
Bayerische Landesanstalt fuer Landtechnik (BLL)
Voettingerstr. 36
D-8050 Freising-Weihenstephan
Federal Republic of Germany

SUMMARY

A lot of farms in germany have a high demand of heat at low temperatures (30° C - 50° C) for heating stables, feeding calves or domestic hot water, which could be provided by solar assisted heating systems. Two self-built low cost collectors (70 m², 54 m²) and a 6.000 l storage tank have been installed. A model collector measured on a test loop has an optical efficiency η_0=0.57 and a heat loss coefficient U_L=5.64 W/(m²K). In the first measuring period in autumn 1983 storage temperatures ranged between 38°C - 45°C. 18.5 % of the solar radiation were delivered to the storage tank, 7.5 % were used for stable heating and 7.7 % for domestic hot water. These poor values are caused by insufficient operation of the regulation unit, thermosyphonic circulations through the collector and an unintentional energy flow from the back-up-system to solar storage tank.

Nevertheless the solar covering in 1983 was 100 % in summer and 90 % in autumn. The total investment was about 24 000,-DM, the costs of maintenance are 250,- DM/a and the energy savings are 1.900,- DM/a.

1. INTRODUCTION

BLL is working on a projekt of measuring the performance of a simple solar water collector system for stable heating and domestic hot water, to work out a thermal and economical analysis and to improve the installation. This projekt is financed by the Comission of the European Communities (CEC Group H) and the german BMFT (Bundesministerium für Forschung und Technologie). All energy flows in the total plant should be monitored, the solar energy input to the storage tank, the contribution of the solar system to the stable heating system and to the domestic hot water, and the fuel consumption of the backup system as well as the electricity input to operate the plant.

2. DESCRIPTION OF THE PLANT

Picture 1 shows a scheme of the plant. The solar system consists of two simple self-built solar collectors (70 m², 54 m²) and a 6 000 l storage tank. Both absorbers are made out of polypropylen plastic tubes (25 mm diameter). Collector I has 15 mm wood fibre boards and 40 mm styropor as insulation at the backside, Collector II has wood fibre boards too but 40 mm polyurethan-foamboards as a insulation. Both collectors have a double glazing; the inner one is a heat resistent (to 140°C) polyester-foil and the outer one partial a high transmissive PVC plate, (to 80°C) partial a polycarbonate plate (to 140°C).

The storage tank is made out of steel and has a 80 mm thick insulation of cork. A 50 m long copperpipe heat exchanger is incorporated in the tank for heating the water used in the stable. The storage water circulates directly through the low temperature floor and radiator heating system. The domestic hot water in the living house is heated via a heat exchanger in the boiler.

The collector circuit does not contain an anti-freeze-mixture, only water is used. Therefore a freezing protection is necessary. If ambient temperature falls below 3°C the collectors are emptied automatically via a temperature controlled valve and refilled by the circulation pump.

In the living house an electric resistance heater in the boiler is used as backup system, the water heating for stable use and the stable heating are besides solar provided by an oil burner.

3. MEASURING EQUIPMENT

In summer 1983 a data acquisition system has been installed. Picture 1 gives an overview of the total plant with the places, where the different sensors for measuring temperatures (T) and massflows ($\dot{m}$) are located. Also a set of meteorological parameters like global radiation in the collector plane, wind-speed and -direction and ambient temperature is registered.

For temperature measurement two by two selected Pt 100 sensors are used. They are connected with a 1 mA constant current source. The linearisation of the Pt 100 resistors is made by software.

The water massflows are measured by hot water flow meters, which give 1 pulse for 1.0 l water.

For a longterm data acquisition a PC-100 microcomputer (Siemens) with an A/D-Converter, a scanner and computerinterface from Analog-Device (μMac-4000) is used. All the temperature and meteorological sensors are scanned continuously, the pulses are counted if there are any. Over a registration interval of 15 minutes these values are integrated or averaged and stored on magnetic tape for further evaluation.

4. RESULTS OF MEASUREMENTS

A model collector (2 m²) of the installed type was measured on a testing loop. An optical efficiency of $_0$=0.57 and a heat loss coefficient of U_L=5.64 W/(m²K) was found out. An efficiency line is drawn in picture 4. A detailed data acquisition started at the beginning of September. Picture 2 shows the curve of solar radiation, storage temperature and efficiencies of the two collector fields from 7.9.1983.

From the data of a first measureing period of 6 weeks the daily stored energy is related to the daily insolated energy. This graph is shown in picture 3.

Picture 5 gives a scheme of the energy distribution to the different consumers on several days in September and October 1983. It could be seen, that on sunny days the demand of the stable heating system and of the domestic hot water system could be provided by solar. The energy flow diagramm in picture 6 shows the distribution over one month. The 14 308 kWh of the insolated energy are set to 100 %. The optical and thermal losses are 81.5 %. This high value is caused by the low density of absorber tubes per square meter, 124 m² of collector area contain about 55 m² of absorber. These values related to the absorber area only give 6 350 kWh insolated energy, 42 % stored energy and 58 % optical an thermal losses, 17.5 % are used for heating the farrowing house and 17.5 % for domestic hot water. In this case the heat losses of the storage are 7 %. Further problems occurred with the temperature difference control unit. Sometimes the pump operates during night and rainy days and so the storage tank was cooled via the collector. Also the backup system does not work only on the heating system of the farrowing house, it heats sometimes the solar storage tank too and the collector then works on a higher temperature level with poor efficiency.

The investment of the total solar system was about 24.000,- DM. The energy savings are about 1.900,- DM/a, the costs for maintenance 250,- DM/a-

5. CONCLUSIONS

This simple solar water heating collector is not of a high efficiency type. If could be improved by increasing the absorbertube density per unit area without extremely increasing the investment. Most of the problems that occurred are caused by incorrect operation of automatic valves or of the control unit. Also the regulation method of the stable heating system has to be improved.

This and other practical experience show that economic applications of the use of solar energy in agriculture are possible on the basis of low cost solar installation, but it is neccessary to be very careful in planing and performing. Also there is a lot of work to improve these systems and to introduce these experiences in further applications.

Acknowledgements: This research is sponsored by the Bundesministerium für Forschung und Technologie der Bundesrepublik Deutschland and the Commission of the European Communities.

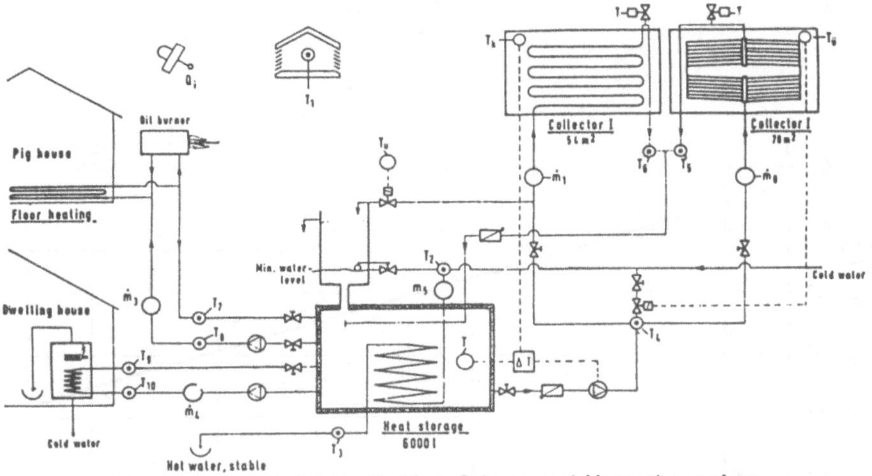

Solar Water-, and Stable-Heating, Scheme and Measuring-points

(Berger, Buchhofen)

LANDTECHNIK WEIHENSTEPHAN

Vogt 983/3a W

Picture 1:

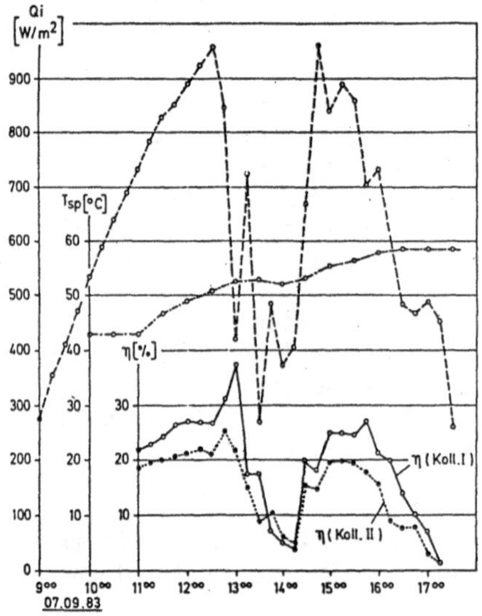

Solar Radiation (Qi), Storage Temperature (T_{sp}) and Collector Efficiencies (η) of 07.09.83

(Solar System Berger, Buchhofen)

Picture 2:

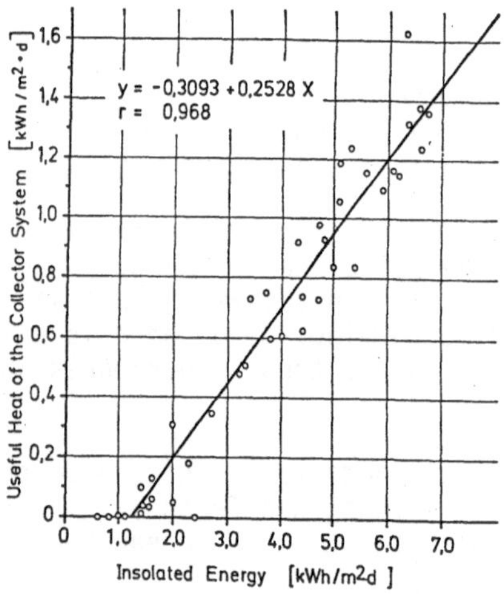

Useful Heat of the Collector System (124 m²)

(from 16.09.83 to 28.10.83, Solar System Berger, Buchhofen)

Picture 3:

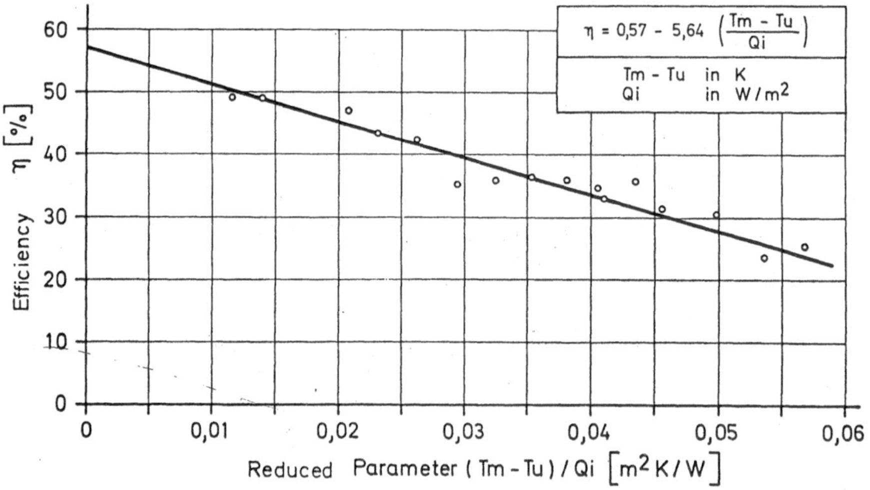

Efficiency Line of a Serpentine Plastic Tube Collector (2 m²)

Picture 4:

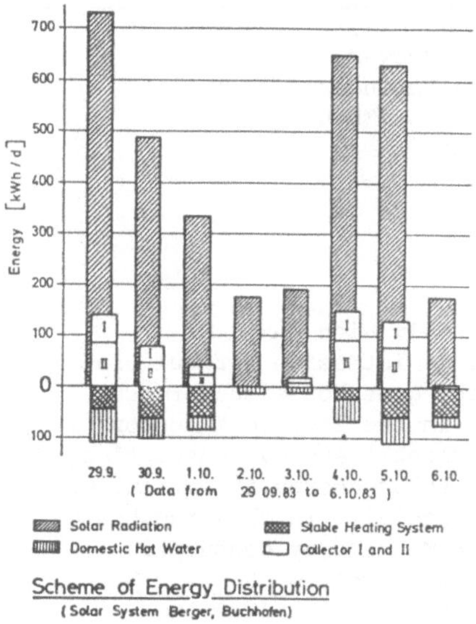
Energy [kWh / d]
700
600
500
400
300
200
100
0
100
29.9. 30.9. 1.10. 2.10. 3.10. 4.10. 5.10. 6.10.
(Data from 29.09.83 to 6.10.83)
Solar Radiation
Domestic Hot Water
Stable Heating System
Collector I and II
Scheme of Energy Distribution
(Solar System Berger, Buchhofen)

Picture 5

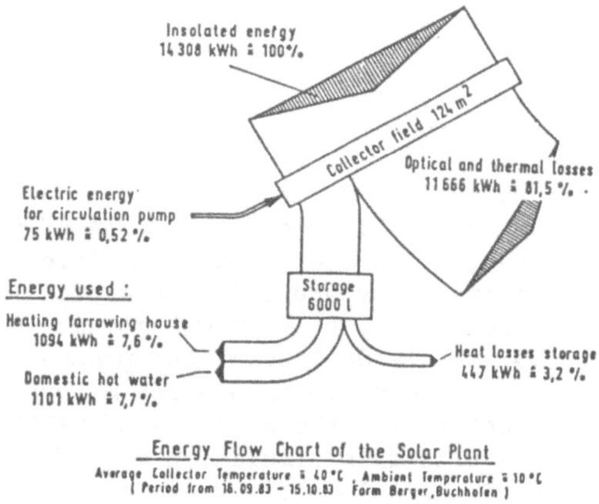
Insolated energy
14 308 kWh ≙ 100 %
Collector field 124 m²
Optical and thermal losses
11 666 kWh ≙ 81,5 %
Electric energy
for circulation pump
75 kWh ≙ 0,52 %
Energy used :
Storage
6000 l
Heating farrowing house
1094 kWh ≙ 7,6 %
Domestic hot water
1101 kWh ≙ 7,7 %
Heat losses storage
447 kWh ≙ 3,2 %
Energy Flow Chart of the Solar Plant

Picture 6

Picture 7: View of the collector fields

MODELLING AND MONITORING SOLAR ENERGY PROJECTS IN THE DUTCH AGRICULTURE

F.J.M. Lenting, G. Brouwer
Van Heugten Consulting Engineers
P.O. Box 305, 6500 AH NIJMEGEN
The Netherlands

Summary

Dutch farming is practised at an extremely high level of industrialization and is one of the most intensive agricultural systems in the world. Activities in the greenhouse sector (cut flowers, ornamental crop growing) and in the livestock sector (poultry, cattle, pigs) go with considerable energy consumption. The study (1) deals with the prospects of solar energy in agriculture and of passive insulation measures, especially in the greenhouse sector. For the analysis of DIRECT THERMAL energy consumption results from simulation models are used to rely on.
This article shortly explains the monitoring technique and performance test procedure of a solar energy project for hot water production in the livestock sector. Monitoring results are compared and discussed with results of the simulation models.

1. INTRODUCTION

Favourable factors for the introduction of solar technologies are the high standards of technical innovations in the highly industrialized Dutch farming and the profitable contribution of agriculture to the country's balance of payments.
Negative factors are the adverse climate and the availability of competitive energy resources in an extensive infrastructure.

The energy-consuming sectors are critically emphasized with respect to the amount of energy, the time of demand and the period, the heating system, the temperature and the type of fuel.

The greenhouse sector is the most consuming one : about 125 PJ in 1981 ; the animal husbanding asks for 7,7 PJ, directly followed by the sector related industries 7,3 PJ.

Based on the situation nowadays the substitution possibility of traditional energy systems by solar energy is investigated.
This analysis uses computerised simulation models and results in easy readable graphics for energy gain estimation, system-dimensioning and economic evaluation. One of them (2) is pointed out in this study.
No competitive energy saving methods are incorporated.

At this moment the use of solar energy is loosing out from an economical point of view in many applications. Insulation, the improvement of the efficiency of equipment, the use of waste heat (milkcooling, power plants) and the introduction of heatpumps are in most cases the first options. The economic feasibility of greenhouse insulating techniques is most attractive. Important prospects are offered by the sectors : production of hot water for animal husbandry (especially at calf fattening farms), fish farming (heated eelbasins) and low temperature heat demands in the food and drink industry.

2. MODELLING METHOD OF APPROACH

To start with, the general characteristics of the Dutch agriculture are critically analyzed in order to trace possibilities for solar energy use. The total energy savings with solar techniques in agriculture amount to about 30 PJ/a in the greenhouse sector and to about 5 PJ/a in the livestock sector. For a number of specific agricultural activities ; greenhouse horticulture, animal husbandry (cows, calves, pigs, poultry), computerized simulation models are developed. As a result from parameter studies by the simulation model energy gain estimation, system dimensioning and economic evaluation can be carried out by the use of easy readable graphics.

In general the investigation went as follows :
For each sector a scheme of the solar energy installation is defined. The analysis of the potential heat demand in holdings with the related physical parameters is incorporated in a computerized simulation model of the complete installation using the hourly weatherdata of a so called reference year.
The optimization process concerns many operational parameters. The economical feasibility, derived from the relationship between investment costs and heat gain, depends on the decreased surplus value of the solar heat gain. This economical analysis is executed for each sector.

3. THE MONITORING PROCESS OF SOLAR SYSTEMS

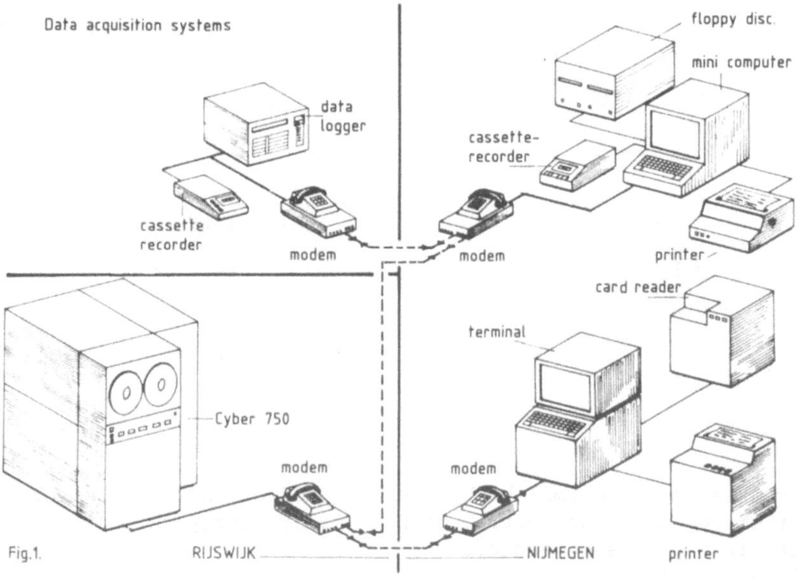

Fig.1.

The data of the projects are obtained with relatively sophisticated data processing systems (manufacturer Acurex type Autodata 10/5). The sensors are more simple, a solarimeter (Dirmhirn) in the collector plane, fluid (turbine) flow meters, temperature sensors (4 leads platinum resistance elements) and a wind velocity meter.

The scanning of analog ous and digital inputs, mostly every 20 seconds and the programmable arithmetic functions of the data acquisition system make it possible to obtain instantaneous and long term performance data of the solar system.
These analysed data are stored on a recording medium (cassette). Programming and transfer of data with data recording is also possible at the computer analysis centre by means of 2 modems and a telephone line ; the modem on site has an automatic answer function.

In this way more than one monitoring system can be controlled, programmed, changed and recorded at the computercentre of our company. This presentation is given in figure 1.

In general this monitoring process is used for the previous described projects. The measuring results are also described in the EC Reporting format.

In the datalogging system the complete Performance Data Sheet can be programmed.

4. THE PRODUCTION OF HOT WATER FOR ANIMAL HUSBANDRY

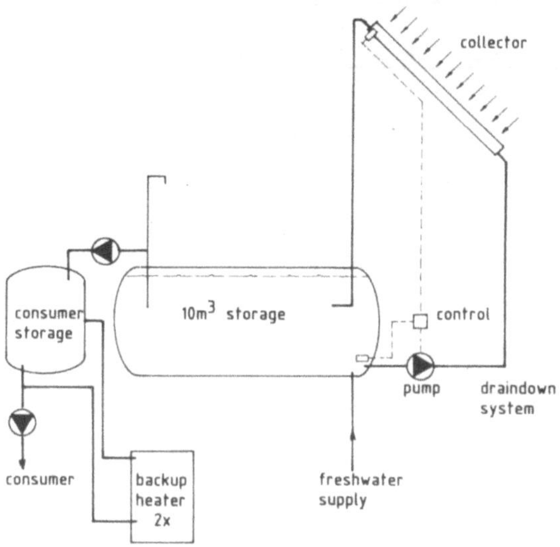

Fig.2.

In the sector of animal husbandry the production of hot water is an important energy use. The contribution of solar energy for hot water production is a promising sector.
The applications are :
* rinsing and cleaning of dairy- and fatting calve equipment ;
* stimulation and cleaning of udders ;
* preparation of feeds ;
* drinking water for cattle ;
* dissolving of milkpowders.

Growing calves need a great amount of hot water, about 5 ltr a calf/day, 80 °C. The hot water demand, mostly electrical heating, for milk-cows is 3 à 4 litr a cow/day. The required temperature is about 40 °C for drinking, the other applications ask for a temperature of about 60 - 70 °C (dissolving and preparation) and 80 °C (rinsing).
In pig holdings there is a need for warm water of about 30 °C in a so called closed holding, i.e. 4 ltr a pig/day.

An example of an actual project is a farm with 800 calfs, 120 m² collector area, 10 m³ storage tank and a hotwater demand of 2.000 ltr twice a day (80 °C).
The scheme of the installation is given in figure 2.
The calculated and predicted heat gains are presented in the diagram of figure 3. Over a whole year the heat gain of the solar system is about 33% of the heatdemand (= 143 GJ).

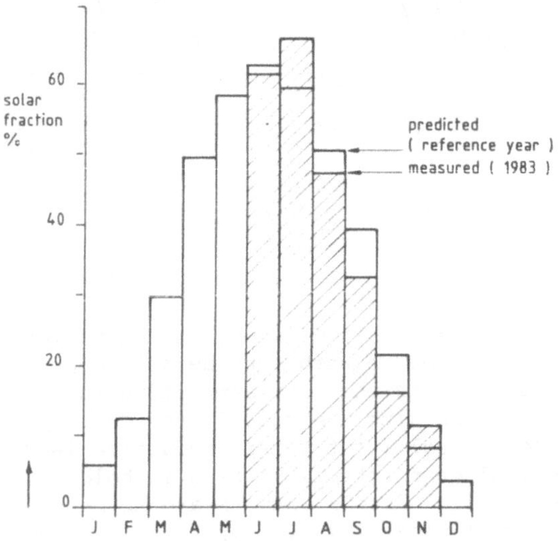

Fig.3.

For this agricultural application of solar hot water systems (80 °C) the following design rules are considered :

- a single glazed collector with a spectral selective coating on the absorber is preferred
- the storage capacity equals once to twice the daily hotwater demand
- the optimal collector area (spectr. sel.) equals : $0{,}025 \times Q \times \sqrt[3]{\frac{I}{Q}}$ ± 25% ; Q = daily hotwater demand (l/day) ; I = storage capacity (l)
- the flow rate through the collector equals 25-50 l/h.m²
- the optimal oriëntation is south, the tilt is 35 - 50 °
- direct heating of the consumed water is preferred (no heat exchanger). A drain-down system protects the system against frost damage.

Some results of a parameterstudy are given in figure 4.

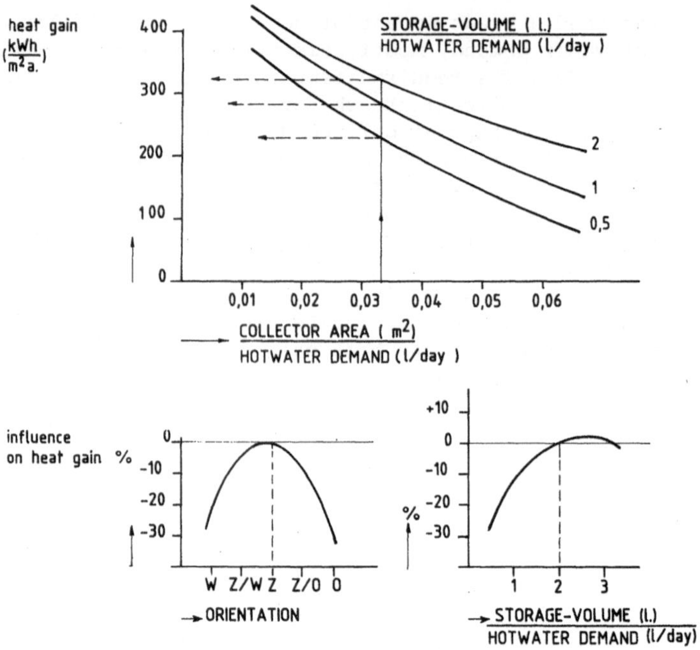

Fig.4.

5. CONCLUSIONS

The highly industrialized and intensive activities in the greenhouse and in the livestock sector go with considerable energy consumption. In greenhouses insulation is the most attractive energy conservation measure ; considerable results are scored, new developments are emerging. In the livestock sector solar hot water production leads to favourable results. The same applies to stable heating in holdings for pig rearing and poultry fattening. Favourable results are to be expected if the farmer's skillfulness (installing s.e. systems) is encouraged, information is provided and results of the ongoing projects are reported. In general price attractive applications can be realized with large installations and with simple low performance collectors. New projects in the different branches of the Dutch agriculture have to be selected for introduction of s.e. systems. Results have to be validated in relation to predictions (simulation models) and actual prospects.

6. REFERENCES

1. G. Brouwer
Assessment study of the application of solar energy in agriculture and most promising related industries in the Netherlands.
Raadgevend Technies Buro van Heugten, febr. 1984.
2. F.J.M. Lenting, G. Latour, G. Brouwer
Systeem analyses voor aktieve benutting van zonnewarmte in land- en tuinbouw.
Raadgevend Technies Buro van Heugten, 1983/1984.

SOLAR RESTORATION OF A NEOCLASSICAL HOTEL IN ATHENS, GREECE

E. Stournas-Triantis*, M. Santamouris**
* University of Ioannina, Ioannina, Greece
** University of Patras, Patras, Greece

Summary

This paper briefly describes a project which combines restoration of hotel "Banghion", a classified neoclassical building at the center of Athens, with restructuring of the hotel's energy system by the use of passive and active solar energy techniques in addition to standard energy conservation measures. The project is therefore expected to provide a solution to two of the hotel's most acute problems at present: a) soaring energy costs, aggravated by a generous proportion of "common" spaces to be heated as well as the rooms, and b) descreasing occupation rates, due to the general ageing and outmodedness of the building and its services. The final solution retained was arrived at by a simulation process whereby nine different "solar roof" configurations were analysed and their performance comparatively evaluated according to selected design parameters. This project has won C.E.E. support as a demonstration project in the solar energy sector.

1. INTRODUCTION

The "Banghion" hotel, located at 18, Omonia square in Athens, Greece, is a historic building of neoclassical style built by the famous architect Ziller in 1880 (Fig. 1). It has been classified as a "work of art" by the Ministry of Culture (1975), and has been in continous operation as a hotel since the beginning-which makes in the oldest hotel in Greece still operating in its original building. Banghion has 54 rooms, a restaurant for 45 persons and very generous circulation and living spaces, organized around a central courtyard (Fig. 2, 3). These spaces are very little used at the moment because of a general lack of excitement and outmodedness of their style and function.

The whole of the building is at the point of needing a general restoration for esthetic and technical reasons. Included in these reasons is the energy problem of the hotel, which has become quite serious in the last few years as it uses an important quantity of conventional fuel. Also serious is the constantly decreasing occupation rate which is accompanied by a falling rate of profits. The main objective of this operation is to regenerate most of the common spaces to their real function by the use of the principles of passive and active solar systems which can at the same time solve the hotel's energy problems and be easily integrated to its style.

Fig. 1. View of Banghion hotel from Omonia square

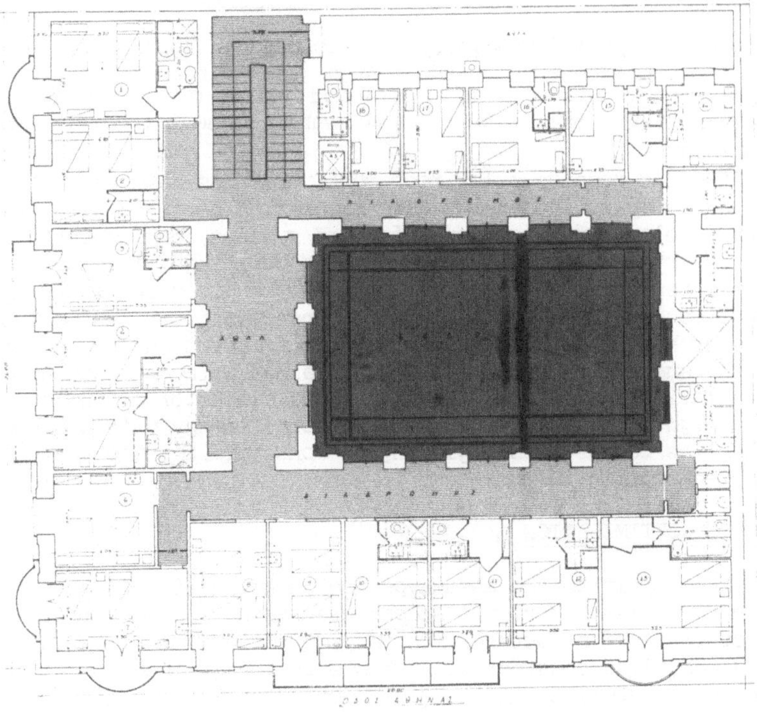

Fig. 2 First floor plan

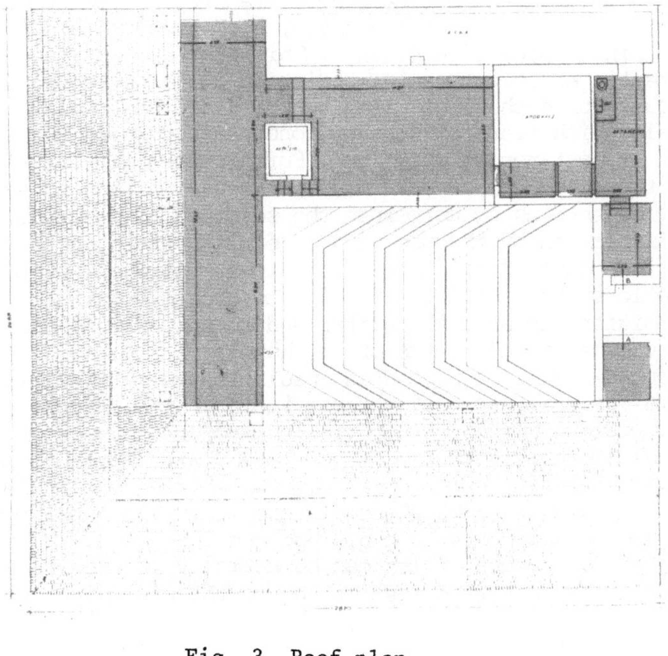

Fig. 3 Roof plan

DESCRIPTION OF THE PROJECT - THE SIMULATION PROCESS

The project's operating principle is to use the "greenhouse effect" in order to collect solar energy at the central courtyard space of the hotel, which will be covered by a specially designed "solar roof" construction, and transfer it from there, mostly by natural means, to heat the surrounding building structure.

Nine different passive solar roof configurations have been studied in order to obtain comparative energy gain figures. A detailed simulation program has been developed for each one of these systems, taking into account the shading effects caused by the surrounding buildings (1). The system examined and the results obtained are graphically exposed in figure 4.

A comparison of these systems' performance can prove that a "shed-type" roof assures highest energy gains during winter months and lowest energy collection in the summer. Results have been obtained with and without presence of night insulation and strongly favor its use since for a single - glazed construction night insulation can increase total energy gain by about 15% in the winter. As far as double glazing is concerned its use seems not to be a very efficient option in this case (2).

FINAL DESIGN OPTIMISATION

In order to optimize the shed-type roof construction to be used for the final design, a series of further steps were taken: 1) The orientation of some parts of the roof structure with respect to early morning and late afternoon shading was first considered. A change in the orientation of the West part of the roof system, turned to face South-East instead of South resulted, with a similar change in the East part, turned to face South-West instead. (fig. 5a) 2) The effect of shading from one "shed-unit" to the next had to be examined as well. It was decided to slightly lower each subsequent shed unit towards the South in order to maximise the area exposed to solar radiation (fig. 5b). 3) In order to increase heat transfer by radiation from the upper to the lower part of the atrium space, a continuous metallic plate was placed along the inner side of each unit. This plate, of a width of 1,3 m. approx., is directly exposed to solar radiation all along each unit and is extended vertically on each side, along the East and West walls forming long "radiator" panels which transfer heat by conduction down to the first floor level, where it is most needed (fig. 5a). Further warm air circulation and ventilation of the space can be obtained by the use of fans as needed.

As far summer shading is concerned the solution adopted was to combine it with night insulation by the use of a moveable curtain on the inside of each shed unit that could fulfill both functions. The whole system will operate with automatic control devices in order to archieve optimal shading and ventilation conditions both in the winter and in the summer, and increase interior space comfort in all parts of the hotel.

Special emphasis will be placed on insulation and double glazing for most hotel rooms as well, especially those facing north. For general energy conservation purposes a special study will be made of the possibility to renovate the classical heating system of the hotel, with specific changes in order to increase its efficiency. This will not mean replacing the whole system by a new one. Insulation will be placed on the inside of hotel rooms on the North and West sides in order not to disrupt the facade. The interior finish will be of special panels, which can be normally painted or wall-papered. As far as hot water is concerned most of the southfacing part of the interior

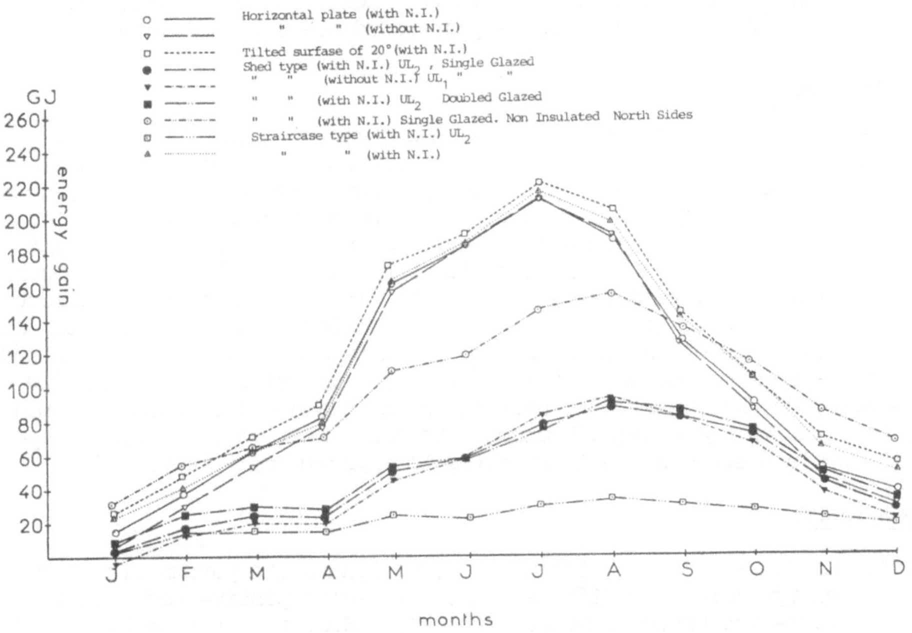

Fig. 4. Comparative evaluation of 9 solar roof systems

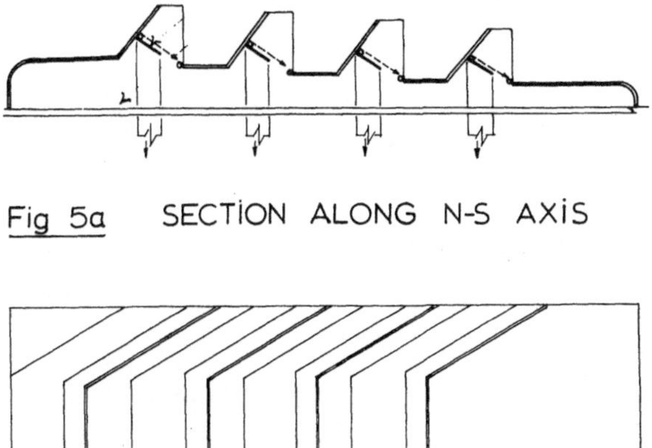

Fig 5a SECTION ALONG N-S AXIS

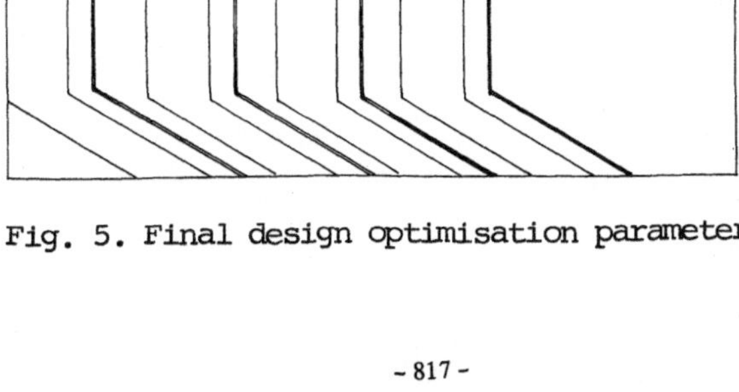

Fig. 5. Final design optimisation parameters

side of the roof (towards the courtyard) will be covered by evacuated solar collectors which will be integrated to it. These collectors will provide for an average of 70-75% of the hotel's year-round needs in hot water.

CONCLUSIONS

This is one of the first efforts in Greece to include solar energy systems in the restoration process of a historic building and it has a pioneering value as such, especially so if the project surmounts some bureaucratic difficulties and passes on to construction and monitoring stages as designed. Although it is quite rare in present-day Athens for a historic building to be not only in good enough condition to be still in use but also in such a position that very little overshadowing from its tall neighbouring structures occurs, this project, if realised, can have obvious applications to other historic buildings of the internal courtyard type as well as to their modern equivalent. Some of the advantages of the systems utilised here are that they are simple, durable, easy to construct and maintain and independent from façade orientation and design. Economic prospects seem good as payback period for the whole system has been estimated to approximately 10 years.

REFERENCES

1. M. Santamouris, E. Triantis, M. Vallindras:"A parametric study of energy gain from different proposed solar passive roof configurations for Banghion hotel, Athens", E.E.C. report (1983).
2. R. Rigopoulos, M. Santamouris, C. Stambolis, M. Valindras:"Windows in Greece-possibilities for solar direct gain utilisation (1984).

SOLAR ENERGY STORAGE FOR ZOOTECHNIC APPLICATIONS

G. RESTUCCIA, S. FRENI and G. CACCIOLA
Istituto CNR di Ricerche sui Metodi e Processi Chimici per la Trasformazione e l'Accumulo dell'Energia - Messina, Italy.

Summary

The main aim of this work is to make an economic evaluation of solar systems applied to zootechnic field. As a typical application, a cattle-breeding farm was choosen, with the energy requirements as following: stable heating during the cold season, heating of washing water and drying of hay during harvest-time. By means of a computer program simulating the hourly operative conditions of the whole system: solar source, storage, thermal load, it was possible to dimension the plant components in order to minimize energy costs. Results are referred to a plant situated in two different locations in Italy, Po Valley (Milan) and Sicily (Messina).

1. INTRODUCTION

In this paper the economics of a solar system, provided with a new storage system, applied to agriculture, have been studied. A cattle-breeding farm heat requirement has been choosen as thermal load since it needs low temperature heat (40-70°C) and exergy at low level. From a thermodynamic point of view, it is non convenient in this case, to burn oil for the load supply, because the heat obtained would be at high temperature and exergy. In this way, efficiencies drop down and the cost of energy goes up. Thus the use of alternative energy sources in agriculture appears to be economically convenient.

2. HEAT REQUIRED FOR CATTLE-BREEDING FARM

The energy requirements for a typical Italian cattle-breeding farm have been considered in this study. It is necessary to supply hot water for stable cleaning and heating, the washing of cow udders and calves' food preparation.

The total consumption has been estimated as a function of the following unitary parameters (1):

a) washing of cow udders: 5 l/day of water at 40°C per animal;
b) stable cleaning: 250 l/day of water at 70°C;
c) food preparation: 7 l/day of water at 40°C per animal;
d) other uses: 300 l/day of water at 60°C.

Further, such a farm must be heated, at 18-20°C, during the cold season. It is useful, also, to use the summer solar energy to dry hay. The total hay consumption is about 300 t/year. All the characteristics taken into consideration are reported in Table I.

3. SOLAR ENERGY RADIATION

This study has been developed for two different locations: Milan (45° lat.) and Messina (38° lat.), respectively in the north and south of Italy. Using available data of the monthly average of daily radiation, the hourly radiation available on flat-plate solar collectors tilted at 50°C and 40°C respectively, has been calculated by a computer program.

A stochastic method with the following assumptions has been used:

- the relationship between the index of the monthly average value of cloudiness and the distribution of its daily values in the course of a month has been utilized to obtain the daily solar radiation;
- the same method has been carried out to obtain the daily hours of sunlight from the monthly average value;
- the hourly value of solar radiation has been assumed constant throughout the day;
- the ambient air temperature has been considered constant from 8.00 a.m. to 8.00 p.m. (diurnal average temperature), from 8.00 p.m. to 8.00 a.m. (night average temperature).

4. SOLAR AND STORAGE SYSTEM

Flat-collectors have been considered coupled with a storage system based on the hydration-dehydration process of an ammonium alum and ammonium nitrate mixture (2); the above mentioned storage system gives the following advantages:

a) stability of the salts in the operative conditions;
b) cheap price of the salts (about 0.15 $/kg);
c) high energy storage density ($9.5 \cdot 10^{-2}$ kwh/kg).

During the charge mode, the storage system runs at 60°C; in the discharge mode the temperature is almost the same.

The hourly heat losses percentage rate, in this case, has been calculated in relation to the storage system characteristics, by the formula:

$$P = 3.38 \times 10^{-2} \left(\frac{1}{E_a}\right)^{1/3} \cdot 100$$

where E_a is the storage capacity (kcal/hr).

5. CALCULATION METHOD

As input data the developed program needs: hourly solar radiation incident on square meter of collector area (calculated by the program described above), working temperature, efficiency of solar collector, hourly behaviour of the thermal load, energy losses of the storage tank. The program develops the energy balances of the whole system, starting from a prefixed value of solar collector area and from an energy storage capacity, taking into account the energy losses and the energy supplied by an auxiliary burner. The program runs iteratively changing the prefixed values till the optimum size of the plant for its economic suitability is found.

The output parameters are:

- solar collector area
- hourly behaviour of the stored energy
- the energy entering into and exiting from the storage system
- the amount of energy that must be supplied by a conventional boiler
- the cost of the energy supplied to the user
- actual cost of the energy in the next twenty years.

6. RESULTS AND CONCLUSIONS

Knowing the requirements of the described utilizer and the metereological data of the two localities (Milan and Messina), different cases have been considered changing both the solar collectors area and the capacity of the storage system. For both locations the best case has been defined as that, which gives the shortest "time-to get the break-even point" (NOTE 1). The best results are:

a) Location:	Messina. Annual energy required	: 167 Mwh/year
	Collector area	: 140 m^2
	Storage capacity	: 500 Kwh
	Annual solar energy utilizable	: 132 Mwh/year
	Percentage of integration by traditional energy	: 21.5%
	Storage system losses	: 825 Kwh/year
	Time to get the break-even point	: 3.5 years
b) Location:	Milan. Annual energy required	: 240 Mwh/year
	Collector area	: 200 m^2
	Storage capacity	: 800 Kwh
	Annual solar energy utilizable	: 139 Mwh/year
	Percentage of integration by traditional energy	: 41.9%
	Storage system losses	: 561 Kwh/year
	Time to get the break-even point	: 6.5 years

The energy demand of the utilizer and the energy inside the storage system are reported in Figure 1 as a time function for both cases described above.

Obviously, results obtained for Messina are better than those relative to Milan; but we can consider both the systems feasible from an economic point of view. As a last reflection, it must be noted that also in solar system applications it is important that the energy demand does not vary in such a way as to require a large storage system.

NOTE 1: "Time to get the break-even point" represents the time that must pass in order that the cost of the energy released by solar system becomes competitive with the cost of traditional energy.

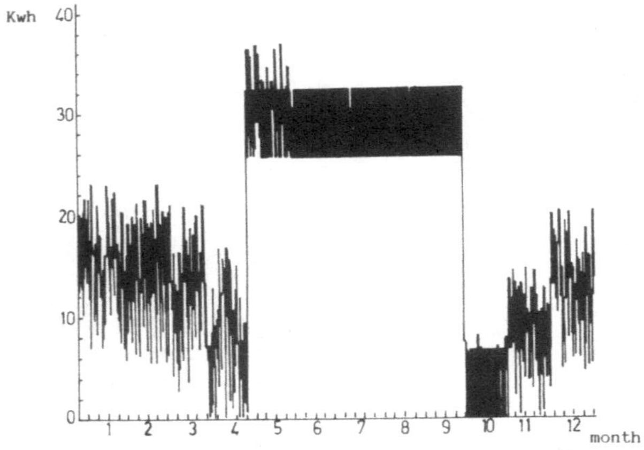

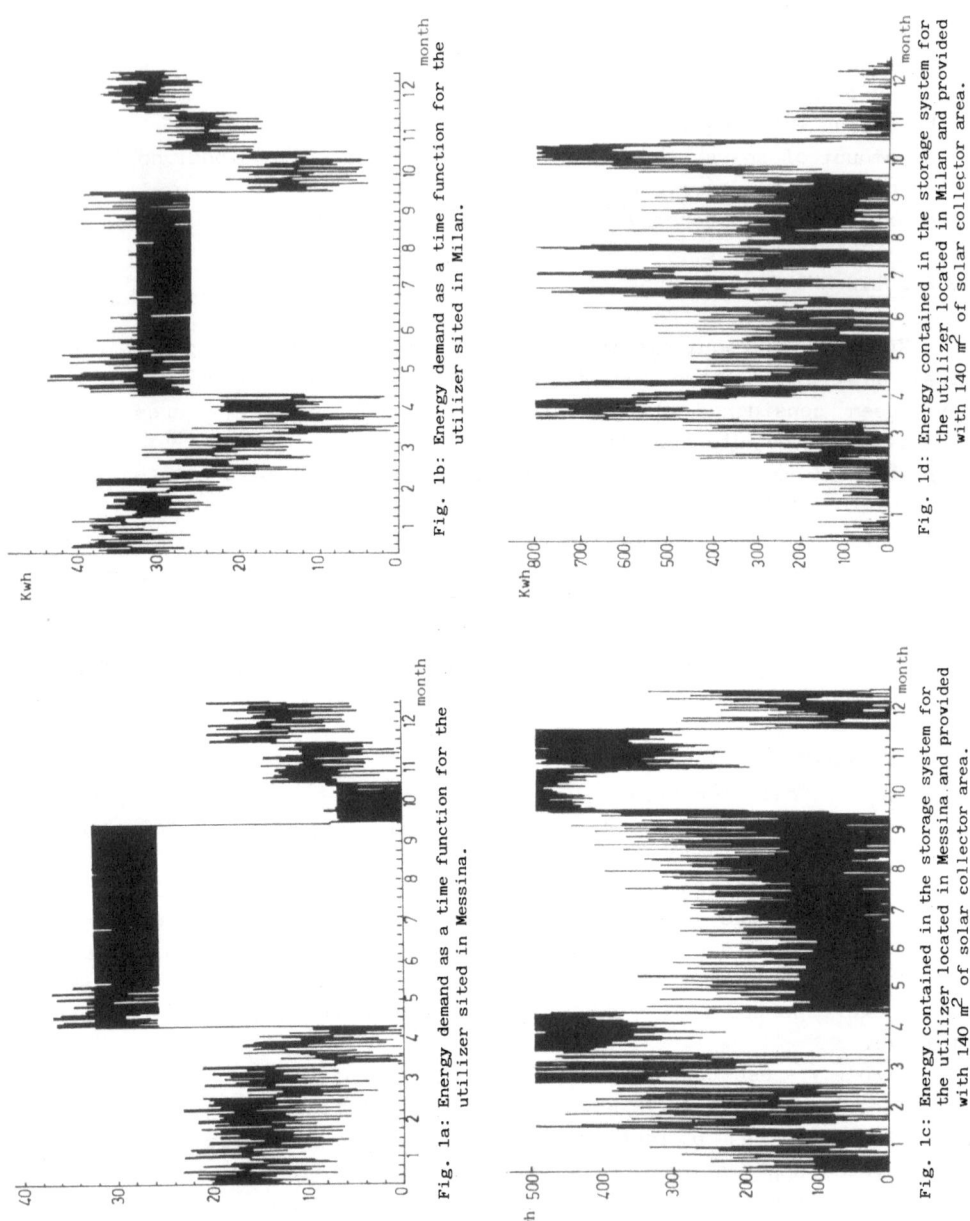

Fig. 1a: Energy demand as a time function for the utilizer sited in Messina.

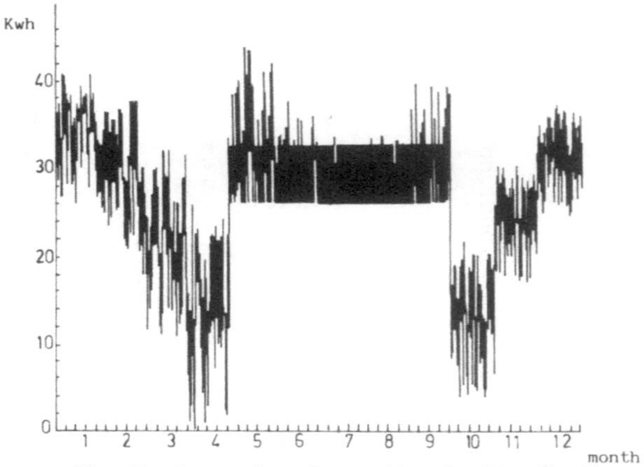

Fig. 1b: Energy demand as a time function for the utilizer sited in Milan.

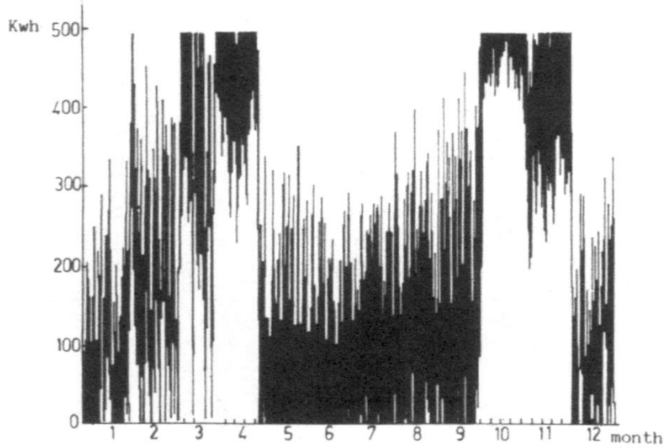

Fig. 1c: Energy contained in the storage system for the utilizer located in Messina and provided with 140 m^2 of solar collector area.

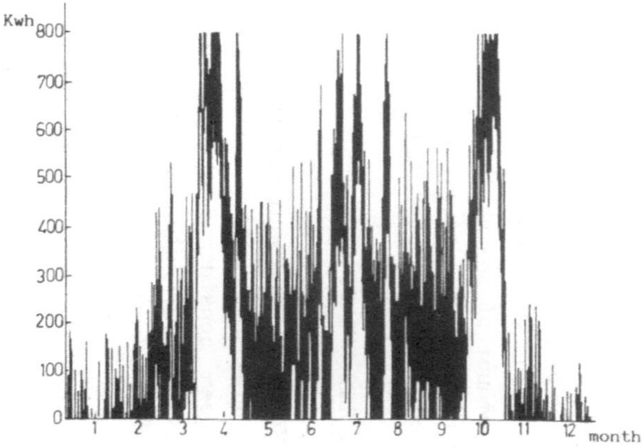

Fig. 1d: Energy contained in the storage system for the utilizer located in Milan and provided with 140 m^2 of solar collector area.

REFERENCES

1. BONDRIA, L., CASTELLI, G. and RIVA, G. "Energia solare in agricoltura: ipotesi di applicazione in una azienda zootecnica", Meccanica Agraria.
2. VACCARINO, C., CIMINO, G., FRUSTERI, F. and BARBACCIA, A. (in press). "A new system for heat storage utilizing salt hydrates". Solar Energy.

TABLE I - Characteristics of the Utilizer.

Stable dimensions:

Lenght	Width	Height
89 m	7.5 m	4.5 m

Energy requirements:

1. Stable heating: $Q = KS(T_i - T_e) = 1.786\,(20 - T_e)$ Kw
   $S = 1536\ m^2$ $T_i = 20$ C (prefixed)
   $K = 1.16 \cdot 10^{-3}$ Kw/m^2C
   (T_i= internal temperature, T_e = external temperature)
2. Water at 40 C: 18.5 kWh/day (530 l/day)
3. Water at 60 C: 17.5 kWh/day (300 l/day)
4. Water at 70 C: 17.4 kWh/day (250 l/day)
5. Hay drying : 81.7 kg/h of dry produce
   Q = 617 Kwh/day
   Period : from May to September
   Initial produce: hay with 50% of humidity
   Final produce : hay with 10% of humidity

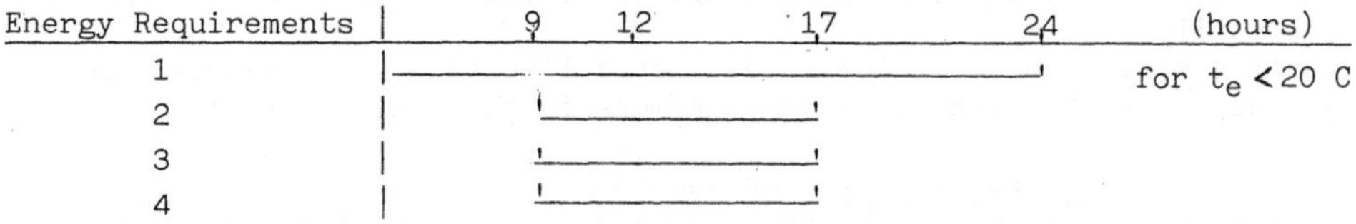

Time table of energy requirements for hot water and heating.

HEAT TRANSFER IN A SOLAR IN-DOOR COOKER

V. Heinzel, J. Holzinger
Universität Karlsruhe
Institut für Reaktortechnik
Postfach 6380
D-7500 Karlsruhe
Germany

Summary

For individual cooking an in-door arranged cooker is preferred. The heat transfer from the solar collector to the cooker has to be made by natural convection. For good heat transfer the cooker is integrated into the system as a double walled pot with the heat transfer medium in the annular. Heat transfer by steam evaporated in the solar receiver tube and condensed at the pot is superior to one phase convection. Experiments showed that the overpressure in the circuit is at boiling condition in the pot low enough, that the circuit can be depressurized for the removal of the pot. Air enclosures can be removed later on easily. A proper designed flow breaker and riser separate the steam and prevents the overflow of water from the boiler receiver to the cooker annular.

1. Posing the problem

A solar cooker with the cooking place in-door allows individual cooking. In order to match variable local conditions the collector and the cooker must be connected flexibly. Natural convection is assumed for heat transfer. In respect of a good efficiency the temperature gap between the collector and the food, to be heated up, should be low. Therefore, the cooking pot will be a part of the system, however, for food preparation and cleaning purpose the pot must be removable.

For power control an adjusting device is necessary. Therefore, a concentrating system is preferred, at which the mirror is movable and the receiver is fixed. The tracking effort should be limited. Therefore a low concentration factor seems to be reasonable. If an east-west orientation or a site near the equator is envisage the collector tube must be arranged horizontally.

As a variety of well designed solar collectors are available the major aspect to be aimed at with this work is the heat transfer.

2. Heat transfer between collector and cooker

The large surface of the pot can be utilized best for heat transfer area, if a double-walled pot is applied with the heat transfer medium passing the annular between the two walls. Two tubes connect the annular with the collector receiver. The installation of the cooker in a groundfloor will still have a height difference to the collector on the ambient ground of about 0.5 m. The natural convection has to work inspite of this small height.

Such a device was simulated with an electric heated receiver tube. All tubes were made from glass. They were enclosed by an insulation, but windows allowed to observe the circulation (Fig. 1). For the first approach the heated tube was inclined with about 10°. The electric resistance heater was attached outside on the tube. With a cooler in or a condenser above the pot the transferred heat could be measured, whereas the heater power was determined with the current measurement. Water was choosen as heat transfer medium. It is everywhere available and is not poisonous.

One phase flow convection is well self-stabilizing. However, the low achievable velocities lead to poor heat transfer coefficients. On the contrary the evaporation in the receiver tube and the heat transfer by steam condensing at the pot could utilize much better heat flow densities. But two-phase convection or boiling and condensation within on circuit is reported to cause pulsations.

2.1 Comparison of the one phase convection and the heat transfer by steam

The first experiments aimed at a comparison of a one phase convection with suppressed boiling and secondly the heat transfer by steam. As expected the one phase convection requires a higher circuit temperature for the equivalent transferred power (Fig. 2) entailing higher heat losses. At boiling condition in the pot and a power released at the receiver of 810 W the overpressure due to the system temperature reaches 0.1 MPa, whereas in case of the heat transfer by steam only 0.027 MPa is necessary. The later is in the range of a modern steam-pot. In addition the low water content in the boiler receiver allows a rapid depressurization, which renders the disconnection and the removal of the pot possible.

0.5 l of boiling water in the pot was replaced by the same quantity but with 20°C. Then the one-phase convection need 482 sec with 810 W of the receiver heater to regain boiling. With the steam transfer boiling was reached again after 258 sec.

Hence the heat transfer shows some remarkable advantages. This is still valid, if it is regarded, that the figure and given numbers are related to the experimental device and will change with the transition to a real collector system.

2.2 Peculiarities of the heat transfer with steam

Steam bubbles when detaching from the heated wall or flowing upwards in an inclined tube stir the fluid. This enhances the heat transfer from the wall to the fluid. With lower power this advantage vanishes. In horizontal tubes the two layer flow may occur with a steam layer on top with a very poor heat exchange. The better surge flow should be aimed at by reducing the tube diameter or a higher steam flow rate.

Figure 3 shows the heat transfer following cold-water inputs to the pot. Accordingly the steam velocities came till 6 m/sec. Therefore, the steam should be separated in the riser soon after the boiler tube exit, so that no water will flow with high velocity through the tubes and hit the pot. Therefore, the diameter of the riser has to be enlarge short behind the receiver exit, so that the original slug flow coming from the boiler tube changes to a film flow, which is finally reversed in a downward flowing film. Steam separation and flow reversal are influenced by the factor of the diameter enlargement, the length of the enlarged tube section, the size of the boiler tube, the power and the steam velocity. An assessment can be made with the formulas for stable film flow. An inclination of the riser i.g. 45° improves the steam separation.

The water level in the system can be varied according to the distance between boiler tube exit and the flow breaker, the riser enlargement. A too short distance reduces the handling tolerance. However, a large water column above the boiler tube acts as cold water trap at start-up and causes cavitation. The optimum depends on the inclination and the diameter of the boiler tube.

After a removal and reinstallation of the pot, the circuit contents air. This air interrupts the heat flow from the boiler to the pot. On the other hand as the heat up-take goes on, steam is produced and causes an overpressure in the again closed system. With the aid of this overpressure the air can be pushed out through a venting valve in the downcomer.

The venting valve should be connected to the downcomer immediately below the pot. Therewith air stays in the remaining downcomer and restricts the heat losses there to the condensate flow. This is demonstrated by the plot of the heat transfer efficiency over the height of the air charge in the downcomer (Fig. 4).

3. Outdoor test

The double-walled pot used hitherto was connected to two single tube collectors. The receiver had boiler tubes from cooper of 3 m length and 15 mm i.D. The receiving area of the tubes were enlarged by lateral fins and covered with a selective foil. Vented glass tubes enclosed the receiver tubes. Each receiver had a parabolic booster mirror with an aperture of 1 m² concentrating the light 4 times.

On Sept. 18th with this cooker 1 l water could be heated up from 28°C till boiling within 12 minutes. The sky was misty at that time. The measurements will be continued.

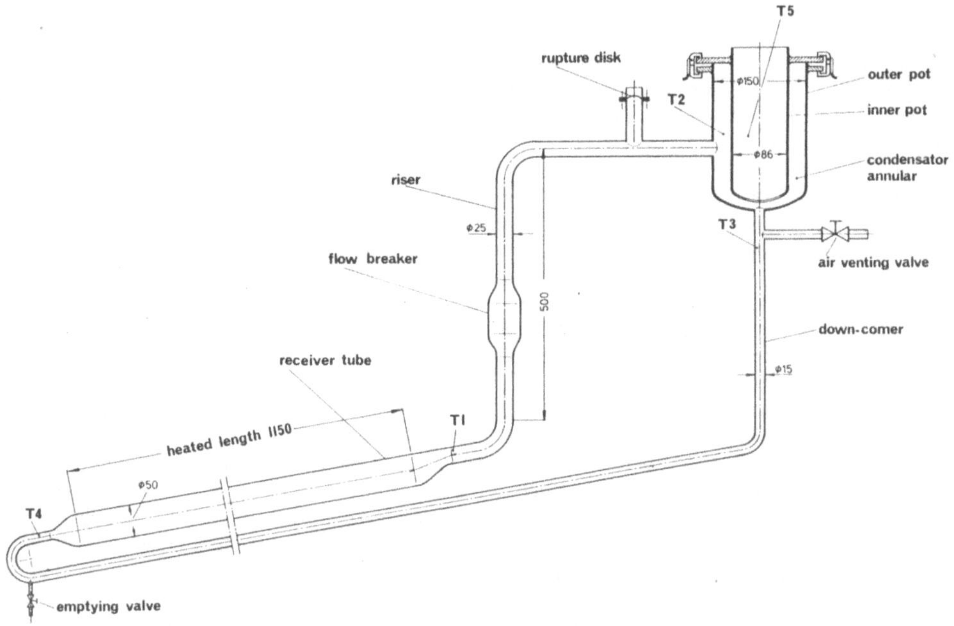

Fig.1: experimental device with electric heating simulating the solar receiver and the double-wall pot (T=thermocouple)

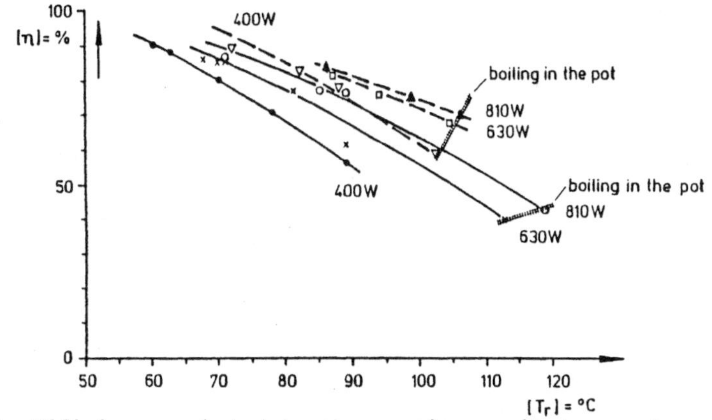

Fig. 2. Efficiency related to the receiver exit temperature

---steam transfer —— one phase flow

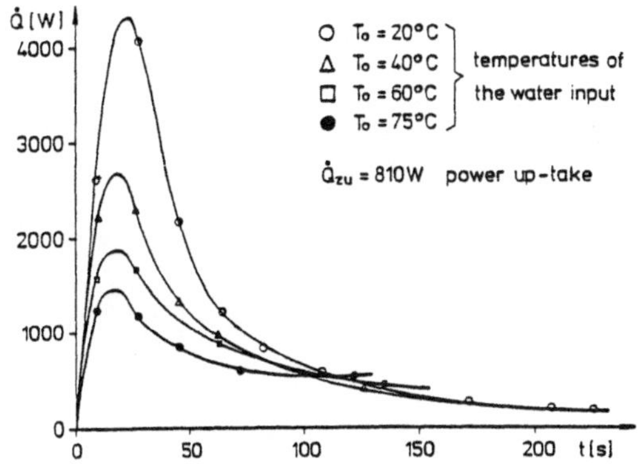

Fig. 3. Heat transfer following an input of 0,5l water with various temperatures

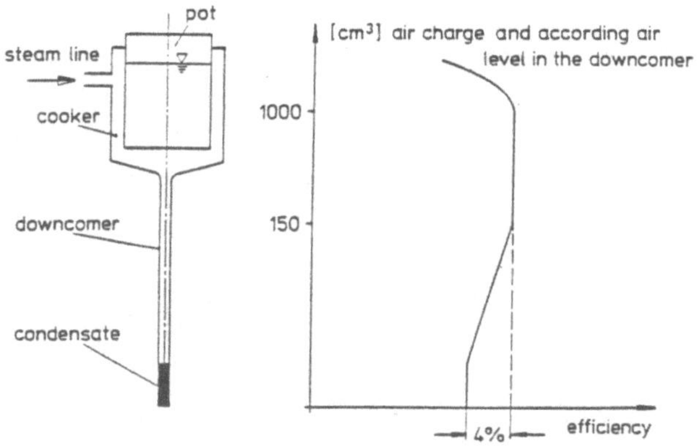

Fig. 4. Heat transfer efficiency depending on the air volume enclosed in the downcomer of the experimental device shown in

SESSION VII - SOLAR HEAT STORAGE, SHORT TERM AND INTERSEASONAL STORAGE CONCEPTS

The sodiumsulphide/water system as a chemical heat pump used for long term energy storage

A general method for determining the overall heat transfer coefficient of thermal storage embedded materials

Determination of the thermal behaviour of a phase change encapsulated material packed bed using air as transfer fluid

Solar heating of a large office building using seasonal aquifer thermal energy storage

Thermal losses analysis of an interseasonal heat storage in the soil: field trials and numerical models

Comparison between two different underground seasonal storage systems using a water vessel and soil heating tubes

Physical model for the study of mass and energy transfers in the non-saturated layer of soil located above a solar energy storage zone

Design and building of a test facility for thermal storage units between -30°C and +90°C

Evaluation of the annual heat losses of an in-ground heat storage

Sizing a latent heat storage with shell and tubes heat exchanger - A new simplified method

Optimal control strategy of solar heating systems using a long term heat storage

Expérimentation d'un système de chauffage utilisant pompe à chaleur et stockage de glace (résultats 82/83)

Experimentation of space heating by heliogeothermal doublet in Aulnay-sous-Bois

Transient heat conduction in a PCM heat storage with metal structures

Economic passive solar warm-air heating and ventilating system combined with short term storage within building components for residential houses

Experimentation on a long term rock-bed thermal storage

Project Wolfsburg-Glockenberg, IEA-task VII: central heating plants with seasonal storage

Study of the thermal behaviour in free temperature evolution of a passive solar inhabited house with high inertia and long period soil heat storage

Thermal storage in phase change materials - Application to a solar air space heating system

Energetic behaviour and economic evaluation of an interseasonal combined solar heat pump system for commercial building space conditioning

Design considerations for short term heat storage systems

THERMAL STORAGE OF SOLAR ENERGY

H. TABOR
The Scientific Research Foundation, Jerusalem, Israel

ABSTRACT

Thermal storage is needed to improve the efficiency and usefulness of solar thermal systems. The paper indicates the main storage techniques, namely, sensible and latent heat which are used for short-term storage. Long term i.e. seasonal storage - which would greatly increase the practical use of solar energy - is more difficult and, for thermal methods, appears limited to sensible-heat storage in water i.e. large tanks, solar ponds or aquifers, or in the ground. The paper stresses aspects of cost and required material characteristics but leaves the question of viability of long-term storage open.

1) PREFACE

Harnessing the sun viably is essentially an exercise in materials technology: the low energy density calls for large area collecting devices that must be low in cost and durable. Similarly the intermittency of solar radiation calls, in most practical cases, for some form of energy storage, again a question of low-cost and durable materials.

2) THE IMPORTANCE OF STORAGE

There are a small number of cases where the need for storing solar energy can be avoided. The first is where the solar energy results in a product that is easy to store - for example, fresh water produced by solar distillation. A second case of no storage needed is where the collected energy is used in real time, either without or with a back-up using conventional energy during periods of inadequate insolation.

It is now generally accepted by the majority of workers in the field that it is only very rarely that a solar device can be designed as a pure "stand alone" installation unless it is hopelessly over-designed. Thus where a firm energy output is mandatory, solar energy may save conventional energy sources - it does not eliminate their use.

For the case where there is no storage - and a substantially constant 24 hour load - the back-up will needs provide most of the energy: this is clear from the time-profile of solar radiation reaching the earth. If we consider a case with no back-up and no energy storage such as a solar pump, this need only operate during periods of insolation, the demand being by storage of the product i.e. water. Fig. 1 taken from Ref. 1 shows that, if a solar power unit without storage is required to pump the same amount of water as a unit operating at constant level for 24 hours a day, the pump, solar heat-engine, piping, etc. all have to be rated at about six times the rating of the continuous unit: and this for clear-day operation: over a year the factor might be of the order of 10:1. Furthermore, the heat-engine, when operating, would do so at widely varying loads, which would result in reduced efficiency. Similarly, Ref. 2 dealing with solar refrigerators, makes the same point: the storage should be on the <u>input</u> side.

These arguments support the contention - expressed many times in the past - that, apart from the need to reduce the capital and maintenance costs of solar devices, the key to wider use of solar energy is a satisfactory and viable energy-storage technology.

The amount of storage capacity needed in any specific application is, in the final analysis, a matter of economics.

3) COST CONSIDERATIONS

We note four principal cost items: (i) the heat storage material itself; (ii) the packaging; (iii) the value of the space occupied by the storage system (which could have been used for other purposes); (iv) the cost and complications of heat exchange between the heat storage system and the heat-transfer fluid (usually water or air) to be heated or cooled. Whilst attention is usually paid to item (i), the other items are equally - and sometimes more - significant. This has to be borne in mind in the final choice of a heat-storage system. For static systems, operating and maintenance (O and M) costs will generally be negligible, but for some dynamic systems, involving pumps, valves, controllers, etc., the O and M costs cannot be ignored.

4) CHARACTERISTICS OF HEAT-STORAGE MATERIALS

The major characteristics to be considered are: (i) heat capacity per unit mass or volume; (ii) complete reversibility; (iii) kinetics; (iv) packaging and corrosivity; (v) hazards - fire and toxicity.

(i) Heat capacity per unit mass or volume. The mass per unit of heat capacity is generally not important (except as it relates to cost). The volume may be important not only because of the value of the space occupied but because, in some cases, the space is simply not available.

(ii) Complete reversibility. This is the ability to maintain full heat capacity after a large number of heating and cooling cycles: some otherwise promising materials for heat-storage have to be rejected because of poor reversibility.

(iii) Kinetics. This refers primarily to the speed with which heat can be added or withdrawn from the material. A low speed may be due to low thermal diffusivity or to slow crystal growth where crystallisation is part of the heat-storage process. A low speed means that more surface area per unit volume of material is needed and this affects the form and cost of the packaging.

Supercooling, which is an undesirable feature that occurs in many liquid-solid systems, can be classified as a kinetic or as a non-reversibility characteristic.

(iv) Packaging. The packaging has two main functions: to contain the heat-storage material and - in many cases - to provide the heat-transfer surface between the heat-storage material and the heat-transfer fluid.

Packaging can be the most critical item in the system, particularly if the heat-storage material has corrosive properties, or varies substantially in volume during thermal cycling.

(v) Hazards. Materials used for domestic applications may be disallowed by the authorities if they represent a serious fire hazard or are toxic.

5) FORMS OF HEAT STORAGE

Four principal forms are (i) sensible heat; (ii) phase change; (iii) heat of solution; (iv) heat of reaction. Most of the effort to date has been concerned with the first two.

(i) Sensible heat. This is the simplest system, though it is not isothermal.

No physical or chemical change occurs in the storage material and the amount of heat stored Q_S is given by the simple equation:

$$Q_S = c_p \int m.d(\Delta T) = c_p M. \overline{\Delta T} \quad \ldots\ldots\ldots\ldots\ldots \quad (1)$$

where the total mass M is divided into parts, each with its own ΔT i.e. the difference between the storage temperature and the lowest useful temperature.

The first form of equation (1) is used when stratification occurs i.e. different parts of the storage medium are at different temperatures. Stratification can result in improved efficiency of the solar collector system, a fact that has been recognised from the earliest thermosyphonic solar water installations using a vertical tank. Stratified storage can also result in lower pumping energy. Prof. Van Koppen and his colleagues have carried the principle of stratification one step further by injecting into the tank the incoming fluid at a level corresponding to its own temperature.

Below 100°C (and occasionally above 100°C where pressure vessels may be used), water is the most likely storage medium as it can also function as the heat-transfer fluid when used with flat-plate collectors for domestic heating and cooling installations. ΔT i.e. temperature swing above the design output temperature would probably have to be limited to 20°C, leading to a heat-storage capacity of 20 kcal/kg or 20 kcal/litre. It is unlikely that any other liquid could compete with water - for top temperatures below 100°C - but the cost of large non-corroding containers is not negligible, particularly in cases where the water is under pressure. (5).

Rockbed or rock-pile storage is a second example of sensible-heat storage usually with air as the heat-transfer medium, (though oils - for improved heat-transfer - have been proposed as the medium for operation at higher temperatures). Here the rocks are stationary i.e. mixing, as in a water tank, cannot occur so that the stratification is less affected by thermal diffusion than in a water tank. Rockbed storage has proven of special interest in house-heating applications because the large containers are easier to construct than large water tanks and air is frequently the heat-transfer medium <u>in</u> the house. However, air is a poorer heat-transfer medium than water: this means that there is a temperature penalty in the collector. Also, for the same storage capacity for a domestic application, the rockbed volume will be about three times that of a corresponding water storage system.

TABLE I compares properties of four sensible-heat storage materials for a T of 20°C. Unfortunately, there seems to be little chance of finding new substances with significantly higher heat capacities than presently known materials.

(ii) Phase-change systems. The variation of temperature in sensible heat storage systems and the relatively large storage volumes needed has led to the interest in phase-change systems.

(a) Solid-liquid systems (heat of fusion). Goldstein (3) considered 300 common inorganic substances, with melting points in the range 30 - 200°C. The heat capacities were all in the range of 9 - 100 kcals/litre: for organic substances, the maximum was 60 kcals/litre, and he concluded, on theoretical grounds, that there was little reason to believe that some new substances would be found with vastly larger capacities per litre. Thus whilst the capacities are a few times that of water used over a limited ΔT of, say, 20°C, many of these materials are so much more expensive than water that they are unlikely to be viable. That they store at a nearly constant temperature is an advantage, particularly where the heat is needed to operate a thermal machine.

Salt hydrates. When heated, these melt to form two phases i.e. a solid phase comprising a lower hydrate - or the anhydrous salt - and a liquid phase comprising a saturated solution of salt in water.

TABLE II shows that the heat capacities of several common hydrates at the melting point is of the order of 160 kcal/litre*, or eight times that of water for $\Delta T = 20^oC$.

The last of these materials, Glauber salts, has been the subject of much study - because of its low cost. But the reversibility is poor - and the studies at the University of Pennsylvania Energy Center suggested that further work on this material be dropped.

Recently, as an alternative to the salt hydrates, materials such as paraffin waxes have been proposed(9): these do not melt at a sharp temperature but the range of 5 - 10°C is often no handicap.

(b) Solid-solid systems. Some solids undergo a reversible transition when heated. Examples are V_2O_4 at 72°C with H = 50 kcal/litre (the material is far too expensive) and F_eS at 138°C, with H = 55 kcal/litre.

Note that all systems involving non-metalic solids require large heat-transfer areas because of poor thermal diffusivity. This means that the storage material must be of small unit size: for solid-liquid systems the materials must be in small packages - to provide large surface area, be non-corrodable, be properly sealed to avoid loss of liquid or vapour, and contain voids or be elastic to allow for volume changes.

(c) Liquid-vapour systems. Latent heats of vaporisation are usually much higher than of fusion but the volume of the vapour is so large that the system is impractical. An alternative is to absorb the vapour in an absorbant. This leads to the 'two-chamber' concept, rather like an absorption refrigerator or heat-pump, with a "hot" chamber, and a "cold" chamber held at substantially ambient temperature. Each chamber contains an absorbant (which may be a volatile liquid or a solid) and is chosen so that the vapour pressure, at a given temperature, is much lower in the hot chamber than in the cold. When the hot chamber is heated, the vapour pressure of the "refrigerant" rises so that it distills over from the hot to the cold side, where it discharges its heat of condensation (and some heat of absorption) to the atmosphere. When the hot side gets cooler, the vapour moves in the opposite direction, condenses in the hot side, heat being liberated. Since the heat to generate the vapour in the cold side came from the environment, we have, in fact, used the environment as the energy store. ** Reasonable storage capacities are possible (Ref. 3) but some of the earlier suggested materials were corrosive or caustic. Some research is underway with more passive materials.

(iii) and (iv) Heats of solution and heats of reaction. These are not very promising areas though there is a basic advantage, in an all liquid system, that heat transfer is facilitated.

---

* Excluding any sensible heat-storage if a temperature swing is allowed.

** If the two vessels are isolated by a valve, and the hot chamber allowed - with the loss of the sensible heat involved - to cool to ambient, there will be no further heat loss and the storage time can be infinitely long.

6) SHORT TERM STORAGE

We have indicated that, in general, back-up energy sources will be used whether a solar system has storage or not. Short-term storage here plays an additional technical function of reducing the wear-and-tear on, and the efficiency of, the auxiliary source.

The solar contribution to total energy needs can be considerably increased by increasing the energy storage capacity i.e. from hours to days, as is well illustrated in the works of the Wisconsin group on f-chart analysis, since the larger storage permits a larger solar collector system without the need to jettison excess heat except perhaps under very unusual conditions.

The storage method used is generally sensible heat; phase-change methods are usually more expensive but are used where space is at a premium. Auxiliary fuel-heating is virtually a necessity to reduce the storage capacity to reasonable proportions.

7) LONG-TERM STORAGE (seasonal)

This is the real challenge. One example: the quantity of energy used for the heating of buildings is very large but the load is in the winter and does not match the solar supply. Thus the motivation to solve the problem of seasonal storage is very great.

Considering phase-change storage and sensible-heat storage, the chances for a viable system of the first kind - for long term storage - are very slim. This arises from the fact that the storage material is only cycled once a year; the quantity needed is two orders of magnitude greater than is needed for diurnal storage, and it is very doubtful if sufficiently cheap materials can be found. For sensible-heat storage, two low-cost materials spring to mind i.e. water and earth and this is the direction in which most long-term thermal storage R & D has been going.

(a) Water. The cost of the container, referred to under short-term storage, can become exhorbitant for large systems if steel or equivalent tanks are considered. In the first place, such tanks cannot be pressurised (which would allow for a wider ΔT and hence less volume per kJ stored) and they cannot be made very deep due to hydrostatic pressure. Remembering the large volume needed, the capital cost of such tanks can rarely be justified, as shown by the studies at Julich on district heating systems (Ref. 5). For individual residences, the volumes are still large and in Canada and elsewhere, such reservoirs under buildings have been proposed to provide seasonal storage. The numbers are rather frightening. For a residence normally burning, say, 5 tons of fuel oil a year - at a combustion efficiency of 80% - this is a load of 40×10^6 kcal per year. * If 60% of this is to be provided by stored solar energy from the summer - with 40% collected during the winter months (these percentages will vary considerably with the climate), this requires a store capacity of 24×10^6 kcals or 800 m^3 of water cycled over a T of 30°C: if the depth of the storage tank were limited to 2 m, its floor area would be 400 m^2. (By comparison, the area of solar collectors needed would be of the order of 60 m^2 !)

In certain cases, natural lakes or ponds could be used - to save the cost of constructing the containers - but great care would need to be taken to assure an impermeable bottom and that the ground does not act as a heat sink. It would also

---

* Assuming a fuel calorific value of 10,000 kcal/kg.

be necessary to cover the lake with a tight, insulated cover to prevent heat losses by evaporation and to the ambient. To avoid the problem of the cover, proposals have been made to use natural aquifers. (10)

One interesting approach to long-term storage of solar energy using water (as the storage medium) is that of the non-convecting solar pond i.e. a large uncovered mass of water used both as the solar collector and the storage system. To keep the surface cool, a temperature gradient must be imposed. Several methods have been proposed, such as dividing the pond into layers by means of a series of horizontal transparent membranes, or by imposing a density gradient by means of a dissolved salt concentration that increases with depth (the "salt-gradient solar pond" - SGSP) or by having the pond as a saturated solution of a salt chosen to have greatly increased solubility with increase in temperature (the "saturated" pond). To date, only the SGSP has been effectively demonstrated in the field and its storage capabilities described (6), (7): as a storage device, all the alternatives behave more or less alike.

The SGSP essentially involves three superimposed layers or zones of water: the top zone, 10 - 40 cm deep, of uniform density and temperature which is mixed by wind forces; a non-convecting zone usually about 1 m deep in which a salt gradient exists, and a bottom mixed zone from which heat is extracted. The non-convecting layer acts as a thermal insulator to the bottom layer where solar radiation is absorbed*, this bottom layer providing the bulk of the storage. Ref. 6 describes the storage capabilities of solar ponds: short term storage is excellent even with a storage zone of 20 cm or more. But even long-term storage is possible by increasing the depth of the storage zone to a few metres. Ref. 7 shows that, in addition to the storage capacity of the mass of the water in the mixed storage zone, the ground under the pond, if not insulated, contributes approximately the equivalent of another 1 m depth of storage, as does the non-convecting layer, which stores heat by diffusion. The cost of seasonal storage is the cost of the extra salt solution in the storage zone plus the cost of building higher walls to the pond.

(b) Earth. The ground has been used as a heat-source for heat-pumps. It can also be used for sensible heat storage but a fundamental problem is the slow rate of diffusion of heat into and out of the ground. Thus whilst the ground may be free, the large surface-area pipes or heat-exchangers needed may make the economic viability questionable. The combination of storage in water <u>and</u> earth i.e. in aquifers (10), is more attractive because the quantity of water can be very large and in some cases, the contact area between water and earth is large. But the aquifer system has several drawbacks: (i) it is very site-sensitive (as, indeed, are solar ponds); (ii) not all the heat put in is recoverable: (iii) there is a considerable drop in temperature in the recovered water. Whilst (ii) and (iii) may not be serious for the storage of waste heat, they <u>are</u> serious for solar heating systems because unrecovered heat means larger collectors whilst a reduction of temperature calls for collectors operating at considerably higher temperatures than the final load requires: aquifers appear more encouraging for storing <u>cold</u> water i.e. as a source or sink for heat-pumps or heat engines.

---

* This refers to the visible part of the spectrum: the IR part, unfortunately, is absorbed in the first few cm of water and never reaches the storage zone.

One generalised comment concerning sensible-heat, long-term storage is that this can only really be considered for very large systems i.e. community heating rather than for individual houses, since the fractional heat loss from the store is proportional to $V^{2/9}$, where V is the volume of the store. For small systems, the losses can be considerable: increasing the store by a factor of 23 times only reduces the fractional loss by a factor of two.

In summary, viable long-term thermal storage is very much dependent upon local site conditions, and a more general solution to long-term storage will probably have to come from other technologies such as chemical or bio-conversion.

TABLE I

Material	Specific heat kcals/kgm	Density kgm/litre	Heat capacity kcals/litre for $\Delta T = 20^oC$	Relative to water
Water	1.00	1.0	20	100%
Concrete or stone	c. 0.2	c. 2.4	9.6	
with 40% voids			5.8	29%
Iron ore Fe_2O_3	0.148	5.3	15.7	
with 40% voids			9.4	47%
Scrap iron	0.112	7.85	17.6	
with 40% voids			10.6	53%

TABLE II

Material	Melting Point C	Density kg/litre	Heat of Fusion kcals/kgm	Heat Capacity kcals per litre
$CaCl_2 6H_20$	30	1.68	75	126
$Na_2 HP0_4 12H_20$	35	1.52	120	182
$Ca(NO_3)_2 4H_20$	51	1.82	90	164
$Na_2 SO_4 10H_20$	32	1.46	105	153

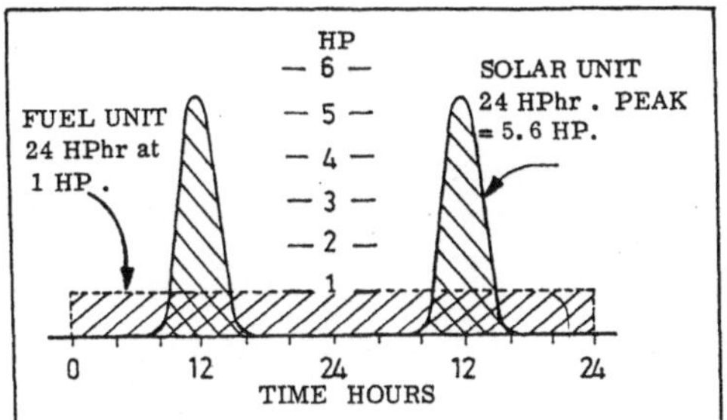

COMPARISON OF SOLAR POWER UNIT, WITHOUT INPUT HEAT STORAGE AND FUEL UNIT, BOTH PROVIDING 24 HPhr. OF OUTPUT PER DAY.

FIG: 1

REFERENCES

1. Tabor, H. "Solar Energy for Developing Regions" UNESCO (Working party on solar energy SC-73/conf 801/2 June 1973.
2. Tabor, H. "The Economics of Solar Refrigerators Derived from Thermodynamic Considerations" COMPLES Bull. No. 15 pp 17-25, 97-103 Dec. 1968
3. Goldstein, M. "Some Physical Chemical Aspects of Heat Storage" UN paper E/CONF 35 5/7 1961 (UN Publication 63-1.39 pp 411-17).
4. Kauffman, K and Chen Chi Pan "Thermal Energy Storage in Sodium Sulfate Decahydrate Mixtures" Report NSF/RANN/SE/GI/27976/TR72/11 (Univ. of Penna. Energy Center).

4a. Kauffman, K and Gruntfest, I. "Congruently Melting Materials for Thermal Energy Storage" Report NCEMP-20 (Nov. 1973) (Univ. of Penna. Energy Center).

4b. Kauffman, K and Chen Chi Pan "Congruently Melting Materials for Thermal Energy Storage in Air Conditioning" Report NSR/RANN/SE/GI/27976/TR73/5 (Univ. of Penna. Energy Center).

5. Scholz. "Seasonal Storage of Low Temperature Heat in Big Water Reservoirs" pp 221-232 in Ref. 8.
6. Tabor, H. and Weinberger, Z "Non-Convecting Solar Ponds" Chapter 10 in Solar Energy Handbook Eds. Kreider and Krieth McGraw-Hill (1981).
7. Tabor, H. "Storage Capability of Solar Ponds" See pages 17-32 in Ref. 8 <u>or</u> "Short and Long-Term Storage in Solar Ponds" CEC Joint Res. Centre Conf on Thermal Energy Storage ISPRA June 1981 Reprinted in Thermal Energy Storage pp 179-195.
8. "Thermal Storage of Solar Energy" Ed. C den Ouden. Martinus Nijhoff being Report of TNO Conf on Thermal Storage of Solar Energy Amsterdam 1980.
9. Mancini, N. "Use of Paraffins for Thermal Storage" pp 99-110 in Ref. 8.

9a. de Jong, A.G. and Hosgendoorn, C.J. "Improvement of Heat Transport in Paraffins for Latent Heat Storage" pp 123-133 in Ref. 8.

10. Chin Fu Tsang "Theoretical Studies on Seasonal Heat Storage in Aquifers" pp 185-196, Ref. 8.

10a. Iris, P. Experimental Study of Seasonal Heat Storage in Aquifers pp 197-207, Ref. 8.

A EUROPEAN TEST PROCEDURE FOR TESTING SOLAR STORAGE DEVICES

E. van Galen
Technisch Physische Dienst TNO-TH
(Institute of Applied Physics TNO-TH)
P.O. Box 155
2600 AD DELFT
The Netherlands

Summary

For an objective comparison of different storage designs, International standards for testing are required. Results obtained by a European group of experts are discussed. It is concluded that the recommended test procedures will lead to new viable designs and system concepts and effective use of R & D funds.

INTRODUCTION

The storage devices developed during the last couple of years, some of which have been put on the market, show a wide diversity in design. To achieve the highest possible output for a given solar installation and climate, researchers have tried and are still trying to improve the design of water stores. Much attention is being paid to size and position of the heat exchanger(s), and measures to avoid energy loss and upkeep thermal stratification.

Priority has also been given to the design of more advanced storage devices using phase change materials. The attention R & D focussed on this type of systems is justified by its two major advantages over sensible heat storage; viz.: higher energy density and isothermal storage. Different technologies for latent heat storage devices were investigated, e.g. bulk storage with direct contact heat exchange or immersed heat exchangers or devices based on macro-encapsulation of the storage material. Each store obviously had its own pros and cons, and because an objective assessment could not easily be made, something had to be done about it.

THE EUROPEAN SOLAR STORAGE TESTING GROUP

The Commission of European Communities established the Solar Storage Testing Group (SSTG), which object was to design test facilities and develop standard test procedures for sensible and latent heat type storage systems.

The objectives of these test procedures are:

- Determination of characteristic storage device parameters, enabling the intercomparison of different storage devices. On the basis of this information design errors can be detected and the right store for a particular application can be chosen.

- Determination of a reliable mathematical model for the storage device tested, describing thermal performances under dynamic conditions and suitable for integration into existing solar system models. Such a model is a necessary tool for calculating the energy savings attributed to the storage device and consequently for an objective cost-performance analysis.

Since - owing to the wide variety in design - the position of sensors for internal measurement could not be described unambiguously, a black box approach was chosen for all test methods from the initial phase. The Institute of Applied Physics TNO had the overall coordination. The other participants were: Technical University of Denmark, Lyngby; Institut für Thermodynamik und Wärmetechnik, Stuttgart; C.S.T.B., Valbonne; ENEA, Rome and Solar Energy Unit, Cardiff.

All participants have used their own storage devices and test facilities (see figure 1 and 2). In that way, each of them could emphasize different aspects of the generalized test methods.

Figure 1: Immersed heat exchanger of prototype latent heat store at SEU, Cardiff.

STORAGE DEVICE CHARACTERIZATION

As a result of the SSTG work, seven parameters were recognized, together characterizing a storage device. These parameters are:

- storage capacity as a function of the temperature;
- storage capacity over the design temperature range;
- heat loss rate at finite flow rate of the heat transfer fluid;
- heat loss rate during a stand-by period;
- heat exchange capacity rate;
- potential for thermal stratification inside storage medium;
- storage efficiency.

Figure 2: Test facility for solar heat stores at the Institute of Applied Physics TNO, Delft.

The last-mentioned may need some explanation. It quantifies the thermal performance of a storage device under well defined conditions. The storage efficiency is by definition the ratio of the actual energy stored in (withdrawn from) the storage device by means of the transfer fluid over a time period t (corrected for heat losses in the case of charging) to the measured capacity of the storage device in the temperature range defined by the temperature step change of the test. The storage efficiency is, therefore, not a real design parameter but it helps to compare the transient thermal performance of storage devices under equal operating conditions.
Test methods were developed for all above-mentioned parameters. The instructions for each of these test methods comprise short descriptions of the method, its objective and importance; demands made upon each test facility; test conditions and measurements to be carried out; and finally, a format for the calculation and presentation of the results.
32 tests are required for a complete characterization of the store according to these recommendations. Especially for complicated designs, efforts like these are counter-balanced by the results, giving direct useful information.

Simpler designs do not require complete characterizations. It is, for example, not necessary to measure the storage capacity as a function of the temperature for an ordinary water store.

A DESIGN GUIDE FOR TEST FACILITIES (see figure 3)

The test facility should be designed to meet the requirements made by the various test methods. Regarding test conditions, the facility should be capable of attaining a constant flow rate, temperature and power, and a well defined positive and negative temperature step. This means that the step should realized rapidly and without to much of an overshoot.

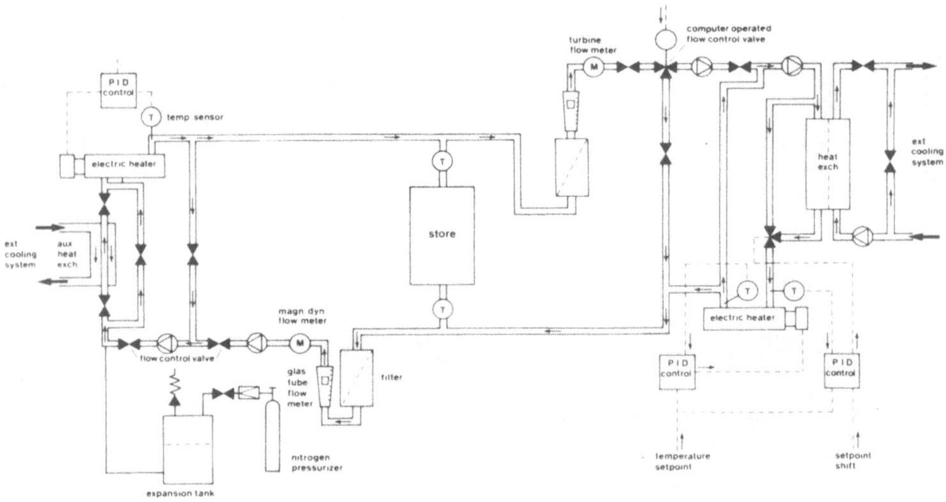

Figure 3: Scheme of recommended test facility.

In the design guide, recommendations have been drawn up, accounting for all these requirements, Moreover, the guide gives recommendations for pipework fittings and instrumentation. Two different test facilities conforming to the requirements are discussed in detail.

STORAGE MODEL DEVELOPMENT

Two different test methods are being developed to determine a mathematical model for a store.

The first method is based on a generalized model developed at SEU Cardiff. There are two unknown parameters, viz. an effective fluid storage material overall heat transfer coefficient UA and the number of nodes in the flow direction, NI. These parameters are to be identified by a curve fitting procedure by which experimental data of a step response test are used. The pair of values, UA and NI, are to be determined as a function of relevant test conditions.

The second method is based on storage efficiency values for charging and discharging. Again, these storage efficiency values are to be determined as a function of relevant test conditions. Interpolation of these values will then result in functions which form the storage model. The effects of stratification and time step on this type of model are being investigated. The advantage of this method is that the storage model is directly derived from test results without requiring other data.

The only alternative for these test based storage models is the development of a detailed theoretical model for each newly designed storage system. In the long run this will be a time consuming and consequently expensive approach. It counts for each model that it must be compatible with detailed models for complete solar systems.

MODEL VALIDATION

Irrespective of the way in which is was developed, each storage model must be validated. Within the framework of the SSTG, different validation procedures are being investigated. Provisional recommendations are made. The recommended validation is based on a five-day dynamic test sequence. Such a dynamic test consists of charging, discharging, alternately charging and discharging, a stand-by period and flow variations for charging and discharging. Figure 4 shows the outcome of these validation test for a theoretical model of a latent heat store. In this case, only minor problems are found directly after stand-by periods during reversal of the flow direction.

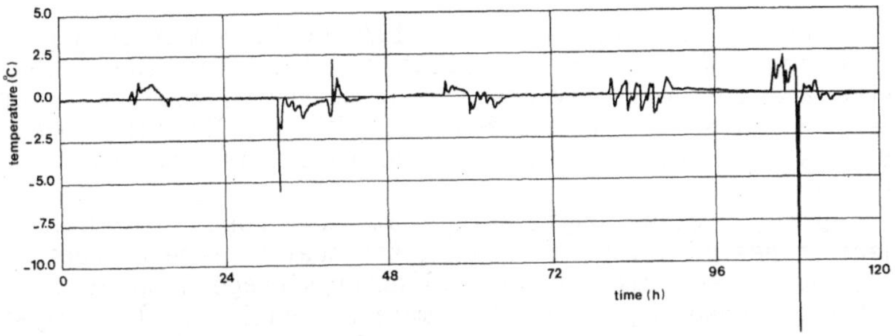

Figure 4: Difference between calculated and measured exit temperature as a method of time for model validation.

REPORTING FORMAT

A standardized reporting format has been designed and printed, containing forms to report the results of the recommended test methods. These format sheets have already proven to be of great use in the solar storage testing programme. In addition special sheets were designed to summarize the characteristics of the storage system and storage material, and to describe the storage system and test facility in short.

CONCLUSIONS

The SSTG succeeded in designing a standardized test procedure for storage device characterization. Good progress was also made in establishing methods to determine models from measurements. Guidelines for test facilities were drawn up and a reporting format was designed. Test results already showed important differences between the various storage technologies.

Some rough conclusions from early tests performed by the SSTG are:

- Macro encapsulation is an excellent method to benefit from isothermal storage, because thermal stratification can easily be achieved.
- The actual storage capacity available is sometimes smaller than the design value as a consequence of inadequate heat exchanger design, for example an asymmetrical distribution.
- The momentary heat transfer power may considerably differ from the expectations.
- A significant heat loss may occur owing to evaporation at temperatures above 50°C when the store is not air-tight.
- For a storage system based on the macro encapsulation principle, air as heat transfer fluid enhances the insulating value during stand-by.

This proves that design errors can be detected by test procedures at an early stage, leading to effective use of R & D funds. Moreover, practical results from standard test procedures will initiate new and viable designs and system concepts.

THE WORK WITHIN THE INTERNATIONAL ENERGY AGENCY ON CENTRAL SOLAR HEATING PLANTS WITH SEASONAL STORAGE

Arne Boysen
Bengt Hidemark Gösta Danielson Arkitektkontor HB
Järntorget 78, S-111 29 Stockholm
Sweden

Summary An agreement was signed in 1979 by 10 member countries of the International Energy Agency (IEA) to determine the technical feasibility and cost-effectiveness of large-scale, seasonal storage solar energy systems for the heating of buildings. The first phase of this Task was finished in 1983, and the result is now available in eight reports.

The reports present engineering data and cost information for five generic types of solar collectors, six heat storage concepts, and heat distribution systems for the collector/storage loop as well as for the storage/house load loop. Mathematical models of all subsystems are combined in a set of computer codes for the total system, which can be used for simulation as well as optimization of the design. Finally ten preliminary designs have been presented, one from each country.

The work is now being continued in Phase II as a parametric analysis in order to evaluate different system configurations. This work is expected to be ready for publication in 1985. Some countries have also built systems, which are now being monitored to provide experimental data for validation of models and for evaluation purposes. A Phase III for an experimental evaluation by an international exchange of such data is also being considered.

For northern countries the amount of solar energy per annum is about the same as for countries much further south - but it is very unevenly distributed over the year. Very little energy during the winter, very much during summer. Storage of a seasonal capacity becomes necessary if solar energy is to substitute the use of fuel. Even so, some auxiliary energy has to be supplied, be it power to a heat pump or fuel to a boiler. These are therefore the main parts of a system (Figure 1).

Figure 1.
Main parts of a Central Solar Heating Plant with a Seasonal Storage.

HP = Heat Pump
B = Boiler

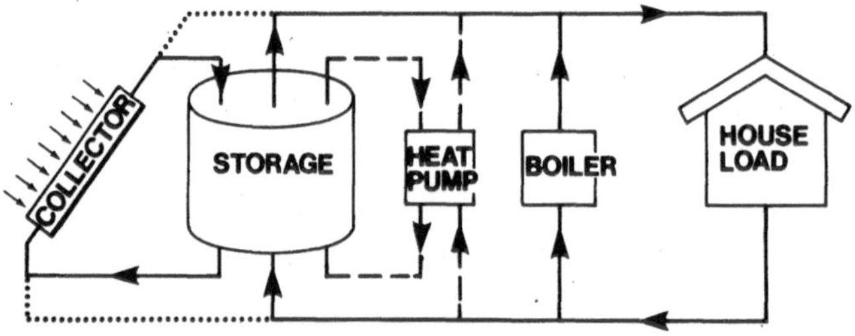

Participants in this project decided at the start to focus the study on systems where energy is stored as sensible heat and distributed as hot water. With these conditions it becomes necessary to have very large storage volumes in order to keep the relative heat losses down. These large volumes have to be combined with large loads - in most cases large number of houses. This concept is indicated by the title of the project - TASK VII - Central Solar Heating Plants with Seasonal Storage, or CSHPSS.

Heat distribution subsystem

In such a system the heat distribution uses the same technique as in a district heating system. Several of the participating countries have long experience in this area, and this was summarized in a short report giving engineering data, cost information and overviews of codes and state-of-the-art in the ten countries (1). The interest for such heat distribution is growing, and even countries like U.S.A. and Canada, where steam is the conventional distribution medium, reported that demonstration projects are underway using what perhaps might be called a European hot-water technique.

In CSHPSS-projects the conditions for the two distribution loops are quite different. In the collector/storage loop the total or close to the total annual energy consumption has to be collected during sunshine hours in about half of the year - which means high peaks and short utilization periods. This favours a concentrated collector field and location of the storage close to the collectors. In the storage/load loop the return temperatures are normally quite low in order to utilize the storage capacity.

Also in countries with long experience of conventional district heating the solar energy application introduces new factors and necessitates a new interest for developing tools for a very careful design, reducing the installation cost, maintenance, and heat losses for the heat distribution.

Heat storage subsystem

The heat storage technique is the key to the problem of the CSHPSS-systems. Seasonal storage means basically that the utilization factor is low, and extremely cost-effective storage concepts must be found. The current R & D is attacking this problem along many lines, testing and evaluating many different concepts. In Task VII six storage alternatives were selected for investigation: Tank, Pit, Cavern, Aquifer, Earth and Rock. For each of these approaches demonstration projects were identified, the state-of-the-art was summarized, and engineering data and cost information was collected.

Tank storage can be considered as a well-known technique. It can be used almost everywhere, has low heat losses and is very flexible in operation. Structural reasons may limit each tank to an approx volume of 100.000 m3, and a maximum height of approx 13 m, but several tanks may be used. FRG has reported a feasibility study of a steel-membrane reservoir for which there is in practice no restriction in size.

For pit storage the waterproof liner is of paramount importance. It must retain the watertightness and strength for many years at elevated temperatures. A maximum temperature of 70-80°C is being recommended. The storage normally has a high surface/volume ratio and insulation becomes necessary to keep the heat losses down. The best geological conditions for building a pit store are easily excavated, stable soil, free of ground water.

Storage in rock caverns is another example of an established technique being applied in a new area. In crystalline rock, large caverns can easily be excavated. Blockfilled caverns have been suggested to reduce the demands on the stability of the mass. The caverns are built without heat-insulating liners, and the surrounding rock participates in the temperature variations. Considerable amounts of heat are needed during the first years to warm up this mass, but then the heat losses are stabilized at a rather low annual level.

Aquifer storage of heat has been demonstrated successfully, but a number of problems still need to be solved. One great advantage of the aquifer storage is that the stored water also is the transfer medium. Thus, the thermal disturbance can be controlled more simply, and large peak loads can be injected or withdrawn compared with earth or rock storage systems.

An earth storage is in principle a layer of subsoil with a heat exchanger in it. At the top, and sometimes around the perimeter, there is an insulation. The heat exchanger is formed by tubes in most cases, vertically or horizontally. Ideally, the earth should be saturated, but high permeability and ground water movements can cause high heat losses. A vertical screen around the storage can limit these losses. Storage temperature is normally low - there is very limited experience for temperatures higher than 40°C.

Rock storage is similar to earth storage with vertical tubes. The heat carrying fluid, normally water, can be circulated through bore-holes in the rock - wells - in open or closed circulation systems. Heat is stored in the rock mass which is perforated by the wells, and buffer storage may be used to reduce peak loads. Deep wells allow storage temperatures above 100°C. Construction of a rock storage utilizes well-known drilling technique.

General engineering data and cost information has been supplied by the participating countries, which also supplied detailed data from 31 projects (Table I) (2), (3).

Table I Storage projects reported from each country

	Tank	Pit	Cavern	Aquifer	Earth	Rock	Sum
Austria					1		1
Canada	2			1			3
CEC	1				1		2
Denmark							-
FRG		3		1			4
Netherlands					1		1
Sweden	4	1	3	2	3	3	16
Switzerland				1	1		2
U.K.		1					1
U.S.A.				1			1
Sum:	7	5	3	6	7	3	31

The proper design of a heat storage has to be based upon simulations. Simulation models were collected, tested and evaluated. Three main families of models were considered, namely one for water tank, pit and cavern storage systems, one for earth and rock storage systems, and finally one for aquifer systems. Of the first family, 7 models were investigated, of the second, 8 models, and of aquifer models 5, in all 20 different models

(Table II). The evaluation led to a selection of three models, all originating from the Lund Institute of Technology (4). Computer codes for these three models are available both in MINSUN - the program developed in Task VII - and TRNSYS.

Table II Evaluated storage models from each country

	Tank, Pit & Cavern	Aquifer	Earth & Rock	Sum
Austria				-
Canada	2		1	3
CEC			1	1
Denmark				-
FRG				-
Netherlands			1	1
Sweden	2	1	2	5
Switzerland			1	1
U.K.			1	1
U.S.A.	3	4	1	8
Sum:	7	5	8	20

Solar collector subsystem

Design data for the solar collector subsystem were developed in a similar way as the data for heat storage. National data and experience were collected, evaluated and generalized. Types of collectors that are suitable for central heating plant application have been identified; analytical models were developed that predict the performance; the performance of large collector arrays was analyzed; and cost equations were defined.

The report (5) identifies five generic types of collectors: Flat plates, shallow solar ponds, evacuated collectors, parabolic troughs and central receiver.

The performance of flat plate collectors has reached maturity - manufacturers are now more concerned with reducing costs than improving efficiency. No significant gains in efficiency are on the whole likely in the next five years, barring unexpected innovations.

For solar ponds the attention has been focused on shallow ponds, in which the absorber is usually a plastic envelope, blackened on the bottom to absorb the radiation, insulated from the earth and covered with an additional plastic or glass glazing.

Evacuated collectors are still being developed to reach higher efficiency. In Task VII data for high performance CPC collectors have been used, anticipating that values reached in laboratories today soon will be reached in commercially available units.

The parabolic troughs that have been considered, are sun-tracking with an EW orientation. For the performance model a hypothetical collector was selected - having an optical efficiency of 0,807 and heat loss coefficients equal to the 1985 goal at the Sandia National Laboratory, U.S.A. These values have already been reached in laboratory tests.

For central receivers the performance model that is used is based on a theoretical analysis.

As the collectors, regardless of type, have to be arranged in large arrays, much attention has been given to what effect this might have on the performance as compared with a single collector module. Very little information has been found in literature and research reports, so the participants initiated a special workshop on this topic, to be held in the U.S.A., in June, 1984. Based on the available experience, on analysis, and on judgement certain performance reduction factors were developed and used in Task VII until better data is available.

There are a number of innovative collector concepts that may be viable for central plant applications that were not examined in any depth. The reason being that the participants decided at the start that the technique used in the study should have achieved a certain maturity, which implied that a mass production of components is established, with a corresponding price level for the collectors.

In reality a design has to be based on data for available components rather than generic types. It was therefore suggested that besides using the jointly developed design information, each participant might want to use alternative data, reflecting national experience and components on the market. No such studies have however been reported.

The MINSUN simulation and optimization program

When the work started, the participants agreed to use the TRNSYS model for simulation of the performance and a Swedish model called MINSUN for optimization. The work has led to some improvements of TRNSYS, but the main work has been done on MINSUN (6), (7). As the work progressed the MINSUN program was further developed to include the five generic collector types that have been mentioned, as well as the six storage concepts, and it can now be used to simulate the thermal behaviour of a central solar energy system as well as determine the optimum size of some of the components. The program can be run in three different modes - Single Simulation, Multiple Simulation and System Optimization.

The simplest application of MINSUN is to perform a Single Simulation for a given, fixed configuration. All parameters of the system are defined by the user. The program simulates the thermal behaviour, does the energy balance and cost calculations, and generates output on the thermal and economic characteristics of the system specified. The thermal characteristics include a daily specification of heat flows among the major subsystems (from collectors, to and from storage, to load, losses, etc.).

In the Multiple Simulation mode MINSUN allows the user to perform several simulations in a single run while systematically varying the parameters defining the system. Only a limited number of result values are kept from each run. This mode is very useful for examining the effects of given input parameters on particular system results. It also uses less computation time than a large number of single simulation runs to get the same outputs.

These multiple runs can be made in two ways. Using the first and simplest of these, the MINSUN set of programs is capable of systematically varying any two (of nine) key design variables and performing single simulations at each point of the grid formed by the two variables. A typical application is to examine system cost as a function of two key variables, say collector area and storage volume. The program automatically spans a specified range for each variable with the requested number of points. Important results such as cost and solar fraction are selected from the simulation results for each grid point and are saved in a separate computer file. These results can then be examined by the user in numeric form or, as intended, plotted using three-dimensional graphics. Then the key results, such as cost or solar fraction, can be examined as a surface over the grid formed by the two variables selected, figure 2.

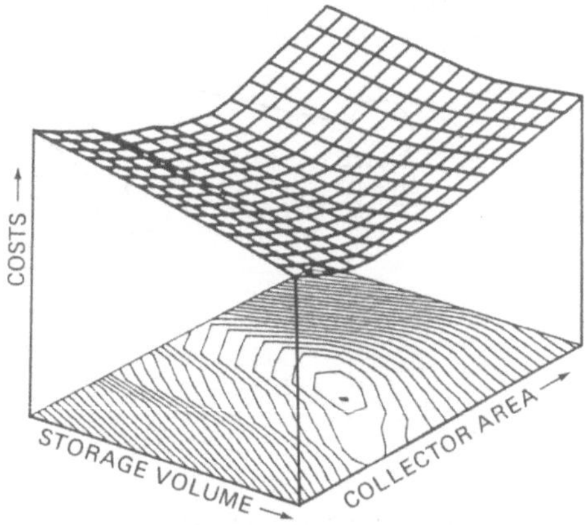

Figure 2.
Sample Cost Surface Plot

Annual cost for various combinations of collector area and storage volume.

Note that the MINSUN package only produced the data. The actual plotting must be done by the individual user.

The second Multiple Simulation option, the MINREP procedure, is slightly more complicated, but much more flexible. Single or iterative changes of any system parameters (not just the nine) can be specified. In addition, any results which appear on the detailed Simulation Summary output to a maximum of twelve variables, can be specified for inclusion in the summary output.

In the System Optimization mode the MINSUN set of programs has the capability to automatically select optimum values for key design variables which minimize overall system levelized annual cost. The variables which can be optimized are the same nine which can be varied in the multiple simulation. The program uses a search procedure to vary the values of the appropriate design variables. It then simulates the thermal behaviour and computes the cost of this system, and compares the cost of this system with that calculated in previous iterations. In this way, the program closes in on the values of the design variables which minimize system cost.

The MINSUN set of programs

As indicated in figure 3 there are two separate main programs: the collector model set and the system simulation and optimization model set. The collector model set requires collector system parameters and other parameters to be set by the user. It then takes hourly solar radiation and temperature data and calculates the amount of energy that would be collected by a collector (per unit area) operating at a given temperature on a daily basis. Several operating temperatures are used and all results are stored for later use.

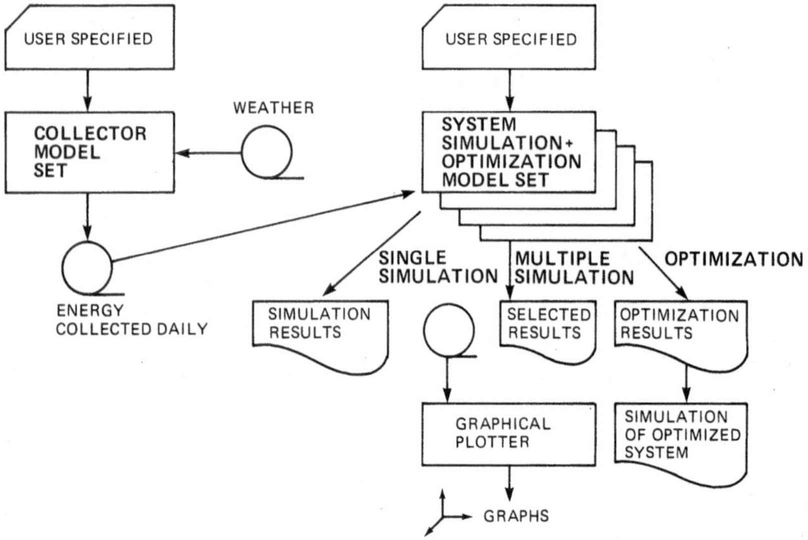

Figure 3. The MINSUN set of programs

The system simulation model requires a large number (approximately 150) of system and other parameters to be specified by the user. It then simulates the thermal performance of a given system on an annual basis. As described above, this system can be applied in one of three modes, for single simulation, for multiple simulation or for optimization. Since the various storage model programs are large, there are separate programs for each storage type as depicted in figure 3.

An overview of the 10 national designs

Figure 4 shows some basic data on the national designs of CSHPSS-system from each of the ten participating countries (8). Each is described in detail in national reports, and five of them (from C, DK, FRG, NL, U.S.A.) are also being presented in separate papers during this conference.

The Austrian report describes a project that was designed and constructed while Phase I of Task VII was in progress. The discussions and the work in Task VII probably contributed to the design of the project, but precisely how much is impossible to say.

The Canadian report also presents a project that was designed and constructed at the time for Phase I. The storage - an aquifer - is primarily used for storing cooling water, and a cooling tower is a dominant feature of the system. In this respect the project does not follow the intentions for the Task VII work. A smaller aquifer stores heat from solar collectors, used mainly for DHW purposes. The project may provide interesting practical experience in the future.

The Danish report is a theoretical design. However, the storage concept, an earth pit, is being tested in full scale within the national energy research program, and as the other subsystems are based on well-known technology, the project may be considered to represent a realistic case.

The ECE project is also a theoretical one. The load is a single building - like the Austrian, the Canadian and the Swiss projects - which does not require a heat distribution system as complex as when a large number of buildings are being served.

The project from FRG is based on an existing site with houses for one or two families. The CSHPSS-system is a theoretical design, supplying heat for space-heating only. Only 23 houses on the site are considered to be connected to the CSHPSS-system, which makes it difficult to design a system with a potential of having a good economy.

In the Netherlands a project has been designed for a group of 96 houses. The houses have been constructed, as has the storage system, while the solar collector subsystem is being simulated with heat from a boiler. Real solar collectors on one house are used to guide the control of the heat supply.

The Swedish presentation is based on a detailed design that was intended to be constructed. At a very late stage the demand for new dwellings in that particular area was drastically reduced, and the project was therefore never realized. The design, however, was based on realistic conditions and on operational experience from several smaller, full-scale testing plants. Of the ten projects this is the only one that approaches a second generation state-of-the-art.

The Swiss project was in most details already designed when the work in Task VII was started. The construction was however delayed in order to utilize the IEA work on design of heat storage. The load is far from being representative of dwellings, and no DHW is being produced.

The U.K. project is purely theoretical, and very much designed to meet the load recommendations that were accepted by all Participants at the start of the IEA work. The design is made for a group of 100 houses, which provides a realistic basis for demonstration of technical features and problems, but is too small to provide the best economy.

The US project, which also is rather small, presents an idea of utilizing existing underground storage tanks in an area undergoing a major redevelopment. The project is therefore characterized by a number of constraints which do not exist in the other nine projects. It is economically favoured by the use of existing storage tanks and tunnels for the piping, which leads to a cost for solar energy which is competitive with conventional heating means.

At the start of the work in Task VII the Participants agreed that the work was to be aimed at hydronic systems for heating of dwellings, with a solar fraction of 80 % or more. This solar fraction was suggested in order to assure designs which made use of considerable storage capacity. The fraction 80 % does not present an economic optimum. On the contrary, it has been claimed by participants. The optimum varies of course from country to country due to different conditions, and probably there are two points that should be investigated, one very close to 100 %, and the other with much less solar.

Heat pumps are being used in seven of the ten designs, in a number of different ways. Also, the use of buffer storage varies. The variation in the choice of solar collectors is in comparison much less. Most striking is the different selection of storage concepts. This may indicate that heat storage is the area that is least well-known.

The overview clearly demonstrates the difficulty in making a general evaluation of the concept of solar heating plants with seasonal storage. The further work in Task VII should however provide some general trends.

However, the ten designs do demonstrate the possibility of adopting this technique to different conditions and considerations. The versatility is such that the potential of development is very great. It is to be hoped that these projects - in many countries the first of their kind - are followed by both design studies and actual plants, which can provide real data and experience.

References

(1) Bruce, T. & Lindeberg, L., 1982, Central Solar Heating Plants with Seasonal Storage - Basic Design Data for the Heat Distribution System, Swedish Council for Building Research, Document D 22:1982, Stockholm, Sweden.

(2) Chuard, P. & Hadorn, J-C., 1983, Central Solar Heating Plants with Seasonal Storage - Heat Storage Systems: Concepts, Engineering Data and Compilation of Projects, Office central fédéral des imprimés et du matériel, Berne, Switzerland.

(3) Hadorn, J-C. & Chuard, P., 1983, Central Solar Heating Plants with Seasonal Storage - Cost Data and Cost Equations for Heat Storage Con cepts, Office central fédéral des imprimés et du matériel, Berne, Switzerland.

(4) Hadorn, J-C. & Chuard, P., 1983, Central Solar Heating Plants with Seasonal Storage - Heat Storage Models - Evaluation & Selection, Office central fédéral des imprimés et du matériel, Berne Switzerland.

(5) Bankston, C., 1984, Central Solar Heating Plants with Seasonal Storage - Basic Performance, Cost, and Operation Information for the Solar Collectors, Argonne National Laboratories, Argonne, U.S.A.

(6) Chant, V. & Biggs, R., 1983, Central Solar Heating Plants with Seasonal Storage, Tools for Design and Analysis, National Research Council, Document CENSOL1, Ottawa, Canada.

(7) Chant, V. & Håkansson, R., 1983, Central Solar Heating Plants with Seasonal Storage - The MINSUN Simulation and Optimization Program - Application and User's Guide, National Research Council, Document CENSOL2, Ottawa, Canada.

(8) Boysen, A. & Dean, E., 1984, Central Solar Heating Plants with Seasonal Storage - Preliminary Designs for Ten Countries, Swedish Council for Building Research, Stockholm, Sweden. (Report in preparation).

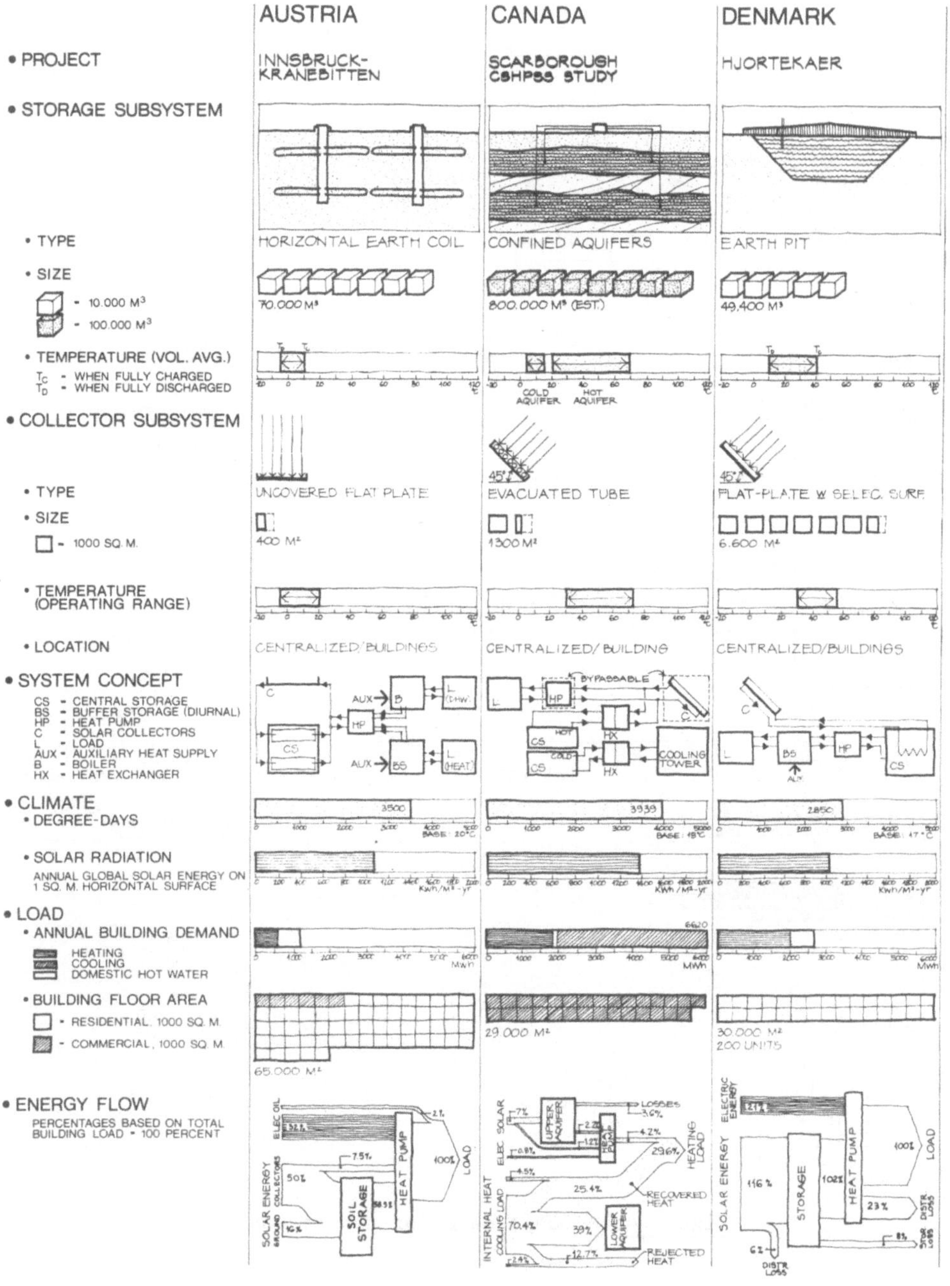

Figure 4a. Basic data for ten national designs

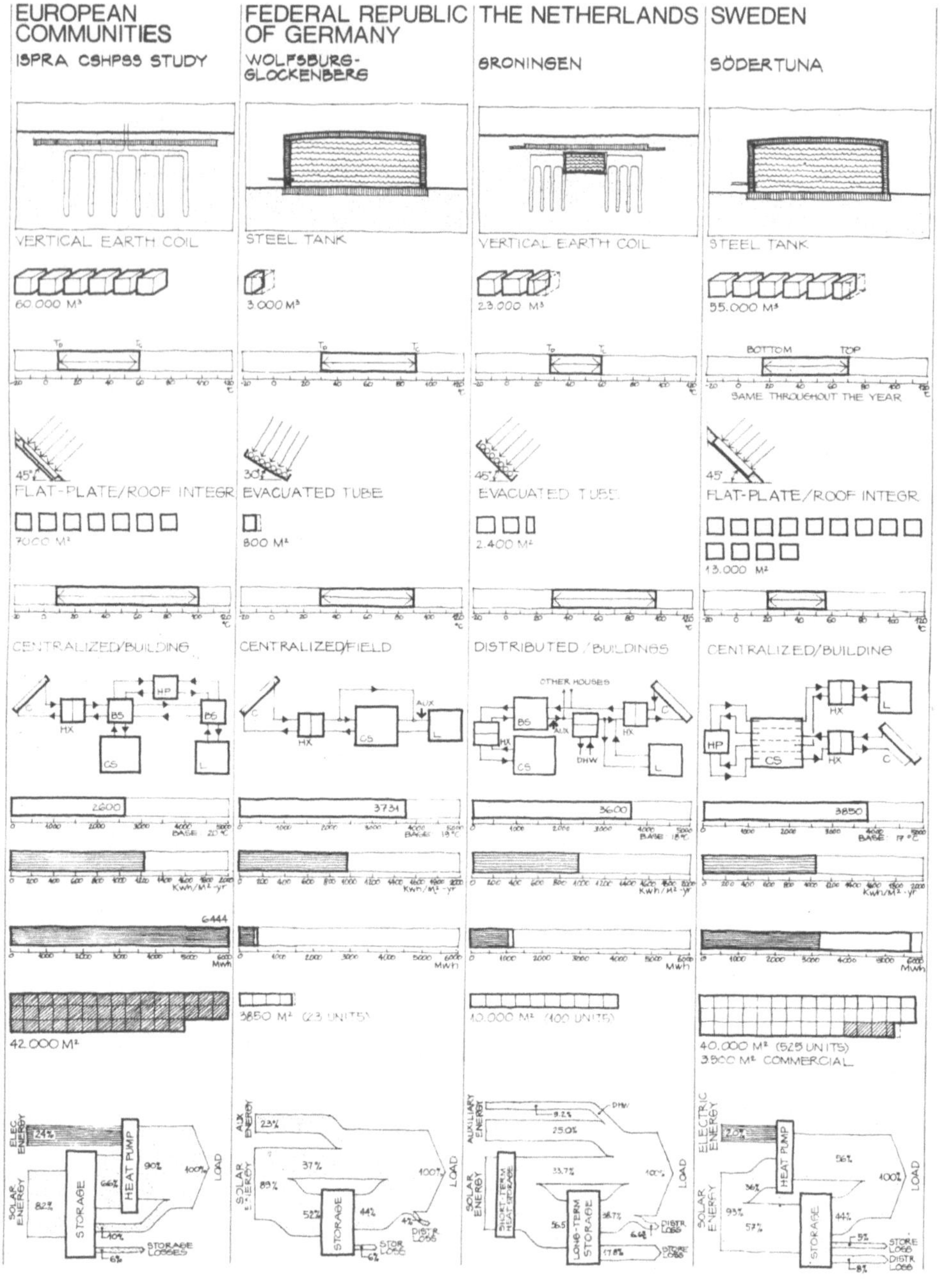

<u>Figure 4b</u>. Basic data for ten national designs

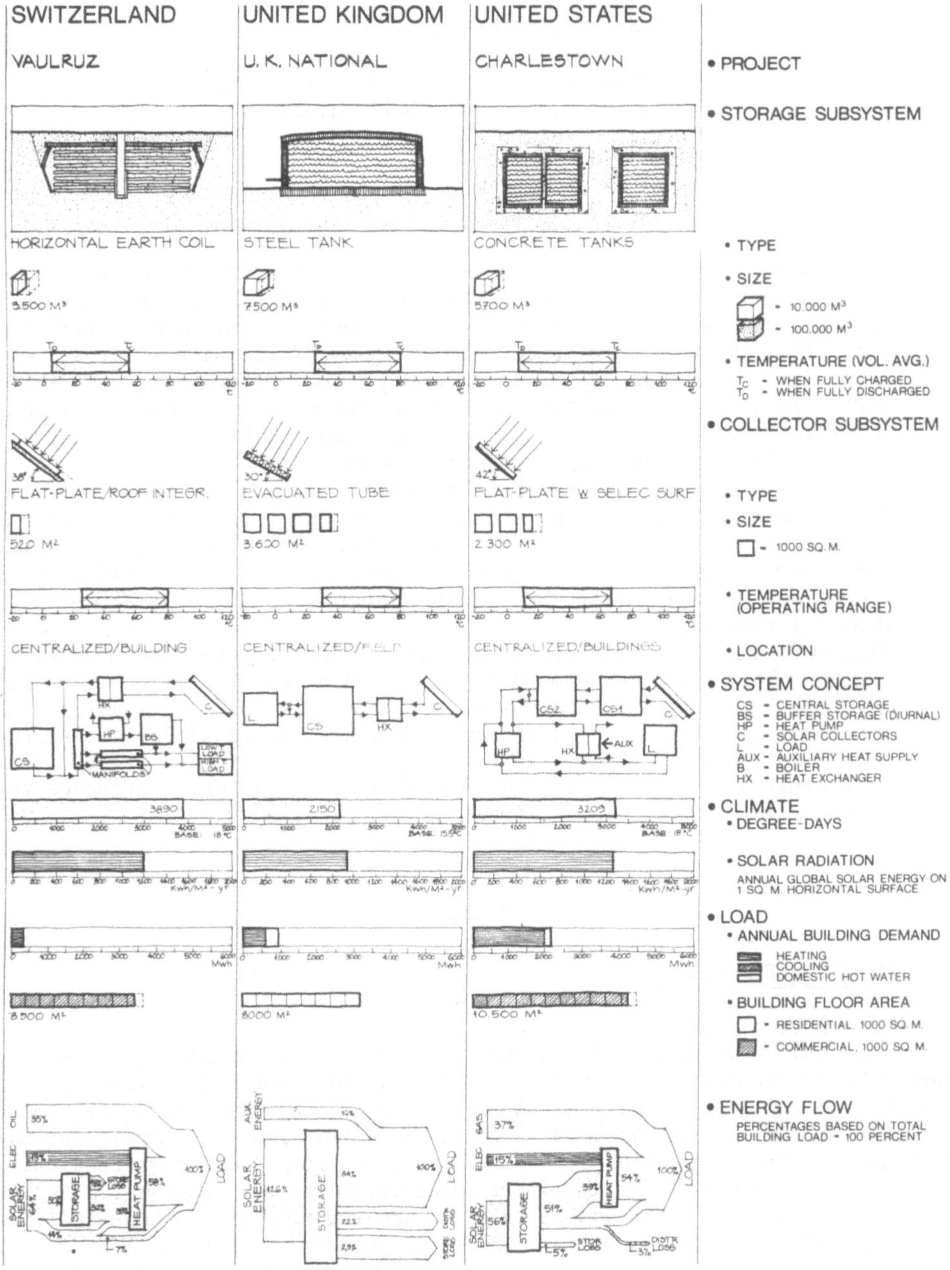

Figure 4c. Basic data for ten national designs

A SEASONAL STORAGE SOLAR HEATING SYSTEM FOR THE CHARLESTOWN, BOSTON NAVY YARD NATIONAL HISTORIC PARK

D.S. BREGER and A.I. MICHAELS
Solar Energy Group
Argonne National Laboratory
Argonne, Illinois 60439 U.S.A.

Summary

This paper concerns the design and analysis of a solar energy system using seasonal heat storage for the National Historic Park in the Charlestown Navy Yard, Boston, Massachusetts. The system uses two existing underground concrete storage tanks filled with water as the heat store and makes use of an existing tunnel network for placement of the heat distribution pipe network. The system includes a heat pump and is designed to supply about 1000 MWH of solar heat to displace the hot water and space heat demand.

The study is part of the U.S. participation in the International Energy Agency Solar Heating and Cooling Program, Task VII on Central Solar Heating Plants using Seasonal Storage. System analysis was performed using the MINSUN computer simulation model developed by this Task specifically for solar seasonal storage systems with district heating.

Analytic results indicated a substantial performance improvement by incorporating a heat pump into the system to extend the useful storage temperature range and reduce the annual average system temperatures. Presently an effort by the U.S. National Park Service and Department of Energy is underway to continue this project into an engineering design phase to determine with greater confidence the system cost and engineering/construction requirements.

1. PROJECT DESCRIPTION

1.1 Site Description

The Boston National Historic Park is an operational part of the decommissioned Charlestown Navy Yard, the remainder of which is the present focus of a major redevelopment project. The Navy Yard is located directly across the Boston Inner Harbor from the downtown area and the Park is open to the public year round and is a major tourist attraction.

The system makes use of two large underground storage tanks built in the 1950's for the Navy Yard, and of existing underground piping tunnels. The central solar system provides district heating for five 2-3 story, historic buildings around the tanks used for residential, administrative, and public purposes. Additional buildings can be readily connected to the system if desired. Figure 1 shows the site plan for the Charlestown Navy Yard, with the buildings to be heated, the storage tanks and the piping tunnels shown in solid black. Area for collector placement is constrained but may be possible on flat roofs of nearby tall buildings shown cross-hatched on Figure 1.

1.2 System Design

The system design is shown schematically in Figure 2 and is based on a seasonal storage concept and a heat distribution network. The available large tanks and relatively constrained area for collectors makes seasonal storage of particular interest. The collector array operates year round and thus the required area (compared to a short-term storage system providing a similar solar fraction) is significantly reduced. The advantage of seasonal storage is strongly dependent on the extent to which the temporal distributions of the system load and the solar insolation are out of phase over the year. In Boston, solar collection remains significant during the winter (especially at the low collector temperatures provided with the heat pump), though the large store provides a use for the large array during the abundant summer insolation period and therefore greater value from the collector array.

Generally, an economic trade-off is implied between collector area and storage capacity. The economies of scale are quite significant for the storage sub-system and, therefore, seasonal storage is of particular interest for large loads. Seasonal storage design is also more favorable for high density loads and loads which are distributed over a rather short period of time during the year which coincides with relatively low solar insolation. The system design concept is characterized as a centralized heating plant (solar and storage) in combination with a low-temperature district heating network.

2. METHODOLOGY

The solar system was designed and analyzed using MINSUN, a computer simulation model developed by the IEA Task VII, specifically for central seasonal storage solar heating plants. The program allows parametric description of each of the sub-systems including three storage types (water tank, direct-earth duct, and aquifer storage), and provides a system simulation on a daily time step. System analysis is performed with an optimization routine or an iterative driving function to change parameters and provide graph-ready output data.

A summary of the most important input parameters for the Charlestown system to the MINSUN model are given in Table 1. Three collector types were considered - a high quality flat plate, a compound parabolic concentrator (CPC) advanced evacuated tube, and a parabolic trough. The system was analyzed with and without a heat pump. The heat pump is a 104 ton (366 kW) unit which operates when the storage temperature drops below the distribution demand temperature and the storage temperature is above about 10°C. Heat is distributed at a minimum of 55°C; distribution temperature then rises by 1.5°C for each degree ambient below 0°C. Distribution return temperature is always at 45°C.

3. RESULTS

With the storage volume fixed in this project, the main design parameters are collector type and area and the use of a heat pump. System performance, shown as annual solar fraction, as a function of collector area is shown in Figure 3, for all three collector types both with and without a heat pump. The heat pump clearly improves performance for all the collectors, though the improvement is most substantial for the flat plate. At the lower temperatures provided by the heat pump the three collector types perform very similarly. The solar fractions are low for typical seasonal storage systems only because the available load is large compared to the storage tank capacity. Despite reducing solar fraction, the larger load does improve system efficiency and total solar energy used by reducing the storage time and losses.

The solar cost ($/MWH) is defined as the annualized system capital cost plus the annualized heat pump operational cost, divided by the solar and heat pump energy supplied to the load. The solar cost as a function of collector area, for all three collector types, with and without the heat pump is given in Figure 4. The heat pump reduces solar cost for all collector types. The flat plate system with heat pump provides comparable performance at lower cost than any other alternative and is thus the clear design choice. A reference system has been selected for detailed performance analysis and consists of 2300 m^2 flat plate collectors operating with a heat pump. This design provides half the 2167 MWH total annual load with solar energy, the remainder supplied by heat pump electric and auxiliary energy.

The daily system performance of the selected reference system is shown in Figure 5 by curves for average storage temperature, cumulative system load, solar energy collected, auxiliary energy used, and heat pump electric input energy. The simulation is for a typical meteorological year in Boston and shows the stored solar energy meeting the space heating load unassisted from October to mid-November. The heat pump with solar is then used solely thru mid-December. The remainder of the heating season is met with a combination of the solar/heat pump and the auxiliary/conventional system. At the reduced storage temperatures produced by the heat pump, solar energy collection remains efficient year round and system heat losses are quite low. The annual system energy flows for the design system are illustrated in Figure 6. Storage temperature stratification is also simulated by the MINSUN program and is important in terms of providing low collector temperatures and higher temperatures for the load or heat pump.

The economic summary of the reference system is shown in Table 2. It is observed that the solar energy cost is competitive with conventional heating at a fuel cost of $0.05/kWh, a real fuel escalation rate of 3.5% and a real discount rate of 4%. The collector sub-system accounts for over 61% of the system capital cost; the storage renovation is 23% and the distribution system is 11%. The storage and distribution portions of total cost are relatively low compared to typical central seasonal storage systems, due to the existence of the tanks and the piping tunnels.

4. PROJECT STATUS

Presently there is interest by the U.S. National Park Service and the Department of Energy to continue this project into an engineering design phase. Funds for this study are presently being sought. The engineering design would provide more detailed system design, cost and performance information on which a decision to implement the system can be based. Site specific evaluation of collector sites, storage renovation, piping networks, and general system layout is necessary. Specification of the major components in conformity with industry availability is desired.

The engineering design is to be preceded by the evaluation of several design alternatives which have been suggested since this study was completed. Of interest is the use of the Boston Harbor or groundwater as a heat source for a water-to-water heat pump system (without storage or augmented by a smaller collector array and storage). This alternative could then be coordinated with the design of one or both storage tanks for winter ice storage for the substantial Boston summer cooling load.

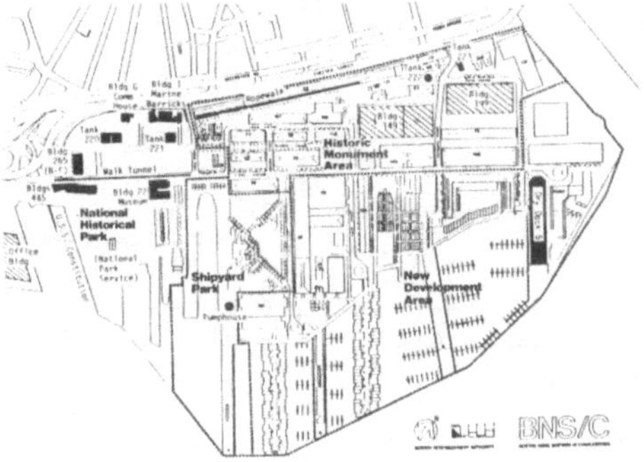

Figure 1. Charlestown Navy Yard Site Plan

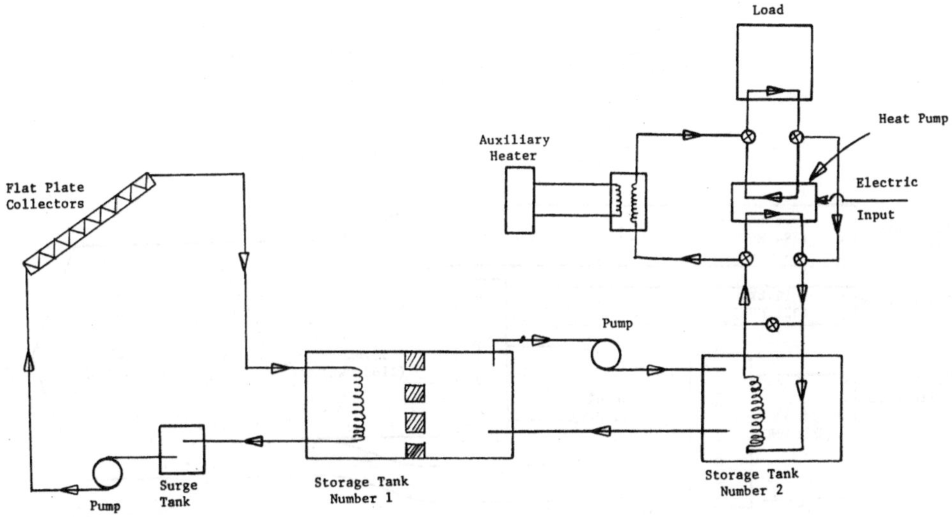

Figure 2. System Design Schematic

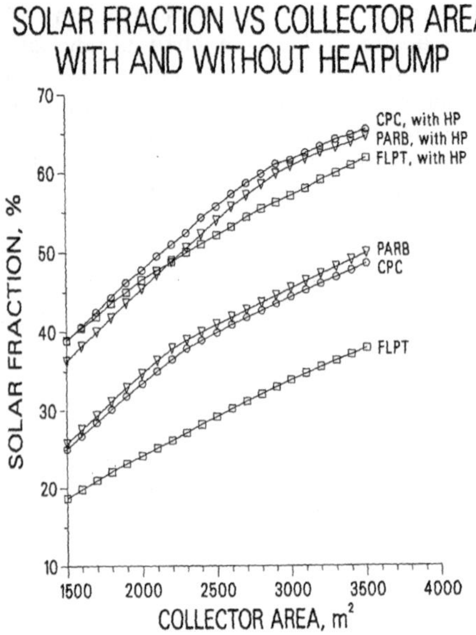

Figure 3. Solar fraction

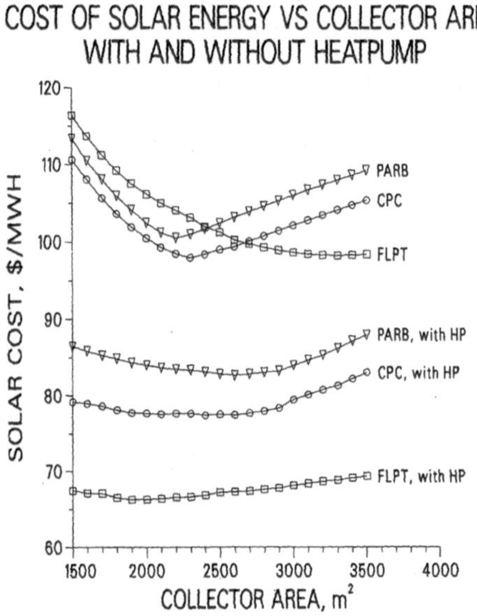

Figure 4. Solar cost

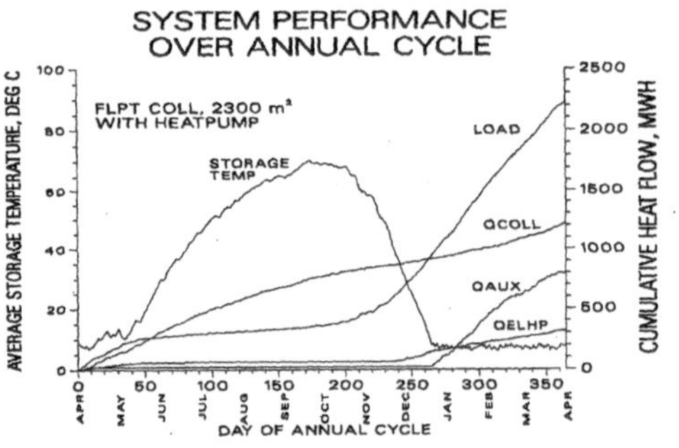

Figure 5. Annual simulation

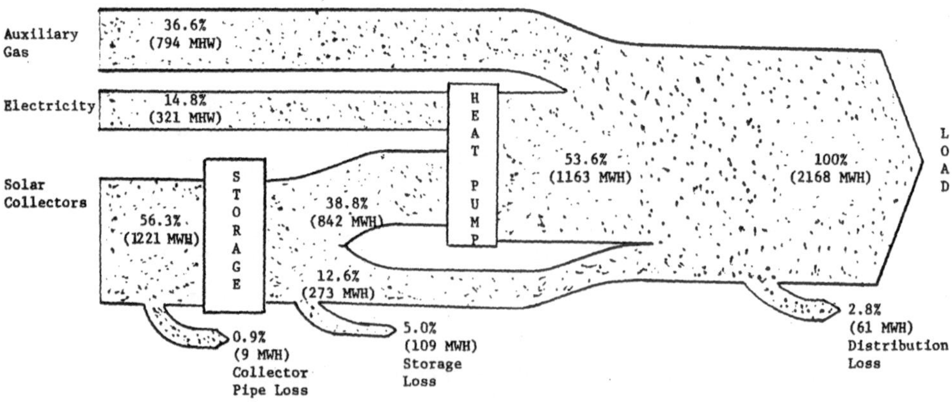

Figure 6. Reference system annual energy flows

Summary of Base Case Input Parmeters

Storage Volume 5700 m^3

Storage Insulation Thickness 0.10 m top, 0.075 m sides, 0.05 m bottom (Thermal conductivity 0.02 W/m/K)

Space Heat Load 2000 MWH

Tap Water Load 19,100 Watts (167 MWH annually)

	Year	FLPT	CPC	PARB
Collector Costs (1982 $/m^2$)	1985	245	370	400

Storage Renovation Cost $6/ft^2 surface area
$846/m^3 insulation (derived to account for total cost)
$216,000 Total

Distribution Cost (Collector and Load) $60/m; $102,480 total

Heat Pump Evaporator heat transfer capacity 80kW/K
Condenser heat transfer capacity 50kW/K
Capital Cost $40,000

Discount Rate (real) 4.0 percent

Energy Escalation Rate (real) 3.5 percent

Aux/Conventional Fuel Cost $0.05/kWh

Electrical Rate $0.07/kWh

Table 1. Summary of MINSUN input data

Economic Summary

System: 2300 m^2 FLPT, with heat pump
Solar Fraction = 50.0%
C.O.P. = 3.6,

Total Capital Cost		$922,000
Collectors	$563,500 (61.1%)	
Storage	$216,000 (23.4%)	
Distribution	$102,400 (11.2%)	
Heat Pump	$ 40,000 (4.3%)	
Operational Cost (first year)		$ 62,130
Heat Pump Electricity	$ 22,450	
Auxiliary Fuel	$ 39,680	
Total Yearly Cost (annualized and 20 year lifetime)		$149,300
Solar Cost		$ 66.60/MWH
Conventional Cost (annualized)		$ 67.61/MWH

Table 2. Economic summary

VALIDATION OF THE GENERALIZED MODEL BASED STORAGE TEST PROCEDURE BY MEANS OF DYNAMIC SIMULATION TESTS

R.H. Marshall
University College Cardiff - Solar Energy Unit
Newport Road, Cardiff, United Kingdom

Summary

Using a test procedure under development by members of the concerted action, Solar Storage Testing Group (SSTG), sponsored by the Commission of European Communities, a solar storage device is to be characterized by some six or more parameters using well defined, typically step response, tests. The question therefore arises as to how well these parameters describe the arbitrary dynamic boundary conditions encountered in service.

The approach taken by this author has been to define a generalized storage model, described elsewhere, and use the step response data to "identify" two key parameters. The dynamic test then serves to confirm or refute the validity of the generalized model. Ultimately, however, it is the link with long term system performance, through the interaction between other system parameters, and the storage parameters that is of interest. A design method known as the SEU Simplified Design Method has already been derived which establishes this link but for which experimental confirmation is lacking.

The work described below therefore has a twofold objective. Firstly, after defining a suitable dynamic test sequence and the criteria for judging the outcome, a typical result for one of three dynamic tests aimed at validating the generalized model is described. Secondly, the comparisons are made between the measured solar fraction and that derived from the SEU Simplified Design Method.

The results indicate that both objectives have been met; that is the generalized model can be deemed to be valid and that the SEU method provides an accurate estimate of system performance.

1. INTRODUCTION TO DYNAMIC TESTING

In performing the many tests aimed at characterizing a storage device deterministic well defined tests, e.g. a step response test, are used. Further, each test begins with a long period of stabilisation to steady state ending with a long period of stabilisation at a new steady state. However, in service the storage device will rarely be exposed to such constant well defined boundary conditions so that the fundamental question to be answered by dynamic testing is:

"how well do the parameters derived from step response tests serve to describe the dynamic operation of the device?"

The question implies, therefore, the existence of a model of the dynamic behaviour of the store where, recall, by definition a model is a calculational procedure for establishing the transient relationship between the inlet temperature, Ti(t), and exit temperature, Te(t), as a function of time. Therefore, the fundamental question reduces to:

"how good is the dynamic model?"

1.1. Dynamic Model(s)

A generalized storage model is assumed which replaces the actual store with a fictitious device described by two coupled partial differential equations, see Marshall (1983a) for details. These coupled equations are used with the step response data to find two values, UA and NI which minimize the error between the model and the experimental results, that is to "identify" the best UA value for a given number of nodes, NI. In establishing the best UA value for a given NI (degree of stratification) it has been found for the Cardiff store that different pairs of UA, NI values as a function of temperature step interval are equally valid, Table 1, and therefore the question arises "which NI value with its associated UA value is to be used?" In what follows we shall answer the questions posed.

1.2 The Dynamic Test Sequence

Selection of an appropriate dynamic test sequence is difficult and outside the current scope of discussion. Certainly far more thought has to be given to the topic. Already two lines of thought have emerged in the SSTG group for the type of sequence to be used:

a) a test sequence which covers the extremes in the boundary conditions which may be encountered, but is unrepresentative of actual operating conditions.

b) a test sequence which employs limited specific but actual boundary conditions.

A real 5-day weather sequence was selected, March 1-5, 1959 (see Marshall (1983c) for complete details) whose statistical properties (mean values and frequency distribution) best matched the long term 20 year average of the transition periods. Therefore performance results can be analysed using the methodology underlying he SEU Simplified Design Method which is a predictor of long term system performance and thus provide a validation of the design method, see Marshall (1982a, 1982b, 1983b). The 5-day dynamic sequence chosen serves: a) to exercise the store using a real weather sequence that is real boundary conditions, b) to permit the validation of the generalized storage model, and c) to permit the validation of the SEU Simplified Design Method.

2. THE SIMULATED SPACE HEATING SYSTEM

Having selected the sequence for the dynamic test the boundary conditions are determined by the system imagined. The system descriptor must take into account any limitations of the test facility, e.g. power levels, flowrates, etc. The schematic of the system is seen as Fig. 1 and the key data for the descriptor are listed in Table 2.

3. TEST DATA

The measured variables (regardless of whether the loop is actually "on" or "off") are

$\dot{m}_c$ - the collector flowrate
$\dot{m}_l$ - the load flowrate
Tst - the top storage temperature
ΔT - the temperature drop across the store
Tas - the ambient about the store

from which the computed variables are found

Tsb - Tst - ΔT, store bottom temperature
T1 - Tsb (no pipe losses, no simultaneous mode) collector inlet temperature
T2 - collector exit temperature

T* - Tst (no pipe losses, no simultaneous mode) emitter inlet temperature

Tr - heat emitter exit temperature

given the weather parameters

G(t) - the irradiance on the tilted collector absorber

Ta(t) - the ambient around the collector

4. THE CONTROL MODES

During the daylight hours, 8:00 - 18:00, the output from a steady state collector equation is computed and if a positive temperature rise of 0.1°C is detected the collector is switched on, i.e. Mode C = 1. For the load side a more complex algorithm has to be used in order to account for a realistic heat extraction profile, the deadband in the room thermostat and the real capacitance of the heated space.

There can occur in our simulations, an interval before the end of an hour when the hourly load has been met and therefore the load loop is "off" corresponding to the reality where the upper thermostat set point has been reached and the building temperature begins its decay. The auxiliary requirement is established as the difference between the demand (QSHL) and that already supplied, QSOUT.

Very briefly four possible modes need be considered;

Mode C = 1; Mode L = 1 simultaneous operation not allowed to occur.

Mode C = 1; Mode L = 0 T2 - T1 > 0.1 8:00 < t < 18:00

Mode C = 0; Mode L = 1 Tr - Trm > 0.1 18:00 < t < 24:00 + 8

Mode C = 0; Mode L = 0 system off.

The actual flow rates within the store are then calculated from the product of the mode multiplied by the measured flow rate.

A compromise between the number of data logged and the accuracy required for the model to be validated has to be made, especially for the frequency with which the control status is checked and updated. A defensible interval for a total 5-day test, i.e. 120 hours, is 0.1 hrs that is 6 minutes (360 seconds). Six minute integrated average values of the primary measured variables, $\overline{Tst}$, $\overline{\Delta T}$, $\bar{\dot{m}}_c$, $\bar{\dot{m}}_l$, $\overline{Tas}$, Mode C and Mode L are formed and logged. From these values the auxiliary values $\overline{Tsb}$, $\overline{T1}$, $\overline{T2}$, $\overline{T^*}$, $\overline{Tr}$ are computed and logged. These two sets of data form the input to the dynamic test using the generalized model. Note that in the "off" mode (Mode C = Mode L = 0) no significance is attached to the measured and computed values as the relationships apply only in the flow mode configuration.

In addition to the 6 minute values of temperatures and flows, the 6 minute values of energy transferred are computed. The totals for each hour of the test are then logged and similarly the daily and the "annual" (5-day total) energy transfers are computed and logged.

5. THE ENERGY TRANSFERS

The integrated hourly, daily and annual (5-day total) average energy transfers considered are (in terms of the sample interval, dt = 360s): the incident irradiation on the collector, QCI; the energy transferred to the collector fluid (Mode C = 1), QCO1 or QCO2; the energy transferred during charging (Mode C = 1), QSIN1 or QSIN2; the energy transferred during discharging (Mode L = 1), QSOUT1 or QSOUT2; energy transferred out of the heat emitter (Mode L = 1), QHX1 or QHX2; reference store losses, QSLOS1 or QSLOS2; auxiliary required (on an hour basis only), QAUX.

The obvious question arises; why are there multiple definitions of similar quantities (e.g. QSIN)? The answer is that the first is the difference between integrated averages and the second is the average integrated difference. As the integration interval tends to zero there will be no difference (barring experimental errors) but a finite interval must be chosen remembering that too small an interval leads to rounding errors when adding up incremental values to a total hourly/daily/annual sum. Secondly there are different experimental errors associated with, for example, (Tst - Tsb) and ΔT. In short, multiple definitions permit an independent check on the measured accuracy and the numerical integration procedures (rectangular rule) for the chosen time interval.

6. VALIDATION OF THE MODEL

There remains only to select criteria for judging the outcome of a dynamic test. Here, the main criterion is taken to mean how well the response for the generalized model agrees with that for the actual device when subject to the same test conditions. During the charge mode, Mode C = 1, the model is driven with the 6 minute integrated values of $\bar{\dot{m}}_c$ and $\overline{Tst}$ and the response is the model's integrated average exit temperature $\overline{Tsb}(t)$ leading to an exit temperature error, ErrTe(t). Similarly when the store is being discharged (Mode L = 1) the model is forced by the measured $\dot{m}_1$ and $\overline{Tsb}$ values leading to the response at the top, $\overline{Tst}(t)$ and the error, ErrTi(t). Provided that the Root Mean Squared errors (RMS Ti, RMS TE) are within an arbitrary limit chosen as 5% of the maximum range of the variables, the model can be said to be validated.

There are, however, many other equally defensible criteria. For example the hourly/daily/annual integrated energy transfers can be compared. In particular, since it is the annual auxiliary energy which must be purchased it can be argued that one need only look at the error in this figure. However, from experience this parameter is too insensitive a measure and there are good reasons for selecting the more sensitive 6 minute averaged temperature profiles as the main criterion for judging the generalized storage model. Recall, it is the temperature level on which the control function must act and these control actions determine if and when energy can be collected and if and when energy can be supplied to the load.

7. RESULTS AND DISCUSSION

Three 5-day dynamic tests have been performed to date and these are denoted by DT1, DT2, DT3. The first, DT1, uses the actual weather data G(t), Ta(t) for Kew, England, March 1-5, 1959. The results obtained were entirely consistent with our expectation, in particular that the temperature swing of the store is low, 20-36°C, even during the transition period in the year (in winter it is lower!). Our storage device is in fact oversized and the melt temperature (52°C) is too high for a space heating duty. In order to test the generalized storage model under more extreme conditions the irradiance values, G(t) were simply doubled. This test is denoted as DT2. For every test the question of repeatability and accuracy must be examined. The third test, DT3, provides some insight into these questions as it is a repeat of DT2 with but minor differences.

For each test both the NI = 9 node generalized storage model and the NI = 3 node model response are analysed using of course, their respective UA values, Table 1. From the step response tests the NI = 9 model was seen to be only slightly preferable to the NI = 3 model, see Marshall (1983a). The complete description and comparison can be found in Marshall (1983c). For the sake of brevity only the DT2 results are discussed herein, as they are typical of all three tests.

7.1 Dynamic Test 1

The two models (NI = 3 and NI = 9) are best compared by looking at the summary, Table 3. Only the NI = 9 node model results in acceptable (less than 5%) errors in both the auxiliary required and the RMS errors in Te and Ti expressed as a percentage of the temperature swing, 20 - 56°C. But the NI = 3 node model leads to lower (and consistent) QSIN errors with a resultant lower RMS Te error while the NI = 9 node model leads to a lower (but less consistent) QSOUT error and a slightly lower RMS Ti error. Recall, again, that if cycling occurs on discharge discrepancies in the energy transfer computations can and do arise. In the work the NI = 9 model is again preferable and can be deemed to be valid.

Referring back to what is meant by "validation" we conclude that on the basis of the maximum instantaneous Ti(t) or Te(t) errors, the RMS Ti and RMS Te errors, and the QAUX error either model can be said to be validated. If the other energy transfers are used as a basis, Table 3, only the NI = 9 and model fulfils the 5% criteria with a caution that the influence of cycling has led to a not negligible discrepancy between the QSOUT1 and QSOUT2 quantities. That is, only the NI = 9 node model can be said to be valid.

A major conclusion is evident from Table 3 for the comparisons between the experimentally determined solar fraction (51.0%), the Cardiff Simplified Design Method (52.1%) and the model(s) (49.3%), NI = 9 and (47.7)% NI = 3.

The results indicate that the SEU Design Method (Marshall 1982a, 1982b, 1983b)) is a good predictor of the measured result that is, the SEU Design Method has been validated.

These same conclusions were also confirmed for the other dynamic tests, DT1 and DT3, Marshall (1983c).

8. CONCLUSIONS

The key conclusions can only be summarized in conjunction with the purpose and choice of the test sequence. Using a realist test sequence we conclude that:

a) the NI = 9 node generalized storage model with its associated UA(T) values meets, in general, both the loose (RMS errors and QAUX errors) and stringent (heat transfer in/out of the store) criteria for all three tests, DT1, DT2 and DT3, that is for the temperature swing 20-56°C. Therefore the NI = 9 node model can be deemed validated.
b) the NI = 3 mode generalized storage model with its associated UA(T) values meets both the loose and stringent criteria for just the first test, DT1, where the temperature swing is from 20-36°C.
c) the results for DT3 agree very well with those for DT2. Therefore the results show good repeatability.
d) the SEU Simplified Method differs from the measured by no more than 1.2 percentage points. The SEU Simplified Method is therefore deemed "validated".

REFERENCES

1. MARSHALL, R.H. (1982a). Modelling of Thermal Storage for Solar Heating Systems. Final Report, U.C. Cardiff, Report 866/SEU 313.

2. MARSHALL, R.H. (1982b). Evaluation of Thermal Storage for Solar Heating Systems. Proc. EC Contractors' Meeting, Meersburg BRD, 14-16 June 1982, eds. Palz, W. & den Ouden, C., D. Reidal Publishing, pp 363-373.

3. MARSHALL, R.H. (1983a). A Generalized Storage Model - Cardiff Subtask 1. Final Report, U.C. Cardiff Report 1021/SEU 400.

4. MARSHALL, R.H. (1983b). Evaluation of Thermal Storage for Solar Heating Systems - Final Report. EC Contractors' Meeting, Brussels 1 - 3 June 1983, eds. Palz, W. & den Ouden, C., D. Reidal Publishing. (In press)

5. MARSHALL, R.H. (1983c). Dynamic Testing - Cardiff Subtask 2, A Final Report. U.C. Cardiff Report 1028/SEU 407.

Temperature Interval	Specific Heat Capacity kJ/kg K	NI = 9 UA(T) W/K	NI = 3 UA(T) W/K
10 - 20	2.25	732	793
20 - 30	3.57	732	793
30 - 40	5.21	1165	1686
40 - 50	6.50	652	717
50 - 60	14.41	1152	1433
60 - 70	2.34	652	717

TABLE 1. Generalized Storage Model Parameters

Collector

$U = 4.5\ W/m^2\ K$ $\eta_o = 0.75$ A_c 15 m^2

$\dot{m}_c = 0.15$ kg/S Cp = 4185 J/kgK

Heat Distribution System

$\frac{\overline{\dot{m}Cp}}{\overline{\dot{m}_1 Cp}}$ $\xi = R\xi = 0.375$ Warm air heating with a fan coil and parallel auxiliary

$\dot{m} = 0.15$ kg/S Cp = 4185 J/kg K

Load

$UA_B = 150$ W/°C $q_1 = UA_B\ (Trm - Ta(t))$

Trm = 20°C $q_L = 0$ 8:00 → 18:00 hrs.

TABLE 2. The System Descriptor

Q-Errors		NI = 9		NI = 3	
	Measured MJ	Model MJ	Model % error	Model MJ	Model % error
QSI1	291.3	294.9	1.2	292.8	0.5
QSI2	269.5	306.6	13.8	284.9	5.7
QSO1	219.4	212.1	-3.3	205.3	-6.4
QSO2	262.2	241.8	-7.8	222.7	-15.0
QLOS1	14.3	14.1	-1.3	13.9	-2.3
QLOS2	12.8	12.5	-2.3	12.7	-1.0
QHX2	219.4	212.1	-3.3	205.3	-6.4
QAUX	210.7	218.0	3.5	224.8	6.7
QSHL	430.0	430.0	0.0	430.0	0.0

Temperature Errors	NI = 9		NI = 3	
RMS TI (224 points)	1.031°C	2.9%	1.191°C	3.3%
RMS TE (301 points)	0.802°C	2.2%	0.581°C	1.6%

Design Method Comparison	NI = 9	% error	NI = 3	% error
Measured	51.0	-	51.0	-
SEU Design Method	52.1	2.2%	52.1	2.2%
Model	49.3	-3.3%	47.7	-6.5%

TABLE 3. Summary of 5-day Performance, DT1

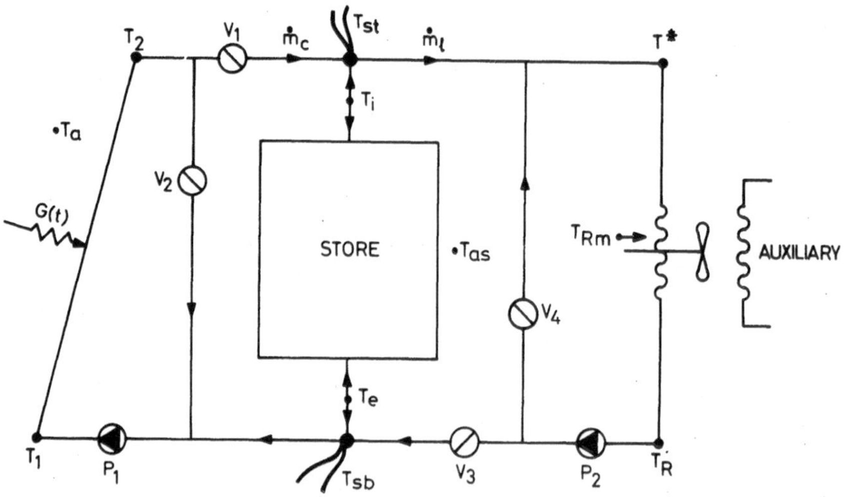

FIGURE 1. Schematic of the Dynamic Test Facility

SALT GRADIENT SOLAR POND
FOR SOLAR HEAT COLLECTION AND LONG TERM STORAGE

P.J. UNSWORTH AND N. AL-SALEH
School of Mathematical and Physical Sciences
University of Sussex, Brighton, Sussex, BN1 9QT, U.K.

Summary

An experimental salt gradient has been built to study the modelling, construction, operation, and monitoring of solar ponds. A new method of heat extraction is described whereby heat is extracted via heat exchangers placed in both the insulation and storage zones of the pond. Extraction from the insulation layer intercepts heat which would otherwise be lost to the surface and permits considerable increase in overall pond efficiency. Efficiency curves for steady state operation are presented. The external surface heat transfer coefficient between a tubular heat exchanger and the non-convecting pond fluid has been measured, as well as the increase in the rate of upward salt diffusion caused by heat extraction from the non-convecting zone.

1. THE EXPERIMENTAL SOLAR POND

The solar pond was constructed in September 1982. It is cylindrical of 4.5 m diameter and 1.2 m high above ground(1).It is formed by bolting together rectangular sheets of galvanised sheet steel. Earth was dug out and replaced by coarse sand to a depth of 0.6 m to give a stable smooth low conductivity floor.Thermo-couples sealed in stainless steel tubes were placed in the ground beneath (and later in the pond) to measure temperatures. The tank was then lined with a flexible liner of black butyl rubber 1 mm thick, sitting directly on the sand. The construction is as used for tanks for fish farming or water treatment.

The pond was filled and a 0.75 m non-convecting insulation layer established by using a diffuser(2)to give a specific gravity varying linearly from 1.00 at the top to 1.08 at the bottom. Beneath the insulation layer is a non-convecting storage layer of depth 0.2 m.

To prevent algae growth, 0.6 ppm of $CuSO_4$ was added during filling, and NaOH was used to give a pH of between 5 and 6 to keep the $CuSO_4$ in solution. After 8 months, clarity started to deteriorate, apparently due to both algae and wind blown dirt and leaves. After 1 year pond acidity had increased so more NaOH was added and $CuSO_4$ increased to 6 ppm. After 18 months, there is no algae problem.

A severe water clouding problem was encountered recently, however. For heat extraction, a copper heat exchanger was used, as copper is regularly used in sea water installations. The clouding came suddenly over a period of a week after 5 months of heat extraction and heat injection experiments. It was due to copper oxides and hydroxides, arising from corrosion of the heat exchanger, which spread around the pond during experiments involving high heat transfer rates between the heat exchanger and the non-convecting zone. The corrosion rate was being accelerated by galvanic action involving the heat exchanger tubing, the salt solution, and other metals in the pond (stainless steel around thermocouples, a lead weight, and brass screws).

Our conclusion from this is that use of metals should, if possible, be avoided, or they should be chosen to minimise corrosion and formation of insoluble products which may cloud the pond.

2. HEAT EXTRACTION FROM THE INSULATION LAYER

Conventionally, heat from solar ponds is extracted from the storage layer (lower convecting zone), which is the hottest part of the pond. If less than maximum temperatures are required, or if the inlet water temperature is low, greater thermodynamic efficiency can be achieved by heat extraction over a suitable region of the insulation layer (non-convecting zone) across which the temperature varies from ambient at the top, to the storage temperature at the bottom. The insulation layer is heated by direct absorbtion of solar radiation passing through the layer, and by the heat flow from the hotter storage layer towards the pond surface where it is lost. By extracting heat via a heat exchanger placed in the insulation layer, this heat flow may be intercepted and utilised.

Two potential problems arise in doing this:

i) To extract heat from the non-convecting zone of the pond, the heat transfer coefficient between the heat exchanger outer surface and the non-convecting pond fluid may be too small.

ii) The temperature difference between the heat exchanger and the pond may induce unwanted convective mixing and increase the upward salt diffusion rate.

Laboratory experiments in a small glass tank were used to observe the convective motion between the heat exchanger tubing and a salt gradient layer. Fluorescene dye and/or Cu_2O powder was used to make the fluid motion visible. Two effects were seen: (a) convective motion was essentially horizontal and laminar (b) the density gradient tended to break up into weakly visible bands of about 1 cm thickness, with alternating directions of outward and return horizontal convective flow. The bands were self healing over an hour or two after heat extraction was stopped. Experiments over a range of heat transfer rates show that the external film heat transfer coefficient h_2 is around 300 $Wm^{-2}C^{-1}$ (see section 3 below).

To test the salt mixing effects of a heat exchanger in the non-convective layer, a vertical array of plastic tubes at 5 cm intervals was installed to enable pond samples to be extracted for laboratory density measurement using a specific gravity bottle, giving results to $\pm$ 3 x 10^{-4} with careful technique. The effects on salt density distribution were monitored over a period of 3 months. To enable greater heat transfer rates and temperature differences to be used, heat was injected into the pond, rather than extracted from it, using a 3 kW auxiliary heater. Heat was transferred continuously at rates up to 2 kW, with breaks of not less than 12 hours before density sampling to allow the pond to return to density equilibrium.

2.1 Salt Mixing Effects

The experimental heat exchanger consisted of 14 m of 8 mm diameter copper tubing, in a 10 stage zig-zag configuration in the insulation layer between 0.25 - 0.75 m below the pond surface. At the highest rates of heat injection, up to 40° C temperature difference existed between the inlet fluid and the pond solution, and simulated the worst conditions for induced salt mixing. The history of pond density was monitored over a period of 76 days, and included two periods of heat transfer, separated by a gap of 36 days to give a comparison of the natural and induced diffusive mixing rates under similar weather conditions (calm winter weather, little or no direct sunlight). Figure 2 shows the overall density change over the whole 76 day period, and over the 36 day period with no heat transfer. The results show that under these worst conditions, the salt diffusion rate was increased by a factor of about six over the normal rate.

3. PERFORMANCE MODELLING

To investigate the possible improvements in operating efficiency and temperature by heat extraction from the insulation layer, we have modelled the pond by a finite difference calculation, looking at steady state behavior to give insight into parametric dependence. In a previous paper (2), we described an analytic method based on the steady state model of Wang and Akbarzadeh (3). In the present work we have modelled fully the behaviour of the heat exchanger, which rules out an analytic solution.

Assuming an infinite pond, the heat flow equation at depth x below the top of the insulation layer is

$$C\frac{\partial T}{\partial t} = -\frac{\partial \Phi}{\partial x} + k_w \frac{\partial^2 T}{\partial x^2} - \dot{Q}$$

where $\Phi(x)$ is the solar flux reaching depth x, and is given by

$$\Phi(x) = \tau\bar{H}\left\{a - b\ln\left[(x+\delta)/\cos r\right]\right\}.$$

$\dot{Q}$ is the heat extraction rate per unit volume, C is the pond heat capacity per unit volume, k_w the pond solution thermal conductivity. $\bar{H}$ is the mean annual solar insulation at ground level, τ the angle of refraction. a and b are constants (4). δ is the depth of the surface mixing layer which is taken to be at the mean ambient temperature.

In the storage layer, heating depends on the balance between the radiation flux $\Phi(d)$ entering the layer (assumed to be completely absorbed) and the sum of the heat losses into the ground and to the insulation layer, and of the heat extraction rate $\dot{Q}_s$ per unit area from the storage zone:

$$C\,d_s\frac{\partial T_s}{\partial t} = \Phi(d) - \frac{(T_s - T_{amb})}{(Di/ki + Dg/kg)} - k_w\left.\frac{\partial T}{\partial x}\right|_{x=d} - \dot{Q}_s$$

d_s is the depth of the storage layer. D_i is the thickness of insulating material under the pond. D_g is the ground thickness below which temperature is taken to be equal to ambient. k_i and k_g are the respective thermal conductivities.

3.1 Heat Exchanger Equations

The transfer rate from the pond to the heat exchanger fluid is

$$\dot{Q} = \dot{m}c\,(T_{fo} - T_{fi}) \tag{1}$$

where T_{fo}, T_{fi} are the fluid outlet and inlet temperatures, $\dot{m}$ the mass flow rate, and c the specific heat. Over a length of tubing dz, the transfer rate is

$$d\dot{Q} = (T_p - T_f)\,U 2\pi\, r_2\, dz$$

where T_p, T_f are the pond and fluid temperatures, r_2 the outer tube radius. U is the overall heat transfer coefficient per unit outer area for the heat exchanger given by

$$\frac{1}{U} = \frac{r_2}{r_1}\frac{1}{h_1} + \frac{r_2}{k} \ln\left(\frac{r_2}{r_1}\right) + \frac{1}{h_2}$$

r_1 is the inner tube radius, k the tube thermal conductivity, and h_1, h_2 the inner and outer surface heat transfer coefficents. h_1 is calculable by standard techniques for water flow in cylindrical pipes(5).

Assuming that the pond temperature varies linearly with distance along the tubing (total length L), the outlet temperature is found to be given by

$$T_{fo} = T_{po} + (T_{fi} - T_{pi})e^{-L/\ell} - \frac{\ell}{L}(T_{po} - T_{pi})(1-e^{-L/\ell}) \qquad (2)$$

where T_{po}, T_{pi} are the pond temperatures at the levels of the inlet and outlet, and $\ell = \dot{m}c/(2\pi r_2 U)$.

Equations (1) and (2) may be used to model the heat transfer over different portions of the heat exchanger to fit in with the finite difference calculation for the pond.

3.2 Pond Efficiency

Figure 1 shows efficiency plots for a solar pond in steady state conditions with a heat exchanger extracting heat from the storage layer only (curve a), from both the storage and insulation layers (curve b) and from the insulation layer only (curve c). Liquid flow rate through the heat exchanger is used to control the amount of heat being extracted in each case. The pond data used is as follows: δ = 0.2 m, d = 1.5 m. The lower convecting layer thickness affects storage time constant, but not steady state performance. D_i = 0.1 m, k_i = 0.035 $Wm^{-2}K^{-1}$, D_g = 5 m, k_g = 1.0 $Wm^{-2}K^{-1}$, k_w = 0.565 $Wm^{-1}K^{-1}$, τ = 0.85 a = 0.36, b = 0.08, r = 39.2°.

For heat extraction from the insulation layer, the level of the heat exchanger inlet should ideally be at a height in the pond where the pond temperature equals the inlet temperature. This can be achieved by being able to raise or lower the inlet by tilting the heat exchanger. Alternatively a number of different inlet points may be provided, selectable by magnetic valves. Similarly, the lower outlet end of the heat exchanger may be adjusted to give the required outlet temperature.

For the curves presented, the inlet temperature was taken to be at ambient temperature, and the heat exchanger to span the whole depth of the insulation layer. It is seen that extraction from the insulation layer alone gives the greatest improvement in efficienty. At the higher operating temperatures Θ/H = 0.3, or 0.4, the efficiency is improved by 70% and 88% respectively. Somewhat reduced improvement is obtained with higher inlet temperatures. However, a number of applications do involve input temperatures not much above ambient e.g. crop drying, swimming pool heating, greenhouse heating, in which case the efficiency improvements are considerable. Calculations and experimental tests are continuing.

4. CONCLUSIONS

An alternative method of heat extraction from solar ponds is described which can give very considerable improvement in pond efficiency - almost double under suitable circumstances. The value of the external surface heat transfer coefficient between the heat exchanger tubing and pond solution, which can convect only in horizontal directions, has been measured for use in heat exchanger design. Under circumstances of very high heat transfer rates and temperature differences, the upward pond salt diffusion rate appears to be increased by sixfold, but this should hopefully decrease to an acceptable figure under reasonable conditions.

REFERENCES

1. UNSWORTH, P.J., AL-SALEH, N., PHILLIPS, V. Solar Energy R. & D. in the European Community: Solar Energy Applications to Buildings, A1, 155, 1981; A2, 354, 1982; A4, 409, 1983. Publ. Reidel, Dordrecht.
2. UNSWORTH, P.J., AL-SALEH, N., PHILLIPS, V. Conference on Long Term Energy Storage in Solar Systems, (C35), U.K.-ISES, 44-55, 1984.
3. WANG, Y.F. and AKBARZADEH, A. Solar Energy 30, 555-562, 1983.
4. BRYANT, H.C. and COLBECK, H. Solar Energy 19, 321-322, 1977.
5. BAYLEY, F.J., OWENS, J.M., TURNER, A.B. Heat Transfer. Publ. Nelson, London, 1972.

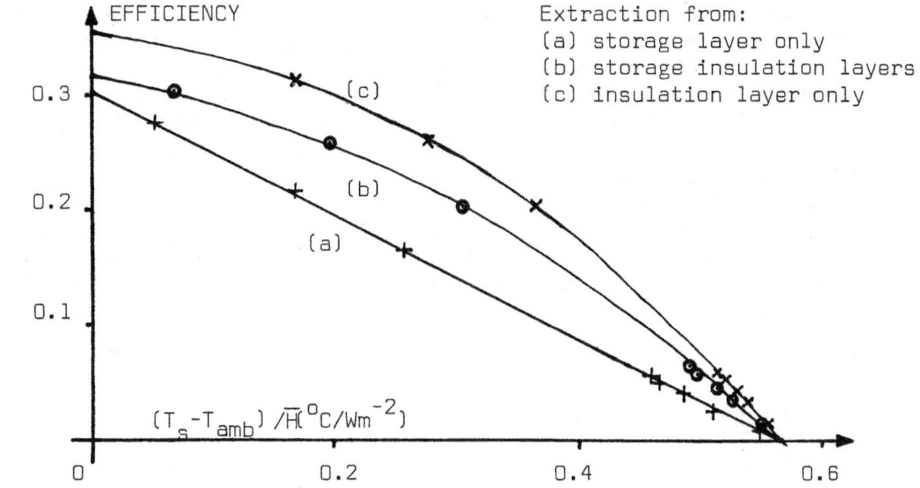

Figure 1. Pond operating efficiencies.

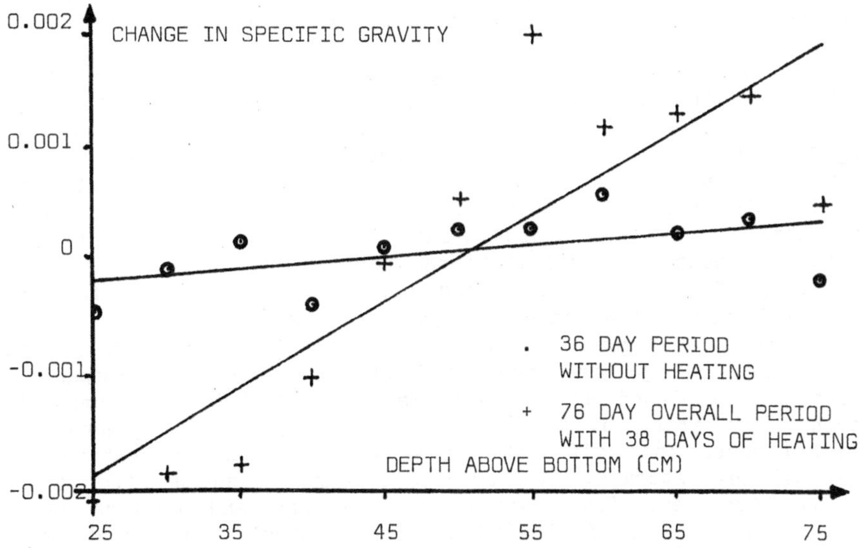

Figure 2. Changes in pond specific gravity

SOLAR HEATING WITH ANNUAL HEAT STORAGE - MODELLING AND PRACTICE

P.D. Lund and S.S. Peltola
Helsinki University of Technology,
Department of Technical Physics,
SF-02150 Espoo 15, Finland

Summary

Central solar heating systems with seasonal heat storage are recognized as one of the most potential forms of solar energy utilization at northern latitudes. Because of the high investment costs associated, special attention has to be paid on system design and component dimensioning in order to guarantee optimal thermal performance. Also, the complexity and dynamics of these systems normally prevent the use of simple rule-of-tumbs for system sizing and numerical mathematical approaches have to be employed. In this paper, computer simulation programs have been used for determining the performance and energy flows of a full-scale district solar heating system with annual heat storage, the Kerava solar village (60°N). Simulated performance has been compared to measured values and a satisfactory agreement was found for most of the energy flows within the heating system. On a yearly basis, a 40-50 percent solar fraction may be expected in the Kerava solar village. The study also reveals the importance of system control on the overall performance.

1. INTRODUCTION

During the last few years, an increasing interest has been shown on large central or district solar heating systems with long-term heat storage. Several full-scale experimental projects have been commenced especially among the IEA-countries (1). Simulation programs for these systems have also been developed especially for the storage-unit (1).

The first Finnish district solar heating system with seasonal storage, the Kerava solar village near Helsinki (60 deg.N), was introduced in the beginning of 1983 and a comprehensive measurement program was started in June. The research program will be accomplished during 1983-86 and its main purpose is to determine the performance and the energy balance of the village as also to assess the technical and economic potential of energy systems of this type. Besides the experimental activity, computational simulation models have been developed for component optimizations, performance studies and to assess the effects of system control and operating strategy on the overall thermal performance of a community solar heating system.

The system configurations simulated in this paper correspond to the Kerava solar village in order to make possible intercomparison of measured and simulated system performances. Two different computational approaches are employed. The NORSOL program (2) is a general-purpose simulation model for different types of central solar heating systems with annual heat storage. This program accounts accurately for transient and convective heat flows within the storage and employs one hour calculation time steps. In the NORSOL, the control of the heating system does not exactly correspond to that in the Kerava solar village. The other FORTRAN program, the KERCONT(2), simulates accurately the control system of Kerava. To prevent prohibited long computer times, the thermal processes in the surroundings of the storage-unit are just generally described. The heat losses are partly calculated by the NORSOL and given as input to the KERCONT. The simulations in KERCONT are performed with a fixed short time step (5-15 min).

2. DESCRIPTION OF THE HEATING SYSTEM IN KERAVA SOLAR VILLAGE

A combined solar heating and heat pump system with annual storage is used in the Kerava solar village. As a comprehensive description of the heating system and operational principles has been reported previously (3), only a brief discussion on the matter is given here.

The village comprises 44 flats altogether with a living area of 3756 m^2. Major part of the triple-glazed windows is placed on the southern facades to increase the passive solar contribution. Table I shows a resume of the technical data.

TABLE I TECHNICAL DATA FOR KERAVA SOLAR VILLAGE

Solar collectors	Heat storage	Residential area
Parameters: $F_R(\tau\alpha)=0.83$ $F_RU_L=6.2$ W/m^2/K	Type: rock cavern Volume: 1500 m^3 (water) 11000 m^3 (rock) (54 boreholes)	Type: terraced houses Floor area: 67,82.5 and 100 (m^2)
Area: 1 100 m^2	Capacity: 250 MWh	Heating system: air heating, heat recovery of exhaust air
Tilt: 70° and 90^0 Expected energy yield: 325-455 MWh	Temp. range: 10-70°C	Expected heat load: 500 MWh/yr

The solar collectors are placed on the roofs and walls of the southern facades of the houses. The heat from the collectors is conducted directly into a suitable water layer of the rock cavern storage corresponding to the fluid temperature in order to maintain the storage stratified. The rock surrounding the water storage is also employed by two circles of boreholes. The holes of the inner circle may be used to charge the rock in the summer if the water temperature in the rock cavern exceeds 60-70°C. During the discharge of the rock, cooled water from the evaporator of the heat pump is circulated through the outer circle of holes to reduce heat losses from the rock storage. The heat pump is electric-driven with a maximum compressor effect of 260 kW and is employed for efficient storage. A conventional electric boiler (400 kW) serves as a back-up. The control and measurements in the village are based on microcomputer utilization and over 140 variables are measured. Typically, one month's measurements give 27000 numbers and the data analysis is performed on a separate computer at the university.

3. THE COMPUTATIONAL MODELLING APPROACHES

As a result of systematic program development at the Helsinki University of Technology since 1979, comprehensive numerical computer

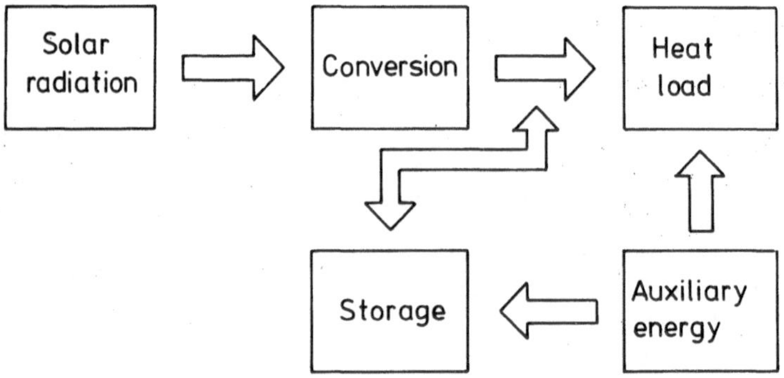

Figure 1. A schematic of the modelling approach.

models have been prepared for solar heating systems with different heat storage means. A leading motive in the development work has been the consideration of the energy system as a whole and to account for transient heat flows within subsystem components which typically require hour-by-hour simulations. A schematic of the modeling principle is shown in Fig. 1. The present program packages comprise the following main subsystems: solar collectors, auxiliary heating systems, heat storage, heat load and a routine for generating weather data. The heat storage has been considered as a central component. The FORTRAN program NORSOL has been coded around a rock cavern with a duct storage in the ground corresponding to that employed in the Kerava solar village. NORSOL (2) has previously been used to calculate the expected energy use of the Kerava solar village (3).

A central component in the NORSOL is the storage-unit through which the major part of heat flows are conducted. In NORSOL, the heat flows originating from the solar collectors, auxiliary heating system and heat load are each fed separately to a suitable layer in the water storage according to their temperature. Thus, the number of required winch mechanisms is four in the NORSOL in order to maintain stratification in the water storage. However, in the Kerava solar village, above-mentioned energy flows are intermixed reducing the number of required winches to one, but causing at the same time quality losses of heat and yielding a lower second-law efficiency. To assess the influence of the different control schemes employed, a FORTRAN program KERCONT (2) was prepared in which the control system has been modelled to represent accurately that of Kerava. Both the NORSOL and KERCONT employ separate subroutines for accurate simulation of the subsystem components. The storage-unit is described by an energy-balance equation. One year's simulation of the thermal performance of the Kerava solar village with the NORSOL requires about 60 min. of CPU-time on a UNIVAC 1100/E61 and with the KERCONT about 30 min., respectively.

4. COMPARISON OF MEASURED AND SIMULATED SYSTEM PERFORMANCE

The measurements in Kerava solar village started in late May 1983 and the period available for comparison is from June 1st to December 31st. During the period in question several system experiments and adjustments were made which have affected some of the energy flows. Consequently the electricity demand was increased even though the heat content of the storage would have been adequate for heat supply. The system simulations were performed assuming undisturbed operation. To make possible the comparison of theory and practice, the results have to be scaled.

A comparison of the simulated and measured energy flows is presented in Table II. It is observed that the measured collector output fits surprisingly well with the simulation results. The discrepancies found are less than 3 percent. On the other hand, the storage utilization has been significantly lower during the measurement period than expected. This is mainly due to the several tests for which auxiliary energy was needed.

TABLE II COMPARISON BETWEEN MEASURED AND SIMULATED SYSTEM PERFORMANCE IN KERAVA SOLAR VILLAGE (JUNE-DECEMBER 1983), ALL UNITS IN MWh

	Storage losses	Auxiliary	Heat load	Solar yield collector	Heat in storage, Dec. 31
Simulated	69.	100.	242.	158.	0.
Measured	138.	246.	262.	154.	32.
Measured*) without tests	95.	169.	262.	154.	10.

*) the extra auxiliary use due to the several tests and system experiments have been deducted from the results.

When the effects of the system experiments are accounted for and excluded from the measured results, the differences are reduced. It should also be noticed that the figures given by the NORSOL include the rock storage utilization which reduces the storage heat losses by 25 MWh per year. The boreholes were not employed during the first measurements period. Later on these have been coupled to the heating system and the discharge of the rock storage has proceeded well. Compared to the NORSOL, the realization of the control system in Kerava would cause an appr. 10-20 MWh increase in the auxiliary demand, or 5 percent of the yearly heat demand. Accordingly, the storage-unit is not employed as effectively as possible. This is well illustrated by Fig. 2, which shows the heat content of the storage for the different modeling approaches of the control system. The expected solar fraction given by NORSOL would be about 50 percent and when accounting for the differences in the control system a 40-45 percent solar contribution would be achievable. If considering the passive yield as a heat input, the solar fraction would still increase well beyond 50 percent.

5. CONCLUSIONS

A comparison between measured and simulated system performance of the Kerava solar village has been made. Even though the available measurement results are confined to a limited period (June-December 1983), some preliminary conclusions may be drawn from the present study.

It was found that for separate system components relatively accurate models can be accomplished. When considering the energy system as a whole taking into account component interactions and the system control, the simulation work becomes more complicated. In particular, the system control and operating strategy were found to have an important influence on the total system performance.

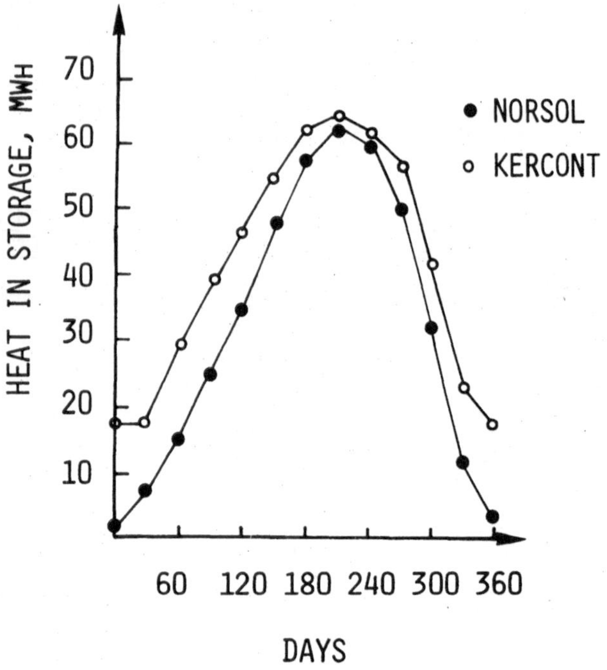

Figure 2. The heat content of the storage. NORSOL employs a 5°C reference temperature and KERCONT 10°C, respectively.

ACKNOWLEDGEMENTS

The financial support of the Ministry of Trade and Industry is gratefully acknowledged.

REFERENCES

1. Swedish Council for Building Research (1983). Proc. of the Int. Conf. on Subsurface Heat Storage, Stockholm, June 6-8, 1983, Vols. 1 and 2.
2. Lund, P.D. (1984). Numerical computer programs for community solar heating systems with different types of storage. Helsinki University of Technology, Report TKK-F-A541.
3. Lund, P.D., Routti, J.T., Mäkinen, R. and Vuorelma, H. (1983). Simulation studies of the expected performance of Kerava solar village. Energy Research 7, 347-357.

SEASONAL STORAGE OF BUILDING WASTE HEAT IN AN AQUIFER

Edward Morofsky
Public Works Canada
Energy Technology
OTTAWA K1A 0M2
Canada

Verne Chant
James F. Hickling
Management Consultants
OTTAWA K1R 7S8
Canada

Thomas LeFeuvre
National Research Council
Solar Energy Program
OTTAWA K1A 0R6
Canada

Summary

The Scarborough Government of Canada Building will have seasonal aquifer thermal energy storage associated with it. The building also has short-term storage tanks for heating and cooling. Possible energy sources for aquifer storage include solar, building waste heat, and ambient air temperatures. The site has two aquifers. The lower aquifer is confined and exhibits excellent hydraulic properties. The upper aquifer is unconfined with inferior hydraulic properties. The natural water temperature in the lower aquifer is not low enough, nor is the regional flow great enough, to effectively cool the building without the injection of chilled water. The building design was modified for aquifer storage at a late stage. This necessitated approximate design calculations. Complete backup systems are in place. Cold storage in the lower aquifer is preferable, due to the naturally lower water temperature and the large building cooling demand. Fewer technical problems are expected with the lesser temperature difference between the injected and natural waters. Thus experimentation will begin with cold water storage, but hot water storage may be attempted later. This paper examines the feasibility of using the upper aquifer for hot water storage, while simultaneously storing cold water in the lower aquifer. Such an option would involve waste heat recovery, solar energy collection, and heat pumps to meet the space heating requirements of the building. Construction costs are based on actual project costs.

Introduction

Construction of the Scarborough G.O.C.B. began in February, 1983. It will be ready for occupancy in December, 1984. During the final design stage in early 1982, two aquifers were discovered at the site separated by a thick clay layer, a lower confined aquifer and an upper unconfined aquifer. Subsequent investigation indicated the suitability of the lower aquifer for storage requiring high injection and withdrawal rates. It was decided to sink four wells at the extremities of the building site into the lower aquifer. The well field configuration resulted in a rectangle of length 130 metres and 65 metres width. The necessary building design adaptations were also made, including more sophisticated controls on the cooling tower, appropriate heat exchangers between the aquifer storage and building systems, and the necessary system controls (Public Works Canada, 1983).

The building has a large cooling load for most of the year (see Fig. 1). It has a balance temperature of $-26\ ^{o}C$ when occupied. The building mechanical system design includes low temperature water radiation heating, heat pump configured chillers, solar energy colletion via evacuated tubes, and short- term energy storage tanks. Cold storage in the lower aquifer involves pumping water to the surface during winter, using the idle cooling tower to chill the water close to freezing, and reinjecting the water into a companion well until needed in summer. The natural temperature of the aquifer (9 °C) is suitable for limited use as a heat sink. The excess heat would normally be dumped by the cooling tower at temperatures ranging from 35 °C in summer to 43 °C in spring and fall when heat is being reclaimed for perimeter heating (Morofsky, 1983).

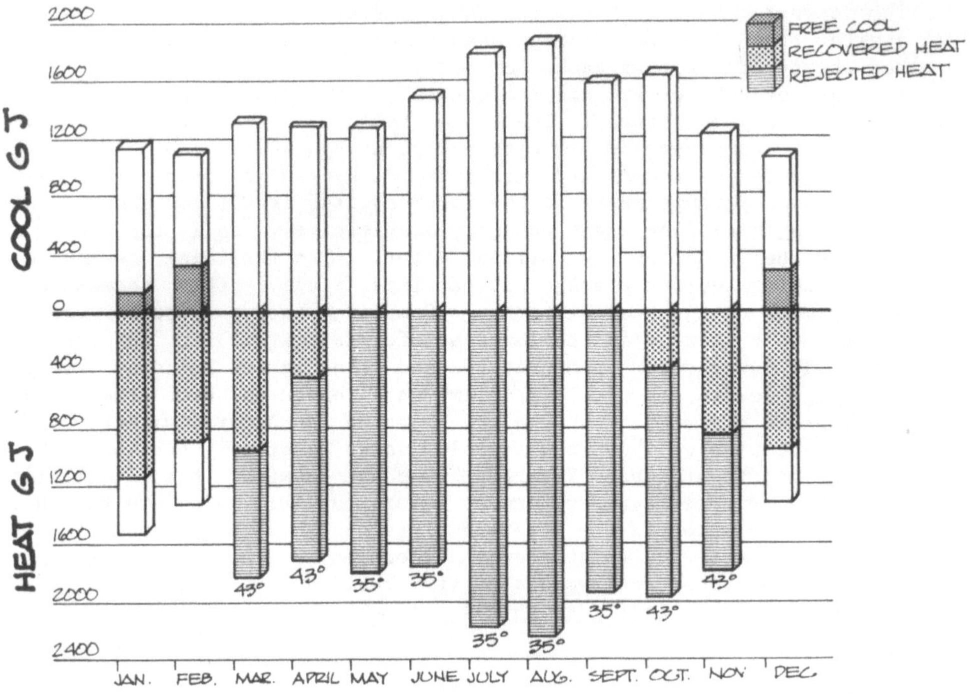

Figure 1. Monthly building heating/cooling demands and rejected heat.

Scarborough, Ontario, Canada is immediately east of the city of Toronto and north of Lake Ontario, at about 44° North latitude and 79° West longitude (see Fig. 2). The Scarborough GOCB has 29,000 square metre of usable floor area on 14 floors of offices and services. The building site (20,000 square metres) is 700 metres south of a major east-west highway (Route 401), on a slight rise, at an elevation of 170 metres above sea level. The building will be founded at about 160 metres elevation. The prevailing groundwater level is 159 metres. It is bounded on the north and southwest by the Scarborough Town Centre and the Scarborough Civic Centre. Intensively developed adjacent areas include an elevated track of the light rapid transit system. The aquifer storage is located directly beneath the building.

Use of the natural water temperature, without reinjection, would not supply a significant fraction of the cooling load. The amount of cooling energy that could be extracted by fully charging the two doublet system during a typical Scarborough winter is estimated to in the range of 6 to 10 TJ. Construction costs for the lower aquifer were $200,000 for the four wells and $150,000 for the associated piping, pumps, and heat exchangers incremental to the original design without aquifer storage.

Figure 2. Location of Scarborough on the map of Canada.

Scope of the study

This study examined the suitability of using the upper aquifer for hot water storage for space heating. The space heating requirement assumed to be met by the waste heat recovery and aquifer system is 1 TJ over the period December through February. For analysis purposes this requirement was distributed as 250 GJ in the months of December and February and 500 GJ in January. Minimum delivery temperature to the load is 41 °C which returns at 35 °C. Heat recovered from the building during May through September at 35 °C would be stored in the aquifer and a heat pump would upgrade storage temperature as required for the load.

Approach

Once the net space heating load was defined and the amounts and temperatures of recoverable heat determined, the study focused on the simulation and design of an aquifer-based seasonal storage scheme using waste heat recovery with a heat pump. It was decided to use the MINSUN system simulation program (Chant and Hakansson, 1983) developed under the International Energy Agency Task VII. With minor modifications this program was able to represent the specific load characteristics (December-February only) and heat recovery from the building. The storage characteristics of importance in the simulation were the energy flows in and out on an annual basis and temperature, especially minimum and maximum. Table 1.1 summarizes the system thermal performance under this concept.

TABLE 1.1 System Thermal Performance

Energy to storage (TJ)	1.6
Net storage loss (TJ)	0.8
Storage to load (TJ)	0.8
Heat pump electric (TJ)	0.2
Total load (TJ)	1.0
Heat pump COP	5.0

Table 1.2 summarizes the cost analysis for this concept and the conventional alternative of using all auxiliary energy. This concept is especially suitable to new building design where waste energy and inexpensive, large-scale storage are available. The cost of this concept can essentially be paid for with energy savings (Chant, LeFeuvre, Morofsky, 1983).

TABLE 1.2 SYSTEM COSTS

	<Bldg Waste> Heat	<Auxiliary> Energy
Capital Cost $000		
Aquifer Storage	120	0
Heat Pump System	17	0
Average Annual Cost $000/a		
Aquifer Storage	9.2	
Heat Pump System	1.3	
Auxiliary Energy	2.4	13.2
Total	12.9	13.2
System Specific Cost (cents/ kWh)	4.7	4.7

<u>Aquifer Characteristics</u>

The upper aquifer is 20 metres thick and occurs 20 metres below ground surface. It is composed of fine sand and has hydraulic properties inferior to those of the lower aquifer. The two aquifers are separated by 10 metres of clay. The lower aquifer will be used to store chilled water at a temperature of about 4.8 ^{o}C. Its natural temperature is about 9 ^{o}C. It has a regional flow of 24 metres per year, a thermal conductivity of 1.56 J/m.s.K, a density of 2 670 kg/m^3, a heat capacity of 2.03 MJ/m^3, and a porosity of 0.28. This storage capacity will enable a large proportion of the summer cooling requirements of the building to be met. We have simulated the use of the upper aquifer for hot water storage to meet the winter space heating requirements of 1.0 TJ. Two wells would be used in the upper aquifer. The hot well would be immediately above the warm well (about 12 ^{o}C) of the cold storage system. This would minimize losses between the two storage systems.

The upper aquifer has not been as extensively investigated as the lower aquifer. The piezometric surface is about 10 metres above the aquifer top, the porosity is in the range 0.3 to 0.35 (compared to 0.28 for the lower), the yield is considerably less than the lower aquifer (about 10 litres per second) resulting from its lower permeability. Regional flow, storativity and transmissivity are unknown. Well costs for the upper aquifer would be less than $10,000 per well based on the reduced depth and flow requirements as compared to the lower aquifer wells.

The marginal energy cost at the Scarborough site is 1.8 cents/kWh with a demand charge of $3.23/kW on a monthly peak basis. Average per unit cost was 3.4 cents/kWh in 1983. An energy cost of 4 cents/kWh was used in this study.

Costs

All costs are in 1983 Canadian dollars and are based on actual work already performed at the Scarborough site. The cost of developing the upper aquifer for hot water storage to meet the building load is expected to be approximately $120,000 ($20,000 for two wells plus $100,000 for piping, pumps and heat exchangers).

Seasonal aquifer storage with heat pump upgrade is an economical option, because waste building heat is available and the building design modifications are relatively inexpensive. In fact, using the preliminary estimates available under this study, the heat recovery, aquifer storage and heat pump extraction concept would be cost competitive with the auxiliary energy supply. This favourable cost depends on the following factors:

- this is a new building whose design and construction take into account using the available aquifer for energy storage and management systems;
- the aquifer storage is available at very low cost, in the order of $1 to $2 per cubic metre equivalent;
- excess heat, although low grade and available at a different time of year than when needed, is essentially free.

Because of these factors, per unit useful energy cost would be cost competitive with auxiliary energy cost and about one half the cost if a greater load were immediately available.

Conclusions

Seasonal heat storage in an aquifer based on waste heat from the building is cost competitive at the Scarborough building. This result is based on having mechanical systems designed for the purpose and large amounts of waste heat available compared to the net heating requirement. Aquifer storage costs were based on actual field experience and may vary at other sites. The efficiency of aquifer storage is based on modelling results and can only be confirmed by operating experience. Additionally, heat storage may not be competitive with cold storage at Canadian sites where conditions do not permit both.

References

Chant, V., LeFeuvre, T. and E. Morofsky (1983) Feasibility analysis of solar energy storage in an aquifer (draft)- Canadian contribution to site specific preliminary design of Task VII - Central Solar Heating Plants with Seasonal Storage, IEA Solar Heating, and Cooling Program, 58 p.

Chant, V. and R. Hakansson (eds.) (1983) the MINSUN simulation and optimization program - application and user's guide, National Research Council of Canada, 80 p. + append., Ottawa.

Fournier, B. (1984) Adapting building systems design to aquifer storage of energy, ASHRAE Trans., June, Kansas City.

Morofsky, E.L. (1983) Overview of Canadian aquifer thermal energy storage field trials, proc. International conf. subsurface heat storage theory and practice, p221-231, Part I, Stockholm, Swedish Council Building Research.

Public Works Canada (1983) Data book for building mechanical systems, Scarborough G.O.C.B., 137 p + 25 system diagrams, March, H. H. Angus.

THE SEASONAL HEAT STORE IN GRONINGEN
Results of experiments
Validation of heat storage models

A.J.Th.M. Wijsman
Institute of Applied Physics TNO-TH
P.O. Box 155
2600 AD DELFT
The Netherlands

G.A.M. van Meurs
Delft University of Technology
P.O. Box 5046
2600 GA DELFT
The Netherlands

SUMMARY

In Groningen a project with seasonal storage of solar heat in the upper layer of the soil is realized. The total solar energy system will involve 96 solar houses connected to a seasonal heat store in the soil of about 23.000 m^3.
The project has two main phases: an experimental phase during which the heat store is tested and a demonstration phase during which the total system is monitored. The store was constructed during the latter half of 1982. In 1983 the store was tested: heat input to the store was effected with a boiler, heat output with a cooler. At the moment the solar houses are under construction. This summer the solar houses will be coupled to the heat store.
In this paper the results of the experiments and those of the validation work on our heat storage models are discussed.

INTRODUCTION

In the Netherlands a conventional solar heating system for a house consists of a collector array mounted onto the south facing roof of the house and a short term (ST) heat store in the house. From seasonal heat storage a twice as high solar contribution is expected. An option for seasonal heat storage is heat storage in the upper layer of the soil, in which case the soil itself acts as storage medium. In principle, the seasonal heat store in the soil consists of a layer of soil (up to a depth of approximately 20 m) with a heat exchanger. The store is not bounded by walls. Only at the top the store is furnished with an insulation foam layer; at the other sides the soil itself acts as insulation.

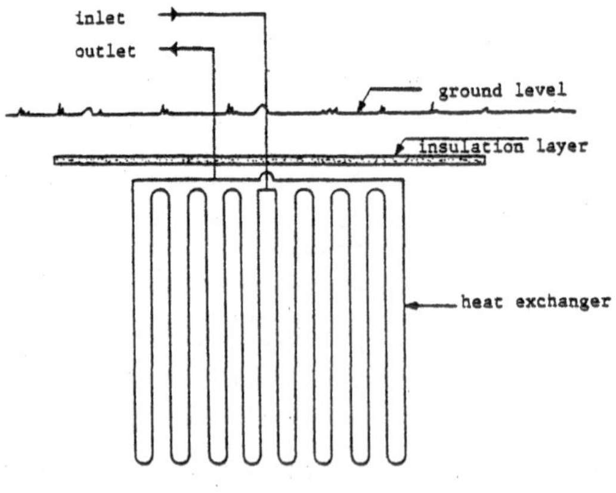

Figure 1: Principle of the heat store in the soil.

The heat exchanger, in which the heat exchange between the transfer medium and the storage medium takes place, consists of vertical tubes, which have been installed in the soil by the specially shaped vibrating lance. For satisfactory performance of a seasonal heat store, it must be large. Therefore, it is essential for seasonal heat storage that a group of solar houses is connected to a central heat store.

After a feasibility study a start was made with the realisation of a project with seasonal heat storage in the soil. The project is located in a new district of the town Groningen in the northern part of the Netherlands. The aim of this project is to gather practical experience with the actual realisation of a large scale seasonal heat storage system and to gather data for the validation of our computer models.

The research work on the project was financed by the Commission of the European Communities and by the Dutch National Solar Energy Programme. The latter also financed the construction.

DESCRIPTION OF THE SOLAR ENERGY SYSTEM WITH A SEASONAL HEAT STORE

The system consists of a group of 96 solar houses, which are situated around a seasonal heat store in the soil.

Figure 2: Solar houses around a seasonal heat store in the soil.

Each house has a solar heating system consisting of about 25 m^2 of high performance solar collectors (total collector area 2400 m^2). The solar heating system contributes to the heat demand for space heating and for domestic hot water. For space heating there is a central auxiliary, for domestic hot water each house has its own back-up system.

The storage system (see fig. 3) is formed by a short term (daily) store and a long term (seasonal) store. The short term store is a 100 m^3 water tank buried in the centre of the seasonal heat store. The volume of the seasonal store is 23,000 m^3 (diameter 38 m, depth 20 m).

The soil in Groningen can be described as water saturated sand with thick layers of clay and some thin layers of peat; the ground water level is about 1 m below the surface. The heat exchanger consists of strings which are vertically inserted in the ground; in one leg of the string the fluid goes downward, in the other leg upward. The strings are of flexible polybutene tube.

The buffertank in the soil made it possible to reduce the size of the soil heat exchanger with a factor of two. To improve and use the radial temperature stratification in the seasonal heat store in the charging mode, the hot transfer fluid enters the heat exchanger in the centre of the reservoir and leaves the heat exchanger at the edge. In the discharging mode the fluid direction is opposite.

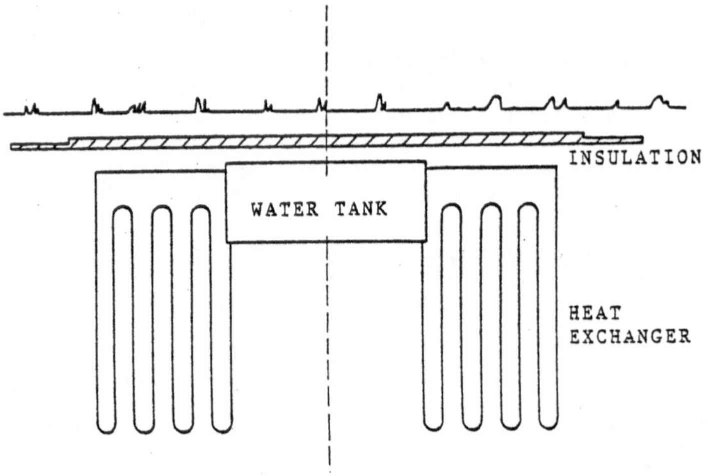

Figure 3: Scheme of the seasonal heat store.

The distribution network has a total length of 1900 m. Because of possible corrosion problems due to oxygen, which diffuses from the ground water through the plastic heat exchanger into the transfer fluid, the soil circuit has been separated from the distribution network by a water-to-water heat exchanger.

The above described system design was a result of extra studies, which brought down the total cost of the system considerably [1]. The construction of the seasonal heat store did not give big difficulties. At the moment the solar houses are under construction.

EXPERIMENTS ON THE SEASONAL HEAT STORE

The heat input/output of the seasonal heat store is realized by a controlled boiler/cooler (see fig. 4) and is in accordance with the thermal behaviour of the 96 solar houses with seasonal heat storage. The thermal behaviour is calculated by a computer simulation model for this system. To monitor the behaviour of the seasonal heat store, the soil has been equipped with temperature sensors, heat flux meters in the top layer, water pressure sensors to detect convection of ground water, tubes for density measurements, observation wells and beacons to detect deformation of the store. The experiments on the heat store started in February 1983 and lasted about one year.

- Description of the experiments

  Three main experiments can be distinguished:

  a. A step response test: during 10 weeks the store was heated up with a constant power delivered by the boiler. After this 10 week period the store was at rest for 2 weeks. Finally during 6 weeks heat was withdrawn from the store by the cooler; the heat output power was constant but limited by a minimum return temperature to the store.
  b. Heat input to the store according to the computer simulation model.

c. Heat output from the store according to the computer simulation model.

Figure 5 gives a timetable of the experiments.

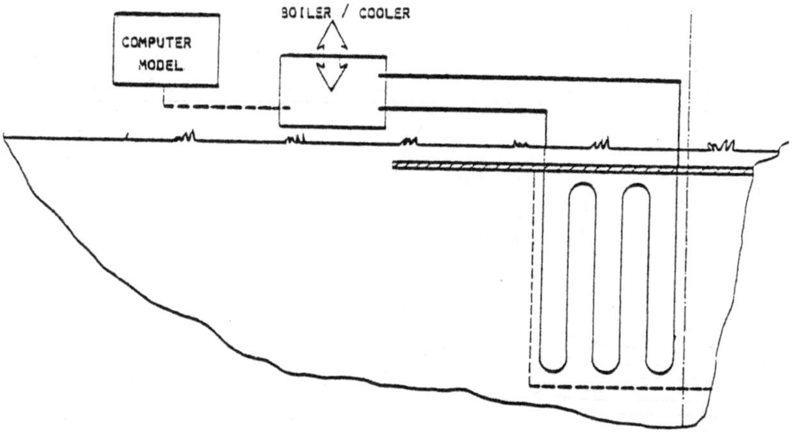

Figure 4: Set-up for the testing of the seasonal heat store.

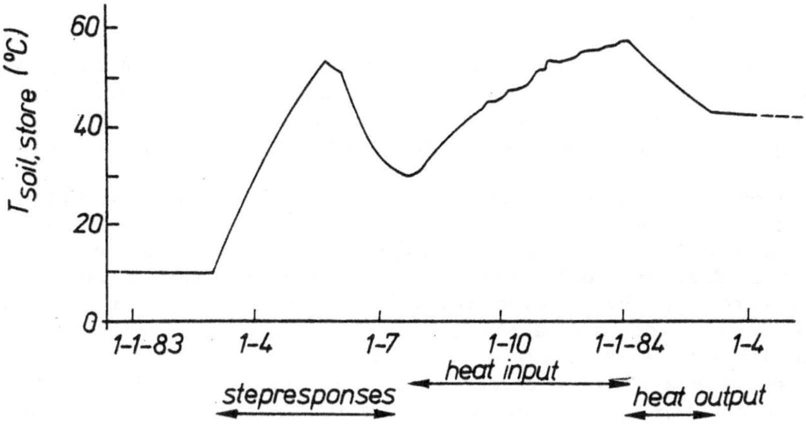

Figure 5: Timetable of experiments.

- Results of the experiments

During the test year the store was heated from 11 °C to 55 °C, cooled down to 35 °C, heated again to 52 °C and finally cooled down to 40 °C. In fig. 6 the temperature profiles in the soil are given at the end of the second heating up period.

The temperature in the centre of the store is 52°C. At a depth of -10 m a bulging of the isotherms can be seen. Between -7 and -13.5 m depth there is a sand layer with permeability $5 * 10^{-12} m^2$ and a heat conductivity of 2.3 W/mK. The bulging out is caused by the higher heat conductivity of this layer. Free convection may also contribute to this effect. The influence of horizontal ground water movement can be seen from an asymmetry in the T-field.

Temperatures have been measured in three radial directions. The deviations are small (within 1°C), so the influence of horizontal ground water movement is small.

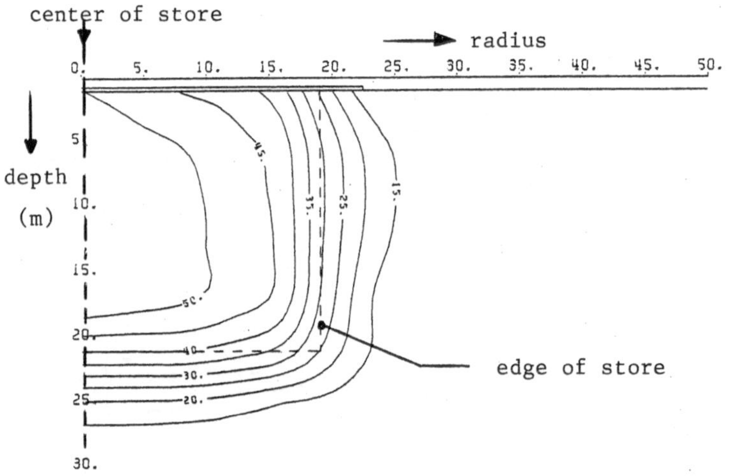

Figure 6: Temperature profile in the soil at the end of the second heating period.

The capacity of the soil heat exchanger is 30 kW/K, which is within 5% from the design value. The heat exchanger between soil circuit and distribution network has a capacity of 245 kW/K.
The soil heat exchanger is divided into three sections to optimize the use of temperature stratification in radial direction. During testing it turned out that the selection function to pick the best section(s) during heat input respectively heat output was not working as required. This problem could be solved by changes in the software of the control computer. Another problem was the position of the temperature sensors for the control of the section selection, these sensors are installed too close to the top insulation layer. The measured temperatures at those positions were a few degrees lower than at a greater depth in the store because of heat losses through the top. During heat output this caused heat output from a more inner section, whereas in the outer section the temperature level was high enough. Another change in the software of the control computer can solve this problem.
The heat losses through the top insulation layer are only according to what is expected during periods of rest. During heat input the heat losses are higher because the interconnections of the soil heat exchanger tubes are just below the top insulation layer. This means a higher temperature at that position. However, the losses are even higher than was to be expected from that temperature. This implies that there must be a second effect: vapour transfer in the 'clay-grain' layer (heat pipe effect). During heat output the losses through the top insulation are lower, for the same reasons.
The short term water tank was functioning well; the heat exchange with the surrounding soil was small (time constant about 5 weeks).
Final remark: the results of the experiments were according to our expectations, only some minor problems occured.

VALIDATION OF COMPUTER MODELS

The models developed describe global and local processes of transfer of heat in the soil.

Two computer models have been developed for the global process. Each of them can predict the temperature distribution in and around the store. The first model (LT-TPD-I) is a simple one (one-dimension) with only heat transfer by diffusion and soil properties are assumed to be homogeneous. In this model the storage region is subdivided into several parts. Each part has its accessory temperature. Heat diffusion is responsible for the heat exchange between these parts. The heat transferred to and from the store is distributed homogeneously over the storage volume.

The second model (LT-TN-I) (a more detailed one) is a finite difference model with heat transfer by diffusion and convection. This model also has the ability to take heterogeneous soil properties into account [2]. The distribution of the exchanged heat over the store, is based on the temperature of the heat transfer fluid along the soil heat exchanger. The influence of free convection and the effect of section selection can be studied with this programme.

For the local process (heat exchange between heat exchanger tubes and surrounding soil) supplementary models were developed. These models calculate the change in the temperature profile around the tubes during the day. The first model (heat transfer by diffusion only) takes average temperatures for store and transfer fluid. In the second model (heat transfer by diffusion and by convection) the tubes connected in series are divided into several parts; each part has its own temperatures for soil and the transfer fluid corresponding to its location in the store.

Validation results:

Earlier work shows that the effect of convection on the global process can be neglected for our specific site in Groningen. The presence of some impermeable horizontal clay layers suppresses free convection [2]. Measurements also show that horizontal ground water movement is small [3]. So it seems justified that convection is neglected during the calculations with the second global model.

The prediction by the computer models of the temperature in the centre of the store and the global temperature distribution are compared to experimental data. The measured heat flows to and from the store were used as input for the models.

The comparison for the centre temperatures (depth 8 m, radius 8 m) is given in fig. 7.

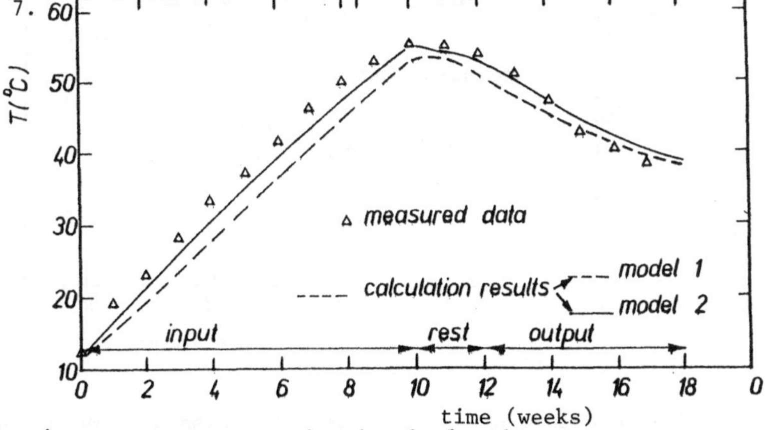

Figure 7: Comparison netween measured and calculated temperature at a depth of 8 m and at a radius of 8 m.

The global radial temperature profile at a depth of 8 m in the reservoir is depicted in fig. 8. The detailed model 2 shows the influence of the cold water tank in the centre.
To get the agreement shown in fig. 7 and 8 both models underwent small changes.

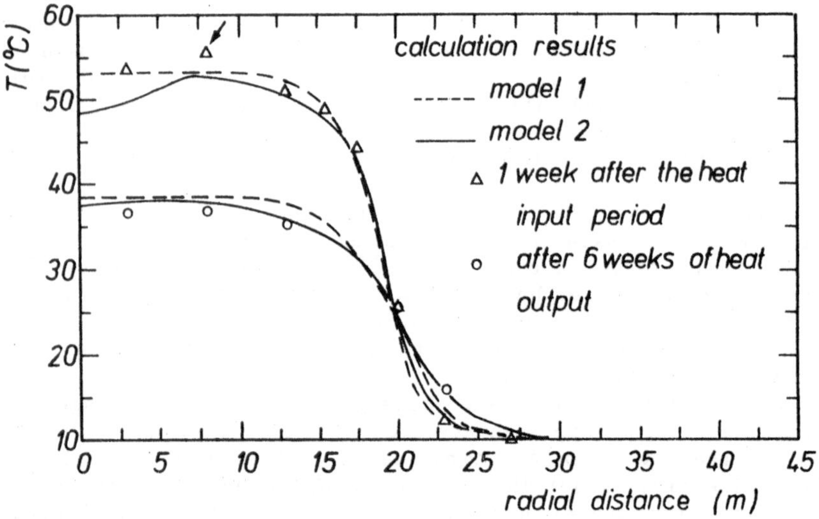

Figure 8: Comparison between measured and calculated temperature profile at a depth of 8 m.

The capacity of the soil heat exchanger is determined by the local heat transfer process around the tubes and the number of tubes. The comparison between heat exchange rate deduced from the experiments and calculations is depicted in fig. 9 as a function time.

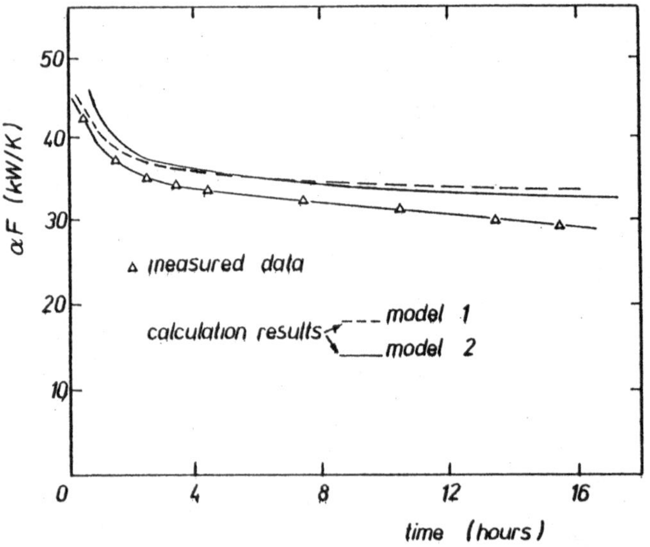

Figure 9: The heat exchange rate as a function of time of the day for the fourth heating day.

The calculated and measured heat flows during heat output are given in figure 10. The measured fluid temperature entering the heat exchanger in the soil was used as input for the models.

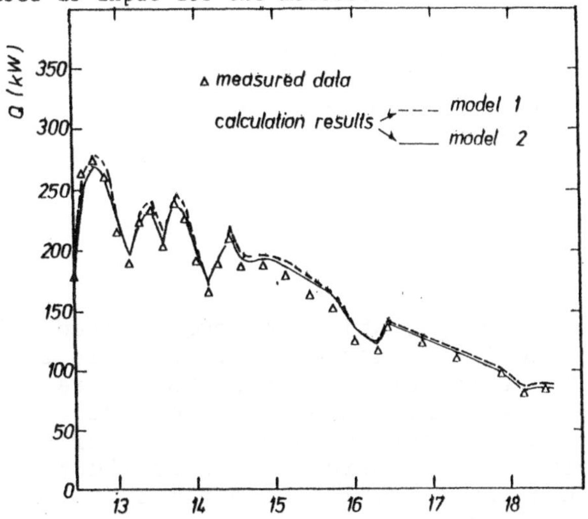

Figure 10: The heat flows during the heat output period.

The agreement is satisfactory. To get this agreement for the simple model the calculated heat output should be based on the mean temperature in the store in stead of the centre temperature.
Final remark: The validation of the models led to some small adaptions. The validated models (both simple and detailed) can predict the behaviour of the seasonal heat store well.

CONCLUSIONS

Seasonal storage of solar heat in the upper layer of the soil is a good option. Based on the test experiences some simplifications in the storage design can be made, which will lead to a further cost reduction. The computer models are suitable for design work. They are a tool to investigate difficulties and problems occuring during this stage of the project.
This summer the solar houses will be coupled to the seasonal heatstore. During two years the total system will be monitored.

References

[1] Wijsman, A.J., The Groningen project: "100 houses with seasonal solar heat storage in the soil using a vertical heat exchanger", paper presented during "Subsurface Heat Storage" conference in Stockholm (June 1983)

[2] Van Meurs G.A.M. and Hoogendoorn C.J., "Influence of natural convection on the heat losses for seasonal heat storage in the soil", paper presented during "Subsurface Heat Storage" conference in Stockholm (June 1983).

[3] De Feyter J.W., "Soil mechanics aspects of seasonal heat storage in Groningen", paper presented during this Amsterdam Congress.

THE SUNCLAY AND KULLAVIK PROJECTS - HEAT STORAGE IN CLAY AT LOW AND HIGH TEMPERATURE

S. OLSSON
AB Andersson & Hultmark
Box 24135
400 22 Göteborg
Sweden

Summary

The SUNCLAY-project, which was constructed 1979-1980, is one of the first projects were clay is used as heat-storage. The temperature of the clay is raised to 14-15°C by low-temperature solar collectors during the summer months. During the winter dieseldriven heatpumps use the clay as heatsource and lower the temperature to 9°C. In this paper results from the operating period are presented. In the Kullavik-project clay is also used to store energy in. The storage is divided into two zones, the high temperature zone (HT) and the low temperature zone (LT). The HT-zone will be warmed up to 55°C and the LT-zone to 25°C. To reach these temperatures solar collectors with one cover-glass (corrugated acryl) are used. This paper presents this project and the results since it was taken into operation in July 1983.

The SUNCLAY-project

The SUNCLAY-project is the heating system of a school in Kungsbacka south of Gothenburg in Sweden. This project has been in operation since April 1981. The heating system consists of 1500 m^2 low temperature solar collectors integrated in the roof, a ground heat storage (87.000 m^3) with vertical plastic pipes as heatexchangers and dieseldriven heatpumps.

The heating installations inside the school are made in a conventional way. Since the start of the operating period measurments have been made on the different pieces of the system. The most interesting measured results during the period 1 July 1982 to 30 June 1983 will be presented in this paper.

The global solar insolation during this period against the solar collectors were 930 kWh/m^2 (3340 MJ/m^2). Produced energy from the solar collectors were measured to 370 kWh/m^2 i.e. the annual efficiency of the collectors were 40%. Typical operating temperatures for the collectors were 24-27°C during the three summer months and 18-20°C during May and September.

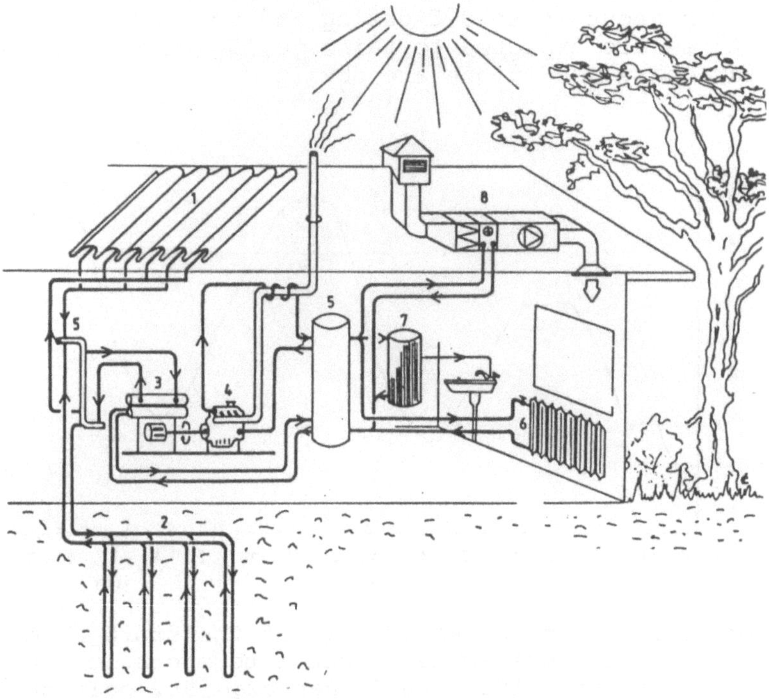

Figure 1. The SUNCLAY-system

During the summer period energy is delivered to the storage from the solar collector. In winter time the heat-pumps use the storage as heat-source.

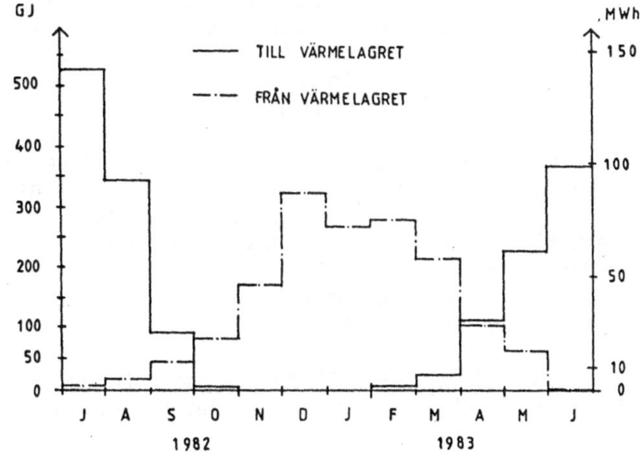

Figure 2. Energy delivered to (-) and from (-·-) the storage

Depending on what season it is the temperature in the storage is different. The mean tempeterature range in the storage was about 14 to 9°C during the current period. These temperatures are expected to raise to 15 and 10° respectively when stationary conditions are obtained.

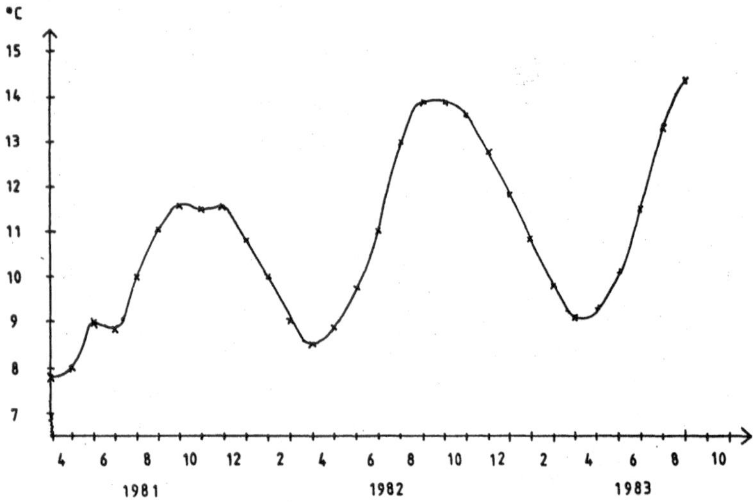

Figure 3. The mean temperature of the storage since the system was taken into operation in April 1981. Until January 1982 only half of the solar collector area (750 m^2) was connected.

Figure 3 shows that the storage reaches its highest temperature in September and its lowest temperature in March-April. The measurments show that the temperature in the clay six meters from the storage is more or less independent of the temperature in the storage.

Due to the temperaturegradient outside the storage the annual heatlosses have been estimated to be 10-15%.

The heat-exchanger in the storage consists of 600 "U"-tubes which are placed vertically to a depth of 35 m. The "U"-tubes consist of polyethylene pipes with an outer diameter of 16 mm.

The results from the measurments show that the capacity of this heat-exchanger is about 1,3 W/m°C as a stationary value and about 2,4 W/m°C as a initial value.

The thermal properties of the clay are:
conductivity - 0,94 W/m°C
heat capacity- 2410 J/kg°C
The density is 1495 kg/m^3.

The morecost of the SUNCLAY-project in 1980 was 1,7 million SEK. With todays oilprices the pay-back period will be eight years.

The Kullavik-project

In Kullavik south of Gothenburg 58 dwellings are supplied with heat and domestic hot water from a heating system which consists of solar collectors, a ground heat storage with two zones, a heatpump and an oilboiler. This system came into operation in July 1983 with 40 dwellings as load. The other 18 dwellings were connected in April 1984.

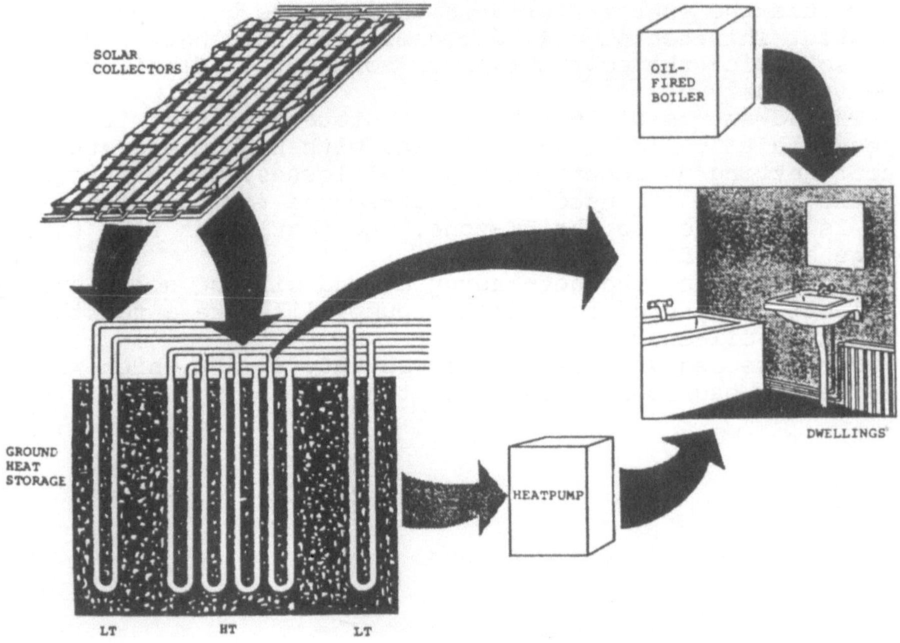

Figure 4. The principle scheme of the Kullavik-project

The solar collectors, which name are Scandinavian IT, are integrated in the roof construction and have a total area of 540 m^2 facing south-east. The collectors have a corrugated acrylic-plate as cover.

The ground beside the dwellings make the heat storage. In the ground, which consists of clay, plastic pipes are placed as heat exchanger. The plastic pipes, with an outer diameter of 32 mm, are placed vertically as "U"-tubes in two groups. In the middle of the storage 130 "U"-tubes are placed in a square with a mutually distance of half a meter. This group of pipes are called the high temperature zone (HT). The low temperature zone (LT) surrounds the HT-zone. In the LT-zone the pipes are placed in three parallell lines with a mutually distance of two meters. Within the lines the mutually distance of the vertical pipes is 1,5 meters.

The distance between the outer pipes of the HT-zone and the inner line of the LT-zone is five meters. The depth of the HT-zone is 8 meters and the LT-zone is 12 meters deep.

During the summer period solar energy is delivered to both the HT-zone and to the LT-zone. In this way HT will reach 55°C and LT 25°C. The total storage volume is 8100 m^3 of which the HT-zone makes 200 m^3.

The HT-zone, which works as a short-term storage during the summer, is connected directly to the heating system. This means that solar energy is used without a heatpump for making hot water in the summer and also partially to heat the dwellings during spring and autumn.

To use the energy stored in the LT-zone a heatpump is connected. This heatpump is designed for about half peak load, so during the coldest winter months the oilboiler supports energy.

The advantage of this two-zone storage is that it makes it possible to build a small storage with high temperatures without getting unacceptable high heatlosses. The heatlosses from the HT-zone are used as a heat-source for the heatpump which is connected to the LT-zone. The heatlosses from this storage are 30-35%.

Two thirds of the net-energy demand will be solar energy and electrical energy for the heatpump will make a fourth. The rest is oil-energy.

The morecost of this system was 800.000 SEK which means that the pay-back period is about ten years.

THERMAL ANALYSIS OF THE BEHAVIOUR OF A LOW-PRESSURE HOT-WATER STORE

R.R. Cohen, P.W. O'Callaghan and S.D. Probert
Applied Energy Group,
School of Mechanical Engineering
Cranfield Institute of Technology
Cranfield, Bedford, UK

Summary

This paper analyses the results of many step input tests on a direct water store. Three common types of inlet port configuration were studied over a wide range of flow conditions. The depth of penetration of the inlet flow is shown to obey a correlation derived using dimensional analysis. The outlet temperature response and thermal effectiveness of the store are demonstrated to be closely determined by this depth of penetration. A one-dimensional, finite-difference computer model, evolved to predict the response behaviour, is described and verified. It is proposed that this model could be usefully employed in annual simulations in place of a conventional segmented tank model. It is also suggested that both the model and the correlations should prove useful for designing stores. They can be used to assess the influence on thermal performance of store volume, aspect ratio and port geometry.

1. INTRODUCTION

Water is currently the most widespread and cost-effective medium for heat storage in diurnal active solar-energy systems, both for space heating and domestic hot-water applications (1). The thermal performance of the store, in particular the degree of stratification achieved, can have a significant influence on the amount of solar energy which is usefully harnessed.

The aims of this investigation are to provide dimensionless correlations which characterise the behaviour of water stores in direct solar-energy systems, and so to evolve a computer model of this behaviour that is simple enough to be incorporated into an annual simulation. The effects of port configuration and flow parameters are described in a form suitable for store designers.

2. APPARATUS AND METHOD

The experimental investigation was conducted using a microprocessor-controlled test facility, as illustrated in figure 1, which delivered water at a specified temperature and flow rate into a storage test module. The microprocessor drove motorised three-way valves so as to minimise deviations between monitored values of temperature and flowrate and specified set points. Control algorithms based on the extrapolation of trends were used to account for various lags in the system. The flow temperature was measured by stabilised and linearised platinum resistance thermometers linked to the microprocessor via a 12 bit ADC. The flow rate was measured using turbine flowmeters, linked to a purpose-built interface which measured the period between pulses from the flowmeter. A two-dimensional lattice of 50 thermocouples, connected to the microprocessor through a Fluke data logger was used to monitor the temperature distribution inside the store.

All the experiments were carried out on a water store made from an upright, cylindrical steel vessel, of diameter 1.36m with an internal volume of $2.85m^3$, which was operated under pump pressure. The store was well insulated with polyisocyanurate foam (of thermal conductivity 0.029 $Wm^{-1}K^{-1}$ at 50°C) to a minimum thickness of 100mm. This served to reduce the difficulties associated with measuring heat losses from the store. A standing heat loss test was performed over a period of 13 days and yielded a rate of heat loss of 4 WK^{-1}. The heat capacity of the complete store was 12.35 MJK^{-1} and so the heat loss time constant was about 36 days.

The experiments performed on the store were of the conventional temperature step-input type and followed the procedure detailed in the ASHRAE Standard 94-77 (2). Three common types of port configuration were evaluated : plain 25mm dia. vertical pipes, plain 25mm dia. horizontal pipes and vertical distributors. The vertical pipes were set in the centre of the domed ends of the store and the horizontal pipes were set at 0.3m (14% of the total vessel height) vertically from the domed ends of the store. The distributors, see figure 2, were designed to absorb the vertical momentum of the inlet flow in order to try to produce perfect plug flow. They comprised 4 horizontal arms, each containing 64, 1.5mm dia. holes, spaced so that each hole "served" an annulus of the same area.

The limits of operation of the test facility constrained the flowrate for a test to the range 0.04 to 0.6 l/sec and the inlet temperature to the range 10°C to 90°C.

3. EXPERIMENTAL RESULTS

A large number of experiments were carried out, with each port configuration, in which the relative magnitude of the momentum force of the inlet flow into the tank and the buoyancy force tending to reverse the momentum was varied. Examination of the temperatures inside the store showed them to be essentially one-dimensional, with the initial depth of penetration of the inlet flow into the tank determining to a large extent the outlet temperature response throughout the experiment.

Dimensional analysis, first applied by Turner (3) to the case of turbulent jets with reversing buoyancy, yielded the following relationship for the height, z, of penetration :

$$z = C \frac{\dot{m}}{\rho_{in}} r_o^{-3/2} (g\beta\Delta T)^{-1/2} \qquad \ldots \qquad (1)$$

where C = constant
$\dot{m}$ = mass flow rate through the inlet (kg s^{-1})
ρ_{in} = density of inlet flow (kg m^{-3})
r_o = radius of inlet (m)
g = gravitational constant (m s^{-2})
β = coefficient of volumetric expansion (K^{-1})
ΔT = temperature difference between inlet flow and tank fluid (K)

Values of z were estimated for each experiment using the monitored temperatures, and are plotted in figure 3 in a manner suggested by equation 1. It can be seen that the relationship holds quite convincingly for all three port configurations although it is clearly most germane to the plain vertical port for which it was originally intended.

The performance of a water store in a direct system, subjected to a step input test, is conveniently described by the storage thermal effectiveness $\eta(\tau)$, after one fill period, τ. This is defined as the energy stored in one fill period divided by the theoretical storage capacity, TSC, i.e.

$$\eta(\tau) = \frac{\int_o^\tau \dot{m}\, C_p \left| T_{in} - T_{ex} \right| dt \pm \int_o^\tau (UA)_{s,a} (T_s - T_a)\, dt}{TSC} \qquad \ldots \qquad (2)$$

where the subscripts in, ex, s and a denote respectively inlet, exit, store and ambient.

The store's behaviour is characterised by the following three limits of operation:

$\eta(\tau)$	=	1	perfectly stratified
$\eta(\tau)$	=	0.63	fully mixed
$\eta(\tau)$	=	0	complete short circuiting

Values of $\eta(\tau)$ were calculated for each experiment and are plotted in figure 4 against the correlating parameter given in equation 1. The decrease in $\eta(\tau)$ at very low values of the correlating parameter (i.e. low flow rates) may well be explained by the fact that the inlet jet is no longer turbulent. This would reduce the rate of mixing between the inlet flow and the water in the tank, thereby allowing the inlet jet to penetrate further into the tank. Another factor causing a decrease in $\eta(\tau)$ at low flow rates is thermal diffusion, which tends to dilute any stratification created in the store. However the relatively low thermal conductivity of water means that diffusion is a comparitively insignificant factor even at the lowest flowrates considered in this investigation.

4. COMPUTER MODEL

A one-dimensional, finite-difference computer model, capable of predicting the outlet temperature response of a direct water store to a step input in inlet temperature, has been evolved. It assumes that the inlet flow will penetrate the tank initially to a depth given by the experimental correlation shown in figure 3. A zone of this depth is calculated analytically to be fully mixed after each time step. Perfect plug flow is assumed for the remainder of the store, which in the model is subdivided into zones of equal volume, V_z, such that

$$V_z = \dot{V}\, dt$$

where $\dot{V}$ is the inlet volume flow rate (in $m^3\ s^{-1}$ units) and dt is the program time step (in seconds).

The size of V_z, and thereby the number of zones employed in the model, determines the degree of resolution of the outlet temperature response and therefore the accuracy of a numerical method of integration to calculate $\eta(\tau)$. Because dt changes linearly with V_z, the resolution, accuracy and computing time all increase according to the inverse square of V_z.

As a step input proceeds, the temperature difference between the inlet flow and the tank fluid steadily reduces thereby moderating the reversing buoyancy effect. The model calculates after each time step the increase in the penetration depth from its initial value due to this effect. When this increase exceeds the depth of one plug-flow zone, the volume of one plug-flow zone is added to the volume of the fully-mixed zone, and the number of plug-flow zones is reduced by unity. This process continues until eventually the whole store is represented by one fully-mixed zone.

The model was applied to the experimental store and used to predict values of $\eta(\tau)$ for various initial depths of penetration. The result, shown in figure 5, is independent of the flow conditions (except in as much as these inherently determine the initial depth of penetration) if diffusion is neglected. If diffusion is included, the model predicts that $\eta(\tau)$ decreases slightly as $\dot{m}$ decreases. This is simply a consequence of the fill period, τ, increasing as $\dot{m}$ decreases, thereby permitting more time for diffusion to occur. The maximum possible predicted effect of diffusion on the experiments reported here has been estimated by running the model with a flow rate of $0.04\,ls^{-1}$, the lowest possible achievable on the test facility. The result, also shown in figure 5, indicates that the effect of diffusion increases with shorter initial penetration depths. This is consistent with the fact that less penetration creates a greater temperature gradient within the store. However, diffusion reduces $\eta(\tau)$ by only 2%, even when the initial penetration is a mere 5% of the store height.

Experimental results for the three port configurations have been plotted on figure 5 and can be seen to provide a good validation of the computer model. Further validation is supplied by figure 6 which shows the measured and predicted outlet temperature response plotted against time for a typical experiment.

5. CONCLUSIONS

(i) Correlations (derived from dimensional analysis, to predict the depth of penetration of a flow into a water store) have been established for three types of inlet port configuration.
(ii) The outlet temperature response of a store to a step input in inlet temperature and the store's thermal effectiveness are demonstrated to be closely determined by the depth of penetration.
(iii) A relatively simple one-dimensional computer model has been evolved which closely predicts the response of a water store to a step input in inlet temperature.
(iv) This model should improve significantly the accuracy of annual simulations if it were inserted in place of a conventional segmented tank model. However its absolute validity in systems producing variable temperatures at the store inlet has yet to be assessed.
(v) Both the model and the correlations should prove useful for designing stores. They can be used to study the sensitivity of the thermal performances of stores to changes in volume, aspect ratio and port geometry.

6. ACKNOWLEDGEMENT

The authors would like to thank the Commission for the European Communities for financial support of this work under contract number ESA/S/039/UK.

7. REFERENCES

1. PALZ, W. AND STEEMERS, T. (1981). "Solar houses in Europe - how they have worked", p.13. Published for the CEC by Pergamon Press.
2. ASHRAE Standard 94-77 (1977). "Methods of testing thermal storage devices based on thermal performance", ASHRAE, 345E, 47th St. NY 10017.
3. TURNER, J.S. (1966). "Jets and plumes with negative or reversing buoyancy". J.Fluid Mech.(1966), vol.26, part 4, pp.779-792.

Figure 1. Photograph of TES test facility.

Figure 2. Photograph of distributor inside water store (and the 2-D thermocouple grid)

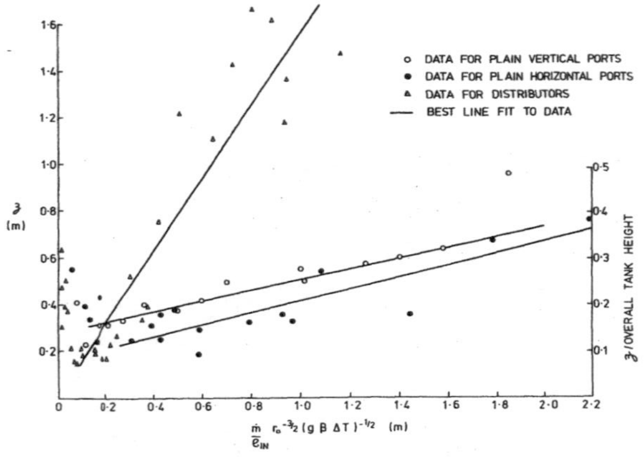

Fig. 3. MEASURED DEPTH PENETRATION VS. CORRELATING PARAMETER FOR THREE PORT CONFIGURATIONS

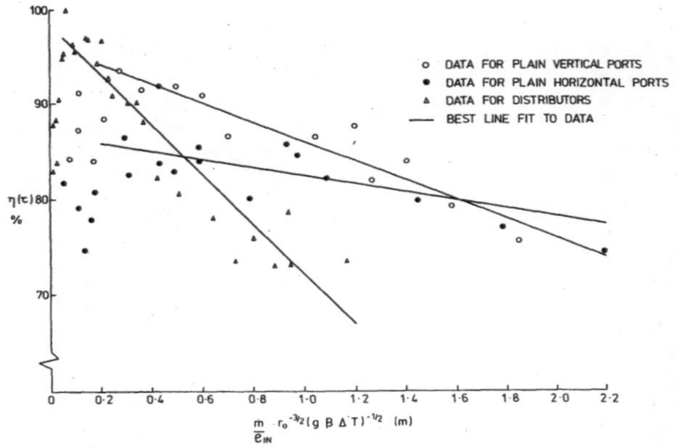

Fig. 4. STORE THERMAL EFFECTIVENESS VS. CORRELATING PARAMETER FOR THREE PORT CONFIGURATIONS

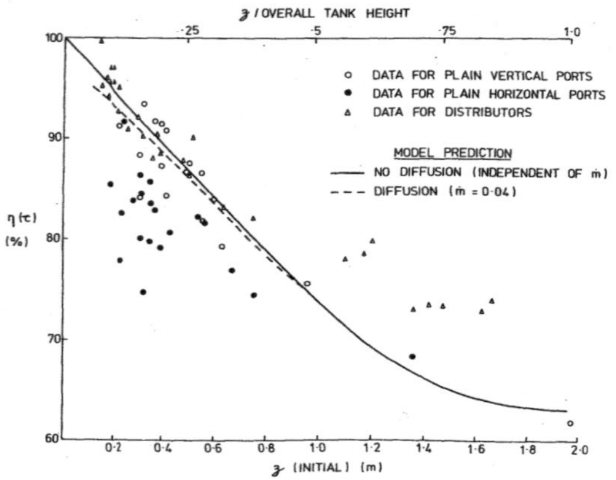

Fig. 5. MEASURED THERMAL EFFECTIVENESS VS. DEPTH PENETRATION COMPARED WITH PREDICTION OF COMPUTER MODEL

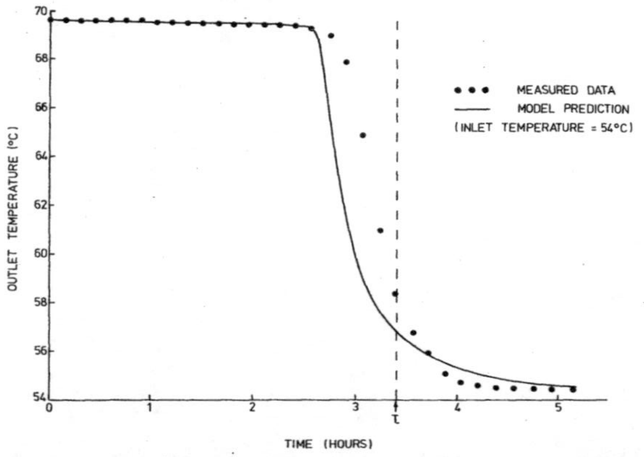

Fig. 6. MEASURED AND PREDICTED OUTLET TEMPERATURE RESPONSE FOR A TYPICAL DISCHARGE EXPERIMENT

THE SOLAR HEAT STORAGE TWIN-SYSTEM OF THE LÜBECK SOLAR HOUSE AS AN EXAMPLE OF A 2nd GENERATION INTERSEASONAL STORAGE CONCEPT

H.WEIK and J. PLAGGE
Fachhochschule Lübeck
D-2400 Lübeck
Fed.Rep. of Germany

Summary

In 1982, a solar heated one-family home with 130 m² heated area was completed at the site of the Fachhochschule Lübeck, the heat storage concept of which equally well utilizes the insolation surplus in summer and the low-intense winter insolation.

This was achieved by a heat pump assisted solar heating system centered around a twin storage consisting of two water tanks of different function: a 4,5 m³ tank, kept at 35 to 50°C, takes care of the heat demand of the low-temperature heating system, and a 20 m³ tank, serving as heat reservoir and - in mid-winter - as heat ('anergy') source for the 2 kW heat pump. In this mode of operation, during which the tank also acts as a latent heat storage because of ice formation, the 34 m² collector field collects the low-intensity winter radiation with high efficiency, thus providing the anergy supply for the heat pump. The solar surplus during late spring is used to melt the ice, and during summer the storage water is heated up to its useful temperature for the new heting season.

The good performance and the low auxiliary energy consumption during the first two years of operation demonstrate the feasibility and the merits of the concept.

1. INTRODUCTION

One of main obstacles for a widestread application of solar energy in buildings and households is the problem of storing solar heat. Contrary to the utilization of solar energy for domestic hot water in summer, where storage is no problem, the energetically more attractive application of solar systems for space heating unfortunately suffers from the half-year phase shift between main solar energy supply and space heat demand. In Lübeck, for instance, 65 to 70 % of the years total insolation occurs during April and September, while 85 % of the total space heating demand for a well insulated one-family home is needed during the October till March period.

Therefore, if one wants to achieve the most effective use of solar energy in our climate, either, one has to accept the expenses for large storage tanks for sensible heat (or to wait for the advent of chemical storage, respectively), or, one has to search for some energetically as well exergetically meaningful combination of a collector - storage system with a heat pump.

Contrary to early Bristish attempts of "solar assisted heat pump systems" (1), it appears to be more attractive to put the emphasis on the collector side of the system instead of the heat pump and to propose rather a "heat pump assisted solar heating system", that comprises suitable solar collectors and a multiple storage with a matching heat pump. The main advantage of such a solar system is the limited size of the heat pump, the compressor of which needs not to exceed a 2 kW driving power (depending upon the choice of the solar components and the size of the house), a magnitude that eventually, with further developments of photovoltaics, can be economically provided by solar generators.

In this contribution, an outline of the above sketched concept will be given, as realized in the solar heated one-family home, completed at the Fachhochschule Lübeck in July 1982. It indicates the central role of storing heat in solar systems, using, however, a conventional 2 kW electric heat pump.

2. OBJECTIVES OF THE PROJECT

All German solar houses so far errected are located in the southern and western regions of the country, where the insolation is fairly high. Therefore, the Lübeck solar house was to demonstrate that even under the unfavourable climatic conditions of nothern Germany (54° N latitude), it is possible and meaningful to utilize the photothermal convertion for space heating and domestic hot water supply by combining architectural design and thermal insulating methods with an efficient solar heating system.

The building complex (Fig. 1) consists of a one-family home of 130 m^2 heated area, and a one-storey, flat roofed laboratory annex of about 40 m^2 area, both having full basements. The two-storey main house is being used as dwelling by a real family of 4 during a four- or five-year test period.

The architectural design criteria for the contruction were, among others,

-- to make the south roof form a slope of 60°, an optimal inclination for winter solar collection in our latitude;

-- to reduce the mean U-value of all enclosure surfaces of the house to 0,35 W/m^2K, by using

- a double-wall construction with a mineral wool layer sandwitched in between, reducing the U-value of the walls to 0,3 W/m^2K;
- triple-glazed windows, with insulating shutters for the night, having an U-value of 2.1/1.6 W/m^2K (day/night).

Based on these constructional data, the calculated total heat demand of the house, including transmission and ventilation heat requirements and the domestic hot water supply during the heating season, but considering also the gains by passive solar collection through the south windows and internal heat from persons and equipment, amounts to some 12000 kWh/heating season (= 43,2 GJ/heating season).

3. THE SOLAR CONCEPT UTILIZING A DOUBLE STORAGE WATER TANK

In a survey computation, based upon the climatic and insolation conditions in Lübeck, the feasibility of a solar space heating system, centered around a double storage water tank and operating with a 2 kW electr. heat pump as the only auxiliary heat source, was examined and proven to be practicable. Subsequently, this fairly elaborate concept was adopted for the main house of the building complex, while for the laboratory annex, which has the size of about one quarter of a normal one-family bungalow, a simple bi-valent solar heating system was chosen, consisting of 8 m² flat plate solar collectors and a 4,5 m³ water storage tank. A small gas boiler is used as auxiliary heat source for the mid-winter period.

The main house gains its solar heat from a 34 m² collector-field on the 60° inclined south roof, consisting of collectors of different types and efficiencies, ranging from a simple do-it-yourself model to a high-efficiency vacuum collector. Each collector group is being individually tested for its performance.

The energy flow chart is shown in Fig. 2, and the system layout - in a simplified form - in Fig. 3. Mainly for demonstrating purposes, the solar heat for the domestic hot water supply is provided by separate collectors (located at the south fassade, cf. Fig. 1), heating up, by natural convection, a 500 liter hot water tank in the attic. During the summer months, the fassade collectors (3 m² active area) are assisted by an 18 m² unglazed 'energy-roof' in the center ot the south roof. Since during the winter half-year, the heat supply from the fassade collectors is unsufficient, supplementary heat for the hot water tank is drawn from the central storage via a heat pipe connection.

During this period, the 'energy-roof' acts as an air heat exchanger, operating into the tank (see below), and delivers, according to laboratory experiments, about 500 kWh low grade heat to the storage system.

The 'heart of the solar system' is the double water storage located in the basement. It consists of two thermally insulated compartments. The smaller one (day tank) of 4.5 m³ volume is kept at a temperature between 35 and 50°C (depending upon the ambient temperature), sufficient to meet the heating requirements for the coldest day of the year. The larger tank (seasonal tank) of about 20 m³ volume (plus 3 m³ water equivalent of the concrete tub as to its sensible heat capacity) is being used as heat reservoir for the day tank, either by direct heat exchange at the beginning of the heating period, or - later in winter - by means of the heat pump.

4. THE STORAGE CYCLE

At the end of summer, the dual storage tank system normally will be fully charged to a temperature of 65 to 75°C. Since the lower temperature limit for direct use in the heating system of the house (floor heating in the ground floor rooms, radiator heating in the first floor rooms) amounts to about 35°C, the useable heat capacity of the storage unit for sensible heat is about 1150 kWh.

During the first 2 or 3 weeks of the heating period, heat gains from the collector field will generally compensate the heat consumption of the house and the heat losses of the tanks. Therefore the large tank will not be discharged until the second decade of October; then its sensible heat will be withdrawn by simple fluid exchange with the small tank. The supply will last roughly 4 weeks.

After that time, the rest of the sensible heat, which ist merely 'anergy' i.e. heat at a temperature level not sufficient for direct use in the heating system, will be extracted by means of the heat pump. Toward the end of this second period of the discharge operation, the temperature of the large tank will drop to 0°C in the upper tank region, after another 1000 kWh stored energy were extracted for the heat requirements of the house during the first decade of December. Heat from the collector field will now flow into the large tank, and this occurs - due to the low temperature level of the storage water - with the highest possible collector efficiency.

The third period of the storage cycle is characterized by the gradual solidification of the water in the large tank, which then acts as phase change (latent heat) storage, providing some 2400 kWh heating supply, including some 1000 kWh driving energy of the heat pump. On particularly sunny days with insolation values > 2.5 kWh/m²d, certain collector types will be automatically switched to deliver their heat directly to the small tank to save driving power of the heat pump.

During April, the heat load of the house and the heat gain from the collector field generally balance. The insolation is high enough to serve the inner tank completely and to shut off the heat pump. In May normally only very little heat demand is necessary for this thermally very well insulated house. The insolation surplus will be used to liquify the large tank, which will then - during the summer months - be heated up to its initial 65 to 75°C temperature for the new heating cycle.

5. CONCLUSIONS

The results of the first complete heating period (Winter 1982/83) confirmed the expectations and the model calculations of the dynamic behavior of the storage system (2; obtained, however, with the earlier storage construction, i.e. both tanks arranged concentrically one into the other). During the water-ice phase change process, the collector field collected the low-intensity winter solar radiation, as expected, resulting in a more or less continous supply of anergy for operating the heat pump, and in April '83, the exergy surplus of solar radiation was used to melt the ice in the large storage tank. May '83 was extremely cool without the normal insolation surplus, but bad sufficient solar harvest to take care of the heat demand of the house. A detailed report has been given elsewhere (3).

The total heat load of the house of the order of some 20 MWh for space heat and hot water supply for the 4-person family was contributed by about 70 % means of passive and active solar techniques leaving a balance of some 6000 kWh electric auxiliary energy.

The excellent performance of the solar space heating system so far indicates not only the feasibility of the twin-storage concept for dwelling houses but also demonstrates its energy saving character, as no fossile fuels were needed in order to meet the total heat requirements.

6. REFERENCES

1. SZOKOLAY, S.V. (1974) Design of an experimental solar heated house at Milton Keynes. Proc. UK-ISES Conf., Low temperature thermal collection of solar energy, Polytechnic of Central London, 23-33
2. WEIK, H. (1983) Efficient energy storage system for solar space heating in dwelling houses. 2nd BHRA Fluid Eng.Intern.Conf. on Energy Storage, Stratford-upon-Avon, 137-150.
3. WEIK, H. (1984) Heizen mit der Sonne - eine Utopie in unseren Breiten?, HR Haustechn. Raundschau 83, 23-31

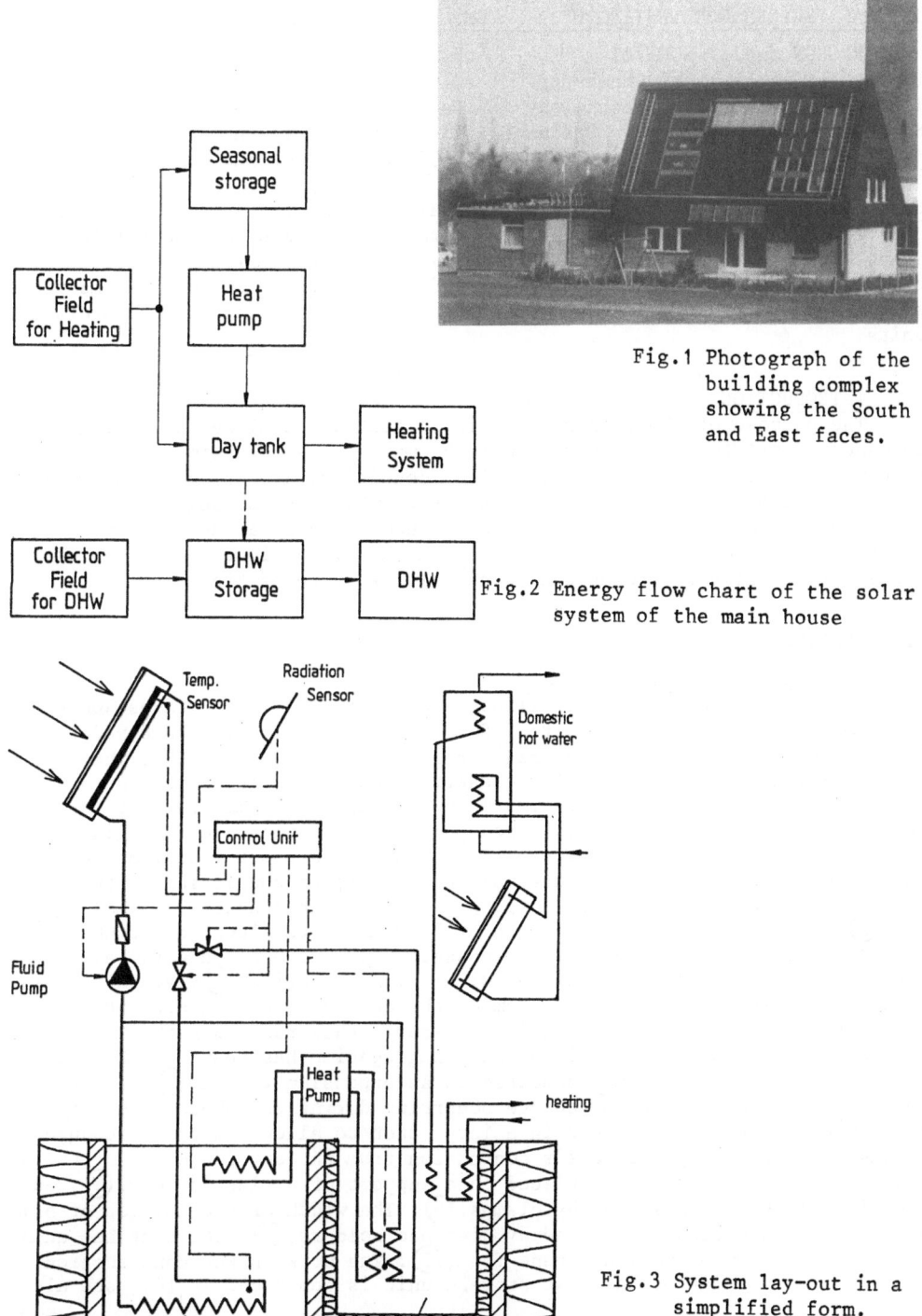

Fig.1 Photograph of the building complex showing the South and East faces.

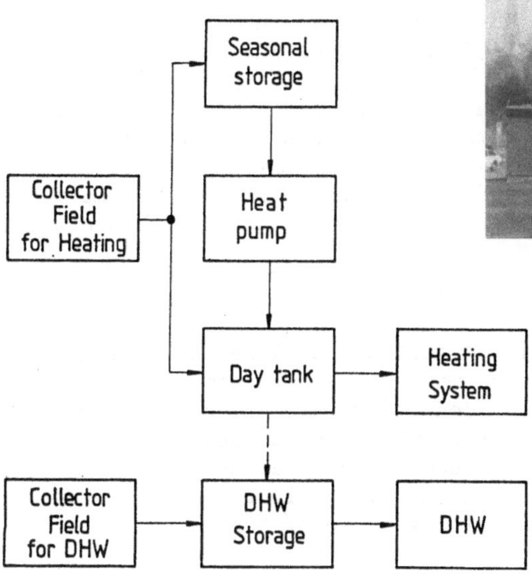

Fig.2 Energy flow chart of the solar system of the main house

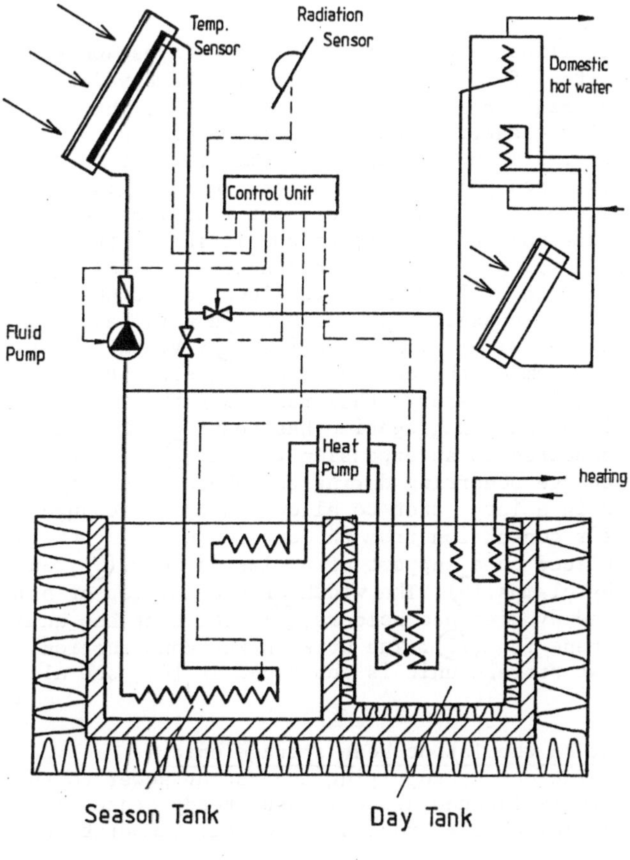

Fig.3 System lay-out in a simplified form.

The depth of the large tank is 2 m, the one of the small tank is only 1.6 m.

THERMAL PERFORMANCE TESTING OF A LATENT HEAT STORAGE UNIT

B. CARLSSON and A. SCHMIDT
Department of Physical Chemistry
The Royal Institute of Technology
S-100 44 STOCKHOLM Sweden

H. OTTOSSON
Division of Energy Systems
Department of Mechanical Engineering
Linköping Institute of Technology
S-581 83 LINKÖPING Sweden

Summary: The results from the experimental testing of a latent heat storage unit containing calcium chloride hexahydrate are presented. Thermal performance data are evaluated from the experimental results by the aid of a quasi-stationary heat transport model and the procedure employed seems well suited both for the rating and for the optimization of latent heat storage units.

1. INTRODUCTION

The purpose of the present work is to test and analyze the thermal performance of a latent heat storage unit by making use of a simple quasi-stationary heat transport model [1,2]. Test conditions are selected, primarily for the contemplated application of short term storage of solar heat. The heat storage unit, which has been dimensioned for a solar heated house [3], was for the purpose of testing built at the Royal Institute of Technology and experimentally tested at the Technical Institute of Linköping.

2. TESTING PROCEDURE

2.1 The latent heat storage unit

The storage unit tested is illustrated in Fig. 1. It contains 400 kg of calcium chloride hexahydrate with 1.5 wt% strontium chloride hexahydrate added and it has a latent heat storage capacity of 70 MJ or 19 kWh. The salt is contained in eight rectangular cassettes, each with the dimension of 1 m in length, 0.8 m in height and 0.05 m in thickness. The material of the cassettes is 2 mm glass-fiber reinforced polyester plastics. For charging of the heat storage unit water is used as the heat transport fluid, and each cassette is equipped with an internal heat exchanger consisting of two coils of cross-linked polyethene tubes (ϕ_{outer} = 15 mm, thickness = 1.5 mm), ① and ②; the upper coil has a length of 7.90 m and the lower one 8.60 m. The internal heat exchangers of the cassettes are coupled in parallel, ⑩ and ⑪ . Each cassette is equipped with a mixing device, ③. Via a perforated tube, extending along the bottom of the cassette, compressed air can be forced through the salt melt and if necessary, thus restoring the salt system to its homogeneous state if a phase segregation has resulted. Each cassette is tightly sealed and connected to an expansion volume, ⑤, in order to compensate for the volume change occurring during the phase change reaction. The cassettes are placed in a large duct to allow for heat exchange during discharging by forced air convection from the outside walls of the cassettes. The cassettes are placed on a support of a plastic foam, ⑧, and are fixed in their positions by clamps, ⑨. The width of the air gap of 5 mm between two cassettes is secured by distance holders, ④. The duct has an insulation of 50 mm of poly-urethane foam, ⑦. The work on the construction and the dimensioning of the heat storage unit is described in [4], see also [3].

2.2 Types of tests carried out on the latent heat storage unit

a) A charging test is carried out by supplying a constant power to the system by the aid of an electric resistance heater. When steady state conditions have been reached during a discharge process (inlet air temperature =

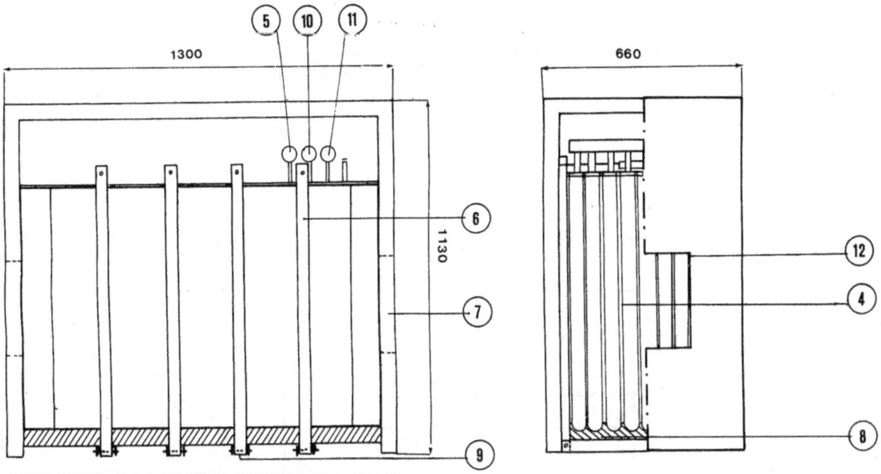

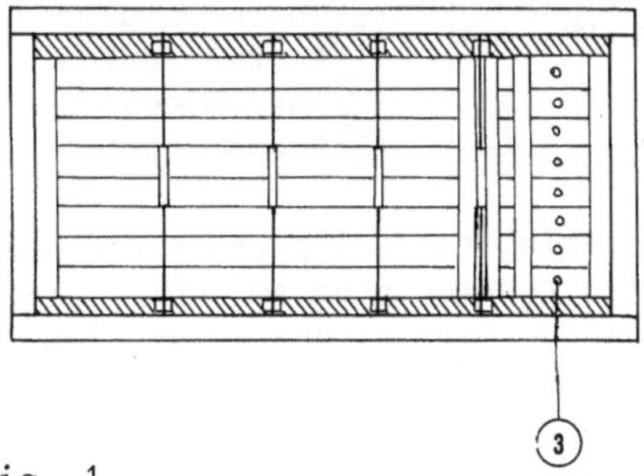

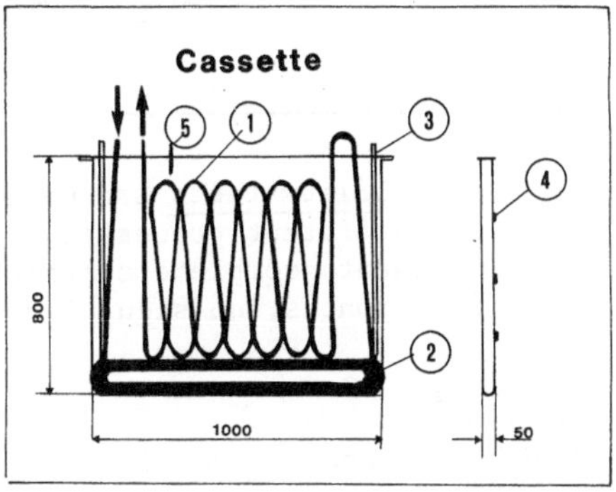

Fig. 1

The latent heat storage unit tested.

20°C), electric power is supplied until the temperature of water of the electric heater reaches 60°C. Afterwards, charging is conducted by keeping this temperature constant until steady state conditions are reestablished. In Fig. 3 the testing results from three charging processes, carried out at different power inputs, are illustrated. The supplied amount of heat is used as abscissa to describe the course of the processes.

b) A discharging test is carried out at a constant flow rate and inlet temperature of the air. After the steady state conditions during charging have been reached, discharging is started and carried out until a new steady state is obtained. In Fig. 4, the results from four different discharging processes are illustrated by using the extracted amount of heat as abscissa. The extracted amount of heat is set equal to zero when the salt temperature at the half height of one cassette, has reached 32°C.

c) Transient tests are conducted both during a charging and a discharging process by changing stepwise the electric power input and inlet air temperature respectively, during the course of the processes. The results obtained are illustrated in Fig. 5 for a charging process and in Fig. 6 for a discharging process.

d) Heat loss measurements are carried out during the steady state period following a charging process. The water inlet and outlet temperatures of the storage unit are then around 58°C and the heat loss rate from the storage unit 106 W. The heat loss coefficient is, consequently, only about 2.8 W/K and the heat losses during the phase change period of the charging and discharging processes are negligible.

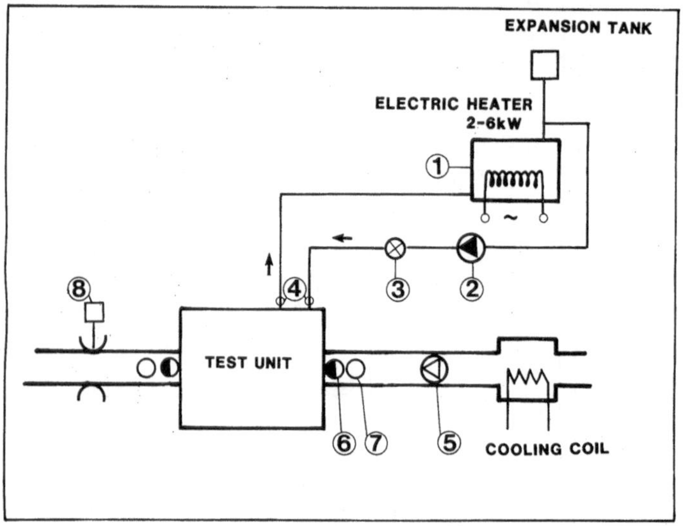

Fig. 2
Experimental set up.
(1) Electric heater
(2) Pump
(3) Water flow rate meter
(4) Thermocouples for temperature difference measurements
(5) Fan
(6) Thermopile of 8 thermocouples
(7) Thermocouple
(8) Nozzle throat

e) Reversibility tests are made by comparing thermal performance data obtained from different processes. During the testing program, which comprised of about twenty complete charging and discharging processes and lasted for about four months, no sign of degradation could be observed.

3. ANALYSIS

3.1 The heat transport model

The heat transport model used for the evaluation of the testing results is based on the assumptions that the heat transfer can be considered stationary and that the heat storage medium be regarded as one homogeneous unit. The heat transfer rate Q is given by the following equations:

$$Q = (\dot{m}c_p)(T_i - T_o) = (\dot{m}c_p)\cdot\varepsilon_M(T_i - T_m) \quad (1)$$

$$\varepsilon_M = 1 - \exp(-U_{su}\cdot A_s/(\dot{m}c_p)) \quad (2)$$

where $(\dot{m}c_p)$ = heat capacity flow rate of the heat transport fluid, T_i and T_o are the inlet and the outlet temperatures of the heat transport fluid, respectively, T_M = mean bulk temperature of the salt melt, ε_M = temperature efficiency for the transport of heat from the heat transport fluid to the salt melt, U_{su} = overall heat transfer coefficient and A_s = effective area of the heat exchanger.

3.2 Predictions of the thermal performance of the unit during the transient tests

As the heat loss rate from the storage unit is of negligible order of magnitude, the testing results in Figs. 3 and 4 can be used directly to predict the results obtained during the transient tests according to eq. (1), if T_M and ε_M are assumed to be dependent solely on the energy content of the storage unit. As is illustrated in Fig. 5, the thermal power input can be predicted from the outlet water temperature and energy content in this simple way with a high accuracy. This is also true for the prediction of the thermal power output obtained from the inlet air temperature and energy content as is illustrated in Fig. 6.

3.3 Evaluation of the thermal performance of the storage unit

The effective heat storage capacity at different discharge rates can easily be evaluated from the data in Fig. 4. The maximal capacities, in the

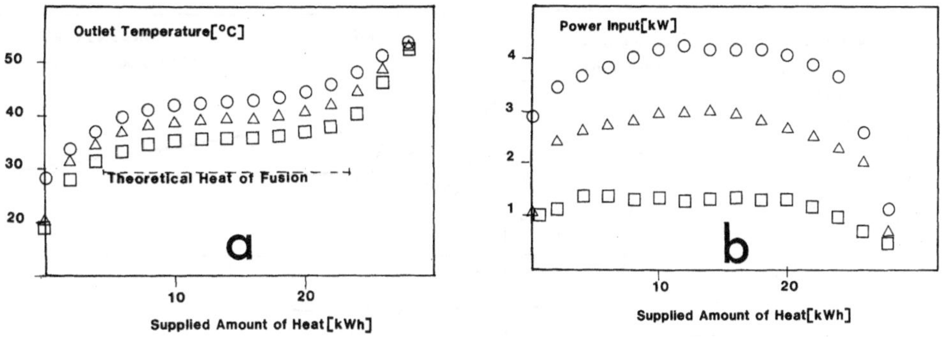

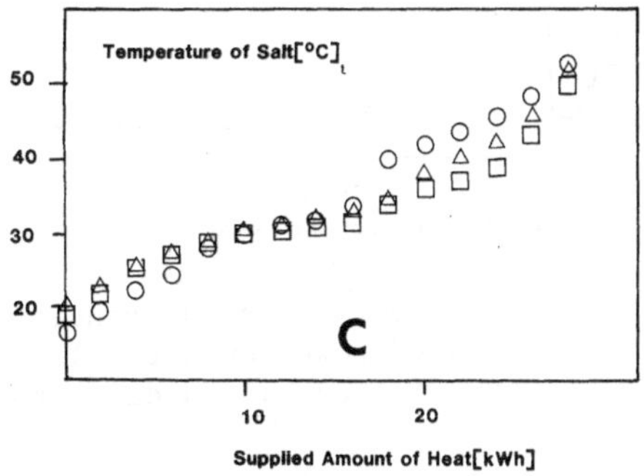

Fig. 3 Charging curves: a) Outlet water temperature, b) power input and c) salt temperature *versus* the supplied amount of heat. The heat capacity flow rate of water is 0.951 kW/K. The salt temperature is measured in one cassette at a height half way to the top of the cassette.

temperature intervals between 20-32°C and 15-32°C, are 26 kWh and 28.5 kWh, respectively. If only power outputs higher than 1.5 kW are taken into account, the corresponding values at an air flow rate of 1000 m^3/h are 14 and 22 kWh, respectively.

The heat exchange characteristics during charging can be analyzed in terms of eq. (1) and (2). With data taken from Fig. 3 at 15 kWh, *i.e.* from the middle of the melting plateau, values of T_M = 32°C and ε_M = 0.30 are obtained. The ε-value corresponds to NTU = 0.36 or to U_{su} = 56 $W/m^2,K$ (A_s = 6.05 m^2)

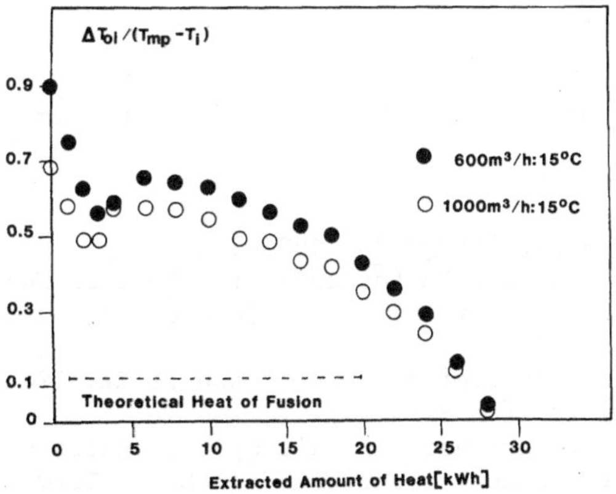

Fig. 4 Discharging curves. The temperature difference between outlet and inlet air (ΔT_{oi}) during four discharging processes at different inlet temperatures and air flow rates (T_i= inlet air temperature and T_{mp} = 29.7°C)

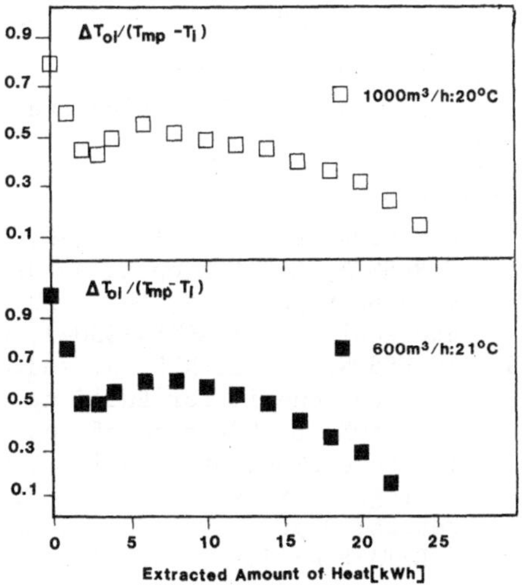

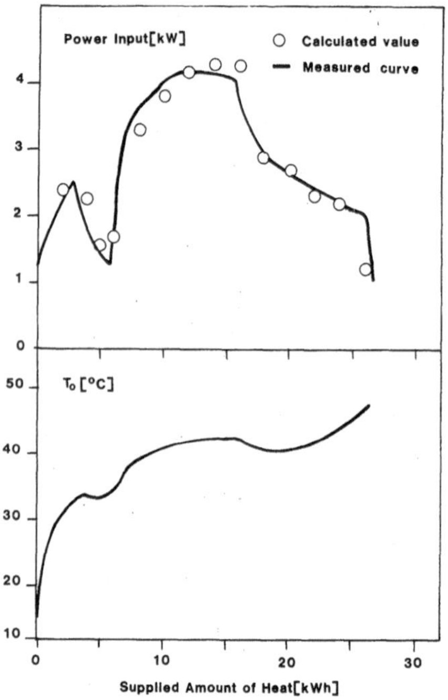

Fig. 5 Charging curve from a transient test. The calculated power input values are obtained by using eq. 1 and data from Fig. 3.

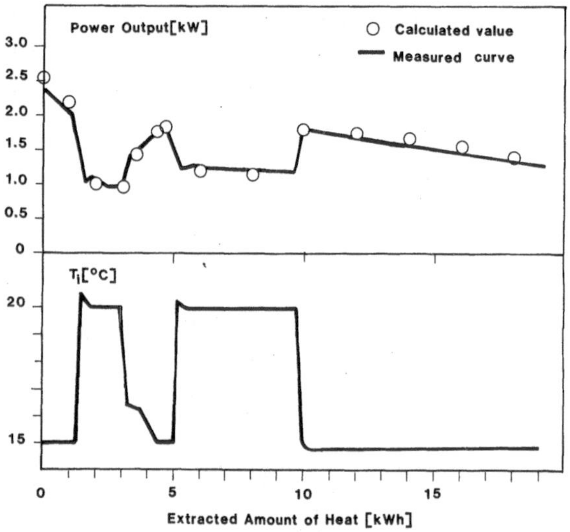

Fig. 6 Discharging curve from a transient test. The calculated power output values are obtained by using eq. 1 and data from Fig. 4.

The heat exchange characteristics during discharging can be analyzed as is indicated in Fig. 4. The ε-values, at an extracted amount of heat = 12 kWh and at the air flow rates of 600 and 1000 m^3/h, are $\varepsilon = 0.59$ and $\varepsilon = 0.50$, respectively, for $T_i = 15^oC$. These values correspond to U_{su} values of 16 W/m^2,K and 21 W/m^2,K ($A_s = 11.1\ m^2$). The order of magnitude of the U_{su} values is determined mainly from the transfer of heat from the outer walls of the cassettes to the streaming air.

ACKNOWLEDGEMENT

This work has been financed by the Swedish Council for Building Research.

REFERENCES

1. CARLSSON, B. and WETTERMARK, G. Solar Energy 24 (1980), 239.
2. CARLSSON, B. "Calculation of the Thermal Performance of a Heat of Fusion Storage Unit", Dept. Phys. Chem., Royal Inst. of Technology, Stockholm, (TRITA-FYK-8304), Oct. 1980.
3. ISAKSSON, P, KELLNER, J. and others "Solveigprojektet" Report R21-1981, Swedish Council for Building Research (in Swedish).
4. CARLSSON, B. "Storage of Low Temperature Heat in Salt Hydrate Melts - Investigations of Materials, Heat Exchange Techniques and Short Term Storage of Solar Heat", Report to the Swedish Council for Building Research; Dept. Phys. Chem., Royal Inst. of Technology, Stockholm, (TRITA-FYK-8306), June 1983 (in Swedish).

SOIL MECHANICS ASPECTS OF SEASONAL HEAT STORAGE IN GRONINGEN

J.W. DE FEIJTER
Delft Soil Mechanics Laboratory, The Netherlands

Summary

The first soil mechanics results, such as soil deformations and groundwater movements, of an operational heat storage in the soil are presented. After extended site investigations and laboratory tests the final design for a heat store in Groningen, located in the northern part of the Netherlands, was completed. During the construction the tubes of the closed loop soil heat exchanger were inserted from the surface to a depth of 20 metres. A lot of sensors were installed for determination of the temperature, the horizontal and free-convection groundwater movement, the gett off of dissolved gasses, the settlement and the horizontal displacement. During the first year of experiments the store was heated up, cooled down and again heated up. The fluctuations in the horizontal groundwater movement and the groundwater table are bigger than assumed in the design stage. The soil deformations such as settlement and horizontal displacement are within the predicted area. Gett off of gasses from the groundwater is not determined. The free convection is as predicted. The groundwater movements justify the take off of the impermeable screen around the store from the basic design. The deformations of the store in the next years have to be much smaller than in the first year therefore the measurements will be continued for at least two years.

1. Introduction

This project is a continuation of the research project 'The use of soil as a storage medium for seasonal storage of solar energy' under EC contract number 516-78-1 ESN. This study was carried out by the Delft Soil Mechanics Laboratory as main contractor and by the Institute of Applied Phisics and Philips as subcontractors. This study provided the theoretical ins and outs of a total solar heating system with long term storage in the soil for a group of solar houses both for the technical and for the economical behaviour. To complete the picture of such a system with actual data on heat balances, cost and geotechnical realization it is necessary to construct and to investigate a heat stoarage in the soil on real scale.

After a long term of negotiations with a lot of involved parties the construction of the store could start in June 1982 and was finished in December 1983. This heat storage project is located in Groningen in the northern part of the Netherlands. 96 Solar houses are coupled to a central heat storage in the soil (diameter: 36 m depth: 20 m volume: 23.000 m³). During emergency situations a boiler house can dèliver the total heat demand of the houses. The experiments were started in February 1983 and are still going on. This paper presents the soil mechanics aspects. The evaluation of the heat balances of the store are presented by Wijsman and Van Meurs in their paper entitled: 'The seasonal heat store in Groningen'.

2. Site investigation

Before designing a heat store one have to determine the soil properties such as soil composition (clay, peat, sand) and homogenity, strength for inserting heat exchanger tubes, permeability for groundwater movement, porosity, heat capacity and thermal conductivity. Therefore site investigations, such as borings, Dutch cone penetration tests and density measurements, and determination of soil properties, such as grain size distribution, permeability and thermal conductivity, in the laboratory are excecuted. Figure 1 summarizes the soil properties of the heat store site.

3. Installation of the soil heat exchanger

The insertion of the heat exchanger strings from the surface was effected by a vibrating lance while at the same time water was being injected. The vibrating lance is a specially designed I-profile and is driven by a modified pile drive vibration equipment (height 30 metres). The flexible tube is at a protected position along the side walls of the lance. At the bottom side the tube is protected by an iron tube bended around the flexible tube; this iron tube stays behind in the soil. In the I-profile 4 pipes for water injection are present. By vibrating and jetting the effective soil stresses are strongly reduced: the saturated soil around the lance is like a liquid (soil liquefaction). So, the friction between the lance and the soil is very small. Within a few minutes the lance is at a depth of 20 metres. Due to vibrating of the lance during pulling back the soil becomes dense again; at the top missing soil is filled up with sand. During one of the insertions the inclination of the vibrating lance was measured in two directions. The deviation at 20 metres depth was less than 2%.

4. Instrumentation in the soil

In the soil temperatures are measured in three directions from the centre of the reservoir. One direction (A) is fully instrumentated. The low groundwater flow rate permits a less dense instrumentation in the other two directions. In figure 2 the total instrumentation is shown.

4.1. The groundwater table and the horizontal groundwater movement

The horizontal groundwater movement in the field may cause heat losses of the store. From the gauging of six observation wells the magnitude and the direction of the groundwater flow will be determined. This six observations wells are installed at the depth of 6 metres: three near the store (30 m from centre) and three in the far field (200 metres from the store centre).

4.2. The free convection groundwater movement

The groundwater flow causes by free convection gives extra heat losses in saturated sandy soils. The very low velocitities (10^{-8} - 10^{-6} m/s) are hardly directly detectable. To get an impression of this phenomenon the groundwater pressure is measured in a vertical plane through the axis of the storage: at depths of 9, 19 and 30 metres there are 4 measuring points.

4.3. The get off of dissolved gasses

In the groundwater several gasses are dissolved by heating the store these gasses can get off. The gas lowers the effective permeability and the heat capacity. This is determined by measuring the density of the soil and the watercontent: gas bubbles will lower these values considerably. In the store three tubes are installed, in which a nuclear density device can be lowered. This backscatter probe for gamma radiation and neutrons measures the mass density and the moisture content. In this way saturation degree, i.e. the ratio of the watervolume and the pore volume can be determined.

4.4. The deformations of the store

Soil deformations can be occur on account of disturbing of the soil stresses and soil fabric during the construction of the store and of the thermal expansion or shrinkage of the soil during the cyclic thermal loading. For the vertical settlement 9 beacons are placed with their base plate at a depth of 1 metre below groundlevel. For the horizontal displacement 4 steel tubes (length 5 m, diameter 0.2 m) were vertically vibrated into the soil at the edge of the store.

5. Results

5.1. Groundwater table

Figure 3 presents the fluctuations in the groundwater level and the local rainfall. The groundwater table is strongly dependent on the rainfall. After finishing the construction of this district the rainwater will mostly be drained by a sewerage.

5.2. Horizontal groundwater movement (hgm)

Table I shows the values of the horizontal groundwater movement at different stages of the experiment.

point of time	time (weeks)	Pore velocity (m/s)	Darcy velocity (m/s)	direction angle A to C (°)
start heating up I	0	$7\ 10^{-8}$	$2.5\ 10^{-8}$	65
3 weeks heating up I	3	$9\ 10^{-8}$	$3.5\ 10^{-8}$	105
end heating up I	10	$13\ 10^{-8}$	$5.0\ 10^{-8}$	70
end cooling down I	20	$19\ 10^{-8}$	$7.5\ 10^{-8}$	35
end heating up II	42	$14\ 10^{-8}$	$5.5\ 10^{-8}$	75

Table I : Velocity vector of the horizontal groundwater movement

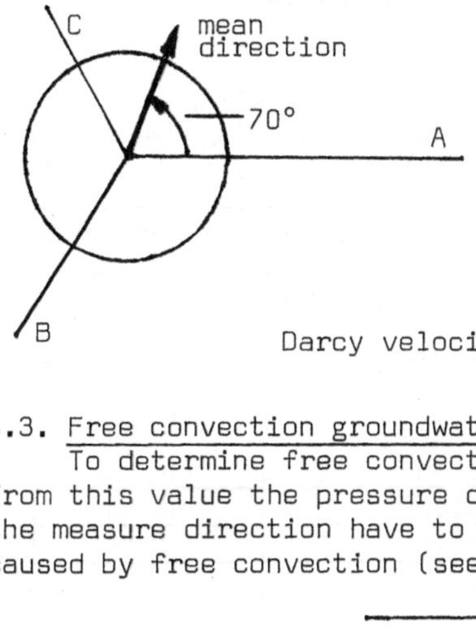

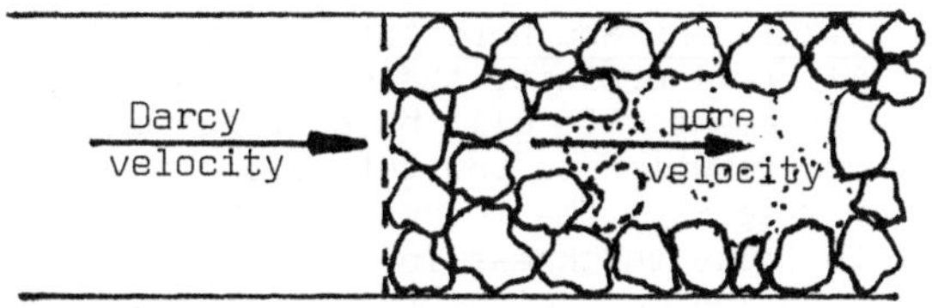

Darcy velocity = porosity * pore velocity

5.3. Free convection groundwater movement (fcgm)

To determine free convection the groundwater pressures were measured. From this value the pressure differences caused by the 'hgm' component in the measure direction have to be extracted to get the pressure difference caused by free convection (see figures below).

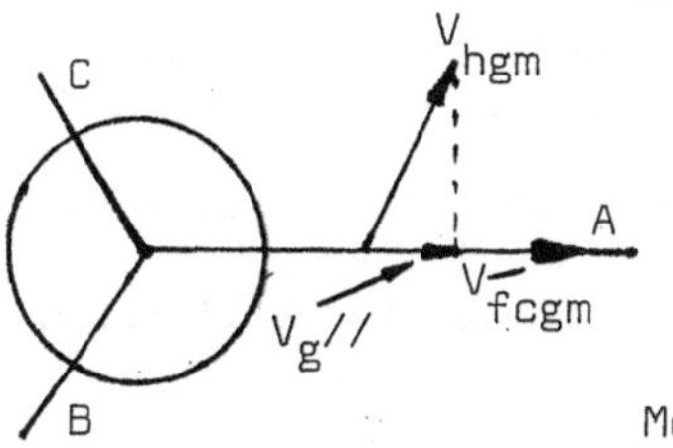

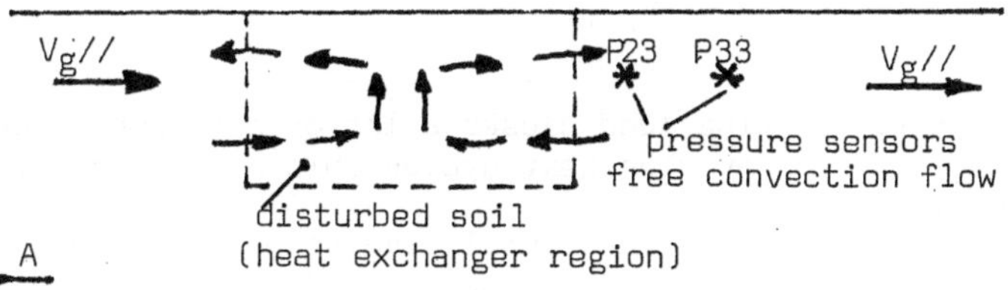

Measured data $P_{23}-P_{33} = \Delta P_{Vg//} + \Delta P$ convection.

With the professinal 3-D computercode LIGHTS for heat transfer by conduction and convection in saturated poreus media also the pressure differences can be calculated: the temperature data in the field are put in an axial symmetric model of the heat store the accessory pressures will be computed. Table II shows the measured and calculated values for the free convection groundwater movement.

point of time	pressure difference (N/m²)				V_{Darcy} (m/s)
	from tot. gm (measured)	from hgm	from fcgm	from fcgm calculated	measured at R = 23 m
3 weeks heating up I	-10	-15	5	10	-
end heating up	150	35	115	130	$7\ 10^{-8}$
end cooling down I	260	130	130	150	$8\ 10^{-8}$
end heating up II	185	30	155	215	$9\ 10^{-8}$

Table II : Free convection groundwater flow

The mean ratio between the measured and calculated data is 0.8. One may conclude that the determinated soil properties and the assumed horizontally homogenity are slightly different than in the field. The total heat losses by groundwater movement will be a little bit less than the heat losses from previous calculations (6% decrease in efficiency), on which was decided to take no preventions against groundwater movement.

5.4. The get off of dissolved gasses

From the measure data follows no changes in the saturation degree. So get off of dissolved gasses did not occur. Therefore no influence on the effective permeability and the heat capacity of the soil.

5.5. Settlement and horizontal displacement

Figure 4 presents the settlement per heating up or cooling down period. The settlement rate will be decrease each next thermal loading cycle. The main contribution to the settlement is the shrinkage of clay in the store. From the borings the thickness of the involved clay layers is estimated. From the clay thickness in combination with the temperature history of a cycle and the mechanical thermal properties of clay one can calculate the settlement. Table III shows the measured and calculated values of the settlements of the first thermal loading cycle at three locations.

distance to centre	settlement calculated	first cycle measured
8 m	68 mm	59 mm
14 m	56 mm	55 mm
18 m	44 mm	41 mm

Table III: Comparison settlements.

point of time	horizontal displacement
end heating up I	-16 mm
end cooling down I	-48 mm
end heating up II	-40 mm

Table IV: Decrease of the diameter.

The first heating load causes a lot of settlement, because of the dewatering of the clay. The next cycles this effect is for the major portion over.

Table IV shows the horizontal displacement: the diameter of the store decreases. The total volume decrease after the second heating up period is about 140 m³. Therefore near the store is a relaxtation of the soil stresses. The stability of constructions in or on the soil could be influenced.

6. Conclusions

From the first year of testing a heat store in the subsoil the next main conclusions on account to the soil mechanics aspects can be drawn:

- The inserting technique for the heat exchanger strings is reliable and rapid in soils with only thin layers of clay.
- The rainfall has a great impact on the groundwater movement.
- The free convection is of the same magnitude as the horizontal groundwater movement.
- Dissolved gasses in the groundwater don't gett off.
- The settlement of the store was predictable for the first heating up and cooling down period.
- The measurements has to be continued till groundwater movement and soil deformations are stabilized.

DEPTH (m-GL)	MEAN SOIL PROPERTIES: SOIL TYPE	PORO-SITY (%)	PERMEA-BILITY (10^{-12} m²)	THERMAL CONDUC-TIVITY (W/m K)	SPECIFIC HEAT CA-PACITY (10^6J/m³K)
0	SAND SILT	45	SMALL	1,9	2,9
-5	SAND (FINE)	29	5	2,2	2,5
	CLAY/PEAT	60	SMALL	1,0	3,3
-10	SAND (FINE)	38	5	2,3	2,8
-15	SAND (FINE) CLAY&PEAT LAYERS	48	< 5	2,0	3,0
-20	SAND (MEDIUM)	36	20	2,1	2,7
	CLAY + SAND LAY	55	SMALL	1.1-1.6	3,1
-25	SAND (MEDIUM)	41	8	2,1	2,8
-30	SAND (FINE)	43	6	2,3	2,9

Fig.1: Soil properties at Groningen

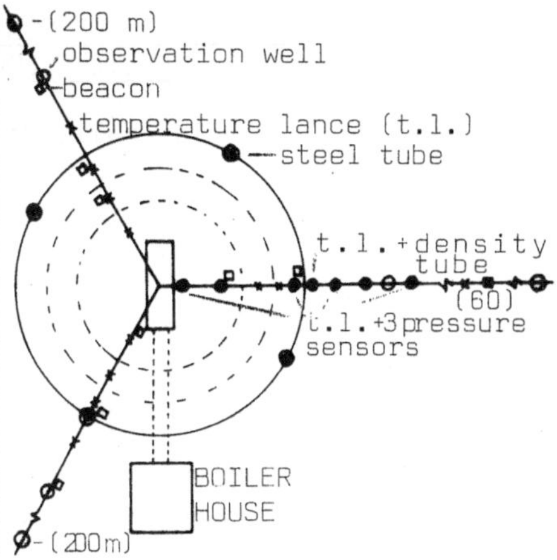

Fig.2.: Instrumentation in the field

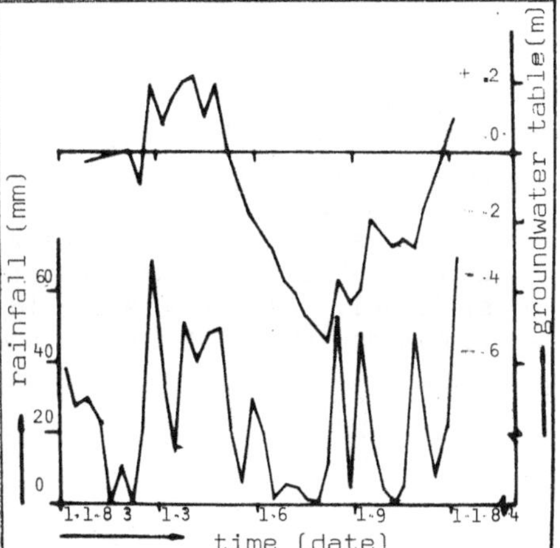

Fig.3.: Groundwater table and ten days rainfall

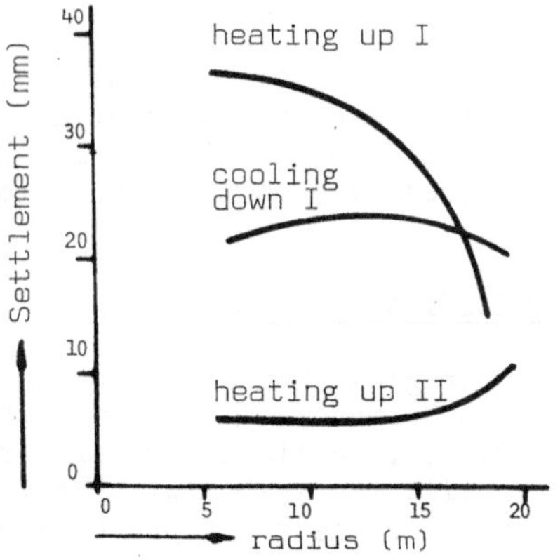

Fig.4.: Settlement per period

LONG TERM STORAGE OF SOLAR ENERGY IN SILICAGEL-WATER SYSTEMS FOR SPACE HEATING

R. Hallermayer, R. Sizmann

Ludwig-Maximilians-Universität München
Federal Republic of Germany

Summary

The thermochemical reaction of silicagel-water is suitable for long term storage of solar energy. There is no thermal degradation and charging temperatures of 75 °C are sufficient. The useful heat from the storage depends strongly on some factors, like temperature and humidity of the inlet air, line temperature of the heating system at discharge. In a computer model we investigated the annual performance of a solar system with water and silicagel storage for autonomous solar space heating. 40 m² high performance collectors, two water tanks with 2 and 10 m³ and 26 t silicagel can match the annual demand of 10.3 MWh of a single-family house. The thermochemical storage supplies 4.6 MWh of the total amount. The energy density is 175 kWh/ton silicagel or 50 % of the heat input into the material. At the discharge process the water storage supplies 1.25 MWh at a temperature level of about 15 °C. Installation of a back-up system can provide heat if there is no additional energy from the water storage. This avoids unfavourable operating conditions and increases the energy density of silicagel. As an example 18 t silicagel supply 3.5 MWh heat energy combined with 1.1 MWh back-up energy.

1. INTRODUCTION

An autonomous solar space heating system in Central Europe requires large solar collector areas and big, well insulated water tanks for storage. Due to the low heating demand, except in winter, thermal losses of sensible heat storage are high and, therefore, the annual performance of the total system is poor. Thermochemical storage materials provide higher energy densities than water and avoid thermal degradation of the stored energy. We report on the performance of a solar system for space heating of a single-family house with silicagel-water as storage and zero back-up energy, investigated by computer simulation.

2. RESIDENTIAL HEATING DEMAND

The investigated house is a so-called low energy house, commercially available in Germany. It has two storeys with 170 m² housing space for one or two families. Walls and floors are well-insulated (see Table I). A forced air ventilation system with heat exchanging is installed, which recovers about 60 % of the spent preheating energy into the fresh air intake.
The residential heat load was computed with hourly weather data of the year

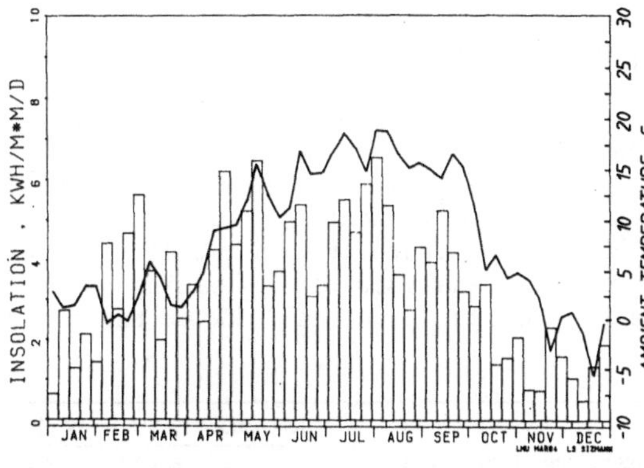

1975 for Weihenstephan (48.24 north, 11.44 east) (1) . The room temperature was kept to 20 °C, except to 17 °C during nighttime. The annual heat demand is 10.3 MWh, with about 70 % from November to February. There is a maximum heat load of 50 W/m^2 at an ambient temperature of - 16 °C. The heating system operates at 35/30°C line temperatures.

Fig. 1 Insolation (bars) on a 48° S tilted surface and ambient temperatures (weekly mean values)

Table I: characteristic data of the house

Building area 9.2 x 13.5 m		124 m^2
Outer walls heat loss coefficient		0.32 W/m^{2}K
Roof (48° tilted)		0.28 W/m^{2}K
Ceilings		0.5,0.3,0.25 W/m^{2}K
Window area 28 m^2	heat loss daytime	2.5 W/m^{2}K
	heat loss nighttime	1.3 W/m^{2}K
Ventilation 0.6 air changes per hour (0.75 in the upper floor)		

3. SILICAGEL-WATER SYSTEM

Thermochemical reactions A + B = AB + heat are suitable for storing solar energy. Advantages are: long term stability by keeping the reactants A and B apart; high storage densities. In an open system with silicagel-water only dry silicagel has to be stored. The water vapour is released into the surroundings. Air is the transfer medium for heat and water vapour (2,3).
In the charging process (drying silicagel) hot, dry air is blown through a bed of the adsorbent. The heat is provided by a solar collector (see Fig. 2). Cold and humid air leaves the storage. Charging temperatures of 60 to 80 °C are sufficient. In the discharging process (heating mode) cold, wet air flows into the adsorbent bed and leaves it warm and dry. For example 15 °C, 100 % humid air is warmed up to 50 °C at the outlet (Figure 3).

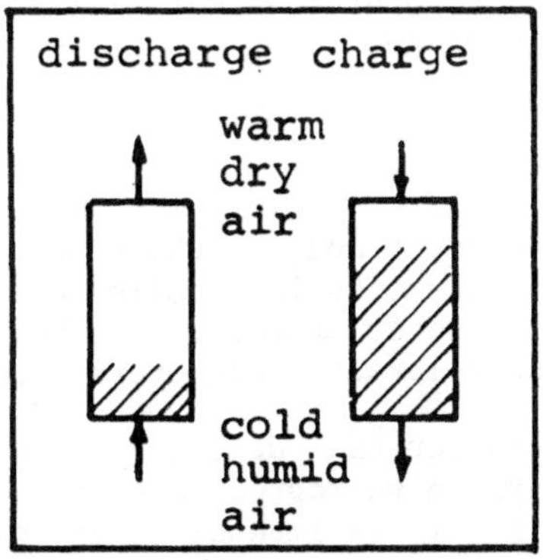

Fig. 2: Operation scheme charging - discharging mode

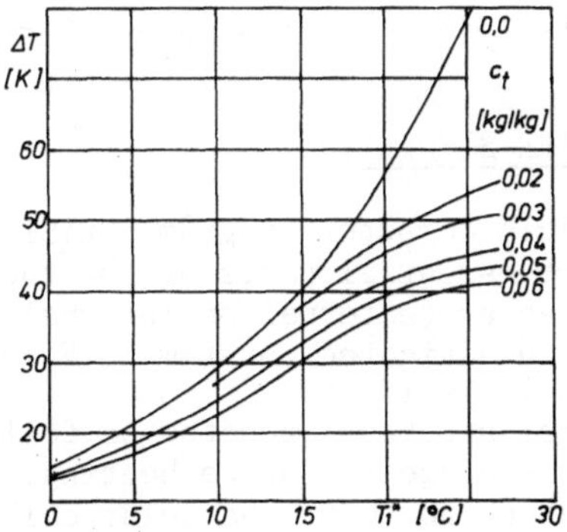

Fig. 3: Maximum temperature raise of the air within silicagel with water content Ct

The dry material can absorb 39% water at the maximum with a heat of adsorption of 1350 kJ/kg. About 70 % of this amount is heat of condensation of the water vapour. Therefore, maximum heat from the storage is extracted only if there is 100 % humidity in the inlet air.
The useful heat from the storage is influenced by the following factors (see Figure 4/5): Temperature T1 and humidity RH1 of the inlet air, water content CI of the dry silicagel, storage efficiency ESP (ratio actual temperature raise outlet-inlet to maximum raise), heating temperature TR, heat exchanger efficiencies E1 and E2, temperature TU and humidity RHU of the ambient air. Fig 5 shows a sensitivity analysis of a heating system with constant air and water flow and heat output power of 5 kW at the fixed operating point.

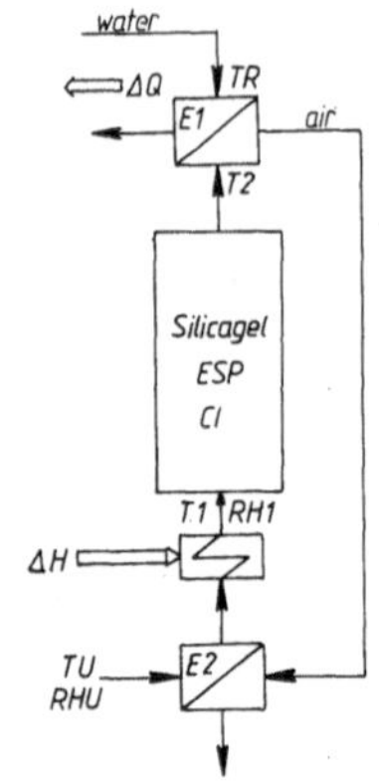

Fig. 4: System in heating mode

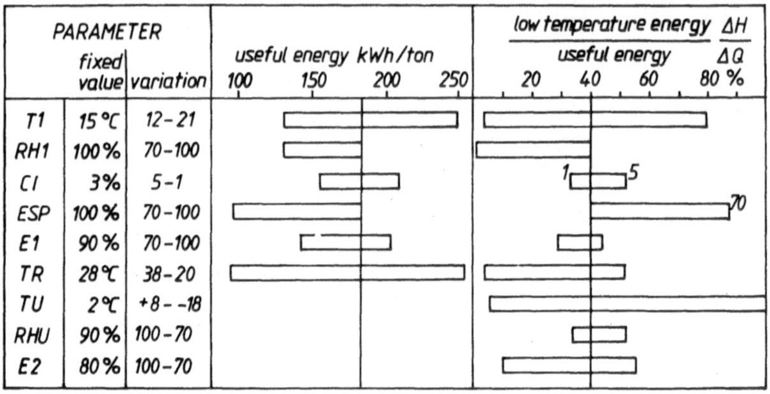

Fig 5: Sensitivity analysis with constant air and water flow

The useful energy depends strongly on temperature and humidity of the inlet air and the heating return line temperature. Where is the lost energy, though there is no thermal degradation? The stored energy of the silicagel warms the air stream for example from 15 °C to 50 °C. The lowest heating temperature is 28 °C, so that only the energy in the air above 28 °C can be used; this is 63 %. The remaining energy can be used to warm and wet the inlet air, but it is lost for space heating. A considerable part of the useful energy has to be provided as additional input energy at a temperature level of about 15 °C to produce humid air for discharging. The availability of this energy is a vital point of using such a storage material and a characteristic of the proposed thermochemical storage.

4. COMPUTER SIMULATION

The utilized computer program SOLYST is a finit-element model, which contains the essential parts of the thermal solar system: solar collectors, heat exchanger (external or internal), stratified water storage. A scheme of the investigated system is Figure 6 and the relevant parameters are listed in Table II.
The control strategy operates as follows: The smaller tank of 2 m³ serves as buffer storage for space heating. It has always to be kept at a temperature of about 50 °C by the solar collectors or if this is impossible at the front line temperature of the heating system by the silicagel storage. The second water tank serves as low energy reservoir for the wet inlet air of the silicagel. It warms the water evaporator and is to be kept at a

Table II: System Parameters

Collector area	40 m^2
Absorptance-transmittance product	0.72
heat loss coefficient	1.8 W/m^2K
water storage	2 and 10 m^3
insulation of the storage	0.4 and 0.1 W/m^2K
external heat exchanger (solar circuit)	2000 W/K
heating temperature (at -5 °C ambient)	35/30 °C

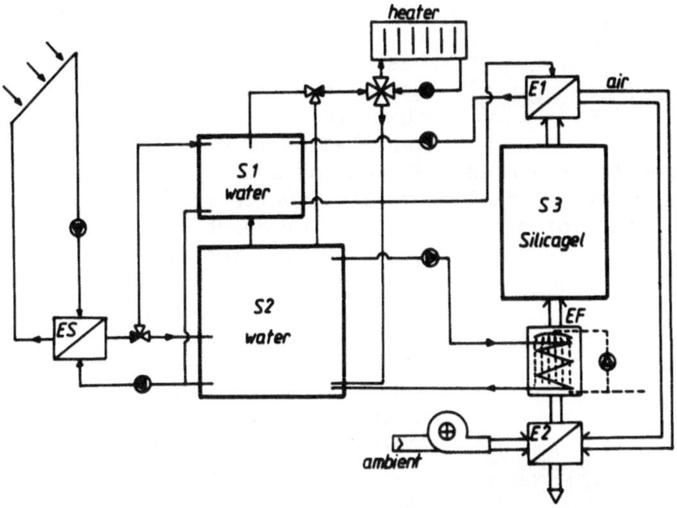

Fig. 6: Scheme of the solar system (heating mode)

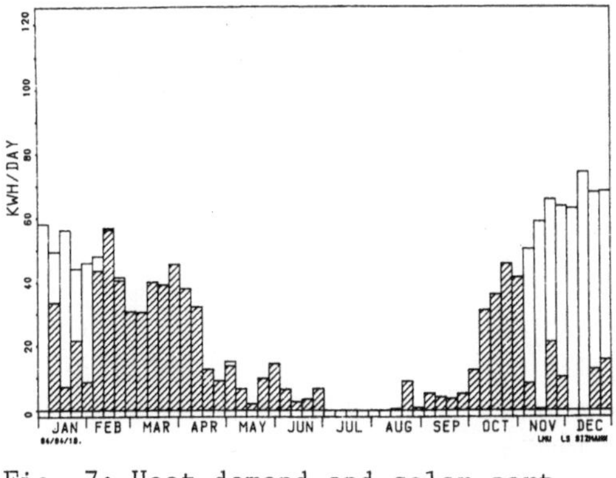

Fig. 7: Heat demand and solar part via water storage (hatched bars)

temperature of 22 °C, if possible. From April to September the solar collectors charge the silicagel storage to 3 % water content at least and stop charging at 26 tons of dry silicagel. In fall the collectors have time (and sunshine) enough to load the bigger water tank to a temperature of about 90 °C. In the heat season the energy comes at first from the water storage until it has no further heat energy. Then the silicagel storage is discharged with the evaporation energy taken from the big tank.

The residential heat demand is matched by the solar system throughout the year. The hatched bars in Fig. 7 are heat energy which appears as space heating via water storage. This amount is 5.7 MWh. The remaining part comes from the silicagel. The collector system has a good performance of about 38 % throughout the year. Figure 8 shows the temperatures of the two water tanks and the energy from the solar collectors into the water tanks. As you can see the water storage operates especially in fall and spring.

Figure 9 shows the useful energy density for the opeerating days of the silicagel storage. The first day occurs in November, the last in February. The lowest amounts are found in December, where the demand is high and the

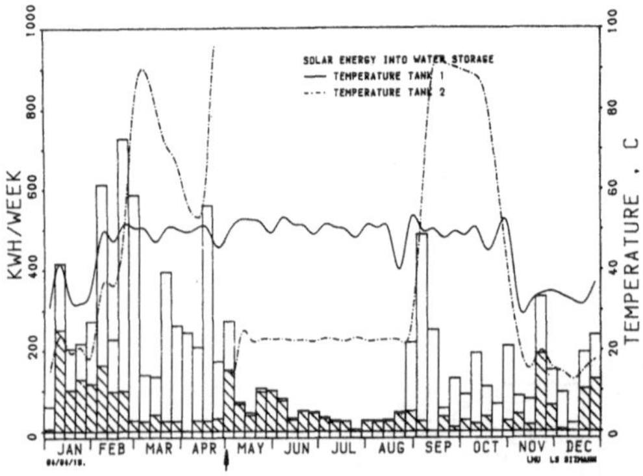

Fig. 8: Water storage temperatures and solar energy into water tank 1 and 2 (hatched bars). Start of simulation is 1st of May.

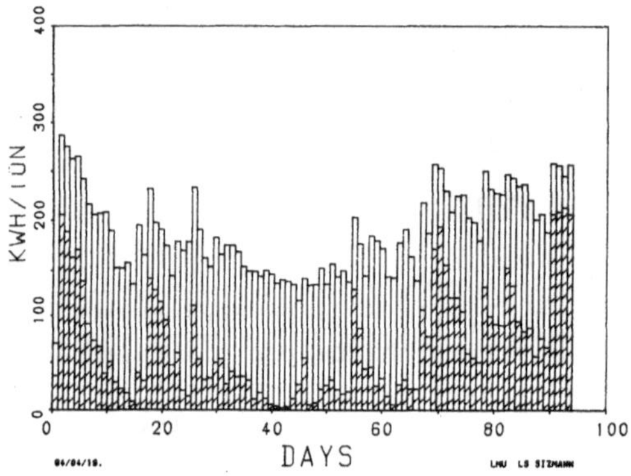

Figure 9: Useful energy from the silicagel and low temperature energy from water tank 2 (hatched); mean values of operating days.

ambient and tank temperatures are low. The total mass of discharged silicagel is 26 tons, necessary for an autonomous solar driven house heating system. The total amount of heat energy from silicagel is 4.6 MWh and the water storage provides another 1.25 MWh to discharge the silicagel.

The electrical energy for water circulation is 400 kWh in the solar circuit. Charging of the silicagel has operation times of about 700 hours and discharging 1100 hours, so that the electrical energy for air ventilation is 850 kWh. For heating 26 tons of silicagel (volume 32 m^3) are used, which leads to an energy density of about 175 kWh/ton. The input energy into the silicagel was 350 kWh/ton, so that 50 % of the stored energy could be used by the system.

The total volume of well insulated water tanks and silicagel is about 60 m^3. There is sufficient space for this volume in every single family house. In contrast to this, an autonomous solar space heating system with water would require 60 m^2 high performance collectors and 80 m^3 tank reservoir with total volume of 140 m^3 including insulation.

REFERENCES

/1/ Deutscher Wetterdienst Offenbach/Hamburg

/2/ E. Lävemann, D. Jung, N. Khelifa, R.Sizmann
Energy balance in a thermochemical storage system using activated silica and water vapour for low temperature heat
4. Int. Solar Forum, Berlin, 1982

/3/ N. Khelifa, E. Lävemann, D. Jung, R.Sizmann
Energy storage in silica-water systems
ISES Congress, Perth, Australia, 1983

The research project is supported by the Federal Ministry of Research and Technology, Federal Republic of Germany.

THERMAL HEAT STORAGE IN SOLAR HEATING SYSTEMS FOR DOMESTIC HOT WATER SUPPLY

S. ØSTERGAARD JENSEN
Thermal Insulation Laboratory
Technical University of Denmark

Summary

Selection of the optimum type storage for solar heating systems for domestic hot water supply. Rules-of-thumb for the optimum storage volume and collector area. Simple method for calculation of heat transfer for heat exchange spirals. Design of heat exchange spirals for storages in solar heating systems for domestic hot water supply.

1.1 Storage type

Three types of heat storage in solar heating systems for domestic hot water supply are to-day used in Denmark:

a) Storage with a built-in heat exchange spiral,
b) Storage with a heating jacket,
c) Storage with a built-in tank for domestic hot water and heat exchange spiral.

Investigations have been carried out concerning which one of the three storage types is the best suited for solar heating systems for domestic hot water supply. Computer calculations of the total performance per year as well as economical considerations have been carried out. Fig. 1 (1) show the results of these calculations. The storages are hight, they have a height/diameter ratio of 4 to secure a good temperature stratification. The heat exchange spiral are placed at the bottom of the storages in case a) and c). For storages of type a) this improve the performance per year by about 10%, instead of letting the spiral be placed in the lower half of the storages (which is normal to-day). When the spiral is placed at the bottom of the storage the temperature of the solar collector fluid will always be as low as possible, which increases the thermal performance. The heating jacket in type b) will have to start at the middle of the storage and end at the bottom. The temperature stratification will thus be increased and at the same time the solar collector will work at the lowest temperature. For the same reasons the built in tank in type c) will have to be high and slim.

Fig. 2 (1) the results of calculations when the storages are installed in a solar heating system with a solar collector area of 4 m^2. The storage volume is 200 l and the daily hot water consumption is 200 l at 45 ^{o}C. The total performance per year has been found by adjusting the net performance per year for the efficiency of the back-up-system and by adding the saved energy consumption by turning the back-up-system off during the summer (May-September). In this case the back-up-system is a modern oil fired boiler with an efficiency of 85% and a thermal loss of 350 W. As can be seen from fig. 2 systems with storages of type a) have the highest performance, but is this type of storage also the one best suited economically?

An investigation show that c)-type storages (in Denmark to-day) are more expensive than the other two types. a) and b)-type storages containing 200 l can be bought at the same price, but the price of b)-type storages with volumes larger than 200 l is higher than that of a similar a)-type storage, while b)-type storages with the volume smaller than 200 l are cheaper than a)-type storages. The economy of a solar heating system with a collector area of 2 m^2, a storage of 100 l and a daily consumption of 100 l has therefore been investigated. In this case the performance of an a)-type system is 12% greater than a b)-type system, but systems with a)-type storages are only 2-3% more expensive than systems with b)-type storages. For the domestic hot water supply a)-type storages are therefore to-day the best suited in Denmark.

1.2 Storage volume

In this chapter the optimum volume of the storage and solar collector area (with selective surface) as function of the daily consumption of hot water at a temperature of 45 °C is discussed. Fig. 3 (1) show the ratio between the cost of the system (Dkr) and the total performance per year (kWh) of solar heating systems for a daily consumption of 200 l at 45 °C.

The total performance per year has been calculated with a modern oil fired boiler as back-up-system. Similar calculations as the above mentioned has been carried out for other daily consumptions and other back-up-systems.

On the basis of three calculations the following equations can be stated:

- Optimum storage volume: daily hot water consumption at 45 °C + 50 l
- Optimum collector area: $\frac{\text{storage volume (l)}}{50}$ (m^2) (with selective surface)

Sensitivity tests have further been made on various user variations in the daily hot water consumption: Change of the consumption pattern per day, per week and per year + the influence of the summer holiday. All these analyses show unanimously that when the starting point is the hot water consumption per day (average per year) the above given equations are not influenced by variations in the consumption pattern.

2.1 Heat exchange spirals: Method of calculation

At the Thermal Insulation Laboratory, Technical University of Denmark, a simple method for calculation of the heat transfer for heat exchange spirals has been developed (2). The method is developed on basis of already existing equations for heat transfer from straight tubes. These equations are not directly usable for heat exchange spirals. Therefore by experiments the equations have been adjusted and validated in such a way, that the experiments are in good agreement with the adjusted theory. The adjustments have been carried out by changes of the influence of the Reynold's, Prandtl's and Grasshof's number.

It is wellknown, that the heat transfer from a heat exchange sprial is a linear function of the temperature of the surrounding fluid (the storage temperature). But as fig. 4 show, the heat transfer is also depending on the motive force, either expressed as the transfered power (W) or as the temperature difference ΔT between the solar collector fluid entering the heat storage and the storage temperature surrounding the spiral.

This equation is very simple:

$$H = a(\Delta T) + b(\Delta T) \cdot T_s \quad \text{W/°C}$$
$$a(\Delta T) = c + d \cdot \ln(\Delta T) \quad \text{W/°C}$$
$$b(\Delta T) = e + f \cdot \ln(\Delta T) \quad \text{W/°C}^2$$
$$\Delta T = T_i - T_s \quad \text{°C}$$

where H is the heat transfer power per ^{o}C temperature difference between the inlet temperature to the spiral and the storage temperature.

T_i is the inlet temperature to the spiral (^{o}C),

T_s is the storage temperature (^{o}C),

c, d, e and f are constants depending on the solar collector fluid and the flow-rate in the spiral, the material, the tube dimension and the length of the spiral.

2.2 Heat exchange spirals: Results

Experiments (2) show, that a flat spiral has the same heat transfer as the spirals used to-day with an extention in the height. Therefore the spirals should be so flat, that they can be located at the bottom of the storage in solar heating systems. This will as mentioned above increase the thermal performance.

The above mentioned simple equations has been used in calculations of the performance per year of solar heating systems for domestic hot water supply (3). As fig. 5 shows the performance from these systems are very little influenced by the flow-rate in the spirals. (In fig. 5 and 6 only the total performance per year is shown. The net performance show the same picture). Only with very short spirals and small flow-rates, the performance is influenced by the flow-rate in the spiral.

Fig. 6 show the total performance per year for solar heating systems for domestic hot water supply designed for daily hot water consumptions of 100, 200, 300 and 450 litres at 45 ^{o}C. The storages and the collector areas are designed according to the Rules-of-thumb from chapter 1.2. The figure show, that there is no real optimum of the performance as a function of the spirals length. But practical considerations concerning the location of the spirals at the bottom of the storages compared with the cost for an increasement in the length of the spirals show, that the spirals should be about 2 m per m^2 collector area (selective surface).

REFERENCES

(1) S. ØSTERGAARD JENSEN and S. FURBO (1984). Storage type and storage volume in solar heating systems for domestic hot water supply. (In danish). Report no. 148. Thermal Insulation Laboratory, Technical University of Denmark.

(2) S. ØSTERGAARD JENSEN (1984). Heat transfer from heat exchange spirals submerged in water. (In Danish). Report 84-10. Thermal Insulation Laboratory, Technical University of Denmark.

(3) S. ØSTERGAARD JENSEN (will be published in the summer 1984). Heat transfer in small well-insulated water storages. (In Danish). Thermal Insulation Laboratory, Technical University of Denmark.

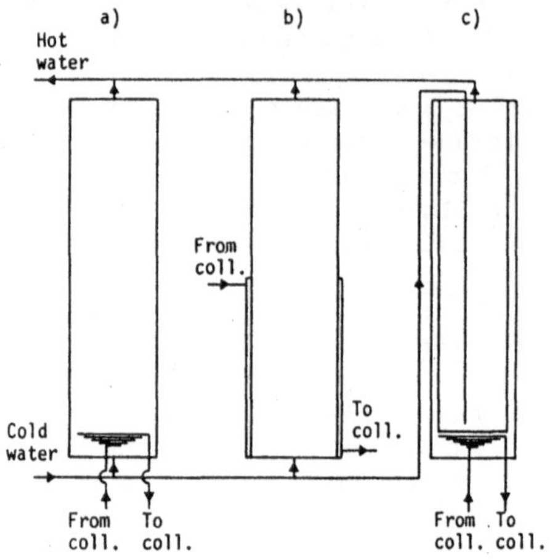

Fig. 1. The three storage types the calculations are made for.

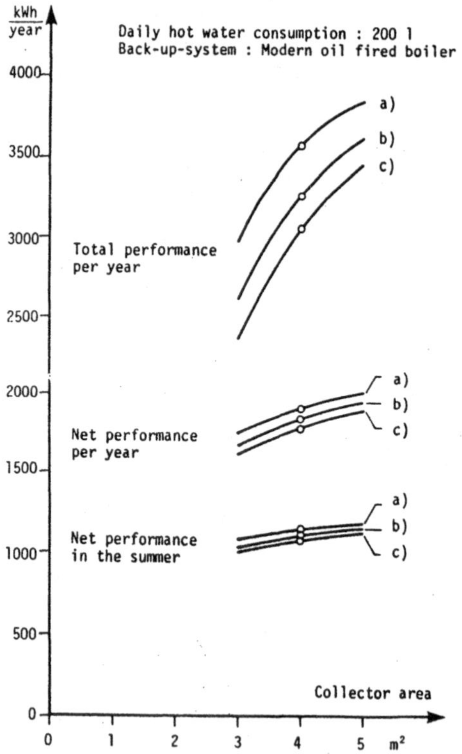

Fig. 2. The performance as a function of the storage type and the collector area.

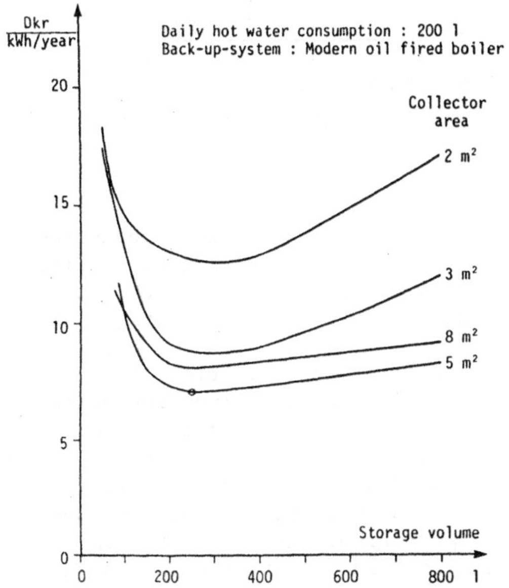

Fig. 3. The ratio between the cost and the total performance per year for different solar heating systems.

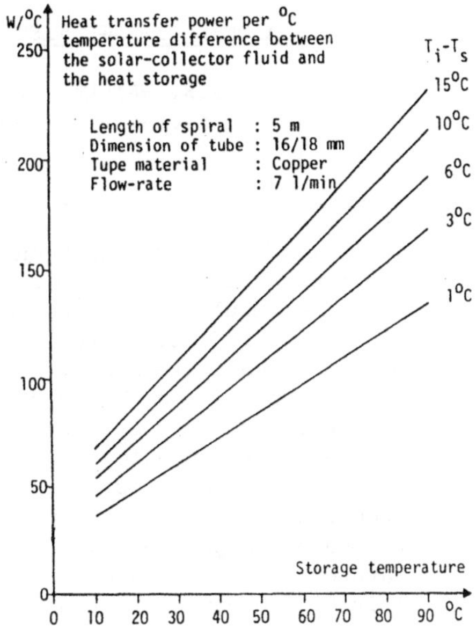

Fig. 4. The heat transfer as a function of the storage temperature and the temperature difference ΔT.

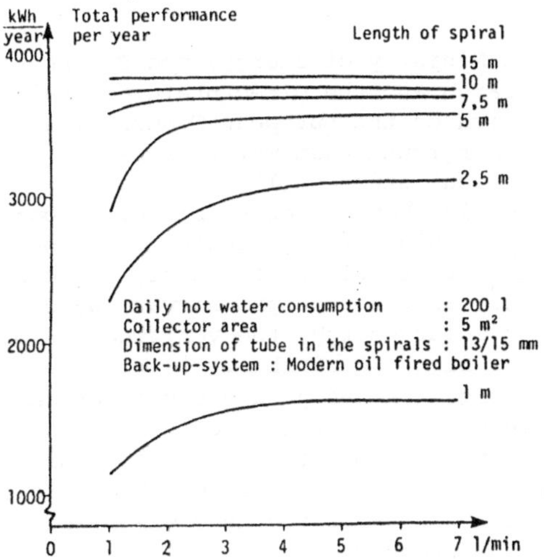

Fig. 5. The performance as a function of the flow-rate in the spiral and the length of the spiral.

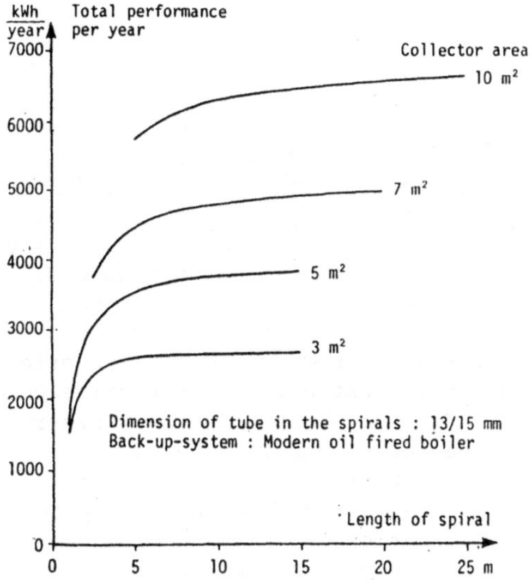

Fig. 6. The performance as a function of the length of the spiral and the collector area.

SEASONAL STORAGE TEST PIT. TESTING OF A 500 M^3 STORE

K.K. HANSEN, P.N. HANSEN and V. USSING
Technical University of Denmark, Thermal Insulation Laboratory

Summary

The aim of the work has been to show, that warm water pits, uninsulated towards the soil, are well suited for seasonal heat storage. Limiting the insulation to the top surface will contribute to a construction price reduction, which is so important, if seasonal storage shall be economically feasible.
The design work was based on the geotechnical investigations of the site selected at the laboratory test grounds on the university campus. The thermal properties of the soil were determined to form the basis of the computed analysis of the heat flows.
Upon construction of the pit, charge and discharge operations varying the water temperature ± 15 °C in a sinoidal pattern with mean temperatures of 45 °C and 65 °C have been performed. Measurement from the initial 210 days reported earlier showed good agreement with analytical and numerical predictions of the efficiency of the pit.
The measurements of 330 subsequent days reported now continue to agree well with numerical simulations. The durability of the high density polyethylene liner has so far given no reason to complaints and the vapour transmission through the floating lid has taken place without measurable accumulation in the insulation.

1. INTRODUCTION

In a joint venture between the Danish Ministry of Energy, the Danish Council for Scientific and Industrial Research and the Directorate General for Science and Development of the Commission of the European Communities the design and construction of a 500 m^3 underground warm water store was undertaken during 1981-82. Previous theoretical studies (1), (2) had indicated that underground warm water stores would show an acceptable efficiency even if the soil interfaces were left uninsulated. As studies and experiments with solar collectors had indicated, that only sufficiently low construction costs for solar collectors as well as for underground warm water stores would secure competitive prices for solar energy, the verification of the above mentioned theoretical studies was highly desirable.

Financed by the Danish Ministry of Energy the design and construction of the 500 m^3 underground warm water store was undertaken. The design did not attempt to make the experimental facility a model of future full scale warm water stores, but attempted to make the verification of the theoretical investigations as easy as possible. Thus the store for instance was placed completely underground in stead of partly underground.

Simarily the store was placed adjacent to an existing building in spite of certain inaccuracies in the verification was to be expected from this location in order to facilitate the installation of the datalogging at a short distance from the point of measurements, which would secure recording at an acceptable accurracy.

2. CONSTRUCTION OF THE STORE

The store with a pyramidal geometry was dug into the ground from the ground level. The store/soil interfaces are made waterproof by use of a 2.5 mm polyethylene plactic liner.

The floating lid is heat insulated by 0.5 m polystyrene and protected agains evaporation from the store and against climatic conditions by use of a 2.5 mm polyethylene plastic bottom liner and a 1.0 mm butyl rubber liner for the topside. The lid was constructed in a "dry dock" next to the pit and was easily floated to the final position. A photo of the final plant is shown in figure 1. The construction was completed in 3 months during the summer of 1982 and are explained in more details in [3].

The ground water level is 40 m below the ground level.

Figure 1. The final plant. The "dry dock" where construction of the lid was undertaken, is partly seen to the left.

3. EXPERIMENTATION

Approximately 80 copper constantan thermocouples are being used to measure temperatures in the water and in the soil under and around the pit. The thermocouples are arranged in chains which are located in two planes. The location of the thermocouples in the ground in one of the planes is shown in figure 2.

The temperatures are scanned automatically once an hour. Every day the data logger system prints all the daily mean temperatures on paper and on tape (for later transfer to our computer center).

The heat input to the water in the store is generated by two gas boilers. The first charge period started in September 1982 with a water temperature of 15°C.

Because of the limited time being available for experimentation in this project, it was decided to accelerate the testing by reducing the charge/discharge cycle from 365 days to 70 days. Further it was decided to have total temperature mixing of the water in the store. The charge and discharge cycles can be seen in figure 3.

4. COMPUTER MODEL

A computer program has been constructed based on the method of finite differences. The program treats the geometry of 1/8 of the store and the surrounding soil as indicated in figure 2. The computer code has been constructed to be truly 3-dimensional. The enmeshment is shown in figure 2. The centre points of the volume elements shown are the meshpoints used in the simulations.

5. SIGNIFICANT RESULTS

The temperature range in the water was originally $30^{\circ}C$-$60^{\circ}C$ and the first period of 210 days of the measurements with this temperature range is reported earlier [3], [4]. Here the calculated temperatures in the 51 thermocouples in the ground coincide very well with the temperatures measured.

From April 1983 and during the summer the temperature range in the water was changed to $50^{\circ}C$-$80^{\circ}C$ while the sinoidal relative temperature change was maintained, see figure 3. In the winter 1983-84 the temperature range once more reverted to 30-$60^{\circ}C$. As indicated on figure 4 the calculated temperatures at the position of the 51 thermocouples in the ground still coincide very well with the temperatures measured.

6. CONCLUSIONS

By the termination of the measurements in the fall of 1984 it is expected that the calculated temperatures in the soil will continue to coincide with the temperatures measured as has been the case so far.

It is also anticipated that no visible reduction of the quality of the high density polyethylene liner which was used as water barrier on the sides and the bottom of the pond as well as on the bottom of the floating lid will be observed.

It is also expected that the efficiency of the storage will have improved as envisaged theoretically corresponding to the number of cycles of operation performed.

7. REFERENCES

1. HANSEN, P.N., Varmetab fra store varmelagre, (Heat losses in big heat stores). 1979. Thermal Insulation Laboratory, Technical University of Denmark.
2. HANSEN, P.N., Analytical description of the heat losses from underground thermal seasonal heat stores. Paper presented at the U.S. Department of Energy Conference "Seasonal Thermal Energy Storage", October 19-21, 1981, Seattle, Washington, U.S.A.
3. HANSEN, K.K., HANSEN, P.N. and USSING, V., Seasonal Heat Storage in Underground Warm Water Stores - Construction and Testing of a 500 m^3 store. Thermal Insulation Laboratory, Technical University of Denmark. Meddelelse nr. 134. 1983.
4. HANSEN, P.N., HANSEN, K.K. and USSING, V., Seasonal Heat Storage in Underground Warm Water Stores - Construction and Testing of a 500 m^3 Store. In Solar Energy Applications to Dwellings - Solar Energy R&D in the European Community. Series A, Vol. 4. D. Reidel Publishing Company. 1983.

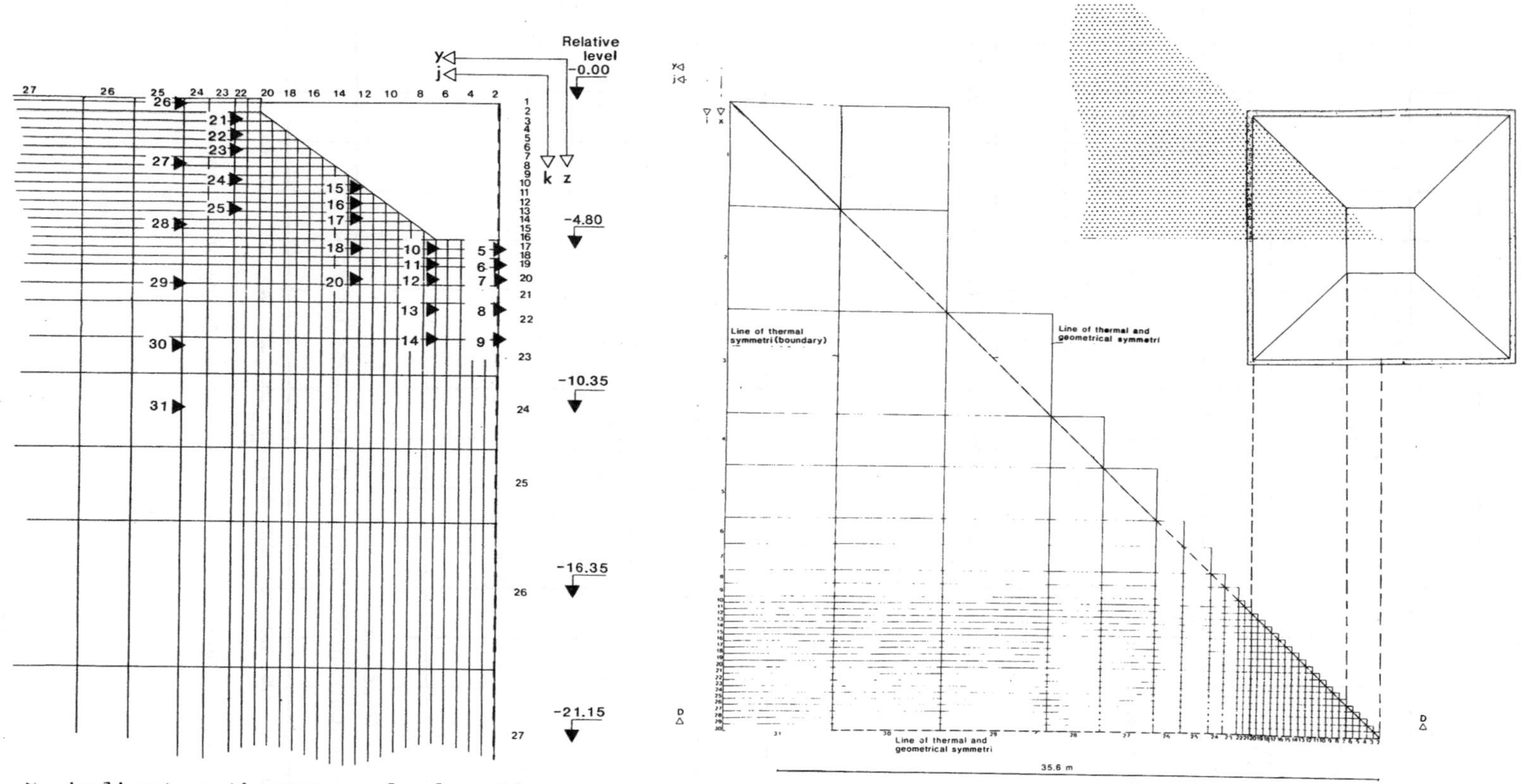

▶ indicates thermocouple locations.

Figure 2. Position of the grid points in vertical plane to the left and horizontal plane to the right.

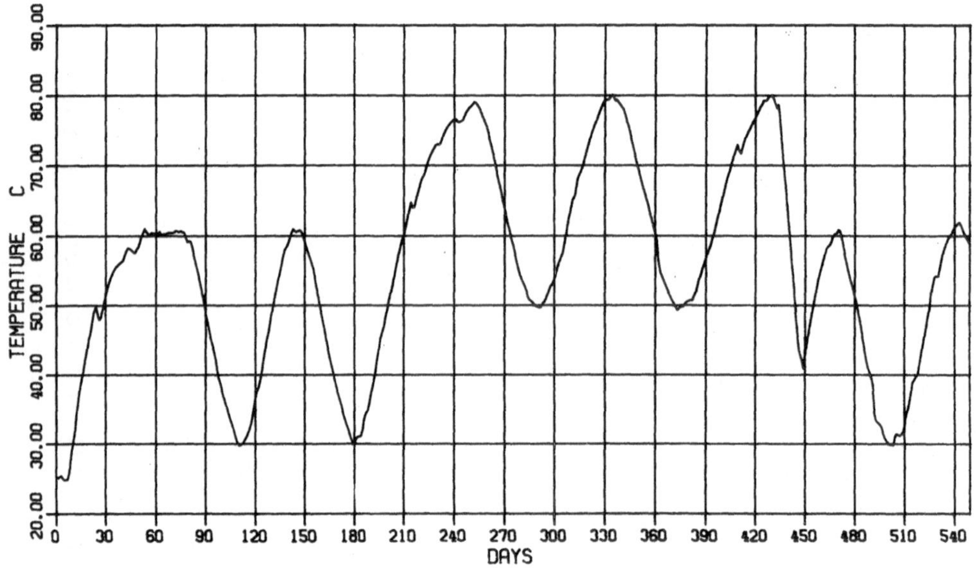

The measured average watertemperature

Figure 3.

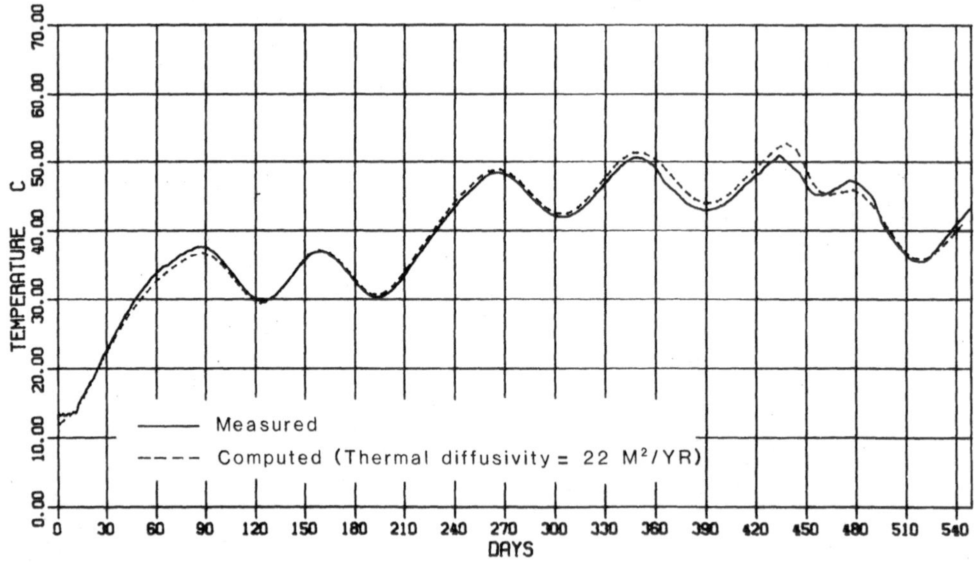

Soiltemperature. Measuring point no. 12 (about 1.5 m below bottom)

Figure 4.

SEASONAL HEAT STORAGE IN UNDERGROUND WARM WATER PIT

(Design of a 30.-50.000 m^3 storage)

K.K. Hansen, P.N. Hansen and V. Ussing
Thermal Insulation Laboratory
Technical University of Denmark

Summary

The design, construction and simulated operation of an underground 500 m^3 warm water store preceded this work. The aim of this project has been to design a 30.-50.000 m^3 underground warm water storage as a pilot project for a large seasonal storage. In order to minimize the construction cost the soil interface of pit has been left uninsulated and the lid has been foreseen to float on the water surface. The design data for the project have been taken from a district heating system based on co-generation expressing serious interest in realization of the design assuming that the construction cost will prove to be competitive with different tank storage solutions. The geometrical and geotechnical problems related to the design have partly been reported earlier. This paper gives a resumé of the major points from the final report to be published later in 1984. The design work was finished jointly by the Commission of European Communities and the Danish Council for Scientific and Industrial Research.

1. INTRODUCTION

A seasonal heat storage is a necessity for heating systems based on solar energy. It has been the purpose of this work to design a 30.-50.000 m^3 storage pond using the experience gained from the construction and simulated seasonal operation of a 500 m^3 test facility at the Technical University of Denmark (1). Studies concerning heat storage ponds in the future energy systems (6) have indicated, that only through sufficiently low costs of collectors as well as of storage ponds can solar energy be competitive with other sources of energy. Only through actual construction of a number of full scale heat storage ponds can low priced ponds be developed.

It was of prime importance to base the pilot design of a 50.000 m^3 heat storage pond on data from a large district heating system, expressing serious interest in an early realization of the design, assuming that the cost estimates proved this type of storage competitive against other designs. Insulated new steel tanks are already in operation and serious studies are being done to renovate existing surplus large fuel oil tanks for storage purposes. Speedy realization of the design appeared only possible if the project was designed to serve as storage in a large district heating system based on co-generation. For such systems large heat storage ponds can be shown to be very advantageous. A series of such ponds may help to obtain the experience in construction allowing reduced construction costs required in seasonal storage.

2. REQUIRED CAPACITY AND CHOICE OF GEOMETRY AND LOCATION

The storage is built into a system planned to have a maximum load of 489 MW in 1999. Based on the 24-hour load variations in the system a 6 hour seperation of the generating facility from the district heating system demands a storage of 2905 MWh. Limiting the temperature of the storage to 95°C and having $\Delta t = 47^{o}$C in the system at maximum load, the required volume can be figured to be 54,539 m^3.

The problems related to choice of geometri and the soil mechanical problems have to some extent been reported earlier (2), (3), (4) and (5).

The storage has been located as close to the electric power plant as possible on an area created by filling in shallow coastal waters with fly-ash. Fig. 1 shows the location of the storage.

3. MAIN FEATURES OF THE CONSTRUCTION

Fig. 2 shows a sectional view of the storage design. By employing a number of bleeder wells and a sheet pile wall the execution of the construction is secured in spite of the fact that the bottom of the pond is 9.2 m below the surface of the sea just south of the pond. The water tightness of the storage is secured primarily by clay. The chemical composition of the water is maintained by a 2mm thick polypropylen sheet covering the bottom, the walls as well as the bottom of the floating lid. The lid is built on the lid bottom liner reinforced by steel wires in the center structure and in the concrete edge beam. The lid consists of insulation, a rubber liner, gravel and top soil placed as the water level is raised. The concrete edge beam and the sheet pile wall are anchored to surrounding ground.

4. CONNECTING THE STORAGE TO THE CO-GENERATION SYSTEM

A principal diagram showing the charge operation for the storage is shown in fig. 3. The storage is connected to the main connection between the power plant and the transmission station of the district heating system. The district heating system is operating at 6 bar. As the storage is planned to function at about 0.9 bar reducing valves (or turbines) and sufficient pumps are used to allow a flow of up to 9.000 m^3/h to and from the storage.

5. OPERATION LOSSES

Based on geotechnical investigations and numerical computations the heat loss in % of maximal theoretical heat content at a ΔT of 44°C is shown in fig. 4 as function of the time elapsed in the fourth year of operation. Underground hot water storage ponds with large volumes can be constructed assuming that suitable layers of clay can be found in the vicinity of the power station or clay can be obtained at a reasonable cost from other locations.

A number of large storages primarily serving load management purposes in a co-generation system will prove to be very economical and will constribute to rapid development of low construction cost for large seasonal storages, which is a condition for use of solar energy on a large scale.

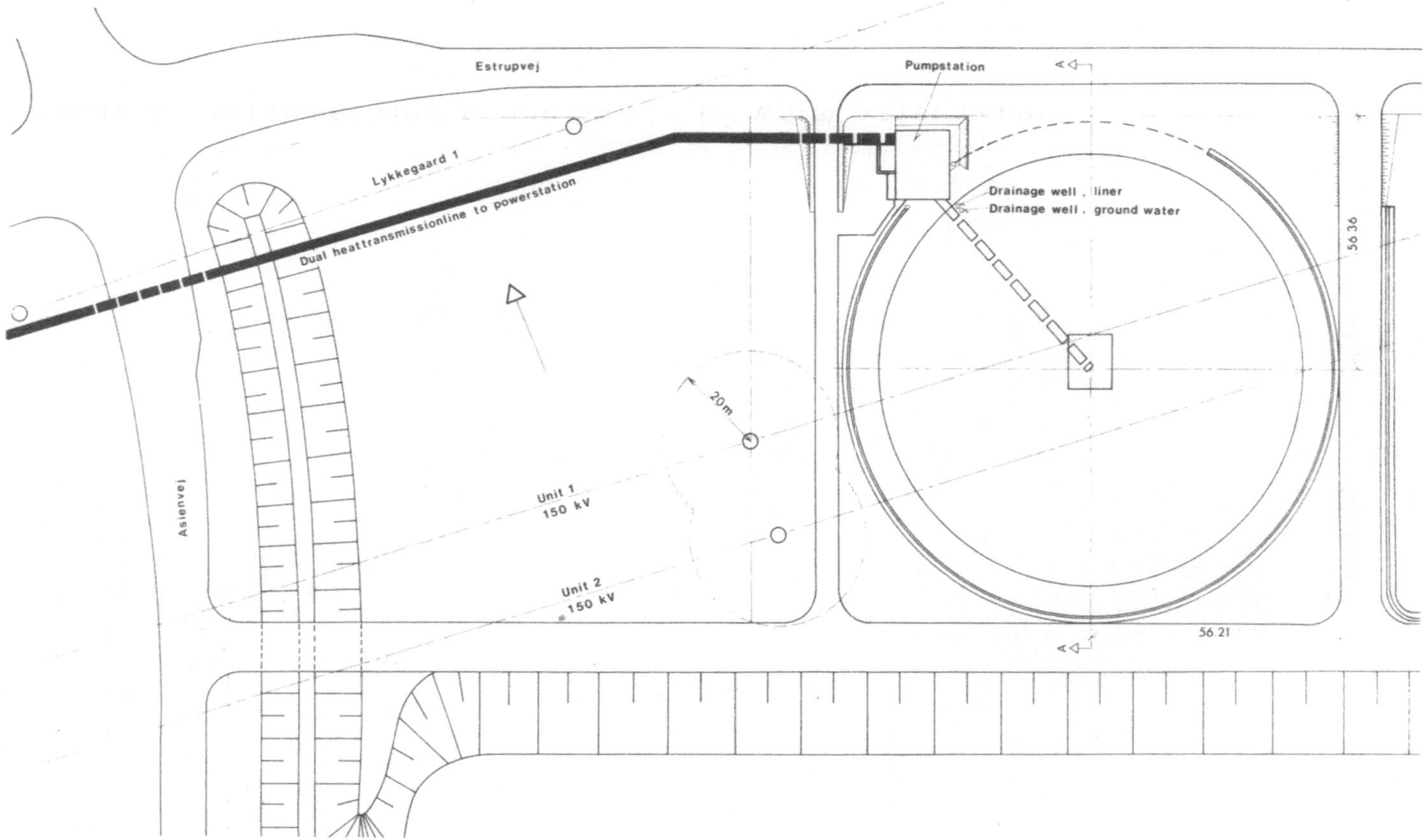

Figure 1. Plan view.

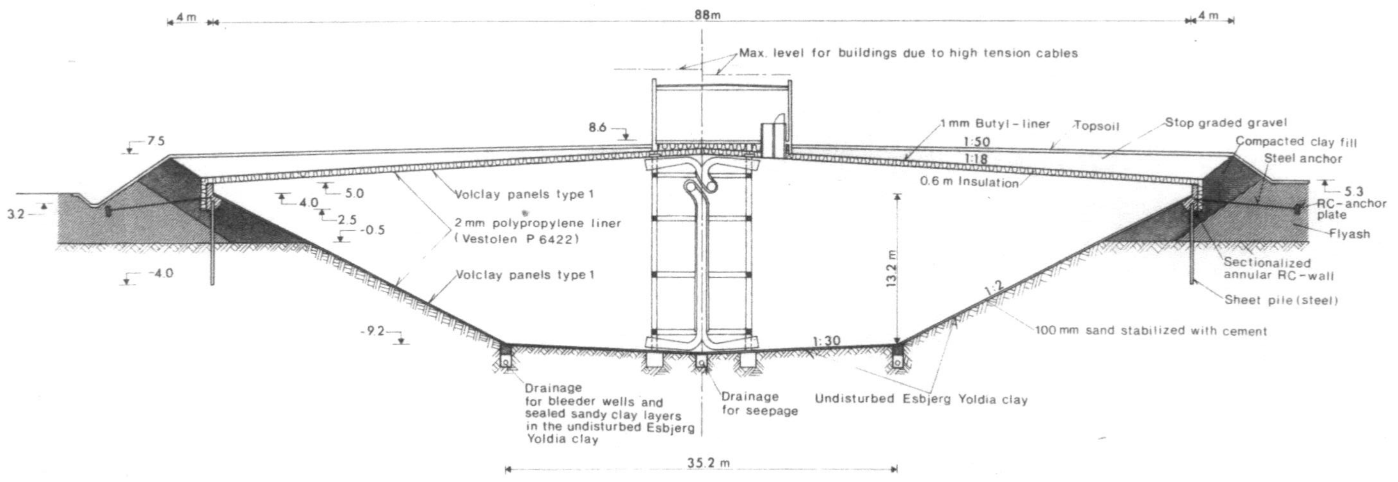

Figure 2. Sectional view of the 50,000 m³ warm water pond.

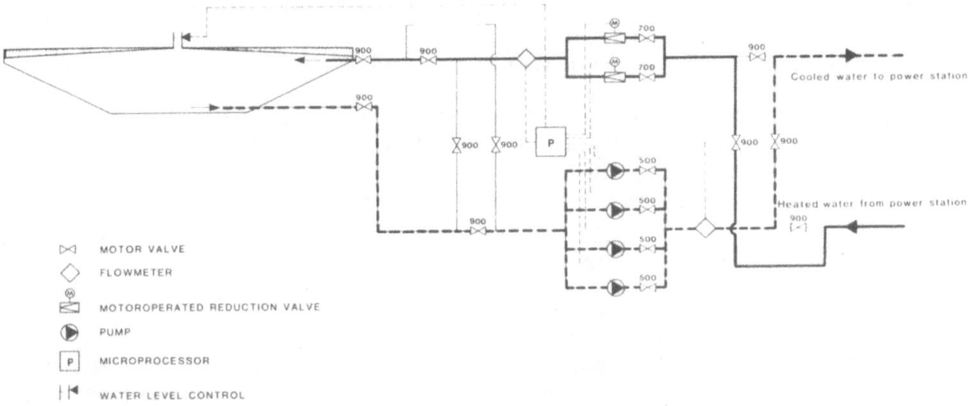

Figure 3. Pumpstation. Principal diagram. Charge operation.

HEAT LOSS IN % OF THE MAXIMAL THEORETICAL HEAT CONTENT BASED ON A TEMPERATURE AMPLITUDE OF 44 °C.
THE TEMPERATURE OF THE STORED WATER IS 73 °C (CONSTANT), AND THE PERIOD IS THE 4. YEAR.

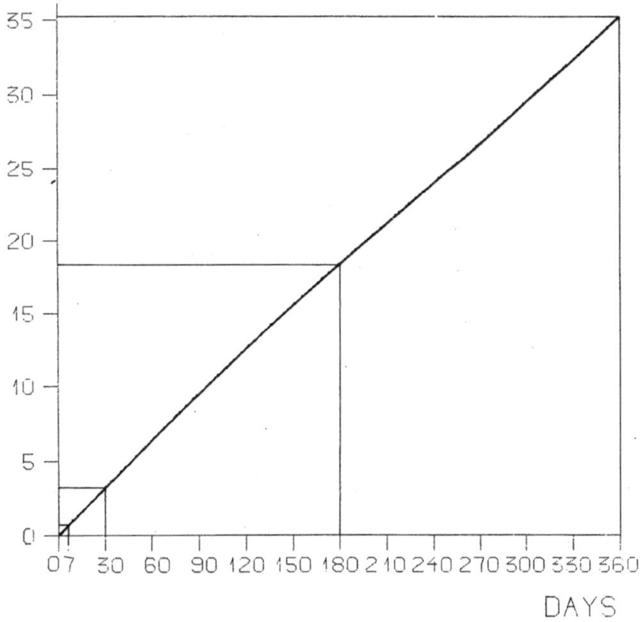

Figure 4. Heat loss calculation.

6. REFERENCES

1. HANSEN, K.K., HANSEN, P.N. and USSING, V. (1983). Seasonal Heat Storage in Underground Warm Water Stores - Construction and Testing of a 500 m^3 Store. Thermal Insulation Laboratory, Technical University of Denmark. Meddelelse nr. 134.
2. HANSEN, K.K., HANSEN, P.N. and USSING, V. (1983). Seasonal Heat Storage in Underground Warm Water Pit(Design of a 30.-50.000 m^3 storage). In Solar Energy Applications to Dwellings - Solar Energy R&D in the European Community. Series A, Vol.4. D. Reidel Publishing Company.
3. Danish Geotechnical Institute: Esbjerg. Heat Storage in Large Reservoirs. Geotechnical report No. 26 and enclosures Nos 1a-11a plus appendix 2C. (1984). Unpublished.
4. HANSEN, K.K., USSING, V. and HANSEN, P.N. (1983). Store Varmelagre. Forholdet mellem dambegrænsningsfladernes areal mod jord og dammens volumen for forskellige geometrier. Laboratoriet for Varmeisolering, Danmarks Tekniske Højskole. Rapport nr. 83-6.
5. Dipco Engineering ApS: Sæsonlagring af varme af store vandbassiner. Laboratoriet for Varmeisolering, Danmarks Tekniske Højskole Meddelelse nr. 91. November 1979.
6. HANSEN, K.K., HANSEN, P.N. and USSING, V. Perspektiver vedrørende damvarmelagre i fremtidens energisystem. Thermal Insulation Laboraty. Rapport No. 83-39.

THE SODIUMSULPHIDE/WATER SYSTEM AS A CHEMICAL HEAT PUMP USED FOR LONG TERM ENERGY STORAGE

O.S. DYRNUM
Thermal Insulation Laboratory, Technical University of Denmark
DK 2800 Lyngby, Denmark

Summary

A chemical heat store using sodiumsulphide as absorbent and water as working medium ulitizing the chemical heat pump principle is designed, constructed and tested. The reaction involved is:

$$Na_2S + 4\frac{1}{2}\ H_2O \rightleftarrows Na_2S;\ 4\frac{1}{2}H_2O + 283\ kJ$$

Test results shows that charging the system last for 200 hours at a charging temperature at 80°C and a condenser temperature at 15°C. Discharging the system was carried out at rates of 25 and 70 W and lasted for 140 and 50 hours respectively, both at a discharging temperature of 50°C.
After 3 cycles the energy turn over raised to 90% of the theoretically value due to maturing and stabilizing of the absorbent material. The system is working well within its limitations, and has a quick response to changes in discharging rates but much of the stored energy is required to heat the system from storage temperature to working temperature and much energy is lost to the surroundings through the surface of the storage tank.

1. INTRODUCTION

The final goal with this work is to construct a seasonal heat store which should be economic and suitable for installing in connection with solar heating systems utilizing chemical bonding energy as the storage medium. The sodiumsulphide/water system was considered suitable on account of its high temperature raise and energy density, and this work is dealing with the design and test of a chemical heat pump based on water absorption in sodiumsulphide ulilizing the process:

$$Na_2S + 4\frac{1}{2}\ H_2O \rightleftarrows Na_2S;\ 4\frac{1}{2}\ H_2O + 283\ kJ$$

2. THE TEST RIG

The test rig is constructed as shown on fig. 1. The sodiumsulphide container is made of plain carbon steel and so is the embedded heat coil as it has shown to be resistant to sodium sulphide. The condenser/evaporator is made of stainless steel.

Two thermostatic baths are simulating the ground coil and the heat delivery and consumption respectively.

The water is circulated in the condenser/evaporator by means of a magnetic coupled gear pump and the system is evacuated by a vacuum pump going to 10^{-4} mbar.

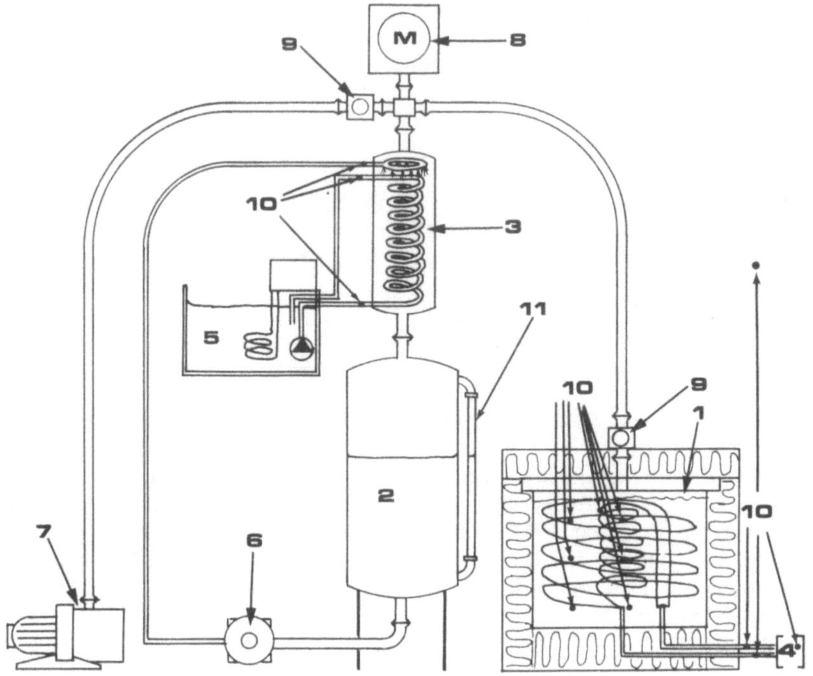

Fig. 1. Schematic view of the test rig.

1. The storage vessel with embedded heat coil
2. The water tank
3. The condenser/evaporator with heat coil
4. Thermostatic bath simulating the heat consumption and delivery
5. Thermostatic bath simulating the ground coil
6. Magnetic coupled gear pump for water cirkulation
7. Vacuum pump
8. Manometer
9. Valves
10. Temperature measuring points
11. Water level glass

3. EXPERIMENTAL

Charging was carried out by heating the sodiumsulphide to 80°C by cirkulating hot oil from a thermostatic bath through the embedded heat coil. The condenser was kept constant at 15°C by means of another thermostatic bath. Temperatures at 14 measuring points was recorded and so was the amount of condensed water as a representation for the energy turn over.

Discharging was carried out by extracting heat from embedded heat coil. The discharging rates was regulated by the flow and the inlet temperatures as the outlet temperature was about constant at 50°C.

Temperatures and amount of evaporated water was likewise recorded.

Cooling the heat store

For calculating the heat losses from the heat store it was heated to 90°C and was let to cool by itself. Hereby a cooling curve was obtained and with knowledge of the thermal capacity the heat losses can be calculated.

9 chargings was carried out
5 dischargings at 25W was carried out
4 discharging at 70W was carried out and
2 cooling curves was recorded.

4. CALCULATIONS

Degree of packing of available volume	59%
Degree of packing of total volume	48%
Volume of space between krystals	34%
Volume of embedded heat coil	18%
Filling of heat store: 104 mol = 25 kg =	17.5 l
Heat of absorption pr. mol $Na_2S \rightarrow Na_2S;\ 4\frac{1}{2}\ H_2O$	283 kJ
Heat of absorption pr. g H_2O	3.50 kJ
Calculated water absorption	8.42 l
Calculated storage capacity	29.5 MJ
Calculated nett storage density	800 kJ/l

5. RESULTS

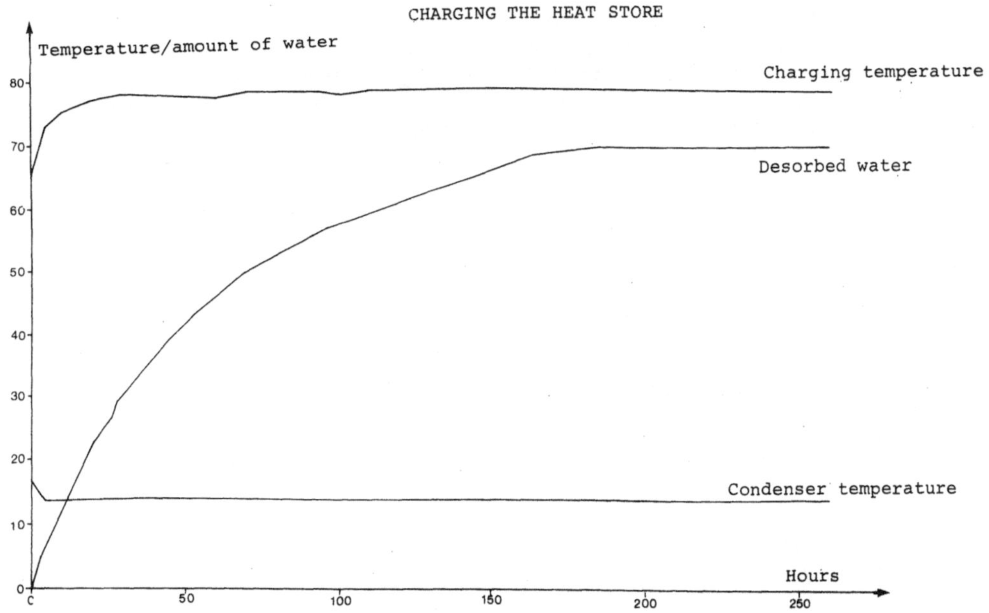

Fig. 2. Charging the heat store. The amount of water is in dl and corresponds to the stored energy - 3.50 kJ pr. g water. After about 25 hours the slope of the curve flattens due to inert gases from a leak. After a new evacuation the slope of the curve is steeper again.

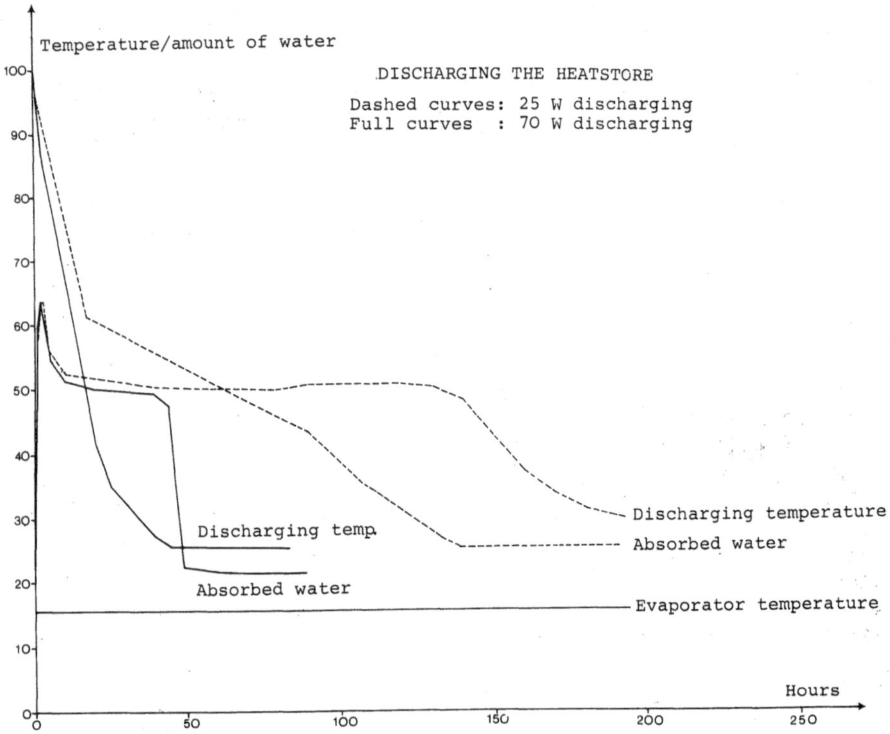

Fig. 3. Discharging the heat store.
Two dischargings are shown, one at a rate of 25W and another at a rate of 70W. The temperature in the storage container is rapidly growing from room temperature to 64°C. About half the energy stored is used for heating the storage container which is seen of the amount of water absorbed.

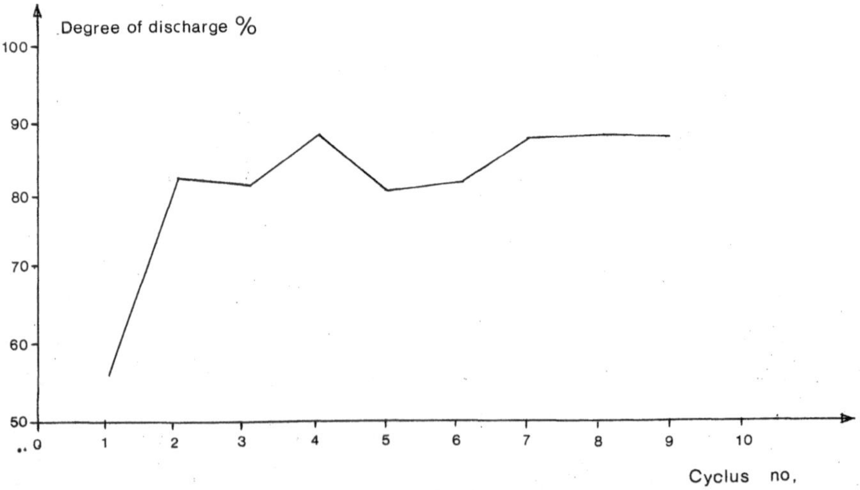

Fig. 4. Degree of unloading represented by the absorbed water in percent of the theoretical calculated water absorption.

6. CONCLUSION

As the system is more complicated and more expensive than an ordinary hot water store it should only be used for long term storage. For this purpose it is advantageous on account of its high energy storage density and freedom from thermal losses during the storage period. For seasonal storage the storage container should be much greater, great enough for a seasonal energy demand, which inplies:

- Less thermal losses during working period due to a smaller surface/volume ratio.
- Better performance as the effect pr. kg storage material is small and the reactions is closer to equilibrium conditions.
- Simpler construction of storage container as small effects pr. kg material implies smaller heat exchangers and therefore a better utilization of the container volume.

Attention should especially be paid to:

- Completely airtightness of the system as the performance is drastic reduced by even small amounts of inert gasses.
- A well thermal insulation of the storage container to avoid thermal losses during long working periods.

A GENERAL METHOD FOR DETERMINING THE OVERALL HEAT TRANSFER COEFFICIENT OF THERMAL STORAGE EMBEDDED MATERIALS

A. SPENA

Dipartimento di Tecniche dell'Edilizia e del Controllo ambientale, Facoltà di Ingegneria, Università degli Studi di Roma "La Sapienza", Via Eudossiana 18, 00184 Roma - Italy

Abstract

The EST (Equivalent Storage Temperature) Method for determining the overall heat transfer coefficient of thermal storage embedded materials is presented. The proposed procedure covers a gap in the existing literature, where very few may be found in the field of testing methods for thermophysical properties of particulate materials to be used in thermal storage packed beds. The method can be applied to either sensible and latent heat storages. An application is studied with a sensible-heat material, using air as transfer fluid. A comparison between experimental and calculated results is made, together with a sensitivity study of the influence of errors in data evaluation.

1. INTRODUCTION

In recent years, packed beds for heat storage have been widely studied, either as complete systems, either with respect to their elementary phenomenologies. But, as a matter of fact, a general criterion is not yet available, which would be able to characterize a storage material from the point of view of its attitude to heat exchange and thermal stratification, regardless to the features of the shell of the unit, and of its thermal insulation. Tis also leads to the absence of any guide for research and development of new materials, together with lack on information about the existing ones.[12]

Present paper, as a first contribution to the fulfilment of the exposed gap, deals with an original method, of general validity and satisfactorily accurated, able to evaluate, in unambiguous and easy fashion, the overall heat transfer coefficient of any storage medium. This coefficient is assumed, following present criterion, as the thermophysical property better suited to characterize the material alone, from the point of view of its idoneity to store heat in packed beds; this also for comparison purposes. As a matter of interest, it must be observed that in literature only studies on the external heat transfer coefficient of the wall of particles can be found, being Biot numbers of the order of less than 0.1, thus neglecting the inner thermal behaviour of the material.[3 4 5]

2. DESCRIPTION OF THE EST-METHOD

The EST-Method basically consists on the use of a mathematical model which enable us, starting from few temperature readings made only in the transfer fluid and easy to obtain, to determine a conventional temperature field inside the storage medium; and then, by means of local energy balances, the value of the overall heat transfer coefficient H, defined as the ratio $H=H_V/\sigma$; where σ is the heat exchange surface per unit volume, and H_V is defined as the ratio between the thermal power Y exchanged, per unit gross storage volume, between transfer fluid and material, and the difference between the local temperature t_f of the fluid and the value of the storage medium temperature t which conventionally make congruent all the measured heat fluxes simultaneously exchanged by the storage material with the remainders parts of the system; namely

$$H_V = \frac{Y}{(t_f-t)}.$$

From the analytical point of view, two different energy balances are performed for the material and the transfer fluid, in unsteady-state conditions; and then the partial differential equation system is converted into a finite-difference algebraic equation system, to be solved as a function of Y and t, being t_f values available from empyrical readings (see fig.1).

In order to provide the necessary data and informations, and to validate either the method and the procedure, a full-scale, air operated experimental apparatus has been built, especially equipped for intensive temperature recordings of both transfer fluid and storage material (fig.2). It presently results the only test-facility of such kind available in Europe. The storage medium is constituted by small water capsules, optimized as possible to give unambiguous mean values of the inner temperature.[6] Air as transfer fluid has been choosen to magnify the role played by the storage medium in terms of storage capacity, and for its critical behaviour as heat transfer partner. In any case, Biot number results of the order of $0.2 \div 0.6$.

3. RESULTS

Top-charge tests of the unit have been performed, with step variation of 50°C of the inlet temperature of the transfer fluid. As an example, fig.3 shows the trends of the major quantities as a function of the reduced space z/L, being $\dot{m}$ the mass flow, τ_r the reduced time, $\vartheta = (t_f-t)$, $T = (t_f+t)/2$; while fig.4 shows the trend of H, with respect to τ_r, as a function of $\dot{m}$. In fig.5 is then reported a linear regression of H vs. ϑ, while fig.6 sho ws H vs.T curves.

In order to evaluate the reliability of the obtained results (particularly in terms of H), information is needed about the ability of the proposed method to obtain congruent values, with respect to both time and space, of the so-called "equivalent" storage medium temperatures. To do this, a match has been made between calculated and measured results, provided that the latters are considered with the corresponding uncertainties; and that the formers, which come from a purely analytical balance of all the not negligible heat fluxes, are carefully taken into account too.[7]

From fig.7 it really appears that any departure between computed and measured values always remains of the order of the corresponding unaccuracies.

In order to investigate to what extent the unaccuracy on estimation of all the input data of the model can influence the results in terms of the $s=H/H_0$ ratio between actual value of H and a reference value corresponding to a set of "true" input data (H_0), a sensitivity study has been performed by varying one quantity at a time, while keeping all the others at their correct values. The most influencing parameters appear to be t_f and the initial values of t at the start of the test (boundary condition), even if consequences of unaccuracies on their measurements are reasonably a function of the local thermal conditions, which in turn vary with respect to the time.

4. CONCLUSIONS

The proposed method satisfactorily allows us to evaluate potential for heat transmission and for temperature stratification of any kind of material available in particles to be embedded in a pile, regardless to the features of the containing shell of the unit and of its insulation. This aim is reached assuming as characteristic parameter the value of the overall heat transfer coefficient of the material. The procedure requires experimental temperature readings made only in the transfer fluid, and can bear some uncertainties when evaluating input data, as the sensitivity study has shown also quantitatively. The match between calculated and measured results shows that any kind of experimental investigation, in spite of its complexity, ambiguity and cost, can provide no more accurated information on storage material thermal behaviour, than the present method allows. Although the procedure falls as temperature differences results of the order of only 1°C, the method appears interesting also for a better comprehension of thermal behaviour in unsteady-state conditions of thermal storage packed beds, as the trends of the most important parameters with respect to both time and space can reliably investigated.

REFERENCES

1. ASHRAE Standard 94-77,ANSI 3 199.1-1977,N.Y,1977.
2. VanGalen,I.E.,Report CEC 203.212,Delft,1984.
3. Handley,D.,Heggs,P.J.,Int. Jou.Heat Mass Tr.Vol.12, 1969.
4. Jeffreson,C.P.,AIChE Jou. Vol.18,1972.
5. Löf,G.O.G.,Hawley,R.W.,Ind. Eng.Chem.40,1061,1948.
6. Fontana,D.M.,Spena,A.,IV Miami Intn.Conf.Altern.En. Sour.,Miami,1981.
7. VanGalen,I.E.,W.D.08,CEC-SSTG,Delft,1983.

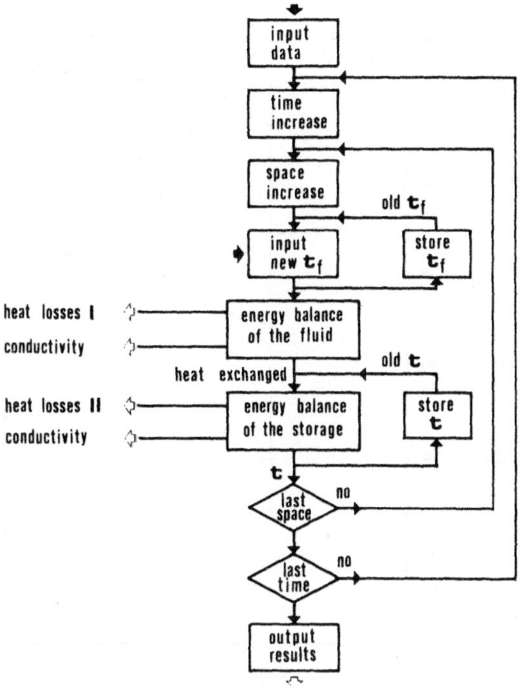

Fig.1

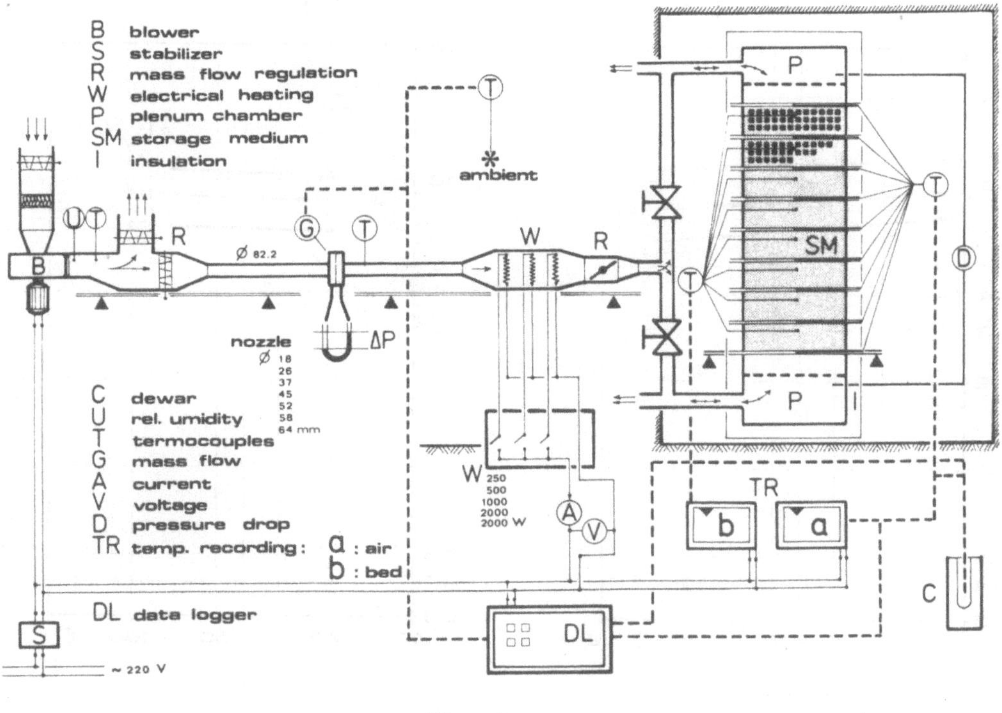

Fig. 2

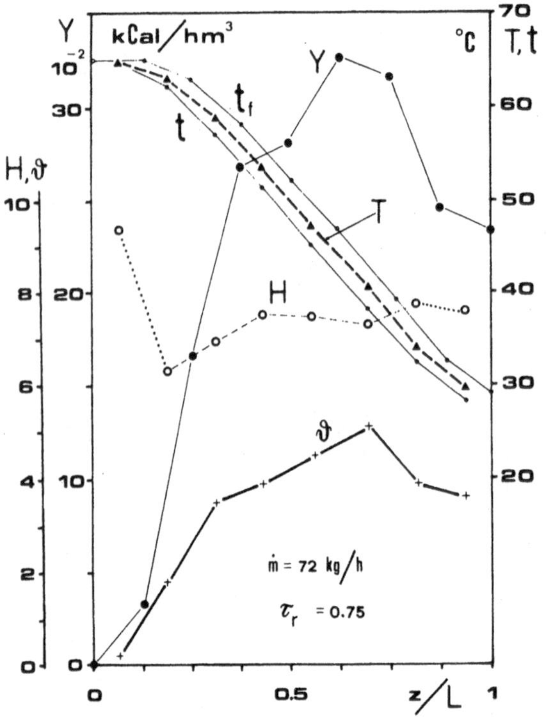

Fig.3

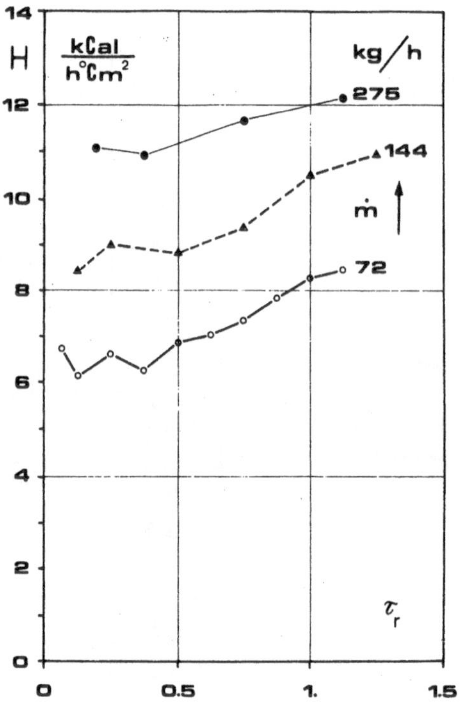

Fig.4

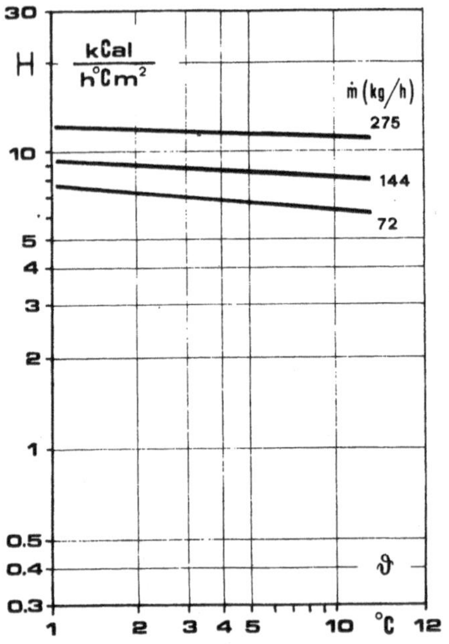

Fig.5

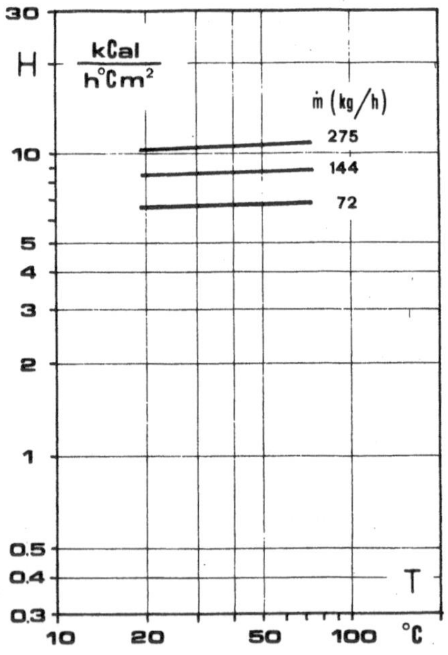

Fig.6

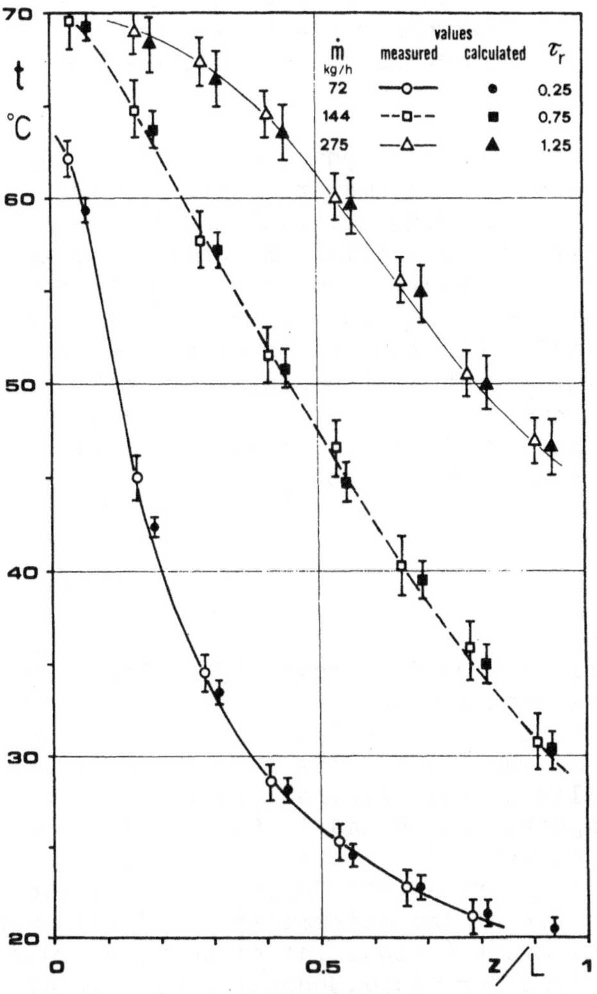

70
60
50
40
30
20
t
°C
values
ṁ
kg/h
measured
calculated
τ_r
72
0.25
144
0.75
275
1.25
0
0.5
z/L
1

Fig.7

DETERMINATION OF THE THERMAL BEHAVIOUR OF A PHASE CHANGE ENCAPSULATED MATERIAL PACKED BED USING AIR AS TRANSFER FLUID

A. SPENA and D. LECOMTE+
Dipartimento di Tecniche dell'Edilizia
e del Controllo Ambientale.
Facoltà d'Ingegneria - Università di Roma
"La Sapienza"-Via Eudossiana,I8-00I84 ROMA

Summary

Measurements made on a phase-change material encapsulated in large size (8cm.diameter) polypropylene spheres are presented.The storage material and the capsules,which are both of current industrial production are tested embedded on a full-scale packed bed unit.From the comparison between experimental results and calculations based on the Schumann model (isothermal spheres),corrective terms to the overall mean heat transfer coefficient,due to intraparticles effect,have been derived for the liquid and the solid phase.The effects of natural convection in the liquid phase are clearly pointed out.

I.INTRODUCTION

Few experiments on full-scale LHS units with air as transfer fluid have been led until now.Aprogram of experimentation using an air based test facility is under realization at University of Rome.The originality of the work undertaken lies in the three following points.Firstly,a LHS unit composed of encapsulating spheres is tested in full scale (200 l).The tested product is a packed bed of large polypropylene spheres (°)filled with a salt hydrate (°°)(Calcium chloride-Tc=27°C).The test program aims at the determination of the macroscopic properties of the store (actual latent heat,supercooling effects) using air as transfer fluid.Secondly,the thermal behaviour of the LHS,far from the current temperature approach widely available in literature,is here characterized by the determination of global heat transfer quantities,namely mean outside and overall coefficients and NTU number.The existing correlations on packed beds can also be validated with a special regard on the effects of phase change on the overall heat transfer coefficient.Finally, experimental validation of simplified models (Schumann model,infinite NTU model) is attempted.

+ Boursier scientifique et technique de la Commission des CE.
° Manufactured by CRISTOPIA(F).
°° Manufactured by The DOW CHEMICAL Co (USA)

Up to now only the two last points have been studied and are developed in the present paper.

2.DESCRIPTION OF THE SYSTEM

The test facility has been thoroughly described in ref.(I) The storage is composed of 8 parallelepipedic matrixes, each of them filled 6I spheres. The spheres are embedded according to a tetragonal network; the geometrical properties of the pile are then well determined. The void fraction is rather high since edge effects are not negligible for such sizes of particles. Copper-constantan thermocouples are placed in the air gap between each matrix, as well as in the inlet and outlet of the accumulator. Thermocouples are also placed within the material and on the surface of one sphere per matrix as seen on fig and . Thermo-physical properties of the store and PCM are listed in Tables I and II.

3.ANALYSIS

The knowledge of the overall heat transfer coefficient greatly conditions the availability of the heat exchanger in a thermal system. It should be emphasized that most correlations on the outside heat transfer coefficient H_e were achieved for packed beds of particles with diameters up to 2cm and are apriori unable to predict with certainty the behaviour of large size particles bed. The most reliable formula, established for rock bed storage with air as working fluid by Löf and Hawley (2) can be written : $H_o = \frac{D}{6(1-\varepsilon)} \, 650 \left(\frac{m_F}{A_F \cdot D}\right)^{0.7}$

In addition, the overall heat transfer coefficient H is dependent on three terms:

$$\frac{1}{H} = \frac{1}{H_o} + R_{wall} + R_i \qquad -1-$$

where $\phi = HS\,(T_F - T_m)$

The first term is a thermal resistance bound to the convection outside the spheres. The second term is the conductive resistance across the capsule. The existence of a third term is derived from the necessity of taking into account intraparticles effects. Thus a mean temperature Tm in the material is considered. Instead of using the isothermal (or Schumann) model (3), we could have evaluated the transient heat flux ϕ by

$$\phi = H_o S\,(T_F - T_S)$$

where T_S is the surface temperature of the spheres. A solution of the heat transfer and mass momentum set of equations inside the spheres is then necessary. For the sake of simplicity a purely conductive model would give:

$$\phi = -\lambda S \,\vec{grad}\, T \cdot \vec{n}\,\Big|_{surface} \qquad -2-$$

with T solution of : $\rho \frac{\partial u}{\partial t} = div(\lambda . \vec{grad}\, T)$

The previous method leads to important numerical calculations. It was used with success by Theunissen and Buchlin (4) with air as working fluid and by Alloncle (5) with water. From the viewpoint of simplicity, formulation -I- is preferible to formulation -2-. But its limits of application must be carefully established. Different states of material are encountered:

-SOLID : a natural approximation would be to assume $R_i=0$ in -I-. However this approximation is only available for weak temperature gradients in the spheres (It was shown (6) that the Biot number should not excess 0.I). But in our case, the Biot number would be in the range 1-3 according to ref.2. A study due to Jeffreson (7) allows to extend the validity of the isothermal model to the range $Bi \leqslant 4$ by introducing the quantity :

$$R_i = Bi/5.H_o$$

-LIQUID : in this case, natural convection will appear in the sphere. We can write : $R_i = 1/H_i$

-SOLIDIFICATION : in that case, the existence of a front makes difficult the evaluation of the mean T_m. In the major part of the solidification process, T_m will be very close to the phase change temperature. As the frozen layer increases, coefficient H decreases and transitory effects become more important like in the SOLID case.

-MELTING : a recent study by Moore (8) has shown the importance of natural convection essentially due to the decantation of the solid phase.

The two equations for heat balance in the fluid and the material are:

$$\dot{m}_F c_F \frac{\partial T_F}{\partial z} = \frac{HS}{\ell}(T_m - T_F) + \frac{(HS)_{loss}}{\ell}(T_a - T_F)$$

$$\rho \frac{\partial u}{\partial t} = \frac{HS}{V}(T_F - T_m)$$

When the material is totally liquid or solid, the equations become in non-dimensionned terms:

$$\partial T_F/\partial z^+ = NTU\,(T_m - T_F) + NTU_{loss}\,(T_a - T_F)$$

$$\partial T/\partial t^+ = NTU\,(T_F - T_m)$$

4. EXPERIMENTAL PROCEDURE AND FIRST RESULTS

Using the method proposed by Spena (1), the overall mean heat transfer coefficients are calculated in the liquid and the solid phase. The experimental curves are a charge (resp. a discharge) in the temperature range 8-22°C (resp. 65-35°C).

Fig.4 shows typical discharge curves in sensible heat .The determination of the NTU is highly dependent on the values calculated for T_m:it is then necessary to know the initial values of T_m with great accuracy;the adopted procedure was to leave the storage with air flowing at the initial temperature for a long period in order to reach the steady-state initial conditions.The error on the calculation of NTU increases with the number of computations i.e as time increases ; to minimize instability factors,smoothed T_f-curves have been chosen instead of the real ones.NTU curves are shown on Fig.5 and 6. Best thermal performances are obtained in liquid phase,which supposes natural convection in the PCM plays a non-negligible part in the transfer rates.Mean values are obtained,the upper one including extreme values,the lower one excluding them.Comparisons between experimental and theoretical values of NTU obtained from ref.2 and ref.7 are satisfactory(see Table III).

5.CONCLUSION

The method has permitted to obtain relatively correct,although a little smaller ,values of the NTU -thus of the overall heat transfer coefficient- compared to the existing correlations.A further developement of this work will be the determination of NTU in charging or discharging mode with latent heat,the evaluation of the mean internal temperature and the proposal of correlations in the phase change case.

-.-.-.-

ACKNOWLEDGEMENTS : The authors wish to thank the Centre d'Energétique of Ecole des Mines de Paris (France) for providing the testing material.

REFERENCES

(1)Spena,A."A method for determining the heat transfer coefficient of thermal storage embedded materials",This conference
(2)Löf,G.O.G.and Hawley R.W,Ind.Eng.Chem.,40,1061 (1948)
(3)Schumann,T.E.W,J.Franklin Inst.,208,405 (1929)
(4)Buchlin J.M. and Theunissen P.H,Colloque S.C.I,8/12/80-Paris
(5)Allloncle,R.Th.Dr.Ing,Toulouse(1982)
(6)Duffie,J.A.and Beckman,W.A,Solar Energy Thermal Processes.
(7)Jeffreson,C.P.,AichE Journal,18,2,409-416 (1972)
(8)Moore,F.E and al,Journal of Heat Transfer,Vol.104,1982,p 19

TABLE I. PHYSICAL PROPERTIES OF TESC81 DOW CHEMICAL

Temperature of fusion		27	°C
Latent heat		192.5	kJ/kg
Specific heat	(sol)	1.423	kJ/kg.°C
	(liq)	2.218	kJ/kg°C
Density	(sol)	1710	kg/m^3
	(liq)	1530	kg/m^3
Thermal conductivity	(sol)	1.09	W/m.°C
	(liq)	0.54	W/m.°C

TABLE II. GEOMETRICAL AND THERMAL PROPERTIES OF THE STORE

Number of matrixes	8	Surface of exchange	9.09m^2
Dimensions	0.48x0.48x0.16	Ratio Surface/Volume	45
Number of spheres	488	Mass of PCM	145.4kg
Sphere diameter	77 mm	Heat capacity capsules	20kJ/°C
Volume of spheres	0.1166m^3	Heat cap. container	37kJ/°C
Total volume	0.2004m^3	Mean thickness of caps.	2mm
Theor.void fraction	0.2595	Heat losses	5.52W/°C
Actual void fraction	0.418	conductivity of capsule	0.20W/m°C
Fluid cross section	0.0962m2	Height of heat exchanger	0.871m

FIG. 1-2. View of the accumulator and location of sensors

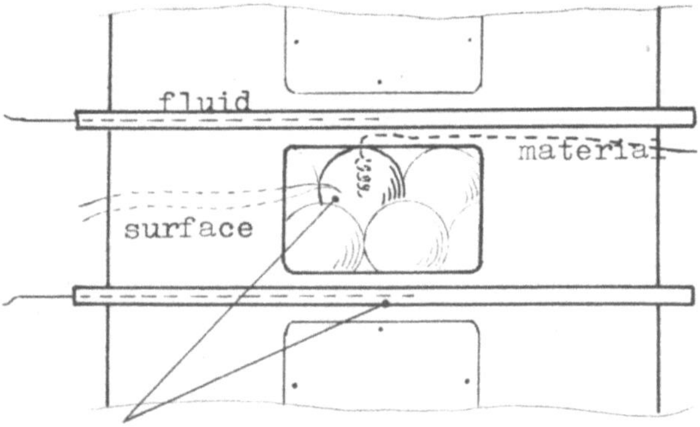

copper-constantan thermocouples

Fig.3 View of a sphere

TABLE III

NTU SOLID	
EXP.	THEO
Upper: 4.00	ref.2: 5.25
Lower: 3.50	ref.2+ref.7: 4.25
NTU LIQUID	
EXP.	THEO.
Upper: 5.00 Lower: 4.65	ref.2: 5.15

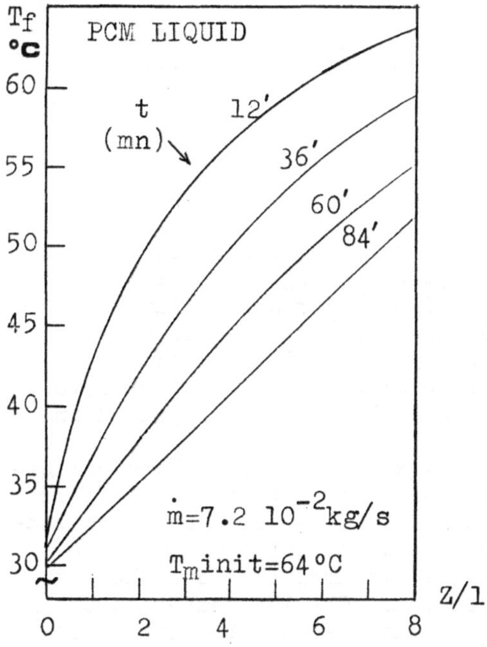

Fig.4. Fluid Temperature v.s z-axis (smoothed).

NOMENCLATURE

A_f: cross-section of flow m^2
Bi: Biot number, $H_oD/2\lambda$
C : specific heat of PCM J/kg°C
C_f: specif. heat of fluid "
D : diameter of spheres m
H : overall heat transfer coefficient $W/m^2 °C$
H_i: inside "
Ho: outside "
l : length of heat exchanger m
$\dot{m}_f$: mass flowrate kg/s
NTU: number of transfer units $HS/\dot{m}_f C_f$
S : heat exchange surface m^2
t : time s
t^+: adimensionnal time, $\rho CV/\dot{m}_f C_f$
T : PCM temperature °C
T_f: fluid temperature °C
U : enthalpy of PCM J/kg
V : volume of PCM m3
z : length variable m
z^+: adimensionnal length, z/l
ρ : density of PCM kg/m3
λ : thermal conductivity W/m°C
ε : porosity.

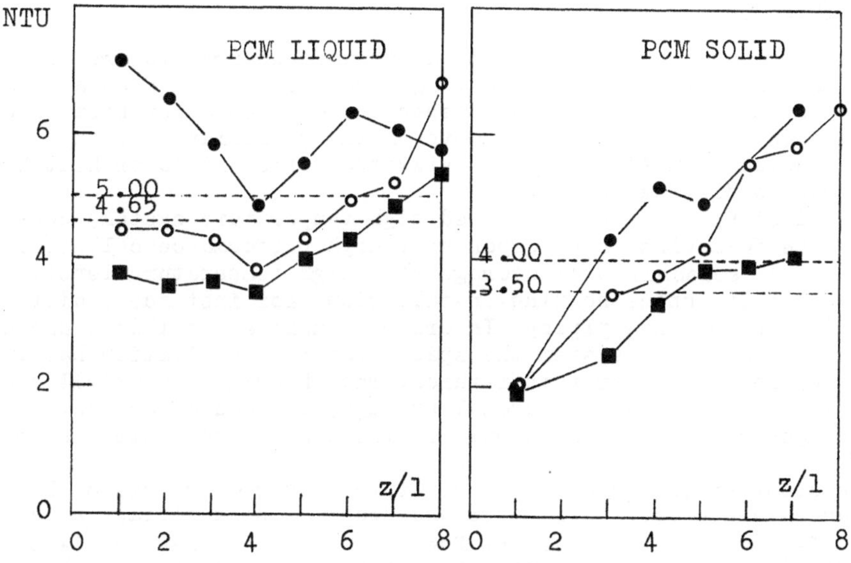

Fig.5 and 6 . Number of transfer units vs z-axis at different times (■ 12 mn, ○ 24 mn, ● 36 mn, ---- average value).

SOLAR HEATING OF A LARGE OFFICE BUILDING USING SEASONAL AQUIFER THERMAL ENERGY STORAGE

A.L. Snijders
Heidemij Adviesbureau B.V.

K. de Wit
Bredero Energy Systems B.V.

Summary

A technical and economical comparison has been made between two different solar systems for the space heating of a large office building. One system is based on high performance solar collectors in combination with aquifer thermal energy storage at a high temperature level, the other system on low performance collectors and aquifer heat storage at a low temperature level. In order to be able to use the heat a heat pump is applied in the latter system.

From a technical point of view the system with high performance collectors offers the possibility to achieve much higher fossil fuel savings than the system with low performance collectors. On the other hand from an economical point of view the high temperature system will not be attractive before the year 2000, even under favourable conditions. The low temperature system however, will be attractive under favourable conditions in the year 1990.

1. INTRODUCTION

Aquifers can be used for seasonal thermal energy storage in large solar systems for space heating. The elements in such a solar heating system cannot be chosen independent of each other because of the influence of each of the elements on the performance of the system as a whole. From this point of view two different solar heating systems can be distinguished.

The first system (the so-called high temperature system) consists of the following main components: high performance collectors, short term and seasonal heat storage at a high temperature level, peak load boiler, space heating installation (for instance radiators) and the interconnecting piping. In order to make an efficient use of the solar heat in this system the space heating installation has to be designed for low water temperatures. Though using a vertical heat exchanger in the soil as a seasonal storage, instead of an aquifer, the Groningen project is an example of the high temperature system (1).

The other system (the low temperature system) consists of low performance collectors, short term and seasonal heat storage at a low temperature level, heat pump, peak load boiler, space heating installation and the interconnecting piping. In this system a heat pump is necessary to use the low temperature solar heat. The Bredero office building which is under construction now at Bunnik, will be provided with the low temperature solar system. As a preparatory study, a technical and economical comparison has been made between the above-mentioned solar heating systems applied to a large office building.

2. METHOD OF APPROACH

In order to get the dimensions and specifications of the components of both systems a design study has been made for a well-insulated office building with a floor area of about 25000m^2. The building has a total heat demand in winter of about 2000 MWh(th), a peak demand of about 1500kW and has no need for cooling in summer. Other starting-points for the design study were:

- the application of a gasengine heat pump in the low temperature system as a consequence of the ratio between gas and electricity prices in the Netherlands.
- the contribution of the peak load boiler has to be less than 50% of the total heat demand.
- the geohydrological conditions for aquifer storage are favourable, i.e. no natural groundwater flow, low permeability (less than 5 Darcy for the high temperature system and less than 25 Darcy for the low temperature system), aquifer thickness about 30 meters and thickness of the confining layers more than 10 meters.

Using the results of the design study a detailed design has been made for both systems. Starting from these detailed designs the technical performance of both systems has been calculated as well as the additional investment and maintenance cost. Also the (expected) development in future of the cost of the main components has been taken into account. At this point for both systems some feed-back loops were necessary to reach at the definite design and cost/performance ratio.

3. TECHNICAL RESULTS

The dimensions and specifications of the main components of both systems are given in table I.

The technical performance of the high and low temperature solar system has been calculated for the reference year 1964 - 1965. The results of these calculations carried out by the Institute of Applied Physics TNO - TH, are summarized in figure 1. The solar contribution amounts to about 53% of the total heat demand for the high temperature system and about 38% for the low temperature system. This results for the high temperature system in a saving of fossil fuels of 54 - 60% (depending on the efficiency of the boiler in the reference situation). For the low temperature system this saving amounts to 36 - 45%. In the latter percentages the efficiency of the gas engine has also been taken into account.

Due to the low water temperature the efficiencies of the solar collectors, the short term storage tank and the aquifer storage are significantly higher in the low temperature system than in the high temperature one. This results amongst others in a much smaller collectors area for the first system, see table I. However it appeared to be necessary to use single covered collectors instead of uncovered collectors in the low temperature system in order to reduce heat losses due to radiation. The efficiency calculated for uncovered collectors amounts to 40% instead of 63%.

Table I. Dimensions and specifications of the main components.

component	high temp. system	low temp. system
collectors		
type	vacuum tube	single covered flat plate
area	$3000 m^2$	$1200 m^2$
outlet temp.	90^oC, flow controlled	25^oC, flow controlled
short term storage		
volume	$70 m^3$	$70 m^3$
aquifer storage		
flow	$55 m^3/h$	$55 m^3/h$
min. recovery temp.	not fixed	14^oC
heat pump		
min. freon evaporation temp.	-	$+ 5^oC$
max. freon condensation temp.	-	$+ 47^oC$
max. power	-	470 kW(th)
space heating instal.		
max. inlet temp.	$42,5^oC$	70^oC
max. outlet temp.	$32,5^oC$	50^oC

4. ECONOMICAL RESULTS

The calculated additional investment for the high temperature system amounts to about Dfl. 1.900.000,- and for the low temperature system to about Dfl. 670.000,-. These amounts are given in Dutch florins of 1984 assuming that the solar collectors and the gas engine heat pump are quantity-produced, resulting in a cost reduction of about 30% as compared to the present cost of these components (fitting included). The investment difference between both systems is mainly caused by both the larger collector area and the higher collector cost pro m^2 in the high temperature system.

Starting from the additional investment, the additional maintenance cost, the additional electricity consumption of circulation pumps and the saving of fossil fuel, the net cash values for both systems have been calculated at different increase rates of the energy prices. The main results of these calculations are given in figure 2 (evaluation period 15 years, inflation rate 5%, interest rate 8%). From this figure it will be clear that the high temperature system will not have a positive net cash value in the near future, even under favourable economical conditions. The low temperature system however, will have a positive net cash value already in 1990 assuming a moderate real increase rate of the energy prices and the collectors quantity-produced.

(1) Wijsman, A.J.Th., The Groningen Project. 100 houses with seasonal solar heat storage in the soil using a vertical heat exchanger. Proc. of the International Conference on subsurface heat storage in theory and practice. Stockholm, June 6 - 8, 1983.

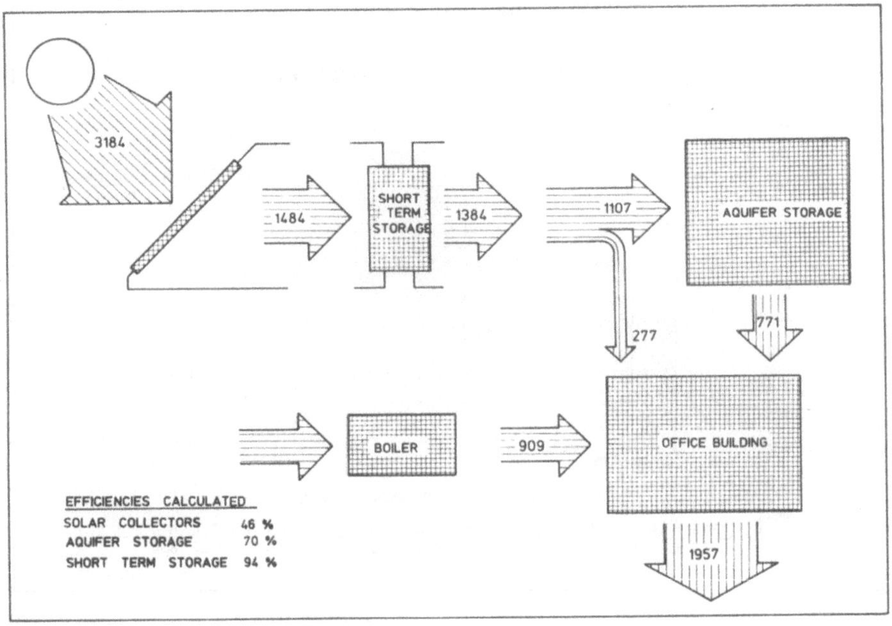

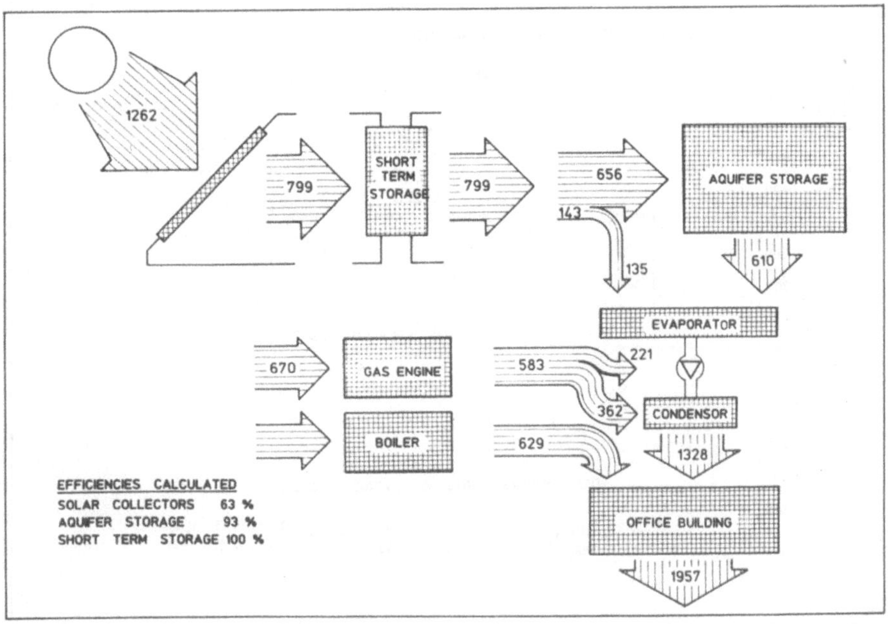

Figure 1. Energy flow diagram in MWh_{th} for the high temperature system (upper side) and the low temperature system.

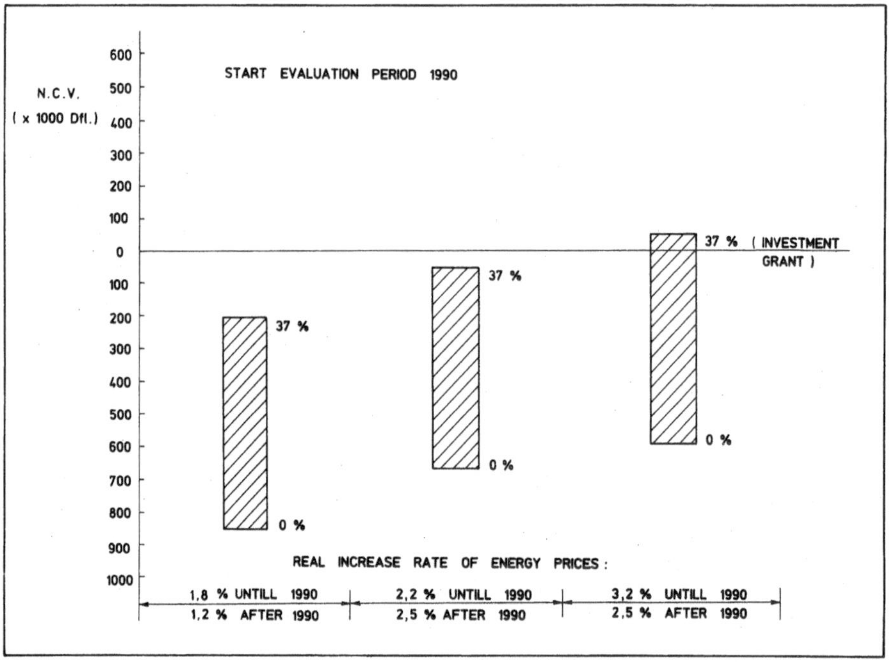

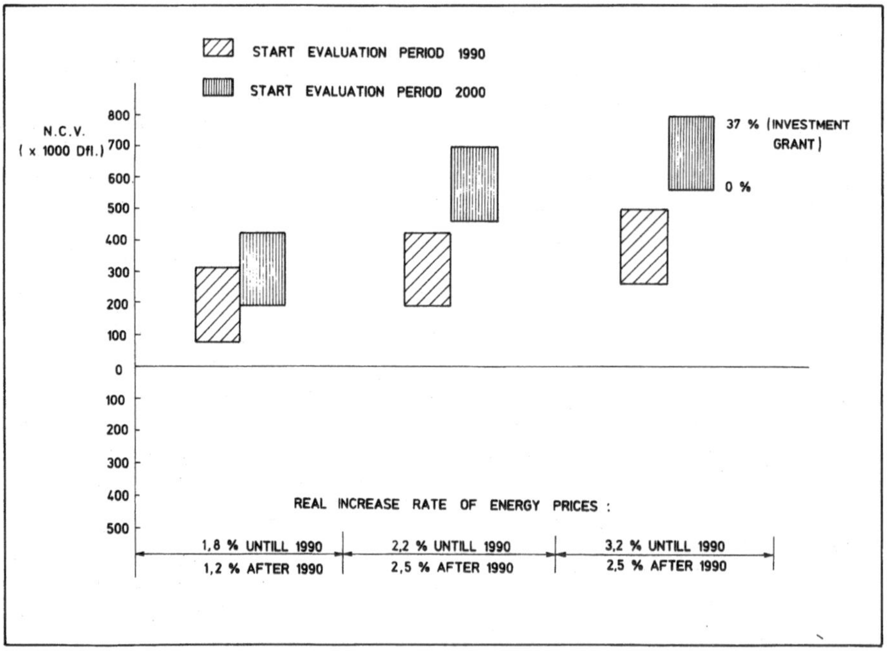

Figure 2. Net cash values in Dfl. of 1984 at different increase rates of the energy prices for the high temperature system (upper side) and the low temperature system. Solar collectors quantity-produced.

THERMAL LOSSES ANALYSIS OF AN INTERSEASONAL HEAT STORAGE IN THE SOIL : FIELD TRIALS AND NUMERICAL MODELS

M. MENOUD and J.P. GAUDET
Institut de Mécanique de Grenoble,
U.S.M.G., BP 68-38, 38402 Saint-Martin-d'Hères, France
and
A. de LA CASINIERE and C. MEYRIER
Centre de Recherches sur les Très Basses Températures,
C.N.R.S., BP 166 X, 38042 Grenoble-Cédex, France

Summary

From field trials achieved between 1980 and 1983 on an interseasonal heat storage in the soil (Grenoble France, 45° North Lat.), a simplified numerical model of the storage behaviour was validated. This model allowed to achieve an exhaustive study of the storage heat loss fluxes during the storing season. Depending on the injection temperature of the energy, which is provided by flat plate solar collectors under the local climate, the amount of heat lost varies from 32 to 37 % of the total entered. Relaxation trials gave heat losses of about 45 kWh/day for 25°C of mean storage temperature.

1. INTRODUCTION

For the both reasons, high costing and practical difficulties, a long term storage in the soil can never be widely insulated. Hence, when high storage temperatures are required, the heat loss fluxes can reach important values.

2. STORAGE AND SOLAR SYSTEM DESCRIPTION

The main elements of the interseasonal storage (see fig. 2 and 3) are steel flat plate panels used as heat exchangers with a cathodic protection against corrosion. These panels are vertically embeded into two parallel trenches 17 m long and 3.5 m depth ; they are 2 m high and have a total of 112 m^2 of exchange surface. In order to insure a good thermal contact, the trenches are filled up with wet sand enclosed into big polyethylene foils. The circulation sense of water in the panels is downward during storing phases and upward during extracting phases. The storing material is the local subsoil ; it is a saturated clay of a high water content level (0.24 m^3/m^3) which remains constant all over the year. An horizontal insulating layer of expansed polyurethane ($7 \times 17 = 119$ m^2) of 0.21 m thickness covers the trenches ; it is buried at 0.60 m below the soil surface and protected against moisture by a polyethylene bag. Fortunately, no drying effect of the soil occurred during heating trials.

The interseasonal heat storage piping is connected with the lower part of a 3 m^3 water tank used as short term intermediary storage. 50 m^2 of flat plate solar collectors, 60^d South tilted, are connected with the same part of this tank which provides hot water to the heating floor loops of a single family house. The auxiliary heating is a gas furnace connected with the upper part of the tank (see fig. 1).

In view to validate the models, 26 temperature probes were buried into the soil in 3 vertical sections A, B, C (see fig. 2 and 3) orthogonal to the trenches, between 0,60 to 4,50 m depth. Several heat flowmeters measure all the energies entering and leaving the short term storage. Other ambiant temperature probes and a solarmeter permit a complete thermal analysis of the solar system. The corresponding data are collected with the aid of a microcomputer.

3. NUMERICAL MODELS

The numerical model, simple enough for needing a microcomputer only, aims mainly at simulate economically and accurately the storage behaviour over periods as large as various years. In return, periods shorter than few days cannot be simulated satisfactorily. The model is bidimensional and uses an alternative difference iterative procedure for calculating meshes energy balances in the soil. The meshes size is constant (0.875 m × 0.400 m ; see fig. 3) as the time step which depends on the water flow value into the buried heat exchangers. The soil surface temperature is given by a sinusoidal law fitted on the real local climate conditions at 2 m above ; an atmosphere generally poorly stratified on the storage site justify a such approximation. The soil is supposed to be homogeneous and isotropic ; the water content remains constant and the convection term is neglected before conduction. Due to a permanent saturation, one admits a contact thermal resistance between wet sand and heat exchangers equal to zero.

This model is able to give the storage outlet water temperature evolution for a fixed storage inlet water temperature ; it can calculate too the meshes mean temperature and heat fluxes, then the heat losses from a given storage volume.

From the above simplified model another model was studied as a subroutine of TRNSYS. Its main features are to be monodimensional only and to allow a sufficiently accurate simulation of short and long term trials, but for previously fixed experimental conditions.

4. FIELD TRIALS

Four main storing field trials were achieved which consisted in series of short daily injections (1 to 2 hours). The water temperature injected into the buried heat exchangers remains constant all over the storing phase :

Storing period (from - to)	heat source	Injection temperature (°C)	Injected energy (kWh)
28/05/80 - 2/07/80	gas furnace	24.6	2400
19/11/80 - 29/12/80	gas furnace	40	4000 to 4500
18/08/82 - 5/10/82	solar collectors	42	1580
17/05/83 - 13/10/83	solar collectors	49	5100

The first model was supposed to be validated when, using the real values of the soil heat capacity and thermal conductivity, the temperatures measured into the soil are consistent with those calculated.

As no transient simulation can be done, the injection is supposed to be continuous but with a flowrate which gives at the end of the trial the same water volume injected.

5. HEAT LOSS STUDY

Three model trials were achieved in order to simulate energy injections provided by 50 m^2 of flat plate collectors, from May 15th to October 15th under the local climate conditions. As expected, the amount of energy injected depends on the temperature injection level (see table 1) ; it is calculated by the aid of a standard method of Electricité de France (1).

The storage volume from which are calculated the losses is arbitrarily choiced to be equal to about 860 m^3. It corresponds, roughly, to the soil zone which temperature is significantly modified by the storing and extracting operations.

$T_{injection}$ (°C)	$Q_{injected}$ (kWh)	Total losses (kWh)	Total losses/$Q_{injected}$ (%)
50	8 800	2 800	31,8
45	10 200	3 600	35,3
42	11 100	4 050	36,5

Table 1 : Injected and lost storage energies from 15/5 to 15/10

A relaxation study over 18 days permits to find an approximately linear law of the heat loss flux values versus the mean storage Temperature $\overline{T}$:

$$\text{Loss flux} = 2.78\ \overline{T}\ (°C) - 24.5\ (kWh/day)$$

Although higher, these values are quite consistent with those given in a previous rough study (2) :

$$\text{Loss flux} = 2.65\ \overline{T}\ (°C) - 29.6\ (kWh/day)$$

$T_{injection}$ (°C)	Stored energy (kWh)	$\overline{T}$ (°C)	Loss flux (kWh/day)
50	6 000	24,3	42,8
45	6 600	25,6	46,9
42	7 050	26,5	48,9

Table 2 : Loss fluxes for various mean injection temperatures

Injection temperature upper than 50°C induces poor stored energy values due to a lower solar collectors efficiency ; below 42°C, the buried heat exchangers surface is too small to store completely the daily disposable energy.

6. CONCLUSION

The both reasons heat loss flux values and collectors efficiency lead to limit at about 45°C the temperature injections ; this implies the presence of an intermediary short term storage. At last, the collectors are to be tilted in order to inject still important amounts of energy at the end of the storing period (Oct. 15th) rather than at the beginning (May 15th) ; such tilting will insure a maximum value of the energy remaining into the store at the beginning of the heating season.

Acknowledgements - This experiment was supported by CNRS-PIRDES (Programme Interdisciplinaire pour la Recherche et le Développement de l'Energie Solaire) and COMES/AFME (Commissariat à l'Energie Solaire/Agence Française pour la Maîtrise de l'Energie).

REFERENCES

1. CHOUARD, P., MICHEL, H. and SIMON, M.F., Bilan thermique d'une maison solaire, méthode de calcul rapide.
2. de LA CASINIERE, A., KUHN, G., GAUDET J.P. and al., Interseasonal storage in the subsoil in Meylan : first experimental results. ICBEM II, AMES-IOWA (USA), 30 May-3 June 1983.

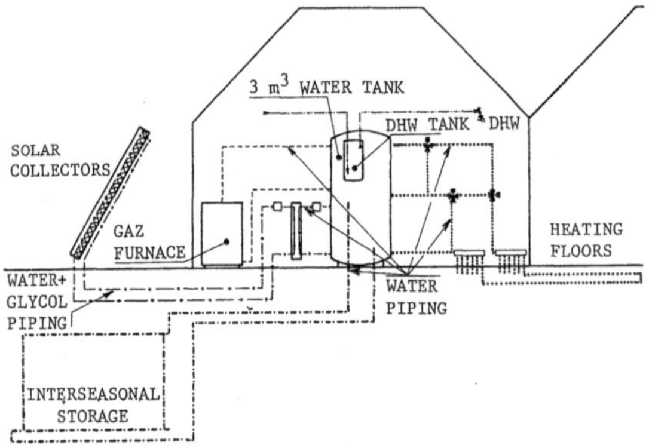

Fig. 1 : Diagram of the complete solar heating system

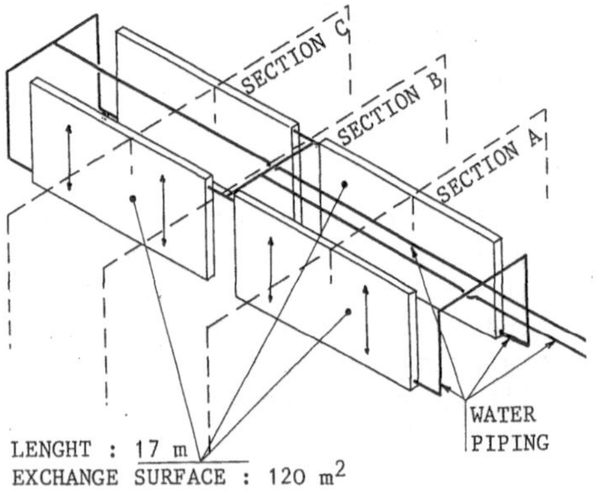

Fig. 2 : Buried heat exchangers of the interseasonal storage

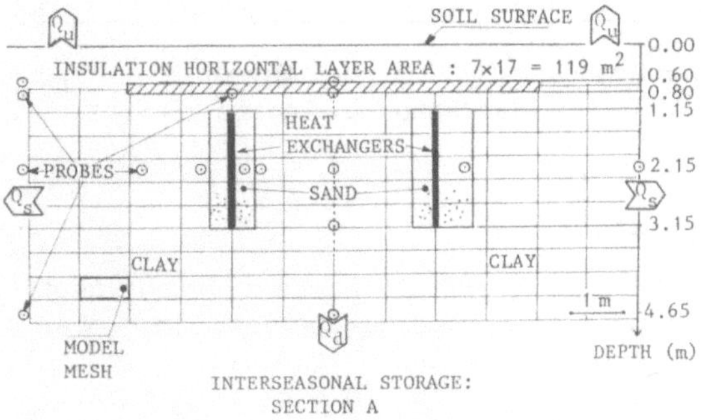

Fig. 3 : Section A of the interseasonal storage

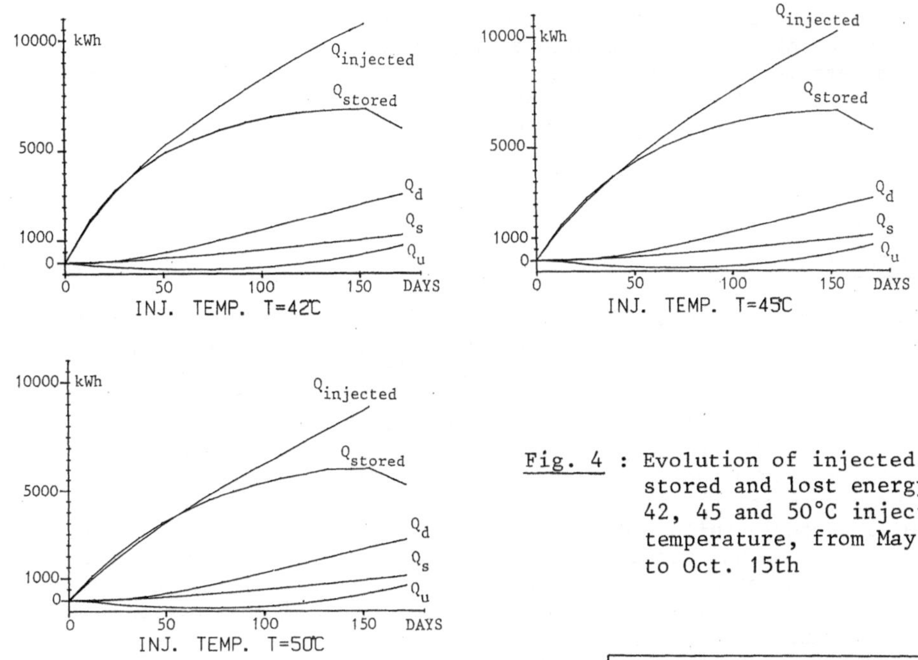

Fig. 4 : Evolution of injected, stored and lost energy for 42, 45 and 50°C injection temperature, from May 15th to Oct. 15th

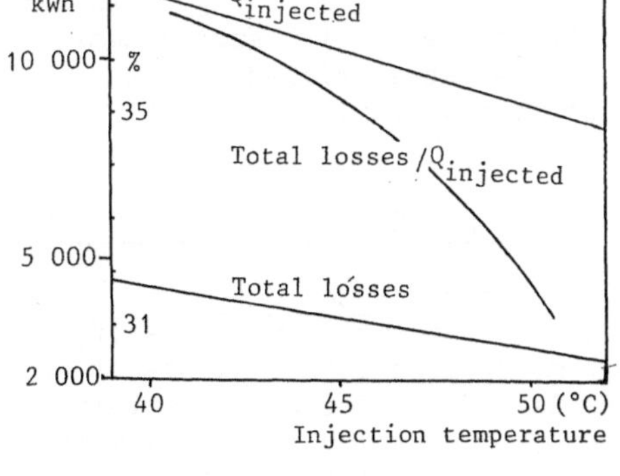

Fig. 5 : Injected and lost energy from May 15th to Oct. 15th, against injection temperature.

COMPARISON BETWEEN TWO DIFFERENT UNDERGROUND SEASONAL STORAGE SYSTEMS USING A WATER VESSEL AND SOIL HEATING TUBES

E.Aranovitch, G.Binda. M.Hardy, L.Rigolini - CEC, JRC-Ispra, Italy
L.Esposito, L. Ingala, G.Rizzi - ENEA, Ispra, Italy

Summary

The cost-effectiveness of space heating solar systems is penalized in most of Europe by the limited number of sunshine hours which are available during the periods for which heating is needed. These limitations can be overcome in part by seasonal storage but the economic constraints are severe. The paper presents results on two underground seasonal storage systems which are being tested and monitored at the Joint Research Centre of Ispra. The first one is an underground 370 m^3 concrete vessel filled with water. The second system uses 3000 m^3 of earth as the main storage medium. The soil is heated by regularly spaced vertical tubes connected to an array of 180 m^2 of solar collectors. Experimental results during charging and discharging periods are presented. A simplified mathematical model and a cost analysis put in evidence the influence of the scale effect and the ranges of values of the main parameters for which these systems may become cost-effective.

1. INTRODUCTION

An evaluation based on the experience of the past years shows that most solar space heating systems are non-economic at this stage in the greater part of Europe. The performance over cost ratio of these systems is strongly affected by the fact that, as they operate only during the cold season, they will utilize only a fraction of the yearly solar energy available.
For this reason, the problem of seasonal energy storage is becoming the object of increasing attention and is often considered the crucial one to be solved if active solar space heating is to have a future. Another factor which will strongly influence the overall performance of a solar system combined with seasonal storage is the temperature of operation which affects the efficiency of both the solar collectors and the store. The advisability of adding to the system a heat pump which would make it possible to lower these temperatures is still largely open to debate because of cost factors. In order to study the complex interactions between solar collectors, heat pump and long term storage, two large underground storage systems have been built at the Joint Research Centre of Ispra. The first system is a buried concrete vessel filled with water. The second one uses the earth itself as a storage medium.

2. MAIN CHARACTERISTICS OF THE TWO SYSTEMS

2.1 Concrete Water Vessel

The internal dimensions of the buried concrete water vessel are 16x4x5x 4.5 m correspong to a volume of 370 m^3 (Fig.1). The effective volume of water is 330 m^3. The water contains an anti-fungus product and the water surface is covered by an oil film to reduce evaporation-condensation phenomena. The thickness of the concrete walls is of the order of 40 cm. Neither lateral nor bottom walls are thermally insulated. Only the cover slab of the vessel is insulated with 30 cm of polystyrene and 25 cm of gravel. In order to drain rain water, the ground surface around the vessel is asphalted. The water in the vessel is heated by 12 electrical resistances of 5 kW each, leading to a maximum heating power of 60 K. In order to simulate solar heating, the electrical power can be controlled with a prefixed multiplication factor by the output of a reference 2 m^2 solar collector. The temperatures in the water and in the surrounding ground are measured by 48 thermocouples placed at various levels and distances.

2.2 Soil Storage with Vertical Heating Tubes

The earth volume of the store is about 3000 m^3 with 36 vertical heating tubes buried with a spacing of 2.5 m. Fig.2. The tubes, of internal diameter 185 mm, are made of steel with the bottom end closed. They are 10 m long, 6 mm thick and filled with water. Inside each of these tubes is placed a one inch pin tube in which circulates the heat transfer fluid coming from the solar collectors. The water surrounding the pin tubes acts as a thermal buffer in order to compensate for the relatively low thermal diffusivity of the earth.

The ground surface of the store, of dimensions 17.5x17.5 m, is thermally insulated with 10 cm of insulating material and a gravel layer. A vapour barrier has the function to limit humidity loss from the store and excessive rain water penetration. The heating tubes have been divided into three concentric zones which can be charged or discharged independently with a prefixed strategy. A daily storage subsystem of 5 m^3 can be inserted between the large store and the array of 180 m^2 of solar collectors inclined at 45^0. The absorber of these collectors is made of synthetic rubber and can be made to operate eventually without glass covers. The circuit for the cold season is completed by a 32 kW water to water heat pump which adds flexibility to the operation of the whole plant.

The temperatures in the ground are measured by 66 thermocouples distributed at different depth levels and locations. Moreover the circuits are instrumented in order to make various energy balance measurements. A meteorological station provides the plant also with the necessary weather data.

3. EXPERIMENTAL RESULTS AND MODELLING

Seasonal ground storage systems require inherently a long time of testing, because on the basis of one cycle per year, it may take several years before their thermal behaviour reaches a repetitive stage. For two years the two systems have been charged and let to discharge naturally through heat losses.

For the water storage vessel which is insulated only at the top part, one of the main points of interest to investigate was the "apparent" heat loss coefficient during and after a certain number of charge and discharge cycles. Theoretical studies had shown that for a non-insulated vessel, if temperatures in the water during the discharge period become inferior to the temperature at the infinite in the ground, one may observe a heat reflux from the surrounding ground into the store. In that case the surrounding ground acts both as a complementary storing medium and as a modulated insulator which will insulate the store at high temperatures and let ground heat penetrate into the store at low temperatures.

In a simplified manner the "apparent" heat transfer coefficient U may be represented by the following equation:

$$MC \frac{dT}{dt} = P - US\ (T-T_\infty) \qquad (1)$$

MC is the heat capacity of the store, T the store temperature, P the power introduced into the store, S the effective heat loss surface, T_∞ the ground temperature at infinite. U the apparent heat transfer coefficient is a function of the store temperature and integratés the heat diffusivity phenomena in the surrounding ground. In Fig.3, the results of more than two years of testing are concentrated, showing that the apparent heat transfer coefficient U follows remarkably well the temperature cycle profiles. It varies by a factor 3 from a minimum value of the order of 0.5 W/m^2K to 1.5 W/m^2K. It should also be noted that the storage efficiency defined as the ratio of the energy retained at the end of the cycle over the energy instroduced increases with the number of cycles passing from 0.22 to 0.43 after 3 cycles.

Studies on the ground storage with vertical heating tubes include both experimental and theoretical work on the single heating tube behaviour as well as on the storage system taken as a whole. Preliminary experiments were carried out on a single tube and the characteristics of the clayish soil which constitutes the storage medium, were determined in the laboratory. The greater part of the theoretical work on the heating tubes was performed under contract by ADES /1/. These studies put in evidence the influence of the tube diameter and the diffusivity of the soil on the charging and discharging processes.

The effective charging capacity of the store was determined by integration of the temperatures measured in various points in the soil. In Fig.4, two and a half years of measurements are presented in a condensed form, giving respectively the energy delivered to the store by the 180 m^2 of solar collectors, the store average temperature and the temperature in the insulation layer above the store. It can be noted that after the second cycle, the "natural" discharge through heat losses will still leave a residual average temperature in the store of the order of 25°C in January. The temperature profiles in the insulation layer indicate that the upper insulation is insufficient. The ratio of 1 m^2 of solar collector for 15 m^3 of storage volume may be slightly increased.

An energy extraction loop with a 32 kW water to water heat pump will now operate with this storage system in order to evaluate and optimize the global system performance.

4. COST ASPECTS AND CONCLUSIONS

In addition to this experimental program, a simplified model has been developed for the technic-economic evaluation of a solar system coupled with seasonal storage and a heat pump /2/. The model takes into account 26 independent variables (design parameters, weather data, cost factors). From a certain number of numerical examples based on reasonable cost factors some economic indications can be put in evidence. As far as the charging process is concerned, assuming cost factors of the order of 40 ecus per equivalent m^3 of water storage, 200 ecus per equivalent m^2 of installed solar collectors and a yearly level of solar radiation of 1300 kWh/m^2 energy can be collected and stored for costs in the range of .09 ecu per kWh. The influence of the scale effect is significant. Costs are reduced by 50% when passing from storage volumes of 100 m^3 to 10,000 m^3. The results of a cost analysis concerning the present two storage systems are shown in Figure 5, showing that storage in the soil itself looks more promising from an economic point of view, but it is also clear that these promises must be validated by extensive and diversified testing.

REFERENCES

1. Valutazione tecnico economica dei sistemi solari con stoccaggio stagionale del calore nell'habitat. C. Mustacchi, L. Lamorgese, ADES (Roma), Cont. no.1506.81.03 ED.ISPI.
2. A simplified model for the technico-economic evaluation of a solar system coupled with seasonal energy storage and a heat pump. E. Aranovitch, CEC, JRC-Ispra, EUR

ACKNOWLEDGEMENTS

The authors wish to acknowledge the valuable contributions of Dr. V. Cena, L. Lamorgese and C. Mustacchi of ADES (Roma) to the preparation and development of the work presented in this paper.

FIGURE 1 Underground concrete water vessel

FIGURE 2 Ground storage with vertical tubes

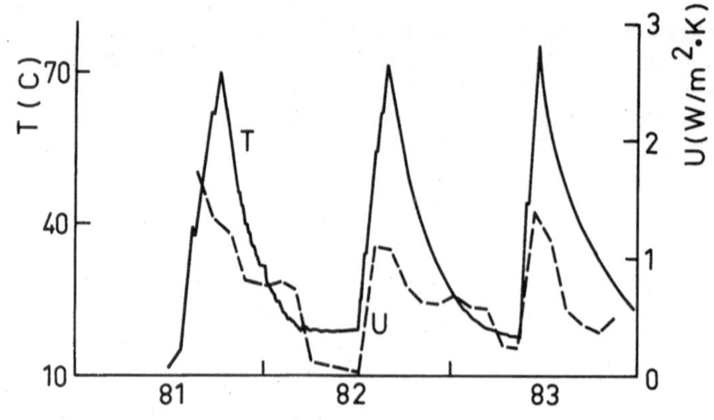

FIGURE 3 Water temperature and heat transfer coefficient

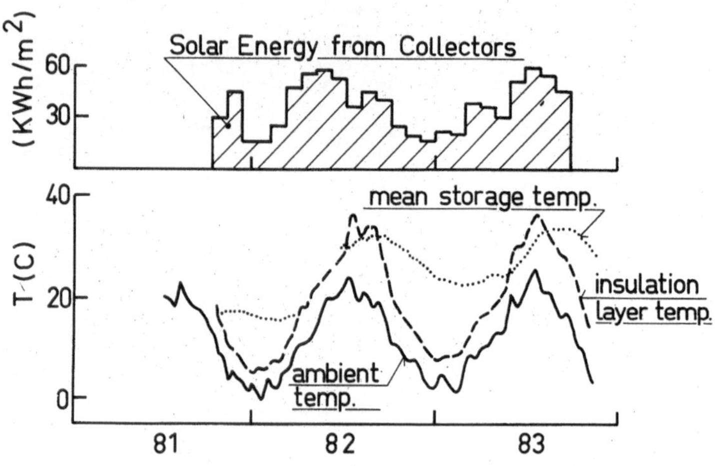

FIGURE 4 Measurements in ground storage

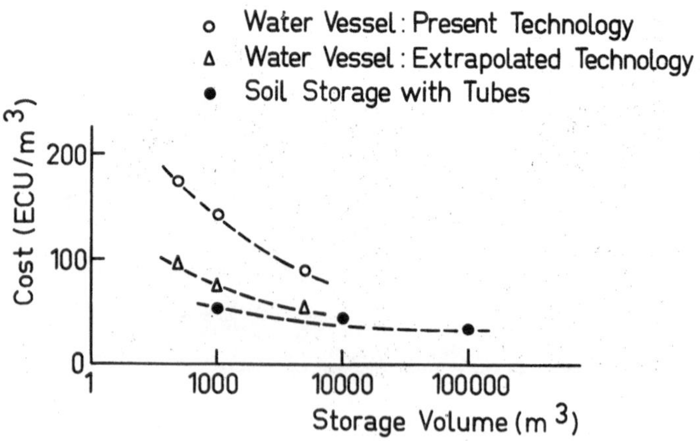

FIGURE 5 Cost factors

PHYSICAL MODEL FOR THE STUDY OF MASS AND ENERGY TRANSFERS IN THE NON-SATURATED LAYER OF SOIL LOCATED ABOVE A SOLAR ENERGY STORAGE ZONE

C. SAIX, J.C. BENET, G. DELLAVALLE, P. JOUANNA
Laboratoire de Génie Civil
Université des Sciences et Techniques du Languedoc
Place Eugène Bataillon
34060 Montpellier Cedex. France

Summary

The efficiency of energy storage in a saturated layer is linked to a great extend to the energy and matter losses through the non-saturated layer of soil lying above the saturated layer used as a storage zone. In order to study this problem we constructed a physical model of sufficiently large size. A test programme has made it possible to simulate heat storage at 60° C. The surface of the soil was subjected to controlled conditions and then natural atmospheric conditions. The physical model serves as a prototype for the study of the behaviour of a natural site. The evolutions of the physical model make it possible to judge the modifications in the state of the non-saturated layer of soil and to come to conclusions regarding possible harmful effects of storage. The physical model has given experimental results that confirm the validity of a local three-phase mathematical model. This confirmation allows this model to be used in the future for making forecasts for other sites.

1. INTRODUCTION - PURPOSE OF THE PHYSICAL MODEL

The purpose of the physical model is the experimental study of mass and energy transfers in the non-saturated layer of soil located above a saturated layer used as a heat storage zone. The physical model constructed simulated the layer of non-saturated soil and the storage zone. The various measurements made in the non-saturated layer of soil made it possible to monitor the evolution of the state of the soil. Interpretation of the experimental results was carried out at two levels: a) Direct interpretation of the measurements considering the model as the prototype for a heat storage installation. The modifications caused by storage are assessed and the various transfers calculated. b) Comparison with the results of e local intgrated mathematical model and verification of the simplifying assumptions relative to this model.

2. PHYSICAL MODEL

2.1 Test zone [1], [2] (Fig. 1)

The storage zone is simulated by a reinforced concrete tank filled with saturated sand. The temperature in the tank is regulated by heating resistances and can vary between 20°C and 60°C. The layer of non-saturated soil is 3. metres thick and formed by compacted silt in 17cm layers. Probes installed in a vertical row in the center of the non-saturated soil enable the development of this layer to be monitored. The information from the various probes are recorded in a nearby laboratory.

2.2 State variables and measuring probes [1], [2], [3]

The state of the soil at any point along the vertical measurement axis is defined by five state variables: apparent dry density of the porous matrix, ρ_1 ; temperature, T; water content, w; total pressure of the gas phase, p^*_g ; partial pressure of water vapour, p^*_v . In order to complete knowledge of phenomena, measurement of thermal conductivity, λ and capillary suction, ψ are also carried out. All these measurements are made by

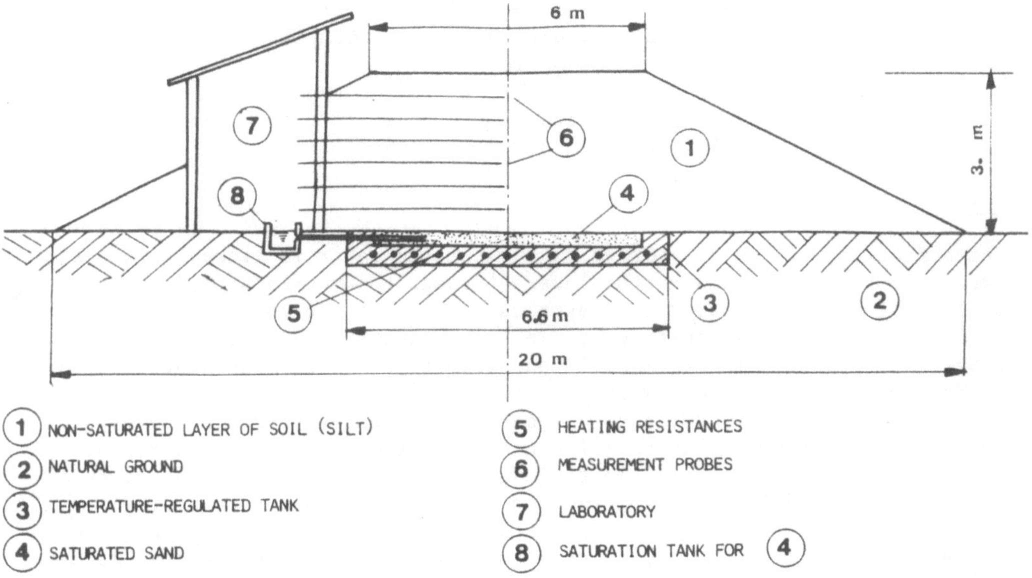

FIG. 1 : DIAGRAM OF THE TEST ZONE

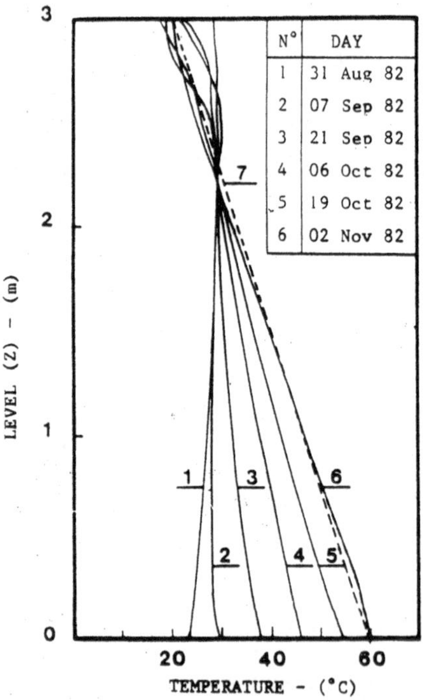

FIG.2 : EVOLUTION OF THE TEMPERATURE PROFILE DURING STORAGE SIMULATION (PHASE B)

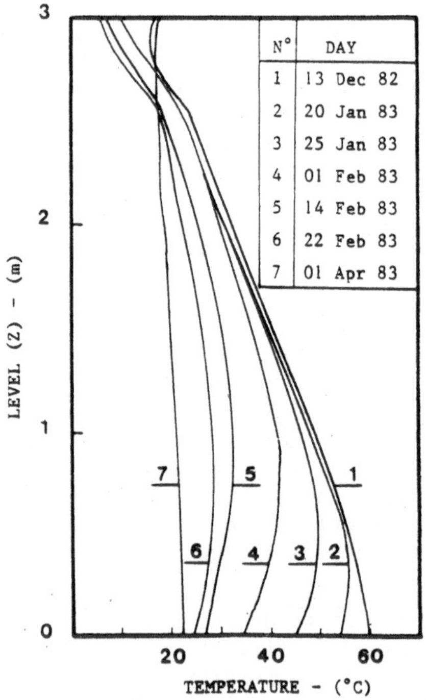

FIG. 3 : EVOLUTION OF THE TEMPERATURE PROFILE AFTER STOPPING OF HEATING (PHASE E)

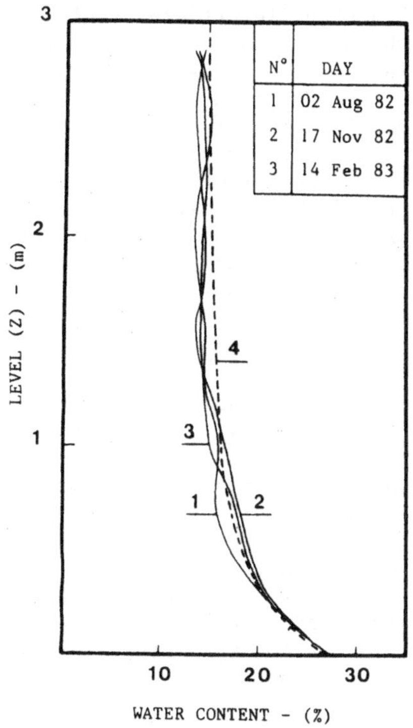

FIG. 4 : WATER CONTENT PROFILES

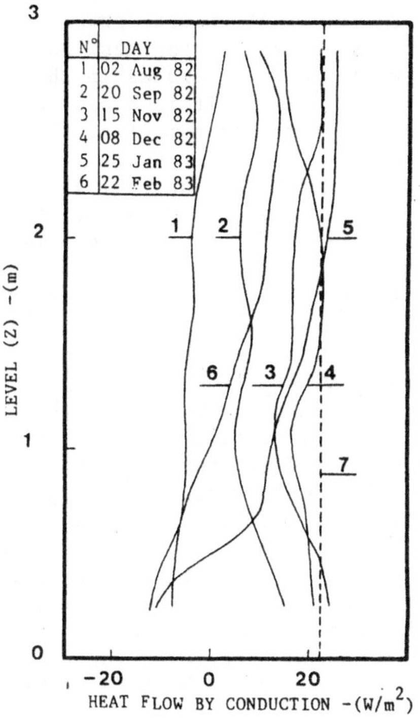

FIG. 5 : PROFILE OF HEAT FLOW BY CONDUCTION

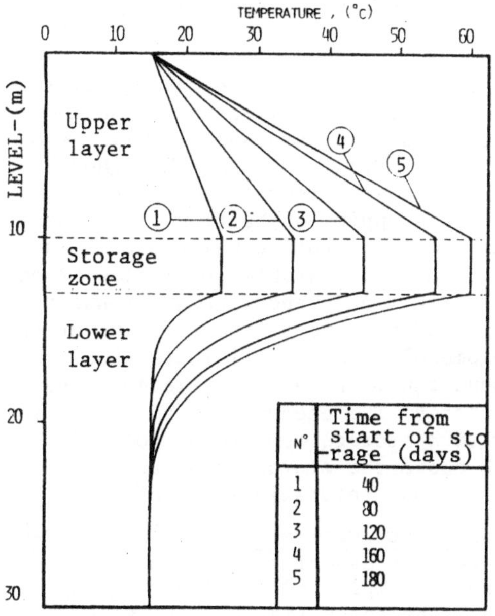

Fig.6-a : Storage phase

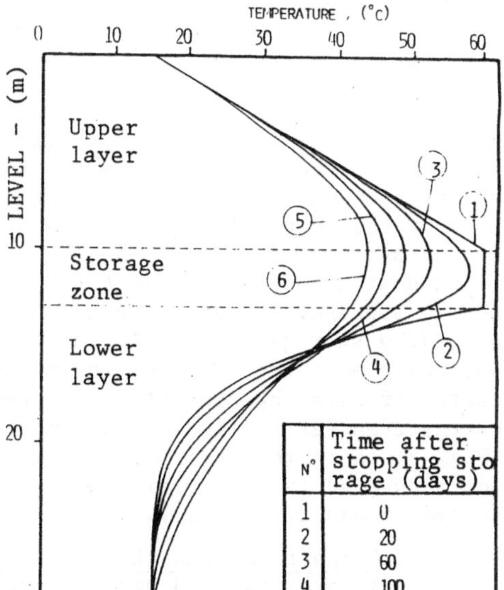

Fig.6-b :After stopping storage

Fig.6 : SIMULATION OF A STORAGE SITE IN THE GROUND : TEMPERATURE PROFILES six probes. Eight of each probe are installed at different depths along the vertical axis of the test zone except the probe measuring p_v^* for which are installed three probes in the surface layer. Probe 1 is for measurement of the settlement. It is based on the principle of communicating vessels. Probe 2 is for measurement of the temperature and thermal conductivity by cylindrical probe in transitory regime. Probe 3 is for measurement of water content by neutronic device. Probe 4 is for measurement of the total pressure of gas phase and consists of a copper tube; the extremity placed in the soil is perforated. Probe 5 is measuring the partial pressure of water vapour in the soil. It is based on the psychrometer principle. Probe 6 is a tensiometer. It measures water pressure through a porous ceramic disc. It has been temperature tested.

3. TEST PROGRAMME

The test programme fixed the evolution of the boundary conditions. It was divided into five phases : Phase A (jul.82-1sep.82), B (1 sept. 82-26 nov. 82), C (26 nov. 82 - 1 déc. 82), D (1déc. 82 - 17 janv. 83), E (17 janv. 83 - 10 mars 83). The purpose of phase A was to attain the state of natural equilibrium in the non-saturated layer of soil; this state was to serve as reference in order to judge the modifications brought about by storage. Phase B simulated the storage of heat, with the temperature increasing in a linear manner from 25°C to 60°C. The purpose of phase C was to attain equilibrium when the storage temperature had stabilized at 60°C. Phase D was to free water vapour transfer in order to judge its extent. The supply of energy at the base was stopped during phase E.

4. RESULTS AND INTERPRETATION

During the programme of tests, probes made it possible to plot the profiles of the various parameters along the vertical in the test zone. The results were interpreted by both direct interpretation and comparison with the results of a mathematical model.

4.1. Direct interpretation

The main results are given in figures 2, 3 and 4 which show examples of temperature and water content profiles. Figure 2 illustrates the modification of the temperature profile as a result of storage. Il will be observed that the temperature variation is practically linear throughout the whole thickness of the non-saturated soil layer. Figure 3 shows the evolution of the temperature profile after the stopping of heating. It is possible to see from figure 4 that the water content profile which became established at the end of phase A remained remarkably stable and little affected by storage. It can be deduced that liquid and vapour phase transfers caused by storage are small. Calculation of transfers from profiles of the variables measured and from known phenomenological coefficients shows that energy transfer by conduction is preponderant. There is little mass and energy transfer in liquid and vapour phases. The energy flow by conduction (Fig. 5) is in the order of 20 W/m^2 with a storage temperature of 60°C, a non-saturated layer of soil 3. m thick above the storage zone and mean thermal conductivity of 1.5 W/m.°C in the non-saturated layer of soil.

4.2 Comparison with the results of a mathematical model [2], [3], [4].

The theoretical and experimental results associated with simplifying assumptions make it possible to envisage a simplified model giving rapidly the temperature profile, the flow of heat by conduction and the water content profile. The results given by this simplified model were first compared with the results of the physical model for validation purposes. A first approximation shows agreement between experimental and theorical results. The simplified model was used to model a storage site consisting of a layer of soil 3m thick located 10 m beneath the surface of the ground. Figure 6-a shows the evolution of the temperature profile in the soil during the 6 - months storage phase when the storage temperature was increased from 15°C to 60°C. Figure 6-b shows the evolution of the temperature profile during the 6 months following the stopping of heating. It can be seen that there was no great dispersion of heat in the soil. Profile N° 6 of figure 6-b shows that the energy stored six months after the stopping of heating formed 80% of the energy present in the soil at the end of the storage process. This is a promising result as regards the efficiency of the procedure.

5. CONCLUSION

The experimental results of the physical model enables the two objectives to be met. Observation of the evolution of this storage installation prototype for a period of eight months has shown that, apart from its temperature, the state of the layer of non-saturated soil is not greatly modified. The preponderant transfer is heat flow by conduction. This was in the order of $20W/m^2$ under the conditions of the experiment. Comparison of the experimental results with the results of an integrated three-phase local mathematical model confirms the qualitative validation of the theorical model. The simplifying assumptions used for integration were confirmed by the experiment; this makes it possible to propose simple models that will be useful for assessing the storage capacity of other sites.

BIBLIOGRAPHY

[1] DELLA-VALLE G., Stockage de chaleur dans les sols. Réalisation d'une aire d'essai. Thèse Doc. Ing. Génie Civil, I.N.S.A. Toulouse 1982.

[2] P. JOUANNA, J.C. BENET, D. CHEVASSUS, G. DELLA-VALLE, P. DURUISSEAUD, D. KOUFOGIANNIS,C. SAIX. Etude du stockage de chaleur dans les sols non saturés. Rapport final du Contrat C.C.E. N° ESA-S-051-F(S). Juin 1983.

[3] J.C. BENET, Contribution à l'étude thermodynamique des milieux poreux non-saturés avec changement de phase. Thèse Doc. es Sciences, U.S.T.L Montpellier 1981.

[4] M.T. GUIRAUD SALES, Transferts d'énergie et d'eau avec hystérésis dans un sol non-saturé au-dessus d'un stockage d'énergie solaire. Thèse Doc. d'Université U.S.T.L. Montpellier 1981.

Acknowledgement : the authors would like to express their thanks to the Commission of the European Communities for its support by means of contract within the framework of the Commission's solar energy R+D programme.

DESIGN AND BUILDING OF A TEST FACILITY FOR THERMAL STORAGE UNITS BETWEEN -30°C and + 90°C

M. BOURDEAU (1), J.L. SALAGNAC (1)
P. ACHARD (2), D. MAYER (2)
(1) Centre Scientifique et Technique du Bâtiment
BP 21
06562 VALBONNE Cédex
France
(2) Ecole Nationale Supérieure des Mines de PARIS
Centre d'Energétique
06560 VALBONNE-France

Summary

This paper presents the reasons that led the Centre Scientifique et Technique du Bâtiment, jointly with the Ecole Nationale Supérieure des Mines de Paris, to build a test facility for thermal storage units, and indicates the main criteria that guided its design. A detailed description of this test facility is given and some comments are made concerning the tests performed up to now.

INTRODUCTION

Complementary to the theoritical and experimental works these two research centers (CSTB and ENSMP) have carried out in the field of thermal energy storage for a few years at SOPHIA ANTIPOLIS, the building of a test facility has proved essential for the following purposes :

- to measure performances of storage units in prescribed conditions ;
- to compare performances of storage units designed for similar uses ;
- to help for the design and improvement of storage units developed either by laboratories or manufacturers ;
- to validate mathematical models.

DESIGN OF THE TEST FACILITY

This test facility has been jointly designed by CSTB and ENSMP, so as to be able to test a wide variety of storage units assigned to heating/cooling systems in building. Therefore, the following requirements had to be met :

- temperature range from -30°C to +90°C ;
- thermal power up to 15 kW (charging and discharging) ;
- flow rates up to 6 m3/h.

Besides, on account of specified tests to be carried out, the test facility was designed so that, on the one hand, step-wise flow rate or temperature changes could be achieved and, on the other hand, the storage inlet temperature could be precisely regulated at a set-point

DESCRIPTION OF THE TEST FACILITY (see figure 1)

This test facility has been built at the end of 1982 according to the previous requirements.

It consists of a cold circuit and of two identical test circuits that can be connected. Such an installation allows to create the most well defined temperature or flow rate step change.

1) Cold circuit

It consists mainly of a 15 kWth heat pump connected to a 1 m3 storage tank.

The fluid coming from the bottom of the tank is cooled in the evaporator of the heat pump and returns to the top of the tank so as to avoid thermal stratification.

The temperature of this loop can be regulated within the range -30°C to +10°C.

The transfer fluid used is a mixture of 50 % water an 50 % monoethylene glycol. The cold tank is used as an expansion tank for the whole test facility.

2) Test circuits

Each test circuit consists of :

- a centrifugal circulation pump (H = 10 m, NPSH = 1 m, Q_{max} = 6 m3/h) ;
- a 300 liter storage tank equipped with a 12 kW electric heater ;
- an inline 3 kW electric heater.

The useful flow rate (through the storage device under test) is adjusted from 0 to 6 m3/h by means of two two-way valves (or by a regulated pneumatic three-way valve on the second circuit).

The inlet temperature is regulated by simultaneous actions on the two electric heaters and on a pneumatic two-way valve that allows cold fluid from the cold circuit to be mixed with the returning fluid from the storage under test.

A set of four two-way valves allows the flow direction through the storage to be reversed without disconnecting the pipes.

3) Particular features

Apart from the regulated components mentioned above, all the valves are manual but could easily be modified to automatic.

Each test circuit has a relatively high thermal capacity, which enables the inlet temperature to be regulated more easily and a useful power to be supplied during a lapse of time if required.

The test loop is equipped with two kinds of flow meters placed in parallel : one orifice plate for higher flow rates and one turbine flow meter for smaller flow rates, giving a better accuracy across the whole range of flow rates.

TESTING PROGRAM

This test facility has been used within the framework of the CEC (SOLAR STORAGE TESTING GROUP) to establish recommendations for storage unit test procedures.

A 300 liters latent storage prototype (designed by ENSMP) with an embedded heat exchanger composed of PVC frames on which flexible tubes are regularly wound, has been tested (see picture 1) according to the

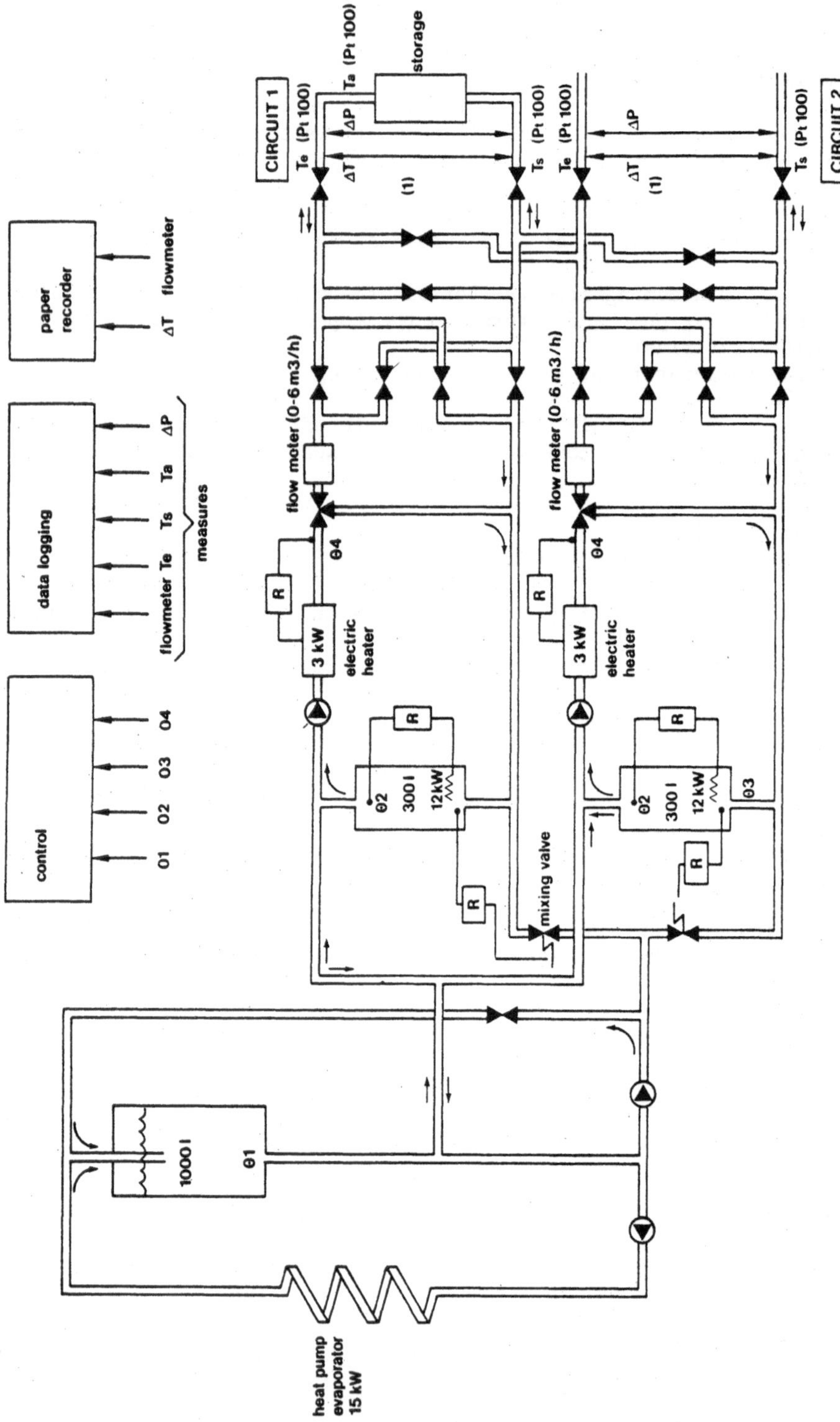

FIGURE 1 The test facility

procedure recommended by the SSTG (1).

The test of a French industrial storage unit (commercialized by CRISTOPIA S.A.R.L. for cooling applications) is now in progress (see picture 2). This storage device is a 1 m3 cylindrical tank containing about 2 500 polypropylene spheres filled with a phase change material.

The major part of the performed tests have given satisfying results, enabling to calculate some characteristic values of the storage devices with a good accuracy (such as the storage capacity, the heat loss capacity rate and the heat exchanger effectiveness), or to estimate various parameters (such as the thermal stratification in the flow direction).

Only a few results turned out to be unreliable, specially those concerning the heat loss capacity rate calculation at high flow rates or near ambient temperature.

CONCLUSION

The various tests already carried out on two storage units with widely different characteristics (one designed for heating applications around 40°C with low nominal flow rate, the other one designed for cooling applications around 0°C with high nominal flow rate) have shown that the test facility can meet the requirements previously listed, for all types of storage devices, even for thermochemical ones. In this latter case, the ability to tune independently for each test circuit the flow rate and the inlet temperature is used so as to provide a "cold" source and a "hot" source.

At least, further improvement of this test facility could be foreseen so as to be able to perform dynamic tests.

REFERENCE

(1) Recommendations for European Solar Storage Test Methods (sensible and latent heat storage devices). Technisch Physische Dienst TNO-TH, Delft, NETHERLANDS.

Picture 1 Storage prototype (designed by ENSMP) under test

Picture 2 Storage unit (commercialized by CRISTOPIA) under test

EVALUATION OF THE ANNUAL HEAT LOSSES OF AN IN-GROUND HEAT STORAGE

J.L. SALAGNAC
Centre Scientifique et Technique du Bâtiment
BP 21
06562 VALBONNE Cédex
France

Summary

The evaluation of the annual heat losses of an in-ground heat storage is of major importance to determine the performances of a solar heating system including such a storage (long term or short term). The volumetric heat loss coefficient G_{SE} ($W\ m^{-3}\ k^{-1}$) has proved to be a significant and sufficient parameter to determine the annual heat losses of a storage unit when the annual average outdoor temperature is taken as the reference temperature (refer to poster P.2.5.). This work aims to replace bulk numerical techniques by a simple method that leads to the determination of the volumetric heat loss coefficient G_{SE} of the heat storage. The walls of the in-ground heat storage are divided in typical elements, the heat losses of which have been predetermined by means of a finite element program. These typical elements take into account the thermal resistance of the storage wall (insulation + building material) and of the surrounding ground, and the geometrical parameters (depth, horizontal dimensions, tilt angle of the lateral walls). This set of typical elements can be increased when new storage configurations are encountered.

THE G_{SE} COEFFICIENT

Let us consider an in-ground insulated heat store (see figure 1). Heat transfer within the ground is purely conductive. The system is at a uniform temperature.

At time t = 0, a stepwise temperature change ΔT occurs at the inner surface of the storage unit.

The storage heat losses decrease with time and reach a steady state (see figure 2).

We then define a volumetric heat loss coefficient

$$G_{SE} = \frac{\text{heat losses under steady state}}{\Delta T \times \text{volume of the store}} \quad (Wm^{-3}\ K^{-1})$$

This coefficient is very useful to determine the annual heat losses of an in-ground storage in a system when a thermal stationnary state in the ground surrounding the storage has been reached.

This can be checked when comparing the results of the two following calculations :

1) due to the linearity of equations the previously mentionned heat losses function (fig 2) can be used in a convolution method to evaluate the storage annual heat losses when the storage temperature is time dependant.

2) the annual heat losses are evaluated from :

$$8760 \times G_{SE} \times V_{SE} \times (\overline{T}_{SE} - \overline{T}_{a}) \text{ (Wh/year)} \qquad (1)$$

$$\text{(h/year)} \times Wm^{-3} \quad K^{-1} \times m^{+3} \times K$$

where $\overline{T}_{SE}$ = average annual storage temperature

$\overline{T}_{a}$ = average annual ambiant temperature.

The plot of these results shows a very good agreement between the two methods (see figure 3).

The G_{SE} coefficient is then a very simple and significant characteristic of an in-ground storage unit.

To evaluate this coefficient we only need the steady state heat losses of the storage (and no transient calculation).

It must be pointed out that this coefficient cannot be used to determine heat losses during a shorter period than the cycle duration of the system in which the storage is used.

This means that, in the case of a seasonal heat storage, monthly or dayly heat losses evaluated from a similar expression to (1) would be false.

EVALUATION OF THE G_{SE} COEFFICIENT

As far as simple geometric shapes are involved (cylinders with vertical or horizontal axis), the G_{SE} coefficient of an uninsulated in-ground storage unit can be determined from existing results once developped for other purposes (electrostatic, heat losses from buried pipes).

When insulation is considered, we can add, as a first approximation, the thermal resistance due to the layer of insulating material to the thermal resistance previously determined. This is done, for example, to evaluate heat losses from a buried insulated pipe.

When other storage shapes are considered (parallelepipeds, truncated pyramids, ...), this method can no more be used and we can either make a complete description of the problem using a suitable method (analogy, 2D or 3D finite element calculation, ...) or try to develop a wider set of typical shapes.

This latter point is being developped and results are already available for a buried storage wall, the tilt angle with the horizontal of which varies from 0 to π radians (see figure 4).

This allows the evaluation of the heat losses of a storage unit as the sum of the heat losses on each face.

The comparison of this global result to the evaluation made from a complete description of the problem shows that a particular attention has to be paid to corners where distorsion of heat flux lines might be important.

Nevertheless, this method of approach looks encouraging and results could be used to evaluate heat losses of other buried structures (basements of houses for instance).

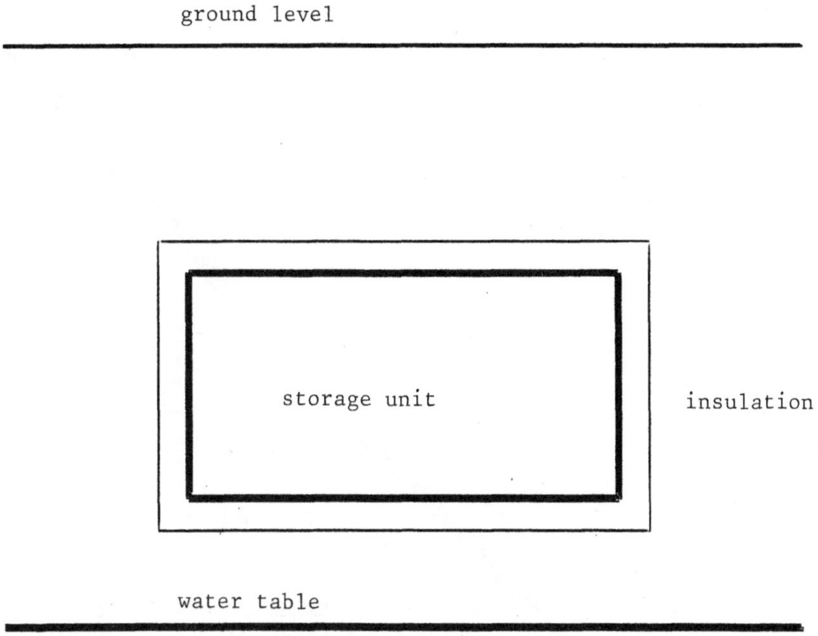

Fig. 1

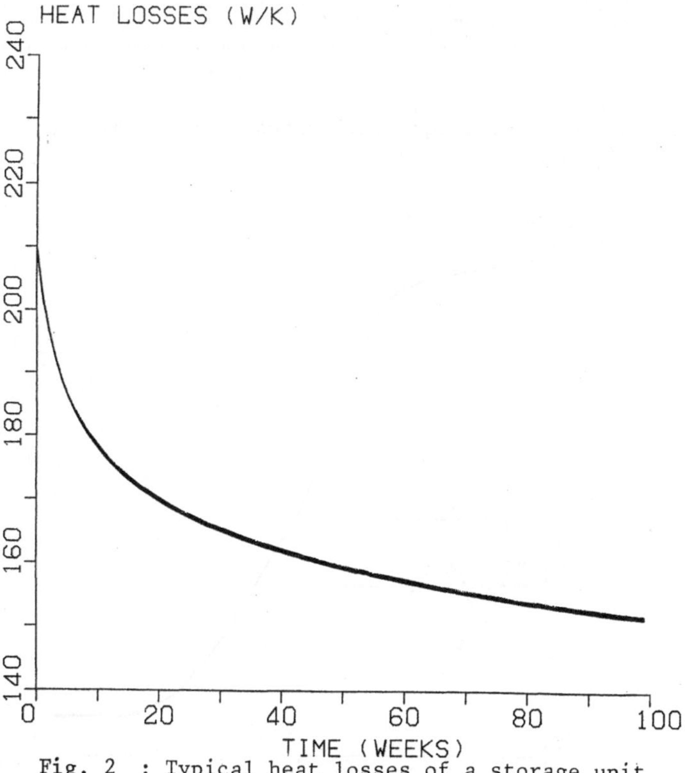

Fig. 2 : Typical heat losses of a storage unit undergoing a temperature step

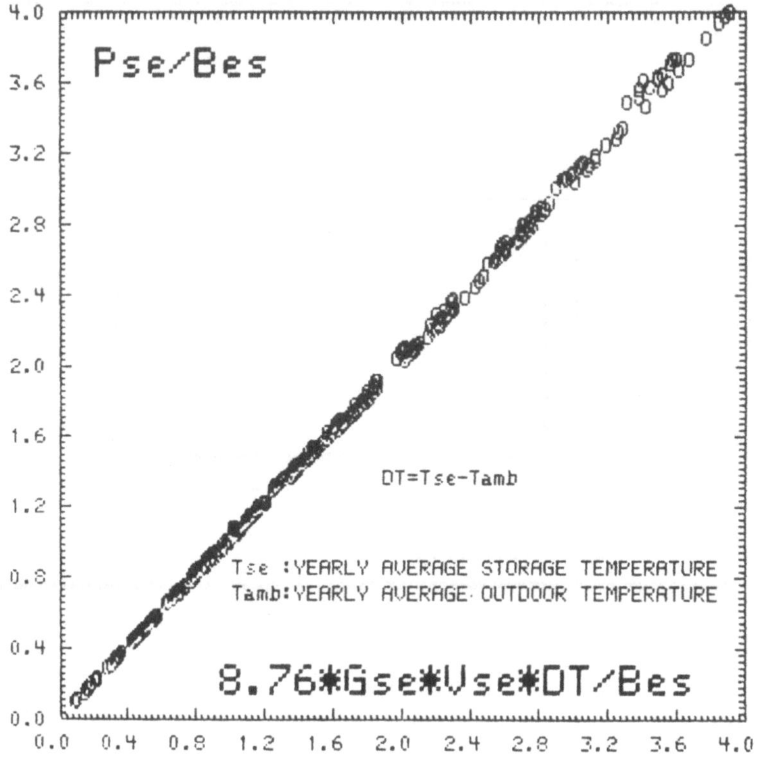

FIG. 3 Evaluation of storage heat losses by means of a volumetric coefficient G_{se}

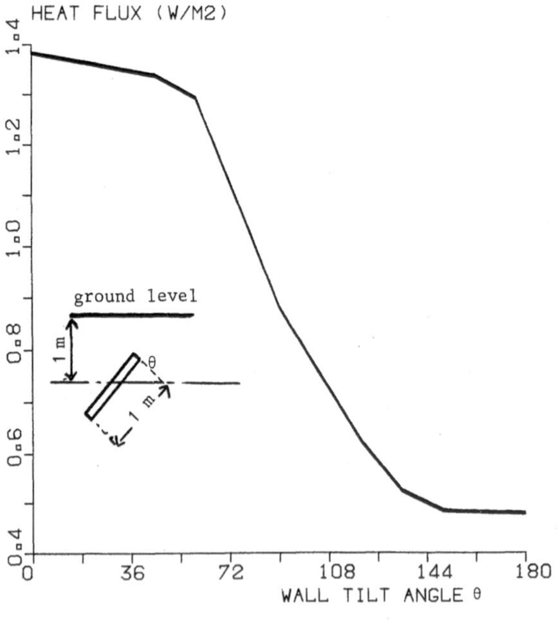

FIG. 4

SIZING A LATENT HEAT STORAGE WITH SHELL AND TUBES HEAT EXCHANGER

A NEW SIMPLIFIED METHOD

D.LECOMTE* - D.MAYER

Ecole Nationale Supérieure des Mines de Paris

Centre d'Energétique-Sophia Antipolis
06565 VALBONNE CEDEX
FRANCE

Summary

The sizing of a latent heat storage device requires to take into account the system in which it will be included.The association of two heat-exchangers in parallel is considered in terms of a minimum power ratio (e.g. the ratio between the minimum power transfered and the maximum transferable power).The sizing problem ,developed in the case of the shell-and-tubes heat-exchanger ,may be expressed in the determination of the length of the tubes and of the space step of the grid in order to deliver a minimal power during a fixed period.This method is very fast and can be used on a pocket calculator.

1.INTRODUCTION

Using a thermal energy storage (TES) in a system implies energy degradation (mainly due to heat-losses) and exergy degradation (mainly due to heat-transfer between the storage and the system). Phase change materials (PCM) show a great theoretical interest since they allow lower temperatures in the thermal systems and smaller volumes .In practice the heat-exchange problem is fundamental when dealing with solid-solid or solid-liquid transitions and in most cases the incompatibility between the storage medium and the heat-transfer fluid (HTF) requires a heat-exchanger (HE).The energy of transition is available at phase change temperature only at the interface between the two phases. Natural convection occurs in the charging mode and therefore the limiting process is the discharge mode which is purely conductive.Sizing a heat-exchanger also needs a sound knowledge of the whole system in which the latent heat storage(LHS) will be included (estimation of the power rates,nature of the other components of the system...).

The aim of this paper is to present a new simplified method in the design of a LHS-HE in a thermal system with a special attention to the shell-and-tubes case.

2.HEAT-EXCHANGERS IN PARALLEL.

Considering two heat-exchangers immersed in two infinite media and connected by a thermal loop ,we assume the steady state conditions are reached in the system (Fig. 1).The two HE allow to transfer a power P between the two media. Knowing the thermal characteristics of HE1 (Global heat transfer coefficient $(UA)_1$) we wish to calculate the characteristics of HE2 as well as the flowrate in the thermal loop.

---

*Post-doctoral fellow of the EEC
Present Address : Dipartimento di Tecniche dell'Edilizia e del Controllo Ambientale.Università di Roma.Via Eudossiana,18.00184 ROMA -ITALIA

The maximum transferable power P between the two media is obtained for an infinite $(UA)_2$ and an infinite flowrate. It can be written:

$$P_{max} = (UA)_1 \; (T_2 - T_1)$$

These two conditions being unrealistic, optimal values of $(UA)_2$ and $\dot{m}$ are to be determined according to thermodynamical and technico-economical criteria /1/. The power actually transfered has the following expression:

$$P = \dot{m}c_F \; (T_2 - T_1) \; \frac{(1-\exp(-(UA)_1/\dot{m}c_F)).(1-\exp(-(UA)_2/\dot{m}c_F))}{1-\exp(-(UA)_1/\dot{m}c_F-(UA)_2/\dot{m}c_F)} \quad (1)$$

Assuming the following adimensionnal terms:

$$NTU_i = (UA)_i/\dot{m}c_F \qquad \text{and} \qquad r = P/P_{max}$$

Equation (1) is reduced to:

$$r = \frac{1}{NTU_1} \; \frac{(1-e^{-NTU_1})\;(1-e^{-NTU_2})}{1-e^{-NTU_1-NTU_2}} \quad (2)$$

3. ANALYSIS

The case of TES in solar heating or cooling systems is particularly relevant since in both full charge or full discharge mode s ,the configuration studied in §2 is reached.

Let us consider a fully loaded (e.g. liquid) LHS which is connected to a heating element.The value of $(UA)_2$ is to be considered as a minimal value and the sizing problem can be stated in the following way : how to determine the surface of HE2 and the volume of the TES so as to perform a minimal power during a fixed period.

Previous works on embedded heat-exchangers /3/ have shown the advantage of the cylindrical geometry with respect to the plane one . For the typical shell and tube heat-exchanger,the two unknowns of the sizing problem can be replaced by the total length of the tubes l and the space-step of the grid p (Fig.2) . A simplified model of the LHS-HE is developed with the following assumptions :

A1 : Heat capacity effects of the tubes are neglected.
A2 : Heat capacity of the HTF are neglected.
A3 : Initial overheating is neglected. The initial temperature of the medium is the phase change temperature : T_c
A4 :Sensible heat in the PCM is neglected (Ste=0).
A5 : No significant interference between adjacent tubes exists as long as the fusion radius has not reached half the space-step.
A6 : The heat exchanger may be approximated by a single tube of length l.
A7 : To account for the extra-volume left appart by the assumption A5 , the space-step is multiplied by a corrective factor such as : $p_c = 1.0501\,p$ (see Ref. /5/ and Fig. 3)
A8 : The bulk temperature of the HTF only depends on time and the z-coordinate (Fig.4) .
A9 : The heat-losses are negligible.
A10 : The Biot number is high enough to neglect the conduction along the z-axis.
A11 : The thermophysical properties of the HTF and the PCM are constant.

A12 : The freezing front position s is independant on z.

The variables of the problem are the front radius s(t) and the HTF temperature $T_F(z,t)$.

The heat-balance of the PCM is written:

$$2\pi s \frac{ds}{dt} \rho L = \frac{T_2 - T_{Fm}(t)}{R} \quad (3)$$

where T_{Fm} is the mean temperature of the HTF (A12)

The heat balance in the HTF gives:

$$\dot{m} c_F \frac{\partial T_F}{\partial z} = \frac{T_2 - T_F(z,t)}{R} \quad (4)$$

where R is the thermal resistance of the solidified shell.

The following dimensionless variables may be introduced:

$\tau = (T_2 - T_1)t / \rho s_o^2 L$	:	dimensionless time.
$n = 2\pi s_o hz / \dot{m} c_F$	:	dimensionless length
$\sigma = s/s_o$	:	dimensionless front.
$Bi = hs/\lambda$	:	Biot number.
$\theta = (T_2 - T_F(z,t))/(T_2 - T_1)$	:	dimensionless HTF temperature.

Equation (3) and (4) become :

$$\sigma \frac{d\sigma}{d\tau} = \frac{Bi\, \theta_m}{1 + Bi\, Ln\sigma} \quad (5) \quad \text{where} \quad \theta_m = \frac{1}{n_o} \int_1^{n_o} \theta(n,\sigma)\, dn$$

$$\frac{\partial \theta}{\partial n} = \frac{-\theta}{1 + Bi\, Ln\sigma} \quad (6)$$

The initial and boundary conditions are : $\sigma(\tau = 0) = 1$

and :
$$\theta(n=0,\tau) = \frac{1 - e^{-NTU_1}}{1 - e^{-NTU_1 - NTU_2(\sigma)}} \quad \text{where } NTU_2(\sigma) = \frac{n_o}{1 + Bi\, Ln\sigma}$$

The temperature θ and θ_m are expressed in terms of the front position :

$$\theta(n,\tau) = \frac{1 - e^{-NTU_1}}{1 - e^{-NTU_1 - NTU_2(\sigma)}}\, e^{-\frac{n}{1 + Bi\, Ln\sigma}} \quad \text{and}$$

$$\theta_m(\sigma) = \frac{1}{NTU_2} \frac{(1 - e^{-NTU_1})(1 - e^{-NTU_2(\sigma)})}{1 - e^{-NTU_1 - NTU_2(\sigma)}}$$

From (5) we obtain the implicit expression of σ in function of τ

$$\tau = \frac{n_o}{Bi} \int_1^{\sigma} \frac{u\, du}{NTU_1\, r(u)} \quad (7)$$

where r(u) is defined by (2)

4. THE SIZING PROBLEM SOLUTION.

The TES is initially liquid ($\sigma=1$). The power ratio is initially at its highest value :

$$r_{init} = \frac{(1 - e^{-n_o})(1 - e^{-NTU_1})}{(1 - e^{-n_o - NTU_1})} \quad (8)$$

when σ reaches the value σ_m after time τ_o, the power ratio has decreased to the value:

$$r_{min} = \frac{(1-e^{-\frac{n_o}{1+Bi.Ln\,\sigma_m}})(1-e^{-NTU_1})}{(1-e^{-\frac{n_o}{1+Bi.Ln\,\sigma_m} - NTU_1})} \qquad (9)$$

The values of NTU_1 and r_{min} are obtained from technico-economical criteria /1/. Bi and τ_o are characteristics of the storage.The two unknowns σ_m and n_o are determined from (7) and (9) by an iterative method. For a constant Biot number corresponding to a reasonnable domain of flowrate, so as to perform turbulent flow in the HE and for a constant minimal power ratio (r_{min}=0.90) ,Figure 5 shows the variations of the ratio n_o/NTU proportionnal to the length of the HE and Figure 6. shows the evolution of σ_m proportionnal to the space-step of the grid , versus the adimensional time :

$$l = \frac{n_o\ (UA)_1}{2\pi\lambda\ Bi} \qquad \text{and} \qquad p = \frac{2\ s_o\ \sigma_m}{1.0501}$$

The number of tubes in parallel are then established so as to maintain pressure drops within reasonnable limits.

5.CONCLUSION.

The method described in the present article has been used in parallel with a fully descriptive finite difference model correlated with experimental data. The values obtained by the method were put into the model and in spite of the rough assumptions, the results were excellent. Details of the procedure are found in Ref./2/. The new simplified model is used for the sizing of a LHS unit which replaces a sensible heat storage in an already defined active solar system.

NOMENCLATURE

c_F	Heat capacity of the HTF	J/kg°C	s_o	Outside radius of pipe	m
h	Heat transfer coef.	W/m2°C	t	Time	s
l	Length of HE2	m	λ	Thermal conductivity	W/m°C
L	Latent heat of PCM	J/kg	ρ	Density of PCM	kg/m3
$\dot{m}$	mass flowrate	kg/s	T_i	Temperature of medium i	°C
NTU_i	number of transfer units of HE_i		R	Thermal resistance of solidified layer	°Cm/W
p	space-step of the grid	m	$(UA)_i$	Global heat transfer coef.	W/°C

REFERENCES.

/1/ D.Lecomte, D.Mayer : "A Design Method for sizing a Latent Heat Storage Heat Exchanger in a thermal system." Submitted to Applied Energy.

/2/ D.Lecomte, "Etude d'accumulateurs-echangeurs thermiques contenant des materiaux à changement de phase à basse temperature (20°C-80°C)" These de Docteur-Ingénieur - ENSMP - 30 Juin 1983.

/3/ P.Achard,D.Lecomte,D.Mayer, "Characterization and modelling of TES units using salt hydrates".,Int. Conf. on Energy Storage,Brighton,UK,April 29-May 1st 1981.

/4/ P.Achard,B.Amann,D.Mayer,"Solar heating test design facility for PCM Bulk Storage." First EC Conference on Solar Heating ,Amsterdam,April 30-May 4 ,1984.

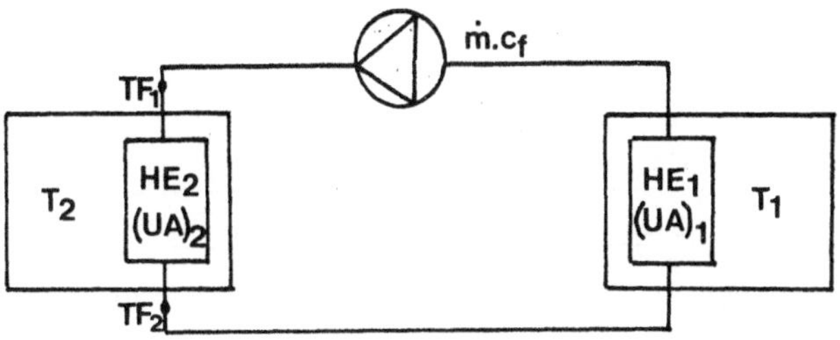

figure 1

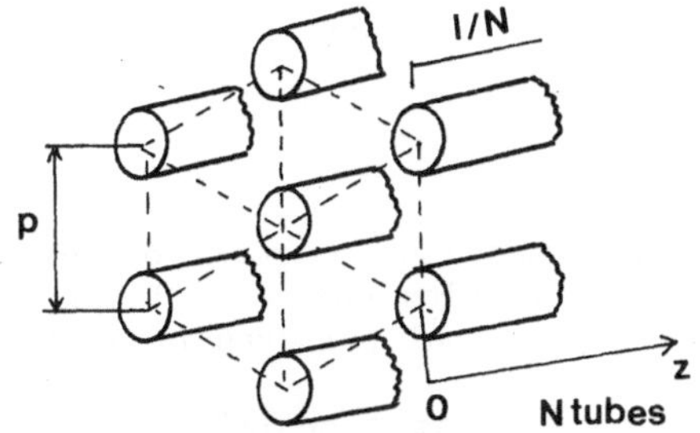

figure 2

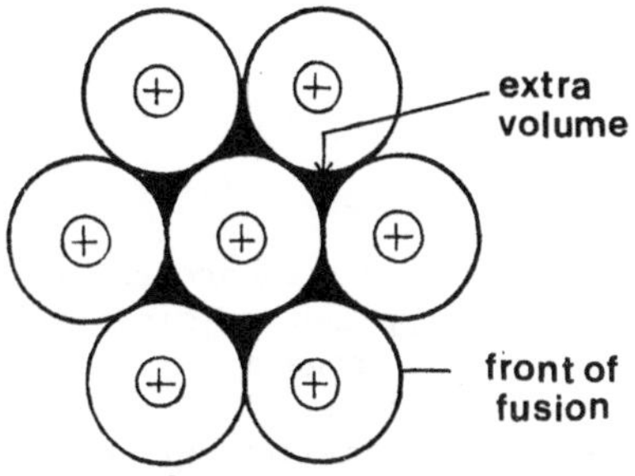

figure 3

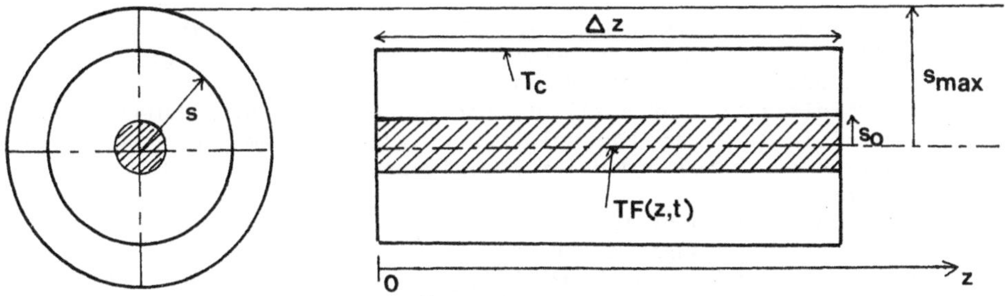

figure 4

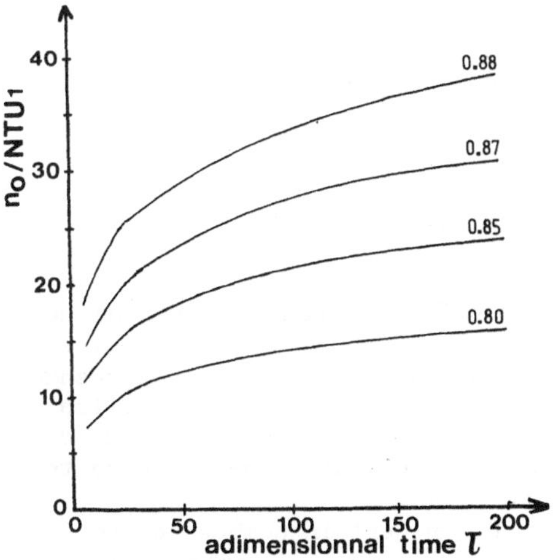

figure 5

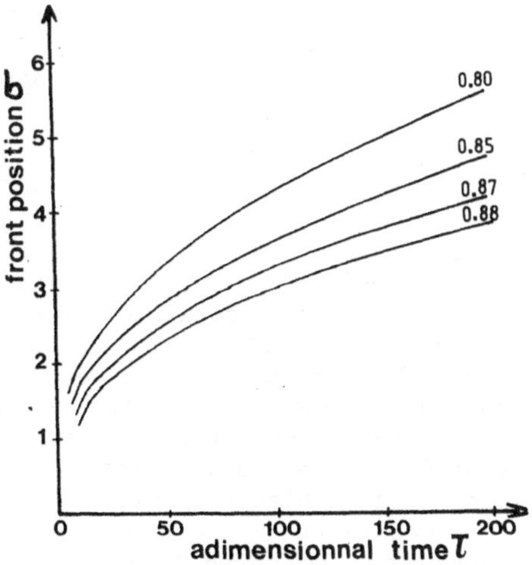

figure 6

OPTIMAL CONTROL STRATEGY OF SOLAR HEATING SYSTEMS USING A LONG TERM HEAT STORAGE

M. BOUCHER - M. POTTIER
GAZ DE FRANCE
2 Rue Curnonsky
75017 PARIS
FRANCE

Y. LENOIR - R. LIDIN
ARMINES - CAI
35 Rue St Honoré
77305 FONTAINEBLEAU
FRANCE

JM. CARDI - P. MORAND
ARMINES - CENERG
Rue Claude Daunesse
06565 VALBONNE
FRANCE

Summary

Some recent researches point out the importance of optimal control theory applied to sizing and management of complex energetical systems with regard to energy saving. The principal purposes of this research are:

- the estimation of the energy gain expected from optimizing the control of a given gas/solar heating system using a long term heat storage in the ground.
- the development of a design tool for these systems able to take optimal strategies into account

The methods now under study aim to:

- the minimization of auxiliary energy consumption. Bellman's Optimum Principle leads to the solution by yearly dynamic programing (storage temperature) and daily dynamic programing (dwelling , temperature)
- the minimization of the maximum required level of auxiliary heating power by a "MIN-MAX" algorithm

Present researches lead to design and control tools taking into account the complementarity of collective and private investments. Computer simulations have been performed to improve control algorithms robustness and to assess the efficiency of such methods with regard to conventional controllers

Computed results show that for the studied system energygains around 20% for auxiliary energy consumption can be obtained through optimized management.

1. INTRODUCTION

Une analyse détaillée des systèmes énergétiques dans le domaine de l'habitat (chauffage, climatisation, production d'eau chaude sanitaire) montre l'importance que revêt le contrôle des transferts thermiques quant aux performances des dits systèmes. Prenant ce fait en considération, des études théoriques et expérimentales ont été menées sur le plan du contrôle optimal d'une part et sur celui de la modélisation des composants énergétiques d'autre part.

Ces études ont permis de mettre au point un certain nombre de procédures de calcul des stratégies optimales de commande pour des systèmes de chauffage solaire comportant un stockage long terme.

2. ASPECTS GENERAUX DU CONTROLE OPTIMAL

Le problème spécifique posé par un système de chauffage solaire est de satisfaire aux contraintes de confort fixées au niveau de l'utilisation en fonction de la réalisation du processus météorologique à court et à long terme. Les systèmes énergétiques étudiés sont constitués de composants réalisant des transferts thermiques et d'éléments de stockage.

Les temps caractéristiques de ces différents composants sont de l'ordre:

- de l'année pour la gestion d'un stockage à long terme
- de la journée pour celle d'un stockage à très court terme
- de quelques heures pour la régulation en temps réel au moyen des composants de transfert thermique (plancher chauffant, murs...)

Les lois de commande mises en oeuvre et testées à ce jour permettent de faire suivre à l'état du stock de chaleur n'importe quelle trajectoire optimale, au sens de la minimisation de la consommation d'énergie commerciale, comprise entre une trajectoire inférieure correspondant à l'absence de contrainte sur la puissance d'appoint disponible et une trajectoire supérieure relative à la puissance d'appoint installée minimale pour passer la saison de chauffage sans transgresser les contraintes de confort.

Les résultats obtenus ont montré que l'optimisation du dimensionnement selon un critère économique prenant en compte les coûts d'investissement tant privés que collectifs, était, en première approximation découplable de celle précédemment décrite de la gestion du stock de chaleur.

Le dimensionnement est alors effectué par une méthode de gradient simple ou à pas de temps variable sur les principaux paramètres techniques de l'installation, celle-ci étant, à chaque pas, simulée, équipée de la puissance d'appoint minimale correspondante (algorithme de MIN-MAX).

Les premières études ont porté sur le système suivant (figure n°1). Dans ce système considéré comme la référence dans la suite de l'étude le stockage long terme joue un rôle de tampon entre les capteurs et l'habitation. Les pompes sont commandées par tout ou rien et le prix de l'énergie auxiliaire est supposé constant tout au long de l'année.

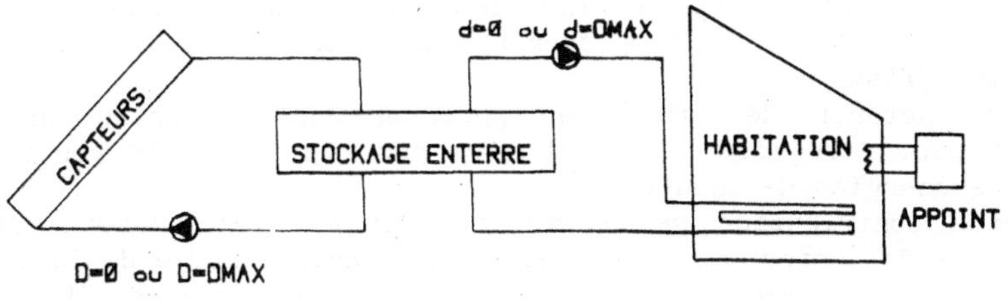

Figure 1: SCHEMA DE PRINCIPE DU SYSTEME DE REFERENCE

3. PRESENTATION DES PROCEDURES DE CONTROLE OPTIMAL

3.1 Régulation optimale instantanée des transferts thermiques

Un premier niveau de commande consiste à maximiser l'énergie solaire captée tout en satisfaisant au besoins instantanés de chauffage. La figure n°2 présente le système sous contrôle optimal. L'optimisation repose sur le choix des positions des vannes (10 cycles hydrauliques) et sur la valeur des débits supposés variables.

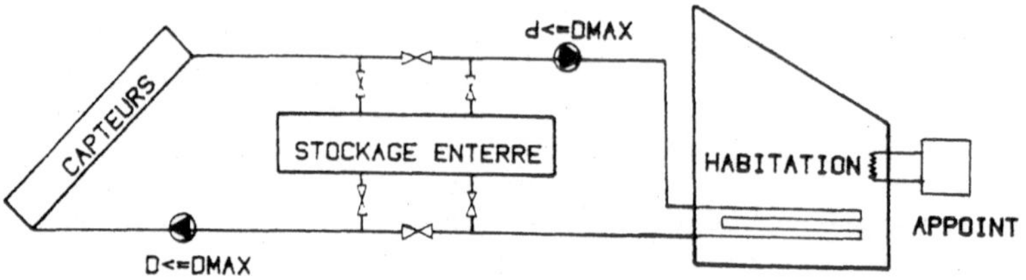

Figure n°2: SCHEMA DE PRINCIPE DU SYSTEME SOUS COMMANDE OPTIMALE

Les variables d'état du système sont:

- la température équivalente de la masse thermique que représente le bâtiment
- la température du stockage long terme supposé non stratifié

3.2 Commande optimale du stockage long terme

L'objet de ce deuxième niveau de commande est de minimiser la consommation en énergie d'appoint sur l'année, la température du bâtiment étant supposées strictement régulée.

Pour ce faire on est conduit à définir un prix relatif, noté, de l'énergie stockée par rapport à celui de l'énergie d'appoint, constant dans le temps et pris égal à 1 comme référence.

Un algorithme itératif, traduction du principe d'optimalité de Bellman, permet de calculer par programmation dynamique annuelle la température du stockage long terme cette fonction de prix, fonction du temps et de la température du stockage, qui minimise l'intégrale sur l'année de l'énergie d'appoint consommée.

Le calcul de la fonction de Bellman λ(temps, température du stock) est effectuée avec les enregistrements météorologiques d'une année de référence (hypothèse, vérifiée par les simulation, de la stationnarité annuelle du processus météorologique vis-à-vis de la gestion optimale du stockage long terme).

3.3 Optimisation des apports de chaleur au bâtiment

Dans le cadre d'une régulation seule, l'appoint n'est utilisé que pour maintenir la température intérieure de l'habitat à sa valeur de consigne lorsque l'énergie en provenance du stockage et/ou des capteurs se révèle insuffisante.

Pour obtenir de meilleures performances, l'inertie thermique du bâtiment peut être mise à contribution pour optimiser la gestion à court terme des énergies de chauffage.

Cette contribution se concrétise par des surchauffes limitées pour des raisons de confort, par une valeur de consigne haute de la température intérieure du bâtiment. Afin d'être aussi brèves que possible, ces surchauffes sont réalisées à débit maximum.

La figure n°3 présente le principe de ces surchauffes anticipées du bâtiment qui permettent de retarder l'utilisation de l'énergie d'appoint.

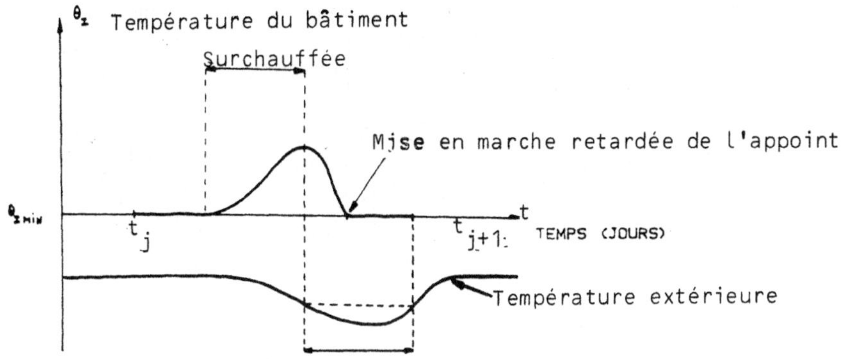

Figure n°3: OPTIMISATION DES APPORTS DE CHALEUR AU BATIMENT

A chaque instant le coût J_j qui est affecté à la fourniture $W_I(t)$ de chaleur au bâtiment prend la forme d'un coût généralisé.

$$J_j = \int_{tj}^{tj+1} (W_a(t) + \lambda J\, W_{II}(t))\, dt$$

λj : valeur de la fonction de Bellman au jour j
$W_I(t)$: besoins en chauffage du bâtiment
$W_a(t)$: énergie d'appoint
$W_{II}(t)$: énergie transmise au bâtiment provenant du stockage long terme
$W_c(t)$: énergie solaire captée directement transmise au bâtiment

La relation suivante est vérifiée à chaque instant t: $W_I(t)=W_a(t)+W_c(t)+W_{II}(t)$

4. PREMIERS RESULTATS DE SIMULATION

Des simulations ont été faites sur différentes années météorologiques pour les stations d'Agen et de Trappes et pour différents dimensionnements du système présenté par la figure n°2 (variation de la surface de captage, du volume de stockage, de l'inertie du bâtiment).

Le premier niveau de commande (régulation optimale instantanée des transferts thermiques) a montré tout le bénéfice retiré par les différents cycles hydrauliques, (notamment la liaison directe entre les capteurs et le bâtiment) ainsi que les débits variables dans les circuits.

La figure n°4 présente les résultats obtenus pour des simulations effectuées avec les données météorologiques d'Agen 1973 et le dimensionnement suivant du système:

- surface de captage 240 m2
- volume du stockage 600 m3
- inertie du bâtiment 0.28 kWh/m3.°

SYSTEME DE REFERENCE — REGULATION OPTIMALE INSTANTANEE DES TRANSFERTS THERMIQUES — COMMANDE OPTIMALE DES APPORTS DE CHALEUR AU BATIMENT

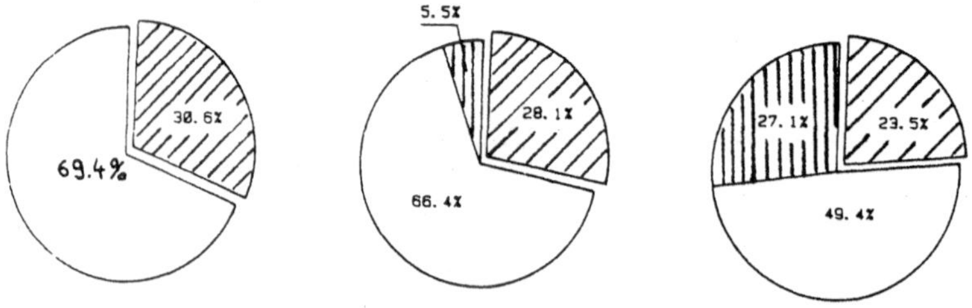

Dans ce cas les économies sont de l'ordre de 20% sur la consommation en énergie d'appoint par rapport au système de référence

- (hachuré) Energie d'appoint
- (blanc) Energie délivrée au bâtiment en provenance du stockage
- (rayé vertical) Energie délivrée au bâtiment en provenance des capteurs (liaison directe)

Figure n°4

5. CONCLUSION

Au vu des résultats acquis une étude liée à l'applicabilité du travail méthodologique fourni en amont doit être engagée en relation avec des projets de réalisation.

Celle-ci devrait permettre dans un délai relativement bref de confirmer les résultats théoriques obtenus et de préciser les technologies de régulation à mettre en oeuvre.

REFERENCES

/1/ Commande optimale des chauffages solaires avec appoint indépendant
Y. LENOIR - L.R GUO CAI de l'ENSMP
Rapport GDF -DETN Février 1982

/2/ Optimisation des systèmes énergétique: évaluation du gain à attendre d'une commande optimisée d'un système de chauffage solaire-gaz
JM. CARDI CENERG de l'ENSMP
Rapport GDF-DETN Juin 1982

/3/ Optimisation statique des installations de chauffage solaire
Y. LENOIR
Revue de Physique Appliquée, Tome 17 Septembre 1982

/4/ Etude d'un système de chauffage solaire à stockage long terme sous commande optimisée
JM. CARDI - P. MORAND
Rapport GDF-DETN Octobre 1983

EXPERIMENTATION D'UN SYSTEME DE CHAUFFAGE UTILISANT POMPE Â CHALEUR ET STOCKAGE DE GLACE (résultats 82/83)

P. BREJON - D. MARCHIO
Centre d'Energétique - Ecole des Mines de Paris
60, bld St Michel 75006 - PARIS

Summary

Le hall expérimental des Ulis a pour but de suivre finement le comportement du système GLASOLTHERM. Ce système utilise des échangeurs climatiques en source froide d'une pompe à chaleur. L'originalité principale du système est d'utiliser un stockage en source froide rendu compact par l'utilisation de la transformation eau→glace. La saison de chauffe 1982-1983 a permis de caler les régulations du système et de vérifier le bon dimensionnement des composants. La période froide ayant eu lieu en février 1983 (température extérieure atteignant -5°C environ) a été entièrement suivie et 25 voies de mesure stockées heure par heure durant toute la séquence. L'analyse de ces relevés conduit aux résultats suivants : COP moyen sur la période 2,86 pour une température extérieure moyenne de 3,6°C ; COP mini journalier obtenu pour -2°C extérieur égal à 2,6 : le stock a été sollicité durant cette séquence pendant une vingtaine de jours. Par ailleurs, une saisie de données plus fréquente (toutes les 2, 3 minutes) sur des séquences précises a complété la bonne connaissance du système sous le rapport des performances des échangeurs climatiques (B=0,75, K=25 W/m^2/°C) et sur les régimes transitoires côté source froide.

L'expérimentation se poursuit en 1983/1984 mais cette fois en continu c'est-à-dire sans réglage ou modification de séquences de régulation.

An experimental building in Les Ulis was constructed to study the behaviour of the GLASOLTHERM system. Climatic exchangers (aero-solar collectors) are used as the cold source for a heat pump. The main original aspect of this system is that it includes a compact cold storage using the water/ice transformation.
The 1982-1983 heating season permitted the system control to be defined and components sizes to be verified.
The coldest period occured during February 1983 (lowest outdoor temperature reached -5°C). During this period, measurements were made and 25 pieces of data stored each hour. Analysis of the data Yielded an average COP of 2.86 for an average outdoor temperature of 3.6°C. Values were averaged over daily periods and the minimum COP was 2.6 wich occured when average ambient temperature was -2°C. The storage had to be used approximately 20 days. During specific sequences, data were aquired more frequently (every 2 or 3 minutes). This allowed calculation of the climatic exchangers characteristics ($\tau\alpha$= 0.75 ; U_L = 25 W/m^2/°C) and observation of transients on cold source side.
The experimental is beeing monitored during 1983/1984 without modification of the control scheme.

1 - LE SYSTEME GLASOLTHERM

Le système GLASOLTHERM satisfait les besoins de chauffage et d'eau chaude sanitaire grâce à une pompe à chaleur puisant ses calories grâce à des échangeurs climatiques récupérant tant le rayonnement solaire que la chaleur de l'air ambiant extérieur. Pour les températures extérieures froides et en l'absence d'ensoleillement on maintient un coefficient de performance supérieur à 2,5 grâce à l'utilisation en source froide d'un stockage de glace. L'énergie solaire sert à charger le stock quand elle est excédentaire ; le stock est déchargé en période froide à une température constante de 0°C dès la prise en glace.

2 - LE HALL EXPERIMENTAL DES ULIS

Ce hall, achevé à l'automne 1982 a pour but de suivre le comportement du système en fonctionnement réel. L'année 82/83 particulièrement a servi au calage des régulations et à la vérification du dimensionnement des composants. En effet, ces 2 aspects avaient été mis au point grâce à un programme de simulation et ont nécessité des modifications au cours de la saison de chauffe. L'année 83/84 fournira des renseignements plus globaux sur le fonctionnement de l'installation. L'expérimentation a également permis d'observer la bonne marche du système en période critique (pointe froide de l'hiver).

Le hall expérimental est représenté sur la figure 1. Il occupe un volume de 2400 m^3 et est chauffé par 2 machines GLASOLTHERM (le bâtiment étant divisé en 2) fournissant chacune 15 kW thermiques par -7°C extérieur. Le dimensionnement correspondant est le suivant : 40 m^2 d'échangeurs climatiques de type FINIMETAL disposés verticalement (voir photo) en façade Sud Sud-Est et ventilés naturellement sur leur face arrière. Le stock de glace est réalisé dans une cuve de 10 m^3.

3 - LA CHAINE D'ACQUISITION DE DONNEES

L'acquisition se fait toutes les heures en régime normal, toutes les minutes éventuellement pour des "loupes" particulières. Pour les enregistrements horaires 25 mesures sont stockées sur cassettes magnétiques grâce à un micro-ordinateur APPLE 2 pilotant une centrale de mesures FLUKE 2400A En régime "loupes" 19 voies sont mesurées et stockées. Ces loupes concernent essentiellement le fonctionnement des échangeurs climatiques et l'analyse du fonctionnement du système côté source froide.

4 - ANALYSE DE LA PERIODE FROIDE DU 12 JANVIER AU 23 FEVRIER 1983

Cette période correspond à la séquence la plus froide de l'hiver, particulièrement au mois de février. La température moyenne est de 2,5°C. La figure 2 illustre l'évolution du coefficient de performance (2) avec la température extérieure (1). Le coefficient de performance est défini ainsi :

$$\mathrm{COP} = \frac{\sum \text{énergies thermiques produites}}{\sum \text{consommations électriques}} \quad \begin{matrix}\text{(chauffage + ECS)}\\ \text{(compresseur, auxiliaires)}\end{matrix}$$

Les auxiliaires sont les suivants : circulation ECS 130 W, pompe de relevage 276 W, circulateur chauffage 119W, circulateur échangeurs climatiques 450W, ventilateur de soufflage d'air neuf 208W, alimentation des vannes électromagnétiques 87W. LE COP s'établit en moyenne sur la période à 2,79. On note également sur la figure la valeur de l'énergie fournie journellement (3). La valeur maximale de l'échelle : 350 kWh correspond à un fonctionnement sur 24h à puissance maximale : 15 kW. Cette valeur est quasiment atteinte le 28ème jour de la période : le 13 février où la machine a fourni en moyenne 14 kW, la température extérieure étant de -2°C. Cette journée est donc représentative d'une journée critique aussi le COP est-il significatif : 2,6.

La figure 3 a pour but d'illustrer l'utilisation du stockage sur la même période. On trouve tout d'abord la valeur de la température du stock (4) ; la période deprise en glace étant très nette : pallier de température à 0°C (du 10 février au 23 février 83). La courbe (6) représente l'énergie amenée au stock. L'aire située au-dessus de zéro correspond aux apports et l'aire située au-dessous aux prélèvements. Il est difficile de connaître la quantité de glace formée sur la période mais on peut avancer l'hypothèse de 4 tonnes environ.
Cette quantité de glace n'est pas représentative d'un fonctionnement normal puisque nous avons volontairement fait fonctionner la machine en fabrication de glace durant cette période. Les renseignements seront plus intéressants sur la saison 83/84 en fonctionnement continu et normal. La courbe (5) correspond aux prélèvements à l'évaporateur qu'ils soient d'origine solaire ou du stock.

5 - PERFORMANCE DES ECHANGEURS CLIMATIQUES

5.1 - Coefficient d'échange

Ce coefficient d'échange par convection a été tout d'abord déterminé en périodes non ensoleillées (essentiellement nocturnes). Ces périodes sont toutes non ventées. Les différentes séquences analysées conduisent à la valeur moyenne de 25,5 $W/m^2/°C$. Nous disposons également de résultats en présence de givre. On obtient un coefficient d'échange de 20,7$W/m^2/°C$. Le givrage se traduit donc par une diminution des performances de 26 % environ confirmant l'ordre de grandeur obtenu sur d'autres types d'échangeurs climatiques.

5.2 - Coefficient d'échange. Facteur optique

La puissance captée s'écrit :

$$\phi_s = BE_n + KS(T_{ext} - \frac{T_e + T_s}{2})$$

E_n étant l'ensoleillement
S la surface
T_{ext}, T_e, T_s les températures extérieure et d'entrée sortie.

On recherche le coefficient d'échange $K(W/m^2/°C)$ et le facteur optique B par une régression à 2 variables sur E_n et $(T_{ext} - \frac{T_e + T_s}{2})$. Nous donnons le résultat de la régression dans le diagramme rendement (ϕ_s/E_n), température réduite $(T_{ext} - \frac{T_e + T_s}{2}) / E_n$ habituel pour les capteurs solaires : figure 4.

FIGURE 1

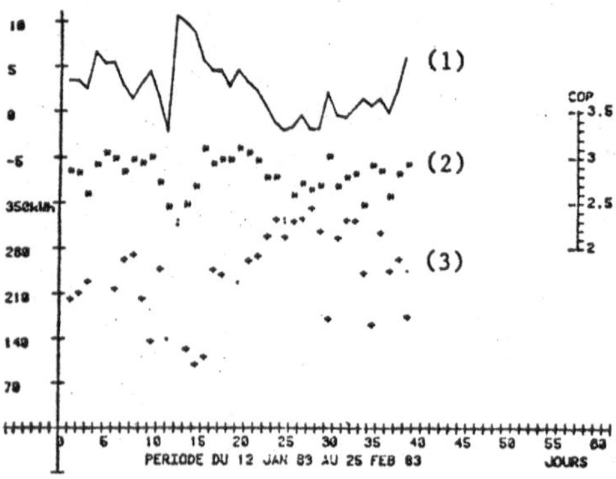

FIGURE 2

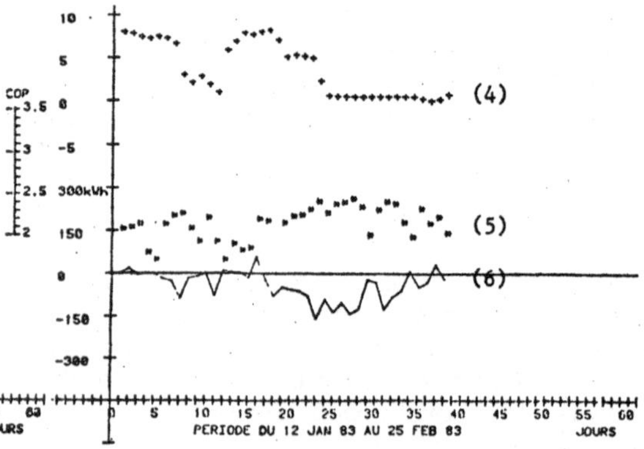

FIGURE 3

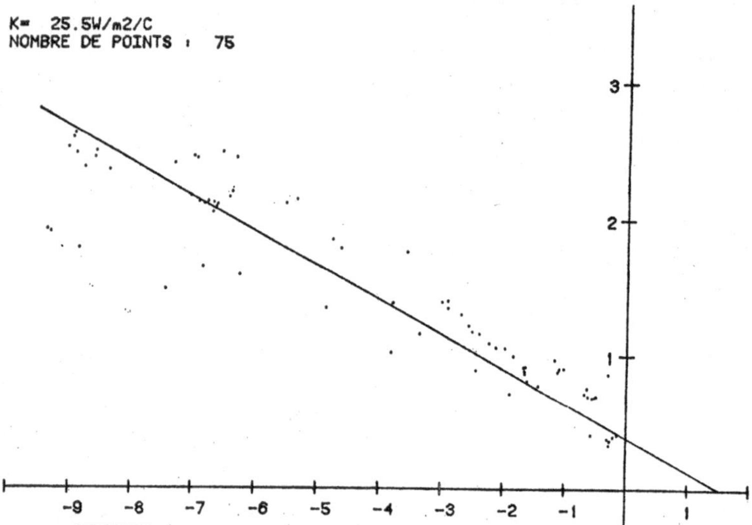

FIGURE 4

en X : 100*(Tm-Text)/En
en Y : RCD*(Ts-Te)/A/En

Dans ce diagramme on remarquera que nous n'avons retenu que les points pour lesquels la température de l'échangeur est inférieure à l'ambiance. On retrouve la valeur de 25,5 W/m^2/°C obtenue lors des essais nocturnes. Les essais aux températures supérieures à l'ambiance conduisent à une valeur plus élevée, les conditions étant différentes (modification du sens de la convection naturelle, influence du rayonnement infra-rouge). Par contre le diagramme précédent conduit à une valeur très imprécise du facteur optique (car correspondant à de faibles ensoleillements). Nous préférons déterminer le coefficient B à partir des points à fort-ensoleillement pour lesquels la température de l'échangeur est supérieure à l'ambiante. On obtient ainsi la valeur de B = 0,77 (échangeurs de couleur grise).

6 - CONCLUSIONS

L'hiver a confirmé le bon dimensionnement des composants et surtout permis de contrôler les performances. La régulation a été largement modifiée dans le but principal de diminuer au maximum le recours au stock de glace. Ces modifications sont opérationnelles depuis le début de l'hiver 83/84 qui fournira les résultats globaux sur une saison de chauffe. Une deuxième expérimentation sur des pavillons habités a également débuté en février 1984 en Touraine.

EXPERIMENTATION OF SPACE HEATING BY HELIOGEOTHERMAL DOUBLET IN AULNAY-SOUS-BOIS

P. IRIS*, E. CORDIER*, J. ADNOT**, R. BONFILS**, D. MARCHIO**
* Centre d'Informatique Géologique } Ecole des Mines de Paris
** Centre d'Energétique } France

Summary

The real scale demonstration project described in the paper deals with the thermal utilisation of shallow aquifer at low temperature (10°C to 15°C in France) which represents a very good cold source for heat pumps. The system of the "heliogeothermal doublet" is proposed to make possible a large development of the thermal use of aquifers (with low natural velocity) in dense urban areas. It is well adapted to captive regional aquifers which are common in large sedimentary basssins. After a feasability study which showed the interest of the project, the sizing of the system has been done in the case of 224 collective appartment in Aulnay-sous-Bois, near Paris ; the aquifer is a sandy aquifer at 80 m depth and a temperature of 13°C which lies under the northern part of the Parisian basin and is well adapted to collective space heating. The system has been tested between March 1983 and September 1983. It began to work October 14th, 1983 at the beginning of the first heating period with inhabitants in the buildings. The paper describes the system and the first results.

1. INTRODUCTION

According to the fact that their temperature generally remains constant, underground aquifers represent an attractive cold source for heat pumps in space heating. But they are finite ressources, particularly on the hydraulical point of view and it can be necessary to inject back the water in the aquifer by a second well, after having extracted the heat from it. Numerical simulations have shown that a thermal perturbation can occur in such a process with risks of dropping of the temperature at the producing well with time, and the progressive cooling of the underground medium around the injection zone. These problems can be accurate in urban areas.

1.1 The heliogeothermal doublet

To avoid these perturbations, it is proposed the system of the heliogeothermal doublet, which is well adapted to captive aquifers with low regional velocity : during winter, water is pumped at its natural temperature and injected back after heat extraction through a second well in the aquifer. During summer, the flow in the wells is reversed, the cold water stored during winter is pumped back and injected back in the producing well at its original temperature after having been heated by climatic convectors which collect ambiant energy both on solar insulation and thermal exchange on ambiant air. This process reduces the size of the cooled zone in the aquifer and permits to keep constant the temperature in the producing well.

1.2 A real scal demonstration project

A real scale demonstration project utilizing this system is now working in Aulnay-sous-Bois, near Paris, in the case of 224 appartments in collective buildings.

It works on a sandy captive aquifer which lies under the whole northern part of the parisian basin at an average depth of 80 m, a temperature of 13°C ; the average flow rate per well is about 80 m^3/h with a thermal power of 600 Kw. Technically and economically this aquifer is an interesting target for middle scale buildings or groups of buildings (300 dwelling or more).

The aim of this project is :

i) to apply the heliogeothermal doublet, to control the sizing of the climatic collectors, to control the thermal behaviour of the aquifer, to control the hydraulical behaviour of the aquifer during injection phasis etc...

ii) to control the technical and economical balance of large scale water heat pumps on aquifer which are not yet very common but should permit important energy savings in acceptable economical conditions.

The project is financially supported by "le Plan Construction", "l'Agence Française par la Maîtrise de l'Energie", "the Commission of the European Communities", and the F.F.F. Company, owner of the buildings.

1.3 Description of the plant

Three water water Electricity driven heat pumps are installed :

- 2 for space heating, 2 x 250 thermal Kw which represents about 60 % of the peak power needed in the buildings. The heat is distributed by radiators and heating floors which maximum temperature is 45°C. During the season, the condensors work between 25°C and 45°C.
- 1 for domestic hot water, 160 thermal Kw. The heat pump is connected to a storage of 25 m^3 (five tanks of five cubic meters each) which is heated at low cost during the night period. During the day, the heat pump heat the distribution network of hot water. The condensor temperature can reach 57°C, the heat is distributed at about 50°C.

Unglazed climatic collectors are placed on the roof : they consist in superposed flat elements (1275 m^2) made of copper thread (mesh) with copper tubes characterized by a high heat exchange coefficient ($K \simeq 45$ $W/m^2.°C$).

The collectors are used to recharge the aquifer (from 4°C to 14°C), to collect energy for domestic hot water during summer (with the help of the heat pump), and they can also work during the warmest days of the heating season together with the aquifer as a cold source for the heat pumps. During the cold days, the aquifer is used alone in order to maintain a correct COP.

1.4 Actual results

The system has been tested from March 15th, 1983 to October 15th, 1983, and began to work since October 15th for the first heating period with inhabitants in the buildings.

At the present time, the main technical problem that we have met concerns the flow rate of the immerged pumps that should vary according to the thermal needs. This regulation did not work correctly : it was due, first to a problem of lubrication of the pump at low flow rate, second, to the frequency variator which drive the velocity of the pump engine and which did not fit well. A solution is to be found to this, but the rate of the producing pump has already been constant (45 m^3/h) for the whole heating season. As consequences, the auxiliary electrical consumption of pumping increased, and a few energy has been collected from the climatic collectors during winter compared to the theoretical calculations.

The rest of the system works as previously.

A good result concerns the injection in the aquifer : the specific flow rate (it represents the flow rate for which the dynamic level of the water in the well at steady state is arising for one meter) remained quite constant till now. Each week, we pump back for one hour in the injection well at about 70 m^3/h in order to backwash the little clay particules that can be injected from the producing well. The specific flow rate in the injection well is about 8 m^3/h/m. From March 1983 to March 1984 about 200.000 m^3 have been injected so.

From October 15th, 1983 to February 29th, 1984 we have the following results :

total energy for space heating = 1349 MWh - heat pumps = 86 % - auxiliary oil = 14 %

brut COP fo the heat pump for space heating (without energy pumping) = 4,1

total energy for domestic hot water = 213,5 MWh - heat pump = 99 % - auxiliary oil = 1 %

brut COP of the heat pump for domestic hot water = 2,76

auxiliary energy of the heat pumps (pumps excepted distribution pumps which should have been also installed in the case of a central oil heating system) = 54 MWh

global COP for the period (taking into account the pumping energy) = 3,32

For the same period the total financial saving is 54 % (1570 FF per appartment compared to 2870 FF that should have been spent with central oil heating system for the same period).

2. CONCLUSION

The demonstration project of space heating of 224 collective appartments with heat pumps on a regional aquifer by the way of the "heliogeothermal doublet" has been tested from March 1983 till October 1983 and works for the first heating period since October 1983.

It works correctly with results very near from the expected ones, excepting some regulations which are not already completely satisfying. The measured average COP till now is about 3,3, with a financial saving of 54 % compared to oil heating system (estimated oil saving per year and per appartment = 1,4 ton).

The overcost of the system is 22.000 French Francs (value 1984, including VAT) and the expected payback is about 9 years (calculated without any particular financial support) ; in a development phasis this payback could decrease to 7 years approximatively which make this system economically feasable.

Next summer, the thermal recharge of the aquifer will start. It will permit to reduce the thermal cold perturbation in the aquifer due to the injection of the cooled water from the evaporators of the heat pumps (4°C). The climatic collectors will be used and we will be able to control their performances (particularly the coefficient K of heat exchange with the exterior air). Measurements are done on the whole installation with an automatic data acquisition system which give hourly or daily averaged values (temperaure at every junction of the system, electricity consumption, flow rates, energy balance, climatic data, pressure in the aquifer, etc...). These results are interpreted with models, especially with the numerical model "systherm" which simulate complex heating systems and which calculated results are compared to measurements on this installation.

Complete results and interpretation of the first complete working cycle will be published in December 1984.

Figure 1 : view of the climatic collectors exchanging with exterior air (natural convection)

Figure 2 : Detail of the climatic collectors : mesh of copper thread with copper tubes

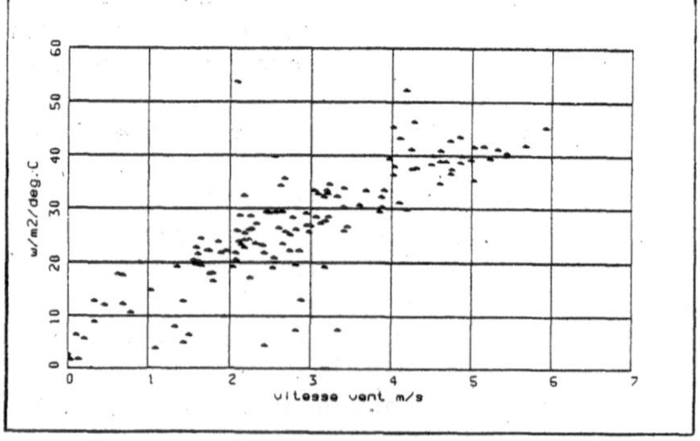

Figure 3 : Heat exchange coefficient in air, as a function of wind velocity (result of measurement)

REFERENCES

1. P. IRIS (1984). Operation pilote de chauffage par pompe à chaleur et héliogéothermie : description et bilans 1983. Rapport AFME-CCE-Plan Construction.
2. R. BONFILS (1984). Opération pilote de chauffage par pompe à chaleur et héliogéothermie : premiers résultats de la simulation de l'installation par le modèle "systherm". Rapport AFME-CCE-Plan Construction.

TRANSIENT HEAT CONDUCTION IN A PCM HEAT STORAGE WITH METAL STRUCTURES

G.C.J. Bart and C.J. Hoogendoorn
Delft University of Technology
Applied Physics Department
Lorentzweg 1, 2628 CJ Delft
The Netherlands

Summary

At the application of phase change materials in heat storage devices, there is a urgent need for a simple description of the heat transfer process with one apparent thermal conductivity, particulary if an embedded metal structure is used to enhance conductivity. Measurements did show that this apparent thermal conductivity indeed could be determined from temperature measurements during cooling down. Now numerical calculations are carried out to investigate the influence of the properties of the phase change material. The numerical results indicate that the description with a single apparent thermal conductivity, although in fact being impossible, yet can give a fair description of the heat transfer process. Discrepancies between measured and calculated conductivity can be explained.

1. INTRODUCTION

To overcome the poor thermal conductivity of organic phase change materials used in a latent heat storage De Jong and Hoogendoorn [1] proposed the application of metal honeycombs or matrix structures. They introduced an apparent thermal conductivity for the so-obtained inhomogeneous material. From measurements on temperature versus time after a stepwise temperature change at the boundary they determined this apparent thermal conductivity and found an improvement with a factor 5 á 6 with adding only 2 volume percent of a thin film metal matrix structure.

In order to be able to predict the behaviour of a phase change material with embedded metal structure we now have developed a numerical model. In this model we also account for the latent heat effects in the phase change material. Use is made of a position dependent thermal conductivity λ, density ρ and specific heat c_p. To account for the latent heat the specific heat has also been taken temperature dependent. Free convection in the liquid phase has been neglected. Up to now the calculations are executed for a fixed percentage (1.6%) of metal structure, this being the most practical case.

2. METHOD OF APPROACH

The temperature T as a function of time t can be obtained by solving the equation of thermal diffusion.

$$\rho c_p \frac{\partial T}{\partial t} = (\nabla . \lambda \nabla T) \qquad (1)$$

This equation has been discretisized on a three dimensional grid within the inhomogeneous rectangular parallellopipidum sketched in figure 1 and after that solved with a fully implicit scheme.

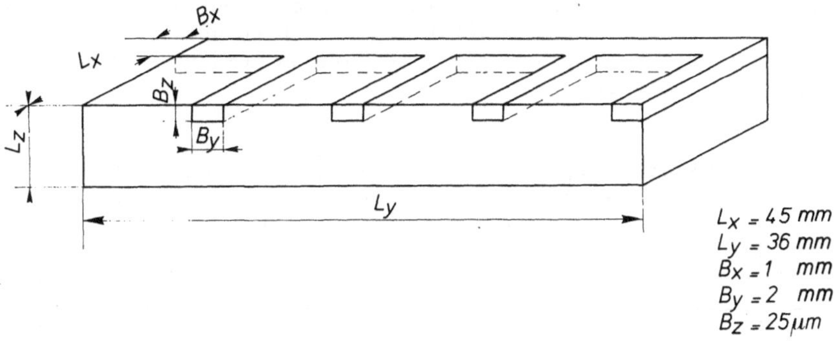

Figure 1. Geometry of the phase change material with embedded metal structure, used in the numerical scheme. Figure not on scale, L_z is determined by the volumepercentage of metal structure.

The plane y=0, left in figure 1, represents a fixed wall and has a temperature step as boundary condition, the other five boundary planes have a no flux condition. The initial temperature has been taken homogeneous. From the calculated time dependent temperature field the extracted heat as a function of time can be obtained. This function will depend on the initial temperature T_o, the boundary temperature T_w, the phase change transition temperature T_s, the constant specific heat outside the melting range c, the latent heat of the material h and the width of the specific heat curve ΔT. A somewhat schematic model of a phase change material with a transition range as shown in figure 2 is used in the numerical model.

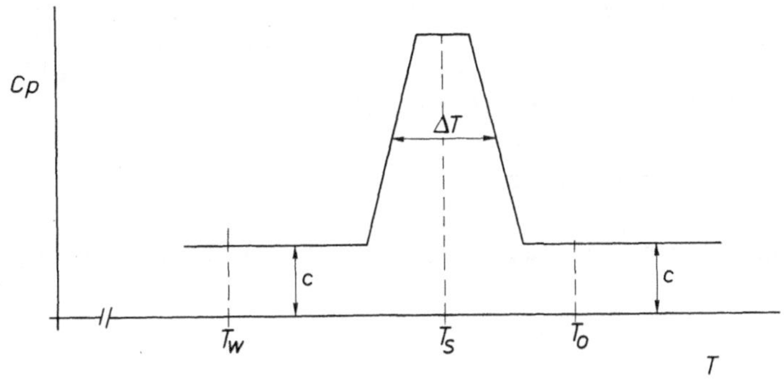

Figure 2. Specific heat versus temperature. The latent heat h equals the extra area under the curve near the transition temperature.

The schematic curve is linear increasing between (T_s-0.75ΔT) and (T_s-0.25ΔT), constant between (T_s-0.25ΔT) and T_s+0.25ΔT) and linear decreasing between (T_s+0.25ΔT) and (T_s+0.75ΔT).
The number of parameters can be reduced by introducing the Stefan-number

$$Ste = c(T_s - T_w)/h \qquad (2)$$

and the temperature ratio

$$\eta = (T_o - T_s)/(T_s - T_w) \tag{3}$$

Troughout the calculations a temperature independent specific heat c=2000 J/kgK and a latent heat h = 180 kJ/kg have been used. The used temperature ranges together with the calculated Stefan number and temperature ratio are shown in table I.

Table I. Parameters used in the numerical calculations.

T_s-T_w (K)	T_o-T_w (K)	1/Ste	η	k
-	20	0.0	-	0.477
5	20	17.6	3	0.122
10	20	8.8	1	0.199
15	20	5.9	1/3	0.265
5	10	17.6	1	0.150
15	30	5.9	1	0.232

In the calculation of the Stefan-number the volume occupied by the metal structure is accounted for.

In the well known Stefan-solution [2] the solidified layer thickness equals $2k\sqrt{at}$, where k is the root of the trancendental equation:

$$\frac{e^{-k^2}}{k\ \mathrm{erf}(k)} - \eta\,\frac{e^{-k^2}}{k\ \mathrm{erfc}(k)} = \frac{\sqrt{\pi}}{Ste} \tag{4}$$

and a is the thermal diffusity.
This equation appears in the analytical solution of the solidification in a semi-infinite region with a PCM with a single transition point instead of a range. If solidification time is short the solution can be used in a limited region. The fraction of extracted heat Φ can be calculated:

$$\Phi = \frac{1}{(1+\eta+1/Ste)\ \mathrm{erf}(k)} \cdot 2\sqrt{\frac{Fo}{\pi}} \tag{5}$$

If there is no transition at all ($Ste \to \infty$) penetration theory ($Fo < 0.2$) gives

$$\Phi_o = 2\sqrt{\frac{Fo}{\pi}} \tag{6}$$

From this it follows that if in case of a phase transition a dimensionless time variable τ is used with

$$\sqrt{\tau} = \frac{\sqrt{Fo}}{(1+\eta+1/Ste).\ \mathrm{erf}(k)} \tag{7}$$

this should result in a coincidention of the calculated results for Φ as far as the approximations do hold.

Representation of the calculated results in a double logarithmic diagram enables us to investigate the behaviour for short times ($\tau<0.2$). Deviations from the theoretical curve with the stationary thermal conductivity(equation (6))can be used to determine the apparent thermal conductivity.

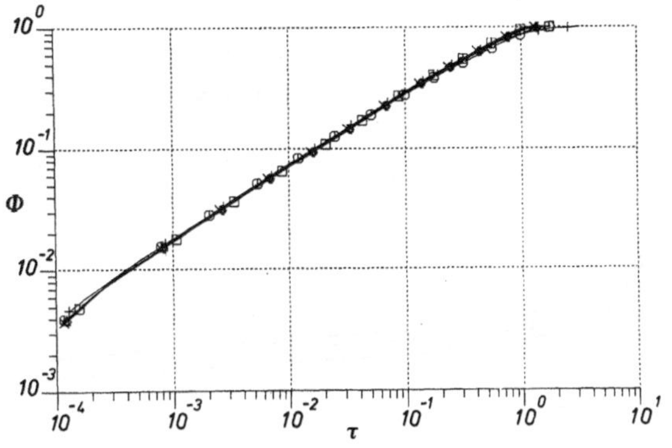

Figure 3. Fraction of extracted heat Φ versus dimensionless time τ.

3. RESULTS

From figure 3 it follows, that if $\Phi < 0.2$ the curves are steeper than the slope predicted by equation (5). For Φ approaching unity the curves first diverge somewhat. The range $0.3 < \Phi < 0.9$ obeys equation (5) pretty good, so a constant apparent thermal conductivity exists here. Moreover, if this conductivity is known in a temperature range outside the transition range, the coincidention of the diagrams when using τ enables us to determine the apparent thermal conductivity in an arbitrary temperature range. The calculations are carried out for ΔT = 2K, enlarging ΔT to 6K, while keeping h constant did not have noticeble effect.

The stationary thermal conductivity seems to be a good approximation of the apparent thermal conductivity if the phase transition is outside the temperature range considered. As yet this calculated value is twice as much as the measured one. This discrepancy can be explained if the geometrical differences (the wavy character of the metal matrix structure and irregularities in the distances between the metal sheets are not found in the numerical model) and the experimental errors are taken into account.

4. APPLICATION IN A PCM-VESSEL.

To test organic PCM's a test unit with a 0.4 m^3 storage vessel has been built. In it simulation tests of heat input and heat extraction can be done. It contains a tube bundle surrounded by the organic PCM embedded in the aluminium matrix structure (2%). Pressurized water flows through the tube bundle for a good heat exchange. Tests are in general done in a 20°C temperature interval around the melting temperature of the material.

Figure 4 gives the dimensionless outlet temperature θ as function of time for a cooling cycle with:

$$\theta = (T_{out} - T_{in})/(T_o - T_{in}) \qquad (8)$$

T_o being to initial vessel temperature.

For this experimental vessel we also calculated the heat extraction by an numerical model in which an apparent thermal conductivity was used. A comparison of the numerical results wity the experimental test data shows a good agreement for an apparent thermal conductivity of about 1 $Wm^{-1} K^{-1}$.

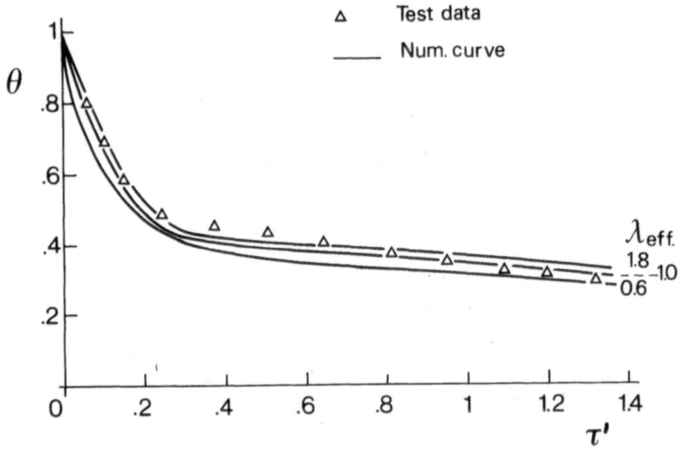

Figure 4. Calculated and measured dimensionless temperature versus a dimensionless time τ'.

REFERENCES

1. De JONG, A.G. and HOOGENDOORN, C.J. (1981). Improvement of Heat Transport in Paraffines for Latent Heat Storages Systems in C. den Ouden (Ed). Thermal Storage of Solar Energy. Martinus Nijhoff Publishers, The Hague.
2. CHURCHILL, S.W. and EVANS, L.B. (1971). Coefficients for Calculation of Freezing in a Semi-Infinite Region. Journal of Heat Transfer, ASME, Vol. 93C, p. 234.

ECONOMIC PASSIVE SOLAR WARM-AIR HEATING AND VENTILATING SYSTEM COMBINED WITH SHORT TERM STORAGE WITHIN BUILDING COMPONENTS FOR RESIDENTIAL HOUSES

K. Bertsch, E. Boy and K.-D. Schall
Fraunhofer-Institut für Bauphysik, Stuttgart
(Fraunhofer Institute for Building Physics, Stuttgart)
Nobelstrasse 12, D-7000 Stuttgart 80
Federal Republic of Germany

Summary

Warm-air heating systems are very suitable for the exploitation of solar energy. A relatively low temperature level combined with the possibility of using existing ventilation systems are cost-reducing factors. Existing building components can often be used for transportation and distribution equipment or as storage elements.

Measurements on natural climate conditions, integrating the heating system in an unoccupied test cell with energy storage in hollow ceilings and inside walls have demonstrated the feasibility of the concept. Different influences have been examined by varying parameters.

The conception and construction of our system allows supplementary installation in existing buildings as well as the integration in new buildings. The user of such systems expects optimal self-regulation. This could be achieved by microprocessor controlling.

The experimental results and price estimation of industrial production of this system are expected to be economic.

The simple construction and a wide range of applications allow conclusions to its marketability.

The reduction of fossil fuel is necessary. Not only the resource situation but also the environmental influences of fossil fuel consumption are important facts to create new possibilities for solar energy using in households.

Our development could be a contribution.

1. INTRODUCTION

As a result of the energy price situation the consciousness for energy saving in buildings has changed in the last few years. Legislative measures in the Federal Republic of Germany lead to higher demands in new building insulation /1/. With the improvement of building insulation the energy demand necessary for the ventilation increases in relation to the heating energy. On the other hand, increasing the tightness of windows leads to ventilation, which is not sufficient any more /2/. Frequently damage to buildings are the result.

In order to avoid condensation as well as for physological and hygienic reasons a minimum of ventilation must be guaranteed. Often this can only be realized by means of additional equipment for the ventilation.

Warm-air heating systems are very suitable for use with solar energy. It is easily possible to integrate heat recovery components in such systems /3/ to /5/.

2. DEVELOPMENT OF A NEW WARM-AIR HEATING SYSTEM

2.1. Description

The economical viability can be obtained by:

- passive use of solar energy,
- use of facade as warm-air collector,
- use of a new transparent insulation system (LDD),
- use of hollow ceilings and inside walls as air-ducts and distribution system,
- use of internal building components as energy storage devices,
- integration of storage material in internal building components.

2.2. Development

In order to develop a prototype (Fig. 1) a testing device was installed. Parameter studies can be carried out.

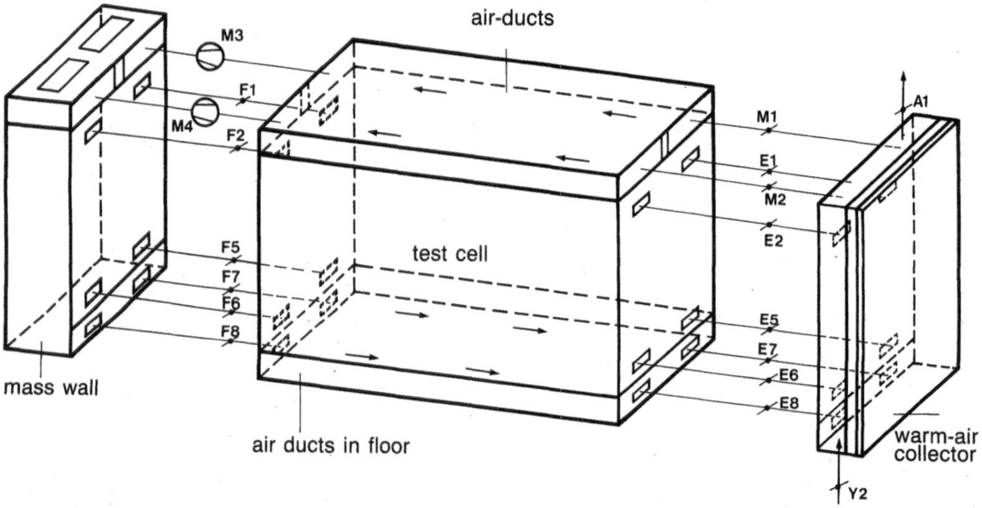

Fig. 1: Test cell arrangements and system outlines

The warm-air collector was built up by using a new transparent insulation system (LDD) on the building facade (Fig. 2). The back wall and the floor are storage wall constructions. Air-ducts are integrated in the ceiling. Fans are installed for air ventilation. Optimal conditions should be reached by measurements. Microprocessor controlling allows an optimal self-regulation.

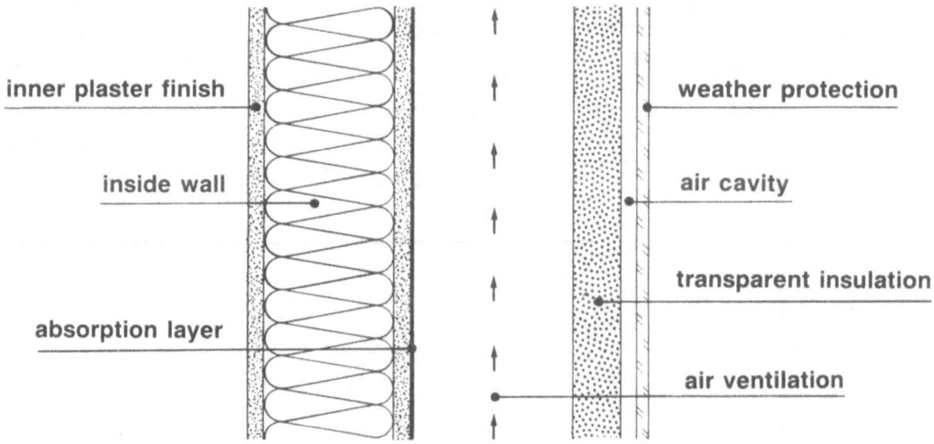

Fig. 2: Air collector system

A picture of the facade integrated transparent insulation system is shown in Fig. 3.

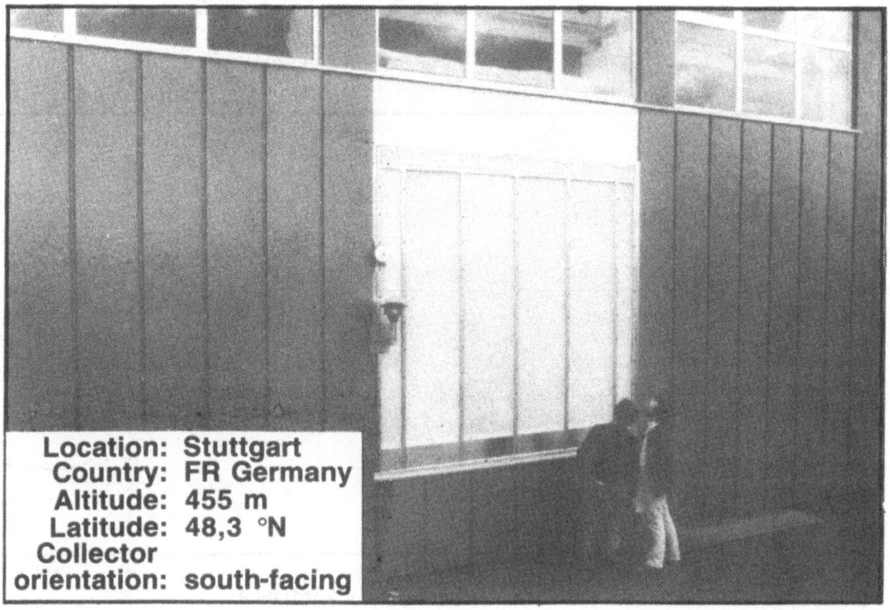

Fig. 3: Facade integrated transparent insulation system

2.3. Measurement

The measurements are not yet finished. In the following some results are given.

Fig. 4 shows the solar irradiation on 23rd February, 1984. On this day the total irradiation on a vertical south area amounted to 4,65 kWh/m^2d. The outdoor temperature is shown in Fig. 5. A mean value of 4,6 °C was determined. This figure presents also the surface temperature of the absorber and the air temperature of the ventilated cavity. Even by an outdoor temperature of - 10 °C the surface temperature of the absorber does not come up to + 8 °C. Temperatures under direct sunshine do not exceed 60 °C. The air temperature in the ventilated cavity exceeded 20 °C for a period of more than 8 hours with maximum temperatures of about 35 °C.

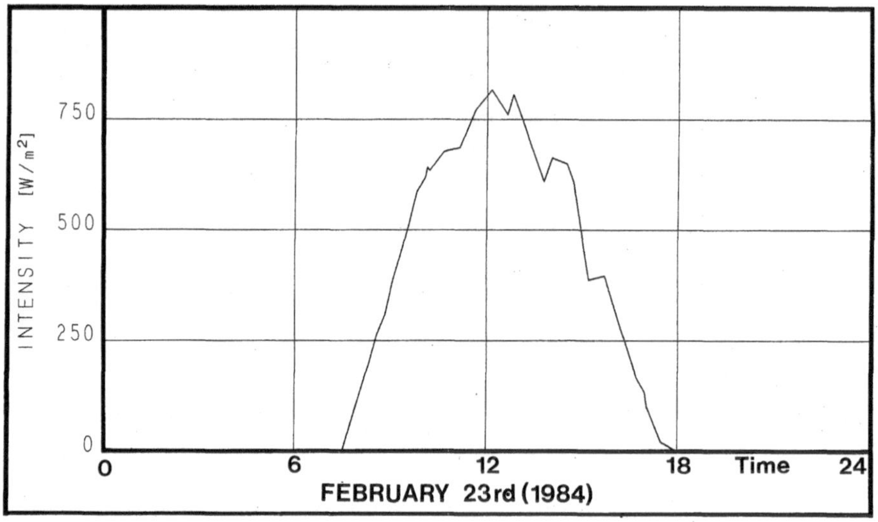

Fig. 4: Solar irradiation on vertical south area

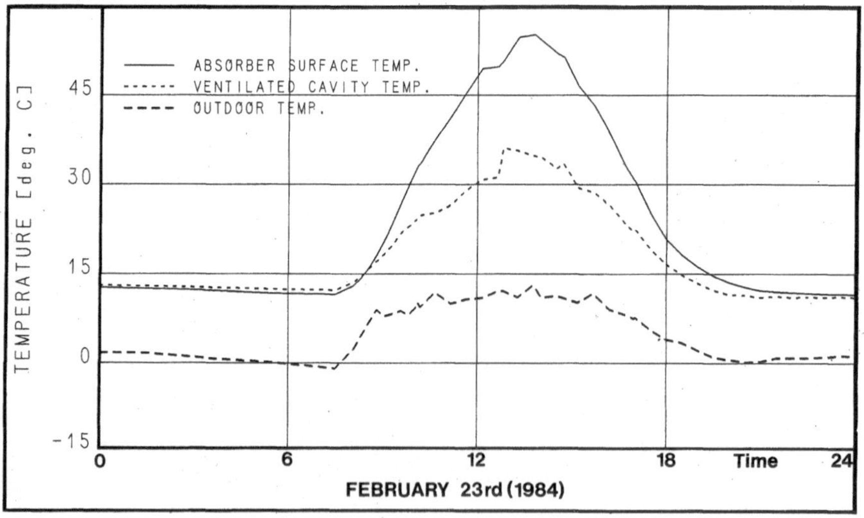

Fig. 5: Outdoor, absorber surface and ventilated cavity air temperature

The storage wall surface temperatures and the room-air temperatures are shown in Fig. 6.

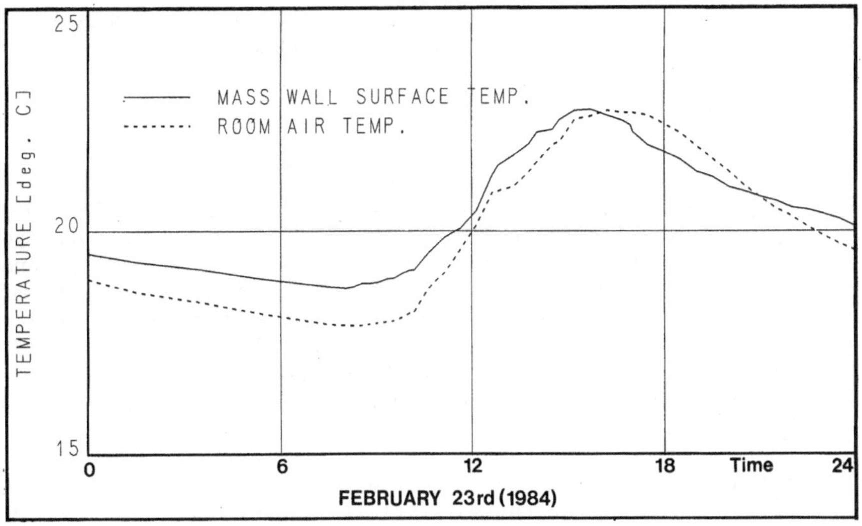

Fig. 6: Storage wall surface and room-air temperature

3. CONCLUSIONS

With the following factors economical solutions are possible:
- a new transparent insulation system as building envelope,
- hollow ceilings and inside walls as air-ducts and distribution system,
- internal building components as energy storage devices,
- additional internal storage material,
- the possibility of combining warm-air heating and the necessary ventilation.

4. REFERENCES

/1/ Wärmeschutzverordnung vom 24. Februar 1982, Bundesgesetzblatt I, 209

/2/ Kolloquium "Wohnungslüftung/-heizung in Zukunft", Bonn 1983

/3/ Bierling, H.J., Kontrollierte Wohnungslüftung - eine Notwendigkeit zur Energieeinsparung und zur Vermeidung von Feuchteschäden, Bauphysik 4, 1 (1982), 8-16

/4/ Künzel, H., Lüftung in Wohnungen, FBW-Blätter 5 (1982)

/5/ Osburg, P., Kontrollierte Lüftung und Wärmerückgewinnung im Ein- und Zweifamilienhaus, sbz 36, 11 (1981), 884-888

EXPERIMENTATION ON A LONG TERM ROCK-BED THERMAL STORAGE

C. DANG-VU* and J. MARTIN**

(*) Equipe R.A.M.S.E.S. (CNRS-PIRSEM) - Centre Universitaire, Bâtiment 208
91405 ORSAY - FRANCE

(**)Centre de Recherche CSTB - Campus scientifique de SOPHIA ANTIPOLIS
06565 VALBONNE - FRANCE

Summary

A large rock-bed thermal storage (160 m^3) was built in 1981 in the basement of C.S.T.B. Laboratory yard, at Sophia Antipolis campus. It has been experimented since mid-82 for various cycles corresponding to heat charge, heat discharge and thermal relaxation. Charge and discharge runs show a good efficiency for the air/rock-bed thermal transfer with a steep thermal gradient along the air-flow direction. Some runs with masks on the input plenum show that the air flow is quickly regularized along the cylinder. Nevertheless large convective effects have been shown for different configurations of thermal relaxation. If the storage was, originally, homogeneously hot, free convective effects induce a strong vertical temperature gradient. If the storage was partially full, with good longitudinal segregation, free convection quickly smoothes the thermal frontier. Though the experimentation shows good operability of a large rock-bed for thermal storage lasting over a few weeks, the conduction losses at the frontier combined with convective effects in the air/pebble medium make the use of such a system unefficient for storage over periods longer than one or two months ; the useful recovery period is shortened to a week in the case of horizontal storage partially full.

1. THE EXPERIMENTAL STORAGE

In 1981, a large experimental rock-bed thermal storage (160 m^3) was built in the basement of CSTB Laboratory yard at Sophia Antipolis Campus. It is a parallelepipedic volume (length : 20 m ; width : 3.2 m ; depth : 2.5 m) ; the sides, the top and the bottom are well insulated (mean value of K = 0.22 $W/m^2.C$). The storage is filled with rounded pebbles the diameter of which varies from 3.5 cm to 6.5 cm. Air is blown lengthwise along the axis with a maximum flowrate of 4000 m^3/hr ; the flow direction can be reversed. The aeraulic loop is connected to a hot water heat exchanger with electric terminal reheat capable of raising the air temperature by 55°C at a flowrate of 1500 m^3/hr. The flowrate is measured with diaphragms at the input and output ends. The storage is instrumented with 112 temperature probes, regularly distributed within the volume as shown on Figure 1. The accuracy of the experimental probes is estimated to be approximately 0.2°C. All measured data are recorded on tape, every two hours. For more details on construction problems see Ref. 1.

2. THE NUMERICAL MODEL

In the aim of understanding the physical phenomena which take place in the air/pebble medium, we developed a three-dimensional finite difference numerical model for conduction in the composite medium with NEUMANN boundary loss conditions on each face. We chose the Alternate Direction Implicit (A.D.I.) method developed by D'YAKONOV (Ref.2) because its splitting formulae are unconditionally stable and particularly because of computer memory core saving.

For the equivalent air/pebble conduction coefficient, we referred to the HENGST-KAGANER modified spherical model (Ref.3, 3') which, in our case, gives a value of $\lambda_e = 0.45$ w/m. The concrete envelope represents about 1/3 of the thermal capacity of the air/pebble storage. Because of core memory problems we replace the concrete by thermocapacity equivalent air/pebble volume, so we have only one medium for the conduction equation. Space and time grid spacings are chosen to be 0.25 m and 30 minutes.

3. EXPERIMENTAL RESULTS AND MODEL PREDICTIONS

During our experimentation several cycles were performed over long periods : heat charge of the storage (one week), heat discharge, or pure thermal relaxation of the hot storage over several weeks ; combined cycles simulating real conditions were also performed.

As to the heat charge we looked carefully for inhomogeneities of the air flow in the storage, already noticed in previous experiments (Ref. 4) ; so we performed special charge and discharge cycles where masks obturated the upper part of the input and output plenum leaving for the air inlet only 1/3 of the storage cross-section, in order to create inhomogeneities in the air flow. Figure 2 shows that after two days all effects due to masks, specially cold boundary and top regions, disappear, and a steep thermal segregation front appears along air-flow direction. Measured pressure drops are 1.6 daPa at 1000 m^3/hr and 10 daPa at 3000 m^3/hr. Another run was performed at 3000 m^3/hr : results are quite similar.

Daily cycles were monitored (6 hours of charge/18 hours of relaxation; 6 hours...) over two months in order to simulate heat charge by solar collectors during the summer. After a relaxation period of 22 days a discharge cycle was monitored with outside air as input. Figure 3 shows the inadequacy of large rock-bed storages for solar heat conservation over long periods.

As to the thermal relaxation over a long period, we compare on Figure 4 experimental data (a smooth curve is drawn through experimental points) and predictions of the numerical model. At the beginning the storage was full ; after a few days of relaxation one can notice a migration of the heat to the top of the storage which is particularly important during the first 15 days of thermal relaxation. We made an attempt to estimate the effects of free convection versus those of conduction :we assumed a conduction coefficient of the air/pebble medium multiplied by three, and introduced a strong asymmetry of the conduction coefficient with the external medium between top and bottom. Figure 5 shows that we can reproduce the observed vertical thermal gradient rather well.

In the case of partially full storage the convective effects seem to be a rather drastic limitation for an efficient heat recovery : convective courants destroy the thermal frontier very quickly (Figure 6).

4. CONCLUSIONS

Rock-bed storage needs a large volume (rock-bed thermal capacity is 1/3 of water capacity), which therefore induces large losses : it seems to be unadequate for interseasonal solar heating. Nevertheless such systems can be of great use for solar heat storage over a few weeks. Free convection effects are observed in the air/pebble medium ; this induces a preference for vertical rather than horizontal storage. In the case of horizontal air/ pebble storage, overdimensioning must be banned because of free convection: heat recovery from partially full storage is very limited.

REFERENCES

1. BOURDEAU, L. (1982). Recherche sur les stockages intersaisonniers par lits de cailloux enterrés, avec l'air comme fluide caloporteur. Rapport interne CSTB - ESP/181/LM (CSTB ; Centre de Sophia Antipolis ; 06565 VALBONNE ; FRANCE).

2. D'YAKONOV, Ya.G. (1963). On the application of disintegration difference operators. Z. Vysisl. Mat. i Mat. Fiz., 3, 385-388

3. HENGST, G. (1934). The thermal conductivity of powered thermalinsulators at high pressure. Ph.D. Dissertation, Technische Hochschule Munchen. KAGANER, M.G. (1966). Thermal insulation in low temperature engineering Izd. Machinostronenie

3'. CRANE, R.A. and VACHON, R.I. (1977). A prediction of the bounds in the effective thermal conductivity of granular material. Int. Journal of Heat and Mass Transfer (20, 711)

4. CLEMENT, D. (1983). Etude expérimentale des systèmes de chauffage solaire pour habitat individuel avec capteurs solaires à air et stockage thermique dans un réservoir de cailloux. Thèse de 3ème cycle, Université PARIS-SUD, Centre d'ORSAY, octobre 83.

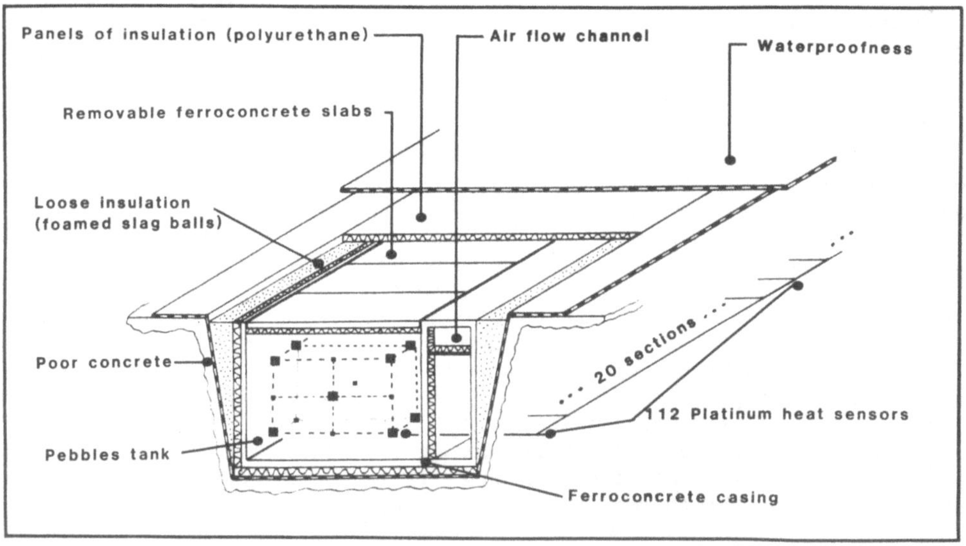

FIGURE 1 Section view of the experimental storage (length : 20 meters; cross section area : 8 m²). The storage is well insulated from the surrounding soil : the conductivity coefficient, K, is 0,22 w/m².°C for lateral sides, 0,25 for bottom side and 0,16 for top side. The location of the 112 thermal probes is shown.

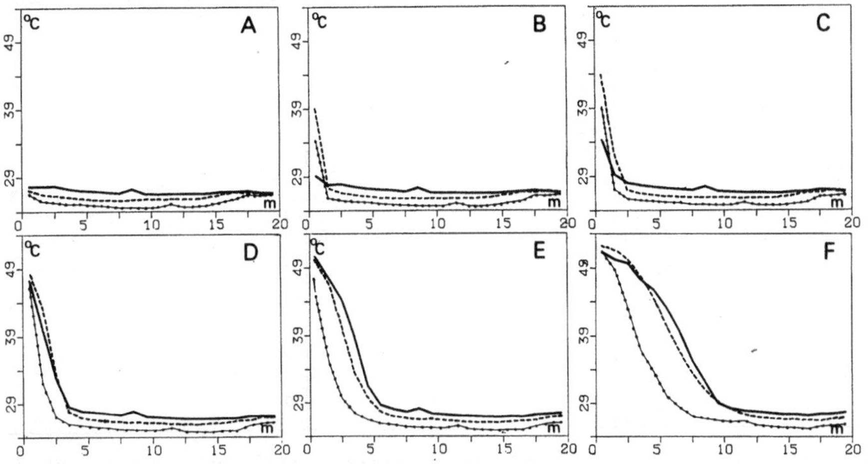

FIGURE 2 The six figures above show the evolution of temperatures in the storage during a test run centered on inhomogeneities in air diffusion. During a six days run of heat charge, with a mean air flow rate of 550 m³/h, the upper part of the input plenum was masked over 2/3 of its surface. One sees the temperature evolution along the storage axis (20 meters) at the beginning of the run (A) ; after 10 hours of charge (B) ; 20 hours (C) ; 40 hours (D) ; 80 hours (E) ; 160 hours (F). The solid line shows the temperature at the top of the storage; the dotted line at the middle; the line with points corresponds to the bottom of the storage.

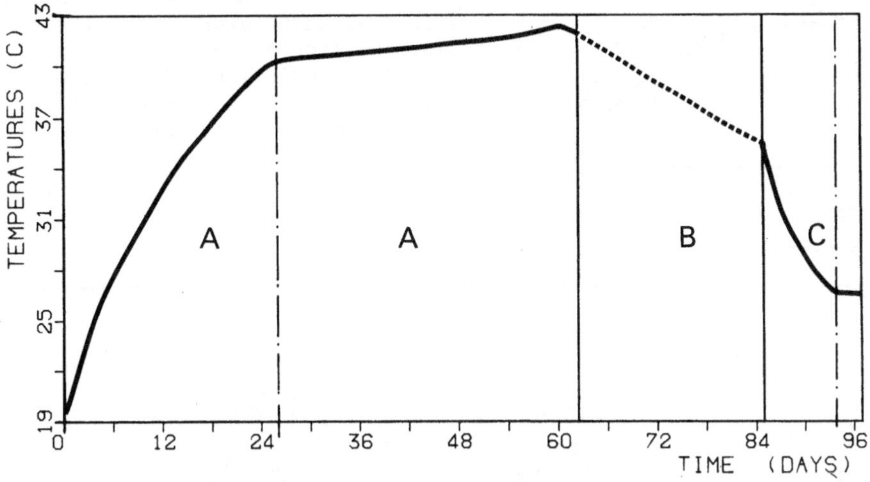

FIGURE 3 Evolution of the mean temperature of the storage during a long test run. During the charge period (A) of two months (63 days) [mean air flow rate : 1500 m³/h; air temperature : 55 °C] daily cycles were simulated : 6 h. of charge; 18 h. of relaxation/ 6 h. etc... After 26 days the storage is charged at 90% level. (B) is a 22 days period of relaxation. (C) is a period of heat discharge where one blows external air in the storage [flow rate : 880 m³/h , mean air temperature : 25 °C]

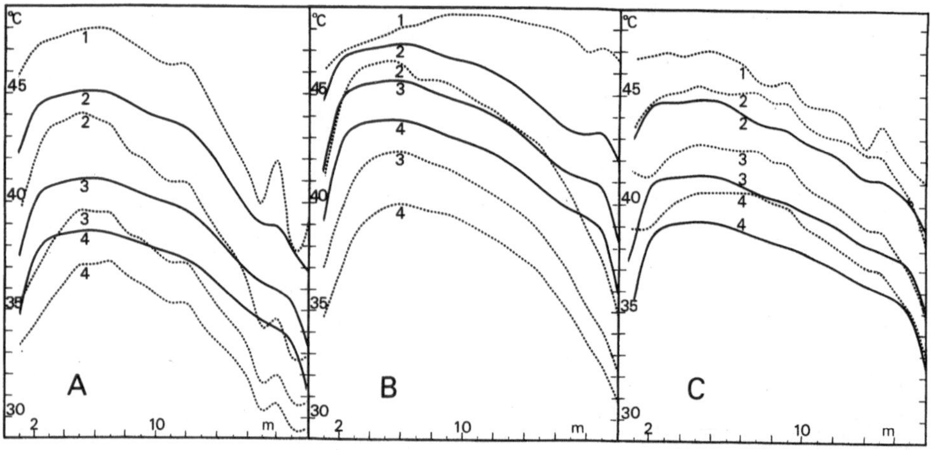

FIGURE 4 (above) : comparison of data (dotted lines) and of model predictions (solid lines) for heat relaxation. (A) : temperatures along bottom axis (ground side); (B) : temperatures along middle axis; (C) : temperatures along top axis (gallery side). Curves (1) show initial temperature; curves (2), (3) and (4) correspond to the maps of temperatures after, respectively, 6, 13 and 18 days.

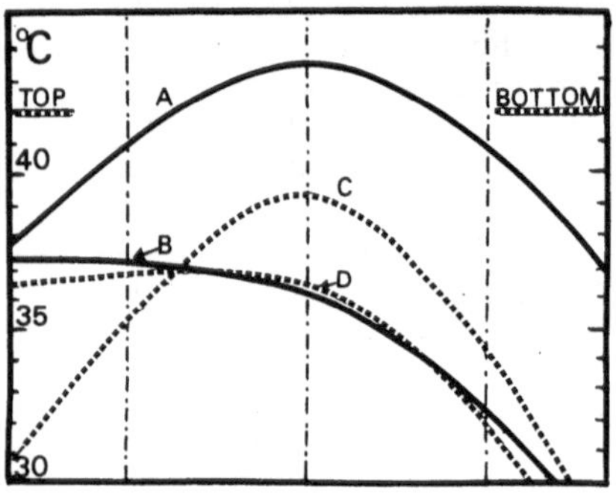

FIGURE 5 Comparison of data (solid lines) and of model predictions (dotted lines) for heat relaxation of the hot storage. Solid lines show the temperatures in a vertical section, in the center region of the storage : (A)shows the initial temperatures, (B) their evolution after 10 days of relaxation; (C) is the prediction to be compared to (B) for a pure conductive model; (D) is the modified conductive model (see § 3)

FIGURE 6 (below) : same conventions as above; the storage is initially partially charged. Curves (2), (3) and (4) correspond to relaxationperiods of, respectively, 5, 12 and 24 days.

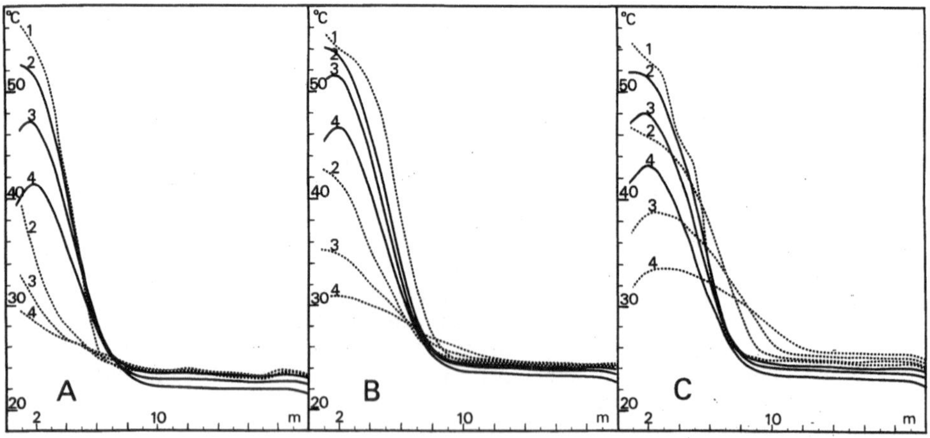

PROJECT WOLFSBURG-GLOCKENBERG, IEA-TASK VII: CENTRAL HEATING PLANTS WITH SEASONAL STORAGE

H. Riemer[+], V. Lottner[++], F. Scholz[+++]
Kernforschungsanlage Jülich GmbH
([+]Programmgruppe Systemforschung und Technologische Entwicklung, [++]Projektleitung Energieforschung, [+++]Institut für Reaktorbauelemente)

Summary

Within the IEA-Task VII: "Central Solar Heating Plants with Seasonal Storage" a solar assisted low temperature district heating system has been designed for a new housing area in Wolfsburg-Glockenberg. 80 % of the total heat load has to be covered by solar energy. It is the aim of the study to optimize the system and to prove the technical and economical feasibility of the concept. The design is based on available technology for which engineering and cost data exist. The system optimization was performed with a simulation programme (SEASON) with hourly data for a whole year. The study shows that solar assisted district heating systems are an interesting option for the utilization of solar thermal. A high solar cover factor can be reached by seasonal storage, however, less economically.

1. Introduction

Solar energy is only able to make an important contribution to the annual heating energy requirements by seasonal storage of low-temperature heat. This is true for climatic/ meteorological conditions of central and northern Europe, i.e. for Germany, too. In contrast to this, however, no technically practicable and, at the same time, economic concept has yet been developed to store low-temperature heat for a long period of time. Parallel to this it is necessary to develop design of solar systems, to test them on their technical realization and efficiency and to optimize them.

The aim of the task VII project is: "Central Solar Heating Plants with Seasonal Storage: Feasibility Study and Design", which was developed as part of the International Energy Agency (IEA) on the basis of the implementing Agreement: Solar Heating and Cooling R & D (duration: Febr. 1, 1980 - June 30, 1983 (1). Phase 1 has been finished by a preliminary draft of a solar plant in subtask 1(e): Preliminary Site Specific System Design. This preliminary design was independently realized by each of the 10 participating countries for a national site. All systems have in common the condition that the solar plant has to cover 80 % of the annual energy demand for the supply area, i.e. by seasonal storage of low temperature heat.

The German partners in the IEA project proposed Wolfsburg-Glockenberg as site, since the houses of that new residential area were equipped with low-temperature central heating which offers thus good conditions for later heat supply by solar plants and by large-scale seasonal heat reservoirs.

The paper at hand only covers part of the results of the investigations under the heading "systems study on the use of solar energy for central heat supply for building complexes" (2). This project, promoted by BMFT, is on the one hand the German contribution to the IEA project subtask 1(e)(3) on the other hand it involves the total spectrum of solar systems with and without seasonal heat reservoirs.

2. Description of the surroundings

2.1 Location and Site

Glockenberg is a new housing area in the district of Fallersleben in Wolfsburg, which lies 80 km east of Hannover, near the border with the German Democratic Republic. The geographical position is latitude 52°26'N, longitude 10°47'E, 88 m above sea level.

The newly-developed area Glockenberg lies in the western part of the town in open, flat countryside. Although the complete area is supplied with district heating, owing to financial considerations only 23 houses have been connected to the low-temperature district heating network (Figure 1). The investigation concentrates on the area supplied by low temperature district heating.

2.2 Climate

In Wolfsburg the lowest outside temperature is assumed to be -15°C for calculations of the heating requirement. With 286 days when heating is required 1586 hours of peak load, one must assume 3700 degree days for an average room temperature of 19°C(4).

The following details are obtained from Brunswick as no sufficient data is available from Wolfsburg itself. Brunswick lies 20 km south of Wolfsburg. The geographical position is longitude 10°27'E and latitude 52°18'N. This data was collected between 1951 and 1970.

The daily mean temperature for each month ranges between -0.2°C and 17°C; the average daily temperature over the whole year is 8.6°C. The relative humidity is 80 % and Brunswick has about 1550 hours of sunshine per year.

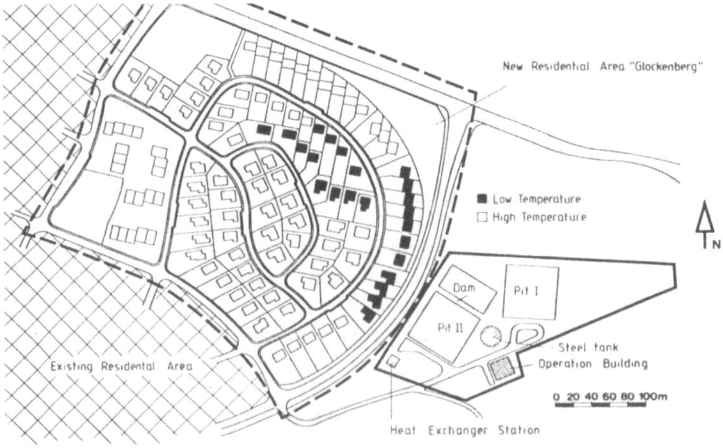

Fig. 1: Site plan with the location of the collector array and steel tank

The total annual global radiation of Brunswich is 990 kWh/m^2 (1980), with 58 % being diffuse radiation (Fig. 2 (5)). Systems calculations as performed here with SEASON model from PHILIPS Research Laboratory require hourly metereologic data which are not recorded by Brunswick on data carriers. A comparison between the total annual global radiation for Brunswick (990 kWh/m^2) and Hamburg (HH) of 1973 (980 kWh/m^2) shows that this is within the range of annual fluctuations, using Hamburg data for model calculations. 1973 is taken as the reference year in the FRG, as far as meteorology is concerned.

3. Description of the System

3.1 Buildings and Load

In Fig. 1 the buildings coloured black (23 altogether) are those connected to the low-temperature district heating network. These are houses for one or two families, some of which are detached and the others are row houses. They all have "improved" heat insulation, by German standards, but passive measures for using solar energy were not implemented. An average house consists of a heated area of 175 m^2. The owners could choose between a combined underfloor/radiator heating system and an entirely underfloor or radiator system. Five houses decided on electric geysers for domestic hot water and the others are supplied with hot water from the district heating system.

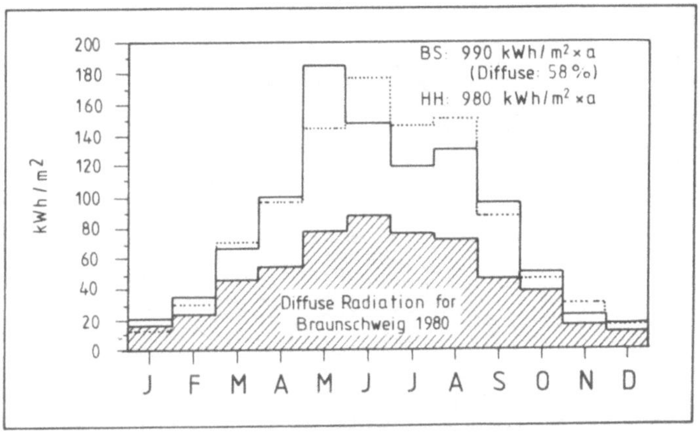

Fig. 2: Monthly radiation on a horizontal surface for Braunschweig (BS) 1980 and Hamburg (HH) 1973. The diffuse radiation for Braunschweig 1980 is additionally marked.

The heating load without hot water production for the individual houses varies considerably between 7.3 and 22.2 kW. The average value is 14 kW, and the minimum, 7.3 kW, is required by a house with a heated area of 104 m^2. The specific heat load per square meter of heated area load per square meter of heated area varies between 68 and 107 W/m^2 with a mean value of 82 W/m^2.

Regarding the yearly heating requirement, one must rely on estimations. For the average house with 175 m^2 heated area snd a specific heat load of 80 W/m^2, an annual heating requirement of 17 MWh is obtained. The hot water demand can be assumed to be about 3000 kWh/a. Hence, the total heating requirement is estimated at 436 MWh/a. With heat losses in the distribution of 5 %, ca. 460 MWh/a must be fed into the network.

3.2 System Design

Current planning is based on a system consisting of collector array of 800 m^2 and a steel tank of 3000 m^3. The system is to run 78 % on solar energy and the remaining 22 % is to be provided by a heat exchanger high-temperature district heating supply (120/60) (Fig. 3).

The collector array of heatpipe collectors is installed close to the reservoir (Fig. 3). The ratio of the collector surface to mounting surface of 800 m^2 requires a surface of 1600 m^2, which is available free of charge at Wolfsburg-Glockenberg. Estimations of the collector piping results in a length of about 180 m with an average nominal diameter of 60 mm. The colllector field is controlled by variable flow, thus achieving the highest possible usable temperature.

A steel tank with a volume of 3,000 m^3 is erected beside the collector array; it serves as a pressureless hot water reservoir up to a maximum temperature of 95 C. Its size is the result of computations presented in Chapters 4. With a form factor of 1, i.e. a ratio of diameter to height of 1, one obtaines a height = diameter of about 15 m. Heat insulation has a k-value of 0.1 $W/m^2 \cdot K$. This corresponds to an insulation thickness of 40 cm when using mineral wool ($= 0.4$ $W/m^2 \cdot K$). The heat losses in a reservoir with annual average temperature of about 60 C can be estimated to be about 60 MWh/A.

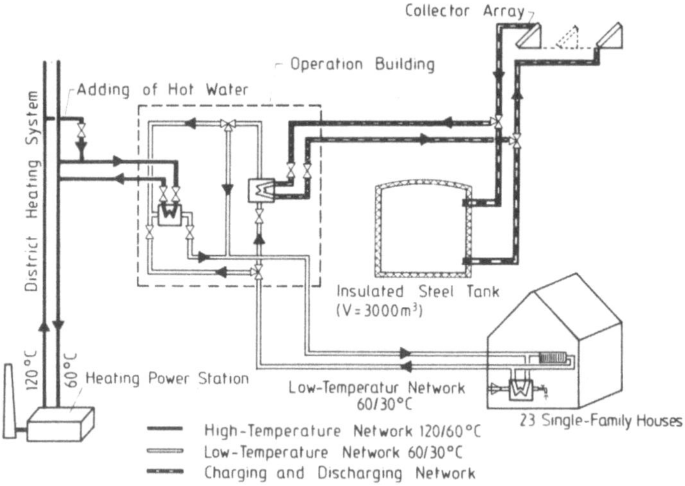

Figure 3: Scheme of the complete system

The distribution grid was installed in 1981/82. The distribution and house connection line has a total length of 1054 m. 70 % of which was installed in the basements of the houses. Using plastic tube material and the favourable distribution technique, especially with respect to installation in the basement, led to a low-price distribution system of about 520,- DM/m. The connection costs thus amount to DM 1060 per kW.

The low temperature grid is supplied via heat exchanger pipe assemblies by the existing conventional district heating grid (520 kW). The heat exchanger which at the present time supplies completely the low-temperature grid will be additionally used as back-up system after the solar part has been realized.

The total system as shown in Fig. 3 can be characterized as a 3-circuit system. The circuits are connected with each other via heat exchangers. No heat exchanger is necessary when applying heatpipe collectors between collector circuit and reservoir circuit.

A heat exchanger is, however, necessary between reservoir/collector circuit and distribution grid (high-temperature and low-temperature grid) because of the different quality of the water. The reservoir collector circuit uses drinking water.

The water heated-up in the collector field is primarily directly transported via heat exchangers to the consumers, i.e. to the 23 houses, for supply of space heating and hot water. If its temperature is not sufficient, heat is extracted from the reservoir as the second step. The heat exchanger couples the high-temperature grid to the low-temperature grid and has the function of a flow heater to ensure the required constant flow temperature of 60 C.

4. Results

4.1 Technical Results

The calculation results described here were performed with the SEASON computer program developed by PHILIPS Research Laboratory in Aachen. The program itself and its possibilities of application are described in detail in (2).

On the basis of 250 houses, extensive calculations were carried out with the SEASON program for the locations of Hamburg and Munich. The following results were derived by a paramter study decreasing the load by a factor of 10, i.e. for 25 houses corresponding to those 23 houses in Wolfsburg-Glockenberg, and taking into consideration scaling regulations for the remaining system. The climate of Hamburg corresponds to that of Wolfsburg-Glockenberg (cf Chapter 2.2).

A cost-optimized system with seasonal central heat reservoir loads to the following design values for the 25 houses: A_c = 800 m^2 (collector area), V_{SP} = 3000 m^3 (storage volume), under the boundary conditions that the system should be realized economically with solar fraction above 75 %. The relative energy price for this system design, as defined as an economic criterion in chapter 4.2, is not only smaller than 100 %, but almost minimal, thus it is optimal.

The system enables 78 % of the annual grid incoming supply of 455 MWh to be covered by solar power. The annual efficiency rate of the system is approximately 43 %.

The collector array annual efficiency is around 47 %. 480 kWh/m^2 a of the collector array were directly fed into the reservoir per year or to the consumer. After having substracted storage losses and other losses in the solar system, about 450 kWh/m^2 a can still be fed into the grid.

Fig. 4 shows the monthly distributions in the course of the year of some heat volumes. Almost 90 % of the demand can be covered in January by solar power, since the annual cycle does not begin with an unloaded reservoir (yearls energy balance of the store nearly zero). About 65 % of the grid supply during the following two months are covered via additional heating which in April is reduced to 43 %, and in May to 23 %. From June on the total demand can be covered from the solar system for the rest of the year.

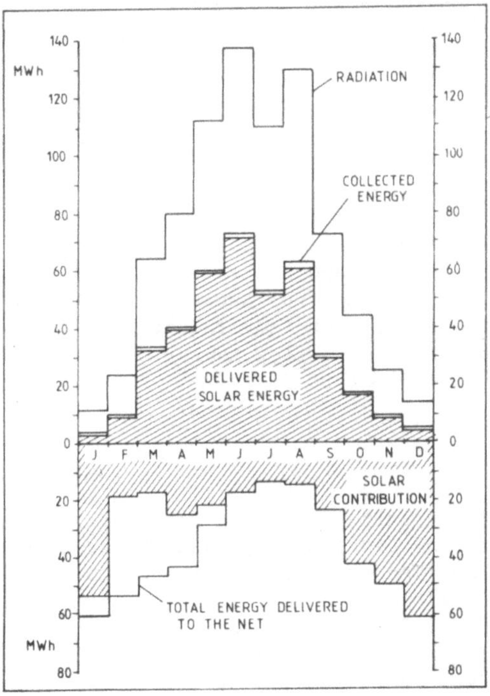

Fig. 4: Yearly cycle of the global radiation, collected energy, energy delivered to the net and downwards the total energy delivered to the net and the solar contribution

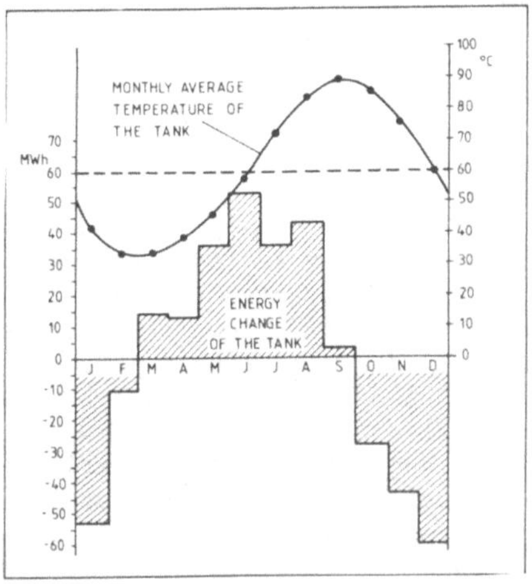

Fig. 5: Yearly cycle of the average temperature of the tank and of the energy change of the tank

Fig. 5 shows the modified reservoir contents and the average reservoir temperature during the course of the year. The shaded area gives an impression of the positive and negative contributions to the reservoir balance. Unloading during 5 months is contrasted with loading for 7 months. The increase in reservoir content already starts in March. The sum of content increase (= sum of content decrease) is 196 MWh/a. The effective contribution to solar grid supply via the reservoir is thus 55 %. 45 % were thus directly supplied to the consumer from the collector array. The reason for such a high contribution is on the one hand the priority in favour of the collector field compared to the reservoir; on the other hand the special control for grid supply which is a mixture of collector, reservoir, and auxiliary heating. The curve of the average reservoir temperature is a sine curve. The reservoir is unloaded in February at about 30 C, and loaded in September at about 90 C. The strongest increases are naturally from May to August when strong insolation is accompanied by low consumption. The annual cycle starts in January with a mean reservoir temperature of 42 C.

Table 1 contains the main performance data of the system.

Energy flow	MWh per year
Global solar radiation	823
Solar energy converted	385
Solar energy provided directly	159
Stored energy	226
Solar energy provided (total)	355
Auxiliary	100
Distribution losses	22
Total consumption	433

Table 1: Performance data

4.2 Economics

Economic considerations have been performed on the basis of life-cycle cost analysis which required certain assumptions. They are in detail:

- interest period N (year) which for reasons of simplicity has been considered corresponding to useful life (no remaining values or new investments).
- the price of conventionally produced thermal unit P_{conv} (1) (DM/kWh) supplied to the grid at the time of solar plant start-up (=conventional energy price).
- the annual increase e (%) of that price.
- the cost of maintenance in per cent as a function of investment costs m (%).
- the annual increase i (%) (=inflation rate) of maintenance costs, and
- an average calculated interest rate (discount rate) p (%) which by simplification represents both the rate of credit interest averaged over the interest period N as well as the rate of debit interests.

A relative (dimensionless) energy price is calculated as a profitability criterion. Therefore the annual costs for energy supplied to the grid are first determined. Dividing the annual supplied total energy results in a mixed price for both the solar system AND conventional system. Fixing this price to an average conventional energy price leads to two aggregated cost parameters which implicitly include all economic boundary conditions that have to be varied in a comprehensible manner. The question wether a system has reached the economic optimum or profitability threshold is thus reduced to calculating A_c and V_{SP} to determine that $\bar{p}_{mix.\ rel}$ is minimal or is $\bar{p}_{mix.\ rel} \leq 1$ ($\bar{P}_{mix.\ rel}$ = relative energy price of the total heating plant (average over life cycle)). This approach of the economic analysis is described in (2) in detail.

The aggregated cost parameters were specially chosen so that a solar fraction of more than 75 % for the solar system could still be achieved economically. This holds for a system with a collector area of A_c = 800 m^2 and a storage volume of V_{SP} = 3000 m^3. The system achieves a solar fraction of 78 % at a relative energy price of 95 %. This design value is close to the border of a weakly characterized minimum.

A certain value for aggregated cost parameters is taken for a large number of possible scenarios. A different estimate for at least two influencing parameters can lead to the same value for the aggregated cost paramters, and thereby to the same economic result. The influence of different input parameters is to be studied at further constant values for the aggregated cost parameters. The columns 2 to 5 of table 2 show the inputs data to calculate the relative energy price. The relative energy price of 95 % results of the values of column 1 for these parameters are applied.

			1	2	3	4	5	6
1	RELATIVE ENERGY PRICE	%	95	95	95	95	95	187
2	AMORTIZATION PERIOD	a	20	10	·	·	10	·
3	CONVENTIONAL ENERGY PRICE	DM/kWh	0.10	·	0.075	·	0.075	·
4	ESCALATION RATE OF THE ENERGY PRICE	%	10	·	·	7,5	7,5	·
5	INTEREST RATE	%	7,5	·	·	·	·	·
6	INFLATION RATE	%	5	·	·	·	·	·
7	RATE OF MAINTENANCE COSTS	%	1,5	·	·	·	·	·
8	SPEC. COSTS OF THE COLLECTOR ARRAY	DM/m^2	600	275	430	490	200	900
9	SPEC. COSTS OF THE STORAGE	DM/m^3	90	42	64	74	30	275
10	INVESTMENT COSTS OF THE COLLECTOR ARRAY	TDM	480	165	344	392	160	720
11	INVESTMENT COSTS OF THE STORAGE	TDM	270	126	192	222	90	825
12	TOTAL INVESTMENT COSTS	TDM	750	291	536	614	240	1545
13	SPEC. SYSTEM COSTS	DM/m^2	938	364	670	768	300	1930
14	INVESTMENT COSTS PER HOUSE	DM	32600	12650	23300	26700	10400	67000

Table 2 : Economic analysis of the system

Very optimistic assumptions have to be made for influencing paramters, especially with respect to energy price increase and low specific costs for the collector array and the reservoir in order to obtain the above result. Assuming 10 years instead of 20 for the interest period, the costs of collector array and reservoir would have to be lower than 300 DM/m^2, or 50 DM/m^3 (column 2). This price is modified in columns 3 and 4 with respect to the conventional heat price and the annual increase. The effects on the investable specific collector and reservoir costs are in the same order of magnitude in the case of a conventional heat price reduced by 25 %. A price increase in the conventional heat price to the amount of the calculation interest rate leads to a reduction in investable specific costs of 18 %.

If all three cases of columns 2, 3 and 4 were to occur at once, the upper cost limit would be 200 DM/m^2 for the collector array, and 30 DM/m^3 for the reservoir. The specific system costs would have to be lower than 300 DM/m^2 in this case (column 5).

Finally if one presumes collector array and storage costs which could be achieved today, one obtains a relative energy price of 187 % with the remaining assumptions from column 2 (Column 6). This energy price is twice that presumed here. Even in the case of very optimistic assumptions about energy price developments, this shows that the collector array costs would have to be reduced by 30 % and the reservoir costs by as much as 60 %.

5. Conclusions

Favourable conditions for the realization of a system with seasonal storage are found in Wolfsburg-Glockenberg:

- a low-temperature district heating network with low, constant forward temperature of 60°C and a return-temperature of 30°C.
- an existing high-temperature district heating supply with heat transfer station, which can function as a back-up system.
- sufficient space available free of charge for the collector array and heat store.

The low temperatures in the network mean a high efficiency of 43 % for the solar system and a solar coverage of 78 %. A 100 % coverage is already possible from June till the end of the year. The system design with a collector area of 800 m^2 and a store capacity of 3000 m^3 was preliminarily optimised based on economic points of view. In this the economic boundary conditions were chosen so that an economic solution was possible for a solar coverage of more than 75 %. However, an analysis of the boundary conditions shows that in spite of good technical results the overall system is too expensive by a factor of 2. Should realization be desirable then state aid is necessary.

REFERENCES

1. BOYSEN, A. (1981). Central solar heating plants with seasonal storage, Review of Task VII of the IEA Solar Heating and Cooling Programme. Proceedings of International Conference on Seasonal Thermal Energy Storage and Compressed Air Energy Storage, Vol. 1, p. 46 - 50, Seattle.

2. BERGMANN, G. et al (1984). Systemstudie zur Nutzung der Sonnenenergie für die zentrale Wärmeversorgung von Gebäudekomplexen. Final report of the project 03E-4453 A funded by the Ministery of Research and Technology.

3. RIEMER, H., LOTTNER, V. and SCHOLZ, F. (1983). Project: Wolfsburg-Glockenberg. National report of the Federal Republic of Germany for IEA Task VII: Central Solar Heating Plants with Seasonal Storage, Subtask 1 e

4. VDI 2067, sheet 2 (1979). Economy calculation of heat consuming installations - Heating installations. VDI-Gesellschaft Technische Gebäudeausrüstung.

5. DEUTSCHER WETTERDIENST, Meteorologisches Observatorium Hamburg (1981). Ergebnisse von Strahlungsmessungen in der Bundesrepublik Deutschland sowie von speziellen Meßreihen am Meteorologischen Observatorium Hamburg. No. 5 1980.

STUDY OF THE THERMAL BEHAVIOUR IN FREE TEMPERATURE EVOLUTION OF A PASSIVE SOLAR INHABITED HOUSE WITH HIGH INERTIA AND LONG PERIOD SOIL HEAT STORAGE

J.M. BALEYNAUD, M. PETIT and A. TROMBE
Laboratoire de Génie Civil I.N.S.A.-U.P.S. de Toulouse
Equipe de Recherche Associée au C.N.R.S. n° 998

Summary

In the frame of the applied researches developed by the building thermal equipment service of the civil engeneering laboratory of the I.N.S.A. and U.P.S. of TOULOUSE, an experimental steady was performed on the "GUY's house". The original feature of that building is that both the designer and the owner wished to give it a complete energetic autonomy by the use of original elements for energy recovering, such as the solar system "greenhouse-long period soil heat storage".

1. HOUSE DESCRIPTION

The "GUY house", built in RAMONVILLE SAINT-AGNE near TOULOUSE is a bioclimatic type house which has been occupied since 1979. Due to its architecture almost the whole windowed area is positionned on sunny walls (S-SE-SW) while shaded walls are protected by earth slope which increases the thermal inertia.

1.1 Building description (see figure 1)

According to the house conceiving main elements converge towards a high thermal mass, the floors are on platforms, the carrier walls made of heavy reinforced concrete and lean on a non-insulated ground slope, the partition walls are made of concrete too and the terrace roof is insulated on the outside.

1.2 Description of the solar system

It consists in a greenhouse and a soil heat storage under the house, whether coupled or not depending on the season.

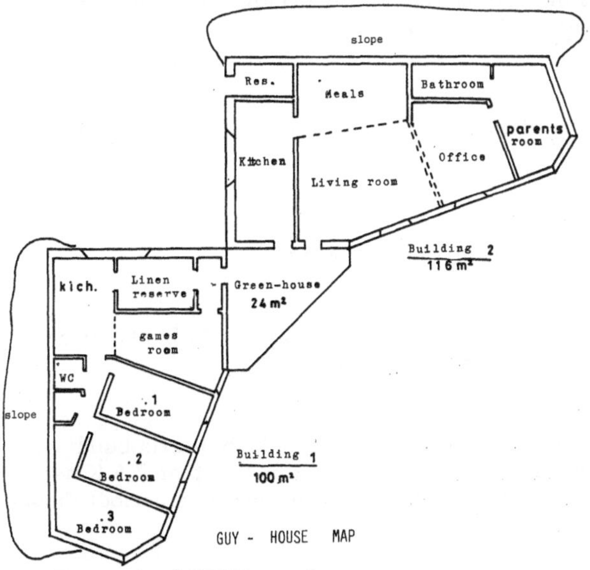

Figure 1 - "GUY house" map

1.21 Greenhouse description

It's the link between the two building bodies (see figure 1) and is composed of a greenhouse which steep part is an air collector.

1.22 Heat storage description

It consists in the clayey soil mass right under the house living rooms (see figure 2).

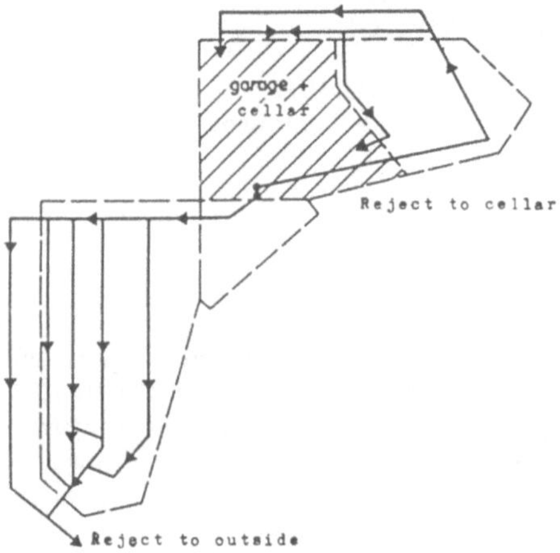

Figure 2 - Storage map, repartition of hot air circulation coils

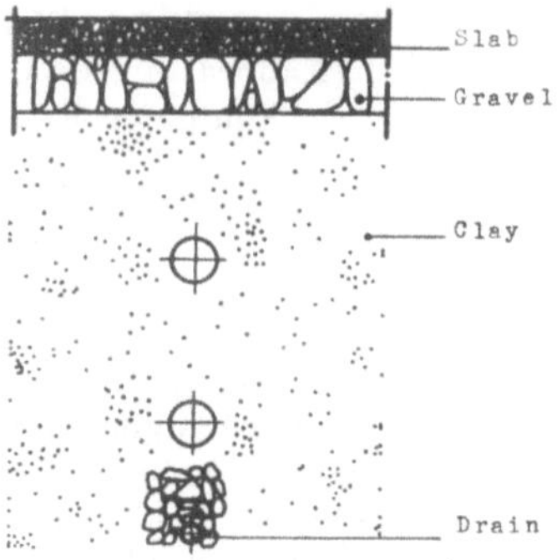

Figure 3 - Storage view

1.23 Set working

Thanks to a forced ventilation system, enslaved to a regulation the air pre-heated in the greenhouse-collector system can be used in two different ways, according to the season :

- in cold season to heat the mains rooms of the house,
- in hot season and mainly in the end of summer the hot air is blown into the buried tubed to load the storage by increasing earth temperature.

2. EXPERIMENTATION

It was led during one year and was performed with appropriate measure and acquisition devices.

3. MAIN RESULTS

Our experimentation led us to show out results on two ways.

3.1 General thermal results

They are corresponding whether to global thermal balances, or to daily evolutions of main parameters typical of the house thermal behaviour.

3.11 Thermal balances (see tables IV)

The results are confirming the house conceiving choice for a high inertia building, we remark that the amplitude of average ambient temperature is less than 7,5 K between june 81 and march 82 (outside amplitude was over 15 K in the same period) though the house is in free temperature evolution.

3.12 Thermal evolutions

That study is limited to the "chidren part" and to two characteristic rooms, bedrooms 1 and 3 of that very building body (figure 1) which volumes are equivalent but positionnings different :

- bedroom 1 is centered in the body and a little linked to outside air,
- bedroom 3 is an outlying room having higher thermal losses.

We could hold 3 caracteristic periods of the house behaviour.

According to the curves drawn on figures 5 and 6 we can hold the following remarks :

	Whole house (8I)				"Children part" (8I-82)			
Month	March	April	May	June	October	November	February	March
Measure period	6|03 -26/03	30|04 -5/05	5|05 -30/05	1|06 -15/06	19|10 -30/10	3|11 -27/11	4|02 -29/02	4|03 -14/03
Free gifts (kwh)	1848	2887	2397	1115	410	683	530	202
Theoretic consumption relative to I8° (kwh)	1872	3057	1235	-729	493	1462	1353	664
Average inside air temperature (°C)	15	16,6	17,4	21,4	18,4	15,5	14,2	13,9
Outside temperature	11,9	12,2	15,0	22,2	10,1	7,3	8,4	6,9

Table IV - Thermal balances

- any beeing the period, average and resulting temperatures never get more than 1 K different,
- the inside temperatures are linked to outside ones, for example apart from the half season before winter where average inside temperature is about 20°C, it decreases to 14°C on november 27 th and finally to 12,5 in february in the same movement as outside.

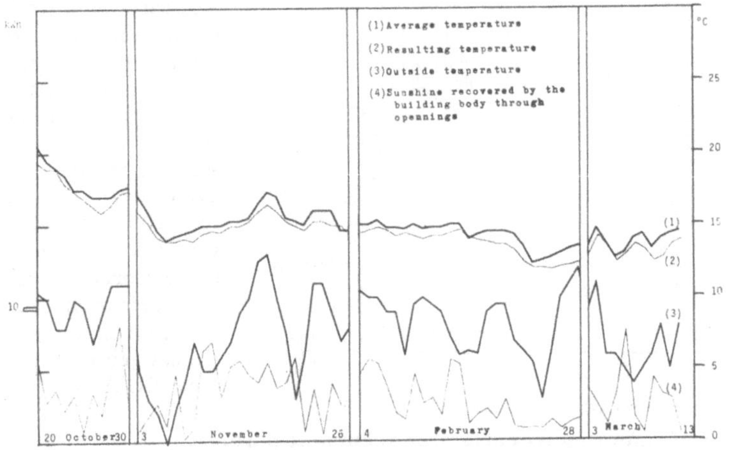

Figure 5 - Average and resulting temperature evolution of "children part"

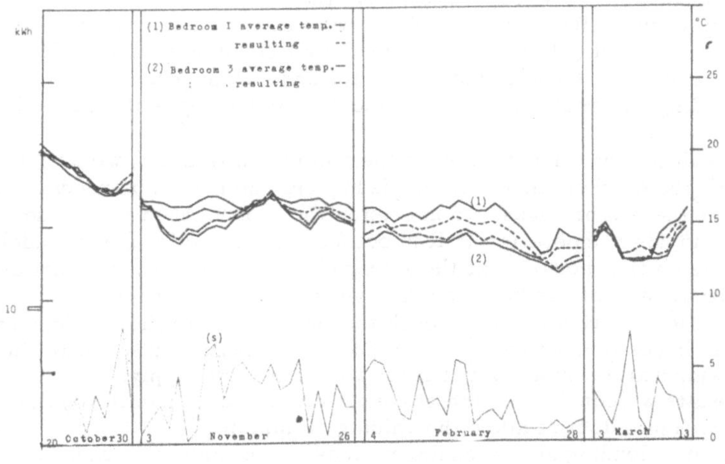

Figure 6 - Comparison of average and resulting temperatures in bedrooms 1 and 3

According to these observed evolutions, we may assert that the very high house inertia increased by the use of a heavy mass of ground allows the stabilization of house temperatures in the order of those measured 1,5 m deep in the soil in the same period.

Moreover, we can remark now, that the working of the long period soil heat storage is more in the way of reducing the thermal losses through the ground than in the way of a winter heating of "children" rooms.

We can remark too that bedroom 3 is much more sensitive to outside temperature evolutions than bedroom 1, which can be explained by the positionnings of these rooms.

3.2 Partial thermal results

They are corresponding to the study of the house long period soil heat storage. Due to the fact that measures performed in ground show both soil and exchanger working, each will be studied in its time to show out weaknesses and possible improvements on each.

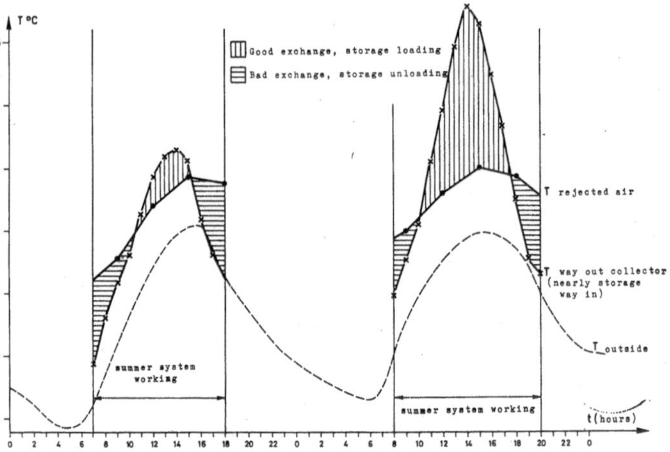

Figure 7 - Air-ground exchanger working

3.21 Exchanger study

Figure 7 shows air temperature evolutions on way in and out from exchanger for two days, throughout we see a regulation defect (early start, late stop) causing a storage unloading. We can see that a timer avoiding fan working before 10 and after 17h30 would have given 47 % more stored energy. A comparison of temperatures right in and out of exchanger would give such a result. Apart from regulation, we could recycle storage air instead of rejecting it, caring for keeping it pure.

3.22 Soil heat storage study

We can value the stored energy through the measures of air temperatures on the way out of collector and out of storage, presuming the whole heat carried by air will be transmitted to the tubes. We define a global output of the summer system as the ratio of stored energy on collector sunshine, this output could be valued to 25 %. Notice that this yield depends on those of both collector and exchanger.

The storage loading was realized since the end of may till september, 22000 kWh were stored.

We show the evolution of temperatures in the storage on figure 8, we notice that this evolution is all the quicker since sunshine is higher. The maximum recorded temperature is 26°C at the end of august 1,5 m deep in the ground, then this value decreases according to outside temperature (8 K in october in storage for 6 K outside).

We can conclude on that period that the storage has soon lost a large part of its energy, whether on benefit of the house or through thermal losses (figure 8).

Figure 9 shows the heat recovering through the floor in the kitchen of "children part". As this room is positionned to the north, no direct sunshine may get to the floor and increase its temperature, moreover it is far from any heat supply.

From these results we could value the phase difference to 1½ month for the thermal wave, however this figure doesn't allow the valuing of damping.

We can notice that loading and unloading periods are in the same order 120 days.

We could value the recovered energy and extrapolate it to the whole house, then we define a global yield of the system as the ratio energy recovered-stored energy that

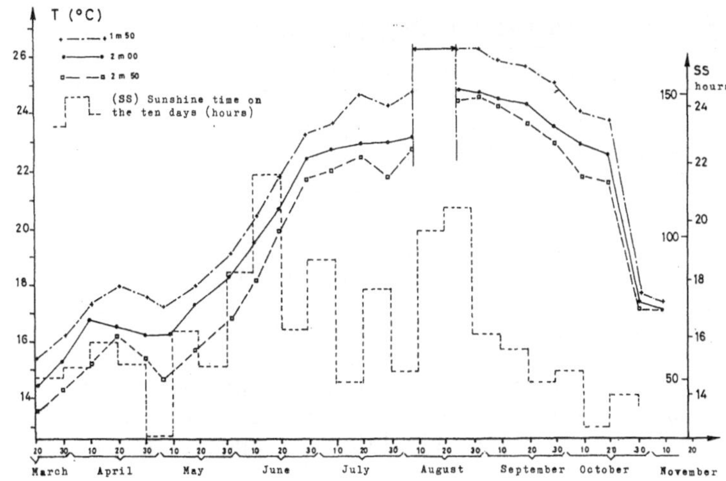

Figure 8 - Evolution of temperatures in storage at different depths

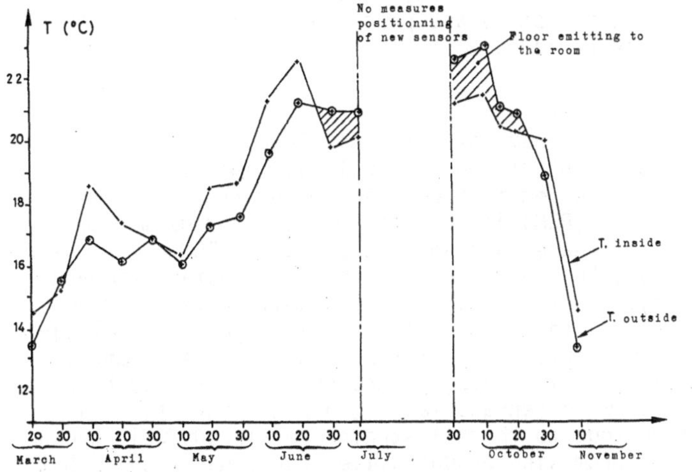

is valued to 20 % which means that 80 % were lost on storage surroundings. Then we define a general yield of the system as the ratio of recovered energy and energy on collector plane valued to 5 %.

For the building of such a house we could advice a peripheral insulation of the storage in the ground and to bury tubes deeper in the ground.

Figure 9 - Storage supply in the kitchen (no direct sunshine on the floor)

4. CONCLUSION

The experimentation performed on the "GUY house" was long and delicate. Many difficulties appeared : building occupancy, difficulties in the exact positionning of sensors where we wished, no extra heating and so on, however we could :
- study the thermal comfort degree of a bioclimatic house,
- determine the global performance level of that type of long period soil heat storage.

5. REFERENCES

FABRE, J. (1982). Etude en site réel d'une maison solaire passive à stockage de chaleur longue durée dans le sol. Thèse Docteur Ingénieur, Toulouse le 8 juillet.

FABRE, J. , TROMBE, A. and JAVELAS, R. (1983). Modélisation du stockage thermique dans le sol de la maison GUY. Revue Générale de Thermique, n° 254 février.

THERMAL STORAGE IN PHASE CHANGE MATERIALS.
APPLICATION TO A SOLAR AIR SPACE HEATING SYSTEM

P.-H. Theunissen & J.-M. Buchlin
von Karman Institute for Fluid Dynamics
B - 1640 Rhode Saint Genèse, Belgium

SUMMARY

Different designs of PCM storages are experimented. In a first step, the experimental observations support an analytical description of the charge and discharge process of an encapsulated PCM storage.
In a second step, a more sophisticated numerical model is used to predict the transient behavior of the encapsulated PCM unit during its charging, discharging and relaxation periods. The resulting computer code is TRNSYS compatible. Numerical simulations compare well with experiments.
Using the TRNSYS program, the performance of solar air space heating systems relying on either PCM or RB (Rock Bed) storage units are investigated and compared on both daily and seasonal basis. PCM decreases by a factor 4.5 the storage volume in respect to the RB.
The conclusions of the thermal study are processed by a micro-economic model to determine payback periods.

1. INTRODUCTION

This study takes place in the frame of Solar Air Space Heating. The volume of conventional Rock Bed (RB) storages as well as their associated costs is known to be the major problem to the spreading of those systems. The use of Phase Change Material (PCM) instead of rock as storage medium is proposed to solve or at least to ease this volumic problem. PCM offers high calorific density (latent heat, L_F) concentrated around a constant temperature (fusion temperature, T_F). Furthermore, the PCM family covers a wide range of fusion temperature and latent heat [1], in which specific application can pick up its own optimum material. Drawbacks of those substances rely mainly in their stability problem [2,3] in the requirement they have for special heat exchanger hardware (encapsulation, direct contact, ...). The change in properties of PCM due to thermal strain are already well known and almost solved for the most promising products, i.e. $CaC\ell_2 6H_2O$ [3,4]. But further work has still to be done in the thermal design of a PCM storage : heat exchange geometry, model of a PCM storage unit, integration of PCM storage into a solar air heating system. Moreover, such a thermal investigation has to end up on an economic analysis to compare the PCM with its challenger, the RB.

Following this framework, the research presented here will consider the PCM storage as a stand alone module by experimenting and modelizing two encapsulated PCM storage units; investigate and compare the thermal performance of space heating system with a PCM or RB storage unit; perform an economic analysis to find out which of these two systems is the most interesting for a given set of price and economic conditions.

This research was carried over in a Ph.D. program and the few results presented in the following lines are described in much more details in [5].

2. PCM STORAGE UNITS

2.1 Heat exchanger definition

A PCM storage unit is made up of three components (Fig. 1) :
-the phase change material itself; -the heat exchange fluid (air in this

study); -the heat exchanger matrix which is the hardware aimed to contain tightly the PCM and provide a close thermal contact between PCM and air. The PCM storage units differ from each other mainly by the matrix geometry. Although a lot of those are proposed to optimize HX effectiveness, the present study will be restricted to two specific encapsulated solutions :
-a Spherical Encapsulated (SE) unit, composed of a volume, V, of .175 m^3 filled with PCM capsules having an equivalent diameter, D_e, of .058 m. Void and PCM fractions, ε and ε_M, are respectively .41 and .48;
-a Tubular Encapdulated (TE) unit, made up of a volume of .25 m^3 filled with PCM rods perpendicular to the main stream, having a diameter, D_e, of .032 m. ε and ε_M are respectively .477 and .339.
The PCM used in both units is the Calcium Chloride Hexahydrate, $CaC\ell_2$-$6H_2O$, stabilized under recommendations of [3]. The fusion temperature, T_F, of the $CaC\ell_2 6H_2O$ is 29°C and thus well suited for space heating system. Furthermore, 29°C is low enough to keep the solar collector to a good efficiciency.

2.2 Experimental investigation

Both geometries had to be experimented in charge and discharge. An indoor solar loop facility was set up and used for this purpose. It is composed of a solar collector area, fed by an artificial sun, an auxiliary heater, a thermal load, a blower, a set of dampers and a storage system. The test facility control as well as the data acquisition and reduction task was taken over by microprocessors.

Charge (discharge) process is simulated by blowing through the storage an air stream, the temperature of which is higher (lower) than the fusion temperature.

Figure 2 pictures the thermal response in charge condition of the TE unit under an air flow, $\dot{m}_s$, of .28 kg/m^{2}s having an inlet temperature of 65°C. Each box of this time sequence shows the temperature profile of the air (top curve) and the PCM (bottom curve) through the unit (0 : inlet; 1 : outlet). Charge process can be viewed as taking place in two successive stages : - a quick uniformization of the PCM temperature to its fusion temparture followed by the melting of the first layer of PCM ($t < t^\star = 4366$);
- motion of the fusion through the bed at a constant velocity, v_F.
This description is strongly backed by Fig. 3 giving the experimental position, x_F, of the interface between the liquid and fusion zones in the unit. The same patterns of thermal response were observed for both TE and SE units in both charge and discharge mode.

Interpretation of experiments suggest to describe a charge (discharge) with the following simple model :

$$x_F = \max\{0, v_F(t - t^\star)\}$$

$$t^\star = \frac{\rho L_F \, \varepsilon_M}{HS_m |T_i - T_F|} \qquad v_F = \frac{\rho L_F \, \varepsilon_M}{\dot{m}_s C p_a |T_i - T_F|}$$

Symbol definitions are given in the nomenclature.
Experiments show that TE and SE units react almost in the same way despite of their geometrical differences.

2.3 Model of an encapsulated PCM storage

The proposed model is based on the description of a storage unit drawn on Fig. 1. Two different options are investigated for the PCM capsules :
- latent heat modelization by either an enthalpic or ablative approach (enthalpic : L_F spread around T_F; ablative : L_F concentrated on T_F);
- bulk (0D) or one dimensional (1D) temperature modelization of the PCM capsule.

For reason of space, this paper will be limited to the enthalpic OD approach. The bulk of the OD enthalpic model is composed of the following thermal balances :

PCM : $$\rho C_p(T) \frac{\partial T}{\partial t} = \frac{2H(1+\omega)}{D_e} \cdot (T_a - T)$$

air in convection : $$\dot{m}_s C_{pa} \frac{\partial T_a}{\partial x} = -S_m \cdot H\ (T_a - T) + H_e S_e (T_e - T_a)$$

air in relaxation : $$-k_{eff} \frac{\partial^2 T}{\partial x^2} = H_e S_e (T_e - T);\ T = T_a$$

These transient equations are solved with a time marching implicit method. The resulting computer code, valid for all the options presented above, is written TRNSYS compatible.

As illustrated in Figs. 3 and 4 for the TE unit, the OD enthalpic simulation gives a fairly good prediction of experiments during charge process. On the opposite, for discharge, the OD ablative model has to be recommended to reach the same agreement quality. The reason has to be found in the behaviour difference of the PCM between melting and freezing [2]. In spite of this, and for convenience, the following system discussions are based on the enthalpic description.

3. PCM STORAGE AND AIR SPACE HEATING SYSTEM

In Brussel weather conditions, the annual heat load, Q_L, of a house can be expressed as the sum of internal and solar passive gains, Q_F, solar active contribution, Q_T, and the auxiliary energy delivered by the backup furnace, Q_{aux} : $Q_L = Q_F + Q_T + Q_{aux}$.

It is convenient to scale the active solar system performance with the solar fraction, F, defined by : $F = Q_T/(Q_T+Q_{aux})$.

After inserting the PCM model into TRNSYS, doing some 300 annual simulations of a space heating system in Brussels, trying both PCM and RB units and using the solar fraction definition, Fig. 5 finally compares the merits of PCM and RB solutions. For a given solar fraction, F, use of PCM instead of RB is shown to decrease storage volume by a factor 4.5 in the small volume range. For larger volume, PCM reaches performances unavailable for RB (less lateral area, less heat losses). For both PCM and RB, small F ($<$20%) are only function of the collector area, A_c, while higher F are much more sensitive to the storage volume.

For a collector area of 50 m^2, Fig. 6 presents a comparison between RB and PCM over a wider range of storage volume, V, including seasonal storage. PCM is seen improving the solar fraction after 1 m^3/m^2 of collector while RB, for realistic insolation levels does not even catch the seasonal frequency of the sun.

4. ECONOMIC ANALYSIS

To assess the soundness of the PCM storage concept for an Air Space Heating system, an economic analysis has to be performed. The baselines of this approach are defined in [6]. Payback periods, N_u, for the systems presented in Fig. 5 are computed neglecting any loan parameters and the discount rate. Each solar system (A_c,V) is characterized by its reduced investment, N_1, i.e., the ratio between the cost of the solar system and the fuel bill of the house without solar system :

$$N_1 = \frac{C_T \eta_{aux}}{(Q_L - Q_F) P_{aux}}$$

The payback period, N_u, is computed using prices and trends which were in the average range of the economical conditions in 1983. In particular, encapsulated PCM was priced at 650 Belgian francs by Mega Joule of latent heat, a mean value of the cost of this product in the US at this time. Within this conditions (Fig. 7) :

- a storage volume decreases the payback period, N_u, by an average of 5 years;
- RB units lead to smaller N_u than the PCM ones, although the difference is not significant taking into account the uncertainty around economic data.

The optimal system is 20 m^2 (40) of collector and 4 m^3 (1) of storage for RB (PCM) unit and have a payback time of 19 years.
For seasonal storages (Fig. 8) the PCM storage does not show up as the best solution in spite of its leading position in the thermal comparison with the RB. This conclusion is true whatever insulation level is considered.

5. CONCLUSIONS

First,TE and SE PCM storage units are studied as stand alone modules. No significant difference of thermal behaviour is found between these two geometries. Interpretation of experiments leads to a linear description of transient behaviour of the PCM bed. A numerical description is proposed and its results are in good agreement with experiments. Second, solar air space heating system is modelized with TRNSYS. Annual simulations show that for daily storage a volume reduction of 4.5 can be expected by switching from RB to PCM. PCM offers significant solar fraction improvement for seasonal storage over RB. And last, RB and PCM turn out to give the same payback time for daily storage while in seasonal storage, PCM does not justify its use in the picked up economic conditions.

REFERENCES

1. ABHAT, A.: Low temperature latent heat thermal energy storage : heat storage materials. Solar Energy, Vol. 30, No 4, 1983, pp 313-332.
2. ABHAT, A.; ABOUL, S.; MALATADIS, N.A.: Heat of fusion storage systems for solar heating applications. in "Thermal Storage of Solar Energy", C. den Ouden, ed. M. Nijhoff Publ., 1980.
3. CARLSON, B.; STYMNE, B.. WETTERMARK, G.: An incongruent heat of fusion system - $CaCl_2 6H_2O$ - made congruent through modification of the chemical composition of the system. Solar Energy, Vol.23, No 4, 1979, pp 343-350.
4. Pour le Stockage Thermique à Basse Température, Casotherm 281, Solvay & Cie. BR 1280-B-1, 25-183, 1983.
5. THEUNISSEN, P.-H.: Accumulation thermique dans les matériaux à changement de phase. Application au chauffage solaire de l'habitat. Thèse de Doctorat, ULB-VKI, September 1983.
6. DUFFIE, J.A. & BECKMAN, W.A.: Solar engineering of thermal processes. New York, Wiley-Interscience, 1980.

NOMENCLATURE

A_c	collector area
C_p	PCM specific heat
C_{pa}	air specific heat
C_T	initial cost of the solar space heating system
D_e	diameter of the PCM shell
F	solar fraction
H	PCM-air heat exchange coefficient
H_e	storage heat losses coefficient
k_{eff}	effective conductivity of the PCM matrix
L_F	latent heat
$\dot{m}_s$	mass flow
N_u	payback period
N_1	reduced investment
P_{aux}	price of auxiliary power
Q_{aux}	annual auxiliary energy consumption
Q_F	annual internal and passive solar gains
Q_L	annual heat loss of the house
Q_T	annual active solar gains
S_m	wetted area of the PCM matrix
S_e	heat losses area of the storage unit
T	PCM temperature
T_e	ambiance temperature
T_F	PCM fusion temperature
T_i	air inlet temperature
t	time
$t^{\star}$	melting time of PCM first layers
V	storage volume
v_F	velocity of fusion interface
x_F	axial position of the fusion interface
ε	void fraction
ε_M	PCM fraction
η_{aux}	efficiency of auxiliary heater
ρ	specific mass of PCM
ω	shell shape coefficient (1 for TE; 2 for SE)

SOLAR COLLECTORS		
Efficiency factor	:	.8
Absorption	:	.95
Number of cover	:	1
Heat loss coefficient	:	7 W/m²°C
STORAGE		
Shape factor, L/D	:	2
Heat loss coefficient, H_e	:	.25 W/m²°C
Ambient temperature, T_e	:	10°C
HOUSE		
Heat loss coefficient	:	250 W/°C
Set temperature	:	19°C
Furnace efficiency, η_{aux}	:	.7
PRICE		
Solar collectors	:	7500 BEF/m²
PCM	:	690 BEF/MJ
RB	:	2500 BEF/m³
Constant costs	:	170 000 BEF
Backup energy	:	351 BEF/MJ
Backup energy inflation	:	14%/year

TABLE 1 - SET OF DATA USED IN ANNUAL SIMULATIONS

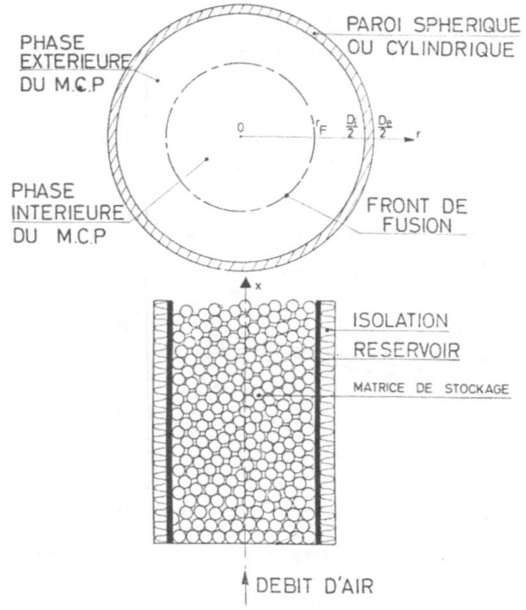

FIG. 1 - DRAWING OF AN ENCAPSULATED PCM STORAGE

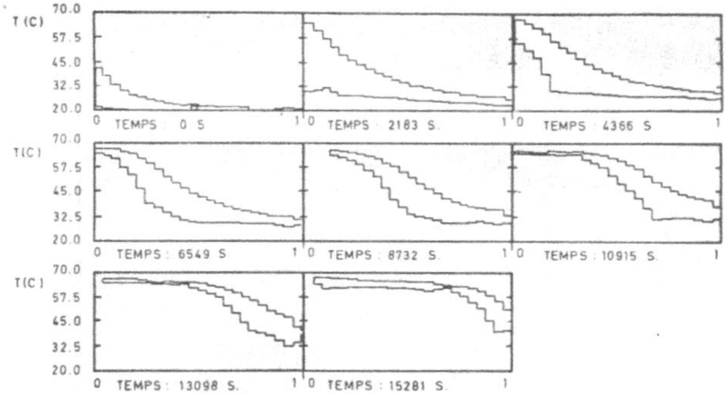

FIG. 2 - THERMAL EVOLUTION OF THE TE PCM UNIT IN CHARGE; $CaC\ell_2 6H_2O$; $\dot{m}_s$ = .28 kg/m²s

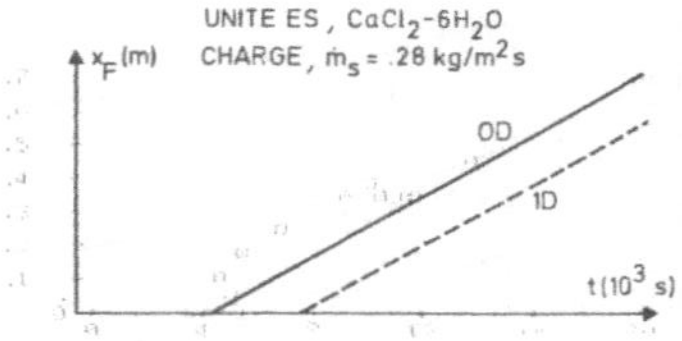

FIG. 3 - MOTION OF THE FUSION INTERFACE, x_F, VS TIME EXPERIMENT (POINTS), ENTHALPIC MODEL OD (——) AND 1D (---) TE PCM UNIT, $CaC\ell_2 6H_2O$; $\dot{m}_s$ = .28 kg/m²s

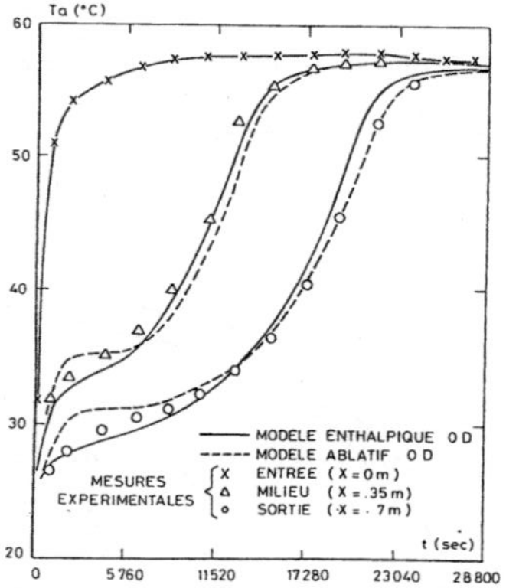

FIG. 4 - AIR TEMPERATURE AT THE INLET (x), MIDDLE (Δ) AND OUTLET OF THE TE PCM UNIT VERSUS TIME. $CaC\ell_2 6H_2O$, $\dot{m}_s$ = .28 kg/m²s

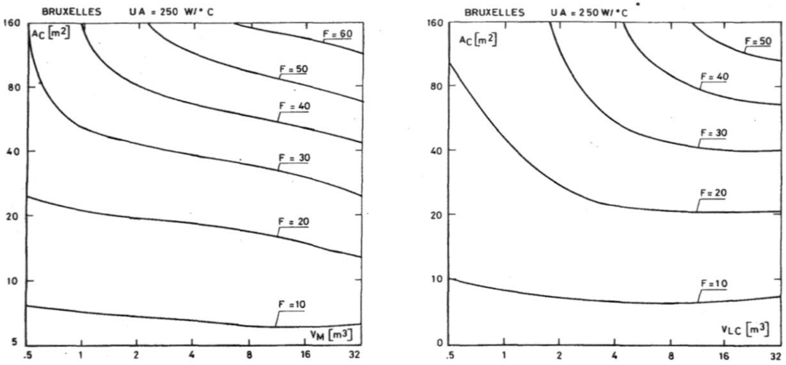

(A) PCM (B) RB

FIG. 5 - ISOSOLAR FRACTION, F, VERSUS COLLECTOR AREA, A_C, AND STORAGE VOLUME, V

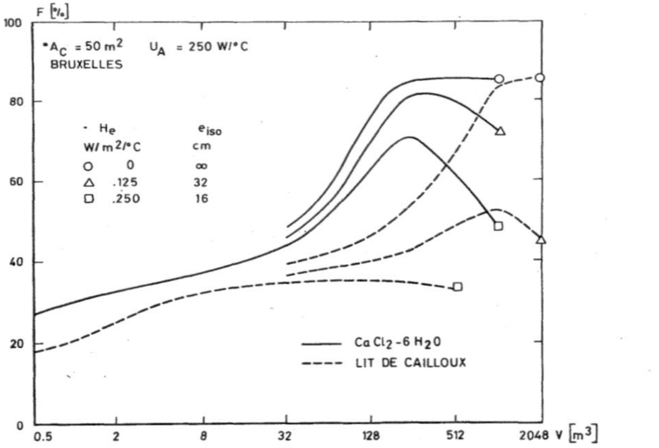

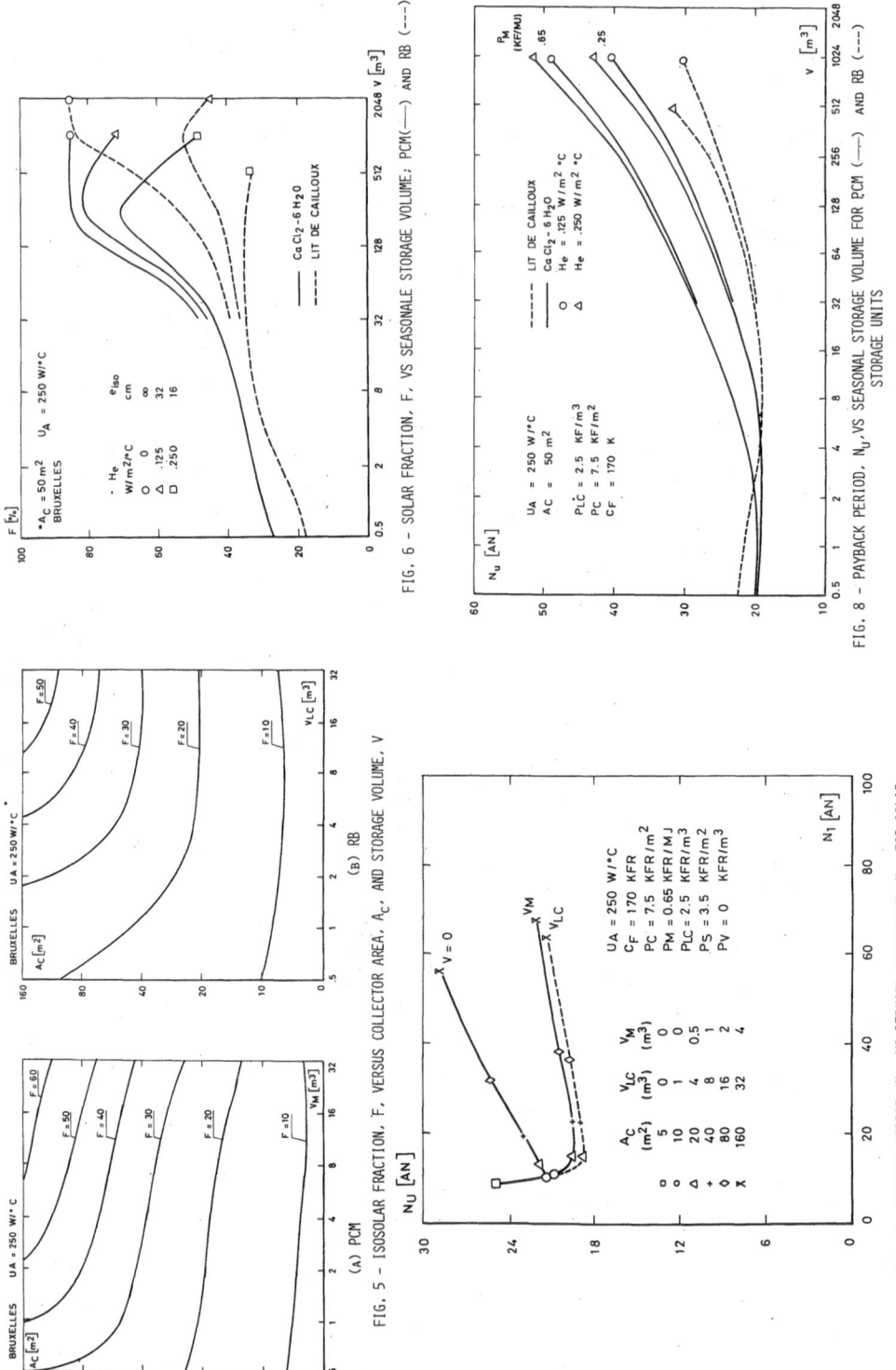

FIG. 6 - SOLAR FRACTION, F, VS SEASONALE STORAGE VOLUME; PCM(——) AND RB (---)

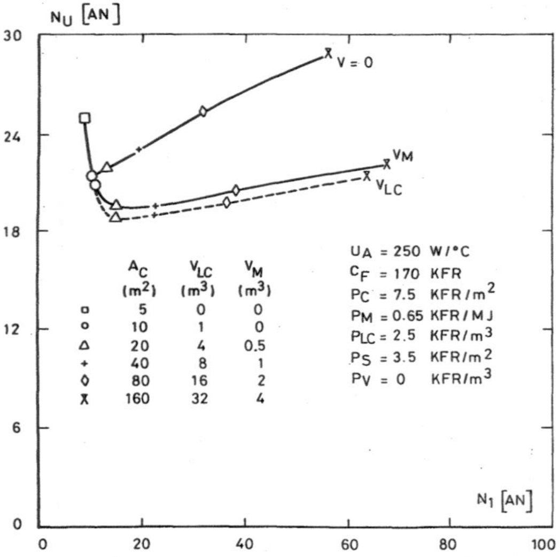

FIG. 7 - PAYBACK PERIOD, N_U, VS REDUCED INVESTMENT, N_1, FOR SOLAR SPACE HEATING SYSTEM W/OUT & W/STORAGE VOLUME PCM(——) AND RB(---)

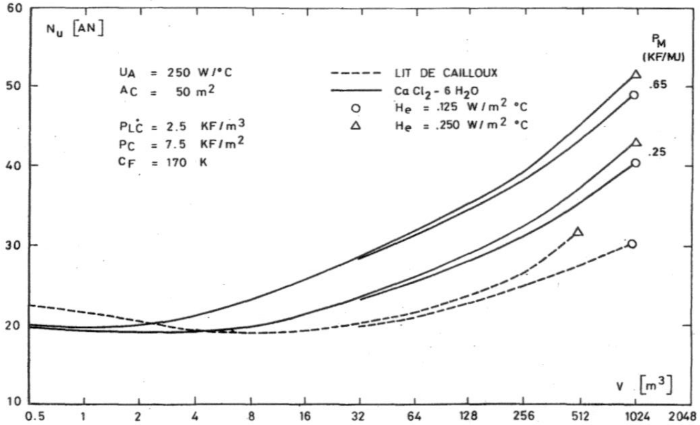

FIG. 8 - PAYBACK PERIOD, N_U, VS SEASONAL STORAGE VOLUME FOR PCM (——) AND RB (---) STORAGE UNITS

ENERGETIC BEHAVIOUR AND ECONOMIC EVALUATION OF AN INTERSEASONAL COMBINED SOLAR HEAT PUMP SYSTEM FOR COMMERCIAL BUILDING SPACE CONDITIONING

D. Borgese - A. Paganuzzi - F. Parrini - R. Viadana
ENEL (Italian Electricity Generating Board) CRTN - Milan

S. Bruca
PHOEBUS S.p.A. - Catania

Summary

ENEL has been engaged for a long time in solar energy experimental researches aimed at producing electricity and low temperature heat for domestic uses. Main targets of these studies are both the primary energy saving evaluation and the estimate of its effects on the electric energy demand, when it represents the back-up energy source. This paper deals with the selection among different interseasonal sensible heat storage solar system, of the most convenient technical solution for the space conditioning of a new ENEL commercial building in Turin. The optimal solution sized to satisfy a temperate winter load was singled out by means of a suitable computer model and consists of the following main items: flat plate solar collector field, seasonal water storage, electric heat pump, daily water storage. The outlined solution provides the following advantages: optimal exploitation of solar collectors due to the low temperature storage, significant primary energy saving, displacement to off-peack hours of electric energy heat-pump consumption; recovery of heat removed from the building when air cooling is operating. The system will be carried out within two years and monitorized to provide a suitable knowledge of both technical and energetic behaviour of major items installed.

1. INTRODUCTION

ENEL has been engaged for years in the study of possible solar energy utilizations for the production of both electric energy and low temperature heat for domestic uses.

Main purposes of these studies are: the attainable primary energy saving evaluation and the estimate of the effects on the demand of electric energy which often represents the back-up energy source. For this purpose many solarized plants with daily heat storage for space heating and domestic hot water production have been accomplished and monitorized by ENEL with CNR collaboration. In this paper, the particular system configuration chosen for space conditioning of an ENEL commercial building in Turin and the size of its major items are presented. The configuration has been singled out among different solar plant systems with long term sensible heat storage.

2. BUILDING

Site climatic conditions are described in table I.

Location	: Turin (Northern Italy) Altitude:239m s.l.m.
Latitude	: 45°N Longitude: 7,4°E
Annual average ambient temp.	: 11°C (jul-aug:21°C/oct-apr:5,4°C)
Wind speed	: 0,8 m/s (annual average)
Relative humidity	: 77,5% (annual average)
Annual global irradiation	: 4920 MJ/m^2 (oct-apr: 1865 MJ/m^2)
Degree days	: 2810 (base temperature 19°C)

Table I: Site climate

The two-storied building is assigned to commercial use.

It will be carried out with particular energy conservation measures as: double glazing with external shields, very high outside insulation (roof, walls) and heat recovery on exhausted air.

Thermal characteristics of the building are reported in table II.

Volume : 6479 m^3 No occupants : 120
Outside wall surface/volume ratio : 0,27 m^{-1}
Glazing surface/net outside wall surface ratio: 0,3

	Heating	Cooling
Design ambient temperature :	-9°C	32°C
Design internal temperature:	20°C	26°C
Ventilation rate :	0,75 h^{-1}	1,5 h^{-1}
Design load :	110 kW	65 kW
Heat loss coefficient :	4853 W/K	
Heat recuperator efficiency:	55% at 6°C	
Internal thermal load :	sensible 21 kW + latent 7 kW	

Table II: Building

3. THEORICAL EVALUATION OF DIFFERENT PLANTS

Among several types of solarized plants with long term water thermal storage, four particularly meaningful configurations have been selected and their simplified diagrams are reported in fig. 1. Solution (A) doesn't require the installation of any auxiliary system and thus it is sized to entirely provide for heating demand. Hot temperature thermal storage causes the reduction of collector efficiency and the increase of storage losses, and, consequently, the increase of the cost of produced energy. In solution (B) a water-water heat pump is placed between the storage tank and heat distribution system. In this way storage can be kept at a low temperature (10°C to 45°C) so improving collectors efficiency and lowering storage losses. In the third solution (C), the heat pump, placed between the collector system and the storage, allows the collectors optimal exploitation and thus a consequent reduction of installed surface. In this configuration storage water reaches quite high temperatures with the previously outlined consequences. In the last examined solution (D) one heat pump is placed between primary loop and the storage and the other between the storage and heat distribution system. This configuration combines the advantages of the two former configurations to the detriment of a greater complexity of the plant. Each of the outlined system solutions has been sized to accomplish the building winter heat requirement by making use of a mathematical model.

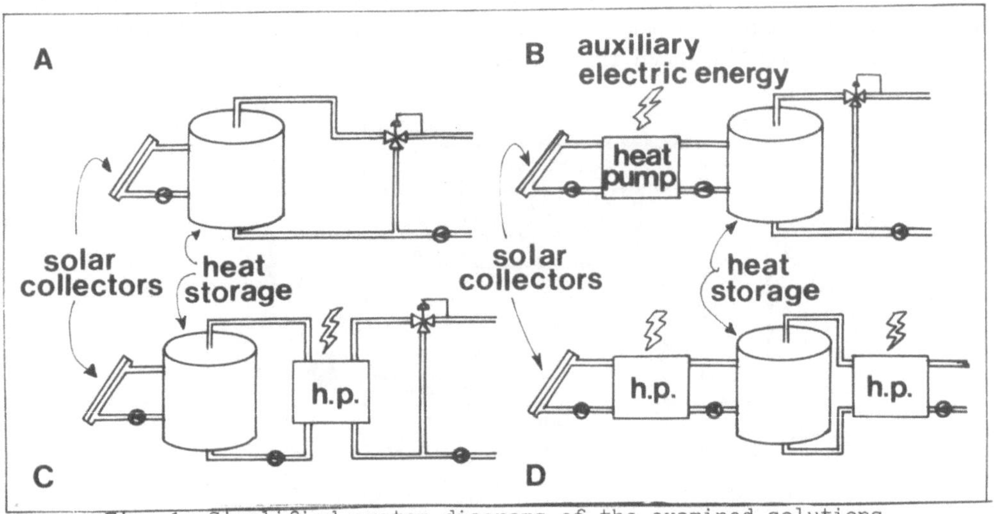

Fig. 1: Simplified system diagrams of the examined solutions

In table III are reported the sizes of main items of each configuration, their most meaningful energetic figures and the cost of supplied energy. Solution (B) was chosen because, in comparison with solution (D), it allows, at the same cost of produced energy, a lower resort to auxiliary source and results easier to carry out.

	A	B	C	D
Collector area (m^2)	400	85	70	40
Collector average efficiency	0,17	0,42	0,54	0,52
Storage volume (m^3)	1400	850	1400	850
Storage max water temperature (°C)	85	50	80	50
Space heating flow temperature (°C)	45	45	45	45
Storage heat losses (GJ)	190	68	108	68
Space heating load (GJ)	198	198	198	198
Electric energy back-up (GJ)	-	55	94	90
Supplied energy specific cost (ECU/MJ)	0,14	0,051	0,072	0,05
1 ECU = 1380 lit.				

Table III: Examined systems main figures

4. DESCRIPTION OF THE PROPOSED SOLUTION

System configuration at solution (B) has been further on refined through the installation of a daily storage, placed between the heat pump and heat distribution system, sized to allow the utilization of electricity for the heat pump only during off-peack hours. The reversibility in the use of the heat pump allows, during summer months space cooling, the recovery of heat removed from the building, thus lowering the size of collector area. The diagram of the proposed system solution in its winter running configuration is shown in fig. 2.

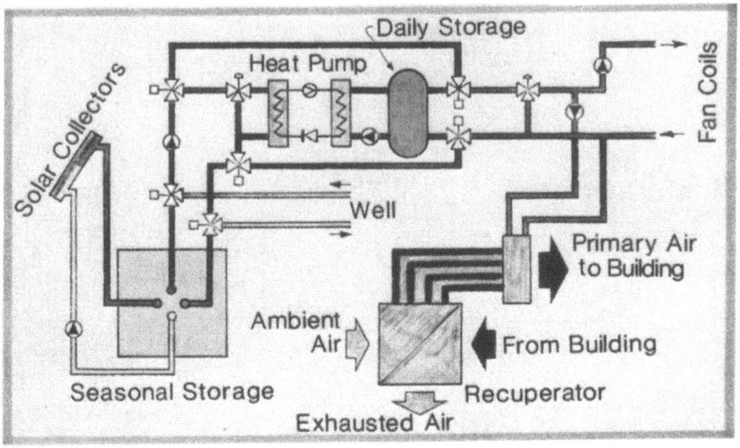

Fig. 2: Simplified diagram of the plant (winter running)

The sizes of major items of the plant are reported in table IV. In order to insure a better exploitation of solar section and to guarantee the necessary reliability of the plant, heat pump can also be fed with well water.

COLLECTOR	Area: 50 m^2	Orientation: 0°	Tilt: 45°
STORAGES		Long term	Short term
Volume		820 m^3	30 m^3
Max water temperature		50°C	60°C
Heat loss coefficient		69,8 W/K	12 W/K
SPACE CONDITIONING		Heating	Cooling
Heat pump type		water-water	
Design COP		3,6	2,8
Nominal electric power		30 kW	23 kW
Emitters		fan-coils	
Design flow temperature		45°C	7°C

Table IV: Major items size

5. ENERGETIC AND ECONOMIC EVALUATIONS

Energetic fluxes concerning the building-plant system have been estimated by means of a mathematical model and are reported in the Sankey diagram (fig. 3). It is possible to observe that:

- collector average efficiency is 42%;
- storage efficiency (supplied energy/input energy) is 65%;
- heat pump average COP is 3,6 in heating and 2,66 in cooling configuration;
- solar fraction is 42% of the actual heat requirement;
- heat recovery when air cooling is operating is 24%;
- electric energy absorbed by heat pump is 34% of the requirement.

Daily storage dimensions are sufficient to guarantee the building energetic requirement, even in cold weather situations, by making use of electricity only during off-peack hours.

The cost of supplied energy is quite high (0,05 ECU/MJ, about 150% of the traditional production cost) even because of the experimental characteristics of the proposed system.

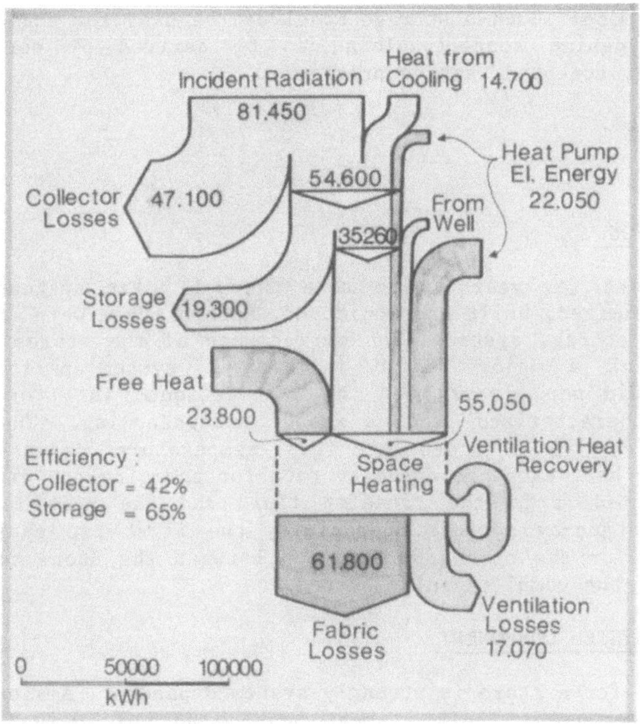

Figure 3 Sankey diagram (computer evaluation on annual basis)

DESIGN CONSIDERATIONS FOR SHORT TERM HEAT STORAGE SYSTEMS

G.J..van den Brink
Technisch Physische Dienst TNO-TH
(Institute of Applied Physics TNO-TH)
P.O. Box 155
2600 AD DELFT
The Netherlands

Summary

A heat storage system can be applied if heat supply and heat demand are out of phase. The main characteristics of the heat store are the temperature dependent storage capacity, the heat exchange capacity rate from storage material to heat transfer medium for both charging and discharging and the heat loss rate. The optimization of the different storage parameters is strongly dependent on the complete system design. The effect of the store on system performance is caused by an interaction with both heat supply and heat demand. A complete characterization of the store is necessary for developing a storage model. Such a storage model can be used for the optimization of the design concept. This can be derived by establishing a generally accepted testing programme.

1. INTRODUCTION

These last few years quite some thermal solar systems have been designed, optimized, built and monitored. Most of them were equipped with thermal heat storage systems. The performance of the stores appeared to be in general a major factor in overall system performance. The performance did not always meet the expectations. The store has to be completely characterized for a good understanding. The important characteristics of a store are the temperature dependent storage capacity, the heat exchange capacity rate for charging and discharging of the storage medium to the transfer fluid and the heat losses during working and stand-by periods. Especially the first two characteristics are the base for the strong interaction between the store and the other components of the complete solar system.

2. STORE AS SYSTEM COMPONENT

The need for a store is strongly system dependent. A short term heat store shall be applied only if heat supply and heat demand are out of phase and the time between supply and demand does not exceed a few days at the most. A strong interaction is found between the store and the other system components like collector and heat distribution system.
System performance is dependent on the storage capacity. A too small storage capacity reduces the solar contribution. The temperature in the collector loop increases and the heat losses cause a decrease of

collector efficiency. The store reaches its maximum allowed temperature too early and too little heat can be stored. Discharging of the store is easier due to a higher storage temperature. A too large storage capacity results in a lower temperature of the collector loop and therefore in an increase of collector efficiency. Problems can occur during discharging of the store if the required distribution temperature is not reached.

A temperature dependent storage capacity influences system performance as well. The higher the temperature for the transition of phase, the lower the collector efficiency shall be due to a higher operation temperature. For discharging a high transition temperature results in an increase of the amount of heat that can be discharged. The lower the temperature for the transition of phase, the higher the collector efficiency shall be, but the lower the amount of heat is which can be discharged due to a lower temperature difference between the storage medium and the storage inlet temperature of the heat transfer fluid.

The system performance is influenced as well by the heat exchanger capacity rate for charging and discharging. Too low a value of the heat exchanger capacity rate for charging the store can cause a higher temperature level in the collector loop. A higher temperature results in increased collector losses and a decreased yearly collector efficiency and system performance. Too low a value of the heat exchanger capacity rate for discharging influences the yearly system performance in a negative way. In most systems an auxiliary heater is required. A too low heat exchanger capacity rate for discharging results in the auxiliary heater coming on earlier and therefore in a lower heat contribution from the store. If the storage device is not discharged completely before the next charging cycle, the influence is a later switch on of the collector flow, higher collector heat losses and a lower collector efficiency. The influence of a higher heat exchanger capacity rate on system performance has the reverse effect.

The heat losses reduce the yearly system performance, although part of the heat losses can, conceivably, contribute to the heat demand. The heat loss rate at zero flow rate can be quite different from the heat loss rate at finite flow rate, especially for storage systems using air as heat transfer medium, where a "self insulating" effect can be found.

3. DESIGN CONCEPTS

The actual decision to use a thermal storage device or not and the choice for the type of store is dependent on a combination of technical, economical and political aspects, such as construction, safety, energy savings, investment, interest, polution by fossile fuels, subsidies etc. Hereafter only some technical aspects shall be discussed.

3.1 Storage capacity

The storage capacity is dependent on the volume of the storage material and on the type of storage material. Heat can be stored in:

- a sensible heat store using e.g. rock, concrete, water;
- a latent heat store using e.g. a salt hydrate 1 , paraffin wax 2 , adsorbing material 3 ;
- a hybrid heat store using both a sensible and a latent heat type storage material.

The main part of the heat stores installed in systems are of the sensible type storage devices. Rock is most in use for systems using air as heat transfer medium. Most systems with a liquid as heat transfer medium use water as storage material. The sensible heat storage systems show a temperature change for changes in enthalpy.

A lot of research has been carried out on latent heat type storage devices, using a phase change material (p.c.m.) or an adsorbent as storage medium. The transition of phase for a phase change material can be solid-solid, solid-liquid or liquid-gas. The advantage of using a latent heat type storage material is that in a small temperature range a relatively large amount of heat can be stored, as compared to e.g. water and that the storage is more or less isothermal. The advantage is dependent on the temperature range of operation of the store. The smaller the temperature range of operation, the larger the advantage of a phase change material is, due to a larger volume reduction.

Different types of latent heat storage materials are available, such as salt hydrates, organic phase change materials (e.g. paraffin waxes) and adsorbing materials. Each material has its advantages and disadvantages. Most salt hydrates shows ageing problems, due to e.g. segregation. Therefore, their reliability is usually small over longer periods. Supercooling is a well-known disadvantage. Special cold zones can be applied to prevent the supercooling. The storage capacity per unit of volume is usually relatively high in comparison with organic materials. The advantage of organic materials is the much higher reliability. The transition temperature range of organic materials, especially of paraffin waxes is usually larger than for salt hydrates. The main disadvantage of using the heat of adsorption for storage of heat is that extra heat exchangers and ventilators have to be applied, resulting in a significant increase of the costs.

The hybrid heat storage systems use both latent heat and sensible heat, e.g. a latent heat storage device with a water reservoir for domestic hot water.

3.2 Heat exchange and encapsulation

Different types of heat exchange from storage medium to heat transfer fluid are possible. In the simplest situation no heat exchanger is applied at all. The transfer fluid and storage material are the same, usually water. Quite a variety of heat exchange concepts are investigated in the past for latent and sensible heat type storage devices, which can roughly be distinguished as:

- Saturated aqueous solutions with an external heat exchanger 4 .
- Saturated aqueous solutions with direct contact heat exchange to an immiscible liquid 5 .
- Saturated aqueous solutions with an immersed heat exchanger 6 .
- Saturated aqueous solutions with multi phase direct contact heat exchange 7 .
- Bulk storage of stable materials with an immersed heat exchanger 8 , 9 .
- Macro encapsulation of stable materials with direct contact heat exchange 10 , 11 .
- Micro encapsulation with direct contact heat exchange 12 .
- Bulk storage of stable materials with direct contact heat exchange 13 .

Based on a technical discussion 14 of the various principles of heat exchange, it can be concluded here, that systems based on saturated aqueous solutions with an external heat exchanger and with direct contact heat exchange are not economically viable, while from a technical point of view systems based on micro/macro encapsulation and bulk storage with direct heat exchange, although still under developments are promising. So far, most experience was gained with systems based on either bulk storage with an immersed heat exchanger or micro/macro encapsulation with direct heat exchange.

3.3 Heat loss

An important design concept is the insulation of the store. The required insulation thickness is dependent on the time to bridge over from charging to discharging of the store. Special care has to be taken to prevent heat losses through cold bridges. Well-known cold bridges are the piping and support frame of the store. A good choice of the material can reduce the heat losses significantly (compare copper and stainless steel piping). The natural convection in the piping by density differences in the transfer fluid due to temperature differences (thermosyphon action) can be minimized by a correct choice of the piping configuration, e.g. by application of an S-turn. Some storage devices show a self-insulating effect during stand-by periods. For storage devices using air as transfer medium in which the air is in direct contact with the outer wall of the store, a large influence can be found. During a stand-by period the air can act as an insulator, resulting in lower heat loss rates for stand-by periods than for finite flow rate conditions.

4. TECHNICAL TARGET SPECIFICATIONS

All target specifications for a heat storage system are determined by the application for which the store should be designed.

- The first specification should define the temperature range, in which the store will be operated. Generally, only a few storage materials will be available for such a particular temperature range, while at the same time the temperate range will restrict the possible storage technologies.

- An important specification concerning the maximum required heat transfer power, both for charging and discharging is given below. When the same heat exchanger is used for charging and discharging, the mode with the lowest driving force will determine the heat exchanger design. Not every storage principle will be able to meet the requirements made by high heat transfer powers. Note that the effective thermal conductivity of certain storage materials can be increased significantly by adding a metal matrix 2 .

- The thermal capacity of the store, which is, in a given case, the time in which a required power can be supplied should be determined on the basis of an economical optimization. The whole system design must be considered then.

- The maximum heat loss rate allowed during operation, or a stand-by period can be specified to increase the energy recorvery ratio of the store.

- The pressure drop across the store should be restricted to a value that can be met by the pumps or ventilators available in connected loops.

5. ECONOMIC TARGET SPECIFICATIONS

Again, the application of the store determines the economic target specifications. Based on the data describing the charging and discharging processes, which are temperature levels and the charging and discharging powers as a function of time, the maximum energy conservation as a result of heat storage can be determined over a certain period of time.

For cyclical processes, there are many examples of these processes in industry, the storage capacity which is part of the economical optimization on the investment costs allowed can easily be determined. The investment costs allowed directly follow from the price of the energy saved per cycle, multiplied by the total number of cycles operated during the pay-back period. The pay-back period for industrial applications is not longer than 3 years. The store itself, required pheripheral equipment, necessary building volume, transport charges and installation costs should be included in these costs.

It is much more difficult to specify the costs allowed for non-cyclical processes.

The storage capacity can not be determined unambiguously and besides the available capacity is used for another part in every charging/discharging cycle. Solar energy applications are good examples for these non-cyclical processes. The investment costs allowed should result from detailed computer simulations for the complete installation. These computations establish the energy saved in consequence of the installation of the store.

A direct comparison of the costs made for a sensible and a latent heat storage system for particular applications does no justice to what is seen as one of the two major advantages of latent heat storage, namely the isothermal storage, that may enhance the efficiencies of conncected components.

A first estimation for the costs allowed can be found, if it is assumed that the same thermal capacity in the operating temperature range is required for competing storage technologies. A volume decrease proportional to the ratio of the energy density of the latent heat store to the energy density of the water store is possible at least. An additional volume decrease owing to the isothermal storage is then kept out of it. A smaller latent heat store for the same price as the sensible heat store will then be the target.

It is obvious, however, that a more accurate and fair-minded appraisal is made by a comparison of the yearly energy savings of the competing systems, by means of simulation models. A simulation model for the store can be based on the technical target specifications, assuming that such a store can be built, but using test data of an existing store would be even better [15].

6. CONCLUSIONS

The main characteristics of a store are the temperature dependent storage capacity, the heat exchange capacity rate for charging and discharging and the heat losses. The strong influence of the characteristics of a store on system performance is shown in a qualitative way.

An overview of the different design concepts is given.

The technical and economic target specifications are strongly system dependent. Therefore no general conclusion on the economical feasibility can be made. Each application must be seen separately. Testing programmes and storage models to be used for system optimization are essential.

REFERENCES

1. Brink, G.J. van den, E. van Galen, C. den Ouden, "Development of a thermal storage system based on encapsulated phase change materials V", Technisch Physische Dienst TNO-TH, report no. 803.219 II, Delft, November 1981.
2. Brink, G.J. van den, E. van Galen, "Thermal energy storage system using organic phase change materials with improved thermal conductivity for storage temperatures between 35 and 120°C", Technisch Physische Dienst TNO-TH, report no. 103.206, Delft, February 1984.
3. Verdonschot, J.K.M., C. den Ouden, "Development of a thermal storage system based on the heat of adsorption in hygroscopic materials", Technisch Physische Dienst TNO-TH, report no. 803.220, Delft, May 1980.
4. Bell, "Low grade heat storage using sodium acetate solution". Proceedings of International Conference on Energy Storage, Brighton, U.K., 1981.
5. Fouda, A.E., et al, "Solar storage systems using salt hydrate latent heat and direct contact thermal heat-I", Solar Energy, Vol. 25, 1980, pp 437-444.
6. Furbo, S., "Heat Storage with an incongruently melting salt hydrate as storage medium based on the extra water principle". Proceedings of an International TNO-symposium "Thermal storage of Solar Energy", Amsterdam, Netherlands, 1980, Martinus Nijhoff Publ.
7. Carlsson, B. et al, "The use of refrigerants for reflux boiling direct contact heat exchanging in heat of fusion storage using salt hydrates". Proceedings of the "6th Miami International Conference on Alternative Energy Sources", Miami Beach, 1983.
8. Marshall, R., "Experimental experience with the ASHRAE/NBS procedures for testing a phase change thermal storage device". Proceedings of International Conference on Energy Storage, Brighton, U.K., 1981.
9. Hoogendoorn, C.J. et al, "Organic phase change materials for solar heat storage". Proceedings of Solar World Congress Perth, Perth, Western Australia, 1983.
10. Galen, E. van, "An optimization method and experimental results of a phase change storage system based on sodium acetate trihydrate". Proceedings of the EC-Contractors' Meeting held in Athens, Greece, 1981, Reidel Publ. Company.
11. Cohen, R.R. et al, "The development and optimization of cost effective thermal energy storage systems for solar space heating by means of a micro processor controlled test facility". Proceedings of the EC-Contractors' Meeting held in Brussels, Belgium, 1983, Reidel Publ. Company.
12. Clen, J. et al, "Pelletization and roll encapsulation of phase change materials". Proceedings of the DOE Thermal and Chemical Storage Annual Contractors' Review Meeting, McLean, Virginia, 1980.
13. Ival, O. et al, "Development of an optimum process for electron beam cross-linking of high density polyethylene pellets". Proceedings of the DOE Thermal and Chemical Storage Annual Contractors' Review Meeting, McLean, Virginia, 1980.
14. Galen, E. van, "Design considerations for latent heat stores". Proceedings of IEA-workshop on latent heat storage, Stuttgart, March 1984.
15. Galen, E. van, "Recommendations for European Solar Storage Test Methods (sensible and latent heat storage devices)", Technisch Physische Dienst TNO-TH, report no. 203.212, Delft, March 1984.

Impression of the Dutch Evening in the Breughel House

INDEX OF AUTHORS

Impression of the Dutch Evening in the Breughel House